HADRON SPECTROSCOPY

Previous Proceedings in the Series of Conferences on Hadron Spectroscopy

Conference		Held in	Publisher	ISSN/ISBN
HADRON05	11th	Rio de Janeiro, Brazil	AIP Conf. Proceedings Vol. 814	0-7354-0307-4
HADRON03	10th	Aschaffenburg, Germany	AIP Conf. Proceedings Vol. 717	0-7354-0197-7
HADRON01	9th	Protvino, Russia	AIP Conf. Proceedings Vol. 619	0-7354-0067-9
HADRON99	8th	Beijing, China	North-Holland Nuclear Physics A, Vol. 675	0375-9474
HADRON97	7th	Upton, New York, USA	AIP Conf. Proceedings Vol. 432	1-56396-765-0

HADRON SPECTROSCOPY

Eleventh International Conference on Hadron Spectroscopy

Rio de Janeiro, Brazil 21 – 26 August 2005

EDITORS

Alberto Reis
Centro Brasileiro de Pesquisas Físicas - CBPF

Carla Göbel
Pontifícia Universidade Católica do Rio de Janeiro

José de Sá Borges
Instituto de Física - UERJ

Javier Magnin
Centro Brasileiro de Pesquisas Físicas - CBPF

SPONSORING ORGANIZATIONS
Financiadora de Estudos e Projetos (FINEP)
Conselho Nacional de Desenvolvimento
Científico e Tecnológico (CNPq)
Banco do Brasil

AIP
75 Years of Service

Melville, New York, 2006
AIP CONFERENCE PROCEEDINGS ■ VOLUME 814

Editors

Alberto Reis
Centro Brasileiro de Pesquisas Físicas - CBPF
Rua Dr. Xavier Sigaud, 150
22290-180 Rio de Janeiro, Brazil
E-mail: alberto@cbpf.br

Carla Göbel
Pontifícia Universidade Católica do Rio de Janeiro
Departamento de Física
Rua Marques de São Vincente, 225
22453-900 Rio de Janeiro, Brazil
E-mail: gobel@fis.puc-rio.br

José de Sá Borges
Instituto de Física - UERJ
Bloco A, Sala 3016
Rua São Francisco Xavier, 524
20599-900 Rio de Janeiro, Brazil
E-mail: saborges@uerj.br

Javier Magnin
Centro Brasileiro de Pesquisas Físicas - CBPF
Rua Dr. Xavier Sigaud, 150
22290-180 Rio de Janeiro, Brazil
E-mail: jmagnin@cbpf.br

L.C. Catalog Card No. 2005938523
ISBN 0-7354-0307-4
ISSN 0094-243X

Printed in the United States of America

IN MEMORIAM

César Lattes

1924 - 2005

Contents

PLENARY SESSIONS

PARALLEL SESSIONS

BARYONS

EXOTICS

HADRON STRUCTURE

HEAVY QUARKS

LIGHT QUARKS

NEW HADRONS

QUARKONIA

REACTIONS

RHIC

SCALARS

SUMMARY TALKS

Preface

The XI International Conference on Hadron Spectroscopy, HADRON05, was organized by the Centro Brasileiro de Pesquisas Fisicas, CBPF, in Rio de Janeiro. Conferences in this series take place every two years, and started in 1985. The previous conference was held in Aschaffenburg, Germany, in 2003. The scope of this conference series is hadron spectroscopy and related aspects of hadron dynamics.

The present conference was held at the Instituto Militar de Engenharia (plenary sessions) and at CBPF (parallel sessions), from 21 to 26 August 2005, and was attended by 192 scientists from institutions all over the world.

After many decades, hadron physics is still a very active field, permanently renewed by new discoveries. A pleasant surprise has been the Tevatron and the B-factories as new players in hadron spectroscopy. One of the highlights of this conference was the discussion on the nature of the recently discovered narrow states. In this meeting many new results from high statistics samples from the charm factories were presented. There has been progress in our understanding of the scalars. An extensive list of high statistics searches for the pentaquark was also presented. In addition, there were interesting contributions on heavy-light quark systems, heavy and light baryons, hadron structure and hadron classification schemes.

We would like to thank the International Advisory Committee for the help in building the scientific program and the effort to guarantee the high scientific level observed throughout HADRON05. We thank the speakers for their excellent presentations and for their contributions to these proceedings. We would like to thank the Local Organizing Committee for their dedication to make HADRON05 a successful conference. We are indebted to the staffs at CBPF and IME. The financial support of our sponsors, Banco do Brasil, FINEP and CNPq, are gratefully acknowledged.

The next conference in this series will be held in Frascati, Italy, in 2007. It is a great pleasure to wish it full success, expecting there exciting new results.

International Advisory Committee

C. Amlser	(Zurich)
P. Barnes	(Los Alamos)
E. Berger	(Argonne)
T. Bressani	(Torino)
S. Chung	(BNL)
F. Close	(Oxford)
W. Dunwoodie	(SLAC)
A, Dzierba	(Indiana)
F. Gross	(JLab)
C. Guaraldo	(Frascati)
E. Klempt	(Bonn)
H. Koch	(Bochum)
H. Lipkin	(Weizmann)
U. Lynen	(GSI)
B. Meadows	(Cincinnati)
V. Metag	(Giessen)
S. Nagamiya	(KEK)
E. Oset	(Valencia)
S. Paul	(Munich)
D. Peaslee	(Maryland)
E. Predazzi	(Torino)
K. Seth	(Evanston)
A. Skrinsky	(Novosibirsk)
K. Takamatsu	(KEK)
U. Wiedner	(Uppsala)
A. Zaitsev	(Protvino)
B. Zou	(Beijing)

Local Organizing Committee

A. Reis	(CBPF)	(Chair)
J. Miranda	(CBPF)	
J. Magnin	(CBPF)	
I. Bediaga	(CBPF)	
C. Göbel	(PUC-Rio)	
J. Borges	(UERJ)	
T. Kodama	(UFRJ)	
M. Nielsen	(USP)	
F. Navarra	(USP)	
G. Krein	(IFT)	
C. Vasconcellos	(UFRGS)	

Conference Secretariat

Z. Quadros	(CBPF)
S. Ferreira	(CBPF)
C. Monteiro	(CBPF)

Sponsors

FINEP
CNPq
Banco do Brasil

César Lattes

1924 - 1985

Cesare Mansueto Giulio Lattes, mostly known as César Lattes, is the most celebrated of Brazilian physicists. He was born in the city of Curitiba, in June 11, 1924. His scientific career begun in the mid 40's, at University of São Paulo, under the supervision of Gleb Wataghin. Soon after the end of the World War II he went to Bristol with Giuseppe Occhialini. Lattes started working with Cecil Powell on cosmic rays, using photographic emulsions. In the spring of 1947 the Bristol group found clear evidence of the existence of a new particle, the pion. Lattes, Muirhead, Occhialini and Powell published their findings in 'Nature' in the issue of 24 May, 1947. A few months later, Lattes brought some photographic plates to Mount Chacaltaya, in the Bolivian Andes, near the capital La Paz, to na old meteorological station built at na altitude of 5,220 m. Examination of the plates from Bolivia led Lattes, Occhialini and Powell, in their October 1947 paper to 'Nature ', to postulate a two-body decay of the primary meson, which they called pion, to a secondary meson, the muon, and one neutral particle.

Lattes spent the years of 1948-1949 at Berkeley, working with Eugene Gardner. At the age of 24, Lattes managed, for the first time, to produce pions artificially, using the new synchro-cyclotron. The discovery of the pion boosted physics research in Brazil. In 1949 Lattes returned to Brazil and founded the Centro Brasileiro de Pesquisas Fisicas, CBPF, the first Brazilian physics research institute.

The research field with the most tradition in Latin America is the study of cosmic rays. Lattes was the pioneer in this field. In January 1952 he installed the Chacaltaya Laboratory of Cosmic Physics, associated with the University of San Andres. The result was a fruitful collaboration between Brazil and Bolivia. In the 50 years since its foundation, the Chacaltaya laboratory has been used extensively by the international scientific community. Scientists from Argentina, England, India, Italy, Japan, and the US, among others, have made contributions there to the knowledge of cosmic radiation. The most effective and long standing collaboration was with Japan, started in 1962. In 1972 Lattes and his collaborators made the first observation of the not yet understood Centauro events.

Lattes was the Scientific Director of CBPF until 1955. He spent the years of 1956 and 1957 in the United States. There he was invited to substitute Enrico Fermi as the head of Fermi's Institute of the University of Chicago. He declined the invitation and came back to Brazil to build the Laboratory of High Energy Cosmic Rays, at the University of São Paulo. He also took part in the group that created the University of Campinas, being responsible for the formation of the Physics Institute.

Lattes won many international prizes, but was always a very simple person. He was particularly proud of having his name given to public schools, streets, squares, in dozens of Brazilian cities.

Cesar Lattes passed away in March 5, 2005.

César Lattes in Chacaltaya.

Lattes and Gardner working in Berkeley.

PLENARY SESSIONS

QCD Effective Field Theories for Heavy Quarkonium

Nora Brambilla

Dipartimento di Fisica dell'Università di Milano and INFN

Abstract. QCD nonrelativistic effective field theories (NREFT) are the modern and most suitable frame to describe heavy quarkonium properties. Here I summarize few relevant concepts and some of the interesting physical applications (spectrum, decays, production) of NREFT.

Keywords: QCD, Effective Field Theories, Quarkonium
PACS: 12.38.-t,12.38.Aw

INTRODUCTION

The study of heavy quark-antiquark bound states touches upon several important physics areas, within and beyond the Standard Model of Particle Physics. These multi-scale systems probe all the energy regimes of QCD, from the hard region, where an expansion in the coupling constant is legitimate, to the low energy region, where nonperturbative effects dominate. Heavy quark-antiquark states are thus an ideal and to some extent unique laboratory where our understanding of nonperturbative QCD and its interplay with perturbative QCD may be tested in a controlled framework. This has been so historically, since quarkonium has been deeply related to the development of QCD, and it is so even more today for the two following main reasons. The first reason is that in the last few years a wealth of new experimental results has become available. The diversity, quantity and accuracy of the data currently being collected is impressive and includes: data on quarkonium formation from BES at BEPC, E835 at Fermilab, KEDR (upgraded) at VEPP-4M, and CLEO-III, CLEO-c; clean samples of charmonia produced in B-decays, in photon-photon fusion and in initial state radiation, from the B-meson factory experiments, BaBar at SLAC and Belle at KEK, including the unexpected observation of large amounts of associated $(c\bar{c})(c\bar{c})$ production; the CDF and D0 experiments at Fermilab measuring heavy quarkonia production from gluon-gluon fusion in $p\bar{p}$ annihilations at 2 TeV; ZEUS and H1, at DESY, studying charmonia production in photon-gluon fusion; PHENIX and STAR, at RHIC, and NA60, at CERN, studying charmonia production, and suppression, in heavy-ion collisions. In the near future, even larger data samples are expected from the BES-III upgraded experiment, while the B factories and the Fermilab Tevatron will continue to supply valuable data for several years. Later on, new facilities will become operational (LHC at CERN, Panda at GSI, much higher luminosity B factories at KEK, a Linear Collider, etc.) offering fantastic challenges and opportunities in this field. A comprehensive review of the experimental and theoretical status of heavy quarkonium physics may be found in the Cern Yellow Report prepared by the Quarkonium Working Group (http://www.qwg.to.infn.it) [1]. See also the experimental review talks [2] at this conference.

CP814 *Hadron Spectroscopy: Eleventh International Conference*
edited by A. Reis, C. Göbel, J. Sá Borges, and J. Magnin

The second reason is the remarkable theoretical progress of the last few years. Effective field theories (EFT), such as Nonrelativistic QCD (NRQCD)[3, 4], provided new tools and definite predictions concerning, for instance, heavy quarkonium production and decays. New effective field theories for heavy quarkonium, as potential NRQCD (pNRQCD)[7, 8] and velocity NRQCD (vNRQCD)[5], have been recently developed and are producing a wealth of new results. The lattice implementation of such effective theories has been partially carried out and many more results with drastically reduced systematic uncertainties are expected in the near future. An extensive review of the latest development in nonrelativistic EFTs can be found in [9].

Therefore, on one hand the progress in our understanding of NREFTs makes it possible to move beyond phenomenological models and to provide a systematic description from QCD of all aspects of heavy-quarkonium physics. On the other hand, the recent progress in the measurement of several heavy-quarkonium observables makes it meaningful to address the problem of their precise theoretical determination. In this situation heavy quarkonium becomes a very special and relevant system to advance our understanding of strong interaction and our control of some parameters of the Standard Model.

In the following we present the main conceptual ideas and simplifications underlying the EFT framework, a brief summary of the main ingredients of NRQCD and pNRQCD and list several applications to the phenomenology of quarkonium, referring to the original publications for the details.

NONRELATIVISTIC EFFECTIVE FIELD THEORIES

Nonrelativistic bound states

Nonrelativistic bound states are characterized by a small relative velocity v (in the centre-of-mass frame) of the particles inside the bound system. The bound state dynamics has quite distinctive features with respect to the scattering case. If one considers the simpler example of a QED nonrelativistic bound system like positronium (or hydrogen), then $v \simeq \alpha \ll 1$. The fine structure constant α being small in QED, physical observables may be evaluated in perturbation theory. However, the bound state is nonperturbative and in this case one needs to resum infinite sets of Feynman diagrams. Due to the fact that the bound state lives close to threshold, i. e. $v \simeq \alpha$, the bound state pole (at leading order) is obtained by the resummation of all the Coulomb ladder photon contributions, $\simeq (\frac{\alpha}{v})^n$, in the Feynman diagram series (cf. Fig.1). As a consequence by solving the Schrödinger equation $(\frac{p^2}{2m} + V)\phi = E\phi$, with V a Coulomb potential, one generates the dynamical scales of the momentum transfer $p \simeq m\alpha$ and of the bound state kinetic energy $E = \frac{p^2}{m} \simeq m\alpha^2$ (for some review see [6]). Such scales manifest themselves in the scalings of positronium radial ($\sim m\alpha^2$), fine and hyperfine splittings ($\sim m\alpha^4$). This is different from a pure scattering calculation where no scales involving α are generated.

Quarkonium scales

Heavy quarkonium provides a non-relativistic system potentially very similar to a QED bound state, with the difference that the low-energy scales will be sensitive to

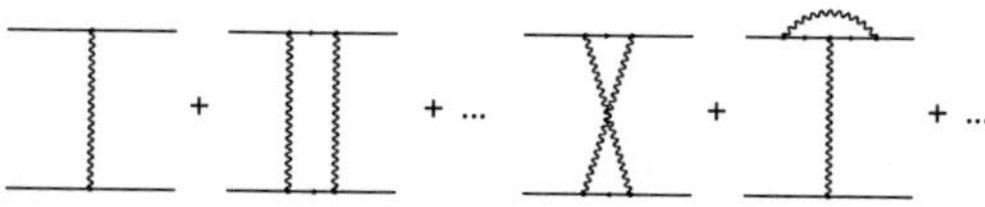

FIGURE 1. QED series of diagram defining the e^+e^- interaction. The resummation of the series of all ladder photon diagrams (of the type of the two first diagrams): $\alpha\left(1+\frac{\alpha}{v}+\ldots\right)$ (with $\alpha \sim v$) gives the (leading order) Coulomb bound state energy pole.

$\Lambda_{\rm QCD}$. Therefore the description of hadrons containing two heavy quarks is a rather challenging problem, which adds to the complications of the bound state in field theory those coming from a nonperturbative QCD low-energy dynamics. A simplification is provided by the nonrelativistic nature of heavy quarkonium, suggested by the large mass of the heavy quarks and manifest in the spectrum pattern. Quarkonium is thus characterized by three energy scales, hierarchically ordered by the quark velocity $v \ll 1$: the mass m (hard scale), the momentum transfer mv (soft scale), which is proportional to the inverse of the typical size of the system r, and the binding energy mv^2 (ultrasoft scale), which is proportional to the inverse of the typical time of the system. In bottomonium $v^2 \sim 0.1$, in charmonium $v^2 \sim 0.3$. In perturbation theory $v \sim \alpha_{\rm s}$. Feynman diagrams will get contributions from all momentum regions associated with the scales. Since these momentum regions depend on $\alpha_{\rm s}$, each Feynman diagram contributes to a given observable with a series in $\alpha_{\rm s}$ and a non trivial counting. For energy scales close to $\Lambda_{\rm QCD}$, the scale at which nonperturbative effects become dominant, perturbation theory breaks down and one has to rely on nonperturbative methods. Regardless of this, the non-relativistic hierarchy $m \gg mv \gg mv^2$ will persist also below the $\Lambda_{\rm QCD}$ threshold.

The wide span of energy scales involved makes also a lattice calculation in full QCD extremely challenging since one needs a space-time grid that is large compared to the largest length of the problem, $1/mv^2$, and a lattice spacing that is small compared to the smallest one, $1/m$. To simulate, for instance, a $b\bar{b}$ state where $m/mv^2 \sim 10$, one needs lattices as large as 100^4, which are extremely time computing demanding.

We may, however, take advantage of the existence of a hierarchy of scales by substituting QCD with simpler but equivalent EFTs. A hierarchy of EFTs may be constructed by systematically integrating out modes associated to energy scales not relevant for the quarkonium system. Such integration is made in a matching procedure that enforces the complete equivalence between QCD and the EFT at a given order of the expansion in v and achieves a factorization between the high energy and the low energy contributions.

What is an EFT?

Let H be a system described by a fundamental Lagrangian $\mathcal{L}$. Suppose that H has two characteristic physical scales: $\Lambda \gg \lambda$. The EFT Lagrangian, $\mathcal{L}_{\rm EFT}$, suitable to describe H at scales lower than Λ is then defined by: a cut off $\Lambda \gg \mu \gg \lambda$; some degrees of freedom that exist at scales lower than μ. $\mathcal{L}_{\rm EFT}$ involves all the operators O_n that may be built from the effective degrees of freedom and are consistent with the symmetries of

$\mathscr{L}$:

$$\mathscr{L}_{\mathrm{EFT}} = \sum_n c_n(\Lambda,\mu)\frac{O_n(\mu,\lambda)}{\Lambda^n} \tag{1}$$

- Once the scale μ has been run down to λ, $\langle O_n \rangle \sim \lambda^n$, so that the EFT is organized as an expansion in the small parameter λ/Λ.
- The EFT is renormalizable order by order in λ/Λ.
- The matching coefficients $c_n(\Lambda,\mu)$ encode the non-analytic behaviour in Λ. They are calculated by imposing that $\mathscr{L}_{\mathrm{EFT}}$ and $\mathscr{L}$ describe the same physics at any finite order in the expansion: this is called matching procedure.
- If $\Lambda \gg \Lambda_{\mathrm{QCD}}$ then the $c_n(\Lambda,\mu)$ may be calculated in perturbation theory.

The EFT has a well defined power counting in a small parameter so that the expansion is systematic. In QCD the EFT approach makes it possible to achieve a rigorous factorization between the high-energy dynamics encoded into matching coefficients calculable in perturbation theory and the nonperturbative QCD dynamics encoded into few well-defined nonperturbative operator matrix elements to be fitted on the data or calculated on the lattice or in QCD vacuum models. Thus, several model independent QCD predictions become possible. EFTs have the additional advantage that they are often close to popular phenomenological models to which experimental results are usually compared to, the main difference being a good control on the systematic errors.

NRQCD

Integrating out degrees of freedom of energy m, which for heavy quarks can be done perturbatively, leads to NRQCD [3, 4]. In the language of the previous section this is the EFT that follows from QCD when $\Lambda = m$. The matching is perturbative. The Lagrangian is organized as an expansion in v and $\alpha_s(m)$:

$$\mathscr{L}_{\mathrm{NRQCD}} = \sum_n c_n(m,\mu) \times O_n(\mu, mv, mv^2, \Lambda_{\mathrm{QCD}})/m^n. \tag{2}$$

The Wilson coefficients c_n are series in α_s and encode the ultraviolet physics that has been integrated out from QCD. The operators O_n describe the low-energy dynamics and are counted in powers of v. However since two scales, the soft and the ultrasoft, still remain dynamical, NRQCD does not have an unambiguous power counting in v. The operators bilinear in the fermion (or in the antifermion) fields are the same that can be obtained from a standard Foldy-Wouthuysen transformation [38] (however corrected with the matching coefficients). In this framework the imaginary part of the NRQCD Hamiltonian, i.e. the imaginary part of the Wilson coefficients of the 4-fermion operators, is responsible for heavy quarkonium annihilations. Being this a field theory, the heavy quarkonium Fock state is given by a series of terms in which the leading one is a $Q\bar{Q}$ in a color singlet state and the first correction, suppressed in v, comes from a $Q\bar{Q}$ in an octet state with a gluon.

Applications of NRQCD

Spectrum. The NRQCD Lagrangian is well suited for lattice calculations. The quark propagators are the nonrelativistic ones and since we have integrated out the scale of the mass, the lattice step used in the simulation may be a factor $1/v$ bigger. Lattice evaluation of heavy systems like bottomonium become thus feasible. The latest results for the spectra (quenched and unquenched) are given e.g. in [10]. The radial splittings are accurate up to $O(\alpha_s v^2)$ while fine and hyperfine splittings are accurate only up to $O(\alpha_s)$, due to the fact that only tree level matching coefficients have been used. A calculation of the NRQCD matching coefficients in the lattice regularization scheme is still missing and would be relevant to improve the precision of the lattice data.

Decays. NRQCD gives a factorization formula for heavy quarkonium inclusive decay widths into light hadrons and electromagnetic decays involving four-fermion matching coefficients and matrix elements of four fermion operators [4]. Singlet operator expectation values may be easily related to the square of the quarkonium wave functions (or derivatives of it) at the origin. Octet contributions remain as nonperturbative matrix elements of operators evaluated over the quarkonium states. In some situations the octet contributions may not be suppressed and become as relevant as the singlet contributions in the NRQCD power counting. In particular octet contributions may reabsorb the dependence on the infrared cut-off of the Wilson coefficients, solving the problem that arised in the color singlet potential model. Systematic improvements are possible, either by calculating higher-order corrections in the coupling constant or by adding higher-order operators. If one goes on in the expansion in v, that seems to be necessary for charmonium, the number of involved nonperturbative matrix elements of octet operators increases in such a way that limits the prediction power [11]. In the matching coefficients large contributions in the perturbative series coming from bubble-chain diagrams may need to be resummed [12].

Production. Before the advent of NRQCD, colour singlet production and colour singlet fragmentation underestimated the data on prompt quarkonium production at Fermilab by about an order of magnitude indicating that additional fragmentation contributions were missing [13]. The missing contribution is precisely the gluon fragmentation into colour-octet $^3S_1$ charm quark pairs. The probability to form a J/ψ particle from a pointlike $c\bar{c}$ pair in a colour octet $^3S_1$ state is given by a NRQCD nonperturbative matrix element which is suppressed by v^4 with respect to to the leading singlet term but is enhanced by two powers of α_s in the short distance matching coefficient for producing colour-octet quark pairs. When one introduces the leading colour-octet contributions, then the data of CDF can be reproduced. Still remains a puzzle the behaviour of the polarization at high p_T (cf. the production chapter in [1]).

PNRQCD

In NRQCD the role of the potential and the quantum mechanical nature of the problem are not yet maximally exploited. A higher degree of simplification may be achieved building another EFT where only the ultrasoft degrees of freedom remain dynamical. pNRQCD [7, 8, 9] is the EFT for heavy quarkonium that follows from NRQCD when $\Lambda = 1/r \sim mv$. We may distinguish two situations: 1) weakly coupled pNRQCD when $mv \gg \Lambda_{QCD}$, where the matching from NRQCD to pNRQCD may be performed in perturbation theory 2) strongly coupled pNRQCD when $mv \sim \Lambda_{QCD}$, where the matching has to be nonperturbative. Recalling that $r^{-1} \sim mv$, these two situations correspond to systems with inverse typical radius smaller than or of the same order as Λ_{QCD}.

Weakly coupled pNRQCD

The effective degrees of freedom are: low energy $Q\bar{Q}$ states (that can be decomposed into a singlet and an octet field under colour transformations) with energy of order $\Lambda_{\rm QCD}, mv^2$ and momentum $\mathbf{p}$ of order mv, plus ultrasoft gluons with energy and momentum of order $\Lambda_{\rm QCD}, mv^2$. All the gluon fields are multipole expanded (i.e. expanded in r). The Lagrangian is then given by terms of the type

$$\frac{c_k(m,\mu)}{m^k} \times V_n(r\mu', r\mu) \times O_n(\mu', mv^2, \Lambda_{\rm QCD})\, r^n. \tag{3}$$

where the potential matching coefficients V_n encode the non-analytic behaviour in r. At leading order in the multipole expansion, the singlet sector of the Lagrangian gives rise to equations of motion of the Schrödinger type. We point out that: each term in the pNRQCD Lagrangian has a definite power counting and there is a systematic procedure to calcolate corrections in v to physical observables; higher order perturbative (bound state) calculations in this framework become much simpler [15, 16]. In particular the EFT can be used for a very efficient resummation of the large logs (typically logs of the ratio of energy and momentum scales) using the renormalization group (RG) adapted to the case of correlated scales [17]; retardation (or non-potential) effects start at the NLO in the multipole expansion and are systematically encoded inside pNRQCD. They are typically related to nonperturbative effects [8, 9]; Poincaré invariance is not lost, but shows up in some exact relations among the matching coefficients [14]. We emphasize that inside the EFT framework the renormalon subtraction may be implemented systematically. The renormalon subtraction allows us to obtain convergent perturbative series and to unambiguously define power corrections[9]. This is one of the main reasons for the success of the several applications listed below (for any details see the quoted references).

Applications of weakly coupled pNRQCD

QCD Singlet Static potential. The singlet and octet potentials are well defined objects to be calculated in the perturbative matching. In [16] a determination of the singlet potential at three loops leading log has been obtained and correspondingly also a determination of α_V showing how this quantity starts to depend on the infrared behaviour of the theory at three loops. The perturbative calculation of the static potential at (almost) three loops and with the RG improvement has been compared to the lattice calculation of the potential and found in good agreement up to about 0.35 fm [29].

***b* and *c* masses.**

Heavy quarkonium is one of the most suitable system to extract a precise determination of the mass of the heavy quarks b and c. Perturbative determinations of the $\Upsilon(1S)$ and J/ψ masses have been used to extract the b and c masses. The main uncertainty in these determinations comes from nonperturbative contributions (local and nonlocal condensates [8]) together with possible effects due to subleading renormalons. Table 1 shows some recent determinations. For a discussion about the errors and the difference among the given results see [9]. A recent analysis performed by the QWG [1] and based on all the previous determinations existing in the literature indicates that the mass extraction from heavy quarkonium involves an error of about 50 MeV both in the bottom

(1% error) and in the charm (4% error) case.

Perturbative quarkonium spectrum.

B_c **mass.** Table 2 shows some recent determinations of the B_c mass in perturbation theory at NNLO accuracy compared with a recent lattice study [28] and the value of the CDF experimental B_c mass. This would support the assumption that nonperturbative contributions to the quarkonium ground state are of the same magnitude as NNLO or even NNNLO corrections, which would be consistent with a $mv^2 \gtrsim \Lambda_{\rm QCD}$ power counting. **Hyperfine splittings.** $c\bar{c}$, $b\bar{b}$, B_c ground state hyperfine splittings have been recently calculated at NLL in [31]. The prediction for η_b mass is $M(\eta_b) = 9421 \pm 10\,(\text{th})\,^{+9}_{-8}\,(\delta\alpha_s)$ MeV. The logs resummation seems to be important. If the experimental error in future measurements of $M(\eta_b)$ will not exceed few Mev, the bottomonium hyperfine separation will become a competitive source of $\alpha_s(M_Z)$ with an estimated accuracy of ± 0.003. **Higher bottomonium resonances** have been investigated in the framework of perturbative QCD most recently in [18, 19, 21]. The surprising result of these studies is that some gross features of the lowest part of the bottomonium spectrum, like the approximate equal spacing of the radial levels, are reproduced by a perturbative calculation that implements the leading-order renormalon cancellation. If this is coincidental or reflects the (quasi-)Coulombic nature of the states will be decided by further studies. A recent NLO calculation of the $1P$ bottomonium fine splittings has been performed in [32]. It seems to indicate either the existence of large NLL/NNLO corrections (as it happens in the hyperfine splittings of the $1S$ levels) or sizeable nonperturbative corrections.

Radiative transitions (M1,E1). A theory of M1 and E1 transitions in heavy quarkonium has been recently formulated using pNRQCD [33]. This may shed some light on recent CLEO results on radiative M1 transitions in the η_b search that have ruled out several models.

Seminclusive radiative decays of $\Upsilon(1S)$. In [35] the end-point region of the photon spectrum in semi-inclusive radiative decays of heavy quarkonium has been discussed using Soft-Collinear Effective Theory and pNRQCD. Including the octet contributions a good understanding of the experimental data is obtained

Top-antitop production near threshold at ILC. In [37] the total cross section for top quark pair production close to threshold in e+e- annihilation is investigated at NNLL in vNRQCD. The summation of logarithms leads to a convergent expansion for the normalization of the cross section, and small residual scale dependence. This makes precise extractions of the strong coupling, the top mass and the top width feasible.

Determinations of α_s. Heavy quarkonia leptonic and non-leptonic inclusive decays rates may provide means to extract α_s. The present PDG determination of α_s from heavy quarkonium pulls down the global α_s average noticeably, due to an error that has been largely underestimated [1]. Using the latest development in the calculation of relativistic corrections and in the treatment of perturbative series in α_s it will be possible to obtain a more appropriate determination of α_s from heavy quarkonium.

Gluelump spectrum. In pNRQCD [8, 34] the full structure of the gluelump spectrum has been studied, obtaining model independent predictions on the shape, the pattern, the degeneracy and the multiplet structure of the hybrid static energies for small $Q\bar{Q}$ distances that well match and interpret the existing lattice data.

Properties of baryons made of two or three heavy quarks. Recently the SELEX experiment has detected first signals from three-body bound states made of two heavy quarks and a light one. These systems are theoretically quite interesting due to the interplay of HQET and NRQCD in the construction of a suitable EFT. Triggered by these experimental data in [36] EFT Lagrangians describing QQQ states and QQq states have been constructed and some model independent predictions on the spectra (hyperfine separations) have been obtained.

TABLE 1. Collection of recently obtained values of $m_b^{\overline{\rm MS}}(m_b^{\overline{\rm MS}})$ and $m_c^{\overline{\rm MS}}(m_c^{\overline{\rm MS}})$ from the $\Upsilon(1S)$ and J/Ψ masses.

Ref.	order	$m_b^{\overline{\rm MS}}(m_b^{\overline{\rm MS}})$ (GeV)
[22]	NNLO	4.24 ± 0.09
[23]	NNLO	4.21 ± 0.09
[24]	NNLO	$4.210 \pm 0.090 \pm 0.025$
[19]	NNLO	$4.190 \pm 0.020 \pm 0.025$
[27]	NNNLO	4.349 ± 0.070
[25]	NNNLO	4.20 ± 0.04
[26]	NNNLO	4.241 ± 0.070

Ref.	order	$m_c^{\overline{\rm MS}}(m_c^{\overline{\rm MS}})$ (GeV)
[18]	NNLO	1.24 ± 0.020

TABLE 2. Different perturbative determinations of the B_c mass compared with the experimental value and a recent lattice determination.

B_c mass (MeV)					
State	[30] (expt)	[28] (lattice)	[20] (NNLO)	[18] (NNLO)	[19] (NNLO)
1^1S_0	$6287 \pm 4.8 \pm 1.1$	$6304 \pm 12^{+12}_{-0}$	6326(29)	6324(22)	6307(17)

Strongly coupled pNRQCD

In this case the matching to pNRQCD is nonperturbative [39]. In the situation where the other degrees of freedom (like those associated with heavy-light meson pair threshold production and heavy hybrids) develop a mass gap of order $\Lambda_{\rm QCD}$, the quarkonium singlet field S remains as the only low energy dynamical degree of freedom in the pNRQCD Lagrangian (if no ultrasoft pions are considered), which reads [39, 40, 8, 9]:

$$L_{\rm pNRQCD} = {\rm Tr}\left\{ {\rm S}^\dagger \left(i\partial_0 - \frac{\mathbf{p}^2}{2m} - V_S(r) \right) {\rm S} \right\}. \tag{4}$$

In this regime we recover the quark potential singlet model from pNRQCD. The matching potential V_S (static and relativistic corrections) is nonperturbative: the real part controls the spectrum and the imaginary part controls the inclusive decays. The potential is calculated in the nonperturbative matching procedure between NRQCD and pNRQCD [39, 9]. The great advantages of this approach include: factorization of hard (in the NRQCD matching coefficients) and soft scales (contained in Wilson loops or nonlocal gluon correlators); the fact that the low energy objects are only glue dependent: this opens a window to confinement investigations, on the lattice [41] or in QCD vacuum models [38]; the existence of a clear power counting indicating leading and subleading terms in quantum-mechanical perturbation theory; the fact that the quantum mechanical divergences (like the ones coming from iterations of spin delta potentials) are absorbed by NRQCD matching coefficients; the definitive disappearance of fake problems like the "Lorentz structure of the potentials"; the fact that one no longer needs to repeat a lattice evaluation for each quarkonium state, but gets in one step the full potential (which, in turn, inserted inside the Schrödinger equation, will produce all the quarkonium masses). The calculations involve only QCD parameters (at some scale and in some scheme).

Applications of strongly coupled pNRQCD

Nonperturbative potentials and Spectrum. The final result for the potential (static and relativistic corrections) appears factorized in a part containing the high energy dynamics (and calculable in perturbation theory) which is inherited from the NRQCD matching coefficients, and a part containing the low energy dynamics given in terms of Wilson loops and chromo-electric and chromo-magnetic insertions in the Wilson loop [39]. The expression obtained for the potential *is* the QCD expression, in particular all the perturbative contributions to the potential at the hard scale are correctly taken into account. This solves the problem of consistency with perturbative one-loop calculations that was previously encountered in the Wilson loop approach. Moreover, further contributions, including a $1/m$ nonperturbative potential, appear with respect to the Wilson loop original results [38, 42]. The full expressions for the potentials is given in [39]. Comprehensive phenomenological applications of these full results are still missing.

Decays. The inclusive quarkonium decay widths in pNRQCD can be factorized with respect to the wave function (or its derivatives) calculated in zero, which is suggestive of the early potential models results: $\Gamma(\mathrm{H} \to \mathrm{LH}) = F(\alpha_s, \Lambda_{\mathrm{QCD}}) \cdot |\psi(0)|^2$. Similar expressions hold for the electromagnetic decays. However, the coefficient F depends here both on α_s and Λ_{QCD}. In particular all NRQCD matrix elements, including the octet ones, can be expressed through pNRQCD as products of universal nonperturbative factors by the squares of the quarkonium wave functions (or derivatives of it) at the origin. The nonperturbative factors are typically integral of nonlocal electric or magnetic correlators and thus depending on the glue but not on the quarkonium state [40]. Typically F contains both the NRQCD matching coefficients at the hard scale m and the nonperturbative correlators at the low energy scale Λ_{QCD}. The nonperturbative correlators, being state independent, are in a smaller number than the nonperturbative NRQCD matrix elements and thus the predictive power is greatly increased in going from NRQCD to pNRQCD. In [40] the inclusive decay widths into light hadrons, photons and lepton pairs of all S-wave and P-wave states (under threshold) have been calculated up to $O(mv^3 \times (\Lambda_{\mathrm{QCD}}^2/m^2, v^2))$ and $O(mv^5)$. A large reduction in the number of unknown nonperturbative parameters has been achieved and, therefore, after having fixed the nonperturbative parameters on charmonium decays, new model-independent QCD predictions have been obtained for the bottomonium decay widths [40].

ACKNOWLEDGMENTS

I would like to thank the Organizers of Hadron05 for the perfect organization, the very enjoable atmosphere and the wonderful location.

REFERENCES

1. N. Brambilla *et al.*, CERN Yellow Report, CERN-2005-005, arXiv:hep-ph/0412158.
2. K. Seth; K. Peters; C. Z. Yuan; G. Boca; K. Trabelsi; D. Muller; A. Maciel, plenary talks at this conference.
3. W. E. Caswell and G. P. Lepage, Phys. Lett. B **167** (1986) 437.
4. G. T. Bodwin, E. Braaten and G. P. Lepage, Phys. Rev. D **51** (1995) 1125 [Erratum-ibid. D **55** (1997) 5853].
5. A. H. Hoang, arXiv:hep-ph/0204299; A. V. Manohar and I. W. Stewart, Phys. Rev. D **62**, 014033 (2000); A. H. Hoang and I. W. Stewart, Phys. Rev. D **67**, 114020 (2003).
6. N. Brambilla, arXiv:hep-ph/0012026.
7. A. Pineda and J. Soto, Nucl. Phys. Proc. Suppl. **64**, 428 (1998);
8. N. Brambilla, A. Pineda, J. Soto and A. Vairo, Nucl. Phys. B **566**, 275 (2000).
9. N. Brambilla, A. Pineda, J. Soto and A. Vairo, arXiv:hep-ph/0410047.
10. A. Gray et al. arXiv:hep-lat/0507013.
11. A. Vairo, arXiv:hep-ph/0311303.
12. G.T. Bodwin and Y.Q. Chen, Phys. Rev. **D64**, 114008 (2001); Phys. Rev. **D60**, 054008 (1999).
13. M. Kramer, Prog. Part. Nucl. Phys. **47**, 141 (2001).
14. N. Brambilla, D. Gromes and A. Vairo, Phys. Lett. B **576**, 314 (2003); Phys. Rev. D **64**, 076010 (2001); N. Brambilla and A. Vairo, Nucl. Phys. Proc. Suppl. **74**, 201 (1999).
15. N. Brambilla, A. Pineda, J. Soto and A. Vairo, Phys. Lett. **B470**, 215 (1999); B. A. Kniehl and A. A. Penin, Nucl. Phys. B **563**, 200 (1999).
16. N. Brambilla, A. Pineda, J. Soto and A. Vairo, Phys. Rev. D **60**, 091502 (1999).
17. A. Pineda, Phys. Rev. D **65**, 074007 (2002); A. Pineda and J. Soto, Phys. Lett. B **495**, 323 (2000).
18. N. Brambilla, Y. Sumino and A. Vairo, Phys. Lett. B **513**, 381 (2001).
19. N. Brambilla, Y. Sumino and A. Vairo, Phys. Rev. D **65**, 034001 (2002).
20. N. Brambilla and A. Vairo, Phys. Rev. D **62**, 094019 (2000).
21. S. Recksiegel and Y. Sumino, Phys. Rev. D **67**, 014004 (2003).
22. M. Beneke and A. Signer, Phys. Lett. B **471**, 233 (1999).
23. A. H. Hoang, Phys. Rev. D **59**, 014039 (1999).
24. A. Pineda, JHEP **0106**, 022 (2001).
25. T. Lee, JHEP **0310**, 044 (2003).
26. C. Contreras, G. Cvetic and P. Gaete, Phys. Rev. D **70**, 034008 (2004).
27. A. A. Penin and M. Steinhauser, Phys. Lett. B **538**, 335 (2002).
28. I. F. Allison, C. T. H. Davies, A. Gray, A. S. Kronfeld, P. B. Mackenzie and J. N. Simone [HPQCD Collaboration], arXiv:hep-lat/0411027.
29. A. Pineda, J. Phys. G **29**, 371 (2003); Y. Sumino, arXiv:hep-ph/0505034.
30. D. Acosta *et al.* [CDF Collaboration], arXiv:hep-ex/0505076.
31. B. A. Kniehl, A. A. Penin, A. Pineda, V. A. Smirnov and M. Steinhauser, Phys. Rev. Lett. **92**, 242001 (2004); A. A. Penin, A. Pineda, V. A. Smirnov and M. Steinhauser, Phys. Lett. B **593**, 124 (2004).
32. N. Brambilla and A. Vairo, Phys. Rev. D **71**, 034020 (2005).
33. A. Vairo, talk at this conference; N. Brambilla, Y. Jia, and A. Vairo, preprint IFUM-841-FT.
34. G. S. Bali and A. Pineda, Phys. Rev. D **69**, 094001 (2004).
35. X. Garcia i Tormo and J. Soto, Phys. Rev. D **72**, 054014 (2005).
36. N. Brambilla, A. Vairo and T. Rosch, Phys. Rev. D **72**, 034021 (2005); S. Fleming and T. Mehen, arXiv:hep-ph/0509313.
37. A. H. Hoang, A. V. Manohar, I. W. Stewart and T. Teubner, Phys. Rev. D **65**, 014014 (2002); Phys. Rev. Lett. **86**, 1951 (2001).
38. N. Brambilla and A. Vairo, arXiv:hep-ph/9904330; N. Brambilla, arXiv:hep-ph/9809263.
39. A. Pineda and A. Vairo, Phys. Rev. D **63**, 054007 (2001) [Erratum-ibid. D **64**, 039902 (2001)]; N. Brambilla, A. Pineda, J. Soto and A. Vairo, Phys. Rev. D **63**, 014023 (2001); Phys. Lett. B **580**, 60 (2004).
40. N. Brambilla, D. Eiras, A. Pineda, J. Soto and A. Vairo, Phys. Rev. **D67**, 034018 (2003); Phys. Rev. Lett. **88**, 012003 (2002).
41. G. S. Bali, Phys. Rept. **343**, 1 (2001).
42. N. Brambilla, P. Consoli and G. M. Prosperi, Phys. Rev. D **50**, 5878 (1994); N. Brambilla, G. M. Prosperi and A. Vairo, Phys. Lett. B **362**, 113 (1995).

Exciting Developments in Hadron Spectroscopy

Kamal K. Seth

Department of Physics and Astronomy, Northwestern University, Evanston, IL, 60208, USA

Abstract. There has been a renaissance in hadron spectroscopy during the last couple of years. Long lost states have been tracked down. Unexpected states are showing up all over, and numerous measurements with unprecedented precision are being reported. A review is presented.

Keywords: Hadron Spectroscopy
PACS: 13.66.Bc,13.20.-V,13.25.-K,13.40.Gp

1. INTRODUCTION

Heavy quark (c, b) spectroscopy is the best way of studying Quantum Chromodynamics, QCD. The $|c\bar{c}>$ charmonium and $|b\bar{b}>$ bottomonium states have several advantages over light quark (u, d, s) hadrons. The relativistic effects are much smaller, the strong coupling constant $\alpha_S \approx 0.2 - 0.3$ is small enough to permit use of perturbative methods to a large extent, and the spectra consist of clearly separated narrow states. This is illustrated in Fig. 1 for charmonium.

Over the last three years there has been a renaissance in heavy–quark spectroscopy with genuine discoveries, important precision measurements, and a host of surprises. These are the subject of this talk, but I also refer you to other related talks presented elsewhere in these proceedings [1].

2. DISCOVERY OF THE LONG–LOST SINGLET STATES OF CHARMONIUM

As is well known, the central $q\bar{q}$ interaction is well represented by the so–called Cornell potential (or its variants) as the sum of a one–gluon exchange Coulombic potential proportional to $1/r$ and a not–well–understood confinement potential, generally taken as Lorentz scalar and proportional to r (Fig. 2). Far less is known about the spin dependence of the $q\bar{q}$ potential which is responsible for the spin–orbit and spin–spin splittings of the $q\bar{q}$ states. The crucial determinant of the spin dependent $q\bar{q}$ interaction is the hyperfine spin–singlet/spin–triplet splitting of the states. Unfortunately, despite numerous and valiant efforts, no spin–singlet states have ever been successfully identified in bottomonium, and we have to depend exclusively on charmonium. In charmonium the singlet S–state $\eta_c(1^1S_0)$ and the triplet S–state $J/\psi(1^3S_1)$ have been known for a long time, with the hyperfine splitting, $M(J/\psi) - M(\eta_c) = 117 \pm 1$ MeV. The other two bound singlet states η_c', or $\eta_c(2^1S_0)$, and $h_c(^1P_1)$ have long eluded successful identification. Well, now they have been finally discovered.

CP814 *Hadron Spectroscopy: Eleventh International Conference*
edited by A. Reis, C. Göbel, J. Sá Borges, and J. Magnin

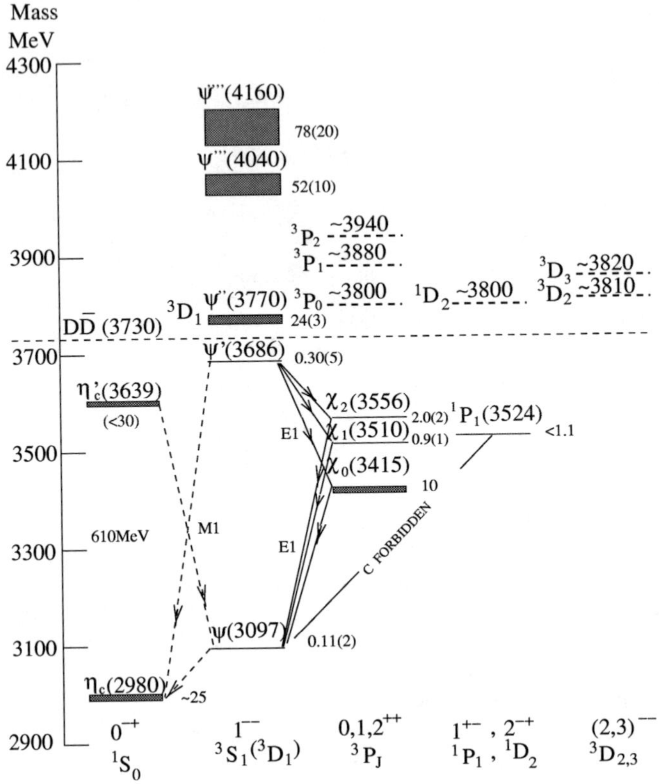

FIGURE 1. Spectra of the states of Charmonium.

2.1 The Radial Excitation of the Charmonium Ground State, η_c', or $\eta_c(2^1S_0)$

As shown in Fig. 2, the S-wave radial excitations, ψ' and η_c' sample the confinement potential almost exclusively. As a result, they provide our only insight into how the hyperfine interaction varies in going from $1S$ states (η_c and J/ψ), with an almost 50–50 exposure to Coulombic and confinement potentials, to $2S$ states (η_c' and ψ'), which lie in the confinement potential territory.

The mass of ψ' is extremely well known, $M(\psi') = 3686.093 \pm 0.034$ MeV, but $M(\eta_c')$ was unknown. None of the earlier attempts to identify η_c', which included Crystal Ball [2], Fermilab E760/E835 [3], DELPHI [4], L3 [5], and CLEO [6] were successful. The breakthrough came from an unexpected source, Belle [7]. Since then several experiments have confirmed the identification of η_c', and their results are summarized in Table 1. Fig. 3 shows the results of CLEO [8] and BaBar [9] for the reaction $\gamma\gamma \to K_S K \pi$. While it is clear that η_c' has been successfully identified, its mass still has substantial uncertainty, and its width remains unmeasured. Using the world average $M(\eta_c') = 3628.3 \pm 2.1$ MeV, we get $\Delta M_{hf}(2S) = 47.8 \pm 2.1$, which means $\Delta M(2S)/\Delta M(1S) = 0.41 \pm 0.02$. Although a potential model calculation can always be found with just about any result, it is a fact

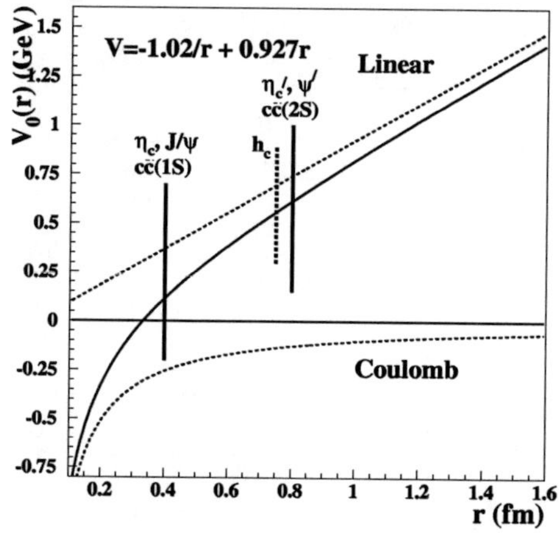

FIGURE 2. Schematic of the QCD $q\bar{q}$ potential (solid line), and its Coulombic and confinement parts (dotted lines) seperately. The approximate locations of the $1S$, $2S$ and $^1P_1$ states are shown.

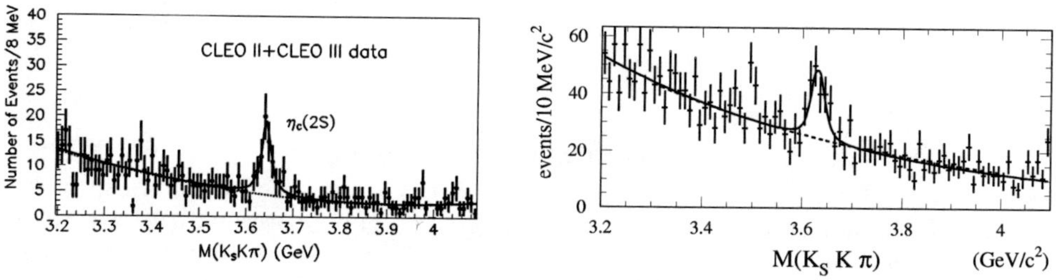

FIGURE 3. Observation of η_c' in the reaction $\gamma\gamma \to K_S K\pi$ by CLEO (left) and Babar (right).

that most predictions were that $\Delta M(2S)/\Delta M(1S) \approx 0.65$. So, here we have a result for theory to digest. On the experimental side we have still much work to do, the first being to find the width of η_c'. Together with $\Delta M_{hf}(2S)$, it will shed light on whether we are observing effects related to the confinement interaction or configuration mixing.

2.2 The Singlet P-wave State of Charmonium, $h_c(1^1P_1)$

The singlet P–wave $h_c(1^1P_1)$ relates to a different aspect of the hyperfine interaction than $\eta_c(2^1S_0)$ does. Here the question is whether or not there is a long range $\vec{s}_1 \cdot \vec{s}_2$ interaction in the $q\bar{q}$ system. In absence of such an interaction, the lowest order pQCD prediction is that the hyperfine, singlet–triplet splitting is finite only for S–wave states, and is identically zero for all higher L, i.e. $\Delta M_{hf}(L \neq 0) = 0$. In particular, $\Delta M_{hf}(1P) = \langle M(^3P_J)\rangle - M(^1P_1) = 0$. The centroid of the $^3P_J(J = 0, 1, 2)$ states has been measured

TABLE 1. Summary of η_c' observations.

Measurement	N(cts)	M–2640 MeV	Width (MeV)
Belle [7] (2002): $B \to K(K_S K\pi)$	39 ± 11	14 ± 10	< 55
Belle [7] (2005): $J/\psi + \lvert c\bar{c}>$	311 ± 42	-14 ± 9	–
BaBar [9] (2004): $J/\psi + \lvert c\bar{c}>$	121 ± 27	5 ± 6	22 ± 14
BaBar [9] (2005): $\gamma\gamma \to K_S K\pi$	112 ± 24	-9 ± 4	17 ± 9
CLEO [8] (2004): $\gamma\gamma \to K_S K\pi$	61 ± 15	$+3 \pm 3$	6 ± 14

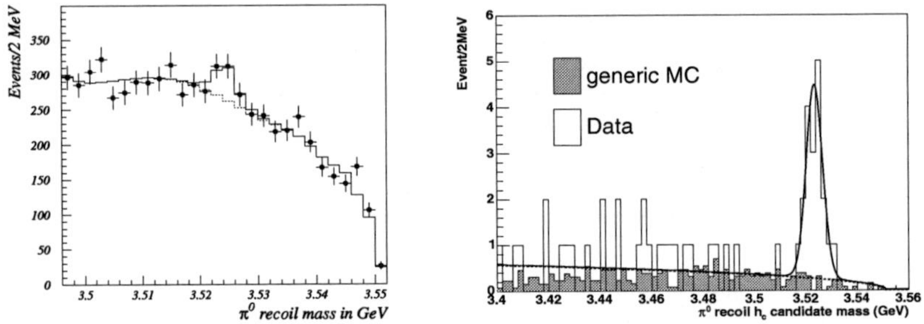

FIGURE 4. Observation of $h_c(1^1P_1)$ in (left) inclusive analysis and (right) exclusive analysis at CLEO [11].

very accurately by the Fermilab E835 experiment, $\langle M(^3P_J)\rangle = 3525.36 \pm 0.06$ [10]. What is needed is a firm identification of $h_c(1^1P_1)$ and a precise measurement of its mass. CLEO [11] has just reported the unambigous (6σ) observation of h_c in a "tour de force" measurement of the reaction $\psi(2S) \to \pi^0 h_c$, $h_c \to \gamma\eta_c$. Both inclusive and exclusive measurements of $M(h_c)$ in recoils against π^0 have been made. In the inclusive measurement, either E_γ or $M(\eta_c)$ are constrained. In the exclusive measurement, η_c is instead identified in seven hadronic decay channels. Both inclusive and exclusive spectra are shown in Fig. 4. The result is $\Delta M_{hf}(1P) = +1.0 \pm 0.6 \pm 0.4$ MeV (for more details, see my parallel session talk). E835 [12] has also just reported the results of the analysis of their 1996+2000 data for the reaction $\bar{p}p \to h_c \to \gamma\eta_c$. They report $\Delta M_{hf}(1P) = -0.4 \pm 0.2 \pm 0.2$ MeV at a significance level of $\sim 3\sigma$.

Two conclusions can be derived from these measurements. The first is that the "naive" pQCD predication $\Delta M_{hf}(1P) = 0$ is not being violated in any substantial manner, as had been feared by many theorists. The second conclusion is that the all–important sign and the small magnitude of $\Delta M_{hf}(1P)$ is not yet pinned down. In the near future, larger ψ' running at CLEO is expected to reduce the errors in the CLEO measurement, and GSI, whenever it becomes operational, may be expected to do even better.

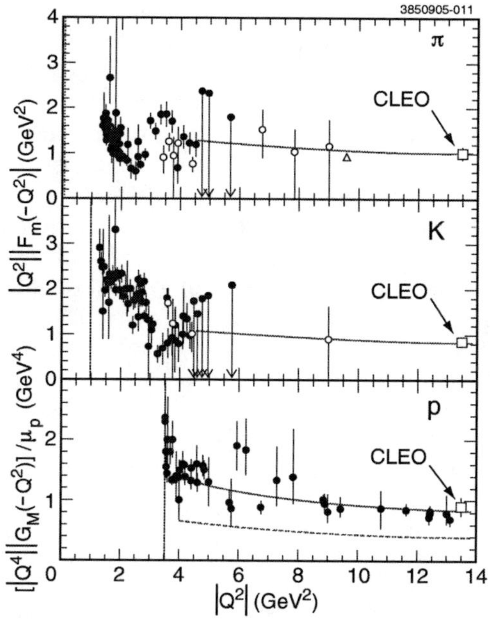

FIGURE 5. CLEO results for the timelike form factors at $Q^2 = 13.48$ GeV^2 of the pion (top), kaon (middle), and proton (bottom).

3. A SAMPLING OF PRECISION CHARMONIUM RESULTS

Spectroscopy does not progress only by discoveries. It makes some of its best gains by precision. Recently, several such gains have been made. At Novosibirsk [13], masses of J/ψ and ψ' have been measured at ± 12 keV and ± 25 keV levels respectively; a precision of ~ 5 parts per million. At Fermilab [10], masses of the χ_J states have been measured with precision of 50–100 parts per million and widths to better than $\pm 10\%$. At CLEO [14, 15], lepton universality has been confirmed at a $\pm 1\%$ level in J/ψ decays to e^+e^- and $\mu^+\mu^-$. At CLEO [15], isospin conservation has been confirmed at a $\pm 2\%$ level in ψ' decay to $\pi^+\pi^- J/\psi$ and $\pi^0\pi^0 J/\psi$, and isospin violation has been observed at a $\pm 0.4\%$ level in ψ' decay to $\pi^0 J/\psi$ and $\eta J/\psi$. Also at CLEO [16, 17], the long standing discrepancy between two–photon widths of the χ_2 state determined by two–photon fusion and by two–photon decay has been resolved by new precision measurements of $\Gamma_{\gamma\gamma}(\chi_{c2})$ and $\mathscr{B}(\chi_{c2} \to \gamma J/\psi)$.

4. TIMELIKE FORM FACTORS OF PION, KAON, AND PROTON

Electromagnetic form factors of hadrons provide deep insight into their quark structure and help define the domain of validity of pQCD. Except for the magnetic form factor of the proton for spacelike momentum transfers, few measurements of spacelike or timelike form factors of any hadrons exist at large enough momentum transfers to shed light on the highly controversial debate on the validity of pQCD at modestly large momentum transfers. The Fermilab [18] discovery that the timelike form factor of the proton is nearly a factor two larger than the spacelike form factor of the proton in the entire region $Q^2 = 5 - 13$ GeV^2 tells us to expect surprises in the measurement of other form factors. Unfortunately, the sparse data which exist for the form factors of pions and kaons (which often could not be separately identified) are essentially limited to $Q^2 \leq 4$ GeV, and have very large (up to 100%) errors. This situation has been remedied recently by a very demanding measurement made at CLEO [19] for $Q^2 = 13.48$ GeV^2. To get an idea of how difficult these measurements are, it is enough to point out that at this momentum, e^+e^- collisions produce about 500 times more muon pairs than pion pairs, and one must distinguish between them. The results of this CLEO measurement with less than ±10% errors are shown in Fig. 5.

5. THE SUPRISING STATES

Let me move on to the unexpected states which seem to be popping up all over during the last two years. The veterans among these are X(3872) and the Pentaquark. The newcomers are X(3943), Y(3943), Z(3931), and V(4260). This proliferation is exciting, but also rather baffling. It arises primarily from the fact that huge integrated e^+e^- luminosities (≥ 300 fb^{-1}) are now available at Belle and BaBar, and very weak resonances are showing up. It will be a while before the dust settles down and we really know what is going on. Let me describe these surprises, and their current status, one by one.

5.1 The Pentaquark

As is probably well known to most of the audience, when the Pentaquark was born in 2003, it caused great interest. Google tells me that there are 31,300 entries for it by now. There were many reported sightings of pentaquarks of all kinds, and even a greater number of reported failures to find the expected signals. Finally, there is the recent JLab report of the absence of the pentaquark signal in a large statistics repeat of their earlier measurement. This reminds me of a similar history of claimed observations of dozens of dibaryons, which all eventually evaporated [20]. My personal, perhaps biased, conclusion is that the pentaquark is now on life–support. May it rest in peace! See the talk by E. Smith [21] for more information.

5.2 The X(3872) Mystery

In 2003 Belle [22] announced the discovery of an unexpected state, X(3872). It was quickly confirmed by CDF [23], DØ [24] and BaBar [25] (see Fig. 6). The decay X(3872)$\rightarrow \pi^+\pi^- J/\psi$ is the dominant decay. The average of the masses measured by the four experiments is $M(\mathrm{X}) = 3871.5 \pm 0.4$ MeV. Note that $M(D^0) + M(\overline{D^{*0}}) = 3870.3 \pm 2.0$ MeV [26]. The best measurement of the width gives $\Gamma(\mathrm{X}) \leq 2.3$ MeV.

The unique decay, the narrow width, and the closeness of its mass to $M(D^0D^{*0})$ have given rise to intense theoretical speculations about the nature of X(3872). Theoretical speculations are that X(3872) is a charmonium state ($1^{++}, 2^{--}, 3^{--}$), a hybrid (1^{++}), a glueball mixed with vector charmonium (1^{--}), a D^0D^{*0} ($1^{++}, 0^{-+}$) molecule? Perhaps

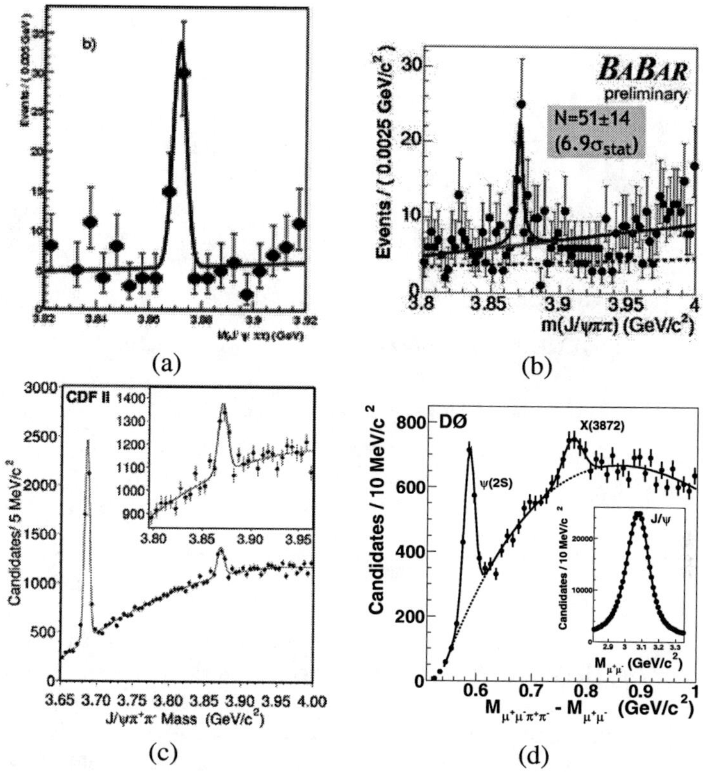

FIGURE 6. Observations of X(3872) by (a) Belle [22], (b) CDF [23], (c)DØ [24], and (d) BaBar [25].

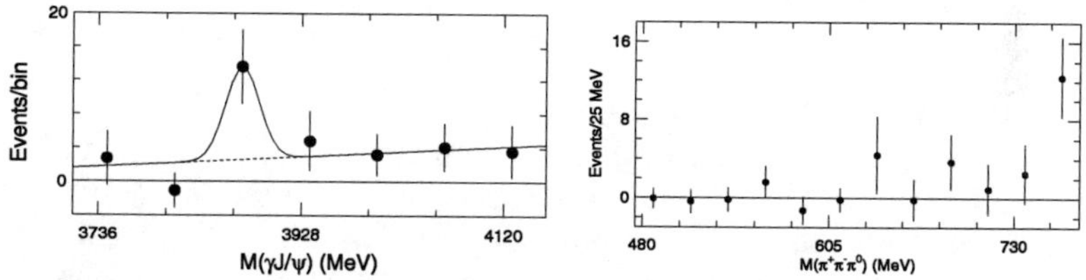

FIGURE 7. Observations by Belle [30] of $X \to \gamma J/\psi$ (left) and $X \to \omega J/\psi$ (right).

the most provocative of these proposals is the molecular model, because no $|q\bar{q}> \cdot |q\bar{q}>$ molecules have ever been found! To sift through these speculations it is necessary to determine $J^{PC}(X)$. Frantic searches are in progress at Belle and BaBar for decays to establish $J^{PC}(X)$.

This state continues to defy understanding. CLEO [27], BES [28], and Babar [29] have put the limit, $\mathscr{B}(X \to \pi^+\pi^- J/\psi)\Gamma(X \to e^+e^-) < 6$ eV, which, depending on how large $\mathscr{B}(X \to \pi^+\pi^- J/\psi)$ is, might weigh in against X being a vector. However,

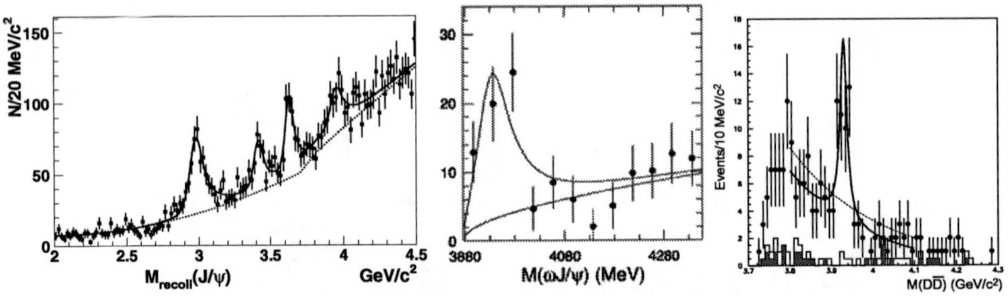

(a) X(3943) (b) Y(3943) (c) Z(3931)

FIGURE 8. Obervations of three new states by Belle.

TABLE 2. Summary of properties of three new states observed by Belle [31, 32, 33].

	M(MeV)	Γ(MeV)	Formed in	Decays in	not in	suggests
X [31]	$3943 \pm 6 \pm 6$	15 ± 10	$e^+e^- \to J/\psi(c\bar{c})$	$D^*\bar{D}$	$D\bar{D}$, $\omega J/\psi$	?
Y [32]	$3943 \pm 11 \pm 13$	87 ± 22	$B \to K(\omega J/\psi)$	$\omega J/\psi$	$D^*\bar{D}$(?)	$c\bar{c}$ hybrid?
Z [33]	$3931 \pm 4 \pm 2$	$20 \pm 8 \pm 3$	$\gamma\gamma$ fusion	$D\bar{D}$		$\chi'_{c2}(2^3P_2)$

DØ finds X $\to \pi^+\pi^- J/\psi$ decays to have all the same characteristics as the vector $\psi(2S)$ decays to $\pi^+\pi^- J/\psi$. Belle [30] has presented arguments against the charge parity of X being negative on the basis of observations of X $\to \gamma J/\psi$ and $X \to \omega J/\psi$. However, as shown in Fig. 7, the two observations consist of 13.6 ± 4.4 and 12.1 ± 4.1 counts (each in one mass bin), respectively. CDF [23] reports the $\pi^+\pi^-$ mass distribution to be consistent with X being a $^3S_1$ vector with $C = -1$, or a $C = +1$ object decaying into $\rho J/\psi$. Finally, potential model calculations do not really rule out $^3D_2(2^{--})$ and $^3D_3(3^{--})$ states at 3872 MeV, In view of all this, I personally believe that all options are still open for X(3872).

5.3 The Saga of X, Y, Z at $M \approx 3940$ MeV

There are three new states reported by Belle with masses which are statistically consistent with being identical [31, 32, 33]. The spectra in which they were observed are shown in Fig. 8, and their characteristics are summarized in Table 2. Each is formed in a different reaction and decays dominantly in a different channel. X(3943) is observed in a recoil mass spectrum in which only $J = 0^\pm$, $C = +$ states $\eta_c(0^{-+})$, $\eta'_c(0^{-+})$, $\chi_{c0}(0^{++})$ are seen. This would suggest $J(\text{X}(3943)) = 0^\pm$, $C = +$. The state Z(3931) with width similar to that of X(3943) is produced in $\gamma\gamma$ fusion, which guarentees $C = +$, but is found to have the $D(\bar{D})$ angular distribution characteristic of $J(\text{Z}(3931)) = 2$. The fitted width of Y(3942) is claimed to be ~ 4 times larger than that of X or Z, and its decays are almost opposite to those of X(3943). All this produces a very confusing picture because it appears to be rather implausible that three distinct states exist within a 10 MeV mass interval. But Nature can always spring surprises! In any case, we can look forward to additional data, particularly from BaBar, to sharpen the picture.

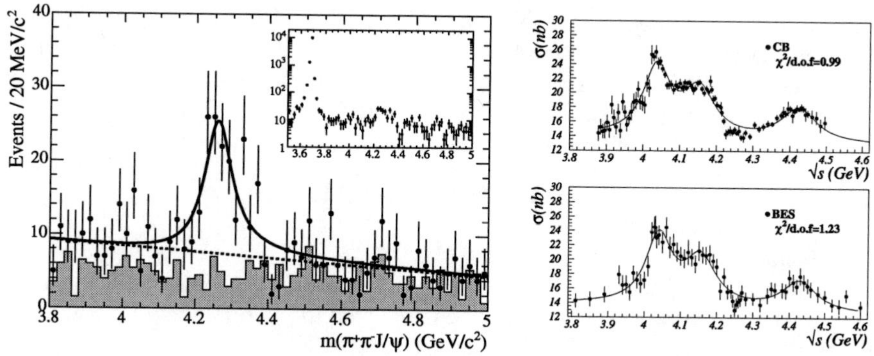

FIGURE 9. (left) Observation by Babar [29] of V(4260) enhancement in ISR, (right) $\sigma(e^+e^- \to \text{hadrons})$ [34] with minimum at $\sqrt{s} = 4260$ MeV.

5.4 Three is Not Enough—Bring in the Fourth, V(4260)

Belle has been running away with too many new resonances. Now comes BaBar. They have analyzed ISR events from 211 fb^{-1} of data [29], and report a convincing (significance 8σ), broad enhancement in the invariant mass $M(\pi^+\pi^- J/\psi)$ spectrum (Fig. 9). Since production via ISR guarentees a vector, and X, Y, and Z have been overused, I take the liberty of christening this state as V(4260). The parameters of this enhancement are $M(\mathrm{V}) = 4259 \pm 8^{+2}_{-6}$ MeV, Γ(V)=$88 \pm 23^{+6}_{-4}$ MeV, $N = 125 \pm 23$ events. From this,

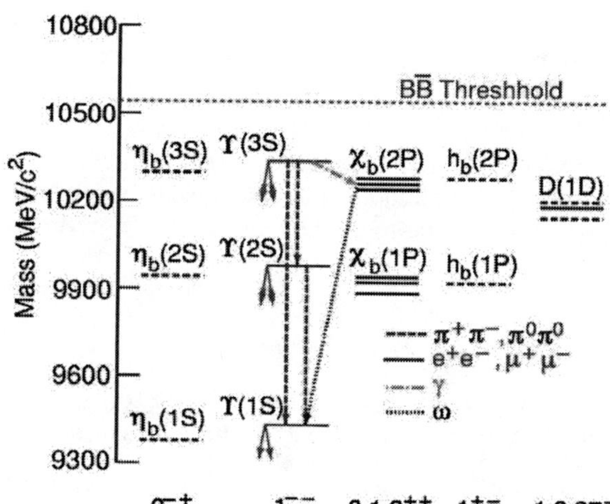

FIGURE 10. Spectra of the bound states of Bottomonium.

BaBar measures $\Gamma(e^+e^- \to \text{V}) \times \mathcal{B}(\text{V} \to \pi^+\pi^- J/\psi) = (5.5 \pm 1.0^{+0.8}_{-0.7})$ eV. They suggest that it might be a previously unobserved 1^{--} resonance. This is quite surprising because no vector around this mass is predicted, and the R measurements actually show a dip in this mass region as illustrated in the adjoining figure from Ref. [34].

6. BOTTOMONIUM

The Bottomonium spectrum is shown in Fig. 10.

The world's largest sample of 21 million $\Upsilon(1S)$, 9 million $\Upsilon(2S)$, and 6 million $\Upsilon(3S)$ comes from CLEO. These data sets have been analyzed to yield interesting new results and improved precision.

- For the first time, a non–$\pi\pi$ hadron transition between bottomonium resonaces has been observed [35], with $\mathcal{B}(\chi'_{b1} \to \omega\Upsilon(1S)) = (1.63 \pm 0.38)\%$ and $\mathcal{B}(\chi'_{b2} \to \omega\Upsilon(1S)) = (1.10 \pm 0.34)\%$.
- The 1^3D_2 state of bottomonium has been identified in a 4–step cascade with $M = 10161.1 \pm 0.6 \pm 1.6$ MeV [36].
- $\Upsilon(1S)$ decays to $X + (J/\psi, \psi(2S), \chi_{c1}, \chi_{c2})$ have been measured [37]. The measured branching fraction is $B(\Upsilon(1S) \to J/\psi + X) = (6.4 \pm 0.4 \pm 0.6) \times 10^{-4}$. The measured branching fraction ratios to $B(\Upsilon(1S) \to J/\psi + X)$ for $\Upsilon(1S) \to (\psi(2S), \chi_{c1}, \chi_{c2}) + X$ are $0.41 \pm 0.11 \pm 0.08$, $0.35 \pm 0.08 \pm 0.06$, and $0.52 \pm 0.12 \pm 0.09$, respectively.
- Precision measurements of $\Upsilon(1S, 2S, 3S) \to \mu^+\mu^-$ have been made, with the result that $\mathcal{B}(\Upsilon(2S) \to \mu^+\mu^-)$ and $\mathcal{B}(\Upsilon(3S) \to \mu^+\mu^-)$ are 56% and 32% larger, respectively, than their current PDG values [38].
- The radiative decays of bottomonium 1P and 2P states have been measured with improved precision [39]. While the branching ratios for $\chi_{bJ}(1P)$ states are found to be in good agreement with the current PDG values, those for $\chi_{bJ}(2P)$ are found to be $\sim 30 - 40\%$ larger.

Postscript for bottomonium: No new data taking at the bound bottomonium resonances is expected at any e^+e^- collider, although much remains to be explored. As an example, none of the spin singlet states of bottomonium, not even the ground state $\eta_b(1S)$, have been identified so far.

REFERENCES

1. For related talks at the conference see plenary session contributions elsewhere in these proceedings by E. Smith (JLab), C. Z. Yuan (BES), K. Trabelsi (Belle), D. Muller (BaBar), and A. Maciel (Tevatron). For parallel session contributions, see M. Negrini, T. Hry'nova, and D. Aston (BaBar), X. Mo, Y. Zhu, X. Shen, and J. Shan (BES), S. Korpar and R. Lesiak (Belle), K. Seth and V. Crede (CLEO), P. Gavillet (DELPHI), and D. Tedeschi and J. Hansen (JLab).
2. Crystal Ball, Phys. Rev. **D 34** (1986) 711.
3. E760 Collaboration, Phys. Rev. **D 52** (1995) 4839; E835 Collaboration, Phys. Rev. **D 64** (2001) 052003.
4. DELPHI Collaboration, Phys. Lett. **B 441** (1998) 479.

5. L3 Collaboration, Phys. Lett. **B 461** (1999) 155.
6. CLEO Collaboration, Phys. Rev. **D 70** (2004) 112002.
7. Belle Collaboration, Phys. Rev. Lett. **89** (2002) 102001; hep-ex/0507019.
8. CLEO Collaboration, Phys. Rev. Lett. **92** (2004) 142001.
9. BaBar Collaboration, Phys. Rev. Lett. **92** (2004) 142002; Phys. Rev. **D 72** (2005) 031101.
10. E835 Collaboration, Nucl. Phys. **B 717** (2005) 34.
11. CLEO Collaboration, Phys. Rev. Lett. **95** (2005) 102003.
12. E835 Collaboration, Phys. Rev. **D 72** (2005) 032001.
13. KEDR Collaboration, Phys. Lett. **B 573** (2003) 63.
14. CLEO Collaboration, Phys. Rev. **D 71** (2005) 111103(R).
15. CLEO Collaboration, Phys. Rev. Lett. **94** (2005) 232002.
16. CLEO Collaboration, Phys. Rev. **D 70** (2004) 112002.
17. CLEO Collaboration, hep-ex/0510033, submitted to Phys. Rev. Lett..
18. E760 Collaboration, Phys. Rev. Lett. **70** (1993) 1212; E835 Collaboration, Phys. Rev. **D 60** (1999) 032002; E835 Collaboration, Phys. Lett. **B 559** (2003) 20, and references therein.
19. CLEO Collaboration, hep-ex/0510005, submitted to Phys. Rev. Lett.
20. "Dibaryons in Theory and Practice", Kamal K. Seth, Invited Paper International Conference on Medium- and High-Energy Nuclear Physics, Taipei 1988.
21. E. Smith, "Review on JLab Results", elsewhere in these proceedings.
22. Belle Collaboration, Phys. Rev. Lett. **91** (2003) 262001.
23. CDF Collaboration, Phys. Rev. Lett. **93** (2004) 072001.
24. DØCollaboration, Phys. Rev. Lett. **93** (2004) 162002.
25. BaBar Collaboration, Phys. Rev. **D 71** (2005) 071103.
26. PDG Collaboration, Phys. Lett. **B 592** (2004) 1.
27. CLEO Collaboration, Phys. Rev. Lett. **94** (2005) 032004.
28. BES Collaboration, Phys. Lett. **B 579** (2004) 74.
29. BaBar Collaboration, Phys.Rev.Lett. 95 (2005) 142001.
30. Belle Collaboration, hep-ex/0505037; hep-ex/0505038.
31. Belle Collaboration, hep-ex/0507019.
32. Belle Collaboration, Phys. Rev. Lett. 94 (2005) 182002.
33. Belle Collaboration, hep-ex/0507033.
34. K. K. Seth, Phys. Rev. **D 72** (2005) 017501.
35. CLEO Collaboration, Phys. Rev. Lett **92** (2004) 222002.
36. CLEO Collaboration, Phys. Rev. **D 70** (2004) 032001.
37. CLEO Collaboration, Phys. Rev. **D 70** (2004) 072001.
38. CLEO Collaboration, Phys. Rev. Lett **94** (2005) 012001.
39. CLEO Collaboration, Phys. Rev. Lett **94** (2005) 032001.

Spectroscopy of D Mesons

Stefano Bianco

Laboratori Nazionali di Frascati dell'INFN
v.E. Fermi 40, 00044 Frascati (Rome) Italy

Abstract. The scenario of heavy quark meson spectroscopy underwent recently a major revolution, after the observation of BABAR and CLEO, confirmed by BELLE, of D_{sJ} L=1 excited states, and by further evidences by SELEX. These experimental results have cast doubts on the incarnations of the ideas of Heavy Quark Effective Theory in heavy quark spectroscopy. I shall review the status of experimental data, discuss implications and sketch an outlook.

Keywords: spectroscopy; hadronic physics; charm quark
PACS: 14.40.Lb; 14.65.Dw; 14.65.Fy; 14.40.Nd

INTRODUCTION

This paper reports on recent experimental results on D meson spectroscopy, discussing the recent events that brought to cast doubts to our current understanding of the overall picture. I shall discuss excited non-strange D mesons, namely the observation of $j_q = 1/2$ broad states, the revolutionary observations of excited strange D_{sJ} mesons which are forcing us to switch the paradigm of HQ spectroscopy, discuss the status of debated $D_{sJ}(2632)$ meson observed by SELEX at Fermilab, finally sketch an outlook and draw conclusions. For a detailed review on charm physics including spectroscopy the reader is referred to Ref.[1], for other charm spectroscopy issues such as charmonium states etc. see other up-to-date reviews such as [2, 3].

Let me pay a tribute to cosmic ray physicists and remind how the — possibly — very first D meson ($D^+ \to K^+\pi^0$) was observed by human eye in nuclear emulsions exposed to cosmic rays in 1971 [4]. The story of charm discovery in nuclear emulsions is narrated in Ref. [5] After 35 years, here is where we are.

HEAVY-LIGHT QUARK SPECTROSCOPY, THE GLOBAL PICTURE

A global interpretation scheme for heavy quark meson spectroscopy is provided by the idea of Heavy Quark Symmetry (HQS). In the infinite heavy-quark mass limit, the heavy-light meson can be described as formed by a the still heavy quark, with all the orbital degrees of freedom being due to the light quark. This means that good conserved quantum numbers are the spin of the heavy quark, and the angular momentum j_q.

Experimentally, for each of the $c\bar{u}$, $c\bar{d}$ and $c\bar{s}$ systems four P-wave and two $n = 2$ radial excitations have been studied. There are four $L = 1$ states, namely two with $j_q = 1/2$ and total spin $J = 0, 1$ and two with $j_q = 3/2$ and $J = 1, 2$. These four states are

CP814 *Hadron Spectroscopy: Eleventh International Conference*
edited by A. Reis, C. Göbel, J. Sá Borges, and J. Magnin

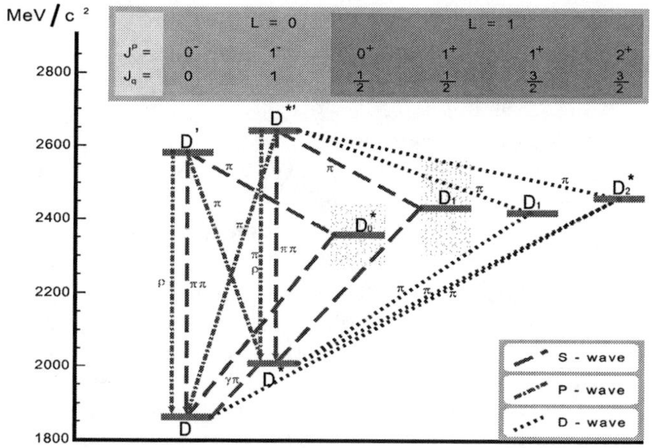

FIGURE 1. *Masses and transitions predicted for the excited non-strange D meson states.*

named respectively D_0^*, $D_1(j_q = 1/2)$, $D_1(j_q = 3/2)$ and D_2^* (Fig.1). Parity and angular momentum conservation force the $(j_q = 1/2)$ states to decay to the ground states via S-wave transitions (broad width), while $(j_q = 3/2)$ states decay via D-wave (narrow width). To be more specific, for the $1/2$ one predicts widths of $\sim$ 100 MeV and for the $3/2$ of about $\sim$ 10 MeV with the exception of the $D_{s1}(j_q = 3/2)(2536)$ which is kinematically forced to a $\sim$ 1 MeV width.

Therefore, the HQS picture has two consequences, which turn to be direct predictions:

1. each L level is split in two J-degenerate doublets, in each doublet one broad and one narrow state;
2. flavour symmetry does exist. In principle in the heavy-quark infinite mass limit one is allowed to use the same chart tfor $c\bar{q}, c\bar{s}, b\bar{q}, b\bar{s}$ mesons, just changing quark labels and absolute mass energy scale.

However, the HQS paradigm was recently put in discussion by BABAR's and CLEO's discovery of D_{sJ} states.

L=1 NON-STRANGE EXCITED D MESONS

All six $L = 1, j = 3/2$ non-strange *narrow* states are well established, with precisions on masses at the 1 MeV level and on widths at the few MeV level. This is due to the fact that excited D states are abundantly produced both at FT experiments, in e^+e^- continuum production, in B decays and at the Z^0 [6].

Table 1 shows the experimental data available for $c\bar{q}$ L=1 mesons, masses and widths, as showing on Ref.[7] updated to 2005, as well as recent measurements not appearing in PDG world averages. In bold I listed measurements that are somehow new or debated.

Let me first of all mention a long-standing dilemma, the $D^{*\prime}$. Called $D^*(2640)^{\pm}$ by PDG, the first L=1 radial excitation was seen by DELPHI seen by DELPHI in the

$D^{*+}\pi^-\pi^+$ final state; it has not been confirmed by any experiment (OPAL, CLEO, ZEUS). Final disproof or confirmation is needed, and should be considered as a relatively easy task considered the level of statistics currently available to contemporary experiments.

The status of the *broad* L=1 states is not clear at all, as well. The assignments of the quantum numbers are largely based on theory expectations for their masses and widths. In 1998 CLEO [8] showed evidence for the $D_1(j_q = 1/2)$ broad state. Two results, by BELLE [9] and photoproduction experiment FOCUS [10], have appeared in 2003 and are now included in the average of PDG 2005. BELLE have studied the $D^{*+}\pi^-$ and $D^+\pi^-$ final states, while FOCUS have studied both isospin channels $D^+\pi^-$ and $D^0\pi^+$. They both claim observation for broad states. Due to the presence of feeddown satellite peaks due to missing neutral kinematics, FOCUS do not claim conclusively that the broad state observed is the D^{*0} predicted by HQS. The mass values found are in disagreement at the $\sim 2\sigma$ level, and consistent with many predictions out of the huge number of papers n the subject. The BELLE mass value is notably close to what predicted a long time ago [11]. More experimental results are needed.

New players in the D meson spectroscopy game could be the experiments at hadron colliders, which have greatly improved charm physics capabilities with impact paramenter trigger which uses silicon vertex detectors. As instance, CDF at the Fermilab Tevatron showed results in 2003 [12] with high statistic peaks of L=1 mesons sitting on huge combinatoric backgrounds, due to high multiplicity of primary interaction vertex. Clearly, hadroproduction is not the best place to look for L=1 mesons. However, more recent unpublished results [13] show great improvements, with $D\pi$ distributions clearly evidencing clean L=1 mesons peaks. We expect interesting news from CDF and D0, possibly at this same conference.

L=1 STRANGE EXCITED D MESONS, OR: NEED TO CHANGE PARADIGMA OF D SPECTROSCOPY ?

Before Spring 2003 we thought we could use the same $c\bar{q}$ chart for $c\bar{s}$, thanks of flavour symmetry of HQS. The narrow D_{s1} and D^*_{s2} states were very well established, and we would expect the two missing broad $c\bar{s}$ states to lie somewhere above the DK and D^*K threshold, respectively.

Instead, surprisingly enough, BABAR finds[14] a prominent peak at 2317 MeVin $D_s\pi^0$ with width compatible to experimental resolution. They also find another narrow peak in $D^*_s\pi^0$, but are not sure whether it is a reflection or not, therefore do not claim observation for a second state. The analysis is complicated by the presence of two reflections from undetected neutrals. Following BABAR announcement, Cleo looks back to circa-1995 data, at the time when they published the first evidence for isospin-violation decay $D^*_s \to D_s\pi^0$. At that time they had much less statistics, now they integrate all events and they also confirm[15] the BABAR state. By availing of a more trained analysis they find and interpret correctly the $D^*_s\pi^0$ state at 2463 MeVas another new state. Belle joins the club by finding evidence[16, 17] of the $D_{sJ}(2463) \to D_s\gamma$ and determines the J^{PC}. A detailed historical account is reported in [1].

TABLE 1. Summer 2005 status of (L=1, n=1) and (L=0, n=2) $c\bar{q}$ mesons (MeV). Statistical and systematical errors added in quadrature, unless noted. Preliminary data from CDF are from [12][13].

j_q	1/2	1/2	3/2	3/2	1/2	1/2
J^P	0^+	1^+	1^+	2^+	0^-	1^-
L,n	1,1	1,1	1,1	1,1	0,2	0,2
	$D_0^*(2400)$	$D_1(2430)$	$D_1(2420)$	$D_2^*(2460)$	D'	$D^{*\prime}[D^*(2640)]$
Decay Mode	$D\pi$	$D^*\pi$	$D^*\pi$	$D\pi, D^*\pi$		$D^*\pi\pi$
			Mass (MeV)			
PDG 0	**2352 ± 50**	2427 ± 36	2422 ± 2	**2461.1 ± 1.6**		
PDG ±	2403 ± 38		2427 ± 5	**2464.9 ± 3.0**		2637 ± 7
CDF prel. ±				**2463.6 ± 2.7 ± x**		
CDF prel. 0			**2421.7 ± 0.9**	**2463.3 ± 1.0**		
			Width (MeV)			
PDG 0	**261 ± 50**	384 ± 117	19 ± 4	**32 ± 4**		
PDG ±	283 ± 42		28 ± 8	29 ± 5		< 15
CDF prel. 0			**20 ± 2**	**49 ± 3**		
			Isospin Mass Splitting(MeV)			
PDG			$4^{+2}_{-3} \pm 3$	**2.4 ± 1.7**		

It seems natural to interpret $D_{sJ}^*(2317)$ and $D_{sJ}^*(2463)$ as 0^+ and 1^+ states, respectively. The decay distributions are consistent with such assignments, yet do not establish them. They together with the mass values would explain the narrow widths: for $D_{s1+}^*(2463) \to DK$ is forbidden by parity, $D_{s0+}^*(2317) \to DK$ and $D_{s1+}^*(2463) \to DK^*$ by kinematics and $D_{s0+}^*(2317) \to D_s^+\pi^0$ and $D_{s1+}^*(2463) \to D_s^{*+}\pi^0$ are isospin violating transition and thus suppressed. Also $D_{s0+}^*(2317) \to D_s^+\gamma$ is forbidden.

There are three puzzling aspects to these states:

- Why have no other decay modes been seen ? In particular CLEO places a low upper bound

$$BR(D_{s0+}^*(2317) \to D_s^{*+}\gamma) < 0.078 \;\; 90\% \; C.L. \qquad (1)$$

Why is it not more prominent, when $D_{s0+}^*(2317) \to D_s\pi^0$ is isospin violating ?

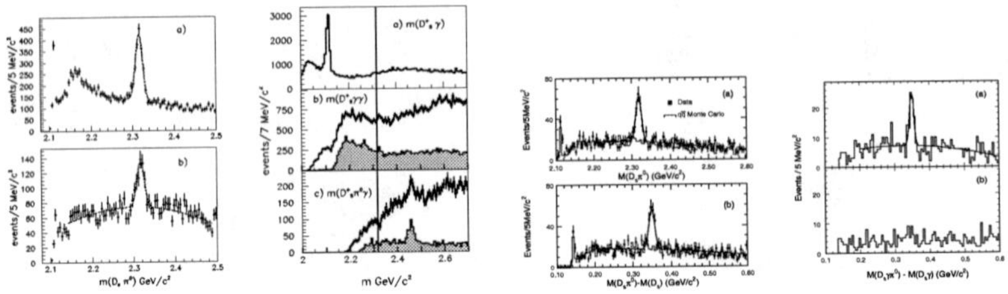

FIGURE 2. *New $D^*_{s0+}(2317)$ and $D^*_{s1+}(2463)$ states observed by BABAR (a,b) [14] and CLEO (c,d) [15].*

- Why are their masses so much below predictions ? One should note that a deficit of ~ 160 and ~ 100 MeV is quite significant on the scale of $M(D^*_{sJ}) - M(D)$. Why is the mass splitting to the previously found narrow states $D_{s1}(2536)$ and $D_{sJ}(2573)$ so much larger than anticipated ?
- A related mystery is the following: where are the corresponding *non*-strange charm resonances ? They should be lighter, not heavier than $D^*_{s0+}(2317)$ and $D^*_{s1+}(2463)$.

Table 2 PDG 2005 entries are dominated by the BABAR measurements. Unpublished results not on PDG are the observation of the 2317 MeVstate by FOCUS, worth to be mentioned because it is the only observation of a DsJ state outside e^+e^- colliders, and some results on 1^+ states.

What is really new is a couple of results from BELLE [20, 19] presented at (northern hemisphere) Summer conferences, most notably measurement of BR,and observation of a nonresonant decay of $D^+_{s1}(2536)$. BELLE have studied the decay $\bar{B^0} \to D_{sJ}(2317)^+K^-$ which is an interesting decays because quark content of final state is totally different from B meson, suggesting non-trivial decay mechanisms: W-exchange, final state interactions, tree diagram if the D_{sJ} has a 4-quark structure. The new measurement improves previous low-statistics results. BELLE find a very large isospin breaking, namely that the rate for $\bar{B}^0$ decays to $D_{sJ}(2317)$ is about three times larger than rate to $D_{sJ}(2460)$. BELLE also study the resonant structure of $D_{s1}(2536)^+ \to D^+\pi^-K^+$ decay, finding a small but non-zero fraction of non-resonant $D^+\pi^-K^+$ component relative to resonant $D^{*+}K^0_s$. Besides, they studied the presence of an S-wave component, which may give informationon mixing between the two newly discovered 1^+ states. I expect BELLE to report on this at this Conference.

THE $D_{sJ}(2632)^+$

Following the discovery of D_{sJ} states by BABAR and CLEO, SELEX (fixed target hadroproduction at Fermilab with Σ^- and π^- beams) looked for signals in strangess-rich channels with a charm meson, such as $D^+_s\eta, D^0K^+$ [21]. They found evidence for a state at 2632 MeV, with a width $\Gamma < 17$ MeV. They also found a very strong isospin breaking, *i.e.*, the D^0K^+ is severely depressed with respect to $D^+_s\eta$.

TABLE 2. Summer 2005 status of (L=1, n=1) and (L=0, n=2) $c\bar{s}$ mesons (MeV). Statistical and systematical errors added in quadrature, unless noted. Preliminary results are from FOCUS[18] and BELLE[19].

j_q	1/2	1/2	3/2	3/2	1/2	1/2
J^P	0^+	1^+	1^+	2^+	0^-	1^-
L,n	1,1	1,1	1,1	1,1	0,2	0,2
Decay Mode	$D^*_{sJ}(2317)$ $D_s^+\pi^0$	$D_{sJ}(2460)$ $D_s^+\gamma, D_s^{*+}\pi^0, D_s^+\pi^+\pi^-$	$D_{s1}(2536)$ $D^*K, D\pi K$	$D^*_{s2}(2573)$ DK	D'_s	$D^{*\prime}_s$
Mass (MeV)						
PDG ±	2317 ± 0.6	2458.9 ± 0.9	2535.35 ± 0.6	2573.5 ± 1.7		
FOCUS prel. ±	2323 ± 2		2535.1 ± 0.3	2567.3 ± 1.4		
$BR(D\pi K)/BR(D^*K)$						
BELLE prel. ±			$2.8 \pm 0.2 \pm 0.4\%$			
Width (MeV)						
PDG ±	<4.6	<5.5	<2.3 90 % cl	15 ± 5		
FOCUS ±			1.6 ± 1.0	28 ± 5		

Given the interest of the SELEX results, quite immediately all other active charm experiments looked for confirmation. Photoproduction experiment FOCUS looked [22] in D^+K_s, D^0K^+, BABAR[23] in $D_s^+\eta, D^0K^+, D^{*+}K_s$, BELLE [24] in $D_s^+\eta, D^0K^+$. All three experiment saw no evidence. Unless a peculiar production mechanism related to the hyperon beam is in place here, one should consider the SELEX evidence not confirmed. Results from hadron beam experiments (CDF/D0, and COMPASS at CERN) would be useful to shed light and revive the case for the $D_{sJ}(2632)$.

CHANGING PARADIGMA OF HQ SECTROSCOPY - A PLETHORA OF IDEAS

Needless to say, the BABAR and CLEO discoveries spurred a plethora of theory papers. My personal list of favourite topics sees the idea [25] in top position: combine HQS and chiral invariance, form doublets by pairing (D_s^+, D_s^{*+}) with $(D_{s0^+}^*(2315), D_{s1^+}(2460))$. By applying chiral dynamics they find that the splitting between doublets should follow the prediction, indeed verified

$$\Delta M \equiv M(D_{s0^+}^*(2315)) - M(D_s)\; M(D_{s1^+}(2460)) - M(D_s^*)\; m_N/3 \tag{2}$$

An interesting comment was made [26] on the relative production rate of 3/2 versus 1/2 states. Sum rules predict dominance of 3/2 states (such as D_1 and D_2^*) versus 1/2 states (such as broad D_0^* and D_1). Since experimentally the opposite is observed, the authors suggest the discrepancy be reconciled with lower mass 1/2 states, compatible to those found by BABAR and CLEO.

As for the SELEX evidence, it was noted[28] how, if the SELEX $D_{sJ}(2632)$ state was confirmed experimentally, the very strong isosping breaking could be explained by a 4q structure $[cd][ds]$.

OUTLOOK AND CONCLUSIONS

We should be aware of the exciting era we are living, at least as far heavy quark spectroscopy is concerned. A lot of new results from all charm experiments active today have urged the need for a critical revision of the basic assumptions in the paradigma used so far.

The discoveries of D_{sJ} states by BABAR and CLEO ask for a critical revision of the HQS paradigma. BELLE entered the game with confirmation of states, new decay modes, and a flurry of new results. Non-strange broad states have now been established, with FOCUS and BELLE confirming the 1998 evidence by CLEO. The PDG average for the newly observed states sums up mass values in mild disagreement, more data is needed and results should come soon. The intriguing evidence of $D_{sJ}(2632)$ by SELEX is not confirmed by FOCUS, BABAR, BELLE. We could be experiencing a peculiar production mechanism connected to the strangeness-rich beam, or simply a statistical fluctuation. There is real opportunity for hadron beam experiments (CDF/D0 at Tevatron, COMPASS at CERN) to say the last word on the issue. As a lot of work is being done presently world-wide, we should expect a wealth of new results in plenary (Mueller, Trabelsi, Maciel) and parallel (Kopar, Poireau, Cumalat) session talks.

Where are we going next ? Of course the list of open problems is fairly large, just to quote some:

- establish the non-strange broad states. In particular all channels with neutrals are unobserved so far;
- measure the widths of D_{sJ} states;
- measure the relative production of 1/2 versus 3/2 states;

- solve the mystery of the existence of the radial excitations.
- investigate the *Terra Incognita*: the beauty L=1 mesons, verify the little data available, mainly dating back to LEP, and check the validity of flavour symmetry, if any.

Most of this shopping list will be addressed by experiments at B-factories. Hadroproduction (CDF/D0) may contribute, as well fixed target (COMPASS at CERN). In the far future I see only SuperBELLE as a player, after the cancellation of the flavour programme in the US. LHC-b seems to be ruled out by the choice of not triggering on charm decays. PANDA [29] will be a major player in charmonium spectroscopy, but seems to me problematic in taming the huge minimum bias background in the search of charm decay verteces. As for the B spectroscopy sector, which is crucial to verify the extent to which one can still apply flavour symmetry, it should be playground of LHC-b at CERN. In any case, the field literally bursts with enigmas, and loads of good data are coming in.

ACKNOWLEDGMENTS

I sincerely thank the Organizers for a perfectly enjoyable meeting. I acknowledge enlightening discussions with D. Asner, I. Bigi, G. Boca, S. Chung, J. Cumalat, A. Maciel, Bigi S. Paul, A. Reis, K. Seth.

REFERENCES

1. S. Bianco, F. L. Fabbri, D. Benson and I. Bigi, Riv. Nuovo Cim. **26N7** (2003) 1 [arXiv:hep-ex/0309021].
2. J. L. Rosner, arXiv:hep-ph/0508155.
3. K.Seth, *these proceedings.*
4. K. Niu, E. Mikumo, Y. Maeda, *Prog. Theor. Phys.* **46** (1971) 1644.
5. The story of the discovery of charm in Japan is narrated in: K. Niu, Proceed. Of the First Intern. Workshop on Nucl. Emulsion Techn., June 12-14, 1998, Nagoya, Japan, preprint DPNU-98-39.
6. P. Roudeau, "Charm, beauty and tau physics at LEP," LAL-97-96 *Talk given at Int. School of Physics, 'Enrico Fermi': Heavy Flavor Physics - A Probe of Nature's Grand Design, Varenna, Italy, 8-18 Jul 1997*
7. S. Eidelman *et al.*, Phys. Lett. **B592**, 1 2004 and 2005 web updates.
8. S. Anderson *et al.* [CLEO Coll.], Nucl. Phys. A **663**, 647 (2000).
9. K. Abe et al., Phts. Rev. **D69** 112002.
10. J.M. Link et al., Phys. Lett. **B586**, 11-20 (2004).
11. T. Matsuki and T. Morii, Phys. Rev. D **56** (1997) 5646 [Austral. J. Phys. **50** (1997) 163] [arXiv:hep-ph/9702366].
12. M. Shapiro, presented at FPCP 2003, Paris.
13. CDF Coll., CDF-Note-7191, available at `www-cdf.fnal.gov`.
14. BABAR Collab., B. Aubert *et al.*, hep-ex/0304021.
15. CLEO Collab., D. Besson *et al.*, hep-ex/0305017.
16. K. Abe *et al.*, arXiv:hep-ex/0307052.
17. K. Abe *et al.* [Belle Collaboration], arXiv:hep-ex/0307041.
18. E.Vaandering (for the FOCUS Coll.), hep-ex/0406044.
19. BELLE Coll., hep-ex/0507030.
20. A. Drutskoy (for the BELLE Coll.), HEP05 (EPS) Lisboa (Portugal) 2005. BELLE-Conf-0516, 0519.
21. A.V.Evdokimov et al., Phys. Rev. Lett. **93**, 242001 (2004).
22. R.Kutschke (FOCUS Coll.), E831-doc-701-v2, available at `www-focus.fnal.gov`.

23. BABAR Coll., hep-ex/0408087.
24. B. Yabsley (BELLE Coll.), hep-ex/0507028.
25. W. Bardeen, E. Eichten, C. Hill, Phys. Rev. **D68** 054024,2003.
26. N. Uraltsev, hep-ph/0406086.
27. F. Jugeau, A. Le Yaouanc, L. Oliver, J-C. Raynal, hep-ph/0504206.
28. L. Maiani, F. Piccinini, A.D. Polosa, V. Riquer, Phys. Rev. **D70** 054009, 2004.
29. G. Boca (for the PANDA Coll.), *these Proceedings*.

Hadron physics at FAIR

K. Peters

Planckstr. 1, 64291 Darmstadt, Germany

Abstract. To deepen our understanding of the properties and structure of matter, new large-scale facilities are under construction or in preparation worldwide that provide for novel research capabilities in many areas of science.

The GSI-Darmstadt laboratory together with the international user community, has developed the concept for such a facility to broadly address frontiers in strong-interaction physics and in the general research with intense ion beams. The new facility builds, and substantially expands, on the present accelerator system. Primary beams of high-energy ions, and secondary beams including antiprotons and beams of short-lived nuclei, together with an intricate system of beam-storage and experimental rings, are its key features.

This report provides a short overview of the science motivation of the hadron community for the facility, as well as of its technical layout and performance characteristics.

Keywords: FAIR, hadron physics, heavy ion reactions, hadron spectroscopy
PACS: 25.75Nq;14.40Gx;14.40Lb;14.40Cs;29.20Lq;29.20Dh

THE FAIR FACILITY

The proposal for the construction of an international Facility for Antiproton and Ion Research (FAIR) was presented by GSI in 2001. The scientific and technical concept for FAIR is based on extensive discussions about the long-term perspectives of research with beams of heavy highly charged ions. Based on these discussions, a first proposal for an 'International Accelerator Facility for Beams of Ions and Antiprotons' was proposed in 2000 followed by a 'Conceptual Design Report' comprising more than 700 pages was presented. The 'Wissenschaftsrat der Bundesrepublik Deutschland' (Research Council of the Federal Republic of Germany) evaluated this proposal together with eight other applications of large-scale facilities for scientific research. The GSI proposal was recommended for funding to the BMBF (Bundesministerium für Bildung und Forschung, Federal Ministry for Education and Research), with the conditions (i) to present a staged plan for the realization of the project and (ii) to raise part of the required funds from foreign sources. The BMBF followed the recommendation of the Wissenschaftsrat and made the following decision in February 2003: The Gesellschaft für Schwerionenforschung (GSI) in Darmstadt shall build the new facilities together with European partners and become a leading European centre for physics. At least one quarter of the investments shall be raised by partners outside Germany.

The main cornerstones for the hadron physics programme at FAIR are the investigation of hadrons in dense baryonic media and the hadron physics being pursued in proton-antiproton annihilations.

CP814 *Hadron Spectroscopy: Eleventh International Conference*
edited by A. Reis, C. Göbel, J. Sá Borges, and J. Magnin

THE CBM EXPERIMENT

The Compressed Baryonic Matter (CBM) experiment is designed to explore the QCD phase diagram in the region of moderate temperatures but very high baryon densities. The phase boundary between quark-gluon matter and hadronic and the location of the critical endpoint is predicted by recent lattice QCD calculations which indicate that for values of μ_B larger than about 400 MeV the phase transition is first order, whereas for μ_B smaller than 400 MeV there is a smooth cross over from the hadronic to the partonic phase.

The CBM research program includes the search for the deconfinement phase transition at high baryon densities, the study of chiral symmetry restoration in superdense baryonic matter, the search for the critical endpoint, and the study of the nuclear equation of state at high densities. The most promising diagnostic probes are particles with open and hidden charm, low-mass vector mesons decaying into dilepton pairs, (multi-)strange particles, eventby- event fluctuations, and collective flow.

The major experimental challenge is posed by the extremely high reaction rates of up to 10^7 events/second. These conditions require unprecedented detector performances concerning speed and radiation hardness. The detector signals are processed by a high-speed data acquisition and an online event selection system. The different detector components are dicussed in the subsequent sections.

Silicon Tracker System

The Silicon Tracking System (STS) serves for track measurement and for determination of primary and secondary vertices. The current STS layout consists of minimum 7 layers and is placed inside a magnetic dipole field which provides the bending power required for momentum determination with an accuracy of $\delta p/p = 1\%$. The STS has to fulfill the following requirements: material budget below 0.3% radiation length per layer to reduce multiple scattering, hit resolution of about 10 μm to achieve a vertex resolution of about 50 μm along the beam axis, radiation hardness up to a dose of 50 MRad corresponding to the dose accumulated in ten years of running, and read-out times of about 25 ns to accommodate reaction rates of 10 MHz. The first 2 stations (distance from the target 5 and 10 cm) will consist of pixel detectors in order to reduce the occupancy to about 1% (for a central Au+Au collision at 25 AGeV). Monolithic Active Pixel Sensors (MAPS) with a pixel size of $40\times40\,\mu m^2$ and a thickness of 100 μm would perfectly fulfill our requirements concerning vertex resolution which is needed to measure the displaced vertices of D mesons. However, the radiation hardness and the read out speed of nowadays MAPS detectors have to be improved by 1–2 orders of magnitude in order to match our requirements. Moreover, we work on track reconstruction methods to resolve the tracks and vertices of up to 50 superimposed events in the MAPS detectors. The ongoing development of low-mass hybrid pixel detectors for future LHC experiments may lead to a substantial reduction of their material budget. These detectors - which are fast and radiation hard - might then serve as a fall back solution for the CBM Vertex Tracker. The next 5 Silicon layers will consist of Silicon microstrip detectors. The current layout foresees a pitch of 25 m and three different strip lengths of 20, 40 and 60 mm. The strips are arranged such that the occupancy is below 2% for a central Au+Au collision at 25 AGeV. The detectors will be double sided with a stereo angle between the strips which has to be optimized by simulations. The ambiguities due to double (or triple) hits having a distance smaller than the strip length result in 2 (or 6) ghost hits which pose a severe problem to the track reconstruction algorithms. It might be necessary to add more Silicon layers with a different orientation of the strips in order to minimize the ambiguities. Another possibility would be to reduce the length of the strips with the consequence of increasing the number of channels. The investigation of this problem is in progress.

Ring Imaging Cherenkov detectors

The RICH detector is designed to provide identification of electrons and suppression of pions in the momentum range of electrons from low-mass vector-meson decays. The actual layout of the RICH detector consists of a radiator (3 gases under study: N_2, 40% He + 60% CH_4, and 50% N_2 + 50% CH_4, length 2.2 m), a mirror (3 mm Be covered with 0.5 mm glass, radius 4.5 m), and a photon detector composed of about 100.000 photomultiplier (PM) channels (6 mm diameter, quantum efficiency 20%). The glass window of the PMs is covered with wave-length shifter (WLS) films in order to increase the absorption of Cherenkov photons. A crucial task is to match the rings to the charged particle tracks. If the track position at the mirror can be determined with an accuracy of 200 μm, and assuming a momentum resolution of 1%, the mismatch of pions to electron rings is less than 10.3 per event. This number will be considerably improved when taking into account particle identification by time-of-flight measurement and by the TRD. Performance studies are in progress, taking into account track reconstruction and more realistic detector properties, in order to optimize the detector layout.

Transition radiation detector

Three Transition Radiation Detector (TRD) stations will serve for particle tracking and for the identification of high energy electrons and positrons (>2000) which are used to reconstruct J/ψ mesons. According to simulations which are based on the experience obtained with the development of the TRD for ALICE and of the TRT for ATLAS, pion suppression factors of up to 200 (for momenta above 2 GeV/c) at an electron efficiency of better than 90% can be achieved.

The major technical challenge is to develop highly granular and fast gaseous detectors which can stand the highrate environment of CBM in particular for the inner part of the detector planes covering forward emission angles. For example, at small forward angles and at a distance of 4 m from the target, we expect particle rates of more than 100 kHz/cm^2 for 10 MHz minimum bias Au+Au collisions at 25 AGeV. In a central collision, particle densities of about 0.05/cm^2 are reached. In order to keep the occupancy below 5% the size of a single cell should be about 1 cm^2.

Within our R&D activities we investigate both the ALICE-TRD and the ATLAS-TRT concept. Both detector versions have to be modified in order to meet the CBM requirements. In the case of the ALICE-TRD, the rate capability of the read-out chambers has to be improved. Various prototypes of fast Multi Wire Proportional Chambers (MWPC) and Gas Electron Multipliers (GEM) have been tested with proton and pion beams up to intensities of 100 kHz/cm^2 and no major deterioration of the performance has been observed. The high-rate performance is required for less than 30% of the active TRD area (which is about 500 m^2 in total). A detector concept for CBM has been studied on the basis of straw tubes (ATLAS-TRT option). Here, the challenge is to build very small straw

tubes which still provide a high efficiency in the regions of highest occupancy. A possible detector concept might be to use an MWPC-based TRD for the inner part of the detectors where the high rates occur, and a straw-based TRT for the outer regions.

Resistive plate chambers

An array of Resistive Plate Chambers (RPC) will be used for hadron identification via TOF measurements. The TOF wall is located about 10 m downstream of the target and covers an active area of about 120 m^2. The required time resolution is about 80 ps. For 10 MHz minimum bias Au+Au collisions the innermost part of the detector has to work at rates up to 20 kHz/cm^2. At small deflection angles the pad size is about 5 cm^2 corresponding to an occupancy of below 5% for central Au+Au collisions at 25 AGeV. With a small-size prototype a time resolution of about 90 ps has been achieved at a rate of 25 kHz/cm^2. Future R&D concentrates on the rate capability, low resistivity material, long term stability, and realization of large arrays with overall excellent timing performance.

Electromagnetic calorimeter

The electromagnetic calorimeter will be used to measure direct photons, neutral mesons decaying into photons, electrons and muons. Simulations and R&D have been started based on the shashlik type of detector modules as used in HERA-B, PHENIX and LHCb. Particular emphasis is put on a good energy resolution and a high pion suppression factor.

THE PANDA EXPERIMENT

The PANDA collaboration plans to build a state-of-the-art universal detector for strong interaction studies at the high-energy storage ring HESR at the international FAIR facility. The detector is designed to take advantage of the extraordinary physics potential, which will be available utilizing high intensity, phase space cooled antiproton beams.

Spectroscopy has been a prime tool for physics in the last century and has played a leading role in the development of quantum mechanics and the standard model of physics. Quantum Chromo Dynamics (QCD) is generally accepted to be the correct underlying theory of the strong interaction. However, our knowledge of the behavior at large distances is still rather primitive. Spectroscopy experiments within hadron physics are the tool to investigate both in the dynamics governing the interaction of fundamental particles and the existence of new forms of matter. The latter could consist of gluonic degrees of freedom, like glueballs and hybrids, previously undiscovered charmonium states, the extension of the nuclear chart into the strangeness dimension, or particles produced inside nuclear matter. Only when we can predict, confirm, and explain the physical states of the theory, we can hardly claim that we understand the strong interaction. Such a deep understanding might have farther- reaching implications in particle physics. For example, it is quite possible that the weak interactions becomes strong at high energies and field theories of the strong interaction may be relevant to the mechanism of the electroweak symmetry breaking. In strong-interaction studies based on QCD we have the ideal laboratory to test our understanding of theory against experimental results.

Significant progress beyond the present understanding can only be made if the statistics and precision of the data and their analysis exceeds past efforts by several orders of magnitude. The PANDA detector will be optimized for the physics goals as discussed above. At the same time, the detection system should be able to accommodate additional physics aspects, like Drell-Yan and CP-violating processes. The PANDA program carries outstanding discovery potential and excellent opportunities for the investigation of additional physics topics.

Placing a detector inside a storage ring allows for optimal use of precious antiprotons but poses severe experimental challenges. The number of antiprotons is limited and the required high-luminosity can only be achieved by an appropriately attuned target. Triggering on decaying charm particles requires precise micro vertex tracking close to the target or lepton identification, dependent on the final state. The PANDA detector has an advanced particle identification system. Charged particle tracking together with a high-resolution electromagnetic calorimeter are mandatory for spectroscopy experiments. The almost full coverage of the solid angle in PANDA is achieved by a combination of a target spectrometer with a forward spectrometer. The complex detector arrangement is to assure the measurement of complete sets of parameters and signatures. The procedure warrants a robust and redundant physics reconstruction which is considered particularly important in a high-luminosity environment.

The PANDA Detector

The design of PANDA is based on previous experience in antiproton experiments and takes advantage of ongoing detector developments performed at the high-energy laboratories for experiments. The general layout of PANDA is based on two magnetic spectrometers. The target spectrometer surrounds the interaction region and has a superconducting solenoid as momentum analyzer. A forward opening of 5–10° (depending on the respective axis) allows high momentum tracks to enter the forward spectrometer with a large-gap dipole magnet. The superconducting solenoidal magnet has a length of

2.5 m, a diameter of 1.9 m and provides an axial field of 2 T. It has to accommodate a gap for pipes of the target device. The target spectrometer is then made up of the following components:

The interaction point is surrounded by a micro vertex detector (MVD), which has five barrel shaped layers plus five disk-shaped detectors in forward direction. The three innermost layers are composed of pixel detectors to achieve best resolution and to be able to easily detect decay vertices displaced from the interaction point. The outer layers are composed of microstrip detectors which are easier to handle. The baseline technology chosen for the pixel detectors are hybrid active pixel sensors as used by several LHC experiments. The electronics still has to be modified to accommodate continuous readout. As alternatives to silicon pixels, GaAs based detectors are considered as well as much thinner monolithic pixel sensors where the problem of radiation hardness would have to be solved.

The MVD is surrounded by a cylindrical tracker. Two options are currently discussed, a straw tube tracker (STT) consisting of 15 double layers of self-supporting straws and a time projection chamber (TPC) with continuous readout. The TPC is the technically more challengingoption since it requires an ungated charge collection based on a GEM readout. However it has the benefit of less material and offers in addition particle identification via dE/dx. On the other hand, the STT is seen as a safe fall-back solution which should still fulfil the basic tracking requirements.

In the forward direction circular or octagonal mini drift chambers are used to track particles with higher momenta before they enter the forward spectrometer.

The next detector is a Cherenkov counter based on the DIRC principle as used in BaBar at SLAC. It consists of quartz rods in which Cherenkov light is internally reflected to an array of photon detectors in the backwards direction. The readout can be either done by imaging a 2D pattern of reflections with a large number of PMTs or APDs or by measuring just one coordinate and the time of light propagation inside the quartz very precisely. In forward direction a disk-shaped Cherenkov counter with quartz radiator and detectors for internally reflected light similar as for the barrel DIRC is planned. Its readout should be located between the solenoid coil and the return yoke to allow the calorimeter end cap to be as close as possible.

An electromagnetic calorimeter is placed outside the DIRC. It consists of a barrel part with 11.360 crystals, a forward end cap with 6.864 crystals and a backward end cap with 816 crystals. As detector material $PbWO_4$ is foreseen, since it is fast and has a reasonable resolution. However, it might require cooling down to -25°C. As alternative BGO with still higher light yield albeit slower signals is considered. Finally, outside the superconducting magnet and its iron return yoke, drift tubes track muons exiting the spectrometer. The necessity of a time-of-flight detector in the target spectrometer is still under discussion. Although the flight path is short, low momentum particles could still be identified.

The heart of the forward spectrometer is a dipole magnet with a large opening and a field integral of 2 Tm. This will provide the required momentum resolution for forward tracks with momenta up to 8 GeV/*c*. One problem of the dipole magnet is the deflection of the beam. If included in the beamline design it can be countered by a magnet chicane. Otherwise the beam could be shielded from the magnetic field, introducing some material in the acceptance.

The detector systems in the forward spectrometer are summarized in the following:

Tracking is provided by mini drift chambers. Before the magnet, they have the same octagonal shape as in the end cap of the target spectrometer. After the magnet, a rectangular shape matches is more suitable for the spread of tracks. The use of straw tube trackers inside the dipole field is considered for better momentum resolution. The option of a third Cherenkov counter, based on gas or aerogel is still under investigation. In addition, a time-of-flight detector is considered for charged particle identification.

An electromagnetic calorimeter based on lead/scintillator sampling and WLS fibre readout (Shashlyk type) is foreseen in the forward spectrometer. It should have 276 channels to cover the acceptance. After the electromagnetic calorimeter, the MIRAC detector from WA80 at CERN is under investigation as a hadron calorimeter after refurbishing its readout.

The last piece of equipment is a muon detector based on drift tubes.

The selected readout and trigger concept foresees continuous digitization of all detector channels. Special trigger hardware is not foreseen. The readout electronics has to be fully pipelined and has to perform autonomously the detection of valid hits as well as intelligent data reduction by clusterization, signal shape analysis, and time reconstruction to transfer only the physically relevant minimum of information. All data are marked by a synchronous timestamp by which event building can be performed at a later stage. The data are fed into a configurable network. Attached computer nodes with the required bandwidth digest the full rate of certain sub-detectors used for the first selection level. Other detectors transfer data only for time slices selected on some level. Parallel selections are possible and several levels will bring the data rate down to the required 100–200 MB/s to mass storage. The clear advantage of this scheme is its flexibility: Any conceivable physics signature of an interesting process measurable by the PANDA detector could be converted into a sequence of selection algorithms. Typical signatures are the decays of J/ψ to leptons, decay vertices of charm or strange hadrons, identified kaons or electromagnetic showers, etc.. The level of complexity is only limited by the required processing power and can be scaled up by attaching more nodes.

The trigger system and the universality of the detector make PANDA a unique tool in hadron physics, ready for future challenges.

PROJECT ORGANIZATION AND STATUS

Organization as an International Project

Since the CDR, considerable progress has been made in both the scientific-technical preparatory work as well as the legal and planning work. A call for letters of intent (LoI) was launched in October 2003. 25 LoIs have been submitted by more than 1750 authors in Spring of 2004. In June of 2004, three programme advisory committees, one on QCD physics (QCD-PAC), one on nuclear structure and astrophysics (NUSTAR-PAC) and another one on atomic physics, plasma physics, applied physics (APPA-PAC) have evaluated the 25 LoIs. The technical advisory committee (TAC) evaluated the ongoing accelerator and infrastructure plans. 23 experiment collaborations obtained the green light to present a Technical Proposal January 15th 2005, the accelerator groups were asked to present Technical Reports detailing the status of the design, planning and costs of the FAIR sub-projects. In close collaboration with the BMBF, a staged plan for the realization has been established.

In February 2004, an international steering committee (ISC-FAIR) was established for FAIR in order to coordinate the international involvement in the organizational and technical development of FAIR. Two working groups have been created to support the ISC-FAIR: a) a working group for Scientific and Technical Issues (STI-FAIR), b) a working group on Administrative and Financial Issues (AFI-FAIR).

The FAIR member countries and a number of countries with observer status are represented in the ISC, and in the STI and AFI working groups, by delegates appointed by their research ministries or funding agencies.

The task of the AFI-FAIR working group includes the development of a suitable legal framework for the new company, the preparation of a Memorandum of Understanding (MoU) as a formal base for the cooperation during the socalled 'preparatory phase', i.e. up to the start of construction (expected in 2007), and the preparation of contracts between the partners for construction, commissioning and operation of the FAIR facility. For an estimate of construction costs, a further working group, the AFI-FCS (Full Cost Structure) working group has been established. Its task is the definition of contributions (deliverables) to the project, the accounting of in-kind contributions etc. After the mode for cost determination has been established, a fixed cost will be established for each work package which forms the base for calculating contributions by the partners. For the legal entity, the current legal status as a German limited liability company (GmbH), expanded by international shareholders, is favoured. The Convention, Articles of Association, and By- Laws are currently being worked out by the AFI-LFI working group (Legal Frame Issues). Up to January 2005, the MoU mentioned above has been signed by 10 countries (Finland, France, Germany, Greece, Italy, Poland, Russia, Spain, Sweden, United Kingdom). Some of the current formal observers (China, Hungary, India, United States of America) may also considering this step in the future. Other countries are invited to join.

Time Schedule for the Stage Plan

A three-stage plan for the FAIR-project construction has been developed together with the BMBF. It follows from the following considerations: First, at the end of each stage, a major advance in research capabilities is achieved and allows the start-up of a corresponding partial research program. Second, civil construction of buildings will be carried out as early as possible to minimize negative impact on facility progress and on the natural environment. In particular, it is aimed to perform all construction requiring considerable earthwork as early as possible, including the tunnels for SIS100/300 and the foundations for the storage-ring buildings as well as these buildings themselves. In this way, construction- caused vibration will be kept to a minimum when the experimental program starts at the end of stage 1. At the end of stage 1, the SIS18 upgrade will provide 2×10^{11} particles/ pulse U^{28+} at 200 AMeV and 4×10^{10} particles/pulse U^{73+} at 1000 AMeV, a repetition rate of 4 Hz at up to 12 Tm and 1 Hz up to 18 Tm, and a bunch compression to 70 ns.

After completion of Stage 1, nuclear structure investigations with a secondary beam of 100-fold increased intensity will be possible after the new superconducting fragment separator (Super-FRS), plasma physics in the existing target area with 200-fold improved power density and atomic physics with radioactive beams. In Stage 2, all buildings will be completed. After completion, the accelerator will deliver 10^{12} particles/pulse U^{28+} at 2.7 AGeV, 10^{11} particles/pulse U^{73+} at 8.3 GeV (or Ne^{10+} up to 14 AGeV), bunch compression up to 50 ns, 2.5×10^{13} protons/pulse up to 29 GeV, up to 10^{11} antiprotons accumulated, stored, and cooled in the HESR at up to 15 GeV, and antiprotons decelerated down to rest in the NESR and HITRAP. Nuclear structure physics will be possible with up to 10000-fold increased secondarybeam intensity, in addition QCD studies with protons and antiprotons and precision spectroscopy with antiprotons and antimatter at rest will form the anticipated research program.

After completion of Stage 3, the full research programme will be possible with, in particular, parallel operation of up to four programmes. Beyond the parameters of Stage 2, plasma-physics research will benefit from up to 2500-fold increased power density. Atomic reaction studies will be possible with highly energetic beams. The accelerator is expected to deliver 2×10^{9} particles/pulse U^{92+} at up to 34 AGeV and stretcher possibility with extraction times of up to several minutes for the nuclear-matter program.

The planning for the construction work will take two more years, following the delivery of the TRs, so that beginning of construction work is anticipated for 2007. The legal procedure for the construction permit and the environmental compensation measures is ongoing. About 0.21 km^2 of forest have to be cleared, of which about 0.11 km^2 will be reforested within the GSI area. From the remaining area, about half will be reforested on adjacent fields. The operation of the FAIR facility has been approved by the Ministery of Environment of the State of Hessen following a thorough evaluation of the proposed radiation safety features worked out by the GSI Radiation Safety group.

ACKNOWLEDGMENTS

I would like to thank the organizers of the conference for their kind invitation and their hospitality in this beautiful surrounding.

Review of Recent Jlab Results

Elton S. Smith

Thomas Jefferson National Accelerator Facility, Newport News, Virginia 23606, USA

Abstract. High quality polarized electron beams at Jefferson Lab make possible precision measurements of hadronic properties in the regime of strongly interacting QCD. We will describe a few programs at Jefferson Lab that are making measurements that link the basic static properties of hadrons to their quark sub-structure. For example, parity-violating electron proton elastic scattering probes the spatial distribution of strange quarks in the nucleon. The nucleon-Delta transition form factors give us information about the deformation of nucleons and Deltas. Finally, new high statistics measurements of photons scattering off proton and deuteron targets are used to set upper limits on the production of exotic baryons with strangeness S=+1. These examples will be used to illustrate the capabilities and focus of the experimental program at JLab.

Keywords: parity violation, nucleon strangeness, exotic baryons, pentaquarks, Delta
PACS: 11.30.Er, 13.40.Gp, 13.60.Fz, 13.60.Le, 13.60.Rj, 14.20.Dh, 14.80.-j

INTRODUCTION TO JLAB

The Continuous Electron Beam Accelerator Facility (CEBAF) at the Thomas Jefferson National Accelerator Facility (Jefferson Lab) is devoted to the investigation of the electromagnetic structure of mesons, nucleons, and nuclei using high energy and high duty-cycle electron and photon beams.

CEBAF is a superconducting electron accelerator [1] with a maximum energy of 6 GeV and 100 % duty-cycle. Three electron beams with a maximum total current of 200 μA can be used simultaneously for electron scattering experiments in the experimental areas. The accelerator design concept is based on two parallel superconducting continuous-wave linear accelerators joined by magnetic recirculation arcs. The accelerating structures are five-cell superconducting niobium cavities with a nominal average energy gain of 5 MeV/m. The accelerator operation has met all design goals, achieving 5.7 GeV for physics running, and delivering three high quality beams with intensity ratios exceeding 10^6:1. The electron beam is produced using a strained GaAs photocathode allowing the delivery of polarized electrons ($P_e \geq 75\%$) simultaneously to all three halls.

Three experimental areas are available for simultaneous experiments, the only restriction being that the beam energies have to be multiples of the single pass energy. The halls contain complementary equipment which cover a wide range of physics topics: Hall A [2] has two high resolution magnetic spectrometers with 10^{-4} momentum resolution in a 10% momentum bite, and a solid angle of 8 msr. Hall B houses the large acceptance spectrometer, CLAS [3]. Hall C uses a combination of a high momentum spectrometer (10^{-3} momentum resolution, 7 msr solid angle and maximum momentum of 7 GeV/c) and a short orbit spectrometer.

To illustrate the physics which is being addressed at Jefferson Lab, we have chosen

CP814 *Hadron Spectroscopy: Eleventh International Conference*
edited by A. Reis, C. Göbel, J. Sá Borges, and J. Magnin

three topics of current interest: measurements of parity violation in electron elastic scattering, the N-Δ transition form factors, and high statistic searches for the Θ^+ exotic strangeness +1 baryon with the CLAS. These programs have the common goal of probing baryon structure which comes from quark pairs beyond the contribution from the standard three valence quarks.

STRANGE QUARKS IN THE NUCLEON

Early deep-inelastic scattering experiments demonstrated that the structure of the proton can not be described by its *uud* valence structure alone. For example, quarks carry only 50% of the proton momentum and are responsible for less than 30% its spin [4]. The fraction and distribution of s and $\bar{s}$ quarks were also determined in ν and $\bar{\nu}$ scattering experiments. For example, NuTeV [5] reports the fraction $(s+\bar{s})/(\bar{u}+\bar{d})$ to be $0.42 \pm 0.07 \pm 0.06$ at a Q^2 of 16 GeV^2. But most of these sea quarks are very short lived ($\sim \hbar/\sqrt{Q^2}$) and arise from perturbative evolution of QCD [6]. The focus of this section is to describe experiments that are sensitive to the long-lived component of $s\bar{s}$ pairs which contribute to the static properties of the nucleon such as its magnetic moment and charge radius.

It is easy to understand how $s\bar{s}$ pairs in the sea can contribute to the proton magnetic moment in a simple hadronic picture [7]. During the process of $uuds\bar{s}$ fluctuations, the quarks will tend to arrange themselves into energetically favorable configurations, the lowest state being a ΛK^+. The K^+ must be emitted in an L=1 state to conserve angular momentum and parity. The probability that the Λ has its spin anti-parallel to the proton is twice as likely as the probability that the spin have the same direction. In this case, the (positive) kaon will have $l_z = +1$ and contribute a positive amount to the magnetic moment of the proton. In addition, the spin of the (negative) s quark in the Λ is anti-aligned with the proton spin so it also gives a positive contribution to the magnetic moment. By convention, however, the usual definition of the strange magnetic moment ($G_M^s(Q^2=0)$) does not include the s-quark charge of $(-\frac{1}{3})$, and therefore this picture actually predicts a *negative* strange magnetic moment. Over the past decade there have been many different models [8] used to estimate both the magnitude and sign of G_M^s, but all have been forced to make non-trivial approximations. Most models predict a negative G_M^s, as in the case of the naive hadronic model above, with a value in the range of -0.8 to 0. But it is clear for the moment that experimental measurements are required to guide our understanding of the contributions of strange quarks to the properties of the proton.

Parity violation in electron scattering is a unique tool for probing the contribution of $s\bar{s}$ sea quarks to the structure of the nucleon. The weak interaction is probed by measuring the electron helicity dependence of the elastic scattering rate off unpolarized targets. Experimental sensitivities of the order of a part per million are required to measure the parity violating asymmetry on a nucleon which can be expressed as

$$A_{PV}^p = \frac{d\sigma_R - d\sigma_L}{d\sigma_R - d\sigma_L} = \frac{G_F Q^2}{4\pi\alpha\sqrt{2}} \left[\frac{A_E + A_M + A_A}{\varepsilon \left(G_M^\gamma\right)^2 + \tau \left(G_M^\gamma\right)^2} \right], \tag{1}$$

where $A_E = \varepsilon G_E^Z(Q^2) G_E^\gamma(Q^2)$, $A_M = \tau G_M^Z(Q^2) G_M^\gamma(Q^2)$ are the electric and magnetic γ-Z interference terms, $\tau = Q^2/4M_N^2$, and ε is the polarization of the virtual photon. The isolation of the electric and magnetic terms can be accomplished by taking advantage of the sensitivity to kinematics (though ε and τ) and taking data at different scattering angles. The last term, A_A, picks out the contribution due to the axial form factor, but the forward-angle measurements are insensitive to this term, so we refer the reader to our references for details [9]. Finally, the flavor decomposition of the form factors can be determined uniquely assuming charge symmetry and by combining parity-violating asymmetries with measurements of the electric and magnetic form factors on the proton and neutron. The scattering amplitude from a zero spin isoscalar target, such as ^4He, is particularly simple because it does not allow magnetic transitions so the asymmetry is sensitive only to the isoscalar electric form factor:

$$A_{PV}^{He} = \frac{G_F Q^2}{4\pi\alpha\sqrt{2}} \left[4\sin^2\theta_W + \frac{G_E^s}{\frac{1}{2}\left(G_E^{\gamma p} + G_E^{\gamma n}\right)} \right], \tag{2}$$

The contribution of the asymmetry which is attributable to strange quarks is obtained by subtracting the calculated asymmetry assuming no strange contribution, including radiative corrections and best estimates of the electromagnetic form factors. The uncertainties in the calculations are estimated and included separately when quoting the uncertainties in the strange form factors. Thus a fairly extensive experimental program is required to isolate all contributions.

There are two major experimental programs at JLab studying parity violation: HAPPEX [14] and G0 [15]. Both collaborations have recently reported results. The precision of these experiments, together with other experimental and theoretical efforts world-wide [9, 16, 17, 18, 19, 20], has now achieved sufficient accuracy to be sensitive to the contribution of strange quarks to the nucleon form factors. The first measurements of parity-violating electron scattering at JLab were carried out by the HAPPEX collaboration. The experiment detected the scattered electrons in each of the Hall A high resolution spectrometers, with sufficient hardware resolution to spatially separate the elastically scattered events from inelastic events into special lead-scintillator absorption counters whose output was integrated during a 30 ms window. Window pairs of opposite helicity were chosen randomly at 15 Hz. A half wave plate in the CEBAF injector was inserted or removed about once a day, switching the helicity of the beam without changing the electronics. HAPPEX has previously reported measurements of parity violation in ep elastic scattering at Q^2=0.48 GeV^2 [21]. The asymmetry errors are completely dominated by statistical uncertainties, and the helicity correlated systematic uncertainties are about 0.1 ppm. Recently it has completed asymmetry measurements at Q^2=0.1 GeV^2 in ep scattering giving $G_E^s + 0.080\, G_M^s = 0.030 \pm 0.025 \pm 0.006 \pm 0.012$ [22] and a measurement of elastic scattering on ^{4}He at a Q^2=0.091 GeV^2 which yields a direct measurement of $G_E^s = -0.038 \pm 0.042 \pm 0.010$, consistent with zero. [23]. These measurements were performed at electron scattering angles of 6°, made possible by the new superconducting septum magnets installed in Hall A [24]. We note that HAPPEX continues to take data and is expected to improve the statistical uncertainty in each measurement by a factor of 2.5–3 this year.

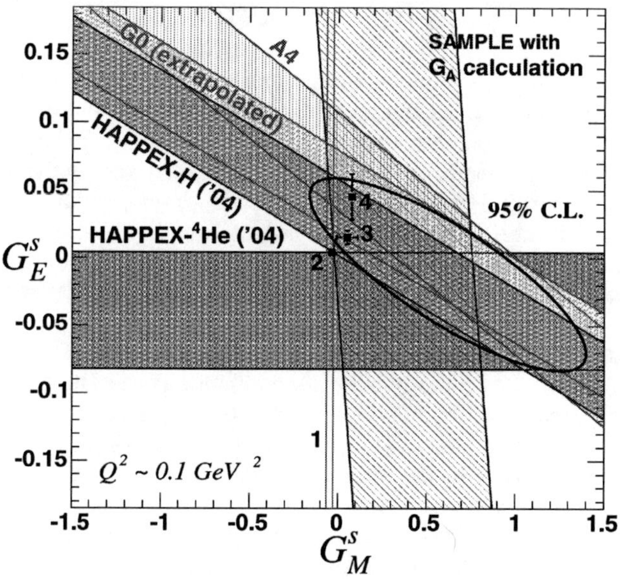

FIGURE 1. Constraints on the G_E^s and G_M^s strange form factors at $Q^2 = 0.1GeV^2$. The curves give the 95% contours. Sample model calculations are also indicated as 1 [10] 2 [11] 3 [12] and 4 [13].

The G0 experiment uses a dedicated detector [25] to determine G_E^s, G_M^s, and G_A^e over a broad kinematical range. To date measurements have been completed for forward angles. The result of their first experimental run has been the determination of the linear combination $G_E^s + \eta G_M^s$ for Q^2 between 0.12 to 1.0 GeV^2 ($\eta = \tau G_M^p / \varepsilon G_E^p$). These measurements are consistent with HAPPEX and A4, but they provide broad coverage in Q^2 to probe the spatial dependence of strange quarks in the nucleon. The three lowest Q^2 points (out of 16 measurements) can be used together with the other experiments (see Fig. 1) to determines a central value for G_E^s and G_M^s separately for $Q^2 = 0.1$ GeV^2.

We close by summarizing the experimental situation, and relating the measured quantities back to our simple hadronic picture for possible strange contributions to the proton magnetic moment. Analysis of the world measurements leads to $G_M^s = 0.55 \pm 0.28$ and $G_E^s = -0.01 \pm 0.03$. To compare to the proton magnetic moment of +2.79 n.m. we multiply G_M^s by $(-\frac{1}{3})$ and, taken at face value, it is approximately 7% of the magnitude of the proton magnetic moment, and opposite in sign. The interpretation of positive values for G_M^s has been investigated within the context of the quark model [26]. This study shows that, even though the quark model is very successful at predicting relations between baryon magnetic moments, the quark model does not lead naturally to positive values for G_M^s. The authors find that positive moments are generated when the $\bar{s}$ is in the ground state and the *uuds* quarks are in excited states. Although these configurations do not seem very natural, they are very analogous to quark-models configurations for pentaquarks, which is the topic of the last section of this paper.

N-Δ DEFORMATION

Within the spin-flavor SU(6) symmetry of the quark model, the Δ resonance is a completely symmetric object with all quark spins aligned. Within this model, the photo-excitation $\gamma^* N \rightarrow \Delta(1232) \rightarrow N\pi$ proceeds via a single quark spin flip in the L=0 nucleon ground state. This magnetic dipole transition, characterized by the M_{1+} multipole, predicts negligible contributions from the electric (E_{1+}) and scalar (S_{1+}) quadrupole multiples. However, strong-interaction dynamics will modify this picture, and several models predict deviations from this simple picture which would indicate deformations of both the nucleon and the Δ [27]. For example, the deformation of the nucleon and Δ could result from D-state admixtures into the baryon wave function. However, D-wave admixtures are expected to be small, but quark-antiquark pairs which are present in the nucleon can also contribute to the quadrupole moment. These latter contribute to the quadrupole moment via a two-body spin tensor in the charge operator, even when the valence quarks are in pure S states. The physical interpretation in this case is that the distribution of $q\bar{q}$ sea quarks in the nucleon deviate from spherical symmetry. These models interpret the negative values of the measured transition quadrupole moments as corresponding to a positive intrinsic quadrupole moment (cigar shape) for the nucleon and a negative moment (pancake shape) for the Δ.

Complete angular distributions of the pion in the reaction $ep \rightarrow ep\pi^0$ at the peak of the Δ allow the experimental determination of the quadrupole moments. The differential cross section of a polarized beam (helicity $h = \pm 1$) and unpolarized target can be written as a function of the transverse and longitudinal structure functions σ_T and σ_L, and the interference terms σ_{TT}, σ_{LT} and $\sigma_{LT'}$:

$$\frac{d^2\sigma}{d\Omega_\pi^*} = \frac{p_\pi^*}{k_\gamma^*}\Big(\sigma_T + \varepsilon_L\sigma_L + \varepsilon\sigma_{TT}\,\sin^2\theta_\pi^*\cos 2\phi_\pi^* + \sqrt{2\varepsilon_L(\varepsilon+1)}\,\sigma_{LT}\,\sin\theta_\pi^*\cos\phi_\pi^* + h\,\sqrt{2\varepsilon_L(\varepsilon-1)}\,\sigma_{LT'}\,\sin\theta_\pi^*\sin\phi_\pi^*\Big), \quad (3)$$

where p_π^*, θ_π^*, and ϕ_π^* are the pion center-of-mass (c.m.) momentum and angles, ε and ε_L are the usual transverse and longitudinal polarizations of the virtual photon, and k_γ^* is the real-photon-equivalent c.m. energy. The moments of interest can be selected as terms in the partial wave expansion of the structure functions using Legendre polynomials [28, 29]. Unique solutions can be obtained by truncating the multipole expansion to terms involving only M_{1+}. Fitting for the coefficients of the expansion allows the extraction of the electric (E_{1+}) and scalar (S_{1+}) quadrupole moments through their interference with the dominant M_{1+}. The systematic errors due to the truncation of higher multipoles are estimated by calculating the effects of higher partial waves using realistic parameterizations of higher mass resonances and backgrounds. At the peak of the Δ and for Q^2 of a few GeV^2, this procedure results in a largely model-independent extraction of the moments.

Early interest in the quadrupole moments stemmed from the perturbative QCD predictions of quark helicity conservation which requires that $E_{1+} \rightarrow M_{1+}$ and S_{1+} approaches a constant at very high Q^2. Therefore, the first experiments at JLab in Hall C [30] determined the quadrupole moments at the highest accessible $Q^2 \sim 3-4$ GeV^2, followed

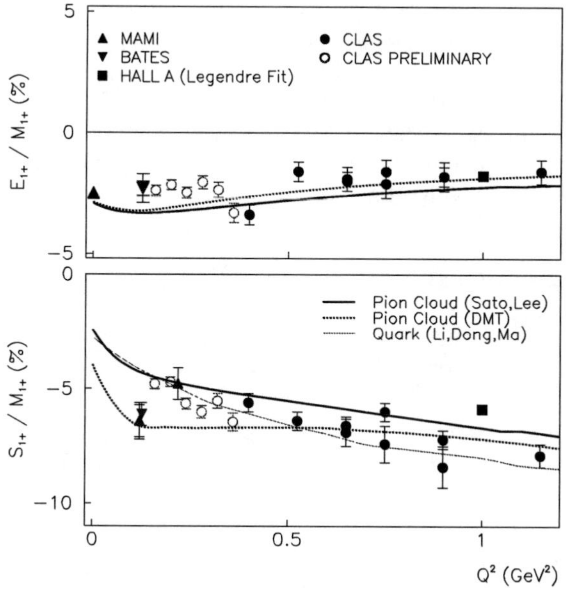

FIGURE 2. Q^2 dependence of the ratios of the electric (E_{1+}) and scalar (S_{1+}) quadrupoles to the dominant dipole moment. The curves are model calculations from references [32], [33] and [34].

by results from CLAS [31] at intermediate momentum transfer. These measurements showed that E_{1+} was not only a few percent of M_{1+}, but also of opposite sign. The asymptotic kinematics of perturbative QCD was clearly well beyond the range of these measurements.

Recently the focus has shifted to understanding the deformation of the nucleon at very low Q^2 which probes chiral symmetry breaking, models of heavy baryon chiral perturbation theory, and the shape of the pion cloud surrounding the nucleon. Data at $Q^2 = 0.12$ GeV^2 from MAMI [35] and Bates [36] show a surprisingly large scalar quadrupole moment of 6.5%, twice that expected from the trend of higher energy measurements and most models. New data from CLAS [37] were taken with an electron beam energy of 1.046 GeV, polarization of $\sim$70% and 10 nA beam current on a 2-cm liquid hydrogen target. The π^0 electroproduction data from this experiment have provided nearly complete angular distributions for $Q^2 = 0.14 - 0.38$ GeV^2 and $W = 1.2 - 1.3$ GeV. The broad acceptance of the CLAS detector allows measurements of the full angular distribution over a broad range of Q^2 and covering the W range of the Δ. The ratios of E_{1+}/M_{1+} and S_{1+}/M_{1+} from this experiment along with other measurements are shown in Fig. 2. Important constraints on the non-resonant backgrounds have been obtained by measuring the helicity-dependent structure function $\sigma_{LT'}$ [38]. Weak non-resonant backgrounds underlying the Δ peak can be enhanced through their interference with the magnetic multipole ImM_{1+}, which is supressed in σ_{LT} due to the vanishing of the real part of the resonant amplitude at the pole.

In summary, there is general agreement on the value of $E_{1+}/M_{1+} \sim$ -2.5%, which is approximately constant at small Q^2. The recent analysis of CLAS indicates that S_{1+}/M_{1+} ratio shows a Q^2-dependence in this same kinematic region, in apparent disagreement with previous measurements that show it is leveling off. Global parameterizations of electroproduction data is underway and will be used along with π^+ data to extract the nucleon parameters with refined accuracy.

SEARCH FOR PENTAQUARKS

Pentaquark states have been studied both theoretically and experimentally for many years [39]. Recent interest was revived by predictions within the chiral soliton model [40] for the existence of an anti-decuplet of 5-quark resonances with spin and parity $J^\pi = \frac{1}{2}^+$. The lowest mass member of the anti-decuplet, now called the Θ^+, is an isosinglet with valence quark configuration $uudd\bar{s}$ with strangeness $S = +1$, was predicted to have a mass of 1.53 GeV/c^2 and a width of $\sim$ 0.015 GeV/c^2. These definite predictions prompted experimental searches to focus attention in this mass region. Shortly after the first observation of an exotic $S = +1$ state by the LEPS collaboration [41], many experiments presented confirming evidence of the initial report, including two observations using the CLAS detector [42, 43]. However, there have been an increasing number of experiments reporting null results, even though in very different reactions and disparate kinematic regions [44].

This paper reports on two searches for a narrow $S = +1$ baryon in the mass range between 1520 and 1600 MeV using the CLAS detector. These new high-statistics searches were conducted using the tagged photon beam facility in Hall B [45] with deuterium [46] and proton [47] targets. The deuterium data can be compared directly with the first reported observation by CLAS [42], which analyzed a data set referred to as "G2a." The G2a data set consisted of two run periods at E_e = 2.478 GeV and 3.115 GeV, corresponding to 0.83 and 0.35 pb^{-1} respectively. The photon energy was tagged beginning below the reaction threshold of 1.51 GeV up to 95% of the electron beam energy. The target was 10 cm long and located on CLAS center, and the CLAS torus field was set at 90% of its maximum value.

The new deuterium data set, referred to as G10, used a primary electron beam energy of E_e = 3.767 GeV creating a tagged photon beam with energies E_γ between 1 and 3.6 GeV. The run was divided evenly between two field settings of the torus magnet (60% and 90% of the full-field current of 3860 A). The hardware trigger required two charged particles in any two sectors and the target was a 24-cm long cell contaning liquid deuterium and located 25 cm upstream of nominal CLAS center. The integrated luminosity for this data set was 38 pb^{-1} for E_γ greater than 1.5 GeV. Evidence for the exotic baryon was searched for using the reaction $\gamma d \to K^- K^+ pn$. The momentum of the charged particles was determined using magnetic analysis and their mass determined using time-of-flight techniques. The analysis selected a detected proton, K^+ and K^- in the final state, all originating from the same beam bucket. The neutron momentum and energy was reconstructed from the measured charge particle tracks and known energy of the incident photon. The exotic Θ^+ baryon was searched for in the decay $\Theta^+ \to K^+ n$,

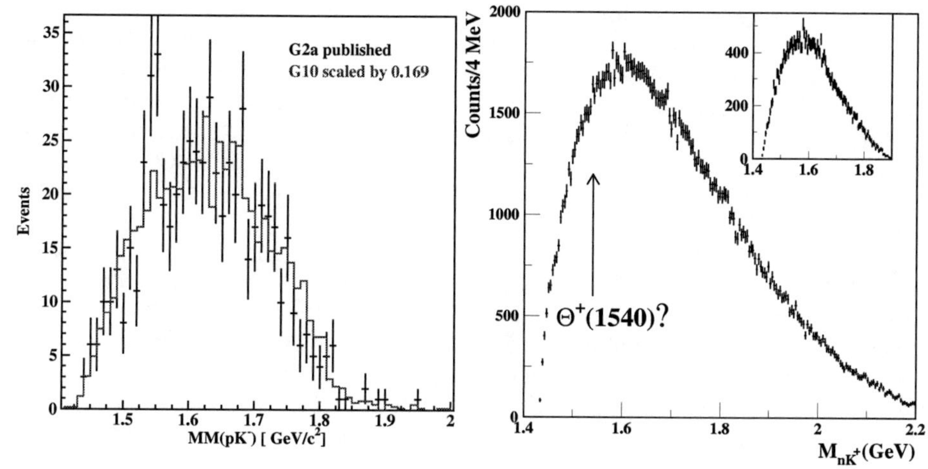

FIGURE 3. (Left) $MM(pK^-)$ distribution of the G10 data (red histogram) compared to the G2a distribution (black points with error bars). Selection cuts have been made on the G10 data set to best represent the experimental conditions of G2a and scaled for direct comparison. (Right)Distribution of the nK^+ mass spectrum for the G11 data. No evidence for narrow structures is apparent. The inset shows the mass distribution with selection cuts to reproduce the SAPHIR analysis [48]. Note: All results from CLAS are preliminary.

where the K^+ uniquely identified the positive strangeness of the baryonic state. The mass of the K^+n system is shown in Fig. 3(left) for both the G2 and G10 data samples. The spectrum is relatively smooth and does not exhibit the peak found in the original smaller data set. The figure overlays the two spectra with selection cuts on G10 to mimic the G2a data set as closely as possible including the photon energy spectrum. If we use the G10 data as a background shape, the peak at 1.54 GeV in the G2 data set is consistent with a 3 σ fluctuation. The new data show no indication of a peak in the $\Theta^+ \to K^+n$ exotic S=+1 channel.

The measurement on the proton (G11) was conducted at lower photon energy and using a different reaction from the published CLAS result [43]. This new experiment used a primary electron beam energy E_e = 4.0 GeV to produce a tagged photon beam in the range of E_γ between 1.6 and 3.8 GeV. The target consisted of a 40-cm long cylindrical cell with liquid hydrogen. The data were taken over a period of 50 days, corresponding to an integrated luminosity of 70 pb^{-1}, which is an order-of-magnitude greater than previous experiments with tagged photons. Evidence for the $S = +1$ exotic baryon Θ^+ was searched for in the reaction $\gamma p \to \overline{K}^0 K^+ n$. The search was conducted for the baryon produced in association with the $\overline{K}^0$ and subsequent decay $\theta^+ \to K^+n$. The positive strangeness of the baryon was uniquely tagged using the K^+ of the decay. The charged kaon, as well as the pions from the $\overline{K}^0 \to \pi^+\pi^-$ decay, were reconstructed in the CLAS detector. The neutron momentum was reconstructed using energy and momentum conservation and the known energy of the incident photon. The missing mass peak of the neutron was reconstructed to be 939 MeV with a width of 10 MeV. The $\Sigma^+ \to \pi^+ n$ and $\Sigma^- \to \pi^- n$ reactions are also present in this data sample and are reconstructed with an accuracy of less than 1 MeV and resolution of 3 MeV. These baryons, as well as the $\Lambda^*(1520)$ were excluded from the present analysis. After all cuts, the reconstructed nK^+ mass spectrum is shown in Fig. 3(right). The spectrum is smooth and structureless, with no indication of a peak near 1540 MeV where the Θ^+ was previously reported [48, 49]. An upper limit to the Θ^+ cross section of less than 1 nb was set based on assuming a t-exchange production mechanism for the reaction.

We conclude with a brief summary of the status of pentaquark measurements with CLAS. There are two published observations [42, 43] of the Θ^+ baryon in photon beam reactions on deteurium and proton targets. The measurement on deuterium has been repeated with higher statistics and the new data do not confirm the original observation. A new experiment to verify the production off the proton in the reaction $\gamma p \to \pi^+ K^- K^+ n$ for photon energies between 3 and 5.5 GeV has been approved [50], but has not yet received beam time. However, a recent high statistics experiment on the proton at lower energy [51] has searched for the narrow exotic baryon in the reaction $\gamma p \to \overline{K}^0 K^+ n$ and found no evidence for the state in the nK^+ mass range between 1520 and 1600 MeV.

ACKNOWLEDGMENTS

I would like to thank Alberto dos Reis and all the organizers of the conference for their gracious hospitality. I would also like to thank K. Paschke, D. Armstrong and D. Beck of the HAPPEX and G0 collaborations for making available the results on parity violation. From the CLAS collaboration I would like to express special thanks to K. Joo and L. Smith for help in preparation of materials on Delta production, and S. Stepanyan, R. De Vita and M. Battaglieri for discussions of the pentaquark search. The Southeastern Universities Research Association (SURA) operates the Thomas Jefferson National Accelerator Facility for the U.S. Department of Energy under contract DE-AC05-84ER40150.

REFERENCES

1. C. Leemann, D. Douglas, and G. Krafft, *Annu. Rev. Nucl. Part. Sci.* **51**, 413 (2001).
2. J. Alcorn, et al., *Nucl. Instr. Meth. A* **522**, 294–346 (2004).
3. B. Mecking, et al., *Nucl. Instr. Meth. A* **503**, 513–553 (2003).
4. E. Hughes, and R. Voss, *Annu. Rev. Nucl. Part. Sci.* **49**, 303 (1999).
5. T. Adams et al. (NuTeV) (1999), [arXiv:hep-ex/9906038].
6. M. Burkardt, and B. Warr, *Phys. Rev. D* **45**, 958 (1992).
7. L. Hannelius, D. Riska, and L. Glozman, *Nucl. Phys. A* **665**, 353 (2000).
8. D. Beck, and B. Holstein, *Int. J. Mod. Phys. E* **10**, 1 (2001), [arXiv:hep-ph/0102053], and references therein.
9. E. Beise, M. Pitt, and D. Spayde, *Prog. Part. Nucl. Phys.* **54**, 289 (2005), [arXiv:nucl-ex/0412054], and references therein.
10. D. Leinweber, et al., *Phys. Rev. Lett.* **94**, 212001 (2005), [arXiv:hep-lat/0406002].
11. V. Lyubovitskij, et al., *Phys.Rev. C* **66**, 055204 (2002), [hep-ph/0207225].
12. R. Lewis, W. Wilcox, and R. M. Woloshyn, *Phys. Rev. D* **67**, 013003 (2003).
13. A. Silva, H.-C. Kim, and K. Goeke, *Phys. Rev. D* **65**, 014016 (2002).

14. K. Aniol et al. (HAPPEX Collaboration), *Phys. Rev. C* **69**, 065501 (2004).
15. D. Armstrong et al. (G0 Collaboration), *Phys. Rev. Lett.* **95**, 092001 (2005), [arXiv:nucl-ex/0506021].
16. F. Maas et al. (A4 Collaboration), *Phys. Rev. Lett.* **93**, 022002 (2004), [arXiv:nucl-ex/0401019].
17. F. Maas et al. (A4 Collaboration), *Phys. Rev. Lett.* **94**, 152001 (2005), [arXiv:nucl-ex/0412030].
18. D. Beck, and R. McKeown, *Annu. Rev. Nucl. Part. Sci.* **51**, 189 (2001).
19. K. Kumar, and P. Souder, *Prog. Part. Nucl. Phy.* **45**, S333 (2000).
20. The 2005 Joint Jefferson Lab/Institute for Nuclear Theory Workshop, Precision ElectroWeak Interactions Workshop (2005), http://www.wm.edu/physics/PreciseEW_2005.html.
21. K. Aniol et al. (HAPPEX Collaboration), *Phys. Lett. B* **509**, 211 (2001).
22. K. Aniol et al. (HAPPEX Collaboration) (2005), [arXiv:nucl-ex/0506011].
23. K. Aniol et al. (HAPPEX Collaboration) (2005), [arXiv:nucl-ex/0506010].
24. P. Brindza, et al., *IEEE Trans. on Applied Superconductivity* **11**, 1294 (2001).
25. P. Roos, *Eur. Phys. J. A* **24**, S2.59 (2005).
26. B. Zou, and D. Riska, *Phys. Rev. Lett.* **95**, 072001–1 (2005), [arXiv:hep-ph/0502225].
27. A. Buchmann, and E. Henley, *Phys. Rev. C* **63**, 015202 (2001), [hep-ph/0101027].
28. V. Burkert, and T. H. Lee, *Int. J. Mod. Phys. E* **13**, 1035 (2004), [nucl-ex/0407020], and references therein.
29. V. Burkert, *Prog. Part. Nucl. Phys.55:108-130,2005* **55**, 108 (2005), and references therein.
30. V. Frolov, et al., *Phys. Rev. Lett.* **82**, 45 (1999), [hep-ex/9808024].
31. K. Joo et al. (CLAS Collaboration), *Phys.Rev. Lett.* **88**, 122001 (2002), [hep-ex/0110007].
32. T. Sato, and T. Lee, *Phys. Rev. C* **63**, 055201 (2001), [nucl-th/0010025].
33. S. Kamalov, et al., *Phys. Rev. C* **64**, 032201(R) (2001), [nucl-th/0006068].
34. Z. Li, Y. Dong, and W. Ma, *J. Phys. G* **23**, 151 (1997).
35. T. Pospischil, et al., *Phys. Rev. Lett.* **86**, 2959 (2001), [nucl-ex/0010020].
36. C. Mertz, et al., *Phys. Rev. Lett.* **86**, 2963 (2001), [nucl-ex/9902012].
37. L. Smith, $\gamma^* p \to \Delta(1232) \to \pi N$ Results from CLAS (2005), contribution to the Japan-US Workshop on *Electromagnetic Meson Production and Chiral Dynamics*, Osaka, Japan, http://www-nuclth.phys.sci.osaka-u.ac.jp/jp-usw/prog.html.
38. K. Joo et al. (CLAS Collaboration), *Phys.Rev. C* **68**, 032201 (2003), [nucl-ex/0301012].
39. P. D. Group, *Phys. Lett. B* **170**, 289 (1986).
40. D. Diakonov, V. Petrov, and M. Polyakov, *Z. Phys. A* **359**, 305 (1997), [arXiv:hep-ph/9703373].
41. T. Nakano et al. (LEPS Collaboration), *Phys. Rev. Lett.* **91**, 012022 (2003), [arXiv:hep-ex/0301020].
42. S. Stepanyan et al. (CLAS Collaboration), *Phys. Rev. Lett.* **91**, 252001 (2003), [arXiv:hep-ex/0307018].
43. V. Kubarovsky et al. (CLAS Collaboration), *Phys. Rev. Lett.* **92**, 032001 (2004), [arXiv:hep-ex/0311046].
44. A. Dzierba, C. Meyer, and A. Szczepaniak, "Reviewing the Evidence for Pentaquarks," in *J. Phys. Conf.*, 9, 2005, p. 192, [arXiv:hep-ex/0412077], and references therein.
45. D. Sober, et al., *Nucl. Instr. Meth. A* **440**, 263–284 (2000).
46. K. Hicks, and S. Stepanyan, spokespersons for Investigation of Exotic Baryons States in Photoproduction Reactions with CLAS (2003), JLab Experiment E-03-113.
47. M. Battaglieri, R. De Vita, and V. Kubarovsky, spokespersons for Spectroscopy of Exotic Baryons with CLAS: Search for Ground and Excited States (2004), JLab Experiment E-04-021.
48. J. Barth et al. (SAPHIR Collaboration), *Phys. Lett. B* **572**, 127 (2003), [arXiv:hep-ex/0307083].
49. M. Ostrick, *Prog. Part. Nucl. Phys.* **55**, 337 (2005).
50. M. Battaglieri, et al., spokespersons for Study of Pentaquark States in Photoproduction off Protons (2004), JLab Experiment E-04-017.
51. M. Battaglieri et al. (CLAS Collaboration), Search for $\Theta^+(1540)$ pentaquark in high statistics measurement of $\gamma p \to \overline{K}^0 K^+ n$ at CLAS (2005), in preparation.

Scalar Mesons in B-decays

Peter Minkowski* and Wolfgang Ochs†

**Institute for Theoretical Physics, University of Bern, CH-3012 Bern, Switzerland*
†Max-Planck-Institut für Physik (Werner-Heisenberg-Institut),
Föhringer Ring 6, D-80805 München, Germany

Abstract. We summarize some persistent problems in scalar spectroscopy and discuss what could be learned here from charmless B-decays. Recent experimental results are discussed in comparison with theoretical expectations: a simple model based on penguin dominance leads to various symmetry relations in good agreement with recent data; a factorisation approach yields absolute predictions of rates. For more details, see [1].

Keywords: Scalar mesons, B-decays, glueball
PACS: 12.39.Mk, 13.25.Hw,14.40.-n

WHY STUDYING SCALARS IN B-DECAYS

There are various reasons for studying scalar particles in B-decays:

1. Dominance of S-wave resonances with little background from crossed channels
In $(B \to 1,2,3)$-decays the masses of (1,2) resonances can extend to $M \lesssim 5$ GeV. Then there is little overlap with resonances in crossed channels (2,3) or (1,3). This is very different from D decays where resonance masses extend only up to ~ 1.5 GeV and in general there is a large overlap. Furthermore, in the final 2-body systems S-wave interactions are dominant.

2. New source of glueballs
The elementary subprocess $b \to sg$ with an isolated gluon is rather well understood theoretically and is described by a penguin diagram. The decay rate has been calculated in next-to-leading order of perturbative QCD as [2]

$$B(b \to sg) = (5 \pm 1) \times 10^{-3}. \tag{1}$$

The gluon may give rise to production of a glueball which could show up as a resonance in the system X of 2-body decays $B \to K^{(*)} + X$. This process adds to the other well known gluon rich processes like: central production in pp collisions, $J/\psi \to \gamma X$ and $p\bar{p}$ annihilation near threshold.

3. Non-charm final states with strangeness
The decays $b \to sq\bar{q}$ are dominated again by the gluonic penguin process whereas the electroweak tree diagrams occur at the level of 20% only. In the leading penguin approximation the decays $b \to su\bar{u}$, $sd\bar{d}$, $ss\bar{s}$ occur with the same fraction and have been calculated to amount to $\sim 2 \times 10^{-3}$ each. In the corresponding hadronic 2-body final states $B \to xy$, if x and y are members of $SU(3)$ multiplets X, Y each, one obtains various symmetry relations [3]. Hopefully, this will ultimately identify the members of the lightest scalar nonet and the mixing properties.

CP814 *Hadron Spectroscopy: Eleventh International Conference*
edited by A. Reis, C. Göbel, J. Sá Borges, and J. Magnin

PROBLEMS OF LIGHT SCALAR MESON SPECTROSCOPY

The interest in light scalar mesons originates from the following expectations:
1. The existence of glueballs
This is a requirement from the first days of QCD and may be the most urgent open problem of the theory at the fundamental level. In lattice QCD, quenched approximation, the lightest glueball appears in the 0^{++} channel with a mass of 1400-1800 MeV [4]. The effect of unquenching is under study but realistic estimates are still difficult, especially because of the large quark masses. An alternative QCD approach is based on QCD sum rules [5] where the lightest glueball is centered around 1000-1400 MeV.

2. Multiplets of $q\bar{q}$ and exotic bound states
There is no general consensus on the members of the lightest $q\bar{q}$ nonet, i.e. the parity partner of π, K, η, η'. In addition, there is the possibility of tetraquarks [6], bound states of di-quarks.

The list of scalar particles provided by the PDG [7] with mass $M \lesssim 1.8$ GeV includes
$I = 0$: $f_0(600)$ (or σ), $f_0(980)$, $f_0(1370)$, $f_0(1500)$, $f_0(1710)$;
$I = \frac{1}{2}$: $\kappa(900)$ (?), $K_0^*(1430)$;
$I = 1$: $a_0(980)$, $a_0(1450)$.

There are two typical scenarios for the classification of these states:
I. One nonet below and one above 1 GeV
The nonet of lower mass includes σ, κ, $f_0(980)$, $a_0(980)$, either $q\bar{q}$ (see, for example, Ref. [8] and Van Beveren [9]) or $qq\bar{q}\bar{q}$ [6] bound states. The higher mass states could then make a $q\bar{q}$ nonet with members $K_0^*(1430)$ and $a_0(1450)$; in the isoscalar sector the three states $f_0(1370)$, $f_0(1500)$ and $f_0(1710)$ could be, as originally proposed in [10], a superposition of the glueball and the two members of the isoscalar nonet.

II. One nonet above 1 GeV
In this scheme the σ and κ with the parameters given are not considered as physical states to be classified along the lines we discuss here. The $q\bar{q}$ nonet is rather formed by $a_0(980)$ (or also $a_0(1450)$), $f_0(980)$, $K_0^*(1430)$ and $f_0(1500)$ [11, 12] whereas two higher mass nonets including $f_0(1370)$ have been proposed in [13]. The $\pi\pi$ S-wave is interpreted as being dominated by a very broad object, centered around 1 GeV, the lower part could be responsible for the $\sigma(500)$ effect. This broad state ($\Gamma > 500$ MeV) has been proposed as representing the isoscalar glueball by various arguments [13, 12].

There are states whose identity is in doubt as can be seen by the large uncertainty in mass and width estimated by the PDG: $\sigma(500)$, $f_0(1370)$ with no single branching ratio or ratio of such numbers accepted by PDG and finally κ or $K^*(800)$ not carried in the main listing of PDG. We will add a few remarks on these problematic states which will be of relevance for our discussion of B decays.

Isoscalar channel: $f_0(600)$ or σ and $f_0(1370)$

Most definitive experimental results on these states can be obtained from the $2 \to 2$ scattering processes $\pi\pi \to \pi\pi, K\bar{K}, \eta\eta$ applying an energy independent partial wave analysis (EIPWA); in this case unitarity provides important constraints in the full energy range. Recently, results on D and B decays as well as $p\bar{p} \to 3$ particles with higher statistics became available. There is no general constraint on the mass dependence of the amplitude which can be affected by various dynamical effects. So far, in these processes no EIPWA over the full energy range has been performed, so an optimal description of data for a particular model parametrization is selected. A promising new approach towards EIPWA in D-decays has been presented at this conference by Meadows [14].

Concerning the $\pi\pi$ interaction there is a general consensus that there exists indeed a broad state with the width of the order of the mass, but the parameters depend on the mass range considered, a feature which is known already since about 30 years.

1. Low mass range $M_{\pi\pi} \lesssim 0.9\ldots1.2$ *GeV.*
In this region the complex $\pi\pi$ amplitude moves along the unitarity circle to its top (phase 90°) where a rapid circular motion follows from $f_0(980)$. An early analysis has been performed by the Berkeley collaboration [15], they found a state, σ, with $M_\sigma = 660 \pm 100$ MeV, $\Gamma_\sigma = 640 \pm 140$ MeV. Recently, results from D-decays by E791 [16], FOCUS [17] and from $J/\psi \to \omega\pi\pi$ by BES [18] have been interpreted in terms of a σ with similar mass, although good fits based on a K matrix parametrization have been obtained without such a state [17]. On the theoretical side, parametrizations of such data using the low mass χPT constraints lead to a low mass pole with $M_\sigma \sim 450$ MeV and $\Gamma_\sigma \sim 450\ldots600$ MeV (see, e.g. Refs. [19, 20] and the reports by Bugg and Pelaes [21]).

2. Extended mass range $500 \leq M \leq 1800$ *MeV*
In case of a broad state the parameters should be determined from the energy interval where its influence is important and this includes the inelastic region above 1 GeV.

All analyses of $\pi\pi$ scattering in this region find again one broad state, but with a higher mass than before, in a range around 1000 MeV and with large width > 500 MeV. The first analysis along these lines goes back again 30 years [22] and in Table 1 we list the pole positions from K matrix fits of various analyses. The fits by Estabrooks [23] refer to the four solutions of an EIPWA of elastic $\pi\pi$ scattering [24] as well as of the $\pi\pi \to K\bar{K}$ reaction. In all solutions of the EIPWA the S-wave amplitude above 1 GeV follows a circular path with some inelasticity in the Argand diagram (*Im T* vs. *Re T*) which can be fitted by a broad resonance. Superimposed is a smaller circle corresponding to a resonance [23] with parameters close to what is known today as $f_0(1500)$. No additional pole, such as $f_0(1370)$, is seen in this analysis. A similar picture is found [12] for the inelastic channels $\pi\pi \to \eta\eta$ and $\pi\pi \to K\bar{K}$ comprising the broad background and $f_0(1500)$ with the interference pattern

$$\begin{array}{lll} \pi\pi \to K\bar{K}: & \text{background - } f_0(1500) & \text{constructive interference} \\ \pi\pi \to \eta\eta: & \text{background - } f_0(1500) & \text{destructive interference} \end{array} \tag{2}$$

This broad state is seen in a variety of processes and has been dubbed $f_0(1000)$ in [25]. Later arguments have been presented that this broad state be a glueball [13, 12]. This state also appears in decay processes although it may happen that the higher mass tails are suppressed for dynamical reasons. As an example, we quote the study by BES [26] of the final state $J/\psi \to \omega K^+K^-$ where the large S-wave background ("σ") extends up to about 2 GeV. A significant flat background has also been observed recently in the gluon rich channel $J/\psi \to \gamma K\bar{K}$ by BES [27].

Apparently, the state $f_0(1370)$ shows up if reactions other than (1)-(3) in Table 1 without unitarity constraints are included in the fits. Whereas $f_0(980)$, "$f_0(1000)$" and $f_0(1500)$ are clearly seen as circles in the Argand diagrams, no such circle has ever been shown to exist for $f_0(1370)$. Before such a behaviour is demonstrated, this state could hardly be considered as established. The strong interference between background and $f_0(1500)$, leads to very different mass spectra, depending on the relative phase, which could easily simulate a "new state" $f_0(1370)$.

TABLE 1. Position of broad state in the T matrix of $\pi\pi$ scattering according to various K matrix fits to data from reactions (1) $\pi\pi \to \pi\pi$, (2) $\pi\pi \to K\bar{K}$, (3) $\pi\pi \to \eta\eta, \eta\eta'$, (4) $p\bar{p}$ annihilation and (5) J/ψ decays

Authors	mass (MeV)	width (MeV)	channels
CERN-Munich [22]	1049	500	1
Estabrooks [23]	750	800-1000	1,2
Au, Morgan & Pennington [25]	910	700	1,2,5
Anisovich and Sarantsev [13]	1530	1120	1,2,3,4

We conclude that there is indeed a broad state in the isoscalar channel with decays into various 2-body final states but there is no standard form for its line shape. Different results on its mass emerge depending on whether the analytic parametrization is fitted to a small or a large mass range (corresponding to either a half resonance circle or an almost full circle) leading either to $\sigma(500)$ or "$f_0(1000)$". There can be little doubt that both results refer to the same state. Studies along path 1 should ultimately extend their parametrization to include higher mass inelastic channels, especially the EIPWA results by Estabrooks, whereas the analyses along path 2 should include the very low mass $\pi\pi$ data as well.

Isospin $I = \frac{1}{2}$ channel: $\kappa(800)$ and $K_0^*(1430)$

The elastic $K\pi$ scattering up to 1700 MeV has been studied some time ago by an experiment at SLAC [28] and the LASS experiment [29]. The S wave phase shifts have been parametrized in terms of $K_0^*(1430)$ with a small inelasticity $< 10\%$ starting only above the inelastic threshold $M_{K\pi} \gtrsim M_{K\eta'} \sim 1450$ MeV and a slowly varying background with an effective range formula. This background phase in the considered range does not exceed about 50° and insofar it is a phenomenon quite different from the background in $\pi\pi$ scattering where the background phase reaches 90° below the first scalar resonance $f_0(980)$. We do not want to enter here into the discussion about a possible state κ but point to different characteristics of the $K\pi$ amplitude in elastic scattering and decay relevant to our later discussion. For a theoretical analysis, see Büttiker et al. [30].

In weak decays like $D \to K\pi\mu\nu$ the $K\pi$ phase equals the one in elastic scattering according to the Watson theorem and this is nicely born out by the data (FOCUS [31]). If rescattering effects are small, then the Watson theorem is still applicable in purely hadronic decays and a recent example for this behaviour is $B \to J/\psi K\pi$ measured by BaBar [32]. On the other hand, in $D \to K\pi\pi$ the $K\pi$ phases determined by E791 [14] follow the trend as in elastic scattering below the inelastic $K\eta'$ threshold $M \sim 1400$ MeV, but with a relative shift of about 70°. The Argand diagram in Fig. 1) shows that the resonance circle related to $K_0^*(1430)$ is much smaller than the circle related to the background, which contrasts to elastic scattering with circles of comparable radii. Therefore the LASS parametrization does not represent the decay amplitude in an energy region beyond 1400 MeV.

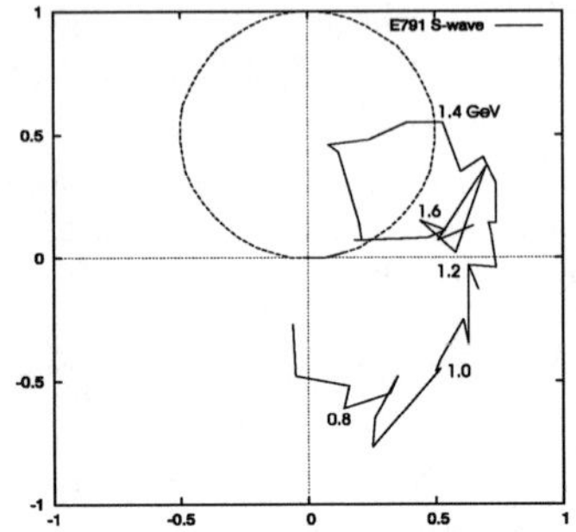

FIGURE 1. Energy independent partial wave analysis of the $I=\frac{1}{2}$ $K\pi$ S-wave using E791 data "c_i" and form factors F_{Di} [14]. Plotted are the rescaled quantities $S_i = c_i F_{Di}\frac{q}{\sqrt{s}}$ (arbitrary units) to be compared with the elastic unitarity circle for $K\pi \to K\pi$ scattering.

B-DECAYS: EXPERIMENTAL RESULTS ON SCALARS

The B branching ratios for the following scalar particles have been measured, for later comparison we present the results corrected for unseen channels, all in units of 10^{-6}.

Isospin I=1: $a_0(980)$

So far only upper limits have been reported by BaBar [33],[1] see Tab. 2.

TABLE 2. B decays into $a_0(980)$ [33] corrected using $B(a_0 \to \eta\pi) = 0.85$; all numbers $\times 10^{-6}$.

$B^0 \to K^0 a_0^0$	< 9.2	$B^+ \to K^+ a_0^0$	< 2.9
$B^0 \to K^+ a_0^-$	< 2.4	$B^- \to \bar{K}^0 a_0^-$	< 4.6

Isospin $I=\frac{1}{2}$: $K_0^*(1430)$

Total decay rates derived from BaBar [35] and Belle data [36] are given in Tab. 3. Belle, in a full Dalitz plot analysis using an isobar model ansatz finds two quite different solutions in $B^+ \to K_0^{*0}\pi^+$ corresponding to different interferences with a coherent background amplitude. Babar is inserting the LASS parametrization for the $K\pi$ phase in a larger energy interval up to $M \sim M_D$ and is then left with only one solution. As discussed above for $D \to K\pi\pi$ the behaviour of the $K\pi$ amplitude above 1400 MeV could be quite different from elastic scattering and a more general ansatz in this mass region seems appropriate. The situation is quite analogous to $f_0(980)$ in $\pi\pi$ interactions where the interference pattern of f_0 and background changes from one reaction to

[1] After the conference results by Belle [34] became available which confirm the tight upper bounds for a_0 production: $B(B \to a_0^- K^+) < 1.6 \times 10^{-6}$ for the $\eta\pi$ channel.

TABLE 3. B decays into $K_0^*(1430)$ corrected for unseen modes using $B(K_0^* \to K\pi) = 0.93$; units in 10^{-6}

	BaBar [35]	Belle [36]	
$B^0 \to K_0^{*0}\pi^0$	12.7 ± 5.0	9.8 ± 2.7	
$B^0 \to K_0^{*+}\pi^-$	36.1 ± 12.2	16.4 ± 5.3	
$B^+ \to K_0^{*0}\pi^+$	37.0 ± 4.4	(I) $45^{+15.3}_{-11.1}$	(II) $8.3^{+3.9}_{-2.6}$

another. It will be therefore important to clarify the existence of two solutions and to possibly exclude one of them by physical arguments.

Isospin I=0: $f_0(1500)$

Both Belle [36] and BaBar [37] see a peak in the K^+K^- Mass spectrum in $B \to (K^+K^-)K$. The mass and width are consistent with $f_0(1500)$. However, no signal in the corresponding $\pi\pi$ decay channel is observed despite the ratio of branching ratios $\Gamma(f_0 \to K\bar{K})/\Gamma(f_0 \to \pi\pi) = 0.241 \pm 0.028$ [7] is favourable for the $\pi\pi$ channel. Therefore both collaborations suggested the existence of a new state, $f_X(1500)$ or $X(1500)$.

In a previous work [3] studying the Belle data [36] we argued that these phenomena are naturally explained by the existence of a broad background which interferes with $f_0(1500)$: constructively in K^+K^- giving rise to the observed peak but destructively in $\pi^+\pi^-$ leading to a vanishing signal. In our analysis we have represented the mass spectra as a superposition of three components $f_0(980)$, $f_0(1500)$ and a broad resonance as background, which fits the data well, see Figs. 2,3. This interference pattern is the same as in inelastic $\pi\pi \to K\bar{K}, \eta\eta$ [12] and elastic $\pi\pi$ scattering [23] (see above).

These signs are consistent with our hypothesis [12] that the background represents a broad glueball (flavour singlet) with mass in the 1 GeV region or above interfering with $f_0(1500)$, which is close to a flavour octet state according to the considerations in [11, 12].

Both collaborations find two solutions for the $B \to K f_0(1500)$ rate corresponding to different interference signs with the background. From the total charmless and the partial fractions we obtain the branching ratios in Tab. 4. According to our model Sol. II is the physical solution.

TABLE 4. B-decay rates ($\times 10^{-6}$) into $f_0(1500)K$ (total rates using $B(f_0(1500) \to K^+K^-) = 0.043$).

$B(B \to f_0(1500)K)$	Belle (B^+)	BaBar (B^0)
Sol. I $(bg - f_0)$	471.8 ± 51.3	223 ± 42
Sol. II $(bg + f_0)$	61.1 ± 14.4	29.9 ± 13.7

Isospin I=0: $f_0(980)$

$f_0(980)$ was the first scalar particle observed in B-decays and the results obtained by the heavy flavor averaging group (HFAG) [38] are presented in Tab. 5. In these decay channels there is again some background contribution as in case of $f_0(1500)$, so we expect two possible solutions corresponding to different interference signs. Note the negative interference in our fits in Figs. 2,3, also observed in $J/\psi \to \phi\pi\pi$ by DM2 [39]. It would be important to study the possibility of a second solution besides the one in Tab. 5.

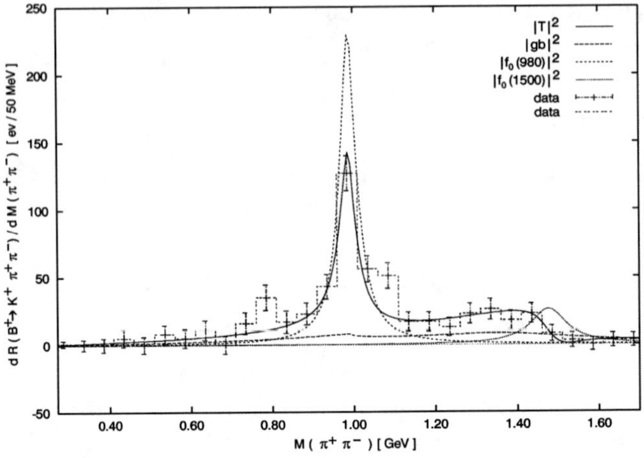

FIGURE 2. $\pi^+\pi^-$ mass spectrum in B-decays (BELLE [36]) in comparison with a model including $f_0(980)$, $f_0(1500)$ and a broad glueball (gb). Also shown are the individual resonance terms. The background (gb) in this fit interferes destructively with both $f_0(980)$ and $f_0(1500)$.

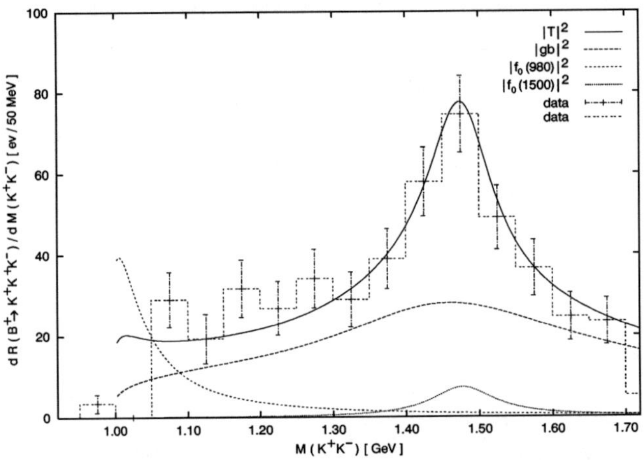

FIGURE 3. K^+K^- mass spectrum in B-decays (BELLE [36]) in comparison with the model amplitude, see Fig. (2). Here $f_0(1500)$ interferes constructively with the background.

Other results on scalars
$f_0(1370)$: A 2σ signal has been observed by Belle [36], not so by BaBar [35].

$\sigma(600)$: No obvious peak near threshold is visible as in $D \to 3\pi$. For our discussion of scalars it would be interesting to obtain the rate for $B \to \sigma K$ (or a limit).

$\kappa(800)$: A $K\pi$ background in $K\pi\pi$ has been observed by Belle [36] but a fit with a κ particle was not successful.

$a_0(1450)$: This state has not been seen yet.

TABLE 5. B-decay rates ($\times 10^{-6}$) into $f_0(980)K$, for $f_0(980) \to \pi\pi$ and total rate using $\Gamma(f_0(980) \to \pi\pi)/\Gamma_{tot} \sim 0.8$.

	$f_0(980) \to \pi\pi$	$f_0(980) \to all$
$B(B^+ \to f_0(980)K^+)$	13.2 ± 1.6	16.5
$B(B^0 \to f_0(980)K^0)$	8.25 ± 1.5	10.3

B-DECAYS INTO SCALARS: THEORETICAL EXPECTATIONS

The theoretical considerations to some extent follow the ideas developed earlier for B-decays into pseudoscalar (P) and vector (V) particles. We outline here two complementary approaches.

Phenomenological amplitudes

The decay rates are expressed in terms of a set of phenomenological amplitudes including the gluonic penguin, the electroweak tree amplitudes and others. Such a scheme has been successfully applied to the decays $B \to PP, PV$ [40, 41].

Here we apply a scheme of this kind [3], but in this exploratory phase for scalars with moderate statistics we restrict ourselves to including only the dominant penguin diagrams and neglect in particular the tree diagrams which give rise to corrections at the 20% level. We then consider in this scheme the three $q\bar{q}$ processes with the same amplitude as well as the gluonic amplitude

$$b \to su\bar{u}, \quad b \to sd\bar{d}, \quad b \to ss\bar{s}, \quad b \to sg. \tag{3}$$

These processes together with the recombination of the spectator quark give rise to 2-body decays $B \to xy$ where x, y are mesons out of the flavour $U(3)$ nonets A and B. Given the members of these multiplets with a particular mixing angle the decay amplitudes can be given in terms of the following parameters: the penguin amplitude p_{AB} with $s \to x$, the exchange amplitude $\beta_{AB}p_{AB}$ for $s \to y$ and $\gamma_{AB}p_{AB}$ for the gluonic amplitude. For a more detailed discussion, see Ref. [3], here we just adress for illustration the decay $B^+ \to \eta' K^{*+}$ for the mixing $\eta' = (u\bar{u}, d\bar{d}, 2s\bar{s})/\sqrt{6}$. The decay amplitude is derived from the penguin amplitudes as in Fig. 4 and reads

$$T = \frac{1}{\sqrt{6}}(1 + 2\beta_{PV} + 4\gamma_{PV})p_{PV} \tag{4}$$

A consequence of this penguin dominance model are various symmetry relations, especially the $I = \frac{1}{2}$ rule: The final state of processes (3) has $I = 0$ and therefore the final state of B decay has the isospin of the spectator, for $B^{\pm}, B^0$ this is $I = \frac{1}{2}$, which is also realized in our amplitudes [3], see also Tab. 6 below.

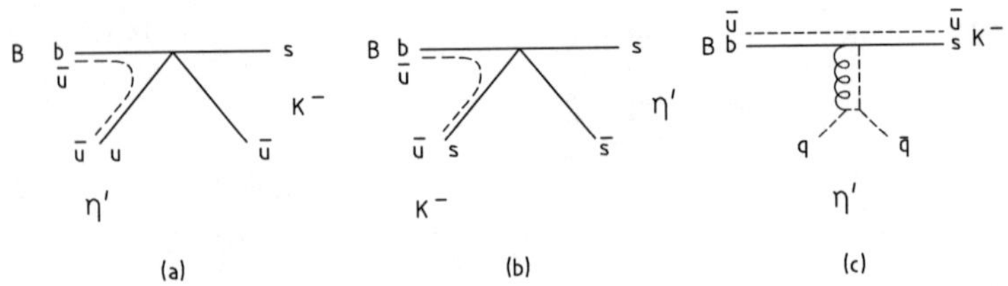

FIGURE 4. Two-body decay $B^- \to K^-\eta'$ (or $B^- \to K^{*-}\eta'$) with three amplitudes from (3): (a) amplitude $p^u_{K^-\eta'}$ with $s \to K^-$, (b) exchange amplitude $p^s_{\eta' K^-}$ with $s \to \eta'$ and (c) amplitude $s_{K^-\eta'}$ for gluonic production of η'.

Application to B-decays into pseudoscalars

As a test of this penguin dominance model we have compared first with data on the decays $B \to PP$, $B \to PV$ [3]. In Tab. 6 this comparison is repeated with new data compiled by HFAG [38]. In col. 2 we show our predictions for 12 decay rates of B^+ in terms of the parameters $p_{PP}, p_{VP}, \gamma_{PP}, \gamma_{VP}, \beta_{VP}$ and the corresponding 12 rates for B^0 obtained after multiplication by the lifetime ratio $\tau(B^0)/\tau(B^+) = 0.921$. From col. 2 various symmetry relations can be obtained, especially the $I = \frac{1}{2}$ rule (favouring charged π or ρ over the neutral decays by a factor 2) for the doublets

$$(K^0\pi^+),\ (K^+\pi^0);\quad (K^+\pi^-),\ (K^0\pi^0);\quad (K^{*0}\pi^+),\ K^{*+}\pi^0);\quad (K^{*+}\pi^-, K^{*0}\pi^0)$$
$$(\rho^+K^0, \rho^0K^+);\qquad (\rho^-K^+, \rho^0K^0)$$

These relations work well, except for one case where the rate for $K^{*0}\pi^0$ is significantly (4.3σ) below the expectation; however, the statistics is very low in this case. Furthermore, there are SU(3) relations between $K^*\pi$ and ϕK, and also between ρK and ωK which work reasonably well.

For a full description we made some simplifying assumptions $\beta_{PV} = -1, \gamma_{PV} = \gamma_{PP}$ which can be removed if necessary with improving statistics. The remaining 3 parameters $p_{PP}, p_{VP}, \gamma_{PP}$ have been determind from 3 input rates. Remarkably, with the data of increased precision obtained in the last year [38] the agreement with the predictions has generally improved in comparison to our earlier results in [3] (2 exceptions with deviations of $> 3\sigma$).

B-decays into scalar particles

After the success of this simple penguin dominance model we take it over to the decays with scalar particles $B \to PS$ and $B \to VS$. We denote the members of the scalar multiplet by a, K^*_{sc}, f_0, f'_0 and define the mixing angle by $f_0 = n\bar{n}\sin\varphi_s + s\bar{s}\cos\varphi_s$, $f'_0 = n\bar{n}\cos\varphi_s - s\bar{s}\sin\varphi_s$, where $n\bar{n} = (u\bar{u} + d\bar{d})/\sqrt{2}$. Then our predictions [3] for scalars are given in Tab. 7. Given the decay branching ratios into scalars one can check any scenario for the multiplet of scalar particles. Hopefully, the symmetries implied by

TABLE 6. Branching ratios for B decays into pseudoscalar (P) and vector (V) particles (cols. 3,4,6,7) and amplitudes (col. 2) as in Eq. (4) with γ_{PP}, γ_{VP} and β_{VP} for gluonic and interchange processes, p_{PP}, p_{VP} set to 1; cols. 3,6: $\gamma_{PP} = \gamma_{VP} = 0.3$, $\beta_{VP} = -1$, $|p_{PP}|^2 = 24.1 \times 10^{-6}$, $|p_{VP}|^2 = 11.4 \times 10^{-6}$. Experimental Data from HFAG, July 2005 [38]

$B^+ \to PP$	$p_{PP} = 1$	$\mathcal{B}_{\text{th}}[10^{-6}]$	$\mathcal{B}_{\text{exp}}[10^{-6}]$	$B^0 \to PP$	$\mathcal{B}_{\text{th}}[10^{-6}]$	$\mathcal{B}_{\text{exp}}[10^{-6}]$
$K^0\pi^+$	1	input p_{PP}	24.1 ± 1.3	$K^+\pi^-$	22.2	18.2 ± 0.8
$K^+\pi^0$	$\frac{1}{\sqrt{2}}$	12.1	12.1 ± 0.8	$K^0\pi^0$	11.1	11.5 ± 1.0
$K^+\eta$	$\frac{1}{\sqrt{3}}\gamma_{PP}$	0.7	2.6 ± 0.5	$K^0\eta$	0.7	< 2.0
$K^+\eta'$	$\frac{1}{\sqrt{6}}(3+4\gamma_{PP})$	input γ_{PP}	70.8 ± 3.4	$K^0\eta'$	65.3	68.6 ± 4.2
$B^+ \to VP$	$p_{VP} = 1$			$B^0 \to VP$		
$K^{*0}\pi^+$	1	input p_{VP}	11.4 ± 1.0	$K^{*+}\pi^-$	10.5	$11.7^{+1.5}_{-1.4}$
$K^{*+}\pi^0$	$\frac{1}{\sqrt{2}}$	5.7	6.9 ± 2.3	$K^{*0}\pi^0$	5.3	1.7 ± 0.8
$K^{*+}\eta$	$\frac{1}{\sqrt{3}}(1-\beta_{VP}+\gamma_{VP})$	20.1	$24.3^{+3.0}_{-2.9}$	$K^{*0}\eta$	18.5	18.7 ± 1.7
$K^{*+}\eta'$	$\frac{1}{\sqrt{6}}(1+2\beta_{VP}+4\gamma_{VP})$	0.8	< 14	$K^{*0}\eta'$	0.1	< 7.6
ρ^+K^0	β_{VP}	11.4	< 48	ρ^-K^+	10.5	$9.9^{+1.6}_{-1.5}$
ρ^0K^+	$\frac{1}{\sqrt{2}}\beta_{VP}$	5.7	$5.11^{+0.82}_{-0.87}$	ρ^0K^0	5.3	$5.11^{+0.82}_{-0.87}$
ωK^+	$\frac{1}{\sqrt{2}}\beta_{VP}$	5.7	5.1 ± 0.7	ωK^0	5.3	5.6 ± 0.9
ϕK^+	1	11.4	$9.03^{+0.65}_{-0.63}$	ϕK^0	10.5	$8.3^{+1.2}_{-1.0}$

TABLE 7. Dominant contributions for B decays into scalar (S) + pseudoscalar (P) or vector (V) particles: penguin amplitudes p_{XY} (normalized to 1 in each sector), exchange and gluonic amplitudes β_{PS}, β_{VS} and $\gamma_{PS}, \gamma_{SP}, \gamma_{VS}$ resp. with scalar mixing angle φ_S; in brackets results for $\sin\varphi_S = 1/\sqrt{3}$ ($\varphi_S \sim \varphi_P$); cols. 3,6: upper sign for B^0, lower sign B^+.

$B^0 \to$ $P+S$	$B^+ \to$ $P+S$	normalization to $p_{PS} = 1$	$B^0 \to$ $V+S$	$B^+ \to$ $V+S$	normalization to $p_{VS} = 1$
K^+a^-	K^0a^+	1	$K^{*+}a^-$	$K^{*0}a^+$	1
K^0a^0	K^+a^0	$\mp\frac{1}{\sqrt{2}}$	$K^{*0}a^0$	$K^{*+}a^0$	$\mp\frac{1}{\sqrt{2}}$
K^0f_0	K^+f_0	$\frac{1}{\sqrt{2}}(1+2\gamma_{PS})\sin\varphi_S$ $+(\beta_{PS}+\gamma_{PS})\cos\varphi_S$ $[\frac{1}{\sqrt{6}}(1+2\beta_{PS}+4\gamma_{PS})]$	$K^{*0}f_0$	$K^{*+}f_0$	$\frac{1}{\sqrt{2}}(1+2\gamma_{VS})\sin\varphi_S$ $+(\beta_{VS}+\gamma_{VS})\cos\varphi_S$ $[\frac{1}{\sqrt{6}}(1+2\beta_{VS}+4\gamma_{VS})]$
K^0f_0'	K^+f_0'	$\frac{1}{\sqrt{2}}(1+2\gamma_{PS})\cos\varphi_S$ $-(\beta_{PS}+\gamma_{PS})\sin\varphi_S$ $[\frac{1}{\sqrt{3}}(1-\beta_{PS}+\gamma_{PS})]$	$K^{*0}f_0'$	$K^{*+}f_0'$	$\frac{1}{\sqrt{2}}(1+2\gamma_{VS})\cos\varphi_S$ $-(\beta_{VS}+\gamma_{VS})\sin\varphi_S$ $[\frac{1}{\sqrt{3}}(1-\beta_{VS}+\gamma_{VS})]$
$\pi^-K_{sc}^{*+}$	$\pi^+K_{sc}^{*0}$	β_{PS}	$\rho^-K_{sc}^{*+}$	$\rho^+K_{sc}^{*0}$	β_{VS}
$\pi^0K_{sc}^{*0}$	$\pi^0K_{sc}^{*+}$	$\mp\frac{1}{\sqrt{2}}\beta_{PS}$	$\rho^0K_{sc}^{*0}$	$\rho^0K_{sc}^{*+}$	$\mp\frac{1}{\sqrt{2}}\beta_{VS}$
ηK_{sc}^{*0}	ηK_{sc}^{*+}	$\frac{1}{\sqrt{3}}(-1+\beta_{PS}+\gamma_{SP})$	ωK_{sc}^{*0}	ωK_{sc}^{*+}	$\frac{1}{\sqrt{2}}\beta_{VS}$
$\eta' K_{sc}^{*0}$	$\eta' K_{sc}^{*+}$	$\frac{1}{\sqrt{6}}(2+\beta_{PS}+4\gamma_{SP})$	ϕK_{sc}^{*0}	ϕK_{sc}^{*+}	1

penguin dominance (isospin, SU(3)) inherent in Tab. 7 will help in selecting the correct assignments of scalar particles. The parameters we have at our disposal are for $B \to PS$: $p_{PS}, \gamma_{PS}, \gamma_{SP}, \beta_{PS}$ and for $B \to VS$: $p_{VS}, \gamma_{VS}, \beta_{VS}$. In our first analysis [3] we used initially, in analogy to the pseudoscalars, $\beta_{PS} = 1$, $\beta_{VS} = -\beta_{PS}$, $\gamma_{VS} = \gamma_{PS} = \gamma_{SP}$.

Considering first the multiplet $\sigma, \kappa, f_0(980)$, $a_0(980)$ along scenario I we note that only $f_0(980)$ has been observed so far. For a meaningful test one would need a measurement of the rates for $B \to K\sigma$ and $B \to \pi\kappa$ which should be possible for a given parametrization.

On the other hand, the decay rates for all members of the multiplet along scenario II $f_0(980)$, $a_0(980)$, $K_0^*(1430)$, $f_0(1500)$ have actually been measured (upper limit for a_0 only). According to our scheme with penguin dominance we should describe these four rates by 3 parameters: p_{PS}, γ_{PS} and β_{PS}.

In a first attempt in 2004 we analysed the data assuming as in case of pseudoscalars $|\beta_{PS}| = 1$. Then we expected for the decay $B(B \to Ka_0^+) \gtrsim 11.0 \times 10^{-6}$ ($\pm 50\%$) The new upper limits from BaBar in Tab. 2 are below this expectation. From Tab. 7 we find $B(B \to K_0^*\pi^\pm)/B(B \to Ka^\pm) = |\beta_{PS}|^2$. The new data require $|\beta_{PS}| \sim 2.7 \ldots 4.6$, or, from averages using the $I = 1/2$ rule $|\beta_{PS}| \gtrsim 2$. The production of a scalar with the spectator is suppressed against production from s-quark.

Until now, there are still considerable experimental uncertainties, especially the ambiguities in the $K_0^*(1430)$ rates and the missing a_0 rate. If we choose $|\beta_{PS}| = 2$ then we find with $B(B \to K_0^*\pi^+) \sim 12 \times 10^{-6}$ (if we include the lower K_0^* mass), for $B(B \to a_0^+K^0) \sim 3 \times 10^{-6}$ and $B(B \to Kf_0(980)) \sim 13 \times 10^{-6}$ four solutions in (β, γ). For $\beta = -2, \gamma = 2$ we find $B(B \to Kf_0(1500)) \sim 25 \times 10^{-6}$ which compares well with Sol. II in Tab. 4. So there is no difficulty in the moment with the multiplet along path II considered. The tests will hopefully become more restrictive with improved data and with measurements of other channels like $K_0^*(1430)\eta$, $K_0^*(1430)\eta'$ and $B \to V+$ scalars.

QCD-improved factorization approximation

In this complementary theoretical investigation one aims at an absolute prediction of rates for scalar particles [42, 43, 44]. This follows the approach applied before to decays $B \to PP$, VP [45]. In the recent work [44] one includes perturbative QCD corrections to the common factorization ansatz but needs to include various non-perturbative objects: formfactors, light cone distribution amplitudes and decay constants where results for scalars are derived from QCD sum rules. In scenario I $\sigma, \kappa, a_0(980), f_0(980)$ are taken as $q\bar{q}$ ground states and $a_0(1450), K^*(1430), f_0(1500)$ as $q\bar{q}$ excited states. In scenario II it is assumed that the low mass multiplet is build of $qq\bar{q}\bar{q}$ states for which no quantitative predictions can be given, whereas the $q\bar{q}$ ground state multiplet includes $a_0(1450), K_0^*(1430)$ and a second multiplet is around 2 GeV.

An early calculation [42] predicted a very small rate for $B^0 \to a_0^+K^-$ which turned out successful. The recent predictions [44] concern decays into $a_0(980)$, $f_0(980)$, also $K^*(1430)$ and $a_0(1430)$. Within scenario I the results on the low mass multiplet are satisfactory whereas the higher mass particles require the low mass solutions with $B(B^- \to K^*(1430)^-\pi^+) < 10 \times 10^{-6}$. In scenario II the K_0^* rates are about twice as large as before, but still smaller than some experimental results.

If this large K_0^* rate is correct, then scenario I is excluded and there are no predictions for the light mesons with $M \lesssim 1$ GeV. It will be important to know the predictions for the other states σ, κ to compare with, likewise predictions for $f_0(1500)$ and the other isoscalar meson.

CONCLUSIONS

1. Experimental results on decays $B \to$ Scalar $+X$
By now $f_0(980), a_0(980), K_0^*(1430)$ and $f_0(1500)$ have been measured in B-decays. Discrete ambiguities are found for $K_0^*(1430)$, $f_0(1500)$ (how about $f_0(980)$?) and emerge naturally in coherent superposions. A clarification is important, possibly these ambiguities can be resolved by physical arguments (comparison with elastic scattering phases, isospin relations fulfilled within $\sim 20\%$).

2. Model with gluonic penguins dominating and $B \to S+X$ amplitudes
This model continues to work well for $B \to PP, VP$ within $\sim 20\%$ or better, especially the $I = \frac{1}{2}$ rule and other SU(3) relations are generally successful. The method has the potential to test the multiplet structure in the scalar sector. Present data within their ambiguities are consistent with a $q\bar{q}$ multiplet $f_0(980), a_0(980), K_0^*(1430), f_0(1500)$. Further tests are possible with $B \to K_0^*(1430)\eta$ (or η') as well as VS rates. The possibility of a light multiplet with σ, κ can be tested once data on $B \to \sigma K, \kappa\pi$ become available.

3. Factorization approach for B-decays into scalar particles
Using QCD sum rules to obtain nonperturbative quantities some absolute predictions have been obtained, a successful one concerns the decay into $a_0(980)$. Further distinctions between different scenarios depend on the magnitude of the ambiguous $K_0^*(1430)\pi$ rate. It will be important to have predictions for the other members of the considered multiplets, especially for $B \to \sigma K, \kappa\pi$, as well as for heavier isoscalars.

4. Broad state: a respectable glueball candidate and the $X(1500), f_X(1500)$ puzzle.
In the $\pi\pi$ channel there is a broad state with $\Gamma \sim M$. It is plausible that $\sigma(600)$ and $f_0(1000)$ refer to the same object. The puzzles with $X(1500), f_X(1500)$ are resolved by taking into account the interference of $f_0(1500)$ with a broad background. The relative signs are explained by taking the background as flavour singlet, in agreement with the glueball hypothesis, and $f_0(1500)$ as a flavour octet state. The same interference phenomenon is known from processes $\pi\pi \to \pi\pi, \eta\eta, K\bar{K}$.

ACKNOWLEDGMENTS

Thanks to the organizers for creating a very lively conference with ample opportunities to in depth discussions!

NOTE ADDED

After this conference a paper by Gronau and Rosner [46] appeared with isospin relations between pairs of B^0 and B^+ 2-body decays as well as 3-body decays also basing on the dominance of penguin amplitudes.

REFERENCES

1. W. Ochs, talk at this conference, http://www.cbpf.br/ hadron05/scientificProg.htm
2. C. Greub and P. Liniger, *Phys. Rev.* **D63**, 054025 (2001).
3. P. Minkowski and W. Ochs, *Eur. Phys. J.* **C39**, 71 (2005), arXiv:hep-ph/0404194.
4. G.S. Bali, in "Fourth Int. Conf. on Perspectives in Hadronic Physics", ICTP, Trieste, Italy, May 2003; arXiv:hep-lat/0308015.
5. S. Narison, *Nucl. Phys.* **B509**, 312 (1998); *Nucl. Phys.* **A675** 54c (2000). E. Bagan and T.G. Steele, *Phys. Lett.* **B243**, 413 (1990); H. Forkel, *Phys. Rev.* **D71**, 054008 (2005).
6. R.L. Jaffe, *Phys. Rev.* **D15**, 267, 281 (1977).
7. S. Eidelman et al. (PDG), *Phys. Lett.* **B592**, 1 (2004).
8. D. Morgan, *Phys. Lett.* **B51**, 71 (1974); M. Roos, N.A. Törnqvist, *Phys. Rev. Lett.* **76**, 1575 (1996).
9. E. van Beveren, presented at this conference.
10. C. Amsler and F.E. Close, *Phys. Lett.* **B353**, 385 (1995); *Phys. Rev.* **D53**, 295 (1996).
11. E. Klempt, B.C. Metsch, C.R. Münz and H.R. Petry, *Phys. Lett.* **B361**, 160 (1995).
12. P. Minkowski and W. Ochs, *Eur. Phys. J.* **C9**, 283 (1999).
13. V.V. Anisovich, Yu.D. Prokoshkin and A.V. Sarantsev, *Phys. Lett.* **B389**, 388 (1996); V.V. Anisovich and A.V. Sarantsev, *Eur. Phys. J.* **A16**, 229 (2003).
14. B. Meadows, these proceedings; E.M. Aitala et al. (E791), hep-ex/0507099.
15. S.D. Protopopescu et al., *Phys. Rev.* **D7**, 1279 (1973).
16. E.M. Aitala et al. (E791), *Phys. Rev. Lett.* **86**, 765 (2001).
17. J.M. Link et al. (FOCUS), *Phys. Lett.* **B585**, 200 (2004).
18. M. Ablikim et al. (BES), *Phys. Lett.* **B598**, 149 (2004).
19. J.A. Oller, E. Oset and J.R. Pelaez, *Phys. Rev.* **D59**, 074001 (1999).
20. G. Colangelo, J. Gasser and H. Leutwyler, *Nucl. Phys.* **B603**, 125 (2001).
21. D. Bugg and J. Pelaez, this conference.
22. B.Hyams et al., *Nucl. Phys.* **B64**,134 (1973); W. Ochs, Ludwig-Maximilians-University Munich, thesis 1973 (unpublished).
23. P. Estabrooks, *Phys. Rev.* **D19**, 2678 (1979).
24. P.Estabrooks and A.D. Martin, *Nucl. Phys.* **B79**, 301 (1974); **B95**, 322 (1975).
25. D. Morgan and M.R. Pennington, *Phys. Rev.* **D48**, 1185 (1993); K.L. Au, D. Morgan and M.R. Pennington, *Phys. Rev.* **D35**, 1633 (1987).
26. M. Ablikim et al. (BES), *Phys.Lett.* **B603**, 138 (2004).
27. J.Z. Bai et al. (BES), *Phys. Rev.* **D68**, 052003 (2003).
28. P. Estabrooks et al., *Nucl. Phys.* **B133**, 490 (1978).
29. D. Aston et al. (LASS), *Nucl. Phys.* **B296**, 493 (1988).
30. P. Büttiker, S. Descotes-Genon, B. Moussallam, *Eur. Phys. J.* **C33**, 409 (2004).
31. C. Goebel (FOCUS), this conference; J.M. Link et al., *Phys. Lett.* **B535**, 43 (2002).
32. B. Aubert et al. (BaBar), *Phys. Rev.* **D71**, 032005 (2005).
33. B. Aubert et al. (BaBar), *Phys.Rev.* **D70**, 111102 (2004).
34. K. Abe et al. (Belle), arXiv: hep-ex/0509003.
35. B. Aubert et al. (BaBar), BaBar-CONF-04/42; arXiv:hep-ex/0408073v3.
36. A. Garmash et al. (Belle), *Phys. Rev.* **D71**, 092003 (2005).
37. B. Aubert et al. (BaBar), BaBar-Conf-05/021; arXiv:hep-ex/0507094.
38. Heavy Flavour Averaging group, P. Chang, R. Harr, F. Lehner, J. Smith; July 2005; http://www.slac.stanford.edu/xorg/hfag/
39. A. Falvard et al. (DM2), *Phys. Rev.* **D38**, 2706 (1988).
40. A.S. Dighe, M. Gronau and J.L. Rosner, *Phys. Lett.* **B367**, 357 (1996); C.W. Chiang, M. Gronau, Z. Luo, J.L. Rosner and D.A. Suprun, *Phys. Rev.* **D69**, 0340 (2004).
41. H.J. Lipkin, *Phys. Lett.* **B415**, 186 (1997).
42. V. Chernyak, *Phys. Lett.* **B509**, 273 (2001).
43. A. Furman, R. Kaminski, L. Lesniak and B. Loiseau, *Phys. Lett.* **B622**, 207 (2005).
44. H.Y. Cheng, C.K. Chua, K.C. Yang, arXiv:ph/0508104.
45. M. Beneke, G. Buchalla, M. Neubert and C.T. Sachrajda, *Phys. Rev. Lett.* **83**, 1914 (1999).
46. M. Gronau and J.L. Rosner, "Symmetry relations in charmless $B \to PPP$ decays"; arXiv:hep-ph/0509155.

Hadron Spectroscopy from BES and CLEOc

Chang-Zheng YUAN

Institute of High Energy Physics, Chinese Academy of Sciences, Beijing, China

Abstract.
Recent results from BES and CLEOc experiments on hadron spectroscopy and charmonium decays using J/ψ, ψ' and ψ'' data samples collected in e^+e^- annihilation are reviewed, including the observation of $X(1835)$ in $J/\psi \to \gamma\pi^+\pi^-\eta'$, study of the scalar particles in J/ψ radiative and hadronic decays, as well as in χ_{c0} hadronic decays, and the study of the "$\rho\pi$ puzzle" in J/ψ, ψ', and ψ'' decays.

Keywords: hadron spectroscopy, charmonium, QCD
PACS: 14.40.Gx, 14.40.Cs, 12.38.Qk, 13.25.Gv

1. INTRODUCTION

BESII [1] running at BEPC and CLEOc [2] running at CESR are the two detectors operating in the τ-charm energy region, and have collected large data samples of charmonium decays including 58 M J/ψ events, 14 M ψ' events, and 33 pb^{-1} data around ψ'' peak at BESII, and 4 M ψ' events, and 281 pb^{-1} ψ'' events at CLEOc. To study the continuum background in the charmonium decays, special data samples at center of mass energy lower than the ψ' mass were taken both at BESII ($\sqrt{s} = 3.65$ GeV) and at CLEOc ($\sqrt{s} = 3.671$ GeV), the luminosity are 6.4 pb^{-1} and 21 pb^{-1} respectively. These data samples are used for the study of the hadron spectroscopy, the D decay properties and the CKM matrix, as well as the charmonium decay dynamics.

In this paper, we focus on the search for the new resonances in J/ψ decays, the properties of the scalars from J/ψ radiative and hadronic decays, and a new approach of studying the scalars using χ_{cJ} decays, and the extensive study of the "$\rho\pi$ puzzle" related physics in J/ψ, ψ' and ψ'' decays.

It should be noted that the CLEOc detector is much better than the BESII detector, especially in the photon detection, this makes its 4 M ψ' events data sample produces results with similar precision as from 14 M ψ' events from BESII.

2. OBSERVATION OF $X(1835)$

The decay channel $J/\psi \to \gamma\pi^+\pi^-\eta'$, with $\eta' \to \pi^+\pi^-\eta$, or $\eta' \to \gamma\rho$, is analyzed using a sample of 58×10^6 J/ψ events collected with the BESII detector [3], to search for the other decay modes of the possible existing $p\bar{p}$ bound state as observed in $J/\psi \to \gamma p\bar{p}$ process at BESII [4]. Figure 1 shows the $\pi^+\pi^-\eta'$ invariant mass spectrum for the combined $J/\psi \to \gamma\pi^+\pi^-\eta'(\eta' \to \pi^+\pi^-\eta)$ and $J/\psi \to \gamma\pi^+\pi^-\eta'(\eta' \to \gamma\rho)$ samples. A clear peak is observed at around 1.835 GeV/c^2. The spectrum is fitted with a Breit-Wigner (BW) function convolved with a Gaussian mass resolution function (with

CP814 *Hadron Spectroscopy: Eleventh International Conference*
edited by A. Reis, C. Göbel, J. Sá Borges, and J. Magnin

$\sigma = 13$ MeV/c^2) to represent the $X(1835)$ signal plus a smooth polynomial background function. The signal yield from the fit is 264 ± 54 events and the statistical significance is 7.7 σ.

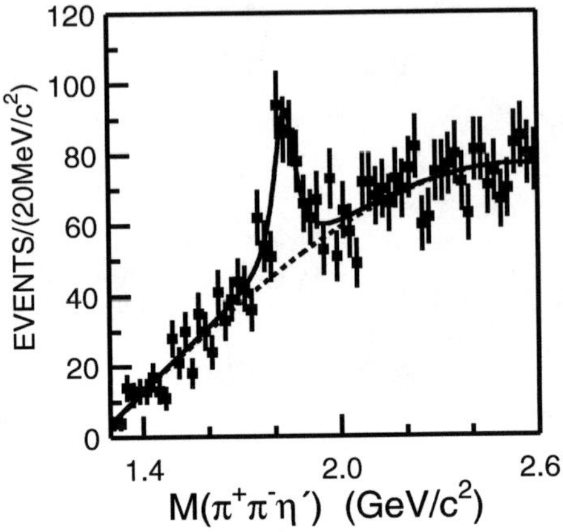

FIGURE 1. The $\pi^+\pi^-\eta'$ invariant mass distribution for selected $J/\psi \to \gamma\pi^+\pi^-\eta'(\eta' \to \pi^+\pi^-\eta, \eta \to \gamma\gamma)$ and $J/\psi \to \gamma\pi^+\pi^-\eta'(\eta' \to \gamma\rho)$ events. Dots with error bars are data, solid line is the fit, and the dashed curve indicates the background function.

The mass of $X(1835)$ is measured to be $M = 1833.7 \pm 6.1(stat) \pm 2.7(syst)$ MeV/c^2, the width is $\Gamma = 67.7 \pm 20.3(stat) \pm 7.7(syst)$ MeV/c^2, and the product branching fraction is $B(J/\psi \to \gamma X) \cdot B(X \to \pi^+\pi^-\eta') = (2.2 \pm 0.4(stat) \pm 0.4(syst)) \times 10^{-4}$. The mass and width of the $X(1835)$ are not compatible with any known meson resonance listed by PDG [5]. In Ref. [5], the candidate closest in mass to the $X(1835)$ is the (unconfirmed) 2^{-+} $\eta_2(1870)$ with $M = 1842 \pm 8$ MeV/c^2. However, the width of 225 ± 14 MeV/c^2, is considerably larger than that of the $X(1835)$ (see also [6], where the 2^{-+} component in the $\eta\pi\pi$ mode of J/ψ radiative decay has a mass 1840 ± 15 MeV/c^2 and a width 170 ± 40 MeV/c^2). Another candidate with mass close to the $X(1835)$ in PDG [5] is the $X(1860)$ observed in the $p\bar{p}$ mass threshold in radiative $J/\psi \to \gamma p\bar{p}$ decays [4], where a mass of 1859^{+6}_{-27} MeV/c^2, and width smaller than 30 MeV/c^2 at 90% C.L. were reported. It has been pointed out that the S-wave BW function used for the fit in Ref. [4] should be modified to include the effect of final-state-interactions (FSI) on the shape of the $p\bar{p}$ mass spectrum [7, 8]. By redoing the S-wave BW fit to the $p\bar{p}$ invariant mass spectrum of Ref. [4] including the Isospin zero, S-wave FSI factor of Ref. [8], BES gets a mass $M = 1831 \pm 7$ MeV/c^2 (in good agreement with the $X(1835)$) and a width $\Gamma < 153$ MeV/c^2 at the 90% C.L. (not in contradiction with the $X(1835)$). There are also theoretical arguments that the two states, $X(1835)$ and $X(1860)$, are indeed one particle, and is a $p\bar{p}$ bound state [9, 10], however, other possible interpretations of the $X(1835)$ that have no relation to the $p\bar{p}$ mass threshold enhancement are not excluded. Further information about the $X(1835)$ and $X(1860)$, especially better precisions on the mass and width measurements and the determination of the spin-parity are strongly desired before concluding the nature of the states.

3. SCALAR PARTICLES IN J/ψ AND χ_{C0} DECAYS

The study of the scalars are very important in two aspects: in experiment, there are still controversies about the resonance parameters of these states; and in theory, it is still hard to incorporate all the experimental results in a self-consistent picture. The reason for the former is somewhat due to the techniques used in extracting the physics information from the experimental data, namely, the partial wave analysis (PWA) was extensively used in the analyses, but sometimes it is rather arbitrary what resonance need to be included in the complicated fit with a few ten and even more than one hundred free parameters. The reason for the latter, in part is due to the fact that the experimental results may not all be reliable, and the complexity in the low energy QCD domain that the higher order terms neglected may not be small, and the mixing of the states in principle is hard to be considered completely.

3.1. Radiative and hadronic J/ψ decays

Using the world largest J/ψ data sample in e^+e^- annihilation experiment, BES studied the scalars decay into pair of pseudoscalars ($\pi^+\pi^-$, $\pi^0\pi^0$, K^+K^- and $K^0_sK^0_s$) in J/ψ radiative decays as well as recoiling against a ϕ or an ω [11, 12, 13, 14]. The full mass spectra and the scalar part in them are shown in Fig. 2.

From the analyses, BES sees significant contributions of σ particle in $\omega\pi^+\pi^-$ and ωK^+K^-, and also hint in $\phi\pi^+\pi^-$. Two independent partial wave analyses are performed on $\omega\pi^+\pi^-$ data and four different parameterizations of the σ amplitude are tried, all give consistent results for the σ pole, which is at $(541\pm 39) - i(252\pm 42)$ MeV/c^2. There is also events accumulation in the low $\pi^+\pi^-$ mass in $\gamma\pi^+\pi^-$ mode, most probably due to the contribution of the σ, however, there is no attempt to analyze the structure at BES, one possible reason is the presentation of the huge background from $\rho^0\pi^0$. Nevertheless, the coupling of the σ with a photon in J/ψ decays is an important piece of information for the understanding of the nature of the particle, a better detector with more statistics may want to measure it. A measurement of the resonance parameters of σ using the BESII $\psi' \to J/\psi\pi^+\pi^-$ data gives similar pole position as measured in $J/\psi \to \omega\pi^+\pi^-$ [15].

Strong $f_0(980)$ is seen in $\phi\pi^+\pi^-$ and ϕK^+K^- modes, from which the resonance parameters are measured to be $M = 965\pm 8(stat)\pm 6(syst)$ MeV/c^2, $g_1 = 165\pm 10(stat)\pm 15(syst)$ MeV/c^2 and $g_2/g_1 = 4.21\pm 0.25(stat)\pm 0.21(syst)$, where M is the mass, and g_1 and g_2 are the couplings to $\pi\pi$ and $K\bar{K}$ respectively if the $f_0(980)$ is parameterized using the the Flatté's formula. The production of $f_0(980)$ is very weak recoiling against an ω or a photon, which indicates $s\bar{s}$ is the dominant component in it.

The $\phi\pi^+\pi^-$ data also show a strong scalar contribution at around 1.4 GeV/c^2, it is due to the dominant $f_0(1370)$ interfering with a smaller $f_0(1500)$ component. The mass and width of $f_0(1370)$ are determined to be: $M = 1350\pm 50$ MeV/c^2 and $\Gamma = 265\pm 40$ MeV/c^2. In $\gamma\pi^+\pi^-$, a similar structure is observed in the same mass region, the fit yields a resonance at mass $1466\pm 6(stat)\pm 16(syst)$ MeV/c^2 with width of $108^{+14}_{-11}(stat)\pm 21(syst)$ MeV/c^2, possibly the $f_0(1500)$, and the contribution from the

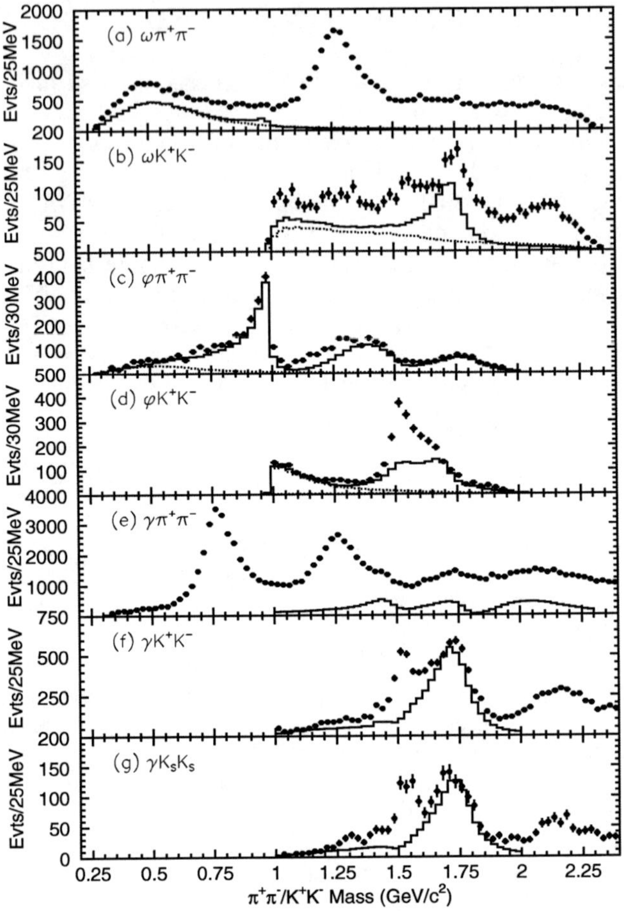

FIGURE 2. The invariant mass distributions of the pseudoscalar meson pairs recoiling against ω, ϕ, or γ in J/ψ decays measured at BESII. The dots with error bars are data, the solid histograms are the scalar contribution from PWA, and the dashed lines in (a) through (c) are contributions of σ from the fits, while the dashed line in (d) is the $f_0(980)$. Notice that not the full mass spectra are analyzed in (e), (f), and (g). Results in (e) are preliminary, otherwise are published.

$f_0(1370)$ can not be excluded. The production of $f_0(1370)$ and $f_0(1500)$ in $\gamma K\bar{K}$ is insignificant.

The K^+K^- invariant mass distributions from $\gamma K\bar{K}$ and ωK^+K^-, the $\pi^+\pi^-$ invariant mass distributions from $\gamma\pi^+\pi^-$, and $\phi\pi^+\pi^-$ show clear scalar contribution around 1.75 GeV/c^2. Two states are resolved from the bump, one is $f_0(1710)$ with $M \sim 1740$ MeV/c^2 and $\Gamma \sim 150$ MeV/c^2 which decays to $K\bar{K}$ mostly, and one possible new state $f_0(1790)$ with $M \sim 1790$ MeV/c^2 and $\Gamma \sim 270$ MeV/c^2 which couples to $\pi\pi$ stronger than to $K\bar{K}$. However, the existence of the second scalar particle needs confirmation: the signal observed in $\phi f_0(1790)$ is rather in the edge of the phase space, and the reconstruction efficiency of the ϕ decreases dramatically as the momentum of the ϕ decreases thus the momentum of the kaon from ϕ decays is very low and can not be detected. Furthermore, there are wide higher mass scalar states above 2 GeV/c^2 as observed in $\gamma\pi^+\pi^-$ (Fig. 2e) and $\gamma K\bar{K}$ [5], whose tails may interfere with the $f_0(1710)$ and produce structure near the edge of the phase space.

The use of these measurements for understanding the nature of the scalar particles can be found in Refs. [16, 17, 18, 19], where the J/ψ decay dynamics and the fractions of the possible $q\bar{q}$ and glueball components in the states are examined.

3.2. Pair production of scalars in $\chi_{c0} \to \pi^+\pi^-K^+K^-$

Partial wave analysis of $\chi_{c0} \to \pi^+\pi^-K^+K^-$ is performed [20] using χ_{c0} produced in ψ' decays at BESII. In 14 M produced ψ' events, 1371 $\psi' \to \gamma\chi_{c0}$, $\chi_{c0} \to \pi^+\pi^-K^+K^-$ candidates are selected with around 3% background contamination.

Fig. 3(a) shows the scatter plot of K^+K^- versus $\pi^+\pi^-$ invariant mass which provides further information on the intermediate resonant decay modes for $(\pi^+\pi^-)(K^+K^-)$ decay, while Fig. 3(b) shows the scatter plot of $K^+\pi^-$ versus $K^-\pi^+$ invariant masses for the resonances with strange quark.

Besides $(\pi\pi)(KK)$ and $(K\pi)(K\pi)$ modes, $(K\pi\pi)K$ mode which leads to a measurement of $K_1(1270)K$ and $K_1(1400)K$ decay processes is also included in the fit. The PWA results are summarized in Table 1. From these results, we notice that scalar resonances have larger decay fractions compared to those of tensors, and such decays provide a relatively clean laboratory to study the properties of scalars, such as $f_0(980)$, $f_0(1370)$, $f_0(1710)$, and so forth. The upper limits of the pair production of the scalar mesons which are less significant are determined at the 90% C.L. to be

$$\mathscr{B}[\chi_{c0} \to f_0(1370)f_0(1370)]\mathscr{B}[f_0(1370) \to \pi^+\pi^-]\mathscr{B}[f_01370) \to K^+K^-] < 2.9 \times 10^{-4},$$
$$\mathscr{B}[\chi_{c0} \to f_0(1370)f_0(1500)]\mathscr{B}[f_0(1370) \to \pi^+\pi]\mathscr{B}[f_0(1500) \to K^+K^-] < 1.8 \times 10^{-4},$$

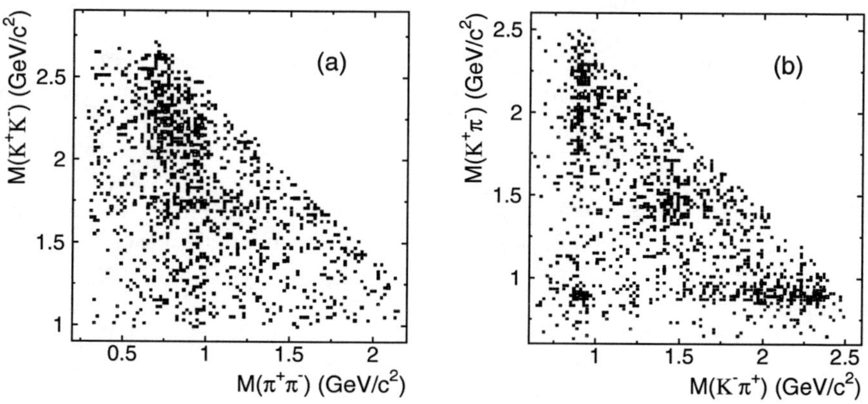

FIGURE 3. The scatter plots of (a) K^+K^- versus $\pi^+\pi^-$ and (b) $K^+\pi^-$ versus $K^-\pi^+$ invariant mass for selected $\psi' \to \gamma\chi_{c0}$, $\chi_{c0} \to \pi^+\pi^-K^+K^-$ events.

$\mathscr{B}[\chi_{c0} \to f_0(1500) f_0(1370)] \mathscr{B}[f_0(1500) \to \pi^+\pi^-] \mathscr{B}[f_0 1370) \to K^+K^-] < 1.4 \times 10^{-4}$,
$\mathscr{B}[\chi_{c0} \to f_0(1500) f_0(1500)] \mathscr{B}[f_0(1500) \to \pi^+\pi^-] \mathscr{B}[f_0(1500) \to K^+K^-] < 0.55 \times 10^{-4}$,
$\mathscr{B}[\chi_{c0} \to f_0(1500) f_0(1710)] \mathscr{B}[f_0(1500) \to \pi^+\pi^-] \mathscr{B}[f_0(1710) \to K^+K^-] < 0.73 \times 10^{-4}$.

TABLE 1. Summary of the $\chi_{c0} \to \pi^+\pi^-K^+K^-$ results, where X represents the intermediate decay modes, N^{fit} is the number of fitted events, and s.s. indicates signal significance.

Decay mode (X)	N^{fit}	**Branching ratio** (10^{-4}) $\mathscr{B}[\chi_{c0} \to X \to \pi^+\pi^-K^+K^-]$	**s.s.**
$f_0(980)f_0(980)$	27.9 ± 8.7	$3.46 \pm 1.08^{+1.93}_{-1.57}$	5.3σ
$f_0(980)f_0(2200)$	77.1 ± 13.0	$8.42 \pm 1.42^{+1.65}_{-2.29}$	7.1σ
$f_0(1370)f_0(1710)$	60.6 ± 15.7	$7.12 \pm 1.85^{+3.28}_{-1.68}$	6.5σ
$K^*(892)^0\bar{K}^*(892)^0$	64.5 ± 13.5	$8.09 \pm 1.69^{+2.29}_{-1.99}$	7.1σ
$K_0^*(1430)\bar{K}_0^*(1430)$	82.9 ± 18.8	$10.44 \pm 2.37^{+3.05}_{-1.90}$	7.2σ
$K_0^*(1430)\bar{K}_2^*(1430) + c.c.$	62.0 ± 12.1	$8.49 \pm 1.66^{+1.32}_{-1.99}$	8.7σ
$K_1(1270)^+K^- + c.c.$, $K_1(1270) \to K\rho(770)$	68.3 ± 13.4	$9.32 \pm 1.83^{+1.81}_{-1.64}$	8.6σ
$K_1(1400)^+K^- + c.c.$, $K_1(1400) \to K^*(892)\pi$	19.7 ± 8.9	< 11.9 (90% C.L.)	2.7σ

The above results supply important information on the understanding of the natures of the scalar states [21].

3.3. Comments on the PWA

PWA is extensively used in extracting physics information from the experimental data, all the information listed above in this section is from PWA. While we know the PWA uses the information in the full phase space for physics study so that it is more powerful than working in one dimension (invariant mass, for example) or a bit higher dimension, it suffers from too many free parameters and other technical problems.

First of all, the experimental data are all affected by the finite resolution in momentum, energy, or direction measurement, this was not considered in current analyses; secondly, the parametrization of the resonance, especially the wide resonance, still have room to improve. Finally, the effect of the imperfect simulation of the detectors is hard to be considered in a fit with a few ten or even more free parameters. These effects may not be significant when the statistics is low, however, as the statistics increases, all these effects will possibly produce significant fake signals. How to handle these, if not now, at least in the near future, should be studied since high luminosity experiments will be soon available.

Studying the papers dealing with the PWA, it is found that two important information are missing in most of the analyses, which are the goodness-of-fit and the correlation coefficients between the fit out parameters.

In many of the analyses, the comparison to the alternative fits are given to show the fit is the best among all the fits tested, however this does not guarantee the fit used in the analysis is reliable, just like a fit to a Gaussian using a 2nd order polynomial is better than using a straight line, but does not mean the fit is acceptable. To give the goodness-of-fit is not easy in case of PWA since the fitting function is always very complicated. A possible way is to supply a simple χ^2 test in one or multi-dimension as has been done in Ref. [20], although not perfect, it shows the reader a feeling how the fit describes the data. Another possible way is try to use toy Monte Carlo to get the expected distribution of the χ^2 or the likelihood after generating many MC samples using the fit out parameters as input, and compare with the one got from the fit to the data — this may be a bit time consuming, however, it is more reliable since the detector effects are considered.

The parameters from the fit are generally correlated, and sometimes, some variables are highly correlated, in this case, only reporting the diagonal error is not enough. This is extremely important when there are two nearby resonances with strong interference, they are anti-correlated and one component will increase (decrease) as the other decrease (increase) to make the total contribution unchanged. Neglecting this in the theoretical analysis will also be dangerous. Another effect of the correlation is the significance of the signal may be affected strongly, that is, a declared significance may not be as high as that determined when there is no correlation, and vice versa.

4. "$\rho\pi$ PUZZLE" IN VECTOR CHARMONIA DECAY

From perturbative QCD (pQCD), it is expected that both J/ψ and ψ' decaying into light hadrons are dominated by the annihilation of $c\bar{c}$ into three gluons or one virtual photon, with a width proportional to the square of the wave function at the origin [22]. This yields the pQCD "12% rule",

$$Q_h = \frac{\mathscr{B}_{\psi' \to h}}{\mathscr{B}_{J/\psi \to h}} = \frac{\mathscr{B}_{\psi' \to e^+e^-}}{\mathscr{B}_{J/\psi \to e^+e^-}} \approx 12\%. \tag{1}$$

A large violation of this rule was first observed in decays to $\rho\pi$ and $K^{*+}K^- + c.c.$ by Mark II [23], known as *the $\rho\pi$ puzzle*, where only upper limits on the branching fractions were reported in ψ' decays. Since then, many two-body decay modes of the ψ' have been measured by the BES collaboration and recently by the CLEO collaboration; some decays obey the rule while others violate it [24, 25].

The extension of the above rule to ψ'' is straightforward, in the same scheme, one would expect

$$Q'_h = \frac{\mathscr{B}_{\psi'' \to h}}{\mathscr{B}_{J/\psi \to h}} = \frac{\mathscr{B}_{\psi'' \to e^+e^-}}{\mathscr{B}_{J/\psi \to e^+e^-}} \approx 1.9 \times 10^{-4}. \tag{2}$$

As the ψ'' data samples are available both at BESII and CLEOc, the search for the decays of ψ'' into light hadrons was performed. Since ψ'' is above the charm threshold, it is expected the dominant decays of it is to charmed meson pairs, however, large fraction of charmless decays of ψ'' is expected if ψ'' is a mixture of S- and D-wave charmonium states and the mixing is responsible for the "12% rule" violation in J/ψ and ψ' decays. The above two rules may be tested by the large data samples of J/ψ, ψ', and ψ'' at both BESII and CLEOc.

4.1. $\psi \to \rho\pi$: current status

The $\rho\pi$ mode of the vector charmonia decays is essential for this study, since this is the first puzzling channel found in J/ψ and ψ' decays. The new measurements, together with the old information, show us a new picture of the charmonium decay dynamics.

4.1.1. $J/\psi \to \pi^+\pi^-\pi^0$

BESII measured the $J/\psi \to \pi^+\pi^-\pi^0$ branching fraction using its J/ψ and $\psi' \to J/\psi\pi^+\pi^-$ data samples [26], and BABAR measured the same branching fraction using J/ψ events produced by initial state radiative (ISR) events at $\sqrt{s} = 10.58$ GeV [27]. Together with the measurement from Mark-II [23], we get a weighted average of the $\mathcal{B}(J/\psi \to \pi^+\pi^-\pi^0) = (2.00 \pm 0.09)\%$.

To extract the $J/\psi \to \rho\pi$ branching fraction, PWA is needed to consider the possible contribution from the excited ρ states, the only available information on the fraction of $\rho\pi$ in $J/\psi \to \pi^+\pi^-\pi^0$ was got at Mark-III. Using the information given in Ref. [28], we estimate $\frac{\mathcal{B}(J/\psi \to \rho\pi)}{\mathcal{B}(J/\psi \to \pi^+\pi^-\pi^0)} = 1.17(1 \pm 10\%)$, with the error from an educated guess based on the information in the paper since we have no access to the covariant matrix from the fit showed in the paper. From this number and the $\mathcal{B}(J/\psi \to \pi^+\pi^-\pi^0)$ got above, we estimate $\mathcal{B}(J/\psi \to \rho\pi) = (2.34 \pm 0.26)\%$. This is substantially larger than the world average listed by PDG [5], which is $(1.27 \pm 0.09)\%$.

4.1.2. $\psi' \to \pi^+\pi^-\pi^0$

$\psi' \to \rho\pi$ was studied both at BESII [29] and CLEOc [25]. After selecting two charged pions and two photons, clear π^0 signals are observed in the two photon invariant mass spectra, the numbers of signals are found to be 229 and 196 from BESII (shown in left plot of Fig. 4 as an example) and CLEOc respectively, and the branching fraction of $\psi' \to \pi^+\pi^-\pi^0$ is measured to be $(18.1 \pm 1.8 \pm 1.9) \times 10^{-5}$ and $(18.8^{+1.6}_{-1.5} \pm 1.9) \times 10^{-5}$ at BESII and CLEOc respectively. The two experiments give results in good agreement with each other. The Dalitz plot of $\psi' \to \pi^+\pi^-\pi^0$ events (right plot of Fig. 4) shows very different signature from that of $J/\psi \to \pi^+\pi^-\pi^0$ decays, there are lots of events in the center part of the Dalitz plot in the former case, while in the latter, almost all the events are in the ρ mass region and there is almost nothing in the center of the Dalitz plot.

To extract the branching fraction of $\psi' \to \rho\pi$, however, BESII uses a PWA including the high mass ρ states and the interference between them, while CLEOc counts the number of events by applying a ρ mass cut. The branching fraction from BESII is $(5.1 \pm 0.7 \pm 1.1) \times 10^{-5}$, while that from CLEOc is $(2.4^{+0.8}_{-0.7} \pm 0.2) \times 10^{-5}$, the difference is large. Although big difference exist between BESII and CLEOc results exist, these do mean the $\psi' \to \rho\pi$ signal exists, rather than the signal is completely missing as has been guessed before. If we take a weighted average neglecting the difference between the two measurements, we get $\mathcal{B}(\psi' \to \rho\pi) = (3.1 \pm 0.7) \times 10^{-5}$.

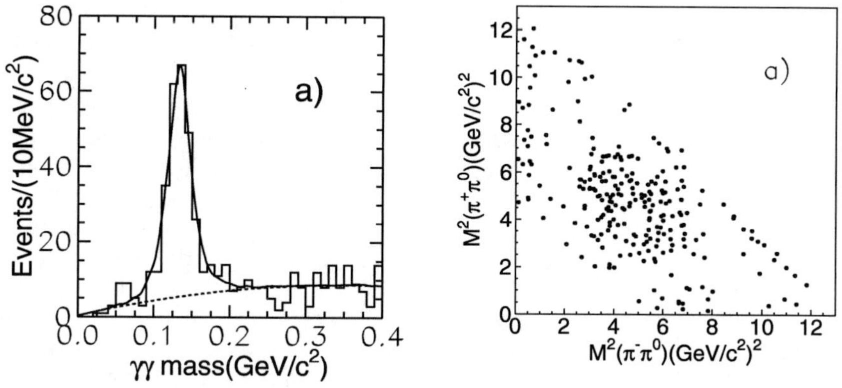

FIGURE 4. Two photon invariant mass distribution (left) and the Dalitz plot (right) after final selection for BESII ψ' data. The histograms are data, and the curves show the best fits.

Comparing $\mathscr{B}(\psi' \to \rho\pi)$ with $\mathscr{B}(J/\psi \to \rho\pi)$, one gets

$$Q_{\rho\pi} = \frac{\mathscr{B}(\psi' \to \rho\pi)}{\mathscr{B}(J/\psi \to \rho\pi)} = (0.13 \pm 0.03)\%.$$

The suppression compared to the 12% rule is obvious.

4.1.3. $\psi'' \to \pi^+\pi^-\pi^0$

It has been pointed out that the continuum amplitude plays an important role in measuring ψ'' decays into light hadrons [30]. In fact, there are two issues need to be clarified in ψ'' decays, that is whether ψ'' decays into light hadrons really exist, and if it exists, how large is it. By comparing the cross sections of $e^+e^- \to \pi^+\pi^-\pi^0$ at the ψ'' resonance peak ($\sqrt{s} = 3.773$ GeV) and at a continuum energy point ($\sqrt{s} = 3.65$ GeV at BESII and 3.671 GeV at CLEOc) below the ψ' peak, both BESII and CLEOc found that $\sigma(e^+e^- \to \pi^+\pi^-\pi^0)$ at continuum is larger than that at ψ'' resonance peak. The average of the two experiments [31, 32] are

$$\sigma(e^+e^- \to \pi^+\pi^-\pi^0)_{\text{on}} = 7.5 \pm 1.2 \text{ pb},$$

$$\sigma(e^+e^- \to \pi^+\pi^-\pi^0)_{\text{off}} = 13.7 \pm 2.6 \text{ pb}.$$

The difference, after considering the form factor variation between 3.65 and 3.773 GeV, is still significant, and it indicates there is an amplitude from ψ'' decays which interferes destructively with the continuum amplitude, which makes the cross section at the ψ'' peak smaller than the pure contribution of continuum process.

For the $\rho\pi$ mode, BESII can only give upper limit of its cross section due to the limited statistics of the data sample, the upper limit at 90% C. L. is found to be 6.0 pb [31] at the ψ'' peak, which is in consistent with the measurement from CLEOc using a much larger data sample: $\sigma(e^+e^- \to \rho\pi)_{\rm on} = 4.4 \pm 0.6$ pb [32]; while the cross section at the continuum is $8.0^{+1.7}_{-1.4} \pm 0.9$ pb measured by CLEOc.

To extract the information on the $\psi'' \to \rho\pi$ branching fraction, BESII developed a method based on the measured cross sections at ψ'' resonance peak and at the continuum [31]. By neglecting the electromagnetic decay amplitude of ψ'', there are two amplitudes contribute to the cross section at the ψ'' peak, the strong decay amplitude of ψ'' and the continuum amplitude. Taking the continuum amplitude as a real number, the ψ'' strong decay amplitude is described as one real number for the magnitude, and one phase between ψ'' strong and electromagnetic decays to describe the relative phase between the two amplitudes. Since only two measurements are available (at ψ'' peak and at continuum), one can only extract ψ'' decay branching fraction as a function of the relative phase. BESII measurement on the upper limit of the $e^+e^- \to \rho\pi$ cross section at ψ'' peak, together with the CLEOc measurement of the continuum cross section restrict the physics region of the branching fraction and the relative phase as shown in Fig. 5(left). From the Figure, we see that the branching fraction of $\psi'' \to \rho\pi$ is restricted within 6×10^{-6} and 2.4×10^{-3}, and the phase is between $-150°$ and $-20°$, at 90% C.L.

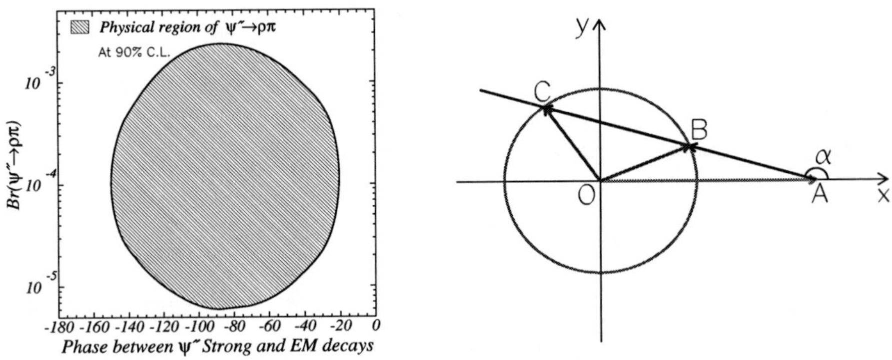

FIGURE 5. Physics region on $\mathcal{B}(\psi'' \to \rho\pi)$ and the relative phase between ψ'' strong and electromagnetic decays from BESII (left); and the illustration of the two solutions in ψ'' decays (right): $\vec{OA}$ represents the continuum amplitudes, $\vec{OB}$ or $\vec{OC}$ represents the peak amplitudes, and $\vec{AB}$ and $\vec{AC}$ are the two solutions for the resonance decay amplitudes, α is the relative phase between continuum amplitude and the ψ'' strong decay amplitude, including the relative phase ϕ and the relative phase between continuum and ψ'' Breit-Wigner amplitudes.

The observation of the $e^+e^- \to \rho\pi$ signal at ψ'' peak and the measurement of the cross section [32] at CLEOc further make the physical region in the branching ratio and relative phase plane smaller: the CLEOc measurement gives a similar out bound of the physical region as BES gives, while also indicates the central part of the physical region in Fig. 5(left) is not allowed by physics. By using a toy Monte Carlo to simulate the CLEOc selection criteria and the interference between resonance and continuum amplitudes, we found that the $\psi'' \to \rho\pi$ branching fraction could be either $(2.1 \pm 0.3) \times 10^{-3}$ or $(2.4^{+3.4}_{-2.0}) \times 10^{-5}$ from the CLEOc measurements, if the relative phase between ψ'' strong and electromagnetic decay amplitudes is $-90°$ as observed in J/ψ and ψ' decays [33].

The reason why there are two solutions for ψ'' decays can be understood as illustrated in Fig. 5b. If we take the continuum amplitude as a real number, it can be shown as a vector along the real axis in the complex plane, the total cross section at ψ'' peak only gives the magnitude of the total amplitude and it is shown as a circle in the plot, the amplitude of the ψ'' decays, represented by a vector connecting the end of the continuum amplitude and the end of the total amplitude may have two cross points with the circle, representing the two solutions of the ψ'' decay amplitudes, and thus the branching fractions. Only in some very special cases, there is only one solution. The angle between the continuum amplitude and the ψ'' decay amplitudes is shown in the plot as α, it is the sum of the relative phase ϕ between the strong and electromagnetic decays of ψ'', and the phase from Breit-Wigner function for the ψ'' resonance. It can be seen that, the two solution only happens when the cross section at continuum is larger than or equal to that at ψ'' peak; otherwise, there is only one solution, as in $\psi' \to \rho\pi$ case. However, as in physics, there is only one ψ'' decay branching fraction, there must be a way to extract the ψ'' decay branching fraction experimentally, this could be made possible if one more data sample is taken at a different energy point, for example, in the shoulder of the ψ'' resonance; furthermore, if one even wants to get the value of the relative phase, one more data point is necessary. In total, to get concrete physics information on the ψ'' decay branching fraction, at least data samples at four different energy points are needed, better have one point far from the resonance, so that it gives absolute measurement of the continuum amplitude.

Based on current data samples, one gets

$$\mathcal{Q}'_{\rho\pi} = \frac{\mathcal{B}(\psi'' \to \rho\pi)}{\mathcal{B}(J/\psi \to \rho\pi)} = (9.0 \pm 1.6)\% \text{ or } (0.10^{+0.15}_{-0.09})\%,$$

to be compared with the pQCD prediction of 0.019% if ψ'' is pure D-wave charmonium.

4.2. Other studies and comments

There are many more new measurements on ψ' decays for the extensive study of the "12% rule" [24, 34, 25], among which the Vector-Pseudoscalar (VP) modes are measured at first priority. The ratios of the branching fractions are all suppressed for these VP modes compared with the 12% rule. The multi-hadron modes and the Baryon-antibaryon modes are either suppressed, or enhanced, or normal, which are very hard to categorize simply. The various models, developed for interpreting specific mode may hard to find solution for all these newly observed modes.

The ψ'' decays into light hadrons were searched for in various ψ'' decay modes, including VP and multi-hadron modes [32, 35]. However, only the comparison between the cross sections at continuum and those at ψ'' resonance peak are given, instead of giving the ψ'' decay branching fractions. In current circumstances, it is still not clear whether the ψ'' decays into light hadrons with large branching fractions, since, as has been shown in the $\rho\pi$ case, there could be two solutions for the branching fraction, and the two values could be very different.

As the luminosity at the ψ'' peak is large enough, current study is limited by the low statistics at the continuum: at CLEOc, the luminosity at continuum is more than an order of magnitude smaller than that at peak, this prevents from a high precision comparison between the cross sections at the two energy points. One conclusion we can draw from the existing data is that the measurements do not contradict with the assumption that the relative phase between ψ'' strong and electromagnetic decay amplitudes is around $-90°$, and the ψ'' decays into light hadrons could be large.

The study of the $\rho\pi$ puzzle between ψ' and J/ψ decays and the charmless decays of ψ'' should not be isolated as they were since J/ψ, ψ' and ψ'' are all charmonium states with very similar quantum numbers, and it is expected ψ' and ψ'' are the mixtures of $2S$ and $1D$ states [36]. In developing models to solve one of the problems, the others should also be considered. There have been a few models developed following this line or can be easily extended to all these three states, like the S- and D-wave charmonia mixing model [36], the $D\bar{D}$ re-annihilation in ψ'' [37], the four-quark component in ψ'' [38], and survival $c\bar{c}$ in ψ' [39], and so on. Experimentally testable predictions are welcome for validating the models.

One further observation is that many of the attempts to interpret the $\rho\pi$ puzzle are based on the potential models for the charmonium which were developed more than 20 years ago, as the B-factories discovered many new charmonium states [40] which are hard to be explained in the potential models, it may indicate even the properties of J/ψ, ψ' and ψ'' are not as expected from the potential models. The further understanding of the other high mass charmonium states may shed light on the understanding of the low lying ones.

5. SUMMARY

There are many new results on hadron spectroscopy from the charmonium decays from BES and CLEOc experiments. While many analyses supply more information on the known states like the light scalar particles to understand the nature of them, there are also new observations which may indicate there are still something unexpected in the low energy regime such as the possible existing baryonium. The decay properties of the vector charmonium states, although have been studied for more than three decades, is still far from being understood, one extreme example is the "$\rho\pi$ puzzle" in J/ψ and ψ' decays. Further studies of all these are expected from the BESIII at BEPCII which will start its data taking in 2007.

ACKNOWLEDGMENTS

The author would like thank the organizer for the invitation to give the talk, and for the successful organization of the conference. I would also like to thank Dr. Xiaohu Mo, Dr. Ping Wang and Dr. Zheng Wang for their contributions and for many helpful discussions. I also thank Dr. Liaoyuan Dong, Dr. Zijin Guo, Dr. Yingchun Zhu and Dr. R. Briere for their help in preparing the talk and this paper.

REFERENCES

1. BES Collaboration, J. Z. Bai *et al.*, Nucl. Instr. Meth. A **344**, 319 (1994); A **458**, 627 (2001).
2. CLEO-c Collaboration, CLEO-c and CESR-c: A New Frontier of Weak and Strong Interactions, CLNS 01/1742.
3. BES Collaboration, M. Ablikim *et al.*, hep-ex/0508025, accepted by Phys. Rev. Lett.
4. BES Collaboration, J.Z. Bai *et al.*, Phys. Rev. Lett. **91**, 022001 (2003).
5. Particle Data Group, S. Eidelman *et al.*, Phys. Lett. B **592**, 1 (2004).
6. BES Collaboration, J.Z. Bai *et al.*, Phys. Lett. **B446**, 356 (1999).
7. B.S. Zou and H.C. Chiang, Phys. Rev. D **69**, 034004 (2003).
8. A. Sibirtsev *et al.*, Phys. Rev. D **71**, 054010 (2005).
9. G.J. Ding and M.L. Yan, Phys. Rev. C **72**, 015208 (2005).
10. S.L. Zhu and C.S. Gao, hep-ph/0507050.
11. BES Collaboration, M. Ablikim *et al.*, Phys. Lett. B **598**, 149 (2004).
12. BES Collaboration, M. Ablikim *et al.*, Phys. Lett. B **603**, 138 (2004).
13. BES Collaboration, M. Ablikim *et al.*, Phys. Lett. B **607**, 243 (2004).
14. BES Collaboration, J. Z. Bai *et al.*, Phys. Rev. D **68**, 052003 (2003).
15. Y.S. Zhu (for the BES Collaboration), this conference.
16. L. Roca *et al.*, Nucl. Phys. A **744**, 127 (2004).
17. D. Bugg, this conference.
18. F. Close and Q. Zhao, hep-ph/0504043.
19. Q. Zhao, B.S. Zou and Z.B. Ma, hep-ph/0508088.
20. BES Collaboration, M. Ablikim, *et al.*, hep-ex/0508050.
21. Q. Zhao, hep-ph/0508086.
22. T. Appelquist and H. D. Politzer, Phys. Rev. Lett. **34**, 43 (1975); A. De Rújula and S. L. Glashow, Phys. Rev. Lett. **34**, 46 (1975).
23. Mark II Collaboration, M. E. B. Franklin *et al.*, Phys. Rev. Lett. **51**, 963 (1983).
24. Many results may be found in Ref. [5]; more recent results may be found in BES Collaboration, J. Z. Bai *et al.*, Phys. Rev. D **69**, 072001 (2004); Phys. Rev. Lett. **92**, 052001 (2004); BES Collaboration, M. Ablikim *et al.*, Phys. Rev. D **70**, 112007 (2004); Phys. Rev. D **70**, 112003 (2004); and Phys. Lett. B **614**, 37 (2005).
25. CLEO Collaboration, N. E. Adam *et al.*, Phys. Rev. Lett. **94**, 012005 (2005).
26. BES Collaboration, J. Z. Bai *et al.*, Phys. Rev. D **70**, 012005 (2004).
27. BES Collaboration, J. Z. Bai *et al.*, Phys. Rev. D **70**, 072004 (2004).
28. MARK-III Collaboration, L. Chen and W.M. Dunwoodie, in *Proceedings of Hadron 91 Conference, College Park, Maryland, 1991*, p. 100, SLAC-PUB-5674 (1991).
29. BES Collaboration, M. Ablikim *et al.*, Phys. Lett. B **619**, 247 (2005).
30. P. Wang, X. H. Mo and C. Z. Yuan, Phys. Lett. B **574**, 41 (2003).
31. BES Collaboration, M. Ablikim *et al.*, hep-ex/0507092, Phys. Rev. D (in press).
32. CLEO Collaboration, G.S. Adams *et al.*, hep-ex/0509011.
33. P. Wang, C. Z. Yuan and X. H. Mo, Phys. Rev. D **69**, 057502 (2004).
34. BES Collaboration, M. Ablikim *et al.*, Phys. Rev. D **71**, 072006 (2005); CLEO Collaboration, R. A. Briere *et al.*, Phys. Rev. Lett. **95**, 062001 (2005); CLEO Collaboration, T. K. Pedlar *et al.*, Phys. Rev. D **72**, 051108(R) (2005). X. H. Mo (for the BES Collaboration), this conference.
35. CLEO Collaboration, G. S. Huang *et al.*, hep-ex/0509046.
36. J. L. Rosner, Phys. Rev. D **64**, 094002 (2001).
37. J. L. Rosner, hep-ph/0405196.
38. M. Voloshin, Phys. Rev. D **71**, 114003 (2005).
39. P. Artoisenet, J. M. Gerard and J. Weyers, hep-ph/0506325; J. M. Gérard and J. Weyers, Phys. Lett. B **462**, 324 (1999).
40. K. K. Seth, this conference.

σ, κ and $f_0(980)$ in E791 and BES II data

D.V. Bugg

Queen Mary, University of London, Mile End Rd., London E1 4NS, UK

Abstract. Both σ and κ are well established from E791 data on $D \to 3\pi$ and $D_s \to K\pi\pi$ and BES II data on $J/\Psi \to \omega\pi^+\pi^-$ and $K^+K^-\pi^+\pi^-$. Fits to these data are accurately consistent with $\pi\pi$ and $K\pi$ elastic scattering when one allows for the Adler zero which arises from Chiral Symmetry Breaking. The phase variation with mass is also consistent between elastic scattering and production data.

Keywords: mesons, resonances
PACS: 13.75Jx, 14.40.Cs, 1440.Ev

At the conference, results on a large range of experiments on the σ and κ were presented. Here, available space limits the discussion to E791 and BES II data. Space also prevents a review of the large background of theoretical papers fitting elastic scattering. An extended review covering all these topics will appear on the web.

THE σ POLE

Early evidence for the σ pole came from elastic scattering data. Markushin and Locher [1] summarise many determinations. Renewed interest was sparked off by E791 data on $D^+ \to (\pi^+\pi^-)\pi^+$ [2]. The $\pi\pi$ mass projection, shown in Fig. 1, has a low mass peak. The fit assumed a conventional Breit-Wigner resonance with $\Gamma(s) \propto \rho(s)$, where $\rho(s)$ is Lorentz invariant phase space $2k/\sqrt{s} = \sqrt{1 - 4m_\pi^2/s}$ and k is centre of mass momentum. This choice of $\Gamma(s)$ will later be shown to be inappropriate, but correcting it to a better form introduces only changes of detail. The pole position is shown in the first entry of Table 1 below. Oller's very interesting refit of these data with ChPT input is also shown

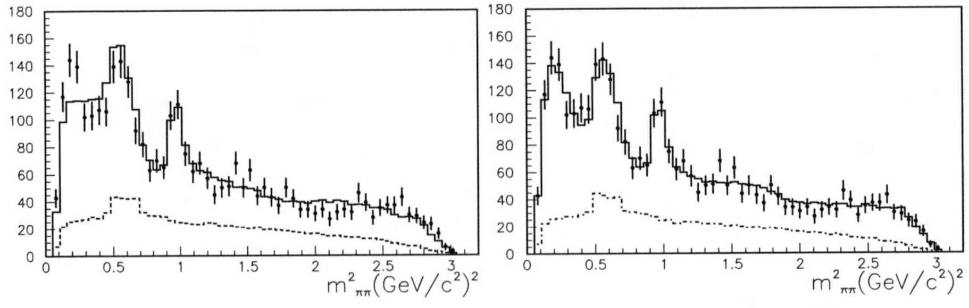

FIGURE 1. The $\pi\pi$ mass projection of E791 data for $D^+ \to \pi^+(\pi^-\pi^+)$, (a) without, (b) with σ in the fit. The dashed histogram shows background.

CP814 *Hadron Spectroscopy: Eleventh International Conference*
edited by A. Reis, C. Göbel, J. Sá Borges, and J. Magnin

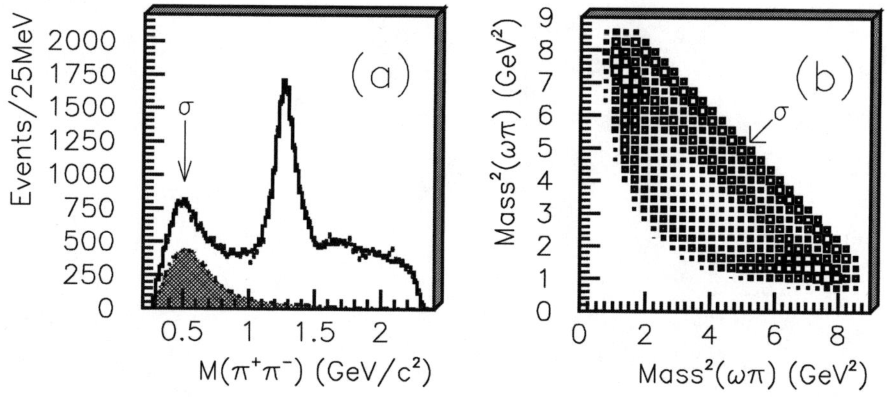

FIGURE 2. (a) The $\pi\pi$ mass projection of BES data; the histogram shows the fit and the hatched area the σ contribution; (b) the Dalitz plot.

there [3].

Fig. 2 shows higher statistics BES II data [4] for $J/\Psi \to \omega\pi^+\pi^-$. Dominant signals are $f_2(1270)$, $b_1(1235)$ and σ, which is clearly visible as a flat band along the right-hand edge of the Dalitz plot in (b). The 0^+ contribution is shown shaded in (a). All four parametrisations which were tried are consistent with an average pole position $\mathrm{M} = (541 \pm 39) - \mathrm{i}(252 \pm 42)$ MeV.

Fig. 3(a) shows the intensity of $\pi\pi$ elastic scattering v. mass. The obvious puzzle is why there is no low mass peak like that in production data of Fig. 2. There is a simple explanation, given in 1965-6 by Adler and Weinberg [5]. They proposed that massless π of zero momentum have zero scattering amplitude. If the $I = 0$ S-wave $\pi\pi \to \pi\pi$ amplitude is expanded as a power series $am_\pi^2 + bk^2$, consistency between s, t and u channels requires that the amplitude is proportional to $(s - 0.5m_\pi^2)$ and has a zero at the Adler point $s_A = 0.5m_\pi^2$. Fig. 3(b) shows the result of dividing (a) by $(s - s_A)^2$. Instantly one sees a resemblance with the σ peak of Fig. 2. So the solution to the puzzle is that the matrix element for $\pi\pi$ elastic scattering is strongly s-dependent: a situation unlike most other resonances.

Let us write the elastic σ amplitude as

$$T_{el}^{00} = [\eta \exp(2i\delta) - 1]/2i = \frac{N(s)}{D(s)} = \frac{N_{el}(s)}{M^2 - s - iN_{tot}(s)}. \quad (1)$$

Here $N_{el}(s)$ is real for $s \geq 0$; the phase variation comes purely from the denominator

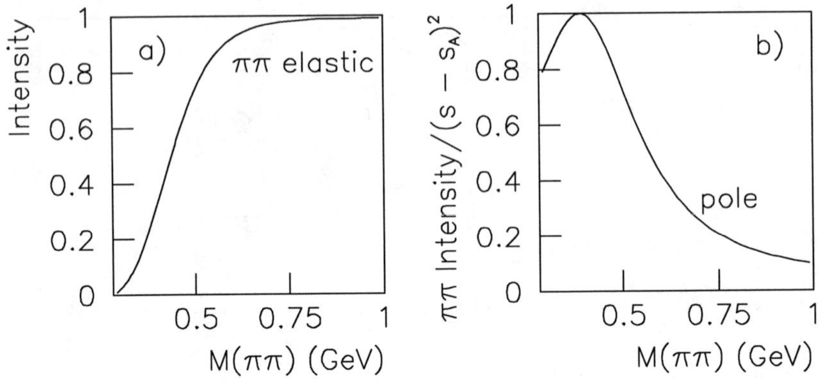

FIGURE 3. (a) The $\pi\pi$ intensity in elastic scattering, (b) with the Adler zero divided out.

$D(s)$. This denominator is universal for all processes involving a $\pi\pi$ pair. For elastic scattering, the Adler zero in $N(s)$ nearly cancels the σ pole for low masses. However, the numerator $N(s)$ is not universal; it is quite different for production processes, where the left-hand cut is distant. Later, it will be shown that E791 data for $D_s \to (K\pi)\pi$ require $N(s)_{prodn} = 1$ within errors. The production amplitude will therefore be written

$$T^{00}_{prodn} = \Lambda / D(s), \tag{2}$$

where Λ is a complex constant.

Fig. 4 illustrates contours of constant intensity for (a) production, (b) elastic scattering. In (b), the Adler zero suppresses the intensity near threshold. The elastic phase shift on the real s axis (where experiments are done) reaches 90° only at $M > 900$ MeV, far above the pole. It is this feature which confuses many people. The phase varies rapidly off the real axis because the width of the resonance increases with s.

Data from BES II, K_{e4} [6] and Cern-Munich [7] are fitted empirically with:

$$N(s) = M(s - 0.5m_\pi^2)\exp[-(s - M^2)/A](1 + \beta s)\rho_{\pi\pi}(s) + M\Gamma_{4\pi}(s). \tag{3}$$

The exponential is required by $\pi\pi$ elastic data to cut off $N(s)$ above 1 GeV. The term $M\Gamma_{4\pi}$ fits inelasticity above 1 GeV, but has little effect on the σ pole. Note from Fig. 2(a) that the σ intensity fitted to BES data is small above 1 GeV. The σ pole is therefore distinct from the broad $f_0(1535)$ fitted by Anisovich and Sarantsev [8]. It is also distinct from the broad pole fitted around 1 GeV by Au, Morgan and Pennington [9]. Contributions from $f_0(980)$, $f_0(1370)$ and $f_0(1500)$ to elastic scattering are included by multiplying their S-matrices with S_σ, to satisfy unitarity; i.e. their *phases* add.

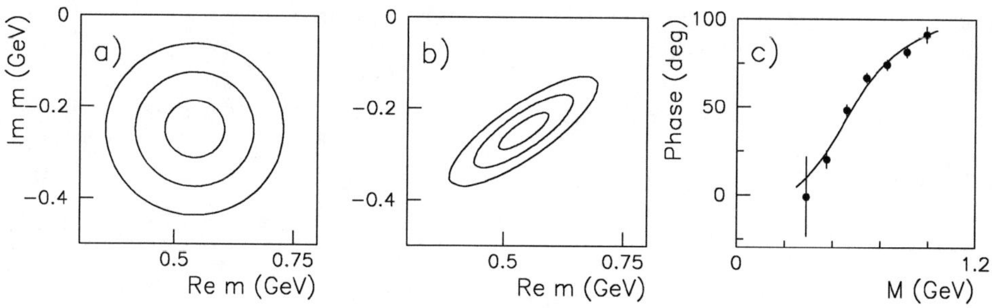

FIGURE 4. Contours of intensity for (a) production, (b) elastic scattering; (c) the phase of the σ amplitude in mass bins 100 MeV wide, compared with the global fit.

TABLE 1. Summary of pole positions.

State	Reference	Data	Pole position (MeV)
σ	[2]	$D \to (\pi^+\pi^-)\pi^+$	$(489 \pm 26) - i(173 \pm 26)$
	[3]	$D \to (\pi^+\pi^-)\pi^+$	$470 - i220$
	[4]	$J/\Psi \to \omega(\pi^+\pi^-)$	$(541 \pm 39) - i(252 \pm 42)$
	[10]	$\pi\pi \to \pi\pi$	$(470 \pm 30) - i(295 \pm 20)$
	[13]	$\pi\pi \to \pi\pi$	$445 - i221$
	[25]	$\pi\pi \to \pi\pi$	$(470 \pm 50) - i(570 \pm 50)$
κ	[28]	$D_s \to (K\pi)\pi$	$(721 \pm 61) - i(292 \pm 131)$
	[30]	$J/\Psi \to K^+\pi^-K^-\pi^+$	$(760 \pm 41) - i(420 \pm 75)$
	[29]	$J/\Psi \to K^+\pi^-K^-\pi^+$	$(841 \pm 82) - i(309 \pm 87)$
	[33]	$K\pi \to K\pi$	$(722 \pm 60) - i(386 \pm 50)$
	here	all	$750^{+30}_{-55} - i(342 \pm 60)$
	[3]	$D_s \to (K\pi)\pi$	$710 - i310$
	[12]	$K\pi \to K\pi$	$770 - i(250 - 425)$
	[14]	$K\pi \to K\pi$	$708 - i305$
	[15]	$K\pi \to K\pi$	$753 - i235$
	[24]	$K\pi \to K\pi$	$(594 \pm 79) - i(724 \pm 322)$
$f_0(980)$	[34]	$J/\Psi \to \phi\pi^+\pi^-$	$(998 \pm 4) - i(17 \pm 4)$
	[12]	$\pi\pi \to \pi\pi$ and KK	994 - i14
$a_0(980)$	[35]	$\bar{p}p \to \eta\pi\pi$ and $\omega\eta\pi^0$	$(1036 \pm 5) - i(84 \pm 9)$

For production, there are hundreds of open channels for D and J/Ψ decays. Within individual channels, unitarity plays a negligible role. Following the standard isobar model, *amplitudes* are added using a complex coupling constant $\Lambda = g\exp(i\phi_0)$ for each amplitude.

The K_{e4} data are available up to 380 MeV and there is then a gap in elastic data until 560 MeV, where Cern-Munich data begin. The σ pole lies in the mass range where there are no elastic data. Although this gap may be bridged by using dispersion relations, the production data are obviously important in filling the gap directly.

In the BES data, $b_1\pi$ contributes 41% of the intensity and $\sigma\omega$ 19%. Strong interferences between them determine the phase variation of the σ with mass. The data have been divided into bins 100 MeV wide from 400 to 1000 MeV. Unfortunately, the b_1 band runs off the corner of the Dalitz plot and does not interfere significantly with the σ below 400 MeV. Fig. 4(c) shows phases for individual bins, in good agreement with the global fit.

Theory

Colangelo, Gasser and Leutwyler [10] have made a precise determination of the σ pole from elastic data and K_{e4} without using production data. This work has the virtue of fitting both the physical region and the left-hand cut.

Oset, Oller, Pelaez and collaborators fit elastic data using 'unitarised' Chiral Perturbation Theory [11-16]. They use ChPT for lowest order and take rescattering from the next order. They fit successfully not only the $I = 0$ S-wave, but also the repulsive $I = 2$ S-wave. This implies that they fit the nearby left-hand cut correctly as well as the physical region.

Schechter's group has also contributed a series of papers on all of σ, κ, $f_0(980)$ and $a_0(980)$ [17–21]. This work examines possible mixing between 2-quark and 4-quark states. Zheng and collaborators at Peking University have developed new types of dispersion relations and have applied them to analysis of data on the σ and κ [22–25].

Van Beveren and Rupp have an interesting but quite different scheme, modelling the spectrum and decays of all mesons from the lightest to charmonium and bottomonium states. This is done with a harmonic oscillator [26] or an arbitrary confining potential [27]. They allow for decays by coupling q-qbar states to outgoing mesons through a transition potential. In this scheme, $f_0(1370)$, $a_0(1450)$, etc. are regular but unitarised $q\bar{q}$ states; σ, κ, $f_0(980)$ and $a_0(980)$ appear as 'extra' states created by coupling of $q\bar{q}$ to decay channels [26,27].

THE κ POLE

E791 data on $D_s \to (K\pi)\pi$ provided the first evidence for the κ pole from production data [28]. A combined fit to these and other data will be considered in detail below.

Next, BES II data on $J/\Psi \to K^+\pi^-K^-\pi^+$ reveal the κ in $J/\Psi \to K^*(890)\kappa$ [29,30]. If one selects a $K^\pm\pi^\mp$ pair within 50 MeV of 890 MeV, the accompanying $K^\mp\pi^\pm$ pair has the mass projection shown in Fig. 5(a) after dividing out phase space. The broad low mass peak below the $K^*(890)$ is the evidence for the κ; one point has a rather large statistical fluctuation. Full details of my fit to these data are given in Ref. [30]. Other strong components are $K^*(890)K_0(1430)$, $K^*(890)K_2(1430)$, $K_0(1430)K_0(1430)$ and $KK_1(1270) + KK_1(1400)$, followed by $K_1 \to K^*\pi$ and ρK. Since K_1 decays populate the low mass $K\pi$ range, it is essential to demonstrate that K_1 decays do not reproduce the κ peak.

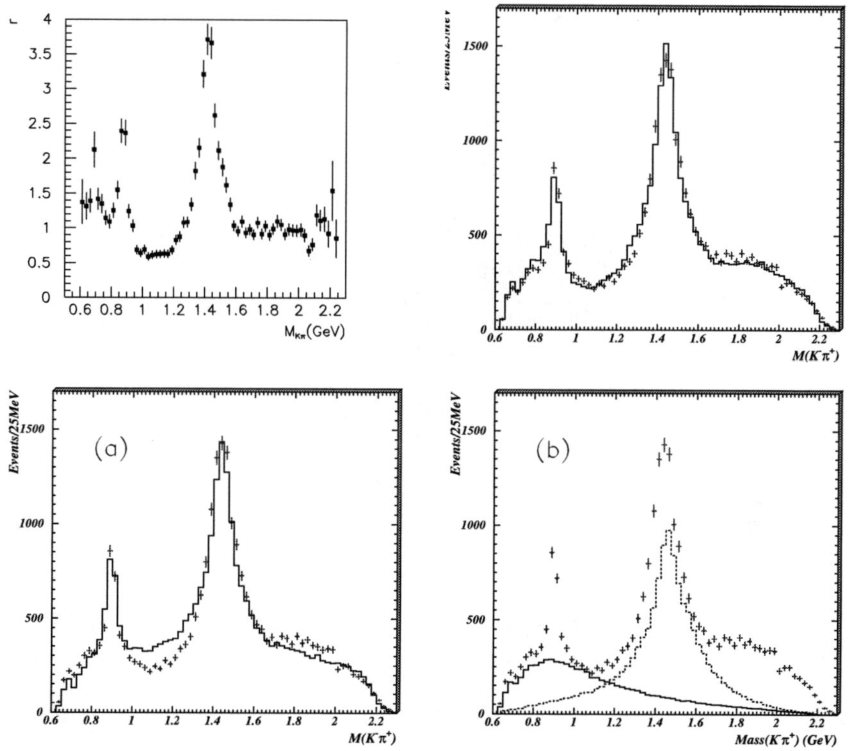

FIGURE 5. BES II data on the κ. Upper left: the $K\pi$ S-wave spectrum in $J/\Psi \to K^*(890)(K\pi)_S$ after factoring out phase space; upper right: the fit to $m(K\pi)$; (a) the bad fit if the Adler zero is omitted; (b) the κ contribution (full histogram) and $K^0(1430)$ (dashed).

A combined fit is made to BES and LASS data [31] for the $K\pi$ $I = 1/2$ S-wave. The fit uses eqn. (1) with

$$N(s) = M(s - s_A)\exp(-\alpha\sqrt{s})\rho_{K\pi}(s) \tag{4}$$

and $s_A = m_K^2 - 0.5m_\pi^2$. The fit to the raw $K\pi$ mass projection is shown in Fig. 5(b). If the factor $(s - s_A)$ is omitted, the poor fit is shown in Fig. 5(c). The κ and $K_0(1430)$ mass projections are shown in Fig. 5(d). There is destructive interference between them. Sensitivity to this interference is one reason for fitting the LASS data simultaneously. A second point is to show that eqn. (4) fits both sets of data successfully.

The phase variation of the κ with mass is well determined in two ways which agree. Firstly, there is large interference between channels $K^*(980)\kappa$ and $KK_1(1270 + 1400)$. Secondly, there is a large interference between κ and $K_0(1430)$, which both contribute to the $K\pi$ S-wave. A bin-by-bin fit has been made where the κ signal is fitted in magnitude and phase in 10 individual bins 100 MeV wide. Results are shown in Fig. 8(c) below.

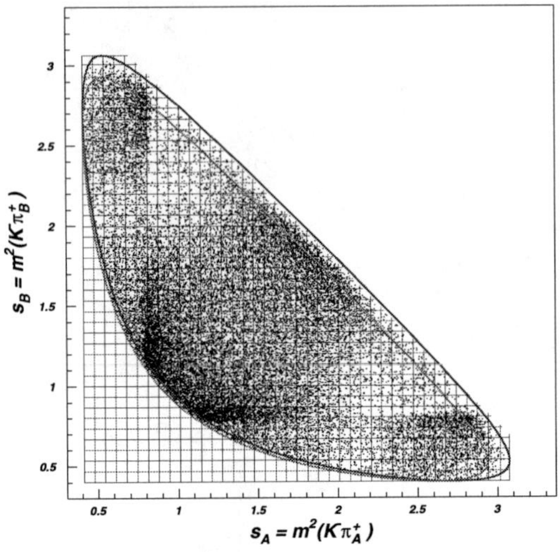

FIGURE 6. The Dalitz plot for E791 data for $D_s \to K\pi\pi$.

An independent analysis of exactly the same data has been reported by the BES group [29]. This fit omits interferences of the κ with both $K_0(1430)$ and the K_1's. It therefore provides no information on the phase of the κ. It also does not fit $K^*(890)$ decays; angular correlations with these decays are crucial in separating channels $K^*(890)K_0(1430)$ and $K^*(890)K_2(1430)$.

RE-ANALYSIS OF E791 DATA

The Dalitz plot for E791 data in shown in Fig. 6. There is obvious interference between $K^*(890)$ and the $K\pi$ S-wave, creating an asymmetry around the K^* band. A new fit has been reported recently where the magnitude and phase of the $K\pi$ S-wave anplitude is fitted separately in 37 mass bins [32]. Magnitudes and phases are shown below in Figs. 7 and 8. No attempt is made to separate κ and $K_0(1430)$.

Both E791 and LASS data contain fairly weak $K_0(1430)$ structures. Results depend the variation of κ intensity under the $K_0(1430)$; i.e. the width of the κ is correlated with the fit to $K_0(1430)$. I have made a combined fit to the data of LASS, BES and E791. The BES data define well the $K_0(1430)$ peak, which is much more conspicuous than in either LASS or E791 data.

In the E791 fit (and also Ref. [28]), the amplitude includes a production form factor $F = \exp(-\alpha q^2)$, where q is the momentum of the κ in the D_s rest frame and $\alpha = 2.08$ GeV^{-2}. I have varied α and Fig. 7 shows results. Panel (a) uses $\alpha = 0$ and gives the best fit. In (b)–(d) α increases in equal steps to the E791 value in (d). Within errors, α optimises at 0. This corresponds to a point-like decay $D_s \to \kappa\pi$ with an

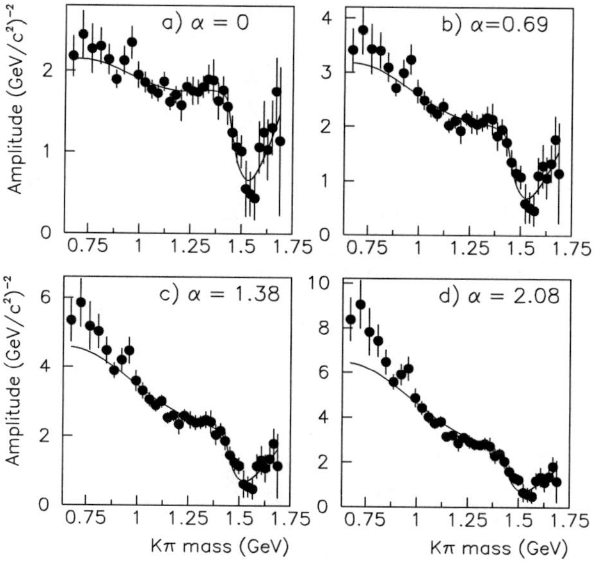

FIGURE 7. The fit to the magnitude of the $K\pi$ S-wave in E791 data for four values of α in the production form factor.

RMS radius < 0.38 fm with 95% confidence. Fits to the phase of the κ also optimise at $\alpha = 0$. The combined fit gives a κ pole at M $= (750^{+30}_{-55}) - i(342 \pm 60)$ MeV. This compares with M $= (722 \pm 60) - i(386 \pm 50)$ MeV for LASS data alone [33] and M $= (760 \pm 20 \pm 40) - i(420 \pm 45 \pm 60)$ MeV for LASS + BES data only [30]. It reduces errors and brings all three sets of data into close agreement.

Fits to the BES and LASS data are shown in Fig. 8. Figs. 8(a) and (b) show the fit to LASS data; (c) and (d) show that the fit to the magnitude and phase of the κ in BES data agrees. Again, the phase variation with mass agrees between elastic scattering and production data. This demonstrates that both can be fitted with the same $D(s)$, and therefore that the data can be fitted with a resonance alone without the need for any background amplitude. The $K_0(1430)$ is a large signal in BES data with well defined centroid and width. The fit to this peak is shown in Fig. 8(e) using 25% $K^*(890)K_2(1430)$ and 75% $K^*(890)K_0(1430)$, as determined from my fit to BES data; (f) shows the good fit to E791 phases using $\alpha = 0$.

$f_0(980)$ AND $a_0(980)$

An excellent new determination of the parameters of $f_0(980)$ is obtained from BES data on $J/\Psi \to \phi\pi^+\pi^-$ and ϕK^+K^- [34]. The $f_0(980)$ appears as a strong peak in $\pi\pi$ and a smaller but clear peak in KK. >From the relative numbers of events in these two

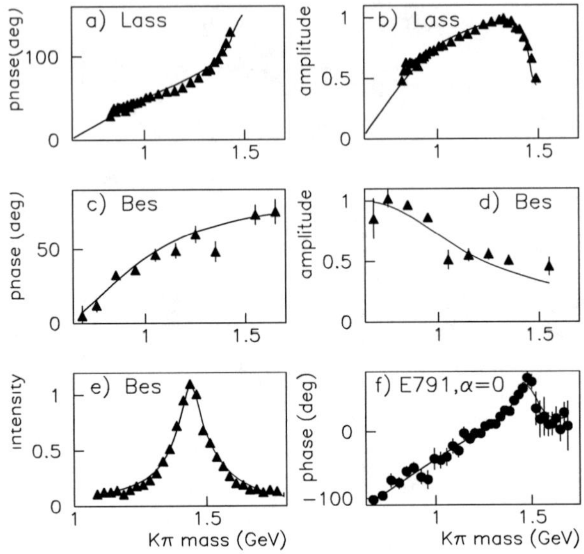

FIGURE 8. Fits to (a) Lass phases for $K\pi$ elastic scattering; (b) magnitudes of the Lass elastic scattering amplitude; (c) the phase of the κ signal in BES II data after dividing into 100 MeV mass bins, (d) BES magnitudes in individual bins, (e) the magnitude of the $K_0(1430)$ peak above background in BES data, (f) E791 phases with $\alpha = 0$.

peaks, the ratio $g^2(KK)/g^2(\pi\pi)$ is well determined. Previously results for this ratio have been erratic. The pole position is shown in Table 1, together with the best available pole position for $a_0(980)$ [35].

INTERPRETATION OF σ, κ, $f_0(980)$ AND $a_0(980)$

This is contentious. The front runner appears to be Jaffe's proposal [36] that they are 4-quark states with $SU(3)$ flavour content $3 \otimes \bar{3}$. His scheme explains naturally the progression of their masses: if $f_0(980)$ and $a_0(980)$ are $s\bar{s}(u\bar{u} \pm d\bar{d})$ states, their near degeneracy is explained naturally, and the large mass difference between σ and $a_0(980)$. There is support for this interpretation from Lattice QCD calculations, where 4-quark combinations dominate at large r and 2-quark combinations at small r [37]. The decay presumably occurs by fission at small r. There is again support for this picture from Ref. [33], where it is shown that the Fourier transform of the σ amplitude v, momentum has an RMS radius of 0.4 fm.

However, the story cannot be quite so simple. The ratio $r = g^2(f_0(980) \to KK)/g^2(a_0(980) \to KK) = 2.7 \pm 0.5$ disagrees with the ratio 1 predicted from Jaffe's model. The likely explanation is that the pole position of $f_0(980)$ is fortuitously very

TABLE 1. Main decay channels of η and η' mesons with related BR's.

η decay channel	BR(%)	ΔBR(%)	η' decay channel	BR(%)	ΔBR(%)
$\gamma\gamma$	39.42	0.26	$\eta\pi^+\pi^-$	44.3	1.5
$\pi^0\pi^0\pi^0$	32.56	0.28	$\gamma\pi^+\pi^-$	29.5	1.0
$\pi^+\pi^-\pi^0$	22.6	0.4	$\eta\pi^0\pi^0$	20.0	1.2
$\pi^+\pi^-\gamma$	4.68	0.11	$\omega\gamma$	3.03	0.31
			$\gamma\gamma$	2.12	0.14

which, although three times lower than the design goal, provides nowadays $\sim$ 8 pb^{-1}/day. Most of the results shown here are based on 450 pb^{-1} of collisions recorded in 2001−2002 data-taking period. After a long shutdown for machine upgrading, we have resumed a stable data-taking on June 2004. In this last run, the machine has already delivered 1916 pb^{-1} of which 1812 recorded by our experiment. We intend to run until the end of 2005 to reach a total integrated luminosity of 2.5 fb^{-1}. The entire data set will correspond to a production of $\sim$ 8 billions of ϕ mesons. Although the branching ratios, BR, are relatively low, the radiative ϕ decays to light scalar and pseudoscalar mesons make DAΦNE a clean η, η', f_0 and a_0 factory. Moreover, apart from the ϕ meson itself, a large production of ρ (ω) vector mesons of 10^9 (10^7) events is also accessible through $\phi \to \rho\pi$ ($e^+e^- \to \omega\pi^0$) processes. In the following, we will restrict our discussion only on light scalar and pseudoscalar mesons and we will review the prospects of our measurement on the hadronic cross section.

KLOE is composed of an hermetic fine sampling lead/scintillating fiber electromagnetic calorimeter, EMC, surrounding a large He-based fully stereo drift chamber, DCH. The whole detector is embedded in a superconducting coil providing a 0.52 T axial field. An excellent momentum resolution, $\sigma(\mathrm{P_T})/\mathrm{P_T} \sim 0.4\%$ at 90°, a good energy resolution, $\sigma_{\mathrm{E}}/\mathrm{E} = 5.7\% \,/\sqrt{\mathrm{E/GeV}}$, a time resolution of $\sigma_{\mathrm{T}} = 54$ ps$/\sqrt{\mathrm{E/GeV}}$ and a high efficiency to low energy photons are the main detector characteristics. We finally remark that, although the detector was not optimized for hadronic physics, its performances and the large dimensions make the overall acceptance and angular resolution excellent for fully neutral decays. A large improvement on the energy resolution is therefore obtained by means of kinematic fitting when imposing the total 4-momentum conservation.

THE LIGHT PSEUDOSCALAR MESONS

The η and η' mesons are produced through radiative ϕ decays with BR of 1.3 % and 6×10^{-5} respectively. These events can be identified by looking at the monochromatic radiative recoil photon. In the η case, the recoil energy is 363 MeV and, in almost all cases, it is very well separated by the end-point of the energy spectrum of all other photons/clusters in the event. This is not always true for the η' since the recoil photon has an energy peaked at $\sim$ 60 MeV. On the entire sample we expect 80×10^6, 450 k events for the η and η' mesons respectively.

TABLE 2. Rare η decay channels with related BR or 90% C.L. limits.

η decay channel	BR(%)	Kind	η decay channel	BR(%)	Kind
$e^+e^-\gamma$	$(6.0 \pm 0.8)\times 10^{-3}$	Dalitz	$\pi^+\pi^-$	$< 1.3\times 10^{-5}$	P,CP
$\mu^+\mu^-\gamma$	$(3.1 \pm 0.4)\times 10^{-4}$	Dalitz	$\pi^0\pi^0$	$< 4.3\times 10^{-5}$	P,CP
$e^+e^-e^+e^-$	$< 6.9\times 10^{-5}$	D.Dalitz	$\pi^0\pi^0\pi^0\pi^0$	$< 6.9\times 10^{-7}$	P,CP
$\pi^+\pi^-e^+e^-$	$(4.0^{+14.0}_{-2.7})\times 10^{-4}$	D.Dalitz	$\pi^0e^+e^-$	$< 4\times 10^{-5}$	C
$\pi^0\gamma\gamma$	$(4.0\pm 1.9)\times 10^{-4}$	CHPT	$\pi^0\mu^+\mu^-$	$< 5\times 10^{-6}$	C
$\gamma\gamma\gamma$	$< 1.6\times 10^{-5}$	C	$\mu^+e^- + \mu^-e^+$	$< 6\times 10^{-6}$	LF

The η meson

The main decay channels

All the main decay channels of the η (left column of Tab. 1) have been extensively studied at KLOE. Fit to the Dalitz-plots of $\eta \to 3\pi$ decays have been performed to extract information on the dynamics of such processes. The fit results are reported in detail in another contribution to these proceedings [3]. The $\eta \to \gamma\gamma$ events have been looked either for normalization studies [4] or to measure the η mass. In Fig. 1.left the distribution of $M_{\gamma\gamma}$ for the events with three photons final state is shown. Well resolved π^0 and η peaks emerge over a negligible background. We expect to measure the η mass with precision better than 50 keV to help clarifying the existing experimental situation. [3] The identification of the $\eta \to \pi^+\pi^-\gamma$ channel is less clean due to some contamination of Bhabha and $\phi \to \pi^+\pi^-\pi^0$ events. However, a good background rejection (S/B $\sim$ 400) is obtained when requiring tight cuts on the recoil photon energy and on the recoil mass of the $\pi^+\pi^-$ system. The interest of such a decay relies on the measurement of the di-pion spectrum which can be sensitive to the box anomaly [6]. Optimization of the cuts is in progress to minimize the bias on this spectrum.

Summarizing, apart from the specific dynamics studies, we can perform a tagged measurement, practically background free, of the four main η decay channels. Statistics is not an issue and we should be able to reduce the existing errors on the BR's which will also help in reducing the systematics on the measurement of the η'-gluonium contents (see η' section).

Rare η decays

The large statistics and the clean environment allow us to look also for rare η decays. For the single conversion (Dalitz) decay we expect to get $\sim$ 500 k (25 k) events in the $e^+e^-\gamma$ ($\mu^+\mu^-\gamma$) final state. This allows to study the di-lepton spectrum and probe the

---

[3] The GEM collaboration has measured [5] in the reaction $p + d \to ^3He + X$ the η mass to be: $M_\eta = (547.31 \pm 0.03_{stat} \pm 0.03_{syst})$ MeV/c^2. This number strongly differs from the previous precise determination by NA48 of: $M_\eta = (547.84 \pm 0.03_{stat} \pm 0.04_{syst})$ MeV/c^2.

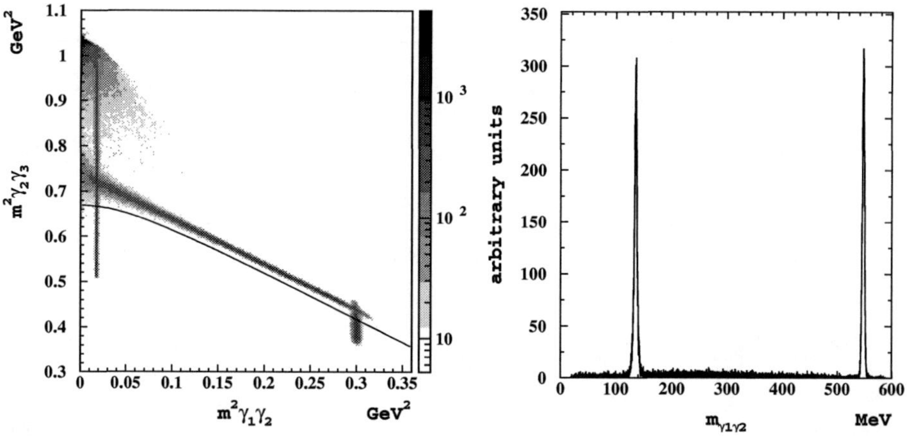

FIGURE 1. Three photons final state for a sample of 50 pb^{-1}: (Left-plot) Dalitz-plot distribution where the photon index is ordered by increasing energy. (Right-plot) projection of $M_{\gamma_1\gamma_2}$ for the events below the solid line drawn in the Dalitz.

transition dynamics of the $\eta - \gamma$ vertex since this spectrum is related to the point-like QED expression via a form factor as follows: $d\Gamma/dq^2 = d\Gamma/dq^2|_{QED} \cdot |F_{\eta\gamma}(q^2)|^2$.

The double Dalitz decays can also be accessed. We expect to get $\sim$ 30 k (6 k) decays on $\pi^+\pi^-e^+e^-$ $(e^+e^-e^+e^-)$. The $\pi^+\pi^-e^+e^-$ final state can be used to test CP violation beyond the CKM mechanism. If we define, in the η center of mass system, ϕ_η as the angle between the $\pi^+\pi^-$ and e^+e^- planes, the following CP violating asymmetry $A_{\phi_\eta} = \frac{N sin\phi cos\phi > 0 - N sin\phi cos\phi < 0}{N sin\phi cos\phi > 0 + N sin\phi cos\phi < 0}$ can be constructed and measured. Standard model prediction are vanishingly small [7].

The most intriguing search on rare η decays we are carrying out on the 2001−2002 collected data concerns the BR measurement of $\eta \to \pi^0\gamma\gamma$. The experimental difficulties of such a search relies on the relative abundance of the $\eta \to 3\pi^0$ decay and on the not negligible fraction of clusters from different photons merged in a single object. It is not a surprise that, in the PDG, you find this BR decreasing by three order of magnitudes in 40 years. The new generation of experiment have a much better acceptance and/or high performance detectors and significantly reduce the $\eta \to 3\pi^0$ and other source of background. KLOE gives [3] a preliminary estimate of the BR of $(8.4 \pm 3.0_{stat+syst}) \times 10^{-5}$. This result has a central value three times smaller than the previous measurement and is in reasonable agreement with p^6 CHPT calculations. The 2.5 fb^{-1} we are collecting will allow us to reduce the statistical error of $\sim$ 2 and to perform at the same time a fit to the invariant mass of the di-photon system which can test VDM and CHPT calculations [8].

In the right column of Tab. 2 a list of the rarest decays of the η related to violation of C, P, CP, LF are shown together with their measured BR or 90 % C.L. limits. Our sensitivity is shown by the fact that the best word limits on $\eta \to 3\gamma/\pi^+\pi^-$ have been measured by KLOE [9, 10] with the 450 pb^{-1} collected in 2001−2002; these two decays are background dominated and we do not expect to improve more than a factor $\sqrt{2}$

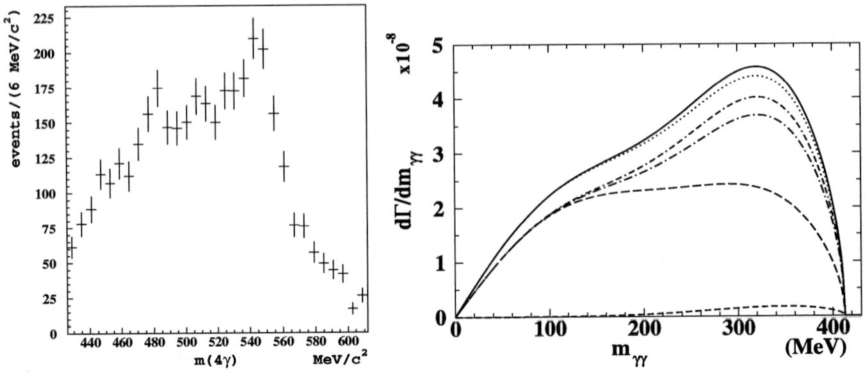

FIGURE 2. Search for $\eta \to \pi^0\gamma\gamma$ decay. (Left-plot) distribution of $M_{4\gamma}$ for the four photons coming from the η. Central values are obtained from the 450 pb^{-1} already analized. Statistical errors have been reduced of a factor two to extrapolate to the entire 2.5 fb^{-1}. (Right-plot) example of di-photon spectrum for the invariant mass of the two bachelor photons from the η [8].

their limits on the entire sample. Most of the other channels are not on the reach of our experiment either due to a limitation on the rate or to the overwhelming existing background (especially from kaons). The only channels we can hope to improve are the C violating decay to $\pi^0 e^+ e^-$ and the Lepton Flavour violating decay to μe. For these channels, the electron identification will greatly help reducing the background contribution.

The η' meson

A tagged search of the main decay channels of the η' (right column of Tab.1) is problematic due to the difficulty of isolating with high purity the recoil radiative photon. However, in the entire 2.5 fb^{-1} sample we will get $\sim$ 450 k η' mesons and, by measuring ratios of BR, we expect to decrease the existing errors on the main decay channels.

Apart from its own interest, this is relevant for our measurement of the η'-gluonium content which is related to the mixing angle on the flavour basis ϕ_P. This quantity can be extracted by the ratio $R_\eta = BR(\phi \to \eta')/BR(\phi \to \eta)$ which is proportional to $\cot^2\phi_P$ [11]. Using the first 17 pb^{-1} collected in 2000, KLOE has determined this angle with an error of 1.5° [12] thus limiting the η'-gluonium content below 15% at 90% C.L. This search was carried out looking for the decay chain $\phi \to \eta'\gamma$ with $\eta' \to \eta\pi^+\pi^-$ and $\eta \to \gamma\gamma$. The new search on 450 pb^{-1} [3] uses instead the decay chain $\phi \to \eta'\gamma$ with $\eta' \to \eta\pi^+\pi^-(\pi^0\pi^0)$ and $\eta \to \pi^0\pi^0\pi^0(\pi^+\pi^-\pi^0)$. An almost background-free sample of $\sim$ 3400 events (see Fig. 3) is obtained. The preliminary result shows a statistical error already twice smaller than the systematic one, which is dominated by the knowledge of the η and η' intermediate BR's, in particular by the BR($\eta' \to \eta\pi\pi$). We intend to use the entire 2.5 fb^{-1} to repeat the measurement by looking at the three already used decay channels and extend the search to all other ones, thus minimizing the effect of the

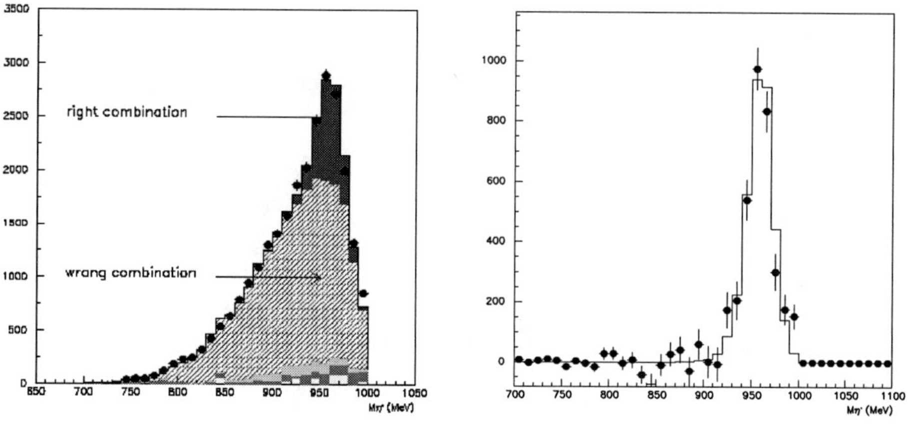

FIGURE 3. Search for $\phi \to \eta'\gamma$ decay with $\eta' \to \eta\pi^+\pi^-(\pi^0\pi^0)$ and $\eta \to \pi^0\pi^0\pi^0(\pi^+\pi^-\pi^0)$: (left-plot) distribution of $M_{\eta'}$ for all possible photon combinations before background subtraction and (right-plot) for the right combination after background subtraction.

intermediate BR.

Moreover, apart from the absolute normalization, a high-statistics study of the Dalitz-plot density for the η' decay to $\eta\pi\pi$ will not only allow to study the box anomaly [6] in this system, but also to extract the contribution of a possible intermediate state due to a $\sigma(600)$ meson [13]. Other interesting searches of the η' Dalitz-decay and the measurement of the $\pi - \eta$ mixing [14], via the ratio of isospin forbidden (3π) with isospin allowed $(\eta\pi\pi)$ final states $R_1 = \frac{\Gamma(\eta'\to\pi^0\pi^0\pi^0)}{\Gamma(\eta'\to\eta\pi^0\pi^0)}$ $(R_2 = \frac{\Gamma(\eta'\to\pi^0\pi^+\pi^-)}{\Gamma(\eta'\to\eta\pi^+\pi^-)})$, are possible.

THE LIGHT SCALAR SECTOR

Through radiative ϕ decays, also the lightest scalar meson with isospin 0, $f_0(980)$, and 1, $a_0(980)$, are accessible at KLOE. The study of their production cross-sections and of the mass (Dalitz-plot) distributions are important ingredients to discriminate their nature [15]. An expanded description of the measurement performed in KLOE with the 450 pb^{-1} collected in 2001-2002 is reported in another contribution to these proceedings [16]. Here, we just summarize the work done and describe the prospects for the entire data-set.

A clear evidence of $f_0(980)$ is observed in the $\pi^+\pi^-$ invariant mass, $M_{\pi\pi}$, spectrum of $\sim 6.5 \times 10^5$ events with a photon at large polar angle. The distribution is clearly dominated by $e^+e^- \to \pi^+\pi^-$ with an additional photon from initial (ISR) or final (FSR) state radiation. The contribution of the scalar term and of its negative interference with FSR are needed to adequately fit the data. The scalar term is parametrized either via the kaon loop, KL [15, 17] or with a simpler no-structure, NS [18], approach. Both parametrizations provide a good description of the spectrum. The KL approach

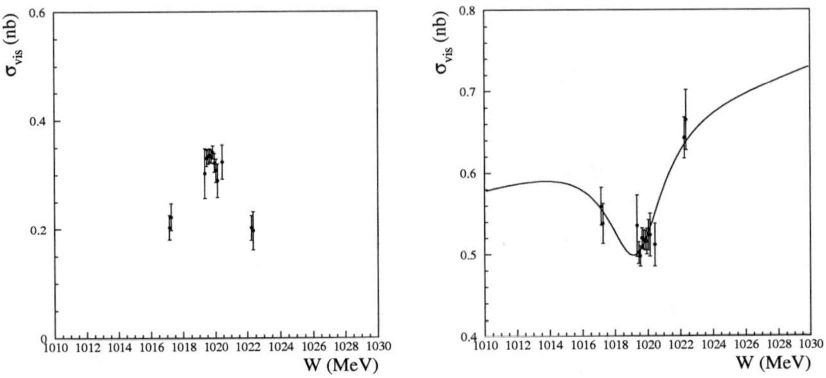

FIGURE 4. Sample of five prompt photons. Dependence of the visible cross-section on W for: (left-plot) the $\phi \to S\gamma$ and (right-plot) the $e^+e^- \to \omega\pi^0$ processes when excluding any interference term.

correctly describes also the dependence of the integrated cross-section as a function of the center of mass energy W. In this analysis, the difficulty on extracting the couplings is related to the large cross-section ratio (5 to 1) between ISR+FSR and the scalar term in the $f_0(980)$ region. However, we determine a strong coupling of f_0 to kaons. An interesting independent observation of the signal comes from the dependence of the charge asymmetry on $M_{\pi\pi}$.

Our search for f_0 to $\pi^0\pi^0$ has already been published with the 17 pb^{-1} collected in 2000 by looking at a sample of 5 prompt photons [19]. While the $M_{\pi^0\pi^0}$ region above 700 MeV is almost background free, the one below 700 suffers of a large background from $e^+e^- \to \omega\pi^0$ and of a possible contribution from the scalar $\sigma(600)$. In the published analysis, we applied an exclusive selection to reject all events with $M_{\pi^0\gamma}$ in within 3 σ's from the ω mass. We are now extending this study to the 450 pb^{-1} collected in 2001, 2002 while changing the analysis scheme to consider a possible interference between the $\phi \to S\gamma$ (scalar) and the $e^+e^- \to \omega\pi^0$ (VDM) processes. S represents either the f_0 or the $f_0 + \sigma(600)$ scalar mesons. If we do not assume any interference between the scalar and VDM terms, the W dependence of their visible cross-sections can be studied. In Fig. 4 the $\sim$ 400 k events collected allow us to clearly distinguish the resonant (not-resonant) behaviour of the scalar (VDM) term. In order to take into account their possible interference, we are fitting, for a given W value, the $M_{\pi^0\gamma} - M_{\pi^0\pi^0}$ plane with both the KL and NS approaches. Work is in progress to extract the relevant couplings from these fits.

The $a_0(980)$ is studied looking at the $\eta\pi^0$ final state with two different decay chains: (1) fully neutral with $\eta \to \gamma\gamma$ and (2) charged and neutral with $\eta \to \pi^+\pi^-\pi^0$. The last one is practically background free. The KL model nicely fits the mass distributions.

The 2.5 fb^{-1} sample will allow to complete the measurement on the scalar sector by improving the precision on the couplings and on the W-dependences of the cross sections. This will be especially true for the a_0 where 25 k events completely background free will be available for the decay chain (2). We are also planning to make the first search for $\phi \rightarrow S\gamma \rightarrow K_S K_S \gamma$, since we expect to collect between 100−500 of such events.

THE MEASUREMENT OF σ_{HAD}

The precise measurement of the muon anomaly, a_μ, at BNL with an accuracy of 0.5 ppm [20] has renewed the interest in the determination of the cross section for e^+e^- annihilation into hadrons, σ_{had}, since via dispersion relation the hadronic contribution, a_μ^{had}, can be obtained by integrating it. The region below 1 GeV contributes $\sim$ 60% of this integral. The novelty of the measurement at DAΦNE is the capability of extracting this cross section running at a fixed center of mass energy, $W \sim M_\phi$, while measuring the cross section for the $\pi^+\pi^-\gamma$ final state. ISR lowers the center of mass energy of the pion pair thus making available the entire energy range from $2m_\pi$ to W. The determination of the $\pi\pi\gamma$ cross section and the knowledge of the radiator function, which describes the ISR in this reaction, allows us to extract σ_{had}.

In ref. [21], we have reported the first measurement of this cross section using 140 pb^{-1} of collisions recorder in 2001 while requiring the radiative photons to be emitted at small polar angle ($\theta_\gamma < 15°$). In this sample, *small_angle*, we ask for two charged tracks coming from the IP with polar angles in the range 50°−130° without any explicit photon detection. To discriminate the photon direction we cut directly on the di-pion production angle $\theta_{\pi\pi}$. [4] The ISR process dominates over final state radiation, FSR, and constitutes an indistinguishable background in our approach. The good momentum resolution of the DCH, and the combination of tracking with the excellent timing performances and granularity of the EMC, allow us to discriminate pion pairs from electron and muon pairs. Residual background from $\mu^+\mu^-\gamma$ and $\pi^+\pi^-\pi^0$ events is reduced and estimated by looking at the kinematic variable M_{trk}. [5] Luminosity is measured with 0.5%(the.), 0.3%(exp.) precision by using large angle Bhabha events.

The $\pi^+\pi^-\gamma$ cross section divided by the radiator function, derived by the MC code Phokhara [22], allows us to extract σ_{had} in the range $0.35 < (S_\pi = M_{\pi\pi}^2) < 0.95$ GeV^2. The obtained cross-section is FSR-inclusive. After correcting it by vacuum polarization we can insert it in the dispersion integral to find: $a_\mu^{\pi\pi} = (388.7 \pm 0.8_{stat} \pm 3.5_{syst} \pm 3.5_{th}) \cdot 10^{-10}$. The experimental errors are dominated by the one related to the filter for machine background (0.6%). All other relevant error sources (acceptance, trigger, tracking, vertexing and background subtraction) each contributes of 0.3% to the error. The theory error is instead dominated by luminosity (0.6%) and by the knowledge of the radiator function (0.5%). To compare with the published results from CMD-2 [23], we have integrated a_μ in the range $0.37 < S_\pi < 0.93$ GeV^2 obtaining a good agreement. This result confirms a rather strong disagreement between the a_μ^{had} determination obtained using τ decay data and disagrees by $\sim 3\ \sigma$ with the direct determination of a_μ.

In the last months, new preliminary measurements of σ_{had} have been shown by the

---

[4] For a single photon emitted the relation $\theta_\gamma = 180° - \theta_{\pi\pi}$ holds.

[5] This variable is derived by the pion momenta, applying 4-momentum conservation in the hypothesis of a final state with two particles with the same mass and one single photon.

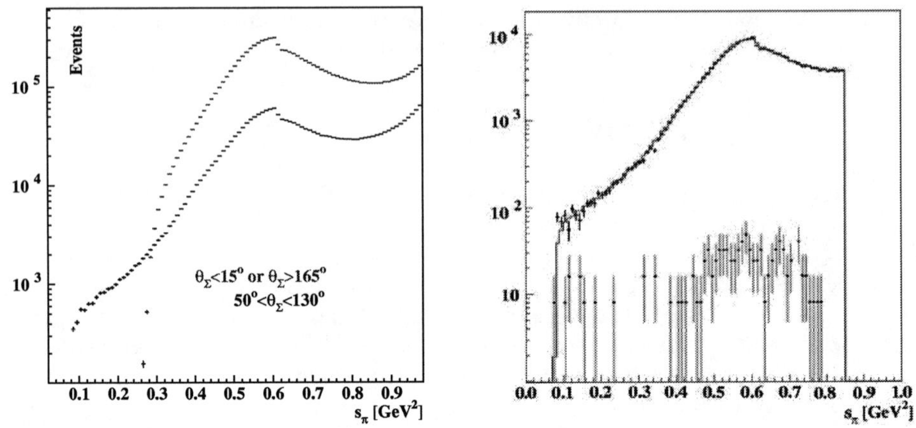

FIGURE 5. (Left-plot) relative rate vs S_π for *small_angle* (red points) and *large_angle* (blue points) selections. (Right-plot) *large_angle*: S_π spectrum data (black points), MC expectation for the signal (red line), residual $\pi^+\pi^-\pi^0$ background (green points).

SND [24, 25] and CMD-2 [26, 27] collaborations indicating a much better agreement with τ-data. This somehow adds another input, apart having a statistical gain of 20, on redoing the analysis on the entire KLOE data sample. Since 2001, we have improved a lot on the triggering, reconstruction and analysis side and most of the experimental systematics quoted in our measurement will be reduced. Furthermore, the large available statistics suggests also a change on the analysis strategy. We can now perform a rate normalization based on $\mu^+\mu^-\gamma$ as follows: $\sigma_{\pi^+\pi^-} = (\sigma_{\pi^+\pi^-\gamma}/\sigma_{\mu^+\mu^-\gamma}) \times \sigma_{\mu^+\mu^-}$. The advantage of such a normalization is related to the first order cancellation of many theory and experimental systematics in the ratio.

The region close to threshold ($S_\pi < 0.35$ GeV^2) still contributes of $\sim 20\%$ to a_μ^{had} for $S_\pi < 1$ GeV^2 and it is poorly experimentally known. However, this region is kinematicaly depressed by the requirements of the *small_angle* analysis. For this reason, we are performing a *large_angle* analysis by requiring the radiative photons to have $\theta_\gamma > 50^\circ$. In this sample an explicit photon detection is necessary and the relative amount of FSR is large (from 10% to 40%, depending on S_π). This will allow us to test the model of scalar QED [28] used in the MC to simulate the radiative corrections. As shown in the scalars section, the presence of the $f_0(980)$ in this sample is well established and we should handle its interference in the spectrum. From the experimental point of view, the largest difficulty is to treat the residual $\pi^+\pi^-\pi^0$ background which is much larger than in the *small_angle* analysis. In Fig. 5, a preliminary data-MC comparison for the S_π spectrum, at the end of the analysis chain, is shown for the 140 pb^{-1} of 2001. This proves that we are already able to handle this background. However, the interference with the $f_0(980)$, and the possibility to simplify our analysis strategy, pushed us to ask for an extension of DAΦNE running to collect ~ 200 pb^{-1} of collisions 20 MeV below the ϕ peak. This should minimize the contribution of the scalars and of the $\pi^+\pi^-\pi^0$ background.

This low energy running will also grant some interesting sample for studying $\gamma\gamma$ physics which can be relevant to improve the actual knowledge of the η-width or to search for, in a completely new way, the $\sigma(600)$ meson.

CONCLUSIONS

KLOE is collecting 2.5 fb^{-1} of clean e^+e^- collisions at $W \sim M_\phi$. Many interesting searches and measurements for low energy hadronic physics are planned.

REFERENCES

1. S. Guiducci *et al.*, Proc. of the 2001 Particle Accelerator Conference (Chicago, Illinois, USA), P. Lucas S. Webber Eds., 2001, 353.
2. KLOE Collaboration, LNF-92/019 (IR) (1992) and LNF-93/002 (IR) (1993).
3. F. Ambrosino for the KLOE Collaboration, these proceedings.
4. S. Giovannella and S. Miscetti, KLOE Note 177 (2002).
5. GEM Collaboration, M. Abdel-Bary *et al.*, Phys. Lett. B 619 (2005) 281.
6. M. Benayoun *et al.*, Eur. Phys. J. C 31 (2003) 525.
7. D. Gao, Mod. Phys. Lett. A 17 (2002) 1583.
8. E. Oset *et al.*, Phys. Rev. D 67 (2003) 073013.
9. KLOE Collaboration, A. Aloisio *et al.*, Phys. Lett. B 591 (2004) 49.
10. KLOE Collaboration, F. Ambrosino *et al.*, Phys. Lett. B 606 (2005) 276.
11. A. Bramon, R. Escribano and M. D. Scadron, Eur. Phys. J. C 7 (1999) 271.
12. KLOE Collaboration, A. Aloisio *et al.*, Phys. Lett. B 541 (2002) 45.
13. A. H. Fariborz and J. Schechter, Phys. Rev. D 60 (1999) 034002.
14. D.J.Gross *et al.*, Phys. Rev. D 19 (1979) 2188.
15. N. N. Achasov and V. N. Ivanchenko, Nucl. Phys. B 315 (1989) 465.
16. C. Bini for the KLOE Collaboration, these proceedings.
17. N. N. Achasov and V. V. Gubin, Phys. Rev. D 56 (1997) 4084.
18. G. Isisdori, L. Maiani and S. Pacetti, private communications.
19. KLOE Collaboration, A. Aloisio *et al.*, Phys. Lett. B 537 (2002) 37.
20. G. W. Bennet *et al.*, Phys. Rev. Lett. 92 (2004) 161802.
21. KLOE Collaboration, A. Aloisio *et al.*, Phys. Lett. B 606 (2005) 12.
22. H. Czyz et al., Eur. Phys. J. C 33 (2004) 333.
23. R. R. Akhmetshin et al., Phys. Lett. B 527 (2002) 161.
    R. R. Akhmetshin et al., Phys. Lett. B 578 (2004) 285.
24. M. N. Achasov *et al.*, hep-ex/0506076.
25. S.Serednyakov on behalf of the SND Collaboration, Proceedings of the HEP2005 International Europhysics Conference on High Energy Physics, July 21st-27th 2005, Lisboa, Portugal.
26. I. B. Logashenko on behalf of the CMD-2 Collaboration, Proceedings of the HEP2005 International Europhysics Conference on High Energy Physics, July 21st-27th 2005, Lisboa, Portugal.
27. A. Sibidanov on behalf of the CMD-2 Collaboration, these proceedings.
28. H. Czyz *et al.*, Eur. Phys. J. C 3327 (2003), 563.

Multi-body Heavy Meson Decay

David Asner

Carleton University, Department of Physics, 1125 Colonel By Drive, Ottawa, ON K1S 5B6, Canada

Abstract. Charm meson decay dynamics have been studied extensively over the last decade. Recently, B meson decay dynamics have also been analyzed. We describe the Dalitz-plot analysis technique which has been applied by many experiments to three-body D and B decays. We discuss results from the E791, FOCUS, CLEO, BABAR and BELLE experiments. These studies probe a variety of physics including charmless B-decay, doubly-Cabibbo suppressed decays, CP violation and T violation, the properties of established light mesons, charm spectroscopy and the properties of $\pi\pi$, $K\pi$ and KK S-wave states.

Keywords: charm,bottom,Dalitz
PACS: 13.25.Ft, 13.25.Hw, 11.80.Et

INTRODUCTION

Charm meson decay dynamics have been studied extensively over the last decade. Recent studies of multi-body decays of charm mesons probe a variety of physics including doubly-Cabibbo suppressed decays [1, 2], searches for CP violation [2, 3, 4, 5], T violation [6], D^0–$\overline{D}^0$ mixing [7], the properties of established light mesons [8, 9, 10, 11], the properties of $\pi\pi$ [1, 10, 12] $K\pi$ [13, 14], and KK [15] S-wave states, and the dynamics of four-body final states [16].

Recently, B meson decay dynamics have been studied. Multi-body decays of B mesons also probe a variety of physics including, charmless B-decays [17, 18, 19, 20, 21, 22, 23, 24], measurements of Cabibbo-Kobayashi-Maskawa (CKM) angle γ/ϕ_3 [25, 26, 27, 28], searches for direct CP violation [21, 22, 29], charm spectroscopy [30, 31], the properties of established light mesons [19, 22, 23], the properties of KK [19, 23] and $K\pi$ [19, 21, 22] S-wave, and the three-body production of baryons [20, 32]. Time-dependent Dalitz-plot (TD) analyses have been used to determine the CKM angle α/ϕ_2 [33]. A TD analysis of $B^0 \to D^{*\pm}K^0\pi^{\mp}$ [34] is sensitive γ/ϕ_3. Future studies could improve sensitivity to new physics in TD analyses of $b \to s$ penguin decays [23].

Weak nonleptonic decays of B and charm mesons are expected to proceed dominantly through resonant two-body decays in several theoretical models [35]; see Ref. [36] for a review of resonance phenomenology. These amplitudes are typically calculated with the Dalitz plot analysis technique [37], which uses the mininum number of independent observable quantities. For three-body final states when the parent particle is a scalar, the decay rate [38] is

$$\Gamma = \frac{1}{(2\pi)^3 32\sqrt{s^3}} |\mathcal{M}|^2 dm_{12}^2 dm_{23}^2, \tag{1}$$

where m_{ij} is the invariant mass of $i-j$ and the coefficient of the amplitude includes all

CP814 *Hadron Spectroscopy: Eleventh International Conference*
edited by A. Reis, C. Göbel, J. Sá Borges, and J. Magnin

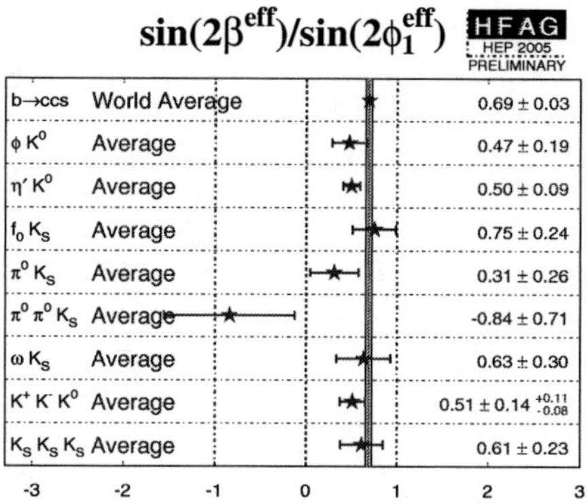

FIGURE 1. Comparison of results for $\sin 2\beta$ determined from $b \to c\bar{c}s$ and several $b \to s\bar{s}s$ processes.

kinematic factors. The scatter plot in m_{12}^2 versus m_{23}^2 is called a Dalitz plot. If $|\mathcal{M}|^2$ is constant, after efficiency correction and background subtraction, the allowed region of the plot will be populated uniformly with events. Any variation in $|\mathcal{M}|^2$ over the Dalitz plot is due to dynamical rather than kinematical effects. It is straightforward to extend the formalism beyond three-body final states. For N-body final states, phase space has dimension $3N-7$. Other cases of interest include one vector particle or a baryon/anti-baryon pair in the final state. For the former case phase space has dimension $3N-5$ and for the latter $3N-4$.

CHARMLESS *B*-DECAY

Several charmless three-body final states have been analyzed by Babar and Belle: $B^+ \to K^+\pi^-\pi^+$ [18, 19, 22], $B^+ \to K^+K^-K^+$ [19], $B^+ \to \pi^+\pi^-\pi^+$ [24], $B^0 \to K^+\pi^-\pi^0$ [21], $B^0 \to K_S^0\pi^+\pi^-$ [17], $B^0 \to K_S^0K^+K^-$ [23] and $B^0 \to \pi^+\pi^-\pi^0$ [33]. These final states are also accessible in D decay and the resonant substructure is modelled following Ref. [3]. The available phase space is much large in B decay and more sophisticated descriptions of the non-resonant contribution are under development. The study of the $\pi\pi$, $K\pi$, KK S-wave at CLEO-c from hadronic charmonium decay, which spans most of the phase space of interest, may provide important input. As in D decay, no strong evidence for direct CP violation has been observed.

Charmless three-body decays of the B meson are a rich laboratory providing information on the weak sector and the dynamics of strong interactions. Decays to the final state $K_S^0K^+K^-$ are dominated by the $b \to s\bar{s}s$ amplitudes and are of particular interest due their sensitivity to physics beyond the Standard Model. This Dalitz-plot is dominated by a non-resonant KK S-wave component and in the low mass KK region a significant P-wave contribution from ϕK_S^0.

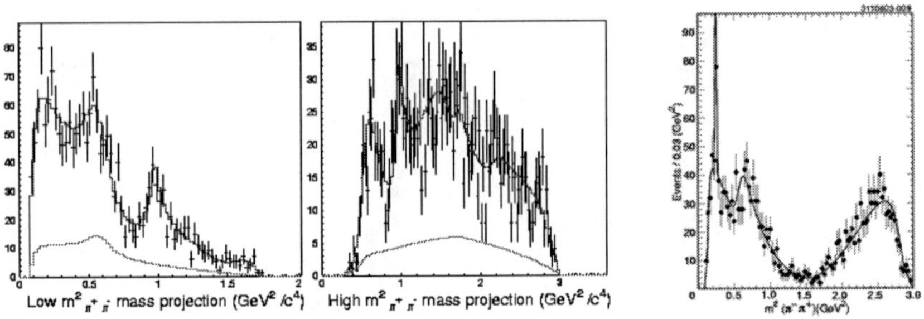

FIGURE 2. $\pi^+\pi^-$ projection of Left: $D^+ \to \pi^+\pi^-\pi^+$ (low mass) and Center: $D^+ \to \pi^+\pi^-\pi^+$ (high mass) from FOCUS [10]. Right: $D^0 \to \pi^+\pi^-\pi^0$ from CLEO [5]. The D^0 final state is consistent with the absence of $\pi\pi$ S-wave while this amplitude dominates the D^+ final state. Data (points) and total fit (curve) are shown. The background contribution for $D^+ \to \pi^+\pi^-\pi^+$ is also shown

The CKM parameter $\sin 2\beta$ can be determined from the time-dependent CP asymmetry in several $b \to s$ penguin processes. These are compared, in Fig. 1, to the precision measurement from $b \to c\bar{c}s$ which is dominated by $B^0 \to \psi K_S$.

TIME-DEPENDENT DALITZ-PLOT ANALYSIS

$B^0 \to \pi^+\pi^-\pi^0$ and $\sin 2\alpha$: The $B^0 \to \pi^+\pi^-\pi^0$ Dalitz plot is dominated by ρ intermediate resoanances. Although $\rho\pi$ is not a CP eigenstate, the time-dependent analysis of this Dalitz plot allows simultaneous determination of the strong transition amplitudes and phases and the CKM angle α. Using 213 million $\Upsilon(4S) \to B\bar{B}$ decays BABAR finds $\alpha = (113^{+27}_{-17} \pm 6)^o$ [33] and 2.9σ evidence for direct CP violation.
$D^0 \to K^0_S\pi^+\pi^-$ and Search for $D^0-\overline{D}^0$ Mixing: The resonant substructure of $D^0 \to K^0_S\pi^+\pi^-$ can be described by a combination of ten quasi two-body intermediate states which include both CP-even and CP-odd eigenstates and one doubly Cabibbo suppressed channel. The time-dependent analysis of this Dalitz plot allows simultaneous determination of the strong transition amplitudes and phases and the mixing parameters x and y without phase or sign ambiguity. Using 9 fb^{-1}, CLEO II.V constrains $(-4.5 < x < 9.3)\%$ and $(-6.4 < y < 3.6)\%$ [7].
Search for New Physics in $b \to s$ Penguin Processes: A time-dependent Dalitz-plot analysis of the $b \to s$ penguin processes listed in Fig. 1 has not been done. Most $b \to s$ penguin final states, such as $K^+K^-K^0_S$, are not pure CP eigenstates. A time-dependent analysis of this Dalitz plot would allow the simultaneous determination of the strong transition amplitudes and phases and the CKM angle β while reducing the systematic uncertainties associated with selecting ϕK^0_S from $K^+K^-K^0_S$. A precision measurement of $\sin 2\beta$ in $b \to s$ penguin processes is required to establish a discrepancy between $\sin 2\beta$ determined with $b \to c\bar{c}s$ and $b \to s\bar{s}s$ which might be attributed to new physics.

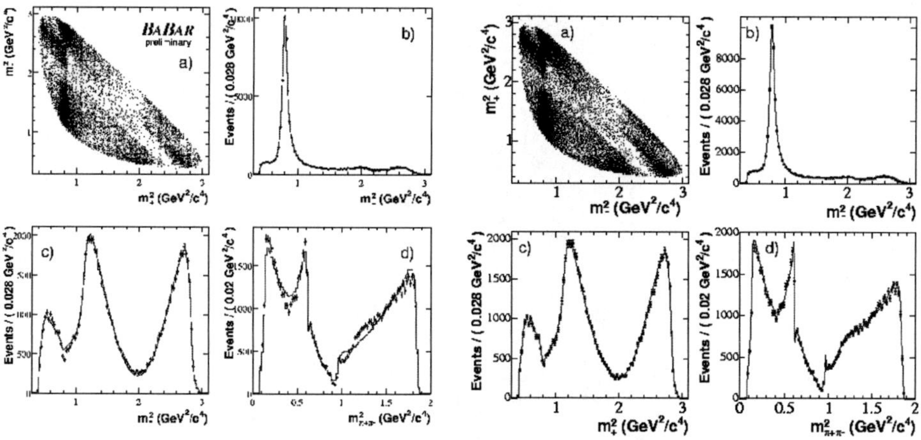

FIGURE 3. Dalitz plot of $D^0 \to K_S^0\pi^+\pi^-$ [28]. Left: Breit-Wigner model [1] does describe the $\pi\pi$ projection of the BABAR data. Right: K-matrix parameterization of $\pi\pi$ S-wave yields much improved fit.

DETERMINING γ/ϕ_3 FROM $B^- \to DK^-, D \to K_S^0\pi^+\pi^-$

Various approaches have been proposed to measure γ/ϕ_3 in $B^\mp \to DK^\mp$ decays [39, 40, 41, 42, 43, 44]. The Dalitz-plot analysis of three-body D^0 decay from $B^\mp \to DK^\mp$ provides the best measurement of γ/ϕ_3. This approach suffers from the Dalitz-plot model uncertainty introduced by the unknown complex phase of D^0 decay. Recently Babar [45] and Belle [46] have reported $\gamma = (70 \pm 31^{+12+14}_{-10-11})^\circ$ and $\phi_3 = (77^{+17}_{-19} \pm 13 \pm 11)^\circ$, respectively, where the third error is the systematic error due to modeling of the Dalitz plot. With larger data samples soon to be available, this uncertainty is expected to be dominant in the γ/ϕ_3 measurement.

The senstivity to γ/ϕ_3 is due to the interference of two amplitudes producing opposite flavors of D mesons. In the case of B^+ decay, the state $D_+ = D^0 + re^{i\theta_+}\overline{D}^0$ where r is the ratio of doubly-Cabbib suppressed to Cabibbo favored amplitude and the relative phase between D^0 and $\overline{D}^0$ is $\theta_+ = \gamma/\phi_3 + \delta$. Analogously for B^-, $D_- = \overline{D}^0 + re^{i\theta_-}D^0$ with $\theta_- = -\gamma/\phi_3 + \delta$

The CLEO-c physics program includes a variety of measurements that will improve the determination of γ/ϕ_3 from the B-factory experiments. The total number of charm mesons accumulated at CLEO-c will be much smaller than the samples already accumulated by the B-factories. However, the quantum correlation of $D\bar{D}$ system from $\psi(3770)$ and $E \sim 4160$ MeV provides a unique laboratory in which to study charm in general and models of multibody charm decay in particular. It has been shown [47] that the Dalitz-plot model uncertainty can be reduced utilizing a sample of D^0 in a CP eigenstate. This sample is being accumulated at CLEO-c from the processes $e^+e^- \to \psi(3770) \to D\bar{D} \to (CP\pm)(K_S^0\pi^+\pi^-)$. It is estimated that a sample of 1 fb^{-1} of integrated luminosity at the $\psi(3770)$ would allow the measurement of γ/ϕ_3 to a precision of a few degrees.

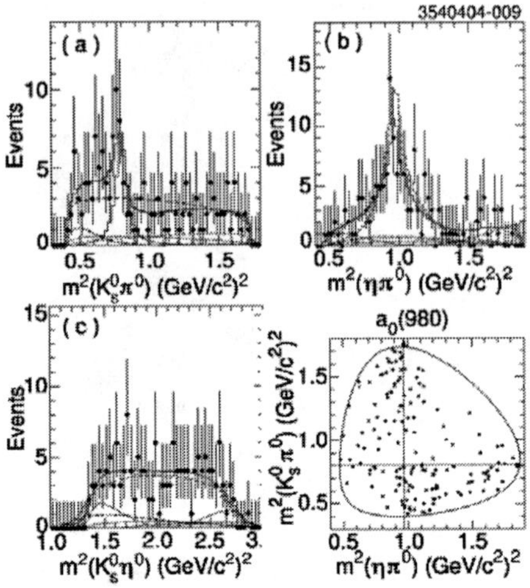

FIGURE 4. The three projections and the $D^0 \to K_S^0 \eta \pi^0$ Dalitz plot. The fit to CLEO data [11] has contributions from $a_0(980)K_S^0$, $K^*(892)\eta$, $K_0^*(1430)\eta$, and $a_2(1320)K_S^0$.

K-MATRIX FORMALISM

Usually the resonances in Dalitz-plot analyses are modeled with a Breit-Wigner and the nonresonant contribution $D \to abc$ is parameterized as an S-wave with no variation in magnitude or phase across the Dalitz plot. Some more recent analyses have used the K-matrix formalism [48] with the P-vector approximation [49] to describe the $\pi\pi$ S-wave.

The decay $D^+ \to \pi^+\pi^+\pi^-$, dominated by $\pi\pi$ S-wave, has been studied by the E687 [8], E791 [12] and FOCUS [10] experiments. FOCUS analyzed its data sample using both the Breit-Wigner formalism and the K-matrix formalism, following the 5 pole, 5 resonance model of Anisovich and Sarantsev [51]. The Breit-Wigner analysis included $\rho(770), f_0(980), f_2(1270), f_0(1500), \sigma(500)$, and a nonresonant contribution. Applying the K-matrix formalism to the S wave and parameterizing the $\rho(770)$ and $f_2(1270)$ with the Breit-Wigner functions also described the FOCUS data well. The K-matrix fit to the FOCUS data is shown in Fig. 2.

The decay $D^0 \to \pi^+\pi^-\pi^0$, dominated by $\pi\pi$ P-wave, has been studied by the CLEO experiment. Only the three $\rho(770)\pi$ resonant contributions are observed. No evidence is found for $\pi\pi$ S-wave with either the Breit-Wigner or a K-matrix parameterization, following the 4 pole, 2 resonance model of Au, Morgan, and Pennington [50]. The K-matrix fit to the CLEO data is shown in Fig. 2.

The decay $D^0 \to K_S^0\pi^+\pi^-$, important for the determination of γ/ϕ_3, has been studied by several experiments. The CLEO model [1] (includes $\rho(770)$, ω, $f_0(980)$, $f_2(1270)$, $f_0(1370)$, $K^*(892)^-$, $K_0(1430)^-$, $K_2(1430)^-$, $K^*(1680)^-$ and the doubly-Cabibbo suppressed amplitude $K^{(}892)^+$ as well as a small non-resonant component) does not provide a good description of the BABAR data. An improved description is obtain in two ways. First, by adding additional Breit-Wigner resonances including two arbitrary $\pi\pi$

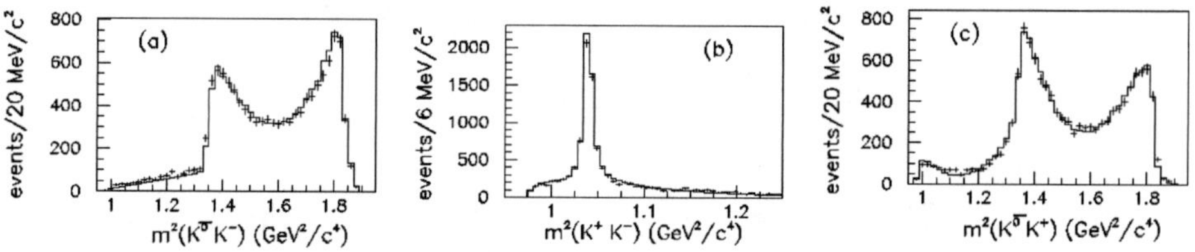

FIGURE 5. Projections of the BABAR data and fit to the $D^0 \to K_S^0 K^+ K^-$ Dalitz plot [15].

resonances, σ_1 and σ_2. Second, following the methodology of FOCUS, BABAR has applied a K-matrix model to describe the $\pi\pi$ S-wave [28]. Both the Breit-Wigner fit and the K-matrix fit are shown in Fig. 3.

STUDYING LOW MASS $\pi\pi$, $K\pi$, KK S-WAVE

The meson content of the 0^{++} of nonet and the quark content of the $f_0(980)$, $a_0(980)$, $f_0(1370)$, $f_0(1500)$, and $f_0(1710)$ mesons are current puzzles in light meson spectroscopy. Measuring branching fractions and couplings to different final states and the comparison of scalar meson production rates among D^0, D^+ and D_s^+ mesons may help resolve these issues.

An interesting example of studies of 0^{++} mesons in D decays follows. A large contribution of $f_0(980)$ to $D^0 \to K_S^0 K^+ K^-$ was reported by ARGUS [52] and BABAR [53]. This is inconsistent with the smaller contribution of $f_0(980)$ observed in $D^0 \to K_S^0 \pi^+ \pi^-$ [1]. One possible explanation is that the $D^0 \to K_S^0 K^+ K^-$ has a large contribution from $a_0(980) \to K^+ K^-$. The CLEO collaboration was motivated to study $D^0 \to K_S^0 \eta \pi^0$ where the dominant contribution is from $K_S^0 a_0(980), a_0(980) \to \eta\pi$ and no contribution from $f_0(980)$. The CLEO data and projection of fit results are shown in Fig. 5. A more recent analysis from BABAR includes a large contribution from $a_0(9800 \to KK$ and little $f_0(980)$ [15].

Indication of a broad $K\pi$ scalar intermediate resonance has been reported by E791 in

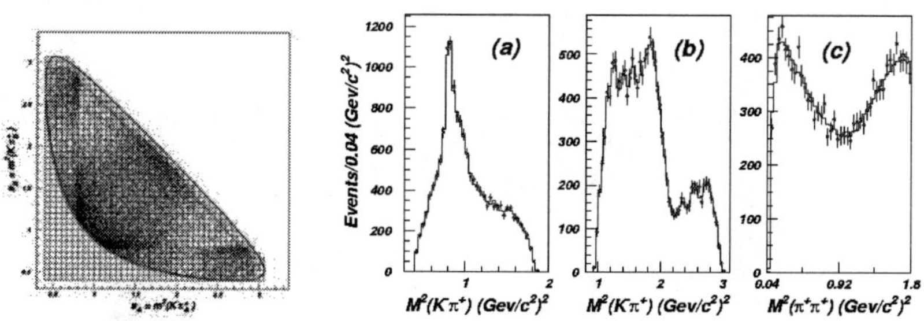

FIGURE 6. Left: Dalitz plot of $D^+ \to K^- \pi^+ \pi^+$ in E791 data [14]. (a) Low mass $K\pi$ (b) high mass $K\pi$ (c) $\pi\pi$ projections.

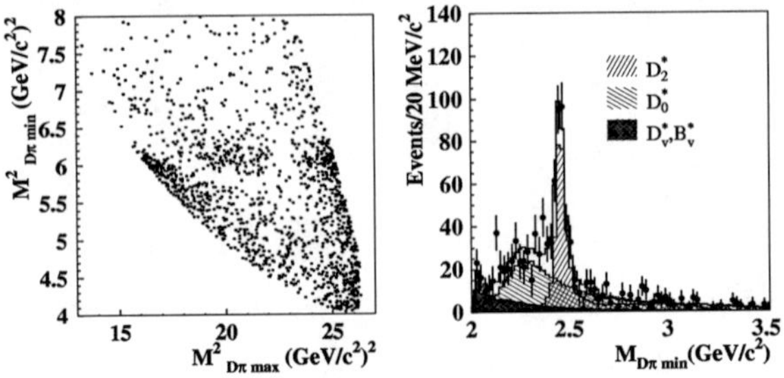

FIGURE 7. Left: Dalitz plot $B^0 \to D^0\pi^+\pi^-$ from Belle [30], . Right: Low mass $D^0\pi$ projection of data and components fit.

the decay $D^+ \to K^-\pi^+\pi^+$ [13]. Fitting the Dalitz plot with $\overline{K}^*(892)^0\pi^+$, $\overline{K}_0^*(1430)^0\pi^+$, $\overline{K}_2^*(1430)^0\pi^+$, and $\overline{K}^*(1680)^0\pi^+$, plus a constant nonresonant component, E791 finds results consistent with earlier results [54, 55] with a nonresonant fit fraction of over 90%. Having reconstructed more events than the other experiments, E791 was led to include an extra low-mass S-wave $\overline{K}\pi$ resonance to account for the poor fit already seen by earlier experiments: A $\kappa(800)$ with $m = 797 \pm 19 \pm 43$ MeV/c^2 and $\Gamma = 410 \pm 43 \pm 87$ MeV/c^2 much improved the fits. The $\kappa(800)$ is now the dominant resonance and the nonresonant fit fraction is reduced from $90.9 \pm 2.6\%$ to $13.0 \pm 5.8 \pm 4.4\%$.

E791 also modeled the $K\pi$ S-wave phase variation as a function of $K\pi$ mass with the $K_0^*(1430)$ resonance only and a nonresonant component following the parameterization of LASS [56]. It was necessary to relax the unitarity constraint to describe the E791 data [13]. The $K\pi$ S-wave phase behavior in this model is consistent with the model that includes the κ resonance.

Finally, E791 performed a model independent partial wave analysis (MIPWA) of the S-save component of the $K\pi$ system. The amplitude and phase for the $K\pi$ S-wave are obtained from $K\pi$ threshold up to 1.72 GeV/c^2. The S-wave amplitudes require no assumptions about its dependence on invariant mass, but do rely on the relatively well understood P-wave and D-wave. The P-wave is described by the $K^*(892)$ and $K^*(1680)$

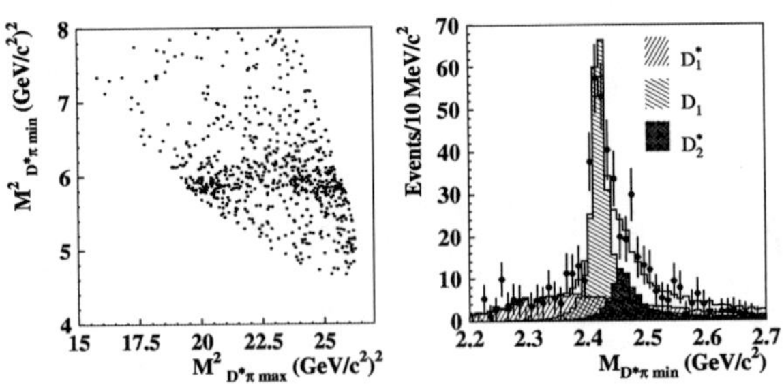

FIGURE 8. Left: Dalitz plot $B^0 \to D^{*+}\pi^+\pi^-$ from Belle [31], . Right: Low mass $D^{*+}\pi$ projection of data and components fit.

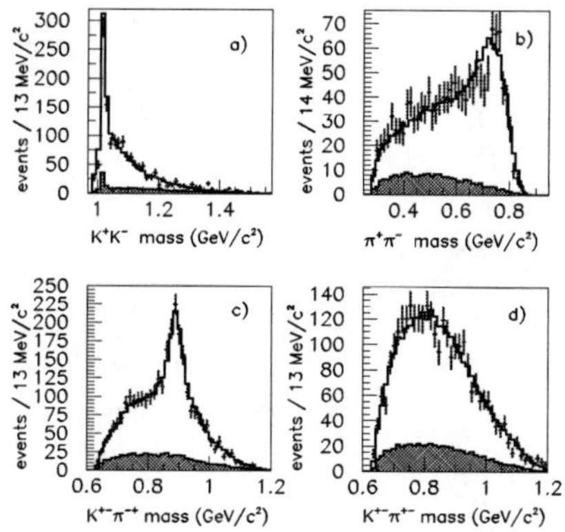

FIGURE 9. Projections of $D^0 \to K^+K^-\pi^+\pi^-$ from FOCUS [16]. a) KK b) $\pi\pi$ c) $K^{\pm}\pi^m p$ d) $K^{\pm}\pi^-\pm$. Data (points), total fit (histogram) and background (shaded) are shown.

and the D-wave is described by the $K_2(1430)$ parameterized as Breit-Wigner resonances. The results of the MIPWA are shown in Fig. 6 and provide a similar description of the data as Ref. [13] which parametrized the S-wave with a κ and $K_0(1430)$ Breit-Wigner resonances and a constant complex non-resonant term.

BEYOND $P \to 3P$ DECAYS

Charm Spectroscopy: The decay $B \to D\pi\pi$ where the D is fully recontructed can be analyzed with the usual Dalitz-plot analysis technique for $P \to 3P$ decays. The Dalitz-plot for this process from Belle [30], is shown in Fig 7. The spin-structure of the narrow D_2^* is evident. In addition to the relative amplitude and phase, the mass and width of the narrow D_2^* and broad D_0^* are determined in the Dalitz-plot fit. The data and projection of the components of the Dalitz-plot fit on to the $m_{D\pi}$ minimum mass is shown in Fig. 7.

The decay $B \to D^*\pi\pi$ where the D^* is fully reconstructed has two additional degrees of freedom, assuming the width of the D^{*+} to be negligible [57]. In addition to the Dalitz-plot variables $(m_{D^{*+}\pi^+}, m_{D^{*+}\pi^-})$ one can choose the variables to be the angle between the pions in the D^{*+} rest frame and the azimuthal angle of the pion from the D^{*+} relative to the $D^{*+}\pi\pi$ decay plane. The Dalitz-plot from Belle [31], is shown in Fig. 8. In addition to the relative amplitudes and phases the mass and width of D_1^0 and $D_1'^0$ are determined. The parameters for D_2^* are fixed from the $B \to D\pi\pi$ analysis. The data and projection of the components of the fit on to the $m_{D^*\pi}$ minimum mass is shown in Fig. 8.

$D \to KK\pi\pi$: The FOCUS experiment has extended the Dalitz-plot formalism to analyze $D^0 \to K^+K^-\pi^+\pi^-$ with four pseudoscalars in the final state. Final states produced through spectator amplitudes in which the W is coupled to a vector (V) or axial-vector

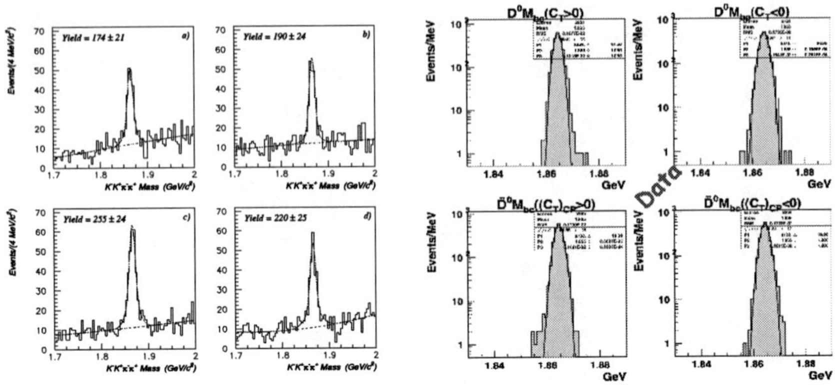

FIGURE 10. Top Row: Invariant mass for $D^0 \to K^+K^-\pi^+\pi^-$ candidates. Bottom Row: Invariant mass for $\overline{D}^0 \to K^+K^-\pi^+\pi^-$ candidates. No indication of T violation is observed in the FOCUS data (left) [6] or CLEO-c data (right).

(A) meson are expected to dominate over those where the W is coupled to a pseudoscaler (P) meson. The dominant contribution to the rate (55%) is from the $D^0 \to AP$ modes, $K_1(1270)^-K^+$ and $K_1(1400)^-K^+$, a large contribution ($\sim 30\%$) from $D^0 \to VV$ modes, mostly $\phi\rho$. The remaining contributions are from $D^0 \to VPP$ and $D^0 \to SPP$, primarily from $K^*(892)^0K^+\pi^-$ and $f_0(980)\pi^+\pi^-$. The projections of the data and fit are shown in Fig. 9.

Search for T-Violation: The FOCUS experiment has searched for T violation four-body charm meson decays using the channels $D^0 \to K^+K^-\pi^+\pi^-$, $D^+ \to K^0_S K^+\pi^-\pi^+$, and $D_s^+ \to K^0_S K^+\pi^-\pi^+$ [6]. The T violation asymmetry is defined as $A_{Tviol} \equiv (A_T - \bar{A}_T)$ where A_T is defined as $A_T \equiv \frac{\Gamma(\vec{v}_1 \cdot (\vec{v}_2 \times \vec{v}_3)>0) - \Gamma(\vec{v}_1 \cdot (\vec{v}_2 \times \vec{v}_3)<0)}{\Gamma(\vec{v}_1 \cdot (\vec{v}_2 \times \vec{v}_3)>0) + \Gamma(\vec{v}_1 \cdot (\vec{v}_2 \times \vec{v}_3)<0)}$ where each v_i is a momentum which are odd under time reversal. They report $A_{Tviol}(D^0) = 0.010 \pm 0.057 \pm 0.037$, $A_{Tviol}(D^0) = 0.023 \pm 0.062 \pm 0.022$, and $A_{Tviol}(D_s^+) = -0.036 \pm 0.067 \pm 0.023$. The FOCUS and CLEO-c $D^0 \to K^+K^-\pi^+\pi^-$ invariant mass distributions are compared in Fig. 10. In 281 pb^{-1} CLEO-c sample is nearly background freewith $\sim$10 times the signal yield of FOCUS. CLEO-c has not reported results for A_{Tviol}.

REFERENCES

1. H. Muramatsu *et al.* [CLEO Collaboration], Phys. Rev. Lett. **89**, 251802 (2002) [Erratum-ibid. **90**, 059901 (2003)].
2. G. Brandenburg *et al.* [CLEO Collaboration], Phys. Rev. Lett. **87**, 071802 (2001).
3. S. Kopp *et al.* [CLEO Collaboration], Phys. Rev. D **63**, 092001 (2001).
4. D. M. Asner *et al.* [CLEO Collaboration], Phys. Rev. D **70**, 091101 (2004).
5. D. Cronin-Hennessy *et al.* [CLEO Collaboration], Phys. Rev. D **72**, 031102 (2005).
6. J. M. Link *et al.* [FOCUS Collaboration], Phys. Lett. B **622**, 239 (2005).
7. D. M. Asner [CLEO Collaboration], Phys. Rev. D **72**, 012001 (2005).

8. P. L. Frabetti *et al.* [E687 Collaboration], Phys. Lett. B **407**, 79 (1997).
9. E. M. Aitala *et al.* [E791 Collaboration], Phys. Rev. Lett. **86**, 765 (2001).
10. J. M. Link *et al.* [FOCUS Collaboration], Phys. Lett. B **585**, 200 (2004).
11. P. Rubin *et al.* [CLEO Collaboration], Phys. Rev. Lett. **93**, 111801 (2004).
12. E. M. Aitala *et al.* [E791 Collaboration], Phys. Rev. Lett. **86**, 770 (2001).
13. E. M. Aitala *et al.* [E791 Collaboration], Phys. Rev. Lett. **89**, 121801 (2002).
14. E. M. Aitala *et al.* [E791 Collaboration], arXiv:hep-ex/0507099.
15. B. Aubert *et al.* [BABAR Collaboration], Phys. Rev. D **72**, 052008 (2005).
16. J. M. Link *et al.* [FOCUS Collaboration], Phys. Lett. B **610**, 225 (2005).
17. K. Abe *et al.* [Belle Collaboration], arXiv:hep-ex/0509047.
18. K. Abe [Belle Collaboration], arXiv:hep-ex/0509001.
19. A. Garmash *et al.* [BELLE Collaboration], Phys. Rev. D **71**, 092003 (2005).
20. K. Abe *et al.* [BELLE Collaboration], arXiv:hep-ex/0409010.
21. B. Aubert *et al.* [BABAR Collaboration], arXiv:hep-ex/0408073.
22. B. Aubert *et al.* [BABAR Collaboration], arXiv:hep-ex/0507004.
23. B. Aubert *et al.* [BABAR Collaboration], arXiv:hep-ex/0507094.
24. B. Aubert *et al.* [BABAR Collaboration], arXiv:hep-ex/0408032.
25. A. Poluektov *et al.* [Belle Collaboration], Phys. Rev. D **70**, 072003 (2004).
26. K. Abe *et al.* [Belle Collaboration], arXiv:hep-ex/0411049.
27. K. Abe *et al.* [Belle Collaboration], arXiv:hep-ex/0504013.
28. B. Aubert *et al.* [BABAR Collaboration], arXiv:hep-ex/0507101.
29. A. Garmash, arXiv:hep-ex/0505048.
30. K. Abe *et al.* [Belle Collaboration], Phys. Rev. D **69**, 112002 (2004).
31. K. Abe *et al.* [Belle Collaboration], arXiv:hep-ex/0412072.
32. K. Abe *et al.* [BELLE Collaboration], arXiv:hep-ex/0409005.
33. B. Aubert *et al.* [BABAR Collaboration], arXiv:hep-ex/0408099.
34. B. Aubert *et al.* [BABAR Collaboration], arXiv:hep-ex/0412040.
35. M Bauer, B. Stech and M. Wirbel, Z. Phys. C **34**, 103 (1987); P. Bedaque, A. Das and V.S. Mathur, Phys. Rev. D **49**, 269 (1994); L.-L. Chau and H.-Y. Cheng, Phys. Rev. D **36**, 137 (1987); K. Terasaki, Int. J. Mod. Phys. A **10**, 3207 (1995); F. Buccella, M. Lusignoli and A. Pugliese, Phys. Lett. B **379**, 249 (1996).
36. J. D. Jackson, Nuovo Cim. **34**, 1644 (1964).
37. R. H. Dalitz, *Phil. Mag.* **44**, 1068 (1953).
38. See the review on Kinematics revised by J. D. Jackson on Page 675-679 of the Review of Particle Physics, Phys.Lett.B592: 1,2004.
39. M. Gronau and D. London., Phys. Lett. B **253**, 483 (1991).
40. M. Gronau and D. Wyler, Phys. Lett. B **265**, 172 (1991).
41. I. Dunietz, Phys. Lett. B **270**, 75 (1991).
42. D. Atwood, G. Eilam, M. Gronau and A. Soni, Phys. Lett. B **341**, 372 (1995).
43. D. Atwood, I. Dunietz and A. Soni, Phys. Rev. Lett. **78**, 3257 (1997).
44. D. Atwood, I. Dunietz and A. Soni, Phys. Rev. D **63**, 036005 (2001).
45. B. Aubert *et al.* [BABAR Collaboration], f arXiv:hep-ex/0504039.
46. K. Abe *et al.* [Belle Collaboration], arXiv:hep-ex/0504013.
47. A. Giri, Y. Grossman, A. Soffer and J. Zupan, Phys. Rev. D **68**, 054018 (2003).
48. S.U. Chung *et al.*, Ann. Physik. **4**, 404 (1995).
49. I. J. R. Aitchison, Nucl. Phys. **A 189**, 417 (1972).
50. K. L. Au, D. Morgan and M. R. Pennington, Phys. Rev. D **35**, 1633 (1987).
51. V. V. Anisovich and A. V. Sarantsev, Eur. Phys. J. A **16**, 229 (2003).
52. H. Albrecht *et al.* [ARGUS Collaboration], Z. Phys. C **33**, 359 (1987).
53. B. Aubert *et al.* [BABAR Collaboration], arXiv:hep-ex/0207089.
54. J.C. Anjos *et al.* [E691 Collaboration], Phys. Rev. D**48** 56 (1993).
55. P.L. Frabetti *et al.* [E687 Collaboration], Phys. Lett. B**331**, 217 (1994).
56. D. Aston *et al.* [LASS Collaboration], Nucl. Phys. B **296**, 493 (1988).
57. A. Anastassov *et al.* [CLEO Collaboration], Phys. Rev. D **65**, 032003 (2002).

Hadron Spectroscopy and Wavefunctions in QCD and the AdS/CFT Correspondence

Stanley J. Brodsky* and Guy F. de Téramond†

*Stanford Linear Accelerator Center, Stanford University, Stanford, California 94309, USA
†Universidad de Costa Rica, San José, Costa Rica

Abstract. The AdS/CFT correspondence has led to important insights into the properties of quantum chromodynamics even though QCD is a broken conformal theory. We have recently shown how a holographic model based on a truncated AdS space can be used to obtain the hadronic spectrum of light $q\overline{q}, qqq$ and gg bound states. Specific hadrons are identified by the correspondence of string modes with the dimension of the interpolating operator of the hadron's valence Fock state, including orbital angular momentum excitations. The predicted mass spectrum is linear $M \propto L$ at high orbital angular momentum, in contrast to the quadratic dependence $M^2 \propto L$ found in the description of spinning strings. Since only one parameter, the QCD scale Λ_{QCD}, is introduced, the agreement with the pattern of physical states is remarkable. In particular, the ratio of Δ to nucleon trajectories is determined by the ratio of zeros of Bessel functions. The light-front quantization of gauge theories in light-cone gauge provides a frame-independent wavefunction representation of relativistic bound states, simple forms for current matrix elements, explicit unitarity, and a trivial vacuum. The light-front Fock-state wavefunctions encode the bound state properties of hadrons in terms of their quark and gluon degrees of freedom at the amplitude level. One can also use the extended AdS/CFT space-time theory to obtain a model for hadronic light-front wavefunctions, thus providing a relativistic description of hadrons in QCD at the amplitude level. The model wavefunctions display confinement at large inter-quark separation and conformal symmetry at short distances. In particular, the scaling and conformal properties of the LFWFs at high relative momenta agree with perturbative QCD. These AdS/CFT model wavefunctions could be used as an initial ansatz for a variational treatment of the light-front QCD Hamiltonian. We also show how hadron form factors in both the space-like and time-like regions can be predicted.

Keywords: Gauge/string duality, Quantum chromodynamics, Hadron mass models and calculations, Light-Front Wavefunctions
PACS: 12.38.-t,12.38.Lg,11.25.Tq,12.40.Yx

INTRODUCTION

A central goal in quantum chromodynamics is to describe the structure and dynamics of hadrons at the amplitude level. The light-front Fock expansion provides a physical description of hadrons as composites of quarks and gluons analogous to the $\psi(\vec{p})$ momentum-space wavefunction description of nonrelativistic bound states of the Schrödinger theory. The light-front wavefunctions $\psi_{n/H}(x_i, \vec{k}_{\perp i}, \lambda_i)$ are functions of the constituent light-cone fractions $x_i = \frac{k_i^+}{P^+} = \frac{(k^0+k^z)_i}{P^+}$, relative transverse momenta $\vec{k}_{\perp i}$, and spin projections $S_i^z = \lambda_i$. They are relativistic and frame-independent, describing all particle number excitations n of the hadrons.

The light-front Fock expansion follows from the quantization of QCD at fixed light-front time $x^+ = x^0 + x^3$. The bound-state hadronic solutions $|\Psi_H\rangle$ are eigenstates of

CP814 *Hadron Spectroscopy: Eleventh International Conference*
edited by A. Reis, C. Göbel, J. Sá Borges, and J. Magnin

the light-front Heisenberg equation $H_{LF}\,|\Psi_H\rangle = M_H^2\,|\Psi_H\rangle$ [1]. The spectrum of QCD is given by the eigenvalues M_H^2. The projection of each hadronic eigensolution on the free Fock basis: $\langle n|\Psi_H\rangle \equiv \psi_{n/H}(x_i, \vec{k}_{\perp i}, \lambda_i)$ then defines the LF Fock expansion in terms of the quark and transversely polarized gluon constituents in $A^+ = 0$ light-cone gauge. The expansion has only transversely polarized gluons. The freedom to choose the light-like quantization four-vector provides an explicitly covariant formulation of light-front quantization and can be used to determine the analytic structure of light-front wave functions and to define a kinematical definition of angular momentum [2]. The front form thus provides a consistent definition of relative orbital angular momentum and J^z conservation: the total spin projection $J^z = \sum_{i=1}^n S_i^z + \sum_i^{n-1} L_i^z$ is conserved in each Fock state. The cluster decomposition theorem [3] and the vanishing of the "anomalous gravitomagnetic moment" $B(0)$ [4] are immediate properties of the LF Fock wavefunctions [5].

Given the light-front wavefunctions $\psi_{n/H}(x_i, \vec{k}_{\perp i}, \lambda_i)$, one can compute a large range of hadron observables. For example, the valence and sea quark and gluon distributions which are measured in deep inelastic lepton scattering are defined from the squares of the LFWFS summed over all Fock states n. Form factors, exclusive weak transition amplitudes [6] such as $B \to \ell \nu \pi$. and the generalized parton distributions [7] measured in deeply virtual Compton scattering are (assuming the "handbag" approximation) overlaps of the initial and final LFWFS with $n = n'$ and $n = n' + 2$. The gauge-invariant distribution amplitude $\phi_H(x_i, Q)$ defined from the integral over the transverse momenta $\vec{k}^2_{\perp i} \leq Q^2$ of the valence (smallest n) Fock state provides a fundamental measure of the hadron at the amplitude level [8, 9]; they are the nonperturbative input to the factorized form of hard exclusive amplitudes and exclusive heavy hadron decays in perturbative QCD. The resulting distributions obey the DGLAP and ERBL evolution equations as a function of the maximal invariant mass, thus providing a physical factorization scheme [10]. In each case, the derived quantities satisfy the appropriate operator product expansions, sum rules, and evolution equations. However, at large x where the struck quark is far-off shell, DGLAP evolution is quenched [11], so that the fall-off of the DIS cross sections in Q^2 satisfies inclusive-exclusive duality at fixed W^2.

The light-front Fock-state wavefunctions encode the bound state properties of hadrons in terms of their quark and gluon degrees of freedom at the amplitude level. They display novel features, such as intrinsic gluons, asymmetric sea-quark distributions $\bar{u}(x) \neq \bar{d}(x)$, $\bar{s}(x) \neq s(x)$, and intrinsic heavy-quark Fock states [12]. Intrinsic charm and bottom quarks appear at large x in the light-front wavefunctions since this minimizes the invariant mass and off-shellness of the higher Fock state. One can use the operator product expansion to show that the probability of such states scales as $1/M_Q^2$ in contrast to $1/M_\ell^4$ fall-off of abelian theory [13]. The remarkable observations of the SELEX experiment of the double-charm baryon Ξ_{ccd} in $pA \to \Xi_{ccd}X$ and $\Sigma^- A \to \Xi_{ccd}X$ at large x_F [14] provides compelling evidence for double-charm intrinsic Fock states in the proton. The coherence of multi-particle correlations within the Fock states leads to higher-twist bosonic processes such as $e(qq) \to e'(qq)'$; although suppressed by inverse powers of Q^2, such subprocesses are important in the duality regime of fixed W^2, particularly in σ_L [15]. In the case of nuclei, one must include non-nucleonic "hidden color" [16] degrees of freedom of the deuteron LFWF.

Measurements of the LFWFs

The E791 experiments at Fermilab [17, 18] has shown how one can measure the valence LFWF directly from the diffractive di-jet dissociation of a high energy pion $\pi A \to q\bar{q}A'$ into two jets, nearly balancing in transverse momentum, leaving the nucleus intact. The measured pion distribution in x and $(1-x)$ is similar to the form of the asymptotic distribution amplitude and the AdS/CFT prediction discussed below. The E791 experiment also find that the nuclear amplitude is additive in the number of nucleons when the quark jets are produced at high $k_\perp$, thus giving a dramatic confirmation of "color transparency", a fundamental manifestation of the gauge nature of QCD [19, 20].

Effects of Final State Interactions

The phase structure of hadron matrix elements is an essential feature of hadron dynamics. Although the LFWFs are real for a stable hadron, they acquire phases from initial state and final state interactions. In addition, the violation of CP invariance leads to a specific phase structure of the LFWFs [21].

Contrary to parton model expectations, the rescattering of the quarks in the final state in DIS has important phenomenological consequences, such as leading-twist diffractive DIS [22] and the Sivers single-spin asymmetry [23]. The Sivers asymmetry depends on the same matrix elements which produce the anomalous magnetic moment of the target nucleon as well as the phase difference of the final-state interactions in different partial waves. The rescattering of the struck parton generates dominantly imaginary diffractive amplitudes, giving rise to an effective "hard pomeron" exchange and a rapidity gap between the target and diffractive system, while leaving the target intact. This Bjorken-scaling physics, which is associated with the Wilson line connecting the currents in the virtual Compton amplitude survives even in light-cone gauge. Thus there are contributions to the DIS structure functions which are not included in the light-front wave functions computed in isolation and cannot be interpreted as parton probabilities [22]. Diffractive deep inelastic scattering in turn leads to nuclear shadowing at leading twist as a result of the destructive interference of multi-step processes within the nucleus. In addition, multi-step processes involving Reggeon exchange leads to antishadowing. In fact, because Reggeon couplings are flavor specific, antishadowing is predicted to be non-universal, depending on the type of current and even the polarization of the probes in nuclear DIS [24].

Another interesting consequence of QCD at the amplitude level is the Q^2-independent "$J=0$ fixed-pole" contribution $M(\gamma^* p \to \gamma p) \sim s^0 F(t)$ to the real part of the Compton amplitude, reflecting the effective contact interaction of the transverse currents [25]. Deeply virtual Compton scattering can also be studied in the timelike domain from $e^+e^- \to H^+H^-\gamma$; the lepton charge asymmetry and single-spin asymmetries allow measurements of the relative phase of timelike form factors and the $\gamma^* \to H^+H^-\gamma$ amplitude.

Nonperturbative Methods for Computing LFWFs

In principle, one can solve for the LFWFs directly from the fundamental theory using methods such as discretized light-front quantization (DLCQ) [26], the transverse lattice [27, 28, 29], lattice gauge theory moments [30], Dyson-Schwinger techniques [31], and Bethe–Salpeter techniques [2]. DLCQ has been remarkably successful in determining the entire spectrum and corresponding LFWFs in one space-one time field theories [32], including QCD(1+1) [33] and supersymmetric QCD(1+1) [34]. The DLCQ boundary conditions allow a truncation of the Fock space to finite dimensions while retaining the kinematic boost and Lorentz invariance of light-front quantization. There are also light-front solutions for Yukawa theory in physical (3+1) space-time dimensions [35, 36] with a limited Fock space. As emphasized by Weinstein and Vary, new effective operator methods [37, 38] which have been developed for Hamiltonian theories in condensed matter and nuclear physics, could also be applied advantageously to light-front Hamiltonian. A review of nonperturbative light-front methods may be found in reference [39].

As we discuss below, one can use the AdS/CFT correspondence to obtain a model for hadronic light-front wavefunctions which display confinement at large inter-quark separation and conformal symmetry at short distances. In particular, the scaling and conformal properties of the LFWFs at high relative momenta agree with perturbative QCD. These AdS/CFT model wavefunctions could be used as an initial ansatz for a variational treatment of the light-front QCD Hamiltonian.

ADS/CFT PREDICTIONS FOR HADRON SPECTRA AND WAVEFUNCTIONS

The AdS/CFT correspondence [40], between strongly-coupled conformal gauge theory and weakly-coupled string theory in the 10-dimensional $AdS_5 \times S^5$ space is now providing a remarkable new insight into the hadron wavefunctions of QCD. The central mathematical principle underlying AdS/CFT duality is the fact that the group $SO(2,4)$ of Poincaré and conformal transformations of physical $3+1$ space-time has an elegant mathematical representation on AdS_5 space where the fifth dimension has the anti-de Sitter warped metric. The group of conformal transformations $SO(2,4)$ in 3+1 space is isomorphic to the group of isometries of AdS space, $x^\mu \to \lambda x^\mu$, $r \to r/\lambda$, where r represents the coordinate in the fifth dimension. The dynamics at $x^2 \to 0$ in 3+1 space thus matches the behavior of the theory at the boundary $r \to \infty$. This allows one to map the physics of quantum field theories with conformal symmetry to an equivalent description in which scale transformations have an explicit representation in AdS space.

Even though quantum chromodynamics is a broken conformal theory, the AdS/CFT correspondence has led to important insights into the properties of QCD. For example, as shown by Polchinski and Strassler [41], the AdS/CFT duality, modified to give a mass scale, provides a nonperturbative derivation of the empirically successful dimensional counting rules [42, 43] for the leading power-law fall-off of the hard exclusive scattering amplitudes of the bound states of the gauge theory. The modified theory generates the hard behavior expected from QCD instead of the soft behavior characteristic of strings.

Other important applications include the description of spacelike hadron form factors at large transverse momentum [44] and deep inelastic scattering structure functions at small x [45]. The power falloff of hadronic light-front wave functions (LFWF) including states with nonzero orbital angular momentum is also predicted [46].

In the original formulation by Maldacena [40], a correspondence was established between a supergravity string theory on a curved background and a conformally invariant $\mathscr{N}=4$ super Yang-Mills theory in four-dimensional space-time. The higher dimensional theory is $AdS_5 \times S^5$ where $R=(4\pi g_s N_C)^{1/4}\alpha_s'^{1/2}$ is the radius of AdS and the radius of the five-sphere and $\alpha_s'^{1/2}$ is the string scale. The extra dimensions of the five-dimensional sphere S^5 correspond to the $SU(4) \sim SO(6)$ global symmetry which rotates the particles present in the supersymmetric Yang Mills supermultiplet in the adjoint representation of $SU(N_C)$. In our application to QCD, baryon number in QCD is represented as a Casimir constant on S^5.

The reason why AdS/CFT duality can have at least approximate applicability to physical QCD is based on the fact that the underlying classical QCD Lagrangian with massless quarks is scale-invariant [47]. One can thus take conformal symmetry as an initial approximation to QCD, and then systematically correct for its nonzero β function and quark masses [48]. This "conformal template" approach underlies the Banks-Zak method [49] for expansions of QCD expressions near the conformal limit and the BLM method [50] for setting the renormalization scale in perturbative QCD applications. In the BLM method the corrections to a perturbative series from the β-function are systematically absorbed into the scale of the QCD running coupling. An important example is the "Generalized Crewther Relation" [51] which relates the Bjorken and Gross-Llewellyn sum rules at the deep inelastic scale Q^2 to the e^+e^- annihilation cross sections at specific commensurate scales $s^*(Q^2) \simeq 0.52\ Q^2$. The Crewther relation [52] was originally derived in conformal theory; however, after BLM scale setting, it becomes a fundamental test of physical QCD, with no uncertainties from the choice of renormalization scale or scheme.

QCD is nearly conformal at large momentum transfers where asymptotic freedom is applicable. Nevertheless, it is remarkable that dimensional scaling for exclusive processes is observed even at relatively low momentum transfer where gluon exchanges involve relatively soft momenta [53]. The observed scaling of hadron scattering at moderate momentum transfers can be understood if the QCD coupling has an infrared fixed point [54]. In this sense, QCD resembles a strongly-coupled conformal theory.

Deriving Hadron Spectra from AdS/CFT

The duality between a gravity theory on AdS_{d+1} space and a conformal gauge theory at its d-dimensional boundary requires one to match the partition functions at the AdS boundary, $z=R^2/r \to 0$. The physical string modes $\Phi(x,r) \sim e^{-iP\cdot x} f(r)$, are plane waves along the Poincaré coordinates with four-momentum P^μ and hadronic invariant mass states $P_\mu P^\mu = \mathscr{M}^2$. For large-$r$ or small-z, $f(r) \sim r^{-\Delta}$, where the dimension Δ of the string mode must be the same dimension as that of the interpolating operator $\mathscr{O}$ which creates a specific hadron out of the vacuum: $\langle P|\mathscr{O}|0\rangle \neq 0$.

The physics of color confinement in QCD can be described in the AdS/CFT approach by truncating the AdS space to the domain $r_0 < r < \infty$, where $r_0 = \Lambda_{QCD} R^2$. The cutoff at r_0 is dual to the introduction of a mass gap Λ_{QCD}; it breaks conformal invariance and is responsible for the generation of a spectrum of color-singlet hadronic states. The truncation of the AdS space insures that the distance between the colored quarks and gluons as they stream into the fifth dimension is limited to $z < z_0 = 1/\Lambda_{QCD}$. The resulting $3+1$ theory has both color confinement at long distances and conformal behavior at short distances. The latter property allows one to derive dimensional counting rules for form factors and other hard exclusive processes at high momentum transfer. This approach, which can be described as a "bottom-up" approach, has been successful in obtaining general properties of the low-lying hadron spectra, chiral symmetry breaking, and hadron couplings in AdS/QCD [55] in addition to the hard scattering predictions [41, 45, 46].

In this "classical holographic model", the quarks and gluons propagate into the truncated AdS interior according to the AdS metric without interactions. In effect, their Wilson lines, which are represented by open strings in the fifth dimension, are rigid. The resulting equations for spin 0, $\frac{1}{2}$, 1 and $\frac{3}{2}$ hadrons on $AdS_5 \times S^5$ lead to color-singlet states with dimension 3, 4 and $\frac{9}{2}$. Consequently, only the hadronic states (dimension-3) $J^P = 0^-, 1^-$ $q\bar{q}$ pseudoscalar and vector mesons, the (dimension-$\frac{9}{2}$) $J^P = \frac{1}{2}^+, \frac{3}{2}^+$ qqq baryons, and the (dimension-4) $J^P = 0^+$ gg gluonium states, can be derived in the classical holographic limit [56]. This description corresponds to the valence Fock state as represented by the light-front Fock expansion. Hadrons also fluctuate in particle number, in their color representations (such as the hidden-color states [16] of the deuteron), as well as in internal orbital angular momentum. The higher Fock components of the hadrons are manifestations of the quantum fluctuations of QCD; these correspond to the fluctuations of the bulk geometry about the fixed AdS metric. For spinning strings orbital excitations of hadronic states correspond to quantum fluctuations about the AdS metric [57]. It is thus also natural to identify higher-spin hadrons with the fluctuations around the spin 0, $\frac{1}{2}$, 1 and $\frac{3}{2}$ classical string solutions of the AdS_5 sector [56].

As a specific example, consider the twist-two (dimension minus spin) gluonium interpolating operator $\mathcal{O}_{4+L}^{\ell_1 \cdots \ell_m} = F D_{\{\ell_1} \ldots D_{\ell_m\}} F$ with total internal space-time orbital momentum $L = \sum_{i=1}^{m} \ell_i$ and conformal dimension $\Delta_L = 4 + L$. We match the large r asymptotic behavior of each string mode to the corresponding conformal dimension of the boundary operators of each hadronic state while maintaining conformal invariance. In the conformal limit, an L quantum, which is identified with a quantum fluctuation about the AdS geometry, corresponds to an effective five-dimensional mass μ in the bulk side. The allowed values of μ are uniquely determined by requiring that asymptotically the dimensions become spaced by integers, according to the spectral relation $(\mu R)^2 = \Delta_L(\Delta_L - 4)$ [56]. The four-dimensional mass spectrum follows from the Dirichlet boundary condition $\Phi(x, z_o) = 0$, $z_0 = 1/\Lambda_{QCD}$, on the AdS string amplitudes for each wave functions with spin < 2. The eigenspectrum is then determined from the zeros of Bessel functions, $\beta_{\alpha,k}$. The predicted spectra [56] of mesons and baryons with zero mass quarks is shown in Figs. 1 and 2. The only parameter is $\Lambda_{QCD} = 0.263$ GeV, and 0.22 GeV for mesons and baryons, respectively.

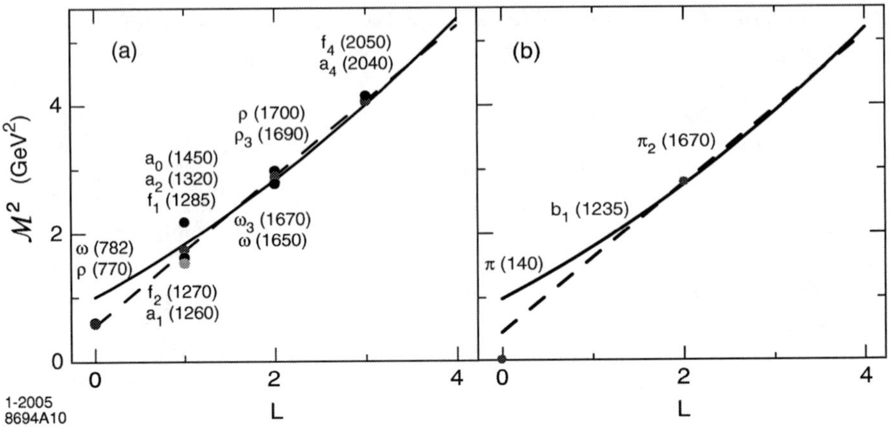

FIGURE 1. Light meson orbital states for $\Lambda_{QCD} = 0.263$ GeV: (a) vector mesons and (b) pseudoscalar mesons. The dashed line is a linear Regge trajectory with slope 1.16 GeV^2.

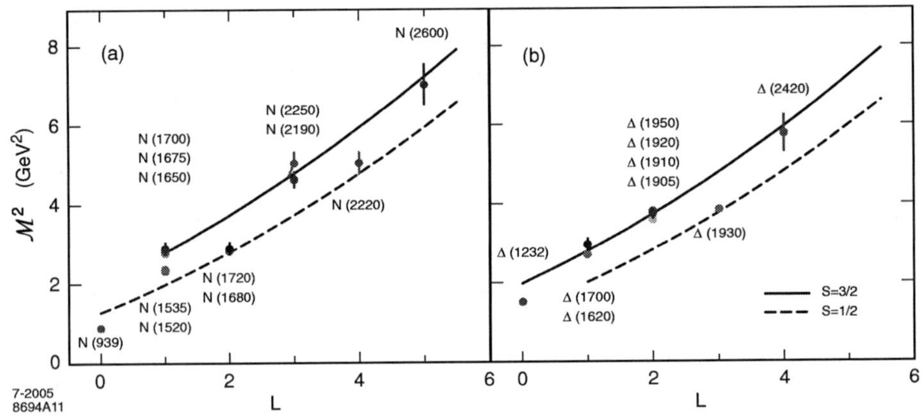

FIGURE 2. Light baryon orbital spectrum for $\Lambda_{QCD} = 0.22$ GeV: (a) nucleons and (b) Δ states.

Dynamics from AdS/CFT

Current matrix elements in AdS/QCD are computed from the overlap of the normalizable modes dual to the incoming and outgoing hadron Φ_I and Φ_F and the non-normalizable mode $J(Q,z)$, dual to the external source

$$F(Q^2)_{I\to F} \simeq R^{3+2\sigma} \int_0^{z_o} \frac{dz}{z^{3+2\sigma}}\, \Phi_F(z)\, J(Q,z)\, \Phi_I(z), \tag{1}$$

where $\sigma_n = \sum_{i=1}^n \sigma_i$ is the spin of the interpolating operator $\mathcal{O}_n$, which creates an n-Fock state $|n\rangle$ at the AdS boundary. $J(Q,z)$ has the value 1 at zero momentum transfer as the boundary limit of the external current; thus $A^\mu(x,z) = \varepsilon^\mu e^{iQ\cdot x} J(Q,z)$. The solution to the AdS wave equation subject to boundary conditions at $Q = 0$ and $z \to 0$ is [45] $J(Q,z) = zQK_1(zQ)$. At large enough $Q \sim r/R^2$, the important contribution to (1) is from the region near $z \sim 1/Q$. At small z, the n-mode $\Phi^{(n)}$ scales as $\Phi^{(n)} \sim z^{\Delta_n}$, and we recover the power law scaling [42], $F(Q^2) \to \left[1/Q^2\right]^{\tau-1}$, where the twist $\tau = \Delta_n - \sigma_n$, is equal to the number of partons, $\tau_n = n$. A numerical computation for the proton magnetic form factor in the space and time-like regions, for the model described here, gives the predictions shown in Fig. 3. The results correspond to a $L = 0$ proton state. It is interesting to compare the holographic predictions with a model-independent analysis of nucleon form factors using dispersion relations [58].

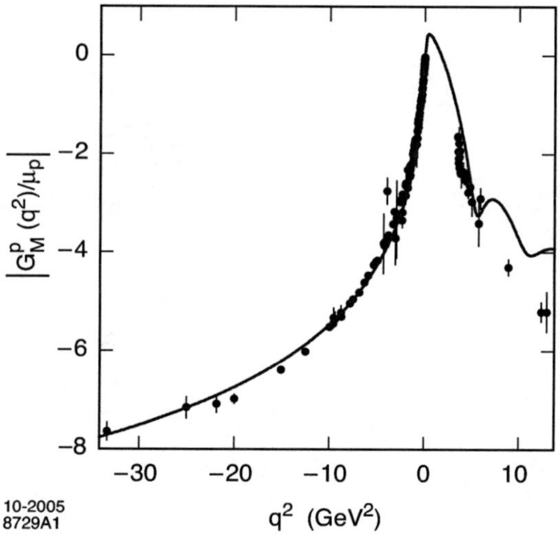

FIGURE 3. Space-like and time-like structure of the proton magnetic form factor in AdS/QCD for $\Lambda_{QCD} = 0.155$ GeV. The data are from the compilation given in ref. [58]. The prediction in the domain $0 < q^2 < 4M_p^2$ represents an analytic continuation into the unphysical region. The results should be modified for values $Q \sim \Lambda_{QCD}$, where the simple form for the cavity current $J(Q,z) = zQK_1(zQ)$ is not valid.

ADS/CFT PREDICTIONS FOR LIGHT-FRONT WAVEFUNCTIONS

The AdS/QCD correspondence provides a simple description of hadrons at the amplitude level by mapping string modes to the impact space representation of LFWFs. It is useful to define the partonic variables $x_i \vec{r}_{\perp i} = x_i \vec{R}_\perp + \vec{b}_{\perp i}$, where $\vec{r}_{\perp i}$ are the physical position coordinates, $\vec{b}_{\perp i}$ are frame-independent internal coordinates, $\sum_i \vec{b}_{\perp i} = 0$, and $\vec{R}_\perp$ is the hadron transverse center of momentum $\vec{R}_\perp = \sum_i x_i \vec{r}_{\perp i}$, $\sum_i x_i = 1$. We find for a two-parton LFWF the Lorentz-invariant form

$$\widetilde{\psi}_L(x, \vec{b}_\perp) = C\, x(1-x)\, \frac{J_{1+L}\left(|\vec{b}_\perp| \sqrt{x(1-x)}\, \beta_{1+L,k} \Lambda_{\rm QCD}\right)}{|\vec{b}_\perp| \sqrt{x(1-x)}}. \tag{2}$$

The $\beta_{1+L,k}$ are the zeroes of the Bessel functions reflecting the Dirichlet boundary condition. The variable $\zeta = |\vec{b}_\perp|\sqrt{x(1-x)}$, $0 \leq \zeta \leq \Lambda_{QCD}^{-1}$, represents the invariant separation between quarks. In the case of a two-parton state, it gives a direct relation between the scale of the invariant separation between quarks, ζ, and the holographic coordinate in AdS space: $\zeta = z = R^2/r$. The ground state and first orbital eigenmode are depicted in the figure below. The distribution in x and $(1-x)$ measured in the E791 experiment for diffractive dijet production $\pi A \to$ Jet Jet A is consistent with the AdS/CFT prediction [17, 18].

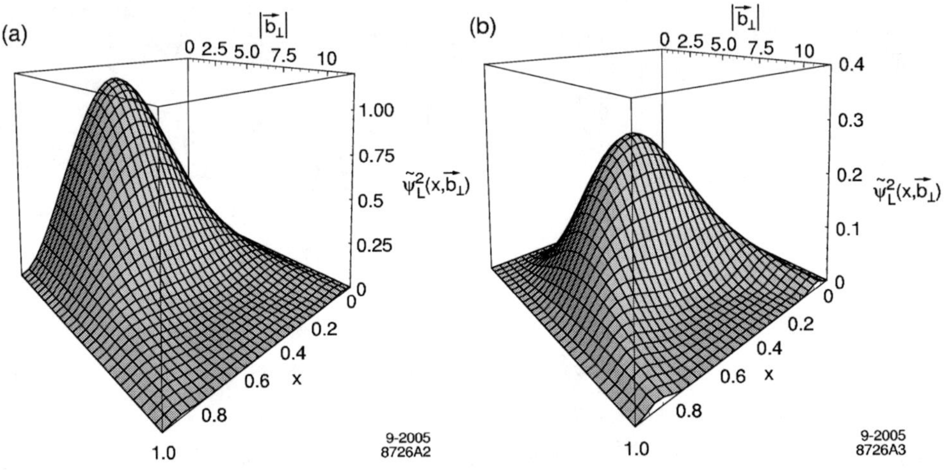

FIGURE 4. Prediction for the square of the two-parton bound-state light-front wave function $\widetilde{\psi}_L(x,\vec{b}_\perp)$ as function of the constituents longitudinal momentum fraction x and $1-x$ and the impact space relative coordinate $\vec{b}_\perp$: (a) $L=0$ and (b) $L=1$.

CONCLUDING REMARKS

The holographic model is quite successful in describing the known light hadron spectrum and hadronic form factors. Since only one parameter, the QCD scale Λ_{QCD}, is introduced, the agreement with the pattern of masses of the physical hadronic states and the space and time-like proton form factor data is remarkable. In particular, the ratio of Δ to nucleon trajectories is determined by the ratio of zeros of Bessel functions. As we have described, non-zero orbital angular momentum and higher Fock-states require the introduction of a Casimir operator derived from quantum fluctuations. It is interesting to note that the predicted mass spectrum $M \propto L$ at high orbital angular momentum, in contrast to the quadratic dependence $M^2 \propto L$ found in traditional string theory. The only mass scale is Λ_{QCD}. Only dimension-3, $\frac{9}{2}$ and 4 states $\bar{q}q$, qqq, and gg appear in the duality at the classical level, thus explaining the suppression of $C = +$ odderon exchange.

We have also shown how one can use the extended AdS/CFT space-time theory to obtain a model for the form of hadron LFWFs. The model wavefunctions display confinement at large inter-quark separation and conformal symmetry at short distances.

In particular, the scaling and conformal properties of the LFWFs at high relative momenta agree with perturbative QCD [59]. These AdS/CFT model wavefunctions could be used as an initial ansatz for a variational treatment of the light-front QCD Hamiltonian. The dominance of the quark-interchange mechanism in hard exclusive processes also emerges naturally from the classical duality of the holographic model.

ACKNOWLEDGMENTS

Presented by SJB at the 11th International Conference On Hadron Spectroscopy, HADRON05, 21-26 Aug 2005, Rio de Janeiro, Brazil. This work was supported by the Department of Energy contract DE–AC02–76SF00515.

REFERENCES

1. S. J. Brodsky, H. C. Pauli and S. S. Pinsky, Phys. Rept. **301**, 299 (1998) [arXiv:hep-ph/9705477].
2. S. J. Brodsky, J. R. Hiller, D. S. Hwang and V. A. Karmanov, Phys. Rev. D **69**, 076001 (2004) [arXiv:hep-ph/0311218].
3. S. J. Brodsky and C. R. Ji, Phys. Rev. D **33**, 2653 (1986); J. Namyslowski, proceedings of the 9th International Conference on Few Body Problems, Eugene, Oregon, 1980.
4. I. Yu. Kobsarev and L. Okun, ZhETF **43** 1904 (1962) [JETP **16** 1343 (1963)]; I. Yu. Kobsarev and V. I. Zakharov, Ann. Phys. **60** 448 (1970); O. V. Teryaev (hep-ph/9904376) (1999).
5. S. J. Brodsky, D. S. Hwang, B. Q. Ma and I. Schmidt, Nucl. Phys. B **593**, 311 (2001) [arXiv:hep-th/0003082].
6. S. J. Brodsky and D. S. Hwang, Nucl. Phys. B **543**, 239 (1999) [arXiv:hep-ph/9806358].
7. S. J. Brodsky, M. Diehl and D. S. Hwang, Nucl. Phys. B **596**, 99 (2001) [arXiv:hep-ph/0009254].
8. G. P. Lepage and S. J. Brodsky, Phys. Lett. B **87**, 359 (1979).
9. A. V. Efremov and A. V. Radyushkin, Phys. Lett. B **94**, 245 (1980).
10. G. P. Lepage and S. J. Brodsky, Phys. Rev. D **22**, 2157 (1980).
11. S. J. Brodsky and G. P. Lepage, SLAC-PUB-2294, Proc. of the Workshop on Current Topics in High Energy Physics, Cal Tech., Pasadena, Calif., (1979).
12. S. J. Brodsky, P. Hoyer, C. Peterson and N. Sakai, Phys. Lett. B **93**, 451 (1980).
13. M. Franz, M. V. Polyakov and K. Goeke, Phys. Rev. D **62**, 074024 (2000) [arXiv:hep-ph/0002240].
14. J. Engelfried [SELEX Collaboration], Nucl. Phys. A **752**, 121 (2005).
15. S. J. Brodsky, E. L. Berger and G. P. Lepage, SLAC-PUB-3027, Proceedings of the Workshop on Drell-Yan Processes, Batavia, Ill., (1982).
16. S. J. Brodsky, C. R. Ji and G. P. Lepage, Phys. Rev. Lett. **51**, 83 (1983).
17. D. Ashery [E791 Collaboration], arXiv:hep-ex/0205011.
18. E. M. Aitala *et al.* [E791 Collaboration], Phys. Rev. Lett. **86**, 4768 (2001) [arXiv:hep-ex/0010043].
19. G. Bertsch, S. J. Brodsky, A. S. Goldhaber and J. F. Gunion, Phys. Rev. Lett. **47**, 297 (1981).
20. S. J. Brodsky and A. H. Mueller, Phys. Lett. B **206**, 685 (1988).
21. S. J. Brodsky, S. V. Gardner, and D. S. Hwang (in progress).
22. S. J. Brodsky, P. Hoyer, N. Marchal, S. Peigne and F. Sannino, Phys. Rev. D **65**, 114025 (2002) [arXiv:hep-ph/0104291].
23. S. J. Brodsky, D. S. Hwang and I. Schmidt, Phys. Lett. B **530**, 99 (2002) [arXiv:hep-ph/0201296].
24. S. J. Brodsky, I. Schmidt and J. J. Yang, Phys. Rev. D **70**, 116003 (2004) [arXiv:hep-ph/0409279].
25. S. J. Brodsky, F. E. Close and J. F. Gunion, Phys. Rev. D **5**, 1384 (1972).
26. H. C. Pauli and S. J. Brodsky, Phys. Rev. D **32**, 2001 (1985).
27. W. A. Bardeen, R. B. Pearson and E. Rabinovici, Phys. Rev. D **21**, 1037 (1980).
28. S. Dalley, Few Body Syst. **36**, 69 (2005) [arXiv:hep-ph/0409139].
29. M. Burkardt and S. Dalley, Prog. Part. Nucl. Phys. **48**, 317 (2002) [arXiv:hep-ph/0112007].

30. L. Del Debbio, M. Di Pierro, A. Dougall and C. T. Sachrajda [UKQCD collaboration], Nucl. Phys. Proc. Suppl. **83**, 235 (2000) [arXiv:hep-lat/9909147].
31. P. Maris and C. D. Roberts, Int. J. Mod. Phys. E **12**, 297 (2003) [arXiv:nucl-th/0301049].
32. D. J. Gross, A. Hashimoto and I. R. Klebanov, Phys. Rev. D **57**, 6420 (1998) [arXiv:hep-th/9710240].
33. K. Hornbostel, S. J. Brodsky and H. C. Pauli, Phys. Rev. D **41**, 3814 (1990).
34. M. Harada, J. R. Hiller, S. Pinsky and N. Salwen, Phys. Rev. D **70**, 045015 (2004) [arXiv:hep-th/0404123].
35. S. J. Brodsky, J. R. Hiller and G. McCartor, Annals Phys. **305**, 266 (2003) [arXiv:hep-th/0209028].
36. S. J. Brodsky, J. R. Hiller and G. McCartor, arXiv:hep-ph/0508295.
37. M. Weinstein, arXiv:hep-th/0410113.
38. H. Zhan, A. Nogga, B. R. Barrett, J. P. Vary and P. Navratil, Phys. Rev. C **69**, 034302 (2004) [arXiv:nucl-th/0401047].
39. S. J. Brodsky, Few Body Syst. **36**, 35 (2005) [arXiv:hep-ph/0411056].
40. J. M. Maldacena, Adv. Theor. Math. Phys. **2**, 231 (1998) [Int. J. Theor. Phys. **38**, 1113 (1999)] [arXiv:hep-th/9711200].
41. J. Polchinski and M. J. Strassler, Phys. Rev. Lett. **88**, 031601 (2002) [arXiv:hep-th/0109174].
42. S. J. Brodsky and G. R. Farrar, Phys. Rev. Lett. **31**, 1153 (1973).
43. V. A. Matveev, R. M. Muradian and A. N. Tavkhelidze, Lett. Nuovo Cim. **7**, 719 (1973).
44. J. Polchinski and L. Susskind, arXiv:hep-th/0112204.
45. J. Polchinski and M. J. Strassler, JHEP **0305**, 012 (2003) [arXiv:hep-th/0209211].
46. S. J. Brodsky and G. F. de Teramond, Phys. Lett. B **582**, 211 (2004) [arXiv:hep-th/0310227].
47. G. Parisi, Phys. Lett. B **39**, 643 (1972).
48. S. J. Brodsky, Y. Frishman and G. P. Lepage, Phys. Lett. B **167**, 347 (1986).
49. T. Banks and A. Zaks, Nucl. Phys. B **196**, 189 (1982).
50. S. J. Brodsky, G. P. Lepage and P. B. Mackenzie, Phys. Rev. D **28**, 228 (1983).
51. S. J. Brodsky, G. T. Gabadadze, A. L. Kataev and H. J. Lu, Phys. Lett. B **372**, 133 (1996) [arXiv:hep-ph/9512367].
52. R. J. Crewther, Phys. Rev. Lett. **28**, 1421 (1972).
53. G. F. de Teramond and S. J. Brodsky, arXiv:hep-ph/0507273.
54. S. J. Brodsky, arXiv:hep-ph/0408069.
55. G. F. de Teramond and S. J. Brodsky, arXiv:hep-th/0409074; J. Erlich, E. Katz, D. T. Son and M. A. Stephanov, arXiv:hep-ph/0501128; S. Hong, S. Yoon and M. J. Strassler, arXiv:hep-ph/0501197; L. Da Rold and A. Pomarol, Nucl. Phys. B **721**, 79 (2005) [arXiv:hep-ph/0501218]; H. Boschi-Filho, N. R. F. Braga and H. L. Carrion, arXiv:hep-th/0507063.
56. G. F. de Teramond and S. J. Brodsky, Phys. Rev. Lett. **94**, 201601 (2005) [arXiv:hep-th/0501022].
57. S. S. Gubser, I. R. Klebanov and A. M. Polyakov, Nucl. Phys. B **636**, 99 (2002) [arXiv:hep-th/0204051].
58. R. Baldini, S. Dubnicka, P. Gauzzi, S. Pacetti, E. Pasqualucci and Y. Srivastava, Eur. Phys. J. C **11**, 709 (1999).
59. X. d. Ji, J. P. Ma and F. Yuan, Phys. Rev. Lett. **90**, 241601 (2003) [arXiv:hep-ph/0301141].

Recent developments in lattice QCD

Z. Fodor

Department of Physics, University of Wuppertal, Germany and
Institute for Theoretical Physics, Eotvos University, Budapest, Hungary

Abstract. Recent results on finite temperature and/or density lattice QCD are reviewed. Lattice techniques and some previous results are shortly discussed. New results for physical quark masses are presented. The critical endpoint, pressure, energy density, entropy density, speed of sound and quark number susceptibilities are determined.

Keywords: Lattice QCD
PACS: 12.38.Gc, 11.15.Ha, 12.38.Mh

INTRODUCTION

In QCD with increasing temperatures (T) we expect a transition at some $T = T_c$. The dominant degrees of freedom are hadrons in the low temperature phase and colored objects in the high temperature phase. Present lattice results suggest a cross-over and a critical point at some non-vanishing T and chemical potential (μ).

Since we are mostly interested in the physics around T_c, non-perturbative methods are necessary among which lattice QCD is the most systematic one. There are at least two serious difficulties with lattice simulations. The first one is connected to the lightness of the quark masses. The cost of computations increases strongly as the quark masses decrease, therefore most lattice results were obtained with unphysically large quark masses. The second difficulty is connected to the continuum limit. Calculations are always performed at a finite lattice spacing (a). In order to get physical results, we have to take the $a \to 0$ limit. Since e.g. for the equation of state (EoS) the computational costs scale as a^{-13} it is not surprising that up to very recently most results were obtained only at one set of lattice spacings.

The situation is much easier in the case of the pure gauge theory. The first problem does not exist since the quark masses are infinite. There are continuum extrapolated results e.g. for the equation of state, both with unimproved and improved lattice actions and they show nice agreement [1, 2, 3]. There are also numerous EoS results for the full theory with dynamical quarks [4, 5, 6, 7] which will be discussed in the following.

For a long time it was believed that no physical answer can be given to questions with non-vanishing baryonic densities. The reason for that is the infamous sign problem, which spoils any Monte-Carlo method based on importance sampling. Recently, new techniques were developed, which are able to cover small to moderate baryonic chemical potentials at non-vanishing temperatures (chemical potential is used to set the baryonic density).

In this talk recent results on non-vanishing densities and the determination of the EoS when approaching the physical quark mass and continuum limit are presented.

CP814 *Hadron Spectroscopy: Eleventh International Conference*
edited by A. Reis, C. Göbel, J. Sá Borges, and J. Magnin

LATTICE FORMULATION, NONVANISHING TEMPERATURES AND DENSITIES

Thermodynamical quantities can be obtained from the partition function which can be given by a Euclidean path-integral:

$$Z = \int \mathscr{D}U \mathscr{D}\bar{\Psi}\mathscr{D}\Psi \exp[-S_E(U,\bar{\Psi},\Psi)] \tag{1}$$

where U and $\bar{\Psi},\Psi$ are the gauge and fermionic fields and S_E is the Euclidean action. The lattice regularization of this action is not unique. There are several possibilities to use improved actions which have the same continuum limit as the unimproved ones. The advantage of improved actions is that the discretization errors are reduced.

Usually S_E can be split up as $S_E = S_g + S_f$ where S_g is the gauge action containing only the self interactions of the gauge fields and S_f is the fermionic part. The gauge action has one parameter, the β gauge coupling, while the parameters of S_f are the m_q quark masses and μ_q chemical potentials. For the fermionic action the two most widely used discretization types are the Wilson and staggered fermions.

For the actual calculations finite lattice sizes of $N_s^3 N_t$ are used. The physical volume and the temperature are related to the lattice extensions as:

$$V = (N_s a)^3, \quad T = \frac{1}{N_t a}. \tag{2}$$

Therefore lattices with $N_t \gg N_s$ are referred to as zero temperature lattices while the ones with $N_t < N_s$ are finite temperature lattices. Since the gauge coupling β has the largest influence on the lattice spacing, it essentially determines the temperature (increasing β increases T).

For large homogeneous systems the pressure is proportional to the free energy density. Unfortunately the free energy density ($-T/V \log Z$) cannot be measured directly. We can only measure the derivatives of $\log Z$ with respect to the parameters of the action. Then, with an integration we can obtain the pressure. This method is known as the integral method for calculating the pressure. In order to remove the divergent zero-point energy we have to subtract the pressure measured on zero temperature lattices. Further thermodynamical quantities can be derived directly from the pressure. For example the energy density (ε), entropy density (s) and speed of sound (c_s) have the following relation with the pressure:

$$\varepsilon = T(\partial p/\partial T) - p, \quad s = (\varepsilon + p)T, \quad c_s^2 = \frac{dp}{d\varepsilon}. \tag{3}$$

Though QCD at finite chemical potential (μ, which as already mentioned, is used to set non-vanishing baryonic density) can be formulated on the lattice [8], standard Monte-Carlo techniques can not be used at $\mu \neq 0$. The reason is that for non-vanishing real μ the functional measure –thus, the determinant of the Euclidean Dirac operator– is complex. This fact spoils any Monte-Carlo technique based on importance sampling. Several suggestions were studied earlier to solve the problem. Unfortunately, none of them was

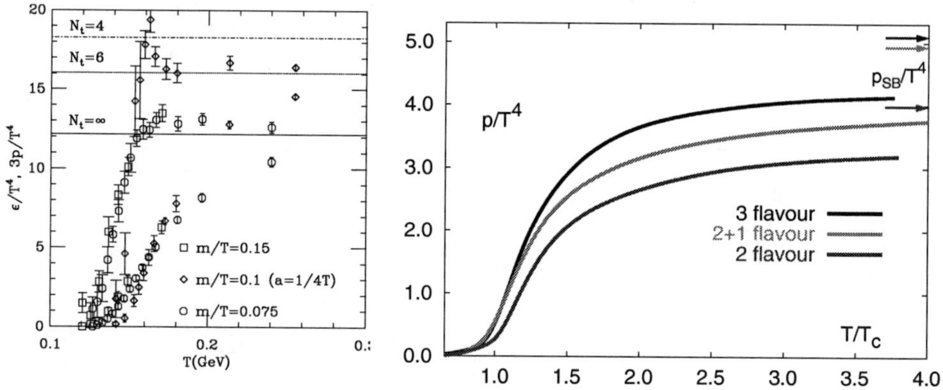

FIGURE 1. *Left:* The pressure (lower symbols) and energy density (upper symbols) with 2 flavors of unimproved staggered fermions on $N_t = 4$ (diamonds) and $N_t = 6$ (squares and circles) lattices for different m_q/T ratios. [5]. *Right:* The pressure 2, 2+1 and 3 flavors of $p4$ improved staggered fermions on $N_t = 4$ lattices [6].

able to give physical answers for non-vanishing densities. About three years ago new techniques appeared, with which moderate chemical potentials could be reached on the lattice.

One of the most popular ideas [9, 10] was to produce an ensemble of QCD configurations at μ=0 and at the corresponding transition temperature T_c (or at any other physically motivated point for which importance sampling works). Then one determined the Boltzmann weights [11] of these configurations at $\mu \neq 0$ and at T lowered to the transition temperatures at this non-vanishing μ. An ensemble of configurations at a transition point was reweighted to an ensemble of configurations at another transition point.

LINE OF CONSTANT PHYSICS

Lattice calculations of the EoS are usually performed with a fixed N_t and then, since in a fixed temperature range N_t is inversely proportional to the lattice spacing, the continuum limit can be approached by increasing N_t. Keeping N_t constant means that the temperature can only be varied by changing the lattice spacing. This is usually achieved by varying the gauge coupling. If we want to keep e.g. the quark masses constant then the dimensionless lattice mass parameters (am_q) have to be tuned accordingly. This defines the line of constant physics (LCP) in the parameter space.

If we keep the mass parameters constant and do not follow the LCP – which is the case in most EoS lattice studies – then we have to face the following unphysical situation. Cooling down two systems, one at $3T_c$ and one at T_c to zero temperature, the quarks in the former case will be 3 times heavier. In this approach not m_q but m_q/T is kept constant.

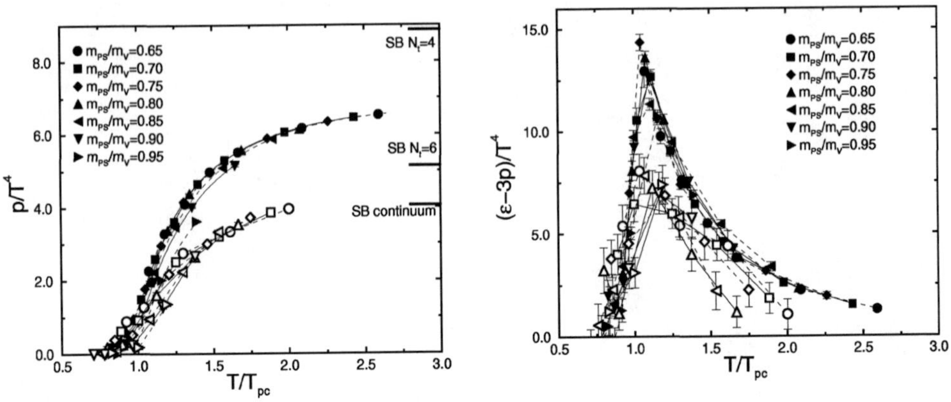

FIGURE 2. The pressure (left) and $\varepsilon - 3p$ (right) obtained from O(a) improved Wilson fermions for several pion masses on $N_t = 4$ (filled symbols) and $N_t = 6$ (open symbols) lattices [7]

PREVIOUS RESULTS ON THE EQUATION OF STATE

There are numerous lattice results for the EoS using dynamical quarks. However, in all cases the quark masses – for computational reasons mentioned in the introduction – were set to higher values than their physical one. The first results were obtained with staggered fermions. Calculations were performed by the MILC collaboration [4, 5] and by Karsch, Laermann and Peikert from Bielefeld [6]. The first calculation with Wilson fermions was done by the CP-PACS collaboration [7].

Staggered results are shown on Fig. 1. No LCP was used in these cases, which means that the curves correspond to constant m_q/T, i.e. increasing quark masses with increasing temperature. Fig. 2 shows the EoS obtained with Wilson fermions for $N_t = 4$ and 6. The lowest quark mass used here corresponds to a pion mass of ≈ 500 MeV. The LCP was used in this analysis.

In the last years small nonzero chemical potentials [9, 10] have also been used to determine the EoS [12, 13, 14, 15].

Recently, at the lattice conference both the MILC collaboration [16] and the RBC-Bielefeld collaboration [17] reported on their ongoing work in QCD thermodynamics.

Although the published results all apply QCD with dynamical quarks they still have several weaknesses.

1. In all cases, unphysical quark masses were used, which results in unphysical pion masses. Since the transition temperature is higher than the physical mass of the pion, but smaller than the pion masses used in these calculations, it might be important to use physical values.
2. The works with staggered fermions did not use the line of constant physics which results in an unphysical dependence of the hadron masses on the temperature.
3. A known problem with staggered fermions is the taste symmetry violation which

causes a non-physical non-degeneracy of the pion masses. This non-degeneracy disappears in the continuum limit, but it is still large for the lattice spacings used in these calculations.

4. The approximate R algorithm [18] was used for the calculations with 2,3 or 2+1 flavors of staggered fermions. This algorithm has an intrinsic stepsize which leads to systematic errors in the results. In order to eliminate this systematics an extrapolation to zero stepsize should be performed. None of the previous works have done such an extrapolation. It should be mentioned that due to the subtraction in the calculation of the pressure the error coming from the typically used finite stepsizes is comparable with the result itself.
5. The discretization errors are still probably large. This is especially true for temperatures around and below T_c where the lattice spacing of $N_t = 4$ lattices can be as large as 0.3 fm.
6. The determination of the physical scale is not always unambiguous. Ref. [6] uses e.g. the string tension which is – strictly speaking – not an existing quantity in full QCD since at large distances the string breaks and a meson pair is produced.

NEW RESULTS WITH PHYSICAL QUARK MASSES, EQUATION OF STATE

In the following the new results obtained in collaboration with Y. Aoki, S.D. Katz, and K.K. Szabó are presented. Details of this work are found in Ref. [20]. We have determined the EoS for two sets of lattice spacings, $N_t = 4$ and 6. We improved on all points listed above.

Lattice action, LCP

The lattice action we used was a combination of the tree level Symanzik-improved gauge action and the stout-improved fermionic action [21]. The stout improvement is known to reduce the taste symmetry violation significantly.

As mentioned above, using an approximate algorithm without performing the necessary extrapolations is dangerous. Instead we used the exact rational hybrid Monte-Carlo (RHMC) algorithm [22, 23].

The quark masses were set to their physical values so that the meson masses agree with their physical values up to a few percent. Moreover, the physical quark masses were kept constant while increasing the temperature, i.e. we followed the LCP.

In order to give the EoS in T/T_c units, we had to find the ratio of the scales at the different simulation points. For this we matched the static quark-antiquark potential for the different points at an intermediate distance. T_c was defined as the turning point of the isospin number susceptibility [20]. The precise determination of T_c, i.e. connecting the scale to physical quantities will be the subject of a subsequent publication.

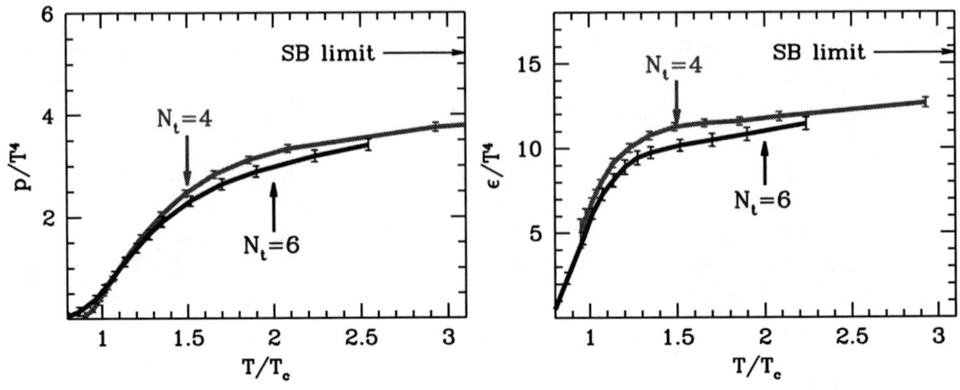

FIGURE 3. *Left:* the pressure p, as a function of the temperature. Both N_t=4 (red, upper curve) and N_t=6 (blue, lower curve) data are obtained along the LCP. They are normalized by T^4 and scaled by c_{cont}/c_{N_t}. In order to lead the eye lines connect the data points. *Right:* the energy density (ε), red (upper) and blue (lower) for N_t=4 and 6 respectively. This result was obtained directly from the pressure.

Results

In order to present the $N_t = 4$ and 6 results on the same plots we rescale all quantities in the following way. At infinite temperatures all quantities should approach their free Stefan-Boltzmann limit (c). This limit is, however different in the continuum (c_{cont}) and on lattices with some fixed N_t (c_{N_t}). Therefore all results are scaled with a factor c_{cont}/c_{N_t} so that they could be compared with the continuum Stefan-Boltzmann limits.

Fig. 3 shows the pressure and the energy density normalized by T^4. For comparison, the Stefan-Boltzmann limit is also shown. Similarly, one can determine the entropy density, speed of sound and quark number susceptibilities [20].

NEW RESULTS WITH PHYSICAL QUARK MASSES, CRITICAL ENDPOINT

A critical point is expected in QCD on the temperature versus baryonic chemical potential plane. Our goal in this section is to determine the location of this critical point.

The lattice action we used was the unimproved staggered action with physical quark masses (it means, that the pion and kaon masses take approximately their physical values).

The partition function of lattice QCD with n_f degenerate staggered quarks is given by the functional integral of the gauge action S_g at gauge coupling β over the link variables

U, weighted by the determinant of the quark matrix M, which can be rewritten [9] as

$$
\begin{aligned}
Z(\beta,m,\mu) &= \int \mathcal{D}U \exp[-S_g(\beta,U)][\det M(m,\mu,U)]^{n_f/4} \\
&= \int \mathcal{D}U \exp[-S_g(\beta_w,U)][\det M(m_w,\mu_w,U)]^{n_f/4} \\
&\left\{ \exp[-S_g(\beta,U)+S_b(\beta_w,U)] \left[\frac{\det M(m,\mu,U)}{\det M(m_w,\mu_w,U)}\right]^{n_f/4} \right\},
\end{aligned}
\tag{4}
$$

where m is the quark mass, μ is the quark chemical potential and n_f is the number of flavors. For non-degenerate masses one uses simply the product of several quark matrix determinants on the $1/4$-th power. Standard importance sampling works and can be used to collect an ensemble of configurations at m_w, β_w and μ_w (with e.g. Re(μ_w)=0 or non-vanishing isospin chemical potential). It means we treat the terms in the curly bracket as an observable –which is measured on each independent configuration– and the rest as the measure. By simultaneously changing several parameters e.g. β and μ one can ensure that even the mismatched measure at β_w and μ_w samples the regions where the original integrand with β and μ is large. In practice the determinant is evaluated at some μ and a Ferrenberg-Swendsen reweighting [11] is performed for the gauge coupling β. The fractional power in eq. (4) can be taken by using the fact that at $\mu = \mu_w$ the ratio of the determinants is 1 and the ratio is a continuous function of the chemical potential. The details of the determinant calculation can be found in Ref. [10].

In the following we keep μ real and look for the zeros of the partition function on the complex β plane. These are the Lee-Yang zeros [19]. Their V$\rightarrow \infty$ behavior tells the difference between a crossover and a first order phase transition. At a first order phase transition the free energy $\propto \log Z(\beta)$ is non-analytic. Clearly, a phase transition can appear only in the V$\rightarrow \infty$ limit, but not in a finite V. Nevertheless, the partition function has Lee-Yang zeros at finite V. These are at "unphysical" complex values of the parameters, in our case at complex β-s. For a system with a first order phase transition these zeros approach the real axis in the V$\rightarrow \infty$ limit (the detailed analysis suggests a $1/V$ scaling). This V$\rightarrow \infty$ limit generates the non-analyticity of the free energy. For a system with crossover the free energy is analytic, thus the zeros do not approach the real axis in the V$\rightarrow \infty$ limit.

Figure 4 shows Im(β_0^∞) as a function of μ enlarged around the endpoint μ_{end}. The picture is simple and reflects the physical expectations. For small μ-s the extrapolated Im(β_0^∞) is inconsistent with a vanishing value, and the prediction is a crossover. Increasing μ the value of Im(β_0^∞) decreases, thus the transition becomes consistent with a first order phase transition.

Setting the scale leads to the final results of the analysis. As we already discussed, the quark masses, used to determine the endpoint, correspond approximately to their physical values. The pion to rho mass ratio, extrapolated to our $T \neq 0$ parameters, is 0.188(2) (its physical value is 0.179), whereas the pion to K mass ratio in the same limit is 0.267(1) (its physical value is 0.277).

Figure 5 shows the phase diagram in physical units, thus T as a function of μ_B, the baryonic chemical potential (which is three times larger then the quark chemical potential). At μ=0 the transition between the hadronic and quark-gluon plasma phases is

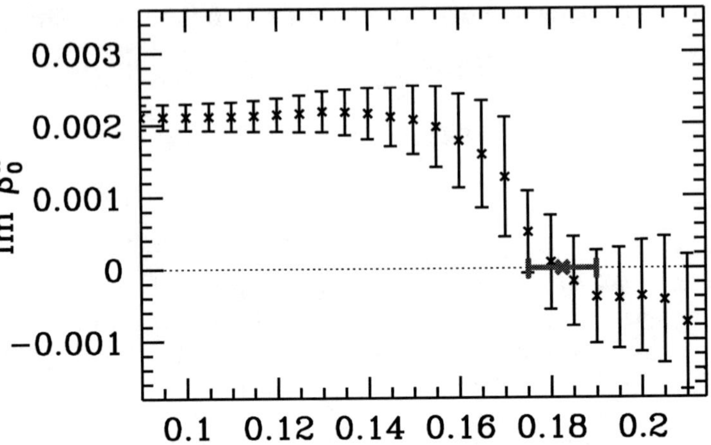

FIGURE 4. Im(β_0^∞) as a function of the chemical potential.

a cross-over. As we increase the chemical potential the transition temperature decreases, but the transition itself remains a cross-over. At a given endpoint chemical potential the transition is a second order one. For even larger chemical potentials the transition temperature further decreases and the transition becomes a first order one.

The curvature of the crossover line separating the QGP and the hadronic phases is given by $T/T_c = 1 - C\mu_B^2/T_c^2$ with C=0.0032(1).

The endpoint is at $T_E = 162 \pm 2$ MeV, $\mu_E = 360 \pm 40$ MeV.

SUMMARY

Previous results using either staggered of Wilson fermions were discussed. They suffer from several weaknesses.

New results on the EoS were presented. Our analysis attempted to improve on previous analyses by several means. We used for the lightest hadronic degree of freedom the physical pion mass. We used two different sets of lattice spacings (N_t=4,6). The system was kept on the line of constant physics (LCP) instead of changing the physics with the temperature. Due to our smaller lattice spacing and particularly due to our stout-link improved fermionic action the unphysical pion mass splitting was much smaller than in any previous staggered analysis. An exact calculation algorithm was applied.

We presented results for the pressure and energy density. In a similar way results for the entropy density, speed of sound and the isospin and strangeness susceptibilities can be obtained.

Although a continuum extrapolation could already be performed with the current data, since the $N_t = 4$ lattices are rather coarse (especially around and below T_c) it would be safer if the EoS on even finer lattices (N_t=8) were obtained. Such an analysis would be

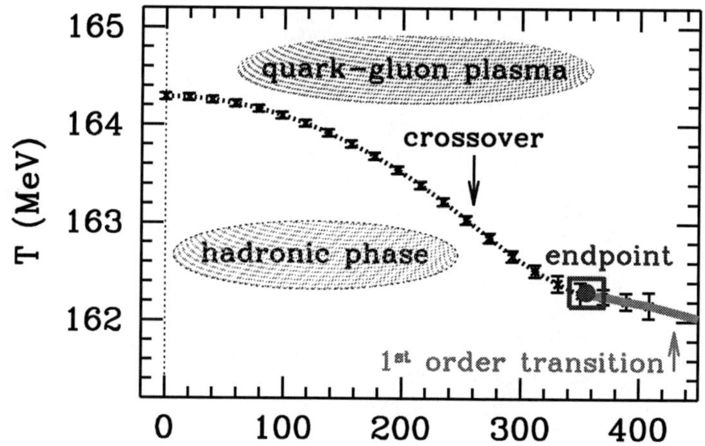

FIGURE 5. The phase diagram in physical units. Dotted line illustrates the crossover, solid line the first order phase transition. The small square shows the endpoint. The depicted errors originate from the reweighting procedure. Note, that an overall additional error of 1.3% comes from the error of the scale determination at T=0. Combining the two sources of uncertainties one obtains $T_E = 162 \pm 2$ MeV and $\mu_E = 360 \pm 40$ MeV.

a major step towards the final results for the equation of state.

We also discussed the overlap-improving multi-parameter reweighting technique, in order to calculate physical observables at non-vanishing temperatures and chemical potentials. A critical point is expected in QCD on the temperature versus baryonic chemical potential plane. Using the above lattice method for $\mu \neq 0$ we studied dynamical QCD with n_f=2+1 staggered quarks of physical masses on $L_t = 4$ lattices. We used physical quark masses in this analysis. Our result for the critical point is $T_E = 162 \pm 2$ MeV and $\mu_E = 360 \pm 40$ MeV. The continuum limit extrapolation is missing also in this case.

ACKNOWLEDGMENTS

This work was partially supported by OTKA Hungarian Science Grants No. T34980, T37615, M37071, T032501. and by the EU Hadron physics project RII3CT-20040506078.

The computations were carried [25] out at Eötvös University on the 330 processor PC cluster of the Institute for Theoretical Physics and the 1024 processor PC cluster of Wuppertal University.

REFERENCES

1. G. Boyd, et al., Nucl. Phys. B **469** (1996) 419 [arXiv:hep-lat/9602007].

2. M. Okamoto *et al.*, Phys. Rev. D **60** (1999) 094510 [arXiv:hep-lat/9905005].
3. Y. Namekawa *et al.*, Phys. Rev. D **64** (2001) 074507 [arXiv:hep-lat/0105012].
4. T. Blum, et al., Phys. Rev. D **51** (1995) 5153 [arXiv:hep-lat/9410014].
5. C. W. Bernard *et al.* [MILC Collaboration], Phys. Rev. D **55** (1997) 6861 [arXiv:hep-lat/9612025].
6. F. Karsch, E. Laermann and A. Peikert, Phys. Lett. B **478** (2000) 447 [arXiv:hep-lat/0002003].
7. A. Ali Khan *et al.*, Phys. Rev. D **64** (2001) 074510 [arXiv:hep-lat/0103028].
8. P. Hasenfratz and F. Karsch, Phys. Lett. B **125** (1983) 308; J. B. Kogut, H. Matsuoka, M. Stone, H. W. Wyld, S. H. Shenker, J. Shigemitsu and D. K. Sinclair, Nucl. Phys. B **225** (1983) 93.
9. Z. Fodor and S. D. Katz, Phys. Lett. B **534** (2002) 87 [arXiv:hep-lat/0104001].
10. Z. Fodor and S. D. Katz, JHEP **0203** (2002) 014 [arXiv:hep-lat/0106002]; JHEP **04** (2004) 050 [arXiv:Xep-lat/0402006];
11. A. M. Ferrenberg and R. H. Swendsen, Phys. Rev. Lett. **61** (1988) 2635; A. M. Ferrenberg and R. H. Swendsen, Phys. Rev. Lett. **63** (1989) 1195.
12. Z. Fodor, S. D. Katz and K. K. Szabo, Phys. Lett. B **568** (2003) 73 [arXiv:hep-lat/0208078].
13. R. V. Gavai and S. Gupta, Phys. Rev. D **68** (2003) 034506 [arXiv:hep-lat/0303013].
14. C. R. Allton, et al., Phys. Rev. D **68** (2003) 014507 [arXiv:hep-lat/0305007].
15. F. Csikor, et al., JHEP **0405** (2004) 046 [arXiv:hep-lat/0401016].
16. C. Bernard *et al.*, arXiv:hep-lat/0509053.
17. C. Jung, arXiv:hep-lat/0510035.
18. S. A. Gottlieb, W. Liu, D. Toussaint, R. L. Renken and R. L. Sugar, Phys. Rev. D **35** (1987) 2531.
19. C. N. Yang and T. D. Lee, Phys. Rev. **87** (1952) 404; T. D. Lee and C. N. Yang, Phys. Rev. **87** (1952) 410.
20. Y. Aoki, Z. Fodor, S.D. Katz, K.K. Szabo, hep-lat/0510084
21. C. Morningstar and M. J. Peardon, Phys. Rev. D **69** (2004) 054501 [arXiv:hep-lat/0311018].
22. M.A.Clark, B.Joo, A.D.Kennedy, Nucl. Phys. Proc. Suppl. **119** (2003) 1015 [arXiv:hep-lat/0209035].
23. M. A. Clark and A. D. Kennedy, Nucl. Phys. Proc. Suppl. **129** (2004) 850 [arXiv:hep-lat/0309084].
24. F. Karsch, E. Laermann and A. Peikert, Nucl. Phys. B **605** (2001) 579 [arXiv:hep-lat/0012023].
25. Z. Fodor, S. D. Katz and G. Papp, Comput. Phys. Commun. **152** (2003) 121 [arXiv:hep-lat/0202030].

Exotic Meson Spectroscopy from BNL E852

Gary S. Adams

Physics Department, Rensselaer Polytechnic Institute, Troy NY 12180

Abstract. BNL experiment E852 has discovered several light-quark exotic mesons. The nature of those states is still under debate but their existence has been established. It is likely that the first few members of the $_1P$ hybrid meson multiplet have been observed.

Keywords: exotic hybrid tetraquark meson
PACS: 13.25-k, 13.85.Hd, 14.40.Cs

Understanding hadron structure is one of the primary goals of modern field theory. However progress has been slow and much of our knowledge of hadrons comes from simple theoretical models. In this context there have been many attempts to identify hadrons that fall outside the basic picture of three-quark baryons and quark-antiquark mesons. For example, recent experiments point to a possible tetraquark meson with hidden charm, the X(3872) [1, 2, 3, 4]. The meson sector is particularly well suited to such studies because there exist quantum-number selection rules for quark-antiquark states. Mesons with forbidden quantum numbers (e.g. $J^{PC} = 1^{-+}$ or 2^{+-}) do not mix with quark-antiquark mesons so they offer a natural testing ground for lattice implementations of quantum chromodynamics (QCD). These spin-exotic mesons may also yield insights into the nature of color confinement [5].

Theoretical predictions indicate that the low-lying spectrum of spin-exotic mesons should be made up of tetraquark states [6] and hybrid mesons [7, 8, 9, 10, 11]. The distinguishing feature of hybrid mesons is that they have explicit gluonic degrees of freedom. This subject has recently been reviewed by Barnes [12].

Experiment E852 at Brookhaven National Laboratory was designed to search for spin-exotic and isospin-exotic mesons. It was carried out at the Multi-Particle Spectrometer facility (MPS). A π^- beam, with laboratory momentum 18 GeV/c, and a liquid hydrogen target were used to measure peripheral meson production leading to multi-meson final states. A description of the experimental apparatus can be found in Ref. [13].

Considerable progress has been made in the experimental identification of isospin-one $J^{PC} = 1^{-+}$ states, π_1 exotic mesons. From E852 there is clear evidence for $\pi_1(1400)$ [13], $\pi_1(1600)$ [14, 15, 16, 17], and $\pi_1(2000)$ [16, 17]. The lower two states have been observed in other experiments as well [18, 19, 20, 21]. There is now sufficient experimental information on these spin-exotic mesons to begin to interpret the mass spectrum.

There is a solid foundation on which to build our interpretation, based on the expected decay properties of the lightest hybrid π_1 meson [9, 22, 23]. It is widely acknowledged that a hybrid π_1 will prefer to decay by emission of one S-wave and one P-wave meson. Thus a light-quark hybrid with $J^{PC} = 1^{-+}$, for example π_1, should be observed in $\pi b_1(1235)$ decay. Just such a decay was observed by the Crystal Barrel experiment,

CP814 *Hadron Spectroscopy: Eleventh International Conference*
edited by A. Reis, C. Göbel, J. Sá Borges, and J. Magnin

however they were unable to determine the resonance parameters for the state [21]. Experiment E852 measured a large signal in the charged mode resulting in the identification of an exotic π_1 meson with mass $M = 1664 \pm 8 \pm 10$ and width $\Gamma = 185 \pm 25 \pm 28$ MeV/c^2 [17]. Moreover, this state was excited primarily by unnatural parity exchange. Unnatural parity exchange is the expected process for hybrid π_1 production with a pion beam because b_1 exchange should dominate. I will label this state $\pi_1(1660)$ to distinguish it from some of the other signals identified as $\pi_1(1600)$. Based on the production and decay properties of $\pi_1(1660)$ it seems likely that it is a $_1P$ hybrid meson. The notation here is from the flux-tube model [10]. It denotes total orbital angular momentum equal to one (P-wave) and a spin-one gluon field.

The mass of the lightest hybrid multiplet has been calculated in many models including the flux-tube model [9], lattice QCD [24], and QCD sum rules [25, 11]. Flux-tube predictions lead one to expect the hybrid spectrum to start around 2 GeVc^2. However later QCD calculations suggest a somewhat lower mass. It is noteworthy that much of the discrepancy between the mass predictions comes from the fact that the QCD string tension is not well known. This leads to the interesting possibility that detailed knowledge of hybrid meson spectroscopy may provide important constraints on this fundamental parameter. The $\pi_1(1660)$ mass is compatible with the latest calculations of the light-quark $_1P$ hybrid mass [25, 11, 24].

Hybrid mesons should also encompass a broad spectrum of states that do not have exotic quantum numbers. Identifying those states is difficult; there are few reliable predictions for their decay properties. However the mass spectrum alone can provide strong indications of unusual states. A case in point is the π_2 spectrum. The $\pi_2(1670)$ is a well established $^1D_2$ quark-model state. Based on the known mass trajectories for other states [26] one expects the first radial π_2 at 2000 MeV/c^2. An excellent candidate for this radial excitation was observed in πf_2 decay [26] and recently confirmed by E852 in $\omega\rho$ decay [17]. The measured masses were $M = 2005 \pm 15$ MeV/c^2 and $M = 1974 \pm 14 \pm 83$MeV/c^2, respectively. An additional high-mass state, $\pi_2(1880)$, was observed in ηa_2 decay [27] with a mass $M = 1880 \pm 20$MeV/c^2 and recently confirmed by E852 in $\omega\rho$ decay [17] with a mass $M = 1876 \pm 11 \pm 67$MeV/c^2. This state, the isovector partner of $\eta_2(1870)$, falls at a mass that cannot be easily accomodated in the quark model. Thus it is a good candidate for a $_1P$ hybrid meson [27]. The observed mass splitting between $\pi_1(1660)$ and $\pi_2(1880)$ is in reasonable accord with the value expected from lattice QCD, although the theoretical predictions have large uncertainties [8].

A high-lying exotic meson has also been observed in experiment E852. The $\pi_1(2000)$ was first observed in πf_1 decay [16] and later confirmed in $\pi b_1(1235)$ [17]. Under the assumption that $\pi_1(1660)$ is a $_1P$ hybrid, the expected mass [10]for a $_1D$ orbitally excited hybrid is about 2060 MeV/c^2, in agreement with the observed $\pi_1(2000)$. The preferred decay modes for $_1D$ hybrids are unknown.

Although further confirmation is needed before $\pi_1(1660)$, $\pi_2(1880)$, and $\pi_1(2000)$ can be definitively identified as hybrid mesons, the available spectroscopic information as well as production and decay properties for these states suggest that the first elements of the light-quark hybrid spectrum have been seen. However several experimental observations indicate that the spectrum of spin-exotic mesons is more complicated than hybrids alone.

The most pressing experimental issue is that several other π_1 decays have been measured in the region of 1600 MeV/c^2. A small $\pi\rho$ signal was observed with clear phase motion in natural-parity exchange [14], but the resonant content of the unnatural parity-exchange signal could not be determined. Thus it is not known if this is the same state identified above as $\pi_1(1660)$ or something else. The situation for π_1 signals observed in πf_1 decay [16] and in $\pi\eta'$ decay [15] is much clearer; they are produced by natural-parity exchange. This indicates that these signals do not come from the $\pi_1(1660)$. I will label this new state $\pi_1(1600)$, although one should take note of the fact that these two natural-parity signals occur at slightly different masses. If these represent two decay modes for the same $\pi_1(1600)$ then one observes a much larger partial width for πf_1 than for $\pi\eta'$ [16]. Both of these signals are small compared to the strength observed for πb_1 decay of $\pi_1(1660)$. One interpretation that has been put forward for the $\pi_1(1600)$ is that it is a dynamically generated object similar to the $f_0(600)$ [28]. This interpretation would also apply to the $\pi_1(1400)$, a state that is clearly too low in mass to be a hybrid meson. It is also possible that $\pi_1(1400)$ is a tetraquark meson. It is widely assumed that such states would be too wide to observe as isolated resonances but no detailed predictions have been made for the spin-exotic π_1. It is even possible that multiple low-lying tetraquark π_1 states could be present [29]. The $\pi_1(1400)$ was first observed in peripheral production as a very small signal interfering with the dominant $a_2(1320)$ wave [30]. Thus one expects that further knowledge about this state will probably come from the large signals observed in annihilation experiments [18, 19, 20]. Hadronic χ decay also offers the possibility of detecting exotic signals without the dominant $a_2(1320)$ seen in peripheral production.

The prospects for future discoveries in exotic meson spectroscopy are very good. Several spin-exotic states have been confirmed. The $\pi_1(1660)$, $\pi_2(1880)$, and $\pi_1(2000)$ states are probably hybrid mesons. Further studies of these states will be helpful. Now that we have a starting point it should be possible to establish the spectroscopy of light-quark ${}_1P$ hybrid mesons. States with $J^{PC} = 1^{--}, 0^{-+}, 1^{-+}, 2^{-+}, 1^{++}, 0^{+-}, 1^{+-}$, and 2^{+-} are expected with masses near those of the $\pi_1(1660)$ and $\pi_2(1880)$. It is particularly important to identify the spin-exotic 0^{+-} and 2^{+-} states. Another avenue of investigation that has received little experimental study is quark flavor. If hybrid mesons exist with typical hadronic widths in the light-quark sector then those same multiplets should be apparent in the strange-quark and charm-quark sectors.

REFERENCES

1. S. L. Olsen (2004), `hep-ex/0412068`.
2. K. Abe (2005), `hep-ex/0505037`.
3. E. Braaten, and M. Kusunoki, *Phys. Rev.* **D69**, 074005 (2004).
4. E. S. Swanson, *Phys. Lett.* **B588**, 189–195 (2004).
5. N. Isgur, *Phys. Rev.* **D60**, 114016 (1999).
6. R. L. Jaffe, *Phys. Rev.* **D17**, 1444 (1978).
7. C. W. Bernard, et al., *Phys. Rev.* **D56**, 7039–7051 (1997).
8. P. Lacock, and K. Schilling, *Nucl. Phys. Proc. Suppl.* **73**, 261–263 (1999).
9. N. Isgur, and J. Paton, *Phys. Rev.* **D31**, 2910 (1985).
10. T. Barnes, F. E. Close, and E. S. Swanson, *Phys. Rev.* **D52**, 5242–5256 (1995).
11. K. G. Chetyrkin, and S. Narison, *Phys. Lett.* **B485**, 145–150 (2000).
12. T. Barnes, *Acta Phys. Polon.* **B31**, 2545–2556 (2000).

13. S. U. Chung, et al., *Phys. Rev.* **D60**, 092001 (1999).
14. S. U. Chung, et al., *Phys. Rev.* **D65**, 072001 (2002).
15. E. I. Ivanov, et al., *Phys. Rev. Lett.* **86**, 3977–3980 (2001).
16. J. Kuhn, et al., *Phys. Lett.* **B595**, 109–117 (2004).
17. M. Lu, et al., *Phys. Rev. Lett.* **94**, 032002 (2005).
18. A. Abele, et al., *Phys. Lett.* **B423**, 175–184 (1998).
19. A. Abele, et al., *Phys. Lett.* **B446**, 349–355 (1999).
20. P. Salvini, et al., *Eur. Phys. J.* **C35**, 21–33 (2004).
21. C. A. Baker, et al., *Phys. Lett.* **B563**, 140–149 (2003).
22. P. R. Page, E. S. Swanson, and A. P. Szczepaniak, *Phys. Rev.* **D59**, 034016 (1999).
23. P. R. Page, *Phys. Rev.* **D70**, 016004 (2004).
24. C. Bernard, et al., *Phys. Rev.* **D68**, 074505 (2003).
25. H. Y. Jin, J. G. Korner, and T. G. Steele, *Phys. Rev.* **D67**, 014025 (2003).
26. A. V. Anisovich, et al., *Phys. Lett.* **B517**, 261–272 (2001).
27. A. V. Anisovich, et al., *Phys. Lett.* **B500**, 222–231 (2001).
28. A. P. Szczepaniak, M. Swat, A. R. Dzierba, and S. Teige, *Phys. Rev. Lett.* **91**, 092002 (2003).
29. S. U. Chung, and E. Klempt, *Phys. Lett.* **B563**, 83–86 (2003).
30. D. R. Thompson, et al., *Phys. Rev. Lett.* **79**, 1630–1633 (1997).

Future Hadron Physics at Fermilab

Jeffrey A. Appel

Fermi National Accelerator Laboratory
PO Box 500, Batavia, IL 60510 USA
appel@fnal.gov

Abstract. Today, hadron physics research occurs at Fermilab as parts of broader experimental programs. This is very likely to be the case in the future. Thus, much of this presentation focuses on our vision of that future – a future aimed at making Fermilab the host laboratory for the International Linear Collider (ILC). Given the uncertainties associated with the ILC - the level of needed R&D, the ILC costs, and the timing - Fermilab is also preparing for other program choices. I will describe these latter efforts, efforts focused on a Proton Driver to increase the numbers of protons available for experiments. As examples of the hadron physics which will be coming from Fermilab, I summarize three experiments: MIPP/E907 which is running currently, and MINERνA and Drell-Yan/E906 which are scheduled for future running periods. Hadron physics coming from the Tevatron Collider program will be summarized by Arthur Maciel in another talk at Hadron05.

Keywords: Fermilab, Tevatron, ILC, hadron, MIPP, MINERνA, Drell-Yan, proton driver.
PACS: 13.85.-t

INTRODUCTION

In order to envisage the future of hadron physics at Fermilab, it will be useful to know the research context at the Laboratory. That can be best done by looking at the recent presentations by Fermilab's new Director, Piermaria Oddone, who began his tenure as Director on July 1, 2005. Pier has had several opportunities to project the vision of Fermilab's future, both inside the Laboratory and to outside committees.[1] I draw heavily on his presentations.

THE VISION OF FERMILAB'S FUTURE

The vision of Fermilab's future is driven by physics; in particular, where we see the largest mysteries. We know next to nothing about the next accessible energy scale. This is demonstrated by the plethora of models for what we might find once physics results are available from the Large Hadron Collider (LHC), the machine which will allow access beyond that at the Fermilab Tevatron. We know very little about the world of neutrinos, having only just discovered that they have mass and not knowing even the scale of all the mixing in the neutrino sector. We are ignorant also about what dark matter and dark energy are.

We have strong clues that the areas where these mysteries reside are the hunting grounds for major new discoveries, discoveries which will change our most basic

CP814 *Hadron Spectroscopy: Eleventh International Conference*
edited by A. Reis, C. Göbel, J. Sá Borges, and J. Magnin

understanding of nature. We also believe there is a deep connection among all these areas.

In addition to the physics context driving the vision of Fermilab's future, there is a strategic context, the U.S. contributions to frontier physics research. If we examine the US domestic accelerator program in 2005, 2010, and 2015, we see the US playing a leading role today in the neutrino frontier, the flavor frontier, and the energy frontier. However, with no new investment, we will be playing a secondary role in all these areas by 2010, and beyond. With investments of new and redirected resources, we can continue to play a leading role in neutrino physics in 2010 and beyond, and return to a leading role at the energy frontier in 2015. We do not presently see how we can play more than a secondary role in flavor physics in the mid-term future, given the plans to turn off the Tevatron and PEPII around the end of this decade.

Highest Priority New Initiative – The ILC

New initiatives at Fermilab will be consistent with our "overarching goals." We want to enable the most powerful attack on the fundamental science questions of our time by providing world class facilities for HEP as part of the global network. We will develop both the science and technology for particle physics and cosmology research. Our specific goals are to make vital contributions to particle physics in the next decade by developing a powerful new tool for discovery at the energy frontier (the International Linear Collider, ILC) and, at the same time, maintaining Fermilab's foremost accelerator-based neutrino program (including the off-axis experiment NOvA, developing a more powerful proton driver, and more).

The energy frontier for Fermilab already includes our efforts at the LHC as the host laboratory for the US LHC Accelerator Research Program and the US CMS collaboration. However, our goal is to also to play a leading role for the ILC. In collaboration with our partners around the world organized via the Global Design Effort (the GDE for short), we will establish all the technical components, costs, engineering designs, and management structures to enable an "early" decision (by 2010) on building the ILC if that is the right decision at that time. We will position the US (and Fermilab) to host the ILC. We will work to position the US (and Fermilab) to play major roles in detector development and physics analysis at the ILC. This is the highest priority initiative for the laboratory.

The ILC research and development roadmap includes milestones for pressing for the early decision on building this energy frontier facility at or near Fermilab. Such a decision might be as early as 2010. On the way, we will see superconducting radio-frequency module assembly and testing, a GDE Reference Design Report (RDR) effort, ILC component industrialization – all with strong Fermilab contributions. Given an LHC discovery by 2010, the scene will be set for a "GO" signal by the science community and by governments. On the other hand, if it is clear from the GDE RDR either that an ILC is not affordable at that point or that a much longer R&D effort is needed, other program choices will be induced. If there is no discovery at the LHC by 2009, again, there will be other program choices to be made. Today, Fermilab's major effort preparing for other program choices focuses on a Proton Driver, an option that would expand our neutrino physics capability substantially.

“Ships of the Line”

While we develop the ILC, we must deliver on our "ships of the line". These include our collider programs, our neutrino efforts, and other elements of our diverse program. Our collider program includes experiments at both the Tevatron, where we have the CDF and DZero experiments, and at the LHC where we play an important role on CMS. Our current neutrino program includes MINOS, which is receiving neutrinos produced by protons from the Fermilab Main Injector, and MiniBooNE, which uses neutrinos coming from our “Booster neutrino beam”. While focusing on these “ships of the line”, we will maintain our scientific vitality with a diverse program that includes smaller accelerator-based experiments, particle astrophysics, theory, and computing and other technology development.

The Tevatron program provides our greatest window into new phenomena until the LHC comes on-line. Between CDF and DZero, there are 1500 collaborators including 600 students and postdocs. The continued success of this program is critically dependent on an ever increasing rate of luminosity integration. The doubling time for luminosity will be a major consideration. This talk will leave a description of the hadron physics at the Tevatron to Arthur Maciel, who will speak on "Hadron Physics at the Tevatron".

Fermilab is the only major US laboratory associated with the CMS experiment at the LHC. As such, Fermilab has a central support role for the US community. This has been the case during construction. Attention is now turning to the huge data and physics discovery challenges. We are establishing an LHC Physics Center (LPC) at Fermilab to coordinate and support these efforts.

The present neutrino program includes two running experiments: MiniBooNE and MINOS. They operate with beams of different energy and with detectors placed at appropriate distances to investigate separate neutrino-mass-difference scales. MiniBooNE has been operating since 2003, and should have its first oscillation results this calendar year. It is looking for ν_e appearance in a ν_μ beam. As Oddone has said, “All hell will break loose if [there is a] positive signal.” At a minimum, the conventional picture of three neutrino generations will disappear. The MINOS program is just starting. It will make the most precise determination of the atmospheric neutrino mass difference by measuring ν_μ disappearance.

TABLE 1. Fermilab’s Currently Running Neutrino Experiments.

Experiment	MiniBooNE	MINOS
Peak Neutrino Energy	1 GeV neutrinos	2 GeV neutrinos
Proton Source Used	Booster	Main Injector
Detector(s)	800 ton oil Cerenkov detector at 0.54 km	5.4 Kiloton far detector at 735 km and 1.0 Kiloton near detector

New Initiatives in Neutrino Physics

We need to measure the full neutrino mixing matrix, and understand it in relation to the quark mixing matrix. However, we do not yet have more than a limit on the value of $\sin^2(2\theta_{13})$. We do not know the neutrino mass hierarchy; i.e., the mass ordering of the mass eigenstates, much less their actual values. We do not know if there is CP violation in the neutrino sector as there is in the quark sector, much less the value of the phase parameter δ. Fermilab is in the best position to make vital contributions to resolve these issues. However, we will need new initiatives in the neutrino program to address the issues. Today, Fermilab has the most powerful beam in the world, at 200 KW, dedicated to the production of neutrinos. Upgrades by 2008 should give 440 KW of primary proton power. When we stop the Tevatron, we'll have more than 600 KW, the same as JPARC will have some time after 2010. We are looking at further improvements; aimed at 1-2 MW of proton beam power.

Our most ambitious efforts to increase the reach of our neutrino program are aimed at developing a Proton Driver, a new source of 8 GeV protons to transfer into our Main Injector ring. We are focused on superconducting rf technology, making use of the R&D for the ILC. The Proton Driver would be a linear accelerator, too, providing a base for extremely flexible operations and a much simplified complex. The Proton Driver energy is only about 2% of an initial ILC. So, proceeding with the Proton Driver could establish useful industrial capability for a full ILC, not to mention provide useful operating experience. Nevertheless, we are also developing alternatives to a full Proton Driver project. Not running the Tevatron means we have the Recycler, Accumulator, and Debuncher rings to think about using. We have VERY EARLY creative ideas on how to take advantage of these opportunities to deliver 1-2 MW at 120 GeV.

Interlinked Roadmap

The interlinked roadmap just described establishes the main trunk and branches for the laboratory. It supports a vital role for the US community in the global program. However, it does require "strong focusing" now, which narrows the program. On the other hand, it establishes the strong base on which breadth and "texture" can flourish as we move forward.

HADRON PHYSICS NOW AND IN THE NEAR TERM

Today, hadron physics at Fermilab occurs as part of broader programs at the laboratory. This is evident from Table 2, which lists the three components of Fermilab's accelerator-based experiments. We anticipate adding NOνA and MINERνA, two neutrino-beam experiments, and a Drell-Yan fixed-target experiment, E906, in coming running periods. The current draft long range schedule (Figure 1) shows these initiatives, as well as continuing commitments to the operation of test beams, and several "OPEN" periods where there are currently no commitments.

Table 2. Fermilab's Accelerator-Based Experiments.	
Hadron Collider Experiments	CDF and DZero at the Tevatron CMS at the LHC
Neutrino Experiments	MiniBooNE and NuMI/MINOS now NuMI/MINERνA, NuMI/NOνA
Fixed Target - SY120	MIPP/E907 now, Drell-Yan/E906 later, kaons? Test beams

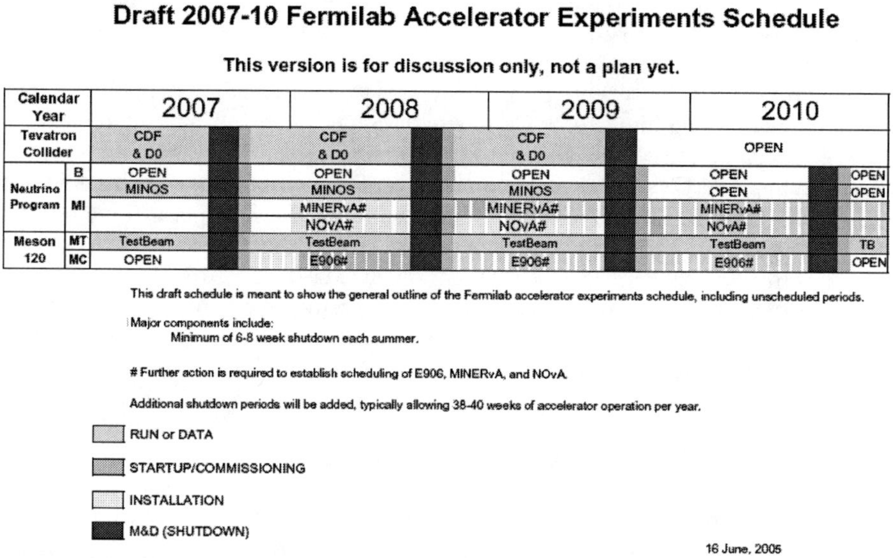

Figure 1. Fermilab Draft Long Range Accelerator Experiment Schedule.

Three experiments are particularly relevant to Fermilab's current hadron physics efforts: MIPP/E907, Drell-Yan/E906, and MINERνA. I will discuss each of these in turn.

Main Injector Particle Production (MIPP/E907)

Using primary protons from the Main Injector, MIPP/E907 is currently taking data on particle production, both by primary protons at 120 GeV/c, and with tagged $\pi^{\pm}$, $K^{\pm}$, and $p^{\pm}$ beams from 5 GeV/c to 90 GeV/c. These beams impinge on a variety of targets: liquid hydrogen and solid targets of various nuclei. The downstream spectrometer is well equipped to measure multiparticle final states, even states with high multiplicity. The detector includes a large Time Projection Chamber, a Ring Imaging Cerenkov counter, two large-aperture magnets, tracking wire chambers, and calorimeters.

The MIPP physics program begins with testing a proposed scaling law of particle fragmentation. This scaling states that the ratio of a semi-inclusive to an inclusive cross section is a function of the invariant mass M^2 only.[2]

$$\frac{\sigma(a+b\rightarrow c+X_{subset})}{\sigma(a+b\rightarrow c+X)} \equiv \frac{\sigma_{subset}(M^2,s,t)}{\sigma(M^2,s,t)} = \beta_{subset}(M^2)$$

European Hybrid Spectrometer data confirms this scaling in 12 reactions, but only at fixed s.[3] MIPP will test this scaling in 36 reactions, as a functions of both s and t, and for various particle types a, b, and c. The scaling law is said to follow naturally if the reaction factorizes into a two step process: (1) formation of a pseudo-resonance and (2) decay of the pseudo-resonance into the final state.

A second part of the physics program of MIPP is to understand, in detail, the particle production model going into Monte Carlo simulation programs. Date will be taken on thin targets and on the actual NuMI thick graphite target. The flux of neutrinos produced in this NuMI target depends directly on the distribution of hadrons leaving the target. MIPP took particle production data with the NuMI target in June and July of this year. Analysis and use of this data should lead to a significant reduction in the systematic error in the comparison of neutrino events in the MINOS near and far detectors, critical to the neutrino oscillation measurements of MINOS. There is a growing appreciation today of this aspect of systematic errors. One may ask whether the current MIPP measurements will be sufficient for future experiments; e.g., NOνA and T2K.

Drell-Yan Experiment E906

The Drell-Yan Experiment E906 is an extension of the previous fixed-target experiment E866 – exploring the sea quarks of the proton, the ratio of down antiquarks to the up antiquarks. The detector and beam energy define the acceptance range in x_{target} and x_{beam}. In fixed target experiments such as E866 and E906,

$$x_F = x_{beam} - x_{target} > 0.$$

The high-x beam partons are essentially only valence quarks. The low- and intermediate-x target sea partons are selected by the Drell-Yan process. These are the partons which can annihilate with the beam valence quarks to produce the high-mass di-muons of the Drell-Yan process. Thus, the fixed-target Drell-Yan process probes target antiquarks.

One may ask whether the proton sea is isospin symmetric. Furthermore, one may ask by what process is the proton sea formed? If the sea arises from perturbative QCD, then

$$\bar{d}(x_t) \equiv \bar{u}(x_t)$$

The study of the ratio of cross sections for deuterium to hydrogen gives the ratio

$$\left.\frac{\sigma^{pd}}{2\sigma^{pp}}\right|_{x_b >> x_t} \approx \frac{1}{2}\left[1+\frac{\bar{d}(x_t)}{\bar{u}(x_t)}\right].$$

E866 data clearly show an asymmetry. E906 will extend the measurements to higher x as shown in Figure 2.

Another aspect of the E906 physics program is to look for the effects of nuclear pions, an expected antiquark enhancement in nuclei relative to nucleons. Early predictions (Berger and Coester[4]) were proven false by Fermilab E772 Drell-Yan data. However, the E772 data has relatively large uncertainties, especially as x increases. E906 will challenge the revised predictions, shown in Figure 3. The new models for the ratio of cross sections on iron and hydrogen shown there still diverge sharply from unity at larger Bjorken-x – and from each other.

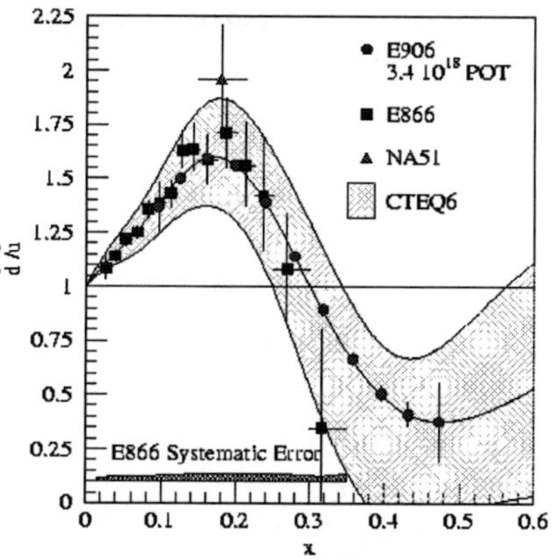

FIGURE 2. Bjorken-x dependence of existing and projected data on the ratio of down antiquarks to the up antiquarks in the proton sea.

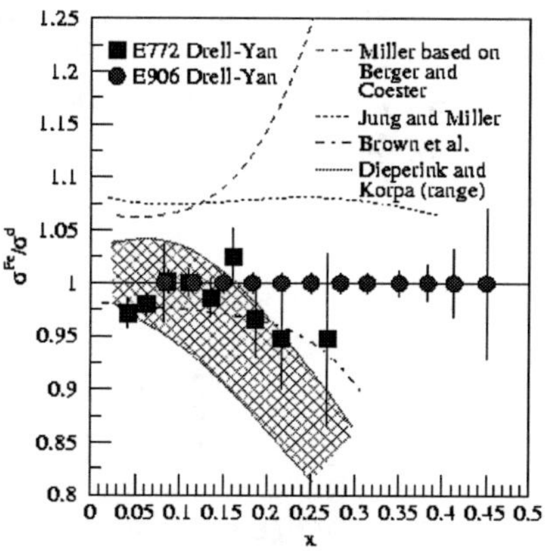

FIGURE 3. Drell-Yan cross section ratio on iron relative to deuterium, showing predictions[4] for the effects of nuclear pions.

MINERνA

MINERνA is a low-energy neutrino-nucleus scattering experiment using the NuMI beam. The name of the MINERνA experiment derives from the capital letters in "Main INjector ExpeRiment ν–A". MINERνA's main physics objective is to make measurements of the neutrino-nucleus scattering processes important for minimizing the systematic errors in neutrino oscillation experiments such as the cross sections for quasi-elastic, resonance and coherent π production. MINERνA will also significantly augment studies by the nuclear physics community at Jefferson Lab such as using the weak current to study the transition region of perturbative to non-perturbative QCD and the use of a variety of nuclear targets (Pb, Fe and C) to investigate nuclear effects in ν-induced interactions a topic of considerable importance for oscillation experiments as well.

In order to distinguish among the various processes, the detector has a high-granularity, fully-active scintillator-strip central core surrounded by electro-magnetic and hadron calorimeters. Triangular-shaped scintillator extrusions yield position precision through light-sharing between adjacent strips. The MINERνA detector will be placed just upstream of MINOS near detector.

A statistical sample of over 16 million charged current events will be collected in a four-year run. The experiment received Stage I approval in April, 2004, and had a successful Director's Review in January, 2005. The funding is anticipated to come from the Department of Energy through Fermilab, with construction and installation complete in the fall of 2008; physics data-taking starting in 2009.

In addition to the oscillation-motivated measurements, MINERνA will measure the poorly known high-x parton distribution functions, strange particle production channels crucial for proton decay experiments and the recently developed generalized parton distribution functions (GPDs) via the weak deeply-virtual Compton scattering process, shown in Figure 4. This will be the first measurement of GPDs with neutrinos, and as such, will allow flavor separation of GPDs. According to a calculation by A. Psaker,[5] MINERνA would accumulate 10,000 weak deeply-virtual Compton scattering events in a 4-year run.

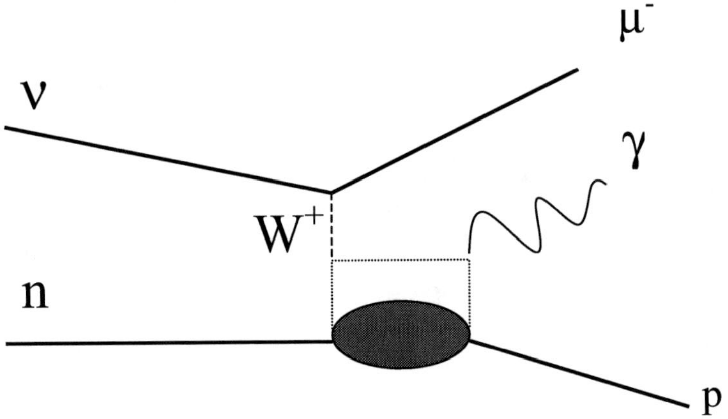

FIGURE 4. Weak deeply-virtual Compton scattering, an exclusive reaction with small momentum transfer t and a high energy photon γ.

A PROTON DRIVER IN FERMILAB'S FUTURE?

There is a need for more intense proton sources for future long-baseline neutrino-physics experiments. This couples to a need to replace the "old" upstream elements in the Fermilab accelerator chain, the Linac and Booster. A common solution to these needs, as mentioned in the discussion of Fermilab's future, would be to build a new proton source to replace the existing Linac and Booster, a Proton Driver.

The Proton Driver Parameters and Technology

The high-level parameters discussed for a Fermilab Proton Driver are listed in Table 3. The favored implementation for the Proton Driver is a superconducting radio-frequency module linac. It has several advantages over a synchrotron solution: (1) better performance, (2) more flexibility of beam types, and (3) a direct connection to the recently announced "cold technology" choice for the ILC. Thus, the development of the technology, any industrialization of the manufacture of components for the Proton Driver, and any operating experience can feed directly into ILC efforts. It is clear how this fits into the vision of the ILC timeline and of Fermilab's future.

Table 3. High-Level Parameters for a Fermilab Proton Driver.	
Beam power at 8 GeV	0.5-2.0 MW
Beam power at 120 GeV	2.0 MW
Resulting increase in beam power of the Main Injector beam.	10 x today's power

The Physics Case for the Proton Driver

The Physics case for a Proton Driver has been documented[6] following a workshop (October 6-9, 2004) and study[7] chaired by Steve Geer. The study and workshop summarized results from several working groups, including those on (1) neutrino interactions, (2) kaons and pions, (3) neutrons, and (4) antiprotons. Much of the physics considered is hadron physics.

FINAL COMMENTS

The DOE has announced that Fermilab will be the home of accelerator-based HEP in the US in the future, even after the Tevatron is turned off. Thus, defining the future of much of HEP hadron physics in the US and beyond will depend on a combination of funding and user demand for facilities - two tightly coupled things.

I have summarized a few dedicated hadron physics experiments which already fit into other, directly funded programs (Tevatron and neutrinos, but also a commitment to a test beam for ILC and other detector R&D). Being part of other programs may be the best model for future hadron physics at Fermilab.

What will be the total available beam? Will we build a Proton Driver? What fraction of the available protons will go to hadron physics? Will a stretcher ring be available (either using the Tevatron or a new ring to be proposed and built?

A robust program depends on robust interest, and eventually, compelling physics-driven experiment proposals.

ACKNOWLEDGMENTS

I would like to acknowledge the help to me as I prepared this presentation of Holger Meyer, Paul Reimer, and Jorge Morfin of MIPP, E906, and MINERνA, respectively.

I thank each of them for their help. Of course, I have also drawn heavily on the presentations of Piermaria Oddone who has already produced a clear statement of the plans for Fermilab's future.

REFERENCES

1. http://vmsstreamer1.fnal.gov/VMS_Site_03/Lectures/EPP2010/050516Oddone/index.htm
2. R. Raja, *Phys. Rev.* D18, 204-209 (1978).
3. R. Raja and Yuri V. Fisyak, "*Batavia 1992, The Fermilab Meeting, DPF 92, vol. 2*" 1012-1015, (1992).
4. E.T. Berger and F. Coester, *Phys. Rev.* **D32**, 1071 (1985) and E. T. Berger, F. Coester, and R.B. Wiringa, *Phys. Rev.* **D29**, 398 (1984). H. Jung and G.A. Miller, *Phys. Rev.* **C41** 659 (1990). G.E. Brown *et al.*, *Nucl. Phys.* **A593**, 295 (1995). A.E.L. Dieperink and C.L. Korpa, *Phys. Rev.* **C55**, 2665 (1997).
5. A. Psaker, JLAB-05-04-305, Dec 2004. *"Fermilab Proton Driver Workshop, Batavia, Illinois"* 6-9 Oct 2004. e-Print Archive: hep-ph/0412321.
6. M.G. Albrow, *et al.*, FERMILAB-FN-0778-AD-E, e-Print Archive: hep-ex/0509019.
7. http://www-td.fnal.gov/projects/PD/PhysicsIncludes/Workshop/index.html

Complex Meson Spectroscopy

Eef van Beveren*, Frieder Kleefeld† and George Rupp†

*Centro de Física Teórica, Departamento de Física, Universidade, P3004-516 Coimbra, Portugal
†Centro de Física das Interacções Fundamentais, Instituto Superior Técnico, Edifício Ciência, P1049-001 Lisboa, Portugal

Abstract. We do meson spectroscopy by studying the behavior of S-matrix poles in the complex-energy plane, as a function of the coupling strength for $^3P_0$ quark-pair creation. Thereto, a general formula for non-exotic hadron-hadron scattering involving arbitrary quark confinement is used, which can be applied to all flavors. We find two distinct types of poles, which we call *confinement* and *continuum* poles, respectively. Together, they suffice to understand the experimental meson spectrum.

Keywords: meson-meson scattering, complex poles, scalar mesons
PACS: 14.40.Ev, 14.40.Lb, 14.40.Cs, 13.25.-k

Introduction. From past studies on atomic spectra, we got so used to the terms "line spectrum" and "states" that we also apply them to mesonic spectra. Such a situation is depicted in Fig. 1a. The decay widths are thought of as higher-order effects, which indicate the life times of unstable states. This is indeed a fruitful strategy for atomic spectra. Starting from QED, one first determines the properties of the force which binds electrons to the nucleus, and next the corresponding bound-state spectra and states. Subsequently, one determines life times of states by considering the electromagnetic

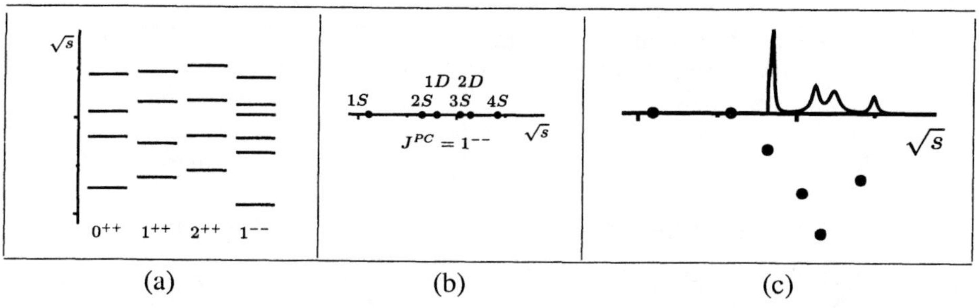

FIGURE 1. (**a**): Spectra as we are used to. (**b**): The same, but restricted to one set of quantum numbers. (**c**): A modern spectrum of bound states below and resonances above threshold, in terms of complex poles of the scattering amplitude (Note that the units along the real and the imaginary axis are not the same in this figure).

(EM) interaction of the photon field with the electrons.

Seen from a different angle, the above strategy amounts to separating the interaction of electrons and the nucleus with the EM field into two distinct parts: the one binding electrons and the nucleus into an atom, and the one describing the scattering of photons off the atom.

Here, we apply the same technique to strong interactions, but with the important

CP814 *Hadron Spectroscopy: Eleventh International Conference*
edited by A. Reis, C. Göbel, J. Sá Borges, and J. Magnin

difference that no manageable scheme exists to disentangle the two parts when setting out from QCD. Hence, they must be dealt with on the same footing. As a consequence, also the central resonance position and the resonance width must be described on the same footing, which can be achieved by referring to the complex pole position of the corresponding scattering amplitude. Such situation is depicted in Fig. 1c.

Furthermore, the term "states" is not very well defined for resonances having large widths compared to their level separations. One may, of course, refer to the state at a certain energy, say the central resonance position. But it is not reasonable to assume that at energies say 100 MeV higher or lower, the state of the system is the same. Moreover, different resonances may be overlapping.

Here, we shall be studying non-exotic meson-meson scattering. Through quark-pair creation, this implies multi-channel scattering, as well as intermediate states of confined systems of quarks and antiquarks. Hence, the advantage of complex spectra is apparent, since the pole structure of the full scattering matrix is the same for all channels, although the structure of peaks and zeros in the various channels can be very different.

Meson-meson scattering amplitude. On rather general grounds [1, 2], it can be shown that a system of confined quarks and antiquarks coupled to free two-meson channels is well described by a partial-wave (ℓ) inverse cotangent matrix of the form

$$K_\ell(p) = \frac{\lambda^2 \sum\limits_{n=0}^{\infty} \dfrac{\mathcal{J}^*_{n\ell}(p)\,\mathcal{J}_{n\ell}(p)}{E(p)-E_{n\ell}}}{\lambda^2 \sum\limits_{n=0}^{\infty} \dfrac{\mathcal{J}^*_{n\ell}(p)\,\mathcal{N}_{n\ell}(p)}{E(p)-E_{n\ell}} - 1} . \tag{1}$$

Here, λ parametrizes the intensity of quark-pair creation, p stands for the relativistic relative on-shell momentum in one of the two-meson channels, and $E(p)$ is the total invariant mass of the system. Furthermore, $E_{n\ell}$ ($n = 0, 1, \ldots$) represent the *confinement spectrum*, which is of course model-dependent, since it cannot be measured directly, while $\mathcal{J}_{n\ell}$ and $\mathcal{N}_{n\ell}$ are convolution matrix integrals over quark-antiquark distributions, meson-meson scattering wave functions, and a transition potential.

If we approximate the latter potential by a radial delta function, then we find for the ij-th matrix element of the partial-wave scattering amplitude

$$[T_\ell]_{ij}(p) = \frac{\lambda^2 \left\{ \sum\limits_{n=0}^{\infty} \dfrac{r_i(n) r_j(n)}{\sqrt{s}-E_{n\ell}} \right\} 2a\sqrt{\dfrac{\mu_i\mu_j}{p_i p_j}}\, p_i p_j\, j_{\ell_i}(p_i a)\, j_{\ell_j}(p_j a)}{1+\lambda^2 \sum\limits_{m=1}^{N} \left\{ \sum\limits_{n=0}^{\infty} \dfrac{|r_m(n)|^2}{\sqrt{s}-E_{n\ell}} \right\} 2ia\mu_m p_m\, j_{\ell_m}(p_m a)\, h^{(1)}_{\ell_m}(p_m a)} . \tag{2}$$

Here, j_ℓ represents the radial Bessel function, $h^{(1)}_\ell$ the radial Hankel function of the first kind, a the delta-shell radius, $\sqrt{s}$ the total invariant mass of the system, and μ_i the reduced mass in the i-th two-meson channel. The coefficients $r_i(n)$ (*vertex functions*), which stem from the convolution integrals, are discussed later.

Cross sections and poles. In Fig. 2 we show how cross sections following from formula (2) vary with increasing λ, for S-wave isodoublet $K\pi$ scattering. In Fig. 2a the nonstrange-strange ($n\bar{s}$) confinement spectrum is well visible for small λ, whereas in Fig. 2c, for the model value of λ, experiment is reproduced.

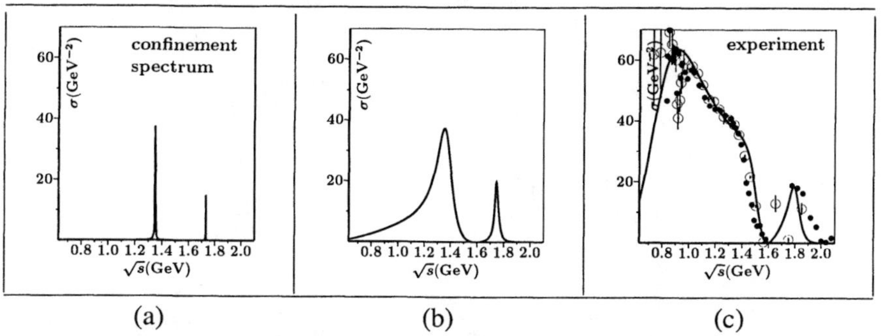

FIGURE 2. Cross section for S-wave isodoublet $K\pi$ scattering. Left: For very small values of λ, one observes the $J^{PC} = 0^{++}$ $n\bar{s}$ confinement spectrum. Middle: When λ takes about half its model value, one notices some more structure for low invariant masses. Right: At the model's value of λ, this structure is dominant and well in agreement with the experimental observations. The data are taken from Ref. [3] (open circles) and Ref. [4] (full circles).

The pole structure for varying λ is depicted in Fig. 3. In Fig. 3a we see how the pole stemming from the ground state of the confinement spectrum moves through the complex-energy ($E = \sqrt{s}$) plane when λ changes. The pole movement shown in Fig. 3b is very different. For small λ the pole disappears into the $K\pi$ continuum, which is the reason why it is not visible in Fig. 2a.

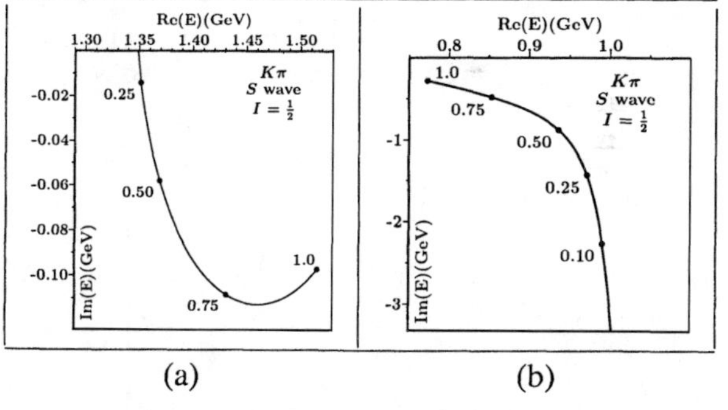

FIGURE 3. Pole positions for S-wave isodoublet $K\pi$ scattering, as a function of the strength of quark-pair creation, parametrized by λ (see Eqs. (1,2)). (**a**): Example of a *confinement pole*, which for $\lambda = 0$ lies on the real axis at total invariant mass $E_{0\ell}$, i.e., the ground state of the $n\bar{s}$ $J^P = 0^+$ confinement spectrum. (**b**): Example of a *continuum pole*, which for $\lambda = 0$ has an infinite negative imaginary part, i.e., the $K\pi$ continuum.

The lowest-lying scattering-amplitude poles for DK in $J^P = 0^+$ and D^*K in $J^P = 1^+$ are shown in Fig. 4. The two poles in Fig. 4a represent the $D_{s0}^*(2317)$ resonance, which, in the absence of isospin-violating processes, comes out on the real axis, plus a broad resonance at about 2.8–2.9 GeV, undetected so far.

The two poles shown in Fig. 4b represent the $D_{s1}(2460)$ and $D_{s1}(2536)$ resonances, which both come out on the real axis.

In Fig. 4a we depict two possible alternatives (one pair of solid lines and one pair of dashed lines). They result from very small variations in the delta-shell radius. Consequently, it is not at all trivial how to connect experimental poles with either confinement or continuum poles.

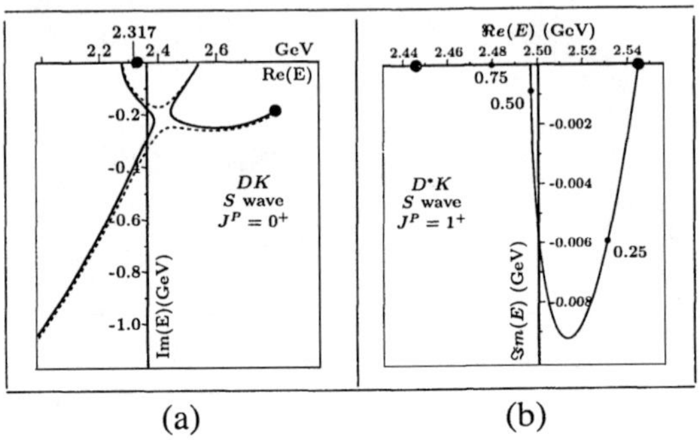

FIGURE 4. The lowest-lying poles for S-wave $J^P = 0^+$ DK (**a**) and $J^P = 1^+$ D^*K (**b**) scattering. The fat dots represent the pole positions for the model value of λ.

The two poles in Fig. 4b both stem from the ground state of the $J^P = 1^+$ charm-strange confinement spectrum, since there are two such states, i.e., $^3P_1$ and $^1P_1$, which come out degenerate in our confinement spectrum. The linear combination 33% $^3P_1$ plus 67% $^1P_1$ decouples from D^*K S-wave scattering, hence becomes a bound state in the scattering continuum (see also Ref. [5]). On the other hand, the orthogonal linear combination of $^3P_1$ and $^1P_1$ couples fully, and turns into a bound state below threshold.

Threshold behavior. One may wonder how the lowest poles in Fig. 4 move along the real axis when λ is increased. This is depicted in Fig. 5, where we continuously vary λ

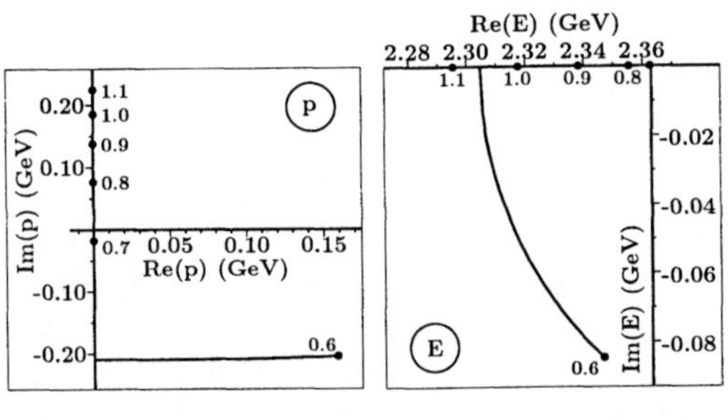

FIGURE 5. Movement of the $D_{s0}^*(2317)$ pole in S-wave DK scattering near the DK threshold, as a function of λ. Left: Movement in the p plane. Right: Movement in the E plane. We have marked the places where λ equals 0.6, 0.7, 0.8, 0.9, 1.0, and 1.1. On the real axis in the E plane, the pole moves first towards threshold and then towards smaller real E for increasing λ.

from a small (0.60) towards a large value (1.2). In the complex-momentum (p) plane, we observe that the pole turns purely imaginary somewhere in the lower half-plane. This is typical for S-wave scattering. For any higher angular momentum this happens at $p = 0$.

The S-matrix pole represents a bound-state when it lies on the positive imaginary axis. On the negative imaginary axis, it represents a virtual bound state. In this example, the pole lies on top of threshold for a value of λ slightly larger than 0.7.

The pole position in the E plane can be obtained from the pole position in the p plane through the relation

$$E = \sqrt{p^2 + m_D^2} + \sqrt{p^2 + m_K^2} \ .$$

Properties. The properties of the scattering amplitude following from the K matrix of Eq. (1) are the following.

- Analytic in the total invariant mass.
- Proper analytical and smooth behavior at all thresholds.
- Unitary for real total invariant mass above and below thresholds.
- Symmetric for any complex value of the total invariant mass.
- Poles in the scattering matrix may be searched for in the same fashion as poles in any analytical function.
- It contains an infinity of interfering resonances.
- It also allows an arbitrary number (N) of coupled meson-meson channels.
- For small λ it exhibits the confinement spectrum $E_{n\ell}$.
- For the model's value of λ, it reproduces the resonances and bound states of meson-meson scattering.
- It contains kinematical Adler-type zeros [6] at almost the same invariant masses as the theoretical Adler zeros [7].
- It also reproduces automatically the low-lying scalar mesons.

The parameters. The parameters λ, a, and ω in formula (2) represent the probability of quark-pair creation, taken equal for up, down, strange and zero for the heavy flavors in the present version of the model, the average distance for quark-pair creation, and the average level spacing of the $q\bar{q}$ confinement spectrum, respectively. As expected, a is of the order of half a fermi, and ω of the order of 200 MeV. In principle, λ, a, and ω should follow from QCD. Furthermore, in our expression

$$E_{n\ell} = m_q + m_{\bar{q}} + \omega\left(2n + \ell + \tfrac{3}{2}\right) \ , \tag{3}$$

for the confinement spectrum, we must specify the constituent quark masses. We employ here the values $m_n = 0.406$ GeV, $m_s = 0.508$ GeV, $m_c = 1.562$ GeV, and $m_b = 4.724$ GeV from Ref. [1].

Moreover, flavor invariance demands that λ^2 and a^2 be scaled by the flavor mass, or the reduced quark mass in the case of different flavors [8, 9]. Consequently, for quark-pair creation, the average distance and probability decrease with increasing quark mass, as one naively would expect.

Sum rules for vertex functions. The vertex functions, represented by $r_i(n)$ in formula (2), where i indicates the channel index and n the radial excitation of the intermediate resonance, contain the full structure of all quantum numbers, with no parameters involved. The former can be determined from the recoupling matrix elements of the harmonic-oscillator wave functions for the effective quarks. The complete procedure is described in Refs. [10], assuming the $^3P_0$ mechanism for quark-pair creation.

Some of the resulting vertex functions are shown in Table 1, for a system with quantum numbers $J^{PC} = 0^{++}$. Each of the two scattering products (mesons) is characterized

TABLE 1. Recoupling coefficients as a function of radial excitation n, for scalar ($J=0, \ell=1, S=1, n$) decay into two mesons. Notice that the recoupling coefficients for $n=0$ add up to one. This means that, in the harmonic-oscillator approach, there are no additional two-meson channels which can couple to the ground state of the confinement spectrum. For the higher radial excitations, the table is still very incomplete.

meson 1 $(nJLS)_1$ (J^{PC})	meson 2 $(nJLS)_2$ (J^{PC})	relative LS	recoupling coefficients $\{r(n)\}^2 \times 4^n$
$(0,0,0,0)$ (0^{-+})	$(0,0,0,0)$ (0^{-+})	0,0	$\frac{1}{24}(n+1)$
$(0,0,0,0)$ (0^{-+})	$(1,0,0,0)$ (0^{-+})	0,0	$\frac{1}{144}(2n+3)(n-1)^2$
$(1,0,0,0)$ (0^{-+})	$(1,0,0,0)$ (0^{-+})	0,0	$\frac{1}{3456}n(2n+1)(2n+3)(n-3)^2$
$(0,0,0,0)$ (0^{-+})	$(0,1,1,1)$ (1^{++})	1,1	$\frac{1}{6}$
$(0,1,0,1)$ (1^{--})	$(0,1,0,1)$ (1^{--})	0,0	$\frac{1}{72}(n+1)$
$(0,1,0,1)$ (1^{--})	$(0,1,0,1)$ (1^{--})	2,2	$\frac{1}{18}(2n+5)$
$(0,1,0,1)$ (1^{--})	$(1,1,0,1)$ (1^{--})	0,0	$\frac{1}{432}(2n+3)(n-1)^2$
$(0,1,0,1)$ (1^{--})	$(0,1,2,1)$ (1^{--})	0,0	$\frac{1}{540}(2n+3)(2n-5)^2$
$(0,1,0,1)$ (1^{--})	$(0,1,1,0)$ (1^{+-})	1,1	$\frac{1}{6}$
$(0,0,1,1)$ (0^{++})	$(0,0,1,1)$ (0^{++})	0,0	$\frac{1}{432}(2n+3)(n-3)^2$
$(0,1,1,1)$ (1^{++})	$(0,1,1,1)$ (1^{++})	0,0	$\frac{1}{144}(2n+3)(n-2)^2$
$(0,1,1,0)$ (1^{+-})	$(0,1,1,0)$ (1^{+-})	0,0	$\frac{1}{144}(2n+3)(n-1)^2$

by its internal quantum numbers n_ℓ, J_ℓ, L_ℓ, and S_ℓ ($\ell = 1,2$), i.e., the internal radial excitation, total angular momentum (spin of the meson), orbital angular momentum of the $q\bar{q}$ system, and intrinsic $q\bar{q}$ spin, respectively. The relative motion of the two mesons is characterized by their relative orbital angular momentum L and total spin S, dictated by the J^{PC} quantum numbers. The relative radial excitation of the two-meson system is directly related to n (see Refs. [11] for details).

When we determine the recoupling coefficients, we find that they decrease rapidly for higher radial excitations n (see Table 1), which implies that the higher terms in the sum over n in formula (2) are suppressed.

***S*-wave scattering.** In Fig. 2c we compare the result of formula (2) to the data of Refs. [3, 4]. We find a fair agreement for total invariant masses up to 1.6 GeV. However, we should bear in mind that the LASS data must have larger error bars for energies above 1.5 GeV than suggested in Ref. [4], since most data points fall well outside the Argand circle. Hence, for higher energies, the model should better not follow the data too precisely.

Now, in order to have some idea about the performance of formula (2) for S-wave $I = 1/2$ $K\pi$ scattering, we argue that, as in our model there is only one non-trivial eigen-phase shift for the coupled $K\pi$+$K\eta$+$K\eta'$ system, we may compare the phase shifts of our model for $K\eta$ and $K\eta'$ to the experimental phase shifts for $K\pi$. We do this comparison in Figs. 6 and 7, where, instead of the phase shifts, we plot the cross sections, assuming no inelasticity in either case. The latter assumption is, of course, a long shot. Nevertheless, we observe an extremely good agreement. In particular, for

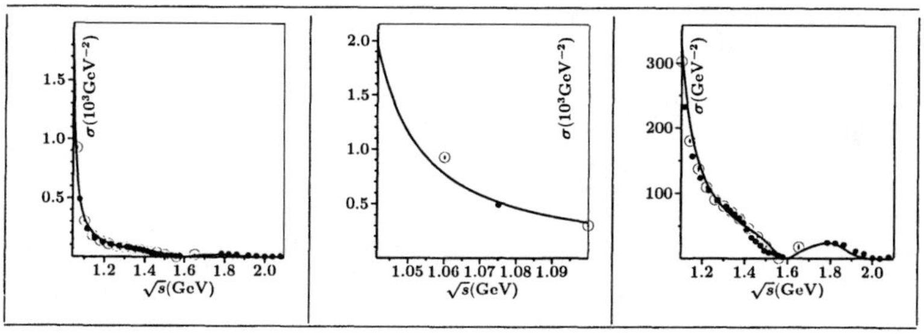

(a) (b) (c)

FIGURE 6. S-wave $K\eta$ "cross section" (see text), as a function of total invariant mass. (**a**): From threshold up to 2.1 GeV. (**b**): Detail for lower energy. (**c**): Detail for higher energy. The data are taken from Ref. [3] (open circles) and Ref. [4] (full circles).

$K\eta'$ (Fig. 7) we become aware of a structure in the data at about 1.9 GeV, indicating the presence of a not-anticipated pole. This is something we would not have easily noticed from the data alone.

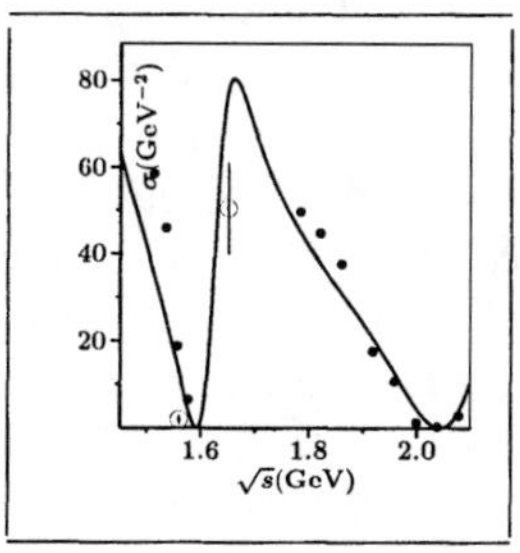

FIGURE 7. S-wave $K\eta'$ "cross section" (see text), as a function of total invariant mass. The data are taken from Ref. [3] (open circles) and Ref. [4] (full circles).

When we inspect formula (2) for poles in the S-wave isodoublet $K\pi$ scattering amplitude, then we find the pole structure as summarized in Table 2, i.e., five poles at energies up to about 2.2 GeV real part. The first pole, at $0.772-0.281i$ GeV, describes the heavily disputed $K_0^*(800)$ structure, whereas the second pole, at $1.52-0.097i$ GeV, represents the well-established $K_0^*(1430)$ resonance.

TABLE 2. T-matrix poles for S-wave $K\pi$ scattering, as obtained from Eq. (2).

Pole (GeV)	$0.772-0.281i$	$1.52-0.097i$	$1.79-0.052i$	$2.04-0.15i$	$2.14-0.065i$
Origin	continuum	confinement	confinement	continuum	confinement

Our model is explicitly flavor-independent, meaning that the only flavor breaking in formula (2) stems from the effective quark masses which determine the ground state of the confinement spectrum (see Eq. 3), and from the masses of the mesons in the scattering channels. Consequently, $\pi\pi$ scattering is not very different from $K\pi$ scattering in our model. We may expect then that each of the two flavor combinations that couple to isoscalar S-wave $\pi\pi$ and KK scattering has a pole structure similar to the one in isodoublet $K\pi$ scattering, with the proviso that $n\bar{n}$-$s\bar{s}$ mixing in the $I=0$ case introduces an extra complication. The to-be-expected poles are given in Table 3, alongside the f_0 structures reported in experiment.

TABLE 3. The left-hand table shows the real parts of the poles in the amplitude for $n\bar{n}$ (upper row) $s\bar{s}$ (lower row) S-wave $\pi\pi$+KK coupled-channel scattering, as expected in analogy with $K\pi$ (see Table 2), whereas the right-hand table shows the observed f_0 states listed in the Tables of Particle Properties [12]. In the upper (lower) rows, we collect those resonances which we expect to be mainly $n\bar{n}$ ($s\bar{s}$).

$n\bar{n}$ (GeV)	0.66	1.42	1.69	1.94	2.04
$s\bar{s}$ (GeV)	0.86	1.62	1.89	2.14	2.24

f_0 (GeV)	0.6	1.37	1.71	2.02	2.33
f_0 (GeV)	0.98	1.50		2.20	

The often read comment that *too many isoscalar states are observed* [13], in order to justify the application of alternative quark, or even quarkless, configurations [14], is not confirmed here.

Most probably, mesons are just mixtures [15] of *quark-antiquark* states [16], *two-meson molecules* [17], *glueballs* [18], *tetraquarks* [19], *hexaquarks, hybrids* [20], and so forth. Here, we have shown that the first two of the latter list of possible components are the most relevant ones. Moreover, a resonance is really a collection of states, all with different masses. Each of these states will have a different composition.

In Table 4 we summarize the scalar poles which we obtained in the past [21, 22, 8]. The first four lines in the third column of Table 4 are refered to in the literature as the scalar-meson nonet [23], given by the isoscalars $f_0(600)$ and $f_0(980)$, the isotriplet $a_0(980)$, and the isodoublet pair $K_0^*(800)$ [24]. Here, they form part of the complex pole spectrum of the general scattering amplitude (1), where they appear as the lowest-lying continuum poles. The first four lines in the fourth column of Table 4 describe the nonet of scalar mesons that stem directly from the $J^P=0^+$ ground states of the confinement spectrum.

Summary and discussion. In the foregoing, we have shown that the most relevant information on (meson) spectra amounts to the knowledge about the complex pole positions of the scattering amplitudes. Furthermore, we presented a general form of amplitudes for non-exotic hadron-hadron scattering with any possible flavor combination,

TABLE 4. S-wave scattering poles for various flavor/isospin combinations with $J^P = 0^+$. In the columns "continuum", "ground state", and "excitation" we indicate the origin of the poles, i.e., either meson-meson continuum, or confinement spectrum (ground state or first radial excitation).

channel	q-$\bar{q}$	continuum	ground state	excitation
		GeV	GeV	GeV
$\pi\pi$	n-$\bar{n}$	$0.47-0.21i$	$1.36-0.13i$	-
$K\pi$	n-$\bar{s}$	$0.77-0.28i$	$1.52-0.10i$	$1.79-0.05i$
$\eta\pi$	n-$\bar{n}$	$0.97-0.028i$	$1.45-0.13i$	-
$\pi\pi$	s-$\bar{s}$	$0.99-0.020i$	$1.51-0.06i$	-
$D\pi$	c-$\bar{n}$	$2.14-0.16i$	$2.58-0.12i$	-
DK	c-$\bar{s}$	2.33	$2.80-0.20i$	-
$B\pi$	n-$\bar{b}$	$6.06-0.29i$	$5.46-0.03i$	$6.03-0.05i$
BK	s-$\bar{b}$	$6.21-0.33i$	5.61	$6.05-0.03i$
BD	c-$\bar{b}$	$7.12-0.43i$	6.64	$7.11-0.03i$

based upon the most prominent properties of strong interactions, namely *quark confinement*, *quark-pair creation*, and *flavor invariance*. The resulting pole spectrum may very well be compared to experiment. It is true that the predicted cross sections and phase shifts are not always in very accurate agreement with the data, as could hardly be expected in view of the extremely wide scope of the model. In part this may also be due to the inaccuracy of some experimental numbers. Nevertheless, the model can be improved as well, as we suggest next.

- At present, we determine the vertices in the framework of the $^3P_0$ mechanism, assuming that all quark masses involved are the same. However, different quark masses can be dealt with, too. Computer programs already exist for several years, but have not yet been implemented.
- The one-delta-shell approximation "unfortunately" works too well. Hence, although it is perfectly known how to implement more complicated transition potentials [25], and even some computer code is ready, the finishing touch could still take a while, unless some extra manpower becomes available.
- The kinematics of meson pairs below threshold is most certainly not being dealt with in the most appropriate way by us. This is a subject which deserves to be studied in more detail.

Acknowledgment. One of us (EvB) wishes to thank the organizers for inviting him to the Hadron05 conference, their warm hospitality, and for organizing this great event. This work was partly supported by the *Fundação para a Ciência e a Tecnologia* of the *Ministério da Ciência, Tecnologia e Ensino Superior* of Portugal, under contract POCTI/FP/FNU/50328/2003 and grant SFRH/BPD/9480/2002.

REFERENCES

1. E. van Beveren, G. Rupp, T. A. Rijken, and C. Dullemond, Phys. Rev. D **27**, 1527 (1983).
2. E. van Beveren and G. Rupp, arXiv:hep-ph/0304105.
3. P. Estabrooks, R. K. Carnegie, A. D. Martin, W. M. Dunwoodie, T. A. Lasinski, and D. W. Leith, Nucl. Phys. B **133**, 490 (1978).
4. D. Aston *et al.* [LASS collaboration], Nucl. Phys. B **296**, 493 (1988).
5. S. Godfrey, Phys. Rev. D **72**, 054029 (2005) [arXiv:hep-ph/0508078].
6. G. Rupp, F. Kleefeld, and E. van Beveren, AIP Conf. Proc. **756**, 360 (2005) [arXiv:hep-ph/0412078].
7. S. L. Adler, arXiv:hep-ph/0505177.
8. E. van Beveren and G. Rupp, Mod. Phys. Lett. A **19**, 1949 (2004) [arXiv:hep-ph/0406242]; E. van Beveren, J. E. G. Costa, F. Kleefeld and G. Rupp, arXiv:hep-ph/0509351.
9. E. van Beveren, C. Dullemond, and G. Rupp, Phys. Rev. D **21**, 772 (1980) [Erratum-ibid. D **22**, 787 (1980)]; F. Kleefeld, AIP Conf. Proc. **717**, 332 (2004) [arXiv:hep-ph/0310320].
10. E. van Beveren, Z. Phys. C **21**, 291 (1984); E. van Beveren and G. Rupp, Eur. Phys. J. C **11**, 717 (1999) [arXiv:hep-ph/9806248]; Phys. Lett. B **454**, 165 (1999) [arXiv:hep-ph/9902301].
11. J. E. Ribeiro, Phys. Rev. D **25**, 2406 (1982); E. van Beveren, Z. Phys. C **17**, 135 (1983).
12. S. Eidelman *et al.* [Particle Data Group Collaboration], Phys. Lett. B **592**, 1 (2004).
13. Claude Amsler, Phys. Lett. B **541**, 22 (2002) [arXiv:hep-ph/0206104].
14. F. E. Close, arXiv:hep-ph/0411396.
15. T. Barnes, F. E. Close, J. J. Dudek, S. Godfrey and E. S. Swanson, Phys. Lett. B **600**, 223 (2004) [arXiv:hep-ph/0407120].
16. M. R. Pennington, J. Phys. Conf. Ser. **18**, 1 (2005) [arXiv:hep-ph/0504262]; arXiv:hep-ph/0509265; M. D. Scadron and M. Nagy, arXiv:hep-ph/0507168; D. K. Hong and C. j. Song, arXiv:hep-ph/0407274.
17. A. H. Fariborz, R. Jora, and J. Schechter, Phys. Rev. D **72**, 034001 (2005) [arXiv:hep-ph/0506170]; L. Roca, E. Oset and J. Singh, Phys. Rev. D **72**, 014002 (2005) [arXiv:hep-ph/0503273]; T. Kunihiro, arXiv:hep-ph/0407355; M. F. M. Lutz and E. E. Kolomeitsev, arXiv:hep-ph/0406015; P. Bicudo, Nucl. Phys. A **748**, 537 (2005) [arXiv:hep-ph/0401106].
18. A. H. Fariborz, AIP Conf. Proc. **688**, 154 (2004); Int. J. Mod. Phys. A **19**, 5417 (2004); T. Teshima, I. Kitamura, and N. Morisita, AIP Conf. Proc. **619**, 487 (2002).
19. M. Napsuciale and S. Rodriguez, Phys. Rev. D **70**, 094043 (2004) [arXiv:hep-ph/0407037]; Phys. Rev. D **71**, 074008 (2005) [arXiv:hep-ph/0411150]; J. R. Pelaez, Mod. Phys. Lett. A **19**, 2879 (2004) [arXiv:hep-ph/0411107]; T. V. Brito, F. S. Navarra, M. Nielsen, and M. E. Bracco, Phys. Lett. B **608**, 69 (2005) [arXiv:hep-ph/0411233].
20. Farida Iddir and Lahouari Semlala, arXiv:hep-ph/0211289.
21. E. van Beveren et al., Z. Phys. C **30**, 615 (1986).
22. E. van Beveren and G. Rupp, Eur. Phys. J. C **22**, 493 (2001) [arXiv:hep-ex/0106077]; arXiv:hep-ph/0312078.
23. M. M. Nagels, T. A. Rijken, and J. J. de Swart, Phys. Rev. D **20**, 1633 (1979); M. D. Scadron, Phys. Rev. D **26**, 239 (1982); A. V. Anisovich and A. V. Sarantsev, Phys. Lett. B **413**, 137 (1997) [arXiv:hep-ph/9705401]; T. Ishida, M. Ishida, S. Ishida, K. Takamatsu, and T. Tsuru, in *Upton 1997, Hadron Spectroscopy*, pp. 385–388 [arXiv:hep-ph/9712230]; D. Black, A. H. Fariborz, F. Sannino, and J. Schechter, Phys. Rev. D **59**, 074026 (1999) [arXiv:hep-ph/9808415]; A. H. Fariborz and J. Schechter, Phys. Rev. D **60**, 034002 (1999) [arXiv:hep-ph/9902238]; M. Ishida, arXiv:hep-ph/9905259; D. Black, A. H. Fariborz, and J. Schechter, in *Kyoto 2000*, p. 115 [arXiv:hep-ph/0008246]; N. A. Törnqvist, arXiv:hep-ph/0201171; I. Bediaga [Fermilab E791 collaboration], arXiv:hep-ex/0208039; M. Ishida, AIP Conf. Proc. **688**, 18 (2004); M. Ablikim *et al.* [BES Collaboration], arXiv:hep-ex/0506055; J. A. Oller, AIP Conf. Proc. **688**, 167 (2004); D. M. Li, K. W. Wei, and H. Yu, Eur. Phys. J. A **25**, 263 (2005) [arXiv:hep-ph/0507121]; Y. B. Dai and Y. L. Wu, Eur. Phys. J. C **39**, S1 (2005) [arXiv:hep-ph/0304075]; D. V. Bugg, arXiv:hep-ex/0510014; arXiv:hep-ex/0510021; G. Erkol, R. G. E. Timmermans, and T. A. Rijken, Phys. Rev. C **72**, 035209 (2005).
24. E. M. Aitala *et al.* [E791 Collaboration], Phys. Rev. Lett. **89**, 121801 (2002) [arXiv:hep-ex/0204018]; C. Göbel, on behalf of the E791 Collaboration, in *Rio de Janeiro 2000, Heavy quarks at fixed target*, pp. 373–384 [arXiv:hep-ex/0012009]; AIP Conf. Proc. **619**, 63 (2002) [arXiv:hep-ex/0110052].
25. C. Dullemond, G. Rupp, T. A. Rijken, and E. van Beveren, Comput. Phys. Commun. **27**, 377 (1982).

Hadron Physics at Belle

Karim Trabelsi

Department of Physics and Astronomy, University of Hawaii at Manoa,
2505 Correa Road, Honolulu, Hawaii 96822, U.S.A.

Abstract. We report recent results on the properties of the $X(3872)$ produced via the $B \to KX(3872)$ decay process in the Belle detector. We compare these properties with the expectations for possible assignments. The discoveries by Belle of other new particles : $Y(3940)$, $X(3940)$ and $Z(3931)$ are briefly reviewed as well as the observation of an isotriplet of excited charmed baryons.

Keywords: hadron spectroscopy, $X(3872)$, $Y(3940)$, $X(3940)$, $Z(3931)$, charmonium
PACS: 43.35.Ei, 78.60.Mq

INTRODUCTION

Recently there have been a number of reports on observation of new charmonium or charmonium-like states. One of the most intriguing puzzles is to explain the nature of charmonium-like state $X(3872)$ decaying into $J/\psi\pi^+\pi^-$ final state and observed for the first time by Belle in B decays [1]. This state exhibits a number of unusual properties and so far has not been assigned to any charmonium state relying on the conventional quark model. Even more unusual is the state observed recently in the $J/\psi\omega$ final state at a mass around 3.94 GeV/c^2 [2]. Although the latter state is above DD^* threshold, it has not been found in either the $D\bar{D}$ or $D^*\bar{D}$ final states; the observed decay to $J/\psi\omega$ is isospin-violating if the parent is a conventional charmonium state. More recently, two other new states ($X(3940)$, $Z(3931)$) were observed by Belle. These experimental results have renewed theoretical interest in the spectroscopy, decay and production of charmonium.

SEARCH FOR AN ASSIGNMENT FOR THE $X(3872)$

The $X(3872)$ is a narrow state that decays into $\pi^+\pi^-J/\psi$, first seen in exclusive $B \to KX(3872)$ decays by the Belle experiment and subsequently seen in inclusive $p\bar{p}$ collisions by the CDF [3] and D0 [4] collaborations, and in exclusive B-meson decays by the BaBar collaboration [5]. Although it looks like a typical charmonium particle (i.e. a $c\bar{c}$ meson), its assignment to any of the as-yet unseen narrow charmonium states has been proven to be problematic.

CP814 *Hadron Spectroscopy: Eleventh International Conference*
edited by A. Reis, C. Göbel, J. Sá Borges, and J. Magnin

New decay modes of the $X(3872)$

Using a 275 million $B\bar{B}$ events sample, Belle searched for $B \to K\gamma J/\psi$ and $B \to K\pi^+\pi^-\pi^0 J/\psi$ decays [6]. The candidates are identified by the energy difference $\Delta E = E_B^{\rm cms} - E_{\rm beam}^{\rm cms}$ and the beam-energy constrained mass $M_{\rm bc} = \sqrt{(E_{\rm beam}^{\rm cms})^2 - (p_B^{\rm cms})^2}$, where $E_{\rm beam}^{\rm cms}$ is the center-of-mass (cms) beam energy and $E_B^{\rm cms}$ and $p_B^{\rm cms}$ are the cms energy and momentum of the B candidate daughters. Events with $M_{\rm bc} > 5.20$ GeV/c^2 and $|\Delta E| < 0.2$ GeV are selected and a signal region : 5.2745 GeV/c^2 $< M_{\rm bc} <$ 5.2855 GeV/c^2 and $|\Delta E| < 0.035$ GeV is defined.

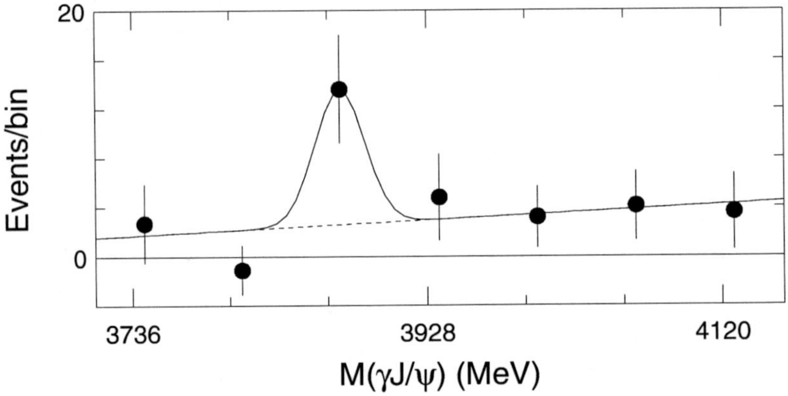

FIGURE 1. The signal yields from fits to the $M_{\rm bc}$ and ΔE distributions in bins of M($\gamma J/\psi$).

For the $K\gamma J/\psi$ combination, a peak near 3510 MeV in the $M(\gamma J/\psi)$ distribution, due to $B \to K\chi_{c1}$; $\chi_{c1} \to \gamma J/\psi$, is used as a calibration mode to determine $M_{\rm bc}$, ΔE and $M(\gamma J/\psi)$ resolutions and shifts in the nominal peak positions. For six 64-MeV-wide M($\gamma J/\psi$) mass bins in the $X(3872)$ region, a simultaneous fit of the $M_{\rm bc}$ and ΔE distributions is performed. The fit uses single Gaussians for the $M_{\rm bc}$ and ΔE signals with widths fixed at the values determined from the $\chi_{c1} \to \gamma J/\psi$ event sample. The results of the fits are plotted in Fig. 1. The signal yield in the 3872 MeV bin is 13.6 ± 4.4 events, corresponding to a significance of 4.0σ. The observation of the $X(3872) \to \gamma J/\psi$ decay mode established its charge-conjugation parity as $C = +1$. The following partial width ratio is then derived :

$$\Gamma(X \to \gamma J/\psi)/\Gamma(X \to \pi^+\pi^- J/\psi) = 0.14 \pm 0.05 \quad (1)$$

This result is inconsistent with the χ'_{c1} (1^{++} charmonium) assignment for $X(3872)$, as in this case a value around 40 would be expected.

For the $K\pi^+\pi^-\pi^0 J/\psi$ combination, the events of the type $B \to K\pi^0\psi(2S)$; $\psi(2S) \to \pi^+\pi^- J/\psi$ are eliminated by vetoing events where $M(\pi^+\pi^- J/\psi)$ is within 3σ of $m_{\psi(2S)}$. To select possible $X(3872) \to \pi^+\pi^-\pi^0 J/\psi$ events, $|\Delta M|$ <0.0165 GeV where $\Delta M \equiv M(3\pi J/\psi) - 3.872$ GeV is required; this corresponds to a $\pm 3\sigma$ requirement. The

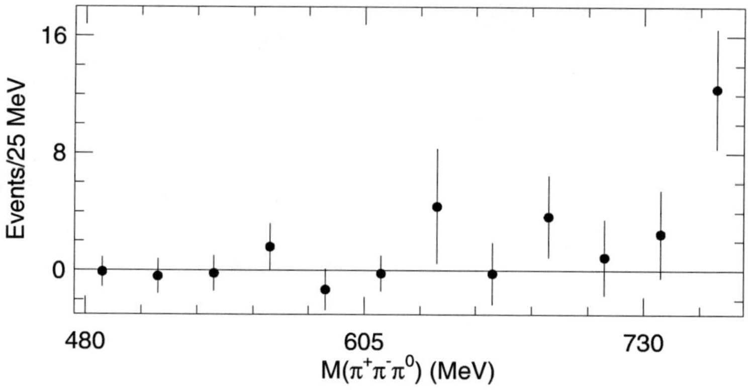

FIGURE 2. The B meson signal yield in bins of 3π invariant mass determined from fits to the $M_{bc} - \Delta E$ distributions.

yields obtained by a simultaneous fit to the M_{bc} and ΔE distributions for these events in 25 MeV-wide $\pi^+\pi^-\pi^0$ invariant mass bins are shown Fig. 2. All of the fitted yields are consistent with zero except for the $M(\pi^+\pi^-\pi^0) > 750$ MeV bins where the fit gives 12.4 ± 4.1 events. To determine the branching fraction for $X \to \pi^+\pi^-\pi^0 J/\psi$, all of the signal events with M($\pi^+\pi^-\pi^0$) > 750 MeV are attributed to $X \to \pi^+\pi^-\pi^0 J/\psi$ decay. The ratio of branching fractions is

$$\frac{\mathscr{B}(X \to \pi^+\pi^-\pi^0 J/\psi)}{\mathscr{B}(X \to \pi^+\pi^- J/\psi)} = 1.0 \pm 0.4(\text{stat}) \pm 0.3(\text{syst}) \tag{2}$$

where the systematic error reflects the uncertainty in the relative acceptance, the level of possible feed-across and non-resonant background, and possible event loss due to the M(3π) > 750 MeV requirement, all added in quadrature. The strong peak between 750 MeV and the kinematic limit of 775 MeV suggests that the process is dominated by the sub-threshold decay $X \to \omega J/\psi$. The partial widths for $3\pi J/\psi$ and $2\pi J/\psi$ are of comparable size, which implies a large violation of isospin symmetry.

J^{PC} assignment

Belle collaboration recently updated [7] with a larger sample (275 million $B\bar{B}$ pairs sample), the study of $B \to X(3872)K$ where $X(3872) \to J/\psi\pi^+\pi^-$. The total signal obtained is 49.1 ± 8.4 events and background in the signal region, estimated from the $M(J/\psi\pi^+\pi^-)$ sidebands, is 11.4 ± 1.1 events. Because of the good signal-to-noise ($\simeq 4$), angular analysis can be performed and provide information on the J^{PC} content of the $X(3872)$ as suggested by Rosner [8]. Among the twelve possible J^{PC}, the ones with negative charge conjugation-parity can be removed because of the observation of $X(3872) \to J/\psi\gamma$ (see previous section).

For 0^{++}, two invariant amplitudes are possible, corresponding to the ρ and J/ψ in relative $S-$ or $D-$waves. Because of the limited phase-space, the $D-$wave can be expected to be strongly suppressed relative to the $S-$wave term and is ignored. The amplitude is then proportional to the scalar product of the ρ and J/ψ polarizations. $\theta_{l\pi}$ is defined as the angle between the l^+ and the π^+ in the $X(3872)$ rest frame. In the limit where the $X(3872), J/\psi$ and ρ rest frames coincide, $dN/d(\cos\theta_{l\pi}) \propto \sin^2\theta_{l\pi}$. Figure 3 shows the $|\cos\theta_{l\pi}|$ distribution, computed in the ρ rest frame, for $X(3872)$ signal region events. The agreement with the $S-$wave 0^{++} Monte Carlo (MC) expectations (histogram) is poor — $\chi^2/d.o.f. = 31.0/9$ — and thus provides evidence against 0^{++} assignment.

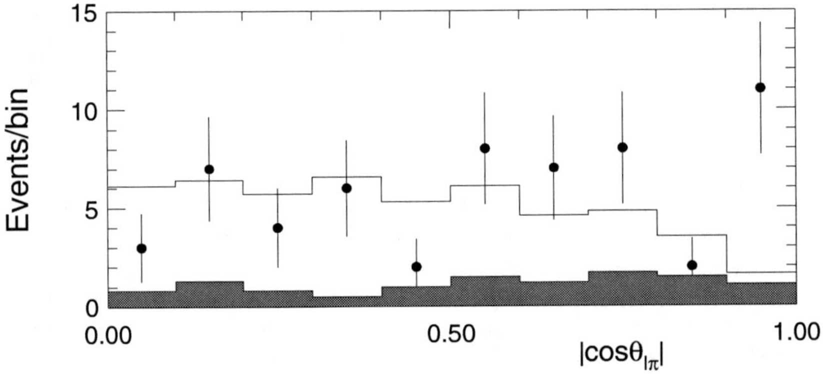

FIGURE 3. The $|\cos\theta_{l\pi}|$ distribution for events in the $X(3872)$ signal region (points with error bars). The open histogram is the expected distribution for a 0^{++} assignment including background. The hatched histogram shows the scaled sideband.

For 0^{-+}, there is only one invariant amplitude corresponding to a ρ and J/ψ in a P-wave. The decay amplitude is proportional to the scalar triple product of the ρ and J/ψ polarizations and their relative momentum. As a result, the polarizations are perpendicular to each other and their relative momentum. If we define θ as the angle between the l^+ and the z axis in the J/ψ rest frame and ψ as the angle between the π^+ and the x axis in the dipion rest frame, the expected distribution for 0^{-+} is $d^2N/d(\cos\theta)d(\cos\psi) \propto \sin^2\theta\sin^2\psi$. The agreement is marginal for $\cos\theta$ ($\chi^2/d.o.f. = 17.7/9$) but poor for $\cos\psi$ ($\chi^2/d.o.f. = 34.2/9$). This latter distribution allows us to reject the 0^{-+} with high confidence.

A similar study is performed in the case 1^{++}, and the fits on the corresponding angular distributions are in good agreement.

Figure 4 shows the $M(\pi^+\pi^-)$ distribution for events in the $X(3872) \to \pi^+\pi^- J/\psi$ signal region, the histogram indicates the sideband determined background, parameterized by the fourth-order polynomial shown in the figure as a dot-dashed curve. The solid curve shows the result of a fit which uses the background function plus an acceptance-weighted ρ BW line-shape with an S-wave cut-off factor at the upper kinematic boundary; the dashed curve shows the fit with a $P-$wave cut-off factor. The S-wave fit ($\chi^2/d.o.f = 43.1/39$) is much better than the P-wave fit ($\chi^2/d.o.f = 71.0/39$)

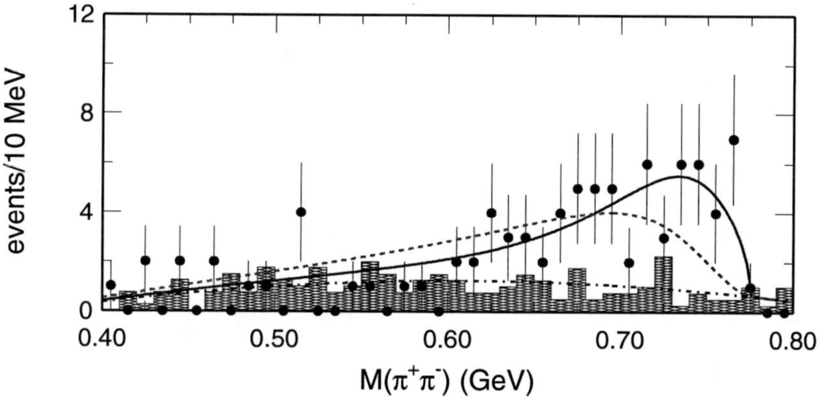

FIGURE 4. $M(\pi^+\pi^-)$ distribution for events in the $X(3872) \to \pi^+\pi^- J/\psi$ signal region.

thus J^{++} is strongly favored over J^{-+}.
The assignment 2^{++} was strongly disfavored by the recent, preliminary observation by Belle [9] of the decay $B \to KX, X \to D^0\bar{D}^0\pi^0$, where the 11.3 ± 3.6 signal events (5.6σ) concentrate close to the threshold for the final state $D^0\bar{D}^0\pi^0$. In case of the 2^{++}, the decay of a spin 2 state to the three pseudo-scalars $D^0\bar{D}^0\pi^0$ would require at least one pair of them to be in a relative D-wave. In such configuration the near threshold production would be strongly suppressed by a centrifugal barrier.
All the above observations strongly favor the assignment of $J^{PC} = 1^{++}$ for the $X(3872)$. This J^{PC} assignment, as well as the narrow width of $X(3872)$, the $\pi^+\pi^-$ shape and the large isospin violation in the $X(3872)$, match the expectations of the models interpreting the $X(3872)$ as $D^0\bar{D}^{*0}$ bound state [10, 11].

DISCOVERY OF THE $Y(3940)$

Using the same data sample than the one used for the $X(3872)$ studies, events of the type $B \to K\pi^+\pi^-\pi^0 J/\psi$ were selected, where both charged and neutral kaons were used. The signal region is defined as 5.2725 GeV/c^2 $< M_{bc} <$ 5.2875 GeV/c^2 and $|\Delta E| < 0.030$ GeV, which is about $\pm 2.5\sigma$ from the central values. The three-pion combinations with 0.760 GeV/c^2 $< M(\pi^+\pi^-\pi^0) <$ 0.805 GeV/c^2 are considered as candidate ω mesons. To suppress events of the type $B \to K_X J/\psi$; $K_X \to K\omega$ where K_X denotes strange meson resonances such as $K_1(1270)$, $K_1(1400)$ and $K_2^*(1430)$ that are known to decay to $K\omega$, the cut M($K\omega$) $>$ 1.6 GeV/c^2 is applied. The B-meson signal yields from the fits to the individual bins are plotted *vs* M($\omega J/\psi$) in Fig. 5, showing an enhancement around M($\omega J/\psi$)=3940 MeV/c^2. The curve used for the fit in Fig. 5(a) is a threshold function of the form $f(M) = A_0 q^*(M)$, where $q^*(M)$ is the momentum of the daughter particles in the $\omega J/\psi$ rest frame. In Fig. 5(b), a S-wave Breit-Wigner function is added to represent the enhancement. The fit yields a signal yield of 58 ± 11 events with a mass

$M = 3943 \pm 11$ MeV and width $\Gamma = 87 \pm 22$ MeV. The statistical significance of the

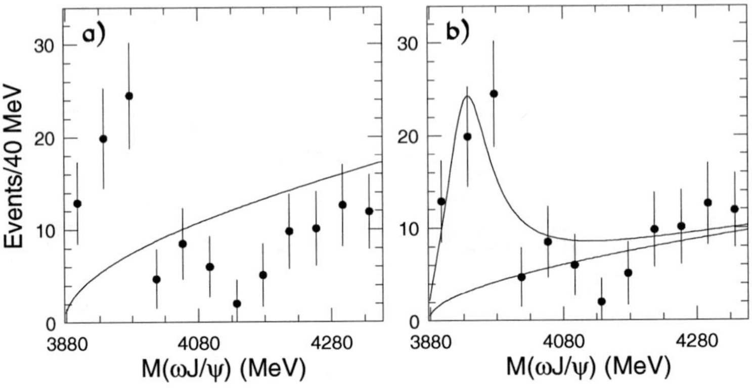

FIGURE 5. $B \to K\omega J/\psi$ signal yields *vs* $M(\omega J/\psi)$. The curves are described in the text.

signal is 8.1 σ.

$$\mathscr{B}(B \to KY(3940))\mathscr{B}(Y(3940) \to \omega J/\psi) = (7.1 \pm 1.3 \pm 3.1) \times 10^{-5} \quad (3)$$

It is expected that a $c\bar{c}$ charmonium meson with this mass would dominantly decay to $D\bar{D}$ and/or $D\bar{D}^*$; hadronic charmonium transitions should have minuscule branching fractions. These properties are similar though to those of some of the $c\bar{c} - gluon$ hybrid charmonium states [12], expected to be produced in B meson decays [13, 14, 15]. Lattice QCD calculations have indicated that partial widths for such decays can be comparable to the width measured, however they predict masses for these states that are between 4300 and 4500 MeV [16, 17], substantially higher than the value measured for $Y(3940)$.

DISCOVERY FOR THE $X(3940)$

Double charmonium production in the e^+e^- annihilation was first observed by Belle [18]. The process $e^+e^- \to J/\psi\eta_c$ was inferred from the η_c peak in the spectrum of masses of the system recoiling against the reconstructed J/ψ in inclusive $e^+e^- \to J/\psi X$ events. The measured Belle cross-section for the process $e^+e^- \to J/\psi\eta_c$ was an order of magnitude larger than those calculated within a non-relativistic QCD approach [19, 20]. To resolve this discrepancy many speculations were discussed in literature proposing that the Belle peak was a signal from other processes : $e^+e^- \to J/\psi J/\psi$ [21] or $e^+e^- \to J/\psi\mathscr{G}$ [22, 23]. The interpretation of recoil mass peak as η_c was confirmed with further Belle studies [24] fully reconstructing the $e^+e^- \to J/\psi\eta_c$ process, validating the correct calibration of the recoil mass scale and angular analysis. The cross sections for the processes $e^+e^- \to J/\psi\eta_c, e^+e^- \to J/\psi\chi_{c0}$ and $e^+e^- \to J/\psi\eta_c(2S)$ were measured with high accuracy.

The study was updated with a larger data sample corresponding to an integrated luminosity of 357 fb^{-1}. The J/ψ reconstruction is identical to previous studies. QED

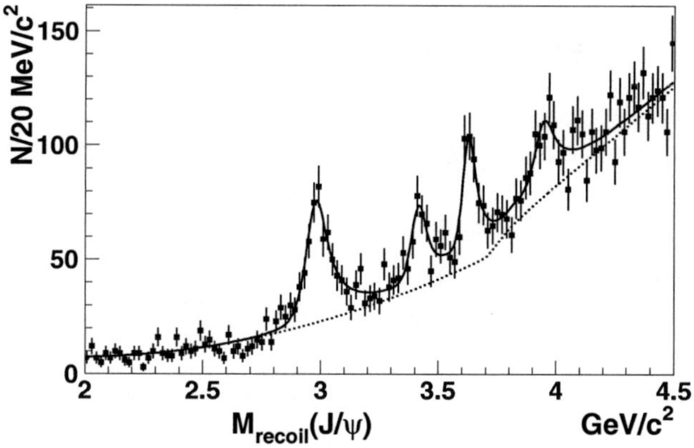

FIGURE 6. Distributions of masses recoiling against the reconstructed J/ψ in inclusive $e^+e^- \to J/\psi X$ events

processes are significantly suppressed by requiring the total charged multiplicity (N_{ch}) in the event to be greater than 4. Background due to J/ψ mesons produced from $B\bar{B}$ events is removed by requiring the center-of-mass momentum $p^*_{J/\psi}$ to be greater than 2.0 GeV/c. The recoil mass is defined as

$$M_{\text{recoil}}(J/\psi) = \sqrt{(E_{\text{cms}} - E^*_{J/\psi})^2 - p^{*2}_{J/\psi}} \tag{4}$$

where $E^*_{J/\psi}$ is the J/ψ cms energy after the mass constraint. The recoil mass spectrum for the data is shown in Fig. 6. Compared to the previous studies, the recoil mass interval is extended above the $D\bar{D}$ threshold. In addition to the three previously observed peaks ($\eta_c, \chi_{c0}, \eta_c(2S)$), another significant peak can be seen around a mass of 3.94 GeV/c^2; this new state is called $X(3940)$.

TABLE 1. Summary of the signal yields, charmonium masses and significances for $e^+e^- \to J/\psi(c\bar{c})_{\text{res}}$

$(c\bar{c})_{\text{res}}$	N	M [GeV/c^2]	σ
η_c	501 ± 44	2.970 ± 0.005	15.3
χ_{c0}	230 ± 40	3.406 ± 0.007	6.3
$\eta_c(2S)$	311 ± 42	3.626 ± 0.005	8.1
$X(3940)$	266 ± 63	3.936 ± 0.014	5.0

In the fit performed, the mass positions of these four states are free parameters and the expected signal lineshapes are determined from MC simulation. The background is parametrized by a second order polynomial and a threshold term ($\sqrt{M_{\text{recoil}}(J/\psi) - 2M_D}$) to account for a possible contribution from $e^+e^- \to J/\psi D^{(*)}\bar{D}^{(*)}$. The results of the fit (masses and yields) are given in Table 1 as well as the significance ($=\sqrt{-2\ln(\mathcal{L}_0/\mathcal{L}_{\text{max}})}$) for each state. As the new state has a mass well above $D\bar{D}$ threshold, a search for $X(3940)$ decaying into $D\bar{D}$ and $D^*\bar{D}$

final states was performed. A search for the decay $X(3940) \to J/\psi\omega$ was also performed to check if $X(3940)$ and $Y(3940)$ could be the same particle. A significant signal is obtained only for $X(3940) \to D^*\bar{D}$ (5.0σ), this decay appears to be the dominant with a measured branching fraction of $(96^{+45}_{-32} \pm 22)\%$. The mass of the new state is measured to be $(3.943 \pm 0.006 \pm 0.006)$ GeV/c^2 with a width smaller than 52 MeV/c^2 at 90% C.L. Upper limits are set for the decays $X(3940) \to D\bar{D}$ and $X(3940) \to J/\psi\omega$ of less than 41% and 26% at 90% C.L., respectively.

DISCOVERY OF THE $Z(3931)$

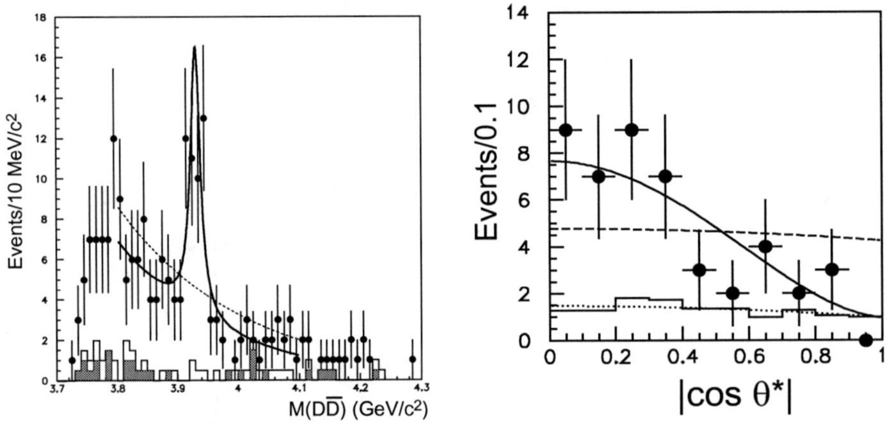

FIGURE 7. The sum of the $M(D\bar{D})$ invariant mass distributions for all four processes (left) : the curves show the fits with (solid) and without (dashed) a resonance component. The histograms show the distribution of the events from the D-mass sidebands. The $|\cos\theta^*|$ distribution (right) in the $3.91 < M(D\bar{D}) < 3.95$ GeV/c^2 (points with error bars) and background scaled from the $M(D\bar{D})$ sideband (solid histogram). The solid and dashed curves are expected distributions for the spin=2 (helicity=2) and spin=0 hypotheses, respectively.

Exclusive $D\bar{D}$ production was searched in the following four combinations of decays :
$\gamma\gamma \to D^0\bar{D}^0$, $D^0 \to K^-\pi^+$, $\bar{D}^0 \to K^+\pi^-$
$\gamma\gamma \to D^0\bar{D}^0$, $D^0 \to K^-\pi^+$, $\bar{D}^0 \to K^+\pi^-\pi^0$
$\gamma\gamma \to D^0\bar{D}^0$, $D^0 \to K^-\pi^+$, $\bar{D}^0 \to K^+\pi^-\pi^+\pi^-$
$\gamma\gamma \to D^+D^-$, $D^+ \to K^-\pi^+\pi^+$, $D^- \to K^+\pi^-\pi^-$

To enhance the two-photon $\gamma\gamma \to D\bar{D}$ production, the requirement $P_t(D\bar{D}) < 0.05$ GeV/c was applied, where P_t is the total transverse momentum in the e^+e^- center-of-mass frame with respect to the incident e^+e^- axis that approximates the direction of the two-photon collision axis. The invariant mass distribution for the combined $D^0\bar{D}^0$ and D^+D^- channels is shown for 10 MeV/c^2 bins in Fig. 7. The fit curve is the result of an unbinned likelihood fit to the data in the region $3.80 < D\bar{D} < 4.10$ GeV/c^2 using a relativistic Breit-Wigner signal function plus a background of the form $\sim M(D\bar{D})^{-\alpha}$ where α is a free parameter. The fit provided a resonance mass of 3931 ± 4(stat) MeV/c^2, a width of 20 ± 8 MeV/c^2 and a total yield of 41 ± 11(stat) events. The statistical significance of the signal is 5.5 standard deviations. The points with error bars in Fig. 7(b) show the event yields in the 3.91-3.95 GeV/c^2 region versus $|\cos\theta^*|$, where θ^* is the angle of a D meson relative to the beam axis in the $\gamma\gamma$ c.m. frame. For a spin-0 resonance this distribution should be flat (dashed curve) whereas for a spin-2 resonance the expected angular distribution is $\propto \sin^4\theta^*$ (solid curve). The data significantly prefer a spin two assignment over spin zero. The product of the two-photon decay width and $D\bar{D}$ branching fraction assuming $J=2$ is $\Gamma_{\gamma\gamma}\mathscr{B}(Z(3920)\to D\bar{D}) = 0.23\pm0.06(\text{stat})\pm0.04(\text{syst})$ keV. The mass, decay angular distributions and $\Gamma_{\gamma\gamma}$ are all consistent with expectations for the χ'_{c2}, the 2^3P_2 charmonium state.

OBSERVATION OF AN ISOTRIPLET OF EXCITED CHARMED BARYONS DECAYING TO $\Lambda_c^+\pi$

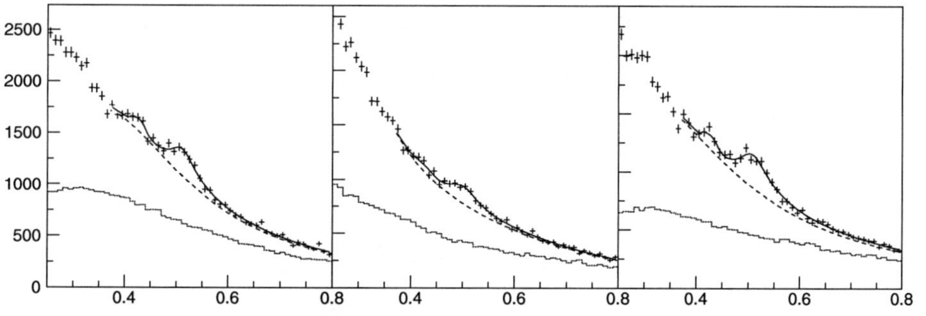

FIGURE 8. $M(\Lambda_c^+\pi) - M(\Lambda_c^+)$ distributions of the selected $\Lambda_c^+\pi^-$ (left), $\Lambda_c^+\pi^0$ (middle) and $\Lambda_c^+\pi^+$ (right) combinations. Data from the Λ_c^+ window (points with error bars) and normalized sidebands (histograms) are shown, together with the fits described in the text (solid curves) and their combinatorial background components (dashed).

Λ_c^+ baryons are reconstructed using the $pK^-\pi^+$ decay mode and combined with a pion candidate. To suppress background, a $x_p > 0.7$ requirement on the $\Lambda_c^+\pi$ pair is applied, where $x_p \equiv p^*/p^*_{max}$, p^* is the momentum of $\Lambda_c^+\pi$ in the cms and $p^*_{max} \equiv \sqrt{E^{cms}_{\text{beam}} - M^2}$, M is the mass of $\Lambda_c^+\pi$. Figure 8 shows distributions of the mass difference $\Delta M(\Lambda_c^+\pi) \equiv M(\Lambda_c^+\pi) - M(\Lambda_c^+)$ for the $\Lambda_c^+\pi^-$, $\Lambda_c^+\pi^0$ and $\Lambda_c^+\pi^+$ combinations in the region above $\Sigma_c(2455)$ and $\Sigma_c(2520)$ resonances. The enhancement near 0.51 GeV/c^2 is interpreted as signals of new excited charmed baryons, forming an isotriplet. These new baryons are denoted $\Sigma_c(2800)^0$, $\Sigma_c(2800)^+$, and $\Sigma_c(2800)^{++}$ respectively. The enhancement near $\Delta M = 0.43$ GeV/c^2 in $\Delta M(\Lambda_c^+\pi^-)$ and $\Delta M(\Lambda_c^+\pi^+)$ spectra is attributed to feed-down from the decay $\Lambda_c(2880)^+ \to \Lambda_c^+\pi^+\pi^-$, already observed by CLEO [25]. The fitting function used is the sum of three components : signal, feed-down, and combinatorial background functions. Assuming the $\Sigma_c(2800)$ to be Σ_{c2} baryons, decaying to $\Lambda_c^+\pi$ in D-wave, the signal is parametrized by a D-wave Breit-Wigner function convolved with the detector resolution of 2 MeV/c^2 (7 MeV/c^2) for final states containing only charged (one neutral) pions. The fit starts at 0.37 GeV/^{2} since at lower mass a feeddown from $\Lambda_c(2765)^+ \to \Lambda_c^+\pi^+\pi^-$ is expected. The significances are 8.6, 6.2 and 10.0 standard deviations for $\Sigma_c(2800)^0$, $\Sigma_c(2800)^+$, and $\Sigma_c(2800)^{++}$ respectively and their masses and widths are given in Table 2.

TABLE 2. Signal yield, mass and width for $\Sigma_c(2800)^0$, $\Sigma_c(2800)^+$, and $\Sigma_c(2800)^{++}$. The first uncertainty is statistical; the second is systematic.

State	Yield/10^3	ΔM, MeV/c^2	Γ, MeV
$\Sigma_c(2800)^0$	$2.24^{+0.79+1.03}_{-0.55-0.50}$	$515.4^{+3.2+2.1}_{-3.1-6.0}$	61^{+18+22}_{-13-13}
$\Sigma_c(2800)^+$	$1.54^{+1.05+1.40}_{-0.57-0.88}$	$505.4^{+5.8+12.4}_{-4.6-2.0}$	62^{+37+52}_{-23-38}
$\Sigma_c(2800)^{++}$	$2.81^{+0.82+0.71}_{-0.60-0.49}$	$514.5^{+3.4+2.8}_{-3.1-4.9}$	75^{+18+12}_{-13-11}

SUMMARY

The properties of the $X(3872)$ are reviewed : new observed modes and angular distributions allow a 1^{++} assignment for this new state, in agreement with the deuson hypothesis. New particles, $Y(3940)$, $X(3940)$ and $Z(3931)$, recently observed by Belle are described in detail. Only the latter seems to have an obvious charmonium assignment (χ'_{c2}). Finally, the observation of an isotriplet of excited charmed baryons decaying to $\Lambda_c^+\pi$ is described.

REFERENCES

1. S.-K.Choi, S.L.Olsen *et al.* (Belle Collaboration), Phys. Rev. Lett. **91**, 262001 (2003)
2. S.-K.Choi, S.L.Olsen *et al.* (Belle Collaboration), Phys. Rev. Lett. **94**, 182002 (2005)
3. D.Acosta *et al.* (CDF-II Collaboration), Phys. Rev. Lett. **93**, 072001 (2004)
4. V.M. Abazov *et al.* (D0 Collaboration), Phys. Rev. Lett. **93**, 162002 (2004)
5. B.Aubert *et al.* (BaBar Collaboration), Phys. Rev. **D 71**, 071003 (2004)
6. K.Abe *et al.* (Belle Collaboration), hep-ex/0505037
7. K.Abe *et al.* (Belle Collaboration), hep-ex/0505038
8. J.Rosner, Phys. Rev. **D 70**, 094023 (2004)
9. P.Pakhlov, proceedings of the HEP2005 International Europhysics Conf. on High Energy Physics.
10. N.A.Tornqvist, Phys. Lett. **B 590**, 209 (2004)
11. E.S.Swanson, Phys. Lett. **B 588**, 189 (2004)
12. D.Horn and J.Mandula, Phys. Rev. **D 17**, 898 (1978)
13. F.E.Close *et al.*, Phys. Rev. **D 57**, 5653 (1987)
14. G.Chiladze, A.Falk and A.A. Petrov, Phys. Rev. **D 58**, 034013 (1988)
15. F.E.Close and A.Godfrey, Phys. Lett. **B 574**, 210 (2003)
16. C.Banner *et al.*, Phys. Rev. **D 56**, 7039 (1997)
17. Z.-H. Mei and X.-Q. Luo, Int. J. Mod. Phys. **A 18** 5713 (2003)
18. K.Abe *et al.* (Belle Collaboration), Phys. Rev. Lett. **89**, 142001 (2002)
19. E.Braaten and J.Lee, Phys. Rev. **D 67**, 054007 (2003)
20. K.Y.Liu, Z.G.He and K.T.Chao, Phys. Lett. **B 557**, 45 (2003)
21. G.T.Bodwin, J.Lee and E.Braaten, Phys. Rev. Lett. **90**, 162001 (2003)
22. S.J.Brodsky, A.S.Goldhaber and J.Lee, Phys. Rev. Lett. **91**, 112001 (2003)
23. K.Hagiwara, E.Kou and C.-F.Qiao, Phys. Lett. **B 570**, 39 (2003)
24. K.Abe *et al.* (Belle Collaboration), Phys. Rev. **D 70**, 071102 (2004)
25. M.Artuso *et al.* (CLEO Collaboration), Phys. Rev. Lett. **86**, 4479 (2001)

Recent Physics Results with the COMPASS Experiment

Stephan Paul
for the COMPASS Collaboration

*Technische Universität München, Physik Department E18
James-Franck Strasse, D-85478 Garching*

Abstract. The COMPASS experiment has obtained first physics results in the field of polarized distribution functions for quarks and gluons using muon scattering off polarized deuterons. The analysis using open charm production and pairs of high p_T hadrons is presented. We also have used a transversely polarized target to address transverse information for quarks inside the nucleon. In addition, a pilot run with incoming pions taken late 2004 will give first information on the pion polarizabilities and hadron resonances. The physics prospects from this run as well as from future data taking in this field are also outlined.

Keywords: Polarized distribution functions, gluon polarization, charm, transversity, Sivers, Collins, ΔG, polarizabilities, exotic hadrons, spectroscopy
PACS: 13.60.-r, 13.88.+e, 14.20.-c, 14.40.-n

INTRODUCTION

The structure of hadrons can be investigated at different length scales revealing different properties and descriptions of hadrons. At small distance scales (large Q^2), deep inelastic scattering reveals the nucleon structure in terms of current quarks and gluons, using structure functions. The hot topic are spin degrees of freedom, the understanding of the nucleons spin in the interplay of its constituents. At medium distance scales the tool of spectroscopy is useful, a domain still dominated by the language of constituent quarks. The search for exotic quantum numbers and glueballs at masses between 1.5 and 2.5 GeV/c^2 should reveal more insights to the phenomenon of confinement. Very large distance scales probe quasi static hadron properties like polarizabilities of mesons. Effective field theories derived from QCD are expected to give a consistent description of the low energy quantities. The COMPASS experiment at CERN offers the unique opportunity of studying the three fields of hadron physics within one experiment using flexible beams and spectrometers setups. As most of the beam time has so far been devoted to experiments using muons (deep inelastic scattering - DIS and hard interactions) most of this report will be devoted to first results on this topic. At the end of last year a first pilot run with pions has been performed aiming at the measurement of polarizabilities and diffractive meson production. A first glimpse to this program will end this report.

CP814 *Hadron Spectroscopy: Eleventh International Conference*
edited by A. Reis, C. Göbel, J. Sá Borges, and J. Magnin

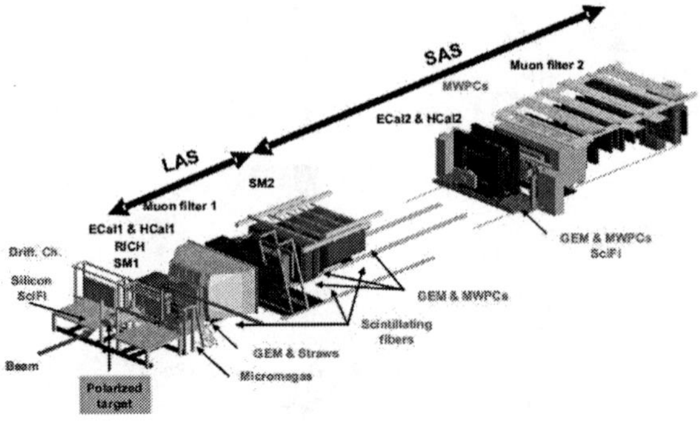

FIGURE 1. Layout of the COMPASS experiment at the CERN SPS

THE COMPASS EXPERIMENT

The COMPASS experiment at the CERN SPS is a modern 2-stage magnetic spectrometer with a flexible setup to allow for a variety of physics programs to be performed with different beams (fig. 1). Common to all measurements is the requirement for high beam intensity and interaction rates with the needs of a high readout speed. The spectrometer is equipped with tracking systems based on silicon detectors for high precision tracking in the target region, micromega and GEM detectors for small area and wire/drift chambers as well as straw tubes for large area tracking. Particle identification is performed in a large acceptance ring imaging cherenkov detector downstream of the first spectrometer magnet and in hadronic (HCAL) and electromagnetic (ECAL) calorimeters in both parts of the spectrometer. The calorimeters are also used for the energy reconstruction of neutral particles. Both parts of the spectrometer are also equipped with muon detectors.

For measurements with a polarized nucleons a polarized target is installed about 3m upstream of the first magnet. It consists of two independent target cells filled with ^6LiD crystals mounted inside a very homogeneous solenoidal field of 2T. Using the technique of dynamic nuclear polarization 50% of the nucleons could be polarized with a degree polarization of more than 50%. The two target cells are thereby operated with opposite polarization to reduce systematic effects on asymmetry measurements owing to time dependent performance of the apparatus. The study of exclusive reactions using hadron beams requires a short solid target inside a veto-detector surrounded by silicon telescopes. The veto-detector is sensitive to charge and neutral particles with small openings in forward and backward direction.

The trigger system is mostly based on scintillator hodoscopes mounted in front of the target and downstream of the second magnet. Using a muon beam deep inelastic scattering (DIS) at large Q^2 is selected as well as low Q^2 events (quasi-real photons) with accompanying hadrons observed in HCAL. For diffractive and Primakoff scattering with hadron beams the target is surrounded by a veto box and small multiplicity events are selected by means of a scintillator and within an online event filter operated inside the event building computers.

PHYSICS RESULTS WITH MUON BEAMS

Spin-structure functions

The low energy description of the nucleons spin is based on the SU(6) wave functions for non-interacting constituent quarks of spin $\frac{1}{2}$ where two quarks are spin-aligned and one quark spin-antialligned making a total spin of $\frac{1}{2}$ for the nucleon. In turn, at small distance scales the nucleon is described in terms of QCD where quarks interact among each other via gluons. The resulting sharing of the nucleons momentum is described via structure functions (number densities) which are a function of the Bjorken variable x denoting the momentum fraction of a constituent and the energy scale Q^2 (which sometimes is identified with a spatial resolution scale). Quarks and gluons are not independent partners and the repartitioning of kinematic variables when changing the energy scale is described by the DGLAP equations. Measurements of the quark helicity within a polarized nucleon has been performed by a number of experiments using DIS on polarized nucleons at various Q^2. The pictorial interpretation of the results are not straightforward but its safe to say that quarks only carry part of the nucleons spin. Owing to the axial anomaly (coupling of quarks and gluons) the singlet axial charge a_0 (which enters the cross section asymmetries) combines spin contributions of quarks $\Delta\Sigma$ and gluons ΔG

$$a_0 = \Delta\Sigma - \frac{3\alpha_a\left(Q^2\right)}{2\pi}\Delta G \tag{1}$$

$$with\ \Delta\Sigma = \Delta u + \Delta d + \Delta s = 2\left\langle S_z^{quark}\right\rangle \tag{2}$$

Using a QCD analysis of the data we can extract values for $\Delta\Sigma \sim 0.2$. The accuracy is limited by the data quality at very low values of x, as $\Delta\Sigma$ is obtained from the integral of the spin-dependent quark structure function $g_1(x)$. The same analysis also allows to extract an estimate of the gluon spin structure-function $\Delta G(x)$. As experiments probe a_0 at different values of Q^2 we can use the evolution equations to obtain an estimate of the gluon contribution of the spin. However, the Q^2 range is small and thus uncertainties on $\Delta G(x)$ are large.

Measurement of A_1 on the deuteron

The high rate capability of COMPASS allows a precision study of the quark asymmetry A_1^d on the deuteron [1]. Using data taken in '02 and '03 we have collected about $3 \cdot 10^7$ DIS-events. The measured asymmetry A_1^d is obtained combining data from periods with opposite spin polarizations within the two target cells (see fig. 2 left). Thus, acceptance effects cancel out. We can translate the asymmetries into the polarized quark distribution functions normalizing the measured asymmetry to the unpolarized cross section determined by the structure function $F_2(x)/2x$ and the ratio of longitudinal to transverse cross

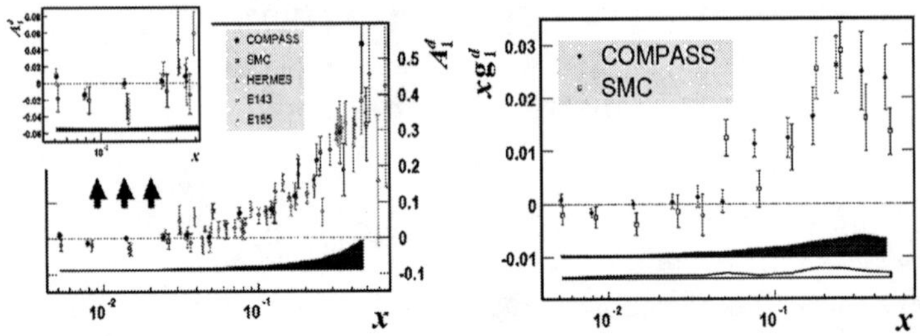

FIGURE 2. Left: Measured asymmetry using DIS events on the deuteron. The data are compared to results from previous experiments. The area shaded in dark red denotes the systematic uncertainties. The inlet depicts the results for small values of x, where COMPASS has the largest impact. Right: Results for $x \cdot g_1(x)$ extracted from the measured asymmetry. Data are taken at large Q^2 and compared to results from the SMC experiment. Bands for systematic uncertainties are given for both data sets (COMPASS: dark red)

section R.

$$g_1^d = \frac{F_2^d}{2x(1+R)} A_1^d \tag{3}$$

The results for $x \cdot g_1(x)$ extracted from the measured asymmetry are shown in the right of fig. 2. The data are taken at large Q^2 and are compared to results from the SMC experiment. Systematic uncertainties are marked by bands and are given for both data sets (COMPASS: dark red). The high precision of the new COMPASS data at low x help to better define the integral over $g_1(x)$ and thus allow to reduce the uncertainties on $\Delta\Sigma$ by a factor 2.

$$\Delta\Sigma = 0.237^{+0.024}_{-0.029} \tag{4}$$

The Gluon Spin Contribution

From the results obtained for the spin contribution of the quarks we conclude that the picture must be far more complicated than assumed at first. We can thus write the nucleons spin formally as a sum of four contribution.

$$\frac{1}{2} = S_z = \frac{1}{2}\Delta\Sigma + \Delta G + L_q + L_G \tag{5}$$

However, the interpretation and distinction between the different terms depends on the reference system and thus is not unique. While contributions to angular momenta are currently only accessible in principle[1] the COMPASS experiment aims at determining the role of the gluon spin to this sum. This requires scattering processes which are a)

[1] In the recent years a new mathematical formulation of angular momentum inside the nucleons in terms of so called generalized parton distributions has been found. However, this requires a large set of exclusive

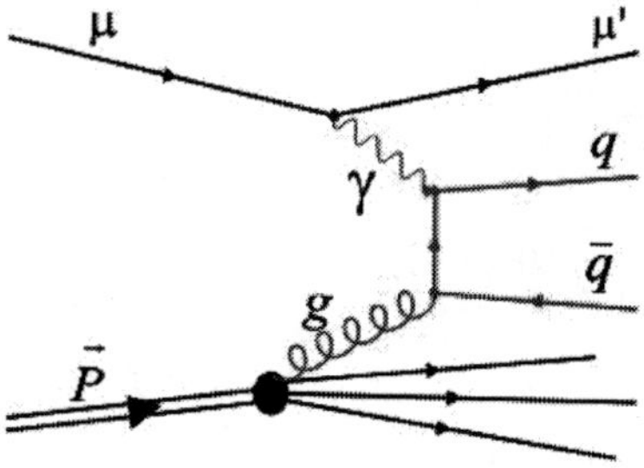

FIGURE 3. The γg-fusion process. It can be tagged using open charm or high p_T-hadron pairs in the final state.

sensitive to gluons and b) depict a sizeable analyzing power. In COMPASS the photon-gluon fusion process (PGF) is used, which can be tagged in two different ways.

1. The production of open charm is a very clean sign for the underlying hard QCD process as charm production in either fragmentation processes or via intrinsic charm is negligible. The energy scale of the process is determined by the minimal value $\sqrt{s} \geq 2m_c \simeq 3GeV/c^2$, thus we can use quasi real photons ($Q^2 \simeq 0$).
2. A second tag is the pair production of light hadrons at large transverse momentum ($p_T \geq 1GeV/c$) which also determines the scale of the underlaying process.

We will discuss the results for both processes studied.

ΔG via production of open charm

The PGF-process is depicted in fig. 3. The yield of charmed hadrons is about 1%, the analyzing power 10%. Using data obtained in 2002 and 2003 we have extracted a sample of D-mesons. This is done in two ways. D^0 candidates are selected via there decay $D^0 \to K\pi$ where the kaons are identified within the RICH. A very clean signal is obtained using reconstructed D^*-decays ($D^* \to D^0\pi_{slow}$). These events are shown on the left of fig. 4. The full D^0 sample with and without D^*-tagging is summarized on the right of fig. 4 From these data we can extract ΔG using NLO calculations for the spin dependence of the $\gamma g-$process.The result is $\frac{\Delta G}{G} = -1.08 \pm 0.73$ at a mean value of $\langle x_G \rangle = 0.15$. We expect to improve the statistical error by about a factor two using 2004 data.

---

measurements over a wide kinematic range for which dedicated experiments and machines will have to be constructed. It turns out that also the measurement of transversity inside the nucleon is connected to angular momentum via the determination of the Sivers function. However, this will be very model dependent.

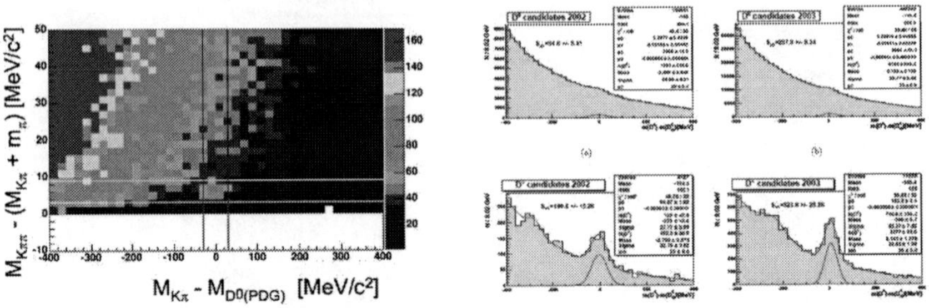

FIGURE 4. Left: $M(K\pi\pi)-M(K\pi)$ versus the $M(K\pi)-M(D^0)$ mass difference. Right: $(K\pi)$ invariant mass plots showing D^0-signals for two different selection schemes and the two data periods (left: 2002, right: 2003). Upper plots:$(K\pi)$ with hard K-identification. Lower plots:$(K\pi)$ combinations tagged by slow pions consistent with $D^* - D^0$ mass difference.

ΔG via production of high p_T hadron pairs

The channel of high p_T hadron pairs shares the virtue of high statistics with the drawback of systematic uncertainties on the role of background processes. The selection requires individual $p_T \geq 0.7$ combined with $p_{1T}^2 + p_{2T}^2 \geq (2.5(GeV/c)^2$ as well as $m(h_1,h_2) > 1.5GeV/c^2$. The data set is divided into $Q^2 > 1(GeV/c)^2$ and $Q^2 \leq 1(GeV/c)^2$, which in our data set corresponds to quasi real photon production. For the high Q^2 sample the background processes are ordinary DIS events with fragmentation or gluon-Bremsstrahlung. However, as the quark asymmetries A_1^d are negligible at small x, these two processes do not dilute the asymmetry from the underlaying PGF process for the analyzing power and the partial cross section are taken from theory (and MC). The result using data from 2002 and 2003 is:

$$\Delta G/G = 0.06 \pm 0.31(stat.) \pm 0.06(syst.) \;\; for\, Q^2 > 1(GeV/c)^2; < x_B > \;1.15 \qquad (6)$$

For quasi-real photons the main background processes in the asymmetry are contributions from resolved photons (fluctuations of the photon into a hadron). As the polarized structure function of the photon is unknown, these processes contribute largely to the systematic errors as the contribution is model dependent and maximal and minimal scenarios are used to obtain a handle on their contribution. It should be noted that all corrections are only calculated in LO. From the measured asymmetry $A_{LL}/D = 0.002 \pm 0.019(stat)$ we derive a gluon polarization of

$$\Delta G/G = 0.024 \pm 0.089(stat) \pm 0.057(syst) \;\; for\, Q^2 \leq 1(GeV/c)^2; < x_B > \;0.95 \qquad (7)$$

Fig. 5 summarizes the COMPASS results for $\Delta G/G$ together with previous measurements from SMC and HERMES. It should be noted that previous determinations have not used the same method of subtracting background which is due to other processes. The small value of the gluon polarization at $x \sim 0.1$ can be compared to various models for the functional dependence $\Delta G(x)/G(x)$ assuming different values for the integral value. It seems that large values of $\Delta G/G$ are very unlikely.

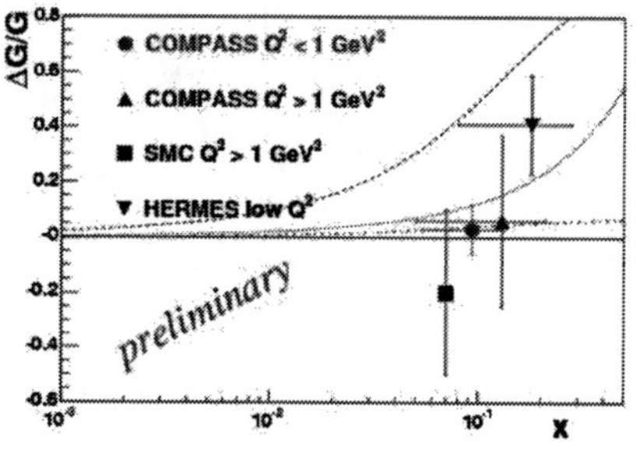

FIGURE 5. World data on direct measurements of the gluon polarization $\Delta G/G$ using high p_T hadron pairs [2]. The results from the COMPASS analysis for both ranges of Q^2 are shown together with previous analysis from SMC [3] and HERMES [4]. Note that the HERMES result does not include a systematic uncertainty linked to the background subtraction from other processes but PGF. Experimental curves taken from [5]

Transverse Spin Effects

In addition to the spin distribution functions G(x) and g_1(x) for longitudinally polarized nucleons the transversity function $\delta q(x)$ describes the distribution of transversely polarized quarks inside a transversely polarized nucleon. This function is more difficult to measure and to interpret, though formally it has to be considered on the same footing as the other structure functions. The reason for the experimentally difficult access to this function is that a physical measurement requires a suitable analyzing function. One possibility is a spin dependent fragmentation function (Collins function) in which a production asymmetry arises for hadron produced with larger transverse momentum (eq. 8). This asymmetry has to be measured w.r.t. to the spin direction of the nucleon and is called 'Collins function' [8]. We expect a modulation of the hadron intensity with $\sin(\phi + \phi_s)$. The plane of reference for all azimuth angles is the lepton scattering plane (see fig. 6 for the definition of angles).

$$A_T \propto \delta q(x) \cdot H_1(z, k_T) \tag{8}$$

HERMES has measured the combination of the two functions for the first time on polarized protons and obtained a non-zero result pointing to both functions being non zero [7]. The results for COMPASS measured on transversely polarized deuterons [6] are depicted in fig. 6. A second effect leading to non-zero asymmetries in reactions on transversely polarized targets is the Sivers effect [9] which correlates transverse momentum of quarks inside the nucleon with the transverse spin direction of the nucleon $\sin(\phi - \phi_s)$. Sivers and transversity/Collins effects can thus be disentangled by their different azimuthal dependence.

The COMPASS results show no sign of any asymmetry which might be explained by a cancelation of the effect by neutron and proton inside the deuteron.

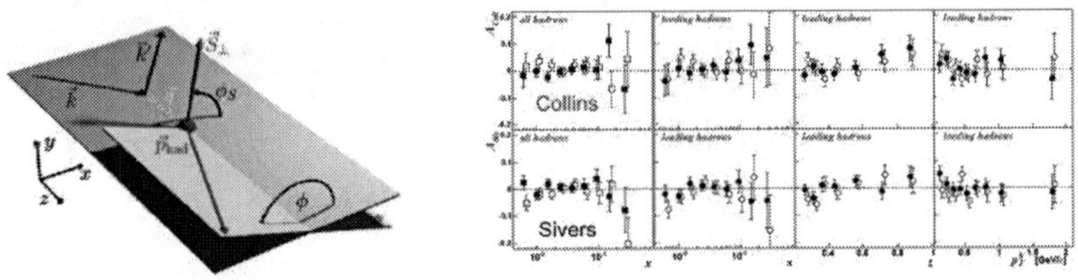

FIGURE 6. Left: Definition of angles in transversity measurements. Right: Azimuthal asymmetries obtained with a transversely polarized target in COMPASS. Upper: Collins angles, lower: Sivers angle.

DIFFRACTIVE PROCESSES AND HADRON STRUCTURE

Using an incoming hadron beam COMPASS aims at studying hadron polarizabilities using Primakoff reactions and light meson spectroscopy in diffractive and central production. In either case high rates and good acceptance should allow to improve over the existing data sets by about a factor 10 in the statistical sample.

A pilot run with Pions

In 2004 COMPASS has performed a short run with low intensity pions ($2 \cdot 10^6 \pi/spill$) impinging on a segmented lead (copper) target of 50% and 25% radiation length (two modes of running). The main purpose was as measurement of the pion polarizability using scattering off virtual photons on a lead nucleus. The Compton scattering process in inverse kinematics (fig. 7) is dominated by the Thompson cross section for point like particles and contains two terms describing the electric and magnetic polarizabilities of the hadron which can be separated by their different angular dependence (see eq. 9, where the cross section is given in the reference frame of the outgoing pion).

$$\frac{d\sigma_{\gamma\pi}(\theta,\omega)}{d\omega d\theta} = \frac{2\alpha^2}{m_\pi^2}\{F_{Th}^{\gamma\pi} + \frac{m_\pi\omega^2}{\alpha}\frac{\bar{\alpha_\pi}^2(1+\cos^2\theta)+\bar{\beta_\pi}^2\cos\theta}{(1+\frac{\omega}{m_\pi}\cos\theta)^3}\} \quad (9)$$

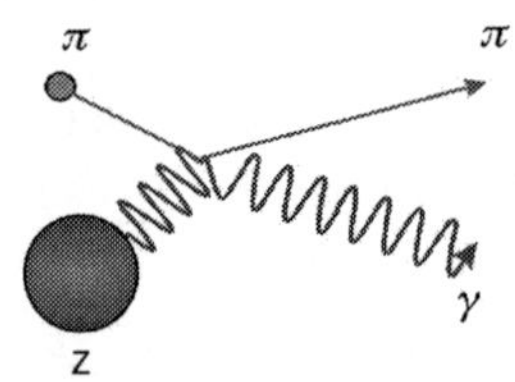

FIGURE 7. Sketch of the Primakoff reaction

$\bar{\alpha_\pi}$ and $\bar{\beta_\pi}$ are the electric and magnetic polarizabilities, respectively. Theoretically polarizabilities can be described in the framework of chiral perturbation theory, an effective field theory derived from QCD where mesons are the relevant degrees of freedom. The low energy expansion parameter is $\frac{m_\pi}{p}$ and higher order terms p^n contain

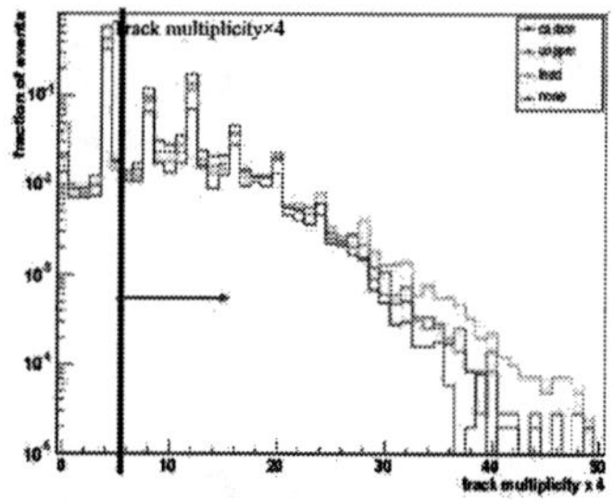

FIGURE 8. Track multiplicities (x4) as observed in the COMPASS silicon telescope downstream of the target using incoming pions. This info is used in a fast online data filter.

low energy constants derived from experiment. By now the calculations are of very high accuracy and constitute a challenge to experiments. While the first experiments using the same technique of Primakoff scattering obtained a value of $\bar{\alpha_\pi} = (6.8 \pm 1.4_{stat.} \pm 1.2_{sys}) \cdot 10^{-4} fm^3$ [10] a more recent extraction using threshold pion production in electron scattering off a proton obtained $\bar{\alpha_\pi} - \bar{\beta_\pi} = (11.6 \pm 0.5_{stat.} \pm 3.0_{sys} \pm 0.5_{model}) \cdot 10^{-4} fm^3$ [11]. These numbers can be compared directly if one assumes $\bar{\alpha_\pi} + \bar{\beta_\pi} = 0$

COMPASS has taken a first data set and within a few days of running about 35000 Primakoff events have been recorded with a 4-momentum transfer $t \leq 0.0015 (GeV/c)^2$, which separates the Primakoff region from the diffractive background dominating a larger values of t. We expect to improve in statistical accuracy by about a factor 4-5 over the previous experiment from Serpukhov.

In addition, we have recorded events with neutral pions in the final state to obtain a new value for the chiral anomaly using the Primakoff reaction $\pi^- \to \pi^- \pi^0$. In the same data set we also have obtained events for diffractive production off a nucleus. Using a CH-target and a multiplicity trigger (fig. 8) we investigated reactions like $\pi^- p \to \pi^- \pi^- \pi^+ p$ as well as $\pi^- p \to \gamma\gamma\pi^- p$. This programm is a continuation of the VES [12] and BNL [13] measurements on diffractive production on exotic mesons, however, with much higher beam energy.

Future of meson spectroscopy using central production

Central production ($pp \to p_{fast} p_{slow} X$) of a state X has proven to be a very useful tool in the search for exotic mesons and gluon rich states [14] (fig. 9) . As the cross section drops very fast with the mass of X we need high beam intensities and acceptance to extend the measurements into the region above 2 GeV/c^2. Using a 270 GeV/c proton beam, a liquid hydrogen target and full electromagnetic calorimetry we will investigate this reaction in 2007. Beam intensities of a few 10^7/spill are required to obtain a major gain in statistical significance as compared to previous experiments. This experimental challenge asks for fast triggering ($p_T - balance$, ECAL) and high resolution.

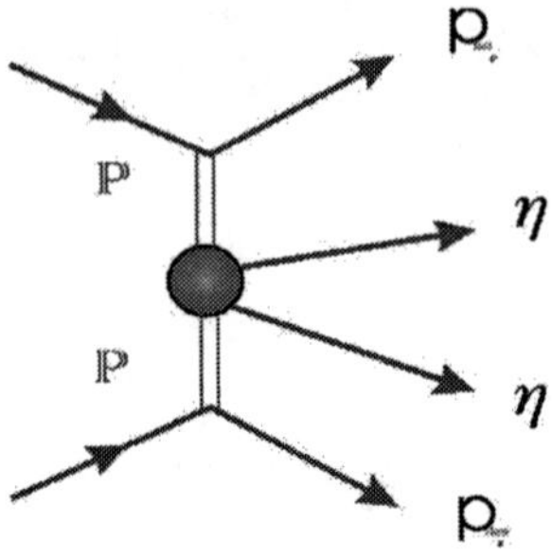

FIGURE 9. Sketch of central production of a state X decaying into $\eta\eta$.

SUMMARY

Using data taken in 2002-2003 COMPASS has made a major step forward in determining the gluon polarization in a direct measurement on polarized deuterons. The results obtained using high p_T hadron pairs points to a very small value for $\Delta G/G$ at $x_G \approx 0.1$ which does not favor expectations derived from QCD analysis of polarized DIS on nucleons. No significant sign for transversal asymmetries has been found so far, neither in transversity nor for the Sivers functions.

First data for the measurement of the pion polarizability has been collected and are currently being analyzed. This marks the beginning of the physics program with hadron beams within COMPASS, which is planned to be continued in 2007 with measurements on central production of hadrons.

REFERENCES

1. E.S. Ageev et al., Phys.Lett.B612:154-164,2005
2. E.S. Ageev et al., submitted to Phys.Lett.B
3. B. Adeva et al., Phys.Rev.D70:012002,2004
4. A. Airapetian et al., Phys.Rev.Lett.84:2584,2000
5. M. Gluck, E. Reya, M. Stratmann, W. Vogelsang, Phys.Rev.D63:094005,2001
6. V.Yu.Alexakhin et al. Phys.Rev.Lett.94:202002,2005
7. A. Airapetian et al., Phys.Rev.Lett.94:012002,2005
8. J. Collins, Nucl.Phys.396:161,1993
9. D. Sivers, Phys.Rev.D.41:83,1990
10. Yu.M. Antipov et al., Phys.Lett.B121:445,1983
11. J. Ahrens et al., Eur.Phys.J.A23:113-127,2005
12. D.V. Amelin et al. Phys.Lett.B356:595-600,1995
13. E.I. Ivanov et al. Phys.Rev.Lett.86:3977-3980,2001
14. D. Alde et al., Phys.Lett.B205:397,1988 and D. Barberis et al., Phys.Lett.B397:339-344,1997

GlueX: The Search for Gluonic Excitations at Jefferson Laboratory

Daniel S. Carman

Ohio University, Department of Physics and Astronomy, Athens, OH 45701

Abstract. One of the unanswered and most fundamental questions in physics regards the nature of the confinement mechanism of quarks and gluons in quantum chromodynamics (QCD). Exotic hybrid mesons manifest gluonic degrees of freedom and their detailed spectroscopy will provide the precision data necessary to test assumptions in lattice QCD and the specific phenomenology leading to confinement. Photoproduction is expected to be a particularly effective manner to produce exotic hybrids, however, existing data using photon beams are sparse. At Jefferson Laboratory, plans are underway by the GlueX Collaboration to use the coherent bremsstrahlung technique to produce a linearly polarized photon beam. A solenoid-based hermetic detector will be used to collect data on meson production and decays with statistics that will exceed existing photoproduction data by several orders of magnitude after the first year of running. In order to reach the ideal photon energy of 9 GeV required for these studies, the energy of the Jefferson Laboratory electron accelerator, CEBAF, will be doubled from its current maximum energy of 6 GeV to 12 GeV. The physics motivating the search and the status of the project are reviewed.

Keywords: Gluonic excitations, meson spectroscopy, Jefferson Laboratory
PACS: 12.38.Aw,12.39.Mk,14.40.-n,25.20.Lj

MOTIVATION FOR THE STUDY OF HYBRID MESONS

The specific goal of the GlueX Collaboration at Jefferson Laboratory [1] is to better understand the detailed nature of confinement. The nature of this mechanism is one of the great mysteries of modern physics, and in order to shed light on this phenomenon, we must better understand the nature of the gluon and its role in the hadronic spectrum. Confinement within the theory of strongly interacting matter, Quantum Chromodynamics (QCD), arises from the postulate that gluons can interact among themselves and give rise to detectable signatures within the hadronic spectrum. These signatures are expected within hadrons known as hybrids, where the gluonic degree of freedom is excited. GlueX is designed to primarily focus on the study of the light quark hybrid mesons and its central purpose is to perform a detailed mapping of the spectrum of these states. It is in these systems where the gluonic degree of freedom is manifest and can provide for a more detailed understanding of the confinement mechanism in QCD.

Gluonic mesons represent a $q\bar{q}$ system in which the gluonic flux-tube contributes directly to the quantum numbers of the state. In terms of the constituent quark model, the quantum numbers of the meson are determined solely from the quark and antiquark. However QCD indicates that this simple picture is incomplete. Lattice QCD calculations predict that states with the flux-tube carrying angular momentum (called hybrids) should exist, as well as purely gluonic states (called glueballs). Modern lattice calculations for mesons show that indeed a string-like chromoelectric flux-tube forms between distant

CP814 *Hadron Spectroscopy: Eleventh International Conference*
edited by A. Reis, C. Göbel, J. Sá Borges, and J. Magnin

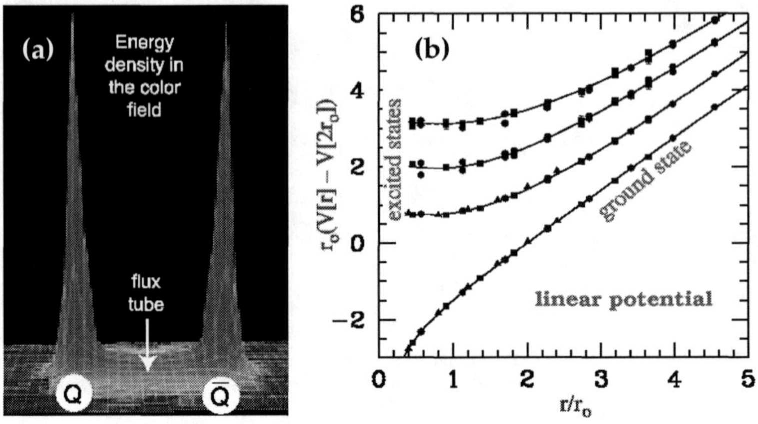

FIGURE 1. (a). A lattice QCD calculation of the energy density in the color field between a quark and an antiquark. This density peaks at the locations of the q and $\bar{q}$ and is confined to a flux-tube stretching between the pair [2]. (b). Lattice calculation of the hybrid potential for the ground and low-lying excited states [3].

static quark charges as shown in Fig. 1a. The non-perturbative nature of the flux-tube leads to the confinement of the quarks and to the well-known linear inter-quark potential from heavy-quark confinement with $dV/dr \sim 1$ GeV/fm (see Fig. 1b). Confinement implies that an infinite energy is required to separate the quarks off to infinity. These calculations predict that the lowest lying hybrid meson states are roughly 1 GeV more massive than the conventional meson states. This provides a reference point for the mass range to which experiments must be sensitive.

HYBRID MESON PROPERTIES

Within the standard non-relativistic constituent quark model, conventional mesons are made up from $q\bar{q}$ pairs with the spin 1/2 quarks coupled to a total spin S=0 or 1. These pairs are then coupled with units of orbital angular momentum L=0,1,2,..., and a possible radial excitation. The relevant quantum numbers to describe these states for a given principal quantum number n are J^{PC}, where $\vec{J} = \vec{L} + \vec{S}$ represents the total angular momentum, $P = (-1)^{L+1}$ represents the intrinsic parity, and $C = (-1)^{L+S}$ represents the C-parity of the state.

The light-quark mesons are built up from u, d, and s quarks and their antiquarks. For each value of S, L, and n, there are nine possible $q\bar{q}$ combinations, thus a nonet of mesons is expected for each value of J^{PC}. The lowest possible mass states for these conventional mesons (i.e. for L=0, n=1) are then $J^{PC} = 0^{-+}$ for S=0 (corresponding to the π, η, η', and K mesons) and $J^{PC} = 1^{--}$ for S=1 (corresponding to the ρ, ω, ϕ, and K^* mesons). The lowest lying S=0 mesons have masses of about 500 MeV and the lowest lying S=1 mesons have masses of about 800 MeV. Nonets of higher mass mesons for a given principal quantum number are then built up by adding in units or orbital angular momentum. Conventional mesons correspond to states with the flux-tube in its ground state, and as such, the gluonic degree of freedom does not contribute.

The hybrid meson quantum numbers can be predicted within the flux-tube model [4]. In its ground state, the flux-tube carries no angular momentum. The lowest excitation is an L=1 rotation which contains two degenerate states (corresponding to clockwise and counterclockwise rotations). Linear combinations of these states give rise to quantum numbers of $J^{PC} = 1^{-+}$ or 1^{+-} for the flux-tube. Adding these quantum numbers to those for the $q\bar{q}$ pair gives the possible J^{PC} for the hybrid mesons, which also are expected to have a nonet of states for each J^{PC}. For $S = L$=0, the possible quantum numbers are $J^{PC} = 1^{--}$ and 1^{++}. Note that exotic hybrid mesons are *not* generated when S=0 as these quantum numbers are possible for conventional $q\bar{q}$ states such as ρ, ω, and ϕ. An important consideration in the study of these states is that non-exotic hybrids may mix with conventional $q\bar{q}$ states making clear experimental identification difficult. Establishing the hybrid nonets will depend on starting with nonets whose quantum numbers are exotic to which ordinary $q\bar{q}$ states cannot couple.

For S=1, L=0 states the possible quantum numbers are $J^{PC} = 0^{+-}$, 0^{-+}, 1^{-+}, 1^{+-}, 2^{+-}, 2^{-+}. Of these states, those with $J^{PC} = 0^{+-}$, 1^{-+}, and 2^{+-} are manifestly exotic. These unusual quantum numbers are not allowed for ordinary $q\bar{q}$ states. They provide the "smoking gun" experimental signature of gluonic excitations. Although GlueX aims to make sense of the full spectrum of light conventional mesonic states, hybrid mesons, exotic hybrid mesons, glueballs, multi-quark states, and molecular states, the exotic states will be the main initial thrust of the experimental program. The GlueX experiment has been designed to perform spectroscopy over a range of meson masses up to roughly 3 GeV.

EVIDENCE FOR HYBRID MESONS AND PERSPECTIVES

There are two avenues to investigate in the review of current evidence for exotic hybrid mesons. On the one hand powerful indicators have been provided by current Lattice QCD (LQCD) calculations. A summary of current LQCD calculations for the lowest lying non-strange $J^{PC} = 1^{-+}$ exotic meson is provided in Table 1. These quantum numbers should correspond to the lightest nonet. These calculations are quenched or partially quenched at the current time with some employing improved lattice actions, and they indicate that the rough mass scale to search in is about 1.8 - 2.1 GeV.

On the experimental side, a number of reasonably high statistics searches for such states have been carried out with beams of hadrons and have given rise to credible evidence for the $J^{PC} = 1^{-+}$ exotics $\pi_1(1400)$, $\pi_1(1600)$, and $\pi_1(2000)$. The $\pi_1(1400)$ was first reported by the VES Collaboration at IHEP [11] and was confirmed by the E852 Collaboration at BNL in the $\pi^- p \to \eta\pi^- p$ reaction [12]. This was followed with additional confirmation by the Crystal Barrel Collaboration at CERN's LEAR facility in antiproton-neutron annihilation [13]. The strength of the exotic signal measured by Crystal Barrel was as large as that of the conventional $a_2(1230)$ meson. In the Particle Data Group (PDG) listings [14], the mass of the $\pi(1400)$ is M=1376±17 MeV and its width Γ=300±40 MeV with observed decays to $\pi\eta$. A second $J^{PC} = 1^{-+}$ exotic hybrid candidate, the $\pi_1(1600)$, was first observed at BNL E852 in $\pi^- p \to \rho p \to$

TABLE 1. Quenched lattice QCD calculations for the mass of the lowest-lying $J^{PC} = 1^{-+}$ non-strange exotic hybrid meson.

Collaboration	Year	Computed Mass (GeV)	Reference
UKQCD	1997	1.87±0.2	[5]
MILC	1997	1.97±0.09	[6]
MILC	1999	2.11±0.13	[7]
SESAM	1998	1.9±0.20	[8]
Mei & Luo	2003	2.013±0.026	[9]
Bernard *et al.*	2004	1.792±0.139	[10]

$\pi^+\pi^-\pi^- p$ [15]. Since that time further evidence for the $\pi_1(1600)$ has been provided through its decays into $\eta'\pi$ [16], $f_1\pi$ [17], and $b_1\pi$ [18]. Evidence in these decay modes has also been provided by VES [19]. The PDG [14] lists the mass and width of this state as M=1596^{+25}_{-14} MeV and Γ as 312^{+64}_{-24} MeV.

The interpretation of these states is presently controversial. The $\pi_1(1400)$ is significantly lighter than theoretical expectations, and its only observed decay mode into $\eta\pi$ is not expected for a hybrid. Hadronic form factors suppress those decays which couple the flux-tube orbital quantum to the relative motion of the final mesons. Thus two-body decays involving one L=0 and one L=1 meson are preferred. It has been suggested that the $\pi_1(1400)$ could represent a meson-meson molecule. Other recent work suggests that this state may actually arise from non-resonant scattering, similar to S-wave $\pi\pi$ scattering at low energy [20]. The same explanation can also be invoked to explain a sizeable fraction of the $\eta'\pi$ signal for the $\pi_1(1600)$ [21]. The mass of the $\pi_1(1600)$ as seen in $\rho\pi$ is still somewhat lower than theoretical expectations, but could well be a hybrid. However a recent analysis of a larger data set from BNL E852 looking for $\pi_1(1600) \rightarrow \rho\pi$ [22] using what is claimed to be an improved partial wave analysis with a larger set of basis states, claims that there is no evidence for a $J^{PC} = 1^{-+}$ meson in studies of two different 3π isospin channels. The VES results [19] provided measurements of the relative branching ratios into $b_1\pi$, $\eta'\pi$, and $\rho\pi$ as 1:1.0:1.6. These predictions are highly at odds with predictions of the flux-tube model. Thus either these three modes are not all due to a hybrid meson or there is a problem with the underlying theory.

Another candidate $J^{PC} = 1^{-+}$ meson in the literature is the $\pi_1(2000)$ which has been seen through its decays into $b_1\pi$ [23] and $f_1\pi$ [24] at BNL E852. This state currently may be the least controversial of the announced exotics in the sense that its mass is in agreement with theoretical expectations and its decays modes are fully in line with expectations for hybrid mesons. However the statistical precision of the existing data is still somewhat limited.

The current landscape makes it clear that a new dedicated spectroscopy experiment with the design and expected statistics of GlueX will be able to help greatly clarify the current ambiguous situation in the mass range below 3 GeV. This will be necessary to come to an understanding of the possible overpopulation currently reported for the 1^{-+}

hybrid nonet, where there should be only one π_1 state.

In the experimental search for exotic hybrid mesons it is important to appreciate why exotics are hard to find. Most past studies have investigated two-body final states. It is clear now that the coupling of hybrids to low multiplicity final states is essentially non-existent. In addition, unambiguous discovery of these new states requires a detailed knowledge of the full meson spectrum and understanding of multiple decay modes. Thus a full picture of conventional mesons, hybrid mesons, exotic hybrid mesons, multiquark states, molecular states, and glueball states must be disentangled. It should also be kept in mind that many of the exotic states are expected to be relatively broad, and thus will be difficult to isolate without sophisticated analysis tools.

A key part of current experimental evidence for exotic mesons relies on sophisticated partial wave analysis (PWA) tools. This type of analysis is required to identify the J^{PC} quantum numbers of the states. In simple terms, a PWA determines the production amplitudes by fitting the decay angular distributions of the final state particles. The fit includes information on the polarization of the beam and target, the spin and parity of the resonance, the spin and parity of any daughter resonances, and any relative orbital angular momentum. The analysis seeks to establish the production strengths, production mechanisms, and the relative phase motion of various production amplitudes.

While the implementation of a partial wave analysis is in principle straightforward, there are difficulties that arise due to the detector system employed as well as ambiguities within the PWA framework itself. Effects such as detector acceptance and resolution can conspire to allow strength from a dominant partial wave to appear as strength in a weaker wave. Intrinsically the PWA can be misleading if the set of basis states chosen is not complete. Both of these problems lead to a phenomenon known as "leakage" where strength that should be attributed to a specific partial wave leaks into other waves. A key part of the efforts of the GlueX Collaboration is to improve the analysis tools to reduce these potential problems. Of course designing the detector system to provide full acceptance for both charged and neutral particle final states is a requisite part of the success of the analysis, as is the ability to acquire large statistics data sets. In the next generation experiment, it is apparent that sophisticated and modern analysis tools leading to detailed spectroscopy will be the answer toward making progress in understanding the nature of confinement.

THE NEXT GENERATION EXPERIMENT

The specifications for the next generation experiment in meson spectroscopy are already quite clear. The essential ingredients are that the detector should be hermetic for both charged and neutral particle final states, with excellent resolution and particle identification capability. This is essential for successful and unambiguous partial wave analysis. The beam energy must be high enough to allow for sufficient phase space for the production of the exotic states, however it should not be too high, otherwise the cross section for background processes will increase and overwhelm the signals that are being sought. The partial wave analysis requires high statistics experiments with sensitivity to production cross sections at the sub-nanobarn level.

In terms of beam properties, it is clear that the next generation experiment should

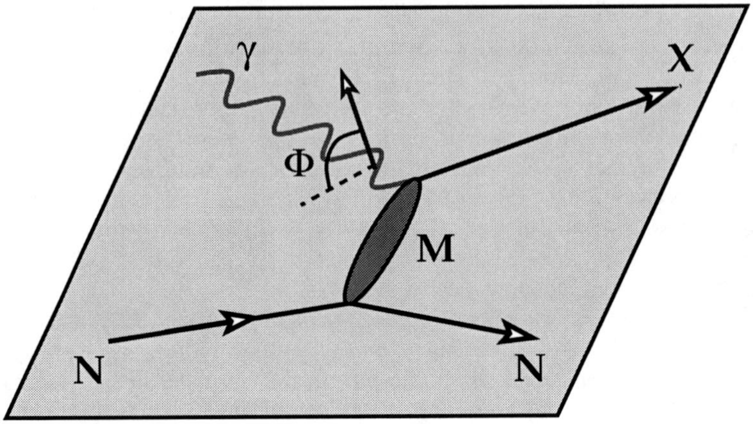

FIGURE 2. Generic meson exchange diagram using a photoproduction reaction with a linearly polarized photon beam showing the angle Φ used to determine the naturality of the meson M in the intermediate state.

be carried out with beams of linearly polarized photons. Photon beams are expected to be particularly favorable for the production of exotic hybrids, as the photon sometimes behaves as a virtual vector meson. When the flux tube in this S=1 system is excited, both ordinary and exotic J^{PC} are possible. In contrast, for an S=0 probe (e.g. pions or kaons), the exotic combinations are not generated. To date, almost all meson spectroscopy experiments in the light quark sector have been done with incident pion, kaon, or proton probes. High flux photon beams of sufficient quality and energy have not been available, so there are virtually no data on the photoproduction of mesons with masses below 3 GeV. Thus up to now, experimenters have not been able to search for exotic hybrid mesons precisely where they are expected. Theoretical calculations indicate that the photoproduction cross sections of light quark exotic mesons should be comparable to those for conventional meson states in this energy range [25].

As mentioned above, the next generation experiment should employ a linearly polarized photon beam. The diffractive production of a meson X (see Fig. 2) takes place via natural parity exchange ($J^P = O^+, 1^-, 2^+, ...$) in the intermediate state, whereas exotic meson production takes place via unnatural parity exchange ($J^P = O^-, 1^+, 2^-, ...$). Experiments that take place with an unpolarized or a circularly polarized photon beam cannot distinguish between the naturality of the exchanged meson M. However with a longitudinally polarized beam, one can distinguish the naturality of the intermediate state by selection based on the angle Φ the polarization vector makes with the hadronic production plane, where the angle is related to the naturality of the exchanged meson. This capability will be essential in helping to isolate the exotic waves in the data analysis.

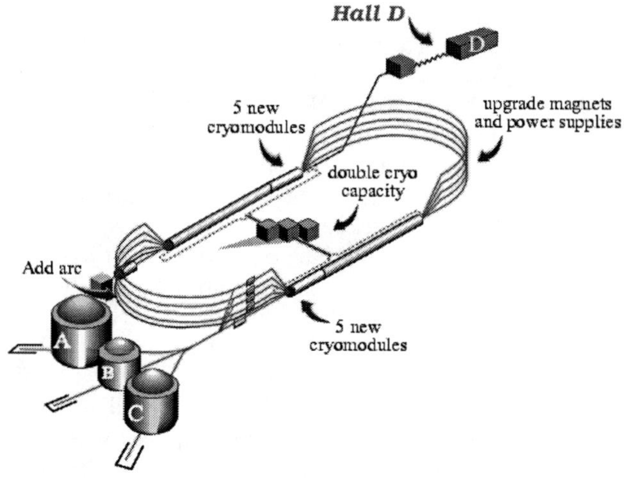

FIGURE 3. The configuration for the proposed 12 GeV energy upgrade of the CEBAF accelerator at Jefferson Laboratory. Also shown is the location of the new experimental Hall D for the GlueX experiment.

THE JLAB UPGRADE & THE DESIGN OF THE GLUEX EXPERIMENT

Jefferson Laboratory represents the world's most powerful electron microscope. At its current operating point, the accelerator can deliver electron beams of up to nearly 6 GeV at currents of $\sim$200 μA and linear polarization up to $\sim$80%. The planned upgrade of this facility is designed to increase the maximum electron beam energy to 12 GeV, to construct a new experimental Hall, called Hall D, for the GlueX experiment, and to lead to equipment enhancements in the three existing Halls A, B, and C. These details are shown in Fig. 3. In April of 2004 the laboratory was awarded CD-0 approval from the U.S. Department of Energy in a first step to approve the upgrade plans. In July of 2005 the laboratory completed the next level of review that will lead to the awarding of CD-1 before the end of 2005. The current schedule has the GlueX experiment on track to begin its physics program in 2011-2012.

Fig. 4 shows a schematic layout of the planned hermetic detector to be installed in Hall D to study the photoproduction of mesons [26]. The photon beam is incident on a liquid-hydrogen target that is located within a large 2 T superconducting solenoidal magnet. The target is surrounded by drift chambers for charged particle tracking. The tracking system includes a straw tube chamber for tracking at central and backward angles and a series of cathode strip chambers for tracking at forward angles. These systems are designed to reconstruct 3–D track points with a resolution of better than 200 μm to enable accurate linking of the track hits along the helical paths followed in the magnetic field. Both of these packages will also provide for particle identification via dE/dX measurements. Surrounding the tracking detectors is a barrel calorimeter for particle identification and timing. Filling the upstream end of the solenoid will be another calorimeter. Downstream of the magnet is a large time-of-flight array, a Cherenkov detector for particle identification (presently planned to be a high pressure gas detector or a DIRC design), and a large lead-glass array that forms an electromagnetic calorimeter. The solenoidal geometry is ideally suited for a high-flux photon beam. The electromagnetic charged particle background (electron-positron pairs) from interactions in the target is contained along the beam line by the axial field of the magnet. The GlueX spectrometer has been designed for large and uniform acceptance for both charged and neutral particle final states, while providing the requisite resolution and particle identification capabilities for spectroscopy. The momentum resolution will be roughly $\Delta p/p \sim 1\%$.

The GlueX photon beam will be produced using the coherent bremsstrahlung technique. Here the 12 GeV electron beam will impinge on a thin ($\sim 20\ \mu$m) diamond crystal and will produce a linearly polarized beam of 9 GeV photons after collimation. In this facility at special settings for the orientation of the crystal, the atoms of the crystal can be made to recoil together from the radiating electron leading to an enhanced emission at particular photon energies, which gives rise to linearly polarized photons. At the target the photon will achieve an average degree of linear polarization in the coherent peak of roughly 40%. In addition the photons emitted in a 0.5 GeV-wide window will be tagged using a focal plane spectrometer. The energy resolution of the photon tagger is 0.1%. This detector system, along with the high duty factor of the electron beam from the accelerator make the search for hybrid mesons quite feasible. Current plans call for the initial operation of GlueX with photon fluxes of 10^7 s^{-1}, increasing to 10^8 s^{-1} when the experiment is in its nominal data acquisition mode. GlueX will collect enough data in its first year of operation to exceed existing photoproduction data by several orders of magnitude.

Hand in hand with the design of the experimental equipment, the GlueX Collaboration is performing detailed Monte Carlo studies of the physics, not only to test and improve the design of the spectrometer, but also to develop the software tools essential for the success of the data analysis. The performance of the detector, the beam flux, and the linear polarization of the photon beam determine the level of sensitivity for mapping the hybrid spectrum. The detector acceptance has been designed to be high and uniform in the relevant meson decay angles in the Gottfried-Jackson frame for *both* charged and neutral particle final states. A program of double-blind Monte Carlo studies have been carried out using the GlueX partial wave analysis software. These studies have shown that low-level exotic signals (at the few percent level) can be successfully pulled out from our data, even with relatively low statistics by GlueX standards (i.e. with only several days of running under nominal conditions). Even with extreme distortions in these simulations, leakage effects due to the spectrometer are no more than 1%. To further certify our PWA codes, consistency checks with Monte Carlo and real data will be made among different final states for the same decay mode.

At the current time the GlueX Collaboration, which has been in existence for more than 7 years, consists of more than 100 physicists from more than 25 institutions. The collaboration is growing and is very driven in its quest to construct the detector on the appropriate time line. Presently our groups are very active in refurbishing existing equipment, as well as with prototyping and designing the systems that will ultimately ensure the long term success of the experiment. The superconducting solenoid to be used as the

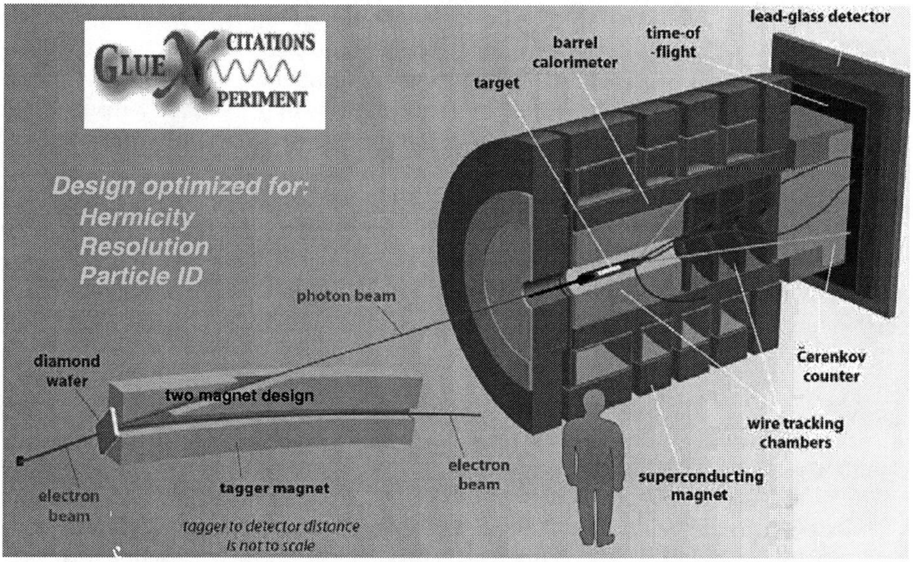

FIGURE 4. Schematic picture of the proposed GlueX detector and coherent bremsstrahlung photon tagging facility. The subsystems within the solenoid are indicated.

basis for the GlueX experiment was originally part of the LASS experiment at SLAC and the MEGA experiment at Los Alamos. Presently the detector is being refurbished at Indiana University. The downstream lead-glass array and time-of-flight wall (which were originally part the BNL E852 experiment) are now being refurbished and reconfigured for GlueX. Research and development programs being lead by collaboration members at Carnegie Mellon University and Ohio University for the straw tube chambers and cathode strip chambers, respectively, are moving toward making decisions for the final tracking system design. The central calorimeter and the upstream calorimeter are nearing the end of their respective design phases at the University of Regina and Florida State University, respectively. Work is ongoing by collaborators at Indiana University and Oak Ridge National Laboratory to perform simulations of various Cherenkov options to allow the collaboration to finalize this technology choice. The collaboration also includes a number of electronics engineers and designers working at Jefferson Laboratory, Indiana University, and the University of Alberta to complete the experiment-specific electronics components (amplifiers, ADCs, TDCs, etc.) that will be necessary to meet to requirements of the GlueX experiment. The design of the tagging system by collaborators from the University of Glasgow, Catholic University, and the University of Connecticut is nearing completion and much research and development of the thin diamond radiators has taken place. A number of subsystem prototypes are presently undergoing beam tests at facilities around the world. All in all there is a lot of ongoing activity within the collaboration.

SUMMARY AND CONCLUSIONS

Understanding confinement requires an understanding of the glue that binds quarks into hadrons. Hybrid mesons are perhaps the most promising laboratory for studying the nature of the glue. Here photoproduction promises to be rich in hybrids, starting with those having exotic quantum numbers where little or no data exist. The planned GlueX experiment that will take place at the energy-upgraded Jefferson Laboratory, will employ photon beams of the necessary flux, duty factor, and polarization, along with an optimized state-of-the-art detector. This experiment will provide for the detailed spectroscopy necessary to map out the hybrid meson spectrum, which is essential for an understanding of the confinement mechanism and the nature of the gluon in QCD.

ACKNOWLEDGMENTS

The author thanks the conference organizers for the opportunity to discuss the GlueX experiment at HADRON05 on behalf of the GlueX Collaboration. He also gratefully acknowledges his fellow GlueX collaborators for their support and many efforts on the experiment to this point in time. This research is supported by the National Science Foundation and the U.S. Department of Energy.

REFERENCES

1. GlueX Collaboration homepage: `http://dustbunny.physics.indiana.edu/HallD/`.
2. G.S. Bali, K. Schilling, and C. Schlichter, Phys. Rev. D **51**, 5165 (1995).
3. C.J. Morningstar and M. Peardon, Phys. Rev. D **60**, 034509 (1999).
4. N. Isgur and J. Paton, Phys. Rev. D **31**, 2910 (1985).
5. P. Lacock *et al.*, Phys. Lett. B **401**, 309 (1997).
6. C. Bernard *et al.*, Phys. Rev. D **56**, 7039 (1997).
7. C. Bernard *et al.*, Nucl. Phys. B (Proc. Suppl.) **73**, 264 (1999).
8. P. Lacock and K. Schilling, Nucl. Phys. B (Proc. Suppl.) **73**, 261 (1999).
9. Z.H. Mei and X.Q. Luo, Int. J. Mod. Phys. A **18**, 5713 (2003).
10. C. Bernard *et al.*, Phys. Rev. D **68**, 074505 (2003).
11. G.M. Beladidze *et al.* (*VES Collaboration*), Phys. Lett. B **313**, 276 (1993).
12. D.R. Thompson *et al. (E852 Collaboration)*, Phys. Rev. Lett. **79**, 1630 (1997).
13. A. Abele *et al. (Crystal Barrel Collaboration)*, Phys. Lett. B **423**, 175 (1998).
14. Review of Particle Physics; S. Eidelman *et al.*, Phys. Lett. B **592**, 1 (2004).
15. G.S. Adams *et al. (E852 Collaboration)*, Phys. Rev. Lett. **81**, 5760 (1998).
16. E.I. Ivanov *et al. (E852 Collaboration)*, Phys. Rev. Lett. **86**, 3977 (2001).
17. J. Kuhn *et al. (E852 Collaboration)*, Phys. Lett. B **595**, 109 (2004).
18. M. Lu *et al.* (*E852 Collaboration)*, Phys. Rev. Lett. **94**, 032002 (2005).
19. Y. Khokhlov *et al.* (*VES Collaboration)*, Nucl. Phys. A **663**, 596 (2000).
20. A.R. Dzierba, Phys. Rev. D **67**, 094015 (2003).
21. A.P. Szczepaniak *et al.*, Phys. Rev. Lett. **91**, 092002 (2003).
22. A.R. Dzierba *et al.*, Preprint hep-ex/0502022, (2005).
23. M. Lu *et al. (E852 Collaboration)*, Phys. Rev. Lett. **94**, 032002 (2005).
24. J. Kuhn *et al. (E852 Collaboration)*, Phys. Lett. B **595**, 109 (2004).
25. A.P. Szczepaniak and M. Swat, Phys. Lett. B **516**, 72 (2001).
26. Hall D Conceptual Design Report Version 4, `http://www.halld.org/`.

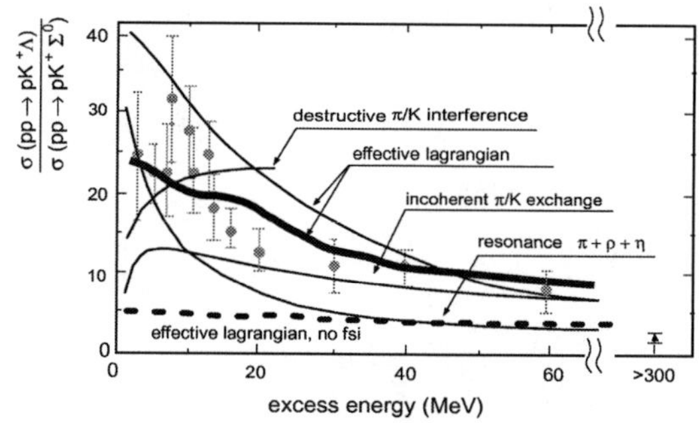

FIGURE 2. The ratio of associated strangeness production as function of the excess energy.

excitation function at zero degree for both reactions. In TOF the complete Dalitz plot is filled. A *fsi* can be seen as enhancement forming a line with constant mass of the $p\Lambda$ system. N^* excitation is seen on the other hand as an enhancement as a line with constant mass of the $K\Lambda$ system. An enhancement with constant mass of the pK^0 system was interpreted by the TOF collaboration as strange pentaquark Θ^+ [9]. The group has in the meantime repeated the experiment in a longer beam time and with an improved efficiency of their detector.

2.3. A precision measurement of the η mass

Compared to other light mesons, the mass of the η is surprisingly poorly known. Though the Particle Data Group (PDG) quotes a value of $m_\eta = 547.75 \pm 0.12$ MeV/c^2 in their 2004 review [10], this error hides differences of up to 0.7 MeV/c^2 between the results of some of the modern counter experiments quoted. The PDG average is in fact dominated by the result of the CERN NA48 experiment, $m_\eta = 547.843 \pm 0.051$ MeV/c^2, which is based upon the study of the kinematics of the six photons from the $3\pi^0$ decay of 110 GeV η-mesons [11]. In the other experiments employing electronic detectors, which typically suggest a mass ≈ 0.5 MeV/c^2 lighter, the η was produced much closer to threshold and its mass primarily determined through a missing-mass technique where, unlike the NA48 experiment, precise knowledge of the beam momentum plays an essential part. GEM performed a high precision determination of the η meson mass. The idea of the experiment is as follows. Three reactions were measured at the same time at a beam momentum, where products of all three reactions are detected simultaneously in the acceptance of the Big Karl spectrometer. The reactions are

$$p+d \to {}^3H + \pi^+ \quad (1)$$
$$p+d \to \pi^+ + {}^3H \quad (2)$$
$$p+d \to {}^3He + \eta. \quad (3)$$

The experiment simultaneously detected forward emitted pions and backward emitted tritons in the c.m. system together with backward emitted ${}^3He$ ions (doubly charged) as

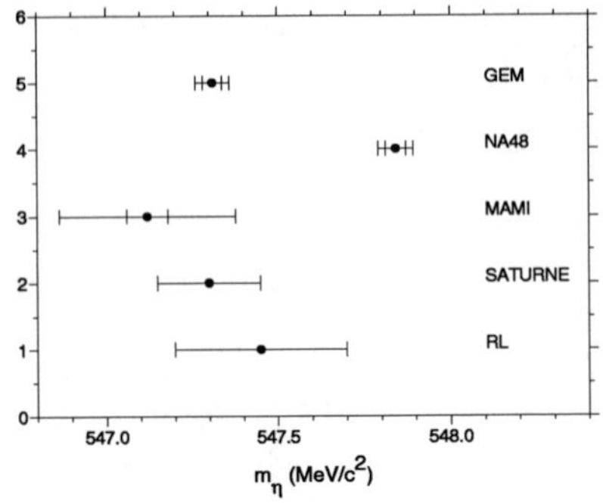

FIGURE 3. The results of the η-mass measurements, in order of publication date, taken from the Rutherford Laboratory (RL) [13], SATURNE [14], MAMI [15], NA48 [11], and GEM [12]. When two error bars are shown, the smaller is statistical and the larger total.

at a proton beam momentum around 1640 MeV/c. Details of the experiment are found in Ref. [12]. The final result of this measurement is

$$m_\eta = 547.311 \pm 0.028\,(\text{stat.}) \pm 0.032\,(\text{syst.})\ \text{MeV/c}^2\,. \tag{4}$$

Our value of the mass of the η meson is compared in Fig. 3 with the results of all other measurements reported in the current PDG compilation [10]. Though significantly smaller than that reported in the NA48 experiment [11], it is in excellent agreement with the other results.

2.4. η and η' production in proton-proton collisions

A comparison of neutral pseudoscalar meson production in pp collisions should shed light on the reaction mechanism and the interactions among the reaction partners. The η and η' mesons are the isospin zero partners of the π^0 which has isospin one. The latter has a very weak interaction with nucleons with respect to the nucleon-nucleon interaction. Furthermore, the influence of an intermediate nucleon resonance, the $\Delta(1332)$, was found only in the Pp partial wave [16]. In contrast to this case, the η-nucleon and even more the η-nucleus interaction is not that small. The $N^*(1535)$ has a strong coupling to the η-nucleon channel. On the other hand no resonance is known to couple to the η'-nucleon channel. The meson-nucleon interaction in all three cases can only be studied in *fsi* because of the short lifetime of the mesons. In Fig. 4 the total cross section for the reactions as a function of the center-of-mass excess energy Q are shown. The sources of data are given in Ref. [17]. The dashed lines indicate a phase-space integral normalized arbitrarily. The solid lines show the phase-space distribution including the $^1S_0$ proton-proton strong and Coulomb interactions. In the case of the $pp \to pp\eta$ reaction, the solid line was fitted to the data in the excess energy range between 15 and 40 MeV. Additional inclusion of the proton- η *fsi* is indicated by the dotted line. The scattering length and the effective range parameter have been

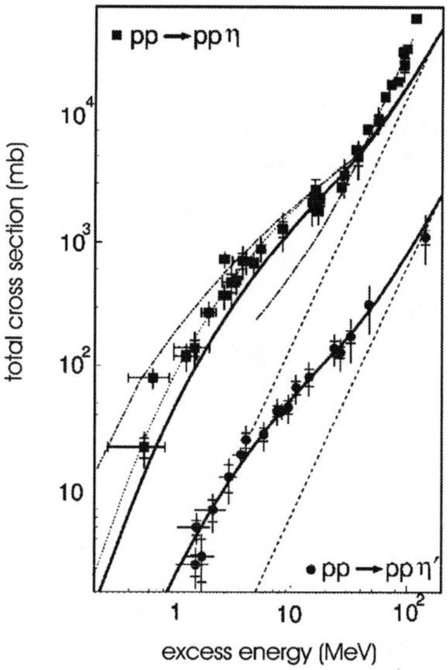

FIGURE 4. Excitation functions of the total cross sections for $pp \to \eta pp$ (circles) and $pp \to \eta' pp$ reactions (squares). The different curves are discussed in the text.

arbitrarily chosen. The dash-dotted line represents the energy dependence taking into account the contribution from the ${}^3P_0 \to {}^1S_0s$, ${}^1S_0 \to {}^3P_0s$, and ${}^1D_2 \to {}^3P_2s$ transitions [18]. Preliminary results for the ${}^3P_0 \to {}^1S_0s$ transition with the full treatment of the three-body effects are shown as a dashed-double-dotted line [19]. The absolute scale of dashed-double-dotted line was fitted with an arbitrary strength in order to demonstrate the energy dependence. While the η' production is nicely reproduced by phase space plus *pp-fsi* this is not the case for η production. Most importantly the Dalitz plot can not be explained by $p\eta$- and *pp-fsi* . The necessity for a rigorous three body calculation was found [17].

2.5. Proton-neutron final state interaction

There has been an extensive search for spin-singlet contribution in *pn-fsi* . Favorite reactions were $dp \to p\{pn\}$ and $dp \to \pi^+\{pn\}$. In both cases the pole (i. e. the deuteron) can also be measured. There is a theorem due to Fäldt and Wilkin which connects the pole to the continuum [20]. Thus the absolute height of the spin-triplet contribution in the continuum is given. The residual cross section is usually attributed to the spin-singlet fraction, however, the resolution and background conditions of most experiments were not sufficient to unambiguously extract the spin-singlet contribution. The characteristic feature of this contribution is a very narrow peak due to an unbound pole at 60 keV, which under these unfavorable conditions could not be seen directly. Therefore, at

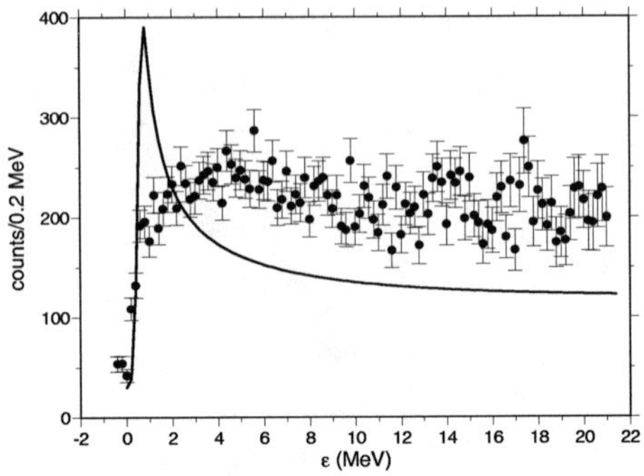

FIGURE 5. Comparison of the measured *pn* excitation energy spectrum on a linear scale with the prediction the singlet cross section folded with the present resolution. The error bars contain a tiny contribution from the uncertainty in the acceptance correction.

COSY and CELSIUS the π^+ and proton were detected in coincidence [21, 22]. While identifying the continuum channel well one loses the relative normalization with the $\pi^+ d$ final state. Therefore, one had to rely on Monte Carlo simulations [21].

At GEM they measured the pions from *pp* interactions with extremely high resolution due to a 2 mm thin liquid hydrogen target and a stochastically extracted beam which was electron cooled at injection energy [23]. A missing mass resolution of $\sigma = 97$ keV was achieved for the deuteron. In addition an almost halo-free beam resulted in a very small background. The Fäldt and Wilkin *fsi* -theorem yielded only 50% of the yield in the *pn* continuum. In Fig. 5, a fit of the spin-singlet *fsi* to the data is shown. Obviously, that calculated cross section can not account for the data. The reason for the discrepancy is not clear at the moment. One explanation is that this is due to *D*-state effects in the *pn* system [23]. Another possibility might be a failure of the Fäldt and Wilkin theorem which is exact only at the pole position.

3. PHYSICS WITH WASA AT COSY

3.1. Pseudoscalar meson mixing

The *QCD* Hamiltonian can be split into to parts

$$H_{QCD} = H_0 + H_m \tag{5}$$

with H_0 the Hamiltonian for massless quarks. In flavor SU(3) the term containing the mass is given as

$$H_m = \int dx^3 \left(m_u u\bar{u} + m_d d\bar{d} + m_s s\bar{s} \right). \tag{6}$$

The latter term breaks chiral symmetry. The neutral pseudoscalar mesons $\tilde{m}$ in ideal mixing are connected to the physical mesons m via

$$\begin{pmatrix} \tilde{\pi} \\ \tilde{\eta} \\ \tilde{\eta}' \end{pmatrix} = \begin{pmatrix} \frac{1}{\sqrt{2}}\left(u\bar{u}-d\bar{d}\right) \\ \frac{1}{\sqrt{6}}\left(u\bar{u}+d\bar{d}-2s\bar{s}\right) \\ \frac{1}{\sqrt{3}}\left(u\bar{u}+d\bar{d}+s\bar{s}\right) \end{pmatrix} = A \begin{pmatrix} \pi \\ \eta \\ \eta' \end{pmatrix}. \quad (7)$$

The matrix A consists mainly of the $\pi^0-\eta$ mixing angle and the $\eta-\eta'$ mixing angle. The former is given by

$$\sin\theta_{\pi\eta} \equiv \frac{\sqrt{3}}{4}\frac{m_d-m_u}{m_s-\hat{m}} \quad (8)$$

with $\hat{m}=(m_d+m_u)/2$. A measurement of the $\pi^0-\eta$ mixing angle would therefore provide information about the up quark and down quark mass difference.

A first attempt to measure this mixing angle was put forward by Magiera and Machner [24]. They proposed to measure the ratio of backward emitted pions from the two reactions $pd \to {}^3He\pi^0$ and $pd \to {}^3H\pi^+$ in the vicinity the η threshold (below and above) in the reaction $pd \to {}^3He\eta$. They argued that in the case of π^0 production this channel could be enhanced below threshold due to $\pi^0-\eta$ mixing. Indeed an effect was found in an experiment leading to $\theta_{\pi\eta}=0.006\pm0.005$ [25]. However, Baru et al. [26] claimed that a possible effect is not solely due to the mixing but also $\eta-{}^3He$ should contribute. The problem might be solved more cleanly by not measuring meson production but studying meson decay instead. This can and will be be done with the WASA detector at COSY.

3.2. The WASA detector

The WASA detector operated until summer of this year at the CELSIUS facility. It is presently dissembled and has been shipped to Jülich where it will be installed at COSY in late fall and spring of next year. Fig. 6 shows a cross section through WASA. WASA consists of a forward part (right) for measurements of charged target-recoil particles and scattered projectiles and a central part (left) designed for measurements of the meson decay products. The forward part consists of eleven planes of plastic scintillators and of proportional counter drift tubes. The central part consists of an electromagnetic calorimeter of CsI(Na) crystals surrounding a superconductive solenoid. Inside of the solenoid a cylindrical chamber of drift tubes and a barrel of plastic scintillators are placed. The arrows indicate a typical reaction $pp \to pp\eta'$ with a subsequent decay of the η' into an η and two charged pions. Finally, the η decays into two γ's.The two protons will be measured in the forward detector, the charged pions in the volume of the magnetic field and the γ's in the CsI crystals.

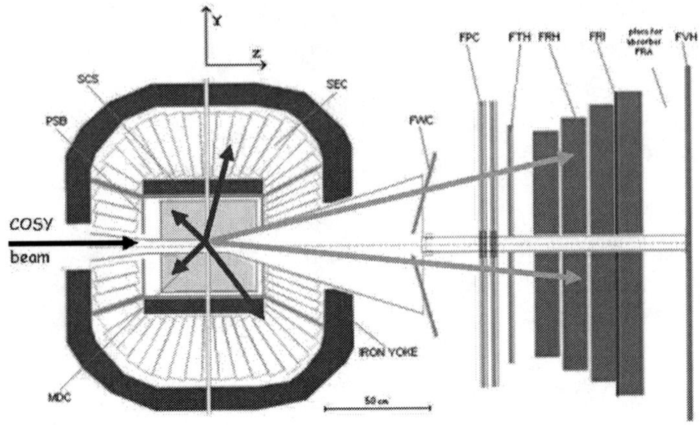

FIGURE 6. Cross section to the WASA detector. The different components are discussed in the text.

3.3. η and η' decays

One possibility of studying the meson mixing angles are the decays of η and η' mesons. We will first concentrate on the isospin forbidden decays of the η' into three pions. Instead of measuring the decays alone it was proposed by Gross, Treiman and Wilczek [27] to measure the ratios of the forbidden decays to the allowed decays into η and two pions:

$$\begin{aligned} R_{ch} &= \frac{\Gamma(\eta' \to \pi^0\pi^+\pi^-)}{\Gamma(\eta' \to \eta\pi^+\pi^-)} = PS_{ch}\sin^2\theta_{\pi\eta} \\ R_{neut} &= \frac{\Gamma(\eta' \to \pi^0\pi^0\pi^0)}{\Gamma(\eta' \to \eta\pi^0\pi^0)} = PS_{neut}\sin^2\theta_{\pi\eta}. \end{aligned} \tag{9}$$

PS denotes the ratio of the three body phase. From the branchings given by the PDG [10] one estimates for the charged pions $R_{ch} < 0.11$ while Gross et al. give an estimate of 1.49×10^{-3}. The upper limit means that no events for this isospin forbidden decay have been observed so far. For the neutral channel the numbers are $(7.4 \pm 1.2) \times 10^{-3}$ while the theory predicts 1.37×10^{-3}. The experimental number is based on two experiments with a count rate of ≈ 150 counts in total. One can expect that with the high luminosity anticipated at COSY more than an order of magnitude more events will be recorded in a couple of weeks.

The same information can in principle also be gained from the study of the decay $\eta \to \pi^+\pi^-\pi^0$. However, here Coulomb corrections and theory input, both with some uncertainties [28, 29], are necessary.

Another approach is the study of the Dalitz plot of the decay $\eta \to 3\pi$. Here we will concentrate on the decay into three neutral pions. The amplitude can be written as

$$\left|A(\eta \to 3\pi^0)\right|^2 = 1 + 2\alpha z \tag{10}$$

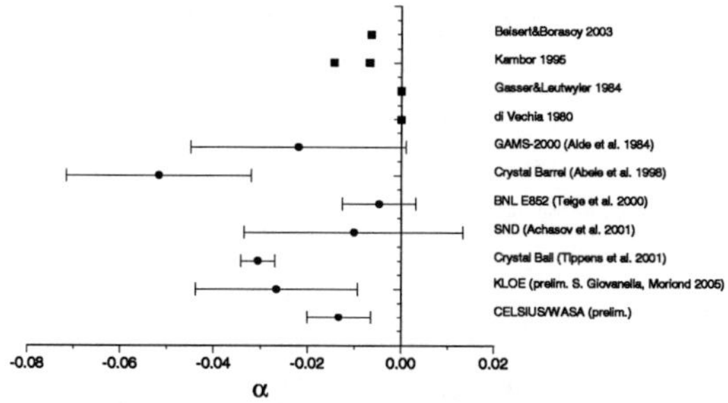

FIGURE 7. The slope parameter α from various measurements and calculations. The references are given in [31] except for the preliminary results.

with $z = \rho^2/\rho_{max}^2$ the relative radial distance. The distance to the center of the Dalitz plot is ρ. A recent measurement of the angle at WASA/CELSIUS [30] yielded a preliminary value of $\alpha = -0.027 \pm 0.015$ (stat.) ± 0.010 (syst.). This value is compared in Fig. 7 with previous measurements and model calculations. The result is based on 37 thousand events. A much richer data sample can be expected at WASA/COSY.

In QCD with $N_f = 3$ there exists a non-Abelian anomaly which breaks the chiral symmetry explicitly. In the effective chiral Lagrangian this anomaly is appropriately reproduced by introducing the Wess-Zumino action [32, 33]. The expansion of the Wess-Zumino-Witten Lagrangian is shown in Fig. 8. Some interactions of the neutral

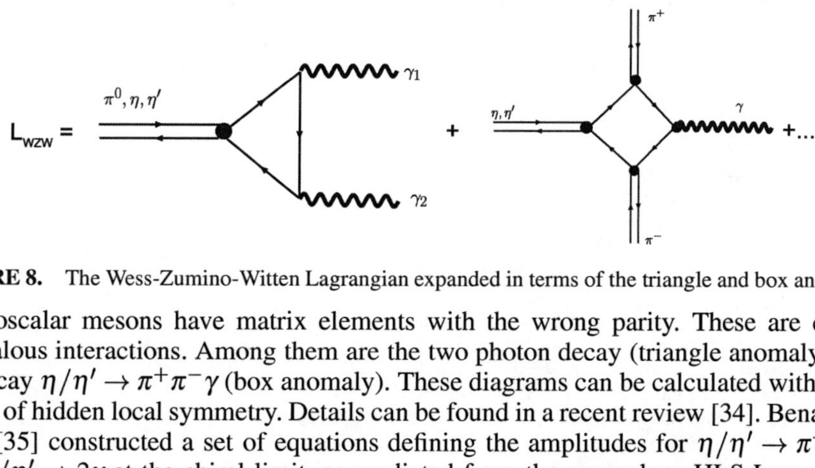

FIGURE 8. The Wess-Zumino-Witten Lagrangian expanded in terms of the triangle and box anomaly.

pseudoscalar mesons have matrix elements with the wrong parity. These are called anomalous interactions. Among them are the two photon decay (triangle anomaly) and the decay $\eta/\eta' \to \pi^+\pi^-\gamma$ (box anomaly). These diagrams can be calculated within the model of hidden local symmetry. Details can be found in a recent review [34]. Benayoun et al. [35] constructed a set of equations defining the amplitudes for $\eta/\eta' \to \pi^+\pi^-\gamma$ and $\eta/\eta' \to 2\gamma$ at the chiral limit, as predicted from the anomalous HLS Lagrangian and appropriately broken. For the decay $\eta' \to \pi^+\pi^-\gamma$ they predict an invariant mass distribution which is the same as the one from $\rho\pi^+\pi^-$ while for $\eta \to \pi^+\pi^-\gamma$ the center of gravity of the distribution is shifted to ≈ 350 MeV/c^2. In these calculations the $\eta - \eta'$ mixing angle enters. The WASA detector at COSY will allow one to measure the corresponding two pion distribution with high statistics.

In the decay studies other rare decays will be measured. Most of them violate $\mathscr{C}$-symmetry. The decay $\eta \to \pi^+\pi^-e^+e^-$ violates $\mathscr{C}\mathscr{P}$-symmetry outside the CKM mechanism. Since this decay is flavor conserving, it is outside the standard model. In a first run at WASA/CELSIUS already 25 candidates were seen compared to 5 events in total in the literature.

ACKNOWLEDGMENTS

The author is grateful to M. Hartmann, B. Hoistadt, A. Koukaz, P. Moskal, E. Roderburg and the members of the GEM collaboration for their discussions concerning their data. He is also thankful to R. Shyam for discussions about his calculations and to J. B. Lieb for a critical reading of the manuscript. The work was supported in part by Deutsche Forschungsgemeinschaft MA 926/3-1.

REFERENCES

1. S. Okubo, *Phys. Rev.*, **D 16**, 2336, G. Zweig, CERN-8182-TH-401, (1964); CERN-8419-TH-412, (1964), I. Izuka, *Prog. Theor. Phys. Suppl.*, **37-38**, 21 (1966).
2. F. Balestra et al., *Phys. Rev.*, **C 63**, 0240004 (2001).
3. S. Abd El-Samad et al., *Phys. Lett.*, **B 522**, 16 (2001).
4. A. Gasparian et al., *Phys. Lett.*, **B 480**, 273 (2000).
5. A. Sibirtsev et al., *Nucl. Phys.*, **A 646**, 427 (1999).
6. R. Shyam et al., *Phys. Rev.*, **C 63**, 022202 (2001).
7. R. Shyam, arXive **hep-ph** (2004) 0406297.
8. A. Sibirtsev et al., *arXive*, **nucl-th**, 0004022 v2 (2000).
9. M. Abdel Bary et al. (COSY-TOF Collaboration), *Phys. Lett.*, **B 595**, 127 (2004).
10. S. Eidelman et al. *Phys. Lett.*, **B 592**, 1 (2004).
11. A. Lai et al., *Phys. Lett.*, **B 533**, 196 (2002).
12. M. Abdel-Bary et al. (GEM Collaboration), *Phys. Lett.*, **B 619**, 281 (2005).
13. A. Duane et al., *Phys. Rev. Lett.*, **32**, 425 (1972).
14. F. Plouin et al., *Phys. Lett.*, **B 276**, 526 (1992).
15. B. Krusche et al., *Z. Physik*, **A 351**, 327 (1995).
16. M. Betigeri et al., *Phys. Rev.* **C 65**, 064001 (2002).
17. P. Moskal et al., *Phys. Rev.*, **C 69**, 025203 (2004).
18. K. Nakayama, J. Haidenbauer, C. Hanhart, J. Speth, *Phys. Rev.*, **C68**, 045201 (2003).
19. A. Fix, H. Ahrenhövel, unpublished.
20. G. Fäldt, C. Wilkin, *Z. Physik*, **A 357**, 241 (1997).
21. A. Betsch et al., *Phys. Lett.*, **B 446**, 179 (1999).
22. V. V. Abaev et al., *Phys. Lett.*, **B 521**, 158 (2001).
23. M. Abdel-Bary et al., *Phys. Lett.*, **B 610**, 31 (2005).
24. A. Magiera, H. Machner, *Nucl. Phys.*, **A 674**, 515 (2000).
25. M. Abdel-Bary et al., *Phys. Rev.*, **C 68**, 021603 R (2003).
26. V. Baru, J. Haidenbauer, C. Hanhart, J. A. Niskanen, *Phys. Rev.*, **C 68**, 35203 (2003).
27. D. J. Gross, S. B. Treiman, F. Wilczek, *Phys. Rev.*, **D 19**, 2188 (1979).
28. H. Leutwylwer, *Phys. Lett.*, **B 378**, 313 (1996).
29. H. Leutwyler, *Phys. Lett.*, **B 374**, 181 (1996).
30. Bo Hoistadt, priv. communication, (2005).
31. W. B. Tippens et al., *Phys. Rev.*, **D 53**, 052001 (2001).
32. J. Wess, B. Zumino, *Phys. Lett.*, **B 37**, 95 (1971).
33. E. Witten, *Nucl. Phys.*, **B 223**, 422 (1983).
34. M. Harada, K. Yamawaki, *Phys. Rep.*, **381**, 1 (2003).
35. M. Benayoun, P. David, L. DelBuono, Ph. Leruste, H.B. O'Connell, *The European Physical Journal*, **C 31**, 525 (2003).

Hadron Physics at BaBar

David Muller
Representing the BaBar Collaboration

Stanford Linear Accelerator Center, Stanford, CA 94309, USA

Abstract. The *BABAR* experiment at SLAC is designed to measure CP violation in the B meson system, however the very high statistics combined with the different e^+ and e^- beam energies, the detector design and the open trigger allow a wide variety of spectroscopic measurements. We are beginning to tap this potential via several production mechanisms. Here we present recent results from initial state radiation, hadronic jets, few body B and D hadron decays, and interactions in the detector material. We also summarize measurements relevant to D_s meson spectroscopy, pentaquarks and charmonium spectroscopy from multiple production mechanisms.

Keywords: Initial State Radiation, Charm, Pentaquarks
PACS: 12.38.Qk, 13.66.Bc, 14.40.Lb

INTRODUCTION

The *BABAR* program at SLAC is designed to measure CP violation in the B meson system by producing coherent $B\overline{B}$ pairs and studying the time dependence of their decays into CP eigenstates. A 9 GeV electron beam and a 3.1 GeV positron beam collide at a center of mass energy of 10.58 GeV, corresponding to the peak of the $\Upsilon(4S)$ resonance. The asymmetric *BABAR* detector covers about 85% of the solid angle in the e^+e^- center of mass (c.m.) frame, and is designed to isolate the low multiplicity B decays that occur in less than 0.001% of events through a combination of excellent charged particle tracking, charged hadron identification and photon detection.

These features are also beneficial for spectroscopy, and the very high luminosity required for CP violation studies yields very large samples of hadrons from hadronic jets and bottom (B) and charmed (D) hadron decays and allows us to study e^+e^- annihilations at lower $\sqrt{s}$ through initial state radiation. Along with high luminosity come high backgrounds, and we are studying interactions of both beam halo and final state particles with the beampipe and detector material. These allow not only a good understanding of the detector itself, but also studies of $e^\pm$ and long-lived hadron interactions with materials such as beryllium, tantalum, silicon, carbon and water.

Here we present an overview of our results in the area of hadron spectroscopy. Details can be found in these proceedings [1] and in the references. The first three sections discuss the three main production mechanisms we have used so far, and present a number of results. We follow with sections summarizing studies of D_s meson spectroscopy, pentaquarks, and charmonium spectroscopy that involve more than one of these mechanisms. We have accumulated over 250 fb^{-1} of data so far and expect at least four times this. These results thus constitute the beginning of an extremely rich program of spectroscopy at *BABAR*, and we look forward to many more results in the future.

CP814 *Hadron Spectroscopy: Eleventh International Conference*
edited by A. Reis, C. Göbel, J. Sá Borges, and J. Magnin

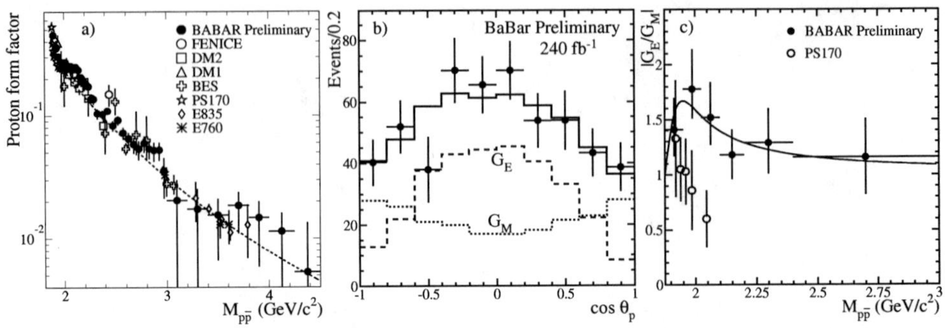

FIGURE 1. a) The effective proton form factor as a function of $\sqrt{s}$ derived from our $e^+e^- \to \gamma_{ISR} p\overline{p}$ data, compared with previous results. b) The proton angular distribution (points) for events in the range 1950–2025 MeV; the solid histogram represents a fit of the sum of electric (dashed) and magnetic (dotted) components. c) The ratio of electric and magnetic form factors as a function of $\sqrt{s}$.

INITIAL STATE RADIATION

An e^+e^- pair can couple directly to any resonance with $J^{(PC)} = 1^{--}$ and if a decay mode of that resonance can be reconstructed then the cross section as a function of $\sqrt{s}$ in the vicinity of the resonance can be used to determine its mass, total width, e^+e^- width and width to that mode. We run 90% of the time at the peak of the $\Upsilon(4S)$ resonance ($\sqrt{s}$ =10.58 GeV), 10% below the $\Upsilon(4S)$ ($\sqrt{s}$ =10.54 GeV) to study non-$B\overline{B}$ backgrounds, and occasionally at other energies. Our total hadronic cross section yields improved values of the $\Upsilon(4S)$ mass, 10579.3±1.2 MeV/c^2, total width, 20.7±3.0 MeV, and e^+e^- width, 0.321±0.034 keV [2].

Annihilations at lower $\sqrt{s}$ can be studied via initial state radiation (ISR) events, in which one of the incoming leptons emits a photon and subsequently annihilates with the other. The cross section for this process is the product of the e^+e^- annihilation cross section at the reduced c.m. energy $\sqrt{s'}$ and a well known radiator function; the latter falls rapidly with increasing photon energy, but the former increases at low $\sqrt{s'}$ and becomes dominated by low multiplicity resonant contributions. By detecting the ISR photon, resonctructing a given final state and requiring overall 4-momentum conservation, we identify clean samples of ISR events and measure cross sections from threshold up to $\sqrt{s'}$ =4.5 GeV. This includes the charmonium region, and we have improved measurements of the J/ψ, and sometimes $\psi(2S)$, branching fraction into each mode studied.

We have previously published [3] measurements of the $\pi^+\pi^-\pi^0$, $\pi^+\pi^-\pi^+\pi^-$, $K^+K^-\pi^+\pi^-$ and $K^+K^-K^+K^-$ final states with better coverage than all previous experiments and comparable or better precision, using 89 fb^{-1} of data. These states can be studied both in terms of cross section and internal structure. The $\pi^+\pi^-\pi^0$ final state is dominated by ω and ϕ resonances, and we have improved the world's knowledge of excited ω states. The $\pi^+\pi^-\pi^+\pi^-$ final state is dominated by the two-body $a_1(1260)\pi$ intermediate state; the $K^+K^-\pi^+\pi^-$ final state shows no significant two-body states, but a rich three-body structure including $K^*(890)K\pi$, $\phi\pi\pi$, ρKK and $K_2^*(1430)K\pi$; the $K^+K^-K^+K^-$ final state shows no substructure, in particular no ϕ signal.

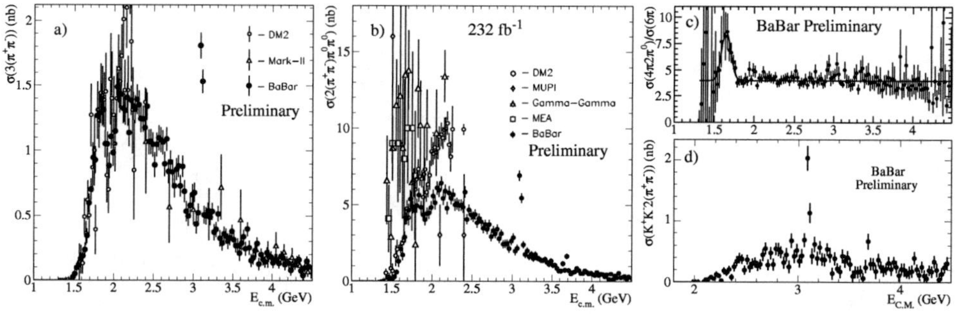

FIGURE 2. The cross section for $e^+e^- \rightarrow$ a) $\pi^+\pi^-\pi^+\pi^-\pi^+\pi^-$ and b) $\pi^+\pi^-\pi^+\pi^-\pi^0\pi^0$ vs. $\sqrt{s}$, compared with previous data. c) The ratio of the cross sections in (a) and (b). d) The cross section for $e^+e^- \rightarrow K^+K^-\pi^+\pi^-\pi^+\pi^-$.

We have a preliminary measurement of the $e^+e^- \longrightarrow p\overline{p}$ cross section; the corresponding effective form factor is shown in Fig. 1a, along with previous data from e^+e^- and $p\overline{p}$ experiments, with which we are consistent. The $\sqrt{s'}$ dependence shows the familiar threshold enhancement, as well as two structures featuring sharp drops at 2.25 and 3 GeV, which illustrate the power of data from a single experiment over a wide range with no point-to-point systematic error. We separate the electric and magnetic form factors using the proton angular distribution, which is shown in Fig. 1b for the range $1950 < \sqrt{s'} < 2025$ MeV; it deviates significantly from uniformity, indicating the electric form factor is substantially larger than the magnetic. The G_E:G_M ratio is shown as a function of $\sqrt{s'}$ in Fig. 1c; the data show a systematic trend toward values above unity at low $\sqrt{s'}$, which is inconsistent with the LEAR data. The line in the figure is a parametrization used to calculate the effective form factor in Fig. 1a.

We also have preliminary studies of the $\pi^+\pi^-\pi^+\pi^-\pi^+\pi^-$, $\pi^+\pi^-\pi^+\pi^-\pi^0\pi^0$ and $K^+K^-\pi^+\pi^-\pi^+\pi^-$ final states. The cross sections for the first two are shown in Figs. 2a,b; large improvements over previous measurements are evident. In the all-charged mode the structure around 1900 MeV seen previously by DM2 and also in diffractive production by FOCUS, is clear and well measured. There is surprisinlgly little substructure in this mode; a simulation with one ρ^0 and four pions distributed according to phase space is adequate to describe all the internal distributions. The $\pi^+\pi^-\pi^+\pi^-\pi^0\pi^0$ mode shows a rather similar cross section, but a much more complex internal structure: we observe signals for ρ^0, $\rho^\pm$, ω and η, and a substantial contribution from the two-body $\omega\eta$ intermediate state, which appears to be resonant. The ratio of the two cross sections is shown in Fig. 2c; the structure near 1650 MeV is due to the $\omega\eta$ contribution, and the ratio is constant with a value near 4.0 over the remainder of the range, which is difficult to explain in light of the very different substructures. More data and modes containing more π^0 are under study.

The $K^+K^-\pi^+\pi^-\pi^+\pi^-$ cross section is measured for the first time and shown in Fig. 2d. There is interesting substructure, with a weak ϕ signal, but a strong $K^*(890)$. More statistics and final states such as $K^+K^-\pi^+\pi^-\pi^0\pi^0$ will be analyzed.

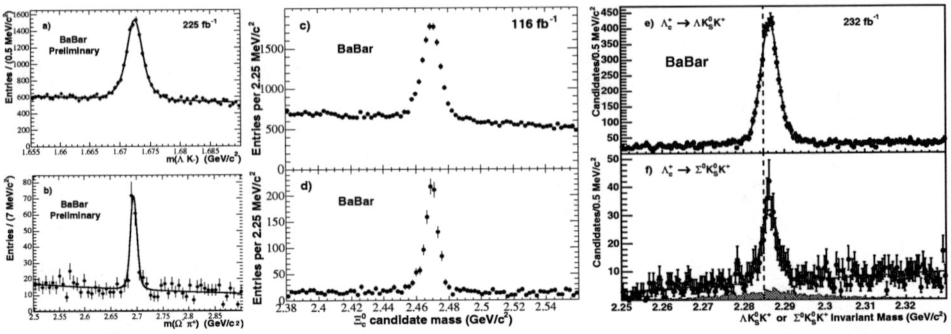

FIGURE 3. Inclusive a) ΛK_S^0 and b) $\Omega^-\pi^+$ invariant mass distributions in the vicinity of the Ω^- and Ω_c^0, respectively. Inclusive d) $\Xi^-\pi^+$ and e) Ω^-K^+ invariant mass distributions in the vicinity of the Ξ_c^0. Inclusive e) $\Lambda K_S^0 K^+$ and e) $\Sigma^0 K_S^0 K^+$ invariant mass distributions in the vicinity of the Λ_c^+.

HADRONIC JETS

Our most prevalent events are of the type $e^+e^- \to q\bar{q}$ at $\sqrt{s}$ =10.58 GeV, in which the quark and antiquark fragment into a number of hadrons, which produce typically ten stable charged tracks and photons. This process is not understood quantitatively, so there is no quantum number information from the production process; however states sufficiently narrow to be seen above background can yield masses, widths and branching fractions, and some quantum numbers can be excluded by observed decay modes.

Our samples of reconstructed hyperons, Λ^0, $\Sigma^{0,\pm}$, $\Xi^{0,-}$ and the Ω^- shown in Fig. 3a, are much larger and cleaner than in pre-B-factory e^+e^- experiments. These signals have numerous applications, such as the study of charmed baryons from $c\bar{c}$ events and B hadron decays. For $\Xi^-\pi^+$ and Ω^-K^+ pairs we show distributions of invariant mass $M(\Xi^-\pi^+)$ and $M(\Omega^-K^+)$ in Figs. 3c and 3d, respectively; both show clear signals for the charmed-strange baryon $\Xi_c(2470)^0$ and we measure the branching ratio $\mathscr{B}(\Xi_c^0 \to \Omega^-K^+)/\mathscr{B}(\Xi_c^0 \to \Xi^-\pi^+) = 0.294 \pm 0.024$ [4]. We also observe the doubly strange, charmed baryon $\Omega_c(2700)^0$ in the $M(\Omega^-\pi^+)$ distribution shown in Fig. 3b. Signals for other decay modes are not yet significant, but we measure [5] (preliminary)

$$\begin{aligned}
\mathscr{B}(\Omega_c^0 \to \Xi^-K^-\pi^+\pi^+)/\mathscr{B}(\Omega_c^0 \to \Omega^-\pi^+) &= 0.31 \pm 0.15 \\
\mathscr{B}(\Omega_c^0 \to \Omega^-\pi^+\pi^-\pi^+)/\mathscr{B}(\Omega_c^0 \to \Omega^-\pi^+) &= 0.16 \pm 0.10.
\end{aligned}$$

We measure the mass of the Λ_c^+ baryon using the $\Lambda K_S^0 K^+$ and $\Sigma^0 K_S^0 K^+$ decay modes. These have low Q values (and therefore low branching fractions), which minimize the systematic error due to uncertainties in the mass scale and resolution of the detector. The invariant mass distributions for these two modes are shown in Figs. 3e and 3f; the former mode contains over 4,000 true Λ_c^+ baryons and the mass measurement is systematics limited; the latter mode has a smaller systematic error, but is statistics limited. The masses measured in the two modes are consistent and the combined result, $M_{\Lambda_c} = 2286.46 \pm 0.14$ MeV$/c^2$ [6], is four times more precise than the current PDG value of 2284.9 ± 0.6 MeV$/c^2$ [7] and higher by 2.5 of the PDG standard deviations.

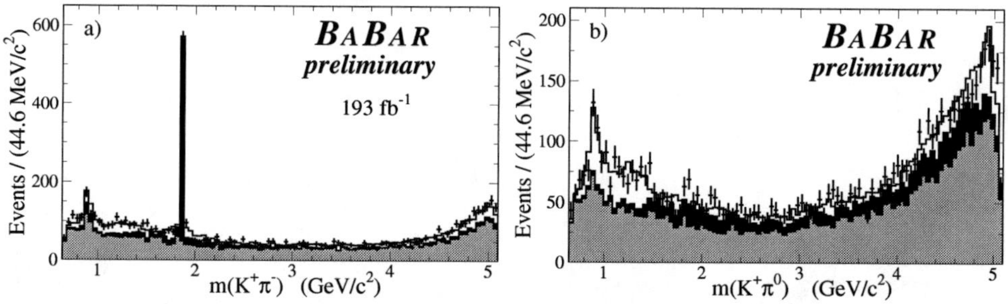

FIGURE 4. The a) $K^+\pi^-$ and b) $K^+\pi^0$ invariant mass distributions in selected $B^0 \to K^+\pi^-\pi^0$ candidates (points with error bars). The gray areas represent the combinatoric background from e^+e^- annihilations into hadrons, and the black areas the contributions from other B meson decays, including the $B^0 \to \overline{D}^0\pi^0$ intermediate state. The histograms represent the result of a fit.

B AND *D* HADRON DECAYS

The weakly decaying B and D hadrons are pseudoscalars, with their (quasi-)two-body decays limited to appropriate quantum numbers and helicity states. These uncommon low-multiplicity decays can be sensitive to CP-violation, and are the goal of our program. We often use "applied spectroscopy", e.g. using angular distributions to reduce backgrounds or project out CP-even and CP-odd components. Recently we improved our measurement [8] of CP violation in $B^0 \to J/\psi K^{*0}$, including a full analysis of the $J/\psi K\pi$ system. We see phase motion around 890 MeV$/c^2$, the direction of which resolves a sign ambiguity in the cosine of the CP-violating phase, and the S-P phase difference agrees with corresponding LASS data from $K^-p \to K^-\pi^+n$ scattering.

Dalitz analyses of three-body modes can supply a wealth of spectroscopic information. We have analyzed several D meson decays with much higher statistics than earlier experiments, and are starting to study decays of much heavier B mesons, both for their own sake and to maximize CP information. We have preliminary results for $B^0 \to K^+\pi^-\pi^0$, $B^+ \to K^+\pi^-\pi^+$ and $B^+ \to \pi^+\pi^-\pi^+$ [9]. The $M(K^+\pi^-)$ and $M(K^+\pi^0)$ projections for the former are shown in Fig. 4. The background from $e^+e^- \to q\overline{q}$ events is large but measurable in off-resonance data. The background from other B meson decays can be more complicated: here we consider $B^0 \to \overline{D}^0\pi^0$ as background, since the $\overline{D}^0$ decays weakly and its mass projections are calculable; the remaining B decay background is suppressed strongly by the selection criteria, but must be modelled.

These data yield qualitative information on the internal structure of this decay. Both $K\pi$ systems are dominated by S-wave components, with the $K_0^*(1430)\pi$ amplitude being 3–4 times larger than that of the $K^*(890)\pi$. In the $\pi^-\pi^0$ system, however, there is a large $K^+\rho^-$ contribution, but no significant S-wave resonance. In the $B^+ \to K^+\pi^-\pi^+$ decay we observe S-wave dominance, with the $K_0^*(1430)^0\pi^+$ amplitude being ~5 times larger than that of the $K^*(890)^0\pi^+$ and the $K^+f_0(980)$ being about twice the $K^+\rho^0(770)$. On the other hand, in the $B^+ \to \pi^+\pi^-\pi^+$ decay we observe large contributions from $\rho(770)^0\pi^+$, $\rho(1450)^0\pi^+$ and $f_2(1270)^0\pi^+$, but no significant S-wave contribution.

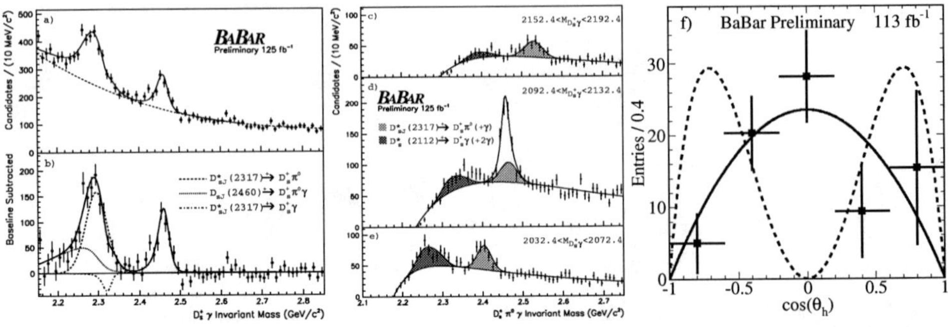

FIGURE 5. The inclusive $D_s(1968)\gamma$ invariant mass distribution a) before and b) after subtraction of combinatoric background. The lines represent the results of a fit to the data with the components indicated. c,d,e) The inclusive $D_s(1968)\pi^0\gamma$ invariant mass distribution in three ranges of the $D_s(1968)\gamma$ submass. The gray areas represent the contributions from the indicated feed-up modes, and the lines indicate results of a fit to the data. f) The helicity angle distribution for $D_{sJ} \rightarrow D_s\gamma$ where the D_{sJ} is produced in a $B^0 \rightarrow \overline{D}D_{sJ}$ decay; the solid (dashed) line represents the expectation for the spin-1 (spin-2) hypothesis.

D_S MESON SPECTROSCOPY

The D_s^+ mesons are bound states of a 'heavy' charmed quark and a 'medium' mass strange antiquark and their spectrum provides an interesting theoretical testing ground. Prior to the B factories, four narrow states were known, with masses of 1968, 2112, 2536 and 2573 MeV$/c^2$, and assigned to the expected narrow states with respective $J^P = 0^-, 1^-, 1^+$ and 2^+, although the spins of the latter three states are unmeasured. The low-mass 0^+ and 1^+ states were expected to be wide and to lie above DK and D^*K thresholds, respectively, so it was surprising when two narrow states, $D_{sJ}(2317)$ and $D_{sJ}(2460)$, were discovered in the isospin violating decay modes $D_s(1963)\pi^0$ and $D_s^*(2112)\pi^0$. These discoveries reinvigorated both experiment and theory in this area, with abundant theoretical speculation that they may not be simple $c\bar{s}$ states, but perhaps 4-quark states, DK 'molecules' or 'hybrid' $c\bar{s}g$ states. It is of great importance to establish their quantum numbers experimentally.

The observation of specific additional states or decay modes can exclude certain hypotheses, and we have done a number of studies [10]. In the $D_s(1968)\pi^{\pm}$ invariant mass spectra we find no evidence for the neutral or doubly charged states that would be expected as isospin partners of any molecular state. In the inclusive $D_s(1968)\pi^-\pi^+$ invariant mass distribution we observe signals at 2460 and 2536 MeV$/c^2$, excluding the 0^+ hypothesis for those states; there is no signal at 2317 MeV$/c^2$.

We have searched for radiative decays to $D_s(1963)\gamma$ and $D_s(1963)\gamma\pi^0$, in which there is considerable cross-feed to be taken into account. In Figs. 5a and 5b we show the inclusive $D_s^+\gamma$ invariant mass distribution before and after subtraction of the combinatorial background, respectively, and in Figs. 5c,d,e we show the inclusive $D_s^+\gamma\pi^0$ invariant mass distribution in three ranges of the $D_s^+\gamma$ submass. There is a signal for $D_{sJ}(2460) \rightarrow D_s^+\gamma$ in Fig. 5a,b, however the structure at lower mass is due to a combination of feeddowns from $D_{sJ}(2460)$ decays and $D_{sJ}(2317) \rightarrow D_s^+\pi^0$ decays where only one of the photons from the π^0 is found. Feed-up from $D_{sJ}(2317) \to D_s^+\pi^0$ and $D_s^*(2112) \to D_s^+\gamma$ with additional photons attached gives rise to the structures indicated in gray in Figs. 5c,d,e that move with the the $D_s^+\gamma$ submass. A signal is seen at 2460 MeV/c^2 only in Fig. 5d, indicating that the $D_{sJ}(2460) \to D_s^+\gamma\pi^0$ decay proceeds entirely through the $D_s^*(2112)\pi^0$ intermediate state. These data confirm the exclusion of the spin-0 hypothesis for the $D_{sJ}(2460)$ and suggest the $D_{sJ}(2317)$ has spin 0.

These studies reveal the limitations of working with production in $e^+e^- \to c\bar{c}$ events. We have also studied D_{sJ} production in exclusve B decays [11], finding evidence for both the 2317 and 2460 states in several $B \to \overline{D}^{(*)}D_{sJ}$ modes. These signals have lower statistics but also much lower background and provide complementary information on masses and branching ratios. In particular we obtain an improved value of $\mathscr{B}(D_{sJ}(2460)^+ \to D_s\gamma)/\mathscr{B}(D_{sJ}(2460)^+ \to D_s^*\pi^0) = 0.27 \pm 0.05$. Since the B and D are both pseudoscalars, the $B \to \overline{D}D_{sJ}$ decays can be used to measure the spin of the D_{sJ} directly. The distribution of the cosine of the angle θ_h between the D_s flight direction and the $\overline{D}$ flight direction in the $D_{sJ}(2460)$ rest frame is shown in Fig 5f, along with the expectations for the spin-1 and spin-2 hypotheses. Statistics are low, but the data show a strong preference for spin-1 over spin-2 (or any higher spin).

PENTAQUARKS

The claims for the exotic baryonic states $\Theta(1540)^+$, $\Xi(1860)^{--}$ and $\Theta_c(3100)^0$ generated strong interest in *BABAR* because of the small widths and the expectation of a large number of additional narrow states. In contrast to fixed target experiments, e^+e^- annihilations produce all members of a given multiplet at about the same rate, giving access to more states. The production rate of such 5-quark states in e^+e^- annihilations is unknown, but ordinary baryons are observed at rates decreasing exponentially with mass, and we observe large samples of baryons with masses above 1540 MeV/c^2, so we should be able to provide useful information as to the nature of these states.

Extensive searches [12] in our inclusive $e^+e^- \to q\bar{q}$ and $\Upsilon(4S)$ data for the $\Theta(1540)^+$, $\Xi(1860)^{--}$ and other states in the antidecuplet generally supposed to contain them yield no signals. We limit their production rates to factors of 5–15 below the corresponding rates for baryons of the same mass, strongly suggesting that these states cannot be ordinary baryons. A preliminary search for the charmed exotic state $\Theta_c(3100)^0$ claimed by the H1 collaboration, using the same pD^{*-} decay mode, is shown in Fig. 6a. There is no evidence for any narrow structure; in particular the region near 3100 MeV/c^2 (inset) is smooth. Few charmed baryon rates have been measured, so we cannot draw the same conclusion as for the non-charmed states; however our D^{*-} sample is two hundred times larger than H1's, so we can exclude that their signal arises from either of the sources to which we are sensitive, hard c-quark fragmentation or B hadron decay.

We search for exotic states in several exclusive B hadron decays, finding no signal. For example, in [13] we study $B^+ \to p\bar{p}K^+$ decays, observing a threshold enhancement in $M(p\bar{p})$, the $\eta_c K^+$, $J/\psi K^+$ and $\psi(2S)K^+$ intermediate states, and evidence for the $\chi_c K^+$ and $p\overline{\Lambda}(1520)$ intermediate states. We see no other structure, such as the excess in $M(pK^+)$ near 1650 MeV/c^2 predicted for an excited Θ multiplet, and we set a stringent

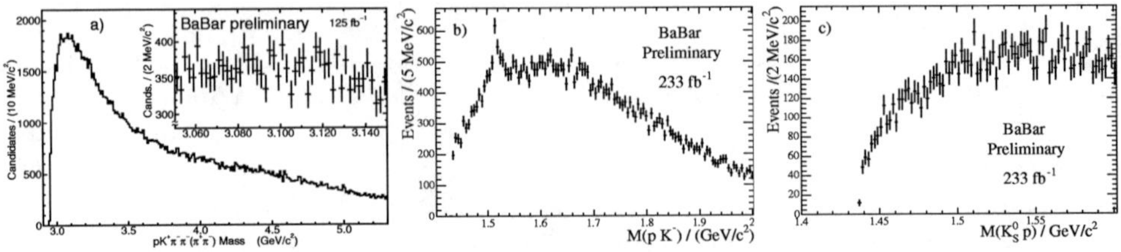

FIGURE 6. a) The inclusive pD^{*-} invariant mass distribution over the full kinematic range and (inset, with suppressed zero) in the vicinity of 3100 MeV/c^2. The b) pK^- and c) pK_S^0 invariant mass distributions for $pK_S^0K^-$ combinations forming a vertex in the beampipe or detector material.

limit on Θ^{*++} production in B decays.

These negative results do not exclude these states' existence, so we search for hadro- and electro-production of $\Theta(1540)^+$ in our beampipe and detector material. Along with high luminosity come high backgrounds from final state particles and off-momentum beam particles interacting in this material. A study of pK_S^0 vertices gives a detailed inner detector map, and we can isolate interactions in Be, Si, Ta, Fe and C. The largest contribution of over 300,000 vertices from $\sim$9 GeV electrons hitting the Be of the beampipe constitutes an electroproduction experiment rather similar to several of the experiments claiming a signal. An observation here, combined with the non-observation in e^+e^- annihilations by the same experiment would be decisive.

We see no signal for the $\Theta(1540)^+$ in the $M(pK_S^0)$ spectrum for any type of interaction. We isolate more exclusive reactions by rejecting events in which another track identified as a proton, deuteron or triton can be attached to the pK_S^0 vertex. In vertices with an associated $\pi^\pm$, we observe $K^*(890)^+ \to K_S^0\pi^+$, $K^*(890)^- \to K_S^0\pi^-$ and $\Lambda \to p\pi^-$, but no structure in the $M(pK_S^0)$ distribution. We consider vertices with an associated K^-; if no other strange particle is produced in such an event, then the K_S^0 must have been a K^0 rather than a $\overline{K}^0$ and the pK_S^0 combination is exotic. The $M(pK^-)$ distribution in such events is shown in Fig. 6b; a signal for $\Lambda(1520)$ is evident. The $M(pK_S^0)$ distribution in the same events, shown in Fig. 6c, shows no signal for $\Theta(1540)^+$.

CHARMONIUM SPECTROSCOPY

The field of charmonium spectroscopy has also been revitalized with the observations of the $\eta_c(2S)$ and h_c states below $D\overline{D}$ threshold, the discovery of the $X(3872)$ state well above $D\overline{D}$ threshold but decaying to $J/\psi\pi^+\pi^-$, and the discovery of states at 3940 MeV/c^2 in three different production processes decaying into non-$D\overline{D}$ modes. Theoretical models predict numerous wide charmonium states above $D\overline{D}$ threshold, but the only experimental data are from the total hadronic cross section in e^+e^-, which show broad structures and the need for at least three states. Again there is speculation regarding molecules, hybrids, etc. In order to improve the experimental situation, further searches and systematic studies of the spectrum in different production processes are needed.

We set a stringent limit on $X(3872)$ production via ISR [3], strongly disfavoring the

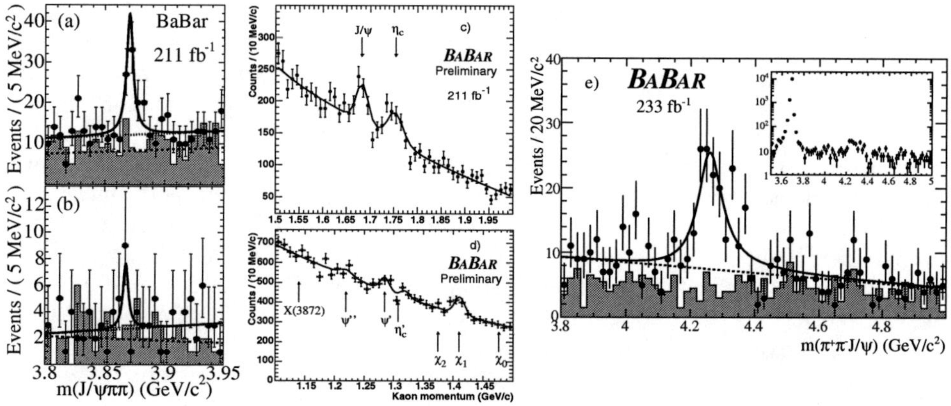

FIGURE 7. Invariant mass distributions for the $J/\psi\pi^+\pi^-$ system in candidate a) $B^+ \to J\psi\pi^+\pi^- K^+$ and b) $B^0 \to J\psi\pi^+\pi^- K_S^0$ decays, in the vicinity of 3870 MeV/c^2. The lines represent the results of a fit. The K^+ momentum spectrum in the rest frame of the system recoiling against a reconstructed B^- candidate, in the ranges corresponding to c) $B^+ \to J/\psi K^+$ and $B^+ \to \eta_c K^+$ decays and d) B^+ decays to a K^+ and a higher charmonium resonance. e) The $J/\psi\pi^+\pi^-$ invariant mass distribution for ISR events in the vicinity of 4300 MeV/c^2 and (inset) on a log scale over a wider range that includes the $\psi(2S)$.

1^{--} hypothesis. We do observe it [14] in the decay $B^+ \to X(3872)K^+ \to J/\psi\pi^+\pi^-K^+$, as shown in Fig. 7a. The analogous neutral decay $B^0 \to X(3872)K_S^0 \to J/\psi\pi^+\pi^-\pi^+\pi^-$ has the potential to discriminate between a number of models: if the $X(3872)$ is a conventional $c\bar{c}$ state, the two decays should have the same rate, although lower experimental acceptance reduces the expected number of the latter to about ten; molecular models predict a suppression of the latter decay; diquark-antidiquark models predict equal rates, but with the observed X mass shifted downward by about 8 MeV/c^2 in the latter case. Our $M(J/\psi\pi^+\pi^-)$ distribution [14] in these decays is shown in Fig. 7b. Current data are consistent with all three of these hypotheses; the best fit, indicated by the line, corresponds to a signal of about seven events with a mass shift of about 3 MeV/c^2.

In a new study of charmonium from B decays, we search for all $B^+ \to K^+X_{c\bar{c}}$ decays by reconstructing the accompanying B^- in the event and measuring the K^+ momentum spectrum in the rest frame of the B^+. We thus measure the absolute branching fractions of the B decays, independent of the $X_{c\bar{c}}$ decay mode, which can be used along with existing data to improve our knowledge of a number of other branching fractions and decay widths. The kaon momentum spectrum is shown in Figs. 7c,d after a selection that enhances quasi-two-body decays and has been optimized separately in the two mass regions shown. There are signals for $X_{c\bar{c}} = J/\psi, \eta_c, \chi_{c1}$, and $\psi(2S)$, as well as enhancements corresponding to $\eta_c(2S)$ and $\psi(3770)$. The observed suppression of χ_{c0} and χ_{c2} is predicted by factorization models. We see no evidence for h_c production, although the uncertainty is large due to its proximity to the χ_{c1}. We see no signal for the $X(3872)$; since we have observed the corresponding exclusive decay, we set a *lower* limit of $\mathscr{B}(X(3872) \to J/\psi\pi^+\pi^-) > 4\%$ at the 90% C.L. No additional states are observed; the region above the $X(3872)$ corresponds to K^+ momenta below 1.1 GeV/c where the background increases rapidly, and is under study.

We verify [15] the Belle result that the exclusive process $e^+e^- \to X_{c\bar{c}}X'_{c\bar{c}}$, where $X_{c\bar{c}}$ and $X'_{c\bar{c}}$ are two charmonium states, has a much higher rate than expected when $X_{c\bar{c}} = J/\psi$ or $\psi(2S)$ and $X'_{c\bar{c}} = \eta_c$, χ_{c0} or $\eta_c(2S)$. In addition we study the exclusive process $e^+e^- \to J/\psi\pi^+\pi^-$ using ISR [16]; here we do not require a detected ISR photon since leptonic J/ψ decays plus missing mass requirements produce a clean signal, but we verify the expected rate and angular distribution of associated photons. The raw number of selected events is shown vs. $M(J/\psi\pi^+\pi^-)$ in Fig. 7e on both linear and (inset) logarithmic vertical scales; in the latter the signal for $\psi(2S)$, a vital control mode, is prominent; in the former there is an excess of events near 4260 MeV/c^2, which we designate the $Y(4260)$. With the current statistics we are unable to determine if this is a single resonance or a more complex structure. If one resonance, its width is $\sim$90 MeV; such a state is expected to decay strongly to $D\overline{D}$ and contribute to the total hadronic cross section, whereas the measured cross section is low in this mass region.

SUMMARY

The excellent detector performance and high statistics in many different production processes makes *BABAR* an exciting place to do spectroscopy. We have already produced results in a number of areas, including e^+e^- annihilations at $\sqrt{s}$ from $\pi^+\pi^-$ threshold up to 4.5 GeV using ISR, charmed hadrons produced in $e^+e^- \to c\bar{c}$ events and B hadron decays, Dalitz analyses of three-body B and D hadron decays, and analyses of interactions in the beampipe and detector material.

With CLEO and Belle, we have revitalized fields of $c\bar{s}$ spectroscopy, with the discovery of the $D_{sJ}(2317)$ and $D_{sJ}(2460)$ states, and charmonium spectroscopy, with the observation of several new states, most recently our $Y(4260)$, and the development of inclusive methods such as $B \to KX_{c\bar{c}}$, $e^+e^- \to X_{c\bar{c}}X'_{c\bar{c}}$, and ISR in specific final states such as $J/\psi\pi^+\pi^-$. We played a strong role in the pentaquark controversy, and look forward to a long and fruitful program in these and more exciting areas.

REFERENCES

1. N. Berger, T. Berger-Hry'nova, V. Poireau, M. Negrini and V. Ziegler, these proceedings.
2. B. Aubert, et al., hep-ex/0405025, submitted to Phys. Rev. Lett.
3. B. Aubert, et al., Phys. Rev. **D70**, 072004 (2004), Phys. Rev. **D71**, 052001 (2004).
4. B. Aubert, et al., Phys. Rev. Lett. **95**, 142003 (2005).
5. B. Aubert, et al., hep-ex/0507011.
6. B. Aubert, et al., hep-ex/0507009, submitted to Phys. Rev. **D**.
7. Particle Data Group, S. Eidelman, et al., Phys. Lett. **B 592**, 1 (2004).
8. B. Aubert, et al., Phys. Rev. **D71**, 032005 (2004).
9. B. Aubert, et al., hep-ex/0408088; hep-ex/0507004 and hep-ex/0507025, submitted to Phys. Rev. **D**.
10. B. Aubert, et al., hep-ex/0408067.
11. B. Aubert, et al., Phys. Rev. Lett. **93**, 181801 (2004).
12. B. Aubert, et al., Phys. Rev. Lett. **95**, 042002 (2005); hep-ex/0408064.
13. B. Aubert, et al., Phys. Rev. **D72**, 051101(R) (2005).
14. B. Aubert, et al., hep-ex/0507090, submitted to Physical Review Letters.
15. B. Aubert, et al., Phys. Rev. **D72**, 031101 (2005).
16. B. Aubert, et al., Phys. Rev. Lett. **95**, 142001 (2005).

Review of Heavy Hadron Spectroscopy

Eric Swanson

Department of Physics, University of Pittsburgh, Pittsburgh PA 15269, USA.

Abstract. The status of several of the newly discovered heavy mesons is discussed.

Keywords: hadrons, spectroscopy
PACS: 14.40.Gx, 14.40.Lb

INTRODUCTION

More than 30 years after the November Revolution, charmonium spectroscopy continues to surprise and challenge. A new era began in April of 2003 when BaBar announced the discovery of the enigmatic $D_s(2317)$. This state continues to perplex us and Fermilab, CLEO, BES, BaBar, and Belle continue to add grist to the mill. This conference report is a brief review of the new heavy hadron spectroscopy.

THE NEW STATES

B_c

CDF recently announced the discovery of the B_c meson. The timing was arranged with the FNAL lattice group to permit them a prediction of its mass. The respective results are 6287(5)(1) MeV[1] and 6304(4) MeV[2]. Thus CDF may claim a state, the lattice may claim a victory, and the B_c appears to carry no surprises. Historians may be interested to know, however, that Godfrey and Isgur[3] predicted this mass with the same accuracy as the lattice, but with a twenty year lead time.

h_c

The h_c has been observed by CLEO[4] with a mass of 3524(1) MeV. This may be compared with typical quark model expectations[5] of 3518 MeV, leading one to suspect that the h_c is also not hiding anything from us. Nevertheless, it tells us something: the h_c lies within 2 MeV of the spin-averaged mass of the χ_{cJ} multiplet: $M_{c.o.g.} = \frac{1}{9}(\chi_0 + 5\chi_1 + 9\chi_2) = 3525.36$. Since the splittings between these states are driven by the $\mathcal{O}(v/c)$ quark interactions, we are learning something about the Dirac structure of confinement (namely that it is an effective scalar).

CP814 *Hadron Spectroscopy: Eleventh International Conference*
edited by A. Reis, C. Göbel, J. Sá Borges, and J. Magnin

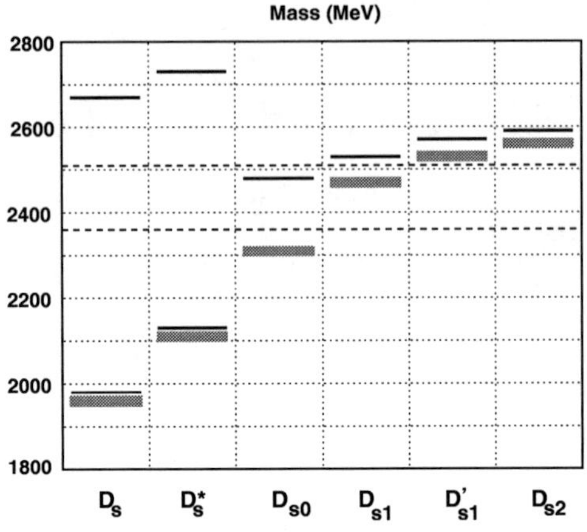

FIGURE 1. The Low Lying D_s Spectrum. Lines are theory from Ref.[3], boxes are experiment. The dashed lines show the DK and D^*K thresholds.

η_c'

The CLEO collaboration[6] has observed the η_c' with a mass of 3638(4) MeV and a width of 19(10) MeV. This is to be compared with the old result of Crystal Ball[7] of 3592(5) MeV. The η_c' is of some interest because its splitting with the ψ' is driven by the hyperfine interaction and hence probes this interaction in a new region. Specifically, the ground state vector-pseudoscalar splitting is $m(J/\psi) - m(\eta_c) = 117$ MeV whereas the excited splitting is now measured to be $m(\psi') - m(\eta_c') = 48$ MeV.

Theoretical expectations for the former range from 108 to 123 MeV in simple (or 'relativised') quark models[5] and thus are within expectations. Alternatively, the latter is predicted to be 67 MeV in the quark model of Eichten, Lane, and Quigg[8]. However, the authors note that including unquenching effects due to open charm meson loops lowers this splitting to 46 MeV which is taken as evidence in favour of their 'unquenched' quark model. However, the simple quark models mentioned above[5] find splittings of 42 - 53 MeV, indicating that it is too early to make definitive conclusions about loop effects. It is worth noting, furthermore, that attempts to unquench the quark model are fraught with technical difficulty[9] and a great deal of effort is required before we can be confident in the results of any model.

$D_s(2317)$ and $D_s(2460)$

These states are roughly 100 MeV below quark model expectations and point to either exotic structure, such as DK molecules[11], or to a deep misunderstanding of heavy-light hadrons. For example, the DK and D^*K continua are both nearby and couple to D_{s0} and D_{s1} in S-wave (see Fig. 1). Is it possible that we have underestimated the importance of coupled channel effects in some systems?

$D_s(2632)$

The $D_s(2632)$ was discovered by the SELEX collaboration at FNAL[12] in the final states D^0K^+ and $D_s\eta$. The measured mass is 2632.6(1.6) MeV and the state is surprisingly narrow with a width of less than 17 MeV at the 90% confidence level. The ratio of the partial widths is measured to be

$$\frac{\Gamma(D_s \to D^0K^+)}{\Gamma(D_s \to D_s\eta)} = 0.16 \pm 0.06. \tag{1}$$

As pointed out by the SELEX collaboration, this is an unusual result since the DK mode is favoured by phase space.

It is unlikely that this state is a $c\bar{s}$ hybrid since the mass of such a state is expected to be roughly 3170 MeV. Possible molecular states include a $D_s^*\eta$ system at 2660 MeV or $D_s^*\omega$ or D^*K^* states at 2900 MeV. However the former is a P-wave which is not favoured for binding, while the latter are too heavy to be plausible.

The remaining possibility is that the $D_s(2632)$ is a radially excited $c\bar{s}$ vector[13] (although it is some 100 MeV lighter than quark model predictions of 2730 MeV). The peculiar decay ratio remains to be explained. Experience with the decay modes of the $\psi(3S)$ to DD, DD^* and D^*D^* points to a possible resolution: transition matrix elements for excited hadrons may have zeroes due to wavefunction nodes. It is possible that such a node is suppressing the DK decay mode. Computation[14] reveals that there is indeed a node but that it occurs at a wavefunction scale which is 20% lower than preferred. Furthermore, the DK mode is always larger than the $D_s\eta$ mode.

We have run out of options and must conclude that the $D_s(2632)$ is an experimental artefact. This conclusion now appears likely because searches by FOCUS[15], BaBar[16], and CLEO[17] have found no evidence for the state.

$X(3872)$

The $X(3872)$ is the poster boy of the new heavy hadrons – it has been confirmed by four experiments[18] at a mass of 3872 MeV and is very narrow, $\Gamma < 2.3$ MeV at 95%. The anomalous nature of the X has led to much speculation: tetraquark[19], cusp[20], hybrid[21], or glueball[22]. But the most popular explanation is that it is a $D\bar{D}^*$ bound state[23, 24]. This model has successfully predicted[24] the quantum numbers of the X ($J^{PC} = 1^{++}$[25]), the decay mode $\pi\pi\pi J/\psi$[26], that the $\pi\pi$ and 3π modes should be comparable[27], and that the $\pi\pi$ invariant mass distribution should be dominated by the ρ[28] while the 3π invariant mass distribution should be dominated by the ω[26].

This string of successes has met with recent experimental challenges: (i) The X has been observed decaying to $\gamma J/\psi$[26] with a strength $Br(X \to \gamma J/\psi)/Br(X \to \pi\pi J/\psi) = 0.14(5)$[29]. This rate is substantially larger than predicted in the model of Ref. [24]. (ii) There are rumours that Belle have seen the mode $X \to D\bar{D}\pi$ and that its rate is ten times larger than that of $\pi\pi J/\psi$[30]. The model predicts this ratio to be 1/20. (iii) BaBar report[31]

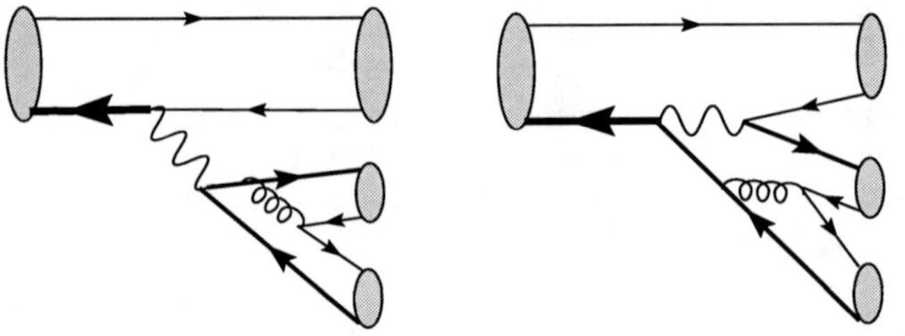

FIGURE 2. Colour Enhanced (left) and Suppressed (right) B Decays.

$$\frac{Br(B^0 \to XK^0)}{Br(B^+ \to XK^+)} = 0.61(36)(6). \tag{2}$$

This is at odds with the molecular picture which predicts a ratio of $\mathcal{O}(\frac{1}{N_c^2}) + \mathcal{O}(\frac{Z_{D^+D^-}}{Z_{D^0\bar{D}^{0*}}}) \approx 10\%$. This prediction is illustrated in Fig. 2. The left panel shows the colour leading diagram which produces an X via the reactions $B^+ \to K^+X(D^0D^{0*})$ and $B^0 \to K^0X(D^+D^{-*})$. Since the latter accesses the X via its suppressed charged mode, one expects $Br(B^0)/Br(B^+) \approx Z_{+-}/Z_{00} \approx 10\%$ (Z is the wavefunction probability). The colour suppressed diagram (right panel) allows B^0 decay to D^0D^{0*} but is weighted by $1/N_c^2$ and hence is also a 10% effect. (iv) Belle[25] and BaBar[31] have measured the product of branching ratios:

$$Br(B^+ \to XK^+)Br(X \to \pi\pi J/\psi) = 1.3(3)[0.85(30)] \cdot 10^{-5}. \tag{3}$$

The new $D\bar{D}\pi$ data imply that $Br(B \to \pi\pi J/\psi) < 0.1$ which implies that $Br(B \to XK) > 10^{-4}$. This is comparable to the rate $Br(B \to \chi_{c1}K)$ and points to a large $c\bar{c}$ component in the X.

All of these new data may be accounted for if the predicted hidden charm interactions of Ref.[24] were over-estimated. This leads to weaker binding which gives rise to a much narrower X with a dominant $D\bar{D}\pi$ mode, weak $\pi\pi J/\psi$ and $\pi\pi\pi J/\psi$ modes, and a radiative transition of the desired magnitude.

$X(3940)$

The $X(3940)$ is seen by Belle[32] recoiling against J/ψ in e^+e^- collisions. The state has a mass of 3943(11)(13) MeV and a width of 87(22)(26) MeV[33]. The X is seen to decay to $D\bar{D}^*$ and not to $\omega J/\psi$ or $D\bar{D}$. It is natural to attempt a $2P$ $c\bar{c}$ assignment for this state since the expected mass of the 2^3P_J multiplet is 3920 - 3980 MeV and the expected widths are 30 - 165 MeV[5]. Finally, if the $D\bar{D}^*$ mode is dominant it suggests that the $X(3940)$ is the χ'_{c1}. The problem with this assignment is that there is no evidence for the χ_{c1} in the same data. This has led Olsen to speculate[34] that the X is the η''_c. Unfortunately this interpretation is also suspect because the η''_c has an expected mass of 4064 MeV, 120 MeV too high.

$Y(3940)$

The $Y(3940)$ is claimed as a resonance in the $\omega J/\psi$ subsystem of the decay $B \to K\pi\pi\pi J/\psi$[35] with a mass of 3940(11) MeV and a width of 92(24) MeV. The state has not been seen in the decay modes $Y \to D\bar{D}$ or $D\bar{D}^*$. Again, the mass and width of the Y suggest a radially excited P-wave charmonium. However, the $\omega J/\psi$ decay mode is peculiar. In more detail, Belle measure $Br(B \to KY)Br(Y \to \omega J/\psi) = 5.0(9)(16) \cdot 10^{-5}$. One expects that $Br(B \to K\chi'_{cJ}) < Br(B \to K\chi_{cJ}) = 4(1) \cdot 10^{-4}$. This implies $Br(Y \to \omega J/\psi) > 12\%$, which is unusual for a canonical $c\bar{c}$ state above open charm threshold.

Thus the Y is something of an enigma, driving the claim of the Belle collaboration that it is a hybrid. This is perhaps premature – certainly more data are required before strong claims can be made.

$Z(3930)$

This state was observed by the Belle collaboration[10] in $\gamma\gamma \to D\bar{D}$ with a mass of 3931(4) MeV and a width of 20(8)(3) and a claimed significance of 5.5 sigma. The $D\bar{D}$ helicity distribution is consistent with J=2. In line with the X and the Y, the Z seems an obvious candidate for the χ'_{c2} (the χ'_{c1} cannot decay to $D\bar{D}$). The predicted mass of the χ_{c2} is 3972 MeV and the predicted width is 80 MeV[5]. However, setting the mass to the measured 3931 MeV restricts phase space sufficiently that the predicted strong width drops to 47 MeV, reasonably close to the measurement. The predicted branching fraction to $D\bar{D}$ is 70%. The largest radiative transition is $\chi'_{c2} \to \psi'\gamma$ with a rate of 180(30) keV. At this stage we have no reason not to believe that the Z is the previously unknown χ'_{c2}.

$Y(4260)$

The $Y(4260)$ was discovered as an enhancement in the $\pi\pi J/\psi$ subsystem of the reaction $e^+e^- \to \gamma_{\rm ISR}\psi\pi\pi$ with a mass of 4259(8)(4) MeV and a width of 88(23)(5) MeV by the BaBar collaboration[36]. Evidentally the state is a vector with $c\bar{c}$ flavour. Of course the low lying charmonium vectors are well known and the only mesonic charmonium vector available is the $\psi(3D)$. However quark model estimates of its mass place it at 4460 MeV, much too heavy for the Y. Of course this statement relies on the quark model itself – Llanes-Estrada[37] has argued that the Y is the $\psi(4S)$ based on the spectrum of a relativistic model. Maiani *al.*[38] claim the Y is a tetraquark $c\bar{c}s\bar{s}$ state which decays predominantly to $D_s\bar{D}_s$. Of the states which we know must exist, the most natural explanation is as a $c\bar{c}$ hybrid[39]. The lightest charmonium hybrid is expected at 4400 MeV, somewhat high, but perhaps acceptable given our lack of experience in this sector.

Lastly, it is tempting to examine molecular interpretations of this state. In particular DD_1 is an S-wave threshold at 4290 MeV – close enough that the enhancement may simply be a cusp effect. If the system does bind, it does so with a novel mechanism since pion exchange does not lead to a diagonal interaction in this channel (unlike the case of the $X(3872)$). Off-diagonal interactions may provide the required novelty.

CONCLUSIONS

The new heavy meson spectroscopy is no mere butterfly collecting – the D_s spectrum and the Xs, Ys, and Zs challenge our understanding of QCD. Can we rise to the challenge? It is clear that the simple constituent quark model must fail somewhere (gluonic degrees of freedom turn on, relativity and chirality become important, and coupled channels become dense) – are we seeing this? Lastly, have we entered a new era of 'mesonic nuclear physics'?

ACKNOWLEDGMENTS

This work was supported by PPARC grant PP/B500607 and the US Department of Energy under contract DE-FG02-00ER41135. I am grateful to Ted Barnes, Denis Bernard, Frank Close, Steve Godfrey, and Steve Olsen for many interesting discussions on the new states. I wish to thank Alberto Reis for the invitation to discuss this fascinating topic in Rio de Janeiro.

REFERENCES

1. D. Acosta *et al.* [CDF Collaboration], arXiv:hep-ex/0505076.
2. I. F. Allison, C. T. H. Davies, A. Gray, A. S. Kronfeld, P. B. Mackenzie and J. N. Simone [HPQCD Collaboration], Phys. Rev. Lett. **94**, 172001 (2005).
3. S.Godfrey and N.Isgur, Phys. Rev. D32, 189 (1985).
4. J. L. Rosner *et al.* [CLEO Collaboration], Phys. Rev. Lett. **95**, 102003 (2005).
5. T. Barnes, S. Godfrey and E. S. Swanson, "Higher charmonia", arXiv:hep-ph/0505002. See also F. E. Close and E. S. Swanson, arXiv:hep-ph/0505206.
6. D.M. Asner *et al.* [CLEO], Phys. Rev. Lett. **92**, 142001 (04).
7. C. Edwards *et al.* [CBALL], Phys. Rev. Lett. **48**, 70 (82).
8. E. J. Eichten, K. Lane and C. Quigg, Phys. Rev. D **69**, 094019 (2004).
9. E. S. Swanson, J. Phys. G **31**, 845 (2005).
10. K. A. Abe *et al.* [Belle], hep-ex/0507033.
11. T. Barnes, F. E. Close and H. J. Lipkin, Phys. Rev. D **68**, 054006 (2003).
12. A. V. Evdokimov *et al.* [SELEX Collaboration], Phys. Rev. Lett. **93**, 242001 (2004).
13. K.-T. Chao, Phys. Lett. **B599**, 43 (04).
14. T. Barnes, F. E. Close, J. J. Dudek, S. Godfrey and E. S. Swanson, Phys. Lett. B **600**, 223 (2004).
15. R. Kutschke, personal communication; E381-doc-701-v2.
16. B. Aubert *et al.* [BABAR Collaboration], arXiv:hep-ex/0408087.
17. R. Galik, talk presented at PIC2004.
18. S.-K. Choi *et al.* [Belle], Phys. Rev. Lett. **91** 262001 (2003); V. M. Abazov *et al.* [D0 Collaboration], Phys. Rev. Lett. **93**, 162002 (2004); D. Acosta *et al.* [CDFII], Phys. Rev. Lett. **93**, 072001 (04); B. Aubert *et al.* [BABAR Collaboration], Phys. Rev. D **71**, 071103 (2005).
19. L. Maiani *et al.*, Phys. Rev. **D71**, 014028 (05).
20. D. Bugg, Phys. Lett. **B598**, 8 (04).
21. B.-A. Li, Phys. Lett. **B605**, 306 (05).
22. K. K. Seth, Phys. Lett. B **612**, 1 (2005).

23. N. Tornqvist, hep-ph/0308277; F.E. Close and P. Page, Phys. Lett. **B578**, 119 (04). For an overview see T. Barnes and S. Godfrey, Phys. Rev. D **69**, 054008 (2004). For an effective field theory approach based on the weak coupling limit see E. Braaten and M. Kusunoki, Phys. Rev. D **72**, 014012 (2005).
24. E. S. Swanson, Phys. Lett. B **588**, 189 (2004); Phys. Lett. B **598**, 197 (2004). For an example of light molecules see K. Dooley, E. S. Swanson and T. Barnes, Phys. Lett. B **275**, 478 (1992).
25. K. Abe *et al.*, arXiv:hep-ex/0505038.
26. K. Abe *et al.*, arXiv:hep-ex/0505037.
27. K. Abe *et al.* [Belle Collaboration], arXiv:hep-ex/0408116.
28. CDF note 05-03-24.
29. BELLE-CONF-0540.
30. G. Bauer, arXiv:hep-ex/0505083; BELLE-CONF-0568; S. Olsen, private communication.
31. D. Bernard, talk at International Conference on QCD and Hadronic Physics, Beijing, June 2005.
32. K. Abe *et al.*, arXiv:hep-ex/0507019.
33. But K. Trabelsi quotes a width of 39(26) MeV at Hadron05.
34. S. Olsen, private communication.
35. S.K. Choi *et al.* [Belle], Phys. Rev. Lett. **94**, 182002 (05); K. Abe *et al.* [Belle Collaboration], Phys. Rev. Lett. **94**, 182002 (2005).
36. B. Aubert *et al.* [BABAR Collaboration], arXiv:hep-ex/0506081.
37. F. J. Llanes-Estrada, Phys. Rev. D **72**, 031503 (2005).
38. L. Maiani, V. Riquer, F. Piccinini and A. D. Polosa, Phys. Rev. D **72**, 031502 (2005).
39. S. L. Zhu, Phys. Lett. B **625**, 212 (2005); F. E. Close and P. R. Page, arXiv:hep-ph/0507199.

Hadron Physics at the Tevatron

Arthur Maciel[1]

Centro Brasileiro de Pesquisas Físicas
Rio de Janeiro, Brazil

Abstract. RunII at Fermilab's *Tevatron* is currently under way and has recently reached a total of $1\ fb^{-1}$ of integrated luminosity delivered to each of its two experiments, CDF and DØ. Selected RunII preliminary results are reviewed, with emphasis on the spectroscopy and decays of heavy mesons with charm and beauty.

Keywords: heavy mesons, spectroscopy, decays
PACS: 13.20.Fc, 13.20.He, 13.25.Ft, 13.25.Hw, 14.40.Lb, 14.40.Nb

INTRODUCTION

The Tevatron, Fermilab's $\overline{p}p$ 1.96 GeV/c^2 C.M. Collider, has recently reached a delivered total of $1\ fb^{-1}$ of integrated luminosity to its two multi-purpose detectors, DØ and CDF, during the current data taking period known as RunII. Initial difficulties with beam stability were efficiently corrected during a shutdown in late 2003, and since then the Tevatron complex has operated remarkably well, with impressive recovery during 2004 and matching its luminosity design goals in 2005 [1]. Both DØ and CDF have been steadily accumulating good quality data throughout RunII with luminosity recording efficiencies now approaching the 90% mark. These detectors were designed to address a wide range of physics topics and measurements, mostly geared towards the high energy frontier, encompassing top and electroweak physics, QCD and the search for phenomena beyond the Standard Model. However, focus here is kept on flavor physics, and most specifically on a few selected topics which we felt were the closest to the interests of this conference. The reader is reminded that the set of results presented here reflects only a very small corner of these detector's heavy flavor programs, and indeed a tiny fraction of their general physics interests [2]. Opening remarks about the detectors and their flavor physics related triggers are followed by a review of the status of B_c measurements at the Tevatron. Recent results on the spectroscopy and decays of doubly excited charm (D^{**}) and bottom (B^{**}) mesons are presented, followed by studies on the $X(3872)$ state.

THE CDF AND DØ DETECTORS

The CDF RunII detector consists of a magnetic spectrometer surrounded by calorimeters and muon chambers, and is described in detail in [3]. The spectrometer contains a silicon microstrip detector (SVX-II) and a cylindrical open-cell wire drift chamber (COT)

[1] For the DØ and CDF Collaborations.

CP814 *Hadron Spectroscopy: Eleventh International Conference*
edited by A. Reis, C. Göbel, J. Sá Borges, and J. Magnin

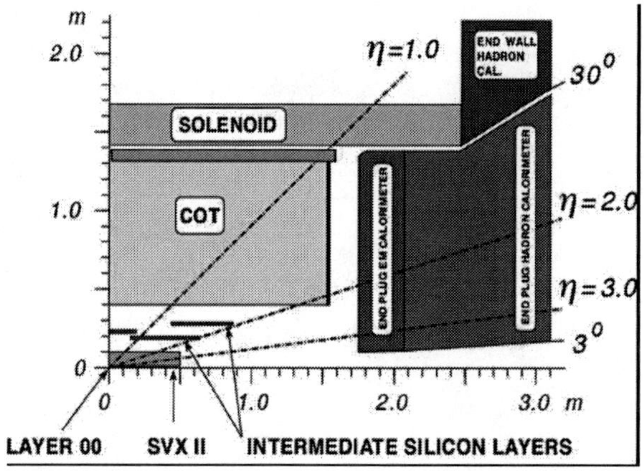

FIGURE 1. A sketch of the CDF inner detectors.

immersed in a 1.4 T solenoidal magnetic field. Charged particles are tracked in the pseudorapidity range $\eta < 1.0$. The SVX-II provides up to five r-ϕ position hits, each with roughly 15 μm precision, at radii between 2.5 and 10.5 cm. Beyond its baseline design, CDF has added an extra (inner) silicon layer (Layer-00) at $r = 1.35$ cm which was used for the first time in the $B_c \rightarrow J/\psi\, \pi$ analysis reported below.

The COT has 96 measurement layers at radii between 40 and 137 cm, and is organized in alternating axial and $\pm 2^o$ stereo superlayers. The solenoid covers $r < 150$ cm and is surrounded by electromagnetic and hadronic calorimeters. Four layers of planar drift chambers (CMU) outside the hadron calorimeter cover the region $\eta < 0.6$ and detect muons of transverse momentum $p_T > 1.4\ GeV/c$ which penetrate the five absorption lengths of inner material. Another four layers of planar drift chambers (CMP, with same η coverage) instrument 60 cm of steel outside the magnet return yoke and improve hit redundancy for higher p_T muons. An extended muon system (CMX) completes the coverage out to $\eta < 1.0$. A sketch of the CDF spectrometer is presented in figure 1.

DØ's central magnetic tracking volume (inexistent in RunI) is a RunII upgrade addition and consists of a silicon microstrip tracker (SMT) surrounded by a central fiber tracker (CFT), both inside a 2 T superconducting solenoidal magnet. Covering $|\eta| < 3$, the SMT has a six barrel/disk longitudinal structure as shown in figure 2. Each barrel module, with radii between 2.7 and 10 cm, consists of four layers of detector ladder assemblies. Layers one and three are single sided and layers two and four are double sided with 2° stereo angle. The single hit resolution is of about 10μm. Barrels are interspersed with twelve F-disks (double sided, $\pm 15^o$ stereo), with four extra H-disks (single sided, $\pm 7.5^o$ stereo) located beyond the barrels to complete the η coverage.

The CFT covers tracking within $|\eta| < 1.6$, reaches an outer radius of 52 cm, and consists of eight thin coaxial cylinders, each supporting a pair of overlapping scintillating fiber doublets. On each cylinder the inner (longitudinal) doublet runs parallel to the beam axis while the outer doublet is mounted at a stereo angle of $\pm 3^o$, alternating with each layer.

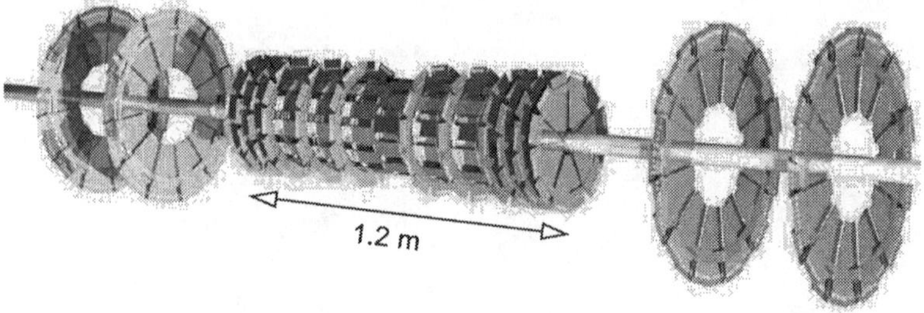

FIGURE 2. The DØ silicon microstrip tracker (SMT).

Outside the coil the next two layers of detection consist of a preshower (scintillator strips) and a liquid argon calorimeter. There follows an outer muon system which covers $|\eta| < 2$ and consists of a layer of multiple wire tracking devices and scintillation counters in front of 1.8 T iron toroids, followed by another two similar detection layers outside the toroids. A detailed description of the DØ detector can be found in [4].

The two detectors are rather different in concept; CDF is built around a spacious and redundant magnetic spectrometer, whereas the DØ project favors the concepts of hermeticity and compactness, towards optimizing missing E_T resolution. With a tracking lever arm of about 1/3 that of CDF, the DØ momentum resolution is rather modest, giving a central $J/\psi(\mu^+\mu^-)$ width near 55 MeV/c^2 whereas that of CDF is below 20 MeV/c^2. A key ingredient to DØ's flavor physics program is its extensive and redundant muon system with an ability to cleanly and efficiently trigger on semi-muonic b decays (or dimuons from J/ψ etc), thus providing a remarkable selection tool for b events in the hostile $\overline{p}p$ collider environment.

Both detectors use a three level trigger system to reduce the event rate of about 2.5 MHz down to an output rate of approximately 50 Hz. Tracking information (from wires or scintillating fibers) is available at level-1 while impact parameters (from silicon) are available at level-2. Here again a basic difference in trigger strategies differentiates the two detectors. While DØ relies on its single muon and dimuon triggers for heavy flavor collection, CDF has a significant amount of extra bandwidth at level-1 for a purely hadronic trigger dedicated to its flavor physics program, with bulk rejection provided at level-2 with an impact parameter trigger (the SVT). Output rates from both detectors at each trigger level are shown in table 1. As part of its RunIIb trigger upgrade, DØ is planning on doubling its event-to-tape rate, with the added 50 Hz being dedicated to B physics measurements.

B_C MESONS

The relevance of the B_c system as a testing ground for theory models has been highlighted in various talks throughout this conference (*e.g.* refs [5] and [6]). The B_c is a unique heavy≠heavy $q\overline{q}$ system that probes the heavy quark potential in a region between the J/ψ and the Υ families, with an expected rich spectroscopy of excited states.

TABLE 1. Approximate trigger accept rates (Hz)

output rates	DØ		CDF	
	Run IIa	Run IIb	*Run IIa*	Run IIb
Level-1	*1500*	2500	*25000*	25000
Level-2	*850*	1000	*300*	750
Level-3	*50*	50+50	*50*	85

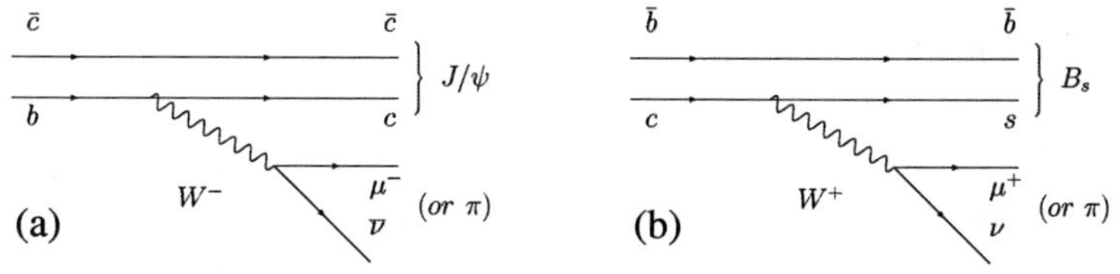

FIGURE 3. Example decays of the B_c. (a) c-quark as spectator, see text. (b) b-quark as spectator, not yet seen, but possibly relevant for the Tevatron B_s mixing program (lifetime and tagging dilution).

Both b and c constituents can decay weakly with comparable probabilities, with a variety of possible final states and associated branching fractions inviting confrontation with calculations. B_c production is also an interesting challenge for QCD based models, with a suggested significant contribution from S wave color octet components in the production mechanism [7]. A recent paper on LQCD [8] proposes several calculation upgrades to heavy meson masses (HQ discretization effects, elimination of the quenched approximation) and predicts the B_c mass with a significant reduction in model uncertainties.

The semileptonic decays of the B_c with a J/ψ in the final state as shown in figure 3(a) have been seen (in both electron and muon modes) in RunI by CDF [9] and in RunII by both CDF [10] and DØ [11]. Mass and lifetime results are all consistent with one another, and since final states are not fully reconstructed (due to a missing neutrino) the B_c mass value is extracted with an uncertainty of about 400 MeV/c^2. Figure 4 shows the DØ (a) and CDF (b) signals (excess over expected backgrounds) in the three muon channel.

Figure 3(a) also shows the hadronic $B_c \to J/\psi(\mu^+\mu^-)\ \pi$ decay. A first evidence for this full reconstruction mode was presented by CDF [10] in December 2004, extracted from 360 pb^{-1} of RunII data. A search range in the $\pi\mu^+\mu^-$ invariant mass is chosen with $\pm 2\sigma$ around the RunI mass value. Signal (modelled from montecarlo) and background (real data) separation is performed using a discrimination figure of merit, and the search region is kept blinded until the final data selection, necessarily very tight on the pion candidate, is established. The resulting $J/\psi\ \pi$ candidate mass spectrum is shown in figure 5(a). This search range is then scanned with a sliding fit window that is 300 MeV wide, and advances in steps of 10 MeV. At each step the scan produces S and B counts by means of a binned likelihood fit of a gaussian plus a linear background within that window. A score function is built from pairs (S,B) and plotted for each step, as shown in figure 5(b). A predetermined threshold for statistical significance was set at 3.5 for the

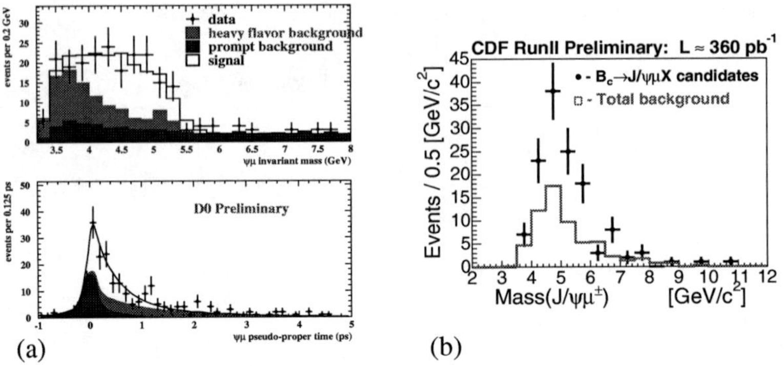

FIGURE 4. Semileptonic B_c signals in the three muon channel; (a) DØ ($210pb^{-1}$) (b) CDF ($360pb^{-1}$).

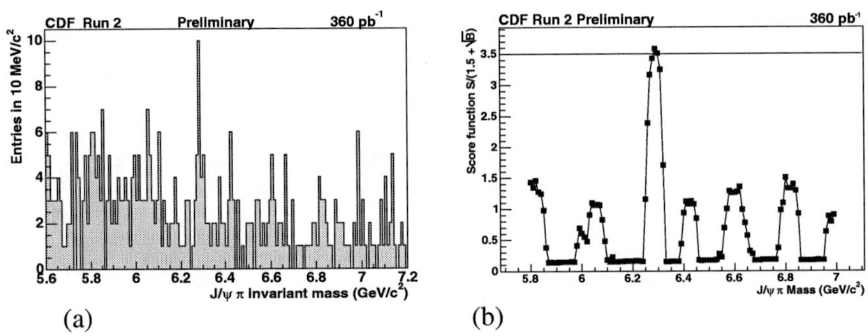

FIGURE 5. (a) The $J/\psi\ \pi$ mass search range after data selection, and (b) the signal score function.

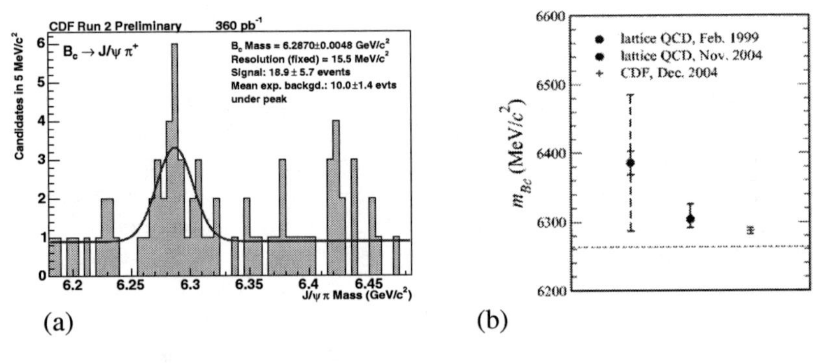

FIGURE 6. (a) The resulting B_c mass fit, and (b) comparison with LQCD calculations (see ref.[8])

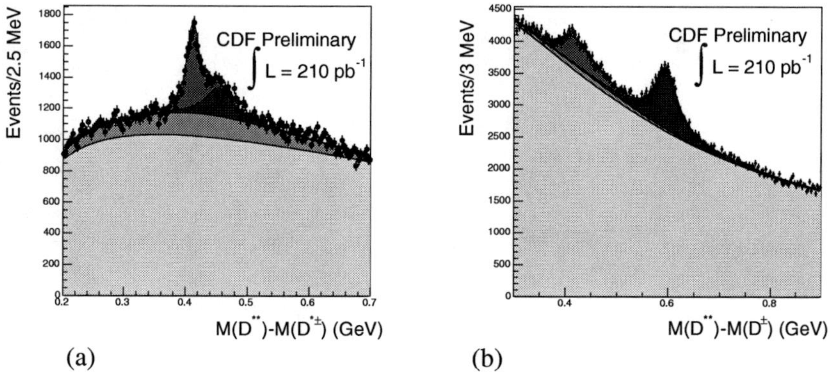

FIGURE 7. Narrow D^{**} states from CDF channels; (a) $D^{*\pm}\pi^{\mp}$, and (b) $D^{\pm}\pi^{\mp}$.

score function, and as shown in figure 5(b) it is clipped only once at around 6.3 GeV/c^2. This is interpreted as signal evidence, and the probability that background fluctuates to this signal is 0.3%. A signal fit dedicated to this neighborhood is shown in figure 6(a) and yields a B_c mass value of $M = 6287.0 \pm 4.8 \pm 1.1\ MeV/c^2$. Figure 6(b) displays this measurement in confrontation with past LQCD calculations, showing the evolution in uncertainty as one departs from the quenched approximation and adds light quark degrees of freedom; see reference [8] and references therein for details.

ORBITALLY EXCITED *D* MESONS

Orbitally excited D mesons, both in the strange and non-strange sectors received a lot of attention throughout this conference, and were featured in many talks, including [12, 13, 14]. In the non-strange sector, the four $L = 1$ excited D mesons are collectively known as D^{**} and are observed as two wide (S wave) states and two narrow (D wave) states [12]. With well defined angular momentum quantum numbers in the massive charm limit, these states offer a unique testing ground for theory models such as HQET (*e.g.* ratios of branching fractions, studies of hyperfine splittings between states of different j_q).

Both CDF [10] and DØ [15] have recent studies on the narrow ($j_q = 3/2$) resonances. Because of their different trigger modes and capabilities, access to these states is rather different among the two experiments. CDF uses its huge band width at Level-1 (see table 1) and the SVT (hadron IP) trigger at L2 for direct collection of hadron-only final states, from which they extract masses and widths of the narrow D^{**} states. In contrast, DØ collects the D^{**}'s that originate from semileptonic B decays by means of a suite of single-muon triggers, and extracts the $B \to D^{**}_{1,2}\ \mu\ X$ branching fractions. A recent LEP study of D^{**} mesons in B decays is available from [16].

Figure 7 shows the CDF data (210 pb^{-1}) in two channels: (a) $D^{*\pm}\pi^{\mp}$, and (b) $D^{\pm}\pi^{\mp}$. Note in (b) the broadenning effect due to π^0 losses in feed down decays from (a), $D^{*\pm} \to D^{\pm}\pi^0$. CDF numbers are extracted from simultaneous fits to the two channels, and are shown in table 2. Backgrounds and broad D^{**} states are treated separately, and

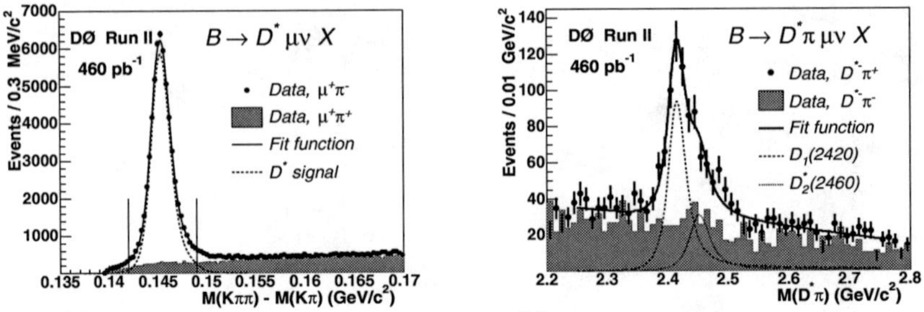

(a) (b)

FIGURE 8. (a) The $J/\psi\ \pi$ mass search range after data selection, and (b) the signal score function.

TABLE 2. CDF results for the narrow D^{**} states.

(GeV/c^2)	*Mass*	*Width*
D_1	$2421.7 \pm 0.7 \pm 0.6$	$20.0 \pm 1.7 \pm 1.3$
D_2	$2463.3 \pm 0.6 \pm 0.8$	$49.2 \pm 2.3 \pm 1.3$

the background excess bands in figures 7 are consistent with contributions from the broad states.

The DØ analysis uses 460 pb^{-1} of data from which about 55K decays $B \to D^*\ \mu\ \nu\ X$ are initially selected, as shown in figure 8(a). The mass spectrum for the narrow D^{**} states is subsequently obtained from the $D^{*\pm}\pi^{\mp}$ channel as shown in figure 8(b). The two states are fit with two relativistic *Breit-Wigner* functions, with widths and mass difference fixed to PDG values. Fit results are also shown in figure 8(b). The DØ numbers for the branching fractions and their ratio (here a first direct measurement) are extracted from fit counts. Inputs (from PDG) to the counting experiment are; $\mathscr{B}(\bar{b} \to B) = (39.7 \pm 1.0)\%$, $\mathscr{B}(D_1 \to D^*\pi) = (100)\%$ and $\mathscr{B}(D_2^* \to D^*\pi) = (30 \pm 6)$ The DØ results are shown in table 3.

In the strange sector (the D_{SJ} family [12]) there were reports at this conference from BaBar [14] and Belle [20]. Very recently, from 485 pb^{-1} of data, DØ [11] has extracted evidence for the decay $B_s \to D_{s1}^{\pm}(2536)\mu\nu X$, where the $D_{s1}(2536)$ is an orbitally excited (c,s) pair in a $J^P = 1^+$ state. The signal is obtained in the $D^{*\pm}\ K_s^0$ channel with an yield of 18±5 events, and is shown in figure 9. Acumulated luminosity during RunII

TABLE 3. DØ results for $\mathscr{B}(B \to D_{1,2}^{**} + \mu\nu X)$.

Decay	$\mathscr{B}$
$B \to \bar{D}_1^0 \ell^+ \nu X$	$(0.33 \pm 0.06)\%$
$B \to \bar{D}_2^{*0} \ell^+ \nu X$	$(0.44 \pm 0.16)\%$
Ratio	1.31 ± 0.29 (stat) ± 0.47 (syst)

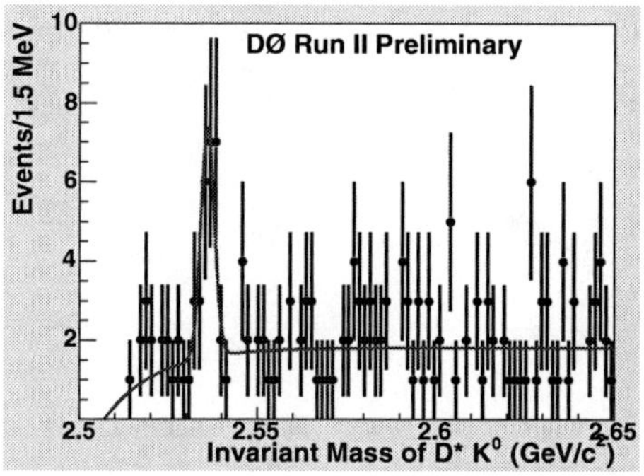

FIGURE 9. DØ evidence ($> 3\sigma$) for the decay $B_s \to D_{s1}^{\pm}(2536)\mu\nu X$.

will enable DØ to proceed towards (as yet inexistent) branching fraction measurements analogous to those presented here in the non-strange ($B \to D^{**}$) sector.

B^{**} MESONS

B^{**} mesons are $(L = 1)$ orbitally excited $b\bar{q}$ states which follow the same pattern of J^P allowed decays as the D^{**} family [12]. In contrast with existing accurate predictions of mass, width and branching fractions in various theory models [17], their spectroscopic properties are not yet well established experimentally, and are not listed in PDG(2004). Previous observations by the LEP experiments and CDF [18] are statistics and resolution limited, unable to separate the two narrow states and to extract individual properties. This separation was achieved by DØ in August 2004 with a study of the narrow B^{**} states decaying to $B^{(*)}\pi$ with exclusively reconstructed B mesons [11], using 350 pb^{-1} of integrated luminosity. The observation of separated sates was subsequently confirmed by CDF [10].

Both collaborations used similar methods to extract the B^{**} signals. Initially a sample of B decays in exclusive channels with a J/ψ ($B^{\pm} \to J/\psi K^{\pm}$, $B^0 \to J/\psi K^{*0}$, $B^0 \to J/\psi K_s$) is selected. A track (with the pion mass hypothesis) is added to the B vertex, and the mass difference $\Delta M = M(B\pi) - M(B)$ is plotted. The expected structure consists of three peaks with identical widths,

$$\begin{array}{lr} (1)\ B_1 \to B^*\pi\ ,\ B^* \to B\gamma & \text{(shifted by } \delta M) \\ (2)\ B_2 \to B^*\pi\ ,\ B^* \to B\gamma & \text{(shifted by } \delta M) \\ (3)\ B_2 \to B\pi & \text{(true position)} \end{array}$$

where $\delta M = M(B^*) - M(B) \approx 46$ MeV is a mass deficit from the undetected γ in $B^* \to B\gamma$. Because of their proximity in mass, peaks (1) and (2) are expected to merge.

The resulting mass difference spectrum for DØ is shown in figure 10, and conforms with expectations. The B_1 and B_2 separation is a simple exercise using peak (3) and the

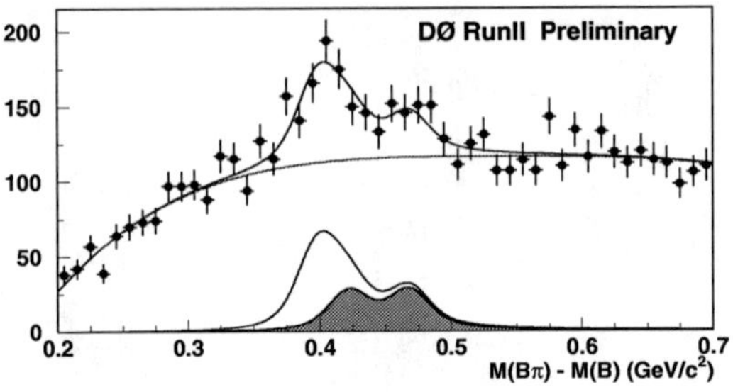

FIGURE 10. The DØ $M(B\pi) - M(B)$ spectrum showing the B_1 and B_2 signals.

identical width constraint. The numbers extracted from the DØ fit are;

$$\begin{aligned} M(B_1) &= 5724 \pm 4(stat) \pm 7(syst) \quad MeV/c^2 \\ M(B_2^*) - M(B_1) &= 23.6 \pm 7.7(stat) \pm 3.9(syst) \quad MeV/c^2 \\ \Gamma_1 = \Gamma_2 &= 23 \pm 12(stat) \pm 9(syst) \quad MeV/c^2 \end{aligned}$$

The CDF numbers are work in progress. Expected increases in statistics throughout RunII will allow precision measurements of their production and decay properties.

THE X(3872)

The ubiquitous $X(3872)$ state, discovered by Belle [19] and widely discussed at this conference (see *e.g.* [20, 21]), was also seen and studied by CDF [22] and DØ [23] in the $J/\psi(\mu^+\mu^-)\pi^+\pi^-$ channel. Using the nearby $\psi(2S)$ as a charmonium template, both collaborations did comparative studies of production and decay properties in an attempt to distinguish it from a $c\bar{c}$ state. No apparent discriminators were found. More recently, motivated by setting constraints to valid J^{PC} values of the $X(3872)$ (narrowed down by Belle [20, 21] by establishing $C = +1$), CDF [10] has examined the dipion mass spectrum in the $\pi^+\pi^-\mu^+\mu^-$ final state. The resulting $M(\pi\pi)$ is displayed in figure 11 and is confronted with QCD multipole expansion predictions for the ${}^3S_1$, ${}^1P_1$ and ${}^3D_J$ hypothesis of a charmonium state. The same plot shows also the pure dipion phase space and the resonant $J/\psi\ \rho$ hypothesis. The resonant ρ shows the best fit to data, closely followed by the $c\bar{c}({}^3S_1)$. This last option was shown not viable [21] since there are no available credible slots. The ${}^1P_1$ and ${}^3D_J$ hypothesis seem quite clearly excluded by the CDF data. As indicated by many speakers in this conference, understanding the nature of the $X(3872)$ state is very much work in progress.

In conclusion, the Tevatron is producing a wealth of heavy flavor physics results, with a competitive program and many unique measurements. It has been a pleasure sharing some of these results in such an exciting and well organized conference.

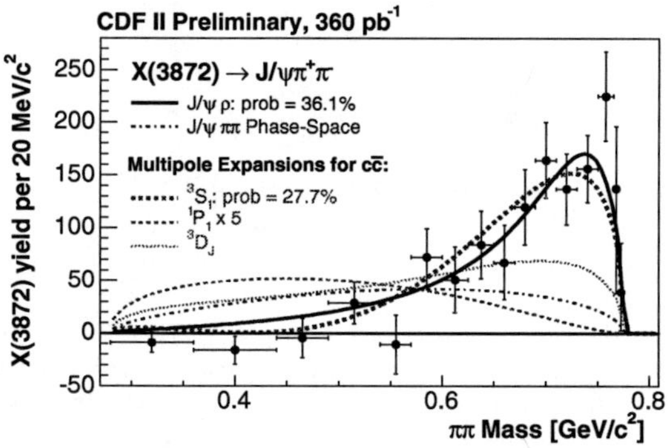

FIGURE 11. The CDF dipion mass spectrum for the $X(3872)$, overlaid with the multipole expansions for the $c\bar{c}$ hypothesis, pure dipion phase space, and the $J/\psi\ \rho$ hypothesis.

REFERENCES

1. Tevatron luminosity plots can be viewed at `http://www-bdnew.fnal.gov/pbar/AEMPlots/today/DataSummaryTables.html`
2. The DØ and CDF collaborations (with links to their physics results pages) can be visited respectively at `http://www-d0.fnal.gov/` and `http://www-cdf.fnal.gov/`
3. D. Acosta *et.al.*, FERMILAB-PUB-04/440-E (2004) and R. Blair *et.al.*, FERMILAB-PUB-96/390-E (1996).
4. "The Upgraded DØ Detector" submitted to Nuclear Instruments and Methods in Physics Research A (NIM-A), Fermilab-Pub-05-341-E (2005), or on the arXiv as physics/0507191.
5. "Effective Field Theories for Heavy Quarkonia", N. Brambilla, these proceedings.
6. "Recent Developments in LQCD", Z. Fodor, these proceedings.
7. C-H Chang *et.al.*, Phys. Rev. D. 71, 074012 (2005).
8. I.F. Allison *et.al.*, Phys. Rev. Lett. 94, 172001 (2005).
9. CDF Collab. Phys. Rev. Lett. 81, 2432 (1998) and Phys. Rev. D. 58, 112004 (1998).
10. `http://www-cdf.fnal.gov/physics/new/bottom/bottom.html`
11. `http://www-d0.fnal.gov/Run2Physics/WWW/results/b.htm`
12. "Spectroscopy of D Mesons", S. Bianco, these proceedings.
13. "Charm Physics at Belle", S. Kopar, these proceedings.
14. "$D_{sJ}^{(*)}$ Production in B Decays" (BaBar), V. Poireau, these proceedings.
15. DØ Collab., Phys. Rev. Lett. in print, and hep-ex/0507046.
16. DELPHI Collab., CERN-EP-PH/2005-015.
17. E. Eichten *et.al.*, Phys. Rev. Lett. 71, 4116 (1993), N. Isgur, Phys. Rev. D57, 4041 (1998), D. Ebert *et.al.*, Phys. Rev. D57, 5663 (1998), erratum Phys. Rev. D59,019902 (1999), A.H. Orsland and H. Hogaasen, Eur. Phys. J. C9, 503 (1999).
18. OPAL Collab., Z.Phys. C66, 19 (1995), DELPHI Collab., Phys. Lett. B345, 589 (1995), ALEPH Collab., Z.Phys. C69, 393 (1996) and Phys. Lett. B425, 215 (1998), CDF Collab., Phys. Rev. D64, 072002 (2001).
19. Belle Collab., Phys. Rev. Lett. 91, 262001 (2003).
20. "Spectroscopy and New Particles" (Belle), T. Lesiak, these proceedings.
21. "Hadron Physics at Belle", K. Trabelsi, these proceedings.
22. CDF Collab., Phys. Rev. Lett. 93, 072001 (2004).
23. DØ Collab., Phys. Rev. Lett. 93, 162002 (2004).

First measurement of the $\pi^+\pi^-$ atom lifetime

Leonid Afanasyev, on behalf of the DIRAC collaboration

Laboratory of Nuclear Problems, Joint Institute for Nuclear Research, 141980 Dubna, Moscow Region, Russia

Abstract. The goal of the DIRAC experiment at CERN (PS212) is to measure the $\pi^+\pi^-$ atom lifetime with 10% precision. Such a measurement would yield a precision of 5% on the value of the S-wave $\pi\pi$ scattering lengths combination $|a_0 - a_2|$. Based on part of the collected data we present a first result on the lifetime, $\tau = \left[2.91\,^{+0.49}_{-0.62}\right] \times 10^{-15}$ s, and discuss the major systematic errors. This lifetime corresponds to $|a_0 - a_2| = 0.264\,^{+0.033}_{-0.020}\, m_\pi^{-1}$.

Keywords: DIRAC experiment, elementary atom, pionium atom, pion scattering
PACS: 36.10.-k, 32.70.Cs, 25.80.E, 25.80.Gn, 29.30.Aj

INTRODUCTION

The aim of the DIRAC experiment at CERN [1] is to measure the lifetime of pionium, an atom consisting of a π^+ and a π^- meson ($A_{2\pi}$). The lifetime is dominated by the charge-exchange scattering process ($\pi^+\pi^- \to \pi^0\pi^0$) and is thus related to the relevant scattering lengths [3]. The partial decay width of the atomic ground state (principal quantum number $n = 1$, orbital quantum number $l = 0$) is [2, 4, 5, 6, 7, 8]

$$\Gamma_{1S} = \frac{1}{\tau_{1S}} = \frac{2}{9}\,\alpha^3\, p\,|a_0 - a_2|^2\,(1+\delta) \tag{1}$$

with τ_{1S} the lifetime of the atomic ground state, α the fine-structure constant, p the π^0 momentum in the atomic rest frame, and a_0 and a_2 the S-wave $\pi\pi$ scattering lengths for isospin 0 and 2, respectively. The term δ accounts for QED and QCD corrections [5, 6, 7, 8]. It is a known quantity ($\delta = (5.8 \pm 1.2) \times 10^{-2}$) ensuring a 1% accuracy for Eq. (1) [7]. A measurement of the lifetime therefore allows to obtain in a model-independent way the value of $|a_0 - a_2|$. The $\pi\pi$ scattering lengths a_0, a_2 have been calculated within the framework of Standard Chiral Perturbation Theory [9] with a precision better than 2.5% [10] ($a_0 = 0.220 \pm 0.005$, $a_2 = -0.0444 \pm 0.0010$, $a_0 - a_2 = 0.265 \pm 0.004$ in units of inverse pion mass) and lead to the prediction $\tau_{1S} = (2.9 \pm 0.1) \times 10^{-15}$ s. The Generalized Chiral Perturbation Theory though allows for larger a-values [11]. Model independent measurements of a_0 have been done using K_{e4} decays [12, 13].

Oppositely charged pions emerging from a high energy proton-nucleus collision may be either produced directly or stem from strong decays ("short-lived" sources) and electromagnetic or weak decays ("long-lived" sources) of intermediate hadrons. Pion pairs from "short-lived" sources undergo Coulomb final state interaction and may form atoms. The region of production being small as compared to the Bohr radius of the atom and neglecting strong final state interaction, the cross section σ_A^n for production of atoms

CP814 *Hadron Spectroscopy: Eleventh International Conference*
edited by A. Reis, C. Göbel, J. Sá Borges, and J. Magnin

with principal quantum number n is related to the inclusive production cross section for pion pairs from "short lived" sources without Coulomb correlation (σ_s^0) [14] :

$$\frac{d\sigma_A^n}{d\vec{p}_A} = (2\pi)^3 \frac{E_A}{M_A} |\Psi_n^C(\vec{r}^* = 0)|^2 \left. \frac{d^2\sigma_s^0}{d\vec{p}_+ d\vec{p}_-} \right|_{\vec{p}_+ = \vec{p}_-} \tag{2}$$

with $\vec{p}_A$, E_A and M_A the momentum, energy and mass of the atom in the lab frame, respectively, and $\vec{p}_+$, $\vec{p}_-$ the momenta of the charged pions. The square of the Coulomb atomic wave function for zero distance $\vec{r}^*$ between them in the c.m. system is $|\Psi_n^C(0)|^2 = p_B^3/\pi n^3$, where $p_B = m_\pi \alpha/2$ is the Bohr momentum of the pions and m_π the pion mass. The production of atoms occurs only in S-states [14].

Final state interaction also transforms the "unphysical" cross section σ_s^0 into a real one for Coulomb correlated pairs, σ_C [15, 16]:

$$\frac{d^2\sigma_C}{d\vec{p}_+ d\vec{p}_-} = |\Psi_{-\vec{k}^*}^C(\vec{r}^*)|^2 \frac{d^2\sigma_s^0}{d\vec{p}_+ d\vec{p}_-}, \tag{3}$$

where $\Psi_{-\vec{k}^*}^C(\vec{r}^*)$ is the continuum wave function and $2\vec{k}^* \equiv \vec{q}$ with $\vec{q}$ being the relative momentum of the π^+ and π^- in the c.m. system[1]. $|\Psi_{-\vec{k}^*}^C(\vec{r}^*)|^2$ describes the Coulomb correlation and at $r^* = 0$ coincides with the Gamov-Sommerfeld factor $A_C(q)$ with $q = |\vec{q}|$ [16]:

$$A_C(q) = \frac{2\pi m_\pi \alpha/q}{1 - \exp(-2\pi m_\pi \alpha/q)}. \tag{4}$$

For low q, $0 \leq q \leq q_0$, Eqs. (2, 3, 4) relate the number of produced $A_{2\pi}$ atoms, N_A, to the number of Coulomb correlated pion pairs, N_{CC} [17] :

$$\frac{N_A}{N_{CC}} = \frac{\sigma_A^{tot}}{\sigma_C^{tot}|_{q \leq q_0}} = \frac{(2\pi\alpha m_\pi)^3}{\pi} \frac{\sum_{n=1}^{\infty} 1/n^3}{\int_0^{q_0} A_C(q) d^3q} = k_{th}(q_0). \tag{5}$$

Eq. (5) defines the theoretical k-factor. Throughout the paper we will use

$$q_0 = 2\text{MeV}/c \quad \text{and} \quad k_{th}(q_0) = 0.615. \tag{6}$$

In order to account for the finite size of the pion production region and of the two-pion final state strong interaction, the squares of the Coulomb wave functions in Eqs. (2) and (3) must be substituted by the square of the complete wave functions, averaged over the distance $\vec{r}^*$ and the additional contributions from $\pi^0\pi^0 \to A_{2\pi}$ as well as $\pi^0\pi^0 \to \pi^+\pi^-$ [16]. It should be noticed that these corrections essentially cancel in the k-factor (Eq. (5)) and lead to a correction of only a fraction of a percent. Thus finite size corrections can safely be neglected for k_{th}.

---

[1] For the sake of clarity we use the symbol Q for the experimentally reconstructed and q for the physical relative momentum.

Once produced, the $A_{2\pi}$ atoms propagate with relativistic velocity (average Lorentz factor $\bar{\gamma} \approx 17$ in our case) and, before they decay, interact with target atoms, whereby they become excited/de-excited or break up. The $\pi^+\pi^-$ pairs from break-up (atomic pairs) exhibit specific kinematical features which allow to identify them experimentally [14], namely very low relative momentum q and q_L (the component of $\vec{q}$ parallel to the total momentum $\vec{p}_+ + \vec{p}_-$). After break-up, the atomic pair traverses the target and to some extent loses these features by multiple scattering, essentially in the transverse direction, while q_L is almost not affected. This is one reason for considering distributions in Q_L as well as in Q when analyzing the data.

Excitation/de-excitation and break-up of the atom are competing with its decay. Solving the transport equations with the cross sections for excitation and break-up, leads to a target-specific relation between break-up probability and lifetime which is estimated to be accurate at the 1% level [19]. Measuring the break-up probability thus allows to determine the lifetime of pionium [14].

The first observation of the $A_{2\pi}$ atom [20] has allowed to set a lower limit on its lifetime [17, 18] of $\tau > 1.8 \times 10^{-15}$ s (90% CL). Here we present a determination of the lifetime of the $A_{2\pi}$ atom, based on a large sample of data taken in 2001 with Ni targets.

THE DIRAC EXPERIMENT

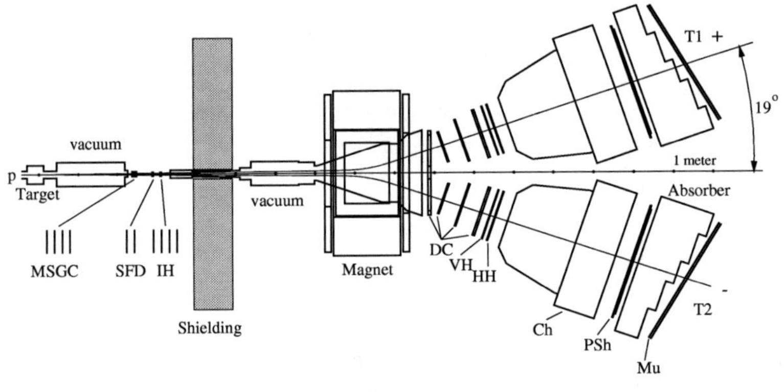

FIGURE 1. Schematic top view of the DIRAC spectrometer. Upstream of the magnet: target, microstrip gas chambers (MSGC), scintillating fiber detectors (SFD), ionization hodoscopes (IH) and iron shielding. Downstream of the magnet: drift chambers (DC), vertical and horizontal scintillation hodoscopes (VH, HH), gas Cherenkov counters (Ch), preshower detectors (PSh) and, behind the iron absorber, muon detectors (Mu).

The DIRAC experiment uses a magnetic double-arm spectrometer at the CERN 24 GeV/c extracted proton beam T8. Details on the set-up may be found in [21]. Since its start-up, DIRAC has accumulated about 15'000 atomic pairs. The data used for this work were taken with two Ni foils, one of 94μm thickness (76% of the $\pi^+\pi^-$ data), and one of 98 μm thickness (24% of the data). An extensive description of the DIRAC set-up, data selection, tracking, Monte Carlo procedures, signal extraction and a first high statistics demonstration of the feasibility of the lifetime measurement, based on the Ni data of 2001, have been published in [22].

The set-up and the definitions of detector acronyms are shown in Fig. 1. The main selection criteria and performance parameters [22] are recalled in the following.

Pairs of oppositely charged pions are selected by means of Cherenkov, preshower and muon counters. Through the measurement of the time difference between the vertical hodoscope signals of the two arms, time correlated (prompt) events ($\sigma_{\Delta t} = 185\,\text{ps}$) can be distinguished from accidental events (see [22]). The resolution of the three components of the relative momentum Q of two tracks, transverse and parallel to the c.m. flight direction, Q_x, Q_y and Q_L, is about 0.5 MeV$/c$ for $Q \leq 4\,\text{MeV}/c$. Due to charge combinatorials and inefficiencies of the SFD, the distributions for the transverse components have substantial tails, which the longitudinal component does not exhibit. This is yet another reason for analyzing both Q and Q_L distributions. Data were analyzed with the help of the DIRAC analysis software package ARIANE [23]. The tracking procedures require the two tracks originate from the intersect of the beam with the target.

ANALYSIS

The spectrometer including the target is fully simulated by GEANT-DIRAC, a GEANT3-based simulation code. The detectors, including read-out, inefficiency, noise and digitalization are simulated and implemented in the DIRAC analysis code ARIANE [23]. The triggers are fully simulated as well.

The simulated data sets for different event types can therefore be reconstructed with exactly the same procedures and cuts as used for experimental data.

The different event types are generated according to the underlying physics.

Atomic pairs: Atoms are generated according to Eq. (2) using measured total momentum distributions for short-lived pairs. The atomic $\pi^+\pi^-$ pairs are generated according to the probabilities and kinematics described by the evolution of the atom while propagating through the target and by the break-up process. These $\pi^+\pi^-$ pairs, starting from their spatial production point, are then propagated through the remaining part of the target and the full spectrometer using GEANT-DIRAC. Reconstruction of the track pairs using the fully simulated detectors and triggers leads to the atomic pair distribution dn_A^{MC}/dQ.

Coulomb correlated $\pi^+\pi^-$ pairs (CC-background): The events are generated according to Eqs. (3,4) using measured total momentum distributions for short-lived pairs. The generated q-distributions are assumed to follow phase space modified by the Coulomb correlation function (Eq. (4)), $dN_{CC}^{gen}/dq \propto q^2 \times A_C(q)$. Processing them with GEANT-DIRAC and then analyzing them using the full detector and trigger simulation leads to the Coulomb correlated distribution dN_{CC}^{MC}/dQ.

Non-correlated $\pi^+\pi^-$ pairs (NC-background): $\pi^+\pi^-$ pairs, where at least one pion originates from the decay of a "long-lived" source (e.g. electromagnetically or weakly decaying mesons or baryons) do not undergo any final state interactions. Thus they are generated according to $dN_{NC}^{gen}/dq \propto q^2$, using slightly softer momentum distributions than for short-lived sources (difference obtained from FRITIOF-6). The Monte Carlo distribution dN_{NC}^{MC}/dQ is obtained as above.

Accidental $\pi^+\pi^-$ pairs (acc-background): $\pi^+\pi^-$ pairs, where the two pions originate from two different proton-nucleus interactions, are generated according to $dN_{acc}^{gen}/dq \propto q^2$, using measured momentum distributions. The Monte Carlo distribution dN_{acc}^{MC}/dQ is obtained as above.

All the Monte Carlo distributions are normalized, $\int_0^{Q_{max}}(dN_i^{MC}/dQ)dQ = N_i^{MC}$, $i = CC, NC, acc$, with statistics about 5 to 10 times higher than the experimental data; similarly for atomic pairs (n_A^{MC}).

The measured prompt distributions are approximated by appropriate shape functions constructed from the simulated distribution discribed above. The fit parameters are n_A^{rec}, N_{CC}^{rec}, N_{NC}^{rec} the reconstructed number of atomic pairs, Coulomb- and non-correlated background, respectively.

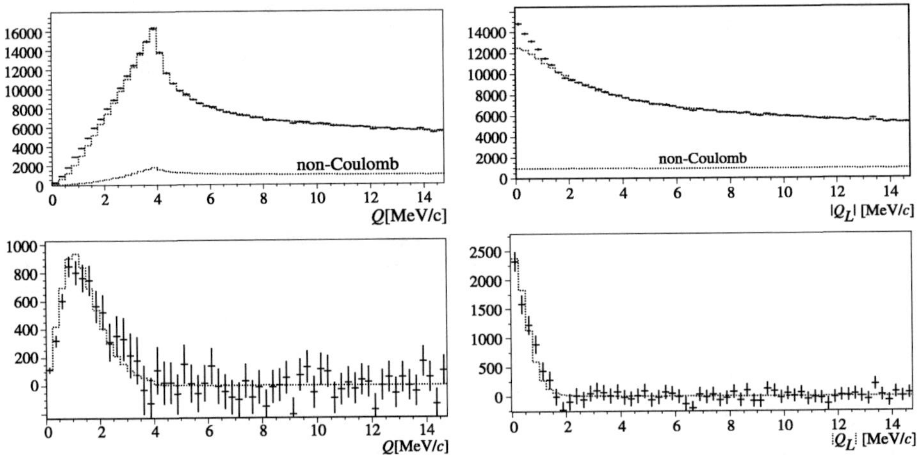

FIGURE 2. Top: Experimental Q and Q_L distributions after subtraction of the prompt accidental background, and fitted Monte Carlo backgrounds (dotted lines). The peak at $Q = 4\,\text{MeV}/c$ is due to the cut $Q_T \leq 4\,\text{MeV}/c$. Bottom: Residuals after background subtraction. The dotted lines represent the expected atomic signal shape. The bin-width is 0.25 MeV/c.

The data taken with 94 and 98 μm thick targets were analyzed separately. The total number of events in the prompt window is $N_{pr} = 471290$.

First, we determine the background composition outside of the atomic pair signal region, i.e. for $Q > 4\text{MeV}/c$ and $Q_L > 2\text{MeV}/c$. For this purpose we require $n_A^{rec} = 0$. As a constraint, the background parameters N_{CC}^{rec} and N_{NC}^{rec} representing the total number of CC- and NC-events, have to be the same for Q and Q_L. Then, with the parameters found, the background is subtracted from the measured prompt distribution, resulting in the residual spectra. For the signal region, defined by the cuts $Q = 4\text{MeV}/c$ and $Q_L = 2\text{MeV}/c$, we obtain the total number of atomic pairs, $n_A^{residual}$ and of Coulomb correlated background events, N_{CC}^{sig}. Results of fits for Q and Q_L together are shown in Table 1.

CC-background and NC- or acc-backgrounds are distinguishable due to their different shapes, most pronounced in the Q_L distributions (see Fig. 2, top). Accidental and NC-background shapes are almost identical for Q and fully identical for Q_L (uniform

TABLE 1. Fit results (94 and 98 μm targets together, background shapes from Monte Carlo (MC)) for the parameters N_{CC}^{rec} (total number of CC-events), N_{NC}^{rec} (total number of NC events) and n_A^{rec} (atomic pairs) and deduced results for the number of atomic pairs from the residuals ($n_A^{residual}$) and the number of CC-background events in the signal region (N_{CC}^{sig}). MC-a: background fit excluding the signal region. MC-b: fit of the entire momentum range including Monte Carlo shape for atomic pairs ("shape fit"). The cuts were at $Q_{cut} = 4\text{MeV}/c$ and $Q_{L,\ cut} = 2\text{MeV}/c$. Q and Q_L-distributions were fitted together. The normalized χ^2 were 0.9 for MC-a and MC-b.

		N_{CC}^{rec}	N_{NC}^{rec}	$n_A^{residual}$	n_A^{rec}	N_{CC}^{sig}
MC-a	Q	374022 ± 3969	56538	6518 ± 373		106500 ± 1130
	Q_L	same	same	6509 ± 330		82289 ± 873
MC-b	Q	374282 ± 3561	56213		6530 ± 294	106549 ± 1014
	Q_L	same	same		same	82345 ± 783

distributions). Thus, the errors in determining the accidental background ω_{acc} are absorbed in fitting the NC background. The correlation coefficient between CC and NC background is -99%. This strong correlation leads to equal errors for N_{CC}^{rec} and N_{NC}^{rec}. The CC-background is determined with a precision better than 1%. Note that the difference between all prompt events and the background is $N_{pr} - N_{CC}^{rec} - N_{NC}^{rec} - \omega_{acc}N_{pr} = 6590$, hence very close to the number of residual atomic pairs ($n_A^{residual}$) as expected. This relation is also used as a strict constraint for fits outside of the signal region (>), $N_{pr}^{>} - N_{CC}^{rec>} - N_{NC}^{rec>} - (\omega_{acc}N_{pr})^{>} = 0$ and, hence, the fit requires only one free parameter, $N_{CC}^{rec>}$.

Second, the atomic pair signal may be directly obtained by minimizing over the full range and including the Monte Carlo shape distribution F_A ("shape fit"). The signal strength has to be the same for Q and Q_L. The result for the signal strength n_A^{rec} as well as the CC-background below the cuts, N_{CC}^{sig}, are shown in Table 1.

The consistency between the analysis in Q with the one in Q_L establishes the correctness of the Q_T reconstruction. A 2D fit in the variables $(Q_L,\ Q_T)$ confirms the results of Table 1.

BREAK-UP PROBABILITY

In order to deduce the break-up probability, $P_{br} = n_A/N_A$, the total number of atomic pairs n_A and the total number of produced $A_{2\pi}$ atoms, N_A, have to be known. None of the two numbers is directly measured. The procedure of obtaining the two quantities requires reconstruction efficiencies and is as follows.

Number of atomic pairs: Using the generator for atomic pairs a large number of events, n_A^{gen}, is generated in a predefined large spatial acceptance window Ω_{gen}, propagated through GEANT-DIRAC including the target and reconstructed along the standard procedures. The total number of reconstructed Monte Carlo atomic pairs below an arbitrary cut in Q, $n_A^{MC-rec}(Q \leq Q_{cut})$ defines the reconstruction efficiency for atomic pairs $\varepsilon_A^{cut} = n_A^{MC-rec}(Q \leq Q_{cut})/n_A^{gen}$. The total number of atomic pairs is obtained from the measured pairs by $n_A = n_A^{rec}(Q \leq Q_{cut})/\varepsilon_A^{cut}$.

Number of produced $A_{2\pi}$ atoms: Here we use the known relation between produced atoms and Coulomb correlated $\pi^+\pi^-$ pairs (CC-background) of Eq. (5). Using the generator for CC pairs, N_{CC}^{gen} events, of which $N_{CC}^{gen}(q \leq q_0)$ (see Eq. 6) have q below q_0, are generated into the same acceptance window Ω_{gen} as for atomic pairs and processed analogously to the paragraph above to provide the number of reconstructed CC-events below the same arbitrary cut in Q as for atomic pairs, $N_{CC}^{MC-rec}(Q \leq Q_{cut})$. These CC-events are related to the originally generated CC-events below q_0 through $\varepsilon_{CC}^{cut} = N_{CC}^{MC-rec}(Q \leq Q_{cut})/N_{CC}^{gen}(q \leq q_0)$. The number of produced atoms thus is $N_A = k_{th}(q_0)N_{CC}^{rec}(Q \leq Q_{cut})/\varepsilon_{CC}^{cut}$ (see Eq. (6)).

The break-up probability P_{br} thus becomes:

$$P_{br} = \frac{n_A}{N_A} = \frac{n_A^{rec}(Q \leq Q_{cut})}{k(Q_{cut})N_{CC}^{rec}(Q \leq Q_{cut})} \qquad \text{with } k(Q_{cut}) = k_{th}(q_0)\frac{\varepsilon_A^{cut}}{\varepsilon_{CC}^{cut}}. \tag{7}$$

The k-factors for different cuts in Q and Q_L and for the two target thicknesses (94μm and 98μm) wer simulated and with the measurements listed in Table 1, the break-up probabilities of Table 2 are obtained. The simultaneous fit of Q and Q_L with the atomic shape results in a single value.

The break-up probabilities from Q and Q_L agree within a fraction of a percent. The values from shape fit and from background fit are in perfect agreement (see Table 1). We adopt the atomic shape fit value of $P_{br} = 0.447 \pm 0.023_{stat}$, because the fit covers the full Q, Q_L range and includes correlations between n_A^{rec} and N_{CC}^{sig}.

Analyzing the data with three allowed hit candidates in the SFD search window instead of two, results in more atomic pairs. The break-up probabilities obtained are 0.440 ± 0.024 and 0.430 ± 0.021 for Q and Q_L, respectively. They are not in disagreement with the adopted value of 0.447. Despite the larger statistics, the accuracy is not improved, due to additional background. This background originates from additional real hits in the upstream detectors or from electronic noise and cross-talk. This has been simulated and leads essentially to a reduced reconstruction efficiency but not to a deterioration of the reconstruction quality. The additional sources of systematic uncertainties lead us not to consider this strategy of analysis further on.

The break-up probability has to be corrected for the impurities of the targets. Thus, the 94 μm thick target has a purity of only 98.4%, while the 98 μm thick target is 99.98 % pure. The impurities (C, Mg, Si, S, Fe, Cu) being mostly of smaller atomic number than Ni lead (for the weighted average of both targets) to a reduction of the break-up probability of 1.1% as compared to pure Ni, assuming a lifetime of $3fs$. Therefore, the measured break-up probability has to be increased by 0.005 in order to correspond to pure Ni. The final result is: $P_{br} = 0.452 \pm 0.023_{stat}$.

SYSTEMATIC ERRORS

Systematic errors may occur through the analysis procedures and through physical processes which are not perfectly under control. We investigate first procedure-induced errors.

TABLE 2. Break-up probabilities for the combined Ni2001 data, based on the results of Table 1 for the cuts $Q_{cut} = 4\text{MeV}/c$ and $Q_{L,cut} = 2\text{MeV}/c$. Errors are statistical.

	$n_A^{residual}$	n_A^{rec}	N_{CC}^{sig}	P_{br}
Q	6518 ± 373		106500 ± 1130	0.442 ± 0.026
Q_L	6509 ± 330		82289 ± 873	0.445 ± 0.023
Q & Q_L		6530 ± 294	106549 ± 1004	0.447 ± 0.023

The break-up probability will change, if the ratio $N_{CC}^{rec}/N_{NC}^{rec}$ depends on the fit range. If so, the Monte Carlo distributions do not properly reproduce the measured distributions and the amount of CC-background may not be constant. The measured ratio is reasonably constant within errors, with the smallest errors for a fit range of $Q = Q_L = 15\text{MeV}/c$. At this point the difference between Q and Q_L fits leads to a difference in break-up probability of $\Delta P_{br}^{CC} = 0.023$.

To investigate the used atomic pair signal shape influence, we studied two extreme models for atom break-up: break-up only from the $1S$-state and break-up only from highly excited states. The two extremes result in a difference in break-up probability of $\Delta P_{br}^{shape} = 0.008$.

Sources of systematic errors may also arise from uncertainties in the genuine physical process. We have investigated possible uncertainties in multiple scattering as simulated by GEANT by changing the scattering angle in the GEANT simulation by $\pm 5\%$. As a result, the break-up probability changes by 0.002 per one percent change of multiple scattering angle. In fact we have measured the multiple scattering for all scatterers (upstream detectors, vacuum windows, target) and found narrower angular distributions than expected from the standard GEANT model [24]. This, however, may be due also to errors in determining the thickness and material composition of the upstream detectors. Based on these studies we conservatively attribute a maximum error of +5% and -10% to multiple scattering.

Another source of uncertainty may be due to the presence of unrecognized K^+K^- and $\bar{p}p$ pairs that would fulfill all selection criteria [25]. Such pairs may be as abundant as 0.5% and 0.15%, respectively, of $\pi^+\pi^-$ pairs as estimated for K^+K^- with FRITIOF-6 and for $\bar{p}p$ from time-of-flight measurements in a narrow momentum interval with DIRAC data. Their mass renders the Coulomb correlation much more peaked at low Q than for pions, which leads to a change in effective $\pi^+\pi^-$ Coulomb background at small Q, thus to a smaller atomic pair signal and therefore to a decrease of break-up probability. The effect leads to a change of $\Delta P_{br}^{\bar{K}K,\ \bar{p}p} = -0.04$. We do not apply this shift but consider it as a maximum systematic error of P_{br}. Admixtures from unrecognized e^+e^- pairs from photon conversion do not contribute because of their different shapes.

Finally, the correlation function Eq. (3) used in the analysis is valid for pointlike production of pions, correlated only by the Coulomb final state interaction (Eq. (4)). However, there are corrections due to finite size and strong interaction [16]. These have been studied based on the UrQMD transport code simulations [26] and DIRAC data on $\pi^-\pi^-$ correlations. The parameters of the underlying model are statistically fixed with data up to 200 MeV/c relative momentum. For $Q \leq 30\,\text{MeV}/c$, the DIRAC data

TABLE 3. Summary of systematic effects on the measurement of the break-up probability P_{br}. Extreme values have been transformed into σ assuming uniform distributions.

source	extreme values	σ
CC-background	+0.012 / -0.012	± 0.007
signal shape	+0.004 / -0.004	± 0.002
multiple scattering	+0.01 /-0.02	$^{+0.006}_{-0.013}$
K^+K^- and $\bar{p}p$	$+0\ /-0.04$	$^{+0}_{-0.023}$
finite size	$+0\ /-0.03$	$^{+0}_{-0.017}$
Total		$^{+0.009}_{-0.032}$

are too scarce to serve as a test of the model. The corrections lead to a change of $\Delta P_{br}^{finite-size} = -0.02$. Due to the uncertainties we conservatively consider 1.5 times this change as a maximum error, but do not modify P_{br}.

The systematics are summarized in Table 3. The extreme values represent the ranges of the assumed uniform probability density function (u.p.d.f.), which, in case of asymmetric errors, were complemented symmetrically for deducing the corresponding standard deviations σ. Convoluting the five u.p.d.f. results in bell-shaped curves very close to a Gaussian, and the $\pm\sigma$ (Table 3, total error) correspond roughly to a 68.5% confidence level and can be added in quadrature to the statistical error.

The final value of the break-up probability is

$$P_{br} = 0.452 \pm 0.023_{stat} \left.{}^{+0.009}_{-0.032}\right\}_{syst} = 0.452\ {}^{+0.025}_{-0.039}. \tag{8}$$

LIFETIME OF PIONIUM

The lifetime may be deduced on the basis of the relation between break-up probability and lifetime for a pure Ni target (Fig. 3). This relation, estimated to be accurate at the 1% level, may itself have uncertainties due to the experimental conditions. Thus the target thickness is estimated to be correct to better than $\pm$ 1 μm, which leads to an error in the lifetime (for $P_{br} = 0.45$) smaller than $\pm$ 0.01 fs, less than 1% of the expected lifetime and thus negligible. The result for the lifetime is

$$\tau_{1S} = \left[2.91\ \left.{}^{+0.45}_{-0.38}\right\}_{stat}\ \left.{}^{+0.19}_{-0.49}\right\}_{syst}\right] \times 10^{-15}\ \text{s} = \left[2.91\ {}^{+0.49}_{-0.62}\right] \times 10^{-15}\ \text{s}. \tag{9}$$

The errors are not symmetric because the $P_{br} - \tau$ relation is not linear, and because finite size corrections and heavy particle admixtures lead to possible smaller values of P_{br}. The accuracy achieved for the lifetime is about $+17\%$, almost entirely due to statistics and -21%, due to statistics and systematics in roughly equal parts. With full statistics (2.3 times more than analysed here) the statistical errors may be reduced accordingly. The two main systematic errors (particle admixtures and finite size correction) will be studied in more detail in the future program of DIRAC.

Using Eq. (1), the above lifetime corresponds to $|a_0 - a_2| = 0.264\,^{+0.033}_{-0.020}\, m_\pi^{-1}$.

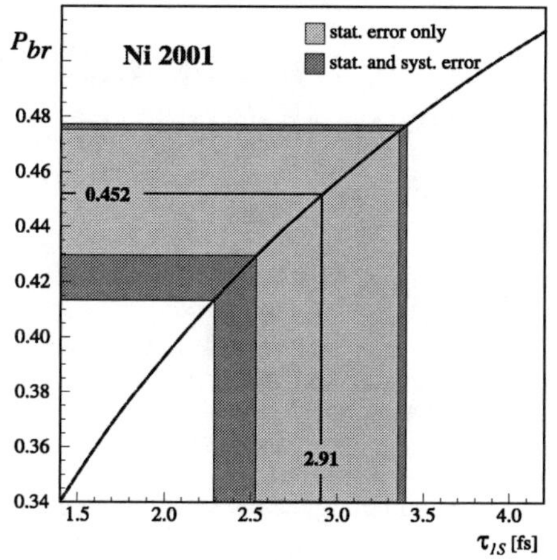

FIGURE 3. Break-up probability P_{br} as a function of the lifetime of the atomic ground state τ_{1S} for the combined 94 and 98 μm thick Ni targets. The experimentally determined P_{br} with statistical and total errors translates into a value of the lifetime with corresponding errors.

REFERENCES

1. B. Adeva et al., DIRAC proposal, CERN/SPSLC 95-1, SPSLC/P 284 (1995).
2. J. Uretsky and J. Palfrey, Phys. Rev. 121 (1961) 1798.
3. S. Deser et al., Phys. Rev. 96 (1954) 774.
4. S.M. Bilenky et al., Yad. Phys. 10 (1969) 812; (Sov. J. Nucl. Phys. 10 (1969) 469).
5. H. Jallouli and H. Sazdjian, Phys. Rev. D58 (1998) 014011; Erratum: ibid., D58 (1998) 099901.
6. M.A. Ivanov et al. Phys. Rev. D58 (1998) 094024.
7. J. Gasser et al., Phys. Rev. D64 (2001) 016008; hep-ph/0103157.
8. A. Gashi et al., Nucl. Phys. A699 (2002) 732.
9. S. Weinberg, Physica A96 (1979) 327; J. Gasser and H. Leutwyler, Phys. Lett. B125 (1983) 325; ibid Nucl. Phys. B250 465, 517, 539.
10. G. Colangelo, J. Gasser and H. Leutwyler, Nucl. Phys. B603 (2001) 125.
11. M. Knecht et al., Nucl. Phys. B457 (1995) 513.
12. L. Rosselet et al., Phys. Rev. D15 (1977) 547.
13. S. Pislak et al., Phys. Rev. Lett, 87 (2001) 221801.
14. L.L. Nemenov, Yad. Fiz. 41 (1985) 980; (Sov. J. Nucl. Phys. 41 (1985) 629).
15. A.D.Sakharov, Z.Eksp.Teor.Fiz. 18 (1948) 631.
16. R. Lednicky, DIRAC note 2004-06, nucl-th/0501065.
17. L.G. Afanasyev, O.O. Voskresenskaya, V.V.Yazkov, Communication JINR P1-97-306, Dubna, 1997.
18. L.G. Afanasyev et al., Phys. Lett. B338 (1994) 478.
19. L. Afanasyev et al., J. Phys B37 (2004) 4749.
20. L.G. Afanasyev et al., Phys. Lett. B308 (1993) 200.
21. B. Adeva B et al., Nucl.Instr.Meth. A515 (2003) 467.
22. B. Adeva et al. J. Phys. G30 (2004) 1929.
23. D. Drijard, M. Hansroul and V. Yazkov, The DIRAC offline user's Guide, and http://dirac.web.cern.ch/DIRAC/offlinedocs/Userguide.html
24. A. Dudarev et al., DIRAC note 2005-02.
25. O.E. Gortchakov and V.V. Yazkov, DIRAC note 2005-01.
26. S. A. Bass et al., Prog. Part. Nucl.Phys. 41 (1998) 225.

Hadronization – the Unsung Hero rather than the alleged Villain in the Tale of CP Violation

I.I. Bigi

Dept. of Physics, University of Notre Dame du Lac, Notre Dame, IN 46556, U.S.A.
e-mail: ibigi@nd.edu

Abstract. The novel successes scored by the Standard Model of High Energy Physics in the last few years concerning heavy flavour dynamics do not weaken the case for 'New Physics' around the TeV scale. They do suggest however that one cannot *count* on that New Physics impacting heavy flavour decays in a numerically massive way. Yet studying this impact will be essential in diagnosing the features of the New Physics. In particular the decays of beauty hadrons have to be analyzed with considerable precision on the experimental as well as theoretical side. While hadronization effects often represent the main bottleneck in our understanding in the short run, they will provide powerful and discriminating tools in the long run, when applied comprehensively and judiciously. The expertise required to exhaust the discovery potential in B decays does exist in the hadron physics community or can be developed without needing a new breakthrough – yet a greater effort has to be made to communicate it to the heavy flavour community.

Keywords: Search for New Physics, **CP** violation, Hadronization
PACS: 11.30.Er, 12.15.Hh, 13.25.Hw

PROLOGUE

There are three reasons why I am grateful for the invitation to participate in this conference and speak to you. Terry Goldman expressed the third reason the other day: Rio de Janeiro gives a new and very pleasant meaning to the word 'winter conference'. The second reason will be explained in my talk. My first reason is of a general nature. I am experiencing high energy physics as a truly noble human activity and view it as the most global one, where I cannot see any drawback to globalization – contrary to other manifestations of globalization. However as far as its geographic distribution is concerned, I am exaggerating: marking on a globe where high energy physics research is done, you will find it happens almost exclusively in the northern hemisphere. This lack of balance concerns me, and I hope for a wider distribution in the future. We owe a lot of gratitude to our colleagues, who maintain and further build high energy physics in the southern hemisphere. I view CBPF here in Rio as the flagship of fundamental physics in the southern hemisphere, appreciate the work Alberto Reis and his colleagues are doing in 'spreading the gospel' and promise my support to them.

CP814 *Hadron Spectroscopy: Eleventh International Conference*
edited by A. Reis, C. Göbel, J. Sá Borges, and J. Magnin

INTRODUCTION

Around the turn of the millenium – in the years 1998 – 2001 – a 'quantum jump' in our knowledge (though *not* our understanding) of fundamental physics has occurred:

- "the completion of an era": the establishment of *direct* **CP** violation in K_L decays;
- "the beginning of a new journey": the first observation of **CP** violation in a system other than the $K^0 - \bar{K}^0$ complex, namely in $B_d \to \psi K_S$;
- "messages from the heavens": the discovery of neutrino oscillations;
- "the birth of a Standard Model of Cosmology": a consistent picture concerning the composition of our Universe [1] in terms of visible and dark matter and 'dark energy' has emerged; it describes the distribution of the cosmic microwave radiation and its fluctuations. The evidence for 'dark energy' can best be commented by repeating Rabi's famous quote about the muon:"Who ordered that?"

The first three items certainly and the fourth one probably impact our views of microscopic dynamics profoundly, although it is still too early to say how. Concerning the first two items I would like to emphasize without an explanation (that can be found in my Varenna lectures [1]) that the SM has scored not 'merely' more or even new successes, but actually novel ones, i.e. of a qualitatively new kind: the *predicted* 'Paradigm of large **CP** violation in B decays' [2] has been verified through the discovery of indirect and direct **CP** violation in $B_d \to \psi K_S$, $B \to K\pi$ and presumably also in $B_d \to \pi^+\pi^-$ [3, 4]; these large **CP** asymmetries are commensurate with **T** violation with*out* assuming the full power of **CPT** invariance [5].

Yet none of these novel successes of the SM weaken the case for it being incomplete and for New Physics being 'nearby', i.e. around the TeV scale. We have 'heavenly' data showing the incompleteness of the SM: baryogenesis, neutrino oscillations, 'dark matter' and the most puzzling feature of all, 'dark energy'. On the theoretical side we still have not solved the Strong CP Problem [7].

More generally these successes do not shed light on any of the mysterious features of the SM.
(i) What are the dynamics driving the electroweak symmetry breaking of $SU(2)_L \times U(1) \to U(1)_{QED}$. How can we tame the instability of Higgs dynamics with its quadratic mass divergence? I find the arguments compelling that point to New Physics at the $\sim$ TeV scale – like low-energy SUSY; therefore I call it the 'confidently predicted' New Physics or *cpNP*.
(ii) The family structure with the quantization of the electric charges of quarks and leptons as expressed for example through

$$Q_e = 3Q_d \, . \tag{1}$$

This would be naturally explained through Grand Unification at very high energy scales implemented through, e.g., $SO(10)$ gauge dynamics. I call this the 'guaranteed New

[1] If indeed there exist other 'universes' besides ours, the term *uni*verse is no longer appropriate. With our 'universe' being just one among countless others, also the name 'cosmos' is misplaced. For this Greek word refers to an order that derives its beauty from careful arrangement.

Physics' or *gNP*.

(iii) *Finite family replication*: we infer from the observed width of Z^0 decays that there are three (light) neutrino species. The hierarchical pattern of CKM parameters as revealed by the data is so peculiar as to suggest that some other dynamical layer has to underlie it. I refer to it as 'strongly suspected New Physics' or *ssNP*. Saying we pin our hopes for explaining the family replication on Super-String or M theory is a scholarly way of saying we have hardly a clue what that *ssNP* is.

Comprehensive studies of heavy flavour dynamics might provide insights into the *gNP* and *ssNP*, items **(ii)** and **(iii)**, but I would not count on it. *However – and this is the central element of my message – they will be crucial in identifying the cpNP.*

In a nutshell my message reads as follows: We expect confidently that New Physics surfaces at the TeV scale. Yet we have to aim beyond 'merely' establishing the existence of New Physics – our goal has to be to identify its salient features. TeV scale dynamics is likely to have some nontrivial impact on B decays, and the discovery potential in B decays is essential for figuring out the *cpNP*, *not* a luxury. Yet due to the past 'unlikely' success of the CKM description one can*not count* on *massive* manifestations of New Physics, at least not in B decays typically. Therefore we need high accuracy both on the experimental and theoretical side in heavy flavour studies. This requires a better *quantitative* understanding of *hadronization* to exhaust the discovery potential in B decays. The necessary expertise exists or can be acquired with*out* a new theoretical breakthrough.

By the past 'unlikely' success of the CKM description I mean the following. As far as we know large fractions of the observables ΔM_K, ε_K and ΔM_B could be due to physics beyond the SM and even most of ε'. Likewise the constraints from the data on the CKM parameters translate into seemingly broad bands in plots of the CKM unitarity triangle [6]. The problem with this statement is that it is not even wrong – it just misses the main point: it is already very remarkable and highly nontrivial that all these constraints can be incorporated in a meaningful way in such plots. This becomes more obvious when plotting these and other observables from the strange, charm and beauty transitions on a commensurate energy scale. They cover several orders of magnitude between 10^{-8} and 10^{-16} MeV. That the CKM expectations are always within 50 % or so of the data over this range is most remarkable and could not be expected a priori – in particular when coupled with the fact that these predictions are based on values for the CKM and other mass parameters that would seem frivolous at best, if they were not forced upon us by independent measurements. The huge top quark mass of about 175 GeV (which is the only quark mass not only to approach the vector boson masses, but even to exceed them considerably) provides a case in point. There could easily have been inconsistencies. In other words: for CKM theory to provide a satisfactory description of the data, its parameters had to reside in a very peculiar corner of the general parameter space – yet nature has put them exactly into that 'neighbourhood' [7]. Therefore some of us had been rather confident that the **CP** asymmetry measured in $B_d \to \psi K_S$ would be close to the CKM prediction, whatever the latter would be – as indeed is the case.

The observation of this asymmetry in 2001 by both BABAR and BELLE [4] has established the CKM paradigm as a *tested* theory that no longer deserves the patronizing label of an 'ansatz'. Furthermore a 'demystification' of **CP** violation has occurred: we have seen that if the dynamics are sufficiently rich as to be able to support **CP** violation, the latter can be large (although we still do not know how to give meaning to the notion of 'maximal' **CP** violation). This process of demystification will have been completed, if **CP** violation is found anywhere in the lepton sector – a point I will return to later. We also know that the standard CKM theory is utterly irrelevant for baryogenesis, that the observed baryon number of the Universe shows the need for New Physics [8].

In summary: we have to conduct dedicated and comprehensive searches for New Physics. You have heard that before, have you not? And it reminds you of a quote by Samuel Beckett:

"Ever tried? Ever failed?
No matter.
Try again. Fail again. Fail better."

Only an Irishman can express profound skepticism concerning the world in such a poetic way. Beckett actually spent most of his life in Paris, since Parisians like to listen to someone expressing such a world view, even while they do not share it. Being in the service of Notre Dame du Lac, the home of the 'Fighting Irish', I cannot just ignore such advice. Yet my response is: "Cheer up – we know there is New Physics (see above) – we will not fail forever!"

My friend Antonio Masiero likes to say: "You have to be lucky to find New Physics." True enough – yet let me quote someone who just missed by one year being a fellow countryman of Masiero, namely Napoleon, who said: "Being lucky is part of the job description for generals." Quite seriously I think that if you as an high energy physicist do not believe that someday somewhere you will be a general – maybe not in a major encounter, but at least in a skirmish – then you are in the wrong line of business.

In the following I will focus on **CP** violation for rather pragmatic reasons; they are one of the most sensitive probes for New Dynamics, since **CP** asymmetries can be *linear* in a New Physics amplitude rather than quadratic and thus exhibit an enhanced sensitivity to New Physics.

PRELUDE: SINGING THE PRAISE OF HADRONIZATION

Hadronization and nonperturbative dynamics in general are usually viewed as unwelcome complication, if not outright nuisances. A case in point was already mentioned: while I view the CKM predictions for ΔM_K, ΔM_B, ε_K to be in remarkable agreement with the data, significant contributions from New Physics could be hiding there behind the theoretical uncertainties. While this is factually correct, it misses a deeper truth. *Without* hadronization bound states of quarks and antiquarks will not form; without the existence of kaons $K^0 - \bar{K}^0$ oscillations obviously cannot occur. It is hadronization that provides the 'cooling' of the (anti)quark degrees of freedom, which allows subtle quantum mechanical effects to add up coherently over macroscopic distances. Otherwise one would not have access to a super-tiny energy difference $\mathrm{Im}\mathcal{M}_{12} \sim 10^{-8}$ eV, which is very sensitive to different layers of dynamics, and indirect **CP** violation could not manifest itself. The same would hold for B mesons and $B^0 - \bar{B}^0$ oscillations.

To express it in a more down to earth way:

- Hadronization leads to the formation of kaons and pions with masses exceeding greatly (current) quark masses. It is the *hadronic* phase space that suppresses the **CP** *conserving* rate for $K_L \to 3\pi$ by a factor ~ 500, since the K_L barely resides above the three pion threshold.
- It awards 'patience'; i.e. one can 'wait' for a pure K_L beam to emerge after starting out with a beam consisting of K^0 and $\bar{K}^0$.
- It enables **CP** violation to emerge in the *existence* of a reaction, namely $K_L \to 2\pi$, rather than an asymmetry; this greatly facilitates its observation.

For these reasons alone we should praise hadronization as the hero in the tale of **CP** violation rather than the villain it is all too often portrayed.

'KING KONG SCENARIOS' FOR NEW PHYSICS SEARCHES

This scenario can be portrayed as follows: "It is unlikely that one will encounter King Kong; yet once it happens there will be no doubt that one has come across something wildly out of the ordinary." The point of analogy is the following. When, say, a certain transition has been observed to proceed although it is predicted not to – like $K_L \to 2\pi$ in 1964 – or when predicted and observed rates differ by orders of magnitude, we can speak of a 'qualitative' discrepancy, which establishes the existence of New Physics right away, though not its nature. This has happened with the decays of strange hadron – and it might happen again.

CP Violation in Charm Decays

I can be very brief here, since a detailed exposition has been given in the 'Cicerone' [9] and we have heard a nice talk by D. Asner [10].

The relative dullness of the weak SM phenomenology for charm can be harnessed as a laboratory for studying hadronization. Yet charm decays still have the potential to reveal New Physics as well. One should note that New Physics in general and flavour changing neutral currents in particular can affect down-type – d, s and b – and up-type quarks – u, c and t – quite differently; charm is the *only* up-type quark allowing the full range of probes for New Physics:

- top quarks do *not* hadronize [11];
- $\pi^0 - \pi^0$ oscillations are not possible, since neutral pions are their own antiparticle; with so few decay modes for pions **CP** asymmetries are basically ruled out by **CPT** symmetry.

My basic contention is: charm transitions provide unique portals for obtaining novel access to New Physics with the experimental situation being a priori favourable (apart from the fact that its leading decays are Cabibbo allowed). Searches for **CP** violation constitute the most powerful and promising tool in the long run for several reasons: (i)

Baryogenesis requires new sources of **CP** violation. (ii) *Within the SM* the weak phase is highly diluted in once Cabibbo suppressed modes, namely $\mathcal{O}(\sin^4\theta_C) \sim 10^{-3}$, and zero in Cabibbo favoured and doubly suppressed modes (except for $D^\pm \to K_S\pi^\pm$ [15]). (iii) A **CP** asymmetry can be *linear* in the New Physics amplitude and thus is more sensitive to it. (iv) Final state interactions are very active in the charm region. (v) The branching ratios for **CP** eigenstates are large. (vi) There is one fly in the ointment, though: $D^0 - \bar{D}^0$ oscillations are slow at best [9].

B-factories clearly can make a tremendous contributions to **CP** studies in the charm sector. The challenge in particular to LHCb is: "Can you?" To which degree can LHCb search for an asymmetry in $D^{*+} \to D^0(t) + \pi^+ \to [K^+\pi^-]_D + \pi^+$ vs. $D^{*-} \to \bar{D}^0(t) + \pi^- \to [K^-\pi^+]_D + \pi^-$ as a function of the time of decay t?

CP Violation in Leptodynamics

There is a compelling impetus for searching for **CP** violation in leptodynamics, namely to complete the aforementioned 'demystification' of **CP** violation and to get a firmer handle on baryogenesis as driven by primary leptogenesis [12].

Neutrino Oscillations

When searching for a **CP** asymmetry in neutrino oscillations, you do not have to worry about hadronization. Yet one has to disentangle genuine **CP** violation from enhancements due to the environment being made up of matter rather than antimatter. This will likely turn out to be a challenging task. I would not be surprised if our colleagues engaged in this enterprise will rue previous requests to be free of hadronization. This reminds me of an ancient Greek saying: "When the gods really want to harm you, they fulfill your wishes."

Electric Dipole Moments

Truly impressive experimental sensitivity has been achieved concerning the electric dipole moments of electrons and neutrons:

$$d_e = (0.07 \pm 0.07) \cdot 10^{-26}\,\text{ecm}\,,\; d_N < 0.63 \cdot 10^{-25}\,\text{ecm} \tag{2}$$

Even so these upper bounds are still many orders of magnitude above CKM predictions

$$d_e^{CKM} < 10^{-36}\,\text{ecm}\,,\; d_N^{CKM} < 10^{-30}\,\text{ecm} \tag{3}$$

On the other hand typical benchmarks for New Physics scenarios are

$$d_e^{NP}\,,\; d_N^{NP} \sim (10^{-28} - 10^{-26})\,\text{ecm} \tag{4}$$

a range that should be reached in the foreseeable future.

CP *Violation in τ Decays – the Next Hero Candidate*

The most promising channels for exhibiting **CP** asymmetries are $\tau \to \nu K \pi$, since due to the heaviness of the lepton and quark flavours they are most sensitive to nonminimal Higgs dynamics, and they can show asymmetries also in the final state distributions rather than integrated rates [13].

There is also a *unique* opportunity in $e^+e^- \to \tau^+\tau^-$: since the τ pair is produced with its spins aligned, the decay of one τ can 'tag' the spin of the other τ. I.e., one can probe *spin-dependent* **CP** asymmetries with *unpolarized* beams. This provides higher sensitivity and more control over systematic uncertainties.

I feel these features are not sufficiently appreciated even by proponents of Super-B factories. It has been recently pointed [14] out that based on known physics one can actually predict a **CP** asymmetry:

$$\frac{\Gamma(\tau^+ \to K_S \pi^+ \overline{\nu}) - \Gamma(\tau^- \to K_S \pi^- \nu)}{\Gamma(\tau^+ \to K_S \pi^+ \overline{\nu}) + \Gamma(\tau^- \to K_S \pi^- \nu)} = (3.27 \pm 0.12) \times 10^{-3} \tag{5}$$

due to K_S's preference for antimatter.

THE NEED FOR PRECISION – EXTRACTING $V(CB)$ AS A LESSON

New Physics scenarios typically will not create massive deviations from the SM predictions in B decays. Thus precision is essential on the experimental as well as theoretical side. This is not a noble, yet unreachable goal. It can be achieved when one combines a *robust* theoretical framework with comprehensive and detailed data. I will briefly illustrate it with a mature example, namely the extraction of $V(cb)$ from inclusive semileptonic B decays.

The robust theoretical framework there was provided by heavy quark expansions, which treat *non*perturbative effects through an expansion in inverse powers of the heavy quark mass m_Q. It provides an analytical algorithm genuinely based on QCD and receiving support from various sum rules; in the future it can incorporate findings from lattice QCD as well [16].

The total semileptonic width of B mesons as well as the lepton spectra encoded through lepton energy and hadronic mass *moments* can be expressed through heavy quark parameters $V(cb)$, beauty and charm quark masses $m_{c,b}$, and B meson expectation values of local heavy quark operators [17, 18]. With a mere handful of such parameters one can describe a host of observables. Those provide a large number of *overconstraints*. Obtaining a consistent fit is thus far from assured a priori. Achieving it provides an impressive demonstration that even systematic uncertainties are under control [19, 20] . As an extra bonus one has found that the fitted values of these heavy quark parameters obey certain theoretical constraints although those were *not* imposed on the fit; furthermore they even agree with other determinations of these parameters. For example the (kinetic) b quark mass can be determined in B production just above threshold – $\Upsilon(4S) \to b\bar{b}$ – which is determined by the *strong and electromagnetic* forces, and in the *weak* decays of B hadrons:

$$m_b^{kin}(1\text{ GeV})|_{\Upsilon(4S)\to b\bar{b}} = 4.57 \pm 0.08\text{ GeV } vs.\ m_b^{kin}(1\text{ GeV})|_{B\to l\nu X_c} = 4.61 \pm 0.068\text{ GeV} \tag{6}$$

With them being fully consistent, it makes sense to average them. A comprehensive analysis [21] yields (with the relative error given in paranthesis)

$$\begin{aligned} m_b^{kin}(1\text{ GeV}) &= (4.59 \pm 0.04)\text{ GeV } (\sim 1\%) \\ m_c^{kin}(1\text{ GeV}) &= (1.14 \pm 0.06)\text{ GeV } (\sim 5\%) \\ |V(cb)| &= (41.58 \pm 0.67)\cdot 10^{-3}\ (\sim 1.6\%) \end{aligned} \tag{7}$$

One should note here that the present uncertainty on $|V(us)|$ is about 1.1 %. The numbers in Eq.(7) show that percent level precision is not utopian even when nonperturbative dynamics is involved.

CASE STUDIES FOR HADRONIZATION AS A DIFFICULT ALLY

ΔM_K, ε_K and ε' – weak observables of central importance – are affected by hadronization effects like $\pi\pi$ phase shifts, the $\eta - \eta'$ wave functions, the role of the σ resonance etc., yet in a way that has never been reliably quantified. Here I will list four case studies of a more recent provenance, which concern B decays and where hadronization can actually be employed as a powerful tool – if applied judiciously.

First I would like to remind you that spectroscopy has been the subject of philosophical reflections over the centuries, the fruits of which can be distilled into the 'three Razors of Spectroscopy':

Ockham's Razor:
"Entities should not be multiplied unnecessarily!"

Peter Minkowski's Razor:
"No experiment can make two discoveries in the same data set!"

Stefano Bianco's Razor:
"You can learn a lot by cutting judiciously!"

These rules are not necessarily binding; however you can ignore them only at your own peril. The first Razor is almost self-evident, although open to interpretation, in particular a posteriori. The second Razor can be based on many examples from history. For the third one, which is the only one with a positive content, I will give an example below.

Case I: The '1/2 > 3/2 Puzzle'

Beyond what charm spectroscopy can teach us about QCD there are (at least) three more motivations for understanding it: (i) To determine the total width for $B \to l\nu X_c$ and keep the error on it small, one needs accurate and realistic modeling of the hadronic system in the final state. (ii) One needs it to gauge the errors on the measurements of the exclusive modes $B \to l\nu D/D^*$ due to a feed down from higher resonances. (iii) With heavy quark symmetry decoupling the spin of the heavy quark, one can label heavy flavour hadrons by their total spin and the angular momentum s_q carried by the light degrees of freedom [22]. To the degree that charm can be treated as a heavy flavour, one has two meson groundstates – D & D^* – and four excited states formed by P wave $c\bar{q}$ configurations, two of which carry $s_q = 3/2$ and are predicted to be narrow with the other two carrying $s_q = 1/2$ and being broad. With heavy quark theory one can derive sum rules from QCD that relate heavy quark parameters – quark masses, B expectation values of heavy quark operators – to subclasses of semileptonic B decay rates, in particular $B \to l\nu D(s_q = 1/2 \,\mathrm{or}\, 3/2)$. If the P wave states saturate those sum rules – a very natural, albeit not proven assumption – then the production of the narrow $s_q = 3/2$ charm meson states should dominate over that of the broad $s_q = 1/2$ ones in semileptonic B decays [23, 9]. This prediction is in apparent conflict with DELPHI studies [19, 24]. They show that the final states not being just D or D^* are made up more by broad than narrow charm states. One possible conjecture for resolving this discrepancy is that a significant fraction of the broad component is made up by *radial* excitations rather than P wave $D(s_q = 1/2)$.

This is not an academic problem and actually goes beyond even precise determinations of $|V(cb)|$ and $|V(ub)|$: the SM predicts $b \to l\nu q$ to proceed by pure $(V-A)_q \times (V-A)_l$ currents. A $(V-A)_q \times (V+A)_l$ coupling is most unlikely even with New Physics, since the required right-handed neutrino is unlikely to be sufficiently light due to Majorana masses; yet New Physics could conceivably induce a $(V+A)_q \times (V-A)_l$ current. Its main impact would be on the *shape* of the lepton spectrum making it softer and thus changing its moments. However such an impact had to be disentangled from hadronization effects. Modeling incorrectly the charm final state could then either fake a signal for $(V+A)_q \times (V-A)_l$ currents or hide one.

Case II: ϕ_1 from $B_d \to 3$ Kaons

Analysing **CP** violation in $B_d \to \phi K_S$ decays is a most promising way to search for New Physics. For the underlying quark-level transition $b \to s\bar{s}s$ represents a pure loop-effect in the SM, it is described by a single $\Delta B = 1$& $\Delta I = 0$ operator (a 'Penguin'), a reliable SM prediction exists for it [25] – $\sin 2\phi_1(B_d \to \psi K_S) \simeq \sin 2\phi_1(B_d \to \phi K_S)$ – and the ϕ meson represents a *narrow* resonance.

Great excitement was created when BELLE reported a large discrepancy between the predicted and observed **CP** asymmetry in $B_d \to \phi K_S$. Based on more data taken, this discrepancy has shrunk considerably: the BABAR/BELLE average for 2005 yields [4]

$$\sin 2\phi_1(B_d \to \psi K_S) = 0.685 \pm 0.032 \tag{8}$$

compared to

$$\sin 2\phi_1(B_d \to \phi K_S) \left\{ \begin{array}{ll} 0.50 \pm 0.25^{+0.07}_{-0.04} & \mathrm{BABAR} \\ 0.44 \pm 0.27 \pm 0.05 & \mathrm{BELLE} \end{array} \right. ; \tag{9}$$

BABAR's as well as BELLE's numbers are below the prediction, albeit by one sigma only. It is ironic that such a smaller deviation, although not significant, is actually more believable than the large one originally reported by BELLE.

This issue has to be pursued with vigour, since this reaction provides such a natural portal to New Physics. One complication has to be studied, though, in particular if the observed value of $\sin 2\phi_1(B_d \to \phi K_S)$ falls below the predicted one by a moderate amount only. For one is actually observing $B_d \to K^+K^-K_S$. If there is a single weak phase like in the SM one finds

$$\sin 2\phi_1(B_d \to \phi K_S) = -\sin 2\phi_1(B_d \to `f_0(980)' K_S)\ , \tag{10}$$

where '$f_0(980)'$ denotes any *scalar* K^+K^- configuration with a mass close to that of the ϕ, be it a resonance or not. A smallish pollution by such a '$f_0(980)'K_S$ – by, say, 10% *in amplitude* – can thus reduce the asymmetry assigned to $B_d \to \phi K_S$ significantly – by 20% in this example.

In the end it is therefore mandatory to perform a *full time dependent Dalitz plot analysis* for $B_d \to K^+K^-K_S$ and compare it with that for $B_d \to 3K_S$ and $B^+ \to K^+K^-K^+$, $K^+K_SK_S$ and also with $D \to 3K$. This is a very challenging task, but in my view essential. There is no 'royal' way to fundamental insights. [2]

An important intermediate step in this direction is given by one application of *Bianco's Razor*, namely to analyze the **CP** asymmetry in $B_d \to [K^+K^-]_M K_S$ as a function of the cut M on the K^+K^- mass.

Case III: ϕ_2 from $B_d \to$ *Pions*

Unlike the preceding case two different operators drive $B \to$ *pions*, namely one obtained from a tree and one from the Penguin diagram. With only the first one depending on the weak phase ϕ_2 it is highly nontrivial to extract its size from these decays. For the **CP** asymmetry depends on the relative weight of those two operators, which is shaped by strong dynamics over which we have less than full theoretical control.

A theoretically clean way to disentangle the impact from the Penguin operator is to measure the isospin two final state $B \to [\pi\pi]_{I=2}$ [26]. By measuring all rates $B_d \to \pi^+\pi^-, \pi^0\pi^0$, $B^\pm \to \pi^\pm\pi^0$ this can be achieved in principle, but maybe not in practice with sufficient accuracy; $B_d \to \pi^0\pi^0$ provides the bottle neck. Less challenging experimentally are the modes $B \to \rho\pi$, but one pays a theoretical price there: the actual final state state consists of three pions, where the ρ cannot be seen as a narrow resonance. Furthermore there are other contributions to the three-pion final state like $\sigma\pi$. It hardly matters actually whether the σ is a bona fide resonance or some other dynamical enhancement. What matters is that merely cutting on the di-pion mass will not produce a $\rho\pi$ final state with sufficient purity; furthermore the σ structure cannot be described adequately by a Breit-Wigner shape. As analyzed first in [27] and then in more detail

---

[2] The ruler of a Greek city in southern Italy once approached the resident sage with the request to be educated in mathematics, but in a 'royal way', since he was very busy with many obligations. Whereupon the sage replied with admirable candor:There is no royal way to mathematics.

in [28] ignoring such complications can induce a systematic uncertainty in the extracted value of ϕ_2. The case of $B \to \rho\rho$, which experimentally is even better than the previous one, is even worse theoretically. We have to aim at a total uncertainty – experimental plus theoretical – of not exceding 5%. To achieve this ambitious goal in $B \to 3\pi$ we have to bring the full machinery of a time dependent Dalitz plot analysis to bear augmented by our understanding of chiral dynamics.

Case IV: ϕ_3 from $B^\pm \to D^{neut} K^\pm$

As first mentioned in 1980 [29], then explained in more detail in 1985 [30] and further developed in [31], the modes $B^\pm \to D_{neut} K^\pm$ should exhibit direct **CP** violation driven by the angle ϕ_3, if the neutral D mesons decay to final states that are *common* to D^0 and $\bar{D}^0$. Based on simplicity the original idea was to rely on two-body modes like $K_S\pi^0$, K^+K^-, $\pi^+\pi^-$, $K^\pm\pi^\mp$. One drawback of that method are the small branching ratios and low efficiencies.

A new method was pioneered by BELLE and then implemented also by BABAR, namely to employ $D_{neut} \to K_S\pi^+\pi^-$ and perform a full Dalitz plot analysis. This requires a very considerable analysis effort – yet once this initial investment has been made, it will pay handsome profit in the long run. For obtaining at least a decent description of the full Dalitz plot population provides considerable cross checks concerning systematic uncertainties and thus a high degree of confidence in the results. BELLE and BABAR find [4]:

$$\phi_3 = \begin{cases} 68^o \pm 15^o(stat) \pm 13^o(syst) \pm 11^o(\text{model}) & \text{BELLE} \\ 70^o \pm 31^o(stat) \pm 12^o(syst) \pm 14^o(\text{model}) & \text{BABAR} \end{cases} \tag{11}$$

I view it still a pilot study, yet a most promising one. It exemplifies how the complexities of hadronization can be harnessed to establish confidence in the accuracy of our results. I consider this to be the way of the future.

ON HEP'S FUTURE LANDSCAPE – A CALL TO ACTION & AN APPEAL

To me there is persuasive evidence of a theoretical as well as experimental nature that the SM is incomplete pointing to New Physics driving the electroweak phase transition: the confidently predicted New Physics *cpNP* at the TeV scale. This is the justification – an excellent one in my view – for the LHC. The goal has to go beyond establishing the presence of some New Physics – we have to identify its salient features as well. One should keep in mind that SUSY – for me the most likely candidate for the *cpNP* – is not a true theory yet or even a class of theories: it is largely an organizing principle given our current lack of understanding its breaking.

It has been recognized that the LHC is primarily a discovery machine sweeping out huge regions of 'terra incognita'; this lead to the proposal of a Linear Collider as a more surgical and focussed probe for the *cpNP* – yet the same justification applies to flavour factories! [3] While comprehensive studies of heavy flavour transitions are not very likely to shed light on the *ssNP* behind the flavour puzzle of the SM (although they might), they will be essential in elucidating salient features of the *cpNP*. For New Physics at the TeV scale can affect flavour transitions significantly, and analyzing them is thus complementary to the program of the LHC and Linear Collider. It is actually essential to obtain all the information experimental research can give us on Nature's Grand Design. *Dedicated and comprehensive studies of heavy flavour dynamics are thus a necessity, not a luxury.*

New Physics having a significant impact on B, τ and charm decays does not mean it will be numerically massive and thus obvious. We must succeed in adding the element of 'high accuracy' to that of 'high sensitivity'.

I view a Super-B factory as crucial for us achieving that high accuracy and decoding the *cpNP*[32, 33, 34]. It requires close collaboration between experiment and theory. A central message of my talk is to treat hadronization and nonperturbative dynamics in general as our ally – albeit a complex and sometimes quirky one – rather than as a nuissance. I would like to combine this with a personal appeal: The expertise required to attain an essential goal – namely to exhaust the discovery potential in heavy flavour transitions by harnessing low-energy hadronization – does exist or can be acquired with *no* need for a breakthrough. However it tends to reside in a community all too often disjoint from the heavy flavour community – *this has to change!* We need input from studies of $\pi\pi$, $K\bar{K}$, $K\pi$ etc. final state interactions at low energies, in nonleptonic as well as semileptonic charm decays and in $B \to multi$neutrals, and we have to refine and extend our understanding of chiral dynamics. These are very complex tasks; when tackling them we should remember the example of Swiss watches for guidance: those became famous by being reliable and sturdy, not necessarily elegant.

One final remark: The pion has been discovered in 1947. Its dynamics have been studied intensely for the last sixty years. This conference here in magnificent Rio has shown that it is not a closed chapter; a great deal still has to be learnt. This demonstrates much better than many words, how momentous the discovery of the pion was, how profound a paradigm shift it created. We thus owe a great deal of gratitude to our late colleague Cesar Lattes for his work, as summarized in the following Memorial.

Acknowledgments: I am grateful to Alberto Reis and his colleagues for organizing such an enjoyable meeting. Being able to stage it in Rio with its breathtaking vistas gave them huge advantage, of course. This work was supported by the NSF under grant PHY03-55098.

REFERENCES

1. I.I. Bigi, lectures given at '**CP** Violation: From Quarks to Leptons', International School of Physics "Enrico Fermi", Varenna, July 2005, to appear in the Proceed.
2. I.I. Bigi, A.I. Sanda, *Nucl. Phys.* **B193** (1981) 85; *Nucl. Phys.* **B281** (1987) 41.

[3] The Linear Collider actually doubles as a top quark factory.

3. The BaBar Collab., B. Aubert *et al.*, *Phys. Rev. Lett.* **87** (2001) 091801; the BELLE Collab., K. Abe *et al.*, *Phys. Rev. Lett.* **87** (2001) 091802.
4. K. Abe, Invited Lecture given at Lepton Photon 2005, Uppsala (Sweden), June 2005; F. Forti, *ibidem.*
5. I.I. Bigi, hep-ph/0411138, to appear in the Proc. of 'Time and Matter', Venice (Italy), August 11 - 17, 2002, World Scientific.
6. See, e.g., CKM Fitter, http://www.slac.stanford.edu/xorg/ckmfitter.
7. For a detailed discussion, see: I. I. Bigi, A. I. Sanda, *CP Violation*, Cambridge University Press, Cambridge, 2000.
8. For a nice discussion see: A. Dolgov, lectures given at '**CP** Violation: From Quarks to Leptons', International School of Physics "Enrico Fermi", Varenna, July 2005, to appear in the Proceed.
9. S. Bianco *et al.*, *La Rivista del Nuov. Cim.* **26,** # 7 - 8 (2003).
10. D. Asner, these Proceedings.
11. I.I. Bigi *et al.*, *Phys.Lett.* **B181** (1986) 157.
12. For a recent review see: W. Buchmüller, R. D. Peccei, T. Yanagida, hep-ph/0502169.
13. J.H. Kühn, E. Mirkes, *Phys.Lett.* **B398** (1997) 407.
14. I.I. Bigi, A.I. Sanda, *Phys.Lett.* **B625** (2005) 47.
15. I.I. Bigi, H. Yamamoto, *Phys.Lett.* **B349** (1995) 363.
16. I.I. Bigi, M. Shifman, N.G. Uraltsev, *Annu.Rev.Nucl.Part.Sci.* **47** (1997)591; N. Uraltsev, in: "At the Frontier of Particle Physics/Handbook of QCD", M. Shifman (Ed.), World Scientific, 2001; hep-ph/0010328
17. D. Benson *et al.*, *Nucl.Phys.* **B665** (2003) 367.
18. P. Gambino, N.G. Uraltsev, *Eur.Phys.J.* **C34** (2004) 181.
19. M. Battaglia *et al.*, *Phys.Lett.* **B556** (2003) 41; The DELPHI Collab., J. Abdallah *et al.*, CERN-EP-PH/2005-015.
20. The BaBar Collab., B. Aubert *et al.*, *Phys.Rev.Lett.* **93** (2004) 011803, hep-ex/0404017.
21. O. Buchmueller, Henning Flaecher, hep-ph/0507253.
22. E. Swanson, these Proceedings.
23. N. Uraltsev, hep-ph/0409125.
24. S. Bianco, these Proceed.
25. See, for example: Y. Grossman *et al.*, *Phys.Rev.* **D68** (2003) 015004.
26. M. Gronau, D. London, *Phys.Rev.Lett.* **65** (1990) 3381.
27. A. Deandrea, A.D. Polosa, *Phys.Rev.Lett.* **86** (2001) 216.
28. S. Gardner, Ulf-G. Meissner, *Phys.Rev.* **D65** (2002) 094004.
29. A. Carter, A.I. Sanda, *Phys.Rev.* **D23** (1981) 1567.
30. I.I. Bigi, A.I. Sanda, *Phys.Lett.* **B211** (1985) 213.
31. M. Gronau, D. Wyler, *Phys.Lett.* **B265** (1991) 172; I. Dunietz, *Phys.Lett.* **B270** (1991) 75.
32. SuperKEKB Task Force, 'Letter of Intent for KEK Super B Factory', KEK Report 2004-4 (2004).
33. J.L. Hewett, D. G. Hitlin (eds.), 'The Discovery Potential of a Super-B factory', SLAC-R-709 (2004).
34. I.I. Bigi, A.I. Sanda, hep-ph/0401003.

PARALLEL SESSIONS

BARYONS

Baryonic Resonances from the Interactions of the Baryon Decuplet and Meson Octet

E. Oset, S. Sarkar and M. J. Vicente Vacas

Departamento de Física Teórica and IFIC, Centro Mixto,
Institutos de Investigación de Paterna - Universidad de Valencia-CSIC
Apdo. correos 22085, 46071, Valencia, Spain

Abstract. We study S-wave interactions of the baryon decuplet with the octet of pseudoscalar mesons using the lowest order chiral Lagrangian. We find two bound states in the $SU(3)$ limit corresponding to the octet and decuplet representations. These are found to split into eight different trajectories in the complex plane when the $SU(3)$ symmetry is broken gradually. Finally, we are able to provide a reasonable description for a good number of 4-star $\frac{3}{2}^-$ resonances listed by the Particle Data Group. In particular, the $\Xi(1820)$, the $\Lambda(1520)$ and the $\Sigma(1670)$ states are well reproduced. We predict a few other resonances and also evaluate the couplings of the observed resonances to the various channels from the residues at the poles of the scattering matrix from where partial decay widths into different channels can be evaluated.

Keywords: $3/2^-$ dynamically generated resonances
PACS: 13.75.-n,12.39.Fe,14.20.Jn,11.30.Hv

INTRODUCTION

The introduction of unitary techniques in a chiral dynamical treatment of the meson baryon interaction has been very successful. It has lead to good reproduction of meson baryon data with a minimum amount of free parameters, and has led to the dynamical generation of many low lying resonances which qualify as quasibound meson baryon states [1, 2, 3]. In particular, the application of these techniques to the s-wave scattering of the baryon octet and the pseudoscalar meson octet have led to the successful description of many $J^P = \frac{1}{2}^-$ resonances like the $N^*(1535)$, the $\Lambda(1405)$, the $\Lambda(1670)$, the $\Sigma(1620)$ and the $\Xi(1620)$ [4, 5, 6, 7, 8]. Naively one may expect that this scheme is not suitable for studying d-wave resonances due to a large number of unknown parameters in the corresponding chiral Lagrangian. However d-wave resonances could be studied in s wave interactions of the meson octet with the baryon decuplet [9], in which case chiral dynamics is quite predictive. The fact that some well known d wave resonances like the $N^*(1520)$, $N^*(1700)$, $\Delta(1700)$ have very large branching ratios to the $N\pi\pi$ channel though the $N\pi$ channel is favoured by phase space, lends support to this scheme.

CP814 *Hadron Spectroscopy: Eleventh International Conference*
edited by A. Reis, C. Göbel, J. Sá Borges, and J. Magnin

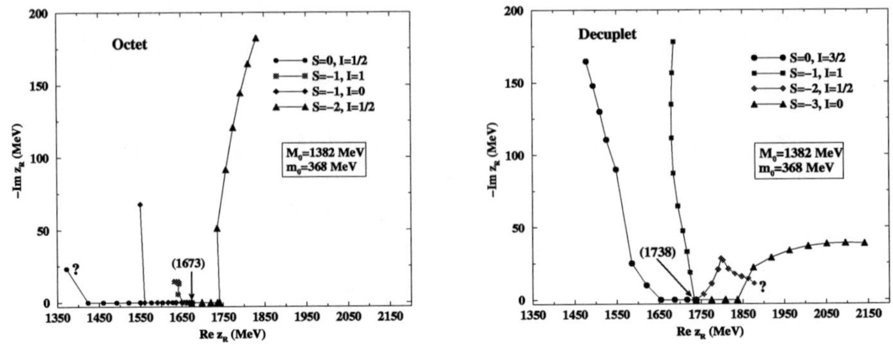

FIGURE 1. Trajectories of the poles in the scattering amplitudes for different values of the $SU(3)$ breaking parameter.

BARYON DECUPLET-MESON OCTET INTERACTION: RESULTS AND DISCUSSION

We have performed a systematic study [10] of the s-wave interaction of the baryon decuplet with the meson octet taking the dominant lowest order chiral Lagrangian, which accounts for the Weinberg Tomozawa term. In this talk we will present the main results of this work.

The tree-level scattering amplitude involving the baryon decuplet and the pseudoscalar octet is used as the kernel of the Bethe Salpeter equation to obtain the transition matrix fulfilling exact unitarity in coupled channels. In this approach, the only free parameter is the subtraction constant in the dimensionally regularized meson baryon loop function for which we took a natural size value. We have looked in detail at the $\frac{3}{2}^-$ resonances which are generated dynamically by this interaction, by searching for poles of the transition matrix in the complex plane in different Riemann sheets. The search was done systematically, starting from an $SU(3)$ symmetric situation where the masses of the baryons are made equal and the same is done with the masses of the mesons. In this case we found attraction in an octet, a decuplet and the 27 representation, while the interaction was repulsive in the 35 representation. In the $SU(3)$ symmetric case all states of the $SU(3)$ multiplet are degenerate and the resonances appear as bound states with no width. As we gradually break $SU(3)$ symmetry by changing the masses, the degeneracy is broken and the states with different strangeness and isospin split apart generating pole trajectories in the complex plane which lead us to the physical situation in the final point, as seen in fig. 1. This systematic search allows us to trace the poles to their $SU(3)$ symmetric origin, although there is a mixing of representations when the symmetry is broken. In addition we also find poles which only appear for a certain amount of symmetry breaking and thus have no analog in the symmetric case.

The search for poles is an ideal technique to identify resonances in addition to visual inspection of the amplitudes. On the other hand, in ref. [9] the speed plot technique is used to identify resonances. However, the speed plot, involving derivatives of the amplitudes, produces artificial peaks at the thresholds of new channels, because the cusp

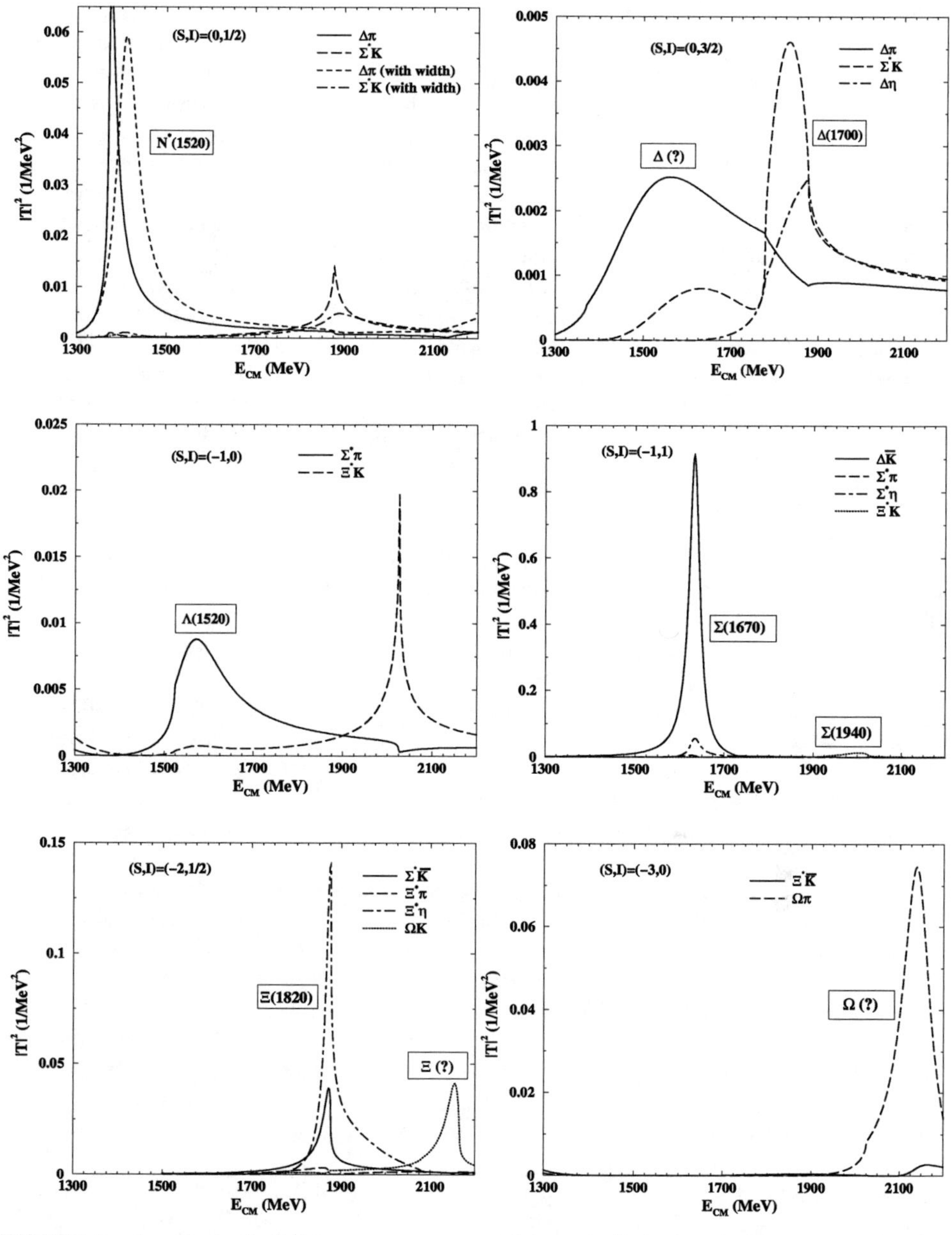

FIGURE 2. Amplitudes for $3/2^-$ resonances generated dynamically in decuplet baryon - pseudoscalar octet interactions.

of the amplitudes, and should not be used there. Neglecting this fact, some peaks which are actually reflections of cusps in the speed plot are misidentified as resonances in [9] while we do not find poles, not even making reasonable changes in the input data.

We have also evaluated the residues of the poles from where the couplings of the resonances to the different coupled channels were found and this allowed us to make predictions for partial decay widths into a decuplet baryon and a meson. There is very limited experimental information on these decay channels but, even then, it represents an extra check of consistency of the results which allowed us to more easily identify the resonances found with some resonances known, or state that the resonance should correspond to a new resonance not yet reported in the Particle Data Book (PDB). In particular, in view of the information of the pole positions and couplings to channels we could associate some of the resonances found to the $N^*(1520)$, $\Delta(1700)$, $\Lambda(1520)$, $\Sigma(1670)$, $\Sigma(1940)$, $\Xi(1820)$ resonances tabulated in the PDB. We could also favour the correspondence of a resonance found in $S=-3$, $I=0$ to the $\Omega(2250)$ on the basis of the quantum numbers, position and compatibility of the partial decay width found with the total width of the $\Omega(2250)$. The scattering amplitudes for different values of strangeness and isospin as a function of the C.M. energy are shown in fig. 2.

We also found several extra resonances, well identified by poles in the complex plane which do not have a correspondent one in the PDB. Some of them are too broad, which could justify the difficulty in their observation, but two other resonances, the $\Delta(1500)$ with a width around 300 MeV and the $\Xi(2160)$ with a width of about 40 MeV stand much better chances of observation. The first one because of its large strength in the $\Delta\pi$ channel and the second one because of its narrowness.

In addition, our study produces couplings of the resonances to baryon-meson channels which could facilitate the identification when further information on these branching ratios is available. We have found that in some cases the dynamically generated resonances couple very strongly to some channels with the threshold above the resonance energy, which makes them qualify as approximately single channel quasibound meson-baryon states.

Lastly, we mention the very interesting case concerning the ΔK resonance [11] found in $S=1$, $I=1$ which shows up as a pole in the unphysical Riemann sheet and leads to ΔK cross sections much larger than the corresponding ΔK cross sections in $I=2$. We suggest that these two cross sections could be seen in ΔK production in $pp \to \Lambda\Delta K$ and $pp \to \Sigma\Delta K$ reactions and could provide evidence for this new 'exotic pentaquark' state, although its structure is more efficiently taken into account in the meson baryon picture where it has been generated. This state, however, appears as a broad bump above the $K\Delta$ threshold and its strength is also tied to details of the interaction. Hence, the ultimate fate of this resonance is difficult to predict at the present stage.

Acknowledgments: This work is partly supported by DGICYT contracts BFM2002-01868, and the E.U. EURIDICE network contract HPRN-CT-2002-00311. This research is part of the EU Integrated Infrastructure Initiative Hadron Physics Project under contract number RII3-CT-2004-506078.

REFERENCES

1. N. Kaiser, P. B. Siegel and W. Weise, Nucl. Phys. A **594** (1995) 325; N. Kaiser, T. Waas and W. Weise, Nucl. Phys. A **612** (1997) 297.
2. E. Oset and A. Ramos, Nucl. Phys. A **635** (1998) 99.
3. J. A. Oller and U.-G. Meißner, Phys. Lett. B **500** (2001) 263.
4. T. Inoue, E. Oset and M. J. Vicente Vacas, Phys. Rev. C **65** (2002) 035204.
5. D. Jido, J. A. Oller, E. Oset, A. Ramos and U. G. Meissner, Nucl. Phys. A **725** (2003) 181.
6. C. Garcia-Recio, J. Nieves, E. Ruiz Arriola and M. J. Vicente Vacas, Phys. Rev. D **67** (2003) 076009 [arXiv:hep-ph/0210311].
7. C. Garcia-Recio, M. F. M. Lutz and J. Nieves, Phys. Lett. B **582** (2004) 49.
8. E. Oset, A. Ramos and C. Bennhold, Phys. Lett. B **527** (2002) 99 [Erratum-ibid. B **530** (2002) 260];A. Ramos, E. Oset and C. Bennhold, Phys. Rev. Lett. **89** (2002) 252001.
9. E. E. Kolomeitsev and M. F. M. Lutz, Phys. Lett. B **585** (2004) 243.
10. S. Sarkar, E. Oset and M. J. Vicente Vacas, Nucl. Phys. A **750** (2005) 294 [arXiv:nucl-th/0407025].
11. S. Sarkar, E. Oset and M. J. Vicente Vacas, Eur. Phys. J. A **24** (2005) 287 [arXiv:nucl-th/0404023].

Photoproduction of Baryons Decaying into $N\pi$, $N\eta$, $K^+\Lambda$, $K^+\Sigma^0$ and $K^0\Sigma^+$

Andrey Sarantsev

Helmholtz–Institut für Strahlen– und Kernphysik (Bonn) and Petersburg Nuclear Physics Institute (Gatchina)

Abstract. A combined analysis of new π^0 and η photoproduction data from CB-ELSA collaboration was performed including the earlier data from SAID data base, new data on beam asymmetry from the GRAAL collaboration and data on $\gamma p \to K\Lambda$ and $K\Sigma$. Three baryon resonances have a substantial coupling to ηN, the well known $N(1535)S_{11}$, $N(1720)P_{13}$ and the newly observed $N(2070)D_{15}$ resonance. A new P_{11} state with mass 1840 MeV and width 140 MeV was observed contributing to most of the fitted reactions. The data also demand the presence of two D_{13} states at 1870 and 2170 MeV. The properties of observed states and their contributions to different observables are discussed.

Keywords: Baryon structure and spectroscopy, photoproduction of mesons, polarization observables
PACS: 11.80.Et, 13.30.-a, 13.60.Le, 14.20.Gk

INTRODUCTION

The energy levels of bound systems and their decay properties provide valuable information about the constituents and their interactions [1]. In quark models, the dynamics of the three constituent quarks in baryons support a rich spectrum, much richer than the energy scheme experiments have established so The properties of baryon resonances are presently under intense investigations at several laboratories like ELSA (Bonn), GRAAL (Grenoble), JLAB (Newport News, VA), MAMI (Mainz), and Spring8 (Hyogo). The aim is to identify the resonance spectrum, to determine spin, parities, and decay branching ratios and thus to supply a vital information for undestanding of strong interactions at low energies.

The largest part of our knowledge on baryons stems from pion induced reactions. However, if a resonance has a large inelasticity, its production cross section in πN scattering is small and it contributes only weakly to the final state. Thus resonances may conceal themselves from observation in elastic scattering. This effect could be a reason why the number of observed states is much smaller that predicted by quark models [2, 3, 4]. Information on resonances coupled weakly to the πN channel can be obtained from photoproduction experiments and the study of final states different from πN such as multibody final states or final states containing open strangeness.

In the present paper we report results of a combined analysis of photoproduction experiments with πN, ηN, $K\Lambda$, and $K\Sigma$ final states. This work is a first step of a forthcoming analysis of all reactions with production of baryon resonances in the intermediate state.

CP814 *Hadron Spectroscopy: Eleventh International Conference*
edited by A. Reis, C. Göbel, J. Sá Borges, and J. Magnin

TABLE 1. Data used in the partial wave analysis and χ^2 contributions.

Observable	N_{data}	χ^2	χ^2/N	Ref.	Observable	N_{data}	χ^2	χ^2/N	Ref.
$\sigma(\gamma p \to p\pi^0)$	1106	1654	1.50	[5]	$\sigma(\gamma p \to \Lambda K^+)$	720	804	1.12	[12]
$\sigma(\gamma p \to p\pi^0)$	861	2354	2.74	[6]	$\sigma(\gamma p \to \Lambda K^+)$	770	1282	1.67	[13]
$\Sigma(\gamma p \to p\pi^0)$	469	1606	3.43	[6]	$P(\gamma p \to \Lambda K^+)$	202	374	1.85	[13]
$\Sigma(\gamma p \to p\pi^0)$	593	1702	2.87	[7]	$\Sigma(\gamma p \to \Lambda K^+)$	45	62	1.42	[14]
$\sigma(\gamma p \to n\pi^+)$	1583	4524	2.86	[8]	$\sigma(\gamma p \to \Sigma^0 K^+)$	660	834	1.27	[12]
$\sigma(\gamma p \to p\eta)$	667	608	0.91	[9]	$\sigma(\gamma p \to \Sigma^0 K^+)$	782	2446	3.13	[13]
$\sigma(\gamma p \to p\eta)$	100	158	1.60	[10]	$P(\gamma p \to \Sigma^0 K^+)$	95	166	1.76	[13]
$\Sigma(\gamma p \to p\eta)$	51	114	2.27	[11]	$\Sigma(\gamma p \to \Sigma^0 K^+)$	45	20	0.46	[14]
$\Sigma(\gamma p \to p\eta)$	100	174	1.75	[6]	$\sigma(\gamma p \to \Sigma^+ K^0)$	48	104	2.20	[13]

FITS TO THE DATA

In this paper, we report results based on a coupled–channel analysis of various data sets on photoproduction of different final states. Amplitudes for photoproduction of baryon resonances and their decays are calculated in the framework of relativistic tensor operators. The formalism is fully described in [15]; here parameterisations of resonances

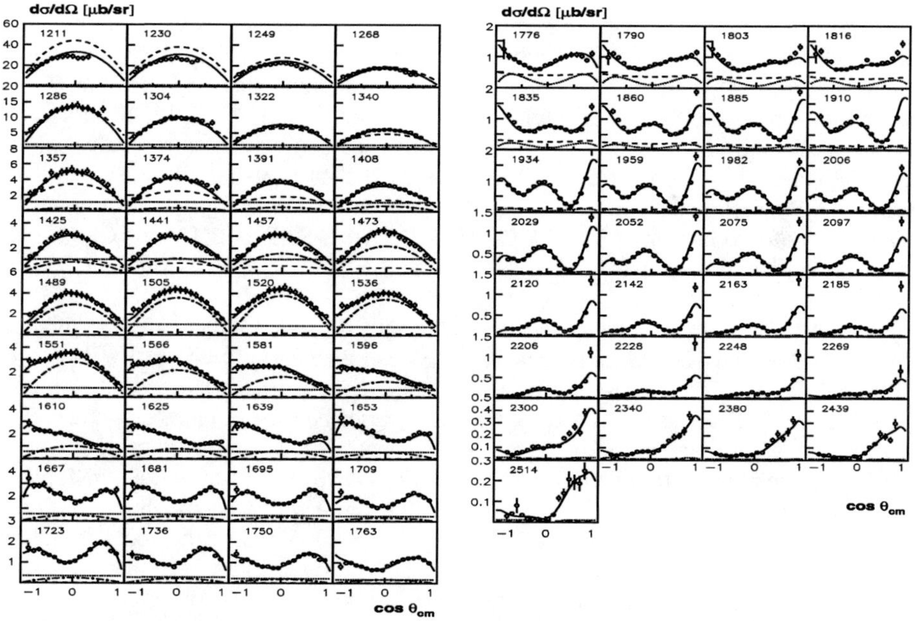

FIGURE 1. Differential cross section for $\gamma p \to p\pi^0$ from CB–ELSA and PWA result (solid line). The left part of the figure shows the contribution of $\Delta(1232)P_{33}$ together with non-resonant background (dashed line), the two S_{11} resonances (dotted line) and $N(1520)D_{13}$ (dash-dotted line); in the right figure, the contributions of $\Delta(1700)D_{33}$ (dashed line) and $N(1680)F_{15}$ (dotted line) are shown.

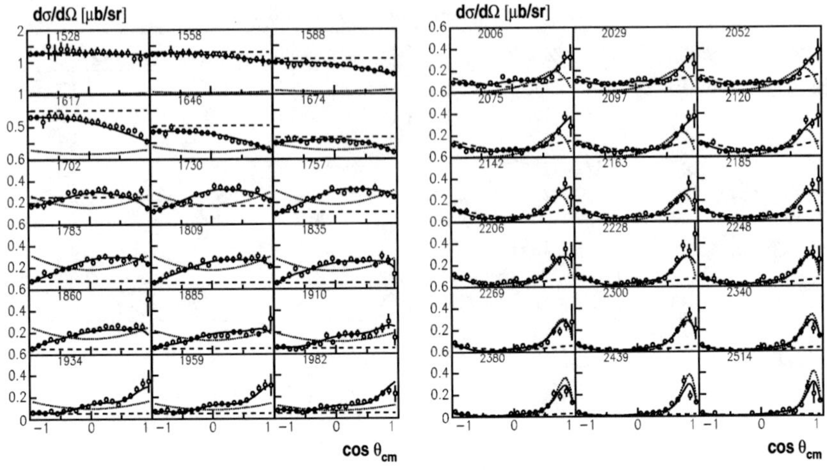

FIGURE 2. Differential cross section for $\gamma p \to p\eta$ from CB-ELSA and PWA result (solid line). In the mass range below 2 GeV the contribution of the two S_{11} resonances is shown as dashed line and of $N(1720)P_{13}$ as dotted line. Above 2 GeV the contributions of $N(2070)D_{15}$ (dashed line) and $\rho(\omega)$ exchange (dotted line) are shown.

used under different conditions are given.

The data comprise CB–ELSA π^0 and η photoproduction data [5, 9]; the Mainz–TAPS data [10] on η photoproduction, beam–asymmetry measurements of π^0 and η [6, 7, 11], and data on $\gamma p \to n\pi^+$ [8]. The high precision data from GRAAL [6] do not cover the low mass region; therefore we extract further data from the compilation of the SAID database [7]. This data allows us to define the ratio of helicity amplitudes for the $\Delta(1232)P_{33}$ resonance.

Data on photoproduction of $K^+\Lambda$, $K^+\Sigma$ and $K^0\Sigma^+$ from SAPHIR [12] and CLAS [13], and beam asymmetry data for $K^+\Lambda$, $K^+\Sigma$ from LEPS [14] are also included in the coupled–channel analysis.

The fit uses 14 N^* resonances coupling to $N\pi$, $N\eta$, $K\Lambda$, and $K\Sigma$, and 7 Δ resonances coupling to $N\pi$ and $K\Sigma$. Most resonances are described by relativistic Breit–Wigner amplitudes. For the two S_{11} resonances at 1535 and 1650 MeV, a four–channel K–matrix ($N\pi$, $N\eta$, $K\Lambda$, $K\Sigma$) is used. The background is described by reggeized t–channel π, ρ (ω), K and K^* exchanges and by baryon exchanges in the s– and u–channels.

The χ^2 values for the final solution of the partial–wave analysis are given in Table 1.

The differential cross sections for the CB–ELSA $\gamma p \to p\pi^0$ data are shown in Fig. 1. The main fit is represented as solid line. The figure also shows the most important individual contributions. The contribution of $\Delta(1232)$ (given as dashed line, on the left panel) dominates the low–energy region, for small photon energies it even exceeds the experimental cross section, thus underlining the importance of interference effects. Non–resonant background amplitudes, given by a pole at $s \sim -1\,\text{GeV}^2$ and by a u–channel exchange diagram, are needed to describe the shape of the $\Delta(1232)$. The pole at negative

s represents the left–hand cuts.

The contribution of the S_{11} K–matrix is depicted as dotted line. The contribution of the $D_{13}(1520)$ shown as dash–dotted line in Fig. 1 (left panel). It is strong in the 1400 – 1600 MeV mass region. At higher energies (Fig. 1, right panel) the most significant contributions come from $\Delta(1700)D_{33}$ (dashed line) and from $N(1680)F_{15}$ (dotted line).

It is important to include data on $n\pi^+$ photoproduction since the combination of the $n\pi^+$ and $p\pi^0$ channels defines the isospin of s-channel baryons.

The CB–ELSA differential cross section for η photoproduction is shown in Fig. 2 and compared to the PWA results. The contribution of the two S_{11} resonances (dashed line, below 2 GeV) dominates the η production region up to 1650 MeV. The most significant further contributions stem from production of $N(1720)P_{13}$ (dotted line, below 2 GeV), of $N(2070)D_{15}$ (dashed line, above 2 GeV) and $\rho\,(\omega)$ exchanges (dotted line, above 2 GeV).

Data on the photon beam asymmetry Σ for $\gamma p \to p\pi^0$ and $\gamma p \to p\eta$ reactions measured by GRAAL [6] are shown in Fig. 3 This data provides essential information on baryon resonances even if their $(p\gamma)$– and/or $(p\pi, p\eta)$–couplings are weak. In addition, the beam asymmetry data are necessary to determine the ratio of helicity amplitudes.

The Figs. 4 and 5 show the differential cross sections $\gamma p \to \Lambda K^+$ (right panels) and $\gamma p \to K^+\Sigma^0$ (left panels) obtained by SAPHIR and CLAS and the results of the best fit described below. The agreement is rather good in both cases if energy renormalization is introduced. Therefore most discrepancies between the two experiments can obviously be ascribed to an overall energy–dependent normalisation error in the region 20%.

The beam polarisation asymmetries are compared to the fit in Fig. 6. The recoil polarisation, obtained from the weak–decay asymmetry of hyperons, provides further constraints to the solution. Data and fit, divided into 24 mass bins, are shown in Fig. 7.

For all reactions the experimental total cross section was determined by summation of the experimental differential cross sections and, in regions where no data exist, predicted values determined by the fit.

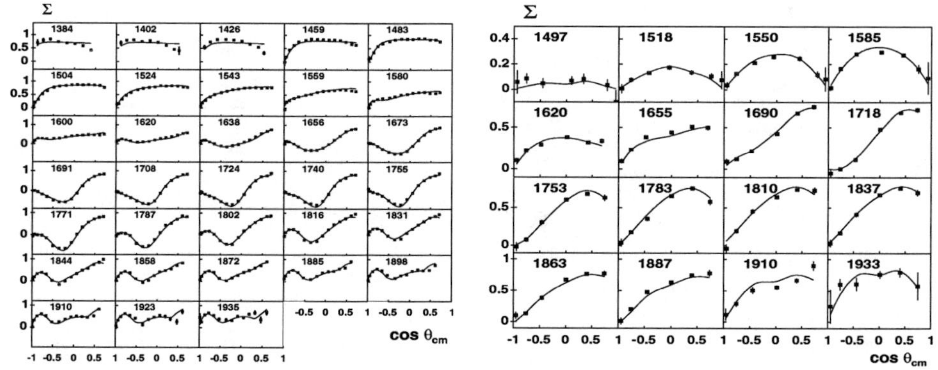

FIGURE 3. Photon beam asymmetry Σ for $\gamma p \to p\pi^0$ (left panel) and for $\gamma p \to p\eta$ (right panel) from GRAAL [11] versus PWA result (solid line).

TABLE 2. Masses, widths and helicity ratio, this analysis.

Resonance	M (MeV)	Γ (MeV)	$A_{1/2}/A_{3/2}$	PDG Rating overall	PDG Rating Nγ
N(1440)P_{11}	1450 ± 50	250 ± 150			
PDG	1440^{+30}_{-10}	350 ± 100		****	***
N(1520)D_{13}	1526 ± 4	112 ± 10	-0.02 ± 0.10		
PDG	1520^{+10}_{-5}	120^{+15}_{-10}	-0.14 ± 0.06	****	****
N(1535)S_{11}*	1530 ± 30	210 ± 30			
PDG	1505 ± 10	170 ± 80		****	***
N(1650)S_{11}*	1705 ± 30	220 ± 30			
PDG	1660 ± 20	160 ± 10		***	****
N(1675)D_{15}	1670 ± 20	140 ± 40	0.40 ± 0.25		
PDG	1675^{+10}_{-5}	150^{+30}_{-10}	1.27 ± 0.93	****	****
N(1680)F_{15}	1667 ± 6	102 ± 15	-0.13 ± 0.05		
PDG	1680^{+10}_{-5}	130 ± 10	-0.11 ± 0.05	****	****
N(1700)D_{13}	1725 ± 15	100 ± 15	0.45 ± 0.25		
PDG	1700 ± 50	100 ± 50	9.00 ± 6.5	***	**
N(1720)P_{13}	1750 ± 40	380 ± 40	1.5 ± 1.1		
PDG	1720^{+30}_{-70}	250 ± 50	-0.9 ± 1.8	***	**
N(1840)P_{11}	1840^{+15}_{-40}	140^{+30}_{-15}	new	new	
PDG	1720 ± 30	100^{+150}_{-50}		***	***
N(1870)D_{13}	1875 ± 25	80 ± 20	1.20 ± 0.45	new	new
N(2000)F_{15}	1850 ± 25	225 ± 40	0.13 ± 1.10		new
PDG	~2000			**	
N(2070)D_{15}	2060 ± 30	340 ± 50	1.10 ± 0.30	new	new
N(2170)D_{13}	2166^{+25}_{-50}	300 ± 65	-1.40 ± 0.80	new	new
PDG	~2080			**	*
N(2200)P_{13}	2200 ± 30	190 ± 50	-0.35 ± 0.40	new	new
Δ(1232)P_{33}◇	1235 ± 4	140 ± 12	0.44 ± 0.06		
PDG	1232 ± 2	120 ± 5	0.53 ± 0.04	****	****
Δ(1620)S_{31}	1635 ± 6	106 ± 12			
PDG	1620^{+55}_{-5}	150 ± 30		****	***
Δ(1700)D_{33}	1715 ± 20	240 ± 35	1.15 ± 0.25		
PDG	1700^{+70}_{-30}	300 ± 100	$1.2^{+0.6}_{-0.4}$	****	***
Δ(1905)F_{35}	1870 ± 50	370 ± 110	>10		
PDG	1905^{+15}_{-35}	350^{+90}_{-70}	$-0.6^{+0.4}_{-0.9}$	****	***
Δ(1920)P_{33}	1996 ± 30	380 ± 40	0.45 ± 0.20		
PDG	1920^{+50}_{-20}	200^{+100}_{-50}	$1.7^{+7.}_{-1.0}$	****	*
Δ(1940)D_{33}	1930 ± 40	200 ± 100	0.20 ± 0.40		new
PDG	~1940			*	
Δ(1950)F_{37}	1893 ± 15	240 ± 30	0.75 ± 0.11		
PDG	1950 ± 10	300^{+50}_{-10}	0.8 ± 0.2	****	****

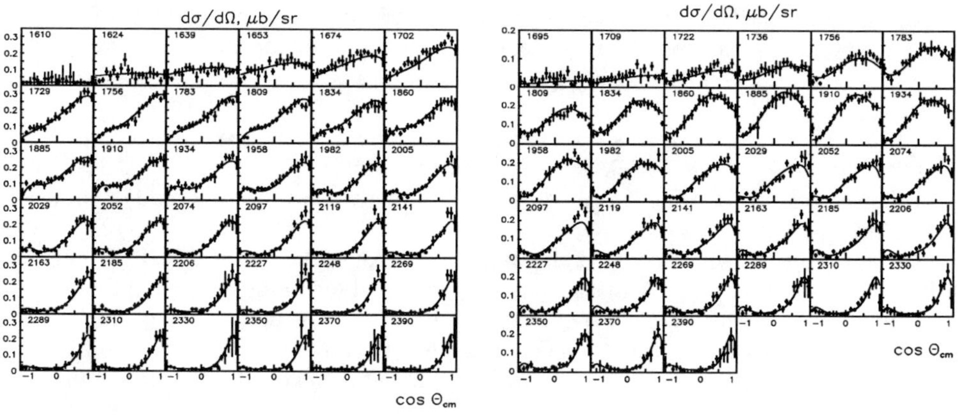

FIGURE 4. The angular distributions as a function of W for $\gamma p \to \Lambda K^+$ (right panel) and $\gamma p \to K^+\Sigma^0$, SAPHIR data.

The total and differential cross sections Σ^+K^0 versus our fit are given in Figs. 8. In this case, no normalisation factor was applied.

The masses and widths of the observed states are presented in Table 2. Additionally, ratios of helicity amplitudes $A_{1/2}/A_{3/2}$ and fractional contributions normalized to the total cross section for the CB–ELSA π^0– and η– photoproduction data are included.

Three new resonances are necessary to describe π^0 and η photoproduction data, N(1870)D_{13}, N(2070)D_{15} and N(2200) with uncertain spin and parity. The best fit is achieved for P_{13} quantum numbers. Two further resonances seen mostly in channels with open strangeness, N(1840)P_{11} and N(2170)D_{13}, have masses which are not consistent with established resonances listed by the PDG. We list them also as new pratices.

Significance of resonance contributions

A systematic study of the significance of new resonances was carried out. For new resonances the quantum numbers were changed to any J^P value with $J \leq 9/2$. In the new fits, all variables were left free for variations including masses, widths, and couplings of all resonances.

The N(2070)D_{15} is the most significant new resonance. Omitting this state from the fit deteriorates χ^2 by 1588 and replacing the J^P assignment from $5/2^-$ to $1/2^\pm$, ..., $9/2^\pm$, the χ^2_{tot} deteriorates it by more than 750. The π^0 photoproduction cross section measured by CB–ELSA are visually not too sensitive to this state. However there is a clear difference between the best fit and other descriptions in the very backward region.

The mass scans of new resonances (χ^2 as a function of the assumed resonance mass) is shown in Fig. 9. In the scan, the mass of the scanned state was fixed while all other fit parameters were allowed to adjust newly.

The N(2200) P_{13} resonance is less significant. Omitting N(2200)P_{13} from the analysis, changes the χ^2 for the CB-ELSA data on η photoproduction by 56, and by 20 for the π^0–photoproduction data. Other quantum numbers results lead to marginally larger χ^2 values.

In Fig. 9 (second panel) we show a mass scan of the new P_{11} state. There is a clear χ^2 minimum at 1810–1850 MeV in the reactions with ΛK^+ and ΣK final states. The minimum in this mass range was also observed in the π^0 and η photoproduction (see Fig. 9 (third panel)).

The N(1840)P_{11} mass is considerably larger than that of the PDG N(1710)P_{11} and is likely to be a different state. A mass scan was performed searching for an additional P_{11} state, fixing the mass of $P_{11}(1840)$ at the optimum value. The description of the CLAS differential cross sections in both final channels was slightly improved if this additional resonance had a large mass and a very wide width, but there was no clear minimum in the χ^2 distributions.

The recoil polarisation for ΣK production still had systematic deviations between data and fit, and ΛK^+ production showed a discrepancy in the 2.2 GeV mass region.

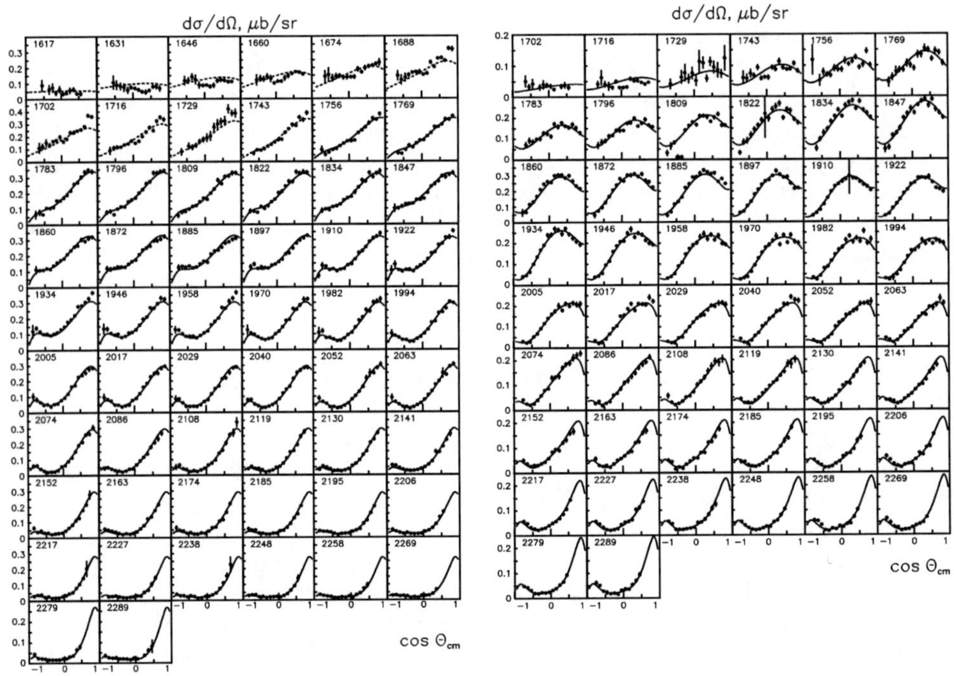

FIGURE 5. The angular distributions as a function of W for $\gamma p \to \Lambda K^+$ (left panel) and $\gamma p \to K^+\Sigma^0$ (right panel) CLAS data. The prediction for the region where data were not used in the fit is shown as dashed lines.

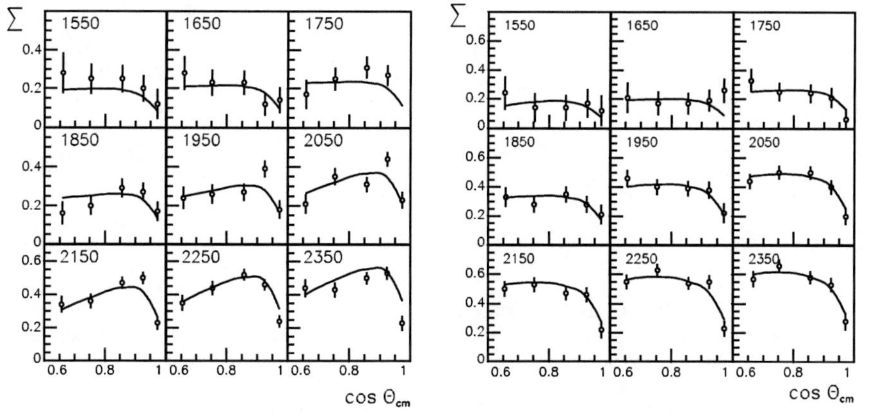

FIGURE 6. The beam polarisation asymmetries as a function of W for $\gamma p \to K^+\Lambda$ (right panel) and $\gamma p \to K^+\Sigma$ (left panel) [14]. The curves are the result of our fit.

Also the recoil polarisation for $\gamma p \to \Lambda K^+$ was not well described in the 2.2 GeV mass region. As before resonances with different quantum numbers were introduced one by one. Only adding contributions from the D_{13} wave improved the picture. Two D_{13} states needed to be introduced; the mass of the lower–mass state optimised for (1875 ± 25) MeV and a width of (80 ± 20) MeV. There is a clear minimum for the ΣK

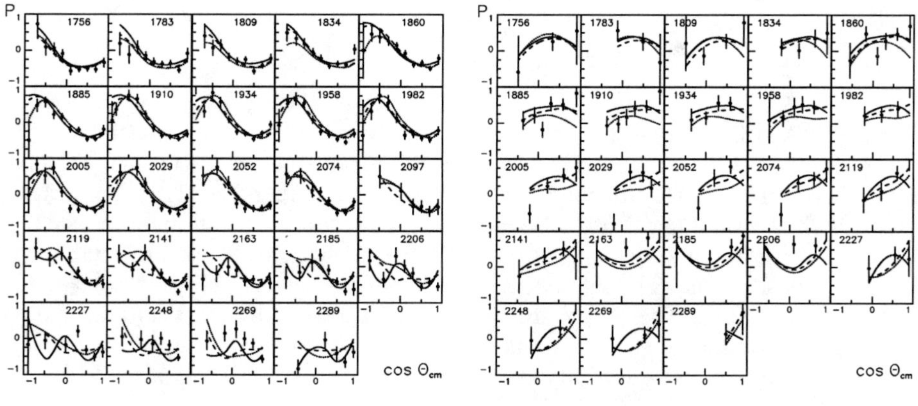

FIGURE 7. left panel: Recoil polarisation of Λ hyperons as a function of W for the center–of–mass kaon angle $\cos(\theta_K^{CM})$. The vertical bars on CLAS data (solid points) combine statistical and systematic errors. The solid line presents our fit. The dashed curves represent results when the N(2170)D_{13} is omitted, for the dotted lines the N(1840)P_{11} was omitted in the fit. Right panel: Recoil polarisations for $K^+\Sigma^0$, CLAS data. The solid curves are results of the fit. The dashed curves correspond to the solution without N(2170)D_{13} and the dotted curves to the solution without (1840)P_{11}.

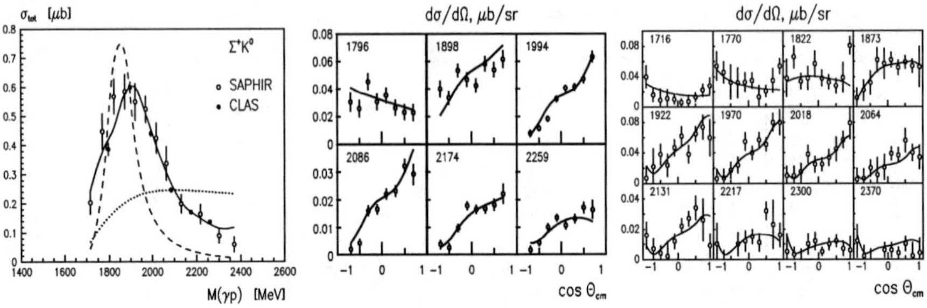

FIGURE 8. Right panel: The total cross sections for $K^0\Sigma^+$ photoproduction. The solid curve represents our fit. The SAPHIR cross section is shown as open circles. For the CLAS data the experimental total cross section (full circles) was determined by summation of the experimental differential cross sections and, in regions where no data exist, predicted values determined by the fit. Dashed line shows the contribution from $P_{11}(1840)$, dotted line shows the contribution from Σ–exchange. Middle panel: Angular distributions for $K^0\Sigma^+$ CLAS data, and right panel: Angular distributions for $K^0\Sigma^+$ SAPHIR data

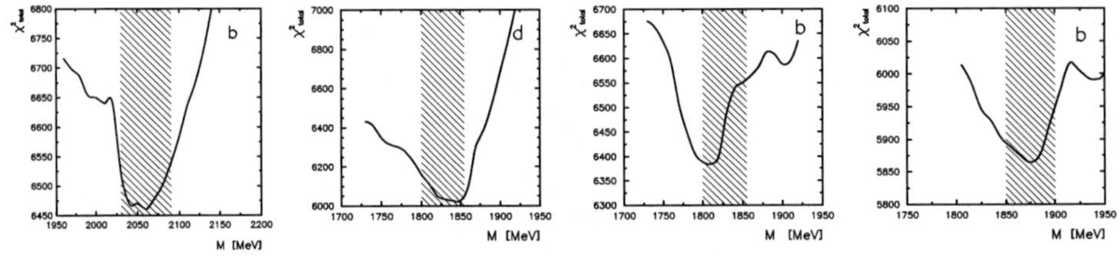

FIGURE 9. The total χ^2 for scan of $D_{15}(2070)$ (first panel), the total χ^2 for all ΛK^+ and ΣK reactions for scan of $P_{11}(1840)$ (second panel), the total χ^2 for all π^0 and η photoproduction reactions for scan of $P_{11}(1840)$ (third panel), the total χ^2 for all ΛK^+ and ΣK reactions for the scan of $D_{13}(1870)$ (fourth panel).

(SAPHIR, CLAS) differential cross sections. A rather shallow minimum is seen for the Λ recoil polarisation. The distribution of sum of χ^2 for all reactions with ΛK^+ and ΣK final states is shown in Fig. 9 (fourth panel). The other state was found at (2166 ± 35) MeV, its width at (280 ± 65) MeV. With this resonance, the recoil polarisation in $\gamma p \to \Lambda K^+$ is described as well. Omitting this contribution from the fit yields the fit which is shown in Figs. 7 as dashed line. The mass scan of individual channels did not show a clear minimum for this state. The χ^2 for the recoil polarisation (differential cross section) show asymmetric minima, shallow on the high–mass (low–mass) side and steeper on the low–mass (high–mass) side. Only their sum gives a pronounced minimum in χ^2.

SUMMARY

We have presented a partial wave analysis of data on photoproduction of πN, ηN, $K\Lambda$, and $K\Sigma$ final states. The data include total and differential angular distributions, beam asymmetry measurements as well as the recoil polarisation in case of hyperon production. New resonances are required to fit the data, N(1840)P_{11}, N(1870)D_{13}, N(2070)D_{15}, N(2170)D_{13}, and N(2200).

Three resonances are found to have very large couplings to Nη, N(1535)S_{11}, N(1720)P_{13}, and N(2070)D_{15}. The dynamical origin of this preference remains to be investigated.

REFERENCES

1. N. Isgur, "Baryons: The promise, the problems, and the prospects," *7th International Conference on the Structure of Baryons, (B.F. Gibson et al. eds.), Santa Fe, New Mexico, 3-7 Oct 1995* World Scientific, Singapore, 1996.
2. S. Capstick and N. Isgur, Phys. Rev. D **34** (1986) 2809.
   S. Capstick and W. Roberts, Prog. Part. Nucl. Phys. **45** (2000) S241.
3. L. Y. Glozman, W. Plessas, K. Varga and R. F. Wagenbrunn, Phys. Rev. D **58** (1998) 094030.
4. U. Löring, K. Kretzschmar, B. C. Metsch and H. R. Petry, Eur. Phys. J. A **10** (2001) 309.
   U. Löring, B. C. Metsch and H. R. Petry, Eur. Phys. J. A **10** (2001) 395, 447.
5. O. Bartholomy *et al.*, Phys. Rev. Lett. **94** (2005) 012003.
6. GRAAL collaboration, article in preparation. We thank the GRAAL collaboration for providing their (preliminary) data to us prior to publication.
7. A. A. Belyaev *et al.*, Nucl. Phys. B **213** (1983) 201.
   R. Beck *et al.*, Phys. Rev. Lett. **78** (1997) 606.
   D. Rebreyend *et al.*,Nucl. Phys. A **663** (2000) 436.
8. K. H. Althoff *et al.*, Z. Phys. C **18** (1983) 199.
   E. J. Durwen, BONN-IR-80-7 (1980).
   K. Buechler *et al.*, Nucl. Phys. A **570** (1994) 580.
9. V. Crede *et al.*, Phys. Rev. Lett. **94** (2005) 012004.
10. B. Krusche *et al.*, Phys. Rev. Lett. **74** (1995) 3736.
11. J. Ajaka *et al.*, Phys. Rev. Lett. **81** (1998) 1797.
12. K. H. Glander *et al.*, Eur. Phys. J. A **19** (2004) 251.
13. J. W. C. McNabb *et al.*, Phys. Rev. C **69** (2004) 042201.
14. R. G. T. Zegers *et al.*, Phys. Rev. Lett. **91** (2003) 092001.
15. A. Anisovich, E. Klempt, A. Sarantsev and U. Thoma, Eur. Phys. J. A **24** (2005) 111.

Investigation of Resonance Structures in the System of Two K_S-Mesons in the Mass Regions around 1595 MeV

E.A. Fadeeva, SERP-E173 experiment [1]

ITEP, Moscow, Russia

Abstract. This report is devoted to previously unknown narrow resonance in the K_SK_S-system that is observed in experimental data coming from 6-m spectrometer (MIS ITEP). The experimental data on the production of K_S pair were obtained in π^-p interaction at 40 GeV by using a neutral trigger. In the K_SK_S-system a maximum of width <3 MeV is observed at mass of about 1595 MeV. The statistical confidence of this state is better than 6 standard deviations. Spin-parity of the state X(1595) is preferably $J^{PC} = 2^{++}$. Seeing its very small width this resonance is likely to be cryptoexotics (see [1],[2] for details).

Keywords: Exotics, Narrow resonances in K_SK_S-system
PACS: 14.40.Cs

INTRODUCTION

The experimental data we analyze had been accumulated by using 6-m spectrometer developed at the Institute for Theoretical and Experimental Physics (ITEP, Moscow) and installed at a 40 GeV π^--meson beam from the accelerator at the Institute for High Energy Physics (IHEP, Serpukhov). A detailed description of the 6-m spectrometer is presented in [3]. The system of two K_S-mesons that was recorded under the experimental conditions of the 6-m spectrometer is produced in the following two reactions:

$$\pi^- p \to K_S K_S n, \tag{1}$$

$$\pi^- p \to K_S K_S + (n + m\pi^0, p + \pi^-, \ldots). \tag{2}$$

Reaction (1) is separated with a trigger facility based on veto counters surrounding the liquid-hydrogen target. Due to imperfect trigger operation, some fraction of events of the reaction (2) is recorded by the setup. The spectrometer records with high efficiency K_S-mesons going to the forward direction and decaying into to charged π-mesons. The precision of the measurement of the effective mass of the K_SK_S-system is better than 5 MeV in mass region around 1600 MeV. The recording efficiency is about 45% for the K_SK_S-system in the mass region around 1600 MeV. It depends on the K_S-meson momenta.

In the analysis of the K_SK_S-system we used the following kinematical variables:

---

[1] I.A.Erofeev, O.N.Erofeeva, V.K.Grigor'ev, Yu.V.Katinov, V.I.Lisin, V.N.Luzin, V.N.Nozdrachev, Yu.P.Shkurenko, V.V.Sokolovsky, V.V.Vladimirsky. ITEP, Moscow.

CP814 *Hadron Spectroscopy: Eleventh International Conference*
edited by A. Reis, C. Göbel, J. Sá Borges, and J. Magnin

the effective mass M_{KK} of the pair of K_S-mesons;
the missing mass squared MM^2 defined as the squared mass of particles that are produced together with the K_SK_S-system and which are not recorded in the spectrometer;
the 4-momentum transferred from the beam to the system being studied, t;
the cosine of the Gottfried-Jackson angle, $cos\theta_{GJ}$; the Treiman-Yang angle, ϕ_{TY}.

The angles are calculated in the rest frame of the pair of K_S-mesons, the beam axis direction in this system being taken for the polar axis. The plane from which the Treiman-Yang angle is reckoned is spanned by the momenta of the beam and of the target proton in this reference frame.

RESONANCE X(1595)

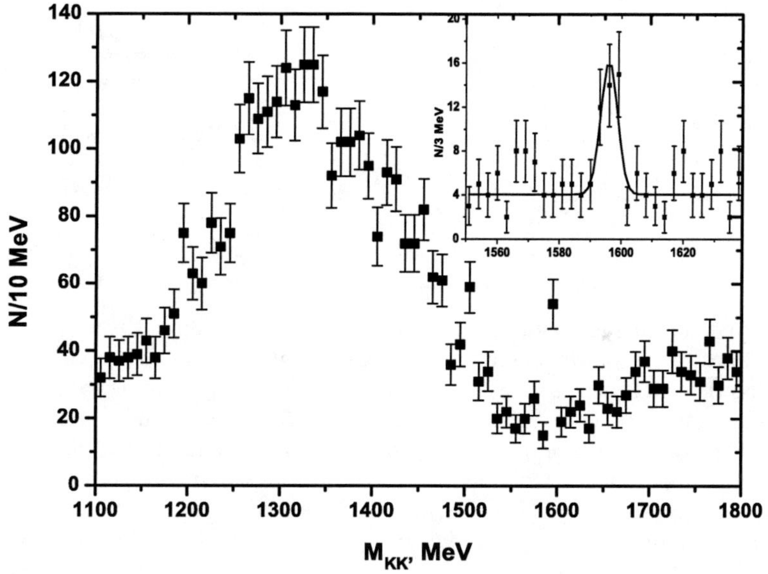

FIGURE 1. Effective-mass spectrum of two K_S-mesons.

The Figure (1) shows the mass spectrum of the K_SK_S-system from 1100 to 1800 MeV and from 1550 to 1640 MeV with the bin width being 10 and 3 MeV, respectively. The resonance feature manifests itself as a maximum in the vicinity of 1595 MeV.

The figure (2) represents the distribution of number of events with respect to the transferred momentum. Momentum-transfer distribution of events from the effective mass region of the K_SK_S-system that are adjacent to the resonance region from the left and from the right (1540< M_{KK} <1590, 1600< M_{KK} <1650 (MeV)) is represented by the curve and dots (the under figure) and of events from the resonance region (1590< M_{KK} <1600 (MeV)) only by the dots (the upper figure). It is seen that these distributions are quite different.

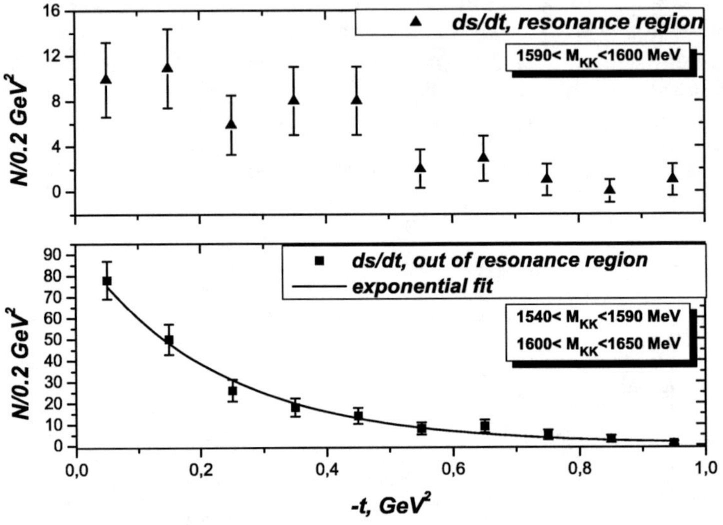

FIGURE 2. Momentum-transfer distribution for X(1595).

In order to determine the parameters of the observed resonance feature and its statistical confidence, the experimental data were fitted by the Maximum-likelihood method (MLM). The main advantage of this method over the histogramming is that the mass and angles are not averaged over the bin width in the fitting procedure, so that the result does not depend on the choice of the reference point and the number of bins into which the mass range under study is divided. Describing the experimental data, we used the probability-density function $F(P;\Omega)$, where P is the set of the parameters (the amplitude, the mass M, the width σ appearing in the Gauss function and the coefficients of the squared amplitudes of the angular distributions). Elements of the phase space Ω are effective mass of two K_S-mesons, the cosine of the Gottfried-Jackson angle $cos\theta_{GJ}$, the Treiman-Yang angle ϕ_{TY}.

Fitting was performed for events falling within the range 1540-1650 MeV of K_SK_S masses. A first-degree polynomial proved to be sufficient for describing the mass dependence of the background. The resonance was approximated by a Gauss function. In order to obtain the most probable values of the parameters we minimized the functional:

$$\int_{\Omega} \varepsilon(\Omega)F(P;\Omega)d\Omega - \sum_{i=1}^{N} \ln F(P;\Omega_i). \tag{3}$$

where $\varepsilon(\Omega)$ is the event-detection recording, N being the number of events. To compare the probabilities of experimental-data description with different parameter set, we calculated χ^2 by the formula:

$$\chi^2 = -2\ln L + const, \tag{4}$$

where $L = \prod_{i=1}^{N} F(P;\Omega_i)$.

TABLE 1. Different sets of minimization for X(1595).

	Background N_{events}				Resonance N_{events}				Parameters M, σ, MeV		$\delta\chi^2$	$N_{st.dev.}$
	S	D_0	D_+	G_+	S	D_0	D_+	G_+	$M \pm \Delta M$	$\sigma \pm \Delta\sigma$		
1	988	288	114	123	–	–	–	–	–	–	100	
2	955	249	82	133	85	–	–	–	$1595.8^{+1.0}_{-1.2}$	3.1±1.0	43.0	6.5
3	991	280	61	122	–	–	57	–	$1595.8^{+1.0}_{-1.2}$	3.1±1.0	27.4	5.2
4	981	224	108	123	–	78	–	–	$1595.8^{+1.0}_{-1.2}$	3.1±1.0	35.6	5.9
5	988	225	67	137	–	59	36	–	$1595.8^{+1.0}_{-1.2}$	3.1±1.0	47.6	6.9

The first line of the table 1 represents the result of fitting without allowing for the Gauss function and without discarding events falling within the resonance region. In the second line we tried to describe the resonance by S-wave. In the third line is given the result of fitting in terms of the D_+-wave. In the forth line is given the result of fitting in terms of the D_0-wave. And, finally, in the fifth line is represented the result of fitting in terms of the D_0- and D_+-waves.

Conclusion

Let us summarize results. At the confidence better than 6 standard deviations, we have obtained an indication of the existence of new resonance feature having
the mass $1595.8^{+1.0}_{-1.2} \pm 3.0$ MeV and the width 2.9±0.8 MeV.

The spin-parity of this resonance is preferably $J^{PC} = 2^{++}$. The product of the cross section for $X(1595)$ formation and the relevant branching ratio $\sigma Br(K_S K_S)$ is estimated at about 8^{+6}_{-3} nbn.

Main feature of this resonance is its very small width (<3 MeV)
and extraordinary cross-section dependence on the transferred momentum.

REFERENCES

1. Landsberg L.G., *Yad. Phys.* **57**, 47 (1994); *Phys. Atom. Nucl.* **57**, 42 (1994).
2. Landsberg L.G., *Usp. Phys. Nauk* **164**, 1129 (1994); *Phys. Uspekhi* **37**, 1043 (1994).
3. Nozdrachev V.N. et al., "The resonance structures of $K_S K_S$ and $\Lambda\bar{\Lambda}$ spectrum at MIS ITEP", in *CP619, Hadron Spectroscopy: Ninth International Conference*, edited by D. Amelin and A.M. Zaitsev, 2001, pp. 155-164.

Unitary coupled channel study of the $\Lambda(1520)$ resonance[1]

L. Roca*, V. K. Magas*, E. Oset* and Sourav Sarkar*

*Departamento de Física Teórica and IFIC, Centro Mixto Universidad de Valencia-CSIC, Institutos de Investigación de Paterna, Aptdo. 22085, 46071 Valencia, Spain

Abstract. We study the $\Lambda(1520)$ resonance in a coupled channel approach involving the $\pi\Sigma(1385)$, $K\Xi(1530)$, $\bar{K}N$ and $\pi\Sigma$ channels. Implementing unitarity in coupled channels, we make an analysis of the relative importance of the different mechanisms which contribute to the dynamical structure of this resonance. From experimental information on some partial wave amplitudes and constraints imposed by unitarity, we get a comprehensive description of the amplitudes and hence the couplings to the different channels. We test these amplitudes in different reactions like $K^-p \to \Lambda\pi\pi$, $\gamma p \to K^+K^-p$, $\gamma p \to K^+\pi^0\pi^0\Lambda$ and $\pi^-p \to K^0K^-p$ and find a fair agreement with the experimental data.

The $\Lambda(1520)$ $(3/2^-, D_{03})$ resonance is capturing renewed attention in the last years. The use of unitary coupled channel techniques in the study of the interaction of the octet of pseudoscalar mesons with the decuplet of baryons has led to the conclusion that the low lying $3/2^-$ baryons are mostly dynamically generated objects [1, 2]. One of these states is the $\Lambda(1520)$ resonance which is generated from the interaction of the coupled channels $\pi\Sigma(1385)$ and $K\Xi(1530)$, and couples mostly to the first channel to the point that, in this picture, the state would qualify as a quasibound $\pi\Sigma^*$ state. Indeed, the nominal mass of the $\Lambda(1520)$ is a few MeV below the average of the $\pi\Sigma^*$ masses. However, the PDG [3] gives a width of 15 MeV for the $\Lambda(1520)$, with branching ratios of 45% into $\bar{K}N$ and 43% into $\pi\Sigma$, and only a small branching ratio of the order of 10% for $\pi\Sigma^*$ and about 85% of the decay into $\pi\pi\Lambda$ is actually $\pi\Sigma^*$. The association of $\pi\pi\Lambda$ to $\pi\Sigma^*$ in the peak of the $\Lambda(1520)$ is a non trivial test since one has no phase space for $\pi\Sigma^*$ excitation and only the width of the Σ^* allows for this decay, hence precluding the reconstruction of the Σ^* resonant shape from the $\pi\Lambda$ decay product. In the work of Ref. [4] the $\bar{K}N$ and $\pi\Sigma$ channels were shown to play a relevant role in building up the resonance. In the present work we tackle this problem by performing a coupled channel analysis of the $\Lambda(1520)$ data with $\pi\Sigma^*$, $K\Xi^*$, $\bar{K}N$ and $\pi\Sigma$, the first two channels interacting in s-wave and the last two channels in d-wave to match the $3/2^-$ spin and parity of the $\Lambda(1520)$ resonance. We test the model making predictions for the $K^-p \to \pi^0\pi^0\Lambda$ and $K^-p \to \pi^+\pi^-\Lambda$ reactions, providing absolute cross sections in good agreement with experiment. Furthermore, we also make predictions for the shape of the K^-p mass distribution for the $\gamma p \to K^+K^-p$ and $\pi^-p \to K^+K^-p$ reactions and for the absolute value of the ratio of the $\gamma p \to K^+\pi^0\pi^0\Lambda$ and $\gamma p \to K^+K^-p$ reactions at the

[1] Contribution to the XI International Conference on Hadron Spectroscopy (HADRON 05), 21-26 August 2005, Rio de Janeiro, Brazil

CP814 *Hadron Spectroscopy: Eleventh International Conference*
edited by A. Reis, C. Göbel, J. Sá Borges, and J. Magnin

FIGURE 1. Diagrammatic representation of the resummation of meson-baryon loops in the unitarization procedure.

peak of the $\Lambda(1520)$ resonance, for which there are no experimental data available.

In Ref.[4] the $\Lambda(1520)$ resonance was studied within a coupled channel formalism including the $\pi\Sigma^*$, $K\Xi^*$ in s-wave and the $\bar{K}N$ and $\pi\Sigma$ in d-waves leading to a good reproduction of the pole position of the $\Lambda(1520)$ of the scattering amplitudes. However, the use of the pole position to get the properties of the resonance is far from being accurate as soon as a threshold is opened close to the pole position on the real axis, which is the present case with the $\pi\Sigma^*$ channel. Apart from that, in the approach of Ref. [4] some matrix elements in the kernel of the Bethe-Salpeter equation were not considered. In the present work we aim at a more precise description of the physical processes involving the $\Lambda(1520)$ resonance. Hence, we introduce the rest of tree level transition potentials relevant for the analysis: $\bar{K}N \to \bar{K}N$, $\bar{K}N \to \pi\Sigma$ and $\pi\Sigma \to \pi\Sigma$. Analogously to Ref. [4], we use for these vertices effective transition potentials which are proportional to the incoming and outgoing momentum squared in order to account for the d-wave character of the channels. Denoting $\pi\Sigma^*$, $K\Xi^*$, $\bar{K}N$ and $\pi\Sigma$ channels by 1, 2, 3 and 4 respectively, the matrix containing the tree level amplitudes is written as:

$$V = \left| \begin{array}{cccc} C_{11}(k_1^0+k_1^0) & C_{12}(k_1^0+k_2^0) & \gamma_{13}\, q_3^2 & \gamma_{14}\, q_4^2 \\ C_{21}(k_2^0+k_1^0) & C_{22}(k_2^0+k_2^0) & 0 & 0 \\ \gamma_{13}\, q_3^2 & 0 & \gamma_{33}\, q_3^4 & \gamma_{34}\, q_3^2 q_4^2 \\ \gamma_{14}\, q_4^2 & 0 & \gamma_{34}\, q_3^2 q_4^2 & \gamma_{44}\, q_4^4 \end{array} \right| , \qquad (1)$$

where $q_i = \frac{1}{2\sqrt{s}}\sqrt{[s-(M_i+m_i)^2][s-(M_i-m_i)^2]}$, $k_i^0 = \frac{s-M_i^2+m_i^2}{2\sqrt{s}}$ and $M_i(m_i)$ is the baryon(meson) mass. The coefficients C_{ij} are $C_{11} = \frac{-1}{f^2}$, $C_{21} = C_{12} = \frac{\sqrt{6}}{4f^2}$ and $C_{22} = \frac{-3}{4f^2}$, where f is $1.15 f_\pi$, with f_π ($= 93$ MeV) the pion decay constant, which is an average between f_π and f_K. The elements V_{11}, V_{12}, V_{21}, V_{22} come from the lowest order chiral Lagrangian involving the decuplet of baryons and the octet of pseudoscalar mesons as discussed in Ref. [2, 1].

In the literature several unitarization procedures have been used to obtain a scattering matrix fulfilling exact unitarity in coupled channels, like the Inverse Amplitude Method [5, 6] or the N/D method [7]. In this latter work the equivalence with the Bethe-Salpeter equation used in [8] was established. In the present work we use the Bethe-Salpeter equation with the V given above as the kernel to obtain the unitarized amplitude T:

$$T = V + VGT \Rightarrow T = [1-VG]^{-1}V \qquad (2)$$

Diagrammatically this means that one is resumming the series expressed in Fig. 1. In Eq. (2) G stands for a diagonal matrix containing the loop functions involving a baryon and a meson [2].

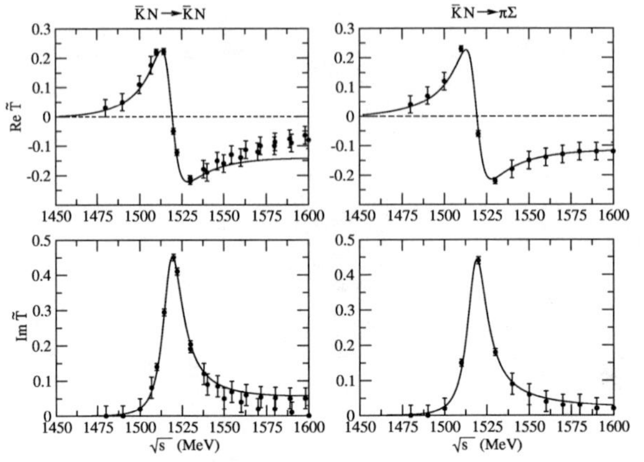

FIGURE 2. Fit to the experimental amplitudes [9, 10]. Left column: $\bar{K}N \to \bar{K}N$; right column: $\bar{K}N \to \pi\Sigma$.

TABLE 1. Parameters obtained from the fit

a_0	a_2	γ_{13} (MeV^{-3})	γ_{14} (MeV^{-3})	γ_{33} (MeV^{-5})	γ_{44} (MeV^{-5})	γ_{34} (MeV^{-5})
-1.8	-8.1	0.98×10^{-7}	1.1×10^{-7}	-1.7×10^{-12}	-0.7×10^{-12}	-1.1×10^{-12}

Since the $\pi\Sigma^*$ threshold lies in the $\Lambda(1520)$ region, the consideration of the width of the Σ^* resonance in the loop function is crucial in order to account properly for this channel. This is achieved through the convolution of the $\pi\Sigma^*$ loop function with the spectral distribution considering the Σ^* width.

In the model described so far we have as unknown parameters γ_{13}, γ_{14}, γ_{33}, γ_{34}, γ_{44} in the V matrix. Apart from these, there is also the freedom in the value of the subtraction constants in the loop functions. We will consider one subtraction constant for the s-wave channels (a_0) and one for the d-wave ones (a_2). Despite the apparent large number of free parameters in the V matrix, it is worth emphasizing that the largest matrix elements are V_{11}, V_{12} and V_{22} [2] which come from a chiral Lagrangian without any free parameters.

In order to obtain these parameters we fit our model to the experimental results on the $\bar{K}N$ and $\pi\Sigma$ scattering amplitudes in d-wave and $I = 0$. We use experimental data from Refs. [9, 10] where $\bar{K}N \to \bar{K}N$ and $\bar{K}N \to \pi\Sigma$ amplitudes are provided from partial wave analysis.

In Fig. 2 we show the result of our fit to the experimental data. The values of the unknown parameters obtained from the fit are given in Table 1. We can see that the value obtained for the subtraction constant for the s-wave channels (a_0) is of natural size (~ -2). Actually, the value of the s-wave loop function obtained using $a_0 = -1.8$ agrees with the result obtained with the cutoff method using a cutoff of about 500 MeV (at $\sqrt{s} \simeq 1520$ MeV). On the other hand, regarding the d-wave loops, the large value obtained for a_2 can be understood comparing also to the cutoff method. If one keeps the momentum dependence of the d-wave vertices inside the loop integral and evaluates the integral with the cutoff method, then also a cutoff of about 500 MeV gives the same

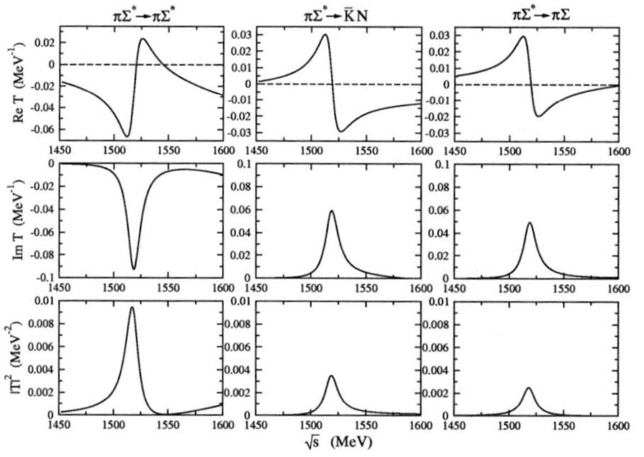

FIGURE 3. Unitary amplitudes involving the $\pi\Sigma^*$ channel. From left to right: $\pi\Sigma^* \to \pi\Sigma^*$, $\pi\Sigma^* \to \bar{K}N$ and $\pi\Sigma^* \to \pi\Sigma$.

result as the dimensional regularization with on-shell factorization and $a_2 \sim -8$.

Next we plot, in Fig. 3, the prediction for the unitarized amplitudes for the different channels involving the $\pi\Sigma^*$. It is worth mentioning that the unitarization procedure does not only provide the amplitude at the peak of the resonance but also obtains the full amplitude at energies away from it. This can have repercussion in some observables in specific physical processes as we will discuss below.

From the amplitudes it is straightforward to obtain the couplings of the $\Lambda(1520)$ to the different channels. Close to the peak the amplitudes can be approximated by

$$T_{ij}(\sqrt{s}) = \frac{g_i g_j}{\sqrt{s} - M_{\Lambda(1520)} + i\Gamma_{\Lambda(1520)}/2} \tag{3}$$

from where the couplings can be easily obtained. Up to a global sign of one of the couplings (we choose g_1 to be positive), the couplings we obtain are $g_1 = 0.91$, $g_2 = -0.29$, $g_3 = -0.54$ and $g_4 = -0.45$. We can see from that the $\Lambda(1520)$ resonance couples most strongly to the $\pi\Sigma^*$ channel. The fact that we are able to predict the value of this coupling is a non trivial consequence of the unitarization procedure that we employ.

The prediction of the amplitudes involving $\pi\Sigma^*$ channels can be checked in particular reactions where this channel could play an important role. First we evaluate the cross section for $K^- p \to \pi\pi\Lambda$ in the lines of Ref. [4] but using the new coupled channel formalism. The mechanisms and the expressions for the amplitudes and the cross sections can be found in Ref. [4] where, apart from the coupled channel unitarized amplitude, other mechanisms of relevance above the $\Lambda(1520)$ peak were included. In Fig. 4 we show our results for $K^- p \to \pi^0\pi^0\Lambda$ and $K^- p \to \pi^+\pi^-\Lambda$ cross section respectively along with experimental data from Refs. [12, 11].

The dashed line represents the contribution from mechanisms other than the unitarized coupled channels, and the solid gives the coherent sum of all the processes. These

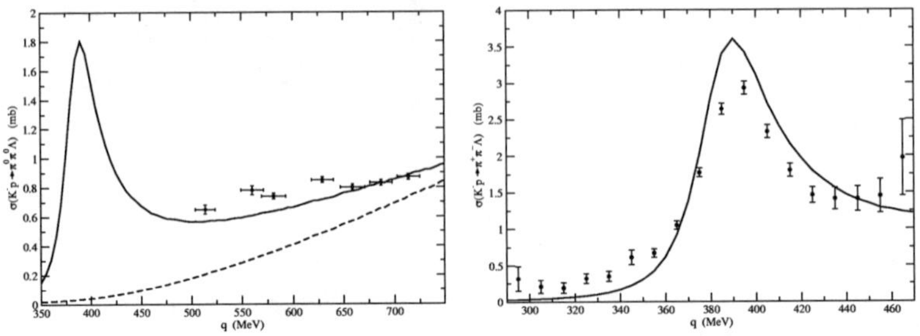

FIGURE 4. Result for the $K^-p \to \pi\pi\Lambda$ cross section.

cross sections depend essentially on the $T_{\bar{K}N\to\pi\Sigma^*}$ amplitude, which is obtained from our coupled channel analysis. It is a non-trivial prediction of the theory since this amplitude has not been included in the fit. Actually, the strength at the peak comes essentially from the unitarity constraint of the theory in analogy with the discussion above regarding the g_1 coupling.

We now test the model evaluating K^-p invariant mass distribution for the $\gamma p \to K^+K^-p$ reaction. In Ref. [13] the basic phenomenological model is explained but there only the $\pi\Sigma^*$ and $K\Xi^*$ channels were considered. In Fig. 5 we present the results obtained with the formalism of the present work along with the experimental data of [14].

We have removed 50 of the arbitrary units in the background, which are apparent below the resonant peak. Given the fact that the absolute strength of the results is a free parameter, the results do no change qualitatively if this background is not removed.

The normalization is arbitrary in the experimental data as well as in our calculation. For the purpose of the present work the shape of the distribution is the most important part and we can see that the agreement is quite fair. It is worth mentioning that in the work of Ref. [13] the reaction was initiated by a $\pi\Sigma^*$ loop. However, with the new channels considered in the present work, initial $\bar{K}N$ and $\pi\Sigma$ loops are also possible. The coupling of these channels to γpK^+ can be different. However we have checked that the shape of the distribution for all these possibilities is very similar around the $\Lambda(1520)$ peak. For higher energies this is also true within about 30%.

In Ref. [13] it was suggested that the ratio

$$R = \frac{d\sigma_{\gamma p\to K^+\pi^0\pi^0\Lambda}/dM_{\pi^0\pi^0\Lambda}}{d\sigma_{\gamma p\to K^+K^-p}/dM_{K^-p}} \tag{4}$$

can provide a test of the coupling of the $\Lambda(1520)$ to $\pi\Sigma^*$ (g_1) since, up to phase space and numerical factors, it is proportional to $(|T_{\pi\Sigma^*\to\pi\Sigma^*}|/|T_{\pi\Sigma^*\to\bar{K}N}|)^2$ which, at the $\Lambda(1520)$ peak position, is $(g_1/g_3)^2$. All considered we obtain a value for the ratio of $R \sim 0.25$, while $(g_1/g_3)^2$ is 2.8.

Next we show, in Fig. 6, the K^-p invariant mass distribution for the $\pi^-p \to K^0K^-p$ reaction. The model used for this purpose is analogous to the one used in the study of the $\gamma p \to K^+K^-p$ reaction. The differences are in the phase space and in the first vertex

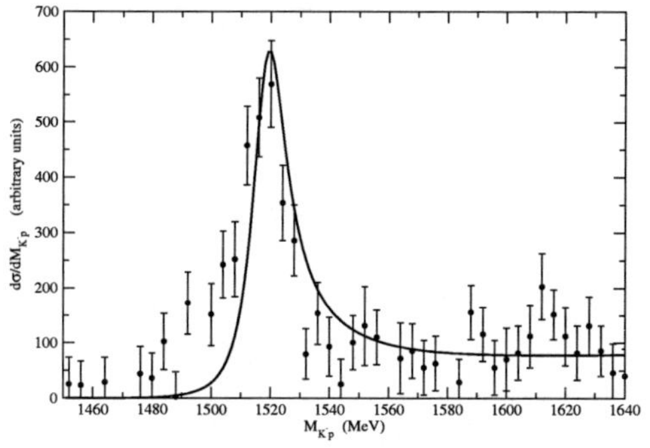

FIGURE 5. K^-p invariant mass distribution for the $\gamma p \to K^+K^-p$ reaction with photons in the range $E_\gamma = 2.8 - 4.8$ GeV. Experimental data from [14].

where we have used a different unknown constant.

In sumary, we have done a coupled channel analysis of the $\Lambda(1520)$ resonance using the $\pi\Sigma(1385)$, $K\Xi(1530)$, $\bar{K}N$ and $\pi\Sigma$ channels. We have used the Bethe-Salpeter equation to implement unitarity in the evaluation of the different amplitudes. The main novelty from previous coupled channel approaches to this resonance is the inclusion of new matrix elements in the kernel of the BS equation and the consideration of the $\Sigma(1385)$ width in the $\pi\Sigma^*$ loop function. The unknown parameters in the $V-$matrix, as well as the subtraction constants of the loop functions, have been obtained by a fit to $\bar{K}N \to \bar{K}N$ and $\bar{K}N \to \pi\Sigma$ partial wave amplitudes. As a consequence of the unitarity of the scheme used, we can predict the amplitudes and couplings of the $\Lambda(1520)$ for all the different channels. The largest coupling is obtained for the $\pi\Sigma^*$ channel. We have then tested the amplitudes obtained in several specific reactions and compared with experimental data at energies close to and slightly above the $\Lambda(1520)$ region. These include the $K^-p \to \Lambda\pi\pi$, $\gamma p \to K^+K^-p$, $\gamma p \to K^+\pi^0\pi^0\Lambda$ and $\pi^-p \to K^0K^-p$ reactions. We have obtained a reasonable agreement with the experimental results that allows us to be confident in the procedure followed to describe the nature of the $\Lambda(1520)$ resonance. For the $K^-p \to \Lambda\pi\pi$ process we could do predictions of the absolute cross sections since the process involves the amplitude for the $K^-p \to \pi\Sigma^*$ transition, which is an output of our theory. In the other two cases the strength of the cross section was left as a free parameter since no theory was done for the coupling of the photon or the pion to the $\Lambda(1520)$ components. Yet, it was found that in all cases the shape of the resonance was rather asymmetric, with a substantial strength above the peak of the resonance in all the reactions, which seems to be an intrinsic property of the $\Lambda(1520)$ resonance, and which in our theoretical framework could be attributed to the d-wave character of the $\bar{K}N$ and $\pi\Sigma$ channels, as well as to the fact that above the resonance peak the phase space for the $\Lambda(1520)$ decay into the $\pi\Sigma^*$ channel grows rapidly. We could see that, in spite of the relevant role played by the $\bar{K}N$ and $\pi\Sigma$ channels, the $\Lambda(1520)$ remains with the largest coupling to the $\pi\Sigma^*$ channel. In a simplified theory in which only the s-wave $\pi\Sigma^*$ and $K\Xi^*$ channels are taken, and their interaction is provided by the chiral Lagrangians, the $\Lambda(1520)$ appears naturally and qualifies as a quasibound $\pi\Sigma^*$ state. The coupling of the

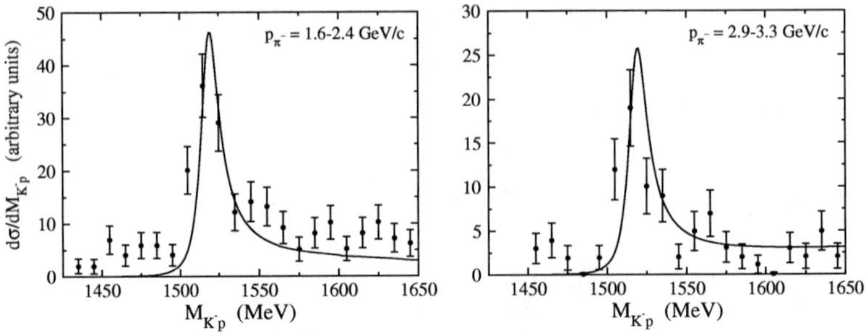

FIGURE 6. K^-p invariant mass distribution for the $\pi^-p \to K^0K^-p$ reaction. Left plot: pions in the range $p_{\pi^-} = 1.6-2.4$ GeV/c; right plot: $p_{\pi^-} = 2.9-3.3$ GeV/c. Experimental data from Ref. [15].

resonance to the $\bar{K}N$ and $\pi\Sigma$ channels in the real world modifies appreciably this picture but the state still keeps memory of the original simplified picture and has still a large coupling to $\pi\Sigma^*$ with a strength of g^2 of the order of three times larger than that of the $\bar{K}N$ or $\pi\Sigma$ channels.

REFERENCES

1. E. E. Kolomeitsev and M. F. M. Lutz, Phys. Lett. B **585** (2004) 243.
2. S. Sarkar, E. Oset and M. J. Vicente Vacas, Nucl. Phys. A **750** (2005) 294.
3. S. Eidelman *et al.* [Particle Data Group], Phys. Lett. B **592**, 1 (2004).
4. S. Sarkar, E. Oset and M. J. Vicente Vacas, Phys. Rev. C **72** (2005) 015206.
5. A. Dobado and J. R. Pelaez, Phys. Rev. D **56** (1997) 3057.
6. J. A. Oller, E. Oset and J. R. Pelaez, Phys. Rev. D **59** (1999) 074001 [Erratum-ibid. D **60** (1999) 099906].
7. J. A. Oller and E. Oset, Phys. Rev. D **60** (1999) 074023.
8. J. A. Oller and E. Oset, Nucl. Phys. A **620** (1997) 438 [Erratum-ibid. A **652** (1999) 407].
9. G. P. Gopal, R. T. Ross, A. J. Van Horn, A. C. McPherson, E. F. Clayton, T. C. Bacon and I. Butterworth [Rutherford-London Collaboration], Nucl. Phys. B **119** (1977) 362.
10. M. Alston-Garnjost, R. W. Kenney, D. L. Pollard, R. R. Ross, R. D. Tripp, H. Nicholson and M. Ferro-Luzzi, Phys. Rev. D **18** (1978) 182.
11. T. S. Mast, M. Alston-Garnjost, R. O. Bangerter, A. Barbaro-Galtieri, F. T. Solmitz and R. D. Tripp, Phys. Rev. D **7** (1973) 5.
12. S. Prakhov *et al.*, Phys. Rev. C **69** (2004) 042202.
13. L. Roca, E. Oset and H. Toki, arXiv:hep-ph/0411155.
14. D. P. Barber *et al.*, Z. Phys. C **7** (1980) 17.
15. O. I. Dahl, L. M. Hardy, R. I. Hess, J. Kirz, D. H. Miller and J. A. Schwartz, Phys. Rev. **163** (1967) 1377.

Evidence for two $\Lambda(1405)$ resonance states

E. Oset*, V. K. Magas* and A. Ramos†

**Departamento de Física Teórica and IFIC, Centro Mixto,*
Institutos de Investigación de Paterna - Universidad de Valencia-CSIC
Apdo. correos 22085, 46071, Valencia, Spain
†Departament d'Estructura i Constituents de la Matéria,
Universitat de Barcelona, Diagonal 647, 08028 Barcelona, Spain

Abstract. The $K^- p \to \pi^0\pi^0\Sigma^0$ reaction is studied within a chiral unitary model. The distribution of $\pi^0\Sigma^0$ states forming the $\Lambda(1405)$ shows, in agreement with a recent experiment, a peak at 1420 MeV and a relatively narrow width of $\Gamma = 38$ MeV. We use these data in combination with those of the $\pi^- p \to K^0\pi\Sigma$ reaction and elements of chiral unitary theory to prove that there are two $\Lambda(1405)$ states instead of one as so far assumed.

Keywords: Two $\Lambda(1405)$ states
PACS: 13.75.-n,12.39.Fe,14.20.Jn,11.30.Hv

The history of the $\Lambda(1405)$ as a dynamical resonance generated from the interaction of meson baryon components in coupled channels has experienced a boost within the context of unitary extensions of chiral perturbation theory ($U\chi PT$) [1, 2, 3, 4, 5], where the lowest order chiral Lagrangian and unitarity in coupled channels generates the $\Lambda(1405)$ and leads to good agreement with the $K^- p$ reactions. The surprise, however, came with the realization that there are two poles in the neighborhood of the $\Lambda(1405)$ both contributing to the final experimental invariant mass distribution [3, 4, 5, 6, 7, 8, 9]. The properties of these two states are quite different, one has a mass around 1390 MeV, a large width of about 130 MeV and couples mostly to $\pi\Sigma$, while the second one has a mass around 1425 MeV, a narrow width of about 30 MeV and couples mostly to $\bar{K}N$. The two states are populated with different weights in different reactions and, hence, their superposition can lead to different distribution shapes. Since the $\Lambda(1405)$ resonance is always seen from the invariant mass of its only strong decay channel, the $\pi\Sigma$, hopes to see the second pole are tied to having a reaction where the $\Lambda(1405)$ is formed from the $\bar{K}N$ channel. This is accomplished by the recently measured reaction $K^- p \to \pi^0\pi^0\Sigma^0$ [10] which allows us to test already the two-pole nature of the $\Lambda(1405)$.

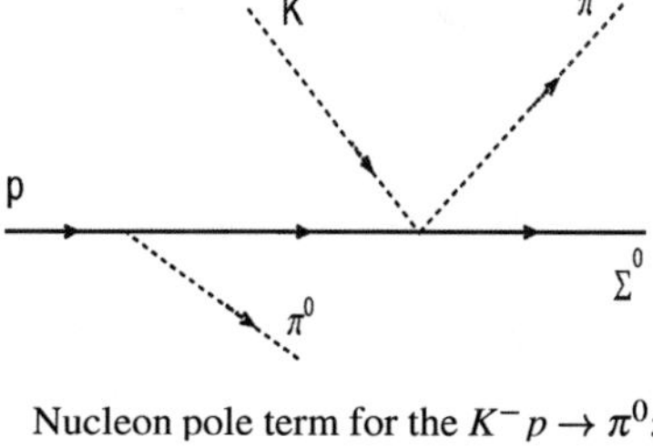

FIGURE 1. Nucleon pole term for the $K^- p \to \pi^0\pi^0\Sigma$ reaction.

CP814 *Hadron Spectroscopy: Eleventh International Conference*
edited by A. Reis, C. Göbel, J. Sá Borges, and J. Magnin

Our model for the reaction $K^- p \to \pi^0 \pi^0 \Sigma^0$ in the energy region of $p_{K^-} = 514$ to 750 MeV/c, as in the experiment [10], considers those mechanisms in which a π^0 loses the necessary energy to allow the remaining $\pi^0 \Sigma^0$ pair to be on top of the $\Lambda(1405)$ resonance. The first of such mechanisms is given by the diagram of Fig. 1. Other mechanisms that involve the meson meson interaction and BBMMM vertices were found negligible in the detailed study of [11].

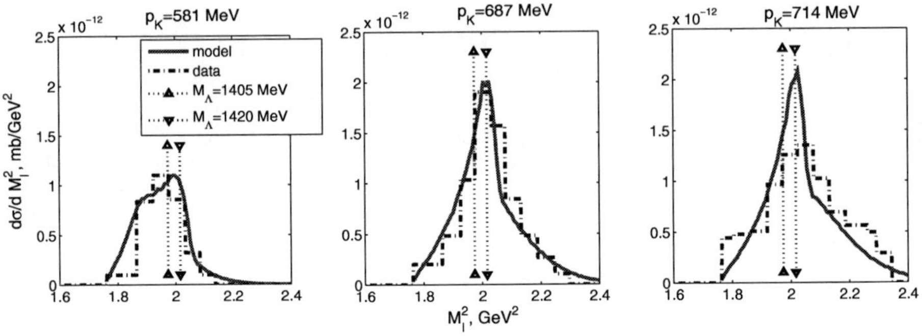

FIGURE 2. The $(\pi^0\Sigma^0)$ invariant mass distribution for three different initial kaon momenta.

The amplitude for the process $K^- p \to \pi^0 \pi^0 \Sigma^0$, considering the diagram of Fig. 1 with a pseudovector πNN coupling (and keeping up to $1/M_N$ terms) is given by:

$$-it_{K^- p \to \pi^0 \pi^0 \Sigma^0} = -\frac{D+F}{2f} \vec{\sigma} \left[\vec{q}\,'(1 + \frac{q^{0\prime}}{2M_N}) + \frac{q^{0\prime}}{M_N} \vec{k} \right] \times$$
$$\frac{M_N}{E_N(\vec{k}+\vec{q}\,')} \frac{1}{E_N(\vec{k}) - q'^0 - E_N(\vec{k}+\vec{q}\,')} t_{K^- p \to \pi^0 \Sigma^0}, \qquad (1)$$

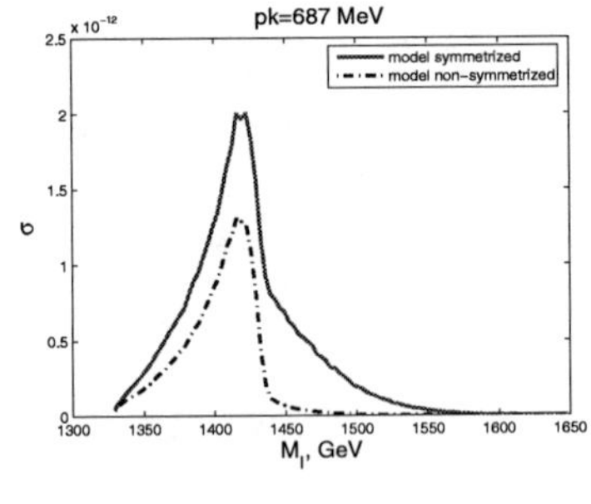

FIGURE 3. Theoretical $(\pi^0\Sigma^0)$ invariant mass distribution for an initial kaon lab momenta of 687 MeV. The non-symmetrized distribution also contains the factor 1/2 in the cross section.

where $q\prime$ refers to the momentum of the first outgoing pion and k is the momentum of the kaon (see Fig. 1). The second pion, from the decay of ther $\Lambda(1405)$, is assumed to have a momentum q_f.

The indistinguishability of the two emitted pions requires the implementation of symmetrization. This is achieved by summing two amplitudes evaluated with the two pion momenta exchanged, $q_f \leftrightarrow q\prime$. In addition, a factor of 1/2 for indistinguishable particles is also included in the total cross section.

Our calculations show that the process is largely dominated by the nucleon pole term shown in Fig. 1. As a consequence, the $\Lambda(1405)$ thus obtained comes mainly from the $K^- p \rightarrow \pi^0 \Sigma^0$ amplitude which, as mentioned above, gives the largest possible weight to the second (narrower) state.

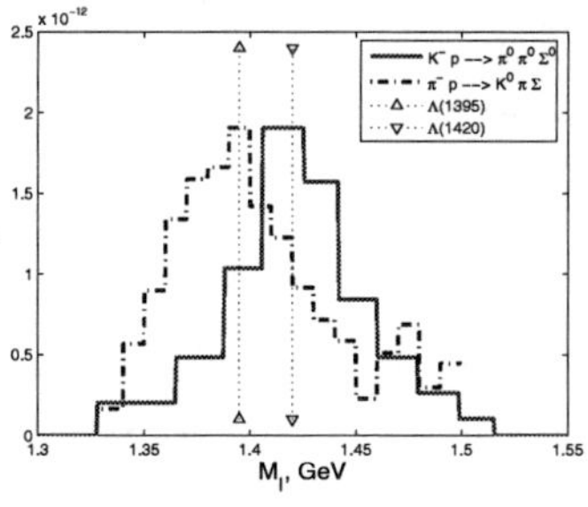

FIGURE 4. Two experimental shapes of $\Lambda(1405)$ resonance. See text for more details.

In Fig. 2 our results for the invariant mass distribution for three different energies of the incoming K^- are compared to the experimental data. Symmetrization of the amplitudes produces a sizable amount of background. At a kaon laboratory momentum of $p_K = 581$ MeV/c this background distorts the $\Lambda(1405)$ shape producing cross section in the lower part of M_I, while at $p_K = 714$ MeV/c the strength of this background is shifted toward the higher M_I region. An ideal situation is found for momenta around 687 MeV/c, where the background sits below the $\Lambda(1405)$ peak distorting its shape minimally. The peak of the resonance shows up at $M_I^2 = 2.02\,\text{GeV}^2$ which corresponds to $M_I = 1420$ MeV, larger than the nominal $\Lambda(1405)$, and in agreement with the predictions of Ref. [6] for the location of the peak when the process is dominated by the $t_{\bar{K}N \rightarrow \pi\Sigma}$ amplitude. The apparent width from experiment is about $40 - 45$ MeV, but a precise determination would require to remove the background mostly coming from the "wrong" $\pi^0 \Sigma^0$ couples due to the indistinguishability of the two pions. A theoretical analysis permits extracting the pure resonant part by not symmetrizing the amplitude. This is plotted in Fig. 3 as a function of M_I. The width of the resonant part is $\Gamma = 38$ MeV, which is smaller than the nominal $\Lambda(1405)$ width of 50 ± 2 MeV [13], obtained from the average of several experiments, and much narrower than the apparent width of about 60 MeV that one sees in the $\pi^- p \rightarrow K^0 \pi \Sigma$ experiment [14], which also produces a distribution peaked at 1395 MeV. In order to illustrate the difference between the $\Lambda(1405)$ resonance seen in this latter reaction and in the present one, the two experimental distributions are compared in Fig. 4. We recall that the shape of the $\Lambda(1405)$ in the $\pi^- p \to K^0 \pi\Sigma$ reaction was shown in Ref. [12] to be largely built from the $\pi\Sigma \to \pi\Sigma$ amplitude, which is dominated by the wider lower energy state.

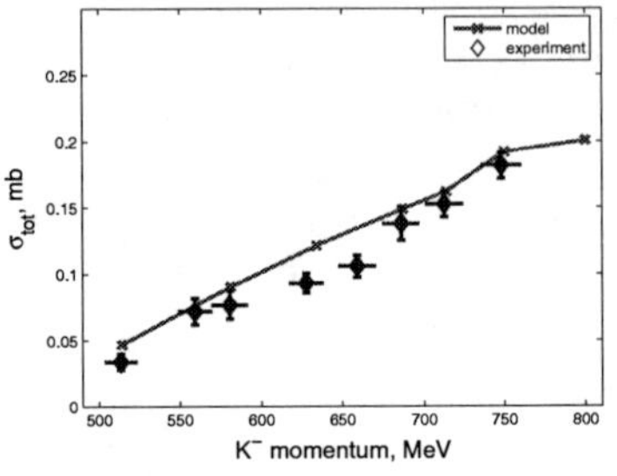

FIGURE 5. Total cross section for the reaction $K^- p \to \pi^0 \pi^0 \Sigma^0$. Experimental data are taken from Ref. [10].

The invariant mass distributions shown here are not normalized, as in experiment. But we can also compare our absolute values of the total cross sections with those in Ref. [10]. As shown in Fig. 5, our results are in excellent agreement with the data, in particular for the three kaon momentum values whose corresponding invariant mass distributions have been displayed in Fig. 2.

In summary, we have shown, by means of a realistic model, that the $K^- p \to \pi^0 \pi^0 \Sigma^0$ reaction is particularly suited to study the features of the second pole of the $\Lambda(1405)$ resonance, since it is largely dominated by a mechanism in which a π^0 is emitted prior to the $K^- p \to \pi^0 \Sigma^0$ amplitude, which is the one giving the largest weight to the second narrower state at higher energy. Our model has proved to be accurate in reproducing both the invariant mass distributions and integrated cross sections seen in a recent experiment [10]. The study of the present reaction, complemental to the one of Ref. [12] for the $\pi^- p \to K^0 \pi\Sigma$ reaction, has shown that the quite different shapes of the $\Lambda(1405)$ resonance seen in these experiments can be interpreted in favour of the existence of two poles with the corresponding states having the characteristics predicted by the chiral theoretical calculations. Besides demonstrating once more the great predictive power of the chiral unitary theories, this combined study of the two reactions gives the first clear evidence of the two-pole nature of the $\Lambda(1405)$.

Acknowledgments: This work is partly supported by DGICYT contracts BFM2002-01868, BFM2003-00856, the Generalitat de Catalunya contract SGR2001-64, and the E.U. EURIDICE network contract HPRN-CT-2002-00311. This research is part of the EU Integrated Infrastructure Initiative Hadron Physics Project under contract number RII3-CT-2004-506078.

REFERENCES

1. N. Kaiser, P. B. Siegel and W. Weise, Phys. Lett. B **362** (1995) 23
2. E. Oset and A. Ramos, Nucl. Phys. A **635** (1998) 99 [arXiv:nucl-th/9711022].
3. J. A. Oller and U. G. Meissner, Phys. Lett. B **500** (2001) 263 [arXiv:hep-ph/0011146].
4. D. Jido, A. Hosaka, J. C. Nacher, E. Oset and A. Ramos, Phys. Rev. C **66** (2002) 025203 [arXiv:hep-ph/0203248].
5. C. Garcia-Recio, J. Nieves, E. Ruiz Arriola and M. J. Vicente Vacas, Phys. Rev. D **67** (2003) 076009 [arXiv:hep-ph/0210311].
6. D. Jido, J. A. Oller, E. Oset, A. Ramos and U. G. Meissner, Nucl. Phys. A **725** (2003) 181 [arXiv:nucl-th/0303062].
7. C. Garcia-Recio, M. F. M. Lutz and J. Nieves, Phys. Lett. B **582** (2004) 49 [arXiv:nucl-th/0305100].
8. T. Hyodo, S. I. Nam, D. Jido and A. Hosaka, Phys. Rev. C **68** (2003) 018201 [arXiv:nucl-th/0212026].
9. S. I. Nam, H. C. Kim, T. Hyodo, D. Jido and A. Hosaka, arXiv:hep-ph/0309017.
10. S. Prakhov *et al.* [Crystall Ball Collaboration], Phys. Rev. C **70** (2004) 034605.
11. V. K. Magas, E. Oset and A. Ramos, Phys. Rev. Lett. **95** (2005) 052301 [arXiv:hep-ph/0503043].
12. T. Hyodo, A. Hosaka, E. Oset, A. Ramos and M. J. Vicente Vacas, Phys. Rev. C **68** (2003) 065203 [arXiv:nucl-th/0307005].
13. K. Hagiwara *et al.* [Particle Data Group], Phys. Rev. D **66**, 010001 (2002).
14. D. W. Thomas, A. Engler, H. E. Fisk, and R. W. Kraemer, Nucl. Phys. B **56**, 15 (1973).

The Nucleon Resonance Program at JLab

Ralf W. Gothe for JLab and the CLAS Collaboration

Department of Physics and Astronomy, University of South Carolina, Columbia, SC 29208

Abstract. The status of the program to study baryon resonances at Jefferson Lab will be exemplified by the latest results on resonance parameters and transition form factors in single and double-pion production as well as kaon-hyperon decays.

Keywords: nucleon resonances and transition form factors.
PACS: 13.60.Rj, 13.60.-r, 14.20.-c, 14.20.Gk.

INTRODUCTION

Two of the four fundamental forces, the weak and especially the electromagnetic one, are reasonably well understood within the underlying framework of perturbative quantum field theories. A complete quantum field theoretical description of the gravitational force is still in its early stages and experimentally the graviton, the field carrier itself, has not yet been measured. In the case of the strong force, we have two experimentally verified perturbative quantum field theories; perturbative Quantum Chromodynamics (pQCD) at small distances which is governed by gluon fields and Chiral Perturbation Theory (ChPT) at larger distances which is governed by pion fields. However, the strong field at intermediate distances, where it generates more than 98% of the total mass of all normal matter, is not understood on similar firm grounds. Laboratories currently provide complementary hadronic or electromagnetic probes in the best suited energy range to perform precision experiments that test the nature of the strong force in this intermediate confinement regime. One of the leading laboratories in this research field is Jefferson Lab (JLab), where some of the most pressing experiments are planned and carried out. Examples reported here will focus on: transition form factor measurements at low Q^2 to investigate the pion fields and to bridge the gap to ChPT, and at high Q^2 to investigate the transition to pQCD in exclusive channels; as well as searching for resonances in sequential decays and for resonances that do not couple to single-pion nucleon states to test the quark flavor symmetry or for highly excited resonances to test chiral symmetry restoration; and pinning down resonance parameters by precision measurements of polarization and double-polarization observables in single- and double-meson production channels.

SINGLE-MESON PRODUCTION

A large portion of the nuclear physics community enthusiastically started to investigate baryon resonances as new optimized detector systems with large solid angle and momentum coverage like CLAS and new high-intensity continuous electron

CP814 *Hadron Spectroscopy: Eleventh International Conference*
edited by A. Reis, C. Göbel, J. Sá Borges, and J. Magnin

beams like at JLab became available. The high versatility of the provided electromagnetic probes that have merely any initial state interaction has been intriguing ever since. The isoscalar or isovector and the electric or magnetic or longitudinal character of the coupling to hadronic matter probe different aspects of the strong interaction; but soon after scientists had to experience the annoying complications introduced by the desired versatility of the electromagnetic probe that mixes all the different coupling amplitudes simultaneously into the measured cross sections. The way out has been successfully demonstrated in the case of the N to $\Delta(1232)$ transition where the small resonant electric (EMR) and scalar quadrupole (SMR) amplitudes could be extracted with respect to the dominant magnetic dipole amplitude at intermediate momentum transfers ($0.2\ (\text{GeV/c})^2 < Q^2 < 1.0\ (\text{GeV/c}^2)$) with absolute systematic uncertainties of typically 0.5 % [1-4]. To obtain such precision results for the extraction of isolated resonance parameters, additional isospin channels and polarization as well as double-polarization observables have been measured to disentangle the individual resonant and non-resonant coupling amplitudes [4-7]. This successful effort has to be extended beyond the first and second resonance region.

Preliminary JLab [4] results at low four momentum transfers down to 0.1 $(\text{GeV/c})^2$ follow the known constant behavior for EMR and the constantly falling behavior with growing Q^2 for SMR. The extrapolation of both ratios to even smaller Q^2 seem to agree with the real photon point EMR results [8] or with Siegert limit respectively, whereas the SMR disagrees with the MAMI [9] and Bates [10] results at $Q^2 = 0.127\ (\text{GeV/c})^2$. A further extension into the momentum transfer region of $0.01\ (\text{GeV/c})^2 < Q^2 < 0.1\ (\text{GeV/c})^2$, where no results are available, is experimentally challenging. How strongly this lack of data impacts the comprehension of the strong interaction in this regime is illustrated in FIGURE 1. One would expect that the total cross section at $Q^2=0\ (\text{GeV/c})^2$ is largest, since the N to $\Delta(1232)$ transition form factors should drop with increasing Q^2, but the total cross section at $Q^2=0.1\ (\text{GeV/c})^2$ is significantly larger than at the real photon point, $Q^2=0\ (\text{GeV/c})^2$ [11]. This can only be explained by a strong longitudinal coupling of the virtual photon to the nucleon and by the fact that the kinematical suppression of this longitudinal coupling drops initially much faster than the transverse transition form factors. On the other side pQCD predicts in the high Q^2-limit, when neglecting higher twist contributions, an EMR of +1 and a Q^2-independent SMR, whereas preliminary results at high Q^2 up to 5 $(\text{GeV/c})^2$ [4] show no indication of this predicted behavior

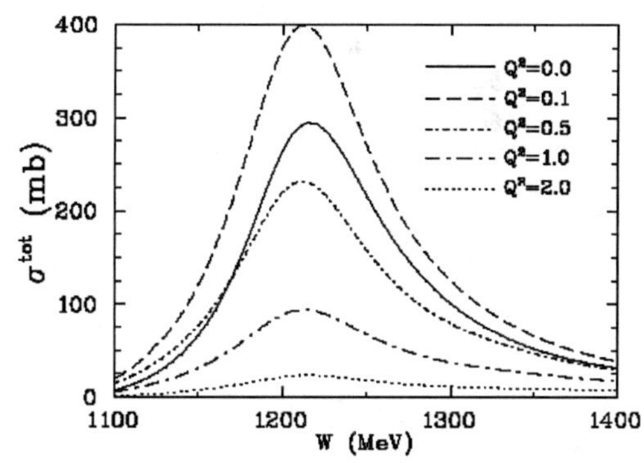

FIGURE 1: Total cross section of the $p\pi^0$ electro-production for different four momentum transfers in GeV [11].

and follow the same overall trend as established in the non-perturbative regime. It is not only important to extend these measurements of the N to Δ(1232) transition form factors to even higher momentum transfers, but also to investigate the high Q^2-evolution of exclusive transition form factors to other higher lying resonances. In the first case we need to push the measurements of exclusive observables and their theoretical description towards the onset of perturbative behaviour. These experiments

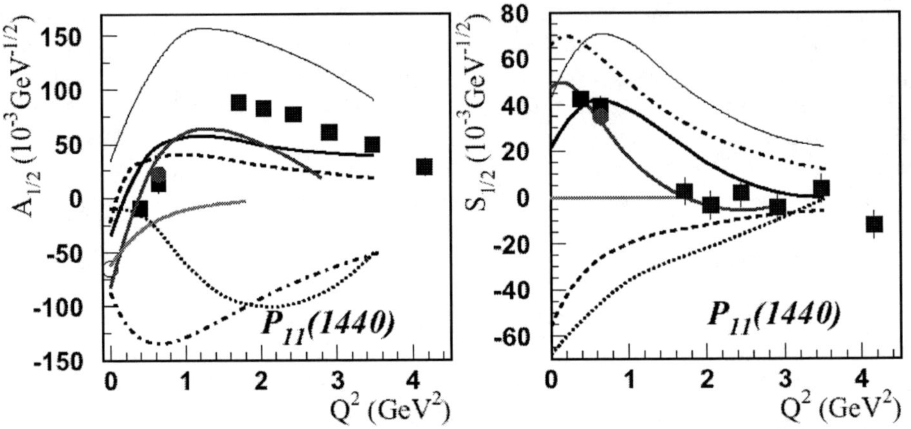

FIGURE 2: All data points are CLAS results in the π photo-production channel (open circle) [12], π and 2π (solid circles) [13], and π (solid squares) low Q^2 [14] and preliminary high Q^2 [15] electro-production channels. The solid bold lines are theoretical model calculations of Cano & Gonzales: qqq+qq$_{bar}$ cloud (red line or gray) [16], Capstick & Keister: light-front RQM (thick black) [17], Li & Burkert: qqqg hybrid (green line or pale gray) [18], and all other lines are further QM calculations [18,19].

require the JLab and CLAS upgrades to 12 GeV. In the latter case the Q^2-dependence of the helicity coupling amplitudes for the higher lying resonances P_{11}, D_{13}, S_{11}, and F_{15} has been measured up to 4 $(GeV/c)^2$ [15]. FIGURE 2 illustrates the quality of the results [12-15] and the shortcomings of the model predictions [16-19]. None of the model calculations are able to describe neither the magnitude nor the Q^2-trend of the helicity amplitudes of all four resonances consistently [4].

MISSING RESONANCES

The term missing resonances refers to the fact that quark models based on the flavor blindness of the strong interaction predict more nucleon excitations than are experimentally found. Besides the two commonly discussed theoretical approaches, di-quark models and coupled channel dynamics, which would reduce the number of predicted baryon states, a possible experiment-related solution to this problem is that these resonances do not couple to the single-pion nucleon channel and therefore have not been seen by now. This problem can initially be tackled from two sides by investigating both the double-pion-nucleon and the kaon-hyperon decay channels.

Many nucleon resonances in the mass region above 1.6 GeV decay predominantly through either $\Delta\pi$ or $N\rho$ intermediate states into $N\pi\pi$ final states [12]. This makes electromagnetic exclusive double-pion production an important tool in the investigation of nucleon resonances and reaction dynamics, as well as in the search for missing baryon states. Unpolarized cross-section measurements of double-pion photo- and electroproduction have been reported by several groups [20] and most recently by the CLAS collaboration [21]. The Q^2-dependent analysis of the $\pi^+\pi^-$ electroproduction data in a phenomenological isobar model [25] indicates the presence of a new $P_{13}(1720)$ resonance that couples strongly to $\Delta\pi$ and only weakly to $N\rho$. Further polarization observables are needed here as in the single-meson production to sufficiently constrain the model dependent interpretation. Polarization degrees of freedom in charged double-pion production have been investigated initially at SLAC [22] and in the context of the Gerasimov-Drell-Hearn Sum Rule at MAMI [23].

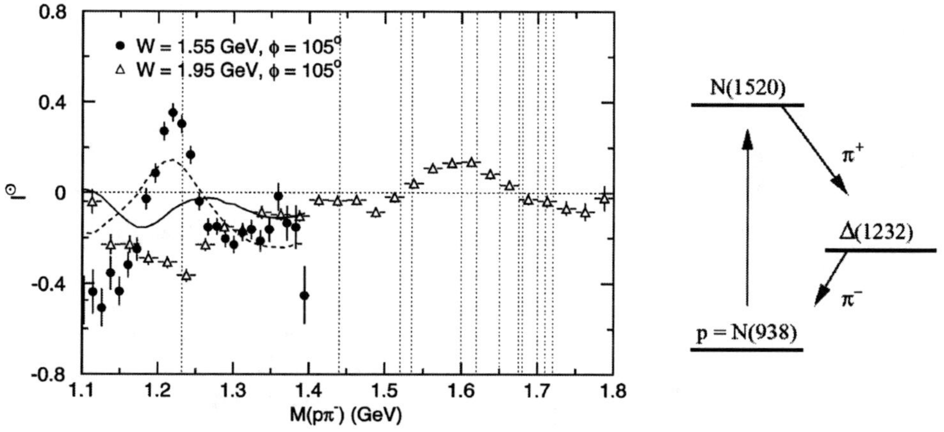

FIGURE 3: Left panel: Helicity asymmetry as a function of the invariant mass $M(p\pi^-)$ for W = 1.55 GeV (filled circles) and 1.95 GeV (open triangles) and a 30°-wide ϕ-angle range centered at ϕ = 105° [24]. The curves are the results of Mokeev *et al.* [25] (solid) and Fix & Arenhövel [26] (dashed) for W = 1.55 GeV only. Note that the result of Fix & Arenhövel is in phase with the data (filled circles) and that of Mokeev *et al.* is not. The vertical lines indicate the masses of the known N and Δ resonances. Right panel: Possible contributing sequential-decay channel.

New beam-helicity asymmetries have recently been reported for the double-pion photoproduction reaction $\gamma p \rightarrow p\pi^+\pi^-$ [24] based on the first measurement of the helicity asymmetry $I^{\circ}=(\sigma^+-\sigma^-)/P_\gamma(\sigma^++\sigma^-)$ in the higher resonance region for center-of-mass energies between 1.35 GeV and 2.30 GeV, where the photon beam is circularly polarized and no target nor recoil polarization is specified. Here, P_γ is the degree of circular polarization of the photon and $\sigma^\pm$ are the cross sections for the two photon-helicity states. FIGURE 3 shows the helicity asymmetry as a function of the invariant mass $M(p\pi^-)$ for two different values of W and a fixed value of ϕ as a typical example, where ϕ is the angle between the scattering and the $\pi^+\pi^-$ plane. The most interesting features are the changes that occur as M(pπ^-) traverses the $\Delta(1232)$ resonance, where for W = 1.55 GeV, a maximum is seen in the region of this resonance. We see a similar trend in the region of the higher-mass resonances around 1.60 GeV but no maximum around the $\Delta(1232)$ resonance for W = 1.95 GeV. This intriguing behavior indicates how valuable helicity asymmetry and other polarization observables can be for studying baryon spectroscopy, and how important sequential decays of resonances are in this energy range, where yet-unobserved resonances are predicted [27].

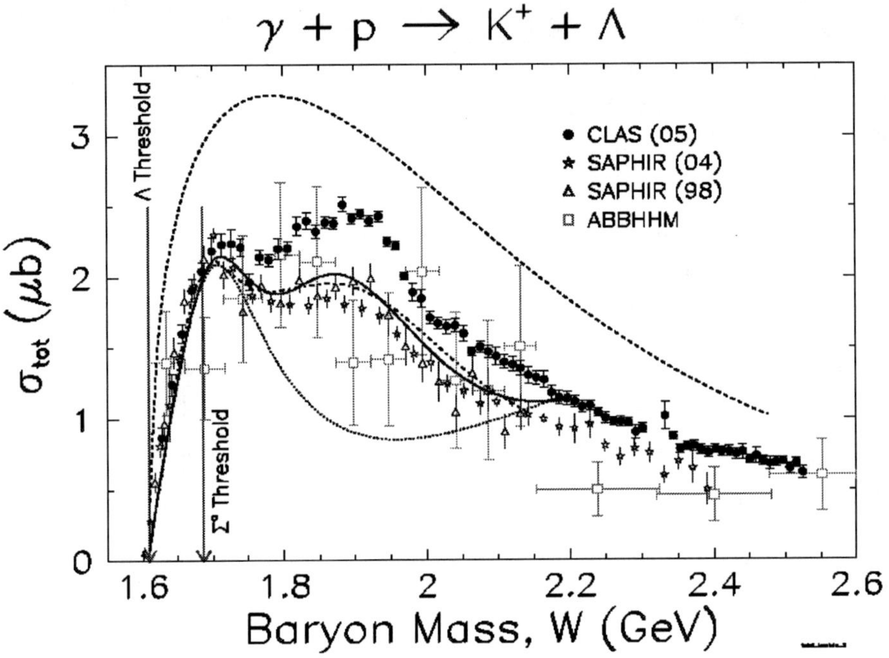

FIGURE 4: Total cross section for $K^+\Lambda$ photoproduction. The CLAS data points (solid circles) are shown with the combined statistical and fitting uncertainties [28] and compared to the results from SAPHIR (stars [29], triangles [30]) and the ABBHHM collaboration (squares) [31]. The lines are from a Regge model (dashed) [32] Kaon-MAID [33] with (solid) and without the $D_{13}(1895)$ (dotted), and an effective Lagrangian approach (dash-dotted) [34].

Very recently high-statistics photoproduction cross section data for $K^+\Lambda$ and $K^+\Sigma^0$ final states, with large kinematical coverage in both 1.6GeV<W<2.5GeV and -0.85< $\cos\theta_K$*<0.95, have become available [28] that allow to look for new resonances decaying into these kaon-hyperon channels. The example shown in FIGURE 4 addresses the discussion whether a new $D_{13}(1895)$ as predicted by Capstick and Roberts or by Loering, Metsch and Petry [33], or a new $P_{13}(1900)$ but no $D_{13}(1895)$ as suggested by Shklyar, Lenske and Mosel is needed [35], or appropriate tuning of the background contributions is sufficient [34] to understand the cross section peak at 1.9 GeV, which is now even more pronounced than previously reported by SAPHIR [29,30]. The additional forward-to-backward angular dependence of the $K^+\Lambda$ cross sections reveals changes in both position and shape of this peak, which cannot be explained by a single new resonance in the s or u-channel.

ACKNOWLEDGMENTS

We thank the staff of the Accelerator and the Physics Divisions at Thomas Jefferson National Accelerator Facility who are making experiments like these possible. Major support is provided by the Department of Energy and the National Science Foundation.

REFERENCES

1. K. Joo *et al.*, Phys. Rev. Lett. **88**, 122001 (2002) 1-5.
2. R.W. Gothe, Proceedings of NSTAR 2002, World Scientific (2003) 220-227.
3. G. Laveissiere *et al.*, Phys. Rev. **C69**, 045203 (2004) 1-15.
4. Preliminary JLab Results, L.C. Smith, M. Ungaro, K. Park, Private Communication (2005).
5. K. Joo *et al.*, Phys. Rev. **C70**, 042201 (2004) 1-6.
6. J.J. Kelly *et al.*, Phys. Rev. Lett. **95** 102001 (2005) 1-6.
7. H. Egiyan *et al.*, submitted to Phys. Rev. **C** (2005) 1-19.
8. R. Beck *et al.*, Phys. Rev. **C61**, 035204 (2000) and G. Blanpied *et al.*, Phys. Rev. Lett. **79**, 4337 (1997).
9. T. Pospischil *et al.*, Phys. Rev. Lett. **86** (2001) 2959-2962.
10. N.F. Spaveris *et al.*, Phys. Rev. Lett. **94** 022003 (2005) 1-4.
11. R. Arndt *et al.*, nucl-th/0301068 or Proceedings of NSTAR2002, Scientific (2003).
12. S. Eidelman *et al.*, Phys. Lett. **B 592**, (2004) 1.
13. I.G. Aznauryan *et al.*, Phys. Rev. **C72**, 045201 (2005) 1-9.
14. I.G. Aznauryan *et al.*, Phys. Rev. **C71**, 015201 (2005) 1-9.
15. K. Park *et al.*, to be submitted to Phys. Rev. Lett. (2005).
16. F. Cano and P. Gonzalez, Phys. Lett. **B 431** (1998) 270-276.
17. S. Capstick and B.D. Keister, Phys. Rev. **D51** (1995) 3598.
18. Z. Li and V. Burkert, Phys. Rev. **D46** (1992) 70.
19. E. Pace, G. Salme and S. Simula, Few Body Syst. Suppl. **10** (1999) 407; S. Simula, Proceedings on NSTAR 2001, Mainz, Germany (2001) 135; M. Warns, H. Schroeder, W. Pfeil, and H. Rollnik, Z. Phys. **C45** (1990) 627; M. Aiello *et al.*, J. Phys. **G 24** (1998) 753.
20. ABBHHM Collaboration, Phys. Rev. **175**, 1669 (1968) ; ABBHHM Collaboration, Phys. Rev. **188**, 2060 (1969) ; A. Braghieri *et al.*, Phys. Lett. **B363**, 46 (1995); W. Langärtner *et al.*, Phys. Rev. Lett. **87**, 052001 (2001) ; M. Wolf *et al.*, Eur. Phys. J. **A9**, 5 (2000) ; F. Härter *et al.*, Phys. Lett. **B401**, 229 (1997) ; Y. Assafiri *et al.*, Phys. Rev. Lett. **90**, 222001 (2003).
21. M. Rapani *et al.*, Phys. Rev. Lett. **91**, 022002 (2003); S.A. Philips, Ph.D. Thesis, The George Washington University (2002); M. Bellis, Ph. D. Thesis, Rensselaer Polytechnic Institute (2004).
22. J. Ballam *et al.*, Phys. Rev. **D5**, 545 (1972).
23. J. Ahrens *et al.*, Phys. Lett. **B551**, 49 (2003).
24. S. Strauch *et al.*, Phys. Rev. Lett. **95**, 162003 (2005) 1-5.
25. V.I. Mokeev *et al.*, Yad. Fiz. **64**, 1368 (2001) or Phys. At. Nucl. **64**, 1292 (2001).
26. A. Fix and H. Arenhövel, nucl-th/0503042 (2005).
27. S. Capstick and W. Roberts, Phys. Rev. D **49**, 4570 (1994).
28. R. Bradford *et al.*, submitted to Phys. Rev. **C** (2005) 1-26.
29. K.-H. Glander *et al.*, Eur. Phys. J. **A19** (2004) 251-273.
30. M.Q. Tran *et al.*, Phys. Lett. **B445**, 20 (1998); M. Bockhorst *et al.*, Z. Phys. **C63**, 37 (1994).
31. R. Erbe *et al.*, Phys. Rev. **188**, 2060 (1969).
32. M. Guidal *et al.*, Nucl. Phys. **A627**, 645 (1997); M. Guidal *et al.*, Phys. Rev. **C61**, 025204 (2000).
33. T. Mart *et al.* at www.kph.uni-mainz.de/MAID/kaon/kaonmaid.html.
34. B. Saghai, *nucl-th/0105001* (2001).
35. V. Shklyar *et al.*, Phys. Rev. **C72**, 015210 (2005) 1-8.

Spectroscopy of doubly charmed baryons

J. Vijande*, H. Garcilazo†, A. Valcarce* and F. Fernández*

*Grupo de Física Nuclear
Universidad de Salamanca, E-37008 Salamanca, Spain
†Escuela Superior de Física y Matemáticas
Instituto Politécnico Nacional, Edificio 9,
07738 México D.F., Mexico

Abstract. We study the mass spectrum of baryons with two and three charmed quarks. For double charm baryons the spin splitting is found to be smaller than standard quark-model potential predictions. This splitting is not influenced either by the particular form of the confining potential or by the regularization taken for the contact term of the spin-spin potential. We consistently predict the spectra for triply charmed baryons.

PACS: 14.20.Lq, 12.40.Yx, 12.39.Jh

SELEX Collaboration, the charm hadroproduction experiment at Fermilab, reported the first observation of a candidate for a double charm baryon, Ξ_{cc}^{+}, [1]. The data are compatible with a narrow state with a mass of 3520 MeV/c^2 decaying through a weak Cabibbo-allowed process into $\Lambda_c^+ K^- \pi^+$. This observation has been confirmed through the measurement of a different weak decay mode that also involves a final state with baryon number and charm, $\Xi_{cc}^{+} \to pD^+K^-$ [2]. SELEX data [3] suggest the existence of four ccq states (q being a light quark) in a mass region of 350 MeV. FOCUS and E791 Collaborations at Fermilab have performed and extensive search for these states in the same reactions, reporting no doubly charmed candidates [4]. SELEX claims that this result is consistent with their findings [2]. Nonetheless, a big effort is being done by BABAR and BELLE Collaborations in an attempt to confirm the expected existence of doubly charmed baryons.

The doubly and triply charmed baryons provide a new window for understanding the structure of all baryons. As pointed out by Bjorken [5] one should strive to study these systems because their excitation spectrum should be close to the perturbative regime. For their size scales the quark-gluon coupling constant is small and therefore the leading term in the perturbative expansion is enough to describe the system. Moreover, the spectroscopy of baryons containing two heavy quarks is of interest because of similarities both to a quarkonium state, $Q\overline{Q}$, and to a heavy-light meson, $\overline{Q}q$. On the one hand, the slow relative motion of the tightly bound color antitriplet $(cc)_{\bar{3}}$ diquark in ccq is similar to quarkonium. On the other hand, for ccq the radius is dominated by the low mass q orbiting the tightly bound cc pair, and therefore is large. As a consequence, the relative $(cc)-(q)$ structure may be described similar to $\overline{Q}q$ mesons, where the cc pair plays the role of the heavy antiquark.

Moreover, heavy-quark baryons are ideal systems to probe QCD dynamics in close connection to the structure of heavy-light mesons and the general structure of hadronic systems. While from the point of view of the interacting potential the analysis of light

CP814 *Hadron Spectroscopy: Eleventh International Conference*
edited by A. Reis, C. Göbel, J. Sá Borges, and J. Magnin

hadronic systems becomes complicated by nonperturbative effects, heavy-quark systems are rather simple. Heavy-quark current masses are clear signals of the explicit breaking of chiral symmetry and as a consequence there are no Goldstone-boson exchange contributions, the interacting potential being controlled by the perturbative short-ranged one-gluon exchange (OGE) and confinement. From a theoretical point of view, in the doubly charmed system one expects a $J = 1/2$ ground state isodoublet, termed $\Xi_{cc}^{+,++}$ in PDG notation. The cc color antitriplet diquark has spin one. The spin of the third quark is either parallel, $J = 3/2$, or anti-parallel, $J = 1/2$, to the diquark. The $J = 3/2$ state has been predicted to be heavier than the $J = 1/2$ state by around 80 MeV/c^2 [6]. For the ccc system the Pauli principle demands a $J = 3/2$ ground state.

Having in mind that the role of models in QCD is to build the simplest physical picture that connects the phenomenological regularities with the underlying structure, in this work we use a model that describes correctly the light and strange baryon [7] and the meson spectra [8] to study the spectra of doubly and triply charmed baryons. This procedure, that will have important consequences as we will see in the following, allows to make parameter-free predictions for the masses of doubly and triply charmed baryons. For this purpose we first make use of the model of Ref. [8] designed to describe the meson spectra from the light pseudoscalar mesons to bottomonium. The model is based on the assumption that the light quark constituent mass is a consequence of the spontaneous breaking of chiral symmetry generating Goldstone boson exchanges between light quarks. Besides, it contains a confining term and a minimal OGE potential. As mentioned above, for doubly and triply charmed baryons chiral symmetry is explicitly broken and therefore the interacting potential gets reduced to the one-gluon exchange and confinement. Following de Rújula *et al.* [9] the OGE is a standard color Fermi-Breit interaction containing a coulomb term plus a spin-spin interaction. The nonrelativistic reduction of the OGE diagram in QCD for point-like quarks presents a contact term that, when not treated perturbatively, leads to collapse [10]. This is why the structure of the OGE is maintained but the δ function is regularized in a suitable way. Such regularization takes into account the finite size of the constituent quarks and should be therefore flavor dependent [11] $r_0(\mu) = r_0 \cdot (\mu_q/\mu_{q_1q_2})$, where μ_q is the reduced mass of two light quarks and $\mu_{q_1q_2}$ is the reduced mass of the two quarks under consideration.

Regarding confinement, lattice calculations in the quenched approximation derive, for heavy quarks, a confining interaction linearly dependent on the interquark distance,

$$V_{CON}^{L}(\vec{r}_{ij}) = \frac{8}{3} a_c \, r_{ij} \,. \qquad (1)$$

This form of strict confinement has been widely used for light and heavy quarks when studying the meson and baryon spectra within a quark model framework. The consideration in the lattice of sea quarks apart from valence quarks (unquenched approximation) suggests a screening effect on the potential when increasing the interquark distance [12]. Creation of light $q\bar{q}$ pairs out of vacuum in between the quarks becomes energetically preferable resulting in a complete screening of quark color charges at large distances. A specific parametrization of these effects was given in the form of a screening multiplica-

TABLE 1. Relative energy ccq and qqq spectra, in MeV, for the model of Ref. [8]. For the qqq system 'Theor.' stands for the results of the model of Ref. [8] and 'Exp.' for the experimental result.

System	$J^P(1/2^+)^*$	$J^P(3/2^+)$	$J^P(1/2^-)$
ccq	287	25	206
qqq (Theor.)	500	290	469
qqq (Exp.)	491−531	291−295	581−616

tive factor in the potential [13],

$$V_{CON}^{S}(r_{ij}) = \frac{8}{3}a_c'\, r_{ij}\left(\frac{1-e^{-\mu\, r_{ij}}}{\mu\, r_{ij}}\right). \qquad (2)$$

Screened confining potentials have been analyzed in the literature obtaining significant improvement both for the baryon [14] and for the heavy-meson spectra [15].

To get the baryon spectrum we have solved exactly the Schrödinger equation by the Faddeev method. For the particular case of three identical particles we have also calculated the spectra by means of the hyperspherical harmonic expansion method. Further details can be found in Ref. [16]. Our results for the excitation energy of the low-lying ccq states for the model of Ref. [8] are presented in Table 1 together with the corresponding results for the qqq system and their experimental values. As can be observed, the first radial excitation lies above the first negative parity state. This is the same situation found in the light baryon spectra where OGE quark-model based potentials predict the first negative parity excitation below the so-called Roper resonance, the first radial excitation. For double charmed baryons (DCBs) this is, however, expected to correspond to the experimental situation opposite to the light baryon case. The reverse of the ordering between the positive and negative nucleon excited states has been explained in terms of the combination of two different effects: relativistic kinematics and the pseudoscalar Goldstone-boson exchange interaction between light quarks. The nucleon Roper resonance is particularly sensitive to the form of the kinetic energy operator [17, 18], its energy being decreased with respect to the first negative parity state when a relativistic kinetic energy operator is used. Once this energy difference has been diminished, the pseudoscalar Goldstone boson exchange potential produces the desired inversion [18, 19]. In DCBs kinetic energy relativistic effects are expected to be much smaller due to the presence of two heavy quarks that makes the system analogous to a hydrogen-like atom [6]. On the other hand, chiral symmetry is explicitly broken and the interaction between heavy-light or heavy-heavy quarks does not present a Goldstone-boson pseudoscalar term. These two effects combined should recover the normal ordering between positive and negative parity states in the case of doubly charmed baryons.

Another striking result appearing in the obtained spectra concerns the $J^P = 1/2^+ - J^P = 3/2^+$ spin splitting, it appears to be much smaller than the one predicted based on potential models directly obtained from the $c\bar{c}$ spectra [6]. Such a small spin splitting seems to be related with our scale-dependent quark-gluon coupling constant and the fact that the $\Delta - N$ mass difference is correctly reproduced.

TABLE 2. Different parameter sets used in the text.

		A	B	C	D	E
	m_q (MeV)			313		313
	m_c (MeV)			1752		1550
	r_0 (fm)	0.25			0.45	0.25
Linear confinement	a_c (MeV fm^{-1})	–	55	–	115	–
Screened confinement	a'_c(MeV)	160	–	300	–	160
	μ(fm^{-1})	0.7	–	0.7	–	0.7

As has been discussed above the interacting potential for doubly and triply charmed baryons may depend on i) the specific form used for the confining term, and ii) the regularization taken for the contact term of the one-gluon exchange potential. To judge the model dependence of our results on these items, we have recalculated the baryon spectra with four different set of parameters that are given in Table 2. Two of them, sets B and D, have an infinite linear confining potential as the one of Eq. (1), while sets A and C present a screened confining potential, as the one of Eq. (2). For both types of confining potentials we have used two different values of the regularization parameter for the spin-spin interaction, r_0. In all cases the $\Delta - N$ mass difference is asked to be correctly described.

The results for the ccq and qqq systems are given in Table 3. As can be seen the same conclusions as before hold for any of the set of parameters used. All set of parameters give comparable radial and orbital excitation energies, the relative position of the first positive and negative parity excited states being preserved. Besides, the spin-splitting appears to be almost constant independently of the set of parameters used and much smaller than the one predicted on the existing literature (one should have in mind at this point that most existing calculations are of rather exploratory nature, since made when double charm physics was considered far future). This result is independent of the strength of the spin-spin force controlled by the values of α_s and r_0 and also of the form of the confining interaction, in agreement with conclusions of Ref. [14], that observes that only the high-energy excited states seem to be "confinement states", and therefore influenced by the explicit form of the confining potential.

The mass of the ccq $J^P = 1/2^+$ ground state can be tuned to reproduce the measured experimental data by means of the constituent charm quark mass. This has been done by means of the set of parameters E in Table 2, that is exactly the same as the set A

TABLE 3. Relative energy ccq and qqq spectra, in MeV, for the first four sets of parameters of Table 2.

	ccq			qqq		
Set	$J^P(1/2^+)^*$	$J^P(3/2^+)$	$J^P(1/2^-)$	$J^P(1/2^+)^*$	$J^P(3/2^+)$	$J^P(1/2^-)$
A	221	23	165	408	291	403
B	221	22	156	473	292	430
C	347	25	243	591	299	548
D	332	21	217	689	297	580

TABLE 4. *ccq* and *ccc* spectra, in MeV, for the set of parameters E of Table 2.

	ccq				*ccc*			
J^P	$1/2^+$	$(1/2^+)^*$	$3/2^+$	$1/2^-$	$3/2^+$	$(3/2^+)^*$	$1/2^+$	$1/2^-$
	3524	3749	3548	3692	4632	4870	4915	4808

except for the constituent charm quark mass (the same could have been done for any of the other sets of Table 2). The results are shown in Table 4. As can be seen there are no variations in the predicted structure of the excited states, but reproducing in this case the experimental energy of the double charm baryon, Ξ_{cc}^{+}, reported by SELEX, 3520 MeV/c^2 [1, 2]. Using the same set of parameters we have also calculated the spectra for the *ccc* system. The results are given in Table 4. In this case it is the $J^P = 3/2^+$ the lighter state. As in the *ccq* system the first negative parity state appears below the first radial excitation. The spin splitting increases up to values close to the results obtained for the light-quark baryons. Therefore, the small spin-splitting found in this work is a characteristic feature of double-heavy baryons.

ACKNOWLEDGMENTS

This work was partially funded by Dirección General de Investigación Científica y Técnica (DGICYT) under the Contract No. FPA2004-05616, by Junta de Castilla y León under the Contract No. SA-104/04, and by COFAA-IPN (Mexico).

REFERENCES

1. M. Mattson *et al.*, Phys. Rev. Lett. **89**, 112001 (2002).
2. A. Ocherashvili *et al.*, hep-ex/0406033.
3. M.A. Moinester, talk presented at the Hadron03 Conference, Aschaffenburg, Germany, (2003).
4. P. Sheldon, Frascati 2002, Charm, Beauty and CP, 79 (2002); FOCUS Collaboration, http://www-focus.fnal.gov/xicc/xicc_focus.html.
5. J.D. Bjorken, FERMILAB-CONF-85/69, Int. Conf. on Hadron Spectroscopy, College Park, (1985).
6. S. Fleck and J.M. Richard, Prog. Theor. Phys. **82**, 760 (1989); J.L. Rosner, Comm. Nucl. Part. Phys. **21**, 369 (1995); M.L. Stong, hep-ph/9505217; R. Roncaglia, D.B. Lichtenberg, and E. Predazzi, Phys. Rev. D **52**, 1722 (1995); C. Itoh, T. Minamikawa, K. Miura, and T. Watanabe, Phys. Rev. D **61**, 057502 (2000).
7. A. Valcarce, H. Garcilazo, and J. Vijande, Phys. Rev. C **72**, 025206 (2005).
8. J. Vijande, F. Fernández, and A. Valcarce, J. Phys. G **31**,481 (2005).
9. A. de Rújula, H. Georgi, and S.L. Glashow, Phys. Rev. D **12**, 147 (1975).
10. R.K. Bhaduri, L.E. Cohler, and Y. Nogami, Phys. Rev. Lett. **44**, 1369 (1980).
11. J. Weinstein and N. Isgur, Phys. Rev. D **27**, 588 (1983).
12. G.S. Bali, Phys. Rep. **343**, 1 (2001), and references therein.
13. K.D. Born, E. Laermann, N. Pirch, T.F. Walsh, and P.M. Zerwas, Phys. Rev. D **40**, 1653 (1989).
14. J. Vijande, P. González, H. Garcilazo, and A. Valcarce, Phys. Rev. D **69**, 074019 (2004).
15. P. González, A. Valcarce, H. Garcilazo, and J. Vijande, Phys. Rev. D **68**, 034007 (2003).
16. J. Vijande, H. Garcilazo, A. Valcarce, and F. Fernández, Phys. Rev. D **70**, 054022 (2004).
17. J. Carlson, J. Kogut, and V.R. Pandharipande, Phys. Rev. D **27**, 233 (1983).
18. H. Garcilazo and A. Valcarce, Phys. Rev. C **68**, 035207 (2003).
19. L.Ya. Glozman and D.O. Riska, Phys. Rep. **268**, 263 (1996).

Charm Baryon Studies at *BABAR*

Veronique Ziegler (On behalf of the *BABAR* Collaboration)

Dept. of Physics and Astronomy, University of Iowa, 203 VAN, Iowa City, IA 52242, USA

Abstract. We present a precision measurement of the mass of the Λ_c^+ and studies of the production and decay of the Ω_c^0 and Ξ_c^0 charm baryons using data collected by the *BABAR* experiment. To keep the systematic uncertainty as low as possible, the Λ_c^+ mass measurement is performed using the low Q-value decays, $\Lambda_c^+ \to \Lambda^0 K_S^0 K^+$ and $\Lambda_c^+ \to \Sigma^0 K_S^0 K^+$. Several hadronic final states involving an Ω^- and a Ξ^- hyperon are analyzed to reconstruct the Ξ_c^0 and the Ω_c^0.

Keywords: Charm baryons
PACS: 13.30.Eg, 14.20.Lq

INTRODUCTION

High luminosity colliders and flavor factories provide excellent opportunities to study the properties of charm baryons with high precision. Using the large charm hadron samples available at *BABAR* the Λ_c^+ mass is measured with a precision significantly improved over the existing PDG value[1]. In addition, a study of the production (from $e^+e^- \to c\bar{c}$ and from B decays) of the Ξ_c^0 charmed baryon and of its decay into $\Omega^- K^+$ and $\Xi^- \pi^+$ is carried out, and a measurement, also improved in precision over previous results[2], for the ratio the branching fractions $\mathscr{B}(\Xi_c^0 \to \Omega^- K^+)/\mathscr{B}(\Xi_c^0 \to \Xi^- \pi^+)$ is obtained. Similarly, the decay of the Ω_c^0 baryon reconstructed into: $\Omega_c^0 \to \Omega^- \pi^+$, $\Omega_c^0 \to \Xi^- K^- \pi^+ \pi^+$, and $\Omega_c^0 \to \Omega^- \pi^- \pi^+ \pi^+$ is investigated and the ratios of branching fractions relative to the $\Omega^- \pi^+$ decay mode are measured. A first clear evidence for Ω_c^0 production form B decays is shown. Charge conjugate reactions are implied throughout. The *BABAR* detector is described elsewhere[3].

MEASUREMENT OF THE Λ_C^+ MASS

Since systematic uncertainties related to track reconstruction, involving energy-loss correction and magnetic field strength scale with Q-value,[1] the Λ_c^+ is reconstructed into its low Q-value final states $\Lambda_c^+ \to \Lambda^0 K_S^0 \pi^+$, $\Lambda^0 \to p\pi^-$, $K_S^0 \to \pi^+\pi^-$ (Q= 178 MeV/c^2) and $\Lambda_c^+ \to \Lambda^0 K_S^0 \pi^+$, $\Sigma^0 \to p\pi^-$, $K_S^0 \to \pi^+\pi^-$ (Q= 101 MeV/c^2), for which the systematic uncertainties in the mass measurement are considerably smaller compared to dominant high Q-value modes such as $\Lambda_c^+ \to pK^-\pi^+$. The large Q-value modes are used as control samples to study systematic uncertainties. This measurement was done on 232 fb^{-1} of *BABAR* data.

[1] The Q-value for the decay $a \to b + c + ...$ is defined as $Q = m(a) + m(b) + m(c)$

CP814 *Hadron Spectroscopy: Eleventh International Conference*
edited by A. Reis, C. Göbel, J. Sá Borges, and J. Magnin

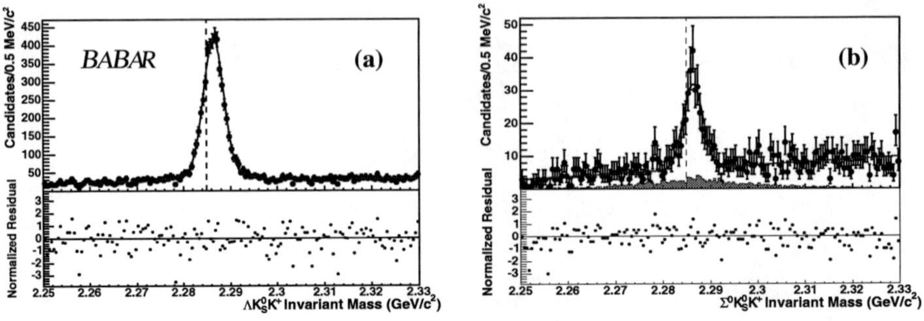

FIGURE 1. Invariant mass distributions for reconstructed Λ_c^+ candidates, for (a) $\Lambda^0 K_S^0 K^+$, and (b) $\Sigma^0 K_S^0 K^+$. The dashed line indicates the present PDG value.

Since the hyperons and the K_S^0 are long-lived particles, the signal-to-background ratio is improved by requiring that their decay vertexes be clearly displaced from that of the parent particle. Particle ID selectors for the proton and the kaon, based on $\mathrm{d}E/\mathrm{d}x$ measurements given by the Silicon Vertex Tracker (SVT) and the Drift Chamber as well as Cherenkov angle information, have been employed. Only intermediate states with an invariant mass within 3σ of the central PDG value are selected.[2] The invariant mass is then constrained to the nominal value. These criteria are also required in the selection of the other charm baryons described in this letter. In addition, to reduce combinatorial background coming mostly from B decays, the momentum of the Λ_c^+ candidate in the e^+e^- center-of-mass frame (p^*) is required to exceed 2 GeV/c. Clear Λ_c^+ signal peaks for both modes are shown in Figures 1 (a) and (b). The presence of a peaking background (absorbed into the signal and background) in $\Lambda_c^+ \to \Sigma^0 K_S^0 K^+$ is due a wrong photon vertexed with a real Λ^0 in the candidate selection. The main systematic uncertainty on the Λ_c^+ mass arises from the energy-loss corrections in charged particle tracking; it is studied using the uncorrected invariant-mass distributions of control samples $\Lambda^0 \to p\pi^-$ and $K_S^0 \to \pi^+\pi^-$. The long lifetimes of these hadrons enable us to study their fitted mass values as a function of the their radial decay distance from the interaction point. Figure 2 shows that the largest K_S^0 fitted mass deviations from the PDG mean occur at low decay radii, indicating the under-estimation of the amount of detector material assumed in the track reconstruction algorithm. The effect of increasing the SVT density assumed in the track reconstruction algorithm by 20% removes the radial dependence of the K_S^0 fitted mass. The dips seen in the distribution are reconstruction artifacts also seen in the reconstruction of simulated data. Another way to investigate the the energy-loss correction is to study the fitted mass as a function of the particle momentum using the high Q-value Λ_c^+ control samples. Similarly, increasing the density of the SVT by 20% removes the momentum dependence of the Λ_c^+ fitted mass. The other major sources of systematic uncertainties are the magnetization of the *BABAR* PEP II

[2] σ corresponds to the invariant mass resolution.

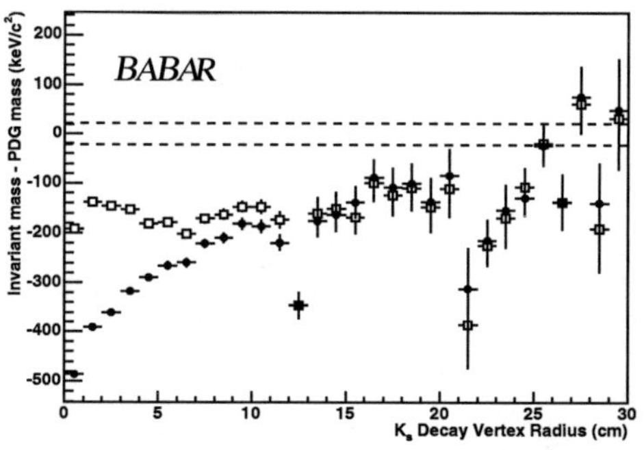

FIGURE 2. The fitted mass values for $K_S^0 \to \pi^+\pi^-$ candidates minus the PDG value as a function of the decay radius. The solid circles are for track reconstruction with the normal amount of detector material. The open squares are for track reconstruction assuming 20% more material in the SVT. The dashed lines show the $\pm 1\sigma$ uncertainty on the PDG value for the masses.

permanent magnets[3] by the solenoidal field and the dependence of the fitted mass value on the azimuthal angle ϕ of the hadron candidate momentum at the origin. The low Q-value fitted Λ_c^+ mass values are corrected for the under-estimated energy loss by increasing the material density of the SVT by 20%, and the systematic uncertainties in these modes are determined using the dominant high Q-value modes.

Taking into account the energy-loss corrections and correlated systematic uncertainties, we obtain a combined mass measurement for the two low Q-value modes: $m(\Lambda_c^+) = 2286.46 \pm 0.14$ MeV/c^2.

PRODUCTION AND DECAY OF THE Ξ_c^0 BARYON

The production and decay of the Ξ_c^0 baryon is studied using 116 fb^{-1} of *BABAR* data. The intermediate states are reconstructed through the decays: $\Xi^- \to \Lambda\pi^-$, $\Omega^- \to \Lambda K^-$ To measure the ratio of branching fractions, the requirement that $p^* > 1.8$ GeV/c is imposed on the Ξ_c^0 candidates improving the signal purity. Furthermore, in order to select a region of high and uniform reconstruction efficiency, the polar angle of the Ξ_c^0 candidate with respect to the collision axis in the center-of-mass frame (θ^*) is limited to the range $-0.8 \leq \cos\theta^* \leq 0.8$ for the $\Xi^-\pi^+$ mode and $-0.8 \leq \cos\theta^* \leq 0.6$ for Ω^-K^+. Taking the ratio of the efficiency-corrected yields for each mode, we obtain:

$$\frac{\mathscr{B}(\Xi_c^0 \to \Omega^- K^+)}{\mathscr{B}(\Xi_c^0 \to \Xi^-\pi^+)} = 0.294 \pm 0.018(stat.) \pm 0.016(syst.),$$

where the dominant systematic uncertainty comes from fits to the mass spectra.

The p^* spectrum of the Ξ_c^0 baryons (allowing for $p^* \leq 1.8$ GeV/c) gives an insight into the production mechanisms of $c\bar{c}$ and $B\bar{B}$ events. Figure 3 (d) shows the background-

[3] made of Samarium Cobalt (permeability $\mu \approx 1.13$)

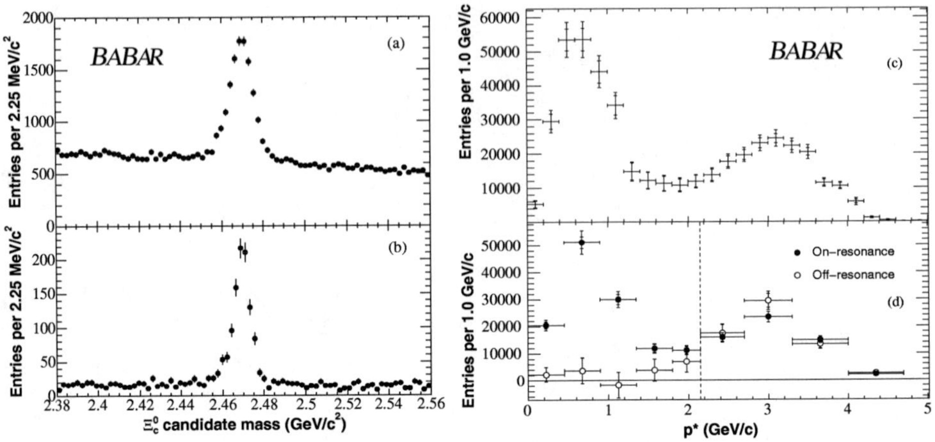

FIGURE 3. Left-hand side: Invariant mass distributions for Ξ_c^0 candidates in data for (a) $\Xi^-\pi^+$, and (b) Ω^-K^+. Right-hand side: The p^* spectrum of Ξ_c^0 candidates decaying via $\Xi^-\pi^+$ shown for the on-resonance data sample (c), and for the on-resonance and off-resonance data samples together (d). The vertical line at 2.15 GeV/c shows the kinematic cutoff for Ξ_c^0 production in B decays at *BABAR*. The inner error bars give the statistical uncertainty and the outer error bars give the sum in quadrature of the statistical and systematic uncertainties.

subtracted, efficiency-corrected p^* distribution for $\Xi_c^0 \to \Xi^-\pi^+$ simultaneously for the off-resonance (collected at approximately 40 MeV/c^2 below the $B\bar{B}$ production threshold), corresponding to Ξ_c^0 production from $e^+e^- \to c\bar{c}$ only, and for the on-resonance (at the $B\bar{B}$ production threshold) data. The absence of a low p^* peak in off-resonance data substantiates the assumption that the low p^* peak observed in the on-resonance data sample distribution originates from Ξ_c^0 produced from B decays. A similar behavior is seen for the Ω^-K^+ decay mode. Comparing the on-resonance and scaled off-resonance samples for $p^* \leq 2.15$ GeV/c gives the yield of Ξ_c^0 produced in B decays and a value:

$$\mathscr{B}(B \to \Xi_c^0 X) \times \mathscr{B}(\Xi_c^0 \to \Xi^-\pi^+) = (2.11 \pm 0.19(stat.) \pm 0.25(syst.)) \times 10^{-4}.$$

The cross section from $e^+e^- \to c\bar{c}$ is obtained from the yield of Ξ_c^0 produced in $c\bar{c}$ events using the scaled off-resonance data set (for $p^* \leq 2.15$ GeV/c) and the on-resonance data set (for $p^* > 2.15$ GeV/c).

$$\sigma(e^+e^- \to \Xi_c^0 X) \times \mathscr{B}(\Xi_c^0 \to \Xi^-\pi^+) = (388 \pm 39(stat.) \pm 41(syst.))\ \text{fb}.$$

PRODUCTION AND DECAY OF THE Ω_c^0 BARYON

The Ω_c^0 charmed baryon is reconstructed into its $\Omega^-\pi^+$ and $\Omega^-\pi^-\pi^+\pi^+$ final states using 225 fb^{-1} of *BABAR* data. And the data sample used to reconstruct $\Omega_c^0 \to \Xi^-K^-\pi^+\pi^+$

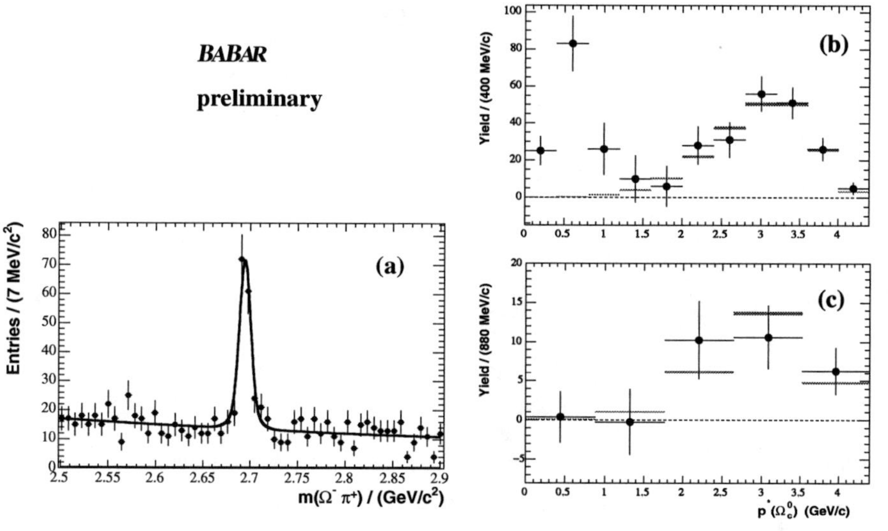

FIGURE 4. Left-hand side: Invariant mass distribution for reconstructed $\Omega_c^0 \to \Omega^-\pi^+$ shown in (a). Right-hand side: The p^* spectrum of Ω_c^0 candidates decaying via $\Omega^-\pi^+$ shown for the on and off-resonance data samples together in (b), and separately for the off-resonance data sample only in (c). The dots correspond to the data (statistical error bars only) and the solid horizontal bar indicate the predictions for the p^* distribution of Ω_c^0 baryons produced from $e^+e^- \to c\bar{c}$.

corresponds to an integrated luminosity of 230 fb^{-1}. High values of p^* are necessary to suppress combinatorial backgrounds in the muti-body Ω_c^0 decays; a common minimum p^* of 2.8 GeV/c is required for candidates used in the measurement of ratios of branching fractions. Figure 4 (a) shows the reconstructed invariant mass spectrum for $\Omega_c^0 \to \Omega^-\pi^+$. A clear peak is observed. Taking the ratio of the efficiency-corrected yields for Ω_c^0 signal events, extracted from the invariant mass spectra, we obtain a value for the ratio of branching fractions:

$$\frac{\mathscr{B}(\Omega_c^0 \to \Xi^- K^- \pi^+ \pi^+)}{\mathscr{B}(\Omega_c^0 \to \Omega^- \pi^+)} = 0.31 \pm 0.15(stat.) \pm 0.04(syst.).$$

Since no significant excess of signal events over the background is observed in the $\Omega_c^0 \to \Omega^-\pi^-\pi^+\pi^+$ mode, we calculate a limit at the 90% confidence level (CL) on the ratio of branching fractions.

$$\frac{\mathscr{B}(\Omega_c^0 \to \Omega^-\pi^-\pi^+\pi^+)}{\mathscr{B}(\Omega_c^0 \to \Omega^-\pi^+)} \leq 0.3(90\%\text{CL})$$

Figure 4 (b) and (c) show the uncorrected p^* distribution of Ω_c^0 candidates for the entire data sample and the off-resonance only sample, respectively; clear evidence of Ω_c^0 production from B decays is visible at low p^* and is absent in the off-resonance data, collected below the $B\bar{B}$ production threshold.

SUMMARY

We have presented a precision measurement of the Λ_c^+ mass $m(\Lambda_c^+) = 2286.46 \pm 0.14$ MeV/c^2, using the low Q-value modes $\Lambda_c^+ \to \Lambda^0 K_S^0 K^+$ and $\Lambda_c^+ \to \Sigma^0 K_S^0 K^+$.
We have measured the ratio of branching fractions

$$\frac{\mathscr{B}(\Xi_c^0 \to \Omega^- K^+)}{\mathscr{B}(\Xi_c^0 \to \Xi^- \pi^+)} = 0.294 \pm 0.018(stat.) \pm 0.016(syst.),$$

and studied the production of Ξ_c^0 from $e^+e^- \to c\bar{c}$ and measured:

$$\sigma(e^+e^- \to \Xi_c^0 X) \times \mathscr{B}(\Xi_c^0 \to \Xi^- \pi^+) = (388 \pm 39 \pm 41) \text{ fb}.$$

We have presented an observation of the production of Ξ_c^0 from B decays:

$$\mathscr{B}(B \to \Xi_c^0 X) \times \mathscr{B}(\Xi_c^0 \to \Xi^- \pi^+) = (2.11 \pm 0.19(stat.) \pm 0.25(syst.)) \times 10^{-4}.$$

We obtained a measurement of the ratio of branching fractions

$$\frac{\mathscr{B}(\Omega_c^0 \to \Xi^- K^- \pi^+ \pi^+)}{\mathscr{B}(\Omega_c^0 \to \Omega^- \pi^+)} = 0.31 \pm 0.15(stat.) \pm 0.04(syst.),$$

and an upper limit for

$$\frac{\mathscr{B}(\Omega_c^0 \to \Omega^- \pi^- \pi^+ \pi^+)}{\mathscr{B}(\Omega_c^0 \to \Omega^- \pi^+)} \leq 0.3(90\%\text{CL}).$$

And we have shown a first observation of the decay $B \to X\Omega_c^0 \to \Omega^- \pi^+$.

ACKNOWLEDGMENTS

We are grateful for the excellent luminosity and machine conditions provided by our PEP-II colleagues, and for the substantial dedicated effort from the computing organizations that support *BABAR*.

REFERENCES

1. Particle Data Group, S. Eidelman *et al.*, Phys. Lett. B **592**, 1 (2004).
2. CLEO Collaboration, S. Henderson *et al.*, Phys. Lett. B **283**, 161 (1992).
3. The *BABAR* Collaboration, B. Aubert *et al.*, Nucl. Instrum. Methods A **479**, 1 (2002).
4. D. Brown, E. Charles, and D. Roberts, "The *BABAR* Track Fitting Algorithm", contributed paper to CHEP 2000.
5. The *BABAR* Collaboration, B. Aubert *et al.*, hep-ex/0507009, contributed paper to LP 2005.
6. The *BABAR* Collaboration, B. Aubert *et al.*, Phys. Rev. Lett. **95**, 142003 (2005).
7. CLEO Collaboration, B. Barish *et al.*, Phys. Rev. Lett. **79**, 3599 (1997).
8. The *BABAR* Collaboration, B. Aubert *et al.*, hep-ex/0507011, contributed paper to LP 2005.
9. P. L. Frabetti *et al.* [E687 Collaboration], Phys. Lett. B **300**, 190 (1993).
10. P. L. Frabetti *et al.* [E687 Collaboration], Phys. Lett. B **338**, 106 (1994).
11. S. Ahmed *et al.* [CLEO Collaboration], Int. J. Mod. Phys. A **16S1B**, 505 (2001).

Baryonic B decays

H. Kichimi

Institute of Particle and Nuclear Studies, High Energy Accelerator Research Organization
Oho-1-1,Tsukuba, Ibaraki, 305-0801 Japan

Abstract. We present recent results of Baryonic B decays from Belle, which contain charmed and charmless baryonic decays into two to four body final states. We report the branching fractions, including new observations of the baryonic decays into two charmed baryons in the final states, and the charmless baryonic decay proceeding through $b \to s\gamma$ transition, and the low di-baryon mass enhancement structures observed in the three-body decays, based on $357 fb^{-1}$ data.

Keywords: B meson, di-baryon
PACS: 01.30.Cc,13.25.Hw,14.20.Lq,14.20.Pt

Introduction

The baryonic B decays have been observed in two to four body final states so far. The measured branching fractions for charmed and charmless baryonic B decays are shown in Figure 1. Since the CKM matrix element magnitude $|V_{cb}|$ [1] is substantially larger than $|V_{ub}|$, the charmed baryonic decays due to $b \to c(\bar{u}d)$ tree transition are dominated [2, 3, 4, 5, 6, 7, 8, 9]. There is found a hierarchy of the branching fraction on the multiplicity of the final state: $\sim 2 \times 10^{-5}$ for the two-body decays $\bar{B}^0 \to \Lambda_c^+ \bar{p}$, $B^- \to \Sigma_c^0(2455)\bar{p}$ and $B^- \to \Sigma_c^0(2520)\bar{p}$, $\sim 1 \times 10^{-4}$ for the three-body decays, and $\sim 7 \times 10^{-4}$ for the four-body decays. In the charmless final states, only the three-body decays are observed [10, 11, 12, 13]: $B^+ \to p\bar{p}\pi^+$ through the $b \to u(\bar{u}d)$ tree transition, $B^0 \to p\bar{\Lambda}\pi^-$, and $B^+ \to p\bar{p}K^+$ and $B^+ \to \Lambda\bar{\Lambda}K^+$ through the $b \to s(\bar{u}u)$ or $(\bar{d}d)$ and the $b \to s(\bar{s}s)$ penguin transitions, respectively. No observations for the two-body decays $B^0 \to p\bar{p}$ and $B^0 \to \Lambda\bar{\Lambda}$ are reported. It indicates a large suppression of the two-body charmless decays due the small value of $|V_{ub}/V_{cb}|^2 \sim 10^{-2}$. These measurements are summarized in Ref. [2].

In this report, we discuss new observations of the baryonic B decay modes: the charmless decays $B^+ \to \bar{\Lambda}p\gamma$ due to the $b \to s\gamma$ transition, two-body and three-body decays accompanied with two charmed baryons or J/ψ (a charmonium), $B^{0(+)} \to \Xi_c^{+(0)}\bar{\Lambda}_c^-$ and $\bar{B}^{0(-)} \to \Lambda_c^+\bar{\Lambda}_c^- K^{0(-)}$, and $B \to J/\psi\Lambda\bar{p}$ due to the $b \to c\ (\bar{c}s)$ transition, and the low di-baryon mass enhancements observed in the three-body final states [14, 15, 16, 17]. The analyses are based on the data sample of 357 fb^{-1} accumulated with the Belle detector [18] at the B-factory at KEK (KEKB), the high luminosity 8 GeV e^- and 3.5 GeV e^+ asymmetric collider [19].

CP814 *Hadron Spectroscopy: Eleventh International Conference*
edited by A. Reis, C. Göbel, J. Sá Borges, and J. Magnin

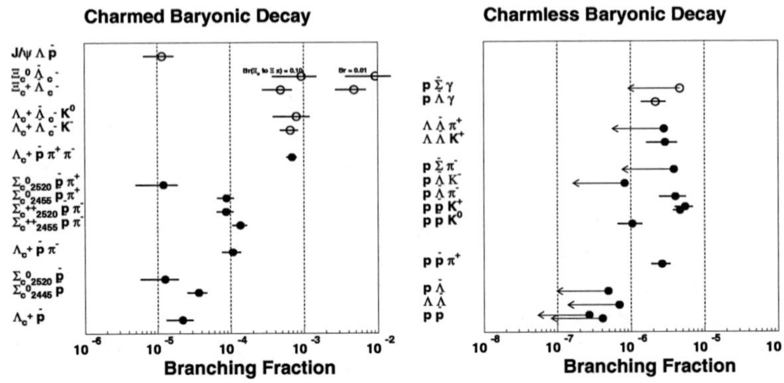

FIGURE 1. Branching fractions for Charmed and Charmless baryonic B decays. The open circles indicate new measurements from Belle.

Observations of $B^{-/0} \to \bar{\Lambda}_c^- \Lambda_c^+ K^{-/0}$ and $B^{+/0} \to \bar{\Xi}_c^{0/-} \Lambda_c^+$ decays

Figure 2 shows the scatter plot of ΔM_B vs M_{bc} for the candidate events of $B^- \to \bar{\Lambda}_c^- \Lambda_c^+ K^-$ decay after event selection. The mass difference ΔM_B is defined as $\Delta M_B = M_{inv}(B) - M_B$, and the beam energy constrained mass M_{bc} is defined as $M_{bc} = \sqrt{E_{beam}^2 - \vec{p}_B^2}$, where E_B and $\vec{p}_B$ are the reconstructed B meson energy and momentum. E_{beam} is the beam energy in the center of mass system (CMS). $M_{inv}(B)$ is the reconstructed mass of B candidate and M_B is the nominal B mass. The parameter ΔM_B is used instead of the beam energy difference $\Delta E = E_B - E_{beam}$, since ΔE shows a correlation with M_{bc}. Λ_c^+ is reconstructed in the decay modes $\Lambda_c^+ \to pK^-\pi^+$, $p\bar{K}^0$ and $\Lambda\pi^+$ by requiring the reconstructed masses to be within 10MeV(2.5σ) of the nominal Λ_c^+ mass. We see a clear signal in the B signal region of $|\Delta M_B| \leq 0.015\mathrm{GeV/c^2}$ and $M_{bc} \geq 5.27\mathrm{GeV/c^2}$. The ΔM_B distribution for $M_{bc} \geq 5.27\mathrm{GeV/c^2}$ and M_{bc} distribution for $|\Delta M_B| \leq 0.015\mathrm{GeV/c^2}$ indicate clear peaks for the B signal, respectively. The curve in the ΔM_B distribution indicates the fit to the data with the MC signal shape and a linear background, while the curve in the M_{bc} distribution shows the fit with the MC signal shape and ARGUS background [20]. The signal yield and significance are determined by 2D fit to the data with the signal shape fixed to the MC signal shape, and with the background and the signal yields floated. The signal yield of $48.5^{+7.5}_{-6.8}$ is obtained with a significance of 15.4σ. The Λ_c^+ and $\bar{\Lambda}_c^-$ mass distributions in the B signal region are investigated to give the yield consistent with the 2D fit. The $B^0 \to \Lambda_c^+ \bar{\Lambda}_c^- K^0$ decay are also observed with a signal yield of $10.5^{+3.8}_{-3.1}$ and a significance of 6.6σ.

Figure 3 shows the ΔE and M_{bc} distributions for $B^+ \to \bar{\Xi}_c^0 \Lambda_c^+$ candidates in the B signal region of $|\Delta E| \leq 0.025\mathrm{GeV}$ and $M_{bc} \geq 5.272\mathrm{GeV/c^2}$. The decay $B^0 \to \bar{\Xi}_c^- \Lambda_c^+$ is also observed. Λ_c^+ is reconstructed in the decay mode $\Lambda_c^+ \to pK^-\pi^+$. The signal yields are obtained by the 2D fit to be $12.4^{+4.2}_{-3.3}$, $16.9^{+4.8}_{-4.0}$ and $8.3^{+3.3}_{-2.5}$ for the decay modes of $B^+ \to \bar{\Xi}_c^0 \Lambda_c^+$ with $\bar{\Xi}_c^0 \to \bar{\Xi}^+ \pi^-$ and $\bar{\Lambda} K^+ \pi^-$, and $B^0 \to \bar{\Xi}_c^- \Lambda_c^+$ with $\bar{\Xi}_c^- \to \bar{\Xi}^+ \pi^- \pi^-$,

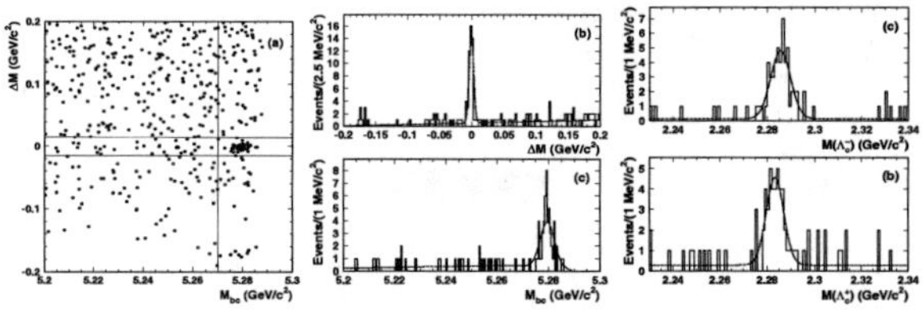

FIGURE 2. Scatter plot for ΔM vs M_{bc}, ΔM vs M_{bc}, and reconstructed $\Lambda_c^{+/-}$ mass distributions.

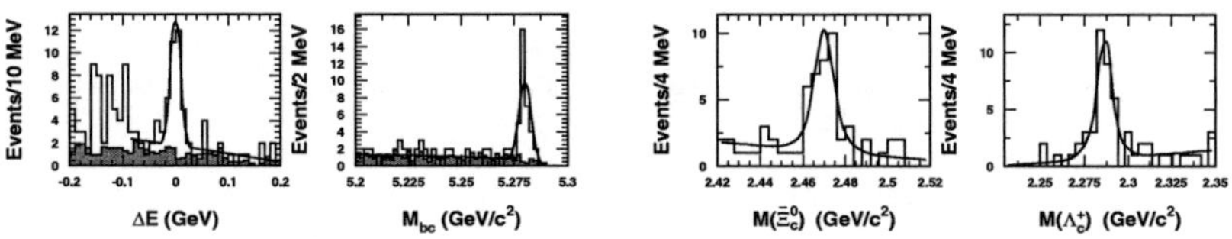

FIGURE 3. ΔE and M_{bc}, and reconstructed $\bar{\Xi}_c$ and Λ_c^+ mass distributions for $B^+ \to \bar{\Xi}_c^0 \Lambda_c^+$.

respectively. The reconstructed $\bar{\Xi}_c^{0/-}$ and Λ_c^+ mass distributions in the B signal region are investigated, which give the yields consistent with those obtained by the 2D fit. As the barnching fractions of the used Ξ decay modes are not known yet, the products of the branching fractions are obtained.

The new observations are compared with other decay modes in Table 1 and Figure 1. The branching fractions of $B^{-/0} \to \bar{\Lambda}_c^- \Lambda_c^+ K^{-/0}$ ($b \to c\bar{c}s$) are comparable to that of the four-body decay $\bar{B}^0 \to \Lambda_c^+ \bar{p} \pi^+ \pi^-$, and significantly larger than the three-body decays proceeding through $b \to c\bar{u}d$. We estimate $Br(B^+ \to \bar{\Xi}_c^0 \Lambda_c^+)$ of $(4.8^{+1.0}_{-0.9} \pm 1.1 \pm 1.2) \times 10^{-3}$ ($\times 10^{-4}$) by assuming the branching fraction of 1% (10%) for the Ξ decay modes, which also are much larger than $Br(\bar{B}^0 \to \Lambda_c^+ \bar{p}) = (2.19^{+0.56}_{-0.49} \pm 0.32 \pm 0.57) \times 10^{-5}$ [7]. These contradict with a naive expectation of the same order of the branching fractions proceeding via $b \to c\bar{c}s$ and $b \to c\bar{u}d$. In contrast, the three-body decays accompanied with J/ψ (charmonium) [17] indicates a comparable or smaller branching fraction than $\bar{B}^0 \to \Lambda_c^+ \bar{p}$. For charmless baryonic decays, we observe a new decay mode $B^+ \to \bar{\Lambda} p \gamma$ [14] proceeding through $b \to s\gamma$ as shown in Table 1 and Figure 1. So far all possible b quark transitions except $b \to c\bar{c}c$ are observed in baryonic B decays, which provide strict constraints to theoretical predictions [21, 22, 23, 24].

Di-baryon low mass enhancement in three-body decays

We study the di-baryon low mass structures in the charmless three-body decays [13] in a view point of b quark fragmentation [25], which predicts a forward peaked structure in a specific helicity frame. In this analysis, B signal yield is obtained by 2D fit to the (M_{bc},ΔE) distribution for each bin of di-baryon invariant mass. Then the branching fraction for each bin is obtained from the fitted B signal yield and the signal efficiency.

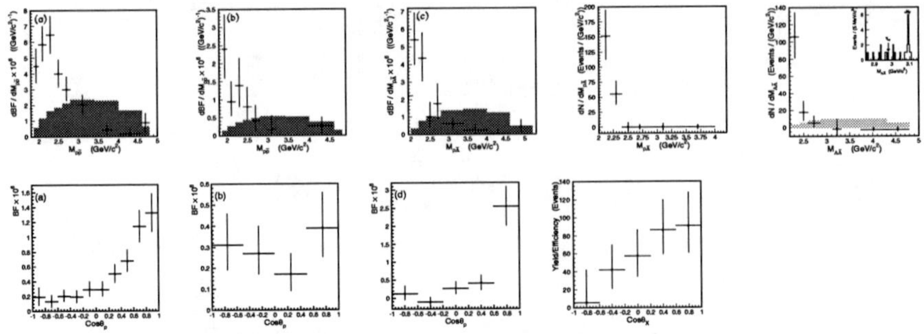

FIGURE 4. Di-baryon mass distributions (differential branching fractions) for (a) $B^+ \to p\bar{p}K^+$, (b) $B^+ \to p\bar{p}K_s^0$, (c) $B^0 \to p\bar{\Lambda}\pi^-$, (d) $B^+ \to p\bar{\Lambda}\gamma$, (e) $B^+ \to \Lambda\bar{\Lambda}K^+$, and the decay angular distributions.

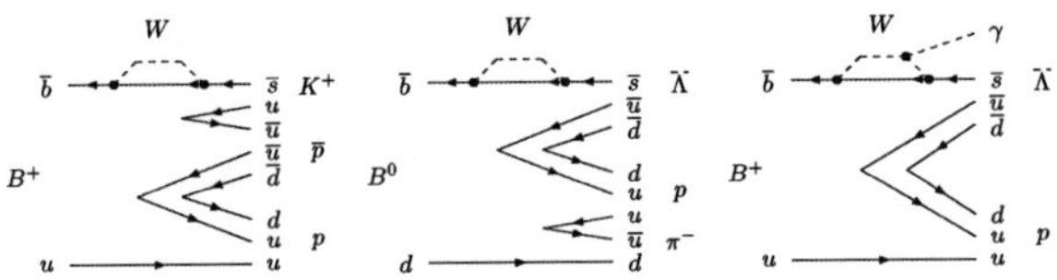

FIGURE 5. Feynman diagrams for (a) $B^+ \to p\bar{p}K^+$, (b) $B^0 \to p\bar{\Lambda}\pi^-$ and (d) $B^+ \to p\bar{\Lambda}\gamma$.

Figure 4 show the di-baryon mass distributions for (a) $B^+ \to p\bar{p}K^+$, (b) $B^+ \to p\bar{p}K_s^0$, (c) $B^0 \to p\bar{\Lambda}\pi^-$, (d) $B^+ \to p\bar{\Lambda}\gamma$, (e) $B^+ \to \Lambda\bar{\Lambda}K^+$, and their decay angular distributions. There is commonly seen an significant structure peaked near the threshold for the $p\bar{p}$, $p\bar{\Lambda}$ and $\Lambda\bar{\Lambda}$ masses. The angular ($cos\theta$) distributions are also obtained by the 2D fit of the B signal for each $cos\theta$ bin. The $cos\theta$ distribution of (a) $\bar{p}$ in the $p\bar{p}$ rest system with respect the K^+ direction, shows a clear forward peak with a forward-backward asymmetry $A = (N_F - N_B)/(N_F + N_B) = 0.59^{+0.08}_{-0.07}$. It is interpreted that the $\bar{b} \to \bar{s}$ transition generates a very energetic K^+ meson in the direction parallel to the incident $\bar{b}$ quark, and the energy flow to the di-baryon is restricted and results in the low mass di-baryon enhancement. On the other hand, in (b) K_s^0 case, it does not show such asymmetric peak structure since the detected K_s^0 has no specific $\bar{s}$ flavor of the $\bar{b}$ quark. The distribution for $B^0 \to p\bar{\Lambda}\pi^-$ (c) shows a forward peak structure, if the angle is defined as the p direction relative to the $\bar{\Lambda}$ direction in the $p\pi^-$ rest frame, while it is flat if the angle is the p direction relative to the π^- direction in the $\bar{\Lambda}p$ rest frame.

The fragmentaion picture [25] gives a qualitative explanation of the low mass enhancements and the asymmetric angular distributions together with Feynman diagrams (a) and (b) in Figure 5. However, it does not necessary predict quantitative distributions of those for the di-baryon system, which distinguishes the fragmentation process from the other mechanisms like final state interaction and threshold enhancement. So, experimental studies of $B^+ \to p\bar{\Lambda}\gamma$ and $B^+ \to \Lambda\bar{\Lambda}K^+$ decays with higher statistics are needed to understand the mechanism in more detail. The low mass enhancement structure of di-baryon is also observed in the three-body charmed baryonic decay $\bar{B}^0 \to \Lambda_c^+\bar{p}\pi^-$ [2, 9].

TABLE 1. Branching fractions for newly observed Baryonic B decays by Belle. See Ref [2] for the previous measurements. The errors are statistical and systematics. The third error in the charmed baryonic decays is due to ambiguity in $Br(\Lambda_c^+ \to pK - \pi^+) = (5.0 \pm 1.3)\%$.

Mode	transition	Branching fraction	Sig(σ)	$L(fb^{-1})$ Ref
$Br(B^+ \to \bar{\Lambda} p\gamma)$	$b \to s\gamma$	$(2.16^{+0.58}_{-0.53} \pm 0.20) \times 10^{-6}$	8.6	140 [14]
$Br(B^+ \to \bar{\Sigma}^0 p\gamma)$		$< 4.6 \times 10^{-6}$ at 90% C.L.		
$Br(B^+ \to \bar{\Xi}_c^0 \Lambda_c^+)$	$b \to c(\bar{c}s)$			357 [15]
$\times Br(\bar{\Xi}_c^0 \to \bar{\Xi}^+ \pi^-)$		$(4.8^{+1.0}_{-0.9} \pm 1.1 \pm 1.2) \times 10^{-5}$	8.7	
$Br(B^0 \to \bar{\Xi}_c^- \Lambda_c^+)$				
$\times Br(\bar{\Xi}_c^- \to \bar{\Xi}^- \pi^+ \pi^-)$		$(9.3^{+3.7}_{-2.8} \pm 1.9 \pm 2.2) \times 10^{-5}$	3.8	
$Br(B^- \to \bar{\Lambda}_c^- \Lambda_c^+ K^-)$	$b \to c(\bar{c}s)$	$(6.5^{+1.0}_{-0.9} \pm 0.8 \pm 3.4) \times 10^{-4}$	15.4	357 [16]
$Br(\bar{B}^0 \to \bar{\Lambda}_c^- \Lambda_c^+ K^0)$		$(7.9^{+2.9}_{-2.3} \pm 1.2 \pm 4.2) \times 10^{-4}$	6.6	
$Br(B^- \to J/\psi \Lambda \bar{p})$	$b \to c(\bar{c}s)$	$(11.6 \pm 2.8^{+1.8}_{-2.3}) \times 10^{-6}$	11.1	253 [17]
$Br(B^- \to J/\psi \Sigma^0 \bar{p})$		$< 1.1 \times 10^{-5}$ at 90% C.L.		
$Br(B^0 \to J/\psi p\bar{p})$		$< 8.3 \times 10^{-7}$ at 90% C.L.		

REFERENCES

1. M. Kobayashi and K. Masukawa, Prog. Theor. Phys. 49, 652 (1973).
2. H. Kichimi, Nucl. Physics B**142** (Proc.Suppl.), 197 (2005).
3. X. Fu *et al.* (CLEO), Phys. Rev. Lett. 79, 3125 (1997). T. E. Coan *et al.* (CLEO), Phys. Rev. D59, 111101 (1999). S. A. Dytman *et al.* (CLEO Collaboration), Phys. Rev. D **66**, 091101(R) (2002).
4. K. Abe *et al.* (Belle Collaboration), Phys. Rev. Lett. **88**, 181803 (2002).
5. K. Abe *et al.* (Belle Collaboration), Phys. Rev. Lett. **89**, 151802 (2002).
6. N. Gabyshev, H. Kichimi *et al.* (Belle Collaboration), Phys. Rev. D **66**, 091102(R) (2002).
7. N. Gabyshev, H. Kichimi *et al.* (Belle Collaboration), Phys. Rev. Lett. **90**, 121802 (2003).
8. K. Abe *et al.*, Belle-conf-0467
9. N. Gabyshev *et al.* (Belle Collaboration), hep-ex/0409005
10. M.-Z. Wang *et al.* (Belle Collaboration), Phys. Rev. Lett. **90**, 201802 (2003).
11. M.-Z. Wang *et al.* (Belle Collaboration), Phys. Rev. Lett. **92**, 131801 (2004).
12. Y.-J. Lee, M.-Z. Wang *et al.* (Belle Collaboration), Phys. Rev. Lett. **93**, 211801 (2004).
13. M.-Z. Wang *et al.* (Belle Collaboration), Phys. Letters B **617**, 141 (2005).
14. Y. -J. Lee, M. -Z. Wang *et al.* (Belle Collaboration), Phys. Rev. Lett. **95**, 061802 (2005).
15. R. Chistov *et al.* (Belle Collaboration), Belle-conf-0536
16. N. Gabyshev *et al.* (Belle Collaboration), hep-ex/0508015
17. Q. L. Xie *et al.* (Belle Collaboration), Phys. Rev. D 72, 051105(R) (2005)
18. A. Abashian *et al.* (Belle Collaboration), Nucl. Instr. and Meth. A **479**, 117 (2002).
19. S. Kurokawa and E. Kikutani *et al.*, Nucl. Instr. and Meth. A **499**, 1 (2003).
20. H. Albrecht *et al.* (ARGUS Collaboration), Phys. Lett. B **229** 304 (1989); B **241** 278 (1990).
21. M. Jarfi *et al.*, Phys. Lett. B237, 513(1990); M. Jarfi *et al.*, Phys. Rev. D43, 1599 (1991); N. Deahpande, J. Trampetic and A. Soni, Mod. Phys. Lett. 3A, 749 (1988).
22. V. Chernyak and I. Zhitnisky, Nucl. Phys. B345, 137 (1990).
23. H. Y. Cheng and K. C. Yang, Phys. Rev. D **66**, 014020 (2002). Phys. Rev. D **66**, 094009 (2002). Phys. Rev. D **67**, 034008 (2003).
24. C. K. Chua, W. S. Hou and S. Y. Tsai, Phys. Rev. D **66**, 054004 (2002).
C. K. Chua and W. S. Hou, Eur. Phys. J. C **29**, 27 (2003).
25. J. L. Rosner, Phys. Rev. D **68**, 014004 (2003).

EXOTICS

Pentaquark Searches With CLAS

David J. Tedeschi for the CLAS Collaboration

Department of Physics and Astronomy, University of South Carolina, Columbia, SC 29208, USA

Abstract. The production of a possible, exotic S=1 pentaquarks off protons and neutrons is being investigated in Hall B at Jefferson Lab. Three dedicated experiments have been carried out to search for pentaquark signals using the CLAS detector in conjunction with a photon beam incident on hydrogen and deuterium targets. The status of on-going analyses and preliminary results from a variety of channels are presented.

Keywords: Pentaquark
PACS: 25.20.Lj, 25.75.Dw

INTRODUCTION

Since the 1960's when Gell-Mann introduced quarks and Nambu introduced color, baryons and mesons have been considered as three-quark and quark-antiquark color singlet configurations. This classification of known particles is largely unchanged today, yet searches for exotic configurations continue to play an important role in testing our understanding of the hadron spectrum.[1] Exotic searches have gained new urgency based on experimental evidence that supported recent theoretical predictions for a new baryon state: the $\Theta^+(1540)$. [2,3,4]

This *S=1* "pentaquark" baryon, composed of 5 quarks ($uudd\bar{s}$), was initially predicted to have a mass of 1540 MeV and a width of 15 MeV.[2] The state is exotic due to its strangeness and enigmatic due to its predicted narrow width. An early discussion by Gao and Ma of pentaquark search possibilities at Jefferson Lab generated little activity.[5] However, since the announcement at PANIC 2002 by the LEPS group for a pentaquark signal in photoproduction data,[6] there has been a renewed effort by the medium-energy and high-energy communities to analyze data for signals of exotic baryons.

At the present time there are both positive and negative results reported in the literature.[7,8] Over the past year many of the new results reported have been negative, casting doubt on the existence of the pentaquark, but most of the new results are in different production channels and kinematic regimes than the positive signals. In response to the unclear experimental situation, the Hall B group has performed three experiments with the CLAS detector optimized to search for exotic particles. One experiment on deuterium (g10) and another on hydrogen (g11) improve the signal statistics for the $\Theta^+(1540)$ by an order of magnitude. Additionally, an effort is underway to search for exotic cascade particles (eg3). This endeavor aims to verify and search for further members of the pentaquark family of which the $\Theta^+(1540)$ is the lowest mass "anti-decuplet" state. These high-statistics experiments will provide

CP814 *Hadron Spectroscopy: Eleventh International Conference*
edited by A. Reis, C. Göbel, J. Sá Borges, and J. Magnin

reliable cross section limits that can be used to address the question of the existence of the exotic pentaquark states in a quantitative way.

PHOTOPRODUCTION OF $\Theta^+(1540)$ ON DEUTERIUM

Data published in 2003 provided the first claim from CLAS for the existence of an exotic $S=1$ baryon.[9] A peak in the invariant mass $m(nK^+)$ spectrum at 1540 MeV with a width consistent with the experimental resolution was reported. The statistical significance of the peak was estimated at 5.8σ when compared with a smooth background, and 4.8σ when compared with a background shape determined from a Monte Carlo simulation. Preliminary results from the new experiment on deuterium (g10) that reproduce the conditions of our published $\Theta^+(1540)$ claim are shown in Figure 1. The missing mass of the pK$^-$ system for the reaction $\gamma+d\rightarrow p+K^++K^-+(n)$ is presented where the black points are the older data published in Reference 9 and the red histogram is the new high-statistics data from CLAS scaled by 0.169. No evidence for a peak at 1.54 GeV/c^2 is seen in the new data. In addition to the order of magnitude increase in statistics, a monumental effort was put into understanding the CLAS calibration and detector corrections. From this data, CLAS has deduced a preliminary upper limit of 4-5 nb on the cross section of the $\Theta^+(1540)$ in this production channel. This result is undergoing final review in preparation for publication.

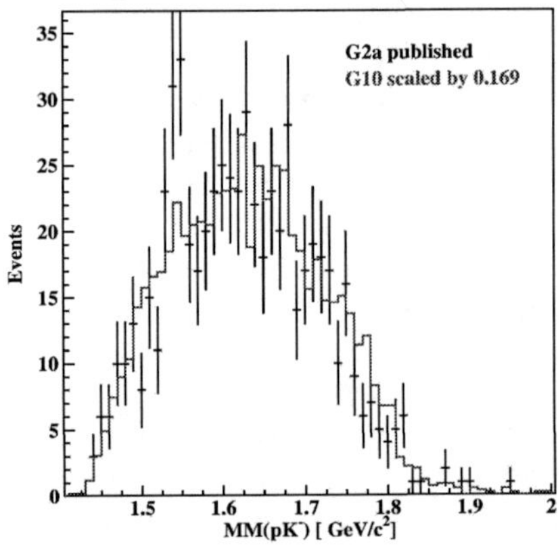

FIGURE 1. Preliminary missing mass of the pK$^-$ system for the reaction $\gamma+d\rightarrow p+K^++K^-+(n)$. The black points are the data published in Reference 9 and the red histogram is the new high-statistics data from CLAS scaled by 0.169.

Searches for the $\Theta^+(1540)$ in complementary channels are also underway using the CLAS g10 data. The high statistics makes possible for the first time the study of the channels $\gamma+d\rightarrow\Lambda+\Theta^+$ and $\gamma+d\rightarrow\Lambda^*(1520)+\Theta^+$. The advantage of these channels is that they do not contain effects from hadronic interferences. On the other hand, understanding the role of the hadronic background is integral to clarify why so many laboratories reported a positive signal. The channel γ+d→p+K^0+K^-+(p) allows a check for exotic pentaquark production on the (quasi-free) neutron by detecting the decay mode $\Theta^+ \to p+K^0$ through analysis of the $m(pK^0)$ invariant mass distribution. The advantage of this channel is that no re-scattering mechanism is needed to produce a pentaquark and detection of the K^- provides a strangeness tag for the K^0. A study of interference effects from meson channels (such as the $a_2(1320)$) that decay to K^0K^- is underway, as these effects have been suggested as the cause of the structure in the mass region of 1.54 GeV/c^2.[10]

PHOTOPRODUCTION OF $\Theta^+(1540)$ ON HYDROGEN

The CLAS g11 run employed photons of energies in the interval 1.6 GeV $<E_\gamma<$3.8 GeV incident upon a 40 cm hydrogen target. Over seven billion triggers corresponding to an integrated luminosity of 70 pb^{-1} were recorded. Each trigger required the detection of at least two charged particles in CLAS to facilitate the pentaquark search in the channels $\gamma+p\to K^++K^0+(n)$ and $\gamma+p\to K^0+p+(K^0)$. In the former reaction, the intermediate state $\Theta^+(1540)$ decays to nK^+ while the latter measures the decay to pK^0. The comparison of these channels, as well as the production data off the neutron will provide an important test of the $\Theta^+(1540)$.[11] The analysis of $\gamma+p\to K^0+p+(K^0)$ is underway and the analysis of $\gamma+p\to K^++K^0+(n)$ is described below.

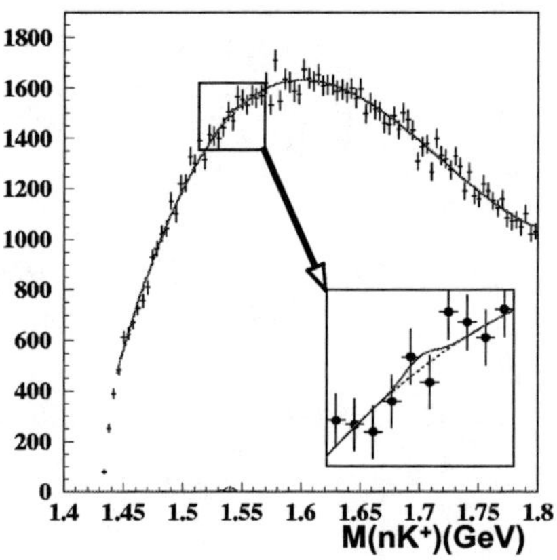

FIGURE 2. Invariant mass of the nK^+ system for the reaction $\gamma+p\to K^++K^0+(n)$. Data are from the analysis of CLAS g11 data: 1.6 GeV < E_γ < 3.8 GeV. Vertical axis is arbitrary yield.

The CLAS search for $\Theta^+(1540)$ production from hydrogen in the reaction $\gamma+p\to K^++K^0+(n)$ has been completed and was first shown at the Spring APS (2005) meeting in Tampa, Florida. The analysis of this channel is a high statistics reproduction of the positive result published by the SAPHIR collaboration.[12] The results of the CLAS measurement are shown in Figure 2. The invariant mass of the nK^+ system is smooth and shows no enhancement in the region of 1.54 GeV/c^2. The

CLAS data put an upper limit of 2 nb on the production cross section in this channel. This number should be contrasted with the limit of 50 nb from SAPHIR after a re-analysis of their published result of 300 nb.[12] It should also be noted that this result does not impact the previously published analysis on hydrogen by CLAS that is an analysis of $\Theta^+(1540)$ in the channel $\gamma+p\rightarrow K^++K^-+\pi^++(n)$ at photon energies 3 GeV $<E_\gamma<$5.4 GeV.[13] It is expected that data to verify this remaining CLAS result will be acquired in 2006/2007, depending on the Jefferson Lab run schedule.

SEARCH FOR CASCADE PENTAQUARKS

The CLAS Collaboration is involved in the search for the Φ^{--} pentaquark state, another member of the exotic anti-decuplet. The eg3 run, successfully completed in early 2005, consisted of a tagged photon beam of energy 4.5 < E_γ < 5.5 GeV incident on a 40 cm long deuterium target. A total of 100 pb^{-1} of data - consisting of 4 billion two and three track triggers - was acquired during the run. A major part of the calibration, namely the timing calibration of tagger, start counter, time-of-flight detector, forward calorimeter, single-photoelectron calibration and the timing calibration for the Cerenkov Counter and RF-signal has been performed. At the present time, the calibrations have been finished.

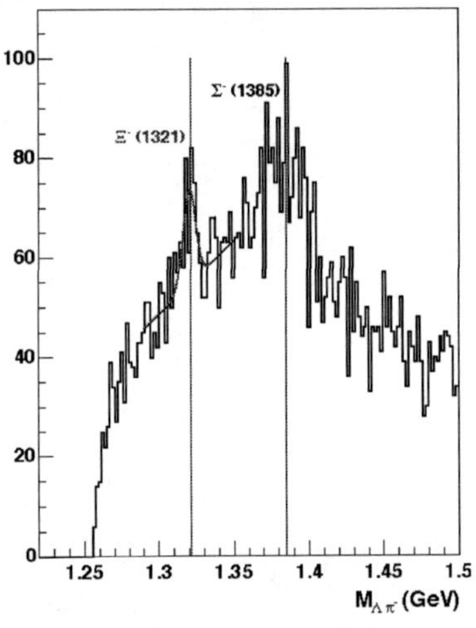

FIGURE 3. Reconstructed $\Lambda\pi^-$ mass from 4% of eg3 data. The lines indicate the mass of the $\Xi(1321)$ and the $\Sigma(1385)$. The fit is a gaussian with a polynomial background to estimate particle yield. Vertical axis is counts.

As a first step in the physical analysis, $p\pi^+\pi^-$ events have been reconstructed. The goal of extracting a well-known and kinematically over determined channel was to test detector systems and reconstruction code for their efficiency and to monitor this efficiency over the whole run period. In the case of the proton target it could be shown that the $\pi^+ \pi^-$ production channel can be isolated even in the absence of the magnetic field. This finding will be useful for future alignment routines for the drift chamber.

Another aspect of the physical analysis is the reconstruction of the $\Xi^-(1321)$ state from its decay products p π^- π^-. The exotic Φ^{--} is expected to decay via three subsequent interactions $\Phi^{--}\rightarrow\Xi^-\pi^-$, $\Xi^-\rightarrow\Lambda\pi^-$, and $\Lambda\rightarrow p\pi^-$. The typical vertex displacement ($d=\gamma\beta\tau c$) of 20 cm allows the reconstruction of intermediate decay vertices in regions with low physical background. This approach improves the signal to background ratio. Additionally, cutting on the invariant mass of the reconstructed Λ's reduces the background under the Σ(1385) peak in the $m(\Lambda\pi^-)$ invariant mass histogram (see Figure 3). We continue to develop this new and important reconstruction algorithm for CLAS to search for the Φ^{--}. Starting with exclusive $p\pi^-\pi^-$ events from 4% of the data, the invariant $\Xi^-(1321)$ mass peak in Figure 3 becomes visible after applying narrow cuts on the Λ mass and decay vertex and after demanding that the missing mass of the system is above the two-kaon threshold. The yield from this preliminary analysis is ~100 $\Xi^-(1321)$'s which translates to ~2500 total $\Xi^-(1321)$'s in the sample. This number is comparable to the yield from high-energy searches and will be used as a benchmark for comparison of the CLAS result to the world data.[14]

ACKNOWLEDGMENTS

These measurements were made possible by the outstanding efforts of the staff of the Accelerator and the Physics Divisions at Jefferson Lab. This work was supported in part by the National Science Foundation under the award NSF-0244982. The Southeastern Universities Research Association (SURA) operates the Thomas Jefferson National Accelerator Facility for the United States Department of Energy under contract DEAC05-84ER40150.

REFERENCES

1. S. Eidelman *et al.*, Phys. Lett. B **592,** 1 (2004).
2. D. Diakonov, V. Petrov, M. Polyakov, Z. Phys. A **359**, 305 (1997).
3. M. Karliner and H. Lipkin, Phys. Lett. B **575**, 249 (2003).
4. R. Jaffe and F. Wilczek, Phys. Rev. Lett. **91**, 232003 (2003).
5. H. Gao and B.-Q. Ma, Mod. Phys. Lett. A **14**, 2313 (1999).
6. T. Nakano *et al.*, Nucl. Phys. A **721**, C112 (2003).
7. A. Dzierba, C. Meyer, A. Szczepaniak, J. Phys. Conf. Ser. **9**, 192 (2005); hep-ex/0412077; A. Dzeirba *et al.*, Phys. Rev. D **71**, 098502 (2005).
8. K. Hicks, Prog. Part. Nucl. Phys. **55**, 647 (2005); hep-ex/0504027.
9. S. Stepanayn, *et al.* (The CLAS Collaboration) Phys. Rev. Lett. **91**, 252001 (2003).
10. Dzierba *et al.*, Phys. Rev. D **69**, 051901 (2004).
11. M. Karliner and H. Lipkin, hep-ph/0506084.
12. J. Barth *et al.,* Phys. Lett. B **572** 127 (2003).
13. V. Kubarovsky *et al.* (The CLAS Collaboration) Phys. Rev. Lett. **92**, 032001 (2004).
14. E.S. Ageev *et al.*, hep-ex/0503033 (March 2005).

High-Resolution Search for Pentaquark Partners in Hall A at Jefferson Lab

J.-O. Hansen, for the Jefferson Lab Hall A Collaboration

Jefferson Lab, Newport News, Virginia 23606, USA

Abstract. We have carried out a high-resolution search ($\sigma = 1.5$ MeV) for narrow exotic resonances ($\Gamma < 10$ MeV) in the mass range $M \approx 1500 - 1850$ MeV in $ep \to e'K^+X$, $ep \to e'K^-X$ and $ep \to e'\pi^+X$ electroproduction at forward angles ($\theta^{CM} \approx 6 - 7°$). Such narrow resonances would be candidates for partner states of the speculative $\Theta^+(1540)$ pentaquark. The experiment employed a 5 GeV CW electron beam incident on a liquid hydrogen target and two high-resolution magnetic spectrometers covering a total center-of-mass solid angle of $\Delta\Omega \approx 30 - 40$ msr. We do not observe a statistically significant signal in any of the three reaction channels. Upper limits on the production cross sections were determined to be between 3 and 16 nb/sr, depending on the channel and the assumed width of the state. In addition, precise electroproduction data of the $\Lambda(1116)$, $\Sigma(1193)$ and $\Lambda(1520)$ resonances were obtained for calibration purposes.

Keywords: pentaquarks, exotics, baryon electroproduction, partner states
PACS: 14.80.-j, 13.60.Rj, 14.20.Jn, 14.20.Gk

INTRODUCTION

Recent claims of the observation [1] of a narrow exotic $S = +1$ baryonic state, the $\Theta^+(1540)$, have generated considerable experimental and theoretical interest. If confirmed, the Θ^+ could be the lowest-mass member of an antidecuplet of pentaquarks, which has been predicted in the framework of the Chiral Quark Soliton Model [2]. Other identifications are also possible. For example, the narrow width of the Θ^+ has been explained in terms of isospin-violating strong decays, which has led to the suggestion of an isotensor multiplet of pentaquarks [3].

If the Θ^+ pentaquark exists then other members of its symmetry group should be observable as well, provided that they are sufficiently narrow. Both the Chiral Quark Soliton Model and the isotensor multiplet model predict relatively narrow ($\Gamma < 30 - 50$ MeV) partner states in the mass region $M \approx 1500 - 2000$ MeV.

In this paper, we report the results of a search for the possible Σ_5^0 and N_5^0 members of the antidecuplet and for the isotensor partner Θ^{++}. The reactions $ep \to e'K^+X$, $ep \to e'\pi^+X$ and $ep \to e'K^-X$ were investigated for evidence of narrow structures in the reconstructed missing mass (M_X). The measurements covered a limited range of forward scattering angles, which did not allow a partial-wave analysis. However, excellent mass resolution was achieved. Precise, high-statistics data of the known $\Lambda(1116)$, $\Sigma(1193)$ and $\Lambda(1520)$ resonances were obtained for calibration purposes. The results presented here are near final.

CP814 *Hadron Spectroscopy: Eleventh International Conference*
edited by A. Reis, C. Göbel, J. Sá Borges, and J. Magnin

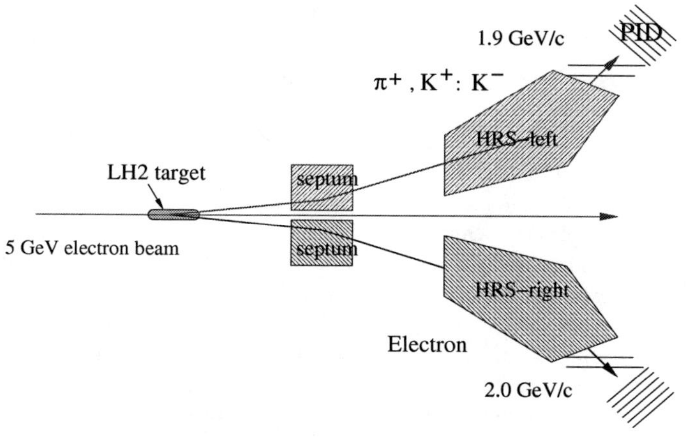

FIGURE 1. Experimental setup.

EXPERIMENT

The experiment ran in Hall A at Jefferson Lab in May 2004 using a 5 GeV continuous-wave electron beam incident on a 15 cm long liquid hydrogen target. Scattered electrons were detected in one of the Hall A High-Resolution Spectrometers (HRS) in coincidence with electroproduced hadrons that were measured in the second HRS. The experimental setup is illustrated in Figure 1. Both HRS were positioned at 6° forward angles, using septa on each arm to reach this small scattering angle. Each spectrometer had an effective acceptance of approximately 4 msr in solid angle and $\pm 4.5\%$ in momentum. For the production runs, to achieve the desired missing mass coverage, the central momentum of the electron HRS was stepped from 1.70 and 2.02 GeV/c, while the central momentum of the hadron HRS was changed between 1.89 and 2.10 GeV/c.

Both spectrometers used a QQDQ magnet arrangement for focusing and a 45° upward bend. The detector packages, placed behind the magnetic elements, were equipped with four planes of vertical drift chambers for tracking and two planes of fast hodoscopes for triggering. Careful calibration of the trigger system yielded a coincidence timing resolution of 600 ps FWHM. The scattering vertex resolution of each spectrometer was 2.5 cm FWHM. The electron HRS employed a CO_2 gas Cherenkov counter for pion rejection. For the kaon measurements, clean particle identification in the hadron HRS was particularly important because of the very high ratio of π/K rates. For this purpose, the hadron HRS was instrumented with two aerogel and a ring-imaging Cherenkov detector as well as a lead glass shower counter. Combined with the coincidence time-of-flight measurement, a total pion rejection factor of 3×10^4 was achieved.

The average momentum transfer carried by the virtual photon was $Q^2 \approx 0.1$ (GeV/c)^2. For the kaon (pion) kinematics, the average center-of-mass scattering angle was $\theta^{CM}_{\gamma^* K} \approx 6°$ ($\theta^{CM}_{\gamma^* \pi} \approx 7°$), and the angular acceptance was $\Delta\Omega^{CM}_K \approx 40$ msr ($\Delta\Omega^{CM}_\pi \approx 30$ msr).

ANALYSIS

The yields measured in each spectrometer were corrected for detector and trigger efficiencies and dead-time. Cuts on the data from the PID detectors were applied to select appropriate particle types, and vertex and coincidence time cuts were used to reduce background from accidentals. The missing mass, M_X, was reconstructed using the measured momenta of the electron and kaon (pion). The resulting spectra were corrected for the missing mass acceptance, which varied strongly within the mass range covered by each kinematic setting because of the limited momentum acceptance of the spectrometers. All spectra were normalized by luminosity.

Calibration data were taken in the $ep \to e'\pi^+X$ channel in the mass range of the neutron, and in the $ep \to e'K^+X$ channel covering the range where the $\Lambda(1116)$ and $\Sigma(1193)$ could be seen. The missing mass resolution was determined to be 3.5 MeV FWHM. Using the well-known masses of the calibration states, the accuracy of our reconstructed missing mass was determined to be better than 3 MeV.

In the search regions, the missing mass data from several kinematics (up to 8) were combined to arrive at the final spectra used for the peak search analysis (see below). The individual data sets overlapped to some extent, allowing a check of the corrections and normalizations. After finalizing the detector analysis, all transitions were found to be smooth, requiring no ad-hoc scaling factors. Background was not subtracted from any of the spectra.

To quantify any possible observations (or non-observations), we performed a systematic search for peaks in the final missing mass spectra of the three reaction channels. Within a 50 MeV wide window, we fitted a linear background to the data. Then a Breit-Wigner peak convoluted with a Gaussian of width $\sigma = 1.5$ MeV was added to this background at the center of the window, the fit was repeated, and the difference of the χ^2 of the two fits, $\Delta\chi^2_{exp}(M_X)$, was recorded. Here, the Gaussian simulated the experimental missing mass resolution, and widths of $\Gamma = 1, 3, 5$, and 8 MeV were assumed for the Breit-Wigner function. This procedure was repeated after shifting the window across the measured missing mass range in steps of 2 MeV. For the Θ^{++} search, χ^2 was replaced with the log-likelihood because of the low statistics of the data. A strong improvement of χ^2 (or log-likelihood, respectively) indicated a candidate for a signal. The statistical significance of the most prominent peak candidates was determined using the standard method of computing the Poisson probability for the background (estimated from sidebands) to fluctuate to the observed signal, expressed in terms of Gaussian sigma.

To obtain upper limits on the production cross section, we carried out a Monte Carlo simulation. In this way, the details of our peak search procedure, the peak shape, experimental resolution, corrections, background fluctuations, etc. were accounted for. The Monte Carlo simulated the presence of a peak, whose cross-section σ_{MC} was varied from 0 to 20 nb/sr. The purpose was to find the correspondence between the Monte Carlo peak strength and the probability to find a certain χ^2 improvement, $\Delta\chi^2(\sigma_{MC})$, due to such a peak when applying the search algorithm described above to the Monte Carlo spectra. We define the quantity $\Delta\chi^2_{90}(\sigma_{MC})$ such that a Monte Carlo peak of cross-section σ_{MC} results in a χ^2 improvement *greater* than $\Delta\chi^2_{90}$ with 90% probability. Thus, if a value $\Delta\chi^2 = \Delta\chi^2_{90}(\sigma_{MC})$ is actually observed in the experiment, then we know that

TABLE 1. Parameters of the largest observed bumps in each reaction channel. The range given for σ_{max} corresponds to the range $\Gamma = 1-8$ MeV of assumed widths used in the peak search.

Channel	Mass value of largest peak (MeV)	Significance	Yield ratio to $\Lambda(1520)$	Cross section limit σ_{max} (nb/sr)
$ep \to e'K^+X$ (Σ_5^0)	1720	2.3σ	3.7%	8–16
$ep \to e'\pi^+X$ (N_5^0)	1855	2.5σ	2.0%	4–9
$ep \to e'K^-X$ (Θ^{++})	1570	3σ	1.4%	3–6

we may have missed a real peak of cross section $\sigma \leq \sigma_{MC}$ with only 10% probability. Hence, we can take the Monte Carlo cross section σ_{MC} for which $\Delta\chi^2_{90}(\sigma_{MC}) = \Delta\chi^2_{exp}$ as the 90% CL upper limit on the production cross section. The resulting values for our searches are given in Table 1.

RESULTS AND CONCLUSIONS

Figure 2 depicts the missing mass spectra obtained for the three reaction channels along with a fit to the most significant candidate peak found in each scan. The parameters of these peaks are given in Table 1. These results were obtained with an assumed width of $\Gamma = 5$ MeV for the Breit-Wigner function; other assumed widths yield very similar results, except that σ_{max} generally increases with increasing peak width, as expected. The table also gives the cross-section ratio of the respective largest peak to the cross section of the $\Lambda(1520)$ as measured in our experiment in the K^+ channel.

There are several known or suspected resonances in our mass region [4], in particular several 3 or 4-star Λ and Σ states in the $ep \to e'K^+X$ channel. Most of these states are either wide ($>$ 50 MeV) or have only been seen in partial-wave analyses. Taken together, they likely add up to a relatively smooth background. Still, we do see several broad structures in this channel, possibly the $\Sigma(1670)$ and $\Sigma(1775)$.

In conclusion, we do not observe statistically significant narrow ($\Gamma < 10$ MeV) structures in any of our three reaction channels. Any signals seen are consistent with background. Our experiment had a relatively small angular coverage, moderate statistics, and was kinematically incomplete, thus our ability to rule out pentaquark partner states or other narrow resonances in our search regions is naturally somewhat restricted.

REFERENCES

1. see, for example, S. Stepanyan *et al.* [CLAS collaboration], Phys. Rev. Lett. **91**, 252001 (2003).
2. D. Diakonov, V. Petrov, and M. Polyakov, Z. Phys. A **359**, 305 (1997).
3. S. Capstick *et al.*, Phys. Lett. B **570**, 185 (2003).
4. *Review of Particle Physics*, S. Eidelman *et al.*, Phys. Lett. B **592**, 1 (2004).

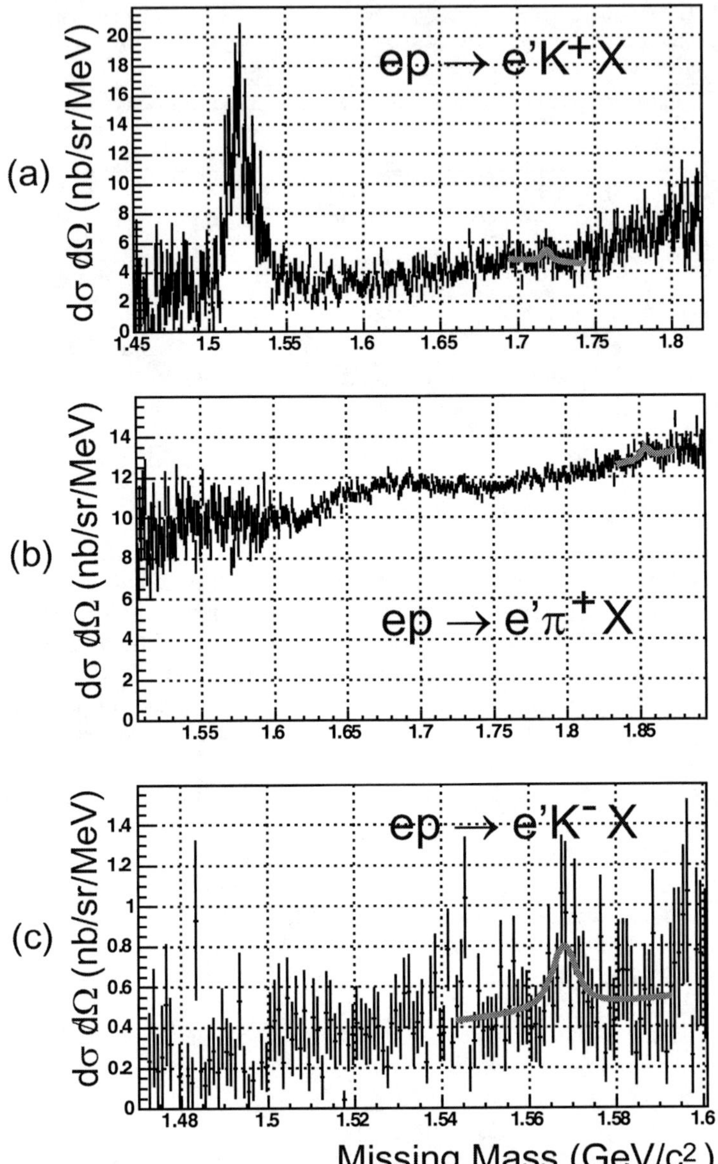

FIGURE 2. Final missing mass spectra obtained for the three reaction channels (a) $ep \to e'K^+X$ (Σ_5^0 search), (b) $ep \to e'\pi^+X$ (N_5^0 search), and (c) $ep \to e'K^-X$ (Θ^{++} search). The fits to the most significant peak are shown in each spectrum; the parameters of these peaks are listed in Table 1. A peak width of $\Gamma = 5$ MeV was assumed for the fits shown; other assumed widths yield similar results. The large peak in the K^+ channel (a) is the $\Lambda(1520)$.

Search for Pentaquarks in the Hadronic Decays of the Z Boson with the DELPHI Detector at LEP

Ph.Gavillet

PH Department, CERN, CH-1211 Geneva, Switzerland
for the
DELPHI Collaboration

Abstract. Recent evidence for pentaquark states has been published, in particular for a strange pentaquark $\Theta^+(1540)$, for a double-strange state called $\Xi(1862)^{--}$ and for a charmed state $\Theta_c(3100)^0$. Such states should be produced in e^+e^- annihilations in Z decays. In this paper a search for pentaquarks using the DELPHI detector is described. Preliminary upper limits at 95% C.L. are set on the production rates per Z decay of such particles and their charge-conjugate state.

Keywords: pentaquark, Z decay
PACS: ?

1. INTRODUCTION

The quark model does not exclude pentaquark bound states of four quarks and one antiquark , e.g. $uudd\bar{s}$. Several models predict the multiplet structure and characteristics of pentaquarks, for example the chiral soliton model, the uncorrelated, correlated quark models, the thermal model, lattice QCD, etc. [1].

Until recently, only bound triplet-quark and quark-antiquark systems have been found. Recent experimental evidence [2], [3], however, may suggest the existence of pentaquark systems. The first possible candidate is[1] the $\Theta^+(1540)$, with mass of 1.54 ± 0.01 GeV/c^2, width smaller than 1 MeV/c^2, and strangeness S=+1, consistent with being made of the quarks ($uudd\bar{s}$). Subsequently, evidence for another exotic baryon doubly charged and with double strangeness, the $\Xi(1862)^{--}$, has been published by the CERN experiment NA49 [4]. More recently, the DESY experiment H1 has reported a signal for a charmed exotic baryon in the pD^{*-} channel [5], the $\Theta_c(3100)^0$ with a mass of 3099 ± 3 (stat) ±5 (syst) MeV/c^2 and a measured width compatible with the resolution. It is interpreted as an anti-charmed baryon with a constituent quark composition of $uudd\bar{c}$.

Pentaquark states might be produced in a significant way in e^+e^- annihilations in Z boson decays and they could be detected and separated from the background using features of particle identification [6]. This paper reports on the results of a search for the $\Theta(1540)^+$, Θ^{++}, $\Xi(1862)^{--}$ and $\Theta_c(3100)^0$ pentaquark states in hadronic Z decays recorded by DELPHI at LEP.

---

[1] The antiparticles are always implicitly included.

CP814 *Hadron Spectroscopy: Eleventh International Conference*
edited by A. Reis, C. Göbel, J. Sá Borges, and J. Magnin

2. EXPERIMENTAL PROCEDURE

The analysis presented here is based on a data sample of over 3 million hadronic Z decays collected from 1991 to 1995 with the DELPHI detector at LEP. The detector is described in detail in [7] and its performance is analyzed in [8].

The charged-particle tracks have been measured in the 1.2 T magnetic field by a set of tracking detectors. The average momentum resolution for charged particles in hadronic final states, $\Delta(1/p)$, is usually between 0.001 and 0.01, depending on their momentum.

A charged particle has been accepted in this analysis if, typically, its momentum p is greater than 300 MeV$/c$, its momentum error $\Delta p/p$ is less than 1 and its impact parameter with respect to the nominal crossing point is within 3-4 cm in the transverse (xy) plane and 3-4 cm/sin θ along the beam direction (z-axis), θ being the polar angle of the track.

Hadronic events are then selected by requiring basically, at least 4 charged particles, 3 GeV as minimum energy of the charged particles in each hemisphere of the event (defined with respect to the beam direction) and total energy of the charged particles of at least 11% of the centre-of-mass energy. The contaminations from $\tau^+\tau^-$ and $\gamma\gamma$ events are respectively less than 0.2% and 0.1% of the accepted events.

Charged particle identification has been provided by the Ring Imaging Cherenkov detectors (RICH) for particles with momenta above 700 MeV$/c$, while the ionization loss measured in the Time Projection Chamber (TPC) has been used for momenta above 100 MeV$/c$. The particle identification tag is based on the combination of both informations as a function of the track momentum. Cuts on the tags have been applied to reject $e^\pm$, $\mu^\pm$ and γ tracks.

The K_s and Λ candidates are detected by their decay in flight into $\pi^+\pi^-$ and $p\pi^-$ respectively. The details of the reconstruction method and the various cuts applied are described in Ref. [9].

To compute efficiencies, $Z \to q\bar{q}$ events were simulated using the JETSET parton-parton generator [10] and then processed through the DELPHI simulation program, DELSIM, which models the detector response. The simulated events passed through DELSIM were then processed by the same reconstruction program as used for the data, DELANA [8]. The amount of simulated events exceeds by a factor 1.5 to 2 the real data.

3. SEARCH FOR STRANGE PENTAQUARKS IN THE PK^0 AND THE PK$^+$ CHANNELS

The state Θ^+ can be detected through its decay into pK^0 pairs; the state Θ^{++} could be observed in its decay into pK$^+$. This analysis studies the invariant mass distributions of pK^0, pK$^+$ pairs in hadronic Z decays. These are compared with the pK$^-$ spectrum, where the $\Lambda(1520)$ is observed.

3.1. Event Selection

The study was restricted to the 1994 and 1995 data taking periods during which the TPC and RICH detectors were optimally set up and functioning especially for particle identification. Beyond the general hadronic cuts described in the previous section, it was additionally required that the track impact parameter to the primary vertex was less than 0.5 mm in the (xy) plane and 1 mm in the z-axis direction to reduce contributions of particle re-interactions inside the detector material. Furthermore there must be at least two tracks inside the angular acceptance $47° < \theta < 133°$ of the Barrel RICH (BRICH).
After applying all track selection criteria, about 1.9 million events remained.

3.2. Analysis of the pK system

We first analysed the $pK^-(\bar{p}K^+)$ invariant mass distribution constructed using identified particles. Figure 1 (Left) shows the pK^- invariant mass spectrum. A clear $\Lambda(1520)$ signal is observed at the expected mass, consistent with published results [11]. This invariant mass distribution was fitted to the sum of a phase space-like term and a normalized Gaussian function accounting for the $\Lambda(1520)$ production. The excess of events in the $\Lambda(1520)$ region is of 306 $\pm$55 events, with a fitted mass and width of 1.520 ± 0.002 GeV/c^2 and 0.010 ± 0.004 GeV/c^2 respectively. The χ^2 per degree of freedom is 1.4. This corresponds to an average $\Lambda(1520)$ production rate per hadronic event of 0.0224 ± 0.0027, compatible with the value reported in PDG [12].

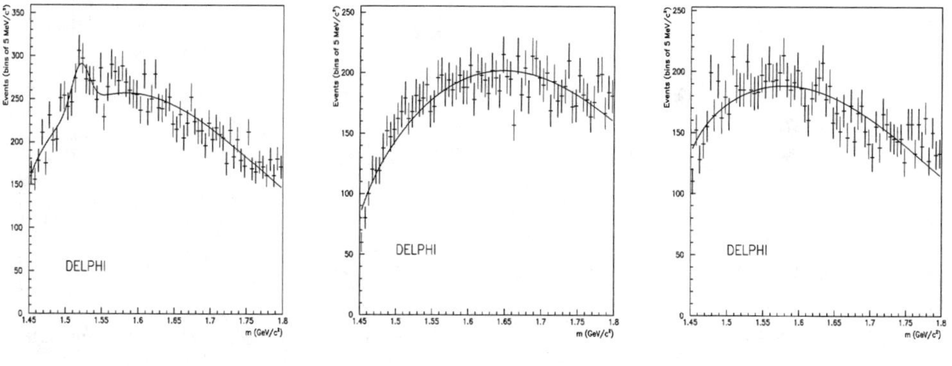

FIGURE 1. Left: (pK^-), Middle: (pK^0), Right: (pK^+) invariant mass spectra. The curves are the results of the fits described in the text

The invariant mass distribution for pK^0 pairs is shown in Figure 1 (Middle). No Θ^+ signal is visible around 1.54 GeV/c^2. We performed the same fit as described above, varying the mass between 1.52 GeV/c^2 and 1.56 GeV/c^2, corresponding to the Θ^+ mass, in bins of 5 MeV/c^2, and imposing the width of 0.010 MeV (the expected experimental resolution). The χ^2 per degree of freedom of the fit is of 1.3. The upper limit at 95% C.L. on the average production rate of the Θ^+, derived from the fit and corrected for inefficiencies is $< N_{\Theta(1540)^+} > < 0.005$.

The invariant mass spectrum for pK$^+$ pairs is shown in Figure 1 (Right). No Θ^{++} peak is visible anywhere. The χ^2 per degree of freedom of the fit to background only is of 1.7. The mass fit was repeated over the range of the Θ^{++} mass estimates, i.e. between 1.50 GeV/c^2 and 1.75 GeV/c^2. The corresponding 95% C.L. upper limit on the average Θ^{++} production rate per hadronic Z boson decay, is $< N_{\Theta^{++}} > \; < 0.006$.

4. SEARCH FOR DOUBLY CHARGED AND DOUBLY STRANGE PENTAQUARKS

4.1. Event selection

In this analysis, in addition to the hadronic selection described in section 2., charged particle tracks from Z decays were required to have a polar angle between 20° and 160°. Furthermore the polar angle of the trust axis, θ_{thrust}, computed for each event, had to satisfy: $|\cos(\theta_{thrust})| < 0.95$.

4.2. Ξ^- Reconstruction

The Ξ^- hyperon was reconstructed through the decay $\Xi^- \to \Lambda\pi^-$. For this, Λ candidates were reconstruced using the standard DELPHI V^0 search algorithm [8] and imposing the invariant mass $M(\mathrm{p}\pi^-)$ to be between 1.10 GeV/c^2 and 1.135 GeV/c^2.

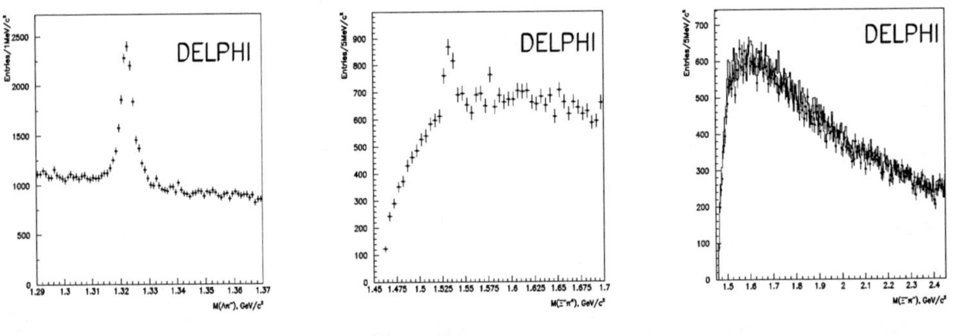

FIGURE 2. Left: ($\Lambda\pi^-$), Middle: ($\Xi^-\pi^+$), Right: ($\Xi^-\pi^-$) invariant mass spectra. The histogram on the Right figure is the MC simulated sample of events

A multivertex fit was then performed on each Ξ^- candidate decaying into $\Lambda\pi^-$. It constrained the sum of the Λ and π momenta to be equal to the Ξ^- momentum and the Λ decay products to give the nominal Λ mass value of 1115.683 ± 0.006 MeV/c^2 [12]. The resulting spectrum of the $\Lambda\pi^-$ invariant mass is shown in Figure 2 (Left).

4.3. Analysis of the $\Xi\pi$ system

Figure 2 (Middle) shows the invariant mass distribution of reconstructed Ξ^- candidate in the mass range between 1.30 GeV/c^2 to 1.34 GeV/c^2 combined with a π^+. A clear $\Xi(1530)$ peak of about 820 ± 50 events is observed.

The mass spectrum of $\Xi^-\pi^-$ combinations is shown in Figure 2 (Right). No significant excess is observed in the 1.86 GeV/c^2 mass region. The histogram is the JETSET7.3 prediction for the $\Xi^-\pi^-$ spectrum without pentaquark. We performed a fit of the $\Xi^-\pi^-$ spectrum to a polynomial background and a Gaussian with a central value of 1.862 GeV/c^2 and a width of 0.015 GeV/c^2 equal to the resolution in this mass region. The fitted number of events is equal to -50 ± 75. The reconstruction efficiency , estimated from a Monte Carlo generated sample of $\Xi(1862)^{--}$ events decaying into $\Xi^-\pi^-$ is $(10.0 \pm 0.5)\%$. This leads to a 95% C.L. estimate of the upper limit of the production rate of a $\Xi(1862)^{--}$ object, of $< N_{\Xi(1862)^{--}} > < 2.8 \times 10^{-4}$.

5. SEARCH FOR CHARMED PENTAQUARKS

5.1. Event Selection

This study was also restricted to the 1994 and 1995 data taking periods corresponding to optimum particle identification. After the standard hadronic event selection listed in section 2. applied, events corresponding to the decay chain $D^{*+} \to D^0 \to K^-\pi^+$ were selected as a first step of the analysis.

Additional cuts were performed to suppress the background:

- $x_E(D^0) \geq 0.15$, where x_E is the energy fraction with respect to the beam energy;
- in the reconstructed D^0 decay, it was required that both the kaon and pion momenta were larger than 1 GeV/c, and that the angles between the K and π momenta were smaller than 90° in the c.m. system;
- the momentum of the bachelor pion had to be between 0.3 GeV/c and 2.5 GeV/c, and the angle between the D^0 candidate and the bachelor π momenta had to be smaller than 90° in the c.m. system;
- the decay length of the D^0 had to be larger than 0 and smaller than 2.5 cm, and different from 0 by at least three standard deviations;
- $\cos\theta_K > -0.9$, where $\cos\theta_K$ is the angle between the D^0 flight direction and the K direction in the D^0 rest frame;
- $1.79\text{GeV}/c^2 < m_{K\pi} < 1.91\text{GeV}/c^2$ and $0.1425\text{GeV}/c^2 < \Delta M < 0.1485\text{GeV}/c^2$, where $\Delta M = m_{K\pi\pi} - m_{K\pi}$;
- the K and π candidates were requested to have at least one hit in the Vertex Detector (VD);
- the K candidates should not have a positive identification as a pion. This cut suppresses about 50% of the combinatorial background surviving all other cuts.

The distribution of $\Delta M = M_{K\pi\pi} - M_{K\pi}$ with $1.79\text{GeV}/c^2 < M_{K\pi} < 1.91\text{GeV}/c^2$ is shown in Figure 3 (Left). One can see a very clear D^* peak over a quite small background.

5.2. Analysis of the D*p system

Figure3 (Middle and Right) shows the effective mass distributions of D*p for total charge zero and for absolute total charge 2 respectively. No narrow resonance peak around 3.1 GeV/c^2 shows up. A 95% C.L. estimate of the upper limit for the production rate, per Z hadronic decay, of a possible $\Theta_c(3100)^0$ state has been worked out, correcting for inefficiencies estimated from a Monte Carlo generated sample of $\Theta_c(3100)^0$ events decaying into D^*p. It is: $< N_{\Theta_c(3100)^0} > \cdot (Br(\Theta_c(3100)^0 \to D^{*+}\bar{p}) \; < 8.8 \times 10^{-4}$.

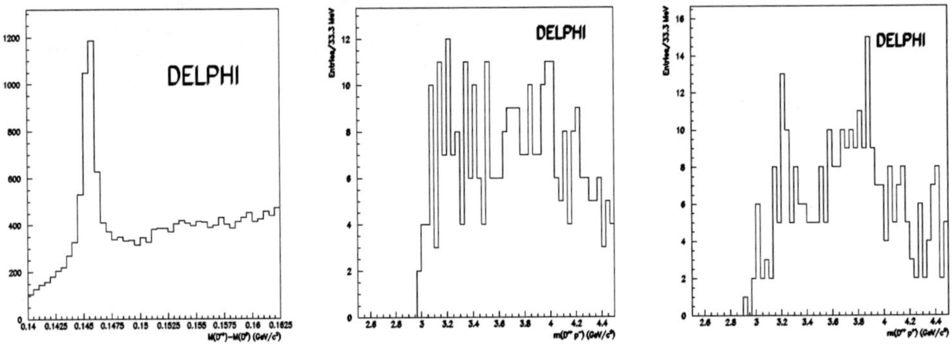

FIGURE 3. Left: (M(D^{*+})-M(D^0)), Middle: (D$^{*+}\bar{p}$), Right: (D$^{*+}p$) invariant mass spectra.

6. CONCLUSIONS

A search for pentaquarks in hadronic Z decays was performed. At 95% C.L., preliminary upper limits were established on the average production rates $< N >$ of such particles and their charge-conjugate state per Z decay:

$$\begin{aligned} < N_{\Theta^+} > &< 0.005 \\ < N_{\Theta^{++}} > &< 0.006 \\ < N_{\Xi(1862)^{--}} > &< 2.8 \times 10^{-4} \\ < N_{\Theta_c(3100)^0} > \cdot (Br(\Theta_c(3100)^0 \to D^{*+}\bar{p}) &< 8.8 \times 10^{-4} . \end{aligned}$$

REFERENCES

1. D. Diakonov et al., Z. Phys. **A359** (1997) 305.
   R. Jaffe and F. Wilczek, Phys. Rev. Lett. **91** (2003) 232003.
   S. Capstick et al., Phys. Lett. **B570** (2003) 185.
   Bin Wu and Bo-Qiang Ma, Phys. Rev. **D69** (2004) 077501.
   J. Ellis et al., JHEP 0405 (2004) **002**.
   R.D. Matheus et al., Phys. Lett. **B578** (2004) 323.
   V. Guzey, Phys. Rev. **C69** (2004) 065203.
2. T. Nakano et al. (LEPS Coll.), Phys. Rev. Lett. **91** (2003) 012002.
   V.V. Barmin et al. (DIANA Coll.), Phys. Atom. Nuclei **66** (2003) 1715.

S. Stepanyan et al. (CLAS Coll.), Phys. Rev. Lett. **91** (2003) 252001.
J. Barth et al. (SAPHIR Coll.), Phys. Lett. **B572** (2003) 127.
A.E. Asratyan et al., Phys. Atom. Nuclei **67** (2004) 682; Yad. Fyz. 67 (2004) 704.
R.A. Arndt, I.I. Strakovsky and R.L. Workman, Phys. Rev. **C68** (2003) 042201.
V. Koubarovsky et al. (CLAS Coll.), Phys. Rev. Lett. **92** (2004) 032001.
3. K. Hicks, "Experimental Search for Pentaquarks", hep-ex/0504027 (April 2005), to be publ. in Progr. Part. Nucl. Phys.
4. C. Alt et al. (NA49 Coll.), Phys. Rev. Lett. **92** (2004) 042003.
5. A. Aktas et al. (H1 Coll.), Phys. Lett. **B588** (2004) 17.
6. S. Schael et al. (ALEPH Coll.), Phys. Lett. **B599** (2004) 1.
7. P. Aarnio et al. (DELPHI Coll.), Nucl. Instr. and Meth. **A303** (1991) 233.
8. P. Abreu et al. (DELPHI Coll.), Nucl. Instr. and Meth. **A378** (1996) 57.
9. P. Abreu et al. (DELPHI Coll.), Z. Phys. **C65** (1995) 587.
10. T. Sjostrand, Comp. Phys. Comm. **82** (1994) 74.
11. P. Abreu et al. (DELPHI Coll.), Phys. Lett. **B475** (2000) 429.
12. S. Eidelman et al., Particle Data Group, Phys. Lett. **B592** (2004) 1.

Searches for Pentaquark Baryons at *BABAR*

Tetiana Berger-Hryn'ova (representing the *BABAR* Collaboration)

SLAC, 2575 Sand Hill Road, Menlo Park, CA 94025, USA

Abstract. This paper presents the results of inclusive searches for the strange pentaquark states $\Theta^+(1540)$, $\Xi_5^{--}(1860)$ and $\Xi_5(1860)^0$ as well as the anti-charm pentaquark state $\Theta_c(3099)^0$ in a dataset of 123.4 fb^{-1} collected on and 40 MeV below the $\Upsilon(4S)$ resonance by the *BABAR* detector at the e^+e^- PEP-II storage rings. No evidence for the pentaquark states is found and upper limits on the rate of $\Theta^+(1540)$ and $\Xi_5^{--}(1860)$ production in e^+e^- annihilation are obtained.

Keywords: pentaquark,baryon,*BABAR*
PACS: 13.25.Hw, 12.15.Hh, 11.30.Er

INTRODUCTION

In the past two years several experimental groups have reported observations of a new, manifestly exotic (B=1, S=1) baryon resonance, the $\Theta^+(1540)$ [1], with an unusually narrow width ($\Gamma < 8$ MeV). Also, the NA49 experiment reported evidence for an additional narrow exotic (B=1, S= -2) state, the Ξ_5^{--}, as well as the corresponding Ξ_5^0 state, with mass about 1862 MeV/c^2 [2]. More recently the H1 collaboration reported [3] a narrow ($\Gamma < 30$ MeV) exotic anti-charmed (B=1, C= -1) resonance, Θ_c^0, with a mass of 3099$\pm$6 MeV/c^2. For these Θ^+, Ξ_5^{--} and Θ_c^0 states the minimal quark content is $(udud\bar{s})$, $(dsds\bar{u})$ and $(udud\bar{c})$, respectively. These results have prompted a surge of pentaquark searches in experimental data of many kinds, mostly with negative results [4]. Several theoretical models [5, 6, 7] have been proposed to describe possible pentaquark structure. They predict that the lowest-mass states containing u, d and s quarks should occupy a spin-1/2 anti-decuplet and octet. The Θ_c^0 pentaquark should be an isospin-zero member of the $\overline{6}$ representation of the 60-plet of SU(4) [8].

The *BABAR* experiment [9] at the SLAC PEP-II e^+e^- collider takes data at center-of-mass energy of $\approx$10.58 GeV. *BABAR* is well suited to search for these states, since it provides excellent pion, kaon and proton identification and good tracking, with the result that excellent mass resolution can be achieved. Charged particle tracks are measured by a five-layer silicon vertex tracker (SVT) and a 40-layer drift-chamber (DCH) located in a 1.5-T solenoidal magnetic field. Charged particles are identified by means of specific ionization (dE/dx) measurements in the SVT and DCH, and from the pattern of Cherenkov photons in the Cherenkov radiation detector (DIRC).

In addition to the inclusive searches for the strange and anti-charm pentaquark states in e^+e^- annihilations presented in this paper, there is also an inclusive search for the Θ^+ in electro- and hadro-production in the material of the *BABAR* detector which shows no evidence for this state [10]. Furthermore an exclusive search in B^+ decay to $p\bar{p}K^+$ final states for the isovector pentaquark candidate Θ^{*++} decaying into pK^+ in the mass range 1.43 to 2.00 GeV/c^2 sets limits on $\mathscr{B}(B^+ \to \Theta^{*++}\bar{p}) \times \mathscr{B}(\Theta^{*++} \to pK^+)$ at the 10^{-7}

CP814 *Hadron Spectroscopy: Eleventh International Conference*
edited by A. Reis, C. Göbel, J. Sá Borges, and J. Magnin

level [11]. A review of these and other *BABAR* hadronic results can be found in these proceedings [12].

INCLUSIVE SEARCHES FOR STRANGE AND ANTI-CHARM PENTAQUARK BARYONS

Although experiments with a baryon in the beam or the target might seem to have some advantage in pentaquark production, e^+e^- interactions are also known for democratic production of hadrons. Mesons and baryons with non-zero charm and strangeness (up to three units) have been observed with production rates that appear to depend on mass and spin, but not quark content (Fig. 1). If pentaquarks are produced similarly, then one might expect a pentaquark rate as high as that for an ordinary baryon of the same mass and spin. The search for inclusive production of the pentaquark states [13] Θ^+, Ξ_5^+, Ξ_5^0, Ξ_5^-, Ξ_5^{--}, Σ_5^+, N_5^0, N_5^+ and Θ_c^0 has been performed with 123 fb^{-1} of data recorded at or slightly below the $\Upsilon(4S)$ resonance [14]. This paper will discuss only the Θ^+, Ξ_5^{--}, Ξ_5^0 and Θ_c^0 searches.

A search for the Θ^+ is carried out in the decay mode $\Theta^+ \to pK_S^0$, where $K_S^0 \to \pi^+\pi^-$. The expected Θ^+ mass resolution is about $2\,\mathrm{MeV}/c^2$. However no peak is seen at the expected mass but a large signal at 2285 MeV/c^2 (with a mass resolution of $6\,\mathrm{MeV}/c^2$ at the Λ_c mass) containing ≈98,000 entries from $\Lambda_c \to pK_S^0$ is observed. This null result for a Θ^+ mass of 1540 MeV/c^2 is quantified by fitting the convolution of a Gaussian and a P-wave Breit-Wigner for the signal line-shape, and a seventh-order-polynomial times a threshold function for the background shape, to the pK_S^0 invariant-mass distribution in the interval from threshold to $1800\,\mathrm{MeV}/c^2$. Since the intrinsic width of the Θ^+ has not been determined so far, width values of $\Gamma = 1\,\mathrm{MeV}$ (for a narrow Θ^+) and $\Gamma = 8\,\mathrm{MeV}$ (best upper limit) are used, and the results quoted for each assumed width. The upper limit, at 95% confidence level, is determined for the number of produced pentaquarks per $e^+e^- \to hadrons$ event, and compared to the production rates of known baryons, assuming $\mathscr{B}(\Theta^+ \to pK_S^0) = 25\%$. The measured upper limit values of 5×10^{-5}/event ($\Gamma = 1\,\mathrm{MeV}$) and 11×10^{-5}/event ($\Gamma = 8\,\mathrm{MeV}$) are between eight and 15 times lower than expected for conventional baryons, as shown in Fig. 1.

A search for the Ξ_5^0 and Ξ_5^{--} resonances was performed using the decay chain $\Xi_5 \to \Xi^-\pi$, $\Xi^- \to \Lambda\pi^-$, $\Lambda \to p\pi^-$. In each case, no peak is seen at the expected mass. In the $\Xi^-\pi^+$ spectrum, prominent peaks for the $\Xi(1530)^0$ and $\Xi_c(2470)^0$ with ≈24,000 and ≈8,000 entries respectively, are seen. No structure is observed in the exotic $\Xi^-\pi^-$ spectrum. A linear function is used for the background, while the signal is modeled as described above. The resolution function is derived from the $\Xi(1530)^0$ and $\Xi_c(2470)^0$ signals in data and simulation, and is described by a Gaussian function with an RMS of $8\,\mathrm{MeV}/c^2$. The fit is performed over a $\Xi^-\pi^-$ mass range from 1760 to $1960\,\mathrm{MeV}/c^2$. As before, two different intrinsic widths of this pentaquark state are used, namely $\Gamma = 1\,\mathrm{MeV}$ (narrow) and $\Gamma = 18\,\mathrm{MeV}$ (best experimental upper limit) in order to determine 95% confidence level upper limit values for the production rate in e^+e^- interactions. The values obtained, 0.74×10^{-5}/event ($\Gamma = 1\,\mathrm{MeV}$) and 1.1×10^{-5}/event ($\Gamma = 18\,\mathrm{MeV}$), are between four and six times lower than those for conventional baryons, as shown in Fig. 1,

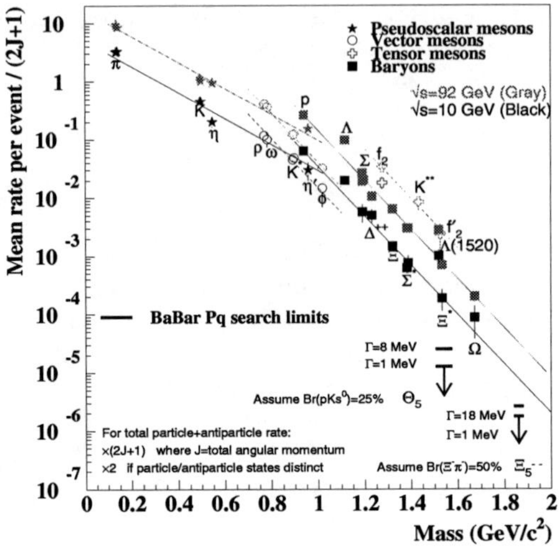

FIGURE 1. Compilation of meson and baryon production rates in e^+e^- annihilation [15] from experiments at the Z^0 (gray) and $\sqrt{s} \approx 10\,\text{GeV}$ (black) as a function of baryon mass. The vertical scale accounts for the number of spin and particle+antiparticle states, and the lines are chosen to guide the eye. The arrows indicate our upper limits on spin-1/2 Θ^+ and Ξ_5^{--} pentaquark states, assuming the branching fractions shown, and are seen to lie below the solid line.

assuming $\mathscr{B}(\Xi_5^{--} \to \Xi^-\pi^-) = 50\%$. It is not possible to determine the total production rate for the Ξ_5^0, as its branching fraction to $\Xi^-\pi^+$ is unknown.

We present preliminary results from a search for Θ_c^0 production performed using the decay mode, $\Theta_c^0 \to pD^{*-}$, where the D^{*-} is reconstructed in the $\bar{D}^0\pi^-$ decay mode, and the $\bar{D}^0$ in the $K^+\pi^-$ and $K^+\pi^-\pi^-\pi^+$ modes; the selection criteria are designed for high efficiency and low bias against production mechanism. The selected candidates satisfy the following criteria: the $\bar{D}^0$ and D^{*-} candidates both have mass values within $20\,\text{MeV}/c^2$ of the respective average peak values, and their mass difference is within $2\,\text{MeV}/c^2$ of the average peak value. These values are chosen to minimize the upper limit on the total cross section after a momentum-dependent correction for efficiency. A total of about 55,000 (73,000) true $D^{*-} \to K^+\pi^-(K^+\pi^-\pi^-\pi^+)\pi^-$ are present in the selected data over a background of 4,000 (62,000).

The invariant mass distributions for the Θ_c^0 candidates in the data are shown in Fig. 2(a) for the two $\bar{D}^0$ decay modes separately. The distributions show no narrow structure, and in particular they are all quite smooth in the region near $3099\,\text{MeV}/c^2$, as shown in the inset. To avoid sensitivity to the details of the production mechanism, the p^* distribution is divided into nine intervals of width $500\,\text{MeV}/c$ from 0 to $4.5\,\text{GeV}/c$, and then a fit to the invariant mass distribution is carried out for each p^* interval.

As for the previous searches a P-wave Breit-Wigner line-shape convolved with the resolution function is used for the signal modeling. The Θ_c^0 invariant mass resolution is obtained from simulation [16], and is represented by a sum of two Gaussian functions

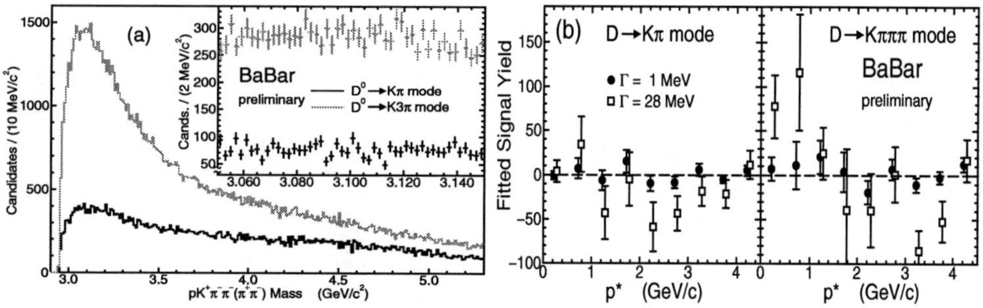

FIGURE 2. (a) Invariant mass distributions of $D^{*-}p$ with $\bar{D}^0$ in the $K^+\pi^-$(black) and $K^+\pi^-\pi^-\pi^+$(gray) modes for combinations satisfying the criteria described in the text. The data are plotted for the full kinematically allowed $D^{*-}p$ range and, in the inset, with statistical uncertainties and a suppressed zero on the vertical axis, for the mass range in which the Θ_c^0 has been reported. (b) The Θ_c^0 yields extracted from the fits to the (left) $pK^+\pi^-\pi^-$ and (right) $pK^+\pi^-\pi^-\pi^+\pi^-$ invariant mass distributions, assuming a mass of 3099 MeV/c^2 and a natural width of $\Gamma = 1$ MeV (circles) or $\Gamma = 28$ MeV (squares).

with a common center. The overall resolution, defined as the FWHM of the resolution function divided by 2.355, averages 2.8 (3.0) MeV/c^2 for the $K^+\pi^-$ ($K^+\pi^-\pi^-\pi^+$) decay modes with a small dependence on p^*, the center-of-mass momentum of the Θ_c^0. The quoted results assume two widths, $\Gamma = 1$ MeV, corresponding to a very narrow state, and $\Gamma = 28$ MeV, corresponding to the width observed by H1. The background is described in each p^* bin by a threshold function. Maximum likelihood fits are performed at several fixed Θ_c^0 mass values in the vicinity of 3099 MeV/c^2. In every case the fit quality is good and the signal obtained is consistent with zero. Results using different mass values are consistent within expected statistical variations. Fixing the mass to the reported value of 3099 MeV/c^2 results in the event yields shown in Fig. 2(b). There is no evidence of a pentaquark signal in any p^* range, and the roughly symmetric scatter of the points about zero indicates low momentum-dependent bias in the background function.

It is not possible to compare the sensitivity of our search with that of H1, due to the presumably different production mechanism, and the fact that there is no cross section measurement in Ref. [3]. It is interesting to note that H1 selects about 3,500 D^{*-} using only the $\bar{D}^0 \to K^+\pi^-$ mode, with a background of about 1,500; of these, about 550 appear in their pD^{*-} mass plot, with mass below 3.6 GeV/c^2, resulting in a Θ_c^0 signal yield of 51±11 events. Thus the observed Θ_c^0 account for roughly 1/70 of their D^{*-} production. In contrast, the *BABAR* search reconstructs about 750,000 D^{*-} above background, 128,000 of which also have an identified proton in the event, but no Θ_c^0 signal is observed.

It is also not possible to compare the production rate of Θ_c^0 to the rates measured for ordinary charmed baryons, as only measurements for $\Lambda_c(2285)$ and $\Sigma_c(2455)$ at much lower mass than Θ_c^0 are available at this point [15]. Furthermore, since $\mathscr{B}(\Theta_c^0 \to pD^{*-})$, is not known it is not possible to set an upper limit on the total production rate of the Θ_c^0.

CONCLUSIONS

A large statistics high-resolution search for the reported pentaquark states Θ^+, Ξ_5^{--}, Ξ_5^0 and Θ_c^0 in e^+e^- annihilations has been performed at *BABAR*. Large signals for known baryon states have been found, but no excess is seen at the reported mass values for the pentaquark states.

ACKNOWLEDGMENTS

The author is grateful for the extraordinary contributions of the PEP-II colleagues in achieving the excellent luminosity and machine conditions that have made this work possible. This work is supported by DOE and NSF (USA), NSERC (Canada), IHEP (China), CEA and CNRS-IN2P3 (France), BMBF and DFG (Germany), INFN (Italy), FOM (The Netherlands), NFR (Norway), MIST (Russia), and PPARC (United Kingdom). Individuals have received support from CONACyT (Mexico), A. P. Sloan Foundation, Research Corporation, and Alexander von Humboldt Foundation.

REFERENCES

1. LEPS Collaboration, T. Nakano *et al.*, Phys. Rev. Lett. **91**, 012002 (2003); SAPHIR Collaboration, J. Barth *et al.*, Phys. Lett. **B 572**, 127 (2003); CLAS Collaboration, S. Stepanyan *et al.*, Phys. Rev. Lett. **91**, 252001 (2003); CLAS Collaboration, V. Kubarovsky *et al.*, Phys. Rev. Lett. **92**, 032001 (2004), Erratum; ibid, 049902; DIANA Collaboration, V.V. Barmin *et al.*, Phys. Atom. Nucl. **66**, 1715 (2003); SVD Collaboration, A. Aleev *et al.*, submitted to Yad. Fiz., hep-ex/0401024 (2004); HERMES Collaboration, A. Airapetian *et al.*, Phys. Lett. **B 585**, 213 (2004); A.E. Asratyan, A.G. Dolgolenko, and M.A. Kubantsev, Phys. Atom. Nucl. **67**, 704 (2004); COSY-TOF Collaboration, M. Abdel-Bary *et al.*, Phys. Lett. **B 595**, 127 (2004); ZEUS Collaboration, S. Chekanov *et al.*, Phys. Lett. **B 591**, 7 (2004).
2. NA49 Collaboration, C. Alt *et al.*, Phys. Rev. Lett. **92**, 042003 (2004).
3. H1 Collaboration, A. Aktas *et al.*, Phys. Lett. **B 588**, 17 (2004).
4. See, *e.g.* K. H. Hicks, Prog Part Nucl Phys **55**, 647 (2005), and references therein.
5. D. Diakonov, V. Petrov and M.V. Polyakov, Z. Phys. **A 359**, 305 (1997).
6. M. Karliner and H. Lipkin, hep-ph/0402260 (2004).
7. R. Jaffe and F. Wilczek, Phys. Rev. Lett. **91**, 232003 (2003).
8. B. Wu and B.-Q. Ma, Phys. Rev. D **70**, 034025 (2004).
9. B. Aubert *et al.*, *BABAR* Collaboration, Nucl. Instr. Meth. A**479**, 1 (2002).
10. J. Coleman, A Search for Pentaquarks at BaBar, Ph.D. Thesis, University of Liverpool (May 2005); *BABAR* -THESIS-05/015; K. Götzen, hep-ex/0510041, presented in QCD05, Montpellier, France, 4-8 Jul 2005.
11. B. Aubert *et al.*, *BABAR* Collaboration, Phys. Rev. D **72**, 051101(R) (2005).
12. D. Muller, these proceedings.
13. Charge-conjugate reactions are included implicitly throughout the paper.
14. B. Aubert *et al.*, *BABAR* Collaboration, Phys. Rev. Lett. **95**, 042002 (2005), B. Aubert *et al.*, *BABAR* Collaboration, hep-ex/0408064.
15. S. Eidelman *et al.*. (Particle Data Group), Phys. Lett. B **592**, 1 (2004).
16. GEANT4 Collaboration, S. Agostinelli et al., Nucl. Instrum. & Methods A **506, 250 (2003).**

PentaQuark Mass and Structure

T. Goldman*, Fan Wang†, Jialun Ping** and Di Qing‡

*Theoretical Division, LANL, Los Alamos, NM87545, USA
†Center for Theoretical Physics, Nanjing University, Nanjing, 210093, China
**Department of Physics, Nanjing Normal University, Nanjing, 210097, China
‡CERN, Ch-1211 Geneva 23, Switzerland

Abstract.
The Quark Delocalization, Color Screening Model has successfully reproduced the binding energy and structure of 9- and 12-quark systems that are small nuclei, namely ^{3}He and ^{4}He, in a relativistic version. In a non-relativistic version it has successfully been applied to 6-quark systems to describe baryon-baryon potentials, phase shifts and cross sections. It is applied here to KN molecular, Jaffe-Wilczek, Karliner-Lipkin and tetrahedral structures proposed for the reported pentaquark state, Θ^+. In this model, we find that the excitation energy above KN threshold can be reproduced in some cases. The negative parity state is always found at a lower mass than the positive parity one. Concerns regarding additional physics effects which may obviate this result and a conjecture regarding two- vs. three-body final state production of the Θ^+ are also noted.

Keywords: pentaquark, quark models, color structure
PACS: 24.85.+p, 12.39.Mk, 12.39.Jh, 12.39.Pn

We have studied the recently reported pentaquark state [1] using various conventional quark models. [2] We found that it is important to take into account mixing of various color structure configurations, which reduces the mass of the state but that a mass as small as that reported experimentally could not be achieved. We report here on the results of using our Quark Delocalization, Color Screening Model (QDCSM) to study the different proposed quark correlation configurations which failed to reproduce the low state mass in the more conventional models. We analyze the molecular (KN, etc.), Jaffe-Wilczek [3] and Karliner-Lipkin [4] quark correlation configurations, including channel coupling. A tetrahedral quark correlation is also considered. Finally, we consider elements missing from all extant quark models and raise a conjecture regarding the relative strength of Θ^+ photoproduction in conjunction with one or with two mesons in the final state.

The philosophy underlying our model is that the physical exchange of quarks between hadrons provides a good representation of the exchange of all of the mesons that would occur in a hadronic model. For the successes of the relativistic and nonrelativistic versions, see Refs.([5, 6, 7, 8, 9, 10, 11, 12]). An extended version of the model, which includes the exchange of pions between quarks adds sufficient attraction to bind the deuteron [13] and reproduces the long-range pion potential. It has little effect on anything else, including the predicted compact d^* state and NΩ dibaryon [8, 9] states. The accuracy of the relativistic version is several MeV in baryons and a few MeV in nuclei. The non-relativistic version achieves 10-20 MeV accuracy in baryons, but can fit the deuteron quite precisely. The accuracy for more compact hadrons, such as dibaryons and pentaquarks is unknown, of course, but we estimate it to be better than 50 MeV on

CP814 *Hadron Spectroscopy: Eleventh International Conference*
edited by A. Reis, C. Göbel, J. Sá Borges, and J. Magnin

the basis of these previous results.

The Hamiltonian and parameters used are those of the QDCSM as fitted to the baryons. The screening scale is 1.0 fm^{-2}, and the strange quark mass is $m_s = 560$ MeV. We note here that this implies a kaon mass of 615 MeV, an excess over experiment not unexpected since the bound state model of a quark and antiquark is not a good representation of a Goldstone boson. This has further implications as it raises the KN threshold by ≈ 120 MeV above its physical value. Since a constituent quark mass is an effective quantity that may vary from state to state, we also carried out our calculations with a strange quark mass of 450 MeV, which reproduces the physical kaon mass.

Under the adiabatic approximation, we calculated the effective potential between a meson containing a strange antiquark and a nucleon, coupled to spin $1/2$, isospin zero and positive or negative parity. (Isospin 1 is fully repulsive.) The delocalization of the wavefunction of a quark from either hadron into the region defined by the other reduces the energy of the state, but it also reduces the color magnetic spin-spin interaction energy. For octet hadrons, where this interaction is net attractive among the quarks involved, this produces a net repulsive effect, raising the energy of the state. (The change in the color confinement interaction energy is also relevant to the balance.) The channels included were KN, K^*N, K_8N$_8$, K_8^*N$_8$ and K_8^*N$_8^*$, where the "8" indicates the quarks were coupled to a color octet instead of a singlet in that cluster ("hidden color" combinations).

Here, in the S-wave, we find no attraction at all whether in the single KN channel, or coupling all of the channels, with or without delocalization when the screening parameter is set to zero (no screening). Note that this occurs even when so-called "hidden color" combinations are included, where the quarks combine to form local color octets rather than singlets. However, when the screening parameter is set to a value consistent with the binding energy for the deuteron in a nuclear calculation, $\mu = 1$ fm^{-2}, a strong minimum of about 70 MeV appears in the effective potential in all cases for a hadron separation of about 0.6 fm. In this case, delocalization is complete at the minimum energy and the quark wavefunctions are symmetric between the two orbitals (one for the meson and one for the nucleon).

After the correction for the physical K, N masses (by using a subtraction procedure), we find a state energy of 1615 MeV, which includes a zero point energy of the fluctuations in separation between the two hadronic "lumps" of 250 MeV. This is less than 100 MeV above the reported mass of the Θ^+. Width calculations are not complete, but appear to be in the normal quark model area of order 10 MeV.

In the P-wave (for the strange antiquark), there is again negligible attraction for $\mu = 0$, and the effective potential develops a ~ 180 MeV minimum for $\mu = 1$ fm^{-2}. However, this occurs at a hadron separation of about 0.4 fm and this much smaller distance effectively prevents delocalization from producing significant reduction of the energy of the state. The mass of this positive parity state is ~ 1820 MeV, and so is too high to match with experiment even taking a threshold correction into account.

Similar results are found with the strange quark mass adjusted in the meson to give the correct mass for the kaon.

In the Jaffe-Wilczek prescription, the two pairs of ud quarks are assumed to be strongly coupled into such diquark pairs. We put each diquark in its own orbital, separated by S and in normal three-body coordinates, the $\bar{s}$-quark is at a distance T at a right angle from the midpoint between the two ud-diquark orbitals. The two diquarks can each be in several spatial-color-isospin-spin states, which we label by integers, $i = 1-4$:

$$1 \equiv [2][2]10;\ 2 \equiv [2][2]01;\ 3 \equiv [2][11]11;\ 4 \equiv [2][11]00;$$

consistent with Pauli statistics. There are then seven four-quark combinations also shown in Table(1), where $i = 1,2,3,4$ and the bar indicate symmetrization and the tilde represents antisymmetrization.

The states are again tto massive when color screening is turned off ($\mu = 0$, conventional quark model), and the positive parity states are also again significantly higher in mass ($\sim 100-200$ MeV). Delocalization and channel coupling have small effects ($\sim 5-10$ MeV reduction in mass). The diquark wavefunctions do not delocalize into each other's orbitals at all, but do develop a relative amplitude of ~ 0.4 intruding into the orbital of the $\bar{s}$-quark which deloclaizes strongly (~ 0.99) into the diquark orbitals. The corresponding values of S and T at the minimum in the energy landscape are $\sim 0.6-0.7$ fm for the lowest energy, negative parity state at ~ 1650 MeV. However, with the correction for the discrepancy with the physical threshold, this closely approximates the reported mass of the Θ^+. The lowest states correspond to $k = 2$ in Table(1).

With the strange quark mass adjusted in the meson to give the correct mass for the kaon, the Θ^+ does appear at a value within the model uncertainty of the experimentally reported value for the S-wave $(1/2^-)$ state.

For the positive parity state, the orbital separations at the energy minimum tend to be $\sim$ 1 fm, although all valules are reduced considerably when color screening is turned off.

TABLE 1. Quantum number correlations for diquark pairs.

k	1	2	3	4	5	6	7
(ij)S_4	$(\overline{13})1$	$(\overline{24})1$	(33)1	$(\widetilde{13})1$	$(\widetilde{24})1$	(33)0	(44)0

In the Karliner-Lipkin configuration, the quarks are correlated into one ud diquark and one three-quark object, containing $u, d$and$\bar{s}$. In this case, we again found the positive parity state to be significantly higher in mass, at 1770 MeV even with channel coupling, and at cluster separation of order 1 fm. The clusters in the negative parity state found an energy minimum at smaller separation, $\sim 0.5-0.6$ fm. Here channel coupling made a marked difference in the delocalization parameters, which were small without coupling, but of order one for both the diquark and triquark clusters with coupling. However, the effect on the final mass of the state was negligible, with a value of ~ 1690 MeV. This seems somewhat high, even after correcting for the model threshold being 120 MeV above the physical KN threshold.

Having already introduced color separations of order a fermi in the cases above, we allowed consideration of a tetrahedral structure that encompasses all of them: Each light quark was assigned an orbital centered on one of the corners of a regular tetrahedron and the strange antiquark orbital was centered at the geometrical center of the tetrahedron. In this case, we considered only the negative parity state and found, with channel coupling of all color configurattions, a state with mass $\sim 1630-1635$ MeV. The minimum energy configuration occurred for a tetrahedron side length of 0.8 fm and with u and d delocalization amplitude parameters of ~ 0.1 (as always, relative to unity in the initial orbital). The strange antiquark was slightly more delocalized, with $\varepsilon \sim 0.2$ at the minimum configuration. Again, if one allow for a $\sim$ 120 MeV correction between the model and physical KN thresholds, the mass of the Θ^+ is accounted for in this molecule-like configuration also. It has the further advantage of being very different in structure from a meson-baryon configuration, which suggests the decay width from this structure will be suppressed relative to normal hadronic expectations.

Further combinations of configurations of the Jaffe-Wilczek and Karliner-Lipkin types in extended channel coupling calculations produced no lower mass states than in the calculation referred to above. Again, color screening was crucial, with states becoming up to 100 MeV more massive without it.

Thus, we have found that "molecular" meson-baryon combinations cannot reproduce the low mass of the Θ^+, even with quark delocalization and color screening. However, for more exotic structures, it seems that it may be possible for the mass of this state to be as low as has been experimentally reported.

New questions have arisen. One is the observation that there is QCD physics omitted in all quark models to date, namely the three- (and four-) gluon coupling vertex which contibutes nothing to three-quark and quark-antiquark states because of a vanishing matrix element in a color singlet system. These presumably do contribute to the higher Fock space component configurations of normal hadrons, but the amplitudes for those are not large [14] so the effects on the state masses may be negligible there.

Another question is whether the multi-body string version of the confinement interaction found in lattice QCD can be approximated by a normal or color-screening two-body confinement interaction, especially the transition interaction between different color structures in multi-quark systems for which we have almost no any reliable information.

A third question is one regarding the production cross section for the state in exclusive reactions. Fig.(1) shows quark line diagrams for photoproduction. Note that if the photon pair produces the strange quark without transferring large momentum to the target nucleon, there is little to choose between $\gamma p \to \overline{K^0}\Theta^+$ and $\to \pi^+K^-\Theta^+$. This is because extra quark-antiquark pairs are fairly easily produced from the vacuum under color separation. However, if the photon strikes a valence quark with high momentum transfer, it is difficult for that quark to re-enter the final state baryon Θ^+ as it must for the two-body final state to occur. Conversely, a fast forward pion is the likely result if the needed quarks appear via soft vacuum pair production to yield the three-body final state. We therefore suggest that a cut on a fast forward pion should enhance the Θ^+ signal in photoproduction experiments. There may already be some evidence to that effect.

We conclude that while the current experimental situation seems murky, theory cannot rule out the existence of the Θ^+ at about the previously reported mass value, despite early QCD sum rule and lattice arguments to the contrary. The study of multi-quark states is important to understanding multigluon coupling effects and coupling between different color structures not previously relevant to quark models of normal hadrons. This will be true whether the existence of the Θ^+ is confirmed or not, and so urge our experimental colleagues on to produce a valuable result, either way. We do predict that if this state exists, its spin-parity will be determined to be $1/2^-$.

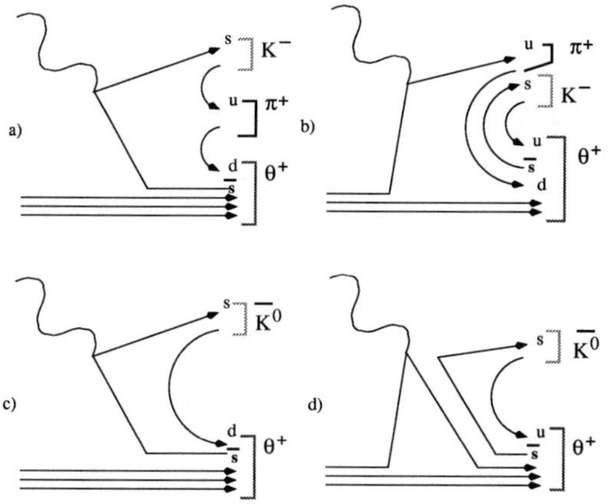

FIGURE 1. Quark line graphs for 2- and 3- body photoproduction of the Θ^+.

ACKNOWLEDGMENTS

This research is supported by the National Research Foundation of China under grants 90103018, 10375030 and by the Department of Energy under contract W-7405-ENG-36.

REFERENCES

1. T. Nakano et al. [LEPS], *Phys. Rev. Lett.*, **91**, 012002 (2003). hep-ex/0301020; V. V. Barmin et al. [DINAN], *Phys. Atom. Nucl.*, **66**, 1715 (2003). hep-ex/0304040; S. Stepanyan et al. [CLAS], *Phys. Rev. Lett.*, **91**, 252001 (2003). hep-ex/0307018; J. Barth et al. [SAPHIR], *Phys. Lett.*, **B572**, 127 (2003); hep-ex/0307083; A. E. Asratyan et al., hep-ex/0309042; V. Kubarovsky et al. [CLAS], hep-ex/0311046; A. Airapetian et al. [HERMES], hep-ex/0312044; A. Allev et al. [SVD], hep-ex/0401024; M. Abdel-Bary et al. [COSY-TOF], hep-ex/0403011; P. Z. Aslanyan et al., hep-ex/0403044; S Chekanov et al. [ZEUS], hep-ex/0403051.
2. Jialun Ping, Di Qing, Fan Wang and T. Goldman, *Phys. Lett.* **B602**, 197 (2004); T. Goldman, *Int. J. Mod. Phys.* **20**, 1963 (2005); Fan Wang, Jialun Ping, Di Qing and T. Goldman, *Phys. Atom. Nucl.*, **68**, 808 (2005); *Yad. Fiz.*, **68**, 841 (2005)-846; nucl-th/0406036.
3. R. L. Jaffe and F. Wilczek, *Phys. Rev. Lett.* **91**, 232003 (2003); hep-ph/0307341.
4. M. Karliner and H. J. Lipkin, *Phys. Lett.* **B595**, 249 (2003); hep-ph/0307243.
5. Kim Maltman, G. J. Stephenson, Jr. and T. Goldman, *Phys. Lett.* **B324**, 1 (1994).
6. T. Goldman, K. R. Maltman, G. J. Stephenson, Jr., and K. E. Schmidt, *Nucl. Phys.* **A481**, 621 (1988).
7. C. J. Benesh, T. Goldman and G. J. Stephenson, Jr., *Phys. Rev. C* **68** 045208 (2003); nucl-th/0307038.
8. T. Goldman, K. Maltman, G. J. Stephenson, Jr., Jia-Lun Ping and Fan Wang, *Mod. Phys. Lett.* **A13**, 59 (1998); nucl-th/9803002.
9. T. Goldman, K. Maltman, G. J. Stephenson, Jr., K.E. Schmidt and Fan Wang, *Phys. Rev. Lett.* **59**, 627 (1987); H.R. Pang, J.L. Ping, F. Wang, T. Goldman and E.G. Zhao, *Phys. Rev. C* **69**, 065207 (2004).
10. Fan Wang, Guang-han Wu, Li-jian Teng and T. Goldman, *Phys. Rev. Lett.* **69**, 2901 (1992); Guanghan Wu, Lijian Teng, J. L. Ping, F. Wang and T. Goldman, *Phys. Rev.* C**53**, 1161 (1996).
11. Jia-lun Ping, Fan Wang and T. Goldman, *Nucl. Phys.* **A657**, 95 (1999); nucl-th/9812068.
12. Guang-han Wu, Jia-lun Ping, Li-jian Teng, Fan Wang and T. Goldman, *Nucl.Phys.* **A673**, 279 (2000); nucl-th/9812079.
13. Jialun Ping, Fan Wang and T. Goldman, *Phys. Rev.* C**65**, 044003 (2002); nucl-th/0012011.
14. Di Qing, Xiang-Song Chen and Fan Wang, *Phys. Rev. D* **58**, 114032 (1998).

Narrow pentaquarks in QCD

F.S. Navarra*, M. Nielsen* and R. Rodrigues da Silva*

*Instituto de Física, Universidade de São Paulo, C.P. 66318, 05389-970 São Paulo, SP, Brazil

Abstract. We study the decay $\Theta \to K^+ n$ within the framework of QCD sum rules and compute the coupling $g_{\Theta nK}$, which is directly related to the pentaquark width. Restricting the decay diagrams to those with color exchange between the meson-like and baryon-like clusters reduces the coupling constant by a factor of four. Whereas a small decay width might be possible for a positive parity pentaquark, it seems difficult to explain the measured width for a pentaquark with negative parity.

It is not clear whether pentaquark states exist or not [1]. All the positive measurements performed so far indicated that it is a very narrow state. Many explanations for this narrow width have been advanced [2]. In this work we calculate the pentaquark decay width within the framework of QCD sum rules (QCDSR) [3]. Several sum rule calculations have been performed for the mass of the pentaquark containing a strange quark [4, 5, 6, 7, 8]. A common problem of all determinations is the large continuum contribution which has its origin in the high dimension of the interpolating currents and results in a large dependence on the continuum threshold. Another problem is the irregular behavior of the operator product expansion (OPE), which is dominated by higher dimension operators and not by the perturbative term as it should be.

In [9, 10] it has been suggested, on a qualitative level, that the narrowness of the pentaquark width can be attributed to the non-trivial color structure of the pentaquark. In this work we will test quantitatively this hypothesis.

The investigation of the pentaquark decay width requires a three-point function which we define as

$$\Gamma(p,p') = \int d^4x d^4y e^{-iqy} e^{ip'x} \langle 0|T\{\eta_N(x) j_K(y) \bar{\eta}_\Theta(0)\}|0\rangle \tag{1}$$

where η_N, j_K and η_Θ are the interpolating fields associated with neutron, kaon and Θ, respectively [11].

We next consider the expression (1) in terms of hadronic degrees of freedom and write the phenomenological side of the sum rule. Treating the kaon as a pseudoscalar particle, the interaction between the three hadrons is described by $\mathscr{L} = ig_{\Theta nK}\bar{\Theta}\gamma_5 Kn$ for $P = +$ and $\mathscr{L} = ig_{\Theta nK}\bar{\Theta}Kn$ for $P = -$.

Writing the correlation function (1) in momentum space and inserting complete sets of hadronic states we obtain an expression which depends on matrix elements connecting the vacuum and the currents associated with the kaon, the neutron and the Θ^+. These numbers, usually called λ_K, λ_N and λ_Θ, can be either extracted from data or obtained from QCDSR calculations. In particular, λ_K is related to the kaon decay constant through

CP814 *Hadron Spectroscopy: Eleventh International Conference*
edited by A. Reis, C. Göbel, J. Sá Borges, and J. Magnin

$\lambda_K = \frac{f_K m_K^2}{m_u+m_s}$. Following the procedure described above we arrive at

$$\Gamma_{phen} = \frac{-g_{\Theta nK}\lambda_\Theta\lambda_N\lambda_K}{(p'^2-m_N^2)(q^2-m_{K+}^2)(p^2-m_\Theta^2)} \times \Gamma_E + \text{continuum} \tag{2}$$

with

$$\Gamma_E = \sigma^{\mu\nu}\gamma_5 q_\mu p'_\nu - im_N \not{q}\gamma_5 + i(m_N \mp m_\Theta)\not{p}'\gamma_5 + i\gamma_5(p'^2 \mp m_\Theta m_N - qp') \tag{3}$$

We shall work with the $\sigma^{\mu\nu}\gamma_5 q_\mu p'_\nu$ structure because, as it was shown in [12], this structure gives results which are less sensitive to the coupling scheme on the phenomenological side.

We now come back to (1) and write the interpolating fields in terms of quark degrees of freedom as

$$j_K(y) = \bar{s}(y)i\gamma_5 u(y) \tag{4}$$

$$\eta_N(x) = \varepsilon^{abc}(d_a^T(x)C\gamma_\mu d_b(x))\gamma_5\gamma^\mu u_c(x) \tag{5}$$

$$\bar{\eta}_\Theta(0) = -\varepsilon^{abc}\varepsilon^{def}\varepsilon^{cfg}s_g^T(0)C\times[\bar{d}_e(0)\gamma_5 C\bar{u}_d^T(0)][\bar{d}_b(0)C\bar{u}_a^T(0)]. \tag{6}$$

Inserting the currents into (1), the resulting expression involves the quark propagator in the presence of quark and gluon condensates, which is known from previous studies. Using it we arrive at a final complicated expression for the correlator represented schematically by the sum of the diagrams (OPE) of Fig. 1.

Let us consider the phenomenological side (2) and, following [13], rewrite it generically as $\Gamma(q^2,p^2,p'^2) = \Gamma_{pp} + \Gamma_{pc1} + \Gamma_{pc2} + \Gamma_{cc}$ where $\Gamma_{pp}(q^2,p^2,p'^2)$ stands for the pole-pole part and reads

$$\Gamma_{pp} = \frac{-g_{\Theta nK}\lambda_\Theta\lambda_N\lambda_K}{(p^2-m_\Theta^2)(p'^2-m_N^2)(q^2-m_K^2)} \tag{7}$$

The continuum-continuum term Γ_{cc} can be obtained as usual, with the assumption of quark-hadron duality [11]. The pole-continuum transition terms are contained in Γ_{pc1} and Γ_{pc2}. They can be explicitly written as a double dispersion integral:

$$\Gamma_{pc1} = \int_{m_{K*}^2}^{\infty} \frac{b_1(u,p^2)\,du}{(m_N^2-p'^2)(u-q^2)} \qquad \Gamma_{pc2} = \int_{m_{N*}^2}^{\infty} \frac{b_2(s,p^2)\,ds}{(m_K^2-q^2)(s-p'^2)} \tag{8}$$

We shall assume that the unknown functions b_1 and b_2 have the following form:

$$b_1(u,p^2) = \widetilde{b_1}(u)\int_{m_\Theta^2}^{\infty} d\omega \frac{b_1(\omega)}{\omega-p^2} \qquad b_2(s,p^2) = \widetilde{b_2}(s)\int_{m_\Theta^2}^{\infty} d\omega \frac{b_2(\omega)}{\omega-p^2} \tag{9}$$

with continuous functions $b_{1,2}(w)$, starting from m_Θ^2. This is our **parametrization A**. The functions $\widetilde{b_1}(u)$ and $\widetilde{b_2}(s)$ describe the excitation spectra of the kaon and the nucleon, respectively. After Borel transform, the pole-continuum term contains one unknown constant factor which can be determined from the sum rules.

We shall now explicitly force the phenomenological side to contain only the pole part of the Θ, both in the pole-pole term and in the pole-continuum terms. This can formally be done by choosing $b_1(\omega) = b_2(\omega) = \delta(\omega - m_\Theta^2)$ in (9). This is our **parametrization B**. In this case we have the Θ in the ground state. Again, in the final expressions this gives additional constants which can be calculated.

The sum rule may be written identifying the phenomenological and theoretical descriptions of the correlation function. In the case of the three-point function considered here, there are two independent momenta and we may perform either a single or a double Borel transform. We shall consider the choices:

$$\textbf{(I)} \quad q^2 = 0 \quad p^2 = p'^2 \qquad \textbf{(II)} \quad q^2 \neq 0 \quad p^2 = p'^2 \tag{10}$$

In (I) we perform a single Borel transform: $p^2 = -P^2$ and $P^2 \to M^2$. In this case we take $m_K^2 \simeq 0$ and single out the $1/q^2$-terms. In (II) we perform two Borel transforms: $p^2 = -P^2$ and $P^2 \to M^2$ and also $q^2 = -Q^2$ and $Q^2 \to M'^2$.

Introducing the notation $G = -g_{\Theta nK}\lambda_\Theta\lambda_N\lambda_K$ and using (I) and (II) we can obtain a set of sum rules. With method (II) we obtain:

$$G e^{-m_K^2/M'^2} \frac{e^{-m_\Theta^2/M^2} - e^{-m_N^2/M^2}}{m_\Theta^2 - m_N^2} + \Gamma_{pc2}(M^2, M'^2) = \int_0^{u_0} du \int_0^{s_0} ds\, \rho_{th}\, e^{-s/M^2}\, e^{-u/M'^2} \tag{11}$$

with $\Gamma_{pc2}(M^2)$ being equal to:

$$= A e^{-m_K^2/M'^2} e^{-m_{N^*}^2/M^2} \quad \text{(param. A)} \qquad = A e^{-m_K^2/M'^2} e^{-m_\Theta^2/M^2} \quad \text{(param. B)} \tag{12}$$

With method (I) we get a quite similar expression. It may be obtained from (11) and (12) setting $m_K = 0$, $u = 0$ and removing the integration over u. In both parametrizations the term Γ_{pc1} is exponentially suppressed and, as discussed in [13], has been neglected. A is an unknown constant and can be determined from the sum rules. In the above expressions ρ_{th} is the double discontinuity computed directly from the theoretical (OPE) description of the correlation function (see [11] for details), s_0 is the continuum threshold of the nucleon defined as $s_0 = (m_N + \Delta_N)^2$ and u_0 is the continuum threshold of the kaon defined as $u_0 = (m_K + \Delta_K)^2$. The hadronic masses are $m_N = 938$ MeV, $m_{N^*} = 1440$ MeV, $m_K = 493$ MeV and $m_\Theta = 1540$ MeV. For each sum rule we can take the derivative with respect to $1/M^2$ and in this way obtain a second sum rule. In each case we have thus a system of two equations and two unknowns (G and A) which can then be easily solved.

In the numerical analysis of the sum rules we use the following values for the condensates: $\langle\bar{q}q\rangle = -(0.23 \pm 0.02)^3\,\text{GeV}^3$, $\langle\bar{s}s\rangle = 0.8\,\langle\bar{q}q\rangle$, $< \bar{s}g_s\sigma.\mathbf{G}s >= m_0^2\,\langle\bar{s}s\rangle$ with $m_0^2 = 0.8\,\text{GeV}^2$ and $\langle g_s^2 G^2\rangle = 0.5\ \text{GeV}^4$. The gluon condensate has a large error of about a factor 2, but its influence on the analysis is relatively small. The couplings constants λ_N and λ_Θ are taken from the corresponding two-point functions and are $\lambda_N = (2.4 \pm 0.2) \times 10^{-2}\,\text{GeV}^3$ and $\lambda_\Theta = (2.4 \pm 0.3) \times 10^{-5}\,\text{GeV}^6$. The coupling λ_K is calculated with $f_K = 160$ MeV, $m_s = 100$ MeV and $m_u = 5$ MeV and is $\lambda_K = 0.37\,\text{GeV}^2$

In Fig. 1, among these OPE diagrams there are two distinct subsets. In the first (from 1a to 1g) there is no gluon line connecting the "petals" and therefore no color exchange.

A diagram of this type we call color-disconnected. In this case the pentaquark has a component similar to a $K-n$ molecule. In the second subset of diagrams (1h, 1i and 1j) we have color exchange and the pentaquark is a genuine 5-quark state with a non-trivial color structure. We may call this type of diagram a color-connected (CC) one. In our analysis we write sum rules for both cases: all diagrams and only color-connected. We work in the Borel window given by $1\,\text{GeV}^2 < M^2, M'^2 < 1.5\,\text{GeV}^2$. We give the values of the coupling extracted at $M^2 = 1.5$ GeV^2 and $M'^2 = 1$ GeV^2 in Table I. We present here our results for the coupling constant $g_{\Theta nK}$ obtained with the color connected diagrams only. In Table I we present a summary of our results for $g_{\Theta nK}$. For the continuum thresholds we have employed $\Delta_N = \Delta_K = 0.5$ GeV.

Table I:$g_{\Theta nK}$ for various cases

case	$\lvert g_{\Theta nK}\rvert$ (CC)	$\lvert g_{\Theta nK}\rvert$ (all diagrams)
I A	0.71	2.59
II A	0.82	3.59
I B	0.84	3.24
II B	0.96	4.48

Considering the uncertainties in the continuum thresholds, in the coupling constants λ and in the quark condensate we get an uncertainty of about 50%. Our final result then reads:

$$\begin{aligned} |g_{\Theta nK}|(\text{all diagrams}) &= 3.48 \pm 1.8\,, \\ |g_{\Theta nK}|(\text{CC}) &= 0.83 \pm 0.42\,. \end{aligned} \tag{13}$$

Including all diagrams, the prediction for Γ_Θ is then 13 MeV (652 MeV) for a positive (negative) parity pentaquark. In the CC case we get a width of 0.75 MeV (37 MeV) for a positive (negative) parity pentaquark. In conclusion, our main results are given in eq. (13). We find that for a positive parity pentaquark a width much smaller than 10 MeV would indicate a pentaquark which contains no color-singlet meson-baryon contribution. For a negative parity pentaquark, even under the assumption that it is a genuine 5-quark state, we can not explain the observed narrow width of the Θ.

This work was supported by CNPq and FAPESP.

REFERENCES

1. F. Close, Nature **435**, (2005) 287; M. Ostrick, Prog. Part. Nucl. Phys. **55**, (2005) 337; D. Barna, Acta Phys. Hung. A **22** (2005) 187; M. Danilov, hep-ex/0509012. K. Hicks, AIP Conf.Proc. 756, (2005) 195.
2. For recent reviews see: M. Karliner, Int. J. Mod. Phys. A **20**, (2005) 199; Shi-Lin Zhu, Int. J. Mod. Phys. A **20**, (2005) 1548; M. Oka, J. Sugiyama and T. Doi, Nucl. Phys. A **755**, (2005) 391; A. Hosaka, hep-ph/0506138; S. H. Lee, H. Kim and Y. s. Oh, J. Korean Phys. Soc. **46**, (2005) 774.

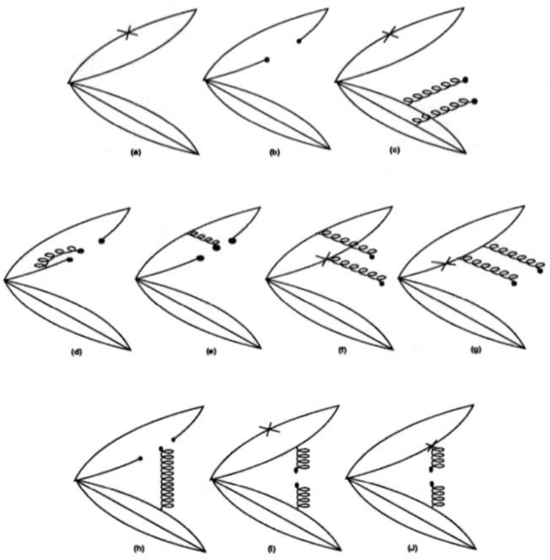

h

FIGURE 1. The main diagrams which contribute to the theoretical side of the sum rule in the relevant structure. a) - g) are the color-disconnected diagrams, whereas h) - j) are the color-connected diagrams. The cross indicates the insertion of the strange mass.

3. M. A. Shifman, A. I. Vainshtein and V. I. Zakharov, Nucl. Phys. B **147**, (1979) 385; Nucl. Phys. B **147**, (1979) 448; L. J. Reinders, H. Rubinstein and S. Yazaki, Phys. Rept. **127**, (1985) 1; S. Narison, World Sci. Lect. Notes Phys. **26**, (1989) 1, H. G. Dosch, M. Jamin and S. Narison, Phys. Lett. B **220**, (1989) 251.
4. S.-L. Zhu, Phys. Rev. Lett. **91**, (2003) 232002.
5. R.D. Matheus, F.S. Navarra, M. Nielsen, R. Rodrigues da Silva and S.H. Lee, Phys. Lett. **B578**, (2004) 323; R.D. Matheus, F.S. Navarra, M. Nielsen, R. Rodrigues da Silva, Phys. Lett. **B602**, (2004) 185.
6. J. Sugiyama, T. Doi, M. Oka, Phys. Lett. **B581**, (2004) 167.
7. M. Eidemüller, Phys. Lett. **B597**, (2004) 314.
8. Y. Kondo, O. Morimatsu, T. Nishikawa, Phys. Lett. **B611**, (2005) 93; Y. Kwon, A. Hosaka and S. H. Lee, hep-ph/0505040.
9. B. L. Ioffe and A. G. Oganesian, JETP Lett. **80**, (2004) 386; A. G. Oganesian, hep-ph/0410335.
10. R.D. Matheus, S. Narison, hep-ph/0412063.
11. M. Eidemuller, F. S. Navarra, M. Nielsen and R. Rodrigues da Silva, Phys. Rev. D **72**, (2005) 034003
12. H. Kim, S.H. Lee and M. Oka, Phys. Lett. **B453** (1999) 199; Phys. Rev. **D60**, (1999) 034007; M.E. Bracco, F.S. Navarra, M. Nielsen, Phys. Lett. **B454**, (1999) 346.
13. B. Ioffe and A. Smilga, Nucl. Phys. **B232**, (1984) 109.

Observation of X(1835) and Multiquark Candidates at BESII

Shan JIN for the BES Collaboration

Institute of High Energy Physics, CAS, Beijing 100049, China

Abstract.
In this talk, I would review the new observation of X(1835), which has properties inconsistent with any know mesons, but consistent with expectations for the state that produces the strong $p\bar{p}$ mass threshold enhancement observed in the $J/\psi \to \gamma p\bar{p}$ process at BESII. I will also talk about some other multiquark candidates which are indicated by $p\overline{\Lambda}$ and $K^-\overline{\Lambda}$ mass threshold enhancements.

Keywords: X(1835), Multiquark State, Threshold Enhancement
PACS: 12.39.Mk, 13.75.Ev, 12.40.Yx, 13.20.Gd

OBSERVATION OF X(1835) IN $J/\psi \to \gamma\pi^+\pi^-\eta'$ AT BESII

An anomalous enhancement near the mass threshold in the $p\overline{p}$ invariant mass spectrum from $J/\psi \to \gamma p\overline{p}$ decays was reported by the BES II experiment [1]. This enhancement was fitted with a sub-threshold S-wave Breit-Wigner resonance function with a mass $M = 1859^{+3+5}_{-10-25}$ MeV/c^2, a width $\Gamma < 30$ MeV/c^2 (at the 90% C.L.) and a product branching fraction (BF) $B(J/\psi \to \gamma X) \cdot B(X \to p\overline{p}) = (7.0 \pm 0.4(stat)^{+1.9}_{-0.8}(syst)) \times 10^{-5}$. This surprising experimental observation has stimulated a number of theoretical speculations [2, 3, 4, 5, 6, 7] and motivated further investigations on baryon-antibaryon mass threshold structures, which led to the subsequent experimental observation of a strong $p\overline{\Lambda}$ mass threshold enhancement in $J/\psi \to pK^-\overline{\Lambda}$ decay [8]. Among various theoretical interpretations of the $p\overline{p}$ mass threshold enhancement, the most intriguing one is that of a $p\overline{p}$ bound state, sometimes called *baryonium* [2, 5, 9], which has been the subject of many experimental searches [10]. The baryonium interpretation of the $p\overline{p}$ mass enhancement requires a new resonance with a mass around 1.85 GeV/c^2 and it would be supported by the observation of the resonance in other decay channels. Possible decay modes for a $p\overline{p}$ bound state, suggested in Ref. [4, 5], include $\pi^+\pi^-\eta'$.

The results reported here are based on a sample of 5.8×10^7 J/ψ decays detected with the upgraded Beijing Spectrometer (BESII) at the Beijing Electron-Positron Collider (BEPC). The analysis is described in detail in Ref. [11]. For the $J/\psi \to \gamma\pi^+\pi^-\eta'(\eta' \to \pi^+\pi^-\eta, \eta \to \gamma\gamma)$ channel, Figure 1(b) shows the $\pi^+\pi^-\eta$ invariant mass distribution , where a clear η' signal is visible. The $\pi^+\pi^-\eta'$ invariant mass spectrum for the selected events is shown in Fig. 1(c), where a peak at a mass around 1835 MeV/c^2 is observed. For the $J/\psi \to \gamma\pi^+\pi^-\eta'(\eta' \to \gamma\rho)$ channel, the $\gamma\pi^+\pi^-$ invariant mass distribution shows an η' signal (Fig. 2(b)). For this channel there is also a distinct peak near 1835 MeV/c^2 in the $\pi^+\pi^-\eta'$ invariant mass spectrum (Fig. 2(c)).

CP814 *Hadron Spectroscopy: Eleventh International Conference*
edited by A. Reis, C. Göbel, J. Sá Borges, and J. Magnin

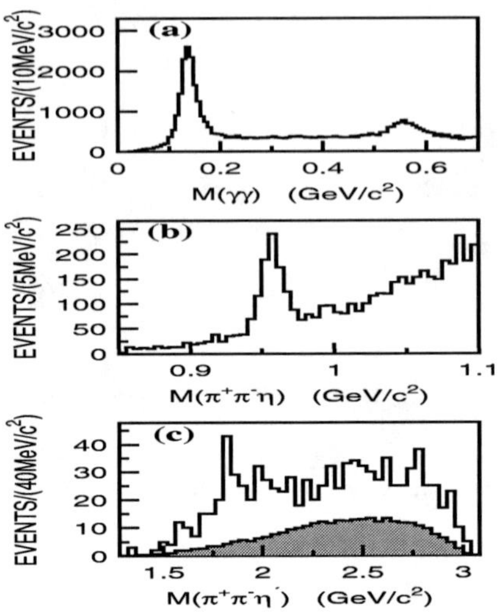

FIGURE 1. Invariant mass distributions for selected $J/\psi \to \gamma\pi^+\pi^-\eta'(\eta' \to \pi^+\pi^-\eta, \eta \to \gamma\gamma)$ candidate events. (a) The invariant mass distribution of $\gamma\gamma$ pairs. (b) The $\pi^+\pi^-\eta$ invariant mass distribution. (c) The $\pi^+\pi^-\eta'$ invariant mass distributions; the open histogram is data and the shaded histogram is $J/\psi \to \gamma\pi^+\pi^-\eta'$ phase-space MC events (with arbitrary normalization).

The combined spectrum (the sum of the histograms in Figs. 1(c) and 2(c))) is fitted with a Breit-Wigner (BW) function convolved with a Gaussian mass resolution function (with $\sigma = 13$ MeV/c^2) to represent the $X(1835)$ signal plus a smooth polynomial background function. The mass and width obtained from the fit (shown in the bottom panel of Fig. 3) are $M = 1833.7 \pm 6.1$ MeV/c^2 and $\Gamma = 67.7 \pm 20.3$ MeV/c^2. The signal yield from the fit is 264 ± 54 events with a confidence level 45.5% ($\chi^2/d.o.f. = 57.6/57$) and $-2\ln L = 58.4$. A fit to the mass spectrum without a BW signal function returns $-2\ln L = 126.5$. The change in $-2\ln L$ with $\Delta(d.o.f.) = 3$ corresponds to a statistical significance of 7.7 σ for the signal.

Using MC-determined selection efficiencies of 3.72% and 4.85% for the $\eta' \to \pi^+\pi^-\eta$ and $\eta' \to \gamma\rho$ modes, respectively, we determine a product BF of

$$B(J/\psi \to \gamma X(1835)) \cdot B(X(1835) \to \pi^+\pi^-\eta') = (2.2 \pm 0.4) \times 10^{-4}.$$

The mass and width of the $X(1835)$ are not compatible with any known meson resonance [12]. We examined the possibility that the $X(1835)$ is responsible for the $p\bar{p}$ mass threshold enhancement observed in radiative $J/\psi \to \gamma p\bar{p}$ decays [1]. It has been pointed out that the S-wave BW function used for the fit in Ref. [1] should be modified to include the effect of final-state-interactions (FSI) on the shape of the $p\bar{p}$ mass spectrum [6, 7]. Redoing the S-wave BW fit to the $p\bar{p}$ invariant mass spectrum of Ref. [1] including the zero Isospin, S-wave FSI factor of Ref. [7], yields a mass $M = 1831 \pm 7$ MeV/c^2 and a width $\Gamma < 153$ MeV/c^2 (at the 90% C.L.); these values are in good agreement with the mass and width of $X(1835)$ reported here. Moreover, according to Ref. [5], the $\pi\pi\eta'$ decay mode is expected to be strong for a $p\bar{p}$ bound state. Thus, the $X(1835)$ resonance

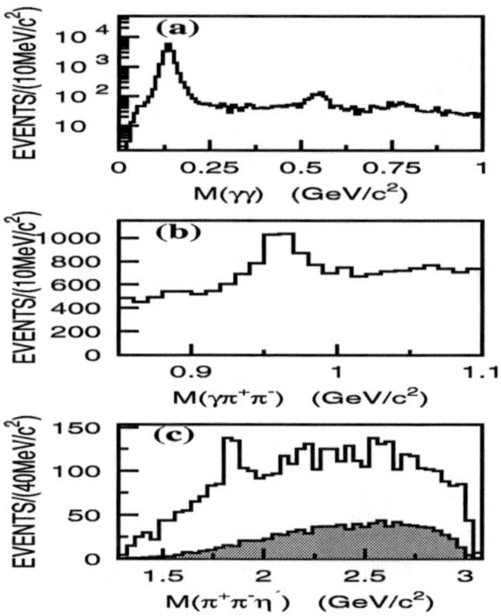

FIGURE 2. Invariant mass distributions for the selected $J/\psi \to \pi^+\pi^-\eta'(\eta' \to \gamma\rho)$ candidate events, (a) The invariant mass distribution for $\gamma\gamma$ pairs. (b) The $\gamma\pi^+\pi^-$ invariant mass distribution. (c) The $\pi^+\pi^-\eta'$ invariant mass distributions: the open histogram is data and the shaded histogram is from $J/\psi \to \gamma\pi^+\pi^-\eta'$ phase-space MC events (with arbitrary normalization).

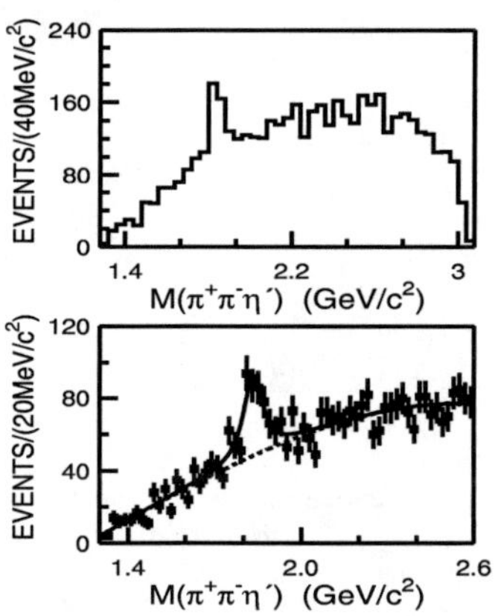

FIGURE 3. The $\pi^+\pi^-\eta'$ invariant mass distribution for selected events from both the $J/\psi \to \gamma\pi^+\pi^-\eta'(\eta' \to \pi^+\pi^-\eta, \eta \to \gamma\gamma)$ and $J/\psi \to \gamma\pi^+\pi^-\eta'(\eta' \to \gamma\rho)$ analyses. The bottom panel shows the fit (solid curve) to the data (points with error bars); the dashed curve indicates the background function.

is a prime candidate for the source of the $p\bar{p}$ mass threshold enhancement in $J/\psi \to \gamma p\bar{p}$ process. In this case, the J^{PC} and I^G of the $X(1835)$ could only be 0^{-+} and 0^+, which can be tested in future experiments. Also in this context, the relative $p\bar{p}$ decay strength is quite strong: $B(X \to p\bar{p})/B(X \to \pi^+\pi^-\eta') \sim 1/3$ [The product BF determined from the fit that includes FSI effects on the $p\bar{p}$ mass spectrum is within the systematic errors of the result reported in Ref. [1].]. Since decays to $p\overline{p}$ are kinematically allowed only for a small portion of the high-mass tail of the resonance and have very limited phase space, the large $p\overline{p}$ branching fraction implies an unusually strong coupling to $p\overline{p}$, as expected for a $p\overline{p}$ bound state [9, 13]. However, other possible interpretations of the $X(1835)$ that have no relation to the $p\overline{p}$ mass threshold enhancement are not excluded.

OBSERVATION OF $P\overline{\Lambda}$ MASS THRESHOLD ENHANCEMENT IN $J/\psi \rightarrow PK^-\overline{\Lambda}$

The detailed analysis is described in Ref. [8]. The $p\overline{\Lambda}$ invariant mass spectrum for the selected events is shown in Fig. 4(a), where an enhancement is evident near the mass threshold. No corresponding structure is seen in a sample of $J/\psi \rightarrow pK^-\overline{\Lambda}$ MC events generated with a uniform phase space distribution. The $pK^-\overline{\Lambda}$ Dalitz plot is shown in Fig. 4(b). In addition to bands for the well established $\Lambda^*(1520)$ and $\Lambda^*(1690)$, there is a significant N^* band near the $K^-\overline{\Lambda}$ mass threshold, and a $p\overline{\Lambda}$ mass enhancement, isolated from the Λ^* and N^* bands, in the right-upper part of the Dalitz plot.

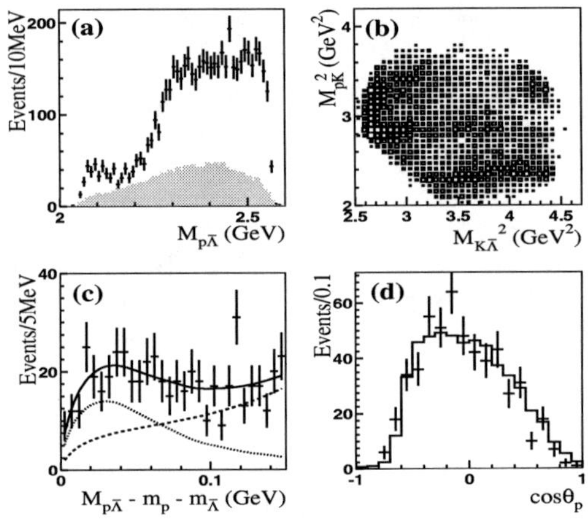

FIGURE 4. (a) The points with error bars indicate the measured $p\overline{\Lambda}$ mass spectrum; the shaded histogram indicates phase space MC events (arbitrary normalization). (b) The Dalitz plot for the selected event sample. (c) A fit (solid line) to the data. The dotted curve indicates the Breit-Wigner signal and the dashed curve the phase space 'background'. (d) The $\cos\theta_p$ distribution under the enhancement, the points are data and the histogram is the MC (normalized to data)

This enhancement can be fit with an acceptance weighted S-wave Breit-Wigner function, together with a function describing the phase space contribution, as shown in Fig. 4(c). The fit gives a peak mass of $m = 2075 \pm 12$ MeV and a width $\Gamma = 90 \pm 35$ MeV. The enhancement deviates from the shape of the phase space contribution with a statistical significance of about 7σ.

The fit yields $N_{res} = 238 \pm 57$ signal events, corresponding to a branching ratio

$$BR(J/\psi \rightarrow K^- X)BR(X \rightarrow p\overline{\Lambda}) = (5.9 \pm 1.4) \times 10^{-5}.$$

Searching for the same enhancement in $K\pi$ and $K\pi\pi$ modes in the $J/\psi, \psi' \to K K\pi, K K\pi\pi$ processes would help to confirm the presence of this anomalous peak and understand its nature, and in particular distinguish whether it is from a conventional K^* meson, a possible multi-quark state, or a resonance in the baryon-antibaryon interaction. If its decay widths to $K\pi$ and $K\pi\pi$ modes are much smaller than to $p\overline{\Lambda}$ modes, the interpretation as conventional K^* meson would be disfavored.

OBSERVATION OF $K^-\overline{\Lambda}$ MASS THRESHOLD ENHANCEMENT IN $J/\psi \to PK^-\overline{\Lambda}$

In the above Dalitz plot (Fig. 4(b)), a clear band, is observed near the $K^-\overline{\Lambda}$ mass threshold. We have performed very preliminary partial wave analysis (PWA) and obtained: The mass of this threshold structure N_X^* is in the range of 1500 MeV to 1650 MeV, the width is about 70 MeV to 110 MeV and its spin-parity favors $1/2^-$. In particular, it has a braching ratio $B(J/\psi \to \overline{p}N_X^*)B(N_X^* \to K^+\Lambda)$ larger than 2×10^{-4}.

Considering the mass of N_X^* is below or very close to $K^-\overline{\Lambda}$ threshold, i.e., the phase space producing $K^-\overline{\Lambda}$ final state is very small, the large branching ratio to $K^-\overline{\Lambda}$ indicates N_X^* has very strong coupling to $K^-\overline{\Lambda}$, suggesting that it could be the $K^-\overline{\Lambda}$ resonant state predicted by chiral SU(3) quark model[14].

ACKNOWLEDGMENTS

This work is supported in part by the National Natural Science Foundation of China under contracts No. 10425523, the 100 Talents Program of CAS under Contract No. U-24, and the Knowledge Innovation Project of CAS under Contract No. KJCX2-SW-N10.

REFERENCES

1. BES Collaboration, J.Z. Bai *et al.*, Phys. Rev. Lett. **91**, 022001 (2003).
2. A. Datta, P.J. O'Donnell, Phys. Lett. **B567**, 273 (2003); M.L. Yan *et al.*, hep-ph/0405087; B. Loiseau, S. Wycech, hep-ph/0502127.
3. J. Ellis, Y. Frishman and M. Karliner, Phys. Lett. **B566**, 201 (2003); J.L. Rosner, Phys. Rev. D **68**, 014004 (2003).
4. C.S. Gao and S.L. Zhu, Commun. Theor. Phys. 42, 844 (2004), hep-ph/0308205.
5. G.J. Ding and M.L. Yan, Phys. Rev. C **72**, 015208 (2005).
6. B.S. Zou and H.C. Chiang, Phys. Rev. D **69**, 034004 (2003).
7. A. Sibirtsev *et al.*, Phys. Rev. D **71**, 054010 (2005).
8. BES Collaboration, M. Ablikim *et al.*, Phys. Rev. Lett. **93**, 112002 (2004).
9. I.S. Shapiro, Phys. Rept. **35**, 129 (1978); C.B. Dover, M. Goldhaber, Phys. Rev. D **15**, 1997 (1977).
10. For recent reviews of this subject, see E. Klempt *et al.*, Phys. Rep. **368**, 119 (2002) and J-M. Richard, Nucl. Phys. Proc. Suppl. **86**, 361 (2000).
11. BES Collaboration, M. Ablikim *et al.*, hep-ex/0508025.
12. Particle Data Group, S. Eidelman *et al.*, Phys. Lett. **B592**, 1 (2004).
13. S.L. Zhu and C.S. Gao, hep-ph/0507050.
14. F. Huang and Z.Y. Zhang, Phys.Rev C **72**, 024003 (2005), nucl-th/0507025.

Search for CHARM pentaquarks in ZEUS at HERA[1]

Yehuda Eisenberg[2]

Weizmann Institute of Science,
Rehovot, Israel

Abstract. Using the full 1996 - 2000 ZEUS data at HERA (121 pb^{-1}) we have searched for the $\Theta_c^0(3100)$ pentaquark in the photoproduction and DIS regime. The search has yielded negative results. The 95% C.L. upper limits on the visible rate $R(\Theta_c^0 \to D^* p/D^*)$ is 0.23% (0.35% for DIS).

Keywords: charmed pentaquarks
PACS: 12.39

INTRODUCTION: PENTAQUARKS

Observation of a narrow exotic baryon with strangeness +1 around 1530 MeV decaying into K^+n was reported during the last 2 years by some low energy experiments [1]. It was suggested that these are due to the $\Theta^+ = uudd\bar{s}$ pentaquark candidate predicted by Diakonov et al. [2], at the top of a $SU(3)$ spin 1/2 anti-decuplet of baryons. Narrow peaks were also seen at a similar mass in the final state $K_S^0 p$, which is exotic if the K_S^0 strangeness is +1. In high energy experiments ZEUS searched for the $\Theta^+(1530) \to K_S^0 p$ [3] in the DIS regime ($Q^2 > 20$ GeV^2) and reported observation of the $\Theta^+(1530)$. See fig.1 for a recent update of the ZEUS results indicating that Θ^+ is mainly produced in the proton fragmentation region. This is contrary to the ZEUS observation [3] that other known states [4] like the $\Lambda(1520)$, are produced equally in positive and negative η. Thus the ZEUS observation of $\Theta^+(1530)$ represents perhaps a very special production mechanism.

It should be noted however that other high energy experiments at LEP, BaBar, CDF and FOCUS searched and have not seen the $\Theta^+(1530)$ pentaquark [5]. Recently some theoretical arguments have been presented [6] suggesting possible reasons why the $\Theta^+(1530)$ can be seen only in very specific cases.

SEARCH FOR CHARM PENTAQUARK DECAYING TO $D^{*\pm}P^{\mp}$

The existence of the strange pentaquark Θ^+ suggests that charmed pentaquarks, $\Theta_c^0 = uudd\bar{c}$, may also exist [7], [8].

---

[1] Supported by the Israel Science Foundation and the U.S.-Israel Bi-national Science Foundation
[2] HADRON 05 conference , on behalf of the ZEUS Collaboration

CP814 *Hadron Spectroscopy: Eleventh International Conference*
edited by A. Reis, C. Göbel, J. Sá Borges, and J. Magnin

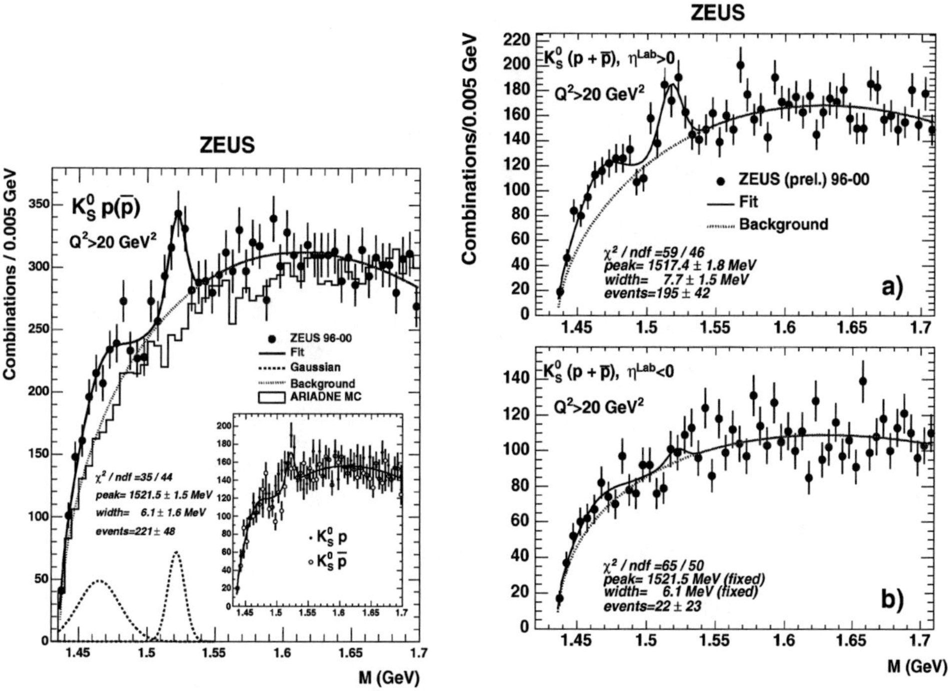

FIGURE 1. Left: $K^0_S(p+\bar{p})$ distribution from original ZEUS paper, for the DIS sample, $Q^2 > 20 GeV^2$. Right: same distributions separate for $\eta > 0$ and $\eta < 0$

The H1 Collaboration reported [9] observation of a narrow signal in the $D^{*\pm}p^{\mp}$ mass spectrum at 3.1 GeV with a width consistent with the detector resolution. The signal was seen in a DIS sample of ≈ 3400 $D^{*\pm} \to D^0\pi^{\pm} \to (K^{\mp}\pi^{\pm})\pi^{\pm}$ with a rate of $\approx 1.5\%$ of the visible D^* production. A less clean signal of a comparable rate was seen also in the H1 photoproduction (PHP) sample.

The Θ^0_c search of ZEUS in the $D^{*\pm}p^{\mp}$ mode was performed with the full HERA-I data [10]. Clean $D^{*\pm}$ signals were seen in the $\Delta M = M(D^{*\pm}) - M(D^0)$ plots (Fig. 2 left). Two $D^{*\pm} \to D^0\pi^{\pm}$ decay channels were used with $D^0 \to K^{\mp}\pi^{\pm}$ and $D^0 \to K^{\mp}\pi^{\pm}\pi^{+}\pi^{-}$. The Θ^0_c search was performed in the kinematic range $|\eta(D^*)| < 1.6$ and $p_T\ (D^*) > 1.35\ (2.8)$ GeV and with ΔM values between $0.144 - 0.147 (0.1445 - 0.1465)$ GeV for the $K\pi\pi$ ($K\pi\pi\pi\pi$) channel. The shaded bands in Fig.2a,b contain a total of ≈ 62000 D^*'s after subtracting wrong-charge combinations with charge ± 2 for the D^0 candidate. Selecting DIS events with $Q^2 > 1$ GeV^2 yielded smaller, but cleaner D^* signals with a total of ≈ 13500 D^*'s (Fig. 2c-d left).

Protons were selected with $p_T(p) > 0.15$ GeV. To reduce the pion and kaon background, a parameterisation of the expected dE/dx as a function of P/m was obtained using tagged protons from Λ decays and tagged pions from K^0_S decays. The χ^2 probability of the proton hypothesis was required to be above 0.15. Fig. 3 shows the

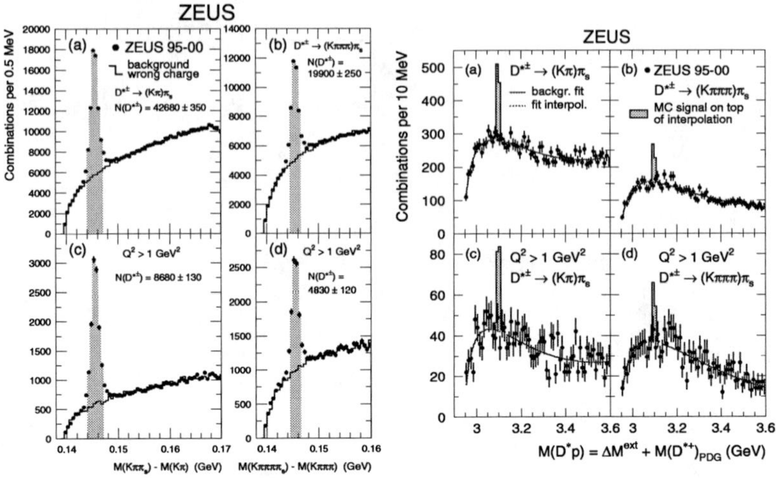

FIGURE 2. Left: ΔM distributions (dots) for (a) $D^* \to K\pi\pi$ and (b) $D^* \to K\pi\pi\pi\pi$ candidates. Events with $Q^2 > 1$ GeV^2 for the two channels, respectively, are shown in (c) and (d). The histograms are for wrong charge combinations. Right: $M(D^{*\pm}p^{\mp})$ distributions (dots) for the same samples. Solid curves are fits to a background function (see text). Shaded historgams are MC Θ_c^0 signals, normalised to $\Theta_c^0/D^* = 1\%$, on top of the background fit.

$M(D^*p) = M(K\pi\pi p) - M(K\pi\pi) + M(D^*)_{PDG}$ distributions for the $K\pi\pi$ channel for the full (left) and the DIS (right) samples, where $M(D^*)_{PDG}$ is the $D^{*\pm}$ mass [4]. In the low-P selection (Fig. 3b), a clean proton sample separated from the π and K dE/dx bands was obtained by taking only tracks with $P < 1.35$ GeV and $dE/dx > 1.3$ mips. In the high-P selection (Fig. 3c) only tracks with $P(p) > 2$ GeV were used. The latter selection was prompted by the H1 observation [9] of a better Θ_c^0 signal-to-background ratio for high proton momenta. No narrow signal is seen in either the $K\pi\pi$ (Fig. 3) or in the $K\pi\pi\pi\pi$ (Fig. 2b,d right,Fig.4a,b left) channel. The $K\pi\pi$ analysis was repeated using very similar selection criteria as in the H1 analysis [9]. No narrow resonance was seen in either the DIS or the PHP event sample [10] (Fig.4a,b right).

95% C.L. upper limits on the fraction of D^* mesons originating from Θ_c^0 decays, $R(\Theta_c^0 \to D^*p/D^*)$, were calculated in a signal window $3.07 < M(D^*p) < 3.13$ GeV for the $K\pi\pi$ and $K\pi\pi\pi\pi$ channels. A visible rate of 1% for this fraction (Fig. 2 right), as claimed by H1 [9], is excluded by 9σ (5σ) for the full (DIS) combined sample. The $M(D^*p)$ distributions were fitted to the form $x^a e^{-bx+cx^2}$, where $x = M(D^*p) - M(D^*) - m_p$ and a,b,c are constants (Fig. 2 right). The number of reconstructed Θ_c^0 baryons was estimated by subtracting in the signal window the background function from the observed number of events, yielding $R(\Theta_c^0 \to D^*p/D^*) < 0.23\%$ and $< 0.35\%$ for the full and DIS combined two channels. The acceptance-corrected rates are, respectively, 0.37% and 0.51%. The 95% C.L. upper limit on the fraction of charm quarks fragmenting to Θ_c^0 times the branching ratio $\Theta_c^0 \to D^*p$ for the combined two channels is $f(c \to \Theta_c^0) \cdot B_{\Theta_c^0 \to D^*p} < 0.16\%$ ($< 0.19\%$) for the full (DIS) sample.

We conclude that the ZEUS data are not compatible with the H1 result of $\approx 1.5\%$ rate

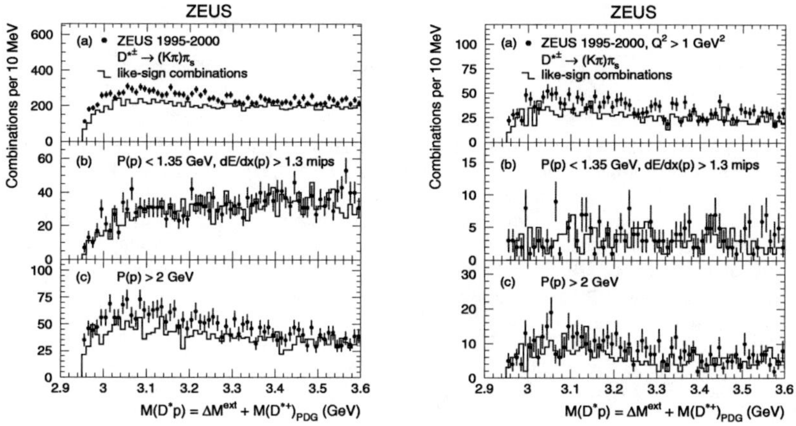

FIGURE 3. Left: $M(D^{*\pm}p^{\mp})$ distributions for the $K\pi\pi$ channel (dots) with (a) all proton candidates, (b) candidates with $P(p) < 1.35$ GeV and $dE/dx > 1.3$, and (c) candidates with $P(p) > 2$ GeV. Histograms show the $M(D^{*\pm}p^{\pm})$ like-sign combinations. Right: Same for DIS events with $Q^2 > 1$ GeV^2.

of $\Theta_c^0 \rightarrow D^*p/D^*$. Such a rate is excluded by more then 9σ for the full data and 5σ for the ZEUS DIS data. Further experiments perhaps with HERA II may help resolve this discrepancy. No other experiment reported the observation of Θ_c^0 [5].

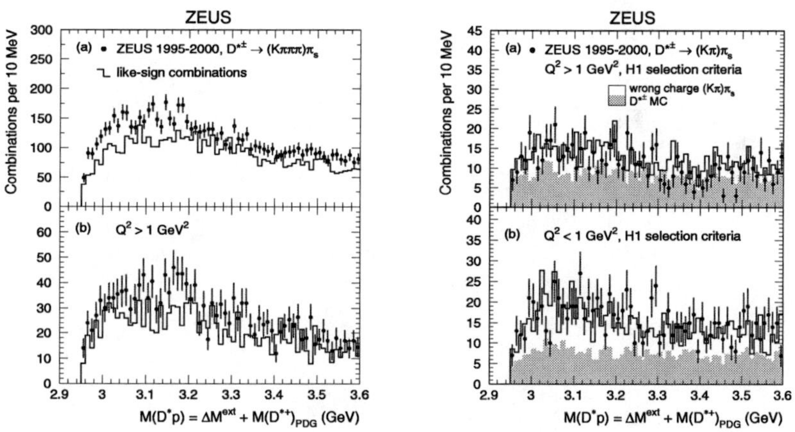

FIGURE 4. Left: $M(D^{*\pm}p^{\mp})$ distributions for the $K\pi\pi\pi\pi$ channel (dots) with (a) full sample, (b) DIS sample, $Q^2 > 1GeV^2$. Right: same distributions for the $K\pi\pi$ channel for DIS ($Q^2 > 1$ GeV^2) and photoproduction using the H1 event selection criteria.

REFERENCES

1. CLAS Coll., V. Kubarovsky et al, Phys. Rev. Lett. 92, 032001 (2004); SAPHIR Coll., J.Narth et al, Phys. Lett. B572,127 (2003). See new CLAS results reported at this conference.
2. D. Diakonov, V. Petrov and M.V. Polyakov, *Z. Phys.* **A359**, 305 (1997).
3. ZEUS Coll., S. Chekanov et al., *Phys. Lett.* **B591**, 7 (2004). See update of ZEUS results reported in Uppsala, XXII Lepton-Photon Interactions Symposium, paper 290, Uppsata July 2005.
4. Particle Data Group, K. Hagiwara et al., *Phys. Rev.* **D66**, 10001 (2002).
5. See reports in this conference: D. Muller, Hadron physics at BaBar ; E. Smith, Review on JLAB Results ; D. Tedeschi, Pentaquarks at CLAS ; J. Cumalat, Results on Baryon Spectroscopy. Aleph Coll.,S. Schael et al, Phys. Lett. B 599,1 (2004); CDF Coll., I. V. Gorelov Preprint (hep-ex/0408025); See also references in K.Hicks,hep-ex/0504027.
6. M. Karliner and H. J .Lipkin, preprint (hep-ph/0506084)
7. A. Jaffe and F. Wilczek, , *Phys. Rev. Lett.* **91**, 232003 (2003).
8. M. Karliner and H.J. Lipkin, Preprint hep-ph/0307343 (2003).
9. H1 Coll., C. Atkas et al., , *Phys. Lett.* **B588**, 17 (2004).
10. ZEUS Coll., S. Chekanov et al., *Eur.Phys. J.* **C38** (2004)29 .

The case of the pseudoscalar glueball

A. Masoni

INFN - Sezione di Cagliari, Cagliari, Italy

Abstract. The $\eta(1405)$, once called the E or ι represents a case in itself in the glueball search. It has been studied through six different production mechanisms by more than thirty experiments for a period spanning four decades. More than 70 experimental results are now quoted within the Review of Particle Properties. The purpose of this review is to focus on the key point: the existence or not of an extra pseudoscalar state . A possible interpretation of the data indicating the presence of a non $q\bar{q}$ state will be presented and discussed. Open questions and alternative interpretations will be considered.

Keywords: Glueballs, light quark spectroscopy, pseudoscalars
PACS: 14.40Cs,12.39.Mk

INTRODUCTION

Glueballs, represent a direct consequence of the non-abelian nature of the QCD and the experimental discovery of such states has fundamental importance for the verification of the theory.

The $\eta(1405)$, once called the E or ι represents a case in itself in the glueball search. More than 30 experiments from 1963 to 2005 studied this case. This review will focus on the main issue: the existence or not of two pseudoscalar states in the 1.4-1.5 GeV mass region. The existence of an extra pseudoscalar state will enforce the exotic nature interpretation of the $\eta(1405)$. Open questions and alternative interpretations will be considered. Following the PDG notation, from 1986[1] until 2002 [2], I will refer to the $\eta(1440)$ to indicate either one pseudoscalar or the two pseudoscalar states. In some cases, when the two pseudoscalar states are separately identified, I will refer to them as $\eta(1405)$ and $\eta(1475)$, following the notation of the 2004 PDG issue[3]. The first observation of the E meson dates back to 1963 in a bubble chamber experiment of the CERN-College de France collaboration[4, 5]. It was the first meson resonance found at CERN and Europe and for this reason was named E, as Europe. However, it was the observation, with a very high branching ratio, in J/Ψ radiative decay [6, 7, 8], that, lead, in 1980 to its glueball interpretation [9, 10, 11, 12]

Later on it became apparent that more than one state was present in the $1.4 \div 1.5$ GeV/c^2 mass region.

In summary, five different experiments, using three different production mechanisms, reported evidence for two pseudoscalars and an axial vector in the E/ι mass region.

These states are now called $\eta(1405)$, $\eta(1475)$ and $f_1(1420)$ [3].

The existence of two pseudoscalar states is a key point for the non $q\bar{q}$ interpretation of the $\eta(1405)$, therefore I will consider the two hypotheses:

1. There are two pseudoscalars, $\eta(1405)$ and $\eta(1475)$. The $\eta(1405)$ main decay

CP814 *Hadron Spectroscopy: Eleventh International Conference*
edited by A. Reis, C. Göbel, J. Sá Borges, and J. Magnin

modes are $a_0(980)\pi$ and direct $\eta\pi\pi$ or $K\overline{K}\pi$. The $\eta(1475)$ decays dominantly through $\overline{K}^{*}K+c.c.$.

2. There is only one pseudoscalar with $a_0(980)\pi$, direct $\eta\pi\pi$ or $K\overline{K}\pi$, and $\overline{K}^{*}K+c.c.$ decay modes

THE EXISTENCE OF AN EXTRA PSEUDOSCALAR STATE

Assuming two resonances, according to hypothesis 1), all the data agree in assigning the $\overline{K}^{*}K+c.c.$ decay mode to the $\eta(1475)$. Therefore this state can be observed in the $K\overline{K}\pi$ final state only.

In the $K\overline{K}\pi$ final state, three states ($\eta(1405)$, $\eta(1475)$ and $f_1(1425)$) are potentially present within a 100 MeV/c^2 mass range and the task of identifying and disentangling them is difficult. However if one retains the high statistics experiments only, one can observe a consistent picture where two pseudoscalars (the $\eta(1405)$ and the $\eta(1475)$) and one axial vector (the $f_1(1425)$) are identified. The mass and width parameters are also in good agreement. The most relevant results, from high statistics experiments, are listed below, according to the production mechanism.

- Antiproton-proton annihilation at rest in liquid hydrogen provides the ideal environment to study pseudoscalar production since the $f_1(1425)$ production is suppressed. It is therefore possible to perform a better study of the $\eta(1405)$ and the $\eta(1475)$ pseudoscalars[13, 14]. By moving from liquid to gaseous hydrogen target both the pseudoscalars and the axial vector have been observed but the $\eta(1405)$ and the $\eta(1475)$ measurement became less precise [15, 16].
- J/Ψ radiative decay data from MarkIII [17] and DM2 [18] report the observation of two pseudoscalars and an axial vector. DM2 results disagree on the decay modes. However the DM2 analysis provided also a second solution with a better agreement with the MARKIII results.
- In peripheral hadroproduction two pseudoscalars and an axial vector are reported by two Brookhaven experiments, E769 in 1989[19] and E852 in 2001[20].
- in $\gamma\gamma$ collision, the L3 experiment[21], observed only one state, decaying trough $\overline{K}^{*}K+c.c.$, compatible with the $\eta(1475)$ while no signal was observed in the $\eta\pi\pi$ final state, this observation is not confirmed by the CLEO-II data[22].

The two photon result from L3 requires a further comment. The simultaneous observation of a signal consistent with the $\eta(1475)$ decaying trough $\overline{K}^{*}K$ and the absence of a signal in the $\eta\pi\pi$ final state provide a strong argument against any possible interpretation in terms of a single pseudoscalar resonance, which cannot be present in one decay mode and absent in the other within the same production mechanism. It is therefore very important to assess the significance of the L3 result. The $\eta(1475)$ signal identification is unmistakable, thanks also to the absence of 1^{++} contribution in the untagged data sample, where spin 1 objects are forbidden by the Yang theorem. The reported upper limit on a $\eta\pi\pi$ signal is 0.095 keV at 95% C.L.. This number can be weighed against the observed width by L3 in $K\overline{K}\pi$. An upper limit on the ratio can be deduced. We can compare this result with the available results on the ratio between the observed branching ratios in $\eta\pi\pi$ and $\overline{K}^*K^{\cdot}$. The data are reported in table 1. The upper limit reported by L3 stays two standard deviations below the lowest value reported in table 1.

We can conclude that the L3 result does not totally exclude the possibility of a signal in $\eta\pi\pi$, due to a single resonance with two decay modes, but it certainly disfavors this interpretation.

More recently CLEO-II studied the process $\eta(1475) \to K_S^0 K^\pm \pi^\mp$ in two photon collision with the declared purpose of enforcing the $\eta(1475)$ observation reported by L3. Actually no pseudoscalar signal was observed but the statistical limit, with respect to L3 was at two standard deviation (without including the systematics), therefore further data are needed to clarify this point.

>From the discussion above it appears that the existence of two pseudoscalars: the $\eta(1405)$ and the $\eta(1475)$, is consistently supported by six experiments, in four different production mechanisms.

Other hypotheses can be considered to interpret the $\eta(1440)$ results in terms of a single resonance. I will review them below.

- The most recent J/Ψ radiative decay to $K\overline{K}\pi$ coming from BES indicate a single resonance[26]. However, their data suffer from severe statistics limitations. The BES $K\overline{K}\pi$ statistics is about six times lower than MARKIII [17]. It is therefore not surprising if two resonances cannot be disentangled.
- According to reference [27] rescattering plays an important role in the $\eta(1440)$ decay modes. They propose a triangle diagram mechanism which can explain some features of the $\eta(1440)$ data. According to their analysis of $\eta\pi\pi$ data from $\overline{p}p$ annihilation at rest there is a significant contribution of $\eta(1440)$ decay trough K^0K or $\overline{K}^*K$ followed by rescattering of the two kaons trough $a_0(980)$ to $\eta\pi$ above the $K\overline{K}$ threshold. This could lead to a mass displacement of the $\eta(1440)$ $\eta\pi\pi$ component with respect to its $\overline{K}^*K$ component. However, as remarked by the authors [27], a combined analysis of $\eta\pi\pi$ and $K\overline{K}\pi$ data with high statistics would be necessary to proceed further. Moreover this hypothesis is disfavored by the L3 $\gamma\gamma$ data [21].

TABLE 1. Summary of results on the ratio of branching ratios between the $\eta\pi\pi$ and the $\overline{K}^*K$ component of the $\eta(1440)$. In $\overline{p}p$ data the results have been obtained combining the $\eta\pi\pi$ result from Crystal Barrel with $\overline{K}^*K$ data from Bubble chamber and Obelix. Statistical and systematic errors are combined

Production mechanism	Experiment	$\frac{\Gamma(\eta\pi\pi)}{\Gamma(\overline{K}^*K)}$
J/Ψ rad.[17, 23]	MARKIII /DM2	1.02 ± 0.34
J/Ψ rad.[23, 18]	DM2	1.27 ± 0.41
$\overline{p}p$ at rest[24, 5]	CRYSTAL BARREL /BUBBLE CHAMBER	3.28 ± 1.02
$\overline{p}p$ at rest[24, 13, 25]	CRYSTAL BARREL / OBELIX	9.17 ± 3.62
$\gamma\gamma$[21]	L3	< 0.45 (95%$C.L.$)

- Another possible interpretation of the $\eta(1440)$ structure in terms of one resonance has been proposed in ref.[28]. A node in the $\eta(1440)$ matrix element can explain the presence of two peaks, at different masses for the $a_0(980)\pi$ and the $\overline{K}^*K + c.c.$ decay modes of the same resonance. According to the calculations of [29], within the $^3P_0$ decay model a single $\eta(1440)$ state can produce a peak, at a mass close to the $\eta(1405)$, in the $a_0(980)\pi$ decay mode and another peak, at a mass close to the $\eta(1475)$, in the $\overline{K}^*K + c.c.$ decay mode. Again this hypothesis is not supported by the $K\overline{K}\pi$ data analysis by Obelix and E852, but even assuming that this effect could forge the MarkIII, Obelix and E792 data, this interpretation, is disfavored by the L3 results in $\gamma\gamma$ collision[21] quoted above.
- The interference effect of a broad background resonance, observed in the analysis of $\eta\pi\pi$ data from $\overline{p}p$ annihilation at rest[30, 27] and in the J/Ψ data from BES [26] could also be responsible for altering the analysis results, indicating two resonances while only one is present. However this hypothesis has been checked an ruled out in the analysis of Obelix data[14, 16] and it is not supported by any $K\overline{K}\pi$ data analysis with the exception of the BES case [26], whose statistics, as remarked above, is so limited that does not allow any solid conclusion on the matter. In this case also the L3 results in $\gamma\gamma$ [21] further disfavor this hypothesis.

CONCLUSIONS

In summary we can say that the existence of two pseudoscalar resonances is supported by a wide range of experiments in different production mechanisms and decay modes. However other interpretations of the data are possible and an improved determination of the $\eta\pi\pi$ and $K\overline{K}\pi$ partial widths for J/Ψ and $\gamma\gamma$ data is mandatory to clarify the picture. A coupled channel analysis of $K\overline{K}\pi$ and $\eta\pi\pi$ data would also be very important to verify alternative interpretations of the data.

Assuming two pseudoscalar resonances, $\eta(1405)$ and $\eta(1475)$, the possible pseudoscalar nonet is shown below.

radial excitation	$n\overline{n}$	$n\overline{n}$	$s\overline{s}$	$n\overline{s}$
0	π	η	η'	K
1	$\pi(1300)$	$\eta(1295)$	$\eta(1475)$	$K(1460)$

Within this framework eight of nine members: the $\pi(1300)$ isotriplet, the $K(1460)$ strange quartet and the $\eta(1295)$ are considered well ascertained[29, 31, 32]. Quite recently the identification of the $\eta(1295)$ as the $\eta(547)$ first radial excitation and its mere existence has been questioned[28]. However we can observe that the $\eta(1295)$ has been identified by four π^-p experiments [20, 33, 34, 35]quoted on the PDG [3]. Moreover possible contributions are also reported in $\overline{p}p$ annihilation [27, 30, 36]and J/Ψ radiative decay (an $\eta(1295)$ signal is quite evident also in the 0^{-+} $\eta\pi\pi$ wave of DM2 data [18]).

If we assume the $\eta(1295)$ being the first radial excitation of the η, the $\eta(1475)$ could be the η' first radial excitation. This state should be mainly $s\overline{s}$, assuming ideal mixing, as suggested by the $\eta(1295)$ and the $\pi(1300)$ mass degeneracy. The $\eta(1475)$, with its $\overline{K}^*K + c.c.$ coupling, is therefore the natural choice. This assignment is further confirmed by the fact that also its width matches the expected width for the radially excited $s\bar{s}$ state [37, 38, 29] even if the non observation in K^-p can cast some doubts on this hypothesis [28].

If one assumes the nonet filled by the states quoted above the $\eta(1405)$ becomes a supernumerary.

The nature of the $\eta(1405)$ has been object of debate for more than twenty years. Since the discovery of a pseudoscalar state in this mass region the glueball hypothesis was claimed. For a long time now, the combined information from $\gamma\gamma$ collision and J/Ψ radiative decay data, has been employed as an indication of a state gluonium nature.

This procedure has been extended and quantified by Close, Farrar and Li [38] who made an estimate of the branching fraction to gluon of a resonance R: $B(R \rightarrow gg)$, from its branching fraction in radiative quarkonium decay. $B(R \rightarrow gg)$ is expected to be $0.1 \div 0.2$ for a $q\bar{q}$ state and 1 for a glueball.

According to this procedure the $\eta(1405)$ branching fraction to gluon is 0.9 ± 0.2 [38].

The $\eta(547)$, $\eta'(958)$ and $\eta(1405)$, gluonium-quarkonium mixing framework has been studied by different authors [38, 39, 40, 41, 42, 43], who compared their models with the data obtained from J/Ψ decay into vector and pseudoscalar mesons and the data on pseudoscalar and vector meson two-photon decays. All of them agree in assigning a dominant gluonium component to the $\eta(1405)$.

However, despite all these elements, indicating a strong gluon content for the $\eta(1405)$, the glueball interpretation is disfavored by lattice calculations, which predict, for the pseudoscalar glueball, a mass above 2 GeV[44, 45].

For this reason alternative hypotheses have been suggested. The $\eta(1405)$ properties are consistent with predictions for a gluino-gluino bound state [46, 47, 48]. Within this framework the 0^{++} glueball would be in a supermultiplet with the pseudoscalar gluino-gluino and the spin 1/2 gluino-gluino bound states.

In addition, it must be considered that, quite recently, Faddeev, Niemi and Wiedner[49] suggested that glueballs can be viewed as closed gluonic fluxtubes. In their article they show that, if glueballs are seen as closed gluonic fluxtubes, i.e. as closed QCD strings, than the 0^{++} $f_0(1500)$ glueball, in a parity-invariant QCD, is naturally related to a 0^{-+} glueball. In this case the $\eta(1405)$ represent the ideal experimental candidate. According to the authors[49] the observed mass difference between the $\eta(1405)$ and the $f_0(1500)$ could mean that an exact parity degeneracy between the Yang-Mills string degrees of freedom become broken in QCD.

Forty years and 30 experiments have not yet been capable to definitely solve the E/iota puzzle. However, a rather consistent picture, where the $\eta(1405)$, is an extra state in the pseudoscalar nonet and shows the characteristics of a strong gluon content emerges from the data. As we saw, other interpretations are possible. A key point will be an improved study of $\eta\pi\pi$ and $K\bar{K}\pi$ final states in $\gamma\gamma$ collision.

ACKNOWLEDGMENTS

It is a real pleasure to thank the organizers of the Hadron 05 Conference for their kind hospitality in such a beautiful setting as Rio. I want to take this opportunity to make a tribute to the memory of Lucien Montanet, who co-signed the first discovery of the E meson, to his contribution to meson spectroscopy, and to his generous and expert advice.

REFERENCES

1. Particle Data Group, M. Aguilar-Benitez et al., Phys. Lett. B 170 (1986)
2. Particle Data Group, K. Hagiwara et al. Phys. Rev. D66 (2002)
3. Particle Data Group, Simon Eidelman, Ken Hayes, Keith Olive et al., Phys. Lett. B 592 (2004)
4. R.Armenteros et al., Proc. of the Siena Int. Conf. on Elementary Particles, vol 1 (1963) 287
5. P. Baillon et al., Nuovo Cim. 3 (1967) 393
6. D.L. Sharre et al.,Phys. Lett B97 (1980) 329
7. C. Edwards et al., Phys. Rev. Lett. 49 (1982) 259
8. C. Edwards et al., Phys. Rev. Lett. 50 (1983) 219
9. C.E. Carlson et al., Phys. Lett. B99 (1981) 353
10. J.F. Donoghue et al., Phys. Lett. B99 (1981) 416
11. K. Ishikawa Phys. Rev. Lett. 46 (1981) 978
12. M. Chanowitz Phys. Rev. Lett. 46 (1981) 981
13. Obelix Collaboration A. Bertin et al. Phys. Lett. B 361 (1995) 187
14. Obelix Collaboration C. Cicalò et al., Phys. Lett. B 462 (1999) 453
15. Obelix Collaboration A. Bertin et al. Phys. Lett. B 400 (1997) 226
16. Obelix Collaboration F. Nichitiu et al., Phys. Lett. B 546 (2002) 261
17. MARKIII Collaboration, Z. Bai et al., Phys. Rev. Lett. 65 (1990) 2507
18. DM2 Collaboration, J.E. Augustin et al., Phys. Rev. D46 (1992) 1951
19. M.G. Rath et al., Phys. Rev. D40 (1989) 693
20. Adams et al., Phys. Lett. B 516 (2001) 264
21. Acciarri et al., Phys. Lett. B 501 (2001) 1
22. R. Ahohe et al. Phys. Rev. D 71 (2005) 072001
23. J.E. Augustin et al., Phys. Rev. D42 (1990) 10
24. Crystal Barrel Collaboration, C. Amsler et al., Phys. Lett. B358 (1995) 389
25. Obelix Collaboration A. Bertin et al. Phys. Lett. B 385 (1996) 493
26. J.Z.Bai et al. Phys. Lett. B 440 (1998) 217
27. A.V. Anisovich et al., Nucl. Phys.A690 (2001) 567
28. E. Klempt, Int. Journ. Mod. Phys. A20 N. 8-9 (2005)
29. T. Barnes, F.E. Close, P. Page and E. Swanson, Phys. Rev. D55 (1997) 4157
30. Crystal Barrel Collaboration, A. Abele et al., Nucl. Phys. B514 (1998) 45
31. S. Godfrey, J. Napolitano, Rev. Mod. Phys. 71 (1999) 1411
32. L. Köpke and N. Wermes, Phys. Rep. 174 (1989) 67
33. S. Fukui et al., Phys. Lett. B267 1991 293
34. D. Alde et al., Phys. Atomic Nucl. 60 (1997) 386; transl. from Yad. Fiz. 60 458
35. J.J. Manak et al., Phys. Rev. D62 (2000) 012003
36. Crystal Barrel Collaboration, C. Amsler et al. Eur. Phys. J C33 (2004) 23
37. T. Barnes, N. Black and P.R. Page, Phys. Rev. D 68 (2003) 054014
38. F.E. Close, G.R. Farrar and Z. Li, Phys. Rev. D55 (1997) 5749
39. N. Morisita, I. Kitamura and T. Teshima, Phys. Rev. D44 (1991) 175
40. I. Kitamura, N. Morisita and T. Teshima, Int. J. Mod. Physiscs, A9 (1994) 5489
41. M. Genovese, D.B. Lichtenberg and E. Predazzi, Z. Phys. C61 (1994) 425
42. W.S. Carvalho, A.C.B. Antunes and A.S. de Castro, Eur. Phys. J C7 (1999) 95
43. D.M. Li et al., Eur. Phys. J C28 (2003) 335
44. G.S. Bali et al., Phys. Lett. B309 (1993) 308
45. C.J. Morningstar and M. Peardon, Phys Rev D60 (1999) 034509
46. M.B. Cakir and G.R. Farrar, Phys. Rev. D50 (1994) 3268
47. G.R. Farrar, Phys. Rev. D51 (1995) 3904
48. G.R. Farrar, Phys. Rev. Lett. (1996) 4111
49. L. Faddeev, A.J. Neemi and U. Wiedner, Phys. Rev. D 70 (2004) 114033

Heavy quarkonium PP transitions and effects of κ and a possible $b\bar{b}q\bar{q}$ state

P.N. Shen*, F.K. Guo*, H.C. Chiang* and R.G. Ping*

*Institute of High Energy Physics, Chinese Academy of Sciences, Beijing, China

Abstract. Heavy quarkonium PP decays are studied with the final state interaction in the framework of Chiral Unitary Approach (ChUA). It is found that $\Upsilon(nS)$ and $\psi(2S)$ $\pi\pi$ decay data, especially the $\Upsilon(3S) \to \Upsilon(1S)\pi^+\pi^-$ decay data, can be well-explained by introducing an additional intermediate state with $J^P = 1^+$, $I = 1$, $M = 10.08GeV$ and $\Gamma = 0.655GeV$, which could be a $b\bar{b}q\bar{q}$ state. By using dimensional regularization method, the singularity in the loop integration in ChUA can be avoid and S-wave $K\pi$ scattering in the energies less than $1.2GeV$ can be well-reproduced by simply employing a lowest order Lagrangian ($O(p^2)$) in ChPA with one adjustable parameter only. As a result, the mass and width of κ are estimated to be $M_\kappa = 757 \pm 33MeV$ and $\Gamma_\kappa = 558 \pm 82MeV$, respectively. It is shown that κ is very important in explaining the broad bump below $1.1GeV$ in the $K^+\pi^-$ invariant mass spectrum of the $J/\psi \to \bar{K}^{*0}(892)K^+\pi^-$ process.

Keywords: heavy quarkonium, decay, scalar meson
PACS: 13.25.Gv, 12.39.Fe, 12.39.Mk

MOTIVATION

Heavy quarkonium decay is a gluon-rich process. It is an idea place to study the nonperturbative effect of Quantum Chromodynamics (QCD) and the properties and the productions of baryons, meson and exotics. In particular, in the heavy quarkonium PP decay, due to the final state interaction (FSI), valuable information about scalar mesons, for instance σ and κ, can be obtained. In recent years, a large amount of data in Υ and ψ PP decays have been accumulated [1], but the difficulty in understanding these data, especially the double peak structure in the invariant $\pi\pi$ mass spectrum (IppMS) and the angular distribution (AGD) in the $\Upsilon(3S) \to \Upsilon(1S)\pi^+\pi^-$ decay, were encountered. For instance, by using QCD multiple expansion method, one can only get a qualitative fit to the IppMS [2]. By including $\pi\pi$ FSI (either in the quark-gluon degrees of freedom or in the hadronic degrees of freedom), although one can obtain a better fit to the double peak character in IppMS, the AGD and even large coupling constant difference in different $\Upsilon(nS) \to \Upsilon(mS)\pi\pi$ decays still can not be explained [3]. Another attempt is introducing an intermediate $b\bar{b}q\bar{q}$ state, but they failed to explain AGD, although fitting to IppMS was satisfactory [4]. In this talk, by choosing a proper theoretical framework, we are trying to solve this problem and further apply this framework to other charmonium PP decay.

CHIAL UNITARY APPROACH AND σ AND κ MESONS

In order to evaluate the heavy quarkonium PP decay accurately, we must adopt a framework which can deal with FSI of PP properly. The framework chosen is so-called

CP814 *Hadron Spectroscopy: Eleventh International Conference*
edited by A. Reis, C. Göbel, J. Sá Borges, and J. Magnin

ChUA [5]. In this approach, the lowest order Lagrangian in the Chiral Perturbation Theory (χPT)

$$\mathscr{L}_2 = \frac{1}{12f^2}\langle(\partial_\mu\Phi\Phi - \Phi\partial_\mu\Phi)^2 + M\Phi^4\rangle$$

is employed to describe the PP interaction at low energies. Then, the lowest order amplitude in χPT is taken as the initial kernel to iterate the coupled channel Bethe-Salpeter equation (BSE), and the relativistic propagators are adopted to describe the intermediate PP state. By factorizing the kernel in BSE in the on-shell approximation, the integral equations can be converted to a set of coupled algebraic equations

$$T = (1 - VG)^{-1}V.$$

In other word, the key point of this approach is that summing over relevant infinite terms in the expansion series needed to provide a proper t-matrix structure. In treating divergence of the propagator in the loop integration in BSE, to avoid the artificial singularity appears in the case of using cut-off method, the dimensional regularization method is employed. As a result, the phase shifts of S-wave I=0 $\pi\pi$ scattering below 1.2 GeV can be well-reproduced in the $\pi\pi + K\bar{K}$ coupled channel calculation (see Fig. 1), and the pole position of σ in the second Riemann sheet is extracted to be $0.469 - i0.203$ GeV [5]. Moreover, the phase shifts of S-wave I=1/2 $K\pi$ scattering below 1.26 GeV can also be well-reproduced in the $K\pi + K\eta + K\eta'$ coupled channel calculation with an $\mathscr{O}(p^2)$ Lagrangian and only one adjustable parameter (see Fig. 1). And the extracted mass and width of κ are $m_\kappa = 757 \pm 33$ MeV and $\Gamma_\kappa = 558 \pm 82$ MeV, respectively [7].

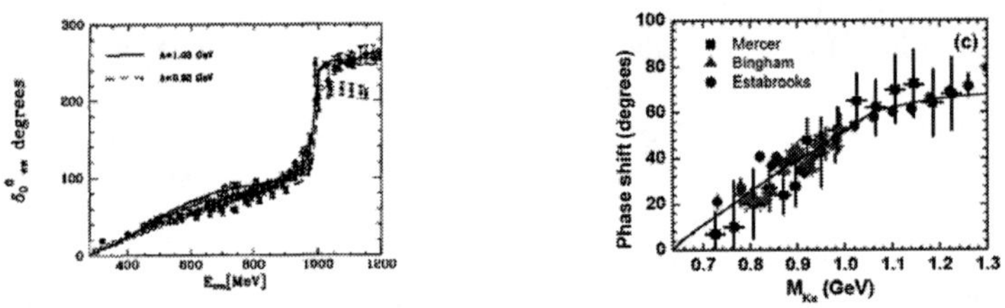

FIGURE 1. Left figure: Phase shifts of the S-wave I=0 $\pi\pi$ scattering in the $\pi\pi + K\bar{K}$ coupled channel calculations [5]. Right figure: Phase shifts of the S-wave I=1/2 $K\pi$ scattering in the $K\pi + K\eta + K\eta'$ coupled channel calculations [7].

$\Upsilon(NS)\ \pi^+\pi^-$ DECAYS AND A POSSIBLE TETRAQUARK STATE

Having such a good theory to deal with FSI, now we are able to study heavy quarkonium decays. The Feynman diagrams for such a decay are shown in Fig. 2. For the direct process shown in the left figure, we employ an effective Lagrangian [3], that incorporates both chiral and heavy-quark expansions appropriately to the lowest order, $\mathscr{L} = \mathscr{L}_0 + \mathscr{L}_{SB}$, where $\mathscr{L}_0$ and $\mathscr{L}_{SB}$ denote the leading term in the derivative expansion obeying chiral symmetry and the term breaking the chiral symmetry. Then, the decay amplitude of the direct process can be written as

$$t = V_0 + V_{0S} \cdot G \cdot 2t^{I=0}_{\pi\pi,\pi\pi}, \quad (1)$$

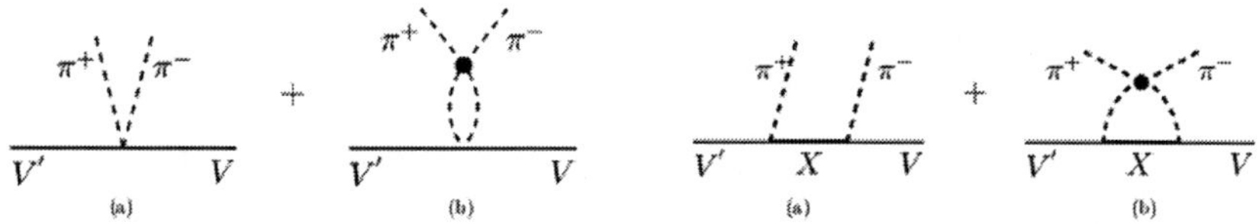

FIGURE 2. Left figure: Diagrams of the direct $V^{'}V\pi^+\pi^-$ decay mechanism with (a) tree diagram, (b) diagram with $\pi\pi$ FSI. Right figure: Diagrams of the sequential X decay mechanism with (a) tree diagram, (b) diagram with $\pi\pi$ FSI.

where V_0 stands for the decay amplitude of the tree diagram

$$V_0 = -\frac{4}{f_\pi^2}(g_1 p_1 \cdot p_2 + g_2 p_1^0 p_2^0 + g_3 m_\pi^2)\varepsilon^* \cdot \varepsilon^{'} = V_{0S} + V_{0D}$$

with V_{0S} and V_{0D} being the S-wave and D-wave components, respectively, G the propagator for the intermediate state, and $2t^{I=0}_{\pi\pi,\pi\pi}$ the contribution from FSI of $\pi\pi$

$$2t^{I=0}_{\pi\pi,\pi\pi} = \langle \pi^+\pi^- + \pi^-\pi^+ + \pi^0\pi^0 | T | \pi^+\pi^- \rangle.$$

The results show that most of the IppMS' for $\Upsilon(nS)$ and $\psi(2S)$ decays can be well described except for the $\Upsilon(3S) \to \Upsilon(1S)\pi^+\pi^-$ decay, and all AGDs for these decays cannot be reproduced. Moreover, the coupling constant g_2 for these decays are quite different even in different sign.

To solve these problem, we include an additional sequential process $\Upsilon(nS) \to X\pi \to \Upsilon(mS)\pi\pi$ shown in the right figure in Fig.2 [6]. The decay amplitude of the tree diagram in the sequential process can be written as

$$V_X^{tree} = g_{nm}\varepsilon^{'}_\mu \varepsilon^*_\nu \left(\frac{-g^{\mu\nu} + p^\mu_{X+} p^\nu_{X+}/m_X^2}{p^2_{X+} - m_X^2 + i m_X \Gamma_X} + \frac{-g^{\mu\nu} + p^\mu_{X-} p^\nu_{X-}/m_X^2}{p^2_{X-} - m_X^2 + i m_X \Gamma_X} \right), \quad (2)$$

and the full t-matrix for the $\Upsilon(nS) \to \Upsilon(mS)\pi\pi$ decay can be given by

$$t = V_0 + V_{0S} \cdot G \cdot 2t^{I=0}_{\pi\pi,\pi\pi} + V_X^{tree} + g_{nm}\varepsilon^{'}_\mu \varepsilon^*_\nu G_X^{\mu\nu} \cdot 2t^{I=0}_{\pi\pi,\pi\pi},$$

where the last term denotes the contribution from FSI in the additional term, and $G_X^{\mu\nu}$ stands for the propagator for the corresponding intermediate state. The resultant IppMS' and AGDs are shown in Fig. 3. The results show that with the additional process included, not only IppMS' but also AGDs for all $\Upsilon(nS)$ $\pi\pi$ decays can be well described. The main reason is that the sequential process provides important contribution in IppMS, and D-wave components of such a process play a decisive role in AGDs in the $\Upsilon(3S) \to \Upsilon(1S)\pi^+\pi^-$ decay. Moreover, the obtained coupling constants for all $\Upsilon(nS)$ $\pi\pi$ decays are fairly close to each other and satisfactory. The resultant coupling constants and the mass and width of intermediate state X are tabulated in Tab. 1. From structure consideration, X should have $b\bar{b}q\bar{q}$ content, for instance $b\bar{b}u\bar{d}$ for X^+, and quantum numbers $J^P = 1^+$ and $I = 1$.

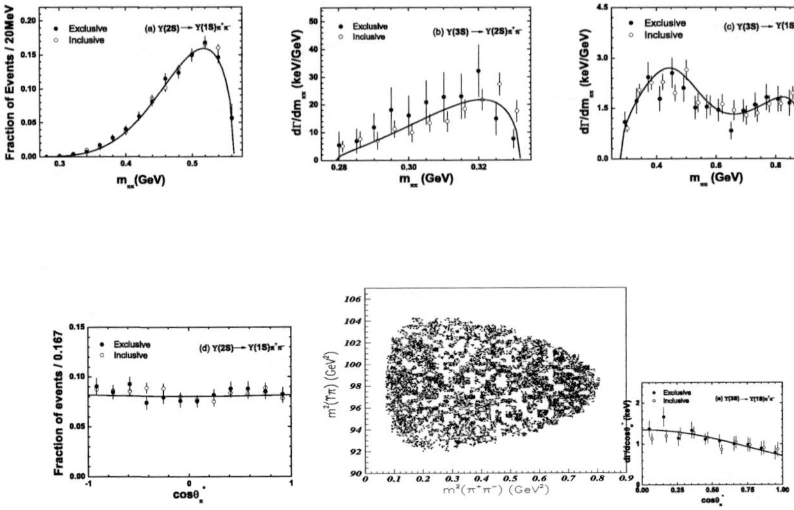

FIGURE 3. Upper line: Figures from left to right: invariant $\pi\pi$ mass spectra in the $\Upsilon(2S) \to \Upsilon(1S)\pi^+\pi^-$, $\Upsilon(3S) \to \Upsilon(2S)\pi^+\pi^-$ and $\Upsilon(3S) \to \Upsilon(1S)\pi^+\pi^-$ decays, respectively. Lower line: Figures from left to right: angular distribution in the $\Upsilon(2S) \to \Upsilon(1S)\pi^+\pi^-$ decay and Dalitz plot and angular distribution in the $\Upsilon(3S) \to \Upsilon(1S)\pi^+\pi^-$ decays, respectively.

$J/\psi \to \bar{K}^{*0}K^+\pi^-$ decay and effect of κ

κ is another interesting scalar meson in hadron physics. Now, we apply ChUA to the $J/\psi \to \bar{K}^{*0}K^+\pi^-$ decay to see the effect of κ. The decay amplitude for the coupled channel case can be written as

$$
\begin{aligned}
t_c = & - g\varepsilon_\mu^{(\lambda)}(p)\varepsilon^{\mu(\lambda')}(p_3)[1+G_{11}(\langle K^+\pi^-|t|K^+\pi^-\rangle + \langle K^+\pi^-|t|\pi^-K^+\rangle) \\
& - \frac{1}{\sqrt{2}}G_{11}(\langle K^+\pi^-|t|K^0\pi^0\rangle + \langle K^+\pi^-|t|\pi^0K^0\rangle) \\
& - \frac{1}{\sqrt{6}}G_{22}(\langle K^+\pi^-|t|K^0\eta\rangle + \langle K^+\pi^-|t|\eta K^0\rangle)],
\end{aligned}
$$

TABLE 1. Resultant coupling constants, and the mass and width of X in $\Upsilon(nS) \to \Upsilon(mS)\pi^+\pi^-$ decays.

Decay	g_1(GeV)	g_2/g_1	g_3/g_1	g_{nm}(GeV^2)	m_X(GeV)	Γ_X(GeV)
$2 \to 1$	0.0886	-0.230	0	-2.316		
$3 \to 2$	0.769	-0.230	0	-0.00418	10.080	0.655
$3 \to 1$	0.00546	-0.230	4.949	4.712		

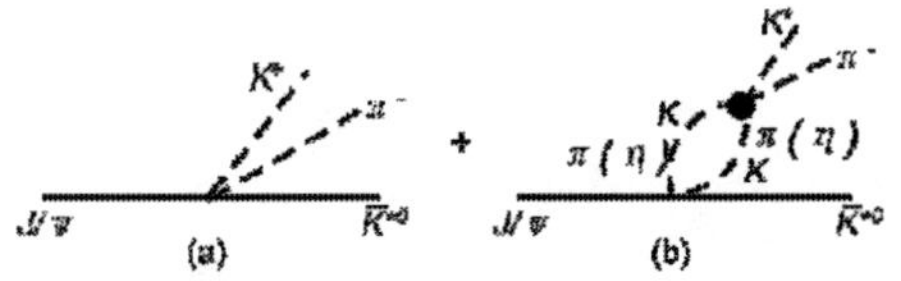

FIGURE 4. Feynman diagram of the $J/\psi \to \bar{K}^{*0}K^+\pi^-$ decay. The left and right figures denote the tree and FSI diagrams, respectively

where $\langle K^+\pi^-|t|P_1P_2\rangle$ ($P_1P_2 = K^+\pi^-$, π^-K^+, etc.) represents the full amplitude of the process $P_1P_2 \to K^+\pi^-$ calculated in ChUT, and $s_{12} = (p_1+p_2)^2 = m_{12}^2$. The resultant invariant mass spectrum of $K^+\pi^-$ is shown in Fig. 5. It is shown that κ is very important in explaining the broad bump below 1.1 GeV in the $K^+\pi^-$ invariant mass spectrum [7].

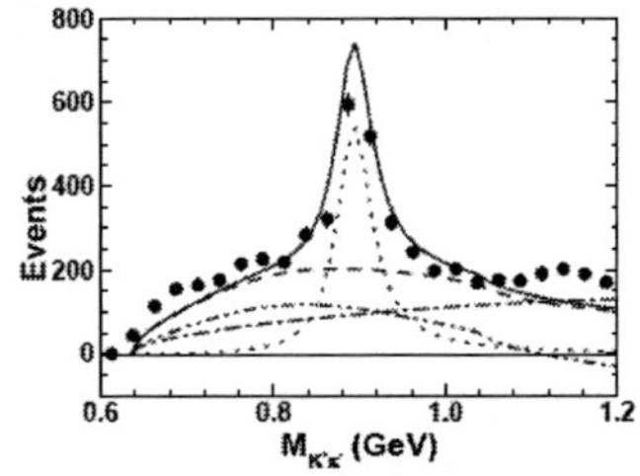

FIGURE 5. Invariant mass spectrum of $K^+\pi^-$ in the $J/\psi \to \bar{K}^{*0}K^+\pi^-$ decay. The dashed-dotted-dotted, dashed-dotted and dotted curves represent the contributions form κ and S-wave $K\pi$ FSI and sideband, respectively.

SUMMARY

ChUA is a powerful approach to treat PP interactions in the energies less than 1.2 GeV. From this approach, the masses and widths of the σ and κ are extracted and coincide with the experimental results. By using this approach and including an additional sequential process, the $\Upsilon(nS) \to \Upsilon(mS)\pi\pi$ decay data can be well explained. The extracted properties of the intermediate X state are $J^P = 1^+, I = 1, M = 10.08$ GeV and $\Gamma = 0.655$ GeV, and the the quark content of X should be $b\bar{b}q\bar{q}$. Applying this approach to the $J/\psi \to \bar{K}^{*0}K^+\pi^-$ decay, one finds that κ is very important in explaining the broad bump at the energy less than 1.1 GeV.

REFERENCES

1. S. Eidelman *et al.*, *Phys. Lett.*, **B592**, 1 (2004).
2. K. Gottfried, *Phys. Rev. Lett.*, **40**, 598 (1978); H. Y. Zhou, and Y. P. Kuang, *Phys. Rev.*,**D44**, 759 (1991).
3. T. Mannel, and R. Urech, *Z. Phys.*, **C37**, 541 (1997); T. A. Lähde, and D. O. Riska, *Nucl. Phys.*,**A707**, 425 (2002); A. Gallegos *et al.*, *Phys. Rev.*,**D69**, 074033 (2004).
4. V. V. Anisovich *et al.*, *Phys. Rev.*,**D51**, R4619 (1995).
5. J. A. Oller and E. Oset, *Nucl. Phys.*, **A620**, 438 (1997); **A652**, 407 (1999); L. Roca *et al.*, *Nucl. Phys.*,**A744**, 127 (2004)
6. F.-K. Guo, P.-N. Shen, H.-C. Chiang, R.-G. Ping, *Nucl.Phys.*, **A761**, 269 (2005).
7. F.-K. Guo, R.-G. Ping, P.-N. Shen, H.-C. Chiang, B.-S. Zou, arXiv: *hep-ph/0509050*.

Exotic baryon searches in 800 GeV/c $pp \rightarrow pX$

M.A. Reyes*, D.C. Christian†, J. Felix*, E.E. Gottschalk†, G. Gutierrez†, E.P. Hartouni**, B.C. Knapp‡, M.N. Kreisler**,§, E. Lopez*, G. Moreno*, I.O. Perez*, M. Sosa*, M.H.L.S. Wang† and A. Wehmann†

*Universidad de Guanajuato, León, Guanajuato, México
†Fermi National Accelerator Laboratory, Batavia, IL
**Lawrence Livermore National Laboratory, Livermore, CA
‡Columbia University, Nevis Laboratory, Irvington, NY
§University of Massachusetts, Amherst, MA

Abstract. We report the results of the search for the pentaquark candidates $\Theta(1540)$ and $\Xi(1862)$ using data from Fermilab experiment E690 in the reaction $pp \rightarrow pX$ at 800 GeV/c. We find no evidence for narrow baryon resonances near 1862 MeV decaying to $\Xi\pi$. Also, we find no evidence of a narrow resonance near 1540 MeV decaying to pK_s^0.

Keywords: Baryon resonances
PACS: 13.85.Rm, 14.20.Jn

A number of different experiments have reported evidence for the existence of the Θ^+, a strangeness +1 baryon that decays to nK^+ or pK^0 [1]. The Θ^+ mass is reported to be approximately 1540 MeV and its width less than ~20 MeV. The Θ^+ has been interpreted as a pentaquark. If this interpretation is correct, then a large number of similar states are expected[2], many of them, besides the Θ^+, having quantum numbers not possible for baryons composed only of three quarks. One such state is a charge -2, strangeness +2 baryon expected to decay to $\Xi^-\pi^-$. In 2003, Wilczek and Jaffe predicted that this state should have a mass of approximately 1750 MeV and should be narrow, with a width only $\sim 50\%$ greater than the width of the Θ^+[3]. Shortly after this prediction, NA49 reported evidence for a $\Xi^-\pi^-$ resonance produced in pp interactions at $\sqrt{s} = 17.2$ GeV with a mass of 1862 MeV and width less than the detector resolution of 18 MeV FWHM[4]. NA49 also reported evidence for states with similar masses and widths decaying to $\Xi^-\pi^+$, $\overline{\Xi}^+\pi^-$, and $\overline{\Xi}^+\pi^+$. However, the experimental case for the existence of pentaquarks is not yet compelling, since many experiments have failed to confirm the existence of the Θ^+ [5, 6], and other have only reported upper limits for the production of a state at 1862 MeV decaying to $\Xi\pi$ [5, 7].

Here we report the results of the search for the pentaquark candidates $\Theta(1540)$ and $\Xi(1862)$ using more than 4.3 billion events collected by the E690 Collaboration during Fermilab 1991 fixed target run. The E690 apparatus consisted of an open-geometry multiparticle spectrometer and a beam spectrometer. The multiparticle spectrometer was used to measure the target system (X) in reactions $pp \rightarrow p_{fast}X$. The E690 apparatus has been described elsewhere[8].

Events were recorded using a trigger that imposed two event selection requirements: (a) an interaction in the target region, indicated by signals from one or more time-of-flight scintillation counters, and (b) a scattered beam particle, detected by counters

CP814 *Hadron Spectroscopy: Eleventh International Conference*
edited by A. Reis, C. Göbel, J. Sá Borges, and J. Magnin

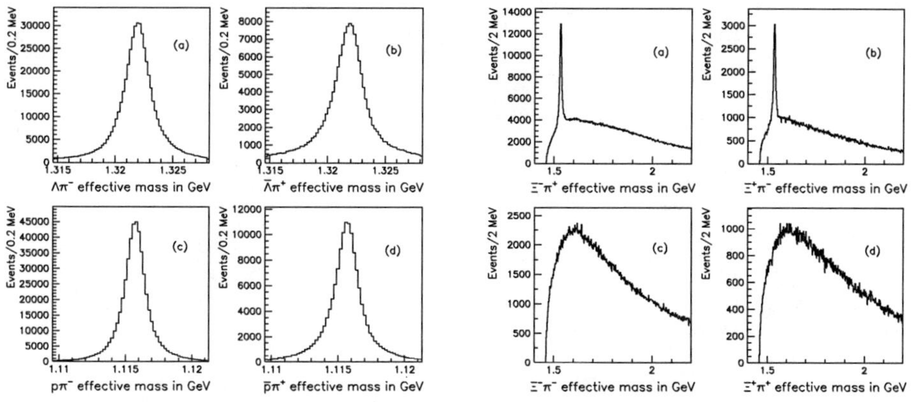

FIGURE 1. Left panel: Hyperon effective mass spectra in events used for this analysis: a)Ξ^-, b)$\overline{\Xi}^+$, c)Λ from Ξ^- decay, and d)$\overline{\Lambda}$ from $\overline{\Xi}^+$ decay. Right panel: Effective mass spectra for: a)$\Xi^-\pi^+$, b)$\overline{\Xi}^+\pi^-$, c)$\Xi^-\pi^-$, and d)$\overline{\Xi}^+\pi^+$.

outside of the beam spot in the forward beam spectrometer. Event reconstruction was performed for the entire data sample. The primary (interaction) vertex was constrained to lie on an incoming beam-track trajectory. For events in which evidence of a "vee" or "cascade" decay was found, the tracks were refit with the constraint that each daughter vertex "point back" to its parent, and the daughter was then "assigned" to the parent. Mass constrained fits were not performed.

For this analysis we selected events, first, for the inclusive reaction

$$pp \to p_{fast}\,\Xi^{\pm}\pi^{\pm}X\,, \quad \Xi \to \Lambda\pi \tag{1}$$

and, second, for the exclusive reaction

$$pp \to p_{slow}\,K_sK^-\pi^+p_{fast} \tag{2}$$

For reaction (1), events were selected if they had a candidate Ξ^- or $\overline{\Xi}^+$ decay. The sample contains approximately 513,000 Ξ^-'s and 153,700 $\overline{\Xi}^+$'s assigned to a primary vertex. The effective mass spectra of the Ξ^- and $\overline{\Xi}^+$ candidates are shown in Fig. 1, together with the effective mass spectra of the corresponding Λ and $\overline{\Lambda}$ candidates. $\Xi\pi$ mass distributions were computed by pairing each Ξ candidate with every charged track (assumed to be a pion) assigned to the same primary vertex. Direct particle identification was not used in this analysis. The effective mass distributions for the four charge combinations are also shown in Fig. 1. The $\Xi^-\pi^+$ and $\overline{\Xi}^+\pi^-$ mass distributions contain prominent $\Xi^0(1530)$ and $\overline{\Xi}^0(1530)$ resonances. None of the four effective mass distributions contains any other obvious resonance.

The plots in Fig. 1 clearly show that if a narrow pentaquark with mass 1862 MeV decaying to $\Xi\pi$ exists, then its production is highly suppressed with respect to the

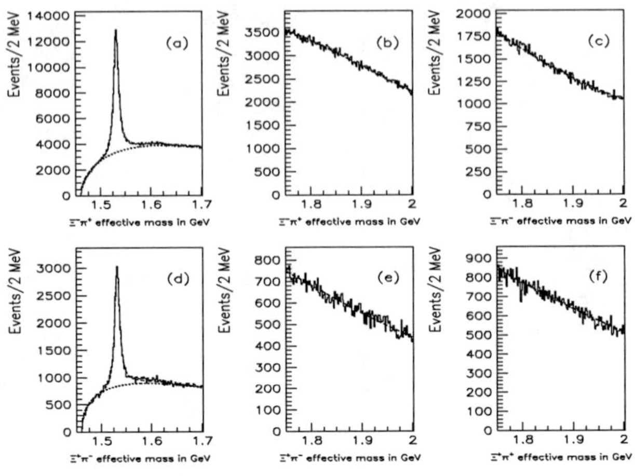

FIGURE 2. Fits to the $\Xi\pi$ effective mass spectra: a) and d) show the region near 1530 MeV. The superimposed curves show the background function and the sum of the background and a relativistic Breit-Wigner resonance, as described in the text. b), c), e), and f) show the region near 1862 MeV. The superimposed curves consist of the background function only.

production of $\Xi^0(1530)$ and $\overline{\Xi}^0(1530)$ in these data. In order to quantify this conclusion, we computed 95% confidence level upper limits[9] for Gaussian signals with mass = 1862 MeV and σ=7.6 MeV (the NA49 mass resolution). We estimated the yield of $\Xi^0(1530)$ and $\overline{\Xi}^0(1530)$ by fitting the $\Xi^-\pi^+$ and $\overline{\Xi}^+\pi^-$ effective mass distributions to the sum of a background function and a signal consisting of a relativistic ($L = 1$) Breit-Wigner function convoluted with a Gaussian with $\sigma = 2.5$ MeV. The Gaussian smearing term represents the spectrometer mass resolution for effective mass close to 1530 MeV. The form of the background function was $F(M) = (M - M_T)^{\alpha} e^{\beta(M-M_T)}$, where M_T is the $\Xi\pi$ threshold mass of 1.46 GeV^2 and α and β are fit parameters [10]. The results of the fits are shown in Fig. 2. The obtained yields were 93728 ± 422 $\Xi^0(1530)$ events, and 22211 ± 219 $\overline{\Xi}^0(1530)$ events (statistical errors only.) The upper limits computed for a Gaussian enhancement at 1862 MeV with $\sigma = 7.6$ MeV are given in Table 1.

TABLE 1. 95% confidence level upper limits for a signal at 1862 MeV. Each limit is also expressed as a fraction of the $\Xi^0(1530)$ or the $\overline{\Xi}^0(1530)$ yield, for baryon and antibaryon modes.

Decay Mode	Limit on Gaussian signal (events) (Mass = 1862 MeV, $\sigma = 7.6$ MeV)	Decay Mode	Limit on Gaussian signal
$\Xi^-\pi^+$	1020 (1.1%)	$\overline{\Xi}^+\pi^-$	310 (0.3%)
$\Xi^-\pi^-$	290 (1.3%)	$\overline{\Xi}^+\pi^+$	288 (1.3%)

Events in the exclusive reaction (2) where selected to satisfy the required topology by imposing energy and momentum conservation. Due to the low multiplicity in this reaction, combinatorics are very limited. In the selection process we imposed a loose

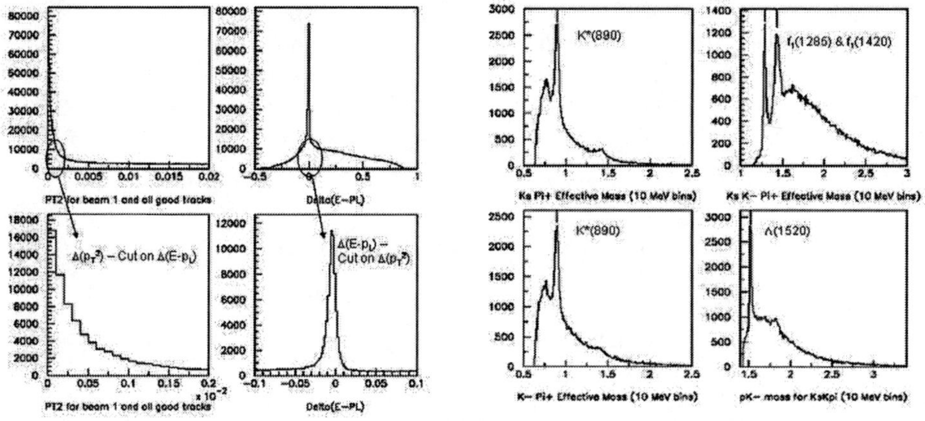

FIGURE 3. Left panel: Events in reaction (2) without cuts (upper plots), and with $\Delta E - p_L$ (lower left) or Δp_T^2 cut (lower right.) Right panel: $K_s\pi^+$, $K_sK^-\pi^+$, $K^-\pi^+$ and pK^- invariant mass distributions.

cut on the conservation of longitudinal momentum (5 GeV/c), but we applied tight cuts on the overall p_T^2 ($p_T^2 < 0.002$ GeV^2/c^2) and $E - p_L$ ($-0.2 < E - p_L < 0.15$ GeV.) We also required that these cuts eliminated events consistent with topology of $p_{slow}K_sK^+\pi^-p_{fast}$, which would not give the adequate strangeness for Θ^+ decays. The energy momentum conservation cuts are depicted in Fig.3 (left panel.) 68,050 events were selected with these cuts. Of these, only 6% had two solutions per event, corresponding to π^+/p_{slow} ambiguity. As it is shown in Fig.3 (right panel), many well known resonances can be seen in these events.

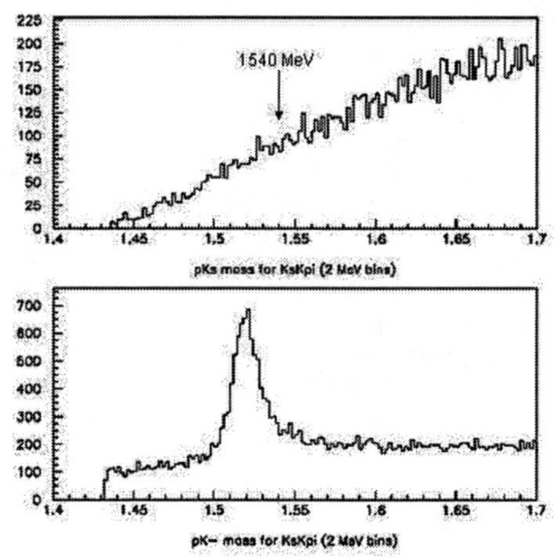

FIGURE 4. pK_s and pK^- invariant mass distributions below 1.7 GeV/c^2 for events in reaction (2).

In Fig.4 we compare the pK_s^0 and pK^- invariant mass distributions. The pK^- effective mass shows about 5000 $\Lambda(1520)$ events above background, with a FWHM of about 14 MeV. However, there is no indication of any exotic resonance structure in the pK_s^0 effective mass distribution below 1.7 GeV/c^2. The Monte Carlo pK_s^0 mass resolution (σ) at 1540 MeV is 1.5 MeV. The estimated yield of a narrow pK_s^0 resonance at 1540 MeV is less than 25 events (95% CL.)

ACKNOWLEDGEMENTS

We acknowledge the superb efforts by the staffs at the University of Massachusetts, Columbia University, Fermilab, and Lawrence Livermore National Laboratory. This work was supported in part by National Science Foundation Grants No. PHY90-14879 and No. PHY89-21320, by the Department of Energy Contracts No. DE-AC02-76 CHO3000, No. DE-AS05-87ER40356 and No. W-7405-ENG-48, and by CoNaCyT of México under Grants 2002-C01-39941, and NSF-J200.810.

REFERENCES

1. T. Nakano *et al.* (LEPS Collaboration), Phys. Rev. Lett **91** (2003) 012002. V. V. Barmin *et al.* (DIANA Collaboration), Phys. Atom. Nucl. **66** (2003) 1715. S. Stepanyan *et al.* (CLAS Collaboration), Phys. Rev. Lett **91** (2003) 252001. J. Barth *et al.* (SAPHIR Collaboration), Phys. Lett. B **572** (2003) 127. A. Airapetian, *et al.* (HERMES Collaboration), Phys. Lett. **B585** (2004) 213. S. Chekanov, *et al.* (ZEUS Collaboration), Phys. Lett. **B591** (2004) 7. M. Abdel-Bary, *et al.* (COSY-TOF Collaboration), Phys. Lett. **B595** (2004) 127. V. Kubarovsky, *et al.* (CLAS Collaboration), Phys. Rev. Lett. **92** (2004) 032001.
2. See, for example, D. Strottman, Phys. Rev. **D20** (1979) 748.
3. R. Jaffe and F. Wilczek, Phys. Rev. Lett **91** (2003) 202003.
4. C. Alt *et al.* (NA49 Collaboration), Phys. Rev. Lett.**92** (2004) 042003.
5. I. Abt, *et al.* (HERA-B Collaboration), Phys. Rev. Lett. **93** (2004) 212003.
6. M.J. Longo *et al.* (HyperCP Collaboration), Phys. Rev. **D70** (2004) 111101(R). Yu.M. Antipov *et al.* (SPHINX Collaboration), Eur. Phys. J. **A21** (2004) 455. J.Z. Bai, *et al.* (BES Collaboration), Phys. Rev. **D70** (2004) 012004. S. Schael, *et al.* (ALEPH Collaboration), Phys. Lett. **B599** (2004) 1.
7. M.I. Adamovich, *et al.* (WA89 Collaboration), Phys. Rev.**C70** (2004) 022201. A. Airapetian, *et al.* (HERMES Collaboration), Phys. Rev.**D71** (2005) 032004. S. Chekanov, *et al.* (ZEUS Collaboration), Phys. Lett. **B610** (2005) 212. E.S. Ageev, *et al.* (COMPASS Collaboration), Eur. Phys. J. **C41** (2005) 469.
8. E.P. Hartouni *et al.*, Phys. Rev. Lett. **72** (1994)1322, M.H.L.S. Wang *et al.*, Phys. Rev. D **56** (1997) 2544.
9. The calculation of 95% confidence level upper limits was done using essentially the same procedure as outlined by G.J. Feldman and R.D. Cousins in Phys. Rev. **D57**, 3873 (1998).
10. D. Aston *et al.*, Phys. Rev. D **32**, 2270 (1985).

The Mass Spectra of the Peculiar Hadrons

Hiroshi Noya* and Hiroshi Nakamura†

*Institute of Physics, Hosei University at Tama, Machida, Tokyo 194-0298, Japan
†2-25-3 Moridai, Atugi, Kanagawa 243-0037, Japan

Abstract.
We calculated the mass spectrum for the hadron by the Diquark Cluster Model and the Two Center Quark Model.

Keywords: hadrons, diquark, spectroscopy
PACS: 14.20.Pt, 14.20.Lq, 14.20.Jn

THE DIQUARK CLUSTER MODEL (DCM) CALCULATION

The Diquark Cluster Model is based on a jj-coupling shell model and the mass formula is the following [1]

$$M = m\, n_T + M(1p_{1/2})\, n(1p_{1/2}) + M(1p_{3/2})\, n(1p_{3/2}) + M(1d_{3/2})\, n(1d_{3/2}) + \Delta_0\,(n_{\phi_0} + n_{\bar{\phi}_0}) + \Delta_1\,(n_{\phi_1} + n_{\bar{\phi}_1}) + \Sigma\Delta_{TS}$$

The mass for u or d quark m and the parameter of the harmonic oscillator ω are $m = \omega = 300$ MeV.

The excitation energy of u or d quark from the $1s_{1/2}$ state are $M(1p_{1/2}) = 150$ MeV for the $1p_{1/2}$ state and $M(1d_{3/2}) = 375$ MeV for the $1d_{3/2}$ state, respectively.
The interaction energies of the diquark cluster in the s state are

$$\Delta_0 = a - 3/4b \text{ for the spin } 0 \quad \Delta_1 = a + 1/4b \text{ for the spin } 1\ ,$$
$$a = 187 \text{ MeV}, \; b = 195 \text{ MeV}.$$

The interaction energies of the diquark cluster in the p state are $\Delta_{00} = \Delta_{11} = \Delta_{01} = 0$, $\Delta_{10} = -60$ MeV.

The parameters for the strange quark are the following.
A strange quark mass is $m' = 476$ MeV, and the excitation energy for the strange quark are $M'(1p_{1/2}) = 178$ MeV for the $1p_{1/2}$ state, and $M'(1d_{3/2}) = 387$ MeV for the $1d_{3/2}$ state, respectively.

The interaction energy of the diquark cluster in the s state with a strange quark is $a' = 171$ MeV, $b' = 123$ MeV, and between u or d quark and s quark interaction energy in the p state diquark cluster are $\Delta'_{1/2,0} = \Delta'_{1/2,1} = 0$.

---

[1] e-mail:hnoya@mt.tama.hosei.ac.jp

CP814 *Hadron Spectroscopy: Eleventh International Conference*
edited by A. Reis, C. Göbel, J. Sá Borges, and J. Magnin

The parameters, except for a', were determined to the analysis of the strange dibaryon [2]. The parameter a' is determined by the mass of $f_0(1710)$.
One of the example of the DCM, we calculated the mass of K = +1 meson as the pentaquark system (Fig.1)

The parameters of the charm quark are the following.
The charm quark mass is m^c= 1645 MeV which is determined by using the charm lambda baryon and the charm sigma baryon masses. The excitation energy for the charm quark are $M^c(1p_{1/2})$ = 123 MeV for the $1p_{1/2}$ state, and $M^c(1d_{3/2})$ = 249 MeV for the $1d_{3/2}$ state, respectively.
The interaction parameters for the s state in the diquark cluster with the u or d quark and where by c quark is a^c = 51.7 MeV, b^c = 123 MeV. The parameter a^c is determined by the mass of Y(3940) [3].

Y(3940) is cq-$\bar{c}\bar{q}$ system insted of $c\bar{c}$ because this particle is able to decay to $c\bar{c}$ (J/ψ) and $q\bar{q}$ (ω). We assume that the X(3872) is the molecular state of DD*. The situation is similar as $a_0(980)$ $1^-(0^{++})$ which is K $\bar{K}$ bound state and above this state the $f_0(1380)$ $0^+(0^{++})$ state exists. This state is the four quark state according to our opinion. We calculated the mass of the anti-charmed baryon [4] states (Fig.2)

THE TWO CENTRE QUARK MODEL (TQM) CALCULATION

The mass formula for the Two Centre Quark Model [5] which was applied to the u and d quark system is
$M = U(k, h) + M_1(k)(i_1^2 + 1/3\ s_1^2 - k) + M_2(h)(i_2^2 + 1/3\ s_2^2 - h).$

We expand this formula to the flavor SU_3 group. The modified mass formula is

$$M = U_f(k, h) + (m' - m)(k_s + h_s) + M_1(k)\{(F_1^f)^2 + 1/3\ s_1^2 - 19/12k\} + M_2(h)\{(F_2^f)^2 + 1/3\ s_2^2 - 19/12h\}$$

$$U_f(k, h) = U(k, h) - k(k-1)\ M_1(k)/12 - h(h-1)\ M_2(h)/12.$$

$U_f(k,h)$ is a flavor, an isospin and a spin independent energy. $M_i(k)$ is an isospin and a spin dependent energy. $(F_i^f)^2$ is the flavor generating operator and each values are in Table 1.

TABLE 1. $(F_i^f)^2$

(Rep.)	(1)	(3)	(6)	(8)	(10)	(15_1)	(15_2)	(21_1)	(21_2)
$F(_1^f)^2$	0	4/3	10/3	3	6	23/3	16/3	40/3	25/3

The parameters for U(k,h), $M_i(k)$ are in Tables 2 and 3.

TABLE 2. U(k,h)

(k,h)	(2,2)	(4,4)	(5,5)	(7,1)	(4,1)	(8,2)	(10,4)	(11,5)
U(k,h)	394	573	640	1917	1094	1958	2050	—

TABLE 3. $M_i(k)$

k	2	4	5	7
$M_i(k)$	21	27	15	5.3

1) Using this mass formula, we calculated for the mass of $q^5 s\, \bar{s}^2$ (S = +1 pentaquark [6]) system.
Lowest state is :$(F_1^f, s_1) = (4/3, 1/2)$, $(F_2^f, s_2) = (10/3, 1)$
I = 0 , S = 1/2, 3/2, M = 1582 (MeV)
1st. excited state is :$(F_1^f, s_1) = (4/3, 3/2)$, $(F_2^f, s_2) = (10/3, 1)$
I = 0 , S = 1/2∼5/2, M = 1597 (MeV)
2nd. excited state is :$(F_1^f, s_1) = (16/3, 1/2)$, $(F_2^f, s_2) = (10/3, 1)$
I = 0 , S = 1/2, 3/2, M = 1642 (MeV)

2) In the case of S = – 2 baryon [7], we calculated the mass of the $q^2\, s^3\, q\bar{s}$ system.
$(F_1^f,s_1) = (25/3,1/2)$, $(F_2^f,s_2) = (10/3,1)$ I = 1/2,3/2, S = 1/2,3/2, M = 1862 MeV
The agreement between the experimental data and calculated one is very good!

REFERENCES

1. 1. Y.Uehara, N.Konno, H.Nakamura and H.Noya, Nucl.Phys. **A606**, 357 (1996)
2. 2. N.Konno, H.Nakamura and H.Noya, Phys. Rev. **37**, 154 (1987)
3. 3. H.Noya and H.Nakamura, AIP Conference **717**, 100 (2004)
4. 4. A.Aktas et al.,Phys.Lett.**B 588**,17 (2004)
5. 5. S.K.Choi et al.,Phys.Rev.Lett.**94**,182002 (2005)
6. 6. T.Nakano et al.,Phys.Rev.Lett.**91**, 012002 (2003)
7. C.Alt et al.,Phys.Rev.Lett. **92**,042003 (2004)
8. S. Eidelman *et al.* (The Particle Data Group), Phys.Lett. **B592** , 1(2004)

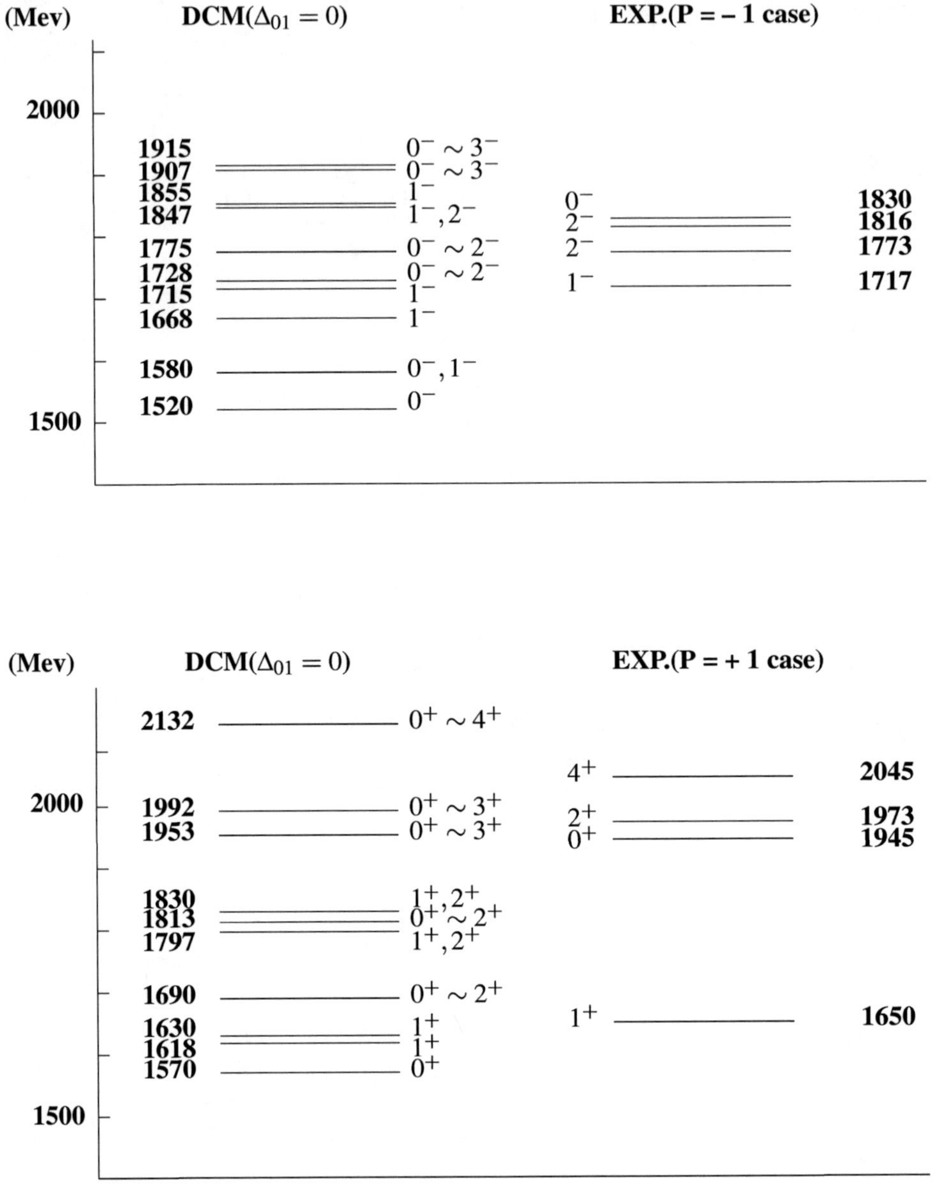

FIGURE 1. The mass spectrum of the $q^2\bar{q}\bar{s}$ system. The experimental data : P.D.G [8]

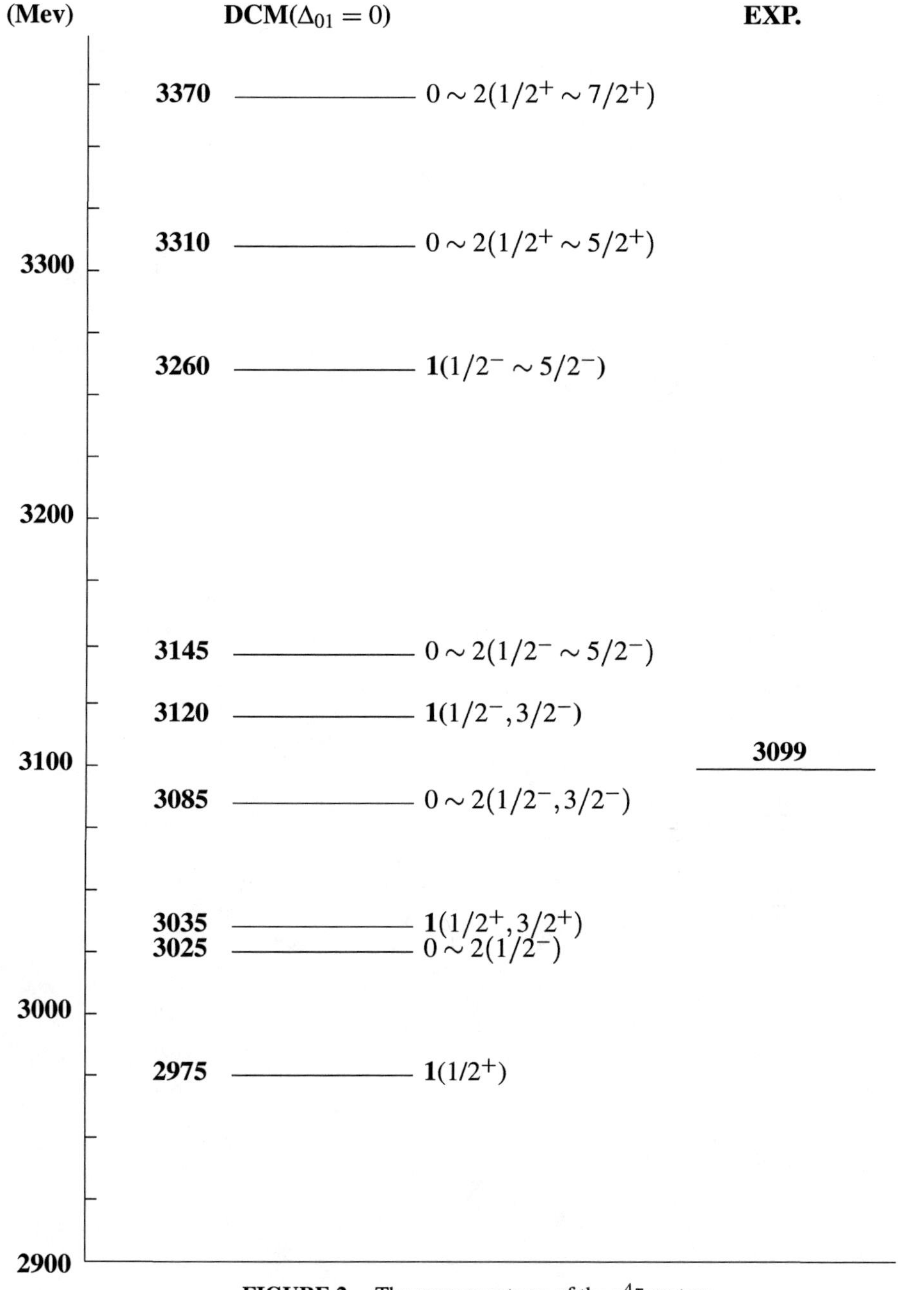

FIGURE 2. The mass spectrum of the $q^4\bar{c}$ system.

HADRON STRUCTURE

Meson structure in soft hadronic reactions at COMPASS

M. Colantoni on the behalf of the COMPASS coll.

Università del Piemonte Orientale e INFN-To

Abstract. The measurement of the electric ($\overline{\alpha}_\pi$) and magnetic ($\overline{\beta}_\pi$) pion polarizabilities supplies a significant test of QCD predictions in particular in the framework of the chiral perturbation theory.

In this perspective we have measured with the COMPASS spectrometer the t-dependence of the cross section for the reactions: $\pi^- + Z \to \pi^- + Z + \gamma$, and $\mu^- + Z \to \mu^- + Z + \gamma$ selecting events corresponding to the Compton $\pi\gamma$ and $\mu\gamma$ scattering respectively. From a fit of the data of the first reaction we can extract $\overline{\alpha}_\pi$ and $\overline{\beta}_\pi$, from those of the second the point-like contribution. This procedure minimizes the systematic errors. Details on the experiment and the present status of the analysis of the data collected in 2004 will be discussed.

Keywords: pion, polarizability, Compton scattering, kaon, chiral anomaly
PACS: 11.30.Rd; 12.39.Fe; 13.40.-f; 13.60.Fz; 13.75.Gx; 14.40.Aq

INTRODUCTION

The pion electric ($\overline{\alpha}_\pi$) and magnetic ($\overline{\beta}_\pi$) polarizabilities probe the response, to an external electromagnetic field, of the $q\overline{q}$ system. Those coefficients are fundamental parameters in pion physics and calculations provide important test for pionic structure as studied in QCD, or from various effective model of the pion (see [1, 2] for a theory review).

The Chiral Perturbation theory (χPT) [3] has been very successful in describing low energy hadronic properties and provides for the pion polarizabilities:

$$\overline{\alpha}_\pi = (2.4 \pm 0.5) \cdot 10^{-4}\ fm^3; \quad \overline{\beta}_\pi = (-2.1 \pm 0.5) \cdot 10^{-4}\ fm^3 \text{ from ref. [4].}$$

An improvement in the theoretical error bars for these values is foreseen in a near future [5].

Taking into account for all different theoretical approaches, the predictions for $\overline{\alpha}_\pi$ fall in the range $(2.4, 8.0) \cdot 10^{-4} fm^3$, whereas for $\overline{\beta}_\pi$ vary between $(-8.0, -2.1) \cdot 10^{-4} fm^3$.

These coefficients have been measured using:

- the crossed channel of $\gamma\gamma \to \pi^+\pi^-$
- the pion photoproduction from nucleons: $\gamma + N \to \gamma + N + \pi$.
- the radiative pion nucleus scattering or Primakoff reaction: $\pi + Z \to \pi + Z + \gamma$.

The results found are listed in Table 1. They are quite uncertain, therefore more precise measurements are needed. This can be done with the COMPASS spectrometer that provides good energy, momentum and angular resolution, high acceptance and stands high beam intensity.

CP814 *Hadron Spectroscopy: Eleventh International Conference*
edited by A. Reis, C. Göbel, J. Sá Borges, and J. Magnin

TABLE 1. Experimental values of $\overline{\alpha}_\pi$, $(\overline{\alpha}_\pi+\overline{\beta}_\pi)$, $(\overline{\alpha}_\pi-\overline{\beta}_\pi)$

Data	**Reaction**	$\overline{\alpha}_\pi$ $[10^{-4} fm^3]$
Lebedev [6]	$\gamma+N\to\gamma+N+\pi$	20±12
PLUTO [7]	$\gamma\gamma\to\pi^+\pi^-$	19.1±4.8±5.7
DM1 [8]	$\gamma\gamma\to\pi^+\pi^-$	17.2±4.6
DM2 [9]	$\gamma\gamma\to\pi^+\pi^-$	26.3±7.4
Mark II [10]	$\gamma\gamma\to\pi^+\pi^-$	2.2±1.6
Serpukhov [11]	$\pi+Z\to\pi+Z+\gamma$	6.8±1.4±1.2

Data	**Reaction**	$(\overline{\alpha}_\pi+\overline{\beta}_\pi)$ $[10^{-4} fm^3]$
Serpukhov [12]	$\pi+Z\to\pi+Z+\gamma$	1.4±3.1±2.8
Data	**Reaction**	$(\overline{\alpha}_\pi-\overline{\beta}_\pi)$ $[10^{-4} fm^3]$
Mami A2 [13]	$\gamma+p\to\gamma+\pi^++n$	11.6±1.5±3.0±0.5

THE PRIMAKOFF REACTION

The radiative pion nucleus scattering or Primakoff reaction is characterized by the sharp dependence on t of the cross section for very small value of t ($t \le 10^{-3}(GeV/c)^2$) ($t=(p^b_\pi - p^s_\pi - p_\gamma)^2$ where p^b_π, p^s_π, p_γ represent respectively the four-momentum of the pion beam, the scattered pion and the real gamma). In this region the contribution of the diffractive interaction background remains small [14].

When given in the anti-laboratory system (alab), that is the rest frame of the incoming pion, the differential cross section is described by the formula:

$$\frac{d^3\sigma}{dtd\omega dcos\theta} = \frac{2\alpha^3 Z^2}{m_\pi^2\omega}\frac{t-t_0}{t^2}|F_A(t)|^2\cdot\left(F^{Pt}_{\pi\gamma} + \frac{m_\pi\omega^2}{\alpha}\cdot\frac{\overline{\alpha}_\pi(1+cos^2\theta)+2\overline{\beta}_\pi cos\theta}{(1+\frac{\omega}{m_\pi}(1-cos\theta))^3}\right) \quad (1)$$

where $t_0 = (\frac{m_\pi\omega}{p_{beam}})^2$, m_π is the pion mass, ω the energy of the virtual photon in the alab system. θ is the angle between the real photon and the virtual photon directions in the alab system, α is the fine structure constant, Z is the charge of the nuclear target, $F_A(t)$ is the electromagnetic form factor of the nucleus ($F_A(t)\approx 1$ for $t<10^{-3}(GeV/c)^2$). Here $F^{Pt}_{\pi\gamma}$ [15] describes the differential Compton cross section for the scattering of photon on a point-like spin-0 particle. The main characteristic of the cross section (Eq. 1) is that it depends on $(\overline{\alpha}_\pi+\overline{\beta}_\pi)$ at forward angles and on $(\overline{\alpha}_\pi-\overline{\beta}_\pi)$ at backward angles.

For high values of Z and higher beam momenta (t_0 gets smaller) the cross section (see Eq. 1) becomes higher. For this reason for the experiment we are reporting here a pion beam momentum of 190 GeV/c has been chosen together with a lead target. For consistency, to check for the Z^2 dependence of the cross section, also carbon (C) and copper (Cu) targets have been used. These additional measurements provide also a cross check of the diffractive background subtraction below the Coulomb region.

Unique feature of COMPASS experiment is the availability of the muon beams that allow for a measurement of $F^{Pt}_{\mu\gamma}$ in the same experimental conditions as in the pion data taking. In this way we have a better control of the systematics errors in the measurement.

A description of the COMPASS spectrometer as used for the 2004 hadron run can be found elsewhere [16]. The performances of the apparatus have been investigated with Monte-Carlo simulations [17, 18, 19, 20, 21]. In particular it was found that the resolution on t, needed to distinguish the EM contributions from the diffractive scattering, was lower than $5 \cdot 10^{-4}(GeV/c)^2$.

THE MEASUREMENT AT COMPASS

Here will be described the *preliminary* analysis of the 2004 data taken with 190 GeV/c pion minus beam and the lead target. In a beam period of about 5 days we collected a total beam flux of more than $10^{11}\pi$.

To detect pions ($|\overrightarrow{p^{s}_{\pi}}|$ 20-110 GeV/c) we added to the COMPASS setup [16] a scintillator hodoscope just in front of the second electromagnetic calorimeter (ECAL2), about 36 m downstream of the target. The energy of the gammas was measured in ECAL2. The coincidence of these two detectors was the trigger selecting Primakoff events and allowing together with the track reconstruction the complete determination of the energy of the $\pi\gamma$ system. For high energy gammas (E_γ>100 GeV) when pion momentum is lower than 20 GeV/c and then the pion can not be detected by the hodoscope, the trigger was given by ECAL2 alone. During the analysis a single pion track was required in the trackers upstream of the hodoscope.

For the data analysis we applied the following main cuts:

1. the reconstructed position of interaction vertex has to fall within ± 5 cm of the position of 3 mm Pb target,
2. total energy $E_{tot} = E_\pi + E_\gamma = E_{beam} \pm 25GeV$,
3. $t < 1.5 \cdot 10^{-3}(GeV/c)^2$.

In fig.1a is reported the reconstructed position of the interaction vertex distribution along the beam axis, from which the target and the nearby detectors can be clearly distinguished. The plot of E_{tot}, obtained selecting only the target events, is given in fig.1b. The peak of the E_{tot} distribution is well centered around the nominal energy of the beam. Applying the E_{tot} cut we can get the t-dependence of the yield given in fig.2. The shape of the distribution is clearly consistent with what is expected for a Coulomb interaction. The mentioned cuts strongly suppress (factor 10) the contamination from the K^- (4.5% of the beam) decay, $K^- \to \pi^- + \pi^0$, where a K^- is misidentified and only one photon coming from the $\pi^0 \to \gamma\gamma$ decay is detected by the trigger. This can be also seen in fig.2 where the remain background from the strong interaction is minimal and will be also reduced from the analysis of the runs taken with an empty target. Other systematic errors will be evaluated from the data collected with the muon beam in the same experimental conditions. The data analysis is still in progress. In the figures is shown less than one tenth of the available statistics. When all the data will be analyzed we expect to have four times more data than in the previous Primakoff experiment [11, 12]

OTHER MEASUREMENTS

We foresee, in the Primakoff program, also the the measurement, for the first time, of the kaon polarizabilities. The pion beam contains about the 4.5% of kaons that will be selected with Cerenkov Differential counters with Achromatic ring focus (CEDAR)[22]. These detectors were successful tested in the 2004 pilot hadron run. We plan to use them in the 2007 hadron runs.

Because of the good spatial resolution of ECAL2, we can detect events with two photons in coincidence to reconstruct the π^0 mass. This allows to study the process $\pi^- + (Z,A) \to \pi^- + (Z,A) + \pi^0$ in parallel to the polarizabilities measurements. The goal is to extract the value of the pion form factor $F_{3\pi}$ in order to check the relation with the chiral axial anomaly term F_π via the low energy theorem [23].

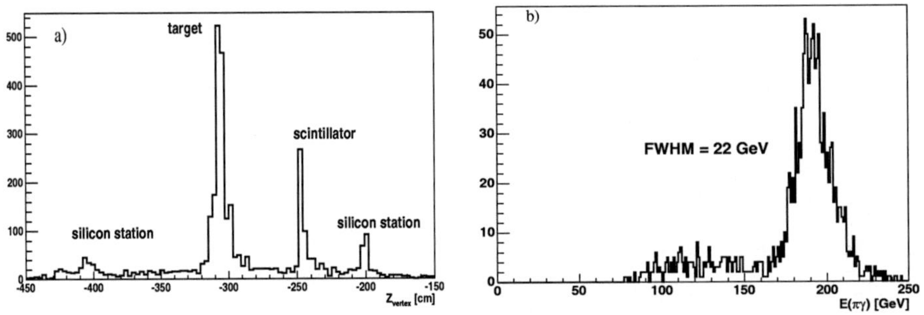

FIGURE 1. a) Vertex distribution along beam axis, b) Total energy of the system $\pi\gamma$

CONCLUSION

The COMPASS 2004 pilot hadron run was successfully completed, collecting an integrated beam flux of more than 10^{11} pions for the polarizabilities measurement. A preliminary analysis shows clearly the signature for the Primakoff reaction, and the statistics expected for this analysis is about 30-40K events that is at least four times larger than in the previous Primakoff experiment [11, 12]. Therefore COMPASS can provide a more precise measurement of the pion polarizabilities. The first measurement of the kaon polarizabilities and important checks of the chiral anomaly hypothesis are also foreseen for the near future.

REFERENCES

1. J. Portoles, M.R. Pennington, *The second DaΦne* Physics Handbook v.2, 579 (1999), hep-ph/9407295
2. C.A. Wilmot, R.H. Lemmer, Phy. Rev. C 65, 035206 (2002)
3. S. Weinberg, Physica **A 96** 327 (1979), J. Gasser and H. Leutwyler, Annals Phys. **158** 142 (1984), J. Gasser and H. Leutwyler, Nucl. Phys. **B250** 465 (1985)
4. U. Bürgi, Phys. Lett. **B377** 147 (1996)

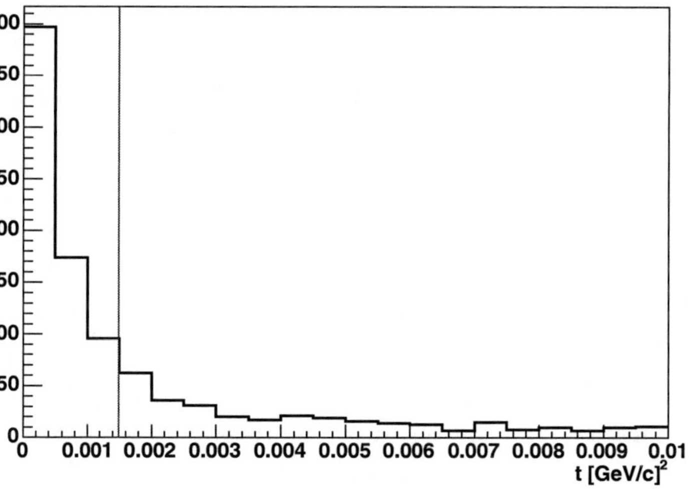

FIGURE 2. Yield plotted vs t

5. J. Gasser, Future Physics @ COMPASS, CERN Yellow Report 2004
6. T.A. Aibergenov et al., Cezch J. Phys B 36, 948 (1986)
7. C. Berger et al., Z. Phys. **C26** 199 (1984)
8. A. Courau et al., Nucl. Phys. **B271** 1 (1986)
9. Z. Ajaltoni et al.,*VII Int. workshop on photon-photon collision* Paris (1986)
10. J. Boyer et al., Phys. Rev **D42**, 1350 (1990), D. Babusci et al., Phys Lett. **B277**,158 (1992)
11. Yu M. Antipov et al., Phys. Lett. **B121**, 445 (1983)
12. Yu M. Antipov et al., Z. Phys. **C26**, 495 (1985)
13. J. Ahrens et al., Eur. Phys. J **A23** 113 (2005)
14. M. Colantoni, Czech. J. Phys. 55 43 (2005) Supp. A
15. M. Buenerd, Nucl. Phys. **A361** 111 (1995)
16. COMPASS Collaboration, *Proposal for a Common Muon Proton Apparatus for Structure and Spectroscopy*, http://wwwcompass.cern.ch/, A. Ferrero Czech. J. Phys. 55 35 (2005) Supp. A
17. M. A. Moinester, Pion polarizabilities and Hybrid meson structure at COMPASS, hep-ex/0012063
18. M. Sans Merce, Ph.D thesis, Ludwig Maximilians University, Munich, 2001
19. R. Kuhn, Diploma thesis, Technical University,Munich, 2001
20. A. Olchevski, M. Faessler, Experimental requirements for COMPASS Initial Primakoff Physics program, COMPASS coll. Meeting 2001.
21. M. Colantoni, Ph.D thesis,Universitá di Torino, Torino, 2002, XL International Winter Meeting on Nuclear Physics, Bormio, Italy, January 21-26, 2002
22. C. Bovet et al., The Cedar counters for particle identificationin the SPS beams: A description and an operation manual, SPS-note-82-40
23. S.L. Adler, Phys. Rev. 177 2426 (1969); J.S. Bell, R. Jackiw, Nuovo Cimento 60A 47 (1969)

Recent Results on the GDH Sum Rule on Neutron and Proton

P. Pedroni

INFN-Sezione di Pavia, via Bassi 6 Pavia, Italy
(for the GDH and A2 collaborations)

Abstract. The Gerasimov-Drell-Hearn (GDH) sum rule is a fundamental theorem that connects ground state properties of the nucleon with its helicity dependent total photoabsorption cross sections. For the first time, this basic theorem has been experimentally verified by the GDH collaboration with circularly polarized photons and longitudinally polarized nucleons at the tagged photon facilities of MAMI (Mainz) ($E_\gamma \leq 800$ MeV) and ELSA (Bonn). (800 MeV $\leq E_\gamma \leq$ 800 MeV).

In this presentation, the experimental approach will be presented together with the final results obtained for the proton and the first data obtained using a deuteron target. Data from the resonance region up to the onset of the Regge regime will be shown. Systematic uncertainties related to the evaluation of the GDH sum rule for the neutron will be discussed.

The behavior of the helicity dependence of the single pion photoproduction channels for $E_\gamma \leq$ 800 MeV, mainly in connection with the excitation of the D_{13} resonance, will be also presented.

Keywords: GDH sum rule; pion photoproduction; Baryon spectroscopy
PACS: 13.60.Le, 13.60.Hb, 25.20.Lj

INTRODUCTION

The Gerasimov-Drell-Hearn (GDH) sum rule [1, 2], relates the total absorption cross section of circularly polarized photons on longitudinally polarized nucleons to the static properties of the nucleon. The two relative spin configurations, parallel or antiparallel, determine the two absorption cross sections $\sigma_{3/2}$ and $\sigma_{1/2}$. The integral over the photon energy ν of the difference of these two cross sections, weighted by the inverse of ν, is related to the nucleon mass, M, and anomalous magnetic moment, κ, as follows:

$$\int_{\nu_0}^{\infty} (\sigma_{3/2} - \sigma_{1/2}) \frac{d\nu}{\nu} = \frac{2\pi^2\alpha}{M^2}\kappa^2 \qquad (1)$$

where ν_0 is the pion photoproduction threshold and α the fine-structure constant.

The GDH sum rule, formulated in the 1960's, is based on very general physics principles (Lorentz-invariance, gauge invariance, unitarity) and an unsubtracted dispersion relation applied to the forward Compton amplitude. Due to its fundamental character this prediction requires an experimental verification which has been awaiting technical developments that have only recently been achieved.

In addition, the measurement of the helicity dependence of the partial $\gamma N \to N\pi(\pi)$ reaction channels gives a new insight into the excitation spectrum of the nucleon. Despite a long history and large experimental effort, our understanding of this spectrum is far from complete. Many predicted states are not sufficiently well established and many properties of the observed states (e.g. coupling constants, branching ratios, helic-

CP814 *Hadron Spectroscopy: Eleventh International Conference*
edited by A. Reis, C. Göbel, J. Sá Borges, and J. Magnin

ity amplitudes) are only poorly known. This is true in particular for resonances above the Δ, where a substantial overlap among the different resonances occurs. The helicity dependent observables give the possibility to access the smaller (and less known) electromagnetic multipoles through their interference with the leading terms.

The aim of the GDH collaboration[1] is to provide an extensive data set of helicity dependent cross sections for all the partial and the total reaction channels both on the proton and on the neutron with a combined use of the MAMI (Mainz) ($m_\pi \leq E_\gamma \leq$ 800 MeV) and ELSA (Bonn) (600 MeV $\leq E_\gamma \leq$ 3 GeV) accelerators. Data takings using polarized proton and deuterium targets have been completed at both facilities and several results have already been pulished ([3] – [11]) In the following, the latest results taken both on the total inclusive and on pion production processes will be presented.

EXPERIMENTAL SETUP

The experiment was carried out at the tagged photon facilities of the MAMI accelerator in Mainz and of the ELSA accelerator in Bonn. In both cases, circularly polarized electrons were produced by bremsstrahlung of longitudinally polarized electrons. The sources of polarized electrons, based on the photoeffect on strained GaAs crystals, delivered electrons with a degree of polarization between $65-75\%$ [12]. The degree of polarization was continuously measured throughout the experiment by Möller scattering in a magnetized vacoflux foil.

Polarized nucleons were available in the frozen spin target [13] that was developed and built by the Bonn and Bochum groups. The system consisted of a horizontal dilution refrigerator and a superconducting polarization magnet, which was used in the polarization phase together with a microwave system for dynamical nuclear polarization (DNP). The polarization was maintained during the measurement in the "frozen spin" mode at temperatures of about 50 mK by an internal superconducting coil (B $\simeq$ 0.4 T) which is integrated into the dilution refrigerator. The target material was butanol (C_4H_9OH). At 2.5 T maximum polarization values close to 90% were obtained for the protons with a typical relaxation time in the "frozen spin" mode of about 200 hours. The holding field was homogeneous enough to allow for continuous NMR monitoring of the target polarization during the experiment.

Two detector concepts were used to meet the special requirements needed for the differen energy ranges covered by the Mainz and Bonn accelerators. In Mainz, the photon induced reaction products were registered by means of the detector DAPHNE[14], made by CEA Saclay and INFN - Sezione di Pavia, which wss complemented at forward angle by the silicon microstrip device MIDAS[15] to increase the solid angle acceptance.

The GDH detector in Bonn [16] has been specifically designed for measurements of total cross section; its concept is to detect at least one reaction product from all possible hadronic processes with almost complete acceptance concerning solid angle

[1] This collaboration is formed by researchers from the Universities of Mainz, Bonn, Bochum, Erlangen, Göttingen, Lund, Nagoya, Tübingen and from INFN-Sezione di Pavia, RUG Gent, CEA Saclay, INR Moscow.

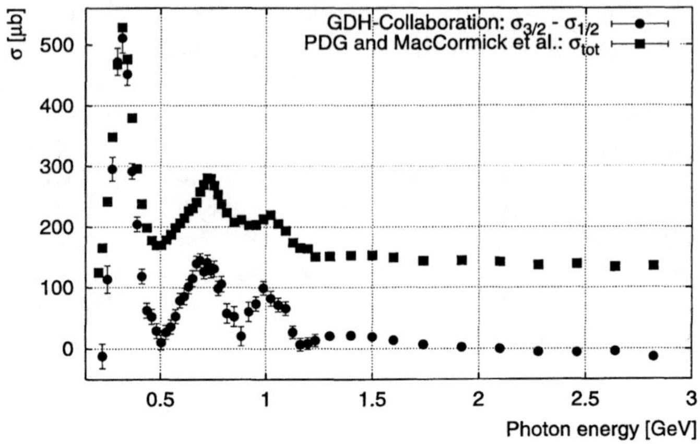

FIGURE 1. The measured helicity difference $\Delta\sigma = \sigma_{3/2} - \sigma_{1/2}$ for the total inclusive cross section on the proton [5, 7, 10] is compared to the unpolarized data [17, 18].

and efficiency. The overall acceptance for any hadronic process is better than 99%. In both cases, electromagnetic background was suppressed in the forward polar region by means of a threshold Cerenkov detector.

RESULTS AND DISCUSSION

Total inclusive cross section on the proton

The helicity difference $\Delta\sigma = \sigma_{3/2} - \sigma_{1/2}$ measured both at Mainz [5] and Bonn [7, 10] is shown in Fig. 1 (circles) and compared to the unpolarized cross section (squares) [17, 18] The three major resonances known from the unpolarized case ($\Delta(1232)$, $D_{13}(1520)$, $F_{15}(1680)$) are present also in the helicity dependent data. Their effect is now more pronounced since the non-resonant background has mostly disappeared.

The value of the GDH integral inside the measured photon energy range (between 200 MeV and 3 GeV) amounts to $254 \pm 5 \pm 12$ μb. Using the existing models [19, 20] for an evaluation of the missing contributions, we obtain a value of $\simeq -30\mu$b for $E_\gamma < 200$ MeV [19, 20] and $\simeq -15\mu$b for $E_\gamma > 3$ GeV [21, 22]. The combination of our experimental result with these predictions yields an estimate which, within the experimental sytematic errors, is consistent with the GDH sum rule value of 205 μb.

Total inclusive cross section on the deuteron

The situation is more complicated for the neutron, for which the sum rule predicts a value of 233 μb. Since no free neutron target is available, experimental data were taken using a polarized deuteron target. A theoretical support is then necessary to extract the

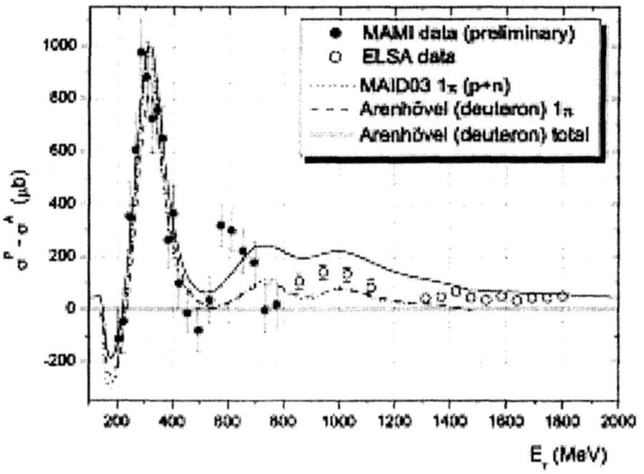

FIGURE 2. The measured helicity difference $\Delta\sigma = \sigma_{3/2} - \sigma_{1/2}$ for the deuteron obtained both in Mainz and Bonn[11]

helicity dependent neutron contribution from the measured data.

The results obtained up to now are shown in Fig. 2. While the published data from Bonn include the full collected statistics, the preliminary data from Mainz were derived using a small part of the total data sample.

In the same figure, the data are compared to the predictions of the MAID multipole model [19] for the free $(p+n)$ case and to the model developed by Arenhovel, Fix and Schwamb [23] with takes into account the deuteron nuclear effect. Apart form the region very close to the pion photoproduction threshold, the results of the MAID model are not very different from the single pion presults predicted using the deuteron nuclear model.

At energies above $\sim$ 500 MeV, nuclear effects should play a minor role, and the free neutron contribution can be simply estimated, in a first approximation, from the difference betweeen the deuteron and proton results. An analysis based on this idea [11] shows that, in this energy region, the neutron data are very similar to the proton ones.

Much more effort is now needed to improve the theoretical calculation and to perform further experimental analyses of exlusive reactions on a deuteron target to evaluate the nuclear effects and to reliably extract the helicity dependent free-neutron data.

Partial reaction channels

In the second resonance region (500 MeV$\leq$ E$_\gamma$ $\leq$900 MeV) the helicity dependent obdervables reveal a high sensitivity to the properties of the D_{13} resonance due to the presence of an interference terms between the electromagneitc multipoles (E_{2-} and M_{2-}) that are responsible for its excitation.

Preliminary data on the total unpolarized and polarized cross section for the $\gamma p \rightarrow n\pi^+$ channels are shown in Fig.3 in comparison with the predictions of MAID [19] and SAID [20] multipole analyses. The differences in the two models are much more

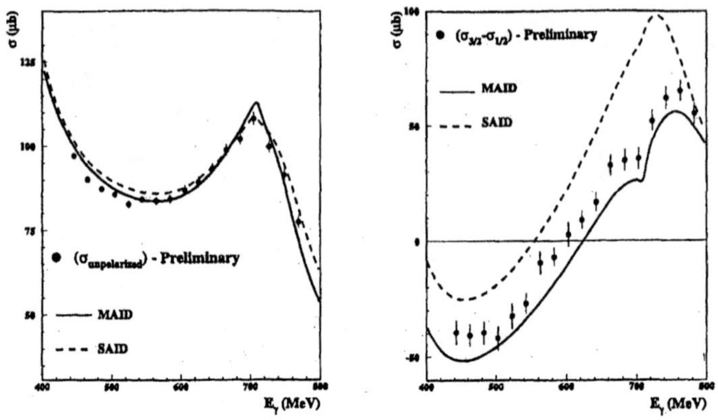

FIGURE 3. The preliminary total unpolarized cross section (left) and the helicity difference $\Delta\sigma = \sigma_{3/2} - \sigma_{1/2}$ for the $\gamma p \to n\pi^+$ channel in comparison with results from MAID and SAID analyses.

pronounced in the helicity dependent case. This originates from significant differences in the balance of the E_{0+} and E_{2-} multipoles, which enter with opposite signs in $\Delta\sigma$. From this example it becomes obvious that these double polarization experiments serve many more purposes than just measuring the GDH integral; in particular they provide a very sensitive tool to study resonance properties.

REFERENCES

1. S. B. Gerasimov, *Sov. J. Nucl. Phys.* **2**, 430 (1966).
2. S.D. Drell, A.C. Hearn, *Phys. Rev. Lett.* **16**, 908 (1966).
3. J. Ahrens *et al.*, *Phys. Rev. Lett.* **84**, 5950 (2000).
4. J. Ahrens *et al.*, *Phys. Rev. Lett.* **87**, 022003 (2001).
5. J. Ahrens *et al.*, *Phys. Rev. Lett.* **88**, 232002 (2002).
6. J. Ahrens *et al.*, *Phys. Lett.* **B 551**, 49 (2003).
7. H. Dutz *et al.*, *Phys. Rev. Lett.* **91**, 192001 (2003).
8. J. Ahrens *et al.*, *Eur. Phys. J. A* **A17**, 241 (2003).
9. J. Ahrens *et al.*, *Eur. Phys. J. A* **21**, 323 (2004).
10. H. Dutz *et al.*, *Phys. Rev. Lett.* **93**, 032003 (2004).
11. H. Dutz *et al.*, *Phys. Rev. Lett.* **94**, 162001 (2005).
12. K. Aulenbacher *et al.*, *Nucl. Instrum. Methods A* **391**, 498 (1997).
13. C. Bradtke *et al. Nucl. Instrum. Methods A* **436**, 430 (1999).
14. G. Audit *et al.*, *Nucl. Instrum. Methods A* **301**, 473 (1991.)
15. S. Altieri *et al.*, *Nucl. Instrum. Methods A* **452**, 185 (2000).
16. K. Helbing *et al.*, *Nucl. Instrum. Methods A* **484**, 129 (2002).
17. M. MacCormick *et al.*, *Phys. Rev. C* **53**, 41 (1998).
18. Particle Data Group, S. Eidelman *et al.*, *Phys. Lett.* **B 592**, 1 (2004).
19. D. Drechsel *et al.*, *Phys. Rev. D* **63**, 114010 (2001).
20. R.A. Arndt *et al.*, *Phys. Rev. C* **66**, 055213 (2002).
21. N. Bianchi and A. Thomas, *Phys. Lett.* **B 450**, 439 (1999).
22. S. Simula *et al.*, *Phys. Rev. D* **65**, 034017 (2002).
23. H. Arenhövel *et al.*, *Phys. Rev. Lett.* **93**, 202301 (2004).

The dynamics of proton structure: Recent results from HERA

Tobias Haas[1]

DESY F1, Notkestrasse 85, 22603 Hamburg, Germany

Abstract. HERA has delivered over 450 pb^{-1} of luminosity to the experiments H1 and ZEUS with both lepton species and polarization of the lepton beam up to 50 %. From these data a large number of highly precise results on inclusive deep inelastic scattering have been obtained. This article reviews a selection of the most important results.

Keywords: deep inelastic scattering, structure functions, proton structure
PACS: 14.20.Dh, 14.65.-q, 14.70.Fm

INTRODUCTION

In the HERA collider at DESY 27.5 GeV/c electrons or positrons collide with 920 GeV/c protons. During the first data taking period from 1992 through 2000 HERA delivered more than 200 pb^{-1} to each one of the experiments H1 and ZEUS. Data from this period will be called HERA I data in the following. The second data period started in 2002. Since then HERA has delivered more than 250 pb^{-1}. Two thirds of these data are taken with an electron beam and one third with a positron beam. In addition the lepton beam is polarized with the polarization reaching 50 % at the end of HERA fills. The data from this data taking period will be called HERA II data in the following.

The availability of large statistics samples with both lepton species and polarization opens up a completely new class of investigations. First results from those will be discussed in the following. In particular in the second section we shall discuss the results on charged current reactions. In the third section results from neutral current reactions at high values of Q^2 will be shown. Using the deep inelastic e^+p and e^-p charged and neutral current scattering cross sections a combined electroweak and QCD analysis can be performed to determine electroweak parameters accounting for their correlation with the parton distributions. These results are shown in the fourth section. Finally some selected results on physics beyond the standard model are shown in the fifth section. A summary is given at the end.

CHARGED CURRENT REACTIONS

Deep inelastic charged current scattering proceeds via the exchange of charged heavy bosons, the $W^{\pm}$, which change the incoming electron or positron into the corresponding

[1] for the H1 and ZEUS Collaborations

CP814 *Hadron Spectroscopy: Eleventh International Conference*
edited by A. Reis, C. Göbel, J. Sá Borges, and J. Magnin

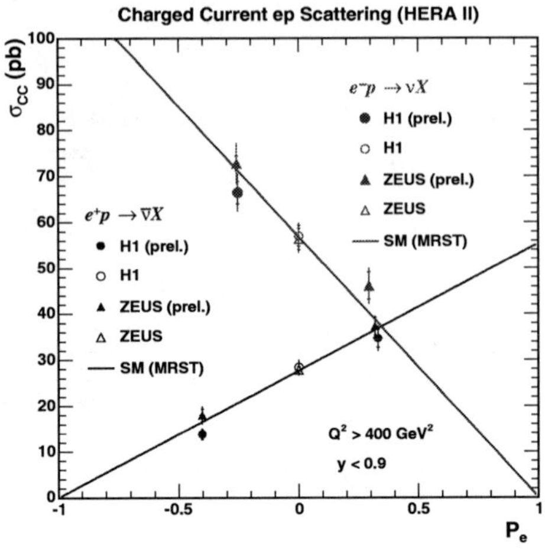

FIGURE 1. Charged current electron proton and positron proton scattering cross section as a function of the polarization, P_e, of the lepton beam. The H1 measurements are shown as the open and closed circles and the ZEUS measurements are shown as the open and closed triangles. The straight lines are the expectation in the Standard Model.

neutrino. The outgoing neutrino then escapes undetected. Charged current events are therefore characterized by large transverse momentum imbalance. The cross section for these processes is much smaller than the neutral current cross section at small values of the momentum transfer but becomes comparable at values of the momentum transfer close to the mass of the heavy bosons squared.

In the Standard Model charged current interactions are maximally parity violating. This means that only left-handed electrons (spin anti-parallel to the momentum) and right-handed positrons (spin parallel to the momentum) interact through the weak interaction. Therefore, the degree of polarization of the beam provides a mechanism to smoothly "switch" the charged current reaction on or off. The first results are shown in Fig. 1.

The average polarization reached for these measuresments was between 30 and 40 %. The lines shown in this figure are the expectation in the Standard Model. The dependence is linear with the cross section being equal to zero for the respective "wrong" helicity. Eventually with more statistics and greater average polarization the lines will be fitted to the data and the extrapolation of the positron data to $P_e = -1$ and the electron data to $P_e = +1$ will provide important contraints on the existance of right-handed weak currents or equivalently the existance of additional very heavy vector bosons.

Another effect that is very clearly visible in the charged current data and also shown in Fig. 1 is the fact that the cross section for unpolarized electron-proton scattering is approximately twice as large as that for positron-proton scattering. The effect is due to the different u and d quark densities in the proton. Due to electric charge conservation, charged current electron-proton scattering proceeds via W^- exchange while positron-

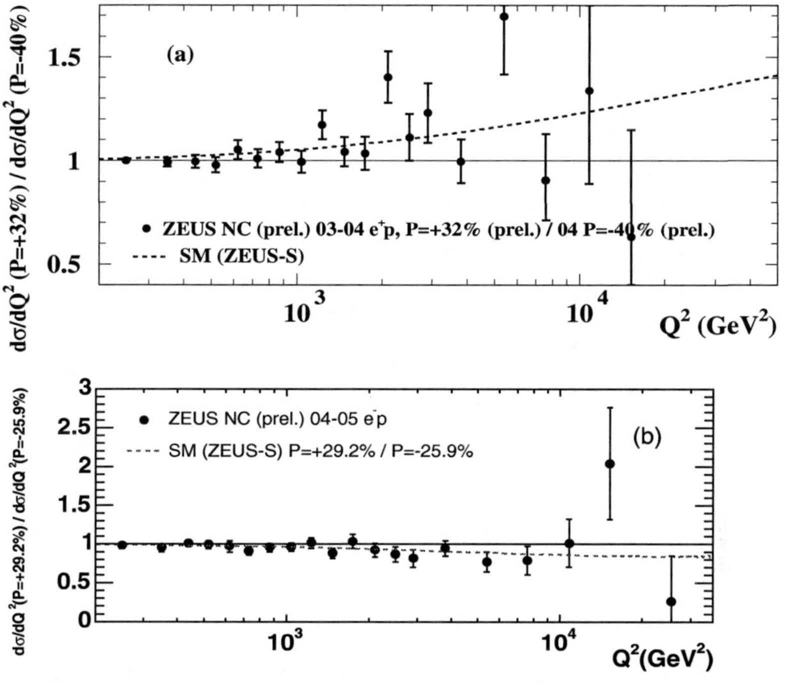

FIGURE 2. Ratios of neutral current cross sections for positive and negative helicity both for (a) positron-proton and (b) electron-proton scattering.

proton scattering proceeds via W^+ exchange. Again, due to electric charge conservation the W^- can only couple to up type quarks while the W^+ couples to down-type quarks. The difference between the two cross sections therefore reflects the difference in the up versus the down type densities in the proton.

NEUTRAL CURRENT REACTIONS AT VERY HIGH Q^2 AND THE EXTRACTION OF THE STRUCTURE FUNCTION F_3

In neutral current deep inelastic scattering the dependence on the lepton beam polarization enters through the interference of photon and Z^0 exchange. At HERA this process is dominated by the purely electromagnetic and therefore parity conserving photon exchange. However, at large values of Q^2 also Z^0 exchange contributes and parity violating effects can be observed. These will however be much smaller and hence more difficult to measure than in the charged current case.

Fig. 2. shows first results on the ratios of cross sections for positive and negative

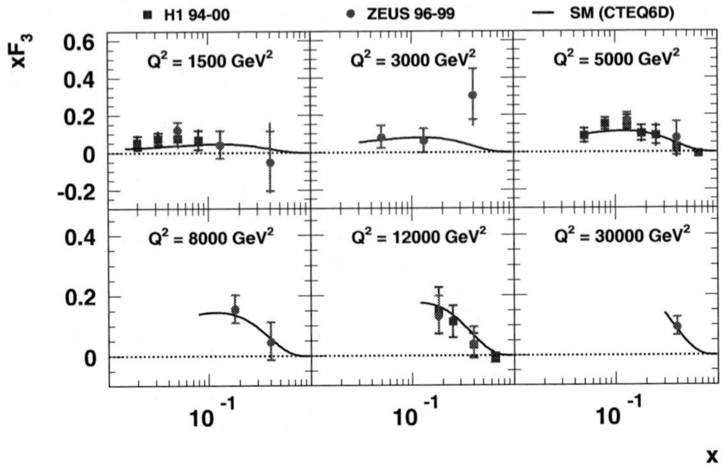

FIGURE 3. The parity violating structure function $xF_3(x)$ as a function of x_{Bj} in different bins of the momentum transfer Q^2. The H1 data are shown as the full squares and the ZEUS data are shown as the full circles.

helicities both for positron and electron beams. In the Standard Model it is expected that the cross section at high Q^2 is greater for positive than for negative helicities for positrons. The reverse is expected for electrons. These expectations are indicated by the dashed lines. The data generally follow the expected trends but the precision is far from sufficient to exclude the case of no polarization dependence at all. Here, future high statistics data are definitely required.

While the measurement of the polarization dependence of the neutral current cross section requires higher statistics than currently available, the dependence on the lepton beam flavour can be extracted with the currently existing statistics. This measurement provides information on the parity violating structure function $xF_3(x)$ which in turn depends on the u and d valence distributions in the proton. These dominate in the region of high x_{Bj} which is particularly relevant for searches beyond the Standard Model. Therefore this measurement is of particular interest for a number of different reasons.

Fig. 3 shows the published results from both H1 and ZEUS [1, 2] based on all the HERA I electron beam data (ca. 16 pb^{-1}) and a fraction of the positron beam data. The much larger sample of electron beam data already available from HERA II will allow a much more precise determination of $xF_3(x)$ soon.

CROSS SECTION FITS AND THE EXTRACTION OF ELECTRO-WEAK PARAMETERS

At HERA, the NC interactions at high Q^2 are sensitive both to γZ^0 interference and Z^0 exchange. Thus the NC data can be used to extract the weak couplings of up- and

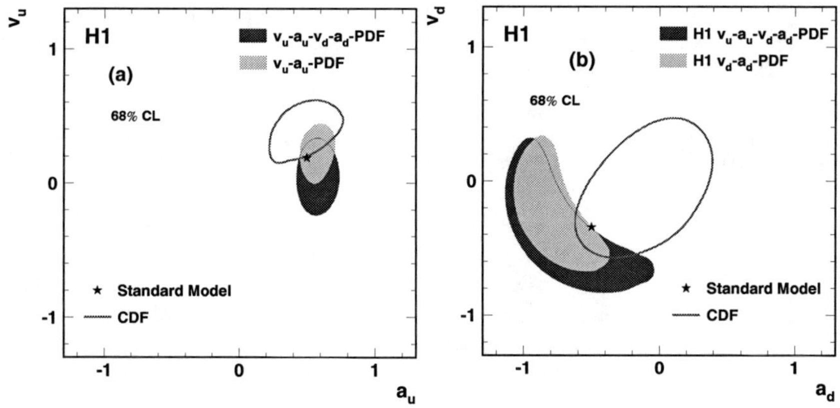

FIGURE 4. Results at 68% confidence level on the weak neutral current couplings of u (a) and d (b) quarks to the Z^0 boson. The dark-shaded contours correspond to results of a simultaneous fit of all four couplings and can be compared with those determined by the CDF experiment (open contours). The light-shaded contours correspond to results of fits where either d or u quark couplings are fixed to their SM values. The stars show the expected SM values.

down-type quarks to the Z^0 boson. At high Q^2 and high x_{Bj}, where the NC $e^{\pm}p$ cross sections are sensitive to these couplings, the up- and down-type quark distributions are dominated by the light u and d quarks. Therefore, this measurement can determine both the light quark densities as well as their couplings. The CC cross section data are used to disentangle the up and down quark distributions.

Fig. 4 shows the vector and axial-vector weak neutral current couplings of u and d quarks to the Z^0 boson as determined at HERA for the first time. The results from this analysis are also compared with similar results obtained recently by the CDF experiment [4]. The HERA determination has comparable precision to that from the Tevatron. These determinations are sensitive to u and d quarks separately, contrary to other measurements of the light quark-Z^0 couplings in νN scattering and atomic parity violation on heavy nuclei. They also resolve any sign ambiguity and the ambiguities between v_u and a_u of the determinations based on observables measured at the Z^0 resonance.

SELECTED RESULTS ON PHYSICS BEYOND THE STANDARD MODEL

The HERA collider operates at the highest CM energies ever reached in lepton proton scattering. So, in addition to being a laboratory for precision measurements, HERA is also potentially a discovery machine with sensitivity in a great number of different areas. This potential has already been explored in many ways with the HERA I data samples.

While no new physics has so far been found, a number of models and scenarios have been excluded and some anomalies have been observed that await clarification with the help of the HERA II data. It is impossible in the context of this article to give an overview over the multitude of ideas and models that have been investigated so far. Therefore using one example, an attempt is made to shine some light both on the richness of the subject and at the same time on the difficulties involved.

The production of real heavy charged $W^{\pm}$ mesons with subsequent decay into leptons is expected at HERA at a very small rate. A significant enhancement of that rate over the Standard Model prediction would be a powerful hint of the existance of new physics. These events are characterized by the presence of an isolated lepton with very high transverse momentum accompanied by a large transverse momentum imbalance. Searches for these events have been performed by both experiments and an excess of events has been observed in the electron channel by the H1 experiment while the ZEUS experiment has observed an excess of events in the tau channel. The results are summarized in table 1

However, the excesses seen both experiments barely exceed the 3σ level and are currently interpreted as statistical fluctuations. Further studies and the evaluation of additional HERA II data will hopefully clarify the situation.

SUMMARY

The lepton proton collider HERA is currently in its second and final data taking period. Already now a very large sample of deep inelastic scattering data has been recorded by the experiments H1 and ZEUS. A few selected results from these data have been presented in this paper. These results have opened up an entirely new kinematic domain in deep inelastic scattering. By the end of HERA operation the experiments expect to collect an integrated luminosity close to 1 fb^{-1}. This will once again more than double the already available data samples and allow to produce measurements of unprecedented precision thus pushing the frontiers of the unknown in subatomic physics quite a bit a further

ACKNOWLEDGMENTS

The author thanks the organizers for inviting him to report at this conference and to travel to a fascinating country and wonderful city.

REFERENCES

1. H1 Collaboration, C. Adloff et al., *Measurement of Charged and Neutral Current Cross Sections in Electron-Proton Collisions at high Q2*, Eur. Phys. J. C 19 (2001), 169–288.
2. ZEUS Collaboration, S. Chekanov et al., *Measurement of high-Q^2 e^-p neutral current cross sections at HERA and the extraction of xF_3 (revised)*, Eur. Phys. J. C 28 (2003) 2, 175-201.
3. H1 Collaboration, C. Adloff et al., *Deep-Inelastic Inclusive ep Scattering at Low x and a Determination of $alpha_s$*, Eur. Phys. J. C21 (2001), 33-61,
   H1 Collaboration, C. Adloff et al., *Measurement of Neutral and Charged Current Cross-Sections in*

TABLE 1. Summary of results of searches for events with isolated leptons and missing transverse momentum at HERA. The number of observed events is compared to the Standard Model predictions (observed/expected). The signal component of the Standard Model expectation, which is dominated by real W production is given as a percentage in parentheses. The quoted errors contain statistical and systematic uncertainties added in quadrature.

H1 preliminary	Electrons 1994-2005 $e^{\pm}$ 211 pb^{-1} obs./exp. (W)	Muons 1994-2005 $e^{\pm}$ 211 pb^{-1} obs./exp. (W)	Taus 1996-2000 $e^{\pm}$ 108 pb^{-1} obs./exp. (W)
All P_T^X	$25/20.4 \pm 2.9$ (68%)	$9/5.4 \pm 1.1$ (82%)	$5/5.8 \pm 1.4$ (15%)
$p_T^X > 25$ GeV/c	$11/3.2 \pm 0.6$ (77%)	$6/3.2 \pm 0.5$ (81%)	$0/0.5 \pm 0.1$ (49%)
ZEUS	**Electrons 1994-2000 $e^{\pm}p$ 130 pb^{-1} obs./exp. (W)**	**Muons 1994-2000 $e^{\pm}p$ 130 pb^{-1} obs./exp. (W)**	**Taus 1994-2000 $e^{\pm}p$ 130 pb^{-1} obs./exp. (W)**
$p_T^X > 25$ GeV/c	$2/2.90^{+0.59}_{-0.32}$ (45%)	$5/2.75^{+0.21}_{-0.21}$ (50%)	$2/0.2^{+0.05}_{-0.05}$ (49%)
$p_T^X > 40$ GeV/c	$0/0.94^{+0.11}_{-0.10}$ (61%)	$0/0.95^{+0.14}_{-0.10}$ (61%)	$1/0.07^{+0.02}_{-0.02}$ (71%)
ZEUS preliminary	**Electrons 2003-2004 $e^{+}p$ 40 pb^{-1} obs./exp. (W)**		
$12 < p_T^X < 25$ GeV/c	$1/0.46 \pm 0.10$ (64%)		
$p_T^X > 25$ GeV/c	$0/0.58^{+0.08}_{-0.09}$ (76%)		

Positron-Proton Collisions at Large Momentum Transfer Eur. Phys. J. C13 (2000) 609-639, H1 Collaboration, C. Adloff et al., *Measurement and QCD Analysis of Neutral and Charged Current Cross Sections at HERA* Eur Phys J C30 (2003) 1-32,

4. CDF Collaboration, D. Acosta et al., *Measurement of the Forward-Backward Charge Asymmetry of Electron-Positron Pairs in p anti-p Collisions at* $\sqrt{s} = 1.96$ TeV Phys. Rev. D 71 (2005)052002.

Studies of heavy flavour production and the hadronic final state in high energy ep collisions

Thomas Kluge [1]

DESY, Notkestr. 85, 22607 Hamburg, Germany
E-mail: thomas.kluge@desy.de

Abstract. An extract of recent results from the H1 and ZEUS Collaborations is shown. Various properties of quantum chromo dynamics are investigated by studying the details of the hadronic final state of high energy electron proton collisions at HERA. The presented results include analyses of jet cross sections and single particle production such as γ and D. Part of the measurements deal with final states involving identified heavy quarks (charm and beauty).

The electron-proton storage ring HERA at DESY operates at a center of mass energy of $\sqrt{s} \approx 320$ GeV, allowing for precision studies of QCD by the H1 and ZEUS Collaborations. High energy electron proton scattering can be divided in two kinematical regimes: photo production γp ($Q^2 < 1$ GeV^2) and deep-inelastic scattering DIS ($Q^2 > 1$ GeV^2). In both cases pointlike interactions of the exchanged virtual photon occur, for γp in addition resolved processes are important, where the photon behaves as a hadron. Heavy quarks are dominantly produced pairwise from fusion of the exchanged photon and a gluon from the proton.

Details about hard scattering and parton density functions (pdfs) can be gained by observation of jets. Both experimentally and theoretically well under control is the measurement [1] of inclusive jet cross sections at high photon virtualities ($Q^2 > 150$ GeV^2) and not too low transverse jet energies ($E_t > 7$ GeV), shown in Fig. 1. The cross section is well described by a perturbative calculation corrected for hadronisation effects, which allows for a determination of the strong coupling constant $\alpha_s(M_Z) = 0.1197 \pm 0.0016\,(\text{exp.})\,^{+0.0046}_{-0.0048}\,(\text{th.})$.

A more direct sensitivity to the strong coupling constant offers the observable $R_{3/2}$ [3], the ratio of the 3- to the 2-jet cross section, which in lowest order is proportional to α_s. Due to the construction as a ratio of two cross sections, part of the uncertainties are expected to cancel. The observable is well described by the theory prediction, shown in Fig. 2. An extraction of the strong coupling constant yields $\alpha_s(M_Z) = 0.1179 \pm 0.0013\,(\text{stat.})\,^{+0.0028}_{-0.0046}\,(\text{th.})\,^{+0.0064}_{-0.0046}\,(\text{th.})$.

The DGLAP equations are used to evolve the pdfs from a starting scale to the scale of the hard scattering. This works well in the high Q^2 phase space of the before mentioned analyses, more problematic are regions of low Bjørken x and $E_t^2 \approx Q^2$, which corresponds to jets in the forward[2] part of the detector. Fig. 1 (right) shows that

---

[1] On behalf of the H1 and ZEUS Collaborations
[2] "Forward" meaning in direction of the proton beam.

CP814 *Hadron Spectroscopy: Eleventh International Conference*
edited by A. Reis, C. Göbel, J. Sá Borges, and J. Magnin

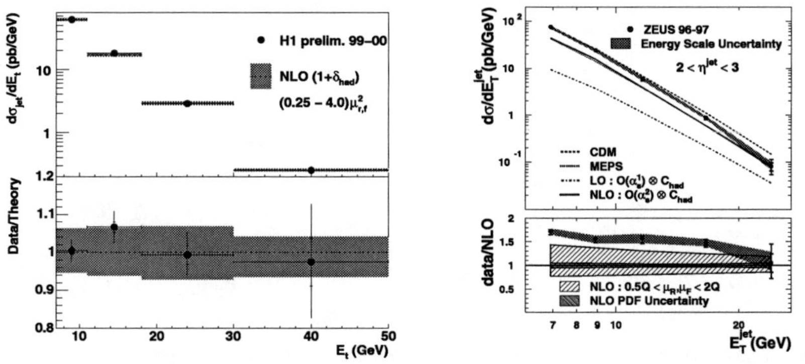

FIGURE 1. Jet cross sections $d\sigma/dE_t$ compared with predictions corrected for hadronisation effects. The inclusive jet cross section for high Q^2 (> 150 GeV^2) is shown on the left hand side plot, the right hand side plot shows the cross section for forward jets at lower Q^2 (> 25 GeV^2). The bands denote the theoretical uncertainty by varying the renormalization and factorization scale by a factor of two.

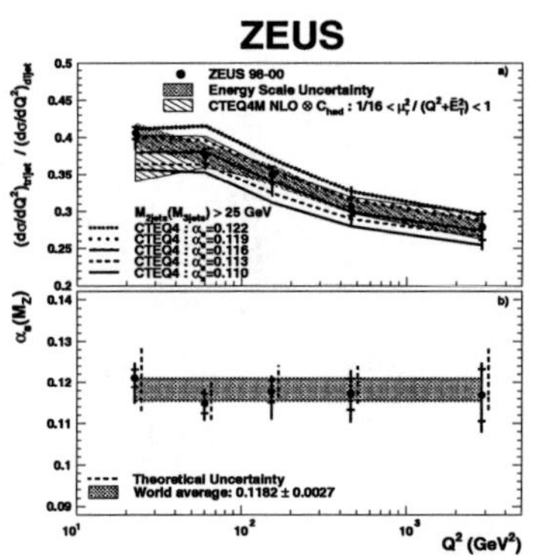

FIGURE 2. The ratio of the inclusive trijet to dijet cross sections as function of Q^2. The predictions of perturbative QCD in next-to-leading order using five sets of pdf are compared to the data. The dashed error bars display theoretical uncertainties.

pQCD calculations at NLO using DGLAP underestimate the data [2]. In comparison, a prediction employing the color dipole model (CDM) does a better job, which contains terms beyond DGLAP. However, it can not be cleary concluded that DGLAP evolution fails, because the effects of higher perturbative orders, resp. the errors on them, are large.

Diffractive ep scattering is still a topic of great interest. In the detector diffraction manifests as events with a rapidity gap, i.e. no proton dissociation occurs. The colourless exchange with the proton can be interpreted in the pomeron picture, employing diffractive pdfs. Such pdfs have been determined by the H1 Collaboration from inclusive data and are used to predict the diffractive differential dijet cross section [4] in DIS, shown on Fig. 3 (left). While the overall good description suggest factorisation in DIS,

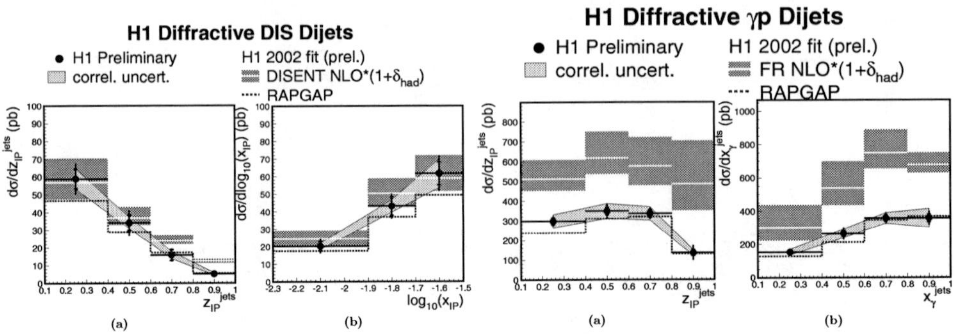

FIGURE 3. Left two plots: Cross section for the diffractive production of two jets in DIS as a function of z_P^{jets} and $\log_{10}(x_P)$. Right two plots: Cross section for the diffractive production of two jets in photo production as a function of z_P^{jets} and x_γ^{jets}.

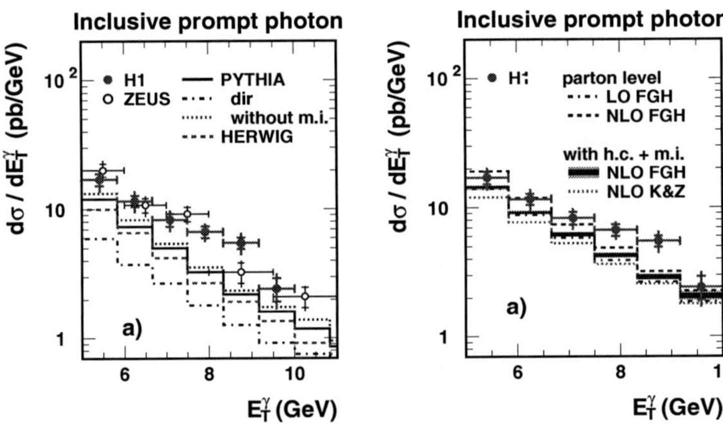

FIGURE 4. Inclusive prompt photon differential cross sections in photo production kinematic range. The data are shown as function of the transverse energy of the photon and compared to predictions from HERWIG and PYTHIA including multiple interactions (left hand side). Also shown is a comparison with NLO pQCD calculations (right hand side).

the predictions for dijets in photo production undershoot the data by far (Fig. 3 right). One may conclude from this observation that factorisation is broken in γp.

As an alternative to the observation of jets, prompt photons emmited by quarks which participate in the hard scattering can be detected without uncertainties due to hadronisation [6]. This benefit is balanced by a reduced cross section and more difficult particle identification compared to jets. Fig. 4 demonstrates that perturbative QCD at NLO describes the shape of the transverse momentum distribution quite well, the normalisation being a bit low. This study is complementary to QCD analyses with direct detection of jets.

Heavy quarks in final states can be identified by reconstruction of heavy mesons as D^*,

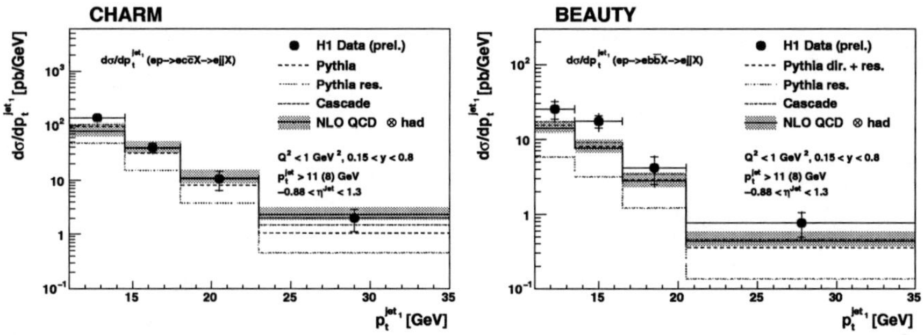

FIGURE 5. Differential charm (left hand side) and beauty (right hand side) cross section as a function of the transverse momentum of the leading jet. The data are compared to a NLO pQCD calculation and predictions from CASCADE and PYTHIA. The contribution in PYTHIA from resolved photon processes is shown separately.

by decay to muns or by lifetime tags. Fig. 5 shows differential cross sections for charm and beauty jets in photo production, where the fraction of heavy jets was determined by lifetime signatures. The distributions of transverse momenta are described well by shape, the normalisation of the beauty cross section is only somewhat smaller predicted by the perturbative calculation at next-to-leading order. This is due to improvements in experiment and theory compared to the past, where larger differences had been observed [5].

Detection of events containing charm can be performed by the reconstruction of a D^* meson by its decay products. Differential cross sections are shown for γp [7] and diffractive DIS [8] in Fig. 6. In both cases the theory predictions are able to describe the cross section.

Many observations as well as non observations of pentaquarks, bound states of five quarks, have been reported. Searches for pentaquarks containing a strange quark decaying to $K_S^0 p(\bar{p})$ were performed by the H1 and ZEUS Collaborations. In the ZEUS analysis a signal is found near 1522 MeV [10], whereby the H1 analysis [9] does not find a signal. Exclusion limits determined by H1 do not contradict the ZEUS observation. Hence more data, which is being recorded now during the HERA II phase, is needed to clarify this issue.

REFERENCES

1. H1 Collaboration, Inclusive Jet Production in Deep Inelastic Scattering at High Q2 at HERA, Submitted to EPS 2005
2. S. Chekanov *et al.* [ZEUS Collaboration], arXiv:hep-ex/0502029.
3. S. Chekanov *et al.* [ZEUS Collaboration], arXiv:hep-ex/0502007.
4. H1 Collaboration, Dijets in Diffractive Photoproduction and Deep-Inelastic Scattering at HERA, Submitted to ICHEP 2004
5. H1 Collaboration, Measurement of Charm and Beauty Photoproduction at HERA using Inclusive Lifetime Tagging, Submitted to ICHEP 2004
6. A. Aktas *et al.* [H1 Collaboration], Eur. Phys. J. C **38**, 437 (2005) [arXiv:hep-ex/0407018].
7. ZEUS Collaboration, Measurement of D* photoproduction at HERA, Submitted to ICHEP 2002

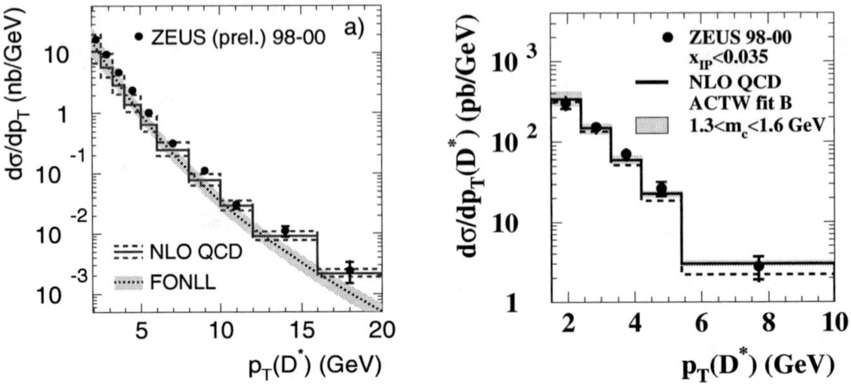

FIGURE 6. Differential cross sections for inclusive $D^{*\pm}$ production as a function of the transverse energy in the kinematic range $Q^2 < 1\text{GeV}^2$ (left). The right hand side plot shows diffractive $D^{*\pm}$ production in DIS ($1.5 < Q^2 < 200\text{GeV}^2$). The data are compared to pQCD calculations at next-to-leading order (NLO).

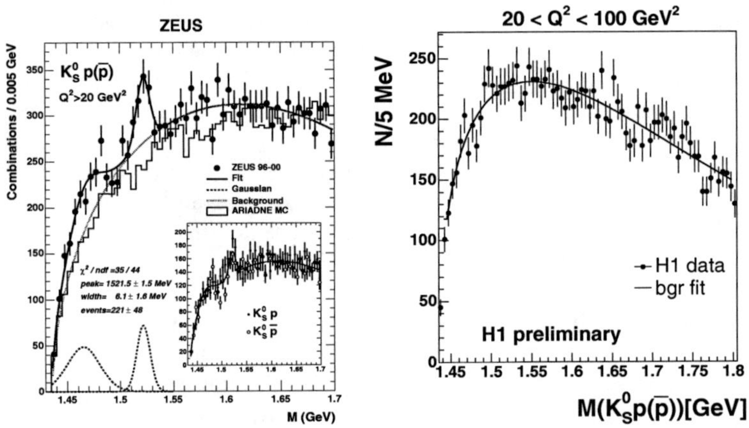

FIGURE 7. Invariant mass spectrum for the $K^0_S(p\bar{p})$ channel in DIS. Data from the ZEUS Collaboration are fitted by a background function plus two Gaussians (Left). The data from the H1 Collboration is fitted by a background function alone(right).

8. S. Chekanov *et al.* [ZEUS Collaboration], Nucl. Phys. B **672**, 3 (2003) [arXiv:hep-ex/0307068].
9. H1 Collaboration, Search for a Narrow Baryonic Resonance Decaying to K0s p(pbar) , Submitted to DIS 2005
10. S. Chekanov *et al.* [ZEUS Collaboration], Phys. Lett. B **591**, 7 (2004) [arXiv:hep-ex/0403051].

Nucleon spin structure studies at COMPASS

Claude MARCHAND†

CEA/DAPNIA/SPhN, CE Saclay, F-91191 GIF-SUR-YVETTE CEDEX, FRANCE
† *On behalf of the COMPASS Collaboration*

Abstract.
One of the main goal of the COMPASS experiment at CERN is the study of the spin structure of the nucleon in DIS, by scattering 160 GeV polarized muon beam on a longitudinally (or transversely) polarized ^{6}LiD target. Besides the scattered muon, the particles produced in the deep inelastic scattering are detected by a two stage magnetic spectrometer equipped with state of the art tracking and particle ID detectors.

The emphasis of COMPASS muon program is the direct determination of the gluon polarization $\Delta G/G$, accessed via asymmetries involving photon-gluon fusion mechanism (PGF). Both open charm production (detecting D^0's), as well as production of hight p_T hadron pairs are used to tag PGF. Preliminary results for $\Delta G/G$ based on the analysis of 2002 and 2003 data are shown. In addition, improved measurement of the deuteron structure function g_1^d at small x, as well as studies of transverse distribution functions in the deuteron by measuring Collins and Sivers azimuthal asymmetries, are reported.

Keywords: nucleon spin, gluon polarization, transversity
PACS: 13.60.Hb; 13.88.+e; 24.85.+p; 25.30.Mr

INTRODUCTION

The spin $1/2$ of the nucleon can be decomposed in the contributions from its constituents as:

$$\frac{1}{2} = \frac{1}{2}\Delta\Sigma + \Delta G + L_{q+g}, \tag{1}$$

where $\Delta\Sigma$ is the contribution from the quarks to the nucleon spin, ΔG the contribution from the gluons (which have spin 1), and L_{q+g} the contribution from orbital angular momentum of quarks and gluons.

The origin of the spin of the nucleon is a subject that draws great interest since 1988, when the EMC experiment found that only a small fraction of the nucleon spin seems carried by the quarks[1] ($\Delta\Sigma = 12 \pm 9 \pm 14\%$). This lead to the so called "spin-crisis", because it is in contrast with the naive quark-parton model, which predicts that 60% of the nucleon spin should come from the contribution of the valence quarks. This result has been confirmed since by a series of deep inelastic scattering (DIS) experiments at CERN[2], SLAC[3] and DESY[4], giving on average $\Delta\Sigma = 20$ to 30%.

The remaining contributions to the nucleon spin must then come from the gluons spin ΔG, and(or) from quark and gluon orbital angular momenta. The direct measurement of the gluon polarization $\Delta G/G$ is presently the main focus of the COMPASS muon program.

CP814 *Hadron Spectroscopy: Eleventh International Conference*
edited by A. Reis, C. Göbel, J. Sá Borges, and J. Magnin

THE EXPERIMENT

COMPASS is a fixed target experiment [5] located at the muon beam line of the CERN SPS, at the exact place of the previous SMC experiment[2]. Data have been collected in 2002, 2003 and 2004, and a muon run is also planned in 2006. Details of the experimental setup can be found in [6, 7], and I only indicate the major factors of improvements with respect to SMC:

- the muon luminosity is about 5 times larger
- a new ^6LiD target polarized up to 50%, with a packing factor of about .4, improving the factor of merit for measuring small asymmetries
- the trigger range in Q^2 has been extended down to quasi-real photo-production to be able to accumulate enough statistics for the very low cross section PGF events
- the vertical acceptance of the two stage spectrometer has been increased to ± 180 mrad, to be able to have full coverage of semi-inclusive DIS events
- new particle ID detectors , among which a RICH detector, and highly segmented and high flux capable tracking detectors to separate charm events from combinatorial backgrounds

POLARIZED DISTRIBUTION FUNCTIONS

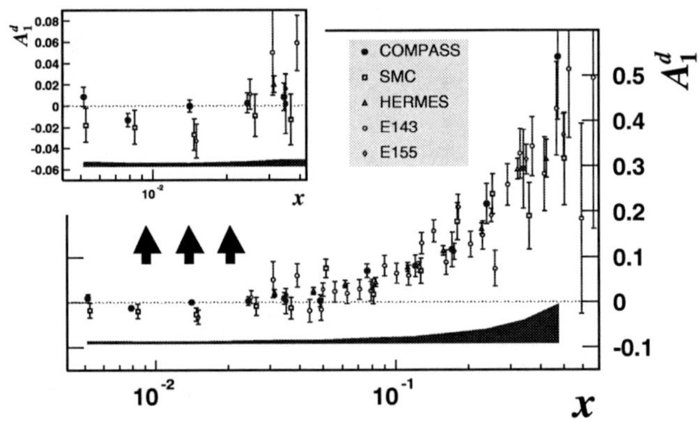

FIGURE 1. The virtual-photon deuteron asymmetry $A_1^d \approx g_1/F_1$ measured by COMPASS. The error bars represent the statistical error and the band the systematic error.

Thanks to the larger luminosity, and extended coverage of the trigger system, we were able to improve the measurement of the polarized asymmetry A_d^1 [8] for inclusive DIS (fig. 1), and hence the structure function g_d^1, in the low x range down to x=0.004.

Incorporating these data in a global QCD fit to world data on polarized DIS asymmetries [1, 2, 3, 4, 9, 10] helps to decrease the statistical error on the contribution from the quarks to the nucleon spin [11]: $\Delta\Sigma = 0.237 \pm 0.025$. This is easily understood by the fact that $\Delta\Sigma$ is derived from the moments of the distribution functions, where the low x data have non negligible weight.

These global QCD fits to the polarized asymmetries, in which DGLAP evolution equations are solved via numerical integration at NLO, allow an indirect determination of the helicity Parton Distribution Functions (PDF) of the quarks and gluons in the nucleon[12, 13, 14]. Although they fix quite well the polarized valence quark distributions, they only poorly constrain the polarized sea quark and gluon distributions, due to the limited accuracy and range in (x,Q^2) of the world data. This situation of course improves over time, with the inclusion of recent precise data from Jlab[9], Hermes[10] and our above cited result on A_d^1. But the obtained x dependence of the gluon distribution will always follow the functional form used in the fit, and remain the main systematic uncertainty of these fits.

GLUON POLARIZATION

Hence, a direct measurement of the polarized gluon distribution functions in semi-inclusive DIS is of crucial importance. This is the main goal of COMPASS muon program, where the gluon polarization $\Delta G/G$ is obtained by measuring the helicity asymmetry of the photon-gluon fusion (PGF) cross-section ($\gamma^* g \to q\bar{q}$). The fraction of PGF events in the data sample can be enhanced by selecting either open charm events or hadron pairs with high transverse momentum.

The cleanest tagging is provided by requiring the $q\bar{q}$ pair to be a $c\bar{c}$ pair, since the presence of c quarks inside the nucleon is negligible. The large mass of the c quark also sets a hard scale to ensure factorization of the "hard" PGF amplitude and the "soft" gluon distribution. The c quark is identified in the form of a D^0 meson which decays to $K\pi$. Due to multiple scattering inside the thick polarized target, the D^0 decay vertex cannot be distinguished from the primary vertex and the D^0 peak sits above a large combinatorial background. This background can be reduced by selecting D^0 coming from the decay $D^* \to D^0\pi_s$. Due to the background and the small cross section, this channel is statistically limited. Using 2002 and 2003 data, we get[15]: $\Delta G/G = -1.08 \pm 0.73(stat)$.

An alternative tagging of PGF is obtained by requiring a pair of hadrons with large p_t. The hard scale is set by the condition $\Sigma(p_T^2) > 2.5 GeV^2$. The measured asymmetry can be written as $A_{\parallel} = R_{PGF}\, a_{LL}^{PGF}\,(\Delta G/G) + A_{bckg}$. Here R_{PGF} is the fraction of PGF events in the sample; a_{LL}^{PGF} is the analyzing power, i.e. the spin asymmetry of the PGF process which can be computed in perturbative QCD. A_{bckg} is the asymmetry of the physics background. It includes the QCD Compton process, $\gamma^* q \to qg$, and at low Q^2 the resolved photon processes where the photon fluctuates to a hadronic state and one of the partons from the photon interacts with one of the partons from the nucleon.

From the 2002 and 2003 data sample at $Q^2 < 1$ GeV^2 and using Pythia to estimate R_{PGF} and A_{bckg}, we obtain [16] :

$$\frac{\Delta G}{G}(x = 0.10) = 0.024 \pm 0.089(stat) \pm 0.057(syst). \qquad (2)$$

At $Q^2 > 1$, the background is easier do deal with, but much less statistics is avalaible, and we obtain:

$$\frac{\Delta G}{G}(x = 0.13) = 0.01 \pm 0.31(stat) \pm 0.06(syst). \quad (3)$$

The high p_T results are presented in fig. 2, together with earlier measurements[17, 18].

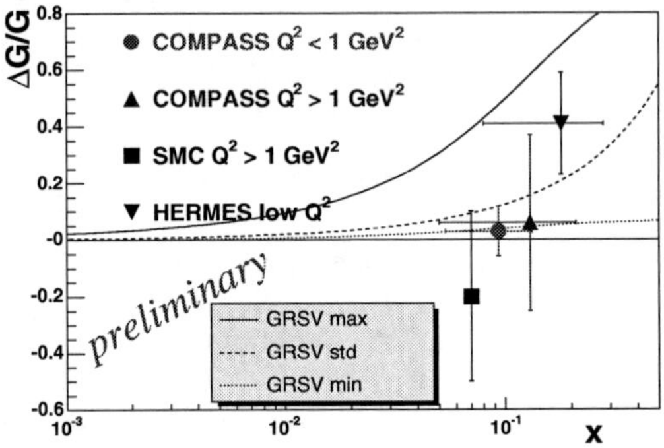

FIGURE 2. Various measurements of $\Delta G/G$ from high p_T events compared to three parametrisations from GRSV[12] corresponding to integrals $\Delta G = \int_0^1 \Delta G(x)dx$ =0.2, 0.6 and 2.5.

TRANSVERSE DISTRIBUTIONS

To completely describe the quark state of the nucleon at twist-two level, three distribution functions are neccessary: the momentum distribution of the partons $q(x)$, the helicity distribution $\Delta q(x)$ and the transverse spin distribution $\Delta_T q(x)$[19].

The transverse distribution $\Delta_T q(x)$ are difficult to measure, as they are chirally-odd and therefore absent in inclusive DIS. They can be accessed by measuring the single spin asymmetries (SSA) in semi-incusive DIS on transversely polarised nucleons, in which a hadron is detected in the final state.

Two kinds of polarized fragmentation have been used so far. The Collins fragmentation function tells how much the transverse spin of the fragmenting quark is reflected in the azimuthal distribution of the produced hadrons. It results in an azimuthal asymmetry in terms of the Collins angle, $\phi_{col} = \phi_h + \phi_s - \pi$, where ϕ_h is the azimuthal angle of the produced hadron and ϕ_s that of the target spin. An azimuthal asymmetry in terms of the Sivers angle, $\phi_{siv} = \phi_h - \phi_s$, is also possible. It is related to the quark momentum distribution in the plane transverse to the nucleon momentum. Our Collins and Sivers azimuthal asymmetries obtained on deuterium from 2002 data are reported in [20]. They are compatible with zero within the statistical error bars, in contrast to recent Hermes results on the proton[21]. This may indicate a possible cancellation between proton and neutron asymmetries.

The so-called interference fragmentation function is due to some interference effects in the production of two hadrons. It results in an azimuthal distribution in terms of the angle $\phi_{RS} = \phi_R + \phi_s - \pi$. The complete description of asymmetries, angles, and the asymmetry $A_{\phi_{RS}}$ obtained from 2002 and 2003 data as a function of the two hadron mass is discussed in ref.[22]. Again, it is compatible with zero within error bars.

CONCLUSION

COMPASS has collected from 2002 to 2004 new and precise informations about the spin structure of the nucleon[8, 16, 20, 22].

The most important result concerns the direct measurement of the gluon polarization. Our most precise result from events with high p_t hadrons at $Q^2 < 1$ GeV^2 clearly rules out values of $\Delta G > 1$, but is not yet precise enough to solve the "spin crisis". The situation will improve by adding 2004 data (as much as 2002 and 2003 combined), and forthcoming 2006 run with improved factor of merit, thanks mainly to improvements in target solid angle and RICH performance. Also, precise direct measurements from polarized p-p collisions will show up soon from RHIC[23].

A complemenatry approach to the "spin crisis" is planned to be addressed after 2010 at COMPASS by the study of generalized parton distributions, from which some information about the angular momentum of the quarks can be extracted.

REFERENCES

1. EMC, J. Ashman *et al.*, Phys. Lett. **B206**, 364 (1988).
2. SMC, B. Aveda *et al.*, Phys. Rev. **D58**, 112001 (1998).
3. E142, Anthony *et al.* Phys. Rev. **D54**, 6620 (1996); E143, Abe *et al.*, Phys. Rev. **D58**, 112003 (1998); E154, Abe *et al.*, Phys. Lett. **B405**, 180 (1997); E155, Anthony *et al.*, Phys. Lett. **B493**, 19 (2000).
4. HERMES, A. Airapetian *et al.*, Phys. Lett. **B442**, 484 (1998).
5. COMPASS proposal, CERN/SPSLC 96-14, SPSLC/P297, 1996
6. G.K. Mallot, Nucl. Instr. Meth. **A518** (2004) 121
7. S. Paul, these proceedings.
8. COMPASS, E.S. Ageev *et al.*, Phys. Lett. **B612**, 154 (2005).
9. JLAB, X. Zheng *et al.*, Phys. Rev. Lett. **92**, 012004 (2004).
10. D. Reggiani, DIS05, AIP Conf. Proc. **792**, 913 (2005).
11. J. Hannappel, DIS05, AIP Conf. Proc. **792**, 917 (2005).
12. M. GLuck, E. Reya, M. Stratmann, W. Vogelsang, Phys. Rev. **D63**, 094005 (2001).
13. AAC, M. Hirai, S. Kumano and N. Saito, Phys. Rev. **D69**, 054021 (2004).
14. E. Leader, A.V. Sidorov and D.B. Stamenov, JHEP **0506**, 033 (2005).
15. S. Hedicke, Czech. J. Phys **56**, (2005);
C. Quintans, Proceedings of QCD05 conference, Montpellier, (2005).
16. C. Bernet, DIS05, AIP Conf. Proc. **792**, 949 (2005).
17. HERMES, A. Airapetian *et al.*, Phys. Rev. Lett. **84**, 2584 (2000).
18. SMC, B. Adeva *et al.*, Phys. Rev. **D70**, 012002 (2004).
19. V. Barone, A. Drago and P.G. Ratcliffe, Phys. Rep. **359**, 1 (2002).
20. COMPASS, V. Y. Alexakhin *et al.*, Phys. Rev. Lett. **94** 202002 (2005).
21. HERMES, A. Airapetian *et al.*, Phys. Rev. Lett. **94**, 012002 (2005).
22. R. Joosten, DIS05, AIP Conf. Proc. **792**, 957 (2005).
23. A. Deshpande, DIS05, AIP Conf. Proc. **792**, 1001 (2005).

HEAVY QUARKS

D Hadronic Branching Fractions and $D\bar{D}$ Cross Section at ψ'' from CLEO-c

V. Credé representing the CLEO Collaboration

Florida State University, Tallahassee, FL 32306, USA

Abstract. We report on the determination of absolute hadronic branching fractions of charged and neutral D mesons using a double tag technique. The analysis is based on 55.8 pb^{-1} of e^+e^- collisions recorded at the $\psi(3770)$ resonance with the CLEO-c detector at CESR. Among measurements for three D^0 and six D^+ modes, we have obtained reference branching fractions $\mathscr{B}(D^0 \to K^-\pi^+) = (3.91 \pm 0.08 \pm 0.09)\,\%$ and $\mathscr{B}(D^+ \to K^-\pi^+\pi^+) = (9.5 \pm 0.2 \pm 0.3)\,\%$, where the uncertainties are statistical and systematic, respectively. It should be pointed out that final state radiation is included in these branching fractions by allowing for additional, unobserved, photons in the final state. We have also extracted the cross sections $\sigma(e^+e^- \to D^0\bar{D}^0) = (3.60 \pm 0.07^{+0.07}_{-0.05})$ nb and $\sigma(e^+e^- \to D^+D^-) = (2.79 \pm 0.07^{+0.10}_{-0.04})$ nb using a determination of the integrated luminosity.

INTRODUCTION

Absolute measurements of hadronic charm meson branching fractions play a central role in the study of the weak interaction because they serve to normalize many D and B meson branching fractions. These are then used to determine elements of the Cabibbo–Kobayashi–Maskawa (CKM) matrix. For example, the determination of the matrix element $|V_{cb}|$ from the decay of a B meson into $B \to D^*l\nu$ requires knowledge of the D meson branching fractions [1]. Two of these modes, $D^0 \to K^-\pi^+$ and $D^+ \to K^-\pi^+\pi^+$, are particularly important because essentially all other D^0 and D^+ branching fractions have been determined from ratios to one of these branching fractions.

Previously, the MARK III collaboration measured hadronic branching fractions at the $D\bar{D}$ threshold using a double tagging technique which relied on fully reconstructed $\psi(3770) \to D\bar{D}$ decays [2, 3]. This technique required the knowledge of the luminosity or the production cross section of $e^+e^- \to D\bar{D}$. We apply a very similar technique using a CLEO-c data sample roughly six times larger than that of MARK III. The resulting precision is comparable to the current PDG world averages.

DETECTOR AND DATA RECONSTRUCTION

The data sample presented here was produced in e^+e^- collisions at the **C**ornell **E**lectron **S**torage **R**ing (CESR) and collected with the CLEO-c detector. It consists of 55.8 pb^{-1} of integrated luminosity collected on the $\psi(3770)$ resonance, at a center-of-mass energy of $E_{cm} = 3773$ MeV. At this energy, no additional hadrons accompanying the $D\bar{D}$ pairs are produced.

The charged particle tracking system, consisting of a six-layer vertex drift chamber

CP814 *Hadron Spectroscopy: Eleventh International Conference*
edited by A. Reis, C. Göbel, J. Sá Borges, and J. Magnin

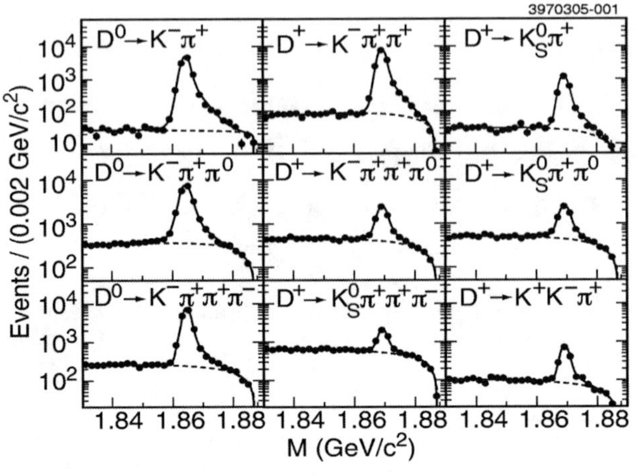

FIGURE 1. Semilogarithmic plot of ST yields and fits for D and $\overline{\mathrm{D}}$ combined for each mode. Data are shown as points with error bars and the solid lines show the total fits. Dashed lines are the background shapes. The high mass tails on the signal are due to initial state radiation.

and a 47-layer central drift chamber, operates in a 1.0 T magnetic field oriented along the beam axis. The momentum resolution of the tracking system is approximately 0.6 % at a particle momentum of 1 GeV. Photons are detected in an electromagnetic calorimeter, composed of 7800 CsI (Tl) crystals. The solid angle coverage for charged and neutral particles is about 93 % of 4π. We make use of two particle identification devices to separate charged kaons from pions. These are the central drift chamber using the energy loss dE/dx and a cylindrical ring-imaging Cherenkov (RICH) detector surrounding the central drift chamber.

In the event reconstruction, charged tracks are accepted based on the track fit quality and whether the track originates from the interaction point. Pions and kaons are identified by consistency with the expected dE/dx and RICH information, when available. $\mathrm{K^0_S}$ candidates are selected from pairs of oppositely-charged and vertex-constrained tracks. The invariant mass is within 12 MeV/c^2 of the known $\mathrm{K^0_S}$ mass whereas for π^0 candidates, the $\gamma\gamma$ invariant mass is within $\sigma \approx 5-7$ MeV/c^2 of the known π^0 mass, depending on poton energy and location. D meson candidates are identified by their invariant mass and total energies. We accept D candidates with M greater than 1.83 GeV/c^2 and with mode-dependent $\Delta E \equiv E_{\mathrm{D}} - E_0$ requirements of approximately 3σ, where E_{D} is the sum of the D candidate daughter energies and E_0 the beam energy.

D TAGGING AND YIELD DETERMINATION

The basic strategy in our analysis is the full reconstruction of one D meson and then to search for the second D meson in the remainder of the event. If one D meson is fully reconstructed, the event is called a single-tag event (ST), otherwise it is called a double-tag event (DT). DT events are obviously a subset of single-tag events.

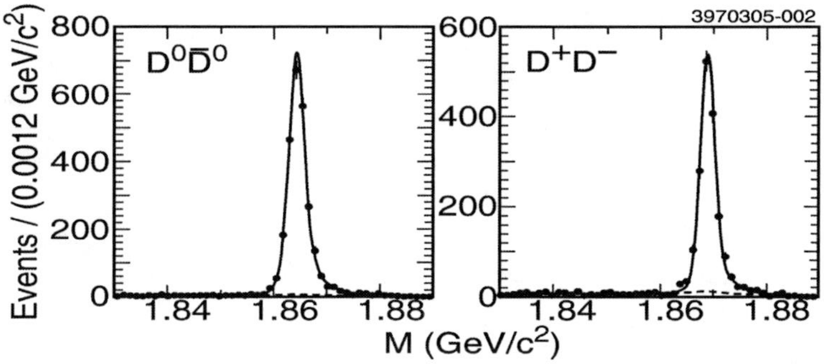

FIGURE 2. The DT yields and fits projected on the D^0 and D^+ axis summed over all modes. The solid lines show the total fit, the dashed lines the background shapes. We find total yields of 2484 ± 51 for D^0 and 1650 ± 42 for D^+. DT modes are very clean. Thus, the statistical yield uncertainties are close to $\sqrt{N_{ij}}$.

We extract ST and DT yields from M distributions by performing unbinned maximum likelihood fits in one and two dimensions, respectively. Figure 1 shows plots of ST yields. The signal shape includes the effects of beam energy smearing, initial state radiation, the line shape of the $\psi(3770)$, and reconstruction resolution. The background in single-tag modes is described by an ARGUS function [4], which models combinatorial contributions. Figure 2 shows DT yields. For DT modes, backgrounds can be uncorrelated, which means that either the D or the $\overline{\mathrm{D}}$ is misreconstructed. However, backgrounds can also be correlated, which means that all the final state particles in the event are correctly reconstructed but are mispartitioned among the D and $\overline{\mathrm{D}}$ meson. In fitting the two-dimensional $M(\mathrm{D})$ versus $M(\overline{\mathrm{D}})$ distribution, we model the uncorrelated background by a pair of functions, where one dimension is an ARGUS function and the other is the signal shape. We model the correlated background by an ARGUS function in $\widehat{M}$ and a Gaussian in the orthogonal variable, which is $[M(\overline{\mathrm{D}}) - M(\mathrm{D})]/2$.

FITTING TECHNIQUE

The number of observed ST events in a specific channel (D and $\overline{\mathrm{D}}$ yields are measured independently and denoted by N_i and $\overline{N}_i$) is the number of produced $\mathrm{D}\overline{\mathrm{D}}$ pairs multiplied by the branching fraction of a D meson into this channel ($\mathscr{B}_i$) multiplied by the efficiency (ε_i) to reconstruct the reaction: $N_i = N_{\mathrm{D}\overline{\mathrm{D}}} \mathscr{B}_i \varepsilon_i$. A further branching fraction ($\mathscr{B}_j$) has to be considered for DT events: $N_{ij} = N_{\mathrm{D}\overline{\mathrm{D}}} \mathscr{B}_i \mathscr{B}_j \varepsilon_{ij}$. While the yields stem from fits to reconstructed D meson peaks as described in the previous section, the efficiencies are determined from Monte Carlo simulations. The goal of the analysis is to find the branching fractions and the number of produced $\mathrm{D}\overline{\mathrm{D}}$ pairs by simultaneously fitting all charged and neutral modes. The branching fraction fitter [5] takes all the systematic uncertainties as input, along with ST and DT yields and efficiencies, crossfeed probabilities among the modes, background branching fractions and efficiencies, and statistical uncertainties on all of these measurements. The estimated crossfeed and background contributions produce yield adjustments of $\mathscr{O}(1\ \%)$. Their dependence on the fit param-

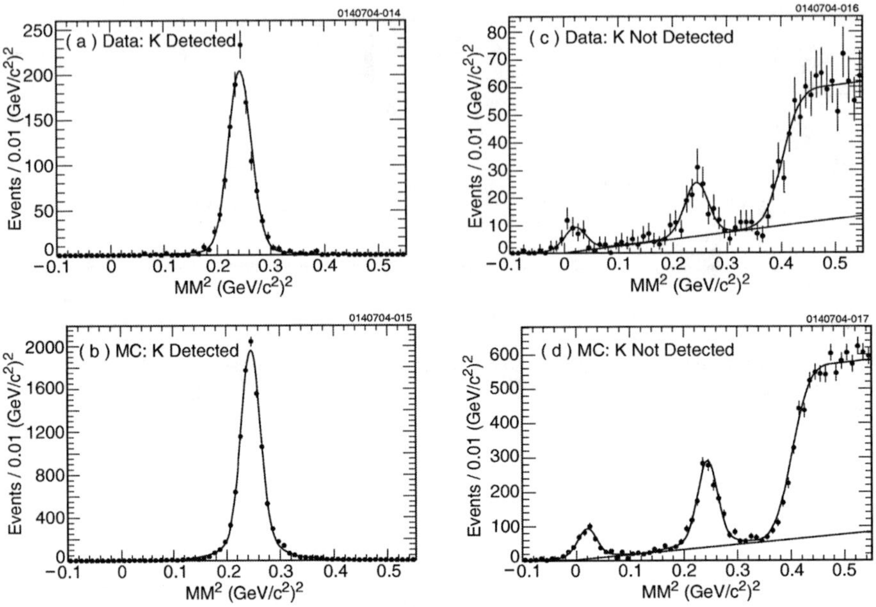

FIGURE 3. The top row of plots are for data and the lower row of plots are for Monte Carlo. The left side shows fits to the missing mass square distribution when the kaon was found. The right side shows the distributions when the kaon was not found.

eters is taken into account both in the yield subtraction and in the χ^2 minimization. We validated the analysis technique, including the branching fraction fit, by studying simulated $D\overline{D}$ events in a sample 50 times the size of our data sample. We reproduced the input branching fractions with biases due to our procedures to be less than one-third of the statistical errors on the data, and consistent with zero.

SYSTEMATIC UNCERTAINTIES

We determine the efficiency for reconstructing tracks, K_S^0 decays, and π^0 decays in both data and simulated events using a missing mass technique. The reactions $\psi(3770) \to D\overline{D}$ with the subsequent decays $D \to K_S\pi\pi$ or $D^0 \to K^-\pi^+$, $\psi(2S) \to J/\psi\pi^+\pi^-$, as well as $\psi(2S) \to J/\psi\pi^0\pi^0$ are fully reconstructed, leaving out one particle, for which the efficiency is to be determined. The missing mass of this combination peaks at the mass of the omitted particle, whether it is detected or not. Then, the desired efficiency is the fraction of this peak with this particle correctly reconstructed. Figure 3 shows the procedure for charged kaons. For tracks and K_S^0 candidates, we find good agreement between efficiencies in data and simulated events. For π^0 candidates, we correct the simulated efficiencies by 3.9 %, which is the level of disagreement with data found in our study. The relative uncertainties in these determinations, 0.7 % per track, 3.0 % per K_S^0, and 2.0 % per π^0, are the largest contributions to the systematic uncertainties.

TABLE 1. Fitted branching fractions and $D\bar{D}$ pair yields, along with the fractional FSR corrections. Uncertainties are statistical and systematic, respectively. The Particle Data Group [6] lists the average branching fractions $\mathcal{B}(D^0 \to K^-\pi^+) = (3.85 \pm 0.09)$ % and $\mathcal{B}(D^+ \to K^-\pi^+\pi^+) = (9.1 \pm 0.7)$ %

Parameter	Fitted Value	Δ_{FSR}
$N_{D^0\bar{D}^0}$	$(2.01 \pm 0.04 \pm 0.02) \times 10^5$	-0.2 %
$\mathcal{B}(D^0 \to K^-\pi^+)$	$(3.91 \pm 0.08 \pm 0.09)$ %	-2.0 %
$\mathcal{B}(D^0 \to K^-\pi^+\pi^0)$	$(14.9 \pm 0.3 \pm 0.5)$ %	-0.8 %
$\mathcal{B}(D^0 \to K^-\pi^+\pi^+\pi^-)$	$(8.3 \pm 0.2 \pm 0.3)$ %	-1.7 %
$N_{D^+D^-}$	$(1.56 \pm 0.04 \pm 0.01) \times 10^5$	-0.2 %
$\mathcal{B}(D^+ \to K^-\pi^+\pi^+)$	$(9.5 \pm 0.2 \pm 0.3)$ %	-2.2 %
$\mathcal{B}(D^+ \to K^-\pi^+\pi^+\pi^0)$	$(6.0 \pm 0.2 \pm 0.2)$ %	-0.6 %
$\mathcal{B}(D^+ \to K^0_S\pi^+)$	$(1.55 \pm 0.05 \pm 0.06)$ %	-1.8 %
$\mathcal{B}(D^+ \to K^0_S\pi^+\pi^0)$	$(7.2 \pm 0.2 \pm 0.4)$ %	-0.8 %
$\mathcal{B}(D^+ \to K^0_S\pi^+\pi^+\pi^-)$	$(3.2 \pm 0.1 \pm 0.2)$ %	-1.4 %
$\mathcal{B}(D^+ \to K^+K^-\pi^+)$	$(0.97 \pm 0.04 \pm 0.04)$ %	-0.9 %

RESULTS

Table 1 shows the results of the data fit. The χ^2 of the fit is 28.1 for 52 degrees of freedom, corresponding to a confidence level of 99.7 %. All nine branching fractions are consistent with, and most are higher than, the current PDG averages [6]. FSR photons are not explicitly reconstructed, but because FSR is simulated in the samples used to calculate the efficiencies, our measurements represent inclusive branching fractions for signal processes with any number of photons radiated from the final state particles. If no FSR were included in the simulations, then all the branching fractions would change by Δ_{FSR} in Table 1.

We obtain the $e^+e^- \to D\bar{D}$ cross sections by scaling $N_{D^0\bar{D}^0}$ ($N_{D^+D^-}$) by the luminosity, which we determine to be $\mathcal{L} = (55.8 \pm 0.6)\ \text{pb}^{-1}$. Thus, at $E_{cm} = 3773$ MeV, we find a peak cross section of $\sigma(e^+e^- \to D\bar{D}) = (6.39 \pm 0.10^{+0.17}_{-0.08})$ nb, where the uncertainties are statistical and systematic, respectively. Our measured cross sections are in good agreement with BES [7] and higher than those of MARK III [3]. We find branching fractions in agreement with, but somewhat higher, than those in the PDG summary. We note that, unlike the branching fractions used in the PDG averages, our measurements are corrected for FSR. Not doing so would lower our branching fractions by $(0.6 - 2.2)$ %.

REFERENCES

1. See review by M. Artuso and E. Barberio in Ref. [6].
2. R. M. Baltrusaitis *et al.* [MARK-III Collaboration], Phys. Rev. Lett. **56**, 2140 (1986).
3. J. Adler *et al.* [MARK-III Collaboration], Phys. Rev. Lett. **60**, 89 (1988).
4. H. Albrecht *et al.* [ARGUS Collaboration], Phys. Lett. B **241**, 278 (1990).
5. W. M. Sun, arXiv:physics/0503050, accepted by Nucl. Instrum. Methods Phys. Res., Sec. A.
6. S. Eidelman *et al.* [Particle Data Group], Phys. Lett. B **592**, 1 (2004).
7. M. Ablikim *et al.* [BES Collaboration], Phys. Lett. B **603**, 130 (2004) [arXiv:hep-ex/0411013].

Open and hidden charm production in 920 GeV proton-nucleus collisions

Pietro Faccioli (for the HERA-B Collaboration)

Dipartimento di Fisica dell'Università di Bologna and INFN Sezione di Bologna - Bologna, Italy
E-mail: faccioli@bo.infn.it

Abstract.
Using the collected sample of 200 million minimum-bias proton-nucleus interactions at 41.6 GeV centre-of-mass energy, HERA-B is measuring the magnitudes of several inclusive particle productions. This contribution reports on preliminary measurements of the D-meson and J/ψ production cross sections and compares them with the current experimental situation.

Keywords: Charm, charmonium, cross sections, hadroproduction
PACS: 13.85.Ni, 13.85.Qk, 14.40.Lb, 14.40.Gx, 24.85.+p

The measurement of charm production in fixed-target collisions gives the opportunity to test perturbative QCD in a borderline context of applicability, where the perturbative expansion is applied to the lightest of the heavy quarks. New, accurate data complementing the current picture of proton-nucleus measurements with different production energies and kinematical conditions can help in determining the importance of higher-order terms and the magnitude of non-perturbative effects, such as nuclear dependence and particle-antiparticle asymmetries. The DESY experiment HERA-B is performing a new measurement of D-meson cross sections, based on 200 million minimum bias $p-A$ events collected during the 2002-2003 data taking. By exploiting the high resolution of the vertex detector and tracking systems and the excellent identification capabilities provided by the Cherenkov detector for kaons and pions, clean D^0, $D^{\pm}$ and $D^{*\pm}$ signals are extracted for each of three target materials (C, Ti, W). The 920 GeV/c momentum of the HERA proton beam corresponds to a production energy of 41.6 GeV in the centre-of-mass of the reaction, the highest among the currently available hadroproduction results.

The same data and experimental conditions are the basis for a new measurement of the inclusive J/ψ cross section, where the J/ψ's are reconstructed in both leptonic decay channels using the information provided by the electromagnetic calorimeter and muon detector. Though initiated by the same partonic couplings (quark-quark annihilation and gluon-gluon fusion) which can give rise to the formation of charmed meson-antimeson couples via hadronization of $c\bar{c}$ pairs, the production of J/ψ can proceed only from a restricted subset of such processes, due to the requirement that the final $c\bar{c}$ bound sate is a physical, colour-neutral particle having the spin-parity properties of a J/ψ. Measurements of J/ψ production can therefore discriminate between QCD-based models differing for the choice of how ample or specific is the class of processes (colour singlet, colour octet, colour evaporation) assumed in the description of the charmonium production mechanism. The present paper discusses only the integral cross section measurement, which takes advantage of the minimum trigger systematics affecting the above-

CP814 *Hadron Spectroscopy: Eleventh International Conference*
edited by A. Reis, C. Göbel, J. Sá Borges, and J. Magnin

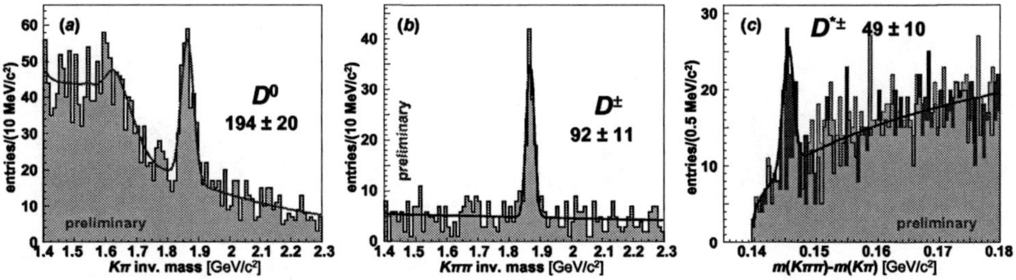

FIGURE 1. *D*-meson signals.

mentioned data sample. Detailed studies of the kinematics of J/ψ production, including the crucial topic of their nuclear dependence, are being performed using a large, dedicated dilepton-triggered data sample, also enabling accurate measurements of the fractions of J/ψ's coming from heavier charmonium states and from B decays; details of these analyses have been reported elsewhere at this conference [1].

***D*-meson production.** – The inclusive production reaction $pA \to DX$ is studied by counting the D mesons detected through the decay channels $D^0 \to K^-\pi^+$, $D^+ \to K^-\pi^+\pi^+$ and $D^{*+} \to D^0\pi^+ \to K^-\pi^+\pi^+$ (charge conjugates are here always implied). An effective background rejection is accomplished with combined requirements on the information provided by the RICH (identification of the decay kaons and pions) and by the vertex detector (clear separation of the reconstructed secondary vertex from the interaction point, small impact parameter of the prolonged D^0 and D^+ tracks). Preliminary signals obtained after a blind, simultaneous optimization of the selection cuts are shown in Fig. 1(a-c) together with the number of reconstructed particles. The bump on the left of the D^0 peak, due to the not reconstructed π^0 of the decay $D^0 \to K^-\pi^+\pi^0$, is well reproduced by the Monte Carlo simulation. The $D^{*\pm}$ signal spectrum, shown in Fig. 1(c), is obtained by requiring that the invariant mass of the detached K and π is within a 50 MeV window around the D^0 mass; the superimposed, brighter spectrum is the background of the events satisfying the opposite requirement. To be mentioned for completeness, $D_s^\pm$ signals are also visible, with lower statistics, in a very preliminary analysis of the decays $D_s^\pm \to \phi\pi^+ \to K^+K^-\pi^+$ and $D_s^\pm \to K_S K^+ \to \pi^+\pi^-K^+$, which will provide the first measurement of the $D_s^\pm$ cross section near this energy.

The cross sections are calculated with the formula $\sigma(D) = N_D^{\rm RD}/(\varepsilon^{\rm MC} \times Br \times \mathscr{L})$, where the number of counted D mesons is divided by the product between the Monte Carlo reconstruction efficiency, the branching ratio of the detection channel and the total luminosity. This latter is determined as the sum over three different target materials (C, Ti, W) assuming no dependence of the per-nucleon cross section on the atomic weight ($\sigma_{pA} = \sigma_{pN} \cdot A^\alpha$ with $\alpha = 1$); the reliability of this assumption is, as will be shown, confirmed by the data. The primary cross section value is limited to the accessible kinematical range in Feynman-x, $-0.1 < x_{\rm F} < 0.05$, corresponding to about 38% of the full spectrum. The overall efficiency is of the order of 1%, with main losses due to detector acceptance ($\sim$ 10%) and detached-vertex requirements ($\sim$ 10%).

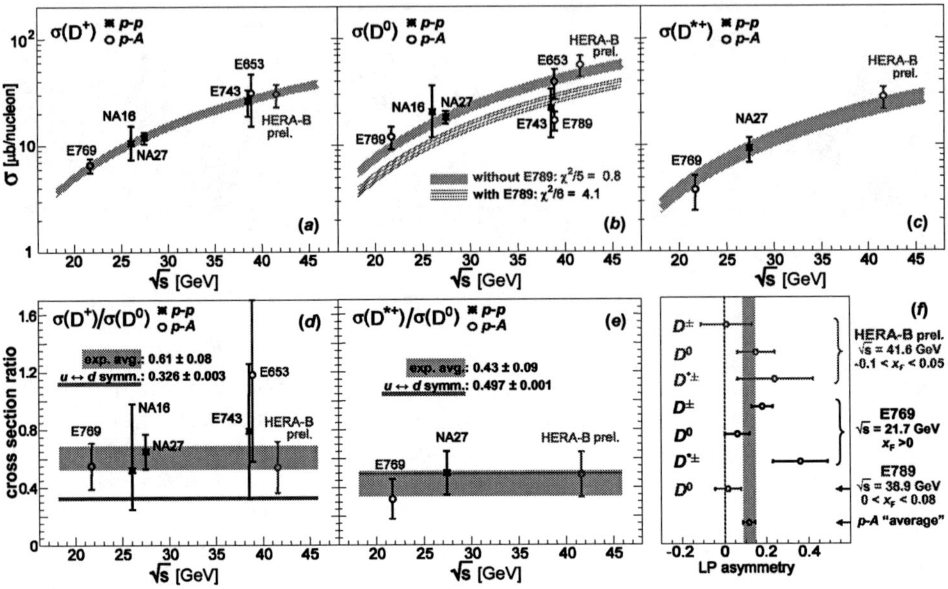

FIGURE 2. *(a-e)* Cross sections of proton-induced charm production and their ratios as a function of energy. *(f)* $p-A$ results on the leading-particle asymmetries.

Preliminary numbers for the cross sections in the visible kinematical range are $\sigma_{D^0} = 21.4 \pm 3.2 \pm 3.6$ μb/nucleon, $\sigma_{D^+} = 11.5 \pm 1.7 \pm 2.2$ μb/nucl and $\sigma_{D^{*+}} = 10.0 \pm 1.9 \pm 1.4$ μb/nucl. Here the first error is statistical, while the second one includes the relevant sources of systematic uncertainties (particle identification and track reconstruction efficiencies, variation of the cuts, signal counting, luminosity, branching ratios). The results are compared in Fig. 2(a-c), after extrapolation to the full kinematical spectrum, to the existing $p-A$ and $p-p$ results [2] as a function of the production energy. Both the D^+ and (with fewer data) the D^{*+} cross sections are in perfect agreement with an energy dependence of the form $\sigma \propto [1 - 1.2/(\sqrt{s})^{0.35}]^{12}$, which is a good parametrization of the perturbative QCD calculation of the $c\bar{c}$ cross section up to $\sqrt{s} = 200$ GeV, performed using different sets of parton densities inside the proton [3]. If the E789 result, significantly lower than the average of the remaining data at $\sqrt{s} \sim 40$ GeV, is left out, also the D^0 measurements remain consistent with the same energy behaviour. The equal energy dependence of the three charm cross sections is actually an obvious qualitative prediction of isospin symmetry, according to which the differences in production rate depend only on the counting of polarization states (enhancing the directly produced D^*'s by a factor of three with respect to the D^0's and D^+'s) and on the $D^* \to D$ feeddown branching ratios. On the other hand, as can be seen in Fig. 2(d-e), the measured cross section ratios indicate a significant violation in magnitude (by a factor of 1.87 ± 0.25) of the isospin symmetry between D^+ and D^0. This effect, still unexplained, is even more pronounced than that observed by pion-nucleus experiments [3] and by ALEPH [4].

With clear enough signals reconstructed in the carbon, titanium and tungsten data separately, the cross sections are determined as a function of the target material and fitted to the functional form $\sigma_{pA} = \sigma_{pN} \cdot A^{\alpha}$ to provide the A-dependence parameter α, previously measured only by E789 for D^0 mesons produced at $\sqrt{s} = 38.9$ GeV in the $0 < x_F < 0.08$ range [2]. The values preliminarily obtained by HERA-B for the three charm species are $\alpha_{D^0} = 0.92 \pm 0.08$, $\alpha_{D^\pm} = 1.02 \pm 0.09$ and $\alpha_{D^{*\pm}} = 1.05 \pm 0.14$. A global D-meson average for the central-x_F production at $\sqrt{s} \sim 40$ GeV, combining the HERA-B and E789 results, is $\alpha_D = 1.01 \pm 0.03$, consistent with the assumed absence of nuclear dependence. This result tends to disfavour the proposed initial state effects as possible explanations of the significant suppression which is observed in the nuclear production of charmonium.

It has been observed in the pion-induced reactions [5] that the production of those D-mesons, called "Leading" particles, which have the light quark in common with the beam particle is increasingly enhanced towards high x_F with respect to the "Non-Leading" D-mesons, resulting in asymmetries $a_{LP} = (\sigma_L - \sigma_{NL})/(\sigma_L + \sigma_{NL})$ of up to 70%. Preliminary HERA-B results for the leading particle asymmetries are compared in Fig. 2(f) with the existing $p-A$ measurements (here the leading particles coincide with the antimesons). To quantify their tendency towards positive values, the measured asymmetries are combined in an informal average, overlooking the different kinematical conditions; the result, $a_{LP}^{pA} = 0.12 \pm 0.03$, indicates the possible presence of a significant effect also in the proton-induced reactions.

J/ψ **production**. – The aim of the analysis is to measure the absolute magnitude of the inclusive J/ψ production, without distinction between its different formation channels. As already said, the kinematical details and feeddown components of charmonium production are the subject of parallel HERA-B analyses performed on dedicated dilepton data [1].

The J/ψ mesons are reconstructed independently in both dilepton decay channels. The data cover a visible kinematical fraction of $63 \pm 1\%$ $(-0.225 < x_F < 0.075)$. The overall efficiencies in this range, after optimized cuts on lepton identification parameters, are 25% and 33% for the dielectron and dimuon channels respectively. Fig. 3(a-b) shows the invariant mass spectra obtained in the full data, containing 57 ± 13 $J/\psi \to e^+e^-$ and 100 ± 12 $J/\psi \to \mu^+\mu^-$ reconstructed decays. The cross section per nucleon is however measured by counting the J/ψ's for each of the three target materials (C, Ti, W) separately, in order to take into account the known A-dependence ($\alpha = 0.96 \pm 0.01$ in this kinematical range [1, 6]). After averaging the electron and muon results for the three material subsamples, fitting them to the usual parametrization, with fixed α, and extrapolating to the full kinematical range, we obtain the result $\sigma_{pN}^{J/\psi} = (643 \pm 71 \pm 48)$ nb/nucleon, where the second error combines the estimated systematic uncertainties due to lepton-ID cuts, track reconstruction, signal counting, luminosity determination, decay branching ratio, kinematical extrapolation and assumed A-dependence.

The result is plotted in Fig. 3(c) together with the existing $p-A$ and $p-p$ measurements for reaction energies from 6.8 to 200 GeV. It is evident, especially if one looks at the inset showing a zoom into the most populated region, that the data are not always consistent with each other. For example, the two visibly high values at $\sqrt{s} = 20.6$ and 23.8 GeV would be obviously incompatible with any phenomenological parametriza-

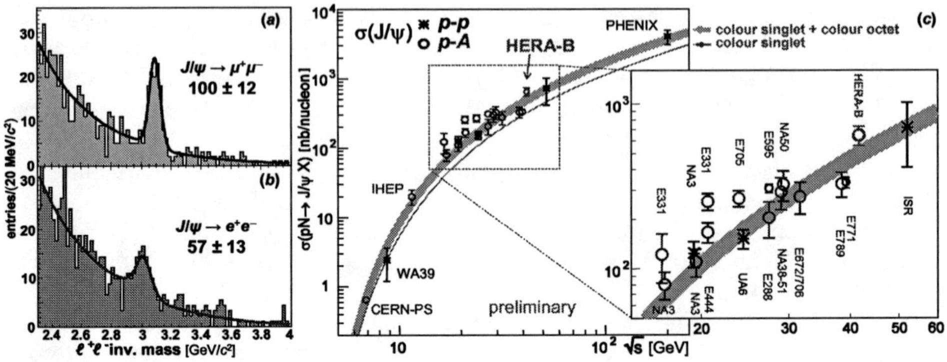

FIGURE 3. (*a-b*) Dilepton invariant mass spectra. (*c*) J/ψ production as a function of energy.

tion of the cross section, denoting a probably too optimistic evaluation of the systematic errors. This said, the curve shown has been obtained by fitting the data, excluding the result here presented, to a parametrization based on NLO NRQCD calculations using different sets of PDF's and taking as additional input the related measurements of ψ' production (29 results in all). Details of the procedure and tables of the experimental inputs can be found in [7]. As can be seen, the HERA-B point, though significantly higher than measurements at slightly lower energy, is in agreement with the curve.

Summary. – The HERA-B analysis on the inclusive production of D^0, $D^{\pm}$ and $D^{*\pm}$ mesons in $p-A$ collisions at $\sqrt{s} = 41.6$ GeV is in its final stage. Compared with the existing experimental situation, the preliminary results confirm an equal trend of increase with energy for all charm species and the absence of nuclear dependence. Isospin symmetry between D^0 and $D^{\pm}$ is violated by about a factor of 2. Small, but significant leading-particle effects, already observed in pion-induced reactions, are also implied by the complex of $p-A$ results. A new measurement of the inclusive J/ψ production cross section has also been presented. Its value, $\sigma_{pN}^{J/\psi} = (643 \pm 71 \pm 48)$ nb/nucleon, is in agreement with a NLO NRQCD parametrization based on all published data.

REFERENCES

1. P. Faccioli and H. Wahlberg, other HERA-B contributions in these Proceedings.
2. E769: G.A. Alves *et al.*, Phys. Rev. Lett. **77**, 2388 (1996); NA16: M. Aguilar-Benitez *et al.*, Phys. Lett. B **135**, 237 (1984); NA27: M. Aguilar-Benitez *et al.*, Z. Phys. C **40**, 321 (1988); E743: R. Ammar *et al.*, Phys. Rev. Lett. **61**, 2185 (1988); E653: K. Kodoma *et al.*, Phys. Lett. B **263**, 573 (1991); E789: M.J. Leitch *et al.*, Phys. Rev. Lett. **72**, 2542 (1994).
3. C. Lourenço and H. Wöhri, J. Phys. G **30**, s315 (2004) and references therein.
4. R. Barate *et al.*, Eur. Phys. J. C **16**, 597 (2000).
5. E.M. Aitala *et al.*, Phys. Lett. B **371**, 157 (1996).
6. M.J. Leitch *et al.*, Phys. Rev. Lett. **84**, 3256 (2000).
7. J. Spengler *et al.*, paper in preparation.

Recent results on D meson decays from the BES Collaboration

Jiangchuan Chen (for BES collaboration)

Institute of High Energy Physics, Beijing, P.R. China

Abstract. Based on a data sample of about 33 pb^{-1} collected at and around the center-of-mass energy of 3.773 GeV with the BES-II detector at the BEPC collider, the decay branching fractions for $D^0 \to K^-\mu^+\nu_\mu$, $D^0 \to \pi^-\mu^+\nu_\mu$ and $D^+ \to \bar{K}^0\mu^+\nu_\mu$ are measured to be $(3.55 \pm 0.56 \pm 0.59)\%$, $(0.38 \pm 0.31 \pm 0.10)\%$ and $(10.3 \pm 2.3 \pm 0.8)\%$, respectively. The ratio of the two partial widths is determined to be $\Gamma(D^0 \to K^-\mu^+\nu_\mu)/\Gamma(D^+ \to \bar{K}^0\mu^+\nu_\mu) = 0.87 \pm 0.24 \pm 0.15$. The decay branching fractions for $D \to K\pi e^+\nu_e$ and $D \to \bar{K}^* e^+\nu_e$ and the decay branching fractions for $D \to \bar{K}^* + X$ and $D \to K^0(\bar{K}^0) + X$ are also measured.

Keywords: Charmed meson, D decay, Semileptonic decay, Inclusive decay
PACS: 12.38.Qk, 13.20.-v, 13.20.Fc, 13.25.Ft, 13.85.Ni

INTRODUCTION

Around the center of mass energy of 3.773 GeV, the $\psi(3770)$ resonance is produced in electron-positron (e^+e^-) annihilation. The $\psi(3770)$ decays predominantly into $D\bar{D}$ pairs. If a $\bar{D}$ meson (through this paper, charge conjugation is implied) is fully reconstructed (this is called a singly tagged $\bar{D}$), the D meson must exist on the recoil side of the singly tagged $\bar{D}$ meson. Using the singly tagged $\bar{D}$ sample, the D decay event can be well selected on the recoil side. Therefore, the absolute branching fractions for D decays can be well measured.

About 33 pb^{-1} data sample were collected at and around the center-of-mass energy of 3.773 GeV with the BES-II [1] detector at the BEPC collider [2]. The singly tagged $\bar{D}$ mesons were obtained by reconstructing four neutral D decay modes and nine charged D decay modes [3, 4]. We obtained the total number of $7584 \pm 198 \pm 341$ reconstructed $\bar{D}^0$ mesons and $5321 \pm 149 \pm 160$ reconstructed D^- mesons.

In this letter we report the measurements of the branching fractions for semileptonic decays of D mesons, determination of the ratio of the partial decay widths $\frac{\Gamma(D^0 \to K^-\mu^+\nu_\mu)}{\Gamma(D^+ \to \bar{K}^0\mu^+\nu_\mu)}$ and measurements of branching fractions for the inclusive Cabibbo-favored $\bar{K}^*(892)$ and Cabibbo-suppressed $K^*(892)$ decays of neutral and charged D mesons [5].

CP814 *Hadron Spectroscopy: Eleventh International Conference*
edited by A. Reis, C. Göbel, J. Sá Borges, and J. Magnin

SEMILEPTONIC D DECAYS

Branching fractions for $D \to K(\pi)\mu^+\nu_\mu$

The isospin symmetry predicts that the partial widths of the exclusive semileptonic decays of the neutral and charged D mesons should be equal. However, the ratio of the two partial widths are $\frac{\Gamma(D^0 \to K^- e^+ \nu_e)}{\Gamma(D^+ \to \bar{K}^0 e^+ \nu_e)} = 1.34 \pm 0.20$ and $\frac{\Gamma(D^0 \to K^- \mu^+ \nu_\mu)}{\Gamma(D^+ \to \bar{K}^0 \mu^+ \nu_\mu)} = 1.2 \pm 0.5$, based on the measured branching fractions and the lifetimes of the D^0 and D^+ quoted from PDG [6]. Recently, BES-II and COLE-c measured the ratio of $\frac{\Gamma(D^0 \to K^- e^+ \nu_e)}{\Gamma(D^+ \to \bar{K}^0 e^+ \nu_e)}$, they obtained the values of the ratio to be $1.08 \pm 0.22 \pm 0.07$ and $1.00 \pm 0.05 \pm 0.04$ respectively [4, 7].

We also measured the branching fractions for $D^0 \to K^- \mu^+ \nu_\mu$ and $D^+ \to \bar{K}^0 \mu^+ \nu_\mu$ and determined the ratio $\frac{\Gamma(D^0 \to K^- \mu^+ \nu_\mu)}{\Gamma(D^+ \to \bar{K}^0 \mu^+ \nu_\mu)}$. Fig. 1 show the distributions of the invariant mass of $mKn\pi$ combinations for the events in which $K^- \mu^+ \nu_\mu$ candidates are observed on the recoil side. After subtracting the background, 87.2 ± 13.8 signal events for $D^0 \to K^- \mu^+ \nu_\mu$ are observed. Similar analysis obtains 9.3 ± 7.5 events for $D^0 \to \pi^- \mu^+ \nu_\mu$ and 28.7 ± 6.4 events for $D^+ \to \bar{K}^0 \mu^+ \nu_\mu$. Those yield the absolute branching fractions to be $BF(D^0 \to K^- \mu^+ \nu_\mu) = (3.55 \pm 0.56 \pm 0.59)\%$, $BF(D^0 \to \pi^- \mu^+ \nu_\mu) = (0.38 \pm 0.31 \pm 0.10)\%$ and $BF(D^+ \to \bar{K}^0 \mu^+ \nu_\mu) = (10.3 \pm 2.3 \pm 0.8)\%$. The upper limit of $BF(D^0 \to \pi^- \mu^+ \nu_\mu) < 0.89\%$ at 90% confidence level is set. The ratio of the two partial widths for the decays of $D^0 \to K^- \mu^+ \nu_\mu$ and $D^+ \to \bar{K}^0 \mu^+ \nu_\mu$ is determined to be $\frac{\Gamma(D^0 \to K^- \mu^+ \nu_\mu)}{\Gamma(D^+ \to \bar{K}^0 \mu^+ \nu_\mu)} = 0.87 \pm 0.24 \pm 0.15$. These results are preliminary.

Branching fractions for $D \to K\pi(K^*)e^+\nu_e$

The exclusive D semileptonic decays $D^+ \to K^- \pi^+ e^+ \nu_e$, $D^0 \to \bar{K}^0 \pi^- e^+ \nu_e$, $D^+ \to \bar{K}^{*0} e^+ \nu_e$ and $D^0 \to K^{*-} e^+ \nu_e$ are studied by BES. From the distribution of the fitted invariant masses for $mKn\pi$ combinations in the events in which the candidates for $D^+ \to K^- \pi^+ e^+ \nu_e$ and $D^0 \to K^0_S \pi^- e^+ \nu_e$ are observed on the recoil side, 31.4 ± 6.2 candidates for $D^+ \to K^- \pi^+ e^+ \nu_e$ and 9.3 ± 3.4 candidates for $D^0 \to K^0_S \pi^- e^+ \nu_e$ are observed.

The distribution of the calculated invariant mass for $K^- \pi^+$ ($K^0_S \pi^-$) of the selected candidates for $D^+ \to K^- \pi^+ e^+ \nu_e$ ($D^0 \to K^0_S \pi^- e^+ \nu_e$) is shown in FIG. 2. A clear $\bar{K}^{*0}$ signal is observed in FIG. 2(a). After subtracting the number of backgrounds, 27.7 ± 6.6 candidates for $D^+ \to \bar{K}^{*0} e^+ \nu_e$ and 6.6 ± 3.1 candidates for $D^0 \to K^{*-} e^+ \nu_e$ are retained.

The absolute branching fractions for these decays are determined to be $BF(D^+ \to K^- \pi^+ e^+ \nu_e) = (3.46 \pm 0.75 \pm 0.28)\%$, $BF(D^0 \to \bar{K}^0 \pi^- e^+ \nu_e) = (2.60 \pm 1.04 \pm 0.27)\%$, $BF(D^+ \to \bar{K}^{*0} e^+ \nu_e) = (4.89 \pm 1.19 \pm 0.38)\%$ and $BF(D^0 \to K^{*-} e^+ \nu_e) = (2.82 \pm 1.39 \pm 0.20)\%$. The ratio of the partial widths of the decays $D^+ \to \bar{K}^{*0} e^+ \nu_e$ and $D^+ \to \bar{K}^0 e^+ \nu_e$ is obtained to be $0.55 \pm 0.17 \pm 0.02$. These results are preliminary.

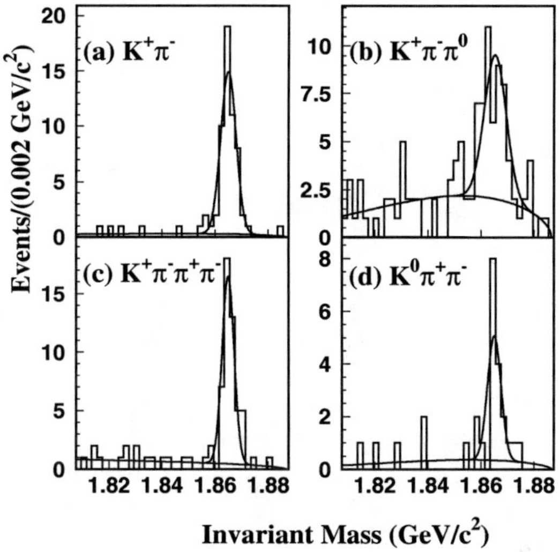

FIGURE 1. Distributions of the fitted invariant masses of $Kn\pi$ combinations for the events in which $D^0 \to K^-\mu^+\nu_\mu$ candidate events are observed on the recoil side of the tagged $\bar{D}^0$. The histograms are the data. The solid lines are the fitted curve. (This result is preliminary.)

INCLUSIVE D DECAYS

Branching fractions for the inclusive Cabibbo-favored $\bar{K}^*(892)$ and Cabibbo-suppressed $K^*(892)$ decays of D mesons

Measurement of branching fractions for the D meson decay modes containing $\bar{K}^*(K^*)$ in the final states can provide useful information about the relative strength of the Cabibbo-favored and Cabibbo-suppressed D decays. Invariant mass spectra for $K^-\pi^+$ combinations observed on the recoil side of the $\bar{D}^0$ tags (the $mKn\pi$ combinations) are shown in Fig. 3 for study of the Cabibbo-favored decay $D^0 \to \bar{K}^{*0}X$, where (a) shows the $K^-\pi^+$ mass spectrum for the events with a tagged $\bar{D}^0$, and (b) shows the normalized $K^-\pi^+$ mass spectrum for the $\bar{D}^0$ sideband events. After subtracting the number of the background events, we obtain 95.9 ± 44.1 signal events for the $D^0 \to \bar{K}^{*0}X$ decay. By similarly analysis, 189.1 ± 36.0, 30.8 ± 13.2 and 12.3 ± 23.3 signal events for $D^+ \to \bar{K}^{*0}X$, $D^0 \to K^{*0}X$ and $D^+ \to K^{*0}X$ decays are obtained, respectively. Branching fractions for the inclusive Cabibbo-favored decay $D \to \bar{K}^{*0}X$ and Cabibbo-suppressed decay $D \to K^{*0}X$ are measured. The results are $BF(D^0 \to \bar{K}^{*0}X) = (8.7 \pm 4.0 \pm 1.2)\%$, $BF(D^+ \to \bar{K}^{*0}X) = (23.2 \pm 4.5 \pm 3.0)\%$ and $BF(D^0 \to K^{*0}X) = (2.8 \pm 1.2 \pm 0.4)\%$. We set an upper limit on the branching fraction at 90% C.L. for the decay $D^+ \to K^{*0}X$ to be $BF(D^+ \to K^{*0}X) < 6.6\%$. [5]

We also measureed the the inclusive Cabibbo-favored and Cabibbo-suppressed $\bar{K}^{*-}/\bar{K}^{*+}X$ decays of D mesons. 28.7 ± 15.3 and 6.4 ± 6.6 inclusive $\bar{K}^{*-}$ events from the D^0 and D^+ Cabibbo-favored decays, and 0.0 ± 3.5 and 2.8 ± 12.6 inclusive

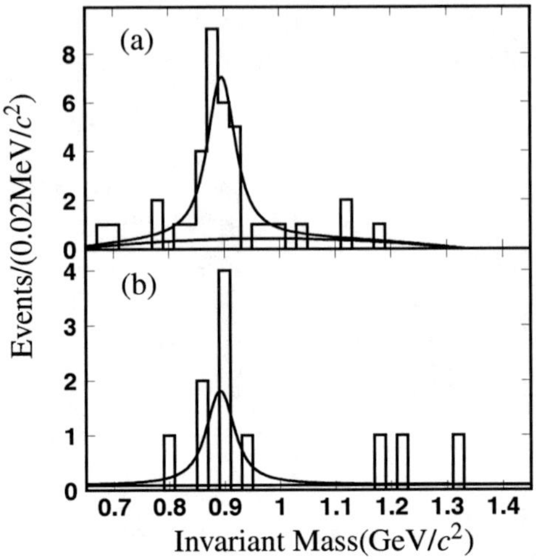

FIGURE 2. Distribution of the calculated invariant mass, (a) $K^-\pi^+$ combination for $D^+ \to \bar{K}^{*0}e^+\nu_e$; (b) $K^0_S\pi^-$ combination for $D^0 \to K^{*-}e^+\nu_e$. (This result is preliminary.)

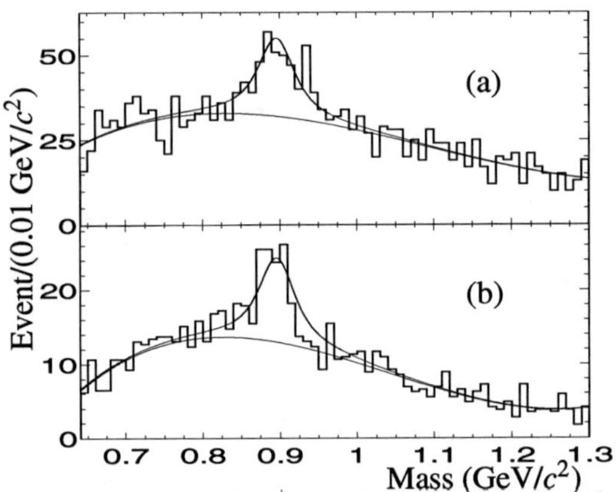

FIGURE 3. Invariant mass spectra for $K^-\pi^+$ combinations observed on the recoil side of the $\bar{D}^0$ tags for study of the Cabibbo-favored decay $D^0 \to \bar{K}^{*0}X$: (a) the mass spectrum for the events with a $\bar{D}^0$ tag, (b) the normalized mass spectrum for the $\bar{D}^0$ sideband events. [5]

K^{*+} events from the D^0 and D^+ Cabibbo-suppressed decays are observed. The inclusive Cabibbo-favored and Cabibbo-suppressed $\bar{K}^{*-}/\bar{K}^{*+}X$ decays of D mesons are measured. The branching fractions for decays $D^{+(0)} \to \bar{K}^{*-}(892)X$ are determined to be $BF(D^0 \to K^*(892)^-X) = (15.9 \pm 8.5 \pm 3.6)\%$ and $BF(D^+ \to K^*(892)^-X) = (5.1^{+5.2}_{-2.2} \pm 1.1)\%$. Upper limits on the branching fraction at 90% C.L. for Cabibbo-Suppressed decays $D \to K^*(892)^+$ are set to be $BF(D^0 \to K^*(892)^+X) < 4.1\%$, $BF(D^+ \to K^*(892)^+X) < 23.0\%$. These results are preliminary.

Branching fractions for the inclusive $K^0/\bar{K}^0$ decays of D mesons

The inclusive decays $D \to \bar{K}^0/K^0X$ are reconstructed on the recoil side of the singly tagged $\bar{D}$, where the $\bar{K}^0/K^0$ is reconstructed through its decay to $\pi^+\pi^-$. We obtain 250.0 ± 26.0 signal events for the decay $D^0 \to \bar{K}^0/K^0X$ and 243.8 ± 22.3 signal events for the decay $D^+ \to \bar{K}^0/K^0X$. The branching fractions for $D \to \bar{K}^0/K^0X$ are determined to be $BF(D^0 \to K^0/\bar{K}^0X) = (47.5 \pm 4.9 \pm 3.0)\%$ and $BF(D^+ \to K^0/\bar{K}^0X) = (62.6 \pm 5.7 \pm 3.3)\%$. These results are preliminary.

SUMMARY

Based on a data sample of about 33 pb^{-1} collected with the BES-II detector at and around the center-of-mass energy of 3.773 GeV, the decay branching fractions for $D^0 \to K^-\mu^+\nu_\mu$, $D^0 \to \pi^-\mu^+\nu_\mu$ and $D^+ \to \bar{K}^0\mu^+\nu_\mu$ are measured. The ratio of the two partial widths is determined to be $\Gamma(D^0 \to K^-\mu^+\nu_\mu)/\Gamma(D^+ \to \bar{K}^0\mu^+\nu_\mu) = 0.87 \pm 0.24 \pm 0.15$. The decay branching fractions for $D \to K\pi e^+\nu_e$ and $D \to \bar{K}^*e^+\nu_e$ and the inclusive decay branching fractions for $D \to \bar{K}^* + X$ and $D \to K^0(\bar{K}^0) + X$ are also measured.

ACKNOWLEDGMENTS

The BES collaboration thanks the staff of BEPC for their diligent efforts. This work is supported in part by the National Natural Science Foundation of China under contracts Nos. 10491300, 10225524, 10225525, the Chinese Academy of Sciences under contract No. KJ 95T-03, the 100 Talents Program of CAS under Contract Nos. U-11, U-24, U-25, and the Knowledge Innovation Project of CAS under Contract Nos. U-602, U-34 (IHEP); and by the National Natural Science Foundation of China under Contract No. 10175060 (USTC), and No. 10225522 (Tsinghua University).

REFERENCES

1. J.Z. Bai et al., *Nucl. Instr. and Methods* **A 458** 627-637 (2001).
2. M. H. Ye and Z. P. Zheng, Proceeding of the XIV International Symposium on Lepton and Photon Interaction, Stanford, California, 1989 (World Scientific, Singapore, 1990)
3. M. Ablikim et al., *Phys. Lett.* **B 597** 39-46 (2004).
4. M. Ablikim et al., *Phys. Lett.* **B 608** 24-30 (2005).
5. M. Ablikim et al., *Phys. Let.* **B 625** 196-202 (2005).
6. S. Eidelman et al. (Particle Data Group), *Phys. Let.* **B 592** 1-1109 (2004).
7. Marina Artuso, 'Charm decays within the Standard Model and beyond', XXII International Smposium on Lepton Photon Interactions at High Energy, Uppsala, Sweden, July 1st, (2005)

Measurements of $\sigma(e^+e^- \rightarrow \text{hadrons})$ and $B(\psi(3770) \rightarrow D\bar{D}, \text{non}-D\bar{D})$

Gang RONG (for BES Collaboration)

Institute of High Energy of Physics, CAS, Beijing, 100049

Abstract. We report measurements of the cross sections for inclusive hadronic event production in e^+e^- annihilation at the energies of 3.650, 3.6648, and 3.773 GeV and measurements of the branching fractions for $\psi(3770) \rightarrow D^0\bar{D}^0, D^+D^-, D\bar{D}$, and for $\psi(3770) \rightarrow \text{non}-D\bar{D}$.

Keywords: Document processing, Class file writing, $\LaTeX\,2_\varepsilon$
PACS: 43.35.Ei, 78.60.Mq

INTRODUCTION

Precision measurements of R values ($R = \sigma^B_{had}/\sigma^B_{\mu^+\mu^-}$, where σ^B_{had} and $\sigma^B_{\mu^+\mu^-}$ are Born order cross sections for inclusive hadron production and for muon pair production in e^+e^- annihilation, respectively.) at low energy region are important for precise test of the Standard Model [1]. On the other hand, the measurements of the R values at the center-of-mass energies around 3.6574 and at 3.773 GeV can be used to determine the branching fractions for $\psi(3770) \rightarrow D^0\bar{D}^0, D^+D^-, D\bar{D}$, and for $\psi(3770) \rightarrow \text{non}-D\bar{D}$. In this paper, we present a more precision measurement of the R values at the center-of-mass energies of 3.650, 3.6648 and 3.773 GeV, and measurements of the branching fractions for $\psi(3770) \rightarrow D^0\bar{D}^0, D^+D^-, D\bar{D}$, and for $\psi(3770) \rightarrow \text{non}-D\bar{D}$ with the data taken with the BES-II detector at the BEPC Collider.

MEASUREMENT OF R VALUES AT 3.650, 3.6648 AND 3.773 GEV

The value of R can be determined by the relation

$$R = \frac{\sigma^{\rm obs}_{\rm had}}{(1+\delta)\,\sigma^{\rm B}_{\mu^+\mu^-}}, \quad \text{with} \quad \sigma^{\rm obs}_{\rm had} = \frac{N^{\rm obs}_{\rm had}}{L\,\varepsilon_{\rm had}\,\varepsilon^{\rm trig}_{\rm had}}, \tag{1}$$

where $N^{\rm obs}_{\rm had}$ is the number of the observed hadronic events, L is the integrated luminosity, $\varepsilon_{\rm had}$ is the efficiency for detection of the inclusive hadronic events, $\varepsilon^{\rm trig}_{\rm had}$ is the trigger efficiency for the events, and $(1+\delta)$ is the radiative correction factor.

The hadronic events are required to have more than 2 good charged tracks, each of which is required to satisfy the following selection criteria: (1) the charged track must be with a good helix fit and the number of dE/dx hits per charged track is required to be greater than 14; (2) the distance of closest approach to the nominal interaction point is less than 2.0 cm; (3) $|\cos\theta| \leq 0.84$, where θ is the polar angle of the charged track; (4)

CP814 *Hadron Spectroscopy: Eleventh International Conference*
edited by A. Reis, C. Göbel, J. Sá Borges, and J. Magnin

2.0 $ns \leq T_{TOF} \leq T_p + 2.0$ ns, where T_{TOF} is the time-of-flight of the charged particle, and T_p is the expected time-of-flight of proton with a given momentum; (5) the charged track must not be identified as muon; (6) $p \leq E_b + 0.1 \times E_b \times \sqrt{(1+E_b^2)}$, where p is the charged track momentum and E_b is the beam energy in GeV; (7) for the charged track, the energy deposited in BSC should be less than 1.0 GeV. In addition, the total energy deposited by an event in BSC should be greater than 28% of the beam energy. Furthermore, the selected tracks must not all point into the same hemisphere in the z direction. Some beam-gas background can also satisfy the selection criteria. The number of the background is estimated from the measured distribution of event vertex points and subtracted from the data. The background due to $e^+e^- \to \tau^+\tau^-$, $e^+e^- \to (\gamma)e^+e^-$, and $e^+e^- \to (\gamma)\mu^+\mu^-$ is estimated by means of Monte Carlo simulation and subtracted from the data. The background from two-photon-exchange processes are expected to be small and are neglected. The systematic uncertainty in measuring the produced hadronic events due to the hadronic event selection criteria is estimated to be about 2.6%.

The integrated luminosities of the data sets are determined from large-angle Bhabha scattering. The uncertainties of the luminosity measurement are estimated to be 1.8%.

The detection efficiency for hadronic events is determined via a special Monte Carlo generator in which the initial state radiative correction to α^2 order and Lorentz boost of the system recoiling against the radiative photons are taken into account. These generated events are simulated with the GEANT3-based Monte Carlo simulation package. The systematic uncertainty in the efficiencies due to the generators is estimated to be 2.0%.

The trigger efficiencies are obtained by comparing the responses to different trigger requirements in the data taken at 3.097 GeV. The trigger efficiencies are measured to be 0.9999 for both the $e^+e^- \to (\gamma)e^+e^-$ and $e^+e^- \to hadron$ events.

The radiative correction is determined by $(1+\delta(s)) = \frac{\sigma^{\rm exp}(s)}{\sigma^{\rm B}(s)}$, where $\sigma^{\rm B}(s)$ is Born cross section for the inclusive hadronic event production and $\sigma^{\rm exp}(s)$ the expected cross section observed experimentally. The latter one can be written as

$$\sigma^{exp}(s) = \int_0^{1-\frac{4m_\pi^2}{s}} dx \cdot \frac{\sigma^B(s(1-x))}{|1-\Pi(s(1-x))|^2} F(x,s), \tag{2}$$

where $F(x,s)$ is a sampling function and $\frac{1}{|1-\Pi(s(1-x))|^2}$ is the correction factor for the effect of vacuum polarization including both the leptonic and hadronic terms from perturbative QED [2] [3].

The R values measured at the three energy points are summarized in table 1, where the first error is statistical, the second is the point-to-point systematic and the third is common systematic error. The common systematic error arises from the uncertainty in luminosity ($\sim$ 1.8%), the uncertainty in selection of hadronic event ($\sim$ 2.6%), the uncertainty in Monte Carlo Modeling ($\sim$ 2.0%)[4], the uncertainty in radiative correction ($\sim$ 1.5%) for the measured cross sections and R values at the three energy points, and the uncertainty in $\psi(3770)$ resonance parameters ($\sim$ 2.7%) for those measured at 3.773 GeV only. Adding these uncertainties in quadrature yields the total systematic uncertainties to be 4.0% and 4.9% for the measured hadronic cross sections and R values

TABLE 1. Radiative correction factors and the measured R values.

Energy (GeV)	$(1+\delta(s))$	R
3.6500	1.287	$2.249 \pm 0.010 \pm 0.025 \pm 0.090$
3.6648	1.264	$2.197 \pm 0.024 \pm 0.024 \pm 0.088$
3.7730	1.208	$3.759 \pm 0.008 \pm 0.041 \pm 0.184$

around 3.6574 and 3.773 GeV, respectively. Averaging the R values measured at 3.650 and 3.6648 GeV by weighting the combined statistical and point-to-point systematic errors yields the average value of R at $\sqrt{s}$ around 3.6574 GeV to be

$$\overline{R}_{uds} = 2.229 \pm 0.030 \pm 0.089, \tag{3}$$

where the first error is the combined statistical and point-to-point systematic errors, and the second is common systematic. This average $\overline{R}_{uds}$ value reflects the Born order cross section for the inclusive light hadronic event production through one-photon process in e^+e^- annihilation. The value is consistent with $R_{uds} = 2.26 \pm 0.14$ obtained by fitting those measured in the energy region between 2.0 and 3.0 GeV [5] and with $\overline{R}_{uds} = 2.21 \pm 0.13$ obtained from fitting to the inclusive hadronic cross sections for both the $\psi(3686)$ and $\psi(3770)$ resonances in the energy region from 3.666 to 3.897 GeV [6].

Using the measured R value at 3.773 GeV listed in table 1 and the $\overline{R}_{uds}$, we obtain the $R_{\psi(3770)}$ due to $\psi(3770)$ decays to be

$$R_{\psi(3770)} = 1.530 \pm 0.052 \pm 0.118, \tag{4}$$

where the first error is the combined statistical and point-to-point systematic errors, and the second is common systematic. In estimation of the systematic uncertainty, we assumed that the same amount of the systematic uncertainties in the measured values of the R and the $\overline{R}_{uds}$ is canceled in subtracting the $\overline{R}_{uds}$ from the R. The corresponding Born order cross section for $\psi(3770)$ production is

$$\sigma^{\mathrm{B}}_{\psi(3770)} = (9.330 \pm 0.317 \pm 0.719) \ \ \mathrm{nb}. \tag{5}$$

BRANCHING FRACTIONS FOR THE DECAYS $\psi(3770) \to D^0\bar{D}^0, D^+D^-, D\bar{D}$ AND FOR $\psi(3770) \to$ NON $- D\bar{D}$

The branching fraction for $\psi(3770) \to D\bar{D}$ can be determined by

$$B(\psi(3770) \to D\bar{D}) = \frac{\sigma^{\mathrm{obs}}_{D\bar{D}}}{(1+\delta)_{D\bar{D}}\ \sigma^{\mathrm{B}}_{\psi(3770)}}, \tag{6}$$

where $\sigma^{\mathrm{obs}}_{D\bar{D}}$ and $\sigma^{\mathrm{B}}_{\psi(3770)}$ are the observed and Born order production cross sections for $D\bar{D}$ and hadronic events, respectively, $(1+\delta)_{D\bar{D}} = (0.768 \pm 0.020)$ [7] is the radiative correction factor for the $D\bar{D}$ production.

BES Collaboration measured $\sigma^{obs}_{D^0\bar{D}^0} = (3.58 \pm 0.09 \pm 0.26)$ nb and $\sigma^{obs}_{D^+D^-} = (2.56 \pm 0.08 \pm 0.26)$ nb [5], which were obtained by analyzing the same data set from which the R value at $\sqrt{s} = 3.773$ GeV is measured. Inserting the $\sigma^{B}_{\psi(3770)}$, $\sigma^{obs}_{D^0\bar{D}^0}$, $\sigma^{obs}_{D^+D^-}$, and $(1+\delta)_{D\bar{D}}$ in Eq. (6), we obtain

$$B(\psi(3770) \to D^0\bar{D}^0) = (50.0 \pm 1.3 \pm 3.9)\%, \tag{7}$$

$$B(\psi(3770) \to D^+D^-) = (35.7 \pm 1.1 \pm 3.4)\%, \tag{8}$$

$$B(\psi(3770) \to D\bar{D}) = (85.7 \pm 1.7 \pm 5.2)\%, \tag{9}$$

which result in the non-$D\bar{D}$ branching fraction of $\psi(3770)$ to be

$$B(\psi(3770) \to \mathrm{non}-D\bar{D}) = (14.3 \pm 1.7 \pm 5.2)\%, \tag{10}$$

where the first error is statistical and the second systematic arising from some uncanceled systematic uncertainties. The uncanceled relative systematic uncertainties are $\sim 6.2\%$, $\sim 8.4\%$ and $\sim 3.3\%$ for the $\sigma^{obs}_{D^0\bar{D}^0}$, $\sigma^{obs}_{D^+D^-}$ and $\sigma^{B}_{\psi(3770)}$, respectively. The systematic error also includes the uncertainty of $\sim 3.2\%$ arising from the combined statistical and point-to-point systematic errors of both the measured R at 3.773 GeV and the $\overline{R}_{uds}$ measured at $\sqrt{s}$ around 3.6574 GeV.

ANALYSIS OF LINE-SHAPES FOR INCLUSIVE HADRONIC AND $D\bar{D}$ EVENT PRODUCTION

A better way to measure the inclusive branching fraction for $\psi(3770) \to \mathrm{non}-D\bar{D}$ is to analyze the energy dependent cross sections for both the inclusive hadron and $D\bar{D}$ event production in the range covering both $\psi(3686)$ and $\psi(3770)$ resonances. To this end, BES Collaboration performed a cross section scan experiment in the energy region between 3.660 and 3.873 GeV.

The observed cross section for $D^0\bar{D}^0$ (or D^+D^-) production is determined based on the number $N_{D^0_{\mathrm{tag}}}$ (or $N_{D^+_{\mathrm{tag}}}$) of the reconstructed D^0 (or D^+) events by

$$\sigma^{obs}_{D^0\bar{D}^0(\mathrm{or}\ D^+D^-)} = \frac{N_{D^0_{\mathrm{tag}}}\,(\mathrm{or}\ N_{D^+_{\mathrm{tag}}})}{2 \times L \times B \times \varepsilon}, \tag{11}$$

where L is the integrated luminosity of the data set used in the analysis, B is the branching fraction for the decay mode in question, and ε is the efficiency determined from Monte Carlo simulation for reconstruction of this decay mode.

Using the method discussed in Section II and Reference [5] [6], we obtained the observed cross sections for both the inclusive hadron and $D\bar{D}$ event production at the energies at which the data were collected. The observed cross section results are displayed in Fig. 1, where Fig. 1(a) and Fig. 1(b) show the observed cross sections for inclusive hadronic event production, and Fig. 1(c) and Fig. 1(d) show the observed cross sections for $D^0\bar{D}^0$ and D^+D^- production, respectively. The error bars represent the combined statistical and point-to-point systematic uncertainties.

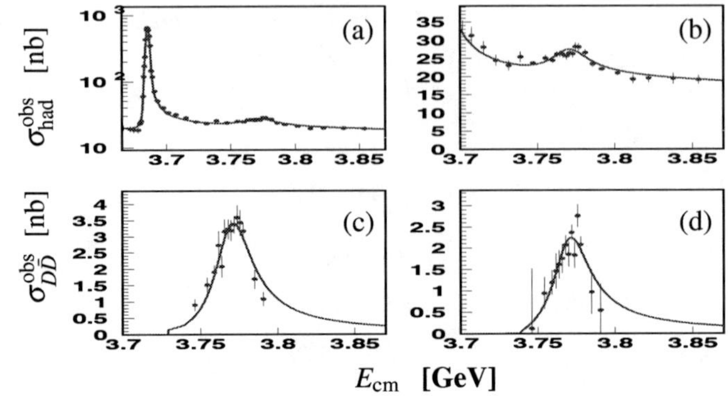

FIGURE 1. The observed cross sections for inclusive hadronic event, $D^0\bar{D}^0$ and D^+D^- event production v.s. the nominal center-of-mass energy.

The determination of the branching fractions for $\psi(3770) \to D\bar{D}$ is accomplished by simultaneously fitting the observed cross sections for $\psi(3686)$, $\psi(3770)$, $D^0\bar{D}^0$ and D^+D^- to functions that described $\psi(3686)$, $\psi(3770)$ resonance shapes, $D^0\bar{D}^0$ and D^+D^- production shapes and the dominant no-resonant hadronic background, and included radiative corrections for resonances themself, and the J/ψ.

For J/ψ and ψ' resonances, we take the standard form

$$\sigma^B(E_{cm}) = \frac{12\pi\Gamma^0_{ee}\Gamma_h}{(E^2_{cm} - M^2)^2 + \Gamma^2 M^2}, \tag{12}$$

where M and Γ are the mass and total width of the resonance, and $\Gamma_{ee} = \Gamma^0_{ee}|1 - \Pi(s)|^{-2}$ and Γ_h are the partial widths to e^+e^- channel and to the inclusive hadronic final state, respectively. We also consider the effects of the BEPC energy spread on the observed cross sections in the fit. For the $\psi(3770)$ resonance, we use a p-wave Breit-Wigner shape with an energy-dependent total width to fit the observed hadronic cross sections and $D\bar{D}$ ($D^0\bar{D}^0$ and D^+D^-) cross sections. The $\psi(3770)$ resonance shape is taken as

$$\sigma^B_{\psi(3770)}(s) = \frac{12\pi\Gamma^0_{ee}\Gamma_{\rm tot}(s)}{(s - M^2)^2 + M^2\Gamma^2_{\rm tot}(s)}, \tag{13}$$

while the $D^0\bar{D}^0$, D^+D^- (or $D\bar{D}$) production shapes are taken as

$$\sigma^B_{D\bar{D}}(s) = \frac{12\pi\Gamma^0_{ee}\Gamma_{\rm D\bar{D}}(s)}{(s - M^2)^2 + M^2\Gamma^2_{\rm tot}(s)}, \tag{14}$$

where $\Gamma_{ee} = \frac{\Gamma^0_{ee}}{|1-\Pi(s)|^2}$ with which the effect of vacuum polarization is absorbed by Γ_{ee}; M and Γ_{ee} are the mass and leptonic width of the $\psi(3770)$ resonance respectively; $\Gamma_{\rm tot}(s)$ and $\Gamma_{\rm D\bar{D}}(s)$ are chosen to be energy dependent, which are defined as

$$\Gamma_{\rm tot}(s) = \Gamma_{D^0\bar{D}^0}(s) + \Gamma_{D^+D^-}(s) + \Gamma_{non-D\bar{D}}(s), \tag{15}$$

with

$$\Gamma_{D^0\bar{D}^0}(s) = \Gamma_0\ \theta(E_{cm} - 2M_{D^0}) \frac{p_{D^0}{}^3}{1+(rp_{D^0})^2} \frac{1+(rp^0_{D^0})^2}{p^0_{D^0}{}^3} B(\psi(3770) \to D^0\bar{D}^0), \quad (16)$$

$$\Gamma_{D^+D^-}(s) = \Gamma_0\ \theta(E_{cm} - 2M_{D^+}) \frac{p_{D^+}{}^3}{1+(rp_{D^+})^2} \frac{1+(rp^0_{D^+})^2}{p^0_{D^+}{}^3} B(\psi(3770) \to D^+D^-), \quad (17)$$

$$\Gamma_{non-D\bar{D}}(s) = \Gamma_0\ [1 - B_{00} - B_{+-}], \quad (18)$$

where p^0_D is the momentum of the D mesons produced at the peak of $\psi(3770)$, p_D is the momentum of the D mesons produced at the c.m. energy $\sqrt{s}$; Γ_0 is the total width of the $\psi(3770)$ at the peak, $B_{00} = B(\psi(3770) \to D^0\bar{D}^0)$ and $B_{+-} = B(\psi(3770) \to D^+D^-)$ are the branching fractions for $\psi(3770) \to D^0\bar{D}^0$ and $\psi(3770) \to D^+D^-$, respectively, which are the fitted parameters, and r is the interaction radius of the $c\bar{c}$, which is left free in the fit; $\theta(E_{cm} - 2M_{D^0})$ and $\theta(E_{cm} - 2M_{D^+})$ are the step functions to account for the threshold of the $D\bar{D}$ production.

The no-resonant hadronic background shape includes a flat term and terms proportional to $p^3_{D^+}$ and $p^3_{D^0}$, which account for continuum hadronic event, non-resonant $D^0\bar{D}^0$ and D^+D^- production, respectively. The J/ψ resonance parameters are fixed at the values given by PDG [9] in the fit.

The curves in Fig. 1 show the fit to the data. The results for this fit are

$$B(\psi(3770) \to D^0\bar{D}^0) = (51.1 \pm 4.2 \pm 3.6)\%,$$

$$B(\psi(3770) \to D^+D^-) = (33.7 \pm 2.9 \pm 3.0)\%,$$

where the first error is statistical and second systematic arising from some uncanceled systematic uncertainties in the measured $\sigma_{\psi(3770)}$ ($\sim 3.3\%$), in $\sigma_{D^0\bar{D}^0}$ ($\sim 6.2\%$), and in $\sigma_{D^+D^-}$ ($\sim 8.4\%$). Considering the correlation between the two branching fractions obtained from the fit, we obtain the branching fraction for $\psi(3770) \to D\bar{D}$ to be

$$B(\psi(3770) \to D\bar{D}) = (84.8 \pm 6.8 \pm 6.2)\%,$$

which results in the non-$D\bar{D}$ branching fraction to be

$$B(\psi(3770) \to non - D\bar{D}) = (15.2 \pm 6.8 \pm 6.2)\%.$$

SUMMARY

From the analysis of the data taken at 3.650, 3.6648 and 3.773 GeV in e^+e^- annihilation, we measured the R values at the energy points, which are $R = 3.759 \pm 0.042 \pm 0.184$ at 3.773 GeV and $\overline{R}_{uds} = 2.229 \pm 0.030 \pm 0.089$ at $\sqrt{s}$ around 3.6574 GeV. Based on the measured R values, we determined the Born cross section for $\psi(3770)$ production at 3.773 GeV to be $\sigma^{\rm B}_{\psi(3770)} = (9.330 \pm 0.317 \pm 0.719)$ nb, the branching fractions for $\psi(3770)$ decays to be $B(\psi(3770) \to D^0\bar{D}^0) = (50.0 \pm 1.3 \pm 3.9)\%$, $B(\psi(3770) \to$

$D^+D^-) = (35.7 \pm 1.1 \pm 3.4)\%$, $B(\psi(3770) \to D^0\bar{D}^0) = (85.6 \pm 1.7 \pm 5.2)\%$, which result in the non-$D\bar{D}$ branching fraction of $\psi(3770)$ to be $B(\psi(3770) \to \mathrm{non} - D\bar{D}) = (14.3 \pm 1.7 \pm 5.2)\%$. From an analysis of the resonance line-shapes for the inclusive hadronic, $D^0\bar{D}^0$ and D^+D^- event production, we also determined the branching fractions for $\psi(3770) \to D^0\bar{D}^0, D^+D^-, D\bar{D}$ and $\psi(3770) \to \mathrm{non} - D\bar{D}$. The results are consistent with those determined based on the measured R values at 3.773 GeV and around 3.6574 GeV. These branching fractions for $\psi(3770)$ decays are measured in first time.

ACKNOWLEDGEMENTS

The BES collaboration thanks the staff of BEPC for their hard efforts. This work is supported in part by the National Natural Science Foundation of China under contracts Nos. 19991480,10225524,10225525, the Chinese Academy of Sciences under contract No. KJ 95T-03, the 100 Talents Program of CAS under Contract Nos. U-11, U-24, U-25, and the Knowledge Innovation Project of CAS under Contract Nos. U-602, U-34(IHEP); by the National Natural Science Foundation of China under Contract No.10175060(USTC),and No.10225522(Tsinghua University).

REFERENCES

1. M. Davier, S Eidelman, A. Hocker, Z. Zhang, Eur. Phys. J. C **27** (2003)497.
2. E. A. Kuraev and V. S. Fadin, Yad Fiz. **41**, 377 (1985); [Sov. J. Nucl. Phys. **41**, (1985)466].
3. F.A. Berends and G.J. Komen, Phys. Lett. **B 63** (1976)432; Andrej B. Arbuzov, Eduard A. Kuraev *et al.*, JHEP01 (1997) 006.
4. J.Z. Bai *et al.* (BES Collaboration), Phys. Rev. Lett. **88** (2002)130.
5. J.Z. Bai *et al.* (BES Collaboration), Phys. Lett. **B 603** (2004)130.
6. M. Ablikim *et al.* (BES Collaboration), Contribution paper to Lepton Photon 2005 (LP05-452).
7. Gang Rong, Dahua Zhang and Jiangchuan Chen, arXiv:hep-ex/0506051.
8. R.H. Schindler *et al.*, Phys. Rev. **D21** (1980)2716.
9. S. Eidelman *et al.* (Particle Data Group), Phys. Lett. **B 592** (2004)1.

Time-Independent Measurements of D^0-$\overline{D}^0$ Mixing and Relative Strong Phases Using Quantum Correlations at CLEO-c

David Asner - for the CLEO Collaboration

Carleton University, Department of Physics, 1125 Colonel By Drive, Ottawa, ON K1S 5B6, Canada

Abstract. Due to quantum correlations in the $C=-1$ and $C=+1$ $D^0\overline{D}^0$ pairs produced in the reactions $e^+e^- \to D^0\overline{D}^0(\pi^0)$ and $e^+e^- \to D^0\overline{D}^0\gamma(\pi^0)$, respectively, the time-integrated $D^0\overline{D}^0$ decay rates are sensitive to interference between amplitudes for indistinguishable final states. The size of this interference is governed by the relevant amplitude ratios and can include contributions from D^0-$\overline{D}^0$ mixing. We make use of fully and partially reconstructed $D^0\overline{D}^0$ pairs in both C eigenstates. and we estimate experimental sensitivities based on a plausible CLEO-c dataset.

Keywords: charm, mixing, correlated production
PACS: 14.40.Lb, 13.20.Fc, 12.15.Mm

INTRODUCTION

Neutral flavor oscillation in the D meson system is highly suppressed within the Standard Model. The time evolution of a particle produced as a D^0 or $\overline{D}^0$, in the limit of CP conservation, is governed by four parameters: $x = \Delta m/\Gamma$ and $y = \Delta\Gamma/2\Gamma$ which characterize the mixing matrix, $\delta_{K\pi}$ the relative strong phase between Cabibbo favored (CF) and doubly-Cabibbo suppressed (DCS) amplitudes to the $K\pi$ final state and R_D the DCS decay rate relative to the CF decay rate. The mixing rate R_M is defined as $\frac{1}{2}(x^2+y^2)$ [1]. Standard Model based predictions for x and y, as well as a variety of non-Standard Model expectations, span several orders of magnitude [2]. It is reasonable to expect that $x \approx y \approx 10^{-3}$ in the Standard Model.

The parameters x and y can be measured in a variety of ways. The most precise limits are obtained by exploiting the time-dependence of D decays [1]. Time-dependent analyses are not feasible at CLEO-c; however, the quantum-coherent $D^0\overline{D}^0$ state provides time-integrated sensitivity to x, y at $\mathcal{O}(1\%)$ level and $\cos\delta_{K\pi} \sim 0.1$ [3, 4] in 1 fb^{-1} of $\psi(3770) \to D\overline{D}$. Although CLEO-c does not have sufficient sensitivity to observe Standard Model charm mixing the projected sensitivity in 1 fb^{-1} at $\psi(3770)$ compares favorably with current experimental results; see Fig. 1.

Two classes of analyses are ongoing at CLEO-c - 1) "targeted" analyses of individual mixing and doubly-Cabibbo suppressed parameters typically based on one double tag rate; 2) a "comprehensive" analysis which simultaneously determines the same parameters by examining various single tag and double tag rates. Greater sensitivity to the mixing and doubly-Cabibbo suppressed parameters is anticipated in the comprehensive analysis as several single tag and double tag rates provide sensitivity to each parameter.

CP814 *Hadron Spectroscopy: Eleventh International Conference*
edited by A. Reis, C. Göbel, J. Sá Borges, and J. Magnin

Targeted Analyses

Measurements of each of the four parameters that describe *CP* conserving mixing can in principle be determined individually by a series of "targeted" analyses. The techniques and projected sensitivities in 1 fb^{-1} are described briefly in the subsections that follow. Greater detail can be found in Ref. [3].

Mixing Rate R_M: The measurement of R_M can be determined unambiguously by considering the decays $\psi(3770) \to (K^-\pi^+)(K^-\pi^+)$ and $\psi(3770) \to (\ell^{\pm}(KX))(\ell^{\pm}(KX))$. The hadronic final state cannot be produced from DCS decay due to a quantum statistics argument - the C-odd initial state cannot produce the symmetric final state required by Bose statistics if both the D^0 and $\overline{D}^0$ decay into the same final state $(K\pi)$. A fit utilizing six constraints make this channel effectively background free. The double semileptonic channel has fewer constraints due to the presence of two neutrinos. The number of "right-sign" $(K^-\pi^+)(K^+\pi^-)$ and $(\ell^{\pm}(KX))(\ell^{\mp}(KX))$ events in the two channels combined produced in 1 fb^{-1} is expected to be $\sim 20{,}000$ corresponding to $\sqrt{2R_M} < 1.7\%$ @95% Confidence Level (C.L.).

Relative Strong Phase $\cos\delta_{K\pi}$: Final states where one D decays to $K\pi$ and the other to a *CP* eigenstate can be used to probe $\cos\delta_{K\pi}$. The *CP*-even eigenstates accessible to CLEO-c include K^+K^-, $\pi^+\pi^-$ and $K^0_S\pi^0\pi^0$ and the *CP*-odd eigenstates include $K^0_S\pi^0$, $K^0_S\eta$, $K^0_S\omega$, $K^0_S\phi$, and $K^0_S\eta'$. The *CP* content of the $K^0_S\pi^+\pi^-$ and $\pi^+\pi^-\pi^0$ Dalitz plots will also be utilized. Sensitivity to $\cos\delta_{K\pi} \sim 1/(2\sqrt{R_D N})$ where N, the total number of *CP* tagged $K\pi$, will be $\sim$9200 and R_D is already well measured, thus leading to an expected precision of ± 0.09 in $\cos\delta_{K\pi}$.

Mixing Parameter y: Tagging one of the D mesons as a *CP* eigenstate, y can be determined by measuring the flavor specific branching ratios of the other meson. The flavor tag width is independent of the *CP* quantum number however the branching ratio is inversely proportional to the total width. Consequently, charm threshold experiments have time-integrated sensitivity to y. Neglecting factors related to DCS decays which are described in detail in Ref. [19], we can construct the ratio $\frac{2\Gamma[\psi \to (D \to CP)(D \to \text{flavor})]}{\Gamma[\psi \to (D \to CP)(D \to X) Br(D \to \text{flavor})]} \sim (1 \pm y)$ where X represents an inclusive set of all states. A positive y would make the above ratio > 1 for $CP+$ tags.

The decay $D^0 \to K^0_S\pi^+\pi^-$ is measured with a Dalitz plot analysis to proceed through intermediate states that are $CP+$ eigenstates, such as $K^0_S f_0$, $CP-$ such as $K^0_S\rho$ and flavor eigenstates such as $K^{*-}\pi^+$ [20]. The presence of mixing through y would introduce an intensity modulation across the Dalitz plot as a function of the *CP* of the contributing amplitudes. The sensitivity at CLEO-c to y with Dalitz plot analyses has not yet been fully evaluated. Preliminary estimates suggests a limit of $y \sim$ few% @95% C.L. is attainable with the CLEO-c data.

Mixing Parameter x: Measurement of y and R_M as outlined above allow the value x to be determined. The upper limit on R_M corresponds to a limit of $|x| < 1.7\%$ @95% C.L..

Doubly Cabibbo Suppressed Rate R_D: Although R_D is determined by the ratio of $(K^-\pi^+)(K^-\ell^+\nu)$ to $(K^-\pi^+)(K^+\ell^-\overline{\nu})$ up to corrections that are second order in x and y, a more precise measurement is attainable using a D^{*+} tag to measure $D^0 \to K^+\pi^-$ relative to $D^0 \to K^-\pi^+$ [1].

Comprehensive Analysis

The comprehensive analysis simultaneously determines mixing and doubly-Cabibbo suppressed parameters by examining various single tag and double tag rates explicitly accounting for QC. Due to quantum correlations (QC) in the $C=-1$ and $C=+1$ $D^0\overline{D}^0$ pairs produced in the reactions $e^+e^- \to D^0\overline{D}^0(\pi^0)$ and $e^+e^- \to D^0\overline{D}^0\gamma$, respectively, the time-integrated $D^0\overline{D}^0$ decay rates are sensitive to interference between amplitudes for indistinguishable final states. The size of this interference is governed by the relevant amplitude ratios and can include contributions from D^0-$\overline{D}^0$ mixing.

We consider in the comprehensive analysis the following categories of final states:

- f or $\bar{f}$: hadronic states that can be reached from either D^0 or $\overline{D}^0$ decay but that are not CP eigenstates. An example is $K^-\pi^+$, which results from Cabibbo-favored D^0 transitions or doubly Cabibbo-suppressed $\overline{D}^0$ transitions. We include in this category Cabibbo-suppressed transitions as well as self-conjugate final states of mixed CP, such as non-resonant $K_S^0\pi^+\pi^-$.
- ℓ^+ or ℓ^-: semileptonic or purely leptonic final states, which, in the absence of mixing, tag unambiguously the flavor of the parent D.
- S_+ or S_-: CP-even and CP-odd eigenstates, respectively.

We calculate decay rates for $D^0\overline{D}^0$ pairs to all possible combinations of the above categories of final states in Ref. [19], for both $C=-1$ and $C=+1$, reproducing the work of Refs. [21, 22]. Such $D^0\overline{D}^0$ combinations, where both D final states are specified, are referred to as double tags (or DT). In addition, we calculate rates for single tags (or ST), where either the D^0 or $\overline{D}^0$ is identified and the other neutral D decays generically. Any DT event also contains two ST decays, and ST rates are obtained by summing the corresponding DT rates.

Using a double tag technique pioneered by MARK III [23, 24], CLEO-c has recently measured $\mathcal{B}(D^0 \to K^-\pi^+)$, $\mathcal{B}(D^0 \to K^-\pi^+\pi^0)$, and $\mathcal{B}(D^0 \to K^-\pi^+\pi^+\pi^-)$ in a self-normalizing way (*i.e.*, without knowledge of the luminosity or $D^0\overline{D}^0$ production cross section) [25]. Measured ST and DT yields and efficiencies are combined in a least-squares fit [26] to extract the branching fractions and the number of $D^0\overline{D}^0$ pairs produced. Quantum correlations were not accounted for in this analysis, and only flavored final states were considered.

In correlated $D^0\overline{D}^0$ decay, the ST and DT rates no longer depend solely on the branching fractions of interest; the rates are modified by x, y, and the magnitudes and phases of various amplitude ratios. Therefore, one must correct the measured ST and DT yields in order to extract the branching fractions. Conversely, as shown in Ref. [19], using these yields, it is possible to determine simultaneously the mixing and amplitude ratio parameters as well as the relevant branching fractions. The least-squares fit described in Ref. [26] can be extended to extract the parameters $\{\mathcal{A}-y,\ R_M,\ r_f,\ z_f\}$ for $C=-1$ or $\{\mathcal{A},\ y,\ r_f,\ z_f,\ w_f x\}$ for $C=+1$, where $\mathcal{A} \equiv \sum_f A_f^2 r_f z_f$, $z_f \equiv 2\cos\delta_f$ and $w_f \equiv 2\sin\delta_f$. For simplicity, we include only one such mode in the analysis: $D \to K^\pm\pi^\mp$, so $f = K^-\pi^+$, and $A_{K\pi}$ ($r_{K\pi} = \sqrt{R_D}$) is the magnitude of the CF (DCS) amplitude to final state $K\pi$. When fitting $C=-1$ and $C=+1$ simultaneously one can also determine x^2, so the free parameters are $\{\mathcal{A},\ y,\ x^2,\ r_f,\ z_f,\ w_f x\}$.

TABLE 1. Estimated uncertainties (statistical and systematic, respectively) for $\Gamma_{D^0\overline{D}^0} = 3.5\times10^6$. For the last two columns, $\Gamma^{C=-1}_{D^0\overline{D}^0} = \alpha\Gamma^{C=+1}_{D^0\overline{D}^0} = 3.5\times10^6$, where α is 1 or 4. The uncertainties on $\cos\delta_{K\pi}$ and $x\sin\delta_{K\pi}$ are half of those on $z_{K\pi}$ and $w_{K\pi}x$, respectively.

Parameter	Value	$C=-1$	$C=+1$	$C=-1$ and $+1$	
				$\alpha=1$	$\alpha=4$
$\mathscr{A}-y(\%)$	0	$\pm1.3\pm0.9$			
$R_M(\%)$	0	$\pm0.03\pm0.03$			
$r_{K\pi}(\%)$	0.06	$\pm0.57\pm0.03$	$\pm0.95\pm0.07$	$\pm0.49\pm0.03$	$\pm0.55\pm0.03$
$z_{K\pi}$	-2	$\pm0.42\pm0.12$	$\pm0.49\pm0.12$	$\pm0.31\pm0.09$	$\pm0.38\pm0.10$
$y(\%)$	0		$\pm1.7\pm0.6$	$\pm1.0\pm0.4$	$\pm1.2\pm0.5$
$\mathscr{A}(\%)$	0		$\pm0.010\pm0.008$	$\pm0.002\pm0.005$	$\pm0.003\pm0.005$
$w_{K\pi}x$	0		$\pm0.043\pm0.012$	$\pm0.025\pm0.007$	$\pm0.033\pm0.010$
$x^2(\%)$	0			$\pm0.06\pm0.06$	$\pm0.06\pm0.06$

The estimated uncertainties on the fit parameters based on 3.5×10^6 $D^0\overline{D}^0$ pairs, in either the $C=-1$ or $C=+1$ configuration are presented in Table 1 using reconstruction efficiencies similar to those found at CLEO-c and assuming no background ($\sigma_N=\sqrt{N}$).

In Figure 1, we plot the projected 95% confidence level (C.L.) contour for $\Gamma^{C=-1}_{D^0\overline{D}^0} = 4\Gamma^{C=+1}_{D^0\overline{D}^0} = 3.5\times10^6$ and $\cos\delta_{K\pi}=-1$. We note that our sensitivity to x depends strongly on the value of $\delta_{K\pi}$, while our sensitivity to y does not, and lowering the value of $|\cos\delta_{K\pi}|$ would reduce the area of the contour. In general, our expected upper limits compare favorably with the current best limits on charm mixing.

REFERENCES

1. D. Asner, D^0-$\overline{D}^0$ Mixing in Review of Particle Physics, *Phys. Lett.*, **B592**, 1 (2004).
2. H. N. Nelson, in *Proc. of the 19th Intl. Symp. on Photon and Lepton Interactions at High Energy LP99* ed. J.A. Jaros and M.E. Peskin, SLAC (1999); S. Bianco, F. L. Fabbri, D. Benson and I. Bigi, Riv. Nuovo Cim. **26N7**, 1 (2003); A. A. Petrov, eConf **C030603**, MEC05 (2003) [arXiv:hep-ph/0311371]; I. I. Y. Bigi and N. G. Uraltsev, Nucl. Phys. B **592**, 92 (2001); Z. Ligeti, AIP Conf. Proc. **618**, 298 (2002) [arXiv:hep-ph/0205316]; A. F. Falk, Y. Grossman, Z. Ligeti and A. A. Petrov, Phys. Rev. D **65**, 054034 (2002); C. K. Chua and W. S. Hou, arXiv:hep-ph/0110106.
3. R. A. Briere *et al.*, CLNS-01-1742 (2001).
4. Gronau, M., Grossman, Y., and Rosner, J. L., *Phys. Lett.*, **B508**, 37–43 (2001).
5. E. M. Aitala *et al.* [E791 Collaboration], Phys. Rev. Lett. **83**, 32 (1999).
6. J. M. Link *et al.* [FOCUS Collaboration], Phys. Lett. B **485**, 62 (2000).
7. S. E. Csorna *et al.* [CLEO Collaboration], Phys. Rev. D **65**, 092001 (2002).
8. B. Aubert *et al.* [BABAR Collaboration], Phys. Rev. Lett. **91**, 121801 (2003).
9. K. Abe *et al.* [Belle Collaboration], arXiv:hep-ex/0308034.
10. E. M. Aitala *et al.* [E791 Collaboration], Phys. Rev. Lett. **77**, 2384 (1996).
11. B. Aubert *et al.* [BABAR Collaboration], Phys. Rev. D **70**, 091102 (2004).
12. C. Cawlfield *et al.* [CLEO Collaboration], Phys. Rev. D **71**, 077101 (2005)
13. K. Abe *et al.* [Belle Collaboration], arXiv:hep-ex/0507020.
14. R. Godang *et al.* [CLEO Collaboration], Phys. Rev. Lett. **84**, 5038 (2000).
15. B. Aubert *et al.* [BABAR Collaboration], Phys. Rev. Lett. **91**, 171801 (2003).
16. K. Abe *et al.* [Belle Collaboration], Phys. Rev. Lett. **94**, 071801 (2005).
17. J. M. Link *et al.* [FOCUS Collaboration], Phys. Lett. B **618**, 23 (2005).
18. D. M. Asner *et al.* [CLEO Collaboration], Phys. Rev. D **72**, 012001 (2005).

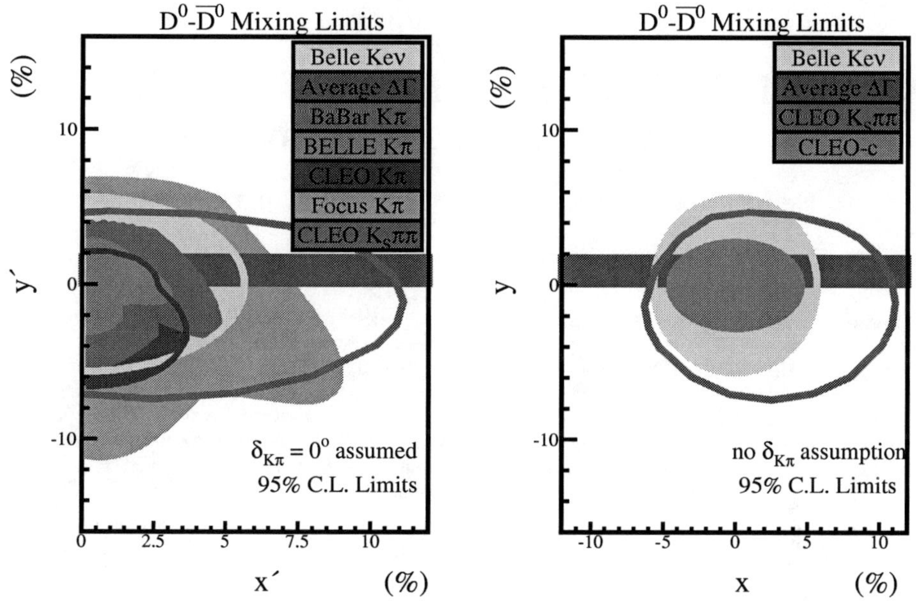

FIGURE 1. **Left**: Current allowed regions in the plane of $y'_{K\pi}$ versus $x'_{K\pi}$. The allowed region for $\Delta\Gamma$ is the average of the results from E791 [5], FOCUS [6], CLEO [7], BABAR [8], and Belle [9]. Also shown is the limit from $D^0 \to K^{(*)}\ell\nu$ from Belle [13]. Semileptonic results from E791 [10], BABAR [11], and CLEO [12] are not shown. The limits from $D \to K\pi$ from CLEO [14], BABAR [15], and Belle [16] allow for *CP* violation in the decay amplitude, in the mixing amplitude, and in the interference between these two processes, while the FOCUS [17] result does not. The CLEO experiment has constrained x and y with a time-dependent Dalitz-plot analysis of $D^0 \to K^0_S\pi^+\pi^-$ [18]. In depicting the y and Dalitz-plot results, we assume $\delta_{K\pi} = 0$; a non-zero value for $\delta_{K\pi}$ would rotate the $\Delta\Gamma$ and $D^0 \to K^0_S\pi^+\pi^-$ confidence regions clockwise about the origin by an angle $\delta_{K\pi}$. All results are consistent with the absence of mixing. **Right:** Current and projected allowed regions in the plane of y versus x. The allowed region for $\Delta\Gamma$ is the average of the results from from E791 [5], FOCUS [6], CLEO [7], BABAR [8] and Belle [9]. Also shown is the limit from $D^0 \to K^{(*)}\ell\nu$ from Belle [13]. Semileptonic results from E791 [10], BABAR [11], and CLEO [12] are not shown. The CLEO experiment has constrained x and y with a time-dependent Dalitz-plot analysis of $D^0 \to K^0_S\pi^+\pi^-$ [18]. All results are consistent with the absence of mixing. We superimpose our projected 95% C.L. contour for $\Gamma^{C=-1}_{D^0\overline{D}^0} = 4\Gamma^{C=+1}_{D^0\overline{D}^0} = 3.5 \times 10^6$ and $\cos\delta_{K\pi} = -1$. Lowering the value of $|\cos\delta_{K\pi}|$ would reduce the area of the contour.

19. D. M. Asner and W. M. Sun, arXiv:hep-ph/0507238, submitted to Phys. Rev. D.
20. Muramatsu, H., *et al.*, [CLEO Collaboration], *Phys. Rev. Lett.*, **89**, 251802 (2002).
21. M. Gronau, Y. Grossman and J. L. Rosner, Phys. Lett. B **508**, 37 (2001).
22. D. Atwood and A. A. Petrov, Phys. Rev. D **71**, 054032 (2005).
23. R. M. Baltrusaitis *et al.* [MARK III Collaboration], Phys. Rev. Lett. **56**, 2140 (1986).
24. J. Adler *et al.* [MARK III Collaboration], Phys. Rev. Lett. **60**, 89 (1988).
25. Q. He *et al.* [CLEO Collaboration], Phys. Rev. Lett. **95**, 121801 (2005).
26. W. M. Sun, arXiv:physics/0503050, submitted to Nucl. Instrum. Methods Phys. Res., Sec. A.

Open and Hidden Beauty Production in 920 GeV Proton-Nucleus Collisions

Hernan Wahlberg on behalf of HERA-B Collaboration

Utrecht University / NIKHEF

Abstract. HERA-B is a fixed-target multi-particle spectrometer experiment at the 920 GeV HERA proton beam at DESY. Approximately 150 million events were recorded with a dilepton trigger during the 2002/2003 HERA run. About 300,000 leptonic J/ψ decays (170,000 in the dimuon channel and 130,000 in the dielectron channel) have been reconstructed in this data sample. This statistics allows a new measurement of the $b\bar{b}$ and of the Υ production cross sections. The b events have been tagged via inclusive bottom quark decays into a J/ψ, by exploiting the longitudinal separation of $J/\psi \to l^+l^-$ decay vertices from the primary proton-nucleus interaction. The Υ cross section is measured in dimuon and dielectron decay channels.

Keywords: p nucleus, inclusive reaction, carbon, titanium, tungsten, beauty, quark, pair production, decay,J/psi, leptonic decay, cross section, magnetic detector, HERA-B, experimental results, DESY HERA Stor, 920 GeV
PACS: 13.85.Ni,13.85.Qk,13.20.He,24.85.+p.

INTRODUCTION

The $b\bar{b}$ production is a field of active interest; the latest theoretical estimations [1, 2] include full NLO calculations and different approaches for higher-order corrections. These calculations present large uncertainties due to the unknown b quark mass and the dependence on the normalization scale. There are three measurements of $\sigma(b\bar{b})$ by fixed target experiments (E789 [3], E771 [4] and HERA-B [5]), but they show incompatible results and present poor statistical precision. For Υ production, several measurements are available but the situation is unclear (conflicting) at $\sqrt{s} \approx 40$ GeV where HERA-B can contribute with a new measurement.

HERA-B EXPERIMENT AND DATA SAMPLE

HERA-B is a fixed target spectrometer [6] that operates at the HERA storage ring at DESY. Interactions are produced between the halo of the 920 GeV proton beam and target wires of different materials (C, Al, Ti, W), arranged in 2 stations separated by 4 cm. The center of mass energy is $\sqrt{s}$ = 41.6 GeV. The spectrometer has a large acceptance (from 15 to 220 mrad in the bending plane and from 15 to 160 mrad in the vertical plane). It has high vertex and track resolution together with good particle identification. In late 2002 and early 2003, HERA-B took most of its data devoted to physics studies. A total sample of 300,000 J/ψ decaying into two leptons was collected.

CP814 *Hadron Spectroscopy: Eleventh International Conference*
edited by A. Reis, C. Göbel, J. Sá Borges, and J. Magnin

OPEN BEAUTY PRODUCTION

To determine the $b\bar{b}$ cross section, we identify the inclusive decay of one b hadron through its decay into a J/ψ + X. Since in proton-nucleon collisions the b quarks are always produced in $b\bar{b}$ pairs, the identification of one b hadron is sufficient to insure the production of the $b\bar{b}$ pair. To distinguish between the prompt J/ψ and those coming from a b hadron decay we take advantage of the long lifetime of the b hadron, which leads to a typical flight path of approx. 9 mm in the HERA-B kinematics, well above the experimental resolution of the longitudinal vertex position of 0.5 mm. Since there are no other long-lived particles that decay into a J/ψ, a J/ψ vertex which is detached from the primary interaction is a unique indication of the presence of a b hadron in the event. The analysis is divided in two steps: first reconstruct the J/ψ particles and second determine the detached ones. To reduce the uncertainties in the efficiencies and luminosity, we normalize the $b\bar{b}$ production yield to the prompt J/ψ yield. We can express the cross section ratio as

$$R_\sigma = \frac{\sigma^A(b\bar{b})}{\sigma^A(J/\psi)} = \frac{n_b}{n_P} \times \frac{1}{\varepsilon_R.\varepsilon_b^{\delta z}.Br(b\bar{b} \to J/\psi X)} \tag{1}$$

where $\sigma^A(J/\psi)$ is the reference prompt J/ψ cross section, n_b and n_P are the measured numbers of b hadrons and prompt J/ψ, respectively; ε_R is the ratio of the J/ψ selection efficiency for the b and the prompt J/ψ sample; $\varepsilon_b^{\delta z}$ is the efficiency of the detached cuts that allows to identify a b decay. The dependency of the production cross sections on the mass number A is parameterized as: $\sigma^A = \sigma_0 \times A^\alpha$, where σ_0 stands for the proton-nucleon cross section, and the parameter α stands for any possible nuclear effect. For open $b\bar{b}$ production no nuclear effect is expected, thus we will assume α = 1. For J/ψ production, the measurement of the E866 collaboration [8] set the value of α at 0.96 $\pm$ 0.01. The branching ratio Br($b\bar{b} \to J/\psi X$) is taken from LEP1 [11] ((2.32 $\pm$ 0.20)%). The efficiencies are determined through a detailed MC simulation [11]. A selection of long tracks including segments in the vertex detector and tracking system, together with particle identification based on muon detector and electromagnetic calorimeter information, is enough to select good lepton candidates. When two lepton candidates with opposite charge are found, a common vertex fit is performed. In Fig. 1 the dilepton mass distributions are shown for candidates with a good reconstructed vertex (dominated by the prompt candidates). To identify a detached pair coming from a b decay we use the longitudinal separation of the J/ψ vertex and the target, the impact to wire for each lepton and the J/ψ impact to wire. Since the cuts are strongly correlated, the best set of cuts is provided by a blind optimization procedure that maximizes the significance of a MC signal (suitably normalized) on background composed of MC and real data. In Fig. 2 the distributions after applying the optimized cuts are shown. Not only those events surviving the cuts downstream of the target wire, which include the signal, but also those in the unphysical region upstream from the target, which must be random combinations between tracks, thus giving an estimate of the equivalent background underneath the real signal. In addition to this, other contributions to the downstream background come from simultaneous semileptonic decays of $b\bar{b}$ and $c\bar{c}$ events. In order to obtain the number of b events in the sample, an unbinned likelihood fit is performed on the mass spectrum.

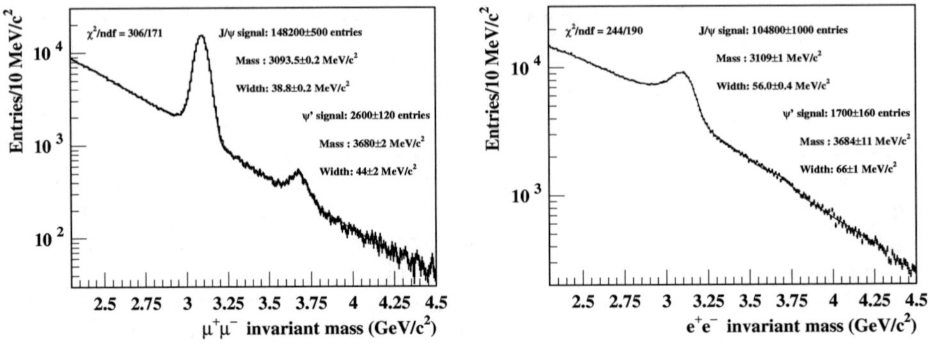

FIGURE 1. Dilepton invariant mass distributions. Left: muons. Right: electrons

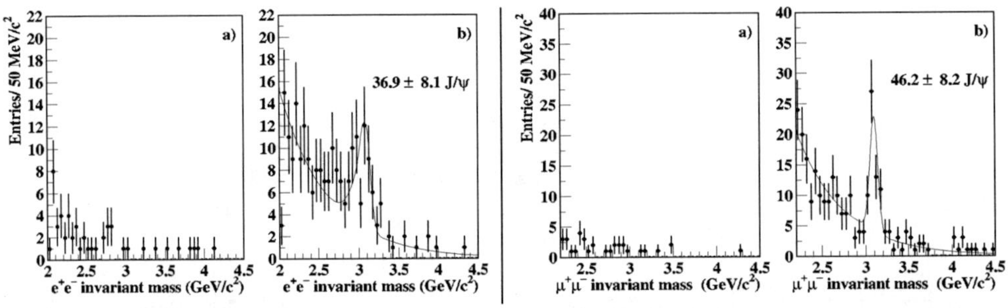

FIGURE 2. Mass spectra of dilepton candidates after detached cuts are applied. Left: electrons. Right: muons. a) events in the unphysical region in front of the target (combinatorial background). b) events downstream from the target (signal).

With these values we obtain a cross section ratio (extrapolated to the complete x_F space) R_σ = $\mathbf{0.027 \pm 0.004}_{stat} \pm \mathbf{0.003}_{sys}$, which is a quantity independent of the reference J/ψ cross section. To be able to compare with other experiments and theoretical results we determine the $b\bar{b}$ cross section to be $\sigma(b\bar{b}) = 14.2 \pm 2.3 \pm 2.4$ nb/nucleon. The HERA-B measurements based on the data taken in 2000 [5] and 2002-2003 were performed with a similar technique and with the same experimental apparatus, thus they can be combined in a final result which yields: $\sigma(b\bar{b})$ = **14.6 ± 2.3 ± 2.4 nb/nucleon** (In the talk, a reference cross section based solely on E771 and E789 results was used. The number given here is based on a more complete, but still preliminary, assessment of the literature:$\sigma(J/\psi) = 493 \pm 20 \pm 43$ nb/nucleon [12]).

Since we do not fully reconstruct the final state of the b hadron, it is important to test if the detached J/ψ sample has some of the characteristics of the b hadrons as, *e.g.*, the long lifetime. On the left of Fig. 3 the proper time distribution of the detached J/ψ candidates is shown. A lifetime fit is performed yielding a value (1.39 ± 0.19 ps) compatible with expectations (1.54 ps [13]). On the right side of Fig. 3, the comparison with theoretical calculations and with the previous measurements are shown. The new result is closer to the lower bound of the theoretical predictions, and seems to be in better agreement with E789 than with E771.

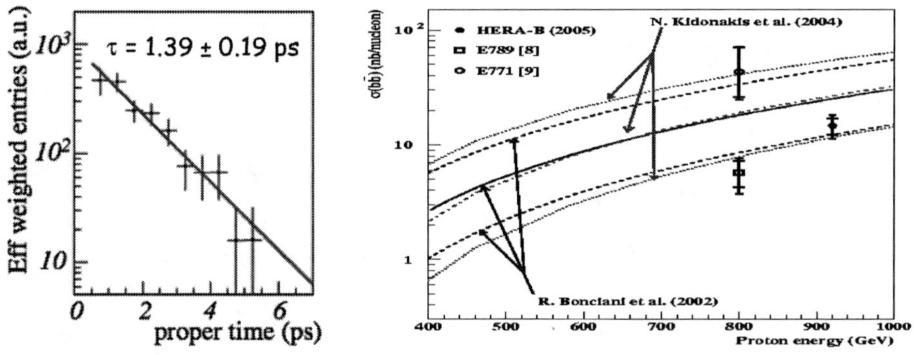

FIGURE 3. Left: Lifetime fit of detached J/ψ candidates. Right: Comparison with other experiments and theoretical predictions.

HIDDEN BEAUTY PRODUCTION

The Υ production is studied using dilepton decays. As for the open beauty case, the production cross section is measured relative to J/ψ production. We can express the differential cross section at central rapidity times the branching ratio as

$$\mathrm{Br}(\Upsilon \to l^+l^-)\left.\frac{d\sigma}{dy}\right|_{y=0}(\Upsilon) = \mathrm{Br}(J/\psi \to l^+l^-)\sigma(J/\psi)\frac{N(\Upsilon)}{N(J/\psi)}\frac{\varepsilon(J/\psi)}{\varepsilon(\Upsilon)}\frac{1}{\Delta y_{eff}} \quad (2)$$

where $\sigma(J/\psi)$ is the J/ψ production cross section, $N(\Upsilon)$ and $N(J/\psi)$ are the numbers of observed Υ and J/ψ decays, respectively, $\varepsilon(J/\psi)/\varepsilon(\Upsilon)$ is the ratio of J/ψ and Υ trigger and reconstruction efficiencies. The Υ production model gives $\Delta y_{eff} = 1.068 \pm 0.002$ at $\sqrt{s} = 41.6$ GeV. The J/ψ cross section values and branching ratios are set as before. Fig. 4 shows the Υ signals for muons (a) and electrons (b). The selection criteria applied are similar to (but tighter than) the ones applied for J/ψ identification. Due to the broad signal and the low number of signal events over background, it is important to have a good description of the background. Random combinatorics (estimated by like-sign pairs from real data) and the Drell-Yan process (estimated from MC simulation) are the main sources of background. These two distributions are shown in Fig. 4 together with the signals. To get a good and stable fit of the signal, the shapes of the background contributions are set together with the relative production ratios of $\Upsilon(1s)/\Upsilon(2s)/\Upsilon(3s)$, fixed according to the E605 results [14]. Combining the muon and electron channels, we obtain as a preliminary result: $\mathrm{Br}(\Upsilon \to l^+l^-) \times \frac{d\sigma}{dy}(\Upsilon)\,|_{y=0}$ = **4.6 ± 1.1 pb/nucleon**, a value in between E605 and E772/E771 measurements [16, 17]. A global fit of the available experimental data is also performed. In order to do this, we used Craigie's parameterization [18] with an additional parameter α for nuclear suppression. In Fig. 4 (c), this parameterization is used to fit the global spectrum, yielding a result compatible with no nuclear suppression.

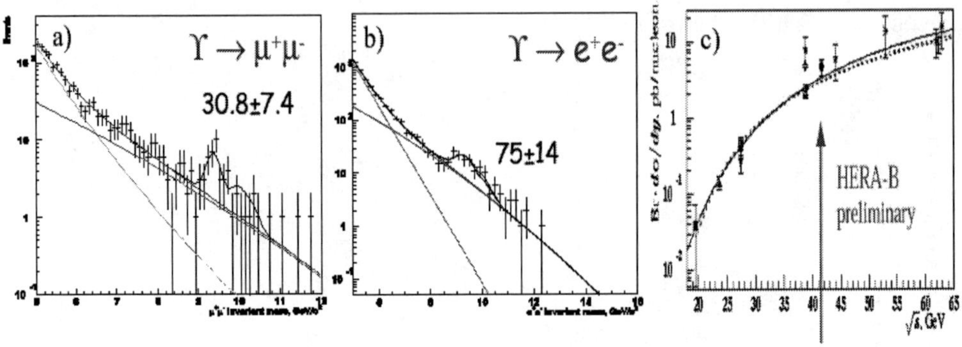

FIGURE 4. Dimuon (a) and dielectron (b) invariant mass spectrums. Background lines show the combinatorial and Drell-Yan contributions. c) Υ production cross section measurements for different experiments at different energy, including the preliminary HERA-B result. The modified Craigie parameterization fit $Br \times \frac{d\sigma}{dy}|_{y=0}(\sqrt{s}) = \sigma_0 \exp\left(-\frac{m_0}{\sqrt{s}}\right) \cdot A^{\alpha-1}$ is also shown.

CONCLUSIONS

The preliminary $b\bar{b}$ cross section measurement gives $14.6 \pm 2.3_{stat} \pm 2.4_{sys}$ nb/nucleon (averaged with previous HERA-B measurement). This result favours the E789 result and is close to the lower bounds of the theoretical predictions. The complete HERA-B sample for b hadron analysis is significantly larger than all previous fixed target experiments. The preliminary Υ differential cross section measurement at central rapidity times the branching ratio yields a value of 4.6 ± 1.1 pb/nucleon. A global fit to the Υ data was performed, which is consistent with no nuclear suppression.

REFERENCES

1. N. Kidonakis *et al.*, *Phys. Rev. D* **6**4 114001 (2001).
2. R. Bonciani *et al.*, *Nucl. Phys.* **B**529 24 (1998).
3. D.M. Jansen *et al.*, *Phys. Rev. Lett.* **74** 3118 (1995).
4. T. Alexopoulos *et al.*, *Phys. Rev. Lett.* **8**2 41 (1999).
5. I. Abt *et al.*, *Eur. Phys.J.* **C**26 345 (2003).
6. I. Abt *et al.*, *DESY-PRC-00-04* (2000).
7. M.J. Leitch *et al.*, *Phys. Rev. Lett.* **7**2 1318 (1994).
8. M.J. Leitch *et al.*, *Phys. Rev. Lett.* **8**4 3256 (2000).
9. M.H. Schub *et al.*, *Phys. Rev. D* **5**2 1307 (1995).
10. T. Alexopoulos *et al.*, *Phys. Rev. D* **55** 3927 (1997).
11. D. E. Groom *et al.*, *Eur. Pys. J.* **C**15 1 (2000).
12. F. Maltoni it et al. J/ψ production cross section at $\sqrt{s} = 41.6$ GeV by means of NRQCD calculation, in preparation.
13. Particle Data Group (K. Hagiwara *et al.*, Phys. Rev. D**6**6 010001 (2002).
14. T. Yoshida *et al.*, *Phys. Rev. D* **39** 3516 (1989).
15. G. Moreno *et al.*, *Phys. Rev. D* **4**3 2815 (1991).
16. T. Alexopoulos *et al.*, *Phys. Lett.* **B**374 271 (1996).
17. P.L. McGaughey *et al.*, *Phys. Rev. D* **5**0 3038 (1994).
18. N.S. Craigie, *Phys. Rept.* **4**7 1 (1978).

Recent Results from the FOCUS Collaboration

John P. Cumalat

Department of Physics, 390 UCB, University of Colorado, Boulder, CO 80309

Abstract. Recent results from the FOCUS Collaboration are presented including a new measurement of the D_s^+ lifetime, new limits on D_s^+ and Λ_c^+ doubly Cabibbo suppressed decays, and the first limits on T Violation in the charm sector.

INTRODUCTION

Data from the FOCUS experiment were collected during the 1996-1997 fixed target run at Fermilab. Bremsstrahlung of electrons and positrons with an endpoint energy of approximately 300 GeV produces photons which interact in a segmented beryllium-oxide target to produce charmed particles. The average photon energy for events which satisfy the event selection trigger is about 180 GeV. Charged decay products are momentum analyzed by two oppositely polarized dipole magnets. Tracking is performed by a system of silicon vertex detectors in the target region and multi-wire proportional chambers downstream of the interaction. Particle identification is performed by three multicell threshold Cerenkov counters, two electromagnetic calorimeters, a hadronic calorimeter, and by sets of muon detectors. A description of the FOCUS spectrometer can be found in references [1-4]. Due to space limitation, I have chosen to only mention results on the D_s^+ lifetime, limits on doubly Cabibbo suppressed decays using genetic programming techniques, and limits on T violation in the Charm sector. In the presentation I also included information on Λ_c^+ branching fractions [5].

D_s^+ LIFETIME MEASUREMENT

The FOCUS experiment has measured the lifetimes of all singly charmed particles. The latest measurement [6] is the lifetime of D_s^+ using $\phi\pi^+$ and $\overline{K^{*0}}K^+$ decays. This measurement is useful as the ratio of the D^0 and D_s^+ lifetimes gives information on the *weak annihilation* contribution to the total decay rate of the D_s^+. In Fig. 1 we present the invariant mass plots and the lifetime distribution plots for the $\phi\pi$ and the K^*K channels. Both D^+ and D_s^+ signals are evident. The lifetimes are determined with a significance of detachment cut of $L > 6\sigma_L$ and using the reduced proper time, namely $t' = (L - 6\sigma_L/\beta\gamma c)$. The use of the reduced proper time ensures that only a small acceptance correction to the lifetime distribution is needed.

The D_s^+ lifetime is measured to be 507.4 $\pm$ 5.5 (stat.) $\pm$ 5.1 (syst.) fs. using 8961 $\pm$ 105 $D_s^+ \to \phi\pi$ and 4680 $\pm$ 90 $D_s^+ \to K^*(892)^0K^+$ decays. The ratio of the D_s^+

CP814 *Hadron Spectroscopy: Eleventh International Conference*
edited by A. Reis, C. Göbel, J. Sá Borges, and J. Magnin

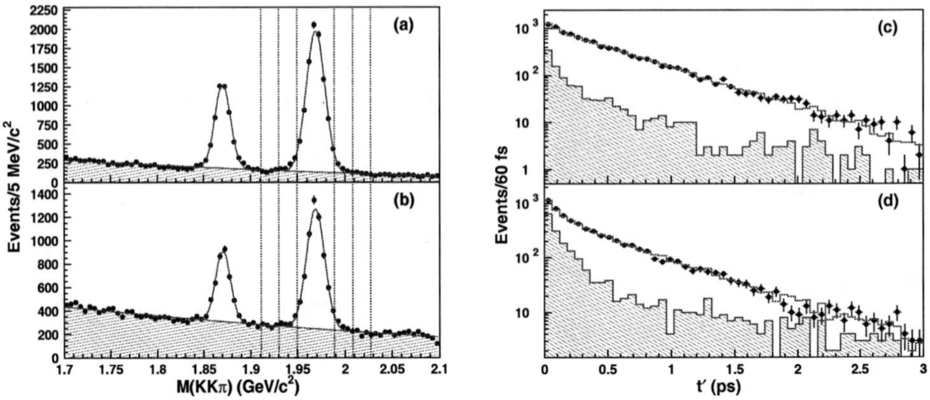

FIGURE 1. The $K^+K^-\pi^+$ invariant mass distributions for (a) $\phi\pi^+$ and (b) $K^*(892)^0K^+$ decay modes. The data is given by the points while the line gives the fit. The vertical lines gives the D_s^+ signal and sideband regions. In (c) $\phi(1020)\pi^+$ and (d) $K^*(892)^0K^+$ the lifetime distribution for all decays in the data signal region (points), and the fit (histograms) are presented. The shaded distribution shows the lifetime distribution of the background component.

TABLE 1. Comparison of D_s^+ lifetime measurements and $\tau(D_s^+)/\tau(D^0)$ ratios. The Particle Data Group (2004) value of the D^0 lifetime is used in the ratios, except for the FOCUS ratio for which the FOCUS measurement is used.

Experiment	$\tau(D_s^+)$ **(fs)**	$\tau(D_s^+)/\tau(D^0)$
E687[7]	$475 \pm 20 \pm 7$	1.158 ± 0.052
E791[8]	$518 \pm 14 \pm 7$	1.262 ± 0.038
CLEO II.5[9]	$486.3 \pm 15.0^{+4.9}_{-5.1}$	1.185 ± 0.039
SELEX[10]	$472.5 \pm 17.2 \pm 6.6$	1.152 ± 0.045
FOCUS (this result)[6]	$507.4 \pm 5.5 \pm 5.1$	1.239 ± 0.017

lifetime to the FOCUS D^0 lifetime [11] is determined to be 1.239 ± 0.014 (stat.) $\pm$ 0.009 (syst.), where the systematic error is reduced by eliminating common systematic contributions. Table 1 compares the FOCUS measurement with recent published results. This measurement represents a significant improvement in precision.

Doubly Cabibbo Suppressed Decays of Λ_c^+ and D_s^+

We have used a genetic programming technique [12] to search for doubly Cabibbo suppressed decays, $\Lambda_c^+ \to pK^+\pi^-$ and $D_s^+ \to K^+K^+\pi^-$. Neither of these decays have been observed, but the naive expectation is that these yields should be around 1/400 of the Cabibbo favored channels. Genetic programming differs from machine learning solutions like neural networks in that the form of the solution is not specified in advance, but is determined by the complexity of the problem. For these searches a fitness is

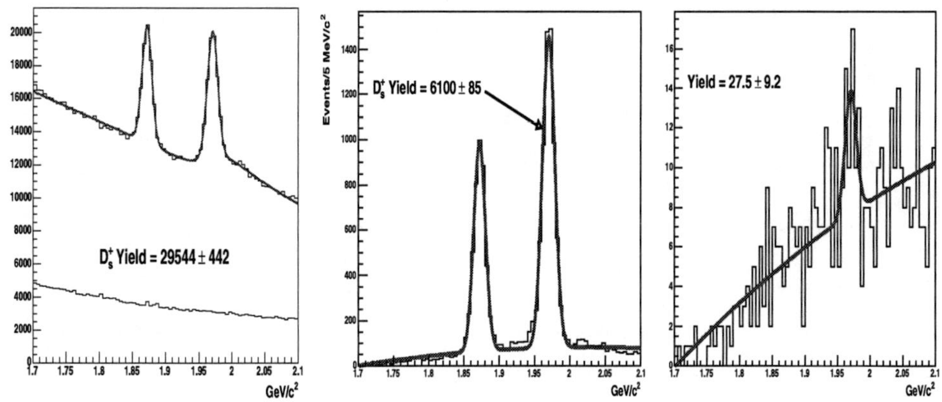

FIGURE 2. The $KK\pi$ invariant mass distributions for Cabibbo favored and doubly suppressed channels used in the $D_s^+ \to K^+K^+\pi^-$ search. The left plot shows the data samples before the genetic programming filter (GPF) is applied. The middle histogram shows the CF decay channel after GPF has been applied while the right histogram shows the DCS sample after the GPF has been applied.

defined using the signal yield in the partner Cabibbo favored channel (CF) and the background (excluding the signal region) in the doubly Cabibbo suppressed decay (DCS). A penalty term is included to encourage filters that have fewer terms. For additional information one can consult reference [13].

For the channel $\Lambda_c^+ \to pK^+\pi^-$ we determine a relative efficiency using a Monte Carlo simulation of $\varepsilon_{CF}/\varepsilon_{DCS} = 1.204 \pm 0.007(stat.)$ and we find the relative branching fraction to be $\frac{\mathrm{BR}(\Lambda_c^+ \to pK^+\pi^-)}{\mathrm{BR}(\Lambda_c^+ \to pK^-\pi^+)} = (0.05 \pm 0.26)\%$, consistent with zero. We quote a 90% CL $< 0.46\%$. The result is based on a CF yield of 3067 $\pm$ 60 events for $\Lambda_c^+ \to pK^-\pi^+$ after the filter was applied.

In Fig. 2 are shown the plots of the DCS search for $D_s^+ \to K^+K^+\pi^-$. The upper distribution in the left histogram shows the CF candidates while the lower distribution shows the DCS candidates before the filter is applied. The middle histogram and right histogram show the CF and DCS samples after the filter selection. The filter retains about 20% of the $D_s^+ \to K^-K^+\pi^+$ signal and reduces the background by roughly a factor of 500. We determine a relative efficiency for CF and DCS channels to be $\varepsilon_{CF}/\varepsilon_{DCS} = 1.154 \pm 0.005(stat.)$ and find a yield of 27.5 $\pm$ 9.2 events. We determine the relative branching fraction to be $\frac{\mathrm{BR}(D_s^+ \to K^+K^+\pi^-)}{\mathrm{BR}(D_s^+ \to K^+K^-\pi^+)} = (0.52 \pm 0.17)\%$. We also quote a 90% CL $< 0.78\%$.

T- VIOLATION

Within the Standard Model, CP violation arises due to the presence of a phase in the (CKM) quark mixing matrix. Triple-product correlations of the form $\vec{p}_1 \cdot (\vec{p}_2 \times \vec{p}_3)$ are odd under time reversal (T). By the CPT theorem, a nonzero value for these correlations

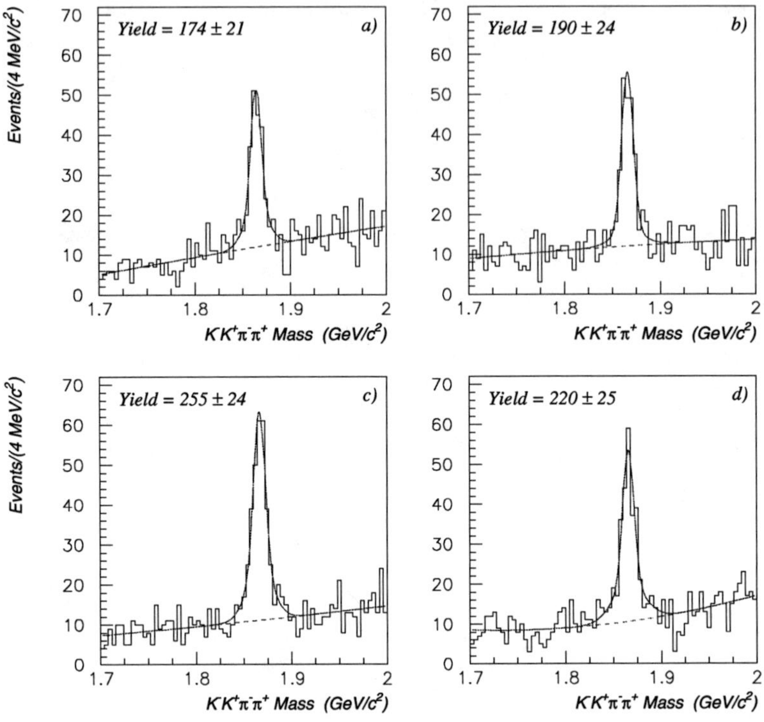

FIGURE 3. (a) D^0 with $C_T > 0$, (b) D^0 with $C_T < 0$, (c) $\overline{D^0}$ with $\overline{C_T} > 0$, (d) $\overline{D^0}$ with $\overline{C_T} < 0$.

would be a signal of *CP* violation. A nonzero triple-product correlation is evidenced by a nonzero value of the asymmetry, A_T, where $A_T \equiv \frac{\Gamma(\vec{p}_1\cdot(\vec{p}_2\times\vec{p}_3)>0)-\Gamma(\vec{p}_1\cdot(\vec{p}_2\times\vec{p}_3)<0)}{\Gamma(\vec{p}_1\cdot(\vec{p}_2\times\vec{p}_3)>0)+\Gamma(\vec{p}_1\cdot(\vec{p}_2\times\vec{p}_3)<0)}$, and Γ is the decay rate for the process. There is a well-known technical complication: strong phases can produce a nonzero value of A_T, even if the weak phases are zero, that is *CP* and *T* violation are not necessarily present. A true *T*-violating signal [14] is found by measuring a nonzero value of $A_{\text{Tviol}} \equiv \frac{1}{2}(A_T - \overline{A_T})$ where $\overline{A_T}$ is the *T*-odd asymmetry measured in the *CP*-conjugate decay process.

Particle and antiparticle are separated by D^{*+} soft pion tag. We form a *T*-odd correlation for the D^0 with the momenta: $C_T \equiv \vec{p}_{K^+} \cdot (\vec{p}_{\pi^+} \times \vec{p}_{\pi^-})$. From the $\overline{D^0}$ sample we form: $\overline{C_T} \equiv \vec{p}_{K^-} \cdot (\vec{p}_{\pi^-} \times \vec{p}_{\pi^+})$. Finding a distribution of C_T different from $-\overline{C_T}$ establishes *CP* violation. While we would prefer to compare the shape of the distributions, due to limited statistics we only compare the number of events with $C_T(\overline{C_T})$ greater than or less than zero. The analysis was encouraged by I. I. Bigi [15].

The mass plots shown in Fig. 3 are the D^0 and $\overline{D^0}$ signals separated by the requirement that $C_T(\overline{C_T})$ be greater or less than zero. The mass plots are fit with a function that includes two Gaussians with the same mean, but with different sigmas to take into account different momentum resolutions in our spectrometer and a second-order polynomial for

TABLE 2. Event yields for D^0, D^+, and D_s^+ signals separated by the requirement that $C_T(\overline{C_T})$ be greater or less than zero.

Final State	Request	D^0 Events	D^+ Events	D_s^+ Events
$K^-K^+\pi^-\pi^+$	$C_T > 0$	174 ± 21		
$K^-K^+\pi^-\pi^+$	$C_T < 0$	190 ± 24		
$K^-K^+\pi^-\pi^+$	$\overline{C_T} > 0$	255 ± 24		
$K^-K^+\pi^-\pi^+$	$\overline{C_T} < 0$	220 ± 25		
$K_S^0K^+\pi^-\pi^+$	$C_T > 0$		122 ± 16	126 ± 17
$K_S^0K^+\pi^-\pi^+$	$C_T < 0$		118 ± 16	147 ± 18
$K_S^0K^-\pi^-\pi^+$	$\overline{C_T} > 0$		145 ± 16	120 ± 17
$K_S^0K^-\pi^-\pi^+$	$\overline{C_T} < 0$		137 ± 16	119 ± 16

the combinatorial background. The D^0 yields are summarized in Table 2 together with the yields from a similar test for D^+ and D_s^+. After correcting for detector and reconstruction efficiencies as given by the Monte Carlo simulation, we find the T-violation asymmetries [16] to be

$A_{Tviol}(D^0)$ = 0.010 $\pm$ 0.057 (stat.) $\pm$ 0.037 (syst.),
$A_{Tviol}(D^+)$ = 0.023 $\pm$ 0.062 (stat.) $\pm$ 0.022 (syst.), and
$A_{Tviol}(D_s^+)$ = -0.036 $\pm$ 0.067 (stat.) $\pm$ 0.023 (syst.).

We have compared the A_{Tviol} measurements with the CP asymmetry measurements for $D^0 \to K^-K^+\pi^-\pi^+$ and $D^+ \to K_S^0K^+\pi^-\pi^+$ using the channels $D^0 \to K^-\pi^+\pi^-\pi^+$ and $D^+ \to K_S^0\pi^+\pi^-\pi^+$ to correct for production differences between particle and antiparticle. We find:

$A_{CP}(D^0)$ = -0.082 $\pm$ 0.056 (stat.) $\pm$ 0.047 (syst.) and
$A_{CP}(D^+)$ = -0.042 $\pm$ 0.064 (stat.) $\pm$ 0.022 (syst.).

Both the A_{Tviol} and the A_{CP} measurements are consistent with zero.

REFERENCES

1. P. L. Frabetti *et. al.*, *Nucl. Instrum. Methods Phys. Res., Sect.* **A320**, 519 (1992).
2. J. M. Link *et. al.*, *Nucl. Instrum. Methods Phys. Res., Sect.* **A516**, 364 (2004).
3. J. M. Link *et. al.*, *Nucl. Instrum. Methods Phys. Res., Sect.* **A484**, 174 (2002).
4. J. M. Link *et. al.*, *Nucl. Instrum. Methods Phys. Res., Sect.* **A484**, 270 (2002).
5. J. M. Link *et. al.*, *Phys. Lett.*, **B624**, 22 (2005).
6. J. M. Link *et. al.*, *Phys. Rev. Lett.*, **95**, 052003 (2005).
7. P. L. Frabetti *et. al.*, *Phys. Rev. Lett.*, **71**, 827 (1993).
8. E. M. Aitala *et. al.*, *Phys. Lett.*, **B445**, 449 (1999).
9. G. Bonvincini *et. al.*, *Phys. Rev. Lett.*, **82**, 4586 (1999)
10. M. Iori *et. al.*, *Phys. Lett.*, **B523**, 22 (2001).
11. J. M. Link *et. al.*, *Phys. Lett.*, **B537**, 192 (2002).
12. J. M. Link *et. al.*, *Nucl. Instrum. Methods Phys. Res., Sect.* **A551**, 504 (2005).
13. J. M. Link *et. al.*, *Phys. Lett.*, **B624**, 166 (2005).
14. W. Bensalem, A. Datta and D. London, *Phys. Lett.*, **B538**, 309 (2002).
15. I. I. Bigi, in Proceedings of KAON2001: International Conference on CP Violation, Pisa, Italy, 12-17 June 2001, p.417 (hep-ph/0107102).
16. J. M. Link *et. al.*, *Phys. Lett.*, **B622**, 239 (2005).

CLEO-c Impact on Lattice QCD

S.A. Dytman - for the CLEO Collaboration

Department of Physics and Astronomy
University of Pittsburgh
Pittsburgh, PA 15260

Abstract. Based on a sample of 60 pb^{-1} collected at the $\psi(3770)$ resonance, we present improved measurements of absolute branching fractions for exclusive D^0 and D^+ semileptonic decays into charged and neutral mesons. We present the first measurements for $D^0 \to \rho^- e^+ \nu$ and $D^+ \to \omega e^+ \nu$. CLEO-c data for $D^+ \to \mu^+ \nu_\mu$ from data taken in 2004-5, 281 pb^{-1} luminosity is now available. We present the updated value of f_D based on this decay.

Keywords: charm,lepton,lattice
PACS: 13.25.Ft,12.38.Gc

INTRODUCTION

The CLEO-c detector differs from CLEO-III only in the replacement of the silicon vertex detector by a compact stereo drift chamber. It is now being used to record data of unprecedented accuracy in the charm energy region. Data reported here were taken during 2 run periods, the first in fall, 2003 and early winter, 2004 and the second a year later. While the results for semileptonic decays use only the first run period (57 pb^{-1}), results for leptonic decays uses the full data set, (281 pb^{-1}). The colliding beams of CESR at total energy of 3-4.5 GeV. The results reported here come from a run at the $\psi(3770)$ resonance which decays most of the time to $D\bar{D}$ and D^+D^- pairs with very low relative momentum. Thus, one D can be detected through its hadronic decay which very cleanly tags the other D. This procedure allows distinct separation of all leptonic and semileptonic decays. This short contribution to the proceedings will discuss results from these events.

The quark mixing parameters in the CKM matrix are fundamental constants needed to fill out the Standard Model in its present form. These constants must come from experiment. Semileptonic D decays to pseudoscalers (e.g. $D \to K\ell\nu$ or $D \to \pi\ell\nu$) have an important role in this effort because they are closely related to V_{us} and V_{ud} through a straightforward calculation,

$$\frac{d\Gamma}{dq^2} = \frac{G_F^2}{24\pi^3}|V_{cq'}|^2 p_P^3 |f_+(q^2)|^2 \tag{1}$$

where G_F is the Fermi coupling constant, q^2 is the four momentum transfer to the meson, p_P is the meson momentum in the D rest frame, and $V_{cq'}$ is the relevant CKM matrix element (V_{cs} for $D \to Kl\nu$ and V_{cd} for $D \to \pi l\nu$). $f_+(q^2)$ is the form factor that measures the probability the quark q' will form the pseudoscaler meson with the spectator quark in the D meson. The branching fraction ($\mathscr{B}$) for a particular decay comes from an integral

CP814 *Hadron Spectroscopy: Eleventh International Conference*
edited by A. Reis, C. Göbel, J. Sá Borges, and J. Magnin

over q^2. With vector mesons in the final state, 3 form factors must be determined and the analysis is more complicated. That analysis will be reported in the future.

Data for $\mathscr{B}$ is sensitive to the product of the desired CKM value and the form factor, f_+. We will depend on an accurate determination of the form factor at q^2=0 from lattice QCD calculations to get ~2% accuracy on V_{cs} and V_{cd}. By providing accurate data for the relative form factor, we will give theorists confidence in their ability to understand the nonperturbative effects that are important for calculating the q^2 dependence. This will in turn give confidence in the corresponding methods to extract V_{ub} from B semileptonic decays.

The leptonic decay has only a W^+ in the intermediate state and is then sensitive to the D decay constant, f_D.

$$\Gamma(D^+ \to \mu^+ \nu_\mu) = \frac{G_F^2}{8\pi} m_\mu^2 M_{D^+} \left(1 - \frac{m_\mu^2}{M_{D^+}}\right)^2 |V_{cd}|^2 f_{D^+}^2 \qquad (2)$$

Thus, a value for f_D requires a value for the appropriate CKM matrix element; we use the PDG value. A preliminary version of this result is published [1]; this was the first observation of this small branching fraction and therefore the first measurement of this fundamental quantity. Important constraints on the CKM unitarity matrix come from B decays and various mixing properties. Many of these require 'exact' calculations from the lattice. One example is ΔM_d, a characterization of B_d mixing, where the best value comes a combination of experiment and lattice calculation. The estimated error on f_{B_d} dominates the error on ΔM_d and a successful calculation of f_D would allow this error to decrease significantly.

LEPTONIC DECAYS

For these events, a total of 158,354±496 D^+ decays were tagged through 6 D^- decay modes representing a total branching fraction of 26.4%. Reconstruction efficiency varied from 20-55%. When the μ^+ is detected, the missing mass for the remainder of the event is shown in Fig. 1. The closest background peak comes from leakage of events for $D^+ \to \pi^+ K_L$. We measure 47.2 signal events and estimate 2.8 background events (from data, cross-checked using Monte Carlo). The corresponding branching fraction is $\mathscr{B}(D^+ \to \mu^+ \nu_\mu) = (4.40 \pm 0.66^{+0.09}_{-0.12}) x10^{-4}$ and the new value for f_D is $222.6 \pm 16.7^{+2.8}_{-3.4}$ MeV [2]. These values are close to those published from a smaller data set. Evidence for $D^+ \to e^+ \nu_e$ was also sought although theory says the branching fraction will be much smaller. An upper limit (90% confidence level) of $2.4x10^{-5}$ has been established.

Confrontation with theory is of great interest and is a recent focus in the lattice community. In fact, some effort was made to announce the experimental [2] and latest lattice results [3] at the same conference this summer. The FNAL-MILC prediction [3] is 201±3±17 MeV. Another recent f_D result from Chiu, et al., is 235±8±14 MeV [4]. With the present estimated experimental error, all calculations are in good agreement with our new result.

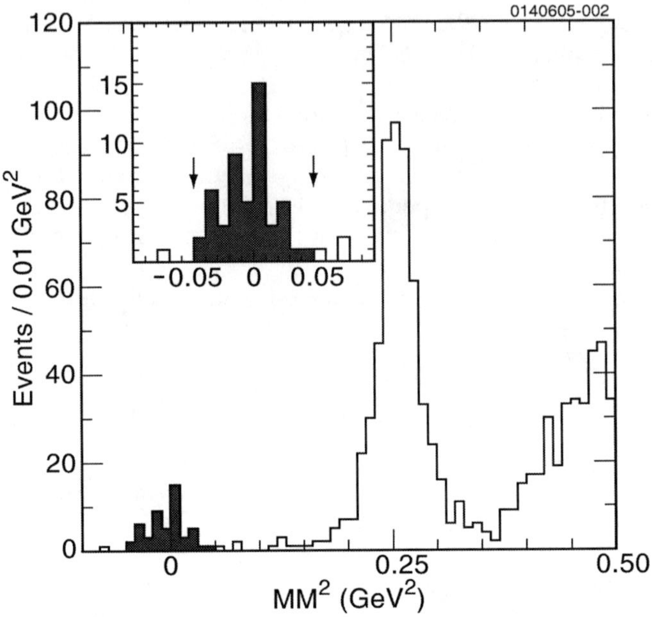

FIGURE 1. Missing mass histogram for $D^+ \to \mu^+ X$. Events at $MM \sim 0$ correspond to the signal for this measurement. The peak at $MM^2 \sim 0.25$ are for the decay $D^+ \to \pi^+ K_L$ where the pion has been misidentified as a muon.

SEMILEPTONIC DECAYS

To measure semileptonic decays, tagging is again involved. The charged D decays are measured with D^- tags similar to the leptonic measurement. D^0 mesons are tagged with 8 $\bar{D}^0$ decay modes adding up to a total of 44.9%. Thus, roughly 25% of all $\psi(3770) \to D\bar{D}$ are tagged, a total of 60041±408 events. In Fig. 2, we show spectra for $E_{miss} - c|\vec{p}_{miss}|$ for D^0 which should peak at 0 for missing neutrinos [5]. The spectra are extremely clean. Similar spectra for D^+ decays [6] are shown in Fig. 3. Details can be found in the preprints. Both $D^0 \to \rho e \nu$ and $D^+ \to \omega e \nu$ decays are seen for the first time.

The yield can then be used to determine the total branching fraction for each semileptonic decay. Results are given in Table 1.

Most of the results are consistent with recent PDG averages. However, $\mathscr{B}(D \to \pi^- e^+ \nu)$ is below the PDG average and $\mathscr{B}(D^+ \to \bar{K}^{*0} e^+ \nu)$ is above. The ratio of the branching fractions for π to K semileptonic decay from D^0 is consistent with the CLEO III result and less than the PDG value [5]. Ratios of D^+ to D^0 semileptonic decays were consistent with isospin symmetry [6].

At the same time, uncorrected form factor results were generated. Resolution studies show that a bin size of 0.03 $(GeV/c)^2$ will be possible. With the luminosity planned for the full CLEO-c data set at $\psi(3770)$, statistics will increase by a large factor and the

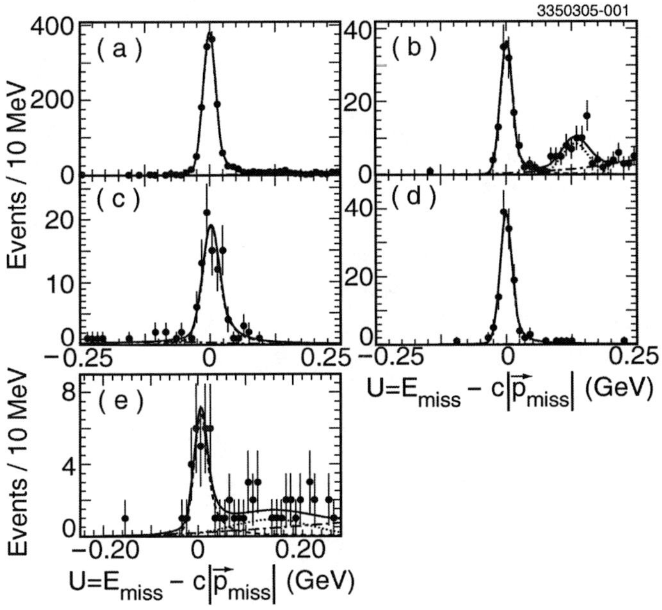

FIGURE 2. D^0 semileptonic decays. (a) $D \to Ke^+\nu$, (b) $D \to \pi e\nu$, (c) $D \to K^*(K^-\pi^0)e\nu$, (d) $D \to K^*(K_S\pi^-)e\nu$, and (e) $D \to \rho^- e\nu$. In all cases, a positron is detected with the $\bar{D}0$ meson fully reconstructed. The solid line represents the total fit to the data, which includes the signal (dashed line), peaking background (dotted line), and non-peaking background (dot-dashed line).

TABLE 1. Branching fraction data from CLEO-c. Values are based on 60 pb^{-1} luminosity.

Mode	CLEO-c result (%)	PDG result (%)
$D^0 \to \pi^- e^+ \nu$	0.26±0.03±0.01	0.36±0.06
$D^0 \to K^- e^+ \nu$	3.44±0.10±0.10	3.58±0.18
$D^0 \to K^{*-}(K^-\pi^0)e^+\nu$	2.11±0.23±0.10	2.15±0.35
$D^0 \to K^{*-}(\bar{K}^0\pi^-)e^+\nu$	2.19±0.20±0.11	2.15±0.35
$D^0 \to \rho^- e^+ \nu$	0.19±0.04±0.01	-
$D^+ \to \pi^0 e^+ \nu$	0.44±0.06±0.03	0.31±0.15
$D^+ \to \bar{K}^0 e^+ \nu$	8.71±0.38±0.37	6.7±0.9
$D^+ \to \bar{K}^{*0} e^+ \nu$	5.56±0.27±0.23	5.5±0.7
$D^+ \to \rho^0 e^+ \nu$	0.21±0.04±0.01	0.25±0.10
$D^+ \to \omega e^+ \nu$	$0.16^{+0.07}_{-0.06}$±0.01	-

form factor results will be impressive. Absolute branching fractions will be measured for D_S mesons in addition to the semileptonic decay channels presented here. The existing accuracy will be dramatically improved beyond previous measurements in every case.

New values for V_{us} and V_{ud} depend on disentangling the product of the CKM matrix element and the form factor in Eq. 1. We will depend on continued improvement in

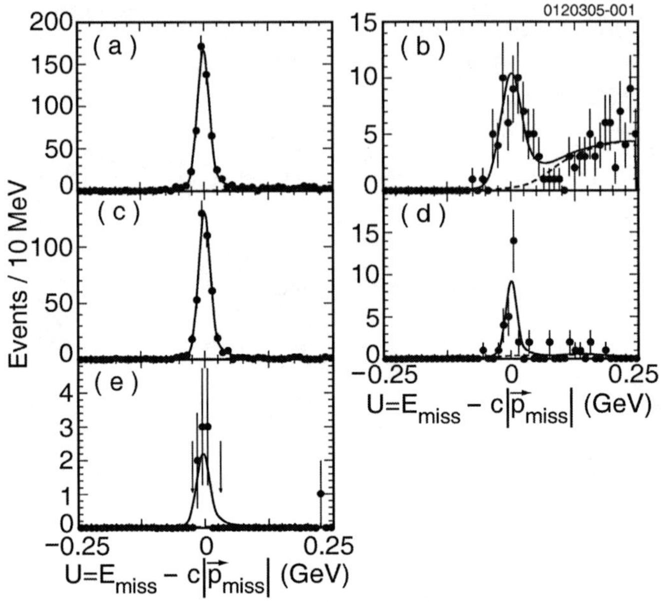

FIGURE 3. D^+ semileptonic decays. (a) $D^+ \to K_S e^+ \nu$, (b) $D^+ \to K^*(K^- \pi^+) e\nu$, (c) $D^+ \to \pi^0 e\nu$, (d) $D^+ \to \rho^0 e\nu$, and (e) $D^+ \to \omega e\nu$. In all cases, a positron is detected. The background contribution is the dashed line only visible in (c) and (d). The arrows show the signal region for $D^+ \to \omega e\nu$ (e).

lattice calculations [7] for the absolute normalization of the form factor.

ACKNOWLEDGMENTS

We gratefully acknowledge the effort of the CESR staff in providing us with excellent luminosity and running conditions. This work was supported by the National Science Foundation and the U.S. Department of Energy.

REFERENCES

1. G. Bonvicini, *et al.* [CLEO Collaboration], *Phys. Rev. D* **70**, 112004 (2004).
2. M. Artuso, *et al.* [CLEO Collaboration], hep-ex/0508057, submitted to *Phys. Rev. Lett.*
3. C. Aubin, *et al.*, hep-lat/0506030.
4. Ting-Wai Chiu, Tung-Han Hsieh, Jon-Yu Lee, Pei-Hua Liu, Hsiu-Ju Chang, hep-ph/0506266.
5. T.E. Coan, *et al.* [CLEO Collaboration], *Phys. Rev. Lett.***95**, 181801 (2005); hep-exp/0506052.
6. G.S. Huang, *et al.* [CLEO Collaboration], *Phys. Rev. Lett.***95**, 181802 (2005); hep-exp/0506053.
7. M. Okamoto, *et al.*, hep-lat/0309107, Nucl. Phys. Proc. Suppl. **129**, 334 (2004).

LIGHT QUARKS

Study of Dalitz plot in the decay $\eta' \to \eta\pi^+\pi^-$

Yu. Khokhlov for the VES Experiment [1]

IHEP, 142281 Protvino, Russia

Abstract.
Dalitz plot of the $\eta' \to \eta\pi^+\pi^-$ decay is studied using data collected in two different exclusive reactions. The world largest statistics made it possible to measure rather precisely the first term of an expansion of the matrix element squared and to determine other coefficients of the non-linear parameterization.

Keywords: Chiral Lagrangian,Meson decay,Dalitz plot
PACS: 12.39.Fe,13.25.Jx

1. INTRODUCTION

Theoretical treatment of decays of light pseudoscalar mesons is based as a rule on the effective chiral Lagrangians with their expansion on outer momenta and masses of the light quarks ([1]). Mainly the octet of (pseudo)goldstone mesons is considered, where the low energy behavior of the theory is determined by so called low energy constants (LEC) of the expansion.

The ninth pseudoscalar meson η', which is predominantly $SU(3)$ singlet state, is distinctive due to so called $U_A(1)$ anomaly - its mass doesn't disappear in the chiral limit of the massless quarks. The η' is accounted in ChPT indirectly through the mixing with the dominantly octet η meson and the changes of the LECs, or is introduced into the theory explicitly, e.g. using the results of the $1/N_c$ expansion.

Phenomenologically the mass of the η' is comparable with the theory scale $\Lambda \approx$ 1.2 GeV and this can degrade the accuracy of the momentum expansion. The early calculations of the decay $\eta' \to \eta\pi^+\pi^-$ within current algebra approach corresponding to the lowest order of the expansion, resulted ([2]) in the width few times less than the experimental value. The analogous calculations in [3] for the Dalitz-plot parameters have overestimated the slope significantly.

It is considered ([4] and references therein) that the re-scattering in the final state and contribution of the meson resonances can play an important role in the decays of the pseudoscalars. The high mass of the η' makes its decays sensitive to the parameters of the intermediate resonances in the relevant approaches.

The study of the form-factors of the $\eta' \to \eta\pi^+\pi^-$ in addition to the data on the decay width can give an information both for the refining of the LECs of the ChPT

[1] V. Dorofeev, R. Dzhelyadin, A. Fenyuk, Yu. Gavrilov, Yu. Gouz, A. Ivashin, V. Kabachenko, I. Kachaev, A. Karyukhin, Yu. Khokhlov, A. Konoplyannikov, V.F. Konstantinov, V. Matveev, V. Nikolaenko, A. Ostankov, B. Polyakov, D. Ryabchikov, A.A. Solodkov, A.V. Solodkov, O. Solovianov, E. Starchenko, A. Zaitsev, A. Zenin

CP814 *Hadron Spectroscopy: Eleventh International Conference*
edited by A. Reis, C. Göbel, J. Sá Borges, and J. Magnin

or the confronting the limits of its applicability, and on characteristics of resonances, especially of scalars. The situation with the latter is still sophisticated theoretically and experimentally and requires more study.

2. EXPERIMENTAL STATUS AND PARAMETERIZATION

From the experimental point of view the measurement of rather small form-factors in $\eta' \to \eta\pi\pi$ suffers from the lack of statistics of a good quality. The preceding experimental works on these form-factors used the so called linear parameterization of the matrix element squared $|M|^2 = |1+\alpha Y|^2 + (cX) + dX^2$, where the X and Y variables are defined in a standard way ([5]). Generally, the purely neutral decay channel and the one with charged pions may have different parameters. The coefficient c, which corresponds to the C-parity violating admixture in the matrix element and equals to zero in the neutral mode by definition, was never measured.

We address the channel studied in this work. The result of [6] is the combined fit of the few contemporary measurements with quite small total statistics of about 1400 decays. One of the fitting parameter in the linear parameterization is in fact the $(Im\alpha)^2$, so the quoted $Im\alpha = 0.0 \pm 0.8$ indicates that fit "got stuck" at the limit upon the parameter. The comparatively recent result of [7] is obtained using much larger statistics but with significantly worse background conditions of the inclusive production of the η'. With quoted rather large errors, two measurements do not contradict to each other, but still are not definitely conclusive even on the significance of the largest term of the expansion.

The present work extends our previous study [5]. We have collected the world highest statistics and have used the more general "non-linear" parameterization

$$|M|^2 = 1 + aY + bY^2 + cX + dX^2 \tag{1}$$

which can not be reduced to the linear one in the parameters region $b < a^2/4$.

3. DATA SAMPLE AND MODELING

The decays of the η' meson used for the analysis were collected with the VES setup in two exclusive reactions : the charge exchange $\pi^- p \to n\eta'$ and the diffraction-like $\pi^- A \to A\pi^-\eta'$, in a single run with a nominal beam momentum 27 GeV/c. The setup description and typical selection procedures can be found elsewhere ([5]).

The same decay mode $\eta \to \gamma\gamma$ was chosen for the analysis in both samples. The number of the η' as determined from the fit with a sum of two Gaussians for the signal peak having FWHM $\approx$ 16 MeV, is about $16 \cdot 10^3$ in the charge exchange ($7 \cdot 10^3$ in diffraction) reaction , while the fraction of the first order polynomial background in the band $\pm$ 21 MeV around the center of the signal is 5% (15%). The events from side bands of appropriate width were used for the background subtraction when filling histograms.

In order to take into account apparatus effects the Monte-Carlo modeling was used. Events for both samples were generated upon the beam and the vertex parameters and the momentum transferred squared $t\prime$ according to the experimental distributions. The decays $\eta' \to \eta\pi^+\pi^-$ (see next section) and $\eta \to \gamma\gamma$ are generated uniformly distributed

TABLE 1. Values of coefficients in non-linear parameterization

Parameter	This experiment	Theory ([10])
a (@Y)	-0.119 ±0.016 (stat.) ±0.021 (syst.)	-0.093
b (@Y^2)	-0.104 ±0.028 (stat.)±0.020 (syst.)	-0.059
c (@X)	-0.018 ±0.011 (stat.)±0.023 (syst.)	
d (@X^2)	-0.066 ±0.018 (stat.)±0.030 (syst.)	-0.003
χ^2/ndf	119./112	

upon the phase space. For the $\eta'\pi^-$ sample in addition the mass $M_{\eta'\pi^-}$ and the polar and azimuthal angles of the η' in each bin on the $M_{\eta'\pi^-}$ are generated according to the solution obtained in the course of the Partial Wave Analysis (PWA) of the system ([8]).

The generated events were traced through the setup model based on the GEANT program. For the γ-detector a simple analytical parameterization of the resolution was implemented instead of an individual shower simulation. The modeled events were subjected to a standard reconstruction procedure and to the same selections as the real data.

4. FITTING PROCEDURE AND SYSTEMATIC UNCERTAINTIES

The fitting procedure consists in minimization of the χ^2 w.r.t. the sought-for parameters a to d of the (1) and two (inessential) normalization factors K_1, K_2:

$$\chi^2(K_1,K_2;\,a,b,c,d) = \sum_{r=1,2}\sum_{i}^{n_{bin}} \frac{(D_{ri} - K_r M_{ri})^2}{\sigma_{ri}^2} \tag{2}$$

Here index r enumerates two reactions considered, D_{ri} stands for number of (weighted) entries in the i-th bin of the two-dimensional histogram in the Dalitz variables for the background-subtracted **d**ata. For the corresponding **M**onte-Carlo histogram $M_{ri} = \sum_{j=1}^{N_r}(1 + aY_{rj} + bY_{rj}^2 + cX_{rj} + dX_{rj}^2)$, where reconstructed X_{rj} and Y_{rj} in the j-th out of N_r generated events determine both the entry weight and the bin number i.

The σ_{ri}^2 are determined by the weights of entries in D_{ri}, neglecting a contribution of finite but large enough MC statistics.

The results of the fit including the χ^2 are given in the table 1. The fig.1 presents the set of distributions upon the Y in the equally sliced X intervals for the real data of the charge exchange reaction, and the fit resulting function superimposed on it.

The two data samples used in the analysis are different in their kinematics and so in systematics. To check their consistency the Dalitz plots for each of them were fitted separately. It turned out that the "variation" of the η' source generally dominates in the overall systematic uncertainty.

The second by importance systematic contribution is due to a variation of the width of the signal band (with the respective tuning of the background bands) of the η', due to the deviation of the MC $\eta\pi\pi$ mass spectrum from the real one, and so the variation of the signal-to-background ratio.

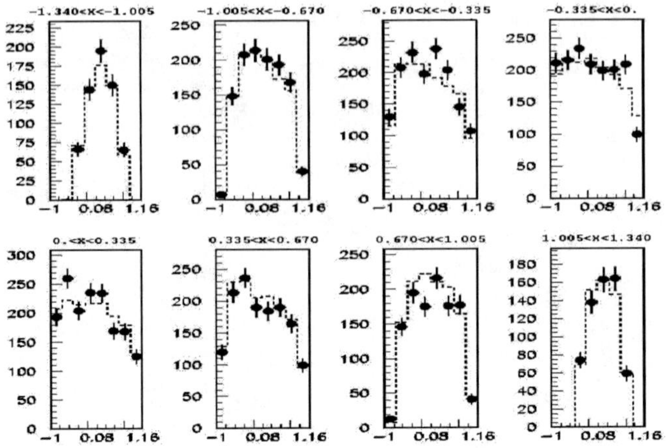

FIGURE 1. Y-distributions in X-slices for charge exchange sample: data (points with errors) and resulting function (line).

Also other potential sources of the systematic error were tested: efficiencies of detectors in the MC model, momentum scale of the magnetic spectrometer, energy scale and parameterization of the resolution of the calorimeter. All of them contributes much less than the major systematic uncertainty.

Finally we take rather conservatively one half of the largest difference over all variations for a given parameter as an estimate of its systematic error (see the table).

For a cross check a simpler model of the setup was tried, consisting in just geometrical acceptances and the parametric specification of the resolutions. This model gives rise to compatible results within errors. Moreover, the full control over the resolution in this model allowed us to check that the distortion of the X,Y variables due to it does not affect the fitted parameters.

5. DISCUSSION OF RESULTS

As a result of the study the slope upon the Y variable for the decay into charged pions is measured with the world best precision (some 22%). It is compatible with the equally precise measurement [9] for the purely neutral decay (assuming $2Re\alpha = a$).

The strong evidence (at 3σ level) to the negative value of the coefficient at the Y^2 is obtained. Nevertheless for the completeness we have tried a fit with the linear model. As one would expect, it gives rise to an unacceptably high value of χ^2/ndf=160/112.

The probing of the weak ($<2\sigma$) indication to the non-zero value of the coefficient at X^2 would require more statistics and better control over the systematics. As for the the C-violating term, it is compatible with zero within our accuracy.

Our results can be confronted with predictions of the theoretical model [10], which are also given in the table and generally are in a qualitative agreement with the experiment. In addition to the tree level contributions of the chiral Lagrangian the model-dependent calculus accounts for an arbitrary number of loops of the S-wave scattering of the final state mesons. The most significant is an attractive interaction in the $\eta\pi$ channel, corresponding, according to authors, to the $a_0(980)$ meson in the resonance terminology. As [10] claims there is still a room for the improvements of the calculations, including the accounting for the higher waves in the scattering.

Another paper [11] is rather close in spirit although using different approach. The consideration of the decay is restricted to tree level diagrams, but scalar resonances are introduced explicitly to the Lagrangian. Again the $a_0(980)$ meson contribution has a major effect upon the form-factors. However, the linear parameterization used in the [11] hinders the direct comparison of our results with their conclusions.

6. CONCLUSIONS

Using the data with highest statistics published so far, the Dalitz plot of the decay $\eta' \to \eta\pi^+\pi^-$ is studied. The inadequacy of the linear parameterization used before is demonstrated.

The most precise measurement of the slope over the Y variable in this decay mode has resulted in the value compatible with the neutral mode data.

The first strong evidence of the essentially negative value of the coefficient at Y^2 is obtained. Also an indication to the negative value of the coefficient at X^2 is obtained.

The C-violating parameter in the decay is measured for the first time and is compatible with zero.

ACKNOWLEDGMENTS

The work is supported by RFBR grants 05-02-08082 and 05-02-17664 and by Presidential grant NSh-1695.2003.2. The participation of Yu.Kh. in the Conference is supported by RFBR grant 05-02-26884.

REFERENCES

1. J. Gasser, H. Leutwyler, Nucl.Phys. **B250** (1985) 465
2. Riazuddin and S. Oneda, Phys.Rev.Lett. **27** (1971) 548
3. J. Cronin, Phys.Rev. **161** (1967) 1483
4. G. Ecker, J. Gasser, A. Pich, E.de Rafael, Nucl.Phys. **B321** (1989) 311
5. D.V. Amelin et al., Phys.At.Nucl. **68** (2005) 372
6. G.R. Kalbfleisch, Phys.Rev. **D10** (1974) 916
7. R.A. Briere et al., Phys.Rev.Lett. **84** (2000) 26
8. V. Dorofeev et al.,*Proceedings of the IX International Conference on Hadron Spectroscopy "HADRON-2001", Protvino, 2001*, AIP Conf.Proc. **619** (2002) 143
9. D.M. Alde et al., Phys.Lett. **B177** (1986) 115
10. N. Beisert, B. Borasoy, Nucl.Phys. **A716** (2003) 186
11. A. Fariborz, J. Schechter, Phys.Rev. **D60** (1999) 034002

Chiral Mass Splitting for $c\bar{s}$ and $c\bar{n}$ Mesons in the $\widetilde{U}(12)$-Classification Scheme of Hadrons

Kenji Yamada*, Tomohito Maeda* and Shin Ishida†

*Department of Engineering Science, Junior College Funabashi Campus, Nihon University, Funabashi 274-8501, Japan
†Research Institute of Science and Technology, College of Science and Technology, Nihon University, Tokyo 101-8308, Japan

Abstract. We investigate the chiral mass splitting of parity-doubled $J = 0,1$ states for $c\bar{s}$ and $c\bar{n}$ meson systems in the $\widetilde{U}(12)_{SF}$-classification scheme of hadrons, using the linear sigma model to describe the light-quark pseudoscalar and scalar mesons together with the spontaneous breaking of chiral symmetry, and consequently predict the masses of as-yet-unobserved $(0^+,1^+)$ $c\bar{n}$ mesons. We also mention some indications of their existence in the recent published data from the Belle and BABAR Collaborations.

INTRODUCTION

Recently, Ishida et al. have proposed the covariant $\widetilde{U}(12)_{SF}$-classification scheme of hadrons[1], which gives covariant quark representations for composite hadrons with definite Lorentz and chiral transformation properties. The $\widetilde{U}(12)_{SF}$-classification scheme has a unitary symmetry in the hadron rest frame, called "static $U(12)_{SF}$ symmetry"[2], embedded in the covariant $\widetilde{U}(12)_{SF}$-representation space, of which tensors can be decomposed into representations of $\widetilde{U}(4)_{DS}\times SU(3)_F$, $\widetilde{U}(4)_{DS}$ being the pseudounitary homogeneous Lorentz group for Dirac spinors. The static $U(12)_{SF}$ contains the Dirac spin group $U(4)_{DS}$ in its subgroups and $U(4)_{DS}$ contains two $SU(2)$ subgroups as $U(4)_{DS}\supset SU(2)_\rho\times SU(2)_\sigma$, where $SU(2)_\rho$ and $SU(2)_\sigma$ are the spin groups concerning the boosting and intrinsic-spin rotation, respectively, of constituent quarks, being connected with decomposition of Dirac γ-matrices, $\gamma\equiv\rho\otimes\sigma$. Thus the static $U(12)_{SF}$ symmetry includes the chiral $SU(3)_L\times SU(3)_R$ symmetry as $U(12)_{SF}\supset SU(3)_L\times SU(3)_R\times SU(2)_\sigma$. This implies that the $\widetilde{U}(12)_{SF}$-classification scheme is able to incorporate effectively the effects of chiral symmetry and its spontaneous breaking, essential for understanding of properties of the low-lying hadrons, into what is called a constituent quark model.

EXPERIMENTAL CANDIDATES FOR THE GROUND-STATE QUARK-ANTIQUARK MESONS

An essential feature of the $\widetilde{U}(12)_{SF}$-classification scheme is to have the static $U(4)_{DS}$ symmetry for light u,d,s quarks confined inside hadrons. The degree of freedom on the ρ-spin, being indispensable for covariant description of spin 1/2 particles, offers a

CP814 *Hadron Spectroscopy: Eleventh International Conference*
edited by A. Reis, C. Göbel, J. Sá Borges, and J. Magnin

basis to define the rule of chiral transformation for quark-composite hadrons. Since we have the ρ-spin degree of freedom, which is discriminated by the eigenvalues of ρ_3, $r=\pm$, in addition to the ordinary σ-spin, the ground states of light-quark $q\bar{q}$ mesons are composed of eight $SU(3)_F$ multiplets with respective J^{PC} quantum numbers, two pseudoscalars $(0_N^{-+}, 0_E^{-+})$, two scalars $(0_N^{++}, 0_E^{+-})$, two vectors $(1_N^{--}, 1_E^{--})$, and two axial-vectors $(1_N^{++}, 1_E^{+-})$ (N and E denoting "normal" and "extra"), where each N (E) even-parity multiplet is the chiral partner of the corresponding N (E) odd-parity multiplet and they form linear representations of the chiral symmetry.

Since the eigenstates only with the ρ_3-eigenvalue $r=+$ are taken for heavy quarks, we have for heavy-light meson systems two heavy-spin multiplets, $(0^-,1^-)$ and $(0^+,1^+)$, which are the chiral partner each other, while for heavy-heavy meson systems we have the same $(0^-,1^-)$-spin multiplets as in the conventional nonrelativistic quark model.

The $\widetilde{U}(12)_{SF}$-scheme assignments for the observed mesons

We try to assign some of the observed mesons to the predicted $q\bar{q}$ multiplets, resorting to their J^{PC} quantum numbers and masses. The observed meson data are taken from the Particle Data Group 2004 edition[3], except for the following mesons:

- $\rho(1250)$. There are several experimental indications of the existence of the $\rho(1250)$ reported by the OBELIX[4] and LASS[5] Collaborations, and others.[1]
- $\omega(1200)$. The existence of $\omega(1200)$ is claimed in the analysis of the $e^+e^- \to \pi^+\pi^-\pi^0$ cross section by the SND Collaboration[6].

We accept the existence of these vector mesons as true[7]. The resulting assignments, though some of them are ambiguous, are shown in Table 1. Here we make some comments on these assignments.

(1) The light scalar mesons $\{a_0(980), \sigma, f_0(980), \kappa\}$ are assined to the (0_N^{++})-nonet as a chiral partner of the π-meson (0_N^{-+})-nonet.
(2) The low-mass vector mesons $\{\rho(1250), \omega(1200), K^*(1410)\}$ are assined to the (1_E^{--})-nonet as a chiral partner of the (1_E^{+-})-nonet $\{b_1(1235), h_1(1170), h_1(1380), K_1(1400)\}$.
(3) The axial-vector mesons $\{a_1(1260), f_1(1285), f_1(1420), K_1(1270)\}$ are assined to the (1_N^{++})-nonet as a chiral partner of the $\rho(770)$-meson (1_N^{--})-nonet.
(4) The recent observed mesons $\{D_{sJ}^*(2317), D_{sJ}(2460)\}$ are assined to the $(0^+,1^+)$ multiplet as a chiral partner of the $(0^-,1^-)$ multiplet $\{D_s, D_s^*\}$[8]. These newly observed mesons, together with the σ-meson nonet, are the best candidates for the hadronic states with $r=-$ whose existence is expected in the $\widetilde{U}(12)_{SF}$ scheme.
(5) It is noted that the normal (N) and extra (E) states with the same J^{PC} generally mix together due to the spontaneous as well as explicit breaking of chiral symmetry and some other mechanism.

---

[1] See the $\rho(1450)$ Particle Listings and the "Note on the $\rho(1700)$" in [3].

TABLE 1. Experimental candidates for ground-state mesons in the $\widetilde{U}(12)_{SF}$-classification scheme.

$q\bar{q}$	P	S_A	$\widetilde{P}$	$\widetilde{S}_B$	V	A	$\widetilde{V}$	$\widetilde{B}$
	1^1S_0				1^3S_1			
	0^{-+}	0^{++}	0^{-+}	0^{+-}	1^{--}	1^{++}	1^{--}	1^{+-}
$n\bar{n}$ $s\bar{s}$	π	$a_0(980)$	$\pi(1300)$		$\rho(770)$	$a_1(1260)$	$\rho(1250)$	$b_1(1235)$
	η	σ	$\eta(1295)$		$\omega(782)$	$f_1(1285)$	$\omega(1200)$	$h_1(1170)$
	$\eta'(958)$	$f_0(980)$	$\eta(1475)$		$\phi(1020)$	$f_1(1420)$		$h_1(1380)$
$s\bar{n}$	K	κ	$K(1460)$		$K^*(892)$	$K_1(1270)$	$K^*(1410)$	$K_1(1400)$
$c\bar{n}$	D				D^*			
$c\bar{s}$	D_s	$D^*_{sJ}(2317)$			D^*_s	$D_{sJ}(2460)$		
$b\bar{n}$	B				B^*			
$b\bar{s}$	B_s				B^*_s			
$c\bar{c}$	$\eta_c(1S)$				$J/\psi(1S)$			
$b\bar{b}$	$\eta_b(1S)$				$Y(1S)$			

CHIRAL MASS SPLITTING FOR THE CHARMED AND CHARMED-STRANGE MESON SYSTEMS

In the $\widetilde{U}(12)_{SF}$-classification scheme heavy-light $(c\bar{q})$ meson fields, aside from the internal space-time wave functions, are given by

$$\Phi(v) = \frac{1}{2\sqrt{2}}(1 - iv\cdot\gamma)(i\gamma_5\mathbf{D} + i\tilde{\gamma}_\mu\mathbf{D}^*_\mu + \mathbf{D}_0 + i\gamma_5\tilde{\gamma}_\mu\mathbf{D}_{1\mu}) \tag{1}$$

with $v_\mu \equiv P_\mu/M$, $\tilde{\gamma}_\mu \equiv \gamma_\mu + v_\mu(v\cdot\gamma)$, where $(\mathbf{D}, \mathbf{D}^*_\mu, \mathbf{D}_0, \mathbf{D}_{1\mu})$ represent the local fields for the $c\bar{q}$ mesons with $J^P = (0^-, 1^-, 0^+, 1^+)$, P_μ (M) is the four-momentum (mass) of meson fields, and flavor indices are omitted for simplicity. To describe the light-quark pseudoscalar and scalar mesons together with the spontaneous breaking of chiral symmetry, we adopt the $SU(3)$ linear sigma model, introducing the chiral field Σ_5 defined by

$$\Sigma_5 = s - i\gamma_5\phi \tag{2}$$

with

$$s = \frac{1}{\sqrt{2}}s^a\lambda^a, \quad \phi = \frac{1}{\sqrt{2}}\phi^a\lambda^a \quad (a = 0, \cdots, 8),$$

where $\lambda^0 = \sqrt{2/3}\,\mathbf{1}$ and s^a (ϕ^a) are the scalar (pseudoscalar) fields. We now write a chiral-symmetric effective Lagrangian which gives the chiral mass splitting between the heavy-light $(0^-, 1^-)$ and $(0^+, 1^+)$ multiplets through the spontaneous breaking of chiral symmetry[11, 8]:

$$\mathscr{L}_{ND} = -g_{ND}\mathrm{Tr}[\Phi\Sigma_5\bar{\Phi}], \tag{3}$$

where g_{ND} is the dimensionless coupling constant of Yukawa interaction in the non-derivative form and the trace is taken over the spinor and flavor indices.

When the chiral symmetry is spontaneously broken, s has the vacuum expectation value, $\langle s\rangle_0 = \mathrm{diag}(a,a,b)$, where a and b are related to the pion and kaon decay constants by

$$a = \frac{1}{\sqrt{2}} f_\pi, \quad b = \frac{1}{\sqrt{2}}(2f_k - f_\pi). \tag{4}$$

Then the mass splitting between the two multiplets is induced and the mass differences $\Delta M_\chi(c\bar{q})$ are given by $\Delta M_\chi(c\bar{n}) = 2g_{ND}a$ and $\Delta M_\chi(c\bar{s}) = 2g_{ND}b$, which leads to the relation

$$\Delta M_\chi(c\bar{n}) = \Delta M_\chi(c\bar{s})\frac{a}{b} = \Delta M_\chi(c\bar{s})\left(\frac{2f_K}{f_\pi} - 1\right)^{-1}. \tag{5}$$

From this relation with the experimental values[3], $\Delta M_\chi(c\bar{s}) = 348.0 \pm 0.8$ MeV and $f_{K^+}/f_{\pi^+} = 1.223 \pm 0.015$, we obtain $\Delta M_\chi(c\bar{n}) = 240.8 \pm 5.4$ MeV and consequently predict the masses

$$M(D_0^*) = 2.11 \pm 0.01 \text{ GeV}, \quad M(D_1) = 2.25 \pm 0.01 \text{ GeV} \tag{6}$$

for the $(0^+, 1^+)$ $c\bar{n}$ mesons, using the measured mass values[3] of the $D(0^-)$ and $D^*(1^-)$ mesons. We hereafter refer to these predicted mesons, respectively, as "$D_0^*(2110)$" and "$D_1(2250)$".

POSSIBLE INDICATIONS OF THE EXISTENCE OF LIGHT SCALAR AND AXIAL-VECTOR CHARMED MESONS

We could ask experimental data whether there was some evidence for the existence of $D_0^*(2110)$ and $D_1(2250)$. Here we check on the recent published data on the $D\pi$ and $D^*\pi$ mass distributions in $B \to (D\pi)\pi$, $(D^*\pi)\pi$ decays from the Belle[9] and BABAR[10] Collaborations.

- **$D\pi$ mass spectrum**: In the Belle data[2] we see an excess of events, a single data point of 20 MeV bin, at a mass of 2.13 GeV near the predicted mass of the $D_0^*(2110)$, and so might regard it as an indication of that resonance, though it is natural to think that its data point should be within a statistical error. On the other hand, it would seem to us that the BABAR data[3] around a mass of 2.1 GeV show a typical pattern of interference between two or more resonances.
- **$D^*\pi$ mass spectrum**: In the Belle data[4] there is also an excess of events, a single data point of 10 MeV bin, at a mass of 2.255 GeV near the predicted mass of the $D_1(2250)$, and so it might be an indication of the resonance. Although it is not clear, the BABAR data[5] around a mass of 2.26 GeV might show a typical pattern

[2] See the $D\pi$ mass distribution in Figure 3 of [9].
[3] See the $D\pi$ mass distribution in Figure 3 (right) of [10].
[4] See the $D^*\pi$ mass distribution in Figure 9 of [9].
[5] See the $D^*\pi$ mass distribution in Figure 3 (left) of [10].

of interference.

If the $D_0^*(2110)$ and $D_1(2250)$ resonances really exist, their widths have to be narrow, $\leq$ 20-30 MeV, judging from the data mentioned above. The dominant decay modes of these resonances are $D\pi$ and $D^*\pi$, respectively, and thus we examine their single pion transitions. To estimate the widths of $D_0^* \to D+\pi$ and $D_1 \to D^*+\pi$ decays, together with $D^* \to D+\pi$, we set up, in addition to the nonderivative interaction $\mathscr{L}_{ND}$ in Eq. (3), the chiral-invariant effective interaction with the derivative form:

$$\mathscr{L}_D = g_D \mathrm{Tr}[\Phi(\partial_\mu \Sigma_5)\gamma_\mu (F_U \bar{\Phi})], \tag{7}$$

where $F_U = \gamma\cdot\partial/\sqrt{\partial\cdot\partial}$ and g_D is the coupling constant with a dimension of $(mass)^{-1}$, which is related to the axial coupling constant g_A by $g_D = g_A/2a = g_A/\sqrt{2}f_\pi$. The pionic decay widths of the D^*, D_0^*, and D_1 states are derived from $\mathscr{L}_{ND}$ and $\mathscr{L}_D$, and the decay widths of D_0^* and D_1 are identical. Using the measured value of $\Gamma[D^{*+} \to D^0 + \pi^+] =$ 65 keV[3] and $g_{ND} = 1.84$ from $\Delta M_\chi(c\bar{n}) = 241$ MeV, the coupling g_D is fixed to 3.96 GeV^{-1} (corresponding to $g_A = 0.521$), and then we obtain

$$\Gamma[D_0^*(2110) \to D+\pi] = \Gamma[D_1(2250) \to D^*+\pi] \approx 30 \text{ MeV}. \tag{8}$$

This value is consistent with the speculated widths of the $D_0^*(2110)$ and $D_1(2250)$.

CONCLUDING REMARKS

We have presented the possible assignments for some of the observed mesons in the covariant $\widetilde{U}(12)_{SF}$-classification scheme. It is necessary and important to examine the strong- and radiative-decay[12] properties of the assigned states in order to establish their assignments. On the basis of these assignments we have also predicted the existence of the low-mass $(0^+, 1^+)$ $c\bar{n}$ mesons with narrow width, which might have been seen in the recent published data on the $D\pi$ and $D^*\pi$ mass distributions from the Belle and BABAR Collaborations.

REFERENCES

1. S. Ishida, M. Ishida, and T. Maeda, *Prog. Theor. Phys.* **104**, 785 (2000).
2. S. Ishida, M. Ishida, K. Yamada, T. Maeda, and M. Oda, hep-ph/0408136.
3. Particle Data Group, S. Eidelman et al., *Phys. Lett.* **B592**, 1 (2004).
4. OBELIX Collaboration, A. Bertin et al., *Phys. Lett.* **B414**, 220 (1997).
5. LASS Collaboration, D. Aston et al., SLAC-PUB-5606 (1994).
6. M. N. Achasov et al., *Phys. Lett.* **B462**, 365 (1999).
7. T. Komada, these proceedings.
8. S. Ishida, in *HADRON SPECTROSCOPY*, edited by E. Klempt et al., AIP Conference Proceedings 717, Melville, New York, 2004, p. 716.
9. Belle Collaboration, K. Abe et al., *Phys. Rev. D* **69**, 112002 (2004).
10. BABAR Collaboration, B. Aubert et al., BABAR-CONF-03/020, SLAC-PUB-10107 (2003), hep-ex/0308026.
11. W. A. Bardeen, E. J. Eichten, and C. T. Hill, *Phys. Rev. D* **68**, 054024 (2003).
    M. A. Nowak, M. Rho, and I. Zahed, *Acta Phys. Polon.* **B35**, 2377 (2004).
12. T. Maeda, K. Yamada, M. Oda, and S. Ishida, these proceedings.

Rare Decays of the η Meson

Zisis Papandreou

Department of Physics, University of Regina, Regina, SK, S4S 0A2, Canada

Abstract. A study of several rare η decays near threshold was carried out at the C-6 (π^-) beamline of the Alternating Gradient Synchrotron at Brookhaven National Laboratory with the Crystal Ball detector, an electromagnetic calorimeter with nearly 4π geometric acceptance that is comprised of 672 optically isolated NaI(Tl) crystals with a thickness of 15.7 radiation lengths.

Results from the search for the CP forbidden decay $\eta \to 4\pi^0$ and the investigation of the quadratic slope parameter in $\eta \to 3\pi^0$ decay will be profiled, among others. The focus of the talk will be on the $\eta \to \pi^0\gamma\gamma$ rare decay: its relative branching ratio was extracted to be $B_1 = (8.3 \pm 2.8 \pm 1.2) \times 10^{-4}$, based on the analysis of 3×10^7 detected η mesons. This leads to a partial width for the eta meson of $\Gamma(\eta \to \pi^0\gamma\gamma) = (0.32 \pm 0.15)$ eV/c^2, a value much lower than past measurements and in line with Chiral Perturbation Theory calculations.

Keywords: eta meson, rare decays, partial width, branching ratio
PACS: 14.40.Aq, 13.20.-v, 12.39.Fe, 12.40.Vv

INTRODUCTION

This Paper will present results from the Crystal Ball (CB) program at Brookhaven National Laboratory (BNL), broken down into two broad categoriers: tests of symmetries and tests of Chiral Perturbation Theory (χPT).

The symmetry properties of any Lagrangian imply the existence of conserved quantities. Alternatively, if the Hamiltonian is invariant under a symmetry, for example under space translations, then the associated momentum operator $\hat{p}$ commutes with the Hamiltonian ($[\hat{p}, H] = 0$), and the momentum's expectation value is conserved. Other classical physics examples are invariance under rotation and time that lead, respectively, to conservation of angular momentum and energy.

The symmetries of interest in subatomic physics are those generated by the parity (P), charge-conjugation (C) and time-reversal (T) operators. Assuming CPT invariance, the combinations of C, P and T invariance violation are divided into four classes that are shown in Table 1. In this Paper, results will be shown from the rare η decay channels $\eta \to 4\pi^0$, $\eta \to 2\pi^0\gamma$ and $\eta \to 3\pi^0\gamma$. The former tests CP violation for classes 3 and 4, whereas the latter two test C for classes 1 and 2.

TABLE 1. The CPT invariance violation combinations are tabulated.

Class	Violated	Valid
1	C, P, CT, PT	T, CP
2	C, T, CP, PT	P, CT
3	P, T, CP, CT	C, PT
4	C, P, T, CP, CT, PT	

CP814 *Hadron Spectroscopy: Eleventh International Conference*
edited by A. Reis, C. Göbel, J. Sá Borges, and J. Magnin

The study of QCD can be broken down into energy regions, based on the applicability of perturbation theory. In the energy region below 0.5 GeV, the symmetries of QCD provide tight constraints on "soft process" features such as structure functions and similar observables. At low energies, χPT provides an accurate description of the strong and electroweak interactions of scalar mesons [1]. Results will be shown for the $\eta \to \pi^0\gamma\gamma$ decay and for the quadratic slope of the $\eta \to 3\pi^0$ that constrain χPT.

The data for all η decay channels were collected simultaneously and the analysis for each channel followed some initial common steps, specializing later. For this reason, the $\eta \to \pi^0\gamma\gamma$ analysis has been selected as the representative channel since it was one of the most difficult to extract and will be presented in detail. As for the other decay channels, only the final results will be shown.

MOTIVATION

In this low-energy region, the branching ratio (BR) of the $\eta \to \pi^0\gamma\gamma$ decay stands out as a stringent test of χPT due to the fact that the first two terms, $O(p^2)$ and $O(p^4)$, in the momentum expansion of χPT are suppressed; the former due to the final state which is composed entirely of neutral particles, and the latter because tree diagrams are absent in the expansion and the one-loop diagrams are suppressed either due to G-parity invariance or because they contain heavy kaons. Indeed, χPT calculations up to $O(p^4)$ result in a partial width $\Gamma(\eta \to \pi^0\gamma\gamma)$ of 3.9×10^{-3} eV/c^2 [2, 3]. However, the $O(p^6)$ terms involve parameters that are not well determined and not trivially related to measurable quantities. Nevertheless, an experimental test of the these terms is valuable in determining the effectiveness of χPT.

With the exception of one theoretical calculation based on quark-box diagrams, all others – and most notably those based on the Vector Meson Dominance scheme and χPT – fall well below the only experimental value listed in the RPP[4]. In that experiment, the final extracted BR was $(7.1\pm1.4)\times10^{-4}$ [5].

Our experiment was carried out at the C-6 beamline of the Alternating Gradient Synchrotron at BNL. The CB detector is an electromagnetic calorimeter with nearly 4π geometric acceptance and is comprised of 672 optically isolated NaI(Tl) crystals with a depth of 15.7 radiation lengths, leaving openings for the entrance and exit of the beam pipe. Centered at the equatorial plane was a 10.57-cm long liquid hydrogen target that was surrounded by a 5-mm-thick, 120-cm long cylindrical plastic scintillator Veto Barrel (VB). Other beamline detectors provided particle tracking and momentum information, vetoed unwanted events, and defined the experimental event triggers.

A total number of 3×10^7 η mesons was produced via the charge-exchange reaction $\pi^-p \to \eta n$ at a nominal beam momentum of 720 MeV/c (the reaction threshold is 684 MeV/c). The η population was determined using the decay $\eta \to 2\gamma$, and cross-checks were carried out with the $\eta \to 3\pi^0 \to 6\gamma$ channel under identical conditions.

In the CB, 98% of a photon's energy is deposited in a cluster defined by a peak crystal and its 12 adjacent neighbors [6]. The calculation of the photon's direction is done under the assumption that events originate at the center of the target, and the position of the geometrical center of the face of each crystal in the cluster is weighted by the square of its energy deposition.

The method to extract the $\eta \to \pi^0\gamma\gamma$ signal relied on a solid understanding of the shape and strength of the background distributions used in the Monte Carlo (MC) simulations. Fortunately, since the background channels were measured simultaneously with the data, sufficient information existed to constrain them to the desired level of confidence. The two background and the signal reactions pertinent to this analysis were:

$$\pi^- p \to 2\pi^0 n \tag{1}$$

$$\pi^- p \to \eta n \to 3\pi^0 n \tag{2}$$

$$\pi^- p \to \eta n \to \pi^0\gamma\gamma n \tag{3}$$

Both background channels were considerably stronger than the signal, by factors of over 1000 each. The $\pi^- p \to 2\pi^0 n$ (four-cluster) channel was handled by extracting its kinematic distributions from the data and using them to calibrate the Monte Carlo. Particularly troublesome, given the granularity and resolution of the CB, was the significant "leakage" of the six-cluster reaction (2) into the four-cluster reaction (3) event sample. The resulting excellent agreement between the MC and data – that allowed the eventual extraction of the signal – is shown with a representative spectrum in Figure 1a.

Once the MC simulations reproduced the data to the level of 1% or better, the signal was extracted from the data in three steps: i) Selection of candidate events through cuts that exclude events outside the kinematic range populated by the $\eta \to \pi^0\gamma\gamma$ signal; ii) Generation of likelihood functions that convolute the probability distributions of several observables to separate the signal from the background events; and iii) Normalization of background MC simulations to the data and subtraction of these events.

Of the 20 observables examined, only six were key ones and used in the final analysis. Three of these were primary observables, i) invariant mass (IM), ii) missing mass (MM), and iii) the energy of the most energetic crystal for each cluster (E_{peak}). The other three were composite quantities, iv) the radial energy spread $R = (E_{\text{cluster}} - E_{\text{peak}})/E_{\text{cluster}}$; R was defined for each cluster as the fraction of its energy that is not contained in the peak crystal, v) the fractional two-pion mass difference $TPD = \min(t_{ij} + t_{kl})$; $i \neq j \neq k \neq l$, where $t_{ij} = (|IM_{ij} - m_{\pi^0}|)/m_{\pi^0}$ and i, j, k and l enumerate the four clusters; TPD is related to the probability that two simultaneous pairs of photons both arose from π^0's, and vi) the transverse momentum in the lab frame $\vec{p_T} = \sum_i \vec{p}_i \times \hat{z}$, where the $\vec{p}_i$ are the three momenta of each of the four reconstructed photons and $\hat{z}$ is a unit vector along the incident beam direction.

After applying suitable cuts for the event selection, the analysis used to separate the signal from background was based on the log-likelihood function used in maximum likelihood fits, and employed the difference in shape between a number of probability distributions to maximize the separation between the signal and background channels. For each event, these probability distributions (see Fig. 1b) are used to evaluate the two likelihood ($L_{m\pi}$) functions given by:

$$L_{2\pi}(i) = \sum \ln\left(\frac{P[X_i = X(\eta \to \pi^0\gamma\gamma)]}{P[X_i = X(\pi^- p \to 2\pi^0 n)]}\right) \tag{4}$$

$$L_{3\pi}(i) = \sum \ln\left(\frac{P[X_i = X(\eta \to \pi^0\gamma\gamma)]}{P[X_i = X(\eta \to 3\pi^0)]}\right). \tag{5}$$

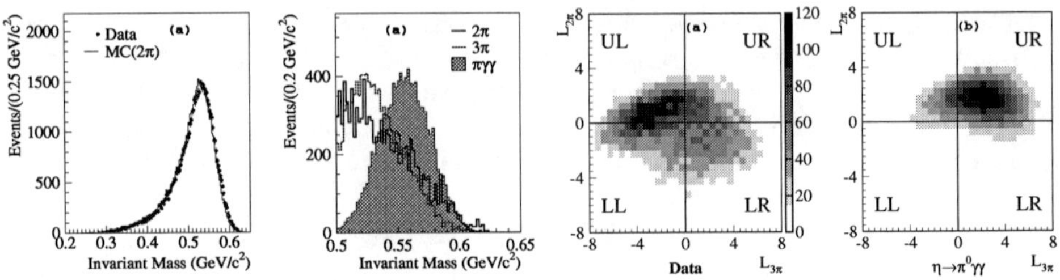

FIGURE 1. (color online) a) Agreement of the $\pi^- p \to 2\pi^0 n$ MC and data. b) Comparison of IM probability distributions for four-cluster events. c) and d) Likelihood distributions, respectively, for the data containing reactions 1-3 , and for the MC of the $\eta \to \pi^0\gamma\gamma$ channel. The regions are: UL - upper left, LL - lower left, UR - upper right and LR - lower right. All cuts have been applied, as described in the text.

The sums are over the set of observables being used: $X \in (IM, MM, PE, R, TPD, p_T)$. The probability functions $P(X)$, are the normalized probabilities that an event is consistent with a particular channel based on observable X.

Figures 1c,d show the $L_{2\pi}$ vs. $L_{3\pi}$ distributions for the data and the signal channel. Reactions (1) and (2) predominantly populate quadrants LR and UL, respectively, however, they spill over into quadrant UR. The signal does not stand out at this stage and the appropriately normalized background MC's were subtracted from the data to leave the number of signal events for each bin. The integration was performed from minimum values of $L_{2\pi}$ and $L_{3\pi}$ through the signal region of quadrant UR. Integration choices for which the significance of the number of found signal events exceeded 3σ were considered "good" regions. The final number of surviving signal events was 120 ± 40.

The normalization fit to the $L_{2\pi}$ vs. $L_{3\pi}$ spectrum and the 3σ cutoff were used to extract the $\eta \to \pi^0\gamma\gamma$ signal. The acceptance-corrected number of signal events was then compared to the number of $\eta \to 3\pi^0$ events to extract the relative branching ratio, $B_1 = BR(\eta \to \pi^0\gamma\gamma)/BR(\eta \to 3\pi^0) = (8.3 \pm 2.8 \pm 2.5) \times 10^{-4}$, with $BR(\eta \to 3\pi^0) = 0.3251 \pm 0.0029$ [4] that leads to $BR(\eta \to \pi^0\gamma\gamma) = (27 \pm 9 \pm 8)\times^{-5}$. The instrumental systematic uncertainties were largely cancelled in taking this ratio, while the analysis systematic uncertainties were investigated by varying the MC tuning parameters, the cut values, and by attempting alternate normalizations. The overall systematic uncertainty in the BR from this analysis was $< 30\%$. This consisted of a $\sim 20\%$ uncertainty from the normalization and variation of the cuts on the analysis parameters, $< 10\%$ uncertainty due to the 3σ cutoff after the background subtraction, $< 10\%$ due to acceptance corrections, and $< 1\%$ uncertainty due to the assumed shape of the $\eta \to \pi^0\gamma\gamma$ distributions between phasespace and χPT predictions. In the end, the main experimental limitations were due to crystal granularity (size) and background suppression using NaI crystals.

The extracted partial width was $\Gamma(\eta \to \pi^0\gamma\gamma) = (0.32 \pm 0.15)\,\mathrm{eV}/c^2$. This result is in agreement with an independent analysis carried out at UCLA [7] but disagrees with the RPP value by nearly two standard deviations. As this work involved 50 times more produced η mesons than the latter, it was possible to employ very aggressive cuts to eliminate virtually all background events and thus to ensure higher confidence in the result. The complexity in extracting this BR explains why a recent measurement [8] only reported an upper limit (0.99 eV/c^2).

Although the present result cannot distinguish among χPT calculations, it is in agreement with them, which allows us to conclude that the chiral series is convergent for this reaction and that SU(3) χPT is indeed applicable at third order. Higher precision measurements will be necessary to constrain any calculations, although it is doubtful whether such levels of precision are achievable with such neutral calorimeters. Indeed, the most recent (preliminary) result is from the KLOE Collaboration [9] using a lead, scintillating fiber calorimeter, and resulted in a $BR(\eta \to \pi^0\gamma\gamma) = (8.4 \pm 2.7 \pm 1.4) \times 10^{-5}$, almost three times lower than our measurement, but consistent within error bars. These two measurements settle the issue, namely that experiment is in agreement with χPT calculations for this *BR*.

Finally, the results for the other decay channels were as follow. The quadratic slope parameter of the $\eta \to 3\pi^0$ decay was measured to be $\alpha = -0.031 \pm 0.004$, in disagreement with χPT, and in line with the inclusion of dynamical effects in the Lagrangian [10]. However, recent results from the KLOE Collaboration [9] point to a much smaller slope ($\alpha = -0.013 \pm 0.005 \pm 0.004$), consistent with χPT. The *BR* of the $\eta \to 4\pi^0$ decay yielded a new upper limit for the test of *CP* violation in classes 3 and 4 of 6.9×10^{-7} [11]. In a similar fashion, the $\eta \to 2\pi^0\gamma$ and $\eta \to 3\pi^0\gamma$ channels tested *C*-invariance in classes 1 and 2 of isoscalar and isovector electromagnetic interactions of light quarks, respectively, at levels of 5×10^{-4} and 6×10^{-5} [12]. The last three measurements were made at the 90% confidence level.

ACKNOWLEDGMENTS

I wish to thank my colleagues in the Crystal Ball Collaboration and as a collaboration we thank SLAC for the loan of the Crystal Ball detector and Brookhaven National Laboratory for their hospitality. I also wish to acknowledge the TRIUMF Scintillator Shop and Dan Kolybaba at the University of Regina for the design and construction of the Veto Barrel. This experiment was supported in part by U.S. NSF & DOE, NSERC of Canada, the Russian Minitry of Science and Technology, the Russian Foundation of Basic Research and a sabbatical support (HMS) by the Volkswagen Stiftung of Germany.

REFERENCES

1. J. Gasser, and H. Leutwyler, *Nucl. Phys.* **B250**, 465 (1985).
2. G. Ecker, *Nucl. Phys.* **B (Proc. Suppl.) 7A**, 1 (1989).
3. L. Ametller, J. Bijnens, A. Bramon, and F. Cornet, *Phys. Lett.* **B276**, 185–190 (1992).
4. K. Hagiwara, et al., *Physical Review D* **66**, 10001 (2002).
5. F. Binon, et al., *Nuovo Cim.* **71A**, 497 (1982).
6. N. Knecht, Measurement of the branching ratio of $\eta \to \pi^0\gamma\gamma$ with the crystal ball detector, Ph.d. thesis, University of Regina (2002), URL `http://www.phys.uregina.ca/sparro/download/`.
7. S. Prakhov, et al., *Phys. Rev. C* **72**, 025201 1–18 (2005).
8. M. N. Achasov, et al., *Nucl. Phys.* **B600**, 3–20 (2001).
9. F. Ambrosino, contribution to these proceedings, KLOE Collaboration (2005).
10. W. Tippens, et al., *Phys. Rev. Lett.* **87**, 192001 1–4 (2001).
11. S. Prakhov, et al., *Phys. Rev. Lett.* **84**, 4802–4805 (2000).
12. B. Nefkens, et al., *Phys. Rev. Lett.* **94**, 041601 1–4 (2005).

Indication of Existence of Low Mass Extra Vector Mesons, $\omega(1250)$ and $\rho(1250)$, expected in the $\tilde{U}(12)$-Classification Scheme of Hadrons

Toshihiko Komada

Department of Engineering Science, Junior College Funabashi Campus, Nihon University, Funabashi 274-8501, Japan

Abstract. Mass distributions of $\pi^+\pi^-\pi^0$ and $\omega\pi^0$ systems obtained by several experiments are analyzed. We obtain indication of extra vector mesons $\omega(1250)$ and $\rho(1250)$ in lower mass region. Those state are expected to exist in the $\tilde{U}(12)$-classification scheme.

INTRODUCTION

Recently the $\tilde{U}(12)$ level-classification scheme of hadrons[1] has been proposed, which is a covariant generalization of non-relativistic scheme based on $SU(6)_{SF}$. In the $\tilde{U}(12)$-classification scheme a new type of relativistic states, called chiralons, are expected to exist in lower mass region for light quark systems. The extra vector meson nonet is predicted as chiralon of the ground state $q\bar{q}$ meson in the classification.

It is noted that the recent data of 3π and $\omega\pi^0$ states in e^+e^- annihilation show hints for the extra states[2]. Studies on them in detail will be interesting. It is also noted that studies in hadroproduction and others have shown indication of low mass extra states[3].

We re-analyzed mass spectra of $\pi^+\pi^-\pi^0$[2, 4, 5, 6] and $\omega\pi^0$[7, 8, 9, 10], and have obtained indication of a low mass ω in and a low mass ρ in respective channels around 1.2 GeV. These seem to be ground states chiralons expected by the $\tilde{U}(12)$ scheme[11].

ANALYSIS OF THE MASS SPECTRA OF $\pi^+\pi^-\pi^0$ AND $\omega\pi^0$. INDICATION OF EXTRA VECTOR MESONS $\omega(1250)$ AND $\rho(1250)$

1. $\omega(1250)$

We re-analyzed the combined mass spectrum of $\pi^+\pi^-\pi^0$ data by SND[2], CMD2[4], ND[5] and DM2[6]. It is considered by the vector meson dominance (VMD) that $e^+e^- \to \gamma \to V \to \rho\pi \to 3\pi$ ($V = \omega, \phi, \omega', \cdots$) contributes dominantly to this process. The $\omega(782), \phi(1020), \omega(1420)$ and $\omega(1650)$ have been taken for vectors so far, while the properties of $\omega(1420)$ and $\omega(1650)$ are not well established yet. On the other hand,

CP814 *Hadron Spectroscopy: Eleventh International Conference*
edited by A. Reis, C. Göbel, J. Sá Borges, and J. Magnin

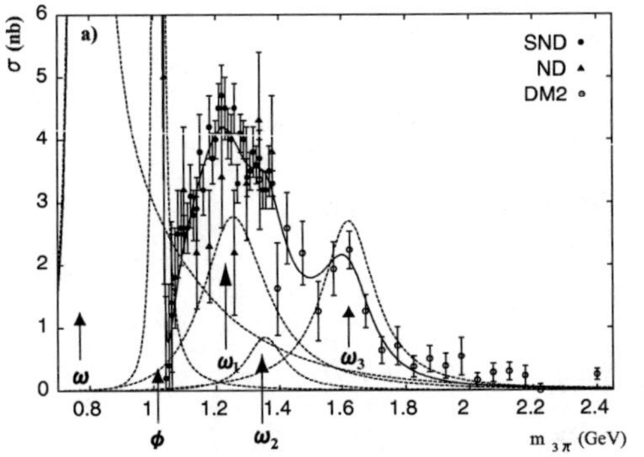

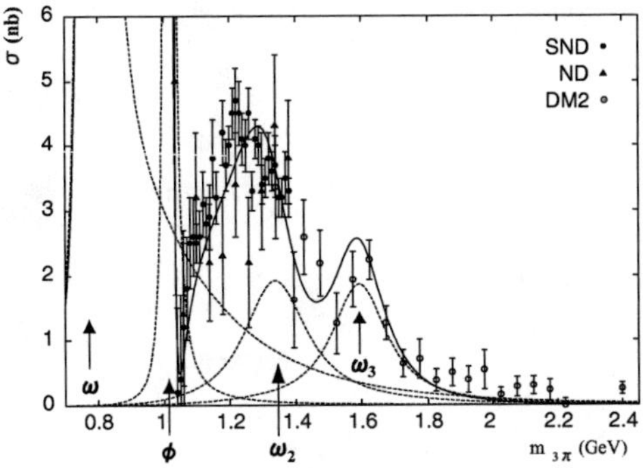

FIGURE 1. Results of the analysis on $\pi^+\pi^-\pi^0$ mass spectrum. a) fit with ω_1. b) fit without ω_1. Data are from SND[2], CMD2[4], ND[5] and DM2[6]. Solid line is result of analysis. Dotted lines are contribution from each amplitude.

TABLE 1. The obtained values of parameters and χ^2/N_F.

	ω_1	ω_2	ω_3
With ω_1	$\chi^2/N_F = 58/51 = 1.14$		
$m(MeV)$	1242 ± 16	1352	1621
$\Gamma(MeV)$	250 ± 39	144	$202^\dagger$
A	-0.152 ± 0.053	-0.040	0.069
Without ω_1	$\chi^2/N_F = 120/54 = 2.21$		
$m(MeV)$	—	1330	1592
$\Gamma(MeV)$	—	$233^\dagger$	$202^\dagger$
A	—	-0.100	-0.059

† Bound for upper limit which is set to be consistent with PDG tables.

there is the huge event accumulation around $m_{3\pi} = 1.2$ GeV in the data. The existence of $\omega(1250)$ is pointed out in analysis of $e^+e^- \to \pi^+\pi^-\pi^0$[2]. $\omega(1250)$ and $\omega(1650)$ were seen and $\omega(1450)$ were not seen in their analysis. In the data, it may be natural to expect a low mass resonance around 1.2 GeV and is interesting to study it in detail. In the present work, a possible contribution of a low mass state is considered, explicitly, in addition to higher vector mesons, $\omega(1420)$ and $\omega(1650)$.

Effective Lagrangians of this process are given as $\mathscr{L}_{e^+e^-\gamma} = ie\bar{\psi}_e\gamma_\mu\psi_e\mathbf{A}_\mu$, $\mathscr{L}_{\gamma V} = (\frac{em_\rho^2}{f_\rho}\rho_\mu^0 + \frac{em_\omega^2}{3f_\omega}\omega_\mu - \frac{\sqrt{2}em_\phi^2}{3f_\phi}\phi_\mu)A_\mu$, ($V = \omega(782), \phi(1020), \omega_1, \omega_2$ and ω_3), $\mathscr{L}_{\omega\rho\pi} = g_{\omega\rho\pi} \in_{\mu\nu\lambda\kappa} \partial_\mu\omega_\nu\partial_\lambda\rho_\kappa \cdot \pi$, $\mathscr{L}_{\phi\rho\pi} = g_{\phi\rho\pi} \in_{\mu\nu\lambda\kappa} \partial_\mu\phi_\nu\partial_\lambda\rho_\kappa \cdot \pi$ and $\mathscr{L}_{\rho\pi\pi} = -f_{\rho\pi\pi}\rho_\mu(\pi \times \partial_\mu\pi)$. ω_1, ω_2 and ω_3 denote a low mass and two higher mass vector states. The cross section is given as

$$\sigma(s) = \frac{4\pi\alpha^2}{s^{\frac{3}{2}}}(\frac{g_{\omega\rho\pi}}{3f_\omega})^2 \left| \frac{m_\omega^2\sqrt{F_\omega(s)}}{m_\omega^2-s-im_\omega\Gamma_\omega} + A_\phi\frac{m_\phi^2\sqrt{F_\phi(s)}}{m_\phi^2-s-im_\phi\Gamma_\phi} + \Sigma_{i=1}^3 A_i\frac{m_{\omega_i}^2\sqrt{F_{\omega_i}(s)}}{m_{\omega_i}^2-s-im_{\omega_i}\Gamma_{\omega_i}} \right|^2 4\pi\Gamma(s),$$

where

$$F_R(s) = \frac{2m_R^2}{m_R^2+s}, A_\phi = \frac{(\frac{-\sqrt{2}g_{\phi\rho\pi}}{3f_\phi})}{(\frac{g_{\omega\rho\pi}}{3f_\omega})} = -\sqrt{2}\frac{f_\omega}{f_\phi}\frac{g_{\phi\rho\pi}}{g_{\omega\rho\pi}}, A_i = \frac{(\frac{g_{\omega_i\rho\pi}}{3f_{\omega_i}})}{(\frac{g_{\omega\rho\pi}}{3f_\omega})} = \frac{f_\omega}{f_{\omega_i}}\frac{g_{\omega_i\rho\pi}}{g_{\omega\rho\pi}}.$$

A_i's are the fitting parameters.

$$\Gamma(s) = \frac{f_{\rho\pi\pi}^2}{3\cdot 16\pi^3}\int_{2m_\pi}^{\sqrt{s}-m_\pi} d\sqrt{s_{12}}|\mathbf{q}|^3|\mathbf{p_1}|^3 \cdot \int_{-1}^{1} d(cos\theta)sin^2\theta\,|\mathscr{F}_{\rho\pi\pi}|^2,$$

$$\mathscr{F}_{\rho\pi\pi} = \frac{\sqrt{F_\rho(s_{12})}}{m_\rho^2 - s_{12} - im_\rho\Gamma_\rho} + \frac{\sqrt{F_\rho(s_{23})}}{m_\rho^2 - s_{23} - im_\rho\Gamma_\rho} + \frac{\sqrt{F_\rho(s_{31})}}{m_\rho^2 - s_{31} - im_\rho\Gamma_\rho},$$

where s_{ij}'s are invariant mass of three possible contribution of π^+ , π^- and π^0. θ is angle between ρ and π in ω at rest. $\mathbf{q}$ and $\mathbf{p_1}$ is 3-momentum of $\mathbf{p}_{\pi_3}$ at $\pi_1\pi_2\pi_3$ C.M.S. and of $\mathbf{p}_{\pi_1}$ at $\pi_1\pi_2$ C.M.S., respectively. Coupling constants are estimated by VMD using experimental values to be f_ω = 5.69 , f_ϕ = 6.23 , $g_{\omega\rho\pi} = 12.47$ and $g_{\phi\rho\pi} = 0.7845976$, respectively. Ranges of mass and width values of $\omega(782), \phi(1020), \omega_2$ and ω_3 in the analysis is restricted to those which are consistent with the PDG tables[14]. Results of analysis with ω_1 and without ω_1 are shown in the fig. 1a) and 1b). The obtained values of parameters and χ^2/N_F are listed in Table 1. The result of fitting with ω_1 is improved by about 8σ from that without ω_1, indicating the existence of the low mass extra vector meson $\omega(1250)$.

2. $\rho(1250)$

We re-analyzed the combined mass spectrum of $\omega\pi^0$ data obtained by CMD2[7], SND[8] and DM2[9] including the $\omega\pi^0$ data from τ decay by CLEO[10]. It is also considered by the VMD that $e^+e^- \to \gamma \to V \to \omega\pi^0$ $(V = \rho, \rho', \cdots)$ contributes dominantly to this process. The $\rho(770)$, $\rho(1450)$ and $\rho(1700)$ have been taken for vectors so far, while the properties of $\rho(1450)$ and $\rho(1700)$ are not well established yet. There is the huge event accumulation recognized around $m_{\omega\pi^0} = 1.25$ GeV in the data. The existence of $\rho(1250)$ is pointed out by sevral experimental group[12, 13] and others[14, 3]. A possible contribution of a low mass state is considered in the present study, explicitly, in addition to higher vector mesons, $\rho(1450)$ and $\rho(1700)$.

Effective Lagrangians are given as $\mathscr{L}_{e^+e^-\gamma} = ie\bar{\psi}_e\gamma_\mu\psi_e A_\mu$, $\mathscr{L}_{\gamma\rho} = \gamma_\rho\rho_\mu^3 A_\mu$, $(\gamma_\rho = \frac{em_\rho^2}{f_\rho})$ and $\mathscr{L}_{\rho\omega\pi} = g_{\rho\omega\pi}\varepsilon_{\mu\nu\lambda\kappa}\partial_\mu\rho_\nu\partial_\lambda\omega_\kappa\pi$. ρ_1, ρ_2 and ρ_3 denote a low mass and two higher mass vector states. The cross section is given as

$$\sigma_0(s) = \frac{4\pi\alpha^2}{s^{\frac{3}{2}}}(\frac{g_{\rho\omega\pi}}{f_\rho})^2 \left| \frac{m_\rho^2\sqrt{F_\rho(s)}}{D_\rho(s)} + \sum_{i=1}^{3} A_i \frac{m_{\rho_i}^2\sqrt{F_{\rho_i}(s)}}{m_{\rho_i}^2 - s - im_{\rho_i}\Gamma_{\rho_i}} \right|^2 P_f(s),$$

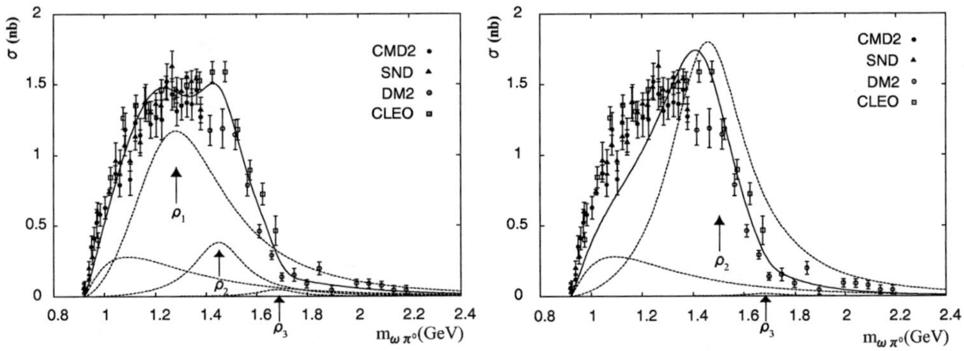

FIGURE 2. Results of the analysis on $\omega\pi^0$ mass spectrum. a) fit with ρ_1. b) fit without ρ_1. Data are from CMD2[7], SND[8], DM2[9] and CLEO[10]. Solid line is result of analysis. Dotted lines are contribution from each amplitude.

TABLE 2. The obtained values of parameters and χ^2/N_F.

Result	ρ_1	ρ_2	ρ_3
With ρ_1	$\chi^2/N_F = 245/72 = 3.40$		
$m(MeV)$	1203 ± 1	$1440^\dagger$	$1680^\dagger$
$\Gamma(MeV)$	506 ± 4	$250^\dagger$	$180^\dagger$
A	-0.445 ± 0.003	-0.079	0.015
Without ρ_1	$\chi^2/N_F = 963/75 = 12.8$		
$m(MeV)$	—	$1440^\dagger$	$1680^\dagger$
$\Gamma(MeV)$	—	$370^\ddagger$	$180^\dagger$
A	—	-0.253	0.009

† Bound for lower limit which is set to be consistent with PDG tables.
‡ Bound for upper limit which is set to be consistent with PDG tables.

where

$$F_R(s) = \frac{2m_R^2}{m_R^2 + s} \quad , \quad P_f(s) = \frac{1}{3}|\mathbf{p}_\omega(s)|^3 \cdot B_{\omega\to\pi^0\gamma} \ .$$

$\mathbf{p}_\omega(s)$ is 3-momentum of ω in ρ at rest. $B_{\omega\to\pi^0\gamma}$ is branting ratio of $\omega \to \pi^0\gamma$ to be 0.085 ± 0.005. Coupling constants are estimated by VMD using experimental values to be $f_\rho = 5.04$ and $g_{\rho\omega\pi} = 12.47$. A_i's are the fitting parameters. Ranges of mass and width values of $\rho(782), \rho_2$ and ρ_3 in the analysis is restricted to those which are consistent with the PDG tables[14]. Results of analysis with ρ_1 and without ρ_1 are shown in the fig. 2a) and 2b). The obtained values of parameters and χ^2/N_F are listed in Table 2. The result of fitting with ρ_1 is significantly improved by about 27σ from that without ρ_1. The results indicate the existence of the low mass extra vector meson $\rho(1250)$.

CONCLUSION

Indication of low mass vector mesons $\omega(1250)$ and $\rho(1250)$ are shown in analyses of mass spectra of $\pi^+\pi^-\pi^0$ and $\omega\pi^0$ data. The obtained values of mass and width are

$$m_{\omega(1250)} = 1242 \pm 16(MeV),$$
$$\Gamma_{\omega(1250)} = 250 \pm 39(MeV),$$

$$m_{\rho(1250)} = 1213 \pm 2(MeV),$$
$$\Gamma_{\rho(1250)} = 507 \pm 5(MeV).$$

The results of the analyses are preliminary. Though the mass distribution data used are interesting and useful for the analyses, they are combined for analyses from those in different experiments performed in different mass regions. We are afraid the structure might be distorted. The recent results[15] of the BABAR collaboration show $\omega(1350)$ and $\omega(1660)$ in the analysis on the 3π of the radiative decay of e^+e^- annihilation, covering the mass region between 1 and 3 GeV. It is strongly desired high statistics data from threshold up to 3 GeV to study the low mass vectors. They might be ingredients of the low mass extra vector meson nonet of ground state which is expected in the $\tilde{U}(12)$ scheme. Further studies are expected to confirm low mass extra vector mesons $\omega(1250)$ and $\rho(1250)$. Existence of those meson are very important for $\tilde{U}(12)$-classification scheme

ACKNOWLEDGMENTS

We would like to express our sincere appreciation to Prof. M. Ishida who initiated this works and gave us helpful and crucial suggestions during the work. We deeply appreciate Prof. K. Yamada and Prof. K. Takamatsu who gave us useful and crucial disscutions and informations. We also thank Prof. S. Ishida, Prof. I. Yamauchi, Prof. T. Tsuru and Prof. T. Maeda. for discussions.

REFERENCES

1. S. Ishida, M. Ishida and T. Maeda, Prog. Theor. Phys. **104** (2000) 785; S. Ishida and M.Y. Ishida , Phys. Lett. **B539** (2002) 249-256.
2. M.N. Achasov et al. (SND Collaboration) , Phys. Lett **B462** (1999) 365-370.
3. A. Donnachie and Yu.S. Karashnikova, Proc. 9th Hadron Conf. 2001, AIP Conf. Proc. **619** (2002) 5; Phys. Rept. **403-404** (2004) 281-301.
4. R. R. Akhmetshin et al. (CMD2 Collaboration), Phys. Lett **B434** (1998) 426.
5. S. I. Dolinsky et al. (ND Collaboration), Phys. Rept. **202** (1991) 99-170.
6. A. Antonelli et al. (DM2 Collaboration), Z. Phys. **C56** (1992) 15.
7. R. R. Akhmetshin et al. (CMD2 Collaboration), Phys. Lett. **B562** (2003) 173.
8. M.N. Achasov et al. (SND Collaboration), Phys. Lett. B **486** (2000) 29.
9. Nuclear Physics B (Proc. Suppi.) **21** (1991) 111-117.
10. K. W. Edwards et al. (CLEO Collaboration), Phys Rev. **D61** (2000) 072003.
11. K. Yamada, in this proceedings.
12. D. Aston et al. (LASS Collaboration), SLAC-PUB-5606 (1994).
13. A. Bertin et al. (OBELIX Collaboration) Phys. Lett. **B414** (1997) 220.
14. Particle Data Group, S. Eidelman et al., Phys. Lett. **B592** (2004) 1.
15. B. Aubert et al. (BABAR Collaboration), Phys. Rev. **D70** (2004) 072004.

ϕ meson radiative decays to η and η'

The KLOE Collaboration presented by F. Ambrosino

Dipartimento di Scienze Fisiche Università "Federico II" e Sezione INFN, Napoli, Italy

Abstract. We present recent results from KLOE on the dynamics of $\eta \to \pi\pi\pi$ decays, as well as searches for rare η decays, including a new determination of the $\eta \to \pi^0\gamma\gamma$ Branching Ratio. We also present a preliminary result on the ratio BR($\phi \to \eta'\gamma$)/BR($\phi \to \eta\gamma$), a quantity related to the $\eta - \eta'$ mixing and the gluonium content of the η'. All results are obtained from the analysis of 450 pb^{-1} integrated luminosity from 2001-2002 runs.

Keywords: ϕ radiative decays; $\eta \to 3\pi$ dynamics; η rare decays; $\eta - \eta'$ mixing
PACS: 14.65.Bt ; 14.40.Aq ; 14.40.Cs

THE η FACTORY

The KLOE[1] experiment [1] is currently taking data on one of the two interaction points of the DAΦNE [2] e^+e^- collider at the Laboratori Nazionali di Frascati of INFN. The machine runs at $\sqrt{s} = M_\phi$. With $\approx 3\mu$b cross section, ϕ production at DAΦNE is by far the most relevant process but elastic scattering. The data set we use in this analysis consists of about 15 billions ϕ decays. Even if the Branching Ratios (BR) of the radiative processes $\phi \to \eta(\eta')\gamma$ are relatively small (about 1.3% for η and 6×10^{-5} for η' respectively) this results in a sample of about 17 millions η and 90 thousands η' decays in a clean e^+e^- environment. Our detector consists of a large, He based, fully stereo drift chamber, and a fine sampling hermetic lead/scintillating fibers calorimeter. A superconducting coil provides a 0.52 T field along the beam axis. Transverse momentum resolution for charged tracks is $\approx 0.4\%$, while energy and time resolution in the calorimeter are respectively $\sigma_E/E = 5.7\%/\sqrt{E(\text{GeV})}$ and $\sigma_t = 54\text{ps}/\sqrt{E(\text{GeV})} \oplus 50\text{ps}$. We improve significantly our resolution by means of kinematic fits where we usually impose only total four-momentum conservation, with no additional constraints on particle masses. The two body nature of $\phi \to \eta(\eta')\gamma$ decays

[1] The KLOE Collaboration:
F. Ambrosino, A. Antonelli, M. Antonelli, C. Bacci, P. Beltrame, G. Bencivenni, S. Bertolucci, C. Bini, C. Bloise, V. Bocci, F. Bossi, D. Bowring, P. Branchini, R. Caloi, P. Campana, G. Capon, T. Capussela, F. Ceradini, S. Chi, G. Chiefari, P. Ciambrone, S. Conetti, E. De Lucia, P. De Simone, G. De Zorzi, S. Dell'Agnello, A. Denig, A. Di Domenico, C. Di Donato, S. Di Falco, B. Di Micco, A. Doria, M. Dreucci, G. Felici, A. Ferrari, M. L. Ferrer, G. Finocchiaro, C. Forti, P. Franzini, C. Gatti, P. Gauzzi, S. Giovannella, E. Gorini, E. Graziani, M. Incagli, W. Kluge, V. Kulikov, F. Lacava, G. Lanfranchi, J. Lee-Franzini, D. Leone, M. Martini, P. Massarotti, W. Mei, S. Meola, S. Miscetti, M. Moulson, S. Müller, F. Murtas, M. Napolitano, F. Nguyen, M. Palutan, E. Pasqualucci, A. Passeri, V. Patera, F. Perfetto, L. Pontecorvo, M. Primavera, P. Santangelo, E. Santovetti, G. Saracino, B. Sciascia, A. Sciubba, F. Scuri, I. Sfiligoi, T. Spadaro, M. Testa, L. Tortora, P. Valente, B. Valeriani, G. Venanzoni, S. Veneziano, A. Ventura, R.Versaci, G. Xu

CP814 *Hadron Spectroscopy: Eleventh International Conference*
edited by A. Reis, C. Göbel, J. Sá Borges, and J. Magnin

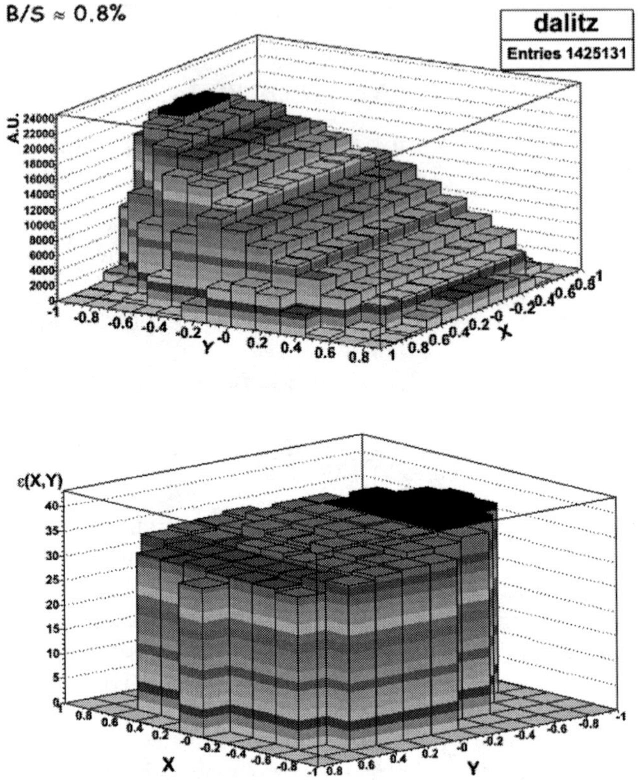

FIGURE 1. *Upper: Dalitz plot of selected $\eta \to \pi^+\pi^-\pi^0$ events. Lower: Efficiency of the selection over the kinematically allowed region.*

helps their identification against background from other ϕ decays, especially for the $\eta\gamma$ channel, where the monochromatic recoil photon has a "high" energy of 363 MeV.

$\eta \to \pi\pi\pi$

The $\eta \to \pi\pi\pi$ decay is due to the isospin breaking part of the strong Lagrangian and its amplitude is proportional to the d-u quark mass difference; a precise study of this decay can lead to a very accurate measurement of $Q^2 = (m_s^2 - \hat{m}^2)/(m_d^2 - m_u^2)$.

Using the 17 millions η mesons produced in 2001/2002, the dynamics of both $\pi^+\pi^-\pi^0$ and $\pi^0\pi^0\pi^0$ final states has been studied through a Dalitz plot analysis.

The η mesons are clearly tagged by detecting the monochromatic recoil photon of the $\phi \to \eta\gamma$ decay (E_{recoil}63 MeV); the background is at the level of few per mill. Efficiency is almost flat all over the kinematically allowed region (see Fig.1).

For the $\pi^+\pi^-\pi^0$ final state, we choose the conventional Dalitz variables $X \propto T_+ - T_-$

TABLE 1. *Fitted parameters of the $\eta \to \pi^+\pi^-\pi^0$ Dalitz plot. The χ^2 probability is 60% for a 147 dof fit.*

a	b	c
$-1.072 \pm 0.006^{+0.005}_{-0.007}$	$0.117 \pm 0.006^{+0.004}_{-0.006}$	$0.0001 \pm 0.0029^{+0.0003}_{-0.0021}$

d	e	f
$0.047 \pm 0.006^{+0.004}_{-0.005}$	$0.006 \pm 0.008^{+0.013}_{-0.000}$	$0.13 \quad \pm 0.01 \quad ^{+0.02}_{-0.01}$

and $Y \propto T_0$, where T is the kinetic energy of the pion. The measured distribution is parametrized as: $|A(X,Y)|^2 = 1 + aY + bY^2 + cX + dX^2 + eXY + fY^3$. As expected from C parity conservation, the odd powers of X are consistent with zero (see Tab. 1) and can be removed from the fit without affecting the determination of the remaining parameters. We clearly observe a non zero quadratic slope in X, and we reach for the first time sensitivity to a cubic term of the expansion; all the cubic terms other than f turn out to be zero in our fit. Using our fitted parameters, the value of Q can be extracted. For example, in Ref. [3] the value Q".8 ± 0.4 is obtained, the error being dominated by the $\eta \to \pi^+\pi^-\pi^0$ width. This value points to a significant violation of the Dashen theorem, and is in agreement with Chiral Perturbation Theory (χ_{PT}) predictions [4] and with other evaluations based on η decays, [5, 6] which have larger errors.

For the $\eta \to \pi^0\pi^0\pi^0$ decay the Dalitz plot density is described by a single parameter α: $|A|^2 \propto 1 + 2\alpha z$. Here $z = \rho^2/\rho^2_{max}$ is the square ratio of the distance of a point from the Dalitz plot center (ρ) to the maximum kinematically allowed distance (ρ_{max}). Being related to the square of the three pion energies in the η rest frame, α is thus a quadratic slope. Photons are paired to π^0's after kinematically constraining the total 4-momentum, thus improving the energy resolution; as a second step a fit constraining also the π^0 mass is performed in order to further improve the resolution. The fitting procedure takes properly into account all the smearing effects and the effect of wrong pairings of photons to π^0's. By fitting a sample with high purity on pairing (98.5%), we get:

$$\alpha = -0.013 \pm 0.005_{\rm stat} \pm 0.004_{\rm syst}. \tag{1}$$

$\eta \to \pi^0\gamma\gamma$ AND OTHER RARE PROCESSES

The $\eta \to \pi^0\gamma\gamma$ decay is an important test of χ_{PT} because of its sensitivity to p^6 in both the BR and the $M_{\gamma\gamma}$ spectrum [7, 8]. The current experimental situation is not completely clear: the most accurate determination of the BR [9] is far from theoretical predictions while more recent measurements [10, 11], with larger relative errors, give a significantly lower value. Moreover, all previous searches were done at hadron machines, using mainly $\pi^- p \to \eta n$. The value of the BR has decreased by three orders of magnitude in the last 40 years, due to the improved separation of the $\eta \to \pi^0\pi^0\pi^0$ background.

KLOE searches for this decay in a much cleaner environment, with different background topologies and experimental systematics. The two orders of magnitude higher

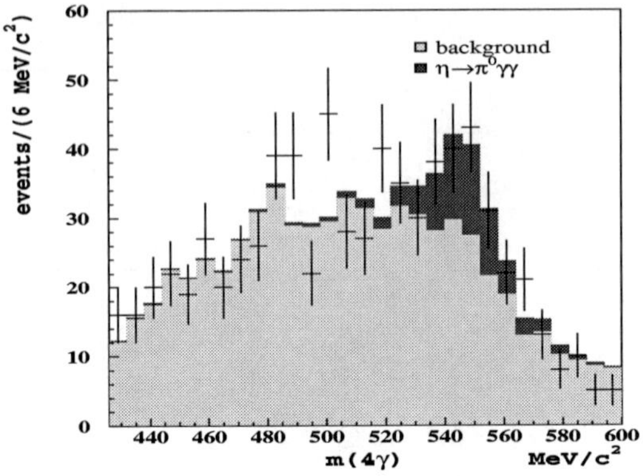

FIGURE 2. *Four photon invariant mass for $\eta \to \pi^0\gamma\gamma$ events. Data (crosses) are fitted with the signal and background contributions evaluated from MC (solid histograms).*

background with the same five photon final state ($e^+e^- \to \omega\pi^0 \to \pi^0\gamma\pi^0$, $\phi \to f_0\gamma \to \pi^0\pi^0\gamma$, $\phi \to a_0\gamma \to \eta\pi^0\gamma$ with $\eta \to \gamma\gamma$) is reduced by vetoing the additional $\omega/\pi^0/\eta$ particles in the event. The remaining background is $\eta \to \gamma\gamma$ with additional clusters from shower fragmentation or machine background and $\eta \to \pi^0\pi^0\pi^0$ with merged/lost photons. We reject them with energy momentum conservation and a likelihood technique to identify merged clusters. The preliminary results obtained fitting the η invariant mass spectrum (Fig. 2) gives a BR in agreement with $\mathcal{O}(p^6)$ χ_{PT} calculations, with a central value which is three times smaller than the previous measurement:

$$\text{BR}(\eta \to \pi^0\gamma\gamma) = (8.4 \pm 2.7_{\text{stat}} \pm 1.4_{\text{syst}}) \times 10^{-5}. \tag{2}$$

Our sensitivity to η rare decays has been also exploited in the search for the C-violating process $\eta \to \gamma\gamma\gamma$, where we obtain an upper limit of 1.6×10^{-5} at 90% C.L. [12] , and in the search for the P and CP violating process $\eta \to \pi^+\pi^-$ where we set a limit at 1.3×10^{-5} at 90% C.L. [13]. In both cases we improve previous exisiting limits by more than an order of magnitude.

$\eta - \eta'$ MIXING (AND GLUONIUM)

The processes $\phi \to \eta(\eta')\gamma$ are well described as magnetic dipole transitions, or, in other words, by a quark spin flip of the decaying vector meson. Neglecting the $\phi - \omega$ mixing, the ϕ is a pure $s\bar{s}$ state. The ratio of these two BR's can then give insight into the strange component of the η and η' mesons, that is into the pseudoscalar mixing angle in the flavour basis. We recall that even if a two mixing angle scenario is unavoidable in describing the $\eta - \eta'$ system, when using the flavour basis the two angles are equal apart from terms which violate the OZI rule, and is thus correct to parametrize the system with *one* mixing angle ϕ_P (this is not true in the SU(3) singlet - octet basis where large effects due to SU(3) symmetry breaking occur). Apart from obvious phase space corrections, and a small correction due to the $\phi - \omega$ mixing, the ratio R= BR($\phi \to \eta'\gamma$)/BR($\phi \to \eta\gamma$)

is proportional to $\cot^2\phi_P$ [14]. Moreover the absolute value of the BR($\phi \to \eta'\gamma$) is itself sensitive to the gluonium content of the η' [15, 16].

The final state we consider here for η' selection is $\pi^+\pi^-7\gamma$ which can be obtained either via the $\phi \to \gamma\eta' \to \gamma\pi^+\pi^-\eta \to \gamma\pi^+\pi^-3\pi^0$ intermediate state or via the $\phi \to \gamma\eta' \to \gamma\pi^0\pi^0\eta \to \gamma\pi^0\pi^0\pi^+\pi^-\pi^0$ one. No other physics channel from the ϕ gives the same final state. In both cases events are selected with small background (less than 10%) and normalized to the number of observed $\phi \to \gamma\eta \to \gamma 3\pi^0 \to 7\gamma$ events, in order to cancel out most of the systematics. The preliminary result we obtain, after observation of about 3,400 signal events is

$$R = (4.8 \pm 0.1_{stat.} \pm 0.2_{syst.}) \times 10^{-3} \quad (3)$$

It is fully compatible with our previous result [17] (which was performed on a different final state) but with much smaller error due to the significantly higher statistics. Systematic error is dominated by the knowledge of intermediate BR's , and in particular by the BR($\eta' \to \eta\pi\pi$).

REFERENCES

1. The KLOE Collaboration (A. Aloisio et al.), LNF-92/019 (IR) (1992) and LNF-93/002 (IR) (1993).
2. S. Guiducci et al., Proc. of the 2001 Particle Accelerator Conference (Chicago, Illinois, USA), P. Lucas S. Webber Eds. 353 (2001).
3. B. V. Martemyanov and V. S. Sopov, *Phys. Rev.* D **71**, 017501 (2005).
4. J. Donoghue and A. Perez, *Phys. Rev.* D **55**, 7075 (1997); J. Bijnens and J. Prades, *Nucl. Phys.* B **490**, 239 (1997).
5. J. Kambor, C. Wiesendanger and D. Whyler, *Nucl. Phys.* B **465**, 215 (1996).
6. A. V. Anisovich and H. Leutwyler, *Phys. Lett.* B **375**, 335 (1997).
7. J. Bijnens and J. Gasser, Proc. of Workshop on Production, Interaction and Decay of the η Meson (Uppsala, October 2001), J. Bijnens, G. Fäldt and B. M. K. Nefkens Eds., T99, 34 (Phys. Scripta, Stockholm, 2002)
8. E. Oset, J. R. Pelaez, L. Roca, *Phys. Rev.* D **67**, 073013 (2003).
9. Alde et al., *Z. Phys.* C **25**, 225 (1984).
10. N. Knecht et al., *Phys. Lett.* B **589**, 14 (2004).
11. Z. Papandreou, *these proceedings*.
12. A. Aloisio et al., *Phys. Lett.* B **591**, 49 (2004)
13. F. Ambrosino et al., *Phys. Lett.* B **606**, 276 (2005)
14. A. Bramon et al., *Eur. Phys. J.* C **7**, 271 (1999)
15. J. L. Rosner, *Phys. Rev.* D **27**, 1101 (1983)
16. E. Kou, *Phys. Rev.* D **63**, 054027 (2001)
17. A. Aloisio et al., *Phys. Lett.* B **541**, 45 (2002)

The low lying axial-vector mesons as dynamically generated resonances[2]

L. Roca*, E. Oset* and J. Singh*

*Departamento de Física Teórica and IFIC, Centro Mixto Universidad de Valencia-CSIC, Institutos de Investigación de Paterna, Aptdo. 22085, 46071 Valencia, Spain

Abstract. We make a theoretical study[1] of the s-wave interaction of the nonet of vector mesons with the octet of pseudoscalar mesons starting from a chiral invariant Lagrangian and implementing unitarity in coupled channels. By looking for poles in the unphysical Riemann sheets of the unitarized scattering amplitudes, we get two octets and one singlet of axial-vector dynamically generated resonances. The poles found can be associated to most of the low lying axial-vector resonances quoted in the Particle Data Book: $b_1(1235)$, $h_1(1170)$, $h_1(1380)$, $a_1(1260)$, $f_1(1285)$ and two poles to the $K_1(1270)$ resonance. We evaluate the couplings of the resonances to the VP states and the partial decay widths in order to reinforce the arguments in the discussion.

The introduction of unitary techniques have allowed to extend to higher energies the predictions of chiral perturbation theory, as well as to tackle new problems barred to a pure perturbative expansion [2, 3, 4, 5]. The unitary extensions of chiral perturbation theory, $U\chi PT$, have brought new light in the study of the meson meson and meson baryon interaction and have shown that some well known resonances qualify as dynamically generated, or in simpler words, they are quasibound states of meson meson or meson baryon. This is, for instance, the case of the low lying scalar mesons σ, $f_0(980)$, $a_0(980)$, $\kappa(900)$ [3, 4, 5] which appear from the interaction of pseudoscalar mesons, which has been succesfully tested in specific decay processes [6, 7, 8]. A suggestive step forward in the direction of studying the meson meson interaction along these lines would be to study the interaction of the octet of pseudoscalar mesons with the octet of vector mesons. Work in this direction has already been done in [9] and, interestingly, most of the low lying axial-vector mesons appear as dynamically generated resonances. In [9] the search for resonances was done by looking at the speed plots of the physical amplitudes. A search for poles of the amplitudes in the unphysical Riemann sheets in the complex plane is a more powerful tool to investigate resonances an a thorough study along these lines is called for. The present work goes in this last direction and we have done a thorough work investigating the following points: 1) The poles of the amplitudes have been systematically searched for in the complex energy plane. 2) The study of the amplitudes around the poles has also allowed us to determine the coupling of the resonances to the different coupled channels and from there the partial decay widths into the different channels. This study has been useful to make an association of the resonances

---

[2] Contribution to the XI International Conference on Hadron Spectroscopy (HADRON 05), 21-26 August 2005, Rio de Janeiro, Brazil

[1] All the details of the model can be found in Ref. [1] on which the present contribution is based.

CP814 *Hadron Spectroscopy: Eleventh International Conference*
edited by A. Reis, C. Göbel, J. Sá Borges, and J. Magnin

found with those of the PDG.

The interaction of two vector and two pseudoscalar mesons can be obtained from the following Lagrangian

$$\mathcal{L}_{VVPP} = -\frac{1}{4f^2} Tr\left([V^\mu, \partial^\nu V_\mu][P, \partial_\nu P]\right), \tag{1}$$

which can be derived from a chiral invariant Lagrangian [10, 1]. From this Lagrangian one can obtain the tree level amplitude which will be used as the kernel of the Bethe-Salpeter equation below.

In the literature several unitarization procedures have been used to obtain a scattering matrix fulfilling exact unitarity in coupled channels, like the Inverse Amplitude Method [2, 4] or the N/D method [5]. In this latter work the equivalence with the Bethe-Salpeter equation used in [3] was established. In the present work we make use of the Bethe-Salpeter equation to resum the diagrammatic series expressed in Fig. 1. The unitarized

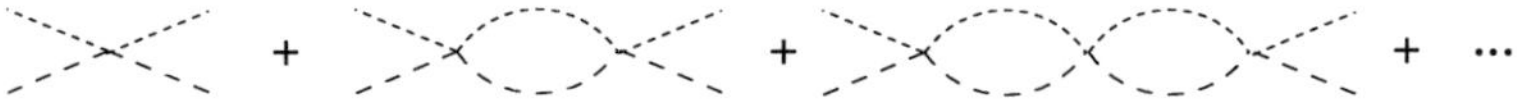

FIGURE 1. Diagrammatic representation of the resummation of loops in the unitarization procedure.

amplitude takes the following form:

$$T = [1 + V\hat{G}]^{-1}(-V)\vec{\varepsilon}\cdot\vec{\varepsilon}\,', \tag{2}$$

where $\hat{G} = G(1 + \frac{1}{3}\frac{q_l^2}{M_l^2})$ is a diagonal matrix with the $l-$th element, G_l, being the two meson loop function containing a vector and a pseudoscalar meson. Differences in the regularization scheme of the loop functions can be found in Ref. [1].

The association of physical resonances to poles of the scattering matrix in unphysical Riemann sheets is a very powerful tool to identify the resonances.

The use of physical vector and pseudoscalar masses, both in the potential V_{ij} as in the loop functions, allows for SU(3) symmetry breaking. Next, we are going to look for poles starting from the exact $SU(3)$ limit and breaking $SU(3)$ symmetry gradually. This can be accomplished by starting using a unique vector meson mass, M_0, and pseudoscalar mass, m_0, (obtained as an average of the physical masses inside each $SU(3)$ multiplet) and changing the masses parametrizing the masses with a parameter x such that the masses used are the $SU(3)$ averaged masses for $x = 0$ and the physical masses for $x = 1$. In Fig. 2 we show the position of the poles for the (S,I) channels $(0,0)$, $(0,1)$ and $(1,1/2)$, (the only (S,I) channels for which we find poles), evaluating the loop functions with the subtraction constant method. In the exact $SU(3)$ limit $(x = 0)$ we find two degenerate octet poles in all the (S,I) channels, and also a singlet pole for the $(0,0)$ channel. As we break $SU(3)$ gradually, by increasing x, two branches for each (S,I) channels of the octet and one for the singlet emerge. The branches end in the physical mass situation, $x = 1$. In the plot we have also written for each pole our guess for the correspondence with physical 1^+ resonances of the PDG. These assignments will be justified in the following discussion.

In Fig. 3 we show the results for the diagonal $|T_{ii}|^2$ matrix elements as a function of the energy with the loop functions evaluated with the subtraction constant method.

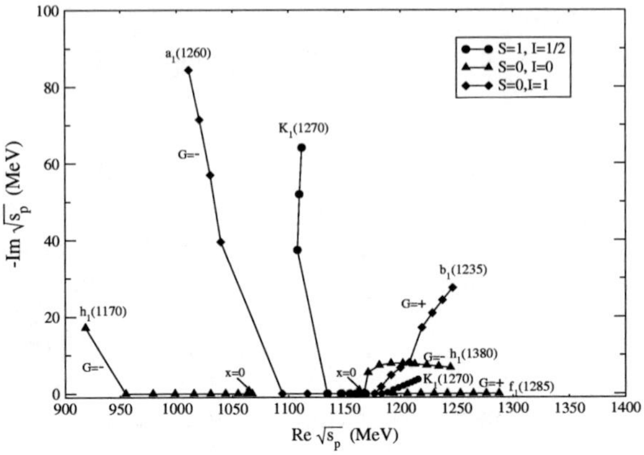

FIGURE 2. Trajectories of the poles in the scattering amplitudes obtained by increasing the $SU(3)$ breaking parameter, x, from 0 to 1.

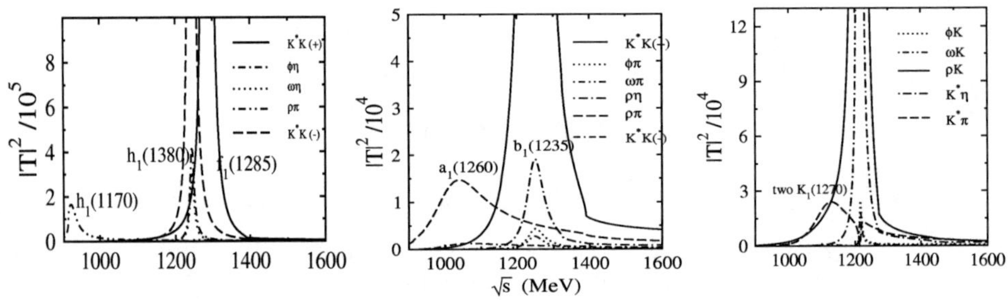

FIGURE 3. Amplitude square for $(S,I) = (0,0)$, $(0,1)$ and $(1,1/2)$ respectively for the different allowed channels.

We can clearly see the reflexion on the real axis of the poles found in Fig. 2. By evaluating the residues at the poles found it is possible to obtain the couplings of the dynamically generated axial-vector resonances to the different channels. With these couplings the partial decay widths of the axial-vector resonances to the different channels can be obtained and are shown in Tables 1, 2 and 3 in comparison with the experimental results quoted in the PDG.

The results that we obtain are compatible with the scarce experimental information in the sense that the dominant decay channels obtained theoretically correspond to the experimental ones.

Regarding the $(S,I) = (0,0)$ channel, the pole found at $1288 - i0$ MeV is reasonable to be assigned to the $f_1(1285)$ resonance. The reason why we do not get width for this resonance, while in the PDG it is said to be 24 MeV, is because there are other decay channels different to VP that we are not obviously considering. In fact the $\bar{K}^*K + c.c.$ is quoted in the PDG as "not seen" which agrees, in a first approximation, with the

TABLE 1. Partial decay widths for $S=0, I=0$. (All the units are MeV.)

	Γ_{tot}^{exp}	Γ_i^{exp}	Γ_i^{th}
$h_1(1170) \to \bar{K}^*K + c.c.$	360 ± 40	$-$	0.5
$h_1(1170) \to \rho\pi$	"	*seen*	77
$h_1(1170) \to \omega\eta$	"	$-$	0
$h_1(1170) \to \phi\eta$	"	$-$	0
$h_1(1380) \to \bar{K}^*K + c.c.$	91 ± 30	*seen*	45
$h_1(1380) \to \rho\pi$	"	$-$	9
$h_1(1380) \to \omega\eta$	"	$-$	25
$h_1(1380) \to \phi\eta$	"	$-$	0
$f_1(1285) \to \bar{K}^*K + c.c.$	24 ± 1	*not seen*	0.3

TABLE 2. Partial decay widths for $S=0, I=1$. (All the units are MeV.)

	Γ_{tot}^{exp}	Γ_i^{exp}	Γ_i^{th}
$b_1(1235) \to \bar{K}^*K + c.c.$	142 ± 9	$-$	7
$b_1(1235) \to \phi\pi$	"	$< 1.5\%$	12
$b_1(1235) \to \omega\pi$	"	*dominant*	25
$b_1(1235) \to \rho\eta$	"	*seen*	8
$a_1(1260) \to \rho\pi$	$250 - 600$	*seen*(60%?)	106
$a_1(1260) \to \bar{K}^*K + c.c.$	"	*seen*(< 10%?)	6

result obtained here as a bound state. The small experimental total width indicates a small coupling to other channels which are kinematically open and barely change the nature of the resonance as a bound system of $K^*\bar{K}$. It is important to note that there is experimentally another resonance, the $f_1(1420)$, which is assigned by the PDG to the 1^{++} nonet. However, in our scheme this resonance has no counterpart. This is because the interaction of two octets give one singlet plus two octets (three poles), therefore there is no way, with the interaction of one vector and one pseudoscalar mesons, to generate dynamically one more pole in this channel and, therefore, there is no room for one more resonance like the $f_1(1420)$.

Regarding the $(S,I) = (0,1)$ channel, the pole at $1247 - i28$ MeV, which is found for the channels with positive $G-$parity, may be assigned to the $b_1(1235)$ resonance. In the PDG the full width is quoted to be 142 MeV and the decay channels are quoted as $\omega\pi$ (dominant), $\rho\eta$ (seen), 4π ($< 50\%$), $K\bar{K}\pi$ ($< 14\%$) and $\phi\pi$ ($< 1.5\%$), (See Table 2). Therefore one could expect that the VP decay channels would account for around 50% of the total width. This is compatible with what we obtain. Note also that the channel quoted as dominant in the PDG is indeed the dominant one in our results. The pole at $1011 - i84$ MeV, which is found for the channels with negative $G-$parity, could be assigned to the $a_1(1260)$ resonance. Note that the values for the mass and width quoted for this resonance in the PDG, 1230 ± 40 and $250 - 600$ MeV respectively, have a large uncertainty. This can justify the discrepancy in the position of the pole we obtain.

Regarding the $(S,I) = (1,1/2)$ channel, we obtain two poles. In the PDG there are two physical $S=1, I=1/2$, resonances with $J^P = 1^+$, which are the $K_1(1270)$ and the

TABLE 3. Partial decay widths for $S = 1$, $I = 1/2$. In the last column, the first element is the value obtained with the coupling to the lower pole and the second element is the result with the coupling to the higher pole. (All the units are MeV.)

	Γ_{tot}^{exp}	Γ_i^{exp}	Γ_i^{th}
$K_1(1270) \to \phi K$	90 ± 20	–	(0,0)
$K_1(1270) \to \omega K$	"	11%	(5,2)
$K_1(1270) \to \rho K$	"	42%	(7,52)
$K_1(1270) \to K^*\eta$	"	–	(0,0)
$K_1(1270) \to K^*\pi$	"	16%	(113,6)

$K_1(1400)$. The resonance found at 1112 MeV couples strongly to the $K^*\pi$ and leads to a partial decay width in this channel of 113 MeV (around 150 MeV should we considered the mass to be 1400 MeV). On the other hand, the 1216 MeV resonance couples dominantly to ρK. In view of this, one would be tempted to assign the 1112 MeV pole to the $K_1(1400)$ resonance and the 1216 MeV pole to the $K_1(1270)$ since this is the experimental case. Yet, the difference of 300 MeV in the mass in the case of the $K_1(1400)$ is not an appealing feature, since this difference is much larger than the $K_1(1400)$ width (174 MeV). There is another possible scenario which we find more appealing. The detailed explanation of the PDG on the determination of the decay widths shows a clear discrepancy between different methods of determination of the $K_1(1270)$ width. This could find an explanation if one assumed that the experimental $K_1(1270)$ resonance is a superposition of two resonances that couple with different strength to a particular channel. This is indeed the case for the two poles that we find, one of them coupling strongly to $K^*\pi$ and the other one to ρK. Different experiments which favor mechanisms that give a bigger weight to one of these channels would lead to very different widths of the resonance.

In summary, the present work substantiates the idea that the low lying axial-vector resonances are dynamically generated, with the exception of the higher mass ones, the $f_1(1420)$ and the $K_1(1400)$, which do not fit easily in our scheme, while at the same time we suggest that the $K_1(1270)$ corresponds actually to two poles, which would have many experimental repercussions.

REFERENCES

1. L. Roca, E. Oset and J. Singh, Phys. Rev. D **72** (2005) 014002.
2. A. Dobado and J. R. Pelaez, Phys. Rev. D **56** (1997) 3057.
3. J. A. Oller and E. Oset, Nucl. Phys. A **620** (1997) 438 [Erratum-ibid. A **652** (1999) 407].
4. J. A. Oller, E. Oset and J. R. Pelaez, Phys. Rev. D **59** (1999) 074001 [Erratum-ibid. D **60** (1999) 099906].
5. J. A. Oller and E. Oset, Phys. Rev. D **60** (1999) 074023.
6. J. E. Palomar, L. Roca, E. Oset and M. J. Vicente Vacas, Nucl. Phys. A **729** (2003) 743.
7. L. Roca, J. E. Palomar, E. Oset and H. C. Chiang, Nucl. Phys. A **744** (2004) 127.
8. E. Oset, J. R. Pelaez and L. Roca, Phys. Rev. D **67** (2003) 073013.
9. M. F. M. Lutz and E. E. Kolomeitsev, Nucl. Phys. A **730** (2004) 392.
10. M. C. Birse, Z. Phys. A **355** (1996) 231.

PWA of 3π Final States and a Search for the $\pi_1(1600)$

Ryan Mitchell

Department of Physics, Indiana University, Bloomington, IN 47405

Abstract. Partial wave analyses (PWA) of the 3π systems in the reactions $\pi^- p \to \pi^+\pi^-\pi^- p$ (the "charged" mode) and $\pi^- p \to \pi^-\pi^0\pi^0 p$ (the "neutral" mode) with an 18.3 GeV/c pion beam were performed using high statistics data from the E852 experiment. Conventional signals, such as the $a_1(1260)$, the $a_2(1320)$, and the $\pi_2(1670)$, were found to be remarkably stable to the choice of waves included in the fit. In contrast, possible evidence for the $\pi_1(1600)$ in the $J^{PC} = 1^{-+}$ exotic wave amplitude disappears when additional decay modes of conventional mesons (especially those of the $\pi_2(1670)$) are included in the PWA fit.

Keywords: Meson Spectroscopy, Hybrid Mesons
PACS: 14.40.Cs, 12.38.Qk, 12.39.Mk

INTRODUCTION

A $q\bar{q}$ bound state is constrained to a series of allowed quantum numbers J^{PC}, where J is the spin of the bound state, P its parity, and C its symmetry under charge conjugation. If, however, the gluonic field between the quark and antiquark is excited (forming a "hybrid" meson) then J^{PC} combinations are allowed that would have otherwise been forbidden. Finding mesons with these "exotic" quantum numbers provides experimental input to our theoretical understanding of quark confinement.

One candidate for a $J^{PC} = 1^{-+}$ exotic meson is the $\pi_1(1600)$ decaying to $\rho\pi$ in a P-Wave. This state was first reported by the E852 Collaboration in 1998 [1, 2] in an analysis based on 250K $\pi^- p \to \pi^+\pi^-\pi^- p$ events collected during the 1994 run of the E852 experiment. The analysis presented here uses 2.6M events (after selection cuts) of the type $\pi^- p \to \pi^+\pi^-\pi^- p$ ("charged" mode) and 3.0M $\pi^- p \to \pi^-\pi^0\pi^0 p$ ("neutral" mode) events collected during the 1995 E852 run.

THE E852 3π DATA SETS

The E852 experiment ran at the Alternating Gradient Synchrotron (AGS) at the Brookhaven National Lab during the years 1994-1995. An 18.3 GeV/c pion beam was incident on a liquid hydrogen target. Charged particles were tracked downstream of the target with a series of proportional wire chambers (PWC) and drift chambers. The positions of recoiling protons were detected with a central cylindrical drift chamber (TCYL). Photons were measured with a downstream lead-glass calorimeter (LGD). Details can be found elsewhere [3].

CP814 *Hadron Spectroscopy: Eleventh International Conference*
edited by A. Reis, C. Göbel, J. Sá Borges, and J. Magnin

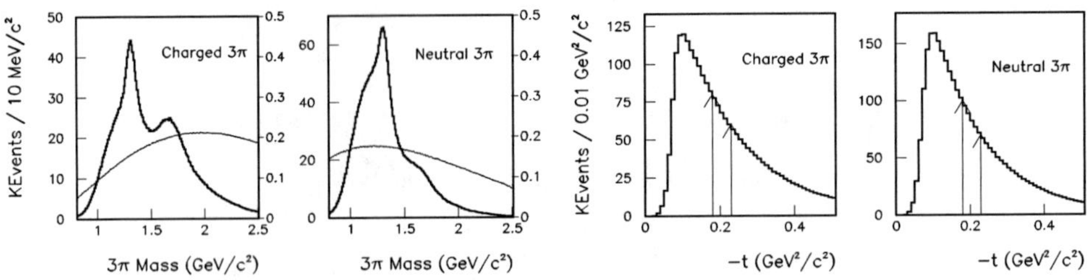

FIGURE 1. Uncorrected mass and $-t$ distributions for the charged and neutral modes. Mass acceptances are superimposed on the mass plots (the scale is given on the right axes).

The trigger for the charged 3π mode required three tracks in the downstream PWCs in coincidence with one track (the recoil) in the central TCYL. The neutral mode trigger[1] required one track in the PWCs, one track in the TCYL, and energy in the LGD inconsistent with a single π^0. Further selection cuts included requiring a single vertex in the target, a confidence level requirement of a kinematic fit, and consistency between the angle of the missing recoil momentum and the position of the TCYL track [4].

Figure 1 shows the uncorrected 3π mass distributions for the two charge modes. Structures corresponding to the $a_1(1260)$, $a_2(1320)$, and $\pi_2(1670)$ are evident. The momentum transfers ($-t$) between the incoming pion and the 3π systems are also shown in Figure 1. The data was divided into 12 different bins of $-t$. Results shown in this paper are from 0.18 to 0.23 GeV^2/c^2, the region in Figure 1 marked by arrows, and the region where there is maximum sensitivity to the largest number of 3π states.

PARTIAL WAVE FIT PROCEDURE

The formalism used in fitting the data was identical to that used in the earlier E852 analysis [2]. The 3π intensity was expanded in each mass bin in a series of partial waves, each of which, in the isobar model, describes a resonance decay to an isobar and a bachelor π and the subsequent isobar decay to $\pi\pi$. Individual partial waves are denoted by $J^{PC}M^{\varepsilon}(isobar)L$, where J^{PC} are the quantum numbers of the resonance, M is its spin projection, L is the orbital angular momentum between the isobar and bachelor π, and ε is the reflectivity (corresponding to the naturality of the exchange particle in the production mechanism).

A set of waves to include in the fit was chosen by the following procedure. First, a parent wave set was defined that included waves with the $\rho(770)$, $\rho_3(1690)$, $f_0(980)$, $f_2(1270)$ and σ isobars, where σ denotes the $\pi\pi$ S-Wave amplitude defined in [2].

---

[1] A global inefficiency of the triggering PWCs resulted in an overall normalization factor between the two charge modes. This overall inefficiency has no effect within a given trigger and was not modeled in the Monte Carlo. To bring the results from the charged and neutral mode triggers into line, all results in the charged mode were scaled by a factor of 1.3.

TABLE 1. Partial waves under consideration in this analysis.

$J^{PC}M^{\varepsilon}(isobar)L$	Wave Set*	$J^{PC}M^{\varepsilon}(isobar)L$	Wave Set	$J^{PC}M^{\varepsilon}(isobar)L$	Wave Set
$0^{-+}0^{+}(\sigma)S$	B	$2^{++}1^{+}(\rho)D$	B	$2^{-+}1^{+}(f_2)D$	B
$0^{-+}0^{+}(f_0)S$	B	$2^{++}0^{-}(\rho)D$	B	$2^{-+}0^{+}(f_2)D$	B
$0^{-+}0^{+}(\rho)P$	B	$2^{-+}0^{+}(f_2)S$	B	$2^{-+}0^{+}(\rho)F$	H
$1^{++}0^{+}(\rho)S$	B	$2^{-+}1^{+}(f_2)S$	B	$2^{-+}1^{+}(\rho)F$	H
$1^{++}1^{+}(\rho)S$	B	$2^{-+}1^{-}(f_2)S$	L	$3^{++}0^{+}(\rho_3)S$	B
$1^{++}1^{-}(\rho)S$	L	$2^{-+}0^{+}(\rho)P$	B	$3^{++}0^{+}(f_2)P$	H
$1^{++}0^{+}(\sigma)P$	H	$2^{-+}1^{+}(\rho)P$	H	$3^{++}0^{+}(\rho)D$	H
$1^{++}0^{+}(f_0)P$	H	$2^{-+}0^{+}(\rho_3)P$	H	$4^{++}0^{+}(\rho_3)D$	H
$1^{++}1^{+}(f_2)P$	H	$2^{-+}1^{+}(\rho_3)P$	H	$4^{++}0^{+}(f_2)F$	H
$1^{++}0^{+}(\rho)D$	L	$2^{-+}0^{+}(\sigma)D$	B	$4^{++}0^{+}(\rho)G$	H
$1^{-+}1^{+}(\rho)P$	B	$2^{-+}0^{+}(f_0)D$	H	$4^{-+}0^{+}(\rho_3)P$	H
$1^{-+}0^{-}(\rho)P$	B	$2^{-+}1^{+}(\sigma)D$	H	$4^{-+}0^{+}(\rho)F$	H
$1^{-+}1^{-}(\rho)P$	B	$2^{-+}1^{+}(f_0)D$	H	Background	B

* Indicates whether wave was used in the high wave set [H], the low wave set [L] or in both [B].

Waves were restricted to those with $J \leq 4$ and $M \leq 1$. Waves were then removed one by one from the fit and the resulting changes in log(Likelihood) were examined. If the change was greater than 40 units, the wave was judged significant and was returned to the wave set[2]. A set of 35 waves survived this procedure. See [4] for details.

A comparison of the resulting wave set (termed the "high wave set") and the wave set used in the original E852 3π analysis (the "low wave set") is given in table 1. The low wave set included 20 waves.

RESULTS

A selection of partial wave intensities resulting from fits with both the high wave set and the low wave set is shown in Figure 2. The $2^{++}1^{+}(\rho)D$ (dominant mode of the $a_2(1320)$), the $1^{++}0^{+}(\rho)S$ (dominant mode of the $a_1(1260)$), and the $2^{-+}0^{+}(f_2)S$ (dominant mode of the $\pi_2(1670)$) intensities show a remarkable stability to the choice of wave set. Furthermore, the relative intensities in the two charge modes are consistent with isospin relations, which state that intensities for waves with isovector isobars (ρ) should be equal, and intensities for waves with isoscalar isobars (f_2) should appear in the charged to neutral ratio of 2:1.

The exotic $1^{-+}1^{+}(\rho)P$ intensity resulting from the low wave fit shows a peaking structure around 1600 MeV$/c^2$ (Figure 2), consistent with the earlier report of the E852 Collaboration. However, the peak is not present in the high wave fit. Note the consistency between the two charge modes.

Phase differences between the 1^{-+}, 2^{-+}, and 2^{++} waves are shown in Figure 3. The phase difference between the 2^{++} and 2^{-+} shows a rise corresponding to the $a_2(1320)$

[2] Exceptions to this method were the 1^{-+} waves, which were judged to be insignificant by the likelihood criteria, but were kept for comparison with the earlier results.

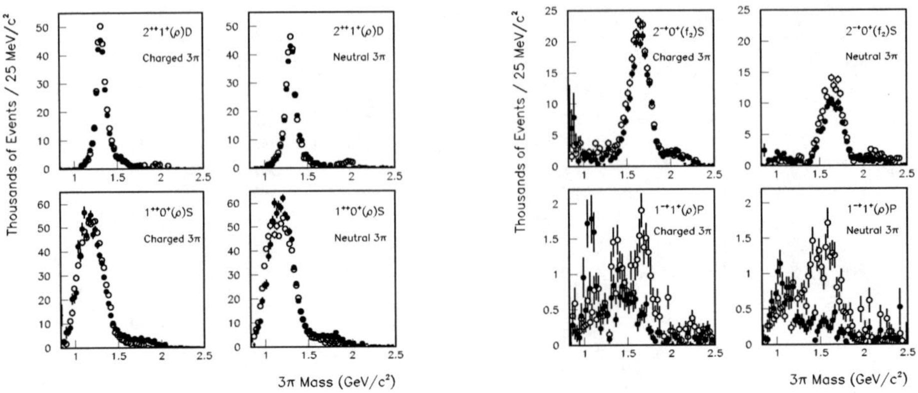

FIGURE 2. Intensities of the $2^{++}1^{+}(\rho)D$, the $1^{++}0^{+}(\rho)S$, the $2^{-+}0^{+}(f_2)S$, and the $1^{-+}1^{+}(\rho)P$ waves resulting from fits with the high (solid circles) and low (open circles) wave sets.

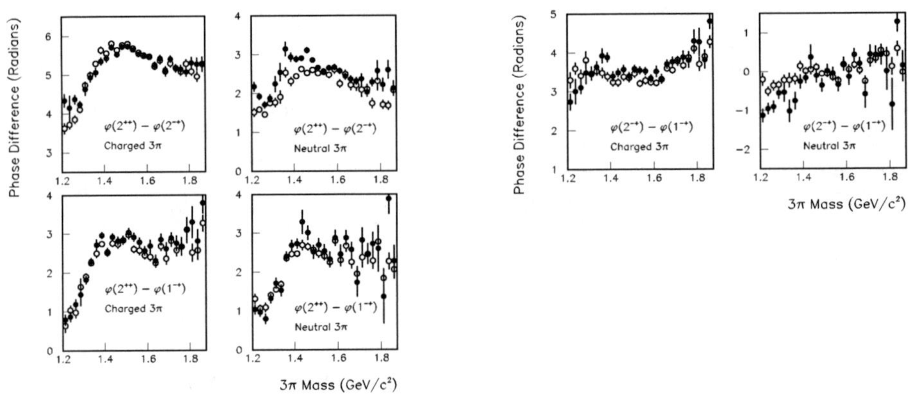

FIGURE 3. Phase differences between the $1^{-+}1^{+}(\rho)P$, the $2^{++}1^{+}(\rho)D$, and the $2^{-+}0^{+}(f_2)S$ waves resulting from fits with the high (solid circles) and low (open circles) wave sets.

resonance and a fall due to the $\pi_2(1670)$. The 1^{-+} phase is found to closely track the 2^{-+} phase, indicating either the presence of a very small $\pi_1(1600)$ with nearly the same mass and width as the $\pi_2(1670)$ or indicating that the small residual background of the 1^{-+} intensity is correlated with the $\pi_2(1670)$, perhaps through leakage.

Intensities of a selection of waves included in the high wave fit but not in the low are shown in Figure 4. The $2^{-+}0^{+}(\rho)F$, the $2^{-+}1^{+}(\rho)F$, and the $2^{-+}1^{+}(\sigma)D$ waves are allowed decay modes of the $\pi_2(1670)$ and show structure consistent with the $\pi_2(1670)$ mass and width. The $1^{++}1^{+}(f_2)P$ wave is a decay mode of the $a_1(1640)$ [5]. All of

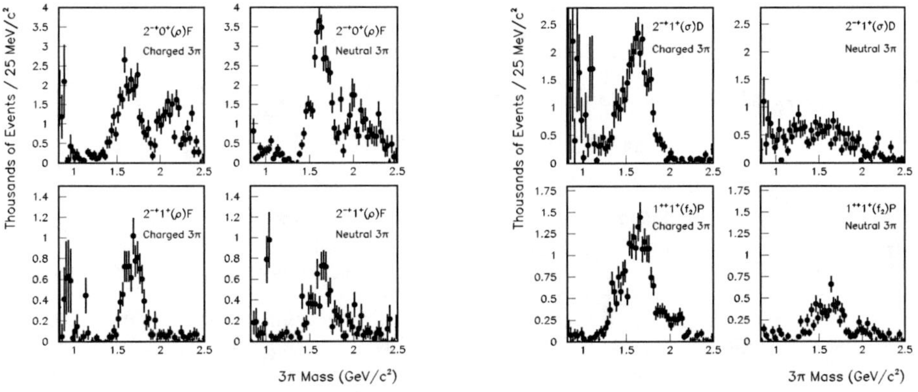

FIGURE 4. Intensities of the $2^{-+}0^{+}(\rho)F$, the $2^{-+}1^{+}(\rho)F$, the $2^{-+}1^{+}(\sigma)D$, and the $1^{++}1^{+}(f_2)P$ waves, which were included in the high wave fit but not in the low.

these waves peak in the 1600 to 1700 MeV/c^2 region. When they are included in the fit, there is no peak in the $1^{-+}1^{+}(\rho)P$ intensity; when they are not included in the fit, a peak appears in the $1^{-+}1^{+}(\rho)P$ wave (Figure 2).

CONCLUSION

The peaking structure around 1600 MeV/c^2 in the exotic $1^{-+}1^{+}(\rho)P$ intensity (the purported $\pi_1(1600)$) is an artifact of the low wave fit. When the wave set includes significant decay modes of the $\pi_2(1670)$ and the $a_1(1640)$, the peak in the exotic intensity disappears, and the small residual intensity shows no convincing evidence for resonance structure.

ACKNOWLEDGMENTS

This work was done in collaboration with Alex Dzierba, Adam Szczepaniak, Maciej Swat, and Scott Teige of Indiana University.

REFERENCES

1. G. S. Adams, et al., *Phys. Rev. Lett.* **81**, 5760 (1998).
2. S. U. Chung, et al., *Phys. Rev. D* **65** (2002).
3. S. Teige, et al., *Phys. Rev. D* **59**, 012001 (1999).
4. A. R. Dzierba, et al. (2005), URL `dustbunny.physics.indiana.edu/3pi_paper`, notes regarding the 3π data processing and selection, PWA formalism, and other details.
5. S. Eidelman, et al., *Phys. Lett. B* **592**, 1 (2004).

Precise measurement of hadronic cross sections with CMD-2 detector

A. L. Sibidanov on behalf of CMD-2 collaboration

Budker Institue of Nuclear Physics, 11 Laverentieva, Novosibirsk 630090, Russia

Abstract. Experiments with the CMD-2 detector on VEPP-2M collider were carried out during 1992 - 2000. The total integrated luminosity of 33 pb^{-1} has been collected. The results of the measurement of electron positron annihilation cross-sections to hadrons in the whole VEPP-2M energy range are presented. Among them the new measurement of the pion formfactor in the whole energy range that confirms our previous result with the same systematic error 0.6% and new data on three, four pion and kaon production on ϕ-meson and above.

Keywords: CMD-2 detector, hadronic cross sections, light vector mesons, R-measurement
PACS: 13.40.Gp, 13.66.Bc, 13.66.Jn, 14.40.-n

INTRODUCTION

Since early 70s up to 2000 the VEPP-2M e^+e^- collider [1] has been running in the Budker Institute of Nuclear Physics(Novosibirsk, Russia) in the c.m. energy range 360-1400 MeV. Experiments with the CMD-2 detector on the VEPP-2M collider were carried out during 1992 - 2000 and a number of significant physical results were obtained. One of the main goal of experiments is a measurement of total hadronic cross section in the whole VEPP-2M energy range. The layout of VEPP-2M collider is shown in Fig. 1.

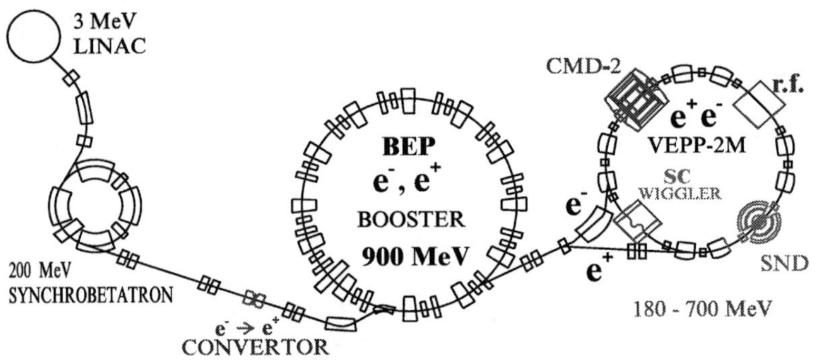

FIGURE 1. The VEPP-2M collider.

The CMD-2 (Cryogenic Magnetic Detector) is a general purpose detector. It consists of tracking system, barrel CsI and end-cap BGO electromagnetic calorimeters and muon range system. Tracking system consists of drift chamber with jet-like cells and double layer proportional Z-chamber inside thin superconductive solenoid with 1 T magnetic field. More detailed description of the detector could be found elsewhere [2].

The features of the detector allows to measure exclusive hadronic cross sections in the whole VEPP-2M energy range. Exclusive approach allows one to keep systematic errors at low level by taking into account efficiencies, acceptance, radiative corrections and so

CP814 *Hadron Spectroscopy: Eleventh International Conference*
edited by A. Reis, C. Göbel, J. Sá Borges, and J. Magnin

on for each particular process. All main hadronic cross sections measured with CMD-2 are presented in Fig. 2.

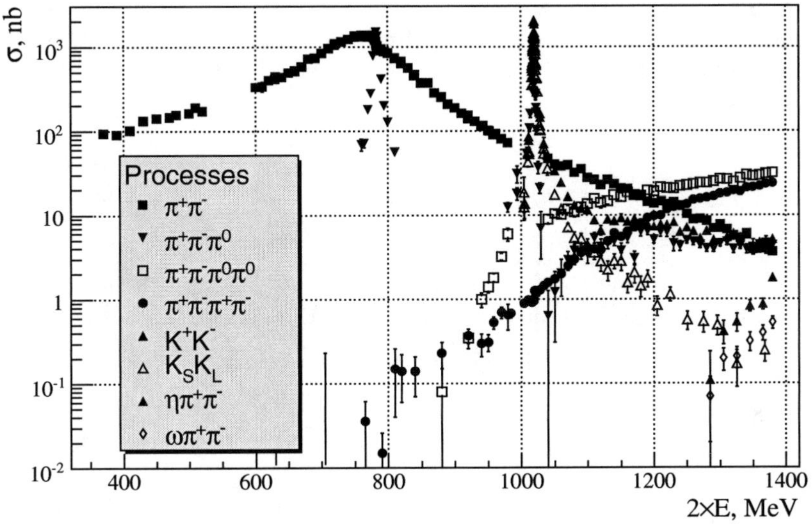

FIGURE 2. All main hadronic cross sections measured at the CMD-2 detector.

Another important goal of CMD-2 experiments is a study of light vector mesons. For this purposes about 2.1×10^7 of ϕ-mesons, 3×10^6 of ω-mesons and 6.5×10^6 of ρ-mesons were collected during data taking.

PION FORMFACTOR

About 10 times more data were analysed since previous publication of CMD-2 high precision measurement of the $e^+e^- \to \pi^+\pi^-$ cross section around ρ-meson [3]. In addition to the region around ρ-meson peak new energy ranges were included in the analysis.

For the pion cross section measurement the whole energy range is naturally subdivided in 3 intervals where different analysis techniques were used. At energies below 600 MeV $e/\mu/\pi$ separation was made using particles momenta. At higher energies, the momentum resolution is not sufficient to distinguish electrons, muons and pions therefore the separation was based on the particles energy deposition in CsI-calorimeter. At energies above 1000 MeV the cross-section of pion production becomes small comparing to other of background. Therefore it becomes necessary to apply different, more rigid, selection criteria on final particles kinematic.

The systematic error for low energy range is estimated at 0.7% with overall statistical error about 2%. For new data ρ-meson region systematic error is about 0.8% which is slightly more then in previous analysis due to different detector conditions but with reduced statistical error. Above ϕ-meson systematic error varies from 1.2% up to 4.2% with c.m. energy.

The parameters of ρ-meson are in agreement with previously published CMD-2 result [3] but with smaller statistical error.

The high precision $e^+e^- \to \pi^+\pi^-$ cross section data were recently published by KLOE [4] and SND [5] detectors. KLOE measured $e^+e^- \to \pi^+\pi^-$ cross section by ISR technique while SND measured $e^+e^- \to \pi^+\pi^-$ cross section in the way similar to CMD-2 by energy scan. The comparison of CMD-2 data with those two results is shown in Fig. 3 and 4. The SND data agrees with those of CMD-2 within the systematic and statistic errors of experiments, while KLOE data have visible difference.

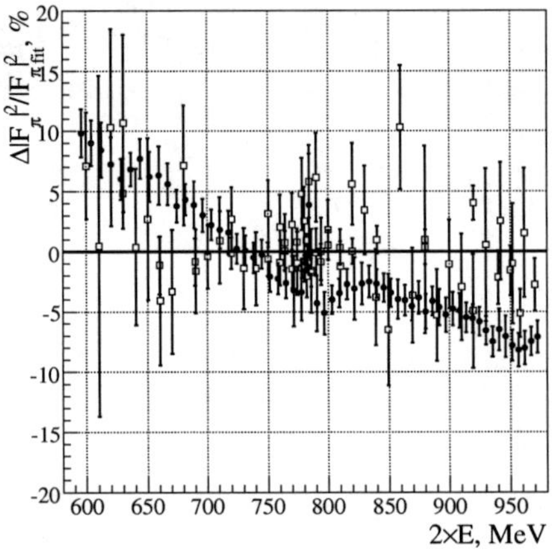

FIGURE 3. Relative difference between KLOE pion formfactor and CMD-2 fit function. Circles - KLOE, squares - CMD-2.

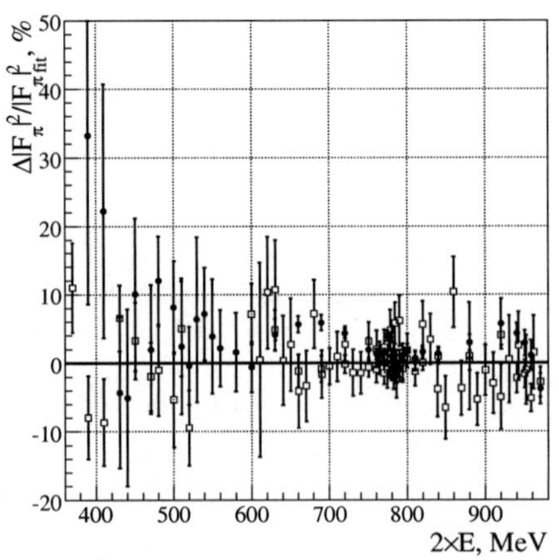

FIGURE 4. Relative difference between SND pion formfactor and CMD-2 fit function. Circles - SND, squares - CMD-2.

THREE PION PRODUCTION

The measurement of three pion cross section above ϕ-meson was based on about 2.6×10^3 selected events in 36 energy points. The total integrated luminosity was about 5.9 nb^{-1}. Systematic uncertainty of this measurement is about 7%. Using 1.15×10^5 events selected around ϕ-meson the following ϕ-meson parameters were obtained:
$\sigma_{\phi\to3\pi} = 641\pm23$ nb,
$m_\phi = 1019.3\pm0.1$ MeV,
$\Gamma_\phi = 4.36\pm0.07$ MeV,
$\mathscr{B}_{ee}\cdot\mathscr{B}_{3\pi} = (4.54\pm0.15)\cdot10^{-5}$.

Global fit was performed to describe cross section behaviour depend on c.m. energy and to obtain parameters of ω-meson excitation states. The fit is shown in Fig. 5. Using three pion data from BaBar [10] above VEPP-2M upper energy range and data from $e^+e^-\to\omega\pi^0$, $e^+e^-\to\omega\pi^+\pi^-$ [11, 12] and $\pi^+\pi^-$ invariant mass spectra from 3π the following parameters of ω' and ω'' were obtained:
$\sigma_{\omega'\to3\pi} = 0.4\pm0.02$ nb,
$m_{\omega'} = 1425\pm10$ MeV,
$\Gamma_\phi = 1100\pm150$ MeV,
$\sigma_{\omega''\to3\pi} = 3.8\pm0.1$ nb,
$\sigma_{\omega''\to\omega\pi\pi} = 2.9\pm0.15$ nb,
$m_{\omega''} = 1681\pm4$ MeV,
$\Gamma_\phi = 355\pm8$ MeV.

FOUR PION PRODUCTION

Four pion final state with neutral pions looks like three pions since it produce also two tracks in tracking system so the analysis of $\pi^+\pi^-2\pi^0$ was done simultaneously with three pions. Systematic uncertainty of this measurement is about 8% which is similar to three pions final state where errors is 7%. The difference in error comes from acceptance of photons from decay of additional neutral pion. About 1.2×10^4 events were selected for analysis.

Four charged pions final state have been already published [6]. The estimated systematic error is about 5%.

The resulting cross sections in comparison with recent precise SND measurements [7] is shown in Fig. 6 and 7. The results of measurements are in good agreement between both detectors.

NEUTRAL KAON PRODUCTION

The measurement of the $e^+e^-\to K_SK_L$ cross section from 1.05 to 1.38 GeV based on 5.7 nb^{-1} integrated luminosity. The systematic uncertainty on the cross section was about 5% below 1.09 GeV and increased to 10% at 1.38 GeV mainly due to changes in selection criteria and background subtraction. About 948 events were selected for analysis. The result has been previously published [8].

Using 3×10^5 K_SK_L events collected during 4 scans of 1996 with integrated luminosity 2.67 nb^{-1} the following ϕ-meson parameters were obtained [9]:

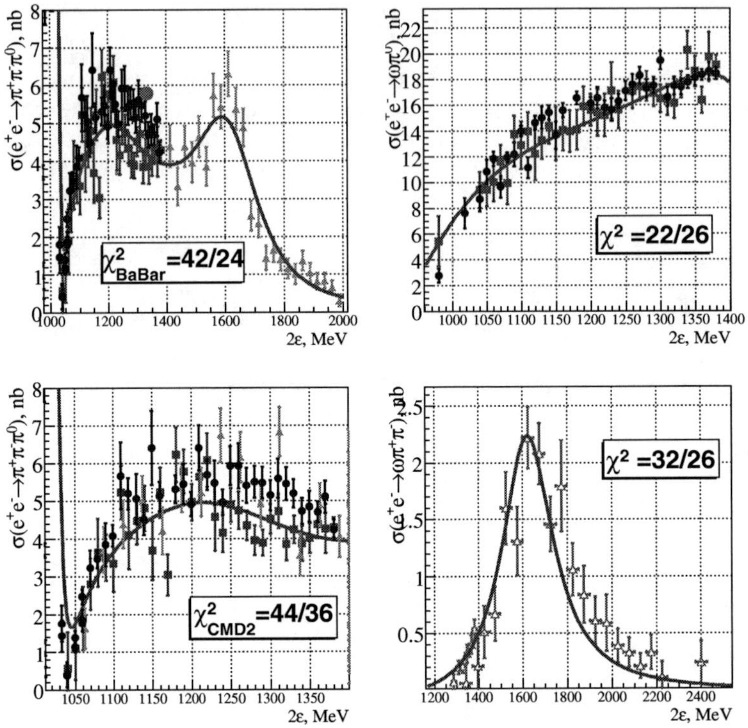

FIGURE 5. Combined fit result with $\pi^+\pi^-\pi^0$ (left upper and lower plots), $\omega\pi^0$ (right upper plot) and $\omega\pi^+\pi^-$ (right lower plot) channels. Squares - CMD-2, circles - SND, full triangles - BaBar, open triangles - DM2.

$\sigma_{\phi\to K_S K_L} = 1376 \pm 6 \pm 23$ nb,
$m_\phi = 1019.483 \pm 0.011 \pm 0.025$ MeV,
$\Gamma_\phi = 4.28 \pm 0.033 \pm 0.025$ MeV,
$\mathscr{B}_{ee} \cdot \mathscr{B}_{K_S K_L} = (9.75 \pm 0.04 \pm 0.17) \cdot 10^{-5}$.

On Fig. 8 the CMD-2 with SND [13] data are shown. It is visible that SND has good agreement with CMD-2 data. The additional amplitude to ρ, ω and ϕ mesons, which one can corresponds to radial excitation of ϕ-meson, is needed to describe neutral kaon cross section as it was mentioned in [8].

CHARGED KAON PRODUCTION

The measurement of the $e^+e^- \to K^+K^-$ cross section from 1.04 to 1.38 GeV based on 5.5 nb^{-1} integrated luminosity. The systematic uncertainty on the cross section was about 8%. About 2.5×10^4 events were selected for analysis.

Using 3.6×10^5 K^+K^- events collected during 2 scans of 1996 with integrated luminosity 1.04 nb^{-1} the following ϕ-meson parameters were obtained [14]:

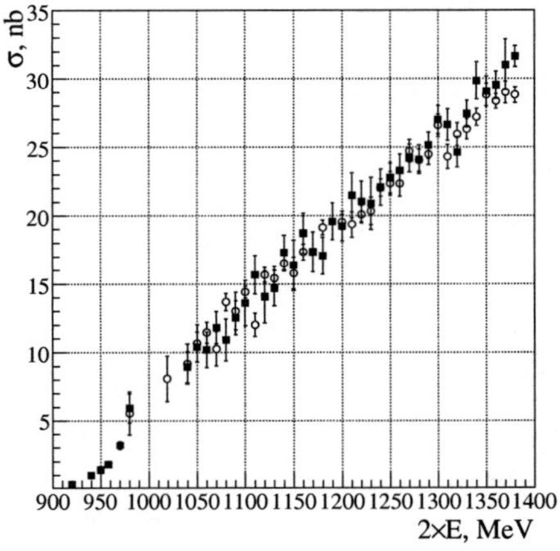

FIGURE 6. Measurement of the $\pi^+\pi^-\pi^0\pi^0$ cross section. Squares - CMD-2, circles - SND.

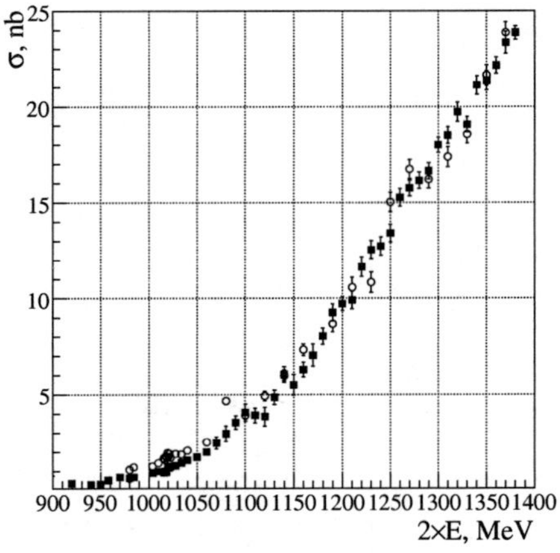

FIGURE 7. Measurement of the $\pi^+\pi^-\pi^+\pi^-$ cross section. Squares - CMD-2, circles - SND.

$\sigma_{\phi\to K^+K^-} = 2044 \pm 10 \pm 57$ nb,
$m_\phi = 1019.448 \pm 0.010 \pm 0.08$ MeV,
$\Gamma_\phi = 4.25 \pm 0.03 \pm 0.03$ MeV,
$\mathscr{B}_{ee} \cdot \mathscr{B}_{K_SK_L} = (14.46 \pm 0.07 \pm 0.4) \cdot 10^{-5}$.
The systematic uncertainty in cross section at ϕ-meson peak is estimated about 2.8%.

On Fig. 9 the CMD-2, SND [13] and OLYA [15] data are shown. It is seen that OLYA results are in good agreement with CMD-2 data at high energies while near ϕ-meson region there is some excess. It can be explained that OLYA might have a contamination of kaon sample due to pions.

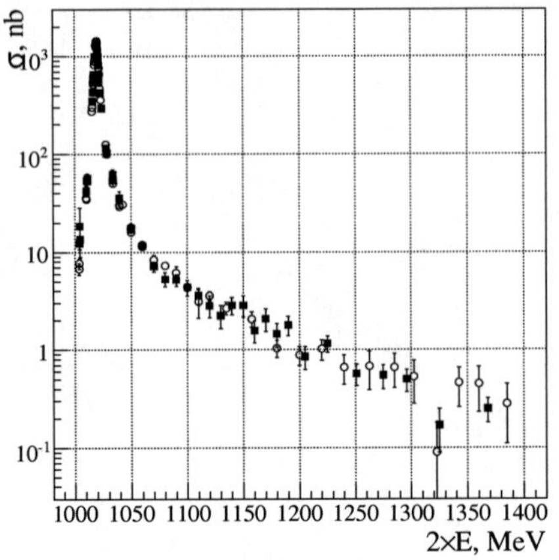

FIGURE 8. Measurement of the $K_S K_L$ cross section. Squares - CMD-2, circles - SND.

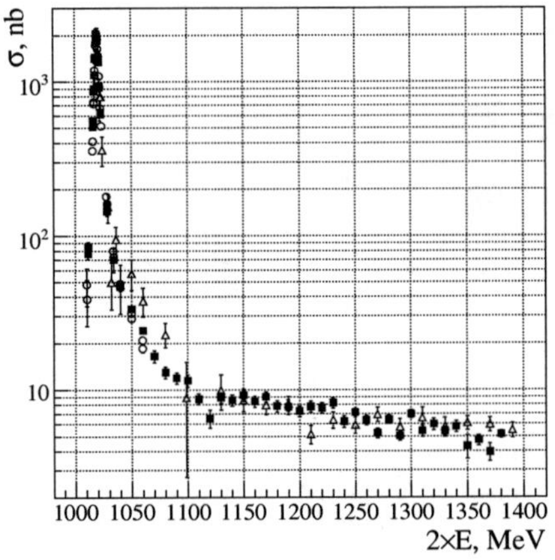

FIGURE 9. Measurement of the K^+K^- cross section. Squares - CMD-2, circles - SND, triangles - OLYA.

ACKNOWLEDGMENTS

The authors are grateful to the staff of VEPP-2M for excellent collider performance, to all engineers and technicians who participated in the design, commissioning and operation of CMD-2.

This work is partially supported by Russian Found of Basic Research, grants N 03-02-16477, 03-02-16280, 04-02-16217, 04-02-16223 and 04-02-16434.

REFERENCES

1. V.V. Anashin *et al.*, Preprint INP 84-114, Novosibirsk (1984).
2. G.A. Aksenov *et al.*, Preprint Budker INP **85-118**, Novosibirsk, 1985;
   E.V. Anashkin *et al.*, ICFA Instr. Bulletin **5** (1988) 18.
3. R. R. Akhmetshin *et al.* [CMD-2 Collaboration], Phys. Lett. B **527**, 161 (2002) [arXiv:hep-ex/0112031].
   R. R. Akhmetshin *et al.* [CMD-2 Collaboration], Phys. Lett. B **578**, 285 (2004) [arXiv:hep-ex/0308008].
4. A. Aloisio *et al.* [KLOE Collaboration], Phys. Lett. B **606**, 12 (2005) [arXiv:hep-ex/0407048].
5. M. N. Achasov *et al.*, arXiv:hep-ex/0506076.
6. R. R. Akhmetshin *et al.* [CMD-2 Collaboration], Phys. Lett. B **595**, 101 (2004) [arXiv:hep-ex/0404019].
7. M. N. Achasov *et al.*, J. Exp. Theor. Phys. **96**, 789 (2003) [Zh. Eksp. Teor. Fiz. **123**, 899 (2003)].
8. R. R. Akhmetshin *et al.*, Phys. Lett. B **551**, 27 (2003) [arXiv:hep-ex/0211004].
9. R. R. Akhmetshin *et al.* [CMD-2 Collaboration], Phys. Lett. B **466**, 385 (1999) [Erratum-ibid. B **508**, 217 (2001)] [arXiv:hep-ex/9906032].
10. B. Aubert *et al.* [BABAR Collaboration], Phys. Rev. D **70**, 072004 (2004) [arXiv:hep-ex/0408078].
11. R. R. Akhmetshin *et al.* [CMD-2 Collaboration], Phys. Lett. B **489**, 125 (2000) [arXiv:hep-ex/0009013].
12. A. Antonelli *et al.* [DM2 Collaboration], Z. Phys. C **56**, 15 (1992).
13. M. N. Achasov *et al.*, Phys. Rev. D **63**, 072002 (2001).
    M. N. Achasov *et al.*, Proc. of the Int. Workshop "e^+e^- collisions from ϕ to J/ψ", Novosibirsk, 1999, p.196.
14. P. A. Lukin [CMD-2 Collaboration], *Prepared for DAFNE 2004: Workshop on Physics at Meson Factories, Rome, Frascati, Italy, 7-11 Jun 2004*
15. P. M. Ivanov *et al.*, Phys. Lett. B **107**, 297 (1981).

Preliminary Results with the Crystall Ball at MAMI

P. Pedroni

INFN-Sezione di Pavia, via Bassi 6 Pavia, Italy
(for the CB@MAMI and A2 collaborations)

Abstract. A new set of experiments on meson photoproduction up to beam energies of 1.5 GeV is being planned for the MAMI-C accelerator in Mainz (Germany).

The experimental apparatus consists of the Crystal Ball multiphoton spectrometer together with the TAPS detector as a forward wall and a central tracker based on the DAPHNE-type cylindrical wire chambers and a central tracker. This configuration provides a geometrical acceptance approaching the full acceptance with good energy and angular resolutions for photons as well as a good angular resolution for charged particles.

Using this set-up, a first round of experiments have already been performed at the existing MAMI-B facility (photon energies up to 820 MeV) to determine basic properties of hadrons like the magnetic moment of the $\Delta^+(1232)$ resonance, branching ratios/upper limits for various rare and forbidden eta-meson decays, test of low-energy theorems for single pion photoproduction.

Keywords: large acceptance detector, pion and η photoproduction
PACS: 13.60.Le, 13.20.-v, 14.40.Aq

INTRODUCTION

A new experimental program has started at the Mainz MAMI microtron facility during the last year. The physics goal is to perform a series of precision experiments on non-perturbative QCD. We intend to make sensitive tests of Chiral perturbation theory, effective Lagrangian models such as the flux-tube quark model and various Lattice QCD computations.

Highlights of the program include:

- threshold strangeness and threshold ω production.
- $\eta, \omega, 3\pi^0, 2\pi^0, \pi^0$ production to measure the radiative couplings of N^* and Δ resonances.
- radiative pion (η) production to determine the magnetic dipole moment of the $\Delta^+(1232)$ ($S_{11}(1535)$) resonance.
- measurements of branching ratios/uper limits for various rare and forbidden η decays to test chiral perturbation theories and C abd CP invariance
- double polarization experiments using linearly or circularly polarized photons on a longitudinally polarized proton/deuteron target. They are a unique probe for an accurate study of the baryon resonance properties.
- coherent π^0 and $2\pi^0$, η and ω production on complex targets to investigate medium modifications.

CP814 *Hadron Spectroscopy: Eleventh International Conference*
edited by A. Reis, C. Göbel, J. Sá Borges, and J. Magnin

The feasibility of all these measurements will become possible by combining a (nearly) 4π detection apparatus with a high quality polarized tagged photon beam.

In the following an overview of the experimental apparatus will be given together with some of the preliminary results obtained up to now during the first round of experiments.

EXPERIMENTAL SETUP

The experimental program outlined in the previous section will be carried out at the tagged photon facility of the MAMI accelerator in Mainz. The maximum photon energy available with the existing MAMI-B machine is $E_\gamma^{max} = 820$ MeV.

The photon energy is determined by the Gasgow-Mainz tagging spectrometer [1] which has an energy resolution of about 2 MeV for an incoming elecron beam of 855 MeV. The resolution can be improved at least by a factor of three using the tagger microscope, an array of small scintillators covering about 80 MeV of the photon bremsstrahlung spectrum.

The maximum photon flux per unit energy bin that can be achieved is $N_\gamma^{max} \sim 5 \cdot 10^5 \mathrm{s}^{-1}\mathrm{MeV}^{-1}$, which is much higher than all other existing tagger photon facilities.

During the past years, a degree of polarization of 60% for linearly and 80% for circularly polarized have been regularly achieved.

A Harmonic Double Sided Microtron (HDSM) is presently under construction to increase the maximum energy of the MAMI-B accelerator. The energy upgraded machine (MAMI-C) will provide $E_\gamma^{max} = 1.5$ GeV, which corresponds to $\sqrt{s} = 1.92$ GeV. This new accelerator is scheduled to come into operation in 2006.

A new Bonn-type forzen spin polarized target, very similar to the one that has successfully worked during the GDH experiment [2], is under construction in Mainz. The target will provide longitudinally polarized protons and deuterons with maximum polarization values of about 85% and 65% respectively. Its horizontal cryostat is ϕ symmetric and allows measurements at $\theta_{lab} < 160°$. This new target will be operational during 2007.

A gas, high pressure polarized ^{3}He target is also in development as a complement of the polarized deuterium target for the evaluation of the helicity dependent reactions on the free neutron.

The experimental detector that will be used is based on three major components (see fig. 1):

- The Crystal Ball (CB) multiphoton spectrometer [3, 4] which is used as a central detector. This device was build at SLAC and used in J/ψ measurements at SPEAR and b–quark physics at DESY [5].
  It consists of 672 optically isolated NaI(Tl) crystals, 15.7 radiation lengths thick. These counters are arranged in a spherical shell with an inner radius of 25.3 cm and an outer radius of 66.0 cm. The detector has an entrance and exit tunnel for the beam and a spherical cavity for the reaction target. The hygroscopic NaI is housed in two hermetically sealed evacuated hemipsheres Each crystal is shaped like a truncated triangular pyramid, 40.6 cm high, pointing towards the center of the Ball. The sides on the inner end are 5.1 cm long ans 12.7 cm on the far end.

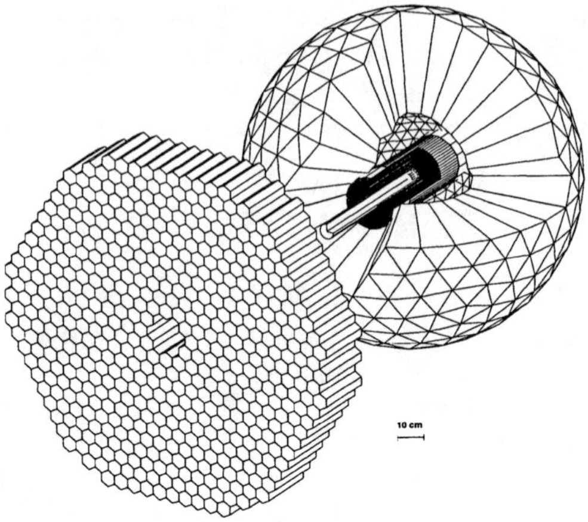

FIGURE 1. Schematic view of the experimental setup used in Mainz. Some of the crystals have been omitted from the Crystall Ball detector to show the position of both the PID-MWPC tracker detector and the target.

Electromagnetic showers in the spectrometer are measured with an energy resolution $\sigma_E \sim 1.7\%(E(GeV))^{0.4}$; the angular resolution for photon showers at energies of 0.05-0.5 GeV is $\sigma_\theta 2^\circ - 3^\circ$ in the polar angle and $\sigma_\phi 2^\circ/\sin(\theta)$ in the azimuthal angle. The Crystal Ball

- The TAPS detector ([6]), which is used as a forward wall. TAPS consists of 509 individual barium fluoride (BaF_2) crystals which are hexagonally shaped with an inscribed diameter of 59 mm and are 250 mm 812 radiation lengths) long.
- A central tracker detector formed by a barrel of 24 narrow scintillator strips (PID) located inside a set of two coaxial multiwire proportional cylindrical wire chambers (MWPC) that were developed for the DAPHNE large-acceptance detector [7]. The chambers provide a track reconstruction effciency of $\sim 80\%$ for relativistic $\pi^\pm$ and $\sim 90\%$ for protons. The resolutions for the polar and the azimuthal angles are respectively $\sigma_\theta \sim 2^\circ$ and $\sigma_\phi \sim 4^\circ$.

The high granularity and the large acceptance (97% of 4π) make this apparatus a unique instrument for measuring reactions with multiphoton final states. In addition, the tracker detector allows the identification and the energy measurement for protons and $\pi^\pm$ that are stopped inside the CB detector.

PRELIMINARY RESULTS

The first production runs using the MAMI-B facility were accomplished from mid-2004 to mid-2005 and mainly aimed at the measurement of i) the photon asymmetry in π^0 photoproduction at threshold; ii) the magnetic dipole moment of the $\Delta^+(1232)$ resonance; iii) the branchng rations for the $\eta \to 3\pi^0$ and $\eta \to \gamma\gamma\pi^0$ decay modes.

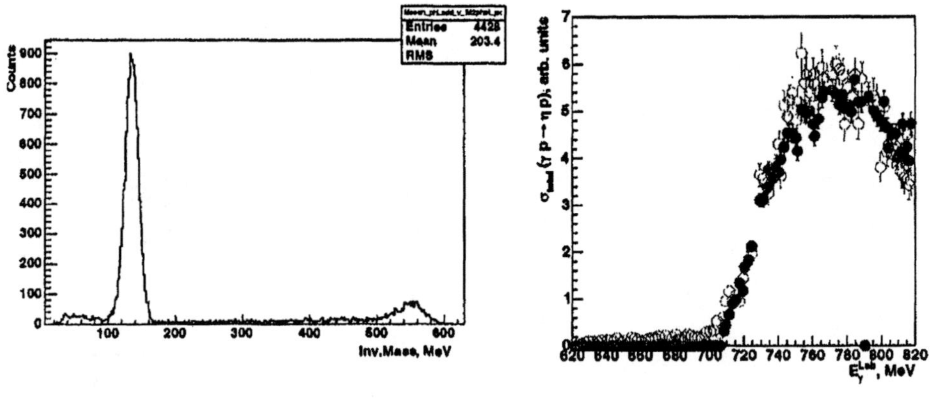

FIGURE 2. Left: Invariant mass of the two-cluster events from beam photons with energy above 700 MeV after requiring the missing mass to be equal to the mass of the proton. The peaks are due to $\pi^0 \to 2\gamma$ and $\eta \to 2\gamma$ decays. Right: The measured yield for the $\gamma p \to p\eta \to p2\gamma$ reaction (solid circles) is compared to measured yield of the events from the $\gamma p \to p3\pi^0$ reactions which are mostly coming from the $\eta \to 3\pi^0$ decay.

Figures 2- 3 illustrate the quality of the data, showing the most characteristic distributions.

The invariant mass of two photons for incident beam photon energies above 700 MeV is shown on the left part of Fig. 2. The two peaks of the spectrum are due to the reactions $\gamma p \to \pi^0(\gamma\gamma)p$ and $\gamma p \to \eta(\gamma\gamma)p$: the 2γ invariant mass is shown for events with the missing mass equal to the mass of the proton. The left part of Fig. 2 shows the measured yield of events from the $\gamma p \to 3\pi^0$ reactions. Below the η production threshold, $3\pi^0$ can be produced by sequential decay of baryon resonances, while above the η threshold essentially all events are produced via the $\eta \to 3\pi^0$ decay mode. The two distributions are normalized at E_γ =730 MeV. The very good agreement between the two differenr experimental shapes is an indication that the detector response is well under control.

The experimental setup used in Mainz is almost perfectly ϕ-symmetric. This feature, combined with ith the good properties of the tagged photon beam in Mainz (low divergence, high stability, high flux) allows polarization measurements with a high statistical and systematical precision.

As an example, the left part of Fig. 3 shows the ϕ dependence of the measured beam photon asymmetry $\Sigma(\phi)$ (not corrected for the beam polarization percentage) for the $\gamma p \to p\pi^0$ reaction. These data, which cover the beam energy interval from 360 to 450 MeV and are integrated over the full θ_π^0 range, represent only a few percent of the obtained statistics. They clearly show the the expected $\cos(2\phi)$ behavior over the full angular range.

In the right part of Fig. 3 are shown the obtained photon asymmetries as a function of $\cos(\theta_{\pi^0}$ in the cms system for some different photon energy in comparison with the MAID [8] model predictions. Also in this case the good agreement with the preliminary

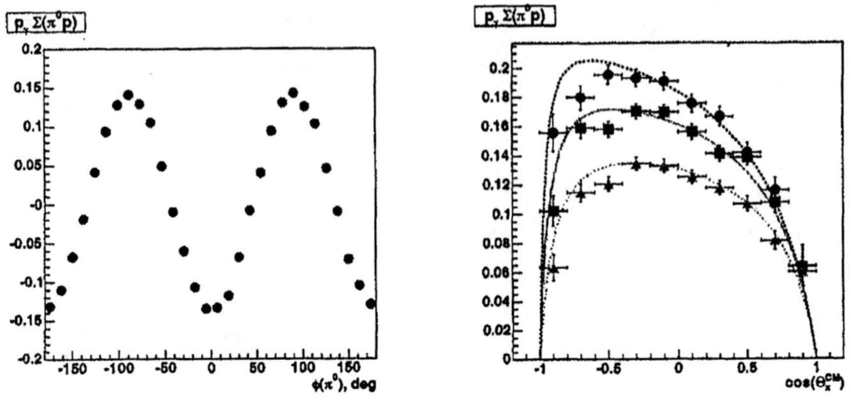

FIGURE 3. Left: ϕ dependence of the beam photon asymmetry for the $\gamma p \rightarrow p\pi^0$ reaction. Right: The preliminary results for the photon asymmetry as a function of cos(θ^0_π in the cms frame for E_γ =375 MeV (triangles), 405 MeV (squares) and 436 MeV (circles). are compared to MAID predictions.

experimental results indicates the high quality of the the collected data

CONCLUSIONS

A unique photonuclear physics program becomes possible when combining a very large acceptance apparatus with tagged and polarized photon beams and with polarized proton and deuterium targets. The new, high-quality experimental facility available at Mainz will allow, among others, the first-ever measurement of the magnetic dipole moments of the $\Delta^+(1232)$ and $S_{11}(1535)$ resonances, precise threshold measurement and a variety of double polarization experiments on the nucleon.

A first round of experiments has been completed using the existing MAMI-B beam ($E_\gamma^{max} = 820$ MeV). The next part of the program will start in 2006 with the new 1.5 GeV MAMI-C machine.

REFERENCES

1. I. Anthony *et al.*, *Nucl. Instrum. Methods A* **301**, 230 (1991).; S. J. Hall et al., *Nucl. Instrum. Methods A* **368**, 698 (1996)..
2. C. Bradtke *et al.*, *Nucl. Instrum. Methods A* **436**, 430 (1999)..
3. A. Starostin *et al.*, *Phys. Rev. C* **64**, 055205 (2001).
4. S. Prakhov *et al.*, *Phys. Rev. C* **64**, 045202 (2004).
5. E.D Bloom and C.W. Peck, *Ann. Rev. Nucl. Sci.* **33**, 143 (1983)..
6. R. Novotny *et al.*, *IEEE Trans. Nucl. Sci.* **38**, 379 (1991).
7. G. Audit *et. al.*, *Nucl. Instrum. Methods A* **301**, 473 (1991)..
8. D. Drechsel *et al.*, *Phys. Rev. D* **63**, 114010 (2001)..

NEW HADRONS

Spectroscopy and new particles

Tadeusz Lesiak [1]

Institute of Nuclear Physics PAN
Radzikowskiego 152, 31-142 Kraków, Poland

Abstract. The properties of three new particles: X(3872), Y(3940) and Z(3931), recently discovered by the Belle collaboration, are briefly reviewed. Negative results of the search for the pentaquark $\Theta(1540)^+$ are also presented.

Keywords: hadron spectroscopy, charmonium, pentaquark
PACS: PACS numbers: 14.40.Lb, 13.25.Ft, 13.25.Gv, 13.20.Jf

INTRODUCTION

In the last two years the Belle collaboration has provided evidence for several new hadrons. This paper presents the observation of three new particles: X(3872), Y(3940) and Z(3931). The other new states discovered recently by Belle were also discussed at this conference in separate talks [1, 2]. The paper also presents the results of searches for the pentaquark state $\Theta(1540)^+$. The Belle detector at the KEKB asymmetric e^+e^- collider [3] is a general purpose spectrometer, described in detail in [4].

PROPERTIES OF THE X(3872)

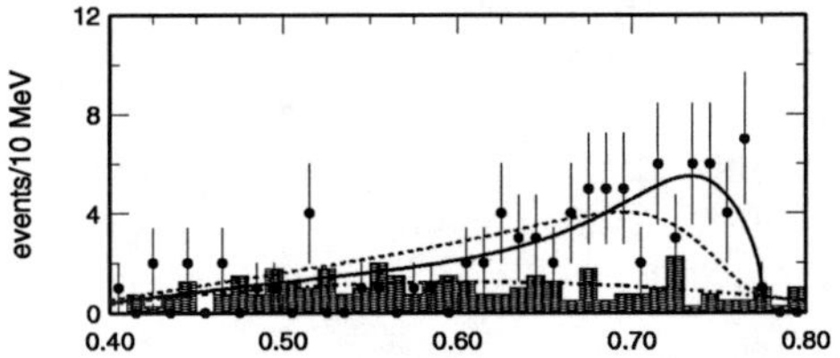

FIGURE 1. The distribution of $\pi^+\pi^-$ invariant mass for events in the $X(3872) \to \pi^+\pi^- J/\psi$ signal region (data points). The shaded histogram corresponds to the background as determined by the X-mass sidebands. The solid (dashed) curve denotes a fit that uses a ρ Breit-Wigner shape with the ρ and J/ψ in a relative $S(P)$-wave. The dash-dotted curve shows a smooth parametrization of the background that is used in the fit.

The state X(3872) was discovered by the Belle collaboration in 2003 [5] by analyzing exclusive decays $B^+ \to \pi^+\pi^- J/\psi K^+$ (charge conjugate modes are included every-

[1] partially supported by the KBN grant No. 2P03B 01324

CP814 *Hadron Spectroscopy: Eleventh International Conference*
edited by A. Reis, C. Göbel, J. Sá Borges, and J. Magnin

where, unless otherwise specified). The B mesons were reconstructed using two kinematical variables: the energy offset $\Delta E = \sum_i E_i - E_{beam}$ and the beam-constrained mass $M_{bc} = \sqrt{E_{beam}^2 - \sum_i(\vec{p}_i)^2}$, where E_i and $\vec{p}_i$ are the center-of-mass (CMS) energies and momenta of the selected B meson decay products and E_{beam} is the CMS beam energy. A very narrow peak in the invariant mass spectrum of the system $\pi^+\pi^-J/\psi$ was observed with a mass of $3872.0 \pm 0.6 \pm 0.5$ MeV/c^2 and a width below 2.3 MeV (90% C.L.).

The observation of X(3872) was very quickly confirmed by the CDF [6], D0 [7] and BaBar [8] experiments. The observed decay mode $X \to \pi^+\pi^-J/\psi$ seemed to favour the explanation of the X(3872) as an excited charmonium state [9, 10]. However, its properties, in particular the very narrow width, did not allow the identification of the X(3872) with any $c\bar{c}$ state. At the same time the coincidence of the X mass with the $D^0\bar{D}^{*0}$ threshold (3871.3 ± 1.0) MeV/c^2) has prompted many theoretical speculations that X(3872) may be a so-called deuson i.e. a loosely bound molecular state of these two mesons [11, 12]. Moreover, the $\pi^+\pi^-$ invariant mass distribution (Fig. 1) was found to peak close to the upper kinematical limit of $M(\pi^+\pi^-)$ as expected for pion-pairs originating from $\rho \to \pi^+\pi^-$ decays.

Recently, the Belle collaboration, using the 253 fb^{-1} data sample collected at the $\Upsilon(4S)$ resonance, has provided the first evidence for two new decay modes: $X \to \gamma J/\psi$ and $X \to \pi^+\pi^-\pi^0 J/\psi$ [13]. They were observed in exclusive B meson decays to the final states $\gamma J/\psi K$ and $\pi^+\pi^-\pi^0 J/\psi K$, respectively. The yield of the decay $B \to \gamma J/\psi K$ plotted in bins of the $\gamma J/\psi$ invariant mass (Fig. 2**a**)) exhibits an excess of 13.6 ± 4.4 events (statistical significance of 4σ). The observation of this decay establishes unambiguously that the charge-conjugation parity of the X(3872) is positive. The partial width ratio $\Gamma(X \to \gamma J/\psi)/\Gamma(X \to \pi^+\pi^-J/\psi)$ amounts to 0.14 ± 0.05. This result is in contradiction with the χ'_{c1} (1^{++} charmonium) assignment for X as in this case a value around 40 would be expected. The second decay mode $X \to \pi^+\pi^-\pi^0 J/\psi$ was found to be dominated by the sub-threshold decay $X \to \omega^* J/\psi$. This is motivated by the fact that the yield of B mesons plotted in bins of the $\pi^+\pi^-\pi^0$ invariant mass (Fig 2**b**)) inside of the signal region from the decay $X \to \pi^+\pi^-\pi^0 J/\psi$ is consistent with zero except for the $M(\pi^+\pi^-\pi^0) > 750$ MeV/c^2. There, the excess of 12.4 ± 4.1 events (4.3σ) is observed. The ratio of branching fractions B($X \to \pi^+\pi^-\pi^0 J/\psi$)/B($X \to \pi^+\pi^-J/\psi$).

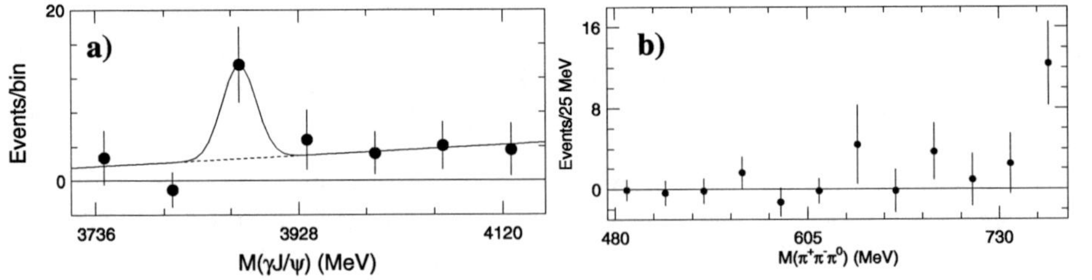

FIGURE 2. The yield of B mesons from the decay **a)** $B^0 \to \gamma J/\psi K$, in bins of the $\gamma J/\psi$ invariant mass and **b)** $B^0 \to \pi^+\pi^-\pi^0 J/\psi K$, in bins of the $\pi^+\pi^-\pi^0$ invariant mass, determined from fits to the ΔE and M_{bc} distributions.

was measured to be $1.0 \pm 0.4 \pm 0.3$, which implies a large violation of isospin symmetry.

The Belle collaboration also attempted to determine the J^{PC} quantum numbers of the X(3872) [14] by studying the angular distributions of the decay $X \to \pi^+\pi^- J/\psi$, as suggested by Rosner [15]. Among the twelve possible J^{PC} assignments, half (0^{--}, 0^{+-}, 1^{--}, 1^{+-}, 2^{--} and 2^{+-}) may be discarded due to their negative charge conjugation-parity. The value 1^{++} is in agreement with the data while the assignments 0^{-+} and 0^{++} are strongly disfavoured by the analysis of angular distributions [14]. The additional two odd-parity possibilities: 1^{-+} and 2^{-+} are discarded as for them the dipion invariant mass spectrum (Fig. 1) is expected to be much softer to compare with the data. On the other side, the distribution of $M(\pi^+\pi^+)$ is in agreement with the 1^{++} hypothesis. The assignment 2^{++} was strongly disfavoured by the recent, preliminary observation by Belle [16] of the decay $B \to KX$, $X \to D^0\overline{D^0}\pi^0$, where the 11.3 ± 3.6 signal events (5.6σ) concentrate close to the threshold for the final state $D^0\overline{D^0}\pi^0$. In the 2^{++} case, the decay of a spin 2 state to three pseudoscalars ($D^0\overline{D^0}\pi^0$) would require at least one pair of them to be in a relative D wave. In such a configuration the near threshold production would be strongly suppressed by a centrifugal barrier.

All the above observations strongly favour the assignment of $J^{PC} = 1^{++}$ to the X(3872). This matches the expectations of models [11, 12] interpreting the X as a $D^0\overline{D^{*0}}$ bound state. This hypothesis also explains the narrow width of the X(3872) and the shape of $\pi^+\pi^-$ and $\pi^+\pi^-\pi^0$ spectra in its corresponding decays and leads to the prediction of large isospin violation in the X decays. Among alternative interpretations are: a 'conventional' charmonium [9, 10, 17], glueball [18], tetraquark [19] or the so called cusp effect [20].

EVIDENCE FOR THE Y(3940)

In 2004 The Belle collaboration provided evidence for another new state, Y(3940), decaying to $\omega J/\psi$[21]. It was again observed in the B^+ meson exclusive decay to the final state $K^+\pi^+\pi^-\pi^0 J/\psi$. A fit to the $\omega J/\psi$ invariant-mass distribution (Fig.3) yielded a signal of 58 ± 11 events (8.1σ) corresponding to a mass of $3943 \pm 11 \pm 13$ MeV/c^2 and the width $87 \pm 22 \pm 26$ MeV. The Y mass coincides with that of another

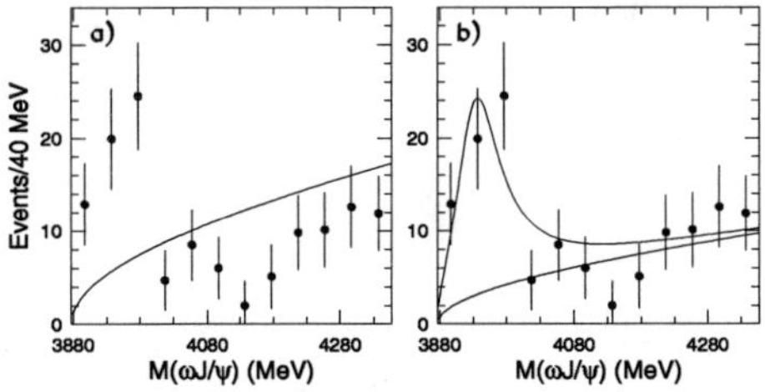

FIGURE 3. $B^+ \to K^+\omega J/\psi$ signal yields vs $M(\omega J/\psi)$. The curve in (a) shows the result of a fit that includes only a phase-space-like threshold function. The curve in (b) corresponds to the result of a fit that includes an S-wave Breit Wigner resonance term.

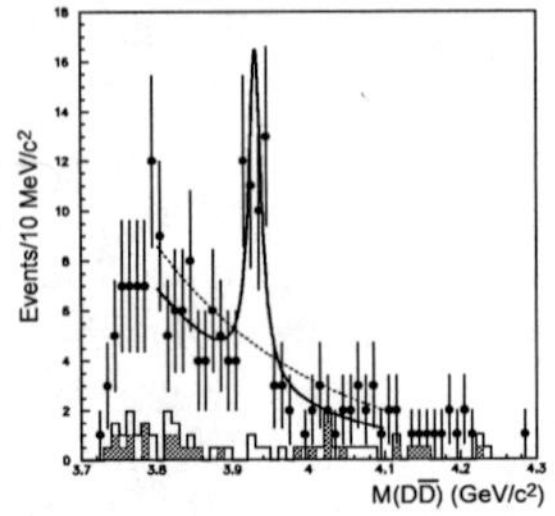

FIGURE 4. Invariant mass distribution of $D\overline{D}$ pairs. The solid (dashed) curve shows the fits with (without) a resonance component. The histograms correspond to the distribution of the events from the D-mass sidebands.

particle, X(3940), also observed by Belle [1, 2]. However, it is unlikely that these two states are the same, since the X(3940) decays to $D\bar{D}^*$ and does not decay to $\omega J/\psi$ and the situation is reversed for the Y(3940), as far as the above-mentioned decays are concerned. The properties of Y(3940) are similar to those expected for the $c\bar{c}-gluon$ hybrid mesons [22].

DISCOVERY OF THE Z(3931)

A recent search by the Belle collaboration for the production of new resonances in the process $\gamma\gamma \to D\overline{D}$ [23] yielded evidence for a new state (Fig. 4) at a mass of $3931 \pm 4 \pm 2$ MeV/c^2 and a width of $20 \pm 8 \pm 3$ MeV. A signal of 41 ± 11 events with a statistical significance of 5.5σ was observed. The properties of this new state match the expectations [24, 10] for the radially excited states χ'_{c0} and χ'_{c2}. A study of angular distributions of the D mesons in the $\gamma\gamma$ rest frame revealed that the data significantly prefer a spin two assignment over spin zero.

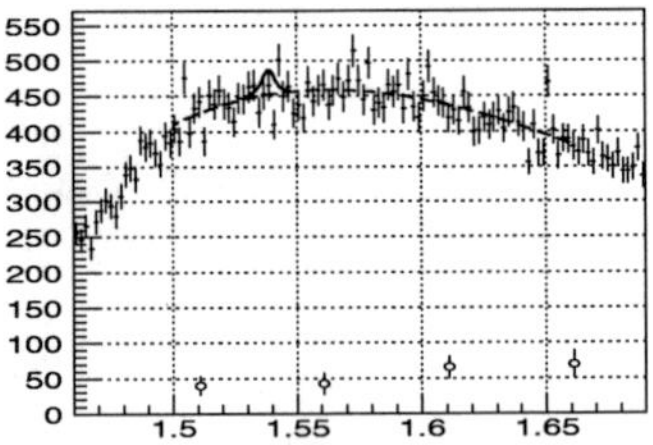

FIGURE 5. Invariant mass spectrum for secondary pK_s^0 pairs and expected yield of the charge exchange reaction per 2 MeV/c^2 (open dots). A fit to a third order polynomial is represented by the dashed curve. The $\Theta(1540)^+$ contribution expected from the DIANA result [26] is presented with the solid line.

SEARCH FOR THE $\Theta(1540)^+$

The Belle collaboration has searched for both inclusive and exclusive production of the $\Theta(1540)^+$ pentaquark using kaon secondary interactions in the material of the detector [25]. An upper limit of 2.5 % (90 % C.L.) was set on the ratio of the $\Theta(1540)^+$ to $\Lambda(1520)$ inclusive production cross section. The search for the exclusive production of the $\Theta(1540)^+$ as an intermediate resonance in the charge exchange reaction $K^+n \to pK_s^0$ yielded an upper limit of $\Gamma_{\Theta^+} < 0.64$ MeV (90 % C.L.) at $m_{\Theta^+} = 1.539$ MeV/c^2. This value is below the current Particle Data Group [27] value of 0.9 ± 0.3.

SUMMARY

The properties of three new particles: X(3872), Y(3940) and Z(3931), recently observed by the Belle collaboration,, were reviewed. For the X(3872) the observation of new decay modes together with angular analysis of the $\pi^+\pi^- J/\psi$ favours the assignment $J^{PC} = 1^{++}$ and is in agreement with the deuson hypothesis. The most plausible interpretations of the Y(3940) and Z(3931) are the $c\bar{c} - gluon$ and χ'_{c2}, respectively. The search for the $\Theta(1540)^+$ yielded a null result giving rise to the limit $\Gamma_{\Theta^+} < 0.64$ MeV (90 % C.L.).

REFERENCES

1. S. Korpar, these proceedings.
2. K. Trabelsi, these proceedings.
3. S. Kurokawa, E. Kikutani, *Nucl. Instrum. Methods A*, **499**, 1 (2003), and other papers incl. in this vol.
4. Belle Collaboration, A. Abashian et al., *Nucl. Instrum. Methods A*, **479**, 117 (2002).
5. Belle Collaboration, S.-K. Choi,S. L. Olsen et al., *Phys. Rev. Lett.*, **91**, 262001 (2003).
6. CDF-II Collaboration, D. Acosta et al., *Phys. Rev. Lett.*, **93**, 072001 (2004).
7. D0 Collaboration, V. M. Abazov et al., *Phys. Rev. Lett.*, **93**, 162002 (2004).
8. BaBar Collaboration, B. Aubert et al., *Phys. Rev. D*, **71**, 071003 (2004).
9. E. J. Eichten, K. Lane and C. Quigg, *Phys. Rev. Lett.*, **89**, 162002 (2002).
10. E. J. Eichten, K. Lane and C. Quigg, *Phys. Rev. D*, **69**, 094019 (2004).
11. N. A. Törnqvist, *Phys. Lett. B*, **590**, 209 (2004).
12. E. S. Swanson, *Phys. Lett. B*, **588**, 189 (2004).
13. Belle Collaboration, K. Abe et al., hep-ex/0505037.
14. Belle Collaboration, K. Abe et al., hep-ex/0505038.
15. J. Rosner, *Phys. Rev. D*, **70**, 094023 (2004).
16. P.Pakhlov, procedings of the HEP2005 International Europhysics Conf. on High Energy Physics.
17. T. Barnes, S. Godfrey, *Phys. Rev. D*, **69**, 054008 (2004).
18. K. K. Seth, *Phys. Lett. B*, **612**, 1 (2005).
19. L. Maiani et al, *Phys. Rev. D*, **71**, 014028 (2005).
20. D.Bugg, *Phys. Lett. B*, **598**, 8 (2004).
21. Belle Collaboration, S.-K. Choi,S. L. Olsen et al., *Phys. Rev. Lett.*, **94**, 182002 (2005).
22. F. E. Close, I. Dunietz et al., *Phys. Rev. D*, **57**, 5653 (1998);
G. Chiladze, A.F. Falk, A. A. Petrov, *Phys. Rev. D*, **58**, 034013 (1998);
F. E. Close and S. Godfrey, *Phys. Lett. B.*, **574**, 210 (2003).
23. Belle Collaboration, K. Abe et al., hep-ex/0507033.
24. S. Godfrey and N. Isgur, *Phys. Rev. D.*, **32**, 189 (1985).
25. Belle Collaboration, K. Abe et al., hep-ex/0507014, subm to *Phys. Lett. B.*
26. DIANA Collaboration, V. Barmin et al., *Phys. Atom. Nucl.*, **66**, 1715 (2003).
27. S. Eidelman et al., *Phys. Lett. B*, **592**, 1 (2004).

Studies of $D_{sJ}^{(*)}$ Production in B Decays and $e^+e^- \to c\bar{c}$ Events

V. Poireau (on behalf of the BaBar Collaboration)

Laboratoire de Physique des Particules, F-74941 Annecy-le-Vieux, France

Abstract. We report a study of $D_{sJ}^*(2317)^+$ and $D_{sJ}(2460)^+$ meson production in B decays. We observe and measure branching fractions for the decays $B^+ \to D_{sJ}^{(*)+}\bar{D}^{(*)0}$ and $B^0 \to D_{sJ}^{(*)+}\bar{D}^{(*)-}$ with the subsequent decays $D_{sJ}^*(2317)^+ \to D_s^+\pi^0$, $D_{sJ}(2460)^+ \to D_s^{*+}\pi^0$, and $D_{sJ}(2460)^+ \to D_s^+\gamma$. In addition, we perform an angular analysis of $D_{sJ}(2460)^+ \to D_s^+\gamma$ decays to test the different $D_{sJ}(2460)^+$ spin hypotheses.

From a dataset of $e^+e^- \to c\bar{c}$ events we measure the masses of the $D_{sJ}^*(2317)^+$ and $D_{sJ}(2460)^+$ mesons and the $D_{sJ}(2460)^+ \to D_s^+\gamma$ and $D_{sJ}(2460)^+ \to D_s^+\pi^+\pi^-$ branching fractions. A search is performed for neutral and doubly-charged partners.

We have also searched for the $D_{sJ}^*(2632)^+$ reported by the SELEX collaboration at FNAL. The resulting $D_s^+\eta$ and D^0K^+ mass spectra show no evidence for the $D_{sJ}^*(2632)^+$ state. In addition, no signal is observed in the $D^{*+}K_s$ mass spectrum.

All the above studies are performed on data samples collected with the BaBar detector at the SLAC PEP-II B factory.

Keywords: Spectroscopy, Hadron, New Resonance
PACS: 14.40.Lb

INTRODUCTION

The unexpected observation of a narrow $D_s^+\pi^0$ resonance with a mass of 2317 MeV/c^2 was recently reported by the BaBar collaboration [1] and confirmed by the CLEO experiment [2]. CLEO observed a second $D_s^{*+}\pi^0$ resonance with a mass close to 2460 MeV/c^2 [2], previously suggested [1] and later confirmed [3] by BaBar. The Belle collaboration confirmed both resonances and found two additional decay modes for the higher-mass state [4], $D_s^+\gamma$ and $D_s^+\pi^+\pi^-$. These resonances are usually interpreted as P-wave $c\bar{s}$ quark states [5, 6, 7, 8, 9], although other interpretations [10, 11, 12, 13, 14] cannot be ruled out, and will be referred to in the following as $D_{sJ}^*(2317)^+$ and $D_{sJ}(2460)^+$ mesons.

In the framework of the heavy quark effective theory, an analogy is made between the $c\bar{s}$ system and a hydrogen atom. The approximation consists to consider a light quark (s quark) spinning around a heavy quark (c quark), such as the heavy quark is regarded as fixed. In this case, the quantum number of the s quark, $j = s_s + L$, where s_s is the spin of the s quark and L the orbital quantum number, is considered as good quantum number. The total quantum number of the $c\bar{s}$ system is defined as $J = j + s_c$, where s_c is the spin of the c quark. In this context, 2 S-wave states are predicted, identified as the $D_s(1968)^+$ and the $D_s^*(2112)^+$ mesons, and 4 P-wave states are predicted, where the states with $j = 3/2$ are identified with the $D_{s1}(2536)^+$ and $D_{s2}^*(2573)^+$ mesons. The agreement

CP814 *Hadron Spectroscopy: Eleventh International Conference*
edited by A. Reis, C. Göbel, J. Sá Borges, and J. Magnin

between the mass prediction of this model [15] and the real masses of the resonances is good. The 2 P-wave states with $j = 1/2$ (with $J^P = 0^+, 1^+$) are predicted with a mass between 2400 and 2600 MeV/c^2, and with a very large width (with a decay to $D^{(*)}K$).

This contradicts with the smaller masses and the narrow widths of the $D_{sJ}^*(2317)^+$ and $D_{sJ}(2460)^+$ mesons. It is then very important to measure further the properties of these 2 new resonances, and in particular to try to measure their quantum number.

The analyses presented below were performed using a 113 fb^{-1} and a 125 fb^{-1} data sample (depending if off-resonance data were used) collected on or near the $\Upsilon(4S)$ resonance with the BaBar detector at the PEP-II asymmetric-energy e^+e^- storage ring. The BaBar detector, a general-purpose, solenoidal, magnetic spectrometer, is described in detail elsewhere [16].

STUDY OF $B \to D_{SJ}^{(*)}\bar{D}^{(*)}$ DECAYS

The new states were first observed in $e^+e^- \to c\bar{c}$ collisions. Their observation in exclusive $B \to D_{sJ}^{(*)}\bar{D}^{(*)}$ decays allows additional properties of the $D_{sJ}^{(*)}$ state to be studied: the helicity angle distribution in B decays can be used to obtain information on the spin J, and the measurement of the different branching fractions can help clarify the nature of these states. This analysis [17] considers the production modes $B^+ \to D_{sJ}^{(*)+}\bar{D}^{(*)0}$ and $B^0 \to D_{sJ}^{(*)+}\bar{D}^{(*)-}$ with the subsequent decays $D_{sJ}^*(2317)^+ \to D_s^+\pi^0$, $D_{sJ}(2460)^+ \to D_s^{*+}\pi^0$, and $D_{sJ}(2460)^+ \to D_s^+\gamma$. After reconstruction of the $\bar{D}^{(*)}$ and $D_s^{(*)+}$ mesons, these candidates are combined with a photon or a π^0 to form B candidates. A B signal region is defined in terms of the beam energy substituted mass, $m_{ES} \equiv \sqrt{s/4 - p_B^{*2}}$, and the difference between the reconstructed energy of the B candidate and the beam energy, $\Delta E \equiv E_B^* - \sqrt{s}/2$, where $\sqrt{s}$ is the total energy in the $\Upsilon(4S)$ center-of-mass frame and E_B^* (p_B^*) is the energy (momentum) of the B candidate in the same frame.Different cuts are defined on these variables depending on the final state, and only one B candidate is selected per event. Signals are observed in the 3 decay modes: 88 ± 17 events with a mass of 2317.2 ± 1.3 MeV/c^2 for the $D_{sJ}^*(2317)^+\bar{D}^{(*)}$ $[D_s^+\pi^0]$ decay mode, 112 ± 14 events with a mass of 2458.9 ± 1.5 MeV/c^2 for the $D_{sJ}(2460)^+\bar{D}^{(*)}$ $[D_s^{*+}\pi^0]$ decay mode and 139 ± 17 events with a mass of 2461.1 ± 1.6 MeV/c^2 for the $D_{sJ}(2460)^+\bar{D}^{(*)}$ $[D_s^+\gamma]$ decay mode. Signals are also observed for each of the twelve $D_s^{(*)+}\bar{D}^{(*)}\pi^0/\gamma$ final states. A significance larger than 4 is observed for 10 of the 12 modes. From the $D_{sJ}^{(*)+}$ event yields in the data, the cross-feed-corrected branching fractions are computed, using the signal efficiency and the relative contributions from cross-feed between the different $D_{sJ}^{(*)+}$ decay modes as obtained from simulated signal events.

A helicity analysis of the $D_{sJ}(2460)^+$ state has been performed, using the decays $B \to D_{sJ}(2460)^+\bar{D}$ with $D_{sJ}(2460)^+ \to D_s^+\gamma$. The helicity angle θ_h is defined as the angle between the $D_{sJ}^{(*)+}$ momentum in the B meson rest frame and the D_s momentum in the $D_{sJ}^{(*)+}$ rest frame. A fit of the $D_s\gamma$ invariant mass is performed for five different $cos(\theta_h)$ regions. A good agreement is found with the $J = 1$ hypothesis.

STUDY OF $D^*_{sJ}(2317)^+$ AND $D_{sJ}(2460)^+$ IN CONTINUUM PRODUCTION

The properties of the $D^*_{sJ}(2317)^+$ and $D_{sJ}(2460)^+$ mesons are studied using $e^+e^- \to c\bar{c}$ events [18]. Searches are performed for the decay to the D_s^+ meson along with one or more π^0, π^+, or γ particles. A search is also performed for neutral and doubly-charged partners.

In this analysis, each D_s^+ candidate is constructed by combining a K^+K^- candidate pair with a π^+ candidate in a geometrical fit to a common vertex. Once a D_s^+ candidate is obtained, a search is performed for all accompanying π^0, γ and $\pi^\pm$ particles. Each final state is restricted to the same minimum center-of-mass momentum (p^*) value of 3.2 GeV/c^2.

The kinematic is complex, and leads to competing contributions and mutual cross-feeds between different modes.

$D_s^+\pi^0$ **final states:** to form $D_s^+\pi^0$ combinations, each D_s^+ candidate is combined with one π^0 candidate (the π^0 momentum is required to be greater than 400 MeV/c). A clear peak signal is seen at the mass of the $D^*_{sJ}(2317)^+$ resonance. Three types of background are present: the combinatorial background, background coming from the contribution of $D_s^*(2112)^+ \to D_s^+\gamma$ (where an unassociated γ forms a fake π^0 candidate), and background coming from the contribution of $D_{sJ}(2460)^+ \to D_s^*(2112)^+\pi^0$. These background contributions must be accurately determined from the simulation in order to extract the properties of the $D^*_{sJ}(2317)^+$. After taking into account these contributions, the results give a $D^*_{sJ}(2317)^+$ mass of 2318.9 $\pm$ 0.3 MeV/c^2, and $D^*_{sJ}(2317)^+$ and $D_{sJ}(2460)^+$ yields of 1275 $\pm$ 45 and 3 $\pm$ 26 mesons (statistical errors only).

$D_s^+\gamma$ **final states:** to form $D_s^+\gamma$ combinations, each D_s^+ candidate is combined with γ candidate with energy greater than 500 MeV/c. A clear peak is seen in the invariant mass distribution for the $D_{sJ}(2460)^+$. There is also a lower mass structure composed of $D^*_{sJ}(2317)^+ \to D_s^+\pi^0$ and of $D_{sJ}(2460)^+ \to D_s^+\pi^0\gamma$, which could be described by the simulation. A $D_{sJ}(2460)^+$ mass of 2457.2 $\pm$ 1.6 MeV/c^2 and yield of 509 $\pm$ 46 mesons is obtained (statistical errors only). The fit, which allows the signal yield to fluctuate to negative values, obtains -107 $\pm$ 84 $D^*_{sJ}(2317)^+$ decays.

$D_s^+\pi^0\gamma$ **final states:** D_s^+, π^0 and γ particles are combined together, with the requirement that the minimum π^0 momentum is 400 MeV/c and that the minimum γ energy is 135 MeV. The $D_{sJ}(2460)^+$ signal can be better isolated by requiring the $D_s^+\gamma$ invariant mass to reside within 2 MeV/c^2 of the $D_s^*(2112)^+$ mass. This procedure isolates clearly the $D_{sJ}(2460)^+$ signal, but also introduces some peaking background in the invariant mass distribution ($D_s^*(2112)^+ \to D_s^+\gamma$ and $D^*_{sJ}(2317)^+ \to D_s^+\pi^0$). A fit is then performed on this distribution, taking into account all contributions: the fit obtains a $D_{sJ}(2460)^+$ mass of 2459.1 $\pm$ 1.3 MeV/c^2 and yield of 292 $\pm$ 29 mesons (statistical errors only). It has also been shown that the decay $D_{sJ}(2460)^+ \to D_s^+\pi^0\gamma$ can be successfully described as proceeding entirely through the channel $D_s^*(2112)^+\pi^0$.

$D_s^+\pi^+\pi^-$ **final states:** to form $D_s^+\pi^+\pi^-$ candidates, each D_s^+ is combined with π^+ and π^- candidates with momentum above 230 MeV/c. The resulting invariant mass distribution has two distinct, narrow peaks, which correspond to the decays of the

$D_{sJ}(2460)^+$ and $D_{s1}(2536)^+$ mesons. The result of the fit of the invariant mass is a $D^*_{sJ}(2317)^+$ yield of 0.6 ± 1.8 decays; a $D_{sJ}(2460)^+$ mass and yield of 2460.1 ± 0.3 MeV/c^2 and 67 ± 11 decays.

$D_s^+\pi^\pm$ **final states:** there has been some conjecture [14, 10] that the $D^*_{sJ}(2317)^+$ may be a four-quark hybrid state. It might be expected, if it was true, that neutral and doubly-charged partners should exist with a similar mass. The $D_s^+\pi^\pm$ system can be used to test this possibility. To form $D_s^+\pi^\pm$, each D_s^+ candidate is combined with $\pi^\pm$ candidates with momentum greater than 300 MeV/c. No resonant structure is observed in the resulting mass distribution.

After taking into account the systematic errors, and combining the different results of this analysis, the $D^*_{sJ}(2317)^+$ and $D_{sJ}(2460)^+$ masses are measured to be 2318.9 ± 0.3 (stat.) $\pm$ 0.9 (syst.) MeV/c^2 and 2459.4 ± 0.3 (stat.) $\pm$ 1.0 (syst.) MeV/c^2, respectively.

SEARCH FOR THE $D^*_{SJ}(2632)^+$

The SELEX collaboration at FNAL has recently reported the existence of a narrow state at a mass of 2632 MeV/c^2 decaying to $D_s^+\eta$ [20]. Evidence for the same state in the corresponding D^0K^+ mass spectrum was also presented. In the present analysis [19], inclusive production of the $D_s^+\eta$, D^0K^+ and $D^{*+}K_s$ systems in $e^+e^- \to c\bar{c}$ collisions is investigated in a search for the $D^*_{sJ}(2632)^+$ state.

Search for $D^*_{sJ}(2632)^+ \to D_s^+\eta$: each D_s^+ candidate is constructed by combining a K^+K^- candidate pair with a π^+ candidate. For events containing a D_s^+ candidate, η candidates are selected in the $\gamma\gamma$ decay mode. The precise cuts for this reconstruction are defined in the original paper [19]. The center-of-mass momentum $p^*(D_s^+\eta)$ of the $D_s^+\eta$ system is required to be at least 2.5 GeV/c to suppress background. In order to establish the presence of an excess of events in the correlated D_s^+ and η production, a two-dimensional subtraction is performed in the $D_s^+ - \eta$ invariant mass plane. After this subtraction is done, and looking at the invariant mass of the $D_s^+\eta$ system, no evidence of a signal has been found.

Search for $D^*_{sJ}(2632)^+ \to D^0K^+$: a D^0 candidate is constructed by combining a $\pi^+ - K^-$ pair in a geometric fit to a common vertex. A good D^0 candidate is combined with a well-identified K^+ track in a fit to a common vertex. The D^0K^+ mass spectrum is obtained after requiring $p^*(D^0K^+) > 4.0$ GeV/c. There is no evidence for structure in the 2.632 GeV/c^2 mass region.

Search for $D^*_{sJ}(2632)^+ \to D^{*+}K_s$: a D^0 candidate with mass within 25 MeV/c^2 of the central mass value is combined with a well-identified π^+ track in a fit to a common vertex. A candidate K_s track is reconstructed by vertexing a well-identified $\pi^+\pi^-$ pair. The center-of-mass momentum of the $D^{*+}K_s$ system is required to be greater than 4 GeV/c. There is no evidence for production of the $D^*_{sJ}(2632)^+$ state in the data.

CONCLUSION

Given the previous results, it is possible to make some assumptions on the spin of the $D^*_{sJ}(2317)^+$ and $D_{sJ}(2460)^+$ resonances. The J^P quantum numbers of the $D^*_{sJ}(2317)^+$

resonance is probably 0^+ since it does not decay to $D_s^*\pi^0$, to $D_s\gamma$ and to $D_s^*\pi^+\pi^-$, as expected for a 0^+ state. The $D_{sJ}(2460)^+$ is probably a 1^+ state since it decays to $D_s\gamma$ and $D_s\pi^+\pi^-$, and does not decay to $D_s\pi^0$ and DK. In addition, the helicity measurement in the exclusive analysis confirms this hypothesis.

Thus, on the experimental point of view, one natural possibility would be to identify the $D_{sJ}^*(2317)^+$ and $D_{sJ}(2460)^+$ with the two missing states 0^+ and 1^+ of the $c\bar{s}$ system.

In summary, the properties of the resonances $D_{sJ}^*(2317)^+$ and $D_{sJ}(2460)^+$ were studied. Masses, spin assignments, decay modes and branching fractions were determined using the data collected by the BaBar detector. In addition, no evidence for the $D_{sJ}^*(2632)^+$ state was found.

ACKNOWLEDGMENTS

The author is grateful for the excellent luminosity and machine conditions provided by our PEP-II colleague, and for the substantial dedicated effort from the computing organizations that support BaBar. The collaborating institutions wish to thank SLAC for its support and kind hospitality.

The author would like also to thank the organizers for this excellent conference on hadron spectroscopy.

REFERENCES

1. B. Aubert *et al.*, Phys. Rev. Lett. **90**, 242001 (2003).
2. D. Besson *et al.*, Phys. Rev. D **68**, 032002 (2003).
3. B. Aubert *et al.*, Phys. Rev. D **69**, 031101 (2004).
4. K. Abe *et al.*, Phys. Rev. Lett. **92**, 012002 (2004).
5. R. N. Cahn and J. D. Jackson, Phys. Rev D **68**, 037502 (2003).
6. W. A. Bardeen, E. J. Eichten and C. T. Hill, Phys. Rev D **68**, 054024 (2003).
7. S. Godfrey, Phys. Lett. B **568**, 254 (2003).
8. Y. B. Dai, C. S. Huang, C. Liu and S. L. Zhu, Phys. Rev D **68**, 114011 (2003).
9. A. Deandrea, G. Nardulli and A. D. Polosa, Phys. Rev D **68**, 097501 (2003).
10. T. Barnes, F. E. Close and H. J. Lipkin, Phys. Rev D **68**, 054006 (2003).
11. E. van Beveren and G. Rupp, Phys. Rev. Lett. **91**, 012003 (2003).
12. H. Y. Cheng and W. S. Hou, Phys. Lett. B **566**, 193 (2003).
13. A. P. Szczepaniak, Phys. Lett. B **567**, 23 (2003).
14. T. E. Browder, S. Pakvasa and A. A. Petrov, Phys. Lett. B **578**, 365 (2004).
15. S. Godfrey, N. Isgur, Phys.Rev. D **32**, 189-231 (1985).
16. B. Aubert *et al.*, Nucl. Instrum. Methods **A479**, 1 (2002).
17. B. Aubert *et al.*, Phys. Rev. Lett. **93**, 181801 (2004).
18. B. Aubert *et al.*, hep-ex/0408067.
19. B. Aubert *et al.*, hep-ex/0408087.
20. A. V. Evdokimov *et al.*, Phys. Rev. Lett. **93**, 242001 (2004).

The $\tilde{U}(12)$-Classification Scheme, Static $U(4)$-Spin Symmetry for Light-Quarks and "Exotic" Hadrons

Shin Ishida[1]

Research Institute of Science and Technology, College of Science and Technology, Nihon University, Tokyo 101-8308, Japan

Abstract. Several years ago we have proposed a manifestly covariant $\tilde{U}(12)_{SF}$-classification scheme of hadrons, which maintains the successful $SU(6)_{SF} \otimes O(3)_L$ framework in Non-Relativistic Quark Model and is also reconcilable with the mechanism of Spontaneously-Broken Chiral Symmetry in Relativistic Field Theoretical Models. The essential point here is to notice the overlooked freedom of the $SU(2)_\rho$-spin for (confined) light quarks, which leads to existence of new type of "exotic" hadrons, called "chiralons" and to a selection rule, ρ_3-line rule, on the spectator quark line.

A series of puzzling new hadrons recently observed such as $X(3872)$-meson family and $\Theta(1540)$ and so on are possibly classified mostly as chiralons.

PRESENT STATUS ON HADRON SPECTROSCOPY

(Conventional Two View-Points on Hadron Classification) The non-relativistic view is based upon NRQM and has the successful $SU(6)_{SF} \otimes O(3)_L$ framework; while the relativistic view resorts to RFTM and realizes the mechanism of SBCS, an indispensable notion in the low-energy hadron physics. On the other hand, both have their own serious difficulties; the former is lacking Lorentz covariance and out of notion chiral symmetry; while the latter is unable to treat internal excitation of hadrons.

(Proposal of $\tilde{U}(12)_{SF}$-Classification Scheme) This[1] is a manifestly-covariant extension of NRQM in conformity with the chirality γ_5-transformation on the light quarks, and accordingly is reconciled with the mechanism of SBCS. The scheme has a unitary symmetry, $U(12)_{SF} \otimes O(3)_L$, in the rest-frame of hadrons, represented in the covariant $\tilde{U}(12)_{SF} \otimes O(3,1)_L$ space; that is, the hadron Wave Function, which is tensors in the above space, has the static symmetry $U(12)_{SF} \supset SU(3)_F \otimes U(4)_S$, embedded in $\tilde{U}(12)_{SF} \supset SU(3)_F \otimes \tilde{U}(4)_{DS}$, where $U(4)_S \supset \boxed{SU(2)_\rho} \otimes SU(2)_\sigma$ and the $U(4)_S(\tilde{U}(4)_{DS})$ being a unitary group(pseudo-unitary Lorentz group) on the 4 components of Dirac spinor(the γ-matrices$\equiv \sigma \otimes \rho$). The basic vectors of the "new" freedom of $SU(2)_\rho$ are given by two eigenstates of ρ_3 with respective eigen value $r = \pm$; "Pauli-spinor" $\Phi_+(X)$ with $j^P = \frac{1}{2}^+$, and "Chiral-spinor" $\Phi_-(X)$ with $j^P = \frac{1}{2}^-$. The chiral spinor $\Phi_-(X)$ with "exotic " quantum numbers leads to existence of "exotic-hadrons"[2].

---

[1] Representing the collaboration group with M.Ishida, T Maeda, M. Oda and K.Yamada

[2] We had defined the two kinds of hadron states: The chiral states are described by tensors with at least one chiral spinor-index, while the Pauli states are by those with only Pauli spinor-index. We call especially

CP814 *Hadron Spectroscopy: Eleventh International Conference*
edited by A. Reis, C. Göbel, J. Sá Borges, and J. Magnin

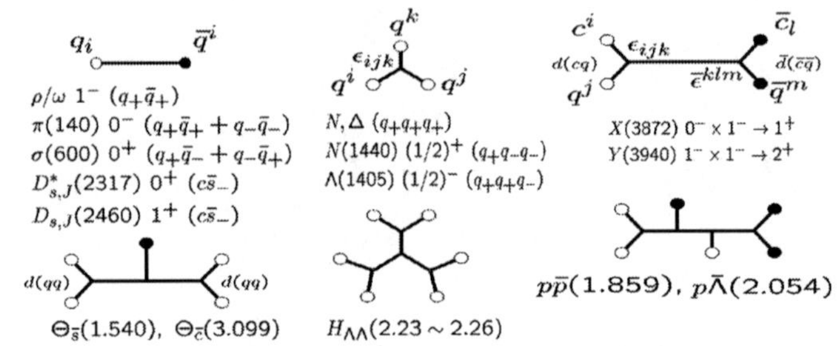

FIGURE 1. Genaral hadrons in $\tilde{U}(12)$-classification scheme

(Observation of New Hadrons), seemingly to be out of the conventional classification scheme, has recently been reported successively. They have features : (F1) Their mass is mostly close to the threshold of their observed or intermediate decay channels, and (F2) their decay width is unexpectedly narrow. The purpose of this work is to point out that these "new hadrons" be promising candidates of chiralons.

OVERVIEW OF HADRONS IN $\tilde{U}(12)$-SCHEME

(Schematic Structures of General Hadrons) in $\tilde{U}(12)$-classification scheme are shown in Fig.1 systematically for conventional $(q\bar{q})$-mesons and (qqq)-baryons and also for multi-quark hadrons resorting on Joined-Spring Quark Model. It had been proposed so as to give color-singlet and triality-zero hadrons a long time ago. In this Fig. we have, in addition to the internal space-time structure, given the structure on flavor, color and the new $SU(2)_\rho$-spin freedom of constituent quarks(, where necessary), and also made some tentative assignment of relevant new hadrons, taking into account theoretical expectations from the static $U(4)$-spin symmetry.

(Existence of Chiral States and Tentative Assignments) The $U(4)_S$-spin WF of hadrons contains a relevant tensor product of basic vectors in $SU(2)_\rho$-space. The basic-vectors are two kinds of Dirac spinors, which are the Fourier conjugates of $\Phi_\pm(X)$, with definite four-volocity $v_\mu = P_\mu/M$ of relevant hadrons(, see later sections). In JSQM(where the symmetry due to quark statistics is somewhat restricted by connected springs) the relevant hadron WF is represented by those of quarks and of di-quarks with the quantum numbers j^P as

$$q_\pm : 1/2^\pm;\quad d(q_+q_+);0^+,1^+;\quad d^\chi(q_+q_-);0^-,1^-;\ \text{and}\ d^{\chi\chi}(q_-q_-);0^+,1^+. \qquad (1)$$

The existence of chiral spinor q_- with $j^P = 1/2^-$ is the origin of exotics and leads to chiral states which show anyhow exotic properties.

WF of $(q\bar{q})$-mesons and (qqq)-baryons The $SU(2)_\rho$-spin WF of $[\pi(140)]/[\sigma(600)]$ nonets comes from their being linear representation of S.B. chiral symmetry. Similarly WF of $D^*_{sJ}(0^+;2317)/D_{sJ}(1^+;2460)$ reflects that they are chiral partners of $D_s(0^-;1968)/D^*_s(1^-;2112)$. The WF of $[\rho]/[\omega]$ being Paulons is due to their electromagnetic properties. The WF for Roper resonance $N(1440)$ and $SU(3)$-singlet $\Lambda(1405)$

the hadrons represented purely by the Pauli/Chiral states as Paulons/Chiralons.

is supposed because of their comparatively lighter mass.
Multi-quark "Exotic hadrons" The $X(3872)$ family($X(3872)$,$Y(3940)$,$\cdots$) are considered to be good candidate of chiralons in the system of $[(cn_-)\cdot(\overline{cn_-})]$ tetra-quarks. The $X(3872)$ had the mass close-to-threshhold of its decay chanel, $J/\psi+\rho,\omega$, and unexpectedly small width $\Gamma < 2.3$MeV. In Fig. we have picked up also the other observed new hadrons with properties of close-to-threshold mass and of the smaller-than-expected width.

OVERLOOKED FREEDOM OF $SU(2)_\rho$-SPIN

(Summary of Covariant Description of Composite Hadrons) It should be first noted that we are not treating a dynamical bound-state problem but presenting a kinematical framework, where the c-number Dirac-spinor as a mathematical tool, called urciton, is simulating the role of physical quarks.
We set up the hadron WF which transforms as a relevant tensor product of the c-number quark field and its Pauli conjugate, $\psi_A(x)$ ans $\bar{\psi}^B(x)$ ($A=(\alpha,a)$etc. $\alpha(a)$ denotes Dirac spinor(flavor) index), and start from the Yukawa-type Klein-Gordon equation to be satisfied by WF

$$[(\frac{\partial}{\partial X_\mu})^2 - \mathscr{M}^2(r_\mu,\partial/\partial r_\mu)]\Phi_{A_1\cdots}^{B_1\cdots}(X,r,\cdots)=0, \tag{2}$$

where the $\mathscr{M}^2$ is depending on relative coordinates of constituents (X_μ/r_μ being C.M./relative coordinates), and assumed to be diagonal on A_1 and B_1etc. In this talk we concern only ground states neglecting the dependence on r_μ of WF. Then WF of hadrons $\Phi(X)$ is given as solutions of the "local" K.G.Eq.(2), and the positive(negative)-frequency Fourier conjugates of WF, become, through the second-quantization, to represent systematically the annihilation(creation) operator of all hadrons(anti-hadrons) composed in the relevant multi-quark system. The Fourier conjugates of Pauli-conjugate WF become similarly the creation(annihilation) operators. This leads to the crossing relation or substitution law of hadrons, an important attribute of hadrons.
The $\tilde{U}(4)_{DS}$ WF of hadrons with $v_\mu = P_\mu/M$ ($P_\mu(M)$;hadron momentum(mass)) is a tensor product of basic vectors, urciton-spinors, in $SU(2)_\rho$-space. They are defined through the two (Pauli and chiral) types of Dirac spinor $\Phi_\alpha(X)=\{\Phi_{+\alpha},\Phi_{-\alpha}(X)\}$ [3], defined by the solutions of K.G. equation as

$$\begin{aligned}(\frac{\partial^2}{\partial X_\mu^2}-M^2)\Phi_\alpha(X) &= [(\gamma_\mu\partial_\mu+M)(\gamma_\mu\partial_\mu-M)\Phi(X)]_\alpha=0, \\ (\gamma_\mu\partial_\mu+|M|)\Phi_+(X) &= 0;\ \ (\gamma_\mu\partial_\mu-|M|)\Phi_-(X)=0\end{aligned} \tag{3}$$

[3] Whether the chiral spinor is required in the actual spectrum is judged only through comparison with experiments. However, for the confined light quarks inside of the lower mass hadrons it is indispensable to describe the relevant chiral symmetry. In describing hadrons with them both "mass conjugate" members of $W(P)$ and of $\bar{W}(P)$ are required, since the hole theory, where $U_-(P)=V_+(P)$ and $V_-(P)=U_+(P)$, is not applicable to the confined quarks independently of the remainders(, Note that particle and anti-particle have mutually conjugate-quantum numbers, for example, $q(\bar{q})$ belongs to $\underline{3}_c(\underline{3}_c^*)$ on the color). On the other hand it is not the case for heavy-quarks. It may be evident from Eq.(4) that the $(\pm)$ sign of $\Phi_\pm$ becomes equal to that of ρ_3-spin in hadron frame.

They are related mutually as $\Phi_{\pm,\alpha}(X) = -\gamma_5\Phi_{\mp,\alpha}(X)$, and the complete set of the basic vectors $W_\alpha(P) = \{U_{\pm\alpha}(P)\}$ and $\bar{W}^\beta(P) = \{\bar{V}_\pm^\beta(P)\}$ are given by Fourier conjugates of $\Phi(X) = \{\Phi_\pm(X)\}$ and, as those of their Pauli conjugate $\bar{\Phi}(X)$.

$$\Phi_{\pm,\alpha}(X) \equiv \sum\nolimits_{P_\mu(P_0>0)} U_{\pm\alpha}(P_\mu)e^{iPX}, \quad \bar{\Phi}_\pm^\beta(X) \equiv \sum\nolimits_{P_\mu(P_0>0)} \bar{V}_\pm^\beta(P_\mu)e^{iPX}, \tag{4}$$

where $(\bar{V}_\pm{}^t)^\beta(P_\mu) = \mp C^{\beta\alpha}U_{\pm\alpha}(P_\mu)$ ($C = \gamma_4\gamma_2 = -i\rho_1\sigma_2$ is charge- conjugation matrix).

MULTI-QUARK EXOTIC HADRONS AND $X(3872)$ FAMILY

(Properties of Multi-Quark Hadrons) The mass of ground states is generally given as: $M = M^{(0)} + \delta M$; $M^{(0)} = \sum_i m_i$(sum of constituent masses), where $M^{(0)}$ is mass in the ideal limit (being the $U(12)_{SF}$-symmetric on light-quarks), while δM represents the effects of broken chiral-symmetry($\delta^\chi M$) and of perturbative QCD($\delta^{OGE}M$). Accordingly the total mass $M^{(0)}$ is degenerate through all types of hadron and/or hadron systems at their thresholds if they as a whole have the same quark-flavor configurations. This leads to F1 one feature of puzzling new hadrons. For an example of $X(3872)$,

$$M_T^{(0)}((cq)(\overline{cq})) = M_\psi^{(0)}(c\bar{c}) + M_M^{(0)}(q\bar{q}) = M_D^{(0)}(c\bar{c}) + M_{\bar{D}}^{(0)}(\bar{c}q) \tag{5}$$

Now we focus our attention on the fission process with small phase volume, as in the above example, among ground state hadrons. In this process the transition matrix element M is considered to come purely from the overlapping $I_{O.L.}$ of WF as $M = g^{(0)} \times I_{O.L.}$ ($g^{(0)}$; a dimensional parameter), and $I_{O.L.}$ consists of a product of those on each quark line as $I_{O.L.} = \Pi_i\, I_{O.L.}[q_i]$. The $I_{O.L.}[q]$ is just the inner product between WF of initial and final quarks. Then, because of the static $U(4)$-spin symmetry for light-quarks, the transiton between urcitons with different r-value is forbidden at threshold, while suppressed strongly in the region close to threhold. This, a kind of selection rule to be called ρ_3-line rule, leads to F2, another feature of new hadrons. The suppression factor is, concerning the $SU(2)_\rho$-space, $\varepsilon(\mathbf{P}) = |\mathbf{P}|/(M + E(|\mathbf{P}|)) < 1$($\mathbf{P}$, E, and M being momentum, energy, and mass of the relevant final hadrons). In this connection the example of $X(3872)$-family are instructive: Firstly the mass of $X(3872)$ is almost equal to the threshold of $(D(1870), \bar{D}^*(2010))$ and is within the threshold region of $(\rho(\omega), J/\psi)$. Both the decay channels are doubly forbidden by the ρ_3-line rule. Experimentally the decay into the latter channel has been obseved with a small width $\Gamma < 2.3$MeV, but not into the former. This fact is understood by the ρ_3-line rule, considering the relation between respective suppression factors, $\varepsilon^4(\mathbf{P})_{D\bar{D}^*} < \varepsilon^4(\mathbf{P})_{J/\psi\cdot\rho(\omega)} \ll 1$. Secondly the mass of $Y(3.943)$ is well above the open-charm thresholds (3.872MeV) of $D\bar{D}^*$ and near to the threshold(3.880) of (ω, J/ψ), of which both decay-channels are similarly double-forbidden by ρ_3-line rule. Experimentally its decay into the $\omega J/\psi$ has been observed with a rather narrow width $\Gamma \approx 90$MeV, while not into the channel $D\bar{D}^*$. This situation may be also understandable from the difference of the suppression factors, $\varepsilon_{D\bar{D}^*} < \varepsilon_{J/\psi\cdot\omega}$.

(Level Structures of Tetra-quark$[(cq)(\overline{cq})]$-System) The quantum numbers of ground-state multiplets are obtained from those of di-quarks and anti-diquarks, $d(cq)$ and $\bar{d}(\overline{cq})$ (see, Eq.(1)). The relevant $SU(2)_\rho$-space WF are classified into three-groups; $T_{\chi\chi}(d^\chi\bar{d}^\chi)$, $T_{P\chi}(d^\chi\bar{d}^P/d^P\bar{d}^\chi)$ and $T_{PP}(d^P\bar{d}^P)$(where $d^\chi \equiv (cq_-)$ and $d^P \equiv (cq_+)$) with $j^P(T) = j^P(d) \otimes j^P(\bar{d})$). In order to obtain the more realistic mass-spectra is necessary

to estimate the symmetry breaking effects $\delta M = \delta^{\chi} M + \delta^{OGE} M$, which are approximated by the respective sum of those of constituents, $\delta M_T = \delta M_d + \delta M_{\bar{d}}$.
Concerning $\delta^{\chi} M \equiv M^{\chi} - M^{P}$, applying the L$\sigma$M for the Yukawa interaction of light quarks inside of $d(cq)$-diquarks and of $D(c\bar{q})$ mesons, we get the relation,

$$\delta^{\chi} M_d(cq) \;=\; -\delta^{\chi} M_D(c\bar{q}) \quad (\delta^{\chi} M_D(c\bar{n}) \propto a,\ \delta^{\chi} M_D(c\bar{s}) \propto b) \tag{6}$$

where $a \equiv \langle n\bar{n}\rangle_0$ and $b \equiv \langle s\bar{s}\rangle_0$ are vacuum eqpectation values of respective scalar densities. Using the experimental values of $\delta^{\chi} M_D(c\bar{s})$ and of the ratio a/b we obtain[2] the values; $\delta^{\chi} M_d(cn) = -242$MeV, $\delta^{\chi} M_d(cs) = -348$MeV. This implies that the masses of $T[(cq)(\overline{cq})]$ -system are in order of $M_T(\chi\chi) < M_T(P\chi) < M_T(PP)$ with the difference $\delta^{\chi} M_d(cq)$, Eq.(6). It is notable that the negative relative-sign between both sides of Eq.(6) reflects the Charge-Conjugation property of thr c-number scalar-density.
Concerning $\delta^{OGE} M$, we set up the effective spin-current interaction between c-quarks and light-quarks as

$$\langle H^J \rangle \approx \langle \bar{c}\sigma_{\mu\nu} c\rangle \langle \bar{q}_{\pm} F_U \sigma_{\mu\nu} q_{\pm}\rangle \propto \langle \sigma^{(c)}\rangle \langle \sigma^{(q)}\rangle; \tag{7}$$

where the middle term is color-gauge invariant and becomes static $U(4)_S$ -symmetric due to the role of unitarizer F_U. This leads to the relation of $\delta^J M \equiv M(J=1) - M(J=0)$,

$$\delta^J M_d(cq) = (1/2)\delta^J M_D(c\bar{q}) = 71\text{MeV} \quad \text{for } q = (n,s) \tag{8}$$

where the positive relative-sign of the first equality in Eq.(7) reflects the C.C. property of the c-number tensor-density and the factor 1/2 comes from the color freedom. The value of Eq.(8) is obtained from the experimental values in the $D(c\bar{q})$ meson system. Here it may be notable that the mass defference between $X(3872)$ and $Y(3940)$ is close to the estimated one Eq.(8). The more detailed investigations on $X(3872)$ family are reported in this conference[3]. Finally, we give brief comments on the related works: The diquark and anti-diquark picture of T-quark system was first discussed in Ref.[4], but without considerations on chiral symmetry. Actually the relevant level structure of [4] is equivalent to that of our T_{PP} system. The $D^0\bar{D}^{*0}$ molecular picture[5] also explains(or starts from) the feature of new hadrons(F1). However, its WF has very loose structure whereas in our case it is, supposed to be, tight due to color confining force.

REFERENCES

1. S. Ishida and M. Ishida, *Phys. Lett.***B539**, 249(2002); S. Ishida, M. Ishida, and T. Maeda, *Prog. Theor. Phys.***104**, 785(2000).
2. S. Ishida et al.,hep-ph/0408136; See, also the precedent work, W. A. Bardeen, E. J. Eichten and C. T. Hill, *Phys. Rev.***D68**, 054024(2003).
3. M. Ishida, in proc. of HADRON'05.
4. L. Maiani, F. Piccinini, A. D. Polosa and V. Riquer, *Phys. Rev.***D71**, 014028(2005).
5. F. E. Close and P. R. Page, *Phys. Lett.***B578**, 119(2004); N. A. Tornqvist, *Phys. Lett.***B590**, 209(2004); C. Y. Voloshin, *Phys. Lett.***B579**, 316(2004); E. Braaten and M. Kusunoki, *Phys. Rev.***D69**, 074005(2004); E. S. Swanson, *Phys. Lett.***B588**, 189(2004); **B598**, 197(2004)

X&Y

A.D. Polosa

Dip. Interateneo di Fisica, Università di Bari, Italy

Abstract. I briefly discuss the 4-quark interpretation of the recently discovered $X(3872)$ and $Y(4260)$ resonances.

Keywords: Multiquark Hadrons
PACS: 12.39.-x, 12.38.-t

$X(3872)$

Quite recently BELLE and BaBar (also CDF and D0) have reported evidence of a new narrow resonance in $B^+ \to K^+ J/\psi\pi\pi(\pi)$, the $X(3872)$ [1], which most likely decays $X \to \rho(\omega)$ with almost equal rate, therefore maximally violating isospin. Even if its mass is slightly above the $D\bar{D}$ threshold, the X(3872) is not observed to decay in this channel. It turns out that it can't because $J^P(X) = 1^+$. The decay $X \to J/\psi\gamma$, with a rate almost ten times smaller than the $J/\psi\rho$, is observed, suggesting $C = +1$, i.e., $J^{PC}(X) = 1^{++}$.

Charmonium and Hybrid. Such a state is a natural candidate for being classified as a new $c\bar{c}$ resonance. Anyway, the experimental assessment of X is in sharp contrast with the 2^3P_1 charmonium assignation, the only possible [2]. What is this narrow X then? The hypothesis next to the charmonium one is the hybrid, i.e., a $c\bar{c}$+constituent(g). Anyway the gluon is flavor blind, it couples to u and d quarks in the same way thus rendering the hybrid state an isospin pure state. Isospin violating decays would not fit this scheme.

Molecule. There are at least two other possibilities for the X: being a $D\bar{D}^*$ molecule (a $4 \div 5$ fm extended object) [3] or a multiquark state (a compact 4-quark state with a size of$\sim$1 fm) [4]. The molecule is a 50% mixture of $I = 0$ and $I = 1$, therefore allowing isospin violating transitions. The D^* has a width of about 70 KeV. Therefore the width of such a molecule is expected to be driven by the D^* width itself. A quite natural decay is then $DD\pi$, which is currently investigated at BELLE and BaBar [5]. Preliminary results agree that this channel is at least $O(1)$ the $J/\psi\rho$ one. On the other hand D and D^* might as well interact via hadron exchange forces giving $J/\psi\rho$ in the final state. In more sophisticated models the X spends part of its time in a mlecular-like state and part in a compact multiquark-like state; once the c quarks are closer to each other the transition to a final state containing a J/ψ is favored.

4-quark. In the four-quark option the X is depicted as a bound state of two colored diquarks $[cq]$ and $[\bar{c}\bar{q}]$. The heavy-light diquarks are supposed to be bound in the $\bar{3}_c$ attractive channel [6]. There is some lattice evidence of the formation of $S = 0$ light-light diquarks, while $S = 1$ seems to be disfavored [7]. In the heavy-light system, spin-spin

CP814 *Hadron Spectroscopy: Eleventh International Conference*
edited by A. Reis, C. Göbel, J. Sá Borges, and J. Magnin

interactions are suppressed by the mass of the heavy constituent, therefore we expect that, *if* the heavy-light diquark is formed, $S = 1$ or $S = 0$ will be equally probable. This opens the way to 4-quark mesons with $J = 0, 1, 2$ and natural and unnatural parities. Depending upon the flavor of the two light constituents, one can build a full nonet of X states where the observed $X(3872)$ is just the neutral one (charged states are expected to occur!).

Diquarks are colored objects. If one tries to pull them apart, at some point will excite the production of a couple of light quarks from the vacuum. This means that the natural decay of such a four quark system is into two charmed baryons. But this is impossible kinematically. Thus some other mechanism must be at work to allow the X decay. If the two heavy quarks escape from their diquark shells and bind to form a J/ψ, then a $X \to J/\psi\rho(\omega)$ decay is easily explained. Also the two diquark structures could be broken by the tunneling of a light anti-quark from the anti-diquark to meet a c quark in the diquark (and viceversa) then forming a $D\bar{D}^* \to DD\pi$ final state.

Isospin forbidden decays are possible if X is not a pure isospin state. A pure $I = 0, 1$ state would be:

$$\begin{aligned} X(I=0) &= \frac{X_u + X_d}{\sqrt{2}} \\ &\text{or} \\ X(I=1) &= \frac{X_u - X_d}{\sqrt{2}}, \end{aligned} \tag{1}$$

where $X_{u,d} = [cu][\bar{c}\bar{u}], [cd][\bar{c}\bar{d}]$. On the contrary if the physical eigenstates are just X_u and X_d, i.e., pure mass eigenstates, maximal isospin violations are possible. Mixtures of X_u and X_d are induced by annihilation diagrams where u quarks transform in d through vacuum. At the scale of the charm quark mass m_c, we expect that annihilation diagrams are suppressed enough to be close to the situation in which X mass eigenstates tend to align to quark masses. We can introduce a θ mixing angle and write [4]:

$$\begin{aligned} X_l &= X_u \cos\theta + X_d \sin\theta \\ X_h &= -X_u \sin\theta + X_d \cos\theta, \end{aligned}$$

expecting a θ to be small. We find that the mass of X_h and X_l must be different by:

$$M(X_h) - M(X_l) \simeq \frac{2(m_d - m_u)}{\cos(2\theta)} \tag{2}$$

and our prediction is $\Delta M \simeq 7$ MeV.

Taking Belle data at face value, we conclude that only one of the two neutral states, $X_{u,d}$ or, more generally $X_{l,h}$, is produced appreciably in B^+ decay: the peak observed is indeed too narrow to describe two resonances about 7 MeV apart. Only one should be produced; our analysis is consistent with the production of X_l. X_h should then be produced in B^0 decays only. The present BaBar measure of this mass difference is $\Delta M = 2.7 \pm 1.3 \pm 0.2$ MeV extracted from $B^0 \to XK^0$ and $B^+ \to XK^+$ [8]. We have to wait for higher statistics analyses to clarify if there are indeed two neutral states separated in mass.

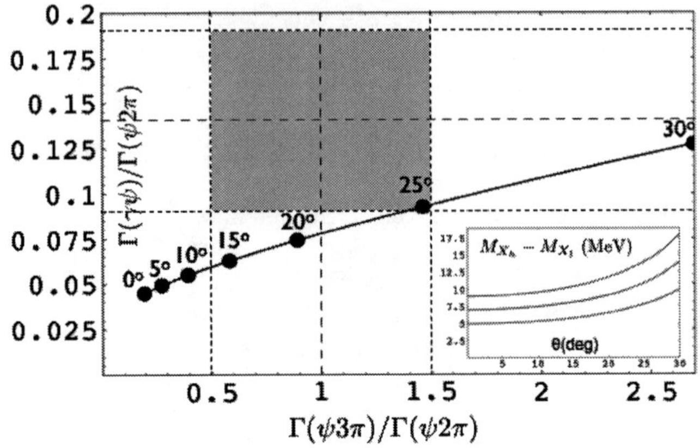

FIGURE 1. The shaded area represents the experimentally allowed region by the decays $X \to J/\psi + \gamma, \rho, \omega$. The solid curve is the 4-quark prediction, related to X_l, for different values of θ. The insert shows the variation of the mass difference $M_{X_h} - M_{X_l}$ versus θ including the theoretical error band.

Differences. In Fig. 1 it is shown the prediction of the X_l widths as a function of θ in the 4-quark model. Data seem to favor a $\theta \sim 20°$. Apparently 4-quarks and molecule interpretations give similar qualitative predictions for the main X decays. Let's stress a bit more their differences. The four quark model predicts the existence of a full nonet of X states. Charged X particles are predicted. Two neutral X particles are predicted with a separation in mass. In this framework the $X(3940)$, seen in $J/\psi\omega$ by BELLE [9], could be interpreted as a 2^{++} 4-quark state. Does it decay also to $J/\psi\rho$?

On the other hand the molecule approach to X does not call for charged X's, X_s's or for two distinct neutral X's. The 4-quark model for X allows a $X \to D\bar{D}^*$ decay width up to two times larger than $J/\psi\rho$. The molecule cannot, the $D\bar{D}^*$ being quite smaller than $J/\psi\rho$ (see e.g. [10]). Experiment will shortly decide between (or beyond) these two models.

$Y(4260)$

Colored objects such as diquarks in a rising confining potential should exhibit a series of orbital angular momentum excitations while in the molecular picture two colorless objects bound by a short range potential should have a very limited spectrum, possibly restricted to S-wave states only.

The $Y(4260)$ has been observed by BaBar as a 1^{--} resonance in $e^+e^- \to \gamma J/\psi\pi\pi$ where γ is an initial state radiation photon. The two pion invariant mass spectrum seems to be peaked at the $f_0(980)$. $Y(4260)$ is not observed to decay into $D\bar{D}$. Is it the first excited state of the X family? [11]

4-quark. We reconstruct the process as $e^+e^- \to \gamma Y \to \gamma J/\psi f_0$. Negative parity of the state calls for one unit of orbital angular momentum in the diquark pair. The

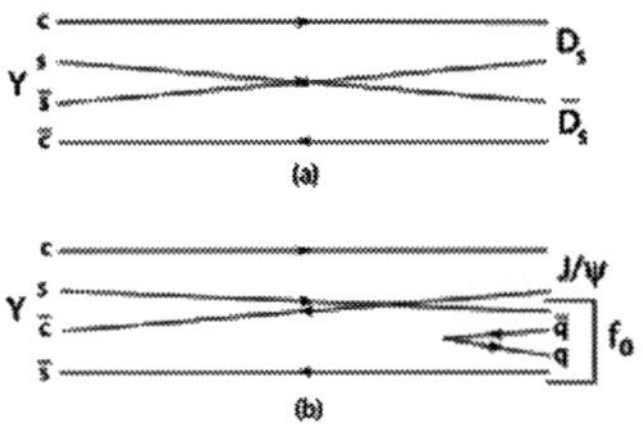

FIGURE 2. Diagram (a) is favored with respect to (b).

final state $f_0 + J/\psi$ suggests, in the 4-quark model of f_0 [12], the quark content $Y \sim \{[cs][\bar{c}\bar{s}]\}_{(\text{P-wave})}$. The most striking prediction of the 4-quark approach is that the observed decay $J/\psi f_0$ should not be the favored one. The most natural decay should be $D_s\bar{D}_s$, as shown in Fig. 2.

Also in the case of Y the 4-quark model predicts a series of states in addition to the observed one. Among them, a state with both diquark and anti-diquark with $S = 1$ can give rise to a $S_{\text{tot}} = 2$ state. This state projects only on spin one $\bar{c}s$ and $\bar{s}c$ states. In the limit in which the spin of the s quark is a good quantum number such state could decay only into $D_s^*\bar{D}_s^*$ with substantial reduction of its decay width. One could then expect another state with a higher mass, by a factor typical of spin-spin interactions, decaying predominantly to $D_s^*\bar{D}_s^*$.

Hybrid. A serious competitor here is the hybrid state $c\bar{c}g$. This is predicted to be subject to selection rules constraining it to decay to $D^{**}\bar{D}$ all other modes being suppressed [13]. Also in this case, like in the case of X, there is a quite stringent experimental test to select among models (or disprove them all).

$B \to J/\psi + \text{ALL}$

The inclusive J/ψ spectrum in B decays has been studied very carefully on the experimental and theoretical side. The Non-Relativistic-QCD predictions agree quite well with the data over most of the range. However for slow J/ψ, i.e., with momenta below 1 GeV, the data show a marked excess over expectations, see Fig 3. No definitive explanation has been furnished yet to explain this excess.

We propose that these soft J/ψ are the footprints of the production of hidden charm four-quark states in $b \to c\bar{c}q$ decays and their subsequent decay into J/ψ plus light-flavor hadrons [14].

SUMMARY

X and Y particles, and their (coming?) partners, could be the first pieces of a new series of mesons with new multiquark body plans (or hybrids?) or could be the first examples of loosely bound states of charmed mesons in the form of molecules. There

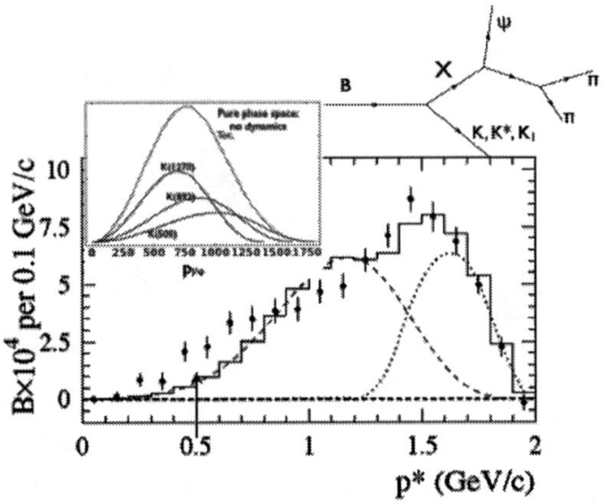

FIGURE 3. The process $B \to J/\psi + \text{all}$ could be dominated by a resonant channel in which the J/ψ comes from an X produced in association to a Kaon, P-wave, or to a K^* or K_1, S-wave. Pure kinematics considerations immediately show that the J/ψ decay momenta should be most likely in the low region, especially when a K^* or K_1 is produced (see the inserted plot). This would justify the J/ψ momentum spectrum (main plot) where a clear discrepancy with non-relativistic QCD is appreciable at low $p_{J/\psi}$ (the region around 500 MeV).

are several crucial experimental tests which will allow to probe the nature of this states and distinguish among models.

ACKNOWLEDGMENTS

I would like to thank L. Maiani for enlightening discussions.

REFERENCES

1. S. K. Choi *et al.* [BELLE Collaboration], Phys. Rev. Lett. **91**, 262001 (2003); D. Acosta *et al.* [CDF Collaboration] Phys. Rev. Lett. **93**, 072001 (2004); V. M. Abazov *et al.* [D0 Collaboration] Phys. Rev. Lett. **93**, 162002 (2004); B. Aubert *et al.* Phys. Rev. **D71**, 071103 (2005).
2. T. Barnes, S. Godfrey, E.S. Swanson, Phys. Rev. **D72**, 054026 (2005).
3. N.A. Törnqvist, Phys. Lett. **B590**, 209 (2004); E. Braaten, M. Kusunoki, Phys. Rev. **D69**, 074005 (2004); E.S. Swanson, Phys. Lett. **B588**, 189 (2004); M. Voloshin, Phys. Lett. **B604**, 69 (2004).
4. L. Maiani, F. Piccinini, A.D. Polosa, V. Riquer, Phys. Rev. **D71**, 014028 (2005).
5. See P. Pakhlov on behalf of BELLE collaboration, EPS-Lisbon-2005
6. R.L. Jaffe, F. Wilczek, Phys. Rev. Lett. **91**, 232003 (2003).
7. C. Alexandrou, Ph. de Forcrand, B. Lucini, [arXiv:hep-lat/0509113]
8. B. Aubert *et al.* [BaBar Collaboration], [arXiv:hep-ex/0507090]
9. K. Abe *et al.*, Phys. Rev. Lett. **94**, 182002 (2005)
10. E.S. Swanson, Phys. Lett. **B588**, 189 (2004)
11. L. Maiani, F. Piccinini, A.D. Polosa, V. Riquer, Phys. Rev. **D72**, 031502 (2005).
12. L. Maiani, F. Piccinini, A.D. Polosa, V. Riquer, Phys. Rev. Lett. **93**, 212002 (2004).
13. E. Kou, O. Pene, [arXiv:hep-ph/0507119]; F.E. Close, P.R. Page, [arXiv:hep-ph/0507199].
14. I. Bigi, L. Maiani, F. Piccinini, A.D. Polosa, V. Riquer, [arXiv:hep-ph/0510307]

Property of cn-$\bar{n}\bar{c}$ System in $\tilde{U}(12)$-Scheme and $X(3872)/Y(3940)$

Muneyuki Ishida

Department of Physics, Meisei University, Hino, Tokyo 191-8506, Japan

Abstract. The properties of four quark cn-$\bar{n}\bar{c}$ states are investigated as cn di-quark and $\bar{n}\bar{c}$ di-antiquark system in $\tilde{U}(12)$-classification scheme of hadrons. We consider the negative-parity di-quark and di-antiquark in ground states, and the properties of $X(3872)$ and $Y(3940)$ are consistent, respectively, with those of the $J^{PC} = 1^{++}$ and 2^{++} states from these negative-parity di-quark and di-antiquark. Their narrow-widths are explained from a kind of conservation law, called ρ_3-line rule. The properties of ground-state cs-$\bar{s}\bar{c}$ system are also predicted in this scheme.

Keywords: $X(3872)$, $Y(3940)$, $\tilde{U}(12)$ group, quark model
PACS: 13.39.Ki, 12.40.Yx

INTRODUCTION

The discovery of $X(3872)$[1] and $Y(3940)$[2] by Belle presented new problems on hadron spectroscopy. Because of their decay-properties, they are considered to be non-$c\bar{c}$ states. We consider the four-quark explanation[3] of $X(3872)/Y(3940)$, where the relevant particles are considered as the cn di-quark and $\bar{n}\bar{c}$ di-antiquark system. This explanation is criticized from the observed decay widths. The cn-$\bar{n}\bar{c}$ system decays to $D\bar{D}^{(*)}$ and $J/\psi\rho$(or $J/\psi\omega$). They are so-called OZI super-allowed processes, which are similar to fission processes in nuclear physics, and proceed without no $q\bar{q}$-pair production or annihilation. Their widths are expected to be so large as a few hundred MeV or more. On the other hand the observed widths of $X(3872)$ and $Y(3940)$ are very small, $\Gamma_X < 2.3$MeV[1] and $\Gamma_Y = 87 \pm 22 \pm 26$MeV[2], respectively.

A few years ago[4], we have proposed a covariant level-classification scheme of hadrons based on $\tilde{U}(12)$ spin-flavor group. It is a relativistic generalization of the $SU(6)_{SF}$-scheme in non-relativistic quark model(NRQM), and the squared-mass spectra of hadrons including light constituent quarks are classified as the representations of $\tilde{U}(12)$[5]. Each spinor index corresponding to light constituent quark in composite hadron WF is expanded by free Dirac spinors $u_{r,s}(v_\mu)$ called urciton spinors. At the rest frame $u_{+,s}(\mathbf{0})$ has the upper two-component which corresponds to the Pauli spinor appearing in NRQM, while the $u_{-,s}(\mathbf{0})$ has the lower two-component representing the relativistic effect. The r index of $u_{r,s}$ represents the eigen-value of ρ_3, and the corresponding freedom is called ρ-spin, while the ordinary Pauli-spin freedom is called σ-spin. In the $\tilde{U}(12)$-scheme, the $u_{-,s}$ is supposed to appear as new degrees of freedom for light constituents, being independent of $u_{+,s}$, and we predict the existence of a number of extra states out of the $SU(6)_{SF}$-framework for meson and baryon systems, called chiral states.

The heavy-light $Q\bar{q}$ mesons in ground states are classified as $\mathbf{12}^*$, which includes the chiral $J^P = 0^+/1^+$ states with positive-parity as well as the negative-parity $0^-/1^-$

CP814 *Hadron Spectroscopy: Eleventh International Conference*
edited by A. Reis, C. Göbel, J. Sá Borges, and J. Magnin

states[5]. The properties of $D_s(2317)/D_s(2460)$ are naturally explained as the chiral states in $c\bar{s}$ system. Applying the $\tilde{U}(12)$-scheme, the heavy-light Qq di-quark forms **12** in ground states, which includes chiral $J^P = 0^-/1^-$ states with negative parity as well as the positive parity $0^+/1^+$ states.

There is a kind of phenomenological conservation rule applicable to chiral states, called ρ_3-line rule[6, 7]. In transitions of chiral states, the ρ_3 along each spectator-quark line is approximately conserved.

There is an interesting possibility that the observed $X(3872)$ and $Y(3940)$ are chiral states, which are made from negative-parity cn di-quark and $\bar{n}\bar{c}$ di-antiquark, and that their narrow-width properties are explained by the ρ_3-line rule. In this talk we investigate this possibility, whether the $X(3872)$ and $Y(3940)$ are explained as cn-$\bar{n}\bar{c}$ states.

DI-QUARK AND DI-ANTIQUARK SPINOR WAVE FUNCTION

The negative-parity cn di-quark($\bar{n}\bar{c}$ di-antiquark), called chiral state, is denoted as $\Gamma^{\chi}_{cn}(\Gamma^{\chi}_{\bar{n}\bar{c}})$, which has negative $\rho_3^n(\rho_3^{\bar{n}})$, while the ordinary positive-parity di-quark(di-antiquark), called Pauli state, is denoted as $\Gamma^{P}_{cn}(\Gamma^{P}_{\bar{n}\bar{c}})$, which is composed of consitituents both with positive ρ_3. Their explicit forms are given in ref.[8].

The Γ_{cn} is related with the WF of $c\bar{n}$-meson $H_{c\bar{n}}$ in heavy quark effective theory, as $\Gamma_{cn} = H_{c\bar{n}}C^{\dagger}$. The WF of positive(negative)-parity di-quark corresponds to the WF of negative(positive)-parity meson. The positive-parity mesons, which are chiral states, become heavier[5] than the negative-parity Pauli mesons through spontaneous chiral symmetry breaking(χSB), which is described by the Yukawa coupling to the scalar σ-meson nonet $s = s^i\lambda^i/\sqrt{2}$ in the framework of $SU(3)$ linear σ model[9]. In Qq di-quark systems, because of the $C^{\dagger}$ factor, the chiral states are expected to have smaller masses than the Pauli states. The mass shift $\delta M^{\chi,P}$ in χSB is given by

$$\delta M^{\chi,P} = g_\sigma s_0 \langle \Gamma^{\chi,P}_{cn}\overline{\Gamma^{\chi,P}_{cn}}\rangle = g_\sigma s_0 \langle H^{\chi,P}_{c\bar{n}} C^{\dagger} C^{\dagger} \overline{H^{\chi,P}_{c\bar{n}}}\rangle = -g_\sigma s_0 \langle H^{\chi,P}_{c\bar{n}} \overline{H^{\chi,P}_{c\bar{n}}}\rangle, \quad (1)$$

where $\overline{\Gamma^{\chi,P}_{cn}} \equiv \gamma_4(\Gamma^{\chi,P}_{cn})^{\dagger}\gamma_4$, and the change of sign comes from $(C^{\dagger})^2 = -1$. The s_0 is vacuum expectation value of s and g_σ is the Yukawa coupling constant. The chiral mass splitting $\Delta M_\chi (\equiv \delta M^P - \delta M^\chi) = 2g_\sigma s_0$ is predicted with 242MeV[5].

Concerning cn-$\bar{n}\bar{c}$ system, we expect three combinations of Γ_{cn}-$\Gamma_{\bar{n}\bar{c}}$: The Γ^P_{cn}-$\Gamma^P_{\bar{n}\bar{c}}$ denoted as Pauli-Pauli states, the Γ^P_{cn}-$\Gamma^\chi_{\bar{n}\bar{c}}$ (or Γ^χ_{cn}-$\Gamma^P_{\bar{n}\bar{c}}$) denoted as Pauli-chiral(or chiral-Pauli) states, and the Γ^χ_{cn}-$\Gamma^\chi_{\bar{n}\bar{c}}$ denoted as chiral-chiral states. Among them the chiral-chiral states have the smallest masses, the chiral-Pauli states are heavier by ΔM_χ, and the Pauli-Pauli states are further heavier by ΔM_χ. The Pauli-Pauli(Pauli-chiral) states decay to Pauli-chiral(chiral-chiral) states through pion emission in S-wave, and the widths are expected to be so large as a GeV. Accordingly, the Pauli-Pauli states and Pauli-chiral states are not observed as resonances, but as non-resonant backgrounds. We focus our interests on chiral-chiral states in the following.

TABLE 1. Masses of ground-state cn-$\bar{n}\bar{c}$ system (in MeV). Decay channels of I=1 states are shown. For I=0, $\rho(\pi)$ is replaced with $\omega(\eta)$.

type	J^{PC}	ΔM_{SS}	Mass	decay channels		
$\Gamma^{0^-}_{cn}$-$\Gamma^{0^-}_{\bar{n}\bar{c}}$	0^{++}	-65	3824	$J/\psi\rho$,	$\eta_c\pi$,	$D\bar{D}$
$\frac{1}{\sqrt{2}}(\Gamma^{0^-}_{cn}$-$\Gamma^{1^-}_{\bar{n}\bar{c}}+\Gamma^{1^-}_{cn}$-$\Gamma^{0^-}_{\bar{n}\bar{c}})$	1^{+-}	-27	3862	$J/\psi\pi$,	$\eta_c\rho$	
$\frac{1}{\sqrt{2}}(\Gamma^{0^-}_{cn}$-$\Gamma^{1^-}_{\bar{n}\bar{c}}-\Gamma^{1^-}_{cn}$-$\Gamma^{0^-}_{\bar{n}\bar{c}})$	1^{++}	-17	<u>3872</u>	$J/\psi\rho$		
$\Gamma^{1^-}_{cn}$-$\Gamma^{1^-}_{\bar{n}\bar{c}}$	0^{++}	-22	3867	$J/\psi\rho$,	$\eta_c\pi$,	$D\bar{D}$
$\Gamma^{1^-}_{cn}$-$\Gamma^{1^-}_{\bar{n}\bar{c}}$	1^{+-}	0.	3889	$J/\psi\pi$,	$\eta_c\rho$,	$D\bar{D}^*$
$\Gamma^{1^-}_{cn}$-$\Gamma^{1^-}_{\bar{n}\bar{c}}$	2^{++}	+43	3932	$J/\psi\rho$,		$D\bar{D}^{(*)}$

OSCILLATOR WF AND GROUND-STATE MASS SPECTRA

We consider spin-spin interaction to predict the mass spectra of ground states. We use the space WF in joint spring quark model(JSQM)[10], where the cn di-quark ($\bar{n}\bar{c}$ di-antiquark) is in color $\mathbf{3}^*(\mathbf{3})$, and they are connected by spring potential. The space-coordinates of c, n, $\bar{n}$ and $\bar{c}$ are denoted, respectively, as $\mathbf{x}_1$, $\mathbf{x}_2$, $\mathbf{x}_3$ and $\mathbf{x}_4$, and the eigen-modes are numerically given by[8]

$$\begin{aligned}\mathbf{s} &= (\mathbf{x}_1-\mathbf{x}_2-\mathbf{x}_3+\mathbf{x}_4)/\sqrt{2},\ \ \mathbf{R}=0.785(\mathbf{x}_1-\mathbf{x}_4)+0.083(\mathbf{x}_2-\mathbf{x}_3),\\ \mathbf{U} &= 0.365(\mathbf{x}_1-\mathbf{x}_4)-0.862(\mathbf{x}_2-\mathbf{x}_3).\end{aligned} \quad (2)$$

We may regard $\mathbf{s}$ as relative coordinate in di-(anti)quark, $\mathbf{R}(\mathbf{U})$ approximately as relative coordinate between c and $\bar{c}$(n and $\bar{n}$). The internal space WF is given by

$$f_s(\mathbf{s})f_R(\mathbf{R})f_U(\mathbf{U}) \;=\; \left(\frac{\beta_s\beta_R\beta_U}{\pi^3}\right)^{\frac{3}{4}} e^{-\frac{\beta_s}{2}\mathbf{s}^2}e^{-\frac{\beta_R}{2}\mathbf{R}^2}e^{-\frac{\beta_U}{2}\mathbf{U}^2}, \quad (3)$$

where $\beta_s=0.129\text{GeV}^2$, $\beta_R=0.254\text{GeV}^2$ and $\beta_U=0.115\text{GeV}^2$.

By using the WF (3), we can estimate the spin-spin interaction, $H_{SS}\propto -\frac{1}{m_im_j}\delta^{(3)}(\mathbf{x}_i-\mathbf{x}_j)\sigma^{(i)}\cdot\sigma^{(j)}$. The proprtional constant is fixed by using $m_{D^*}-m_D$=142MeV as input, and the mass-shifts ΔM_{SS} in ground-states of cn-$\bar{n}\bar{c}$ system are given in Table 1. The mass of $X(3872)$, which plausibly J^{PC}=1^{++}[11], is used as input to fix overall mass-shift. The 2^{++} state is predicted with mass 3932MeV, which is consistent with the experimental value of $Y(3940)$, $M=3943\pm11\pm13$MeV[2]. At the same time, the existence of the other 4 states with $J^{PC}=0^{++}$ and 1^{+-} is predicted in the same mass region.

STRONG DECAY WIDTHS

The relevant decays are quantum-mechanical fission processes, and thus, their decay matrix elements are considered to be obtained only by the overlapping of initial and final WFs. We calculate the overlapping of space WFs by taking the cn-$\bar{n}\bar{c}\to D\bar{D}$ as an example. The initial four quark state and the final D and $\bar{D}$ states are described by a common Hamiltonian, H, as $i\hbar\frac{\partial}{\partial t}|\psi\rangle = H\ |\psi\rangle$; $|\psi\rangle = |c_1n_2$–$\bar{n}_3\bar{c}_4\rangle$, $|D\rangle(=|c_1\bar{n}_3\rangle)$, $|\bar{D}\rangle(=$

$|n_2\bar{c}_4\rangle$). The state-kets are explicitly given by using WF (3) in ref.[8]. The transition amplitude A is given by the overlapping of the state-kets, inserted H. By using the A, the decay width Γ is simply given by Fermi's golden rule.

$$
\begin{aligned}
A &= \langle c_1\bar{n}_3(\mathbf{P}_D)| \cdot \langle n_2\bar{c}_4(\mathbf{P}_{\bar{D}})|\, H \,|c_1 n_2 - \bar{n}_3\bar{c}_4(\mathbf{P})\rangle \\
\Gamma &= \int \frac{d^3\mathbf{P}_D}{(2\pi)^3 2E_D} \frac{d^3\mathbf{P}_{\bar{D}}}{(2\pi)^3 2E_{\bar{D}}} \frac{2\pi}{\hbar} \delta(M - E_D - E_{\bar{D}})\, |A|^2 \,\frac{1}{2MV},
\end{aligned} \tag{4}
$$

where M is the mass of initial four quark state and $V = (2\pi)^3\delta^{(3)}(\mathbf{P}=\mathbf{0})$. The equation (4) predicts huge decay-widths $\Gamma \sim$ a few$\sim$10 GeV for Pauli-Pauli states. On the other hand, for chiral-chiral states, we must take into account another factor coming from spinor WFs, which drastically reduces Γ.

The overlapping of spinor WFs is given for the decays to $c\bar{n}+n\bar{c}$ by

$$
\begin{aligned}
A_s(c\bar{n}+n\bar{c}) &= \langle \Gamma^{\chi}_{\bar{n}\bar{c}} \overline{H}_{\bar{D}}(v_{\bar{D}})(-iv_{\bar{D}}\cdot\gamma)\, {}^t\Gamma^{\chi}_{cn}\, {}^t[iv_D\cdot\gamma \overline{H}_D(v_D)]\rangle \\
&= \langle \Gamma^{\chi}_{\bar{n}\bar{c}}\, B(v_{\bar{D}})\overline{H}_{\bar{D}}\bar{B}(v_{\bar{D}})\, {}^t\Gamma^{\chi}_{cn}\, {}^t[B(v_D)\overline{H}_D\bar{B}(v_D)]\rangle,
\end{aligned} \tag{5}
$$

where $D\bar{D}$ is taken as an example of $c\bar{n}+n\bar{c}$. The $\overline{H}(v)$ is the spinor WF of the final meson with velocity $v_\mu (\equiv P_\mu/M)$, and $\overline{H}$ is the one at the rest frame. They are related by Lorentz booster $B(v), \bar{B}(v) = ch\theta \pm \rho_1 n\cdot\sigma sh\theta$ as $\overline{H}(v) = B(v)\overline{H}\bar{B}(v)$, where the $ch\theta, sh\theta = \sqrt{\frac{E\pm M}{2M}}$. In Eq. (5), the change of ρ_3 on each quark-line occurs only through ρ_1 term in $B(v)$ and $\bar{B}(v)$ being proportional to $sh\theta$, which is much smaller than unity for the relevant decays because of their small phase space. Thus, the ρ_3's along all the spectator-quark lines are separately conserved. This is ρ_3-line rule[6].

In the chiral-chiral states, the n in cn di-quark and the $\bar{n}$ in $\bar{n}\bar{c}$ di-antiquark are in negative ρ_3 state. Thus, their decay amplitudes to the Pauli mesons are doubly suppressed from ρ_3-line rule. In $\tilde{U}(12)$-scheme, the D, η_c, J/ψ, ρ, ω, $D^{(*)}$ and $\bar{D}^{(*)}$, are Pauli mesons, while the WF of Nambu-Goldstone π and η mesons have doubly-negative ρ_3 component. Thus, in the decays of chiral-chiral states, the transition amplitudes to $J/\psi\rho(\omega)$, $\eta_c\rho(\omega)$ and $D\bar{D}^{(*)}$ are strongly suppressed, while the ones to $J/\psi\pi(\eta)$ and $\eta_c\pi(\eta)$ are not suppressed. The decay widths of the latter channels are a few GeV. Thus, when the initial states have no $\pi(\eta)$ decay modes, they have narrow widths, and are expected to be observed as resonant particles. The possible decay channels of the ground six states are shown in Table 1. Among them, only two states with $J^{PC} = 1^{++}$ and 2^{++} have no $\pi(\eta)$ decay modes and are possibly detected as resonances. This theoretical prediction is consistent with the present experimetnal situation: Only two states, $X(3872)$ and $Y(3940)$, are observed as the exotic candidates in the relevant mass region. Furthermore, the decay widths Γ of $Y(3940)$ and $X(3872)$ are predicted with the values consistent with the experiments as shown in Table 2. The predictions for $Y(3940)$ are fairly consistent with the present experiments. By taking the I=0 component $\cos\theta_Y \simeq 1$, the main decay mode of $Y(3940)$ is considered to be $J/\psi\omega$. Experimentally the $Y(3940)$ is observed in $J/\psi\omega$, not in $D\bar{D}$ and $D\bar{D}^*$. The total decay width $\Gamma_Y^{\rm exp} = 87\pm22\pm26$MeV, which is close to our prediction obtained by the sum of $J/\psi, D\bar{D}$ and $D\bar{D}^*$, $\Gamma_Y^{\rm tot}$=88+26+4=118MeV. The Γ_X are also consistent with the present experiment, although they are largely dependent upon the treatment of the effective momenta of final mesons.

TABLE 2. Predicted decay widths of $Y(3940)$, Γ_Y, and $X(3872)$, Γ_X in MeV. For the $\Gamma_Y(D\bar{D}^*)$, $\bar{D}D^*$ is also included. The $\cos\theta_{X,Y}(\sin\theta_{X,Y})$ are I=0(1) component of the flavor WFs of X,Y. See ref.[8] for details.

R	$J/\psi\omega$	$D\bar{D}$	$D\bar{D}^*$		$J/\psi\rho$	$J/\psi\omega$
$\Gamma_Y(R)$	$88.2\cos^2\theta_Y$	25.7	4.2	$\Gamma_X(R)$	$(1.9\sin^2\theta_X)$	$(0.3\cos^2\theta_X)$

TABLE 3. The masses and widths for the ground cs-$\bar{s}\bar{c}$ states with $J^{PC}=1^{++}$, $X^{1++}_{cs-\bar{s}\bar{c}}$, and 2^{++}, $Y^{2++}_{cs-\bar{s}\bar{c}}$. The total width of $Y^{2++}_{cs-\bar{s}\bar{c}}$ is estimated by the sum as $\Gamma^{\rm tot}_{Y^{2++}_{cs-\bar{s}\bar{c}}}$=27+16+12=55MeV.

	$\Gamma_Y(D_s\bar{D}_s)$	$\Gamma_Y(D_s\bar{D}^*_s)$	$\Gamma_Y(J/\psi\phi)$		$\Gamma^{\rm tot}_X$
$Y^{2++}_{cs-\bar{s}\bar{c}}(4145)$	27MeV	16MeV	12MeV	$X^{1++}_{cs-\bar{s}\bar{c}}(4074)$	$<$ 1MeV

CONCLUDING REMARKS

The properties of $X(3872)$ and $Y(3940)$ are consistent with the four quark cn-$\bar{n}\bar{c}$ states with $J^{PC}=1^{++}$ and 2^{++}, respectively, which come from 0^--1^- and 1^--1^- combinations of cn chiral di-quark and $\bar{n}\bar{c}$ chiral di-antiquark in $\tilde{U}(12)$-scheme. Their mass difference is consistently explained by the spin-spin interaction between constituents in JSQM. Their narrow widths are explained from a selection rule, called ρ_3-line rule in $\tilde{U}(12)$-scheme. Only these two states are expected to be observed as resonant particles in ground states, and the other states have huge widths and are observed as non-resonant backgrounds. In order to check this interpretation, we present the predictions for the properties of ground cs-$\bar{s}\bar{c}$ states, obtained by a similar method, in Table 3. There are no simple decay modes for $X^{1++}_{cs-\bar{s}\bar{c}}$, and its full width is expected to be of order keV. The main decay mode of $Y^{2++}_{cs\bar{s}\bar{c}}$ is $D_s\bar{D}_s$. In order to check the four quark nature of this state, it is necessary to observe $J/\psi\phi$ decay.

The author would like to express his sincere gratitude to Prof. S. Ishida and Dr. T. Maeda for useful information and comments.

REFERENCES

1. S. K. Choi et al., Belle Collaboration, Phys. Rev. Lett. **91** (2003) 262001.
2. S. K. Choi et al., Belle Collaboration, hep-ex/0408126.
3. L. Maiani, F. Piccinini, A. D. Polosa and V. Riquer, Phys. Rev. D **71** (2005) 014028.
4. S. Ishida and M. Ishida, Phys. Lett. B 539 (2002) 249.
5. S. Ishida et al., hep-ph/0408136.
6. M. Ishida, hep-ph/0506160, to be published in Phys. Lett. B.
7. S. Ishida, in proc. of Hadron'05, Rio de Janeiro, August, 2005.
8. M. Ishida, hep-ph/0509212.
9. L. H. Chan and R. W. Haymaker, Phys. Rev. D **7** (1973) 402.
   M. Ishida, Prog. Theor. Phys. **101** (1999) 661.
10. S. Ishida, M. Oda, K. Takeuchi and M. Watanabe, Prog. Theor. Phys. **68** (1982) 883.
11. B. Yabsley, hep-ex/0507028.

Recent results on charm spectroscopy

J. Vijande*, F. Fernández* and A. Valcarce*

*Grupo de Física Nuclear
Universidad de Salamanca, E-37008 Salamanca, Spain

Abstract. We present a theoretical framework that accounts for the new D_J and D_{sJ} mesons measured in the open-charm sector. These resonances are properly described if considered as a mixture of conventional $P-$wave quark-antiquark states and four-quark components. The narrowest states are basically $P-$wave quark-antiquark mesons, while the dominantly four-quark states are shifted above the corresponding two-meson threshold, being broad resonances. We study the electromagnetic decay widths as basic tools to scrutiny their nature. The proposed explanation incorporates in a natural way the most recently discovered mesons in charmonium espectroscopy.

PACS: 14.40.Lb, 14.40.Ev, 12.39.Pn.

During the last few years, heavy meson spectroscopy is living a continuous excitation due to the discovery of several new charmed mesons. Two years ago BABAR Collaboration reported the observation of a charm-strange state, the $D_{sJ}^*(2317)$ [1]. It was confirmed by CLEO Collaboration [2] and also by Belle Collaboration [3]. Besides, BABAR had also pointed out to the existence of another charm-strange meson, the $D_{sJ}(2460)$ [1]. This resonance was measured by CLEO [2] and confirmed by Belle [3]. Belle results are consistent with the spin-parity assignments of $J^P = 0^+$ for the $D_{sJ}^*(2317)$ and $J^P = 1^+$ for the $D_{sJ}(2460)$. Thus, these two states are definitively well established, confirmed independently by different experiments. They present unexpected properties, quite different from those predicted by quark potential models. If they would correspond to standard $P-$wave mesons made of a charm quark, c, and a strange antiquark, $\bar{s}$, their masses would be larger, around 2.48 GeV for the $D_{sJ}^*(2317)$ and 2.55 GeV for the $D_{sJ}(2460)$. They would be therefore above the DK and D^*K thresholds, respectively, being broad resonances. However the states observed by BABAR and CLEO are very narrow, $\Gamma < 4.6$ MeV for the $D_{sJ}^*(2317)$ and $\Gamma < 5.5$ MeV for the $D_{sJ}(2460)$.

The intriguing situation of the charm-strange mesons has been translated to the nonstrange sector with the Belle observation [4] of a nonstrange broad scalar resonance, D_0^*, with a mass of $2308 \pm 17 \pm 15 \pm 28$ MeV/c^2 and a width $\Gamma = 276 \pm 21 \pm 18 \pm 60$ MeV. A state with similar properties has been suggested by FOCUS Collaboration at Fermilab [5] during the measurement of masses and widths of excited charm mesons D_2^*. They found significant evidence for a broad excess parametrized as an $S-$wave resonance. Without being able to clearly distinguish the origin of this broad excess, they conclude that their results are in agreement with the Belle results [4]. This state generates for the open-charm nonstrange mesons a very similar problem to the one arising in the strange sector with the $D_{sJ}^*(2317)$. If the $D_0^*(2308)$ would correspond to a standard $P-$wave meson made of a charm quark, c, and a light antiquark, $\bar{n}$, its mass would have to be larger, around 2.46 GeV. In this case, the quark potential models prediction and the measured resonance are both above the $D\pi$ threshold, the large width observed being expected

CP814 *Hadron Spectroscopy: Eleventh International Conference*
edited by A. Reis, C. Göbel, J. Sá Borges, and J. Magnin

TABLE 1. $c\bar{s}$ and $c\bar{n}$ masses (QM), in MeV. Experimental data (Exp.) are taken from Ref. [11], except for the state denoted by a dagger that has been taken from Ref. [4].

$nL\,J^P$	State	QM ($c\bar{s}$)	Exp.	State	QM ($c\bar{n}$)	Exp
$1S\,0^-$	D_s	1981	1968.5±0.6	D	1883	1867.7±0.5
$1S\,1^-$	D_s^*	2112	2112.4±0.7	D^*	2010	2008.9±0.5
$1P\,0^+$	$D_{sJ}^*(2317)$	2489	2317.4±0.9	$D_0^*(2308)$	2465	2308±17±15±28†
$1P\,1^+$	$D_{sJ}(2460)$	2578	2459.3±1.3	$D_1(2420)$	2450	2422.2±1.8
$1P\,1^+$	$D_{s1}(2536)$	2543	2535.3±0.6	$D_1^0(2430)$	2546	2427±26±25
$1P\,2^+$	$D_{s2}(2573)$	2582	2572.4±1.5	$D_2^*(2460)$	2496	2459±4

although not its low mass.

The last step in this series of discoveries has been the observation of a new charm-strange meson, D_{sJ}, announced by SELEX Collaboration at Fermilab [6] with a mass of 2632.5 ± 1.7 MeV/c^2 and a small width, $\Gamma < 17$ MeV. However, up to now no other experiment has been able to confirm the existence of this resonance [7].

The difficulties to identify the D_J and D_{sJ} states with conventional $c\bar{q}$ mesons are rather similar to those appearing in the light-scalar meson sector [8] and may be indicating that other configurations are playing a role. $q\bar{q}$ states are more easily identified with physical hadrons when virtual quark loops are not important. This is the case of the pseudoscalar and vector mesons, mainly due to the $P-$wave nature of this hadronic dressing. On the contrary, in the scalar sector is the $q\bar{q}$ pair the one in a $P-$wave state, whereas quark loops may be in a $S-$wave. In this case the intermediate hadronic states that are created must play a crucial role in the composition of the resonance, in other words unquenching is important. The vicinity of these components to the lightest $c\bar{s}$ ($c\bar{n}$) state implies that they have to be considered, being the first available color-singlet combination: $cn\bar{s}\bar{n}$ ($cn\bar{n}\bar{n}$). This has been shown to be relevant for the proper description of the low-lying scalar mesons, where the coupling of the scalar $q\bar{q}$ nonet to the lightest $qq\bar{q}\bar{q}$ configurations allows for an almost one-to-one correspondence between theoretical states and experiment [9].

In this work we pretend to explore the same ideas for the understanding of the properties of the D_J and D_{sJ} meson states. In non-relativistic quark models gluon degrees of freedom are frozen and therefore the wave function of a zero baryon number (B=0) hadron may be written as $|\mathrm{B}=0\rangle = \Omega_1 |q\bar{q}\rangle + \Omega_2 |qq\bar{q}\bar{q}\rangle + ...$ where q stands for quark degrees of freedom and the coefficients Ω_i take into account the mixing of four-quark and $q\bar{q}$ states. The hamiltonian considering the mixing between both configurations could be described using the $^3P_0$ model, however, since this model depends on the vertex parameter, we prefer in a first approximation to parametrize this coefficient by looking to the quark pair that is annihilated and not to the spectator quarks that will form the final $q\bar{q}$ state. Therefore we have taken $V_{q\bar{q}\leftrightarrow qq\bar{q}\bar{q}} = \gamma$. Further details about the formalism are given in Ref. [9]. The calculations have been performed in the framework of the constituent quark model of Ref. [10].

A thoroughly study of the full meson spectra has been presented in Ref. [10]. The results for the open-charm mesons are resumed in Table 1. It can be seen how the open-charm states are easily identified with standard $c\bar{q}$ mesons except for the cases of the

TABLE 2. Probabilities (P), in %, of the wave function components and masses (QM), in MeV, of the open-charm mesons once the mixing between $q\bar{q}$ and $qq\bar{q}\bar{q}$ configurations is considered. Experimental data (Exp.) are taken from Ref. [11] except for the state denoted by a dagger that has been taken from Ref. [4].

$I=0$					$I=1/2$	
$J^P=0^+$		$J^P=1^+$			$J^P=0^+$	
QM	2339	QM	2421	2555	QM	2241
Exp.	2317.4±0.9	Exp.	2459.3±1.3	2535.3±0.6	Exp.	2308±17±15±28[†]
P($cn\bar{s}\bar{n}$)	28	P($cn\bar{s}\bar{n}$)	25	~ 1	P($cn\bar{n}\bar{n}$)	46
P($c\bar{s}_{1^3P}$)	71	P($c\bar{s}_{1^1P}$)	74	~ 1	P($c\bar{n}_{1P}$)	53
P($c\bar{s}_{2^3P}$)	~ 1	P($c\bar{s}_{1^3P}$)	~ 1	98	P($c\bar{n}_{2P}$)	~ 1

$D^*_{sJ}(2317)$, the $D_{sJ}(2460)$, and the $D^*_0(2308)$. This is a common behavior of almost all quark potential model calculations [12]. In a similar manner, quenched lattice NRQCD predicts for the $D^*_{sJ}(2317)$ a mass of 2.44 GeV [13], while using relativistic charm quarks the mass obtained is 2.47 GeV [14]. Unquenched lattice QCD calculations of $c\bar{s}$ states do not find a window for the $D^*_{sJ}(2317)$ [8], supporting the difficulty of a $P-$wave $c\bar{s}$ interpretation.

The results obtained for the $cn\bar{s}\bar{n}$ configuration with the same interacting potential used for Table 1 are 2731 and 2699 MeV for the $J^P=0^+$ with $I=0$ and $I=1$, and 2841 and 2793 MeV for the $J^P=1^+$ with $I=0$ and $I=1$. For the $cn\bar{n}\bar{n}$ configuration with $I=1/2$ the energy is 2505 MeV. The $I=1$ and $I=0$ states are far above the corresponding strong decaying thresholds and therefore should be broad, what rules out a pure four-quark interpretation of the new open-charm mesons.

As outlined above, for $P-$wave mesons the hadronic dressing is in a $S-$wave, thus physical states may correspond to a mixing of two- and four-body configurations. In the isoscalar sector, the $cn\bar{s}\bar{n}$ and $c\bar{s}$ states get mixed, as it happens with $cn\bar{n}\bar{n}$ and $c\bar{n}$ for the $I=1/2$ case. Let us notice that the interacting potential contains terms mixing $P-$wave $S=0$ and $S=1$ states [10], arising from the different spin-orbit contributions for systems made of quarks of different mass. The mixing angle comes determined by the structure of the interaction. The parameter γ has been fixed to reproduce the mass of the $D^*_{sJ}(2317)$ meson, $\gamma=240$ MeV.

The results obtained are shown in Table 2. Let us first analyze the nonstrange sector. The $^3P_0$ $c\bar{n}$ pair and the $cn\bar{n}\bar{n}$ have a mass of 2465 MeV and 2505 MeV, respectively. Once the mixing is considered one obtains a state at 2241 MeV with 46% of four-quark component and 53% of $c\bar{n}$ pair. The lowest state, representing the $D^*_0(2308)$, is above the isospin preserving threshold $D\pi$, being broad as observed experimentally. The mixed configuration compares much better with the experimental data than the pure $c\bar{n}$ state. The orthogonal state appears higher in energy, at 2713 MeV, with and important four-quark component.

Concerning the strange sector, the $D^*_{sJ}(2317)$ and the $D_{sJ}(2460)$ are dominantly $c\bar{s}$ $J=0^+$ and $J=1^+$ states, respectively, with almost 30% of four-quark component. Such component is responsible for the shift of the mass of the unmixed states to the experimental values below the DK and D^*K thresholds. Being both states below their

TABLE 3. Electromagnetic decay widths, in keV, for the $D^*_{sJ}(2317)$ and $D_{sJ}(2460)$ (QM), compared to the results of two different quark models based only on $q\bar{q}$ states. To compare with the experimental data by CLEO and Belle we have assumed for $\Gamma(D_s^{*+}\pi^0) \approx \Gamma(D_s^+\pi^0) \approx 10$ keV as estimated in Ref. [16].

Transition	Quark models			Experiments	
	QM	Ref. [15]	Ref. [16]	CLEO [2]	Belle [3]
$D^*_{sJ}(2317) \to D_s^{*+}\gamma$	1.6	1.74	1.9	< 0.59	< 1.8
$D^*_{sJ}(2317) \to D_s^{+}\gamma$	0.0	0.0	0.0	< 0.52	< 0.5
$D_{sJ}(2460) \to D_s^{*+}\gamma$	0.06	4.66	5.5	< 1.6	< 3.1
$D_{sJ}(2460) \to D_s^{+}\gamma$	6.7	5.08	6.2	< 4.9	5.5±1.3±0.8

isospin-preserving two-meson threshold, the only allowed strong decays to $D_s^*\pi$ would violate isospin and are expected to have small widths $O(10)$ keV [15, 16]. As a consequence, they should be narrower than the $D_{s2}(2573)$ and $D_{s1}(2536)$, opposite to what it is expected from heavy quark symmetry. The second isoscalar $J^P = 1^+$ state, with an energy of 2555 MeV and 98% of $c\bar{s}$ component, corresponds to the $D_{s1}(2536)$. Regarding the $D^*_{sJ}(2317)$, it has been argued that a possible DK molecule would be preferred with respect to an $I = 0$ $cn\bar{s}\bar{n}$ tetraquark, what would anticipate an $I = 1$ $cn\bar{s}\bar{n}$ partner nearby in mass [17]. Our results confirm the last argument, the vicinity of the isoscalar and isovector tetraquarks, however, the restricted coupling to the $c\bar{s}$ system allowed only for the $I = 0$ four-quark states opens the possibility of a mixed nature for the $D^*_{sJ}(2317)$ while the $I = 1$ $J = 0^+$ and $J = 1^+$ four-quark states appear above 2700 MeV and cannot be shifted to lower energies.

Our results do not show any state around 2600 MeV, the mass region compatible with the $D^+_{sJ}(2632)$ measured by SELEX. Its two possible theoretical partners, the 2^3S_1 state or the 1^3D_1 state, lie above 2700 MeV (2764 and 2873 MeV, respectively). A careful analysis of several theoretical interpretations has been done in Ref. [18], the surprising properties of this state not fitting in any of the scenarios considered. Several experiments have failed trying to confirm the existence of this state [7]. The confirmation or refutation of the $D^+_{sJ}(2632)$ is clearly an important priority for meson spectroscopy.

Apart from the masses, the structure of the $D^*_{sJ}(2317)$ and the $D_{sJ}(2460)$ mesons could be scrutiny also through the study of their electromagnetic decay widths. We compare in Table 3 our results with different theoretical approaches and the experimental limits reported by Belle and CLEO. The main difference is noticed in the suppression predicted for the $D_{sJ}(2460) \to D_s^{*+}\gamma$ decay as compared to the $D_{sJ}(2460) \to D_s^+\gamma$. A ratio $D_{sJ}(2460) \to D_s^+\gamma / D_{sJ}(2460) \to D_s^{*+}\gamma \approx 1-2$ has been obtained assuming a $q\bar{q}$ structure for both states [15, 16] (what seems incompatible with their properties). We find a larger value, $D_{sJ}(2460) \to D_s^+\gamma / D_{sJ}(2460) \to D_s^{*+}\gamma \approx 100$, due to the small 1^3P_1 $c\bar{s}$ probability of the $D_{sJ}(2460)$. A similar enhancement has been obtained in Ref. [19] in the framework of light-cone QCD sum rules in contrast to a previous calculation of the same authors using vector meson dominance [20].

Let us finally mention that the recent charmonium and B_c states measured at different facilities do nicely fit into the predictions of the model used, see Table 4, given confidence to the results obtained in the present work.

TABLE 4. Masses (QM), in MeV, of the recently measured charmonium and B_c states obtained within the model of Ref. [10] used in this work.

Name	Mass	Ref.	$n^{2S+1}L_J$	QM
$X(3940)$	$3943 \pm 6 \pm 6$	[21]	2^1P_1	3923
–	–		2^3P_0	3878
$Y(3940)$	$3943 \pm 11 \pm 13$	[22]	2^3P_1	3915
$X'_{c2}(3940)$	$3931 \pm 4 \pm 2$	[23]	2^3P_2	3936
$Y(4260)$	$4260 \pm 8 \pm 2$	[24]	4^3S_1	4307
$B_c(6287)$	$6287 \pm 4.8 \pm 1.1$	[25]	1^1S_0	6277

ACKNOWLEDGMENTS

This work has been partially funded by Ministerio de Ciencia y Tecnología under Contract No. FPA2004-05616 and by Junta de Castilla y León under Contract No. SA-104/04

REFERENCES

1. BABAR Collaboration, B. Aubert *et al.*, Phys. Rev. Lett. **90**, 242001 (2003).
2. CLEO Collaboration, D. Besson *et al.*, Phys. Rev. D **68**, 032002 (2003).
3. Belle Collaboration, Y. Mikani *et al.*, Phys. Rev. Lett. **92**, 012002 (2004).
4. Belle Collaboration, K. Abe *et al.*, Phys. Rev. D **69**, 112002 (2004).
5. FOCUS Collaboration, J.M. Link *et al.*, Phys. Lett. B **586**, 11 (2004).
6. SELEX Collaboration, A.V. Evdokimov *et al.*, Phys. Rev. Lett. **93**, 242001 (2004).
7. BABAR Collaboration, B. Aubert *et al.*, hep-ex/0408087. Belle Collaboration, B. Yabsley, hep-ex/0507028. FOCUS Collaboration, R. Kutschke, E831-doc-701-v2.
8. G.S. Bali, Phys. Rev. D **68**, 071501(R) (2003).
9. J. Vijande, A. Valcarce, F. Fernández, and B. Silvestre-Brac, Phys. Rev. D **72**, 034025 (2005).
10. J. Vijande, F. Fernández, and A. Valcarce, J. Phys. G **19**, 2013 (2005).
11. S. Eidelman *et al.*, Phys. Lett. B **592**, 1 (2004).
12. S. Godfrey and R. Kokoski, Phys. Rev. D **43**, 1679 (1991). D. Ebert, V.O. Galkin, and R.N. Faustov, Phys. Rev. D **57**, 5663 (1998). M. Di Pierro and E. Eichten, Phys. Rev. D **64**, 114004 (2001). W. Lucha and F.F. Schöberl, Mod. Phys. Lett A **18**, 2837 (2003).
13. J. Hein *et al.*, Phys. Rev. D **62**, 074503 (2000).
14. UKQCD Collaboration, P. Boyle, Nucl. Phys. B (Proc. Supp.) **63**, 314 (1998); Nucl. Phys. B (Proc. Supp.) **53**, 398 (1997).
15. W.A. Bardeen, E.J. Eichten, and C.T. Hill, Phys. Rev. D **68**, 054024 (2003).
16. S. Godfrey, Phys. Lett. B **568**, 254 (2003).
17. T. Barnes, F.E. Close, and H.J. Lipkin, Phys. Rev. D **68**, 054006 (2003).
18. T. Barnes, F.E. Close, J.J. Dudek, S. Godfrey, and E.S. Swanson, Phys. Lett. B **600**, 223 (2004).
19. P. Colangelo, F. de Fazio, and A. Ozpineci, hep-ph/0505195.
20. P. Colangelo and F. de Fazio, Phys. Lett. B **570**, 180 (2003). P. Colangelo and F. de Fazio, and R. Ferrandes, Mod. Phys. Lett. A **19**, 2083 (2004).
21. Belle Collaboration, K. Abe *et al.*, hep-ex/0507019.
22. Belle Collaboration, S.-K. Choi *et al.*, Phys. Rev. Lett. **94**, 182002 (2005).
23. Belle Collaboration, K. Abe *et al.*, hep-ex/0507033.
24. BABAR Collaboration, B. Aubert *et al.*, hep-ex/0506081.
25. CDF Collaboration, M.D. Corcoran, hep-ex/0506061.

Studies of e^+e^- Collisions with a Hard Initial-State Photon at BaBar

Nicolas Berger

SLAC MS 61, Menlo Park, California, U.S.A.

Abstract. We present preliminary BaBar measurements of hadronic cross sections in e^+e^- annihilation using the radiative return technique. The cross sections for $e^+e^- \to p\bar{p}$, $3(\pi^+\pi^-)$, $2(\pi^+\pi^-)2\pi^0$, and $K^+K^-2(\pi^+\pi^-)$ are measured. Measurements of the proton form factor and of the ratio G_E/G_M are also shown.

Keywords: spectroscopy, ISR, radiative return, BaBar

INTRODUCTION

In recent years, precise new results in low-energy hadronic cross sections have been obtained at high-luminosity e^+e^- machines using the radiative return technique. The cross section for $e^+e^- \to \pi^+\pi^-$ has been measured by the KLOE collaboration [1], while those of $e^+e^- \to \pi^+\pi^-\pi^0$, $\pi^+\pi^-\pi^+\pi^-$, $K^+K^-K^+K^-$ and $K^+K^-\pi^+\pi^-$ have been measured by BaBar [2, 3]. These measurements are of particular interest since they provide input to data-driven calculations of hadronic contributions to the muon anomalous magnetic moment, a_μ^{had}, and of the running of the QED coupling constant $\Delta\alpha_{had}^{(5)}$, which appears in global fits to the standard model. As a continuation of this program, preliminary BaBar measurements for the $p\bar{p}$, $3(\pi^+\pi^-)$, $2(\pi^+\pi^-)2\pi^0$, and $K^+K^-2(\pi^+\pi^-)$ final states are presented here. The $p\bar{p}$ and six-hadron results are based on $240\,\text{fb}^{-1}$ and $232\,\text{fb}^{-1}$ of data respectively.

RADIATIVE RETURN

The radiative return method uses the emission of an initial-state photon to probe center-of-mass energies $\sqrt{s'}$ below the nominal collision energy $\sqrt{s}$. The emission of a photon of energy E_γ in the center-of-mass frame corresponds to the production of a recoiling system with $s' = s - 2\sqrt{s}E_\gamma$. Cross-section measurements can therefore be obtained over a wide range of center-of-mass energies without varying thc beam energy, thereby avoiding the usual point-to-point uncertainties associated with discrete measurements.

At BaBar the range $\sqrt{s'} \leq 4.5\,\text{GeV}$ is readily accessible. To reduce backgrounds, events are reconstructed by requiring the ISR photon to be detected within the acceptance of the BaBar electromagnetic calorimeter. Since the recoiling system is produced as a jet opposite the high-energy photon, the event detection efficiency is only weakly dependent on $\sqrt{s'}$, and measurements down to threshold are possible. The efficiency also does not depend strongly on the simulation of the hadronic final state.

CP814 *Hadron Spectroscopy: Eleventh International Conference*
edited by A. Reis, C. Göbel, J. Sá Borges, and J. Magnin

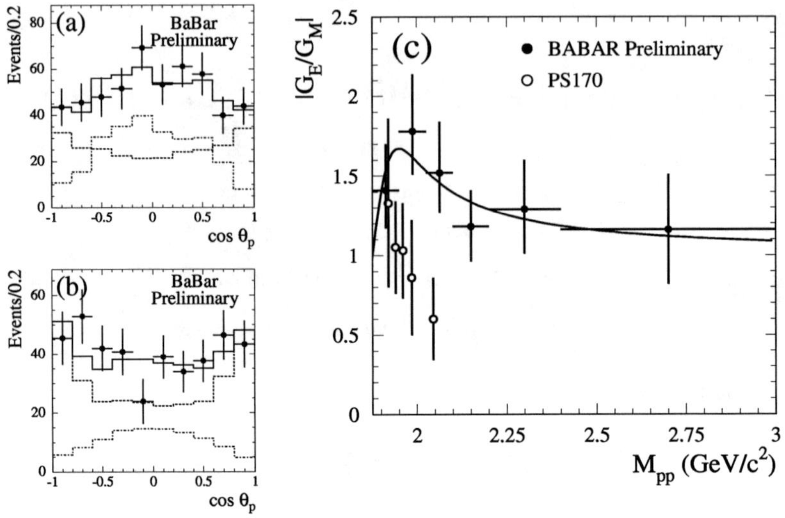

FIGURE 1. $\cos\theta_p$ distributions for the mass bins $[1.877;1.950]\,\mathrm{GeV}/c^2$ (a) and $[2.400;3.000]\,\mathrm{GeV}/c^2$ (b). The dashed (dot-dashed) curve shows the contribution from the simulated $|G_E|^2$ ($|G_M|^2$) cross section term; the solid line is the sum of the two terms and the points correspond to data. (c) Mass dependence of the ratio $|G_E/G_M|$.

At first order the cross section for an ISR process with a photon polar angle acceptance of $\theta_{min} < \theta_\gamma < \pi - \theta_{min}$ in the center-of-mass frame is

$$\frac{d\sigma_{\mathrm{e^+e^-}\to\gamma X}}{dx} = \frac{\alpha}{\pi}\frac{1+x^2}{x(1-x)}\left(\log\frac{1-\cos\theta_{min}}{1+\cos\theta_{min}} - (1-x)cos\theta_{min}\right)\sigma_{\mathrm{e^+e^-}\to X}(s(1-x)) \quad (1)$$

where $x = 1 - s'/s$. A kinematic fit is used to measure s' and reject backgrounds. Cross sections are normalized using the process $\mathrm{e^+e^-} \to \mu^+\mu^-\gamma$.

THE $\mathrm{p\bar{p}}$ FINAL STATE

The cross section for $\mathrm{e^+e^-} \to \mathrm{p\bar{p}}$ is given by

$$\sigma_{\mathrm{e^+e^-}\to\mathrm{p\bar{p}}}(s') = \frac{4\pi\alpha^2 C}{3s'}\sqrt{1-\frac{2m_p^2}{s'}}\left(|G_E(s')|^2 + \frac{2m_p^2}{s'}|G_M(s')|^2\right), \quad (2)$$

where m_p is the proton mass and G_E and G_M are respectively the electric and magnetic form factors of the proton. The factor C accounts for Coulomb interaction in the $\mathrm{p\bar{p}}$ system and has the effect of making the cross section non-zero at threshold.

The detection efficiency for the ISR process is about 17%, with no strong dependence on either the $\mathrm{p\bar{p}}$ invariant mass $m_{\mathrm{p\bar{p}}}$ or the value of $|G_E/G_M|$.

The proton helicity angle θ_p in the $\mathrm{p\bar{p}}$ rest frame can be used to separate the $|G_E|^2$ and $|G_M|^2$ terms. Their respective variations are approximately $\sin^2\theta_p$ and $1+\cos^2\theta_p$, with exact expressions obtained from Monte-Carlo simulation. By fitting the $\cos\theta_p$ distribution to a sum of the two terms the ratio $|G_E/G_M|$ can be extracted. This is done separately in six bins of $m_{\mathrm{p\bar{p}}}$. Preliminary results are shown in Fig. 1, and disagree significantly with previous measurements from LEAR [7].

The cross section for $\mathrm{e^+e^-} \to \mathrm{p\bar{p}}$ is measured and from it we obtain the "effective form factor" $G = \sqrt{|G_E|^2 + 2m_p^2/s'|G_M|^2}$. Preliminary values are shown in Figs. 2(a) and 2(b). Results are good agreement with existing data, and cover the entire energy range from threshold to 4.5 GeV in a single measurement. The data show a clear enhancement at threshold, already observed by LEAR, as well as hints of structures at 2.25 and $3\,\mathrm{GeV}/c^2$. Apart from these structures, the data agree well with a fit to the perturbative QCD expectation [4, 5] $G(m_{\mathrm{p\bar{p}}}) = A/m_{\mathrm{p\bar{p}}}^4 \log(m_{\mathrm{p\bar{p}}}^2/\Lambda^2)$, where A and Λ are constants determined from the fit, which is shown as the dashed curve in Fig. 2(a).

FINAL STATES WITH SIX HADRONS

As an extension of the recent ISR measurements of the cross section of final states with four pions or kaons [3], preliminary results have been obtained for the six-hadron processes $\mathrm{e^+e^-} \to 3(\pi^+\pi^-)$, $2(\pi^+\pi^-)2\pi^0$ and $K^+K^-2(\pi^+\pi^-)$. The corresponding cross section distributions are shown in Figs. 2(c), 2(d) and 2(e) respectively. In the $2(\pi^+\pi^-)2\pi^0$ mode the data disagree strongly with the DM2 results above 1.8 GeV. Apart from this, the data agrees well with existing results, and is considerably more precise. The cross section for $K^+K^-2(\pi^+\pi^-)$ is measured for the first time. Again, the entire energy range from threshold to 4.5 GeV is covered in a single experiment.

A clear dip is visible at about 1.9 GeV in the 6π modes. A similar feature was already seen by DM2 in $\mathrm{e^+e^-} \to 3(\pi^+\pi^-)$ and by FOCUS [8] in the diffractive photoproduction of six charged pions. The cross section distributions are fitted using the parametrization used by FOCUS, a sum of a Breit-Wigner resonance shape and a Jacob-Slansky continuum [6]. For the $3(\pi^+\pi^-)$ ($2(\pi^+\pi^-)2\pi^0$) mode, we obtain values of 1880 ± 30 MeV (1860 ± 20 MeV) for the resonance peak, 130 ± 30 MeV (160 ± 20 MeV) for the resonance width and 21 ± 14 deg (-3 ± 15 deg) for phase shift between the resonance and continuum. The width values differ significantly from the FOCUS result of 29 ± 14 MeV.

In the $3(\pi^+\pi^-)$ channel the resonance structure is surprisingly simple, being well-described by a Monte-Carlo simulation featuring a single $\rho^0 \to \pi^+\pi^-$ resonance and the other four pions distributed according to phase space. However rich resonant structures are observed in the $2(\pi^+\pi^-)2\pi^0$ channel, with signals for $\rho^0 \to \pi^+\pi^-$, $\rho^+ \to \pi^+\pi^0$, $f_0 \to \pi^0\pi^0$ and $f_0(1270) \to \pi^+\pi^-$ in the 2π combinations, and signals for ω and η in $\pi^+\pi^-\pi^0$. A signal for $\mathrm{e^+e^-} \to \omega(\pi^+\pi^-\pi^0)\eta(\pi^+\pi^-\pi^0)$ is also seen, corresponding to a resonant structure in the 6π invariant mass, as shown in Fig. 2(f). Fitting the distribution to a Breit-Wigner shape gives a peak position of 1645 ± 8 MeV and a width of 114 ± 14 MeV, which could correspond to the ω'' state reported in our previous analysis of the $\pi^+\pi^-\pi^0$ final state [2]. Apart from this structure, which is unique to the $2(\pi^+\pi^-)2\pi^0$ mode, the ratio of the $2(\pi^+\pi^-)2\pi^0$ and $3(\pi^+\pi^-)$ cross sections is

TABLE 1. Summary of J/ψ and $\psi(2S)$ branching fraction obtained from six-hadron final states

Mode	Branching fraction (this work)	Branching Fraction (PDG)
$J/\psi \to 3(\pi^+\pi^-)$	$(4.40 \pm 0.29 \pm 0.29) \times 10^{-3}$	$(4.0 \pm 2.0) \times 10^{-3}$
$J/\psi \to 2(\pi^+\pi^-)2\pi^0$	$(1.65 \pm 0.10 \pm 0.18) \times 10^{-2}$	n/a
$J/\psi \to K^+K^-2(\pi^+\pi^-)$	$(5.09 \pm 0.42 \pm 0.35) \times 10^{-3}$	$(3.1 \pm 1.3) \times 10^{-3}$
$J/\psi \to 2(\pi^+\pi^-)2\pi^0$	$(5.3 \pm 1.6 \pm 0.6) \times 10^{-3}$	n/a
$J/\psi \to K^+K^-2(\pi^+\pi^-)$	$(2.1 \pm 1.0 \pm 0.2) \times 10^{-3}$	n/a

remarkably constant over the entire energy range, at a value of $3.98 \pm 0.06(\text{stat.}) \pm 0.41(\text{syst.})$. In the $K^+K^-2(\pi^+\pi^-)$ final state, a signal for $\phi \to K^+K^-$ is seen, with a significant contribution from $J/\psi \to \phi\pi^+\pi^-\pi^+\pi^-$. Signals for $K^{*0} \to K^+\pi^-$ are also present. Finally, the observed J/ψ and $\psi(2S)$ signals can be used to extract branching fraction measurements which, in many cases, considerably improve over current world averages. These results are summarized in Table 1.

CONCLUSION

The program to measure low-energy hadronic cross sections using ISR is well underway at BaBar. In addition to the published results for $\pi^+\pi^-\pi^0$, $2(\pi^+\pi^-)$, $K^+K^-\pi^+\pi^-$ and $2(K^+K^-)$, we now have preliminary results for the $3(\pi^+\pi^-)$, $2(\pi^+\pi^-)2\pi^0$, $K^+K^-2(\pi^+\pi^-)$ and $p\bar{p}$ modes, with analyses ongoing for the final states $\pi^+\pi^-$, K^+K^-, $\pi^+\pi^-\pi^0\pi^0$, η, η', and $D^{(*)}\bar{D}^{(*)}$.

ACKNOWLEDGMENTS

The author is grateful for the extraordinary contributions of the PEP-II colleagues in achieving the excellent luminosity and machine conditions that have made this work possible. This work is supported by DOE and NSF (USA), NSERC (Canada), IHEP (China), CEA and CNRS-IN2P3 (France), BMBF and DFG (Germany), INFN (Italy), FOM (The Netherlands), NFR (Norway), MIST (Russia), and PPARC (United Kingdom). Individuals have received support from CONACyT (Mexico), A. P. Sloan Foundation, Research Corporation, and Alexander von Humboldt Foundation.

REFERENCES

1. S.E. Muller *et al.*, Nucl. Phys. Proc. Suppl. **126**, 335-340 (2004)
2. B. Aubert *et al.*, Phys. Rev. D **70**, 072004 (2004)
3. B. Aubert *et al.*, Phys. Rev. D **71**, 052001 (2005)
4. V. L. Chernyak and A. R. Zhitinsky, JETP Lett. **25**, 510 (1977)
5. P. Lepage and S. Brodsky, Phys. Rev. Lett **43**, 545 (1979)
6. M. Jacob and R. Slansky, Phys. Rev. Lett B**37**, 408 (1971) and Phys. Rev. D **5**, 1847 (1972)
7. G. Bardin *et al.*, Nucl. Phys. B **411**, 3 (1994)
8. E687 Collaboration: P.L. Frabetti *et al.*, Phys. Lett. **B**514, 240-246 (2001)

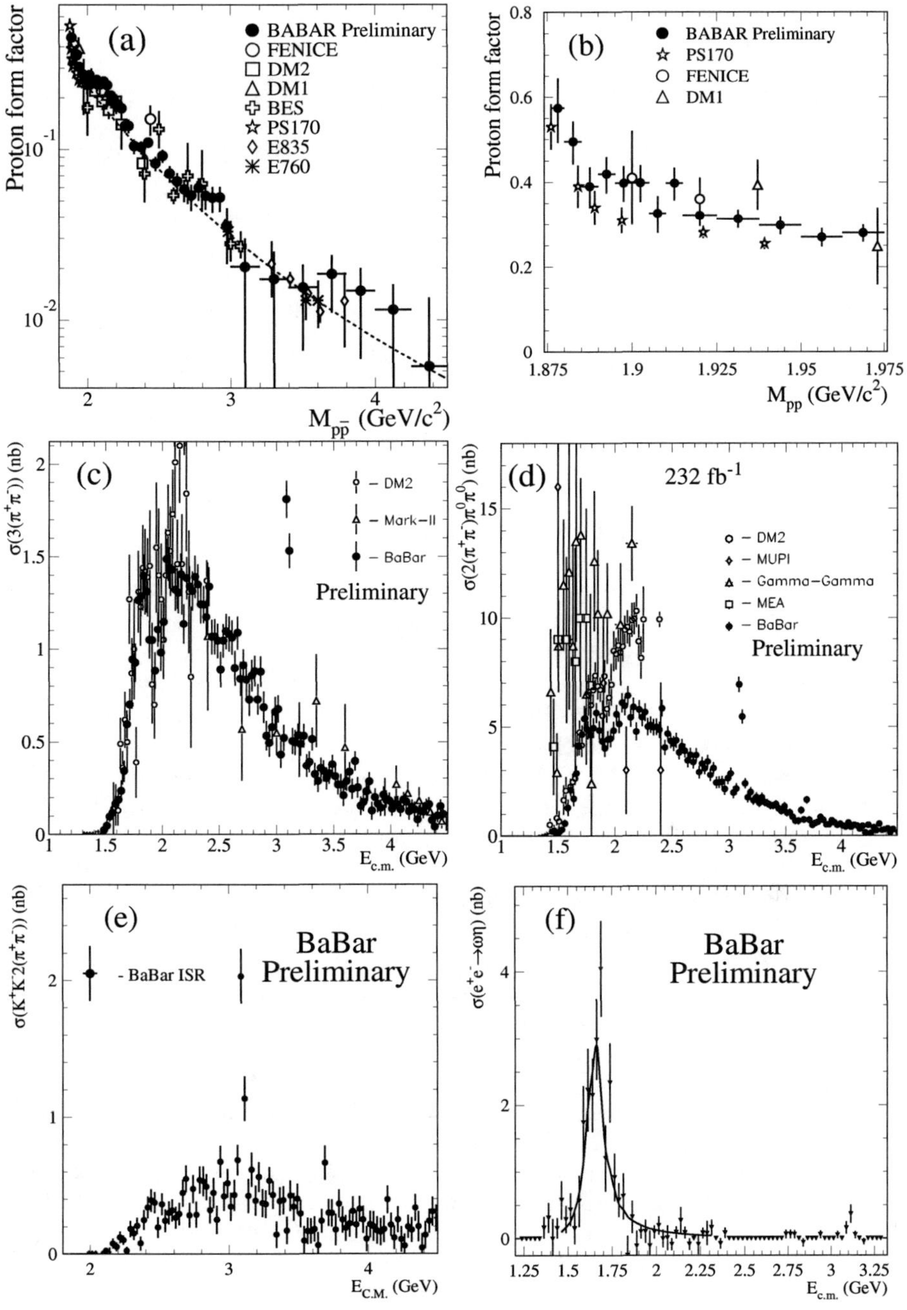

FIGURE 2. (a) Effective proton form factor versus $m_{p\bar{p}}$; (b) Close-up view of the threshold region. Cross section distributions versus $6h$ invariant mass for (c) $e^+e^- \to 3(\pi^+\pi^-)$, (d) $2(\pi^+\pi^-)2\pi^0$, (e) $K^+K^-2(\pi^+\pi^-)$, and (f) $e^+e^- \to \omega(\pi^+\pi^-\pi^0)\eta(\pi^+\pi^-\pi^0)$.

Four-quark state picture of the charmed mesons

M. Nielsen*, R.D. Matheus*, F.S. Navarra*, M.E. Bracco† and A. Lozea†

*Instituto de Física, Universidade de São Paulo, C.P. 66318, 05389-970 São Paulo, SP, Brazil
†Instituto de Física, Universidade do Estado do Rio de Janeiro, Rua São Francisco Xavier 524, 20550-900 Rio de Janeiro, RJ, Brazil

Abstract.
We employ the QCD sum rule approach to investigate the possible four-quark structure of the recently observed charmed scalar mesons $D_0^0(2308)$ (BELLE) and $D_0^{0,+}(2405)$ (FOCUS) and also of the very narrow $D_{sJ}^+(2317)$, firstly observed by BABAR. Using diquak-antidiquark currents, our results indicate that a four-quark structure is acceptable for the resonances observed by BELLE and BABAR: $D_0^0(2308)$ and $D_{sJ}^+(2317)$ respectively, but not for the resonances observed by FOCUS: $D_0^{0,+}(2405)$.

Recently new resonances with open- and hidden-charm have been reported by various experiments. The reported open-charmed states in the scalar sector include $D_0^0(2308)$ [1], $D_{sJ}^+(2317)$ [2], $D_0^0(2407)$ and the $D_0^+(2403)$ [3]. The discovery of these new resonances in the charm sector openned a new challenge for heavy quark system and a systematic analysis is required to understand their internal structure and their properties.

Among the new open-charmed states the very narrow $D_{sJ}^+(2317)$ is particularly interesting, since its mass is approximately 160 MeV below the prediction of the very successful quark model for the charmed mesons [4]. The mass of the scalar meson, $D_0^0(2308)$, observed by BELLE Collaboration is also bellow the prediction of ref. [4] (approximately 100 MeV), while the masses of the states observed by FOCUS Collaboration, $D_0^{0+}(2405)$, are in complete agreement with ref. [4].

Due to its low mass, the structure of the meson $D_{sJ}^+(2317)$ has been extensively debated. In refs. [5, 6, 7] it has been interpreted as a four-quark state. In the light sector the idea that the light scalar mesons (the isoscalars $\sigma(500)$, $f_0(980)$, the isodublet $\kappa(800)$ and the isovector $a_0(980)$) could be four-quark bound states is not new [8, 9]. Indeed, in a four-quark scenario, the mass degeneracy of $f_0(980)$ and $a_0(980)$ is natural, the mass hierarchy pattern of the nonet is understandable, and it is easy to explain why σ and κ are broader than $f_0(980)$ and $a_0(980)$.

In this work we use the method of QCD sum rules (QCDSR) [10] to study the two-point functions of the scalar mesons, $D_{sJ}(2317)$, $D_0(2308)$ and $D_0(2405)$ considered as four-quark states. In a recent calculation [11] the light scalar mesons were considered as S-wave bound states of a diquark-antidiquark pair. As suggested in ref. [12] the diquark was taken to be a spin zero colour anti-triplet. We extend this prescription to the charm sector and, therefore, the corresponding interpolating fields containing zero, one and two strange quarks are:

$$j_0 = \varepsilon_{abc}\varepsilon_{dec}(q_a^T C\gamma_5 c_b)(\bar{u}_d\gamma_5 C\bar{d}_e^T),$$

CP814 *Hadron Spectroscopy: Eleventh International Conference*
edited by A. Reis, C. Göbel, J. Sá Borges, and J. Magnin

$$
\begin{aligned}
j_1 &= \frac{\varepsilon_{abc}\varepsilon_{dec}}{\sqrt{2}}\left[(u_a^T C\gamma_5 c_b)(\bar{u}_d\gamma_5 C\bar{s}_e^T) + u \leftrightarrow d\right], \\
j_2 &= \varepsilon_{abc}\varepsilon_{dec}(s_a^T C\gamma_5 c_b)(\bar{q}_d\gamma_5 C\bar{s}_e^T),
\end{aligned} \tag{1}
$$

where q represents the quark u or d according to the charge of the meson. Since D_{sJ} has one $\bar{s}$ quark, we choose the j_1 current to have the same quantum numbers of D_{sJ}, which is supposed to be an isoscalar. However, since we are working in the SU(2) limit, the isoscalar and isovector states are mass degenerate and, therefore, this particular choice has no relevance here.

The QCDSR for the charmed scalar mesons with n strange quarks are constructed from the two-point correlation function

$$
\Pi(q) = i\int d^4x\, e^{iq.x}\langle 0|T[j_n(x)j_n^\dagger(0)]|0\rangle. \tag{2}
$$

In the OPE side we work at leading order and consider condensates up to dimension six. We deal with the strange quark as a light one and consider the diagrams up to order m_s. To keep the charm quark mass finite, we use the momentum-space expression for the charm quark propagator. We calculate the light quark part of the correlation function in the coordinate-space, which is then Fourier transformed to the momentum space in D dimensions. The resulting light-quark part is combined with the charm-quark part before it is dimensionally regularized at $D = 4$.

We can write the correlation function in the OPE side in terms of a dispersion relation:

$$
\Pi^{OPE}(q^2) = \int_{m_c^2}^{\infty} ds \frac{\rho(s)}{s-q^2}, \tag{3}
$$

where the spectral density is given by the imaginary part of the correlation function: $\rho(s) = \frac{1}{\pi}\mathrm{Im}[\Pi^{OPE}(s)]$.

In the phenomenological side, the coupling of the scalar meson with n strange quarks, S_n, to the scalar current, j_n, can be parametrized in terms of the meson decay constant f_{S_n} as [11]: $\langle 0|j_n|S_n\rangle = \sqrt{2} f_{S_n} m_{S_n}^4$, therefore, the phenomenological side of Eq. (2) can be written as

$$
\Pi^{phen}(q^2) = \frac{2 f_{S_n}^2 m_{S_n}^8}{m_{S_n}^2 - q^2} + \cdots, \tag{4}
$$

where the dots denote higher resonance contributions that will be parametrized, as usual, through the introduction of the continuum threshold parameter s_0 [13]. After making a Borel transform on both sides, and transferring the continuum contribution to the OPE side, the sum rule for the scalar meson S_n can be written as

$$
2 f_{S_n}^2 m_{S_n}^8 e^{-m_{S_n}^2/M^2} = \int_{m_c^2}^{s_0} ds\, e^{-s/M^2}\, \rho_{S_n}(s), \tag{5}
$$

where $\rho_{S_n}(s) = \rho^{pert}(s) + \rho^{m_s}(s) + \rho^{\langle\bar{q}q\rangle}(s) + \rho^{\langle G^2\rangle}(s) + \rho^{mix}(s) + \rho^{\langle\bar{q}q\rangle^2}(s)$, and all the expressions can be found in ref. [14].

We call $D_0^{(0s)}$, $D_0^{(1s)}$ and $D_0^{(2s)}$ the scalar charmed mesons represented by j_0, j_1 and j_2 (in Eq. (1)) respectively.

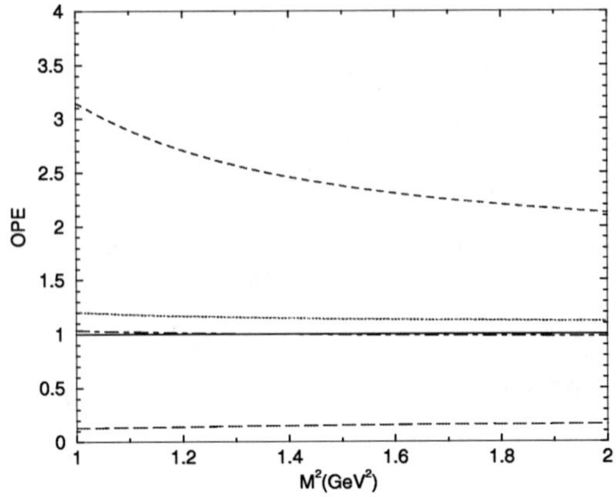

h

FIGURE 1. Borel mass dependence of the relative contributions of the OPE terms: perturbative (long-dashed line), perturbative plus quark condensate (dot-dashed line), previous plus four-quark condensate (dashed line), previous plus mixed condensate (solid line).

In Fig. 1 we show, as a function of the Borel mass, the OPE relative contribution of the: perturbative (long-dashed line), perturbative plus quark condensate (dot-dashed line), previous plus four-quark condensate (dashed line) and previous plus mixed condensate (solid line), for the $D_0^{(1s)}$ meson. We see that there is no good OPE convergence and that the four-quark condensate and the mixed condensate contributions are very big, as compared with the perturbative contribution, and with opposite signal, in such a way that the final result is almost the same as before adding these two contributions. One can argue that this is not a good sum rule, since there is not a good OPE convergence. We notice, however, that this is a common feature of the sum rules for currents with more than three quarks. The same behaviour is obtained for $D_0^{(0s)}$ and $D_0^{(2s)}$ mesons.

We have also studied the percentage of the pole and continuum contributions to the total contribution. We have obtained that in the Borel window $1.0\,\text{GeV}^2 \leq M^2 \leq 1.4\,\text{GeV}^2$ the pole contribution is always bigger than the continuum contribution. Therefore, this is the Borel window that we will consider.

In order to get rid of the meson decay constant and extract the resonance mass, m_{S_n}, we first take the derivative of Eq. (5) with respect to $1/M^2$ and then we divide it by Eq. (5). As can be seen in ref. [14], we get a reasonable stable result for the masses of these resonances, as a function of the Borel mass. Considering the variations on the quark masses and on the continuum threshold [14], in the Borel window considered here our results for the ressonance masses are given in Table I.

Table I: Numerical results for the resonance masses

resonance	$D_0^{(0s)}$	$D_0^{(1s)}$	$D_0^{(2s)}$
mass (GeV)	2.22 ± 0.15	2.32 ± 0.13	2.31 ± 0.14

Comparing the results in Table I with the resonance masses given by BABAR, BELLE and FOCUS: $D_{sJ}^{+}(2317)$, $D_0^0(2308)$ and $D_0^{0,+}(2405)$, we see that we can identify the four-quark states represented by $D_0^{(1s)}$ and $D_0^{(2s)}$ with the BABAR and BELLE resonances respectively. However, we do not find a four-quark state whose mass is compatible with the FOCUS resonances, $D_0^{0,+}(2405)$. Therefore, we associate the FOCUS resonances, $D_0^{0,+}(2405)$, with a scalar $c\bar{q}$ state, since its mass is completly in agreement with the predictions of the quark model in ref. [4].

Besides de masses, another important point to understand the nature of the charmed meson states is their corresponding decay width. One can ask how, in the present approach, it is possible to obtain a extremely narrow width for $D_{sJ}^{+}(2317)$, while the $D_0^0(2308)$ state remain fairly wide? To compute the decay width of the hadronic decay $D_0^{(1s)} \to D_s^+\pi^0$, for example, one has to study the three-point function

$$T_\mu(p,p',q) = \int d^4x\, d^4y\, e^{i.p'.x}\, e^{iq.y} \langle 0|T\{j_{D_s}(x) j_{5\mu}^{\pi}(y) j_1^{\dagger}(0)\}|0\rangle, \tag{6}$$

where $p = p' + q$, and the currents for the two pseudoscalar mesons in the vertex are

$$j_{5\mu}^{\pi} = \frac{1}{\sqrt{2}}(\bar{u}_a\gamma_\mu\gamma_5 u_a - \bar{d}_a\gamma_\mu\gamma_5 d_a), \quad j_{D_s} = i\bar{s}_a\gamma_5 c_a. \tag{7}$$

In the phenomenological side, the three-point function in Eq. (6) is related with the vertex coupling constant, $g_{D_0^{(1s)}D_s\pi}$, which is related with the decay width through the relation:

$$\Gamma(D_0^{(1s)} \to D_s\pi) = \frac{1}{16\pi m_{S_1}^3} g^2_{D_0^{(1s)}D_s\pi} \sqrt{\lambda(m_{S_1}^2, m_{D_s}^2, m_\pi^2)}, \tag{8}$$

where $\lambda(a,b,c) = a^2 + b^2 + c^2 - 2ab - 2ac - 2bc$.

For the light scalar mesons considered as diquark- antidiquark states, the study of their decay width using the QCD sum rule approach was done in ref.[11]. In Table II we show the results obtained for the different vertices studied in ref. [11], as well as the experimental values.

Table II: Numerical results for the coupling constants

vertex	g(GeV)	g^{exp}(GeV)	vertex	g(GeV)	g^{exp}(GeV)
$\sigma\pi^+\pi^-$	3.1 ± 0.5	2.6 ± 0.2	$f_0K^+K^-$	1.6 ± 0.1	
$\kappa K^+\pi^-$	3.6 ± 0.3	4.5 ± 0.4	$f_0\pi^+\pi^-$	0.47 ± 0.05	1.6 ± 0.8

From Table II we see that, although not exactly in between the experimental error bars, the hadronic couplings determined from the QCD sum rule calculation are consistent with existing experimental data. The biggest discrepancy is for $g_{f_0\pi^+\pi^-}$ and this can be understood since the $f_0 \to \pi^+\pi^-$ decay is mediated by one gluon exchange and, therefore, probably in this case α_s corrections could play an important role. In the case of the decay $f_0(a_0) \to K^+K^-$, the coupling can not be experimentally measured due to the lack of phase space.

In the case of the decay $D_0^{(1s)} \to D_s^+\pi^0$, for an isoscalar $D_0^{(1s)}$, in the QCD sum rule approach one only gets a result different from zero for the coupling constant, if one allows a break in the $SU(2)$ symmetry. In this case, the coupling is proportional to the difference of the masses of the u and d quarks, and the difference of the u and d quark condensates. In a preliminary calculation we got

$$g_{D_0^{(1s)}D_s\pi} \sim 0.06\,\text{GeV}, \tag{9}$$

which gives a decay width $\Gamma(D_0^{(1s)} \to D_s\pi) \sim 8$ KeV. It is important to notice that, if we have used a isovector current for the $D_0^{(1s)}$ state instead of an isoscalar current, we would get $\Gamma(D_0^{(1s)} \to D_s\pi) \sim 260$ MeV. Therefore, it seems possible, in this four-quark scenario, to obtain a extremely narrow width for $D_{sJ}^+(2317)$, while the $D_0^0(2308)$ state remain fairly wide.

In conclusion, we have presented a QCD sum rule study of the charmed scalar mesons considered as diquark-antidiquark states. We found that the masses of the BABAR, $D_{sJ}^+(2317)$, and BELLE, $D_0^0(2308)$, resonances can be reproduced by the four-quark states $(cq)(\bar{q}\bar{s})$ and $(cs)(\bar{u}\bar{s})$ respectively. However, the mass of the FOCUS resonance, $D_0^{0,+}(2405)$ can not be reproduced in the four-quark state picture considered here. Therefore, we interpret it as a normal $c\bar{q}$ state, since its mass is in complete agreement with the predictions of the quark model in ref. [4]. We also obtain a mass of ~ 2.2 GeV for a four-quark scalar state $(cq)(\bar{u}\bar{d})$ which was not yet observed, and that should be also rather broad.

This work was supported by CNPq and FAPESP.

REFERENCES

1. BELLE Coll., K. Abe *et al.*, Phys. Rev. **D69**, 112002 (2004).
2. BABAR Coll., B. Auber *et al.*, Phys. Rev. Lett. **90**, 242001 (2003); Phys. Rev. **D69**, 031101 (2004).; CLEO Coll., D. Besson *et al.*, Phys. Rev. **D68**, 032002 (2003); BELLE Coll., P. Krokovny *et al.*, Phys. Rev. Lett. **91**, 262002 (2003); FOCUS Coll., E.W. Vaandering, hep-ex/0406044.
3. FOCUS Coll., J.M. Link *et al.*, Phys. Lett. **B586**, 11 (2004).
4. S. Godfrey and N. Isgur, Phys. Rev. **D32**, 189 (1985); S. Godfrey and R. Kokoshi, Phys. Rev. **D43**, 1679 (1991).
5. H.-Y. Cheng and W.-S. Hou, Phys. Lett. **B566**, 193 (2003).
6. K. Terasaki, Phys. Rev. **D68**, 011501(R) (2003).
7. L. Maiani, F. Piccinini, A.D. Polosa, V. Riquer, Phys. Rev. **D71**, 014028 (2005).
8. R.L. Jaffe, *Phys. Rev.* **D15**, 267, 281 (1977); **D17**, 1444 (1978).
9. for a review see F.E. Close and N.A. Törnqvist, *J. Phys.* **G28**, R249 (2002).
10. M.A. Shifman, A.I. and Vainshtein and V.I. Zakharov, *Nucl. Phys.*, **B147**, 385 (1979).
11. T.V. Brito, F.S. Navarra, M. Nielsen, M.E. Bracco, Phys. Lett. **B608**, 69 (2005).
12. R.L. Jaffe and F. Wilczek, *Phys. Rev. Lett.* **91**, 232003 (2003).
13. B. L. Ioffe, Nucl. Phys. **B188**, 317 (1981); **B191**, 591(E) (1981).
14. M.E. Bracco, A. Lozea, R.D. Matheus, F.S. Navarra, M. Nielsen, hep-ph/0503137.

Solution to Trouble of D_{sJ} Particles [1]

Takayuki Matsuki*, Toshiyuki Morii† and Kazutaka Sudoh**

*Tokyo Kasei University, 1-18-1 Kaga, Itabashi, Tokyo 173, JAPAN [2]
†Faculty of Human Development, Kobe University, Nada, Kobe 657-8501, JAPAN [3]
**Institute of Particle and Nuclear Studies, High Energy Accelerator Research Organization, 1-1 Ooho, Tsukuba, Ibaraki 305-0801, JAPAN [4]

Abstract. Recent discovery of D_{sJ} particles, which are considered to be a great trouble by experimentalists as well as theorists, had already been resolved by our potential model proposed some time ago by two of us (T.M. and T.M.), in which the Hamiltonian and wave functions are expanded in 1/(heavy quark mass) respecting heavy quark symmetry.

Using our model, we explain how narrow states like D_{sJ} can be realized, predict properties of 0^+ and 1^+ states of B and B_s heavy mesons, and interpret how global $SU(3)$ symmetry seems to be recovered for these 0^+ and 1^+ heavy mesons.

Keywords: Heavy Quark Effective Theory; Spectroscopy; Heavy Quarks
PACS: 12.39.Hg; 2.39.Pn; 12.40.Yx; 14.40.Nd

INTRODUCTION

Recent discovery of narrow meson states $D_{sJ}(2317)$ and $D_{sJ}(2457)$ by BaBar [1] and the following confirmation by CLEO and Belle [2] has driven many theorists to explain these states since the former study of these states using a semi-relativistic potential model [3, 4] seems to fail to reproduce these mass values. This triggered a series of study on spectroscopy of heavy mesons again. To understand these states, an interesting explanation is proposed by Bardeen, Eichten, Hill and others [5, 6] who used an effective Lagrangian with heavy quark symmetry combined with chiral symmetry of light quarks but they cannot reproduce absolute mass values of these particles.

This problem of D_{sJ} has been considered by many experimentalists as well as theorists to be a great trouble until recently. It turns out, however, that actually we have already predicted these masses within one percent accuracy at the first order of purturbation in $1/m_Q$ with m_Q heavy quark mass some time ago[7] by using the same potential model as in [3, 4]. Since the data adopted at the time of publication[7] is obsolete, we are now refining calculations up to the second order.[8] The main difference between our treatment of the potential model and others is that we have taken into account negative energy states of a heavy quark in a bound state while others do not. Namely mass is

[1] Talk given by T. Matsuki.
[2] E-mail: matsuki@tokyo-kasei.ac.jp
[3] E-mail: morii@kobe-u.ac.jp
[4] E-mail: kazutaka.sudoh@kek.jp

CP814 *Hadron Spectroscopy: Eleventh International Conference*
edited by A. Reis, C. Göbel, J. Sá Borges, and J. Magnin

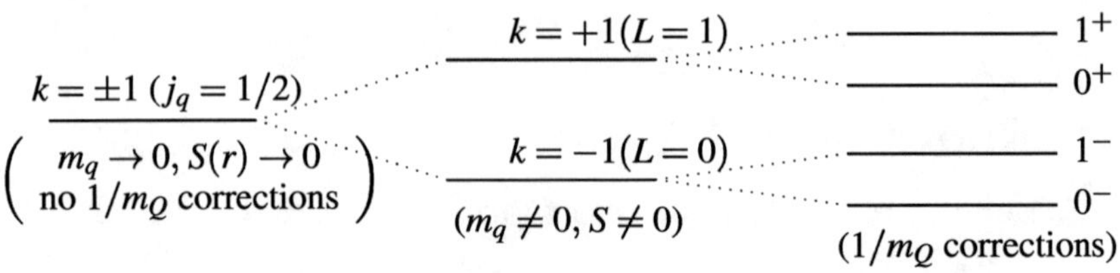

FIGURE 1. Procedure how the degeneracy is resolved in our model.

expressed as an eigenvalue of a Hamiltonian of the heavy-light system, in which a bound state consists of a heavy quark and a light anti-quark and negative energy states appear in the intermediate states of a heavy quark in calculating an energy eigenvalue, that is missed in the former study. These negative-energy state contribution is not so small which is the main reason why people of [3, 4] could not get a right answer.

Figure 1. shows how our model generates masses step by step. At first chiral symmetry is intact with a limit of light-quark mass and scalar potential going to zero. At this stage no heavy quark mass correction is taken into account except for the lowest contribution, m_Q, to a heavy meson mass, i.e., heavy quark mass is finite even at this moment. Next chiral symmetry is broken by introducing light-quark mass and scalar potential. There is still heavy quark symmtetry remained so that some states with partial angular momentum, $L = 0, 1$, between light quark and heavy quark are still degenerate. Finally all degeneracies are resolved by including $1/m_Q$ corrections. The similar scenario is adopted in [5] except for the first stage.

OUR POTENTIAL MODEL

D_{sJ} Particles

Our formulation using the Cornell potential is to expand Hamiltonian, energy, and wave function in terms of $1/m_Q$ and sets coupled euations order by order. The non-trivial differential equation is obtained in the zeroth order, which gives orthogonal set of eigenfunctions, and quantum mechanical perturbative corrections to energy and wave functions in higher orders are formulated. Consistency equation is given by

$$H\Psi_\ell = E^\ell \Psi_\ell,$$
$$(H_{-1} + H_0 + H_1 + \cdots)(\Psi_{\ell 0} + \Psi_{\ell 1} + \cdots) = \left(E_0^\ell + E_1^\ell + \cdots\right)(\Psi_{\ell 0} + \Psi_{\ell 1} + \cdots) \quad (1)$$

where integers of subscripts and superscripts denote order in $1/m_Q$, $H = H_{\rm FWT} - m_Q$, and the Foldy-Wouthuysen-Tani transformation is operated on a heavy quark and the Hamiltonian.[7]

In Tables 1 and 2, J^P stands for total spin and parity, M_0 lowest degenerate mass, c_1/M_0 the first order correction, M_{calc} calculated value of mass, and M_{obs} observed mass.

TABLE 1. *D* meson mass spectra (units are in MeV).

State ($^{2s+1}L_J$)	k	J^P	M_0	c_1/M_0	M_{calc}	M_{obs}
$^1S_0$	-1	0^-	1782	0.477×10^{-1}	1867	1867
$^3S_1$	-1	1^-		1.272×10^{-1}	2009	2008
$^3P_0$	1	0^+	2073	1.024×10^{-1}	2285	2308
"$^3P_1$"	1	1^+		1.687×10^{-1}	2423	2427
"$^1P_1$"	-2	1^+	2123	1.412×10^{-1}	2423	2420
$^3P_2$	-2	2^+		1.616×10^{-1}	2466	2460
$^3D_1$	2	1^-	2329	1.863×10^{-1}	2763	–
"$^3D_2$"	2	2^-		2.021×10^{-1}	2800	–

TABLE 2. D_s meson mass spectra (units are in MeV).

State ($^{2s+1}L_J$)	k	J^P	M_0	c_1/M_0	M_{calc}	M_{obs}
$^1S_0$	-1	0^-	1902	0.351×10^{-1}	1969	1969
$^3S_1$	-1	1^-		1.101×10^{-1}	2112	2112
$^3P_0$	1	0^+	2102	1.088×10^{-1}	2330	2317
"$^3P_1$"	1	1^+		1.762×10^{-1}	2472	2460
"$^1P_1$"	-2	1^+	2242	1.269×10^{-1}	2527	2535
$^3P_2$	-2	2^+		1.462×10^{-1}	2570	2572
$^3D_1$	2	1^-	2349	2.006×10^{-1}	2821	–
"$^3D_2$"	2	2^-		2.170×10^{-1}	2859	–

In Tables 1 and 2, we give our refined first order calculations both of D and D_s heavy mesons with various quantum numbers. Especially 0^+ and 1^+ states should be compared with experiments, i.e., $D_0^{*0}(2308)$, $D_1^{\prime 0}(2427)$,[9] $D_{sJ}(2317)$, and $D_{sJ}(2460)$.[1, 2] The calculated masses, M_{calc}, are within one percent of accuracy compared with the observed masses, M_{obs}. In the calculations, we have used values of parameters listed in Table 3.

0^+ and 1^+ States of B and B_s

In Tables 4 and 5, we give our refined calculations both of B and B_s heavy mesons with various quantum numbers. For B and B_s mesons, unfortunately there are only a few data. We predict the mass of several excited states which have not yet been observed. Among them, calculated masses of 0^+ and 1^+ states for B_s mesons are below BK/B^*K thresholds the same as the case of D_s mesons. Therefore, their decay modes to $B(0^-/1^-)+K/K^*$ are kinematically forbidden, and the dominant modes are the pionic decay:

$$B_s(0^+)\to B_s(0^-)+\pi, \qquad B_s(1^+)\to B_s(1^-)+\pi. \tag{2}$$

The decay widths of these states are expected to be narrow as for those of D_s mesons, since these decay modes violate the isospin invariance.

TABLE 3. Most optimal values of parameters.

Parameters	α_s	a (GeV^{-1})	b (GeV)	$m_{u,d}$ (GeV)	m_s (GeV)	m_c (GeV)	m_b (GeV)
	0.2620	1.937	0.07091	0.00803	0.09270	1.040	4.513

TABLE 4. B meson mass spectra (units are in MeV).

State ($^{2s+1}L_J$)	k	J^P	M_0	c_1/M_0	$M_{\rm calc}$	$M_{\rm obs}$
$^1S_0$	-1	0^-	5255	0.373×10^{-2}	5275	5279
$^3S_1$	-1	1^-		0.994×10^{-2}	5307	5325
$^3P_0$	1	0^+	5546	0.882×10^{-2}	5595	–
"$^3P_1$"	1	1^+		1.453×10^{-2}	5627	–
"$^1P_1$"	-2	1^+	5596	1.235×10^{-2}	5665	–
$^3P_2$	-2	2^+		1.413×10^{-2}	5675	–
$^3D_1$	2	1^-	5802	1.723×10^{-2}	5902	–
"$^3D_2$"	2	2^-		1.870×10^{-2}	5911	–

The similar decay modes are predicted for $B(0^+)$ and $B(1^+)$ mesons.

$$B(0^+) \to B(0^-) + \pi, \qquad B(1^+) \to B(1^-) + \pi, \tag{3}$$

but with broad decay widths, the same as $D(0^+)$ and $D(1^+)$ since isospin invariance is not broken and there is no threshold.

These higher states of B and B_s mesons might be observed in Tevatron/LHC experiments in near future by analyzing above decay modes. We hope that our framework is confirmed in the forthcoming experiments.

Recovery of Global $SU(3)$ for 0^+ and 1^+ Heavy Mesons

When one carefully looks at Tables 1 and 2, one notices that it looks that the global $SU(3)$ flavor symmetry is recovered when $J^P = 0^+$ and 1^+ for $D_{(s)}$ mesons compared with 0^- and 1^-. The magnitude of $SU(3)$ symmetry breakdown for $D_{(s)}(0^-)$ and $D_{(s)}(1^-)$ mesons is given by (units in MeV),

$$M(D(c\bar{u})) \approx M(D(c\bar{d})) \approx M(D(c\bar{s})) - 100, \tag{4}$$

while the symmetry seems to be recovered for $D(0^+)$ and $D(1^+)$ mesons,

$$M(D(c\bar{u})) \approx M(D(c\bar{d})) \approx M(D(c\bar{s})) - (9 \sim 33). \tag{5}$$

For instance, mass difference among members of a multiplet $(D(0^-), D_s(0^-))$ seems to be larger compared with that among those of $(D(0^+), D_s(0^+))$. These equations are both for observed and calculated masses.[7, 8] What causes this recovery of $SU(3)$ symmentry? Is this just an accidental or is there a deep meaning for this? We study this

TABLE 5. B_s meson mass spectra (units are in MeV).

State ($^{2s+1}L_J$)	k	J^P	M_0	c_1/M_0	$M_{\rm calc}$	$M_{\rm obs}$
$^1S_0$	−1	0^-	5375	0.287×10^{-2}	5391	5369
$^3S_1$	−1	1^-		0.898×10^{-2}	5424	–
$^3P_0$	1	0^+	5575	0.945×10^{-2}	5627	–
"$^3P_1$"	1	1^+		1.531×10^{-2}	5660	–
"$^1P_1$"	−2	1^+	5715	1.147×10^{-2}	5781	–
$^3P_2$	−2	2^+		1.322×10^{-2}	5791	–
$^3D_1$	2	1^-	5822	1.866×10^{-2}	5931	–
"$^3D_2$"	2	2^-		2.018×10^{-2}	5940	–

phenomenon in details[10] and have found that this is a general phenomenon occuring for heavy mesons, $(Q\bar{q})$, as well as heavy baryons like (QQq).

This is explained by drawing a figure of mass gap between degenerate states, $M_0(0^-) = M_0(1^-)$ $(L=0)$ and $M_0(0^+) = M_0(1^+)$ $(L=1)$. These degenerate masses are written as $M_0 = m_Q + C_0(m_q)$, i.e., a function of light quark mass m_q. Hence defining

$$\Delta M(m_u) = M_{D0}(0^+) - M_{D0}(0^-), \quad \Delta M(m_s) = M_{D_s0}(0^+) - M_{D_s0}(0^-),$$

these mass gaps have the same values for D and B mesons. This mass gap function has *monotonous decreasing tendency* so that $\Delta M(m_u) > \Delta M(m_s)$ for $m_u < m_s$.

$$\Delta M(m_s) - \Delta M(m_u) = -91 \text{ MeV}, \tag{6}$$

which almost cancels mass difference between $D(c\bar{u})$ $(D(c/\bar{d}))$ and $D(c\bar{s})$ given by Eq.(4). Equation (6) is the origin of recovery of $SU(3)$ invariance, which does not depend on whether the light-heavy system is $Q\bar{q}$ or QQq when $q = u,d,s$. That is, this is a general phenomenon not peculiar to $D(0^\pm)$ and $D(1^\pm)$. Relation between this phenomenon and constituent quark mass will be discussed elsewhere.[10]

REFERENCES

1. BaBar Collaboration, B. Aubert *et al.*, Phys. Rev. Lett. **90**, 242001 (2003).
2. CLEO Collaboration, D. Besson *et al.*, Phys. Rev. D **68**, 032002 (2003); Belle Collaboration, Y. Mikami *et al.*, Phys. Rev. Lett. **92**, 012002 (2004).
3. S. Godfrey and N. Isgur, Phys. Rev. D **32**, 189 (1985); S. Godfrey and R. Kokoski, Phys. Rev. D **43**, 1679 (1991).
4. D. Ebert, V.O. Galkin, and R.N. Faustov, Phys. Rev. D **57**, 5663 (1998).
5. W. A. Bardeen, E. J. Eichten, and C. T. Hill, Phys. Rev. D **68**, 054024 (2003).
6. W. A. Bardeen and C. T. Hill, Phys. Rev. D **49**, 409 (1994); M. A. Nowak, M. Rho, and I. Zahed, *ibid.* **48**, 4370 (1993); A. Deandrea, N. Di Barolomeo, R. Gatto, G. Nrdulli, and A. D. Plosa, *ibid.* **58**, 034004 (1998); A. Hiorth and J. O. Eeg, *ibid.* **66**, 074001 (2002).
7. T. Matsuki and T. Morii, PRD **56**, 5646 (1997); T. Matsuki, K. Mawatari, T. Morii, and K. Sudoh, Phys. Lett. B **606**, 329 (2005).
8. T. Matsuki, T. Morii, and K. Sudoh, work in progress.
9. Belle Collaboration, K. Abe *et al.*, Phys. Rev. D **69**, 1120022 (2004); Talk given by K. Abe at the *International Workshop PENTAQUARK04* held at Spring-8 in July 19-23, 2004.
10. T. Matsuki, T. Morii, and K. Sudoh, work in progress.

Radiative decays of ground-state $q\bar{q}$ mesons in the $\tilde{U}(12)$-classification scheme of hadrons

Tomohito Maeda*, Kenji Yamada*, Masuho Oda† and Shin Ishida**

*Department of Engineering Science, Junior College Funabashi Campus, Nihon University, Funabashi 274-8501, Japan
†Faculty of Engineering, Kokushikan University, Tokyo 154-8515, Japan
**Research Institute of Science and Technology, College of Science and Technology, Nihon University, Tokyo 101-8308, Japan

Abstract. The radiative transitions between ground-states of light $q\bar{q}$ mesons are investigated in the $\tilde{U}(12)$-classification scheme. In this scheme the rich decay-spectra are offered even in transition between ground states due to the appearance of "chiral states", out of the framework of non-relativistic quark model(NRQM) based on the $SU(6)$-symmetry, and it is shown that these states play an important role to improve our previous results, obtained without considering them.

INTRODUCTION

($\tilde{U}(12)$-scheme and ground-state $q\bar{q}$ mesons) Recently we have proposed the $\tilde{U}(12)$-classification scheme of hadrons, based on $\tilde{U}(12)_{SF} \otimes O(3,1)_L$-scheme with static $U(12)_{SF}$-symmetry[1], leading to possible existence of new hadron multiplets(called chiral states or chiralons[1]). The ground-states of light $q\bar{q}$ meson system are assigned as $\mathbf{12} \times \mathbf{12}^* = \mathbf{144}$-representation of $U(12)_{SF}$ group and they are classified into the eight flavor nonets with J^{PC}(see, Table1). It may be noTable that the **144**-plet contains, besides the ordinary pseudo-scalar and vector nonet, the extra-ones, respectively, and also the scalar and axial-vector nonets which appear firstly only as P-wave state in NRQM. The existence of "chiral states" which never appear in the conventional $SU(6)_{SF}$-scheme are due to inclusion of the $SU(2)_\rho$-spin degree of freedom for confined light quarks. In order to check the possibility of existence of these states experimentally, and to confirm the validity of the scheme, it is indespensable to investigate the various properties of known and newly predicted hadrons from the viewpoint of our new framework. Especially, radiative decays give an effective means for this purpose since electromagnetic(EM) gauge interaction is rigoulously well known.
(Radiative dacays and our previous work) In the previous scheme before possible existence of chiralons being noticed, we had been investigating systematically radiative decays of light through heavy quark meson systems by applying the covariant oscillator quark model(COQM)[2]. It had also the $\tilde{U}(12) \otimes O(3,1)$-scheme, but with static $SU(6)_{SF}$-symmetry. In the radiative transition between light-quark meson system final mesons generally move relativistically, so its covariant treatment is absolutely necessary. About this point, the COQM has remarkable features that hadrons are treated in a man-

[1] More precisely, the $U(12)_{SF} (\supset SU(6)_{SF} \otimes SU(2)_\rho)$ is imposed at the hadron rest frame.

CP814 *Hadron Spectroscopy: Eleventh International Conference*
edited by A. Reis, C. Göbel, J. Sá Borges, and J. Magnin

ifestly covariant way and conserved effective EM currents are explicitly given in terms of hadron variable themselves. As a results, in the previous work, it had been shown that qualitative features of experimental data are reproduced *except for some cases.* The new $\tilde{U}(12)$-scheme with static $U(12)_{SF}$-symmetry is extended, keeping the above mentioned covariant properties, to be able to describe the chiral states in addition to the conventional states appearing NRQM(called Pauli states or Paulons).

(Purpose of this work) In this report, focusing on the ground-state $q\bar{q}$-mesons, we shall reexamine our previous analysis in the new scheme, now taking into account an existence of chiral states.

WAVE FUNCTION OF GROUND-STATE $Q\bar{Q}$ MESONS IN THE $\tilde{U}(12)$-SCHEME

(Wave functions) In the $\tilde{U}(12)$-scheme, mesons are unifiedly decsribed by bi-local Klein-Gordon field[2] with two spin-flavor indices as

$$\Phi(x_q, x_{\bar{q}})_A^{(+)B} = Ne^{+iPX}\Phi(x,P)_A^{(+)B} \tag{1}$$

where $A = (\alpha, a)(B = (\beta, b))$ denotes Dirac spinor and flavor indices respectively, $x_q(x_{\bar{q}})$ is a space-time coordinate of the quark(anti-quark) which is related to the center of mass(relative) coordinate for composite meson as $X = (m_q x_q + m_{\bar{q}} x_{\bar{q}})/(m_q + m_{\bar{q}})(x = x_q - x_{\bar{q}})$ $(m_q(m_{\bar{q}})$ being the quark(anti-quark) mass paremater). The bi-local field (1) satisfies the Klein-Gordon type equation

$$\left[\partial_X^2 - \mathscr{M}^2(x, \partial_x)\right]\Phi(X,x)_A{}^B = 0, \tag{2}$$

where $\mathscr{M}^2(x, \partial_x)$ is the $U(4)(\supset SU(2)_\sigma \otimes SU(2)_\rho)$-spin independent squared mass operator including only central confining-force potential. As the basis to expand the above internal wave function $\Phi(x,P)_A{}^B$, we use "*LS*-coupling" type direct products,

$$\Phi(x,P)_A^{(+)B} \approx f(x,P) \otimes W_A^{(+)B}(P), \tag{3}$$

and take the definite-metric type 4-dimentional oscillator function $f(x,P)$ and the extended Bargmann-Wigner(BW) spin WF $W_A^{(+)B}(P)$[3] for the respective parts. Decopmosing the $W_A^{(+)B}(P)$ into irreducible definite J^{PC}-components, their detailed forms for relevant meson system are given as Table1.

[2] We give only the expression of the positeve frequency part in the following.

[3] The complete set of the extended BW spin function consists of totally 16 members in the $\tilde{U}(4)$-space, which are decomposed into the irreducible components of field given in Table1.

TABLE 1. Ground-state light $q\bar{q}$ meson nonets in $\tilde{U}(12)$-scheme. The $v_\mu = P_\mu/M$ denotes 4-velocity of the relevant meson. As mentioned in the text, physical π- and ρ-nonet are corresponding to $P_s^{(N)}$ and $(V^{(N)}+V^{(E)})/\sqrt{2}$ in this Table, respectively.

Meson field:	$P_s^{(N)}$	$P_s^{(E)}$	$V^{(N)}$	$V^{(E)}$	$A_1^{(N)} \equiv A_1^\chi$	$A_1^{(E)} \equiv B_1^\chi$	$S^{(N)} \equiv S$	$S^{(E)}$
J^{PC}:	0^{-+}	0^{-+}	1^{--}	1^{--}	1^{++}	1^{+-}	0^{++}	0^{+-}
$W(P)^{(+)}$:	$\frac{i\gamma_5}{2}$	$\frac{i\gamma_5 v\gamma}{2}$	$\frac{i\gamma_\mu}{2}$	$\frac{-i\sigma_{\mu\nu}v_\nu}{2}$	$\frac{i\gamma_5\tilde{\gamma}_\mu}{2}$	$\frac{\gamma_5\sigma_{\mu\nu}v_\nu}{2}$	$\frac{1}{2}$	$-\frac{v\gamma}{2}$

EFFECTIVE MESON CURRENTS

In order to treat the interaction of $q\bar{q}$-meson with the EM field $A_\mu(X)$, we start from the action of meson in free-space,

$$S_0 = -\int d^4x d^4X (\overrightarrow{F}_U(X)\bar{\Phi}(x,X)\overleftarrow{F}_U(X)) \left(\partial_X{}^2 - \mathcal{M}(x)^2\right) \Phi(x,X), \tag{4}$$

where the factors $F_U(X) = \frac{\gamma_\mu \partial_{X,\mu}}{\sqrt{\partial_X^2}}$ are introduced to guarantee the $U(12)_{SF}$-invariance of action, leading to give the correct sign of mass term and electric charge for the chiral states as well as the Pauli states. By using the conventional minimal substitution method, $\partial_{q,\mu} \to \partial_{q,\mu} - iQeA(x_q)_\mu$ ($Q=$diag$(2/3,-1/3,-1/3)$ being quark charge matrix), we obtain the action for the relevant EM interaction, up to lowest order of coupling constant e_q,

$$S_I^{EM} = \int d^4x_q d^4x_{\bar{q}} \sum_i j_{i,\mu}(x_q,x_{\bar{q}}) A_\mu(x_i) = \int d^4X J_\mu(X) A_\mu \quad (i = q,\bar{q}) \tag{5}$$

where,

$$j_{i,\mu}(x_q,x_{\bar{q}}) = -ie\frac{m_q+m_{\bar{q}}}{m_i}\langle(\overrightarrow{F}_U(X)\bar{\Phi}(x,X)\overleftarrow{F}_U(X))Q(\partial_{i,\mu}^{(-)} + ig_M^{(i)}\sigma_{\mu\nu}^{(i)}\partial_{i,\nu}^{(+)})\Phi(x,X)\rangle \tag{6}$$

Here we have introduced the parameter $g_M^{(i)}$ concerning intrinsic magnetic/electric moment of their "constituents"[2]. Integrating on the relative space-time coordinate for ground-states WF, we obtain the effective meson cuurent in momentum representation,

$$\begin{aligned} J_\mu(P,P') &= J_{q,\mu}(P,P') + J_{\bar{q},\mu}(P,P') & (7)\\ J_{q,\mu}(P,P') &= e\langle(-iv'\gamma)\bar{\Phi}(v')(-iv'\gamma)Q[(P_\mu+P'_\mu) + \frac{M}{m_q}g_M^{(q)}i\sigma_{\mu\nu}q_\nu]\Phi(v)\rangle & \\ J_{\bar{q},\mu}(P,P') &= e\langle(-iv'\gamma)\bar{\Phi}(v')(-iv'\gamma)\Phi(v)Q[(P_\mu+P'_\mu) + \frac{M}{m_{\bar{q}}}g_M^{(\bar{q})}i\sigma_{\mu\nu}q_\nu]\rangle & (8) \end{aligned}$$

where $q_\mu \equiv P_\mu - P'_\mu$ is the four-momentum of photon, $M(M')$ represents the mass of initial(final) meson, and $\langle\cdots\rangle$ denotes taking the trace over Dirac and flavor indices. The first(second) term of respective lines in Eq.(8) is corresponding to the convection(spin)

TABLE 2. Effective transition currents for radiative decays among ground-state mesons. (where $P_s^{(N)} \equiv P$, $S \equiv S^{(N)}$, $(V^{(N)} \pm V^{(E)})/\sqrt{2} \equiv V^{(')}$, and $d \equiv 2(m_q + m_{\bar{q}})$.)

process	current	coupling parameter
$V^{(')} \to P\gamma$	$J_\mu = ie\mu_o \varepsilon_{\mu\nu\rho\alpha} q_\nu \varepsilon_\rho^{V^{(')}} P_\alpha$	$\mu_0 = (\frac{d}{2m_q} g_M^{(q)} \langle PQV^{(')} \rangle + \frac{d}{2m_{\bar{q}}} g_M^{(\bar{q})} \langle PV^{(')} Q \rangle)\ (\frac{1}{\sqrt{2}M})$
$A_1^\chi \to P\gamma$	$J_\mu = 0$ (forbidden)	–
$B_1^\chi \to P\gamma$	$J_\mu = ie\varepsilon_B \varepsilon_\mu^B$	$\varepsilon_B = (\frac{d}{2m_q} g_M^{(q)} \langle PQB \rangle + \frac{d}{2m_{\bar{q}}} g_M^{(\bar{q})} \langle PBQ \rangle)(-qv)$
$S \to V\gamma$	$J_\mu = -e\zeta \tilde{\varepsilon}_\mu^V$	$\zeta = (\frac{d}{2m_q} g_M^{(q)} \langle VQS \rangle + \frac{d}{2m_{\bar{q}}} g_M^{(\bar{q})} \langle VSQ \rangle)(qv')\frac{1}{\sqrt{2}}$
$A_1^\chi \to V\gamma$	$J_\mu = ie\xi_0 \varepsilon_{\mu\nu\rho\sigma} q_\nu \varepsilon_\rho^A \tilde{\varepsilon}_\sigma^V$	$\xi_0 = (\frac{d}{2m_q} g_M^{(q)} \langle VQA \rangle + \frac{d}{2m_{\bar{q}}} g_M^{(\bar{q})} \langle VAQ \rangle)\frac{1}{\sqrt{2}}$

TABLE 3. Formulae for radiative dacy width.

process	Γ
$V^{(')} \to P\gamma$	$\Gamma = \frac{\alpha}{3}\mu_0^2 \lvert q \rvert^3$
$A_1^\chi (B_1^\chi) \to P\gamma$	$\Gamma = \frac{\alpha}{3} {\varepsilon_{A(B)}}^2 \frac{\lvert q \rvert}{M^2}$
$S \to V\gamma$	$\Gamma = \frac{\alpha}{2J+1} \zeta^2 \frac{\lvert q \rvert}{M^2}$
$A_1^\chi \to V\gamma$	$\Gamma = \frac{2\alpha}{3} \frac{{\xi_0}^2}{M^2} \omega(\omega + \omega_3) \lvert q \rvert^3$

current. Taking into account the relation, $P_\mu A_\mu = P'_\mu A_\mu = 0$, the convection current has no contribution in the transition between ground-states. On the other hand, the spin current has a remarkble feature due to the relativistic-covariance of our scheme. Rewriting the spin current term as

$$\sigma_{\mu\nu} iq_\nu A_\mu = \sigma \cdot \mathbf{B} - i\rho_1 \sigma \cdot \mathbf{E}, \tag{9}$$

we see that it contains not only the ordinary magnetic interaction of the constituent, but also electric interaction accompanied by its ρ-spin flip. This "intrinsic electric dipole" transition comes from $\sigma_{i4} iq_i A_4$-component representing the relativistic effect, and gives non-negligible contribution for most transition processes.

(WFs of π- and ρ-nonets) Here it should be noted that generally the $\tilde{U}(4)$-spin WF for P and V are superposition of the (N) and (E) components with the same J^{PC}(see, Table1). However, from the consideration on linear realization of chiral symmetry, the $\tilde{U}(4)$ WF of π-octet should be $i\gamma_5/2$, reflecting its Goldston-boson nature. On the contrary, it seems that the WF of ρ-nonet should take as the same form as in our previous $\tilde{U}(12)$-scheme, being based on its success with $SU(6)_{SF}$-description for ρ-nonet. This is obtained by taking the linear combination of (N) and (E) chiral-representations, so as to make "pure Pauli state".

(Fixing the parameters) From the expression of current Eq.(8), we can derive the formula of the relevant radiative decay width, which are given in Table 2-3. In this work, we take the following values:

$$M, M' : \text{PDG value}, \quad \frac{m_n}{m_s} = \frac{M_\rho}{M_\phi} = 0.76,$$

TABLE 4. Predicted width in comparison with experiments.

process		Γ_{theor}	Γ_{exp}
$V \to P\gamma$	$\rho \to \pi\gamma$	67.6	67.6 ± 7.5
	$K^{*\pm} \to K^{\pm}\gamma$	58.9	50 ± 5
	$K^{*0} \to K^{0}\gamma$	120	116 ± 10
	$\phi \to \eta\gamma$	70.7	58.9 ± 0.5
	$\omega \to \pi^{0}\gamma$	619	757.3
$A_1^{\chi} \to P\gamma$	$(a_1(1260)^{\pm} \to \pi^{\pm}\gamma)$	0	(640 ± 246)
$B_1^{\chi} \to P\gamma$	$b_1(1235)^{\pm} \to \pi^{\pm}\gamma$	227	230 ± 60
$V^{(\prime)} \to P\gamma$	$\rho(1200) \to \pi\gamma$	111	–
	$(K^*(1410) \to K^0\gamma)$	380	(< 52.9)
$A_1^{\chi} \to V\gamma$	$a_1^{\chi}(1000) \to \rho\gamma$	101	–

$$g_M^{(n)}(= g_M^{(s)}) = 1.19(\text{determined from the } \rho \to \pi\gamma), \tag{10}$$

and here we choose, $\omega = (u\bar{u} + d\bar{d})/\sqrt{2}, \phi = -s\bar{s}, \eta = (u\bar{u} + d\bar{d})/2 - (s\bar{s})/\sqrt{2}$.

PREDICTED WIDTH IN COMPARISON WITH EXPERIMENTS

Using the above numerical values, we can predict the width of relevant radiative transitions. Some typical results are shown in Table 4. Concerning $V \to P\gamma$ process, our estimated widths seems to be consistent with the experimental data. In previous scheme, our predicted width for $b_1(^1P_1) \to \pi(^1S_0)$(66keV) was much smaller than experimental one(230 ± 60keV). We had considered at that time that this discrepancy is quite serious for the scheme, since to this process contributes only the convection current, whose form reflects the conservation of EM current. However, it has been shown now that the experiment can be reproduced by assigning $b_1(1235)$ as ground-state "chiralon".

CONCLUDING REMARKS

We have reinvestigated the radiative decay of ground-state light quark mesons in our new scheme. As a result, it is pointed out that the new scheme gives important improvement to our previous predictions. The essential points are that to take the π-on spin WF as $i\gamma_5/2$, to take the ρ-meson spin WF as equal weight superposition of different chiral representations, and to assign $b_1(1235)$ as pure chiral states. It is also shown that $a_1(1260)$ and $K^*(1410)$ are not pure chiral states. So, we have to consider the possibility of mixing them with P-wave Pauli states. Predicted values assuming them pure chiral states, $\rho(1200) \to \pi\gamma$, $a_1^{\chi} \to \rho\gamma$, are $\sim$ 100 keV. It should be checked experimentally.

REFERENCES

1. S. Ishida, M. Ishida and T. Maeda ,Prog. Theor. Phys.104 (2000) ; S. Ishida and M. Ishida , Phys. Lett. **B**539(2002) ; S. Ishida, M .Ishida, T.Maeda, M. Oda and K. Yamada, hep-ph/0408136 ; and also S. Ishida, K. Yamada in this conference talk.
2. S. Ishida, K. Yamada, and M.Oda Phys. Rev. D40(1989)1497

QUARKONIA

Charmonium production in 920 GeV proton-nucleus collisions

Pietro Faccioli (for the HERA-B Collaboration)

Dipartimento di Fisica dell'Università di Bologna and INFN Sezione di Bologna - Bologna, Italy
E-mail: faccioli@bo.infn.it

Abstract.
HERA-B is performing a comprehensive study of charmonium production based on about 300000 J/ψ's, 5000 ψ''s and 15000 χ_c's collected using a dedicated dilepton trigger. By presenting preliminary accounts of production kinematics, nuclear dependence and feeddown components, this report illustrates how these measurements are going to improve the current experimental picture of proton-nucleus results.

Keywords: Charmonium, hadroproduction, nuclear effects
PACS: 13.85.Qk, 14.40.Gx, 24.85.+p

HERA-B is a fixed-target experiment working on the 920 GeV/c proton beam of the HERA ring at DESY in Hamburg. The primary proton-nucleon interactions are induced, at a centre-of-mass energy of $\sqrt{s} \simeq 41.6$ GeV, by collisions between protons of the beam halo and target wires of different materials. A first part of the spectrometer is devoted to tracking and vertex measurements and consists of a silicon vertex detector, a magnet and a tracking system. The second part is focussed on the particle identification and includes a Ring Imaging Cherenkov detector, an electromagnetic calorimeter and a muon detector. The latter two also provide the level-0 of the dilepton and hard photon triggers. A central subject of the physics program is the study of the charmonium states J/ψ, ψ' and χ_c, for which the ongoing analyses can count on a large statistics (about 300000 J/ψ's) acquired in both dilepton decay channels using three different targets (C, Ti, W) and a dedicated trigger.

The availability of charmonium production studies performed over the widest possible range of processes, kinematical configurations and production energies is essential for an extensive verification of perturbative QCD. Moreover, the particular environment in which these processes are studied in a fixed target experiment constitutes by itself an additional reason of interest. In fact, several existing models predict a wide range of possible modifications of the kinematic characteristics of the produced charmonium state as a consequence of its interaction with the surrounding nuclear matter [1]. A solid knowledge of the ordinary effects of scattering from confined nuclear matter constitutes a necessary baseline for the interpretation of the studies of anomalous suppression performed in heavy-ion collisions.

The predictions of the nuclear absorbtion models differ crucially from each other in the negative part of the Feynman-x (x_F) spectrum, where the charmonium state is formed inside the nuclear target. The negative-x_F range is therefore the most favourable testing ground of nuclear dependence models. While the presently available $p-A$

CP814 *Hadron Spectroscopy: Eleventh International Conference*
edited by A. Reis, C. Göbel, J. Sá Borges, and J. Magnin

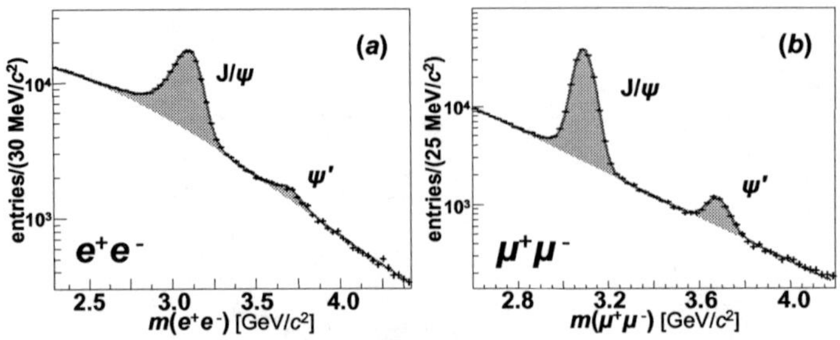

FIGURE 1. J/ψ and ψ' signals in the dilepton invariant mass distributions.

measurements are limited to the positive half of the spectrum, the wide angular coverage of the HERA-B detector (up to 220 mrad) corresponds to a range of acceptance (for the J/ψ) centred, for the first time, at negative x_F ($-0.35 < x_F < 0.15$).

J/ψ **kinematics**. – J/ψ's and ψ''s are reconstructed in their dilepton decay modes (e^+e^- and $\mu^+\mu^-$) by requiring the presence of two trigger tracks of opposite charge, a common vertex and a good lepton identification. The resulting invariant mass distributions, containing respectively ~ 130000 J/ψ's with ~ 2200 ψ''s in the e^+e^- channel and ~ 170000 J/ψ's with ~ 3000 ψ''s in the $\mu^+\mu^-$ channel, are shown in Fig. 1(a-b). Given the large statistics, the J/ψ yield can be determined as a function of the kinematical variables x_F and p_T (transverse momentum) and of the atomic mass of the target nucleus. Preliminary spectra obtained from the dielectron data are compared in Fig. 2(a-b) to fixed-target measurements at slightly lower energy ($\sqrt{s} = 38.8$ GeV) [2]. As can be seen, the new results extend significantly the kinematical coverage towards high p_T and negative x_F values. The p_T distributions are normalized to increasing values with increasing atomic mass A and fitted to the usual parametrization proportional to $[1+(35\pi/256)^2(p_T/\langle p_T\rangle)^2]^{-6}$, which, though purely phenomenological, still describes the data surprisingly well. The average p_T of the produced J/ψ, obtained from each fit, is plotted in the inset as a function of the target atomic mass, in order to give a representation of the known p_T-broadening effect. The increase with A is consistent with a logarithmic dependence, with $d\langle p_T\rangle/d\ln A = 0.025 \pm 0.003$.

The inset in Fig. 2(b) shows the result of a preliminary analysis of the A-dependence of the x_F distribution, based on data acquired with two target wires (C and W) simultaneously inserted into the beam. The plotted quantity (with statistical errors only) is the suppression parameter $\alpha = \ln(\sigma_{pW}/\sigma_{pC})/\ln(A_W/A_C)$ (σ and A indicate, respectively, cross sections and target atomic weights). The result confirms a small, constant suppression near $x_F = 0$, as already observed by E866 and NA50, even if there may be an indication of a decrease in the magnitude of the suppression with increasing production energy. The observation of a flat behaviour of $\alpha(x_F)$ is extended by the new analysis to a significant portion of the negative x_F spectrum. To give an idea of how untrivial this result is, the experimental points are superimposed to theoretical curves calculated (for $\sqrt{s} = 38.8$ GeV) [1] assuming different non-perturbative processes as the dominant sources of nuclear modification of charmonium production (alterations of the parton densities inside a nucleus, energy losses of the incoming parton and/or of the formed

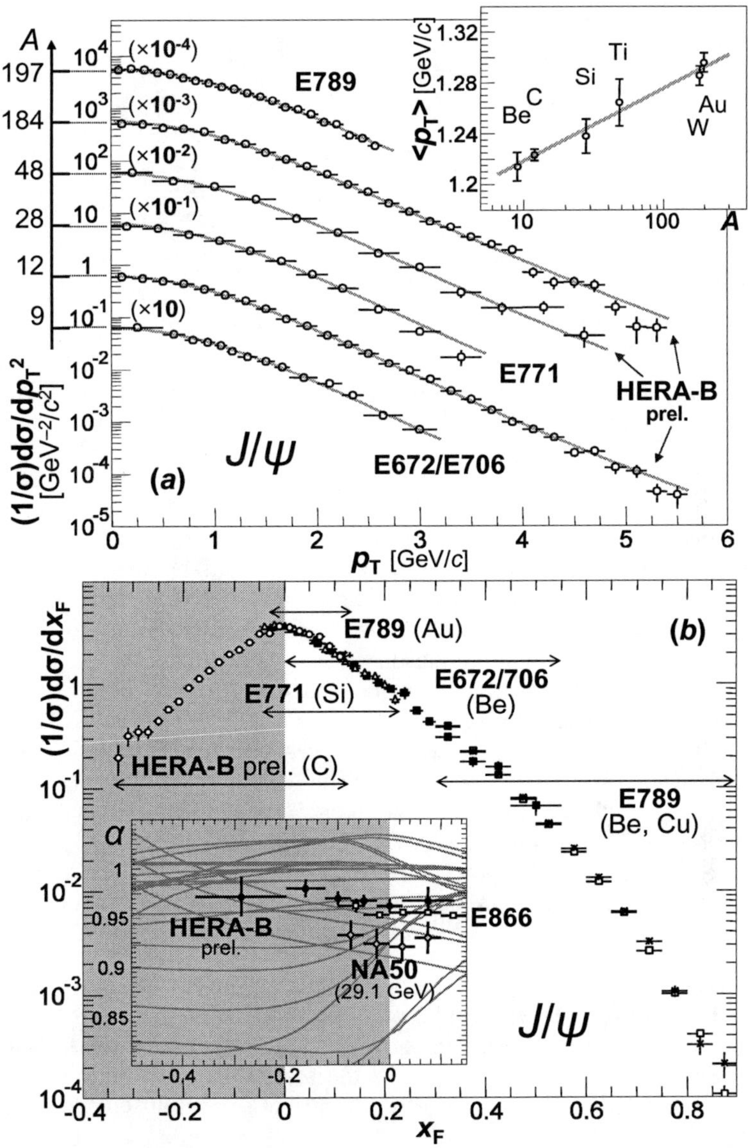

FIGURE 2. J/ψ kinematical distributions: preliminary HERA-B results (e^+e^- channel), compared with those published at $\sqrt{s} = 38.8$ GeV. Insets: (*a*) p_T broadening; (*b*) A-dependence as a function of x_F.

$c\bar{c}$ state, disassociation of the $c\bar{c}$ pair in the interaction with nucleons or other produced particles, occasional hadronization of an intrinsic $c\bar{c}$ component of the scattering proton, etc.). Among the great variety of predicted behaviours, the strongest effects are clearly disfavoured as possible components of the nuclear dependence at negative x_F.

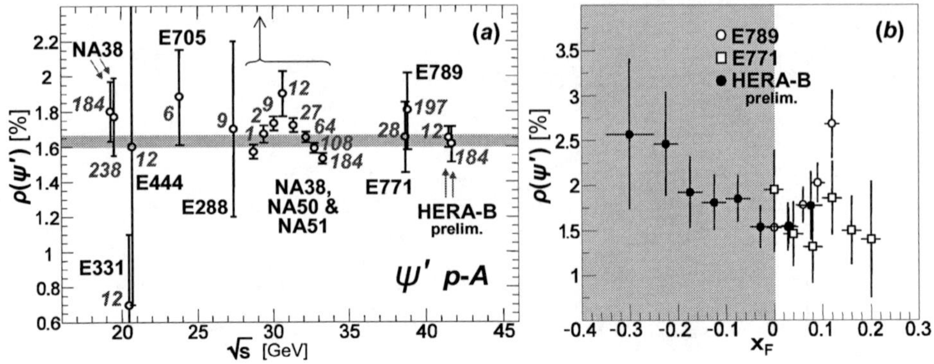

FIGURE 3. ψ' to J/ψ production ratio in $p-A$ collisions: (*a*) as a function of energy and target atomic mass; (*b*) as a function of x_{F}, at $\sqrt{s} = 38.8$ and 41.6 GeV.

ψ' **production.** – The comparison between ψ' and J/ψ production can further contribute to the understanding of nuclear absorbtion processes: different cross section behaviours of the two charmonium states may, on the one hand, be interpreted in terms of the bigger size and lower binding energy of the ψ' with respect to the otherwise very similar J/ψ state; on the other hand, the time elapsing before the originally produced $c\bar{c}$ state evolves into the charmonium meson ψ' or J/ψ may have an attenuating effect on the measured differences. Preliminary results extracted from the combined e^+e^- and $\mu^+\mu^-$ samples for the yield ratio $\rho_{\psi'} = B(\psi' \to e^+e^-)\sigma(\psi')/B(J/\psi \to e^+e^-)\sigma(J/\psi)$ are compared in Fig. 3(a) to the published $p-A$ measurements [3] (obtained in different kinematical windows) as a function of energy. The target atomic mass is indicated next to each data point. Despite the slight (not monotonic?) dependence on the target nucleus shown by the NA38-50-51 results at 29.1 GeV, the data indicate that, within a 4% variation, the ψ'-to-J/ψ ratio is independent of the production energy. The x_F dependence of $\rho_{\psi'}$, preliminarily determined in the e^+e^- channel, is shown in Fig. 3(b), compared with the existing measurements in the positive x_F spectrum at 38.8 GeV.

χ_c **production.** – The fraction R_{χ_c} of the J/ψ's produced via the radiative decay $\chi_c \to J/\psi\gamma$ is an important test of charmonium production models, but the present experimental situation is not clear. As shown in Fig. 4(a) for the muon channel, the χ_c signal is extracted from the spectrum of the mass difference between $\ell^+\ell^-\gamma$ and $\ell^+\ell^-$ combinations. Neglecting the small χ_{c_0} contribution and differences between the unresolved ($\Delta m = 46$ MeV/c^2) states χ_{c_1} and χ_{c_2}, a preliminary analysis based on only 15% of the available $\mu^+\mu^-$ statistics (about 1300 reconstructed χ_c's) provides the result $R_{\chi_c} = 0.21 \pm 0.05_{\mathrm{stat}}$, updating a previous, published value [4]. The full $\mu^+\mu^-$ and e^+e^- data samples, on which the final result will be based, represents the largest χ_c statistics ever analyzed by a hadroproduction experiment. Fig. 4(b) shows the preliminary value together with the expected, final statistical precision, in the context of the published hadroproduction results [5], considered as a function of energy. As can be seen, while the $\pi - A$ data are consistent with a common value of about 40%, the results on proton-induced production form an unfocussed, if not contradictory picture.

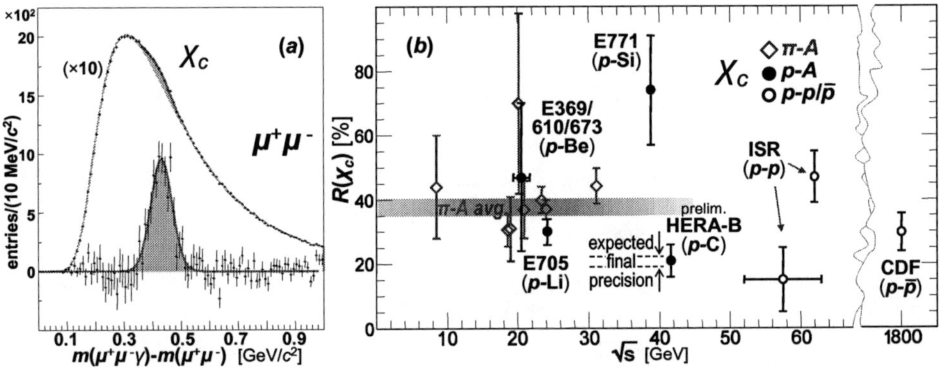

FIGURE 4. (*a*) χ_c signal, before and after background subtraction. (*b*) the preliminary R_{χ_c} value compared to published $p-A$, $p-p/\bar{p}$ and $\pi-A$ results.

Summary. – The preliminary HERA-B results on charmonium production presented here and elsewhere at this conference [6], combined with up-to-date feeddown branching ratios [7], indicate that, of the J/ψ's produced in $p-A$ collisions at $\sqrt{s} \simeq 41.6$ GeV, respectively $(21 \pm 5)\%$ and $(7.0 \pm 0.4)\%$ come from the decays of χ_c and ψ', while the fraction of J/ψ's coming from b-hadrons is $(0.065 \pm 0.011)\%$. The inclusive production cross section is measured as [6] $\sigma_{pN}^{J/\psi} = (643 \pm 71 \pm 48)$ nb/nucleon. In the observed kinematical range $-0.35 < x_{\mathrm{F}} < 0.15$, the nuclear effects consist in a small, roughly constant suppression.

REFERENCES

1. R. Vogt, *Phys. Rev.* C **61**, 035203 (2000); K.G. Boreskov and A.B. Kaidalov, *JETP Lett.* **77**, 599 (2003).
2. E789: M.S. Kowitt *et al.*, *Phys. Rev. Lett.* **72**, 1318 (1994) and M.H. Schub *et al.*, *Phys. Rev.* D **52**, 1307 (1995); E771: T. Alexopoulos *et al.*, *Phys. Rev.* D **55**, 3927 (1997); E672/E706: A. Gribushin *et al.*, *Phys. Rev.* D **62**, 012001 (2000); E866: M.J. Leitch *et al.*, *Phys. Rev. Lett.* **84**, 3256 (2000). NA50: B. Alessandro *et al.*, *Eur. Phys. J.* C **33**, 31 (2004).
3. E288: H.D. Snyder *et al.*, *Phys. Rev. Lett.* **36**, 1415 (1976); E331: J.G. Branson *et al.*, *Phys. Rev. Lett.* **38**, 1331 (1977); E444: K.J. Anderson *et al.*, *Phys. Rev. Lett.* **42**, 944 (1979); E705: L. Antoniazzi *et al.*, *Phys. Rev.* D **46**, 4828 (1992); NA51: M.C. Abreu *et al.*, *Phys. Lett.* B **438**, 35 (1998); NA38: M.C. Abreu *et al.*, *Phys. Lett.* B **444**, 516 (1998); E789, E771 and NA50 already cited in [2].
4. I. Abt *et al.*, *Phys. Lett.* B **561**, 61 (2003).
5. ISR: A.G. Clark *et al.*, *Nucl. Phys.* B **142**, 29 (1978) and C. Kourkoumelis *et al.*, *Phys. Lett.* B **81**, 405 (1979); E705: L. Antoniazzi *et al.*, *Phys. Rev. Lett.* **70**, 383 (1993); E369/610/673: D.A. Bauer *et al.*, *Phys. Rev. Lett.* **54**, 753 (1996); E771: T. Alexopoulos *et al.*, *Phys. Rev.* D **62**, 032006 (2000); CDF: F. Abe *et al.*, *Phys. Rev. Lett.* **79**, 578 (2003). $\pi-A$ *results*: T.B.W. Kirk *et al.*, *Phys. Rev. Lett.* **42**, 619 (1979); Y. Lemoigne *et al.*, *Phys. Lett.* B **113**, 509 (1982); S.R. Hahn *et al.*, *Phys. Rev.* D **30**, 671 (1984); F. Binon *et al.*, *Nucl. Phys.* B **239**, 311 (1984); V. Koreshev *et al.*, *Phys. Rev. Lett.* **77**, 4294 (1996); E705 as cited above.
6. H. Wahlberg and P. Faccioli, other HERA-B contributions in these Proceedings.
7. S. Eidelman *et al.*, *Phys. Lett.* B **592**, 1 (2004).

Quarkonium Spectroscopy at Babar

M. Negrini

On behalf of the Babar collaboration

Dipartimento di Fisica dell'Università di Ferrara / INFN
Via Saragat 1, 44100 Ferrara, ITALY

Abstract. The Babar experiment at the high luminosity storage ring PEP-II offers excellent opportunities in quarkonium spectroscopy. Recent Babar results obtained in this field are reported.

Keywords: Quarkonium, B mesons
PACS: 13.25.Hw, 14.40.Gx

INTRODUCTION

The $X(3872)$ state was first observed by the Belle collaboration [1] in the decay $B^{\pm} \to K^{\pm} J/\psi \pi^+ \pi^-$, where a narrow signal in the $J/\psi \pi^+ \pi^-$ invariant mass is observed at 3872 MeV/c^2. The observation was then confirmed by CDF [2], D0 [3] and Babar [4]. The following observation of other states in the 4 GeV mass region at the B-factories ($Y(3940)$, $X(3940)$, $Z(3930)$ and $Y(4260)$), renewed the interest in quarkonium spectroscopy as well as the need to investigate their properties to shed light on their nature.

The Babar detector [5] is located at the PEP-II B-factory, an asymmetric e^+e^- collider at $\sqrt{s} = 10.58$ GeV. The experiment, designed to perform precision measurements of CP violation in the B meson system, has a much broader physics reach and an extensive program of quarkonium spectroscopy.

Quarkonium states are produced at B-factories through different processes: B decays ($B \to K + c\bar{c}$), two photon production ($\gamma^*\gamma^* \to c\bar{c}$, where two photons emitted by the e^+ and e^- interact to form even C-parity states), double charmonium production ($e^+e^- \to c\bar{c} + c\bar{c}$) and initital state radiation ($e^+e^- \to \gamma_{ISR} c\bar{c}$, where a photon is emitted by one of the $e^{\pm}$ before interaction, lowering $\sqrt{s}$ for the e^+e^- annihilation, and allowing the direct production of states with quantum numbers $J^{PC} = 1^{--}$).

INVESTIGATION OF THE $X(3872)$

The $X(3872)$ was discovered in the decay $B^{\pm} \to K^{\pm} + X(3872)$. The mass value and the fact that $B^{\pm} \to K^{\pm} + c\bar{c}$ is a typical decay mode of the B meson, suggest the possible charmonium nature of the state. However, its mass is not compatible with existing predictions from potential models, and its likely $J/\psi \rho^0$ decay mode would be forbidden for charmonium states because of isospin symmetry. Moreover the mass value is exactly at the threshold for $D^0 \bar{D}^{*0}$ production.

For these reasons, other explanations have been proposed, like a $D\bar{D}^*$ molecule [6, 7] or diquark-antidiquark state [8]. To discriminate among different models, experimen-

CP814 *Hadron Spectroscopy: Eleventh International Conference*
edited by A. Reis, C. Göbel, J. Sá Borges, and J. Magnin

tal results about the quantum numbers, the decay modes, the branching fractions, the angular distributions, and the existence of charged partners are needed.

The decays $B^- \to K^- J/\psi\pi^+\pi^-$ and $B^0 \to K_s^0 J/\psi\pi^+\pi^-$ have been studied using a sample of 211 fb^{-1} of data, corresponding to 232 millions of $B\bar{B}$ pairs recorded by Babar, as reported in [9].

We obtain 61 ± 15 and 8.3 ± 4.5 $X(3872)$ decay events in B^- and B^0 decay respectively, with signal significance of 6.1σ and 2.5σ including systematic uncertainty .

The decay $B^0 \to K^0 X(3872)$ is predicted to be suppressed by one order of magnitude in the $D\bar{D}^*$ molecule hypothesis [10]. Using the number of events and the efficiency we measure $\mathscr{B}^- = \mathscr{B}(B^- \to K^- X(3872), X \to J/\psi\pi^+\pi^-) = (10.1 \pm 2.5 \pm 1.0) \times 10^{-6}$ and $\mathscr{B}^0 = \mathscr{B}(B^0 \to K^0 X(3872), X \to J/\psi\pi^+\pi^-) = (5.1 \pm 2.8 \pm 0.7) \times 10^{-6}$. In the ratio of branching fractions some sources of systematic error cancel and we have $R = \mathscr{B}^0/\mathscr{B}^- = 0.50 \pm 0.30(\mathrm{stat}) \pm 0.05(\mathrm{syst})$.

In the diquark-antidiquark interpretation, the $X(3872)$ state can have quark content $X_u = (cu)(\bar{c}\bar{u})$ and $X_d = (cd)(\bar{c}\bar{d})$, and a mass difference of 7 ± 2 MeV is predicted [8]. The two different mass states should be produced differently in B^- and B^0 decay. From our data we obtained a mass difference $\Delta m = (2.7 \pm 1.3)$ MeV/c^2.

Though no clear interpretation can be obtained from the present data, the method seems promising and more data are needed to discriminate between different models.

Search for $X(3872)$ charged partners

Babar searched for $X(3872)$ charged partners in charged and neutral B decays $B^- \to \bar{K}^0 J/\psi\pi^-\pi^0$ and $B^0 \to K^+ J/\psi\pi^-\pi^0$, by looking at the $J/\psi\pi^-\pi^0$ invariant-mass distribution for 212 fb^{-1} of data; this corresponds to 234×10^6 $B\bar{B}$ events. No charged partner was observed [11], and we determined the upper limits $\mathscr{B}(B^0 \to K^+ X^-, X^- \to J/\psi\pi^-\pi^0) < 5.4 \times 10^{-6}$ and $\mathscr{B}(B^- \to \bar{K}^0 X^-, X \to J/\psi\pi^-\pi^0) < 22 \times 10^{-6}$ at 90% confidence level.

INCLUSIVE CHARMONIUM IN B DECAY

A novel technique for the study of charmonium states consists in the reconstruction of everything except the charmonium state itself. In events in which one B is fully reconstructed, it is possible to determine the rest frame of the recoiling B from the reconstructed B and the beam parameters. The distribution of the K momentum in the recoiling B system presents a series of monochromatic peaks due to two body decays $B \to K + X$, where X is predominantly a charmonium state, superimposed on a continuum background due to K's from secondary or multi-body decays. This technique allows the observation of X states independently of decay mode, and hence the measurement of absolute branching fractions $\mathscr{B}(B \to K + X)$.

The observed $K^\pm$ momentum distribution in the $B^\pm$ rest frame, after a neural network selection, is shown in Figure 1. An unbinned maximum likelihood fit is used to extract the signal yields for nine possible charmonium states. The branching fractions are summarized in Table 1.

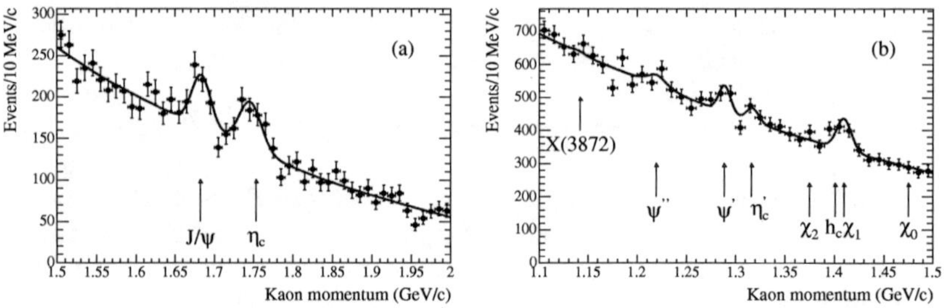

FIGURE 1. The $K^{\pm}$ momentum distribution in the $B^{\pm}$ rest frame. Peaks corresponding to $K^{\pm}\eta_c$ and $K^{\pm}J/\psi$ decay are visible in the high momentum region (a) and to $K^{\pm}\chi_{c1}$, $K^{\pm}\eta_c'$, $K^{\pm}\psi'$, and $K^{\pm}\psi''$ decay in the low momentum region (b).

TABLE 1. Yields and branching fractions for $\mathscr{B}(B^{\pm} \to K^{\pm}X)$. Upper limits are given at 90% CL.

Particle	Yield	$\mathscr{B}$ (10^{-4})	Particle	Yield	$\mathscr{B}$ (10^{-4})
η_c	273 ± 43	$8.4 \pm 1.3 \pm 0.8$	η_c'	98 ± 52	$3.4 \pm 1.8 \pm 0.3$
J/ψ	259 ± 41	$8.1 \pm 1.3 \pm 0.7$	ψ'	139 ± 44	$4.9 \pm 1.6 \pm 0.4$
χ_0	9 ± 21	< 1.8	ψ''	99 ± 69	$3.5 \pm 2.5 \pm 0.3$
$\chi_1 + h_c$	227 ± 40	$8.0 \pm 1.4 \pm 0.7$	$X(3872)$	15 ± 39	< 3.2
χ_2	0 ± 36	< 2.0			

No signal is found for the decay $B^{\pm} \to K^{\pm}X(3872)$, and we obtain an upper limit for the corresponding branching fraction (Table 1). Using the Belle/Babar average $\mathscr{B}(B^{\pm} \to K^{\pm}X(3872)) \cdot \mathscr{B}(X(3872) \to J/\psi\pi^+\pi^-) = (13.3 \pm 2.5) \cdot 10^{-6}$ we can extract a lower limit $\mathscr{B}(X(3872) \to J/\psi\pi^+\pi^-) > 4.2\%$ at 90% CL. No signal of two body decay is observed looking at the $K^{\pm}$ momentum distribution in B^0 decay, therefore we determine the upper limit $\mathscr{B}(B^0 \to K^{\pm}X(3872)^{\mp}) < 5 \times 10^{-4}$ at 90% CL for the production of $X(3872)$ charged partners in B decays.

In the fit to the $K^{\pm}$ momentum distribution, the central value for each signal peak is usually fixed at the value expected from the mass of the corresponding charmonium state. If the peak position is left floating in the fit it is possible to measure the recoiling mass, in this way we obtain $m_{\eta_c} = 2982 \pm 5$ MeV/c^2 and $m_{\eta_c'} = 3639 \pm 7$ MeV/c^2.

DOUBLE CHARMONIUM PRODUCTION

Charmonium states can be observed through the process of double charmonium production $e^+e^- \to \gamma^* \to J/\psi c\bar{c}$. In the recoil of a J/ψ, only $c\bar{c}$ states with even C-parity are expected, although a production of odd C-parity states is possible through $e^+e^- \to \gamma^*\gamma^* \to J/\psi c\bar{c}$.

The invariant mass distribution recoiling against a J/ψ is shown in Figure 2. Clear signals are observed for η_c, χ_{c0} and η_c', while there is no evidence of χ_{c1}, χ_{c2} and odd C-parity states [12]. The measured cross sections agree with the Belle results [13] and

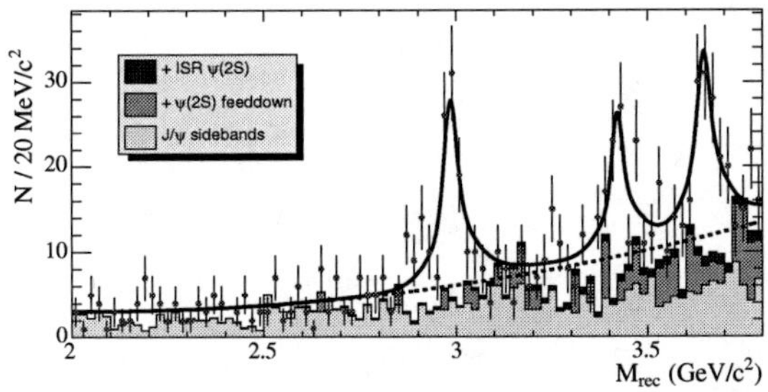

FIGURE 2. The distribution of missing mass recoiling against inclusively-produced J/ψ. The fit result (solid curve) shows clear evidence of η_c, χ_{c0} and η_c' production over background (dashed curve). The histograms represent the contributions from different sources of background.

are larger with respect to NRQCD expectations [14, 15]. Because of the requirement in the event selection of 5 or more tracks to be detected, we measure $\sigma(e^+e^- \to J/\psi c\bar{c}) \times \mathscr{B}(c\bar{c} \to> 2 \text{ charged}) = 17.6 \pm 2.8^{+1.5}_{-2.1}$ fb for $J/\psi\eta_c$, $10.3 \pm 2.5^{+1.4}_{-1.8}$ fb for $J/\psi\chi_{c0}$, and $16.4 \pm 3.7^{+2.4}_{-3.0}$ fb for $J/\psi\eta_c'$.

INITIAL STATE RADIATION

In initial state radiation (ISR) processes, a photon is radiated from either the e^+ or the e^- before the interaction, so the e^+e^- annihilation occurs at an energy smaller than the nominal energy of the machine. This technique is particularly interesting at the B-factories, where the nominal center of mass energy of the interaction is fixed, since it permits an energy scan of e^+e^- annihilation interactions. In the $e^+e^- \to \gamma_{ISR}X$ process, only X states with $J^{PC} = 1^{--}$ are produced.

The ISR photon is not necessarily detected in the analysis, but its presence can be inferred from the small values of the missing mass squared recoiling against the final state and the missing transverse momentum. Moreover, the process $e^+e^- \to \gamma_{ISR}\psi'$ provides a well known benchmark channel.

The reaction $e^+e^- \to \gamma_{ISR}J/\psi\pi^+\pi^-$ has been studied using 233 fb^{-1} of data, and the corresponding $J/\psi\pi^+\pi^-$ invariant mass distribution for the selected events is shown in Figure 3. A broad enhancement at 4260 MeV/c^2, called $Y(4260)$, is clearly observed [16]. An unbinned fit with a Breit-Wigner signal function and a second order polynomial background yields 125 ± 23 $Y(4260) \to J/\psi\pi^+\pi^-$ events, with a mass $M_Y = 4259 \pm 8(\text{stat})^{+2}_{-6}(\text{syst})$ MeV/c^2 and a width $\Gamma_Y = 88 \pm 23(\text{stat})^{+6}_{-4}(\text{syst})$ MeV.

Search for the $Y(4260)$ in B decays

A search for the $Y(4260)$ in B decays has been performed by studying the decay $B^\pm \to K^\pm J/\psi\pi^+\pi^-$ using 211 fb^{-1} of data. Fixing M_Y and Γ_Y to the values measured in the ISR production process, an excess of 128 ± 42 events compatible with $B^\pm \to$

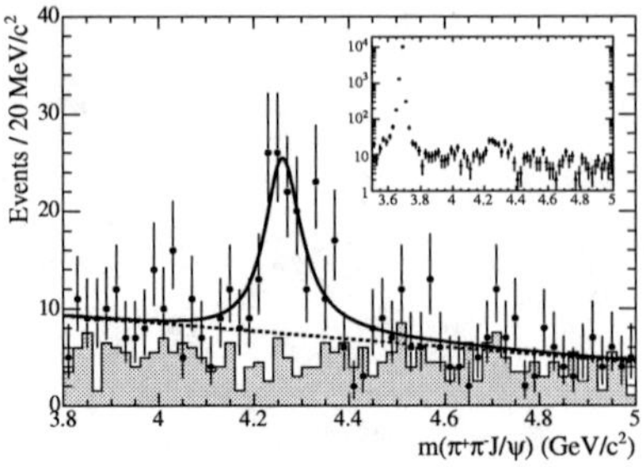

FIGURE 3. The $J/\psi\pi^+\pi^-$ invariant-mass distribution in the range 3.8-5.0 GeV/c^2 for $e^+e^- \rightarrow \gamma_{ISR}J/\psi\pi^+\pi^-$ events; the same distribution for the 3.5-5.0 GeV/c^2 region, including the ψ' peak, is shown in the inset. The solid line is the fit of the distribution with a relativistic Breit-Wigner and a second order polynomial background (dashed line). The histogram represents the background estimated from the J/ψ sideband regions.

$K^{\pm}Y(4260)$, $Y(4260) \rightarrow J/\psi\pi^+\pi^-$ is observed [9]. More data are needed therefore for a study of $Y(4260)$ production in B decay.

SUMMARY

The Babar experiment is performing an extensive study of the spectroscopy and production mechanisms of the new charmonium states. Recent results on the $X(3872)$ state have been presented, and a new state called $Y(4260)$ has been observed recently in ISR processes.

Further investigation and new data will help in the study of quarkonium spectroscopy and in understanding the nature of the states whose interpretation is still unclear.

REFERENCES

1. S. K. Choi *et al.* [Belle Collaboration], Phys. Rev. Lett. **91**, 262001 (2003) [arXiv:hep-ex/0309032].
2. D. Acosta *et al.* [CDF II Collaboration], Phys. Rev. Lett. **93**, 072001 (2004) [arXiv:hep-ex/0312021].
3. V. M. Abazov *et al.* [D0 Collaboration], Phys. Rev. Lett. **93**, 162002 (2004) [arXiv:hep-ex/0405004].
4. B. Aubert *et al.* [BABAR Collaboration], Phys. Rev. D **71**, 071103 (2005) [arXiv:hep-ex/0406022].
5. B. Aubert *et al.* [BABAR Collaboration], Nucl. Instrum. Meth. A **479**, 1 (2002) [arXiv:hep-ex/0105044].
6. E. S. Swanson, Phys. Lett. B **588**, 189 (2004) [arXiv:hep-ph/0311229].
7. N. A. Tornqvist, Phys. Lett. B **590**, 209 (2004) [arXiv:hep-ph/0402237].
8. L. Maiani, F. Piccinini, A. D. Polosa and V. Riquer, Phys. Rev. D **71**, 014028 (2005) [arXiv:hep-ph/0412098].
9. B. Aubert *et al.* [BABAR Collaboration], arXiv:hep-ex/0507090.
10. E. Braaten and M. Kusunoki, Phys. Rev. D **71**, 074005 (2005) [arXiv:hep-ph/0412268].
11. B. Aubert *et al.* [BaBar Collaboration], Phys. Rev. D **71**, 031501 (2005) [arXiv:hep-ex/0412051].
12. B. Aubert *et al.* [BABAR Collaboration], Phys. Rev. D **72**, 031101 (2005) [arXiv:hep-ex/0506062].
13. K. Abe *et al.* [BELLE Collaboration], Phys. Rev. Lett. **88**, 052001 (2002) [arXiv:hep-ex/0110012].
14. E. Braaten and J. Lee, Phys. Rev. D **67**, 054007 (2003) [arXiv:hep-ph/0211085].
15. K. Y. Liu, Z. G. He and K. T. Chao, arXiv:hep-ph/0408141.
16. B. Aubert *et al.* [BABAR Collaboration], arXiv:hep-ex/0506081.

Recent BESII Results of Charmonium Decays

X. H. Mo

Institute of High Energy Physics, CAS, Beijing 100039, China

Abstract. Using 14 million ψ' data sample collected with BES at BEPC, the recent results relevant to χ_{cJ} states are presented, which include measurement of χ_{cJ} resonance parameters, study of $\chi_{cJ} \to VV$ and $K_S^0 K_S^0$ contained final state, and partial wave analysis of $\chi_{c0} \to \pi^+\pi^- K^+ K^-$.

Keywords: χ_{cJ} state, resonance parameter, hadronic decay, PWA
PACS: 14.40.Gx, 12.38.Qk, 13.25.Gv

INTRODUCTION

Precise measurements of the spectrum and the decay properties of charmonia are essential to test potential QCD models and QCD-based approaches. Such an interest is revived after discovery of X(3872), and long expected η_c' and $h_c(^1P_1)$ states. Recently, there are both theoretical and experimental progress concerning $\chi_{cJ}(J = 0,1,2)$ mass. With large ψ' sample and a newly-developed photon conversion technique, BES measured the resonance parameters of χ_{cJ} states [1] which will be given in the second section of this report.

In the third section, we present the results about χ_{cJ} hadronic decays. Compared to J/ψ decay, relatively little is known concerning χ_{cJ} decays, more experimental data on exclusive decays of P-wave charmonia are important for a better understanding of the nature of χ_{cJ} state, as well as testing QCD-based calculations.

The fourth section of this report is devoted to partial wave analysis (PWA) of $\chi_{c0} \to \pi^+\pi^- K^+ K^-$ final state. As we known, PWA is an excellent tool to investigate the intermediate resonant decay modes account for the interference effects when calculating branching ratios.

MEASUREMENT OF χ_{CJ} RESONANCE PARAMETER

Beside statistics, energy resolution is essential to accurate determination of χ_{cJ} resonance parameters, especially its width. So we developed a photon conversion technique to improve our energy resolution greatly. The so-call photon conversion technique is to reconstruct photons from their converted e^+e^- pairs, that is to say, we actually transform the resolution of neutral track into that of charged track and photon energy resolution for $\psi' \to \gamma\chi_{cJ}(J = 0,1,2)$ decays are determined to be 3.78 ± 0.04, 2.58 ± 0.05, and 2.26 ± 0.11 MeV, respectively. Only price we have to pay here is the lose of efficiency. However with 14 M ψ' event, we still expect sufficient statistics.

Fig. 1 shows the observed photon energy spectrum from the ψ' data after the event selection and the corresponding fit result. Subtracting background, we get clear χ_{c0} and

CP814 *Hadron Spectroscopy: Eleventh International Conference*
edited by A. Reis, C. Göbel, J. Sá Borges, and J. Magnin

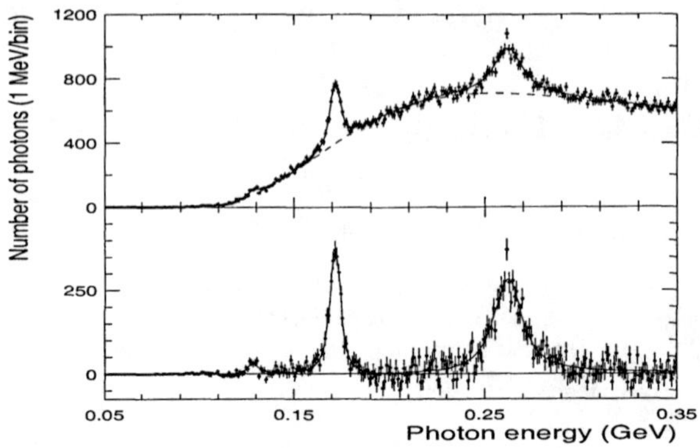

FIGURE 1. Inclusive photon spectrum from photon conversions from 14×10^6 ψ' events. A fit is made to $\psi' \to \gamma\chi_{cJ}$ decays ($J = 0, 1, 2$) plus background described by a threshold function. Points with errors bars are data. The solid line is the fit; the dashed line is the fitted background. Background subtracted results are shown in the lower plot.

χ_{c1} signals and comparatively smaller χ_{c2} signal. So in our final fit, the width of χ_{c2} is fixed to PDG value. The final results are given in table 1 together with comparison with other measurement results. It can be seen our results are consistent with newly-measured results from other experiment groups and have fairly high accuracy.

TABLE 1. BES results for χ_{cJ} resonance parameters and comparison with other experimental results

Exp. Group	χ_{c0}		χ_{c1}		χ_{c2}
	Mass (MeV)	**Width (MeV)**	**Mass (MeV)**	**Width (MeV)**	**Mass (MeV)**
BES (2005)	$3414.21 \pm 0.39 \pm 0.27$	$12.6^{+1.5+1.6}_{-0.9-1.1}$	$3510.30 \pm 0.14 \pm 0.16$	$1.39^{+0.40+0.26}_{-0.38-0.77}$	$3555.70 \pm 0.59 \pm 0.39$
BES (98/99)	$3414.1 \pm 0.6 \pm 0.8$	$14.3 \pm 2.0 \pm 3.0$	3509.4 ± 0.9	–	3556.4 ± 0.7
E835 (2002/2005)	$3415.4 \pm 0.4 \pm 0.2$	$9.8 \pm 1.0 \pm 0.1$	$3510.719 \pm 0.051 \pm 0.019$	$0.876 \pm 0.045 \pm 0.026$	$3556.173 \pm 0.123 \pm 0.020$
E835–BES	-1.2 ± 0.7	-2.8 ± 2.1	0.42 ± 0.22	0.51 ± 0.86	0.47 ± 0.72
CLEO	$3414.07 \pm 0.40 \pm 1.41$	–	$3509.83 \pm 0.20 \pm 0.90$	–	$3555.79 \pm 0.14 \pm 0.66$
CLEO–BES	-0.14 ± 1.5	–	-0.47 ± 0.94	–	-0.09 ± 0.97

STUDY OF χ_{cJ} HADRONIC DECAYS

The decays of χ_{cJ}, in particular χ_{c0} and χ_{c2}, provide a direct window on glueball dynamics in 0^{++} and 2^{++} channels, as the χ_{cJ} hadronic decays may proceed via $c\bar{c} \to gg \to q\bar{q}q\bar{q}$. In addition, exclusive decays of χ_{cJ} are viewed as an ideal laboratory to test the color-octet mechanism.

Table 2 summarizes the results obtained by BES collaboration, including $\chi_{cJ} \to VV$ mode ($V : K^*$, ω) and $\chi_{cJ} \to K_S^0 K_S^0$ contained final states. Most of results are first measurements for χ_{cJ} decays.

TABLE 2. BES results for χ_{cJ} resonance parameters and comparison with other experimental results

Branching ratio (10^{-3})	χ_{c0}	χ_{c1}	χ_{c2}
$K^*(892)K^*(892)$	$1.78 \pm 0.34 \pm 0.34$	$1.67 \pm 0.31 \pm 0.31$	$4.85 \pm 0.56 \pm 0.88$
$\omega\omega$	$2.29 \pm 0.58 \pm 0.41$	–	$1.77 \pm 0.47 \pm 0.36$
$K_S^0 K_S^0$	$3.51 \pm 0.22 \pm 0.47$	< 0.08 (90% C.L.)	$0.89 \pm 0.12 \pm 0.13$
$K_S^0 K_S^0 \pi^+\pi^-$	$6.5 \pm 0.6 \pm 1.0$	$0.80 \pm 0.31 \pm 0.14$	$3.24 \pm 0.61 \pm 0.55$
$K_S^0 K_S^0 K^+K^-$	$1.83 \pm 0.47 \pm 0.33$	$0.31 \pm 0.19 \pm 0.06$	$0.33 \pm 0.24 \pm 0.06$

The detailed analyses of $\chi_{cJ} \to K^*(892)K^*(892)$, $\omega\omega$,$K_S^0K_S^0$-contained final states are presented in Refs. [2, 3, 4], respectively. In addition, in Ref. [4], we also determine the branching fraction of $\psi' \to K_S^0 K_S^0 \pi^+\pi^-$ to be $(2.20 \pm 0.25 \pm 0.33) \times 10^{-4}$.

PARTIAL WAVE ANALYSIS OF $\chi_{C0} \to \pi^+\pi^-K^+K^-$

PWA results

Event selection of $\chi_{cJ} \to \pi^+\pi^-K^+K^-$ is given in Ref. [5], then 5*C*-fit is made with the invariant mass of the $\pi^+\pi^-K^+K^-$ being constrained to the χ_{c0} mass. Finally, selected are 1371 $\psi' \to \chi_{c0}$, $\chi_{c0} \to \pi^+\pi^-K^+K^-$ candidate events, which are utilized as the sample for all afterwards PWA.

Fig. 2(a) shows the scatter plot of K^+K^- versus $\pi^+\pi^-$ invariant mass for selected $\psi(2S) \to \gamma\chi_{c0}$, $\chi_{c0} \to \pi^+\pi^-K^+K^-$ events, which provides further information on the intermediate resonant decay modes for $(\pi^+\pi^-)(K^+K^-)$ decay. For instance, it can be seen that the $f_0(980) \to \pi^+\pi^-$ mainly couples with $f_0(980)$, $f_0(1710)$, and $f_0(2200)$, which decay to K^+K^-, and the concentration of events at large K^+K^- mass is also

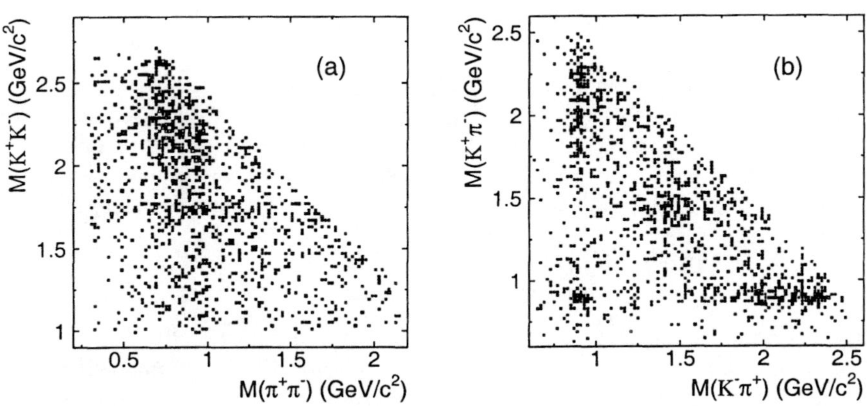

FIGURE 2. The scatter plots of (a) K^+K^- versus $\pi^+\pi^-$ and (b) $K^+\pi^-$ versus $K^-\pi^+$ invariant mass for selected $\psi' \to \gamma\chi_{c0}$, $\chi_{c0} \to \pi^+\pi^-K^+K^-$ events.

associated with $\rho(770)$ decays to $\pi^+\pi^-$.

Fig. 2(b) shows the scatter plot of $K^+\pi^-$ versus $K^-\pi^+$ invariant masses. There are obvious clusters due to $K^*(892)^0$ and around 1.43 GeV/c^2, where there are several known resonances, and evidence for $(K\pi)$ structures at about 1.7 GeV/c^2 and 1.95 GeV/c^2, as well as an enhancement above 2 GeV/c^2. There is also a possible small accumulation outside the $K^*(892)^0\bar{K}^*(892)^0$ cluster, which may be due to the broad S-wave structure κ.

Besides $(\pi\pi)(KK)$ and $(K\pi)(K\pi)$ modes, we also studied $(K\pi\pi)K$ mode which lead to measurement of $K_1(1270)K$ and $K_1(1400)K$ decay processes. All PWA results are summarized in Table 3.

TABLE 3. Summary of numbers used in the branching fraction (or upper limit) calculations and corresponding results, where X represents the intermediate decay modes, N^{fit} is the number of fitted events, and s.s. indicates signal significance.

Decay mode (X)	N^{fit}	**Branching ratio** (10^{-4}) $\mathcal{B}[\chi_{c0} \to X \to \pi^+\pi^-K^+K^-]$	**s.s.**
$f_0(980)f_0(980)$	27.9 ± 6.7	$0.83^{+1.93}_{-1.57}$	5.3σ
$f_0(980)f_0(2200)$	77.1 ± 10.6	$8.42 \pm 1.16^{+1.65}_{-2.29}$	7.1σ
$f_0(1370)f_0(1710)$	60.6 ± 12.4	$7.12 \pm 1.46^{+3.28}_{-1.68}$	6.5σ
$K^*(892)^0\bar{K}^*(892)^0$	64.5 ± 9.9	$8.09 \pm 1.24^{+2.29}_{-1.99}$	7.1σ
$K_0^*(1430)\bar{K}_0^*(1430)$	82.9 ± 12.5	$10.44 \pm 1.57^{+3.05}_{-1.90}$	7.2σ
$K_0^*(1430)\bar{K}_2^*(1430) + c.c.$	62.0 ± 10.7	$8.49 \pm 1.47^{+1.32}_{-1.99}$	8.7σ
$K_1(1270)^+K^- + c.c.$,	68.3 ± 11.0	$9.32 \pm 1.50^{+1.81}_{-1.64}$	8.6σ
$K_1(1400)^+K^- + c.c.$,	19.7 ± 6.9	< 11.9 (90%)	2.7σ

From these $\chi_{c0} \to \pi^+\pi^-K^+K^-$ decay fit results, we notice that scalar resonances have larger decay fractions compared to those of tensors, and such decays provide a relatively clean laboratory to study the properties of scalars, such as $f_0(980)$, $f_0(1370)$, $f_0(1710)$, $f_0(2200)$, $K_0^*(1430)$, and so forth.

Mixing of strange axial mesons

Here is an interesting problem involving $K_1(1270)$ and $K_1(1400)$ that is the mixing of the two strange axial mesons [6]. In the quark model, there are two ground-state axial vector nonets, one is $(^3P_1)$ state or denoted as K_A state, the other is $(^1P_1)$ state or denoted as K_B state. For a strange quark mass is greater than those of up and down quarks, so SU(3) is broken which lead to mixing of K_A and K_B states to give a physical K_1 state, viz.

$$\begin{aligned} K_1(1270) &= K_A \sin\theta_K + K_B \cos\theta_K \,, \\ K_1(1400) &= K_A \cos\theta_K - K_B \sin\theta_K \,. \end{aligned}$$

In SU(3) limit, only K_A couples to the weak current and theoretically if the $I = 1/2$ members of the two octets are degenerate before mixing, SU(3)-symmetry breaking always leads to the maximal mixing $\theta_K = 45°$ which means that there should be equal amount of $K_1(1270)$ and $K_1(1400)$. However, our PWA yields: $K_1(1270) \to K\rho$ signal of 68.3 events; $K_1(1400) \to K^*\pi$ signal of 19.7 events; that is to say, Flavor-SU(3)-violating of $K_1(1270)$-$K_1(1400)$ asymmetry is observed.

Another interesting problem here is about the mixing angle θ_K. According to present BES results, the mixing angle $\theta_K > 57°$ for χ_{c0} decay. But according to previous BES results [7], $\theta_K < 29°$ for ψ' decay and $\theta_K > 48°$ for J/ψ decay. More interesting fact is from τ decay [8]. By virtue of $\tau \to \nu K_1$ measurements and together with relativized quark model estimation [9], the mixing angle should be within a range from $-35°$ to $45°$. All discrepancies displayed here indicate that more theoretical and experimental works are needed.

ACKNOWLEDGMENTS

I would like to acknowledge help from Prof. C. C. Zhang, Dr. Z. J. Guo and Z. Y. Wang, who provided corresponding results presented in this report. Thanks are also due to Prof. F. A. Harris and other colleagues of BES collaboration who gave many suggestions and comments for improvement of my report.

REFERENCES

1. BES Collaboration, M. Ablikim, *et al.*, Phys. Rev. **D71**, 092002 (2005).
2. BES Collaboration, M. Ablikim, *et al.*, Phys. Rev. **D70**, 092003 (2004).
3. BES Collaboration, M. Ablikim, *et al.*, hep-ex/0506045.
4. BES Collaboration, M. Ablikim, *et al.*, hep-ex/0410028.
5. BES Collaboration, M. Ablikim, *et al.*, hep-ex/0508050.
6. H.J. Lipkin, Phys. Rev. **176**, 1709 (1968); H.J. Lipkin, Phys. Lett. **B63**, 224 (1976); M. Suzuki, Phys. Rev. **D47**, 1252 (1993).
7. BES Collaboration, J. Z. Bai, *et al.*, Phys. Rev. Lett. **83**, 1918 (1999).
8. TPC/Two-Gamma Collaboration, D. A. Bauer *et al.*, Phys. Rev. **D50**, R13 (1994).
9. H. G. Blundell *et al.*, Phys. Rev. **D53**, 3712 (1996).

Heavy quarkonium decays and transitions in the language of effective field theories

Antonio Vairo

Dipartimento di Fisica dell'Università di Milano and INFN, via Celoria 16, 20133 Milano, Italy

Abstract. Heavy quarkonium decays and transitions are discussed in the framework of non-relativistic effective field theories. Emphasis is put on the matching procedure in the non-perturbative regime. Some exact results valid for the magnetic dipole couplings are discussed.

Keywords: NRQCD, pNRQCD, heavy quarkonium, decay, radiative transitions
PACS: 12.38.-t, 12.39.Hg, 13.25.Gv

INTRODUCTION

In the last years the B factories, CLEO and BES have produced a large amount of new data for heavy quarkonium observables [1]. These data are not only interesting because they may signal new states or new decay or production mechanisms, but also because heavy quarkonium is a system that to a large extent can rigorously be studied in QCD. Therefore, any new understanding of it may potentially provide new insight on the non-perturbative dynamics of QCD.

Heavy quarkonium, as a non-relativistic bound state, is characterized by a hierarchy of energy scales: m, mv and mv^2, where m is the heavy-quark mass and $v \ll 1$ the heavy-quark relative velocity. Whenever a system is described by a hierarchy of scales, observables may be calculated by expanding one scale with respect to the other. An effective field theory (EFT) is a field theory that makes this expansion explicit at the Lagrangian level. To be more precise, let us call H a system described by a fundamental Lagrangian $\mathscr{L}$ and suppose it characterized by 2 scales: $\Lambda \gg \lambda$. The EFT Lagrangian, $\mathscr{L}_{\rm EFT}$, suitable to describe H at scales lower than Λ, is characterized by (1) a cut off $\Lambda \gg \mu \gg \lambda$; (2) some degrees of freedom that exist at scales lower than μ. The Lagrangian $\mathscr{L}_{\rm EFT}$ is then made of all operators O_n that may be built from the effective degrees of freedom and are consistent with the symmetries of the original Lagrangian $\mathscr{L}$:

$$\mathscr{L}_{\rm EFT} = \sum_n c_n(\Lambda/\mu)\frac{O_n(\mu,\lambda)}{\Lambda^n}. \tag{1}$$

The advantage is that, once the scale μ has been run down to λ, the power counting is homogeneous $\langle O_n \rangle \sim \lambda^n$, so that the EFT is, indeed, organized as an expansion in λ/Λ. Despite the EFT not being renormalizable in the traditional sense, it is renormalizable order by order in λ/Λ. The matching coefficients $c_n(\Lambda/\mu)$ encode the non-analytic behaviour in Λ. They are calculated by imposing that $\mathscr{L}_{\rm EFT}$ and $\mathscr{L}$ describe the same physics at any finite order in the expansion. The procedure is called matching. Finally,

CP814 *Hadron Spectroscopy: Eleventh International Conference*
edited by A. Reis, C. Göbel, J. Sá Borges, and J. Magnin

we note that if $\Lambda \gg \Lambda_{\rm QCD}$, $c_n(\Lambda/\mu)$ may be calculated in perturbation theory, if $\Lambda \sim \Lambda_{\rm QCD}$, the matching must rely on non-perturbative methods.

Several effective field theories for heavy quarkonium that take full advantage of the non-relativistic hierarchy of scales have been developed and used over the last decade. For a recent review we refer to [2]. NRQCD is the EFT that exploits the hierarchy $\Lambda = m \gg \lambda = mv$ [3, 4]. Since $m \gg \Lambda_{\rm QCD}$, the matching coefficients of NRQCD may be calculated in perturbation theory. pNRQCD is the EFT that exploits the hierarchy $\Lambda = mv \gg \lambda = mv^2$ [5, 6]. If $\Lambda_{\rm QCD} \sim mv^2$, then the matching to pNRQCD may be still done in perturbation theory. We call weak coupling this regime, which may be suited to describe ground-state quarkonium. If $\Lambda_{\rm QCD} \sim mv$, then the matching to pNRQCD is non perturbative. We call strong coupling this regime, which may be suited to describe excited quarkonium states.

The fact that EFTs may be built to describe heavy quarkonium in the strong-coupling regime follows from the observation that the non-relativistic hierarchy of scales survives also below $\Lambda_{\rm QCD}$ [7, 8]. The complication of the strong-coupling regime comes from the non-perturbative matching and from new scales that may arise in loops sensitive to $\Lambda_{\rm QCD}$. An example is the three-momentum scale $\sqrt{m\Lambda_{\rm QCD}}$ discussed in [9]. Nevertheless, many advantages remain in treating even strongly-coupled heavy quarkonium in an EFT framework.

In the following we shall sketch a unified framework for the description of inclusive and electromagnetic decays, and radiative transitions in the framework of strongly coupled pNRQCD. For a treatment of inclusive and electromagnetic decay widths in the weak-coupling regime we refer to [2] and references therein. For a treatment of magnetic dipole transitions in the weak-coupling regime we refer to [10].

NRQCD

NRQCD is the EFT that follows from QCD when modes of energy or momentum m are integrated out. The structure of the EFT Lagrangian is like Eq. (1) with $\Lambda = m$ and $\lambda = mv \sim \Lambda_{\rm QCD}$. The scale mv is sometimes called soft. The degrees of freedom of the EFT Lagrangian are quarks, antiquarks and gluons with energy and momentum lower than m (we neglect light quarks).

The NRQCD Lagrangian may be written as

$$\mathcal{L}_{\rm NRQCD} = \mathcal{L}_{2-f} + \mathcal{L}_{4-f} + \mathcal{L}_{\rm light}, \tag{2}$$

where

$$\mathcal{L}_{2-f} = \psi^\dagger \left(iD_0 + \frac{\mathbf{D}^2}{2m} \right) \psi + \frac{c_F}{2m} \psi^\dagger \sigma \cdot g\mathbf{B} \psi - \frac{2c_F - 1}{8m^2} \psi^\dagger \sigma \cdot [-i\mathbf{D} \times, g\mathbf{E}] \psi + \dots$$
$$+ [\psi \to i\sigma^2 \chi^*], \tag{3}$$

$$\mathcal{L}_{\rm light} = -\frac{1}{4} F^{\mu\nu\, a} F^a_{\mu\nu}. \tag{4}$$

ψ is the Pauli spinor field that annihilates a heavy quark of mass m, χ is the corresponding one that creates a heavy antiquark, $iD_0 = i\partial_0 - gT^aA_0^a$ and $i\mathbf{D} = i\nabla + gT^a\mathbf{A}^a$. The term $\mathcal{L}_{4-f}$ stands for the 4-fermion part of the NRQCD Lagrangian (for an explicit expression see, for instance, [4]).

The coefficient c_F is a matching coefficient of the EFT. In Eq. (3), we have made use of reparameterization invariance to reduce the other matching coefficients to this one. It is known at two loops and may be found in [11]. The 4-fermion matching coefficients encode the contribution of the annihilation graphs. As a consequence they develop an imaginary part. We refer to [12] for an updated list of them and for references.

Let us give some definitions concerning the Fock space of NRQCD. If $H_{\rm NRQCD}$ is the Hamiltonian of NRQCD, we call $|\underline{\rm n};\mathbf{r},\mathbf{R}\rangle$ the eigenstates of $H_{\rm NRQCD}$, and E_n the corresponding eigenvalues. $\mathbf{r}$ stands for the relative distance of the two heavy quarks and $\mathbf{R}$ for their centre-of-mass coordinate. Both are good quantum numbers in the static limit. With n we indicate a generic set of conserved quantum numbers. $|\underline{\rm n};\mathbf{r},\mathbf{R}\rangle$ and $E_n(\mathbf{r},\mathbf{R};\nabla_r,\nabla_R)$ satisfy the system of equations:

$$H_{\rm NRQCD}|\underline{\rm n};\mathbf{r},\mathbf{R}\rangle = \int d^3r' d^3R' |\underline{\rm n};\mathbf{r}',\mathbf{R}'\rangle E_n(\mathbf{r}',\mathbf{R}';\nabla_{r'},\nabla_{R'})\delta^3(\mathbf{r}'-\mathbf{r})\delta^3(\mathbf{R}'-\mathbf{R}), \quad (5)$$

$$\langle\underline{\rm m};\mathbf{r},\mathbf{R}|\underline{\rm n};\mathbf{r}',\mathbf{R}'\rangle = \delta_{nm}\delta^3(\mathbf{r}-\mathbf{r}')\delta^3(\mathbf{R}-\mathbf{R}'). \quad (6)$$

INCLUSIVE DECAYS IN PNRQCD

pNRQCD is the EFT that follows from NRQCD when gluons of energy or momentum and quarks of energy larger than mv^2 and quarks of momentum larger than mv are integrated out. The structure of the EFT Lagrangian is like Eq. (1) with $\Lambda = mv \sim \Lambda_{\rm QCD}$ and $\lambda = mv^2$. The scale mv^2 is sometimes called ultrasoft. In the strong-coupling regime, if the gluonic excitations between the two heavy quarks develop a mass gap of order $\Lambda_{\rm QCD}$, then they are all integrated out from the theory. Therefore, the degrees of freedom of the EFT Lagrangian are only singlet quarkonium fields. The Lagrangian of pNRQCD is very simple:

$$\mathcal{L}_{\rm pNRQCD} = \int d^3r \, {\rm Tr}\left\{ {\rm S}^\dagger \left(i\partial_0 + \frac{\nabla_R^2}{4m} + \frac{\nabla_r^2}{m} - V_S \right) {\rm S} \right\}, \quad (7)$$

S is a non-local field, function of $\mathbf{r}$, $\mathbf{R}$ and t, $2\otimes 2$ in spin space and a $3\otimes 3$ singlet in colour space. The trace is taken over colour and spin indices.

All complications go into the potential V_S, which is a non-perturbative function of $\mathbf{r}$ to be determined by a non-perturbative matching procedure. In general V_S contains also an imaginary part inherited from the matching coefficients of the 4-fermion operators of NRQCD. The matching condition reads

$$\langle\underline{0};\mathbf{r}',\mathbf{R}'|H_{\rm NRQCD}|\underline{0};\mathbf{r},\mathbf{R}\rangle = \left(-\frac{\nabla_R^2}{4m} - \frac{\nabla_r^2}{m} + V_S \right) \delta^3(\mathbf{r}'-\mathbf{r})\delta^3(\mathbf{R}'-\mathbf{R}). \quad (8)$$

The matching condition determines V_S as a function of quantities defined in NRQCD (the left-hand side of Eq. (8)). Once V_S has been determined, one may calculate the solutions $\Phi_H(\mathbf{r})$ and E_H of the Schrödinger equation

$$\left(-\frac{\nabla_r^2}{m}+V_S\right)\Phi_H(\mathbf{r})=E_H\Phi_H(\mathbf{r}). \tag{9}$$

Using the optical theorem, the inclusive decay width to light particles (l.p.) is given by

$$\Gamma_{H\to \mathrm{l.p.}}=-2\,\mathrm{Im}\,\langle H(\mathbf{0})|-\mathscr{L}_{\mathrm{pNRQCD}}|H(\mathbf{0})\rangle. \tag{10}$$

$|H(\mathbf{0})\rangle$ stands for a quarkonium state in the rest-frame ($\mathbf{P}$ = 0):

$$|H(\mathbf{0})\rangle=\int d^3r\int d^3R\,\mathrm{Tr}\left\{\Phi_H(\mathbf{r})\mathrm{S}^\dagger(\mathbf{r},\mathbf{R})\right\}|0\rangle, \tag{11}$$

where $|0\rangle$ is the Fock subspace containing no heavy quarks but an arbitrary number of ultrasoft particles. Note that, since Φ_H and $\mathscr{L}_{\mathrm{pNRQCD}}$ have been calculated through the matching procedure, Eq. (10) provides, indeed, a practical tool to calculate the inclusive decay width. Explicit applications of Eq. (10) have been worked out in [13, 14, 15].

RADIATIVE TRANSITIONS IN PNRQCD

Radiative transitions may be described in the same EFT framework that we have discussed so far by enlarging the gauge group to $SU_c(3)\times U_{\mathrm{em}}(1)$. This means that more degrees of freedom have to be taken into account (photons) and more operators added to the EFT Lagrangians. We will concentrate in the following on magnetic dipole transitions [10].

At the level of the NRQCD Lagrangian, magnetic transitions are accounted for by replacing $iD_0\to iD_0-ee_QA_0^{\mathrm{em}}$ and $i\mathbf{D}\to i\mathbf{D}+ee_Q\mathbf{A}^{\mathrm{em}}$ in Eq. (3),

$$\begin{aligned}\mathscr{L}_{2-f}\to\mathscr{L}_{2-f}&+\frac{c_F^{\mathrm{em}}}{2m}\psi^\dagger\sigma\cdot ee_Q\mathbf{B}^{\mathrm{em}}\psi-\frac{2c_F^{\mathrm{em}}-1}{8m^2}\psi^\dagger\sigma\cdot[-i\mathbf{D}\times,ee_Q\mathbf{E}^{\mathrm{em}}]\psi\\&+\frac{c_{W1}^{\mathrm{em}}}{8m^3}\psi^\dagger\{\mathbf{D}^2,\sigma\cdot ee_Q\mathbf{B}^{\mathrm{em}}\}\psi-\frac{c_{W1}^{\mathrm{em}}-1}{4m^3}\psi^\dagger\mathbf{D}^i\sigma\cdot ee_Q\mathbf{B}^{\mathrm{em}}\mathbf{D}^i\psi\\&+\frac{c_F^{\mathrm{em}}-1}{8m^3}\psi^\dagger\left(\sigma\cdot\mathbf{D}\,ee_Q\mathbf{B}^{\mathrm{em}}\cdot\mathbf{D}+\mathbf{D}\cdot ee_Q\mathbf{B}^{\mathrm{em}}\,\sigma\cdot\mathbf{D}\right)\psi+\cdots+[\psi\to i\sigma^2\chi^*],\end{aligned} \tag{12}$$

and $\mathscr{L}_{\mathrm{light}}\to\mathscr{L}_{\mathrm{light}}-\frac{1}{4}F^{\mu\nu\,\mathrm{em}}F_{\mu\nu\,\mathrm{em}}$, where the gauge fields with upperscript "em" are electromagnetic fields and ee_Q stands for the charge of the quark of flavour Q.

The coefficients c_F^{em} and c_{W1}^{em} are new matching coefficients of the EFT associated with the electromagnetic couplings. Again, we have made use of reparameterization invariance to reduce their number. All coefficients are known at least at one-loop level

[16]. In particular, we have[1]

$$c_F^{\rm em} \equiv 1 + \kappa_Q^{\rm em} = 1 + \frac{4}{3}\frac{\alpha_{\rm s}}{2\pi} + \mathcal{O}(\alpha_{\rm s}^2), \tag{13}$$

$$c_{W1}^{\rm em} = 1 + \frac{4}{3}\frac{\alpha_{\rm s}}{\pi}\left(\frac{1}{12} + \frac{4}{3}\ln\frac{m}{\mu}\right) + \mathcal{O}(\alpha_{\rm s}^2), \tag{14}$$

$\kappa_Q^{\rm em}$ is usually identified with the anomalous magnetic moment of the heavy quark.

At the level of pNRQCD, magnetic transitions involving ultrasoft photons are described by adding to the Lagrangian (7) the electromagnetic Lagrangian $-F^{\mu\nu\,{\rm em}} \times F_{\mu\nu\,{\rm em}}/4$ and a term $\mathcal{L}_{\gamma{\rm pNRQCD}}$ responsible for the coupling of the quarkonium to the electromagnetic field:

$$\begin{aligned}
\mathcal{L}_{\gamma{\rm pNRQCD}} = \int d^3r\, {\rm Tr}\Bigg\{ & V_A^{\rm em}\, {\rm S}^\dagger \mathbf{r}\cdot ee_Q\mathbf{E}^{\rm em}{\rm S} \\
& + \frac{V_S^{\frac{\sigma\cdot B}{m}}}{2m}\left\{{\rm S}^\dagger, \sigma\cdot ee_Q\mathbf{B}^{\rm em}\right\}{\rm S} + \frac{V_S^{(r\cdot\nabla)^2\frac{\sigma\cdot B}{m}}}{16m}\left\{{\rm S}^\dagger, \mathbf{r}^i\mathbf{r}^j(\nabla_R^i\nabla_R^j\sigma\cdot ee_Q\mathbf{B}^{\rm em})\right\}{\rm S} \\
& + \frac{V_S^{\frac{\sigma\cdot(r\times r\times B)}{m^2}}}{4m^2\, r}\left\{{\rm S}^\dagger, \sigma\cdot\left[\hat{\mathbf{r}}\times\left(\hat{\mathbf{r}}\times ee_Q\mathbf{B}^{\rm em}\right)\right]\right\}{\rm S} + \frac{V_S^{\frac{\sigma\cdot B}{m^2}}}{4m^2\, r}\left\{{\rm S}^\dagger, \sigma\cdot ee_Q\mathbf{B}^{\rm em}\right\}{\rm S} \\
& - \frac{V_S^{\frac{\sigma\cdot\nabla\times E}{m^2}}}{16m^2}\left[{\rm S}^\dagger, \sigma\cdot\left[-i\nabla_R\times, ee_Q\mathbf{E}^{\rm em}\right]\right]{\rm S} - \frac{V_S^{\frac{\sigma\cdot\nabla_r\times r\cdot\nabla E}{m^2}}}{16m^2}\left[{\rm S}^\dagger, \sigma\cdot\left[-i\nabla_r\times, \mathbf{r}^i(\nabla_R^i ee_Q\mathbf{E}^{\rm em})\right]\right]{\rm S} \\
& + \frac{V_S^{\frac{\nabla_r^2\sigma\cdot B}{m^3}}}{4m^3}\left\{{\rm S}^\dagger, \sigma\cdot ee_Q\mathbf{B}^{\rm em}\right\}\nabla_r^2{\rm S} + \frac{V_S^{\frac{(\nabla_r\cdot\sigma)(\nabla_r\cdot B)}{m^3}}\, 1}{4m^3}\left\{{\rm S}^\dagger, \sigma^i ee_Q\mathbf{B}^{{\rm em}\, j}\right\}\nabla_r^i\nabla_r^j{\rm S} \\
& + \frac{V_S^{\frac{\sigma\cdot(r\times r\times B)}{m^3}}}{4m^3\, r^2}\left\{{\rm S}^\dagger, \sigma\cdot\left[\hat{\mathbf{r}}\times\left(\hat{\mathbf{r}}\times ee_Q\mathbf{B}^{\rm em}\right)\right]\right\}{\rm S} + \frac{V_S^{\frac{\sigma\cdot B}{m^3}}}{4m^3\, r^2}\left\{{\rm S}^\dagger, \sigma\cdot ee_Q\mathbf{B}^{\rm em}\right\}{\rm S} + \cdots \Bigg\}.
\end{aligned} \tag{15}$$

All gauge fields are calculated in the centre-of-mass coordinate **R**. The field S is understood as a singlet also under $U_{\rm em}(1)$ gauge transformations.

In the centre-of-mass of the initial quarkonium state, the power counting goes as follows: $\nabla_r \sim mv$, $r \sim 1/mv$, the electromagnetic fields associated to the external photons go like $\mathbf{E}^{\rm em}, \mathbf{B}^{\rm em} \sim k_\gamma^2$. The centre-of-mass derivative ∇ acting on the recoiling final quarkonium state or emitted photon is of order k_γ, where k_γ is the energy and momentum of the emitted photon.

The coefficients V in Eq. (15) are the matching coefficients of pNRQCD. They encode high-energy contributions to the electromagnetic couplings and are of the same nature as V_S in Eq. (7). In the strong-coupling regime they are determined by non-perturbative matching of 5-points Green functions involving two external quarks, two external antiquarks and an external photon. Let us consider the matching condition for

[1] The coefficients get also QED corrections, but they are numerically negligible.

the $1/m$ operators, it reads

$$\langle \underline{0};\mathbf{r}',\mathbf{R}'| \otimes \langle\gamma| \left(\frac{c_F^{\rm em}}{2m} \int d^3x\, \psi^\dagger \sigma\cdot ee_Q \mathbf{B}^{\rm em} \psi + [\psi \to i\sigma^2\chi^*] \right) |0\rangle \otimes |\underline{0};\mathbf{r},\mathbf{R}\rangle =$$
$$\left(\frac{V_S^{\frac{\sigma\cdot B}{m}}}{2m} + \frac{V_S^{(r\cdot\nabla)^2 \frac{\sigma\cdot B}{m}}}{16m} (\mathbf{r}\cdot\nabla_R)^2 \right) \left(\sigma^{(1)} + \sigma^{(2)}\right) \cdot \langle\gamma| ee_Q \mathbf{B}^{\rm em} |0\rangle \delta^3(\mathbf{r}'-\mathbf{r})\delta^3(\mathbf{R}'-\mathbf{R}). \quad (16)$$

Since corrections to the state $|\underline{0};\mathbf{r},\mathbf{R}\rangle$ involving derivatives or spins are $1/m$ suppressed (see Eq. (3)), $\sigma\cdot ee_Q\mathbf{B}^{\rm em}$ effectively behaves as the identity operator. As a consequence, the electromagnetic matrix element decouples in the left-hand side. From the normalization condition (6) it follows that

$$V_S^{\frac{\sigma\cdot B}{m}} = V_S^{(r\cdot\nabla)^2 \frac{\sigma\cdot B}{m}} = c_F^{\rm em}. \quad (17)$$

This is a rather remarkable result that holds to all orders in the strong-coupling constant and non-perturbatively. It excludes that the $1/m$ magnetic coupling of the quarkonium field is affected by any soft contribution. A fortiori, it excludes large anomalous non-perturbative corrections to this coupling. Similar arguments lead to the following exact results at order $1/m^2$:

$$V_S^{\frac{\sigma\cdot(r\times r\times B)}{m^2}} = \frac{r^2}{2} V_S^{(0)\prime}, \quad V_S^{\frac{\sigma\cdot B}{m^2}} = 0, \quad V_S^{\frac{\sigma\cdot\nabla\times E}{m^2}} = V_S^{\frac{\sigma\cdot\nabla_r\times r\cdot\nabla E}{m^2}} = 2c_F^{\rm em} - 1, \quad (18)$$

where $V_S^{(0)}$ is the static part of the V_S potential. The first equality follows from the fact that Poincaré invariance protects the spin-orbit coupling [17, 18]. The second one remarkably states that to all orders in the strong-coupling constant and non-perturbatively the existence of an effective scalar interaction, which has been often advocated in phenomenological models, is excluded. The third one that those matching coefficients, like the one in Eq. (17), get only hard contributions.

The matching of the $1/m^3$ terms is more complicated. One reason is that at this order kinetic energy and spin-dependent corrections affect the state $|\underline{0};\mathbf{r},\mathbf{R}\rangle$ and $\sigma\cdot ee_Q\mathbf{B}^{\rm em}$ does not behave anymore like the identity operator.

Once the matching has been completed, the transition width is given by:

$$\Gamma_{H\to H'\gamma} = \int \frac{d^3P'}{(2\pi)^3} \frac{d^3k}{(2\pi)^3} (2\pi)^4 \delta^4(P_H - k - P') \overline{\left|\mathscr{A}[H(\mathbf{0}) \to H'(\mathbf{P}')\gamma(\mathbf{k})]\right|^2}, \quad (19)$$

where

$$\mathscr{A}\left[H(\mathbf{0}) \to H'(-\mathbf{k})\gamma(\mathbf{k})\right] \delta^3(\mathbf{P}'+\mathbf{k}) = \langle H'(\mathbf{P}')\,\gamma(\mathbf{k})| - \int d^3R\, \mathscr{L}_{\gamma{\rm pNRQCD}} |H(\mathbf{0})\rangle. \quad (20)$$

The overline stands for the sum over the final-state polarizations and the average over the initial state ones. $P_H = (M_H, \mathbf{0})$ stands for the four-momentum of the initial-state quarkonium of mass M_H. The state $|H(\mathbf{P})\rangle$ is the state (11) boosted by $-\mathbf{P}/M_H$. The Lorentz-boost transformations may be read from [17, 18].

CONCLUSIONS

We have discussed in an unified framework inclusive and electromagnetic decays, and radiative transitions of heavy quarkonium in a regime where the typical momentum transfer is of order $\Lambda_{\rm QCD}$. Noteworthy, also in this situation suitable effective field theories may be constructed, systematic expansions exploited and exact results derived.

It seems rather unlikely that the non-perturbative matching, once completed at order $1/m^3$, will support the formulas traditionally and universally used so far to describe radiative transitions at relative order v^2 and derived from phenomenological assumptions [19, 20]. This may possibly shade some light, for instance, on the radiative transition data for the Υ system recently collected at CLEO [21], whose understanding is problematic in many phenomenological models.

ACKNOWLEDGMENTS

The author acknowledges the financial support obtained inside the Italian MIUR program "incentivazione alla mobilità di studiosi stranieri e italiani residenti all'estero" and the Marie Curie Reintegration Grant contract MERG-CT-2004-510967.

REFERENCES

1. N. Brambilla *et al.*, arXiv:hep-ph/0412158.
2. N. Brambilla, A. Pineda, J. Soto and A. Vairo, arXiv:hep-ph/0410047.
3. W. E. Caswell and G. P. Lepage, Phys. Lett. B **167**, 437 (1986).
4. G. T. Bodwin, E. Braaten and G. P. Lepage, Phys. Rev. D **51**, 1125 (1995) [Erratum-ibid. D **55**, 5853 (1997)] [hep-ph/9407339].
5. A. Pineda and J. Soto, Nucl. Phys. Proc. Suppl. **64**, 428 (1998) [arXiv:hep-ph/9707481].
6. N. Brambilla, A. Pineda, J. Soto and A. Vairo, Nucl. Phys. B **566**, 275 (2000) [arXiv:hep-ph/9907240].
7. N. Brambilla, A. Pineda, J. Soto and A. Vairo, Phys. Rev. D **63**, 014023 (2001) [arXiv:hep-ph/0002250].
8. A. Pineda and A. Vairo, Phys. Rev. D **63**, 054007 (2001) [Erratum-ibid. D **64**, 039902 (2001)] [arXiv:hep-ph/0009145].
9. N. Brambilla, A. Pineda, J. Soto and A. Vairo, Phys. Lett. B **580**, 60 (2004) [arXiv:hep-ph/0307159].
10. N. Brambilla, Y. Jia and A. Vairo, IFUM-841-FT (in preparation).
11. A. Czarnecki and A. G. Grozin, Phys. Lett. B **405**, 142 (1997) [arXiv:hep-ph/9701415].
12. A. Vairo, Mod. Phys. Lett. A **19**, 253 (2004) [arXiv:hep-ph/0311303].
13. N. Brambilla, D. Eiras, A. Pineda, J. Soto and A. Vairo, Phys. Rev. Lett. **88**, 012003 (2002) [arXiv:hep-ph/0109130].
14. N. Brambilla, D. Eiras, A. Pineda, J. Soto and A. Vairo, Phys. Rev. D **67**, 034018 (2003) [arXiv:hep-ph/0208019].
15. A. Vairo, Nucl. Phys. Proc. Suppl. **115**, 166 (2003) [arXiv:hep-ph/0205128].
16. A. V. Manohar, Phys. Rev. D **56**, 230 (1997) [hep-ph/9701294].
17. N. Brambilla, D. Gromes and A. Vairo, Phys. Lett. B **576**, 314 (2003) [hep-ph/0306107].
18. N. Brambilla, D. Gromes and A. Vairo, Phys. Rev. D **64**, 076010 (2001) [arXiv:hep-ph/0104068].
19. H. Grotch and K. J. Sebastian, Phys. Rev. D **25**, 2944 (1982).
20. H. Grotch, D. A. Owen and K. J. Sebastian, Phys. Rev. D **30**, 1924 (1984).
21. M. Artuso *et al.* [CLEO Collaboration], Phys. Rev. Lett. **94**, 032001 (2005) [hep-ex/0411068].

Charmonium at CLEO

Kamal K. Seth

Department of Physics and Astronomy, Northwestern University, Evanston, IL, 60208, USA

Abstract. At CLEO, charmonium spectroscopy is pursued both thorugh e^+e^- annihilation data taken in the Upsilon region and more recently at $\psi(2S)$. A nmber of first observations (η'_c, h_c, π and K form factors) have been made, and numerous high precision measurements have been made in radiative and hadronic decays of charmonium resonances. A brief report of these contributions is presented.

Keywords: charmonium spectroscopy, meson form factor
PACS: 14.40.Gx,13.20.Gd,13.25.Gv,13.40.Gp

1. INTRODUCTION

Charmonium was discovered in 1974. During the next 10 years a great amount of discovery physics was done by the e^+e^- colliders at SLAC, DESY and ORSAY. In the 1990's the Fermilab $p\bar{p}$ experiments E760/E835 brought unprecedented mass resolution and precision to charmonium spectroscopy, and the BES detector also made numerous contributions at the Beijing e^+e^- collider. The BES II detector has since accumulated the world's largest data set with 59 million J/ψ and 14 million $\psi(2S)$. Unfortunately, the Fermilab experiments were limited because their detector did not have the capability to identify charged hadrons, and the BES detector had limited capability for photon detection. In contrast, the CLEO detector, operating at the CESR e^+e^- collider, is a state-of-the-art solenoidal detector with 93% of 4π coverage. It contains an excellent CsI electromagnetic calorimeter with $\sigma_E/E = 1.5\%$ at 5 GeV (4% at 100 MeV), excellent charged particle tracking with $\sigma_E/E \approx 0.6\%$ at 1 GeV/c, and a RICH detector. This allows for precision charmonium spectroscopy using $\sim$ 15 fb^{-1} of e^+e^- data in the bottomonium region ($\sqrt{s} = 9.46 - 11.30$ GeV), ~ 6 pb^{-1} of data taken at $\psi(2S)$ ($\sqrt{s} =$ 3.686 GeV), and ~ 20 pb^{-1} taken just belw $\psi(2S)$ ($\sqrt{s} = 3.670$ GeV).

The list of measurements made at CLEO in the charmonium region is quite impressive. They include discovery measurements and precision measurements as described in the following. To put my presentation in perspective, the charmonium spectrum is shown in Fig. 1 (left).

2. DISCOVERY OF $\eta'_c(2^1S_0)$ AND $h_c(1^1P_1)$

Spin singlet states are notoriously difficult to populate in e^+e^- collisions. For example, $\eta_c(1^1S_0)$ and $\eta'_c(2^1S_0)$ can only be reached by weak M1 radiative transitions from the directly produced vector states J/ψ and $\psi(2S)$, and radiative transition to $h_c(1^1P_1)$ is forbidden by C–parity conservation. However, identification of these states is crucial

CP814 *Hadron Spectroscopy: Eleventh International Conference*
edited by A. Reis, C. Göbel, J. Sá Borges, and J. Magnin

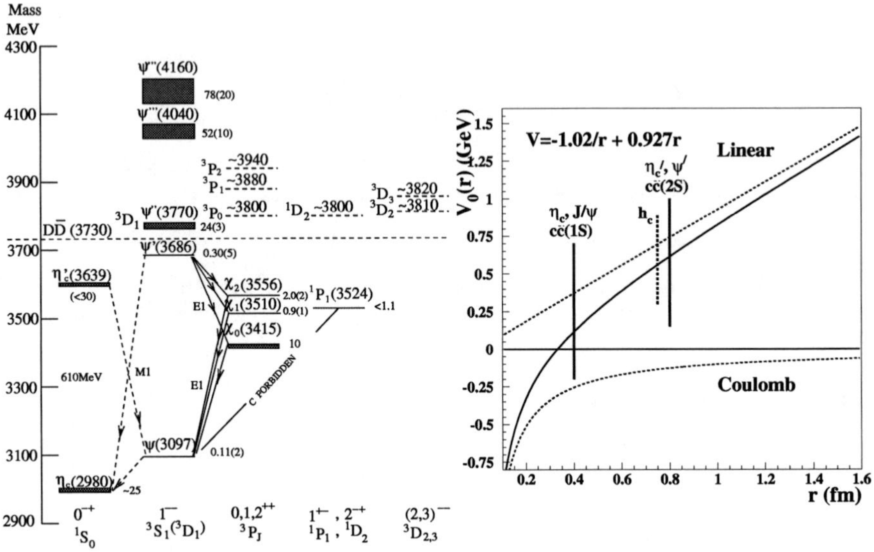

FIGURE 1. (left) Spectra of the states of Charmonium. (right) Schematic of the QCD $q\bar{q}$ potential (solid line), and its Coulombic and confinement parts (dotted lines). The approximate locations of the $1S$, $2S$ and $^1P_1$ states are shown.

for understanding the spin–spin hyperfine interaction in the $q\bar{q}$ system. The nature and development of the hyperfine interaction in going from $1S$ to $2S$ states, and in going from $S-$ to $P-$wave states, can only be detemined by identifying the spin–singlet partners of the well–established spin–triplet states, J/ψ, $\psi(2S)$, and $\chi_J(^3P_J)$. As Fig. 1 (right) illustrates, the $2S$ and $1P$ states sample mainly the confinement part of the $q\bar{q}$ interaction, and determining their hyperfine splitting may be expected to shed light on the nature of the confinement interaction.

2.1 The radial excitation of the charmonium ground state, $\eta_c'(2^1S_0)$

Both $\eta_c(1^1S_0)$ and $J/\psi(^3S_1)$ states are well–established and measured. The hyperfine splitting is known to be $\Delta M_{hf}(1S) = 117 \pm 1$ MeV. In contrast, while $\psi(2^3S_1)$, or ψ', is well established with $M(\psi') = 3686.11 \pm 0.03$ MeV, until now $\eta_c'(2^1S_0)$, or η_c', has remained unidentified. The Crystal Ball's [1] claim of the observation of η_c' with $M(\eta_c') = 3594 \pm 5$ MeV was never confirmed. Several subsequent attempts to find η_c' were unsuccessful. These included $p\bar{p}$ experiments E760 and E835 at Fermilab [2], two–photon fusion experiments at DELPHI and L3 at LEP, and a recent $\psi(2S)$ inclusive photon measurement at CLEO [3].

The breakthrough came, of all the places, from the observation of η_c' in B decays by Belle [4]. It was followed by its observation in $\gamma\gamma$ fusion by us at CLEO [5], and by BaBar [6] (see Fig. 2). The present world average is $\langle M(\eta_c')\rangle = 3638.7 \pm 2.0$ MeV, which leads to the hyperfine splitting $\Delta M_{hf}(2S) = 3686.1 - 3638.7 = 47.4 \pm 2.0$ MeV. This is a factor ~ 2.5 smaller than $\Delta M_{hf}(1S)$, and needs to be understood. Is this due to

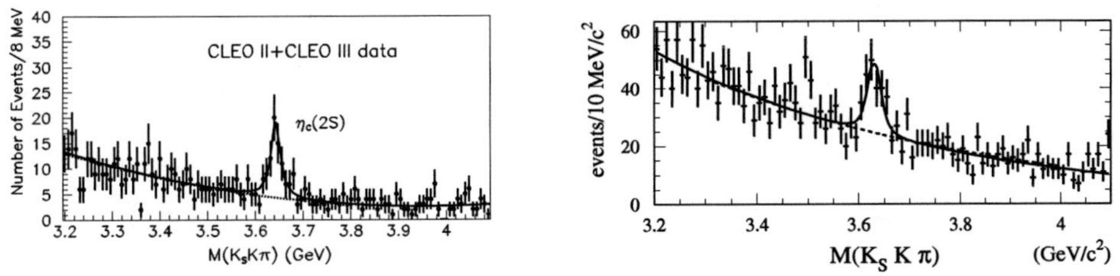

$M(\eta_c') = 3642.9 \pm 3.4$ MeV [5] $\qquad$ $M(\eta_c') = 3630.8 \pm 3.5$ MeV [6]

FIGURE 2. Observation of η_c' in the reaction $\gamma\gamma \to K_S K\pi$ by CLEO (left) and BaBar (right).

configuration mixing or due to some special aspect of the hyperfine $\vec{s}_1 \cdot \vec{s}_2$ interaction in the confinement region? It is worth noting that the width of η_c' is essentially unmeasured so far, and that it may shed light on the possible configuration mixing in η_c'.

2.2 The singlet P state, $h_c(1^1P_1)$, of charmonium

If there ever was a more elusive state than $\eta_c'(2^1S_0)$, it is $h_c(1^1P_1)$, the singlet P-state of charmonium. Its identification is important to determine the hyperfine splitting of P-states. And far greater precision in its mass determination than for $M(\eta_c')$ is required because it is predicted that $\Delta M_{hf}(1P) = M(\langle ^3P_J \rangle) - M(^1P_1) = 0$ for the generally accepted scalar confining potential.

Since the h_c, with $J^{PC} = 1^{+-}$, cannot be populated in a radiative transition from the 1^{--} $\psi(2S)$ state, neither SLAC nor BES experiments have ever claimed to observe $h_c(^1P_1)$. In 1992, the Fermilab $p\bar{p}$ experiment E760 claimed to have identified h_c in the reaction $p\bar{p} \to h_c \to \pi^0 J/\psi$, but larger luminosity runs in 1996 and 2000 have failed to confirm this "observation" [7].

Now, we have made a firm (significance $> 6\sigma$) observation of h_c at CLEO [8]. At CLEO-c data were taken at $\psi(2S)$, with 3.08 million $\psi(2S)$. We have analyzed these data for the population of h_c in the isospin forbidden reaction,

$$\psi(2S) \to \pi^0 h_c \,,\; h_c \to \gamma \eta_c$$

Both inclusive and exclusive analyses were done, and an accurate determination of h_c mass was made in recoils against π^0's whose energy could be measured with precision (see Fig. 3).

Two independent **inclusive analyses**, different in details of event selection, were made. The decay $h_c \to \gamma\eta_c$ was identfied by loosely constraining the photon energy in one analysis, and by loosely constraing the η_c mass in the other analaysis. Completely consistent results were obtained. In the **exclusive analysis**, instead of constraining E_γ or $M(\eta_c)$, seven known decay channels with a total branching fraction of $\sim 10\%$ were measured. Once again, consistent results were obtained.

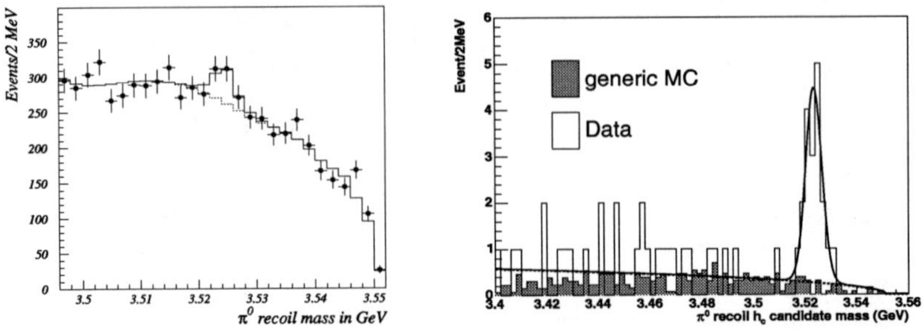

FIGURE 3. Observation of $h_c(1^1P_1)$ in (left) inclusive analysis and (right) exclusive analysis at CLEO [8].

The overall result obtained was $M(h_c) = 3524.4 \pm 0.6 \pm 0.4$ MeV, or

$$\Delta M_{hf}(1P) = \langle M(\chi_{cJ}) \rangle - M(h_c) = +1.0 \pm 0.6 \pm 0.4 \text{ MeV}$$

Two results follow from our unambiguous observation of h_c. The first is that $\Delta M_{hf}(1P)$ is not very different form the pQCD prediction of zero, as was feared by some theorists. The second is that higher statistics data need to be taken at CLEO in order to arrive at a statistically significant result for $\Delta M_{hf}(1P)$.

We note that in a recent publication, E835 has claimed evidence for h_c in the reaction $p\bar{p} \to h_c \to \gamma\eta_c$ in a combined analysis of data from their 1996 and 2000 runs. However, only (7+6=) 13 counts were observed, and the significance of the observation was only $\sim 3\sigma$ [7].

3. TIMELIKE FORM FACTORS OF π, K, AND p

Electromagnetic form factors of composite particles provide deep insight into their structure, and play an important role in determining the momenta above which perturbative QCD can be considered reliable, a question about which there has been great controversy.

We have used the CLEO-c data for e^+e^- collisions at $\sqrt{s} = 3.671$ GeV with a total integrated luminosity of 20.7 pb^{-1} to determine the timelike form factors of the kaon and proton at $|Q^2| = 13.48$ GeV^2 [9].

The excellent charged particle detection capability of the CLEO detector has enabled us to determine

$$\begin{aligned}
Q^2|F_\pi(13.48\,\text{GeV}^2)| &= 1.01 \pm 0.11(\text{stat}) \pm 0.07(\text{syst}),)\ \text{GeV}^2, \\
Q^2|F_K(13.48\,\text{GeV}^2)| &= 0.85 \pm 0.05(\text{stat}) \pm 0.02(\text{syst})\ \text{GeV}^2, \\
Q^4|G_M^p(13.48\,\text{GeV}^2)|/\mu_p &= 0.91 \pm 0.13(\text{stat}) \pm 0.06(\text{syst})\ \text{GeV}^4.
\end{aligned}$$

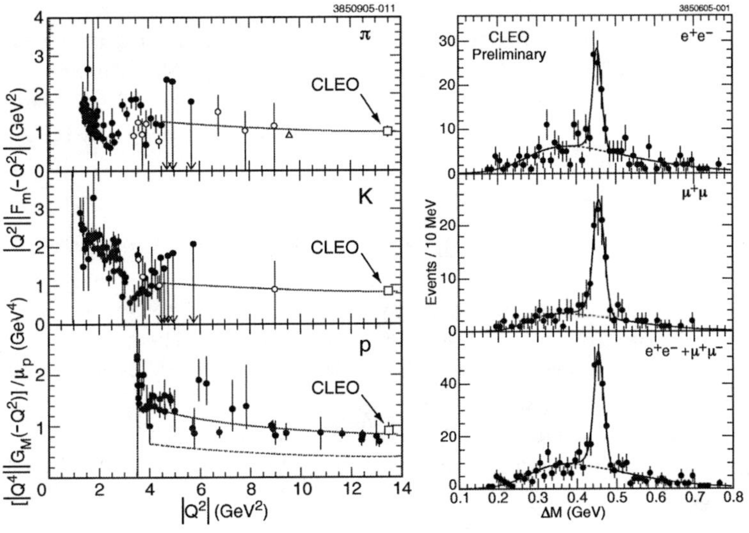

FIGURE 4. (left) CLEO results (open squares) for the timelike form factors at $Q^2 = 13.48$ GeV^2 of the pion (top), kaon (middle), and proton (bottom) [9]. Other results are from the literature. Fig. 4 (right) Distributions of $\Delta M = M(\gamma l^+ l^-) - M(l^+ l^-)$ for the reaction $\gamma\gamma \to \chi_{c2} \to \gamma l^+ l^-$ from CLEO [13].

The proton magnetic form factor result agrees with that measured in the reverse reaction $p\bar{p} \to e^+e^-$ at Fermilab as shown in Fig. 4 (left). The pion and kaon form factor measurements are the first ever direct measurements at $|Q^2| > 4.5$ GeV^2.

4. RADIATIVE DECAYS OF CHARMONIUM

In a recent publication, CLEO [10] has reported the branching ratios, $\mathscr{B}(\psi(2S) \to \gamma\chi_{c0}, \gamma\chi_{c1}, \chi_{c2}, \gamma\eta_c)$, from a careful analysis of the inclusive photon spectrum. It is found that while $\mathscr{B}(\psi(2S) \to \gamma\chi_{c0}, \gamma\chi_{c1}, \gamma\eta_c)$ are in good agreement with earlier results [11], $\mathscr{B}(\psi(2S) \to \gamma\chi_{c2})$ is $\sim 46\%$ larger. This result, when combined with new CLEO [12] measurements of radiative cascades, $\psi(2S) \to \gamma\chi_{cJ} \to \gamma\gamma J/\psi$, leads to new results for the radiative transitions $\mathscr{B}(\chi_{cJ} \to \gamma J/\psi)$. These results are substantially different from the original Crystal Ball results which have long been in use. In particular, it is found that $\mathscr{B}(\chi_{c2} \to \gamma J/\psi) = (19.9 \pm 1.7)\%$, as compared to the Crystal Ball result $\mathscr{B}(\chi_{c2} \to \gamma J/\psi) = (12.4 \pm 1.4)\%$. This new CLEO result has resolved the long–standing discrepancy between the results for the two–photon width of χ_{c2} as measured at e^+e^- colliders, via the production of χ_{c2} in two–photon fusion, and as measured in $p\bar{p}$ annihilation, via the decay of χ_{c2} into two photons. Fig. 4 (right) shows the result of the new CLEO [13] measurement of the reaction

$$\gamma\gamma \to \chi_{c2} \to \gamma J/\psi, \quad J/\psi \to (e^+e^-, \mu^+\mu^-)$$

using ~ 15 fb^{-1} of e^+e^- collision data taken in the bottomonium region. It is found that if the Belle two–photon fusion and E835 $p\bar{p}$ annihilation measurements are analyzed using the new CLEO value for $\mathcal{B}(\chi_{c2} \to \gamma J/\psi)$, their results come in complete agreement with the CLEO result, $\Gamma_{\gamma\gamma}(\chi_{c2}) = 559 \pm 81$ eV. The CLEO result leads to $\Gamma(\chi_{c2} \to \gamma\gamma)/\Gamma(\chi_{c2} \to gg) = (3.65 \pm 0.45) \times 10^{-4}$, and therefore $\alpha_S = 0.29 \pm 0.01$ using pQCD with first order radiative corrections.

5. THE SAGA OF X(3872)

In 2003 Belle [14] announced the discovery of an unexpected narrow state, X(3872), in B–decays. It was quickly confirmed by CDF, DØ, and BaBar. The average of the masses measured by the four experiments is $M(\mathrm{X}) = 3871.5 \pm 0.4$ MeV. Note that this is very close to $M(D^0) + M(\overline{D^{*0}}) = 3870.3 \pm 2.0$ MeV [11].

The unique decay, the narrow width, and the closeness of its mass to $M(D^0 D^{*0})$ have given rise to intense theoretical speculations about the nature of X(3872). Is it a charmonium state (1^{++}, 2^{--}, 3^{--}), or a hybrid (1^{++}), or a glueball mixed with vector charmonium (1^{--}), or a unique 'dimeson', a $D^0\overline{D^{*0}}$ (1^{++}, 0^{-+}) molecule? To sift through these speculations it is necessary to determine $J^{PC}(\mathrm{X})$.

At CLEO [15] we have analyzed ~ 15 fb^{-1} of e^+e^- collision data taken in the bottomonium region for possible production of X(3872) in two–photon fusion and ISR, and have established the following 90% confidence limits:

Two–photon fusion ($J^{PC}(\mathrm{X}) = J^{P+}$): $(2J+1)\Gamma(\mathrm{X} \to e^+e^-) < 0.65$ eV,
ISR ($J^{PC}(\mathrm{X}) = 1^{--}$): $\mathcal{B}(\mathrm{X} \to \pi^+\pi^- J/\psi) \times \Gamma(\mathrm{X} \to e^+e^-) < 8.3$ eV.

With BaBar's recent limit of $\mathcal{B}(\mathrm{X} \to \pi^+\pi^- J/\psi) > 0.042$ [16], the ISR result leads to $\Gamma(\mathrm{X} \to e^+e^-) < 200$ eV. Recall that $\Gamma(\psi(3770)) = 260 \pm 40$ eV.

At CLEO an attempt is underway to measure the mass of D^0 with high precision, so that the proposed interpretation of X(3872) as a $D^0\overline{D^{*0}}$ molecule can be submitted to a critical test.

6. $\psi(2S)$ DECAYS TO BARYON–ANTIBARYON PAIRS

Very few decays of charmonia to baryon–antibaryon pairs have been measured. Using 3.08 million $\psi(2S)$ decays, CLEO [17] has measured the decays $\psi(2S) \to B\bar{B}$ with $B \equiv p$, Λ, Σ^+, Σ^0, Ξ^-, Ξ^0, (Ξ^{*0}) and (Ω^-), many of them for the first time. A rather curious feature of these results is that the branching ratios for all $B\bar{B}$ pairs are nearly the same, $\sim \mathcal{B}(\psi(2S) \to B\bar{B}) \approx 2.7 \times 10^{-4}$. This is surprising, considering that phase space is quite different for the different $B\bar{B}$ pairs.

7. THE $\rho - \pi$ PROBLEM OF $J/\psi(1S)$ AND $\psi(2S)$

A simple prediction of pQCD is that the ratio

$$Q \equiv \frac{\mathscr{B}(\psi(2S) \to \text{hadrons})}{\mathscr{B}(J/\psi \to \text{hadrons})} = \frac{\mathscr{B}(\psi(2S) \to \text{leptons})}{\mathscr{B}(J/\psi \to \text{leptons})} = 0.13 \pm 0.1.$$

BES has studied this over a long time by measuring many hadronic decays of $\psi(2S)$. Recently CLEO [18] has measured many more two–body and many body decays of $\psi(2S)$ using the data for ~ 3 million $\psi(2S)$. The results are that while Q varies between 0.1% and 10% for two–body decays, it is generally larger, varying between 2% and 10% for most multibody decays. For more details, see the talk by C. Z. Yuan in these proceedings.

8. PRECISION MEASUREMENTS OF HADRONIC DECAYS

Spectroscopy does not progress only by discoveries. It makes some of its best gains by precision. Recently, several such gains have been made. At CLEO [12], isospin conservation has been confirmed at a $\pm 2\%$ level in ψ' decay to $\pi^+\pi^- J/\psi$ and $\pi^0\pi^0 J/\psi$, and isospin violation has been observed at a $\pm 0.4\%$ level in ψ' decay to $\pi^0 J/\psi$ and $\eta J/\psi$. Also at CLEO [19] lepton universiality has been confirmed at a $\pm 1\%$ level in J/ψ decays to e^+e^- and $\mu^+\mu^-$.

9. NON–$D\overline{D}$ DECAYS OF $\psi(3770)$

All that we know about $\psi(3770)$ is that it decays dominantly to $D\bar{D}$ ($\sim 100\%$). It is important to measure other possible decays of $\psi(3770)$. $\psi(2S)$ decays to $\pi\pi J/\psi$ nearly 50% of the time. What about $\psi(3770)$? CLEO has invested 281 pb^{-1} of luminosity, and obtained three new results. It is determined that the branching fractions $\mathscr{B}(\psi(3770) \to \pi^+\pi^- J/\psi) = (0.214 \pm 0.033)\%$ and $\mathscr{B}(\psi(3770) \to \pi^0\pi^0 J/\psi) = (0.097 \pm 0.040)\%$ [20]. The branching fraction $\mathscr{B}(\psi(3770) \to \gamma\chi_{c1}) = (0.32 \pm 0.06 \pm 0.04)\%$ was measured, but only upper limits could be set for radiative decays to χ_{c2} and χ_{c0} [21]. No vector+psuedoscalar decays of $\psi(3770)$ were found (except for a hint of $\phi\eta$), and a search of 25 multibody final states also found none [22].

REFERENCES

1. Crystal Ball, Phys. Rev. **D 34** (1986) 711.
2. E760 Collaboration, Phys. Rev. **D 52** (1995) 4839; E835 Collaboration, Phys. Rev. **D 64** (2001) 052003.
3. CLEO Collaboration, Phys. Rev. **D 70** (2004) 112002.
4. Belle Collaboration, Phys. Rev. Lett. **89** (2002) 102001; hep-ex/0507019.
5. CLEO Collaboration, Phys. Rev. Lett. **92** (2004) 142001, and references therein.
6. BaBar Collaboration, Phys. Rev. Lett. **92** (2004) 142002; Phys. Rev. **D 72** (2005) 031101.
7. E835 Collaboration, Phys. Rev. **D 72** (2005) 032001.
8. CLEO Collaboration, Phys. Rev. Lett. **95** (2005) 102003.

9. CLEO Collaboration, hep-ex/0510005, submitted to Phys. Rev. Lett.
10. CLEO Collaboration, Phys. Rev. **D 70** (2004) 112002.
11. PDG Collaboration, Phys. Lett. **B 592** (2004) 1.
12. CLEO Collaboration, Phys. Rev. Lett. **94** (2005) 232002.
13. CLEO Collaboration, hep-ex/0510033, submitted to Phys. Rev. Lett.
14. Belle Collaboration, Phys. Rev. Lett. **91** (2003) 262001.
15. CLEO Collaboration, Phys. Rev. Lett. **94** (2005) 032004.
16. BaBar Collaboration, hep-ex/0510070, submitted to PRL.
17. CLEO Collaboration, Phys.Rev. **D 72** (2005) 051108.
18. CLEO Collaboration, Phys. Rev. Lett. **94** (2005) 012005; Phys. Rev. Lett. **25** (2005) 062001.
19. CLEO Collaboration, Phys. Rev. **D 71**, 111103(R) (2005).
20. CLEO Collaboration, hep-ex/0508023, submitted to Phys. Rev. Lett.
21. CLEO Collaboration, hep-ex/0509030, submitted to Phys. Rev. Lett.
22. CLEO Collaboration, hep-ex/0509011, and hep-ex/0509046, both submitted to Phys. Rev. Lett.

CLEO Results From Υ Decays

V. Credé representing the CLEO Collaboration

Florida State University, Tallahassee, FL 32306, USA

Abstract. Decays of the narrow Υ resonances have been studied with the CLEO-III detector. Among other things, inclusive photon spectra in $\Upsilon(2S)$ and $\Upsilon(3S)$ decays using a large statistics data sample have been obtained. The most precise measurements of electric dipole (E1) photon transition rates and photon energies for $\Upsilon(2\text{S}) \to \gamma\chi_{\text{b}J}(1\text{P})$ and $\Upsilon(3\text{S}) \to \gamma\chi_{\text{b}J}(2\text{P})$ ($J = 0, 1, 2$) are presented. The rare E1 transition $\Upsilon(3\text{S}) \to \gamma\chi_{\text{b}0}(1\text{P})$ has been measured for the first time.

DETECTOR AND DATA RECONSTRUCTION

The CLEO-III detector is a versatile multi-purpose particle detector described more fully in [1, 2, 3, 4]. It is centered on the interaction region of CESR. The charged particle tracking system, consisting of a silicon-strip vertex detector and a 47-layer central drift chamber, operates in a 1.5 T magnetic field oriented along the beam axis. The momentum resolution of the tracking system is approximately 0.6 % at a particle momentum of 1 GeV. Photons are detected in an electromagnetic calorimeter, composed of 7800 CsI (Tl) crystals. The solid angle coverage for charged and neutral particles is about 93 % of 4π. We make use of two particle identification devices to separate charged kaons from pions. These are the central drift chamber using the energy loss dE/dx and a cylindrical ring-imaging Cherenkov (RICH) detector surrounding the central drift chamber.

CLEO-III has the largest world data sample of clean Υ events below the $\text{B}\overline{\text{B}}$ threshold. The data sample has an integrated luminosity of 1.13 fb^{-1} taken at the $\Upsilon(1\text{S})$ energy, $\sqrt{s} = 9.46$ GeV, which corresponds to 21.2 ± 0.2 million $\Upsilon(1\text{S})$ decays. Moreover, $9.3 \cdot 10^6$ and $5.8 \cdot 10^6$ events have been accumulated at the $\Upsilon(2\text{S})$ and $\Upsilon(3\text{S})$, respectively.

Photon Transitions in $\Upsilon(2\text{S})$ and $\Upsilon(3\text{S})$ Decays

CLEO has studied the inclusive photon spectra in $\Upsilon(2\text{S})$ and $\Upsilon(3\text{S})$ decays using a large statistics data sample [5]. Three triplet-P states $\chi_{\text{b}}(1\text{P}_J)$ ($J = 2, 1, 0$) are for example reached via E1 photon transitions from the $\Upsilon(2\text{S})$ state. Measurements of the photon energies determine the P-state masses, whereas transition rates test theoretical predictions for E1 matrix elements. Figure 1 shows spectra for photon transitions in $\Upsilon(3\text{S})$ decays. Photon lines are represented by a Gaussian in addition to a low-energy tail. This so-called Crystal Ball line shape is discussed in more detail elsewhere [6]. The background was carefully studied in the photon spectra observed in off-resonance data and in $\Upsilon(1\text{S})$ data. The $\Upsilon(3\text{S})$ spectrum shows a peak also at higher energies. This peak is due to the overlap of primary, but rare transitions from $\Upsilon(3\text{S})$ to $\chi_{\text{b}J}(1\text{P})$ and

CP814 *Hadron Spectroscopy: Eleventh International Conference*
edited by A. Reis, C. Göbel, J. Sá Borges, and J. Magnin

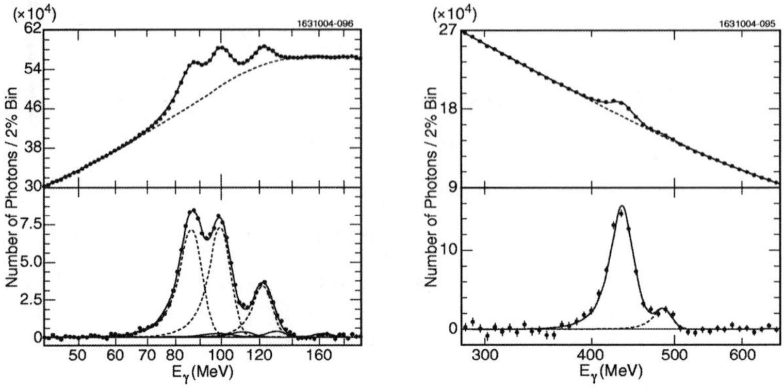

FIGURE 1. The left plot shows a fit to the $\Upsilon(3S) \to \gamma\chi_{bJ}(2P)$ $(J = 2,1,0)$ photon lines in the data. The points represent the data (top plot). Statistical errors on the data are smaller than the point size. The solid line represents the fit. The dashed line represents total fitted background. The background subtracted data (points with error bars) are shown at the bottom. The solid line represents the fitted photon lines together. The dashed lines show individual photon lines. Small solid line peaks in the bottom plot show the $\chi_{bJ}(2P) \to \gamma\Upsilon(1D)$ and $\Upsilon(2S) \to \gamma\chi_{bJ}(1P)$ contributions. Parameters of these peaks in the fit are fixed from the other measurements and also from theoretical predictions. The right plot shows the $\Upsilon(3S)$ spectrum at higher energies, see text for details.

cascades from $\chi_{bJ}(1P)$ to $\Upsilon(1S)$ $(J = 2,1,0)$. In total, there are six photon lines which largely overlap. The latter are fed not only by the direct transitions from the $\Upsilon(3S)$ to the $\chi_{bJ}(1P)$ states, but also by photon or hadronic transitions via the $\Upsilon(2S)$ and $\Upsilon(1D_J)$ states. One line, $\Upsilon(3S) \to \gamma\chi_{b0}(1P)$, has a photon energy sufficiently different from the rest and its amplitude can be measured. The rate has been determined for the first time to be $\mathscr{B}(\Upsilon(3S) \to \gamma\chi_{b0}(1P)) = (0.30 \pm 0.04 \pm 0.10)\,\%$. In summary, the measured energies in photon transitions to $\chi_{bJ}(1P)$ and $\chi_{bJ}(2P)$ are in good agreement with previous measurements and currently, represent the most precise measurements. Branching ratios for $\Upsilon(2S) \to \gamma\chi_{bJ}(1P)$ are also in agreement with previous measurements, but offer much improved experimental errors. Branching ratios for $\Upsilon(3S) \to \gamma\chi_{bJ}(2P)$ are significantly larger than previously measured by CUSB II [7].

Radiative $\Upsilon(1S)$ Decays into two Hadrons: $\Upsilon(1S) \to h^+h^- + \gamma$

Radiative decays of heavy quarkonia, where a photon replaces one of the three gluons from the strong decay, are useful in studying color-singlet two-gluon systems. The two gluons can, among other things, hadronize into a meson, or directly form a glueball. Further information on radiative decays of heavy quarkonia can be found in [8]. Many interesting results have been obtained from the study of radiative J/ψ decays. The tensor states $f_2(1270)$ [9, 10, 11] and $f_2'(1525)$ [12] were observed and helicity production ratio measurements agree with theoretical predictions [13, 14]. In 1996, the BES collaboration claimed to observe the glueball candidate $f_J(2220)$ and recently, an enhancement

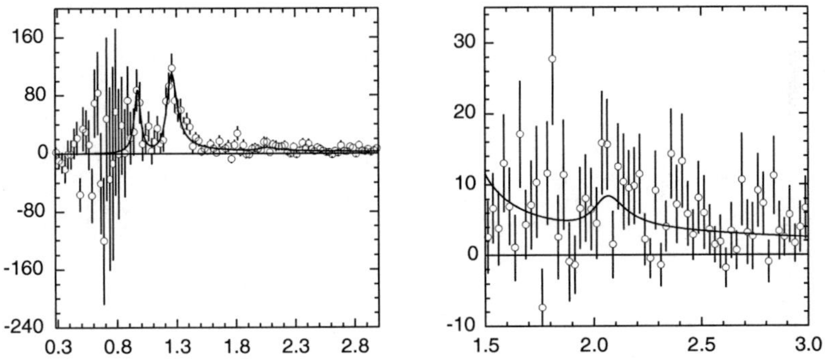

FIGURE 2. Invariant mass of $\pi^+\pi^-$ from $\Upsilon(1S) \rightarrow \gamma\pi^+\pi^-$ and the fit to three spin-dependent relativistic Breit-Wigner functions. The right plot shows the $f_4(2050)$ candidate in the region $1.5-3.0$ GeV/c^2.

at the $p\bar{p}$ threshold was reported by BES [15], a result that is currently being interpreted.

The experimental observation of radiative $\Upsilon(1S)$ decays is challenging because their rate is suppressed by a factor of the order of

$$(\frac{q_b}{q_c})^2(\frac{m_c}{m_b})^2 \approx 0.025 \tag{1}$$

with respect to the rate of J/ψ radiative decays. This factor arises because the quark-photon coupling is proportional to the electric charge, and the quark propagator is roughly proportional to $1/m$ for low momentum quarks. Taking into account the total widths of J/ψ and $\Upsilon(1S)$, the branching fraction of a particular $\Upsilon(1S)$ radiative decay mode is expected to be suppressed by a factor of roughly 0.04.

Figures 2 and 3 show a variety of hadron-antihadron mass distributions. In particular, Figure 2 shows a fit to the continuum subtracted $\pi^+\pi^-$ spectrum. In the left plot, a significant $f_2(1270)$ signal of 944 ± 74 events is clearly visible. It also has two less significant signal candidates: 340^{+140}_{-130} events in the $f_0(980)$ region, and 80 ± 30 events in the $f_4(2050)$ region (right plot) whose significances are 4.3σ and 2.6σ, respectively. The fit to the continuum subtracted K^+K^- spectrum, shown in Figure 3 (left plot), has a significant signal of 312^{+69}_{-61} events identified as the $f_2'(1525)$, and two non-significant signal candidates indicating possible $f_2(1270)$ and $f_0(1710)$ production with 109 ± 36 and 73 ± 29 events whose significances are 3.2σ and 3.3σ, respectively. No recognizable structure is seen in the continuum subtracted $p\bar{p}$ spectrum, which is shown in Figure 3.

In summary, CLEO has confirmed its previous observation of the $f_2(1270)$ in radiative decays and made a new observation of the $f_2'(1525)$, obtaining factors of 0.07 ± 0.01 and $0.08^{+0.04}_{-0.03}$ for the ratio of the $\Upsilon(1S)$ branching fraction with respect to the one measured in J/ψ radiative decays, respectively. These values are about a factor of 2 larger than expected from naive scaling arguments. In both modes, CLEO can confirm by fits to the angular distributions of the photon and charged particles that the two daughter hadrons are produced by a spin-2 parent which is itself produced mostly with helicity 0, in good agreement with predictions. No structure is seen in the $p\bar{p}$ invariant mass distribution. In

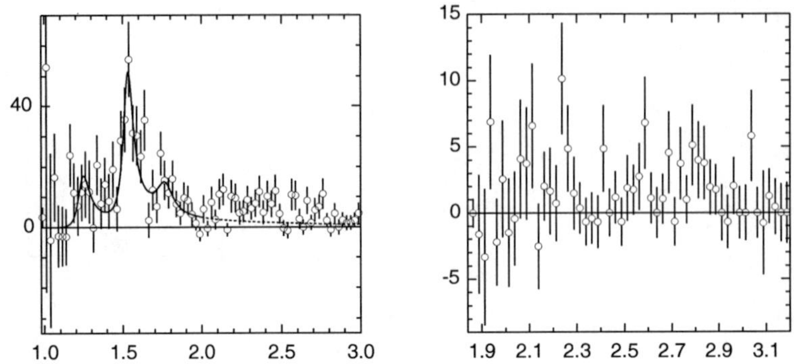

FIGURE 3. The left plot shows the invariant mass of K^+K^- from $\Upsilon(1S) \to \gamma K^+K^-$ and the fit to 3 spin-dependent rel. Breit-Wigner functions. The right plot shows the invariant mass of $p\bar{p}$ from $\Upsilon(1S) \to \gamma p\bar{p}$.

particular, there is no near-threshold enhancement as observed by the BES collaboration. Glueball production is expected to be enhanced in $\Upsilon(1S)$ radiative decays, but it is found that, within the experimental sensitivity, tensor meson states dominate the di-gluon spectrum which are believed to be composed only of quarks.

Direct Photons in Υ Decays and $R_\gamma = \frac{N_{gg\gamma}}{N_{ggg}}$

We have investigated the direct photon spectrum in the decays $\Upsilon(1S) \to \gamma gg$, $\Upsilon(2S) \to \gamma gg$, and $\Upsilon(3S) \to \gamma gg$. The latter two of these are first observations. Our measurement procedures differ from previous ones in the following ways: a) background estimates (primarily from π^0 decays) are based on isospin symmetry rather than an explicit determination of the π^0 spectrum, which permits measurements of the $\Upsilon(2S)$ and $\Upsilon(3S)$ direct photon spectra without having to explicitly correct for π^0 backgrounds from, e.g., χ_b states, b) we estimate the branching fractions with a parametrized functional form (exponential) used for the background, c) we use the high-statistics sample of $\Upsilon(2S) \to \pi\pi\Upsilon(1S)$ to obtain a tagged-sample of $\Upsilon(1S) \to \gamma + X$ events, for which there are no QED backgrounds. We determine values for the ratio of the inclusive direct photon decay rate to that of the dominant three-gluon decay $\Upsilon \to ggg$ ($R_\gamma = \mathscr{B}(gg\gamma)/\mathscr{B}(ggg)$) to be: $R_\gamma(1S) = (2.90 \pm 0.007 \pm 0.22 \pm 0.15)\,\%$, $R_\gamma(2S) = (3.49 \pm 0.03 \pm 0.58 \pm 0.18)\,\%$, and $R_\gamma(3S) = (2.88 \pm 0.03 \pm 0.38 \pm 0.12)\,\%$, where the errors are statistical, systematic, and theoretical model-dependence, respectively. Given a value of Q^2, one can estimate a value for the strong coupling constant $\alpha_s(Q^2)$ from R_γ.

The idea for the study of 2-gluon + photon modes is to search for hadronic events with a nice isolated photon. For this study it is extremely important that background sources are well understood because the direct photon spectrum is the *left-over* spectrum which is the photon spectrum after the subtraction of all other sources. Important background is initial state radiation and photons stemming from π^0 decays.

Observation of $\chi_b(2P) \rightarrow \chi_b(1P)\pi^+\pi^-$

The dominant hadronic transitions among the heavy quarkonia ($Q\overline{Q} = c\overline{c}$, $b\overline{b}$) involve di-pion emission, characterized as the emission of two soft gluons which then hadronize as a di-pion system. These have been studied for transitions among the quarkonia $^3S_1$ states, but have not been observed (yet) in other quarkonia transitions such as $\eta_c^{'} \rightarrow \pi\pi\eta_c$ or the $\chi_b^{'}$ decays. Of particular interest in such transitions is the di-pion invariant mass distribution. While the $m_{\pi\pi}$ spectra for the decays $\psi^{'} \rightarrow \pi\pi J/\psi$, $\Upsilon(3S) \rightarrow \pi\pi\Upsilon(2S)$, and $\Upsilon(2S) \rightarrow \pi\pi\Upsilon(1S)$ all seem to agree nicely with the Yan model [16], the situation is different for the decay $\Upsilon(3S) \rightarrow \pi\pi\Upsilon(1S)$. For this decay, the spectrum of $m_{\pi\pi}$ has two peaks, one at low invariant mass and one at the high endpoint. New interest in $\chi_b^{'}$ decays has also been generated by the CLEO observation of a large branching fraction for the decay $\chi_b^{'} \rightarrow \omega\Upsilon(1S)$ [17]. This is the only observed hadronic decay of the P-wave $b\overline{b}$ states and the only hadronic bottomonium transition that is not $\pi\pi$. Therefore, we have decided to look for another hadronic transition of the $\chi_b^{'}$, namely $\chi_b^{'} \rightarrow \chi_b\pi^+\pi^-$.

We have searched for the di-pion transition $\chi_b^{'} \rightarrow \chi_b\pi^+\pi^-$ in the CLEO-III sample of $\Upsilon(3S)$ decays in the exclusive decay chain, $\Upsilon(3S) \rightarrow \gamma\chi_b^{'}$, $\chi_b^{'} \rightarrow \chi_b\pi^+\pi^-$, $\chi_b \rightarrow \Upsilon(1S)\gamma$, and finally $\Upsilon(1S) \rightarrow l^+l^-$. For comparisons, we have used the well-established channels $\Upsilon(3S) \rightarrow \Upsilon(2S)\pi^+\pi^-$, with either $\Upsilon(2S) \rightarrow l^+l^-$ or $\Upsilon(2S) \rightarrow \gamma\chi_b$. Numerous Monte Carlo simulations were performed to evaluate selection criteria and to predict eventual background. The results from two analyses using events with both and only one soft pion yield 7 observed events with an expected background of 0.6 ± 0.2 events and 17 observed events with an expected background of 2.2 ± 0.6 events, respectively. We refute the null hypothesis, observing a signal greater than 6σ deviation from expected background, and thus claim observation of this di-pion decay. Under reasonable assumptions, we find the two-pion partial width to be $\Gamma_{\pi^+\pi^-} = (0.80 \pm 0.21 \pm^{+0.23}_{-0.17})$ keV. This is the first observation of a $\pi\pi$ transition outside of the $^3S_1$ system.

REFERENCES

1. G. Viehhauser [CLEO-III Collaboration], Nucl. Instrum. Meth. A **462**, 146 (2001).
2. D. Peterson *et al.*, Nucl. Instrum. Meth. A **478**, 142 (2002).
3. A. T. Warburton *et al.*, Nucl. Instrum. Meth. A **488**, 451 (2002) [arXiv:hep-ex/0109015].
4. M. Artuso *et al.*, arXiv:physics/0506132.
5. M. Artuso *et al.* [CLEO Collaboration], Phys. Rev. Lett. **94**, 032001 (2005) [arXiv:hep-ex/0411068].
6. S. B. Athar *et al.* [CLEO Collaboration], Phys. Rev. D **70**, 112002 (2004) [arXiv:hep-ex/0408133].
7. U. Heintz *et al.*, Phys. Rev. D **46**, 1928 (1992).
8. N. Brambilla *et al.*, arXiv:hep-ph/0412158.
9. G. Alexander *et al.* [Pluto Collaboration], Phys. Lett. B **76**, 652 (1978).
10. C. Edwards *et al.*, Phys. Rev. D **25**, 3065 (1982).
11. J. E. Augustin *et al.* [DM2 Collaboration], Z. Phys. C **36**, 369 (1987).
12. J. Z. Bai *et al.* [BES Collaboration], Phys. Rev. D **68**, 052003 (2003) [arXiv:hep-ex/0307058].
13. M. Krammer, Phys. Lett. B **74**, 361 (1978).
14. J. G. Korner, J. H. Kuhn, M. Krammer and H. Schneider, Nucl. Phys. B **229**, 115 (1983).
15. J. Z. Bai *et al.* [BES Collaboration], Phys. Rev. Lett. **91**, 022001 (2003) [arXiv:hep-ex/0303006].
16. T. M. Yan, Phys. Rev. D **22**, 1652 (1980).
17. D. Cronin-Hennessy *et al.* [CLEO Collaboration], arXiv:hep-ex/0311043.

New Results from ψ' decays and $\psi'' \to \rho\pi$ decay

Yongsheng Zhu (for the BES collaboration)

Institute of High Energy Physics, CAS, Beijing 100039, China

Abstract. Based on 14M ψ' events and 17.3 pb^{-1} ψ'' sample collected by BESII detector, following channels are analyzed to study the hadron spectroscopy and to measure the branching fractions: ψ' decays to $\pi^+\pi^-J/\psi$, $p\bar{p}(\pi^0,\eta)$, $3(\pi^+\pi^-)$, $\pi^+\pi^-\pi^0K^+K^-$ and ψ', ψ'' decays to $\pi^+\pi^-\pi^0$, $\rho\pi$ etc. Results of σ resonance and N^* states are also reported.

Keywords: charmonium decay, hadronic decay, σ resonance, PWA
PACS: 13.66.Bc, 13.25.Gv

INTRODUCTION

From perturbative QCD (pQCD), it is expected both J/ψ and ψ' decaying into light hadrons are dominated by the annihilation of $c\bar{c}$ into three gluons or a virtual photon, with a width proportional to the square of the wave function at the origin [1]. This yields the pQCD "12% rule": $Q_h \equiv \frac{B_{\psi'\to h}}{B_{J/\psi\to h}} \approx \frac{B_{\psi'\to e^+e^-}}{B_{J/\psi\to e^+e^-}} \approx 12\%$. A large violation of this rule was observed in decays to $\rho\pi$ and $K^{*+}K^- + c.c.$ [2], known as the $\rho\pi$ puzzle. Since then, many decay modes of the ψ' have been measured by the BES and recently by the CLEO Collaborations; some decays obey the rule while others violate it [3, 4, 5, 6, 7].

Numerous theoretical explanations have been suggested [8, 9] to try solve this puzzle, in which the S- and D- mixing scheme [9] considers both ψ' and ψ'' be the mixtures of S- and D-wave charmonia, while the 12% rule only valid for pure 1S and 2S states. This scheme could qualitatively explain the $\rho\pi$ puzzle, and predict ψ'' decay branching fractions, therefore, it is worth to test experimentally. In general, more and accurate data on hadronic decays of chamonia are desired to test different models.

Excited nucleon states play an important role in the understanding of nonperturbative QCD, however, our knowledge on N^* resonances based on πN and γN experiments is very limited and imprecise. Studies of N^* produced in chrmonium decays provide a new method to probe N^* physics [10, 11, 12]. The possible N^* formation in ψ' decays into $p\bar{p}(\pi^0,\eta)$ may provides additional experimental information.

One of important aspects of strong interaction physics at low energies is the spontaneous breaking of chiral symmetry. The experimental knowledge on the 0^{++} low energy hadrons, for instance, the $f_0(600)$ or σ particle, is important to the understanding of QCD in the non-perturbative region. Recently BES Collaboration determined the pole position of σ resonance as $(541\pm39) - i(252\pm42)$ MeV in $J/\psi \to \omega\pi^+\pi^-$ decay [13]. $\psi' \to \pi^+\pi^-J/\psi$ decay is a good place to study the σ because of its large branching fraction, simple kinematics, and large data sample collected at BESII. Also, the S- wave is dorminant and D- wave is very small in its $\pi^+\pi^-$ system.

In this paper, we report the results on ψ' decays to multi-hadron final states, ψ' and

CP814 *Hadron Spectroscopy: Eleventh International Conference*
edited by A. Reis, C. Göbel, J. Sá Borges, and J. Magnin

ψ'' decay to $\rho\pi$, and on σ pole position via $\psi' \to \pi^+\pi^- J/\psi$ process. The analyses are based on $14.0 \times 10^6 (1 \pm 4\%)$ ψ' events [14], $6.42(1 \pm 4\%) pb^{-1}$ data at 3.650 GeV [15] and $17.3(1 \pm 3\%) pb^{-1}$ data around 3.773 GeV [16] collected by BESII detector [17].

ψ' DECAY TO MULTIHADRON

$\psi' \to p\overline{p}(\pi^0, \eta)$

The candidate events are selected by particle ID for p and $\overline{p}$ using dE/dx and TOF and then performing a 4C kinematic fitting for $\psi' \to p\overline{p}\gamma\gamma$. The π^0 and η signals are obtained by fitting the $M_{\gamma\gamma}$ spectrum. The numbers of $p\overline{p}(\pi^0, \eta)$ events are 256 ± 18 and $44.8^{+8.7}_{-8.4}$, and the branching fractions determined to be $(13.2 \pm 1.0 \pm 1.5) \times 10^{-5}$ and $(5.8 \pm 1.1 \pm 0.7) \times 10^{-5}$, respectively [4]. The corresponding Q_h values, $(12.1 \pm 1.9)\%$ and $(2.8 \pm 0.7)\%$, indicate that the $\psi' \to p\overline{p}\eta$ decay is suppressed wrt 12% rule. On the $p\pi^0/\overline{p}\pi^0$ mass spectrum of $p\overline{p}\pi^0$ events, no enhancement has been seen around 2065 MeV, indicating there is no production of $N^*(2065)$ state, although it is seen in J/ψ decay [12]. A clear peak shows up at 1550 MeV on $p\eta/\overline{p}\eta$ mass spectrum of $p\overline{p}\eta$ events, which is probably due to the $N^*(1535)$ state plus phase space effect. In both channels, there are enhancements appeared at ~ 2 GeV in $M_{p\overline{p}}$ spectrum, but with low significance ($\sim 2.7\ \sigma$).

$\psi' \to 3(\pi^+\pi^-)$

4C kinematic fit is performed to six charged track events with net charge zero to select $\psi' \to 6\pi$ events. $\psi' \to \pi^+\pi^- J/\psi$ background is removed by requiring the invariant mass recoiling against a $\pi^+\pi^-$ pair, $M^{recoil}_{\pi^+\pi^-}$, not falling into (3.06, 3.14) GeV; while $\psi' \to K_sK_s\pi^+\pi^-$ removed by requiring two $\pi^+\pi^-$ pairs falling into (0.47, 0.53) GeV. 671 and 71 candidates are obtained at 3.686 and 3.650 GeV, respectively. The form factor of $e^+e^- \to 3(\pi^+\pi^-)$ is determined to be $|F(s = (3.650GeV)^2)| = 0.19 \pm 0.02$ and $|F(s = (3.686GeV)^2)| = 0.24 \pm 0.02$, respectively, and the branching fraction of $\psi' \to 3(\pi^+\pi^-)$ is $(5.45 \pm 0.42 \pm 0.87) \times 10^{-4}$. $B(J/\psi \to 2(\pi^+\pi^-))$ has been measured by comparing two processes: $\psi' \to \pi^+\pi^- J/\psi, J/\psi \to 2(\pi^+\pi^-)$ and $J/\psi \to \mu^+\mu^-$, in which many common systematic errors cancel out. The determined branching fraction is $(3.53 \pm 0.12 \pm 0.29) \times 10^{-3}$ using the $B(J/\psi \to \mu^+\mu^-)$ of PGD [18]. With these newly measured branching fractions, the Q_h values for $3(\pi^+\pi^-)$ and $2(\pi^+\pi^-)$ final states are determined to be $(14 \pm 8)\%$ and $(13 \pm 3)\%$, respectively [5].

$\psi' \to \pi^+\pi^-\pi^0 K^+K^-$

4C kinematic fit is performed to the four charged tracks with net charge zero plus two photons events to select $\psi' \to \pi^+\pi^- K^+K^- \gamma\gamma$ candidates with pion and kaon identitied by TOF and dE/dx. $\psi' \to \pi^+\pi^- J/\psi$ background is removed by requiring $M^{recoil}_{\pi^+\pi^-}$

not falling into (3.06, 3.14) GeV. ; while $\psi' \to K_s K^\pm \pi^\mp \pi^0$ background removed by requiring $M_{\pi^+\pi^-}$ falling into the region around the kaon mass. By fitting the mass $M_{\gamma\gamma}$ as shown in Fig.1, 698 ± 41 and 35 ± 7 $\psi' \to \pi^+\pi^-\pi^0 K^+K^-$ events are obtained at 3.686 and 3.650 GeV, respectively. Further, we fit the mass $M_{\pi^+\pi^-\pi^0}$ to get $\psi' \to \omega K^+K^-$ signal, the numbers of candidate events are 78 ± 11 and $0^{+1.3}_{-0}$ (at 68.3% C.L.) at 3.686 and 3.650 GeV, respectively. By fitting the peak appearing at 1.71 GeV in $M_{K^+K^-}$ spectrum of ωK^+K^- events from 3.686 GeV data, 18.9 ± 6.2 $\psi' \to \omega f_0(1710)$ events are obtained. Subtracting the contribution from continuum incoherently deduced from 3.650 GeV data, we get branching fractions of ψ' decay to three channels and corresponding Q_h values listed in Table 1, wherein the fisrt two values are consistent with recent CLEO results [7], while the last one is the first measurement.

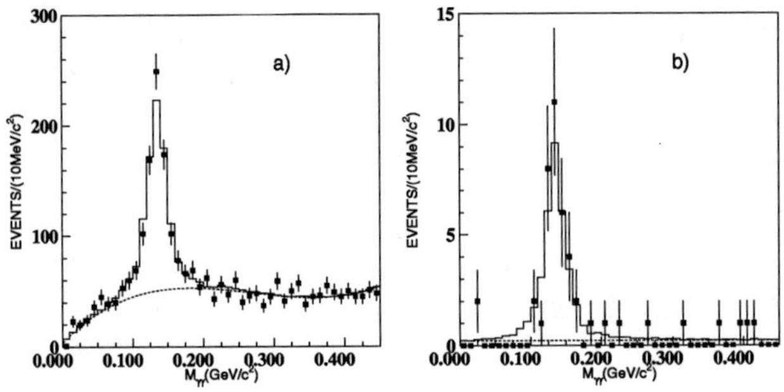

FIGURE 1. $M_{\gamma\gamma}$ at (a) 3.686, (b) 3.650 GeV (Preliminary)

TABLE 1. BRs for ψ' 3 decay modes and their Q_h values (Preliminary)

Channel	$B(\psi')$ 10^{-4}	$B(J/\psi)$ 10^{-4}	$Q_h(\%)$
$\pi^+\pi^-\pi^0 K^+K^-$	$12.4 \pm 1.0 \pm 1.5$	120 ± 28	10.3 ± 2.9
ωK^+K^-	$2.38 \pm 0.37 \pm 0.28$	16.8 ± 2.1	14.2 ± 3.4
$\omega f_0(1710) \to \omega K^+K^-$	$0.59 \pm 0.20 \pm 0.09$	6.6 ± 1.3	8.9 ± 3.8

$\psi', \psi'' \to \rho\pi$

In the Dalitz plot of $\psi' \to \pi^+\pi^-\pi^0$ decay, besides clear ρ bands at the edges, there is a cluster of events in the center, which is very different from the Dalitz plot for $J/\psi \to \pi^+\pi^-\pi^0$ decay, indicating different decay dynamics. The selected events are fitted in the helicity amplitude formalism with an unbinned maximum likelihood method [6]. The branching fractions for $\psi' \to \pi^+\pi^-\pi^0, \rho\pi, \rho(2150)\pi \to \pi^+\pi^-\pi^0$ are $(18.1 \pm 1.8 \pm 1.9) \times 10^{-5}$,$(5.1 \pm 0.7 \pm 1.1) \times 10^{-5}$ and $(19.4 \pm 2.5^{+11.5}_{-3.4}) \times 10^{-5}$, and Q_h values for the first two channels are $(0.86 \pm 0.13)\%$ and $(0.40 \pm 0.09)\%$, respectively, both heavily suppressed compared to the "12% rule", while the $\rho(2150)\pi$ state is the first observation in ψ' decay. It is worth to point out, the BES $B(\psi' \to \pi^+\pi^-\pi^0)$ value is consistent with recent CLEO result, while $B(\psi' \to \rho\pi)$ is 2σ higher than that of CLEO [7]. The reason is that the interference between $\rho(770)$ and $\rho(2150)$ is considered in BES PWA analysis, but not in CLEO analysis, and the continuum subtraction is also different.

The number of selected $\pi^+\pi^-\pi^0$ events at 3.773 and 3.650 GeV shown in Fig. 2 is 11.4 ± 4.7 and 10.0 ± 3.8 with significance 3.1σ and 4.6σ, respectively [19]. The corresponding Dalitz plot for $\pi^+\pi^-\pi^0$ events shows no clear $\rho\pi$ signal at both energies. The fit yields number of $\rho\pi$ events 2.9 ± 2.8 and 4.9 ± 2.7, or 5.1 and 8.3 at 90% C.L. at 3.773 and 3.650 GeV, respectively. The corresponding Born cross sections are listed in Table 2. BES upper-limit (at 90% C.L.) of $\sigma(e^+e^- \to \rho\pi)$ at 3.773 GeV is lower than CLEO cross section $(8.0^{+1.7}_{-1.4} \pm 0.9)pb$ at 3.67 GeV, indicating the existence of $\psi'' \to \rho\pi$ non-zero amplitude when phase and interference taken into account. P. Wang et al pointed out [20], the $\rho\pi$ cross section at ψ'' peak depends on $B(\psi'' \to \rho\pi)$ and relative phase ϕ between strong and EM amplitudes. In this scheme, using the $\rho\pi$ cross section at 3.773 GeV from BES and that at 3.67 GeV from CLEO, one gets (at 90% C.L.) $B(\psi'' \to \rho\pi) \in (6.0 \times 10^{-6}, 2.4 \times 10^{-3})$ and $\phi \in (-150°, -20°)$.

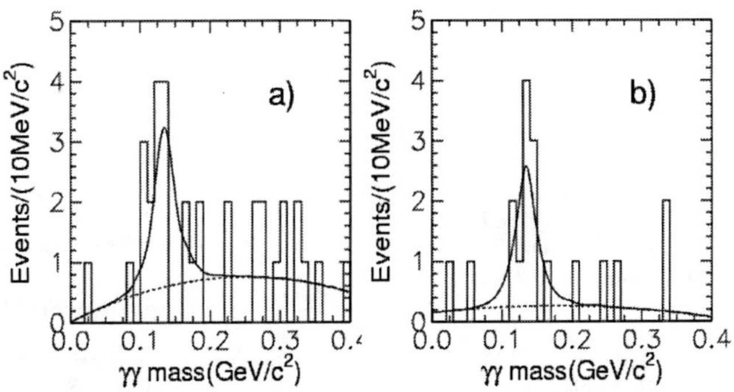

FIGURE 2. $M_{\gamma\gamma}$ at (a) 3.773, (b) 3.650 GeV

TABLE 2. Born cross sections of $e^+e^- \to \pi^+\pi^-\pi^0, \rho\pi$ (upper limit @90% C.L.)

Energy	$\sigma(e^+e^- \to \pi^+\pi^-\pi^0)$	$\sigma(e^+e^- \to \rho\pi)$
3.650 GeV	$19.3 \pm 7.3 \pm 3.7$ pb	<25 pb
3.773 GeV	$8.4 \pm 3.5 \pm 2.0$ pb	<6.0 pb

σ PRODUCTION IN $\psi' \to \pi^+\pi^- J/\psi$

About 40,000 $\pi^+\pi^-\mu^+\mu^-$ events with $M_{\mu^+\mu^-}$ constrained to the J/ψ mass are selected for this study. In the $M_{\pi^+\pi^-}$ distribution of selected events, the severe suppression near $\pi^+\pi^-$ threshold, distinctly different from phase space, suggests the σ production. PWA are performed for the $\psi' \to \pi^+\pi^- J/\psi$ process. The $\pi^+\pi^-$ system has three contributions including σ resonance, D-wave(2^+) and a contact term, which is the destructive background required by chiral symmetry. Three types of BW parameterization are used to describe σ resonance:(a) a form from PDG [18] with constant width, (b) a form used by E791 [21] with the width containing kinematic factor $\Gamma(s) = \rho\Gamma = \sqrt{1 - \frac{4m_\pi^2}{s}}\Gamma$, (c) PKU ansatz [22], which removes the spurious singularity hidden in form (b): $\Gamma(s) = \rho\frac{s}{m^2}\Gamma = \sqrt{1 - \frac{4m_\pi^2}{s}}\frac{s}{m^2}\Gamma$.

All three BWs fit data well through a strong destructive interference between σ and the contact term. The D-wave has a contribution as small as $(0.3 \sim 1)\%$ level. The pole positions for three BWs are (preliminary): $(553 \pm 15 \pm 47) - i(254 \pm 23 \pm 54)$, $(559 \pm 6 \pm 26) - i(179 \pm 7 \pm 17)$ and $(554 \pm 13 \pm 66) - i(240 \pm 4 \pm 20)$ MeV, respectively. The second scheme gives a smaller σ width, however, this model creates a virtual state in the real energy below the $\pi\pi$ threshold, hence, its prediction may be questionable. In contrast,the pole positions determined by constant width BW and PKU ansatz agree with each other, and is well consistent with that of the recent result obtained from $J/\psi \to \omega\pi^+\pi^-$ decay ($(541 \pm 39) - i(252 \pm 42)$ MeV) [13].

REFERENCES

1. T. Appelquist and H. D. Politzer, Phys. Rev. Lett. **34**, 43 (1975); A. De Rújula and S. L. Glashow, Phys. Rev. Lett. **34**, 46 (1975).
2. MarkII Collab., M. E. B. Franklin *et al.*, Phys. Rev. Lett. **51**, 963 (1983).
3. Many results can be found in PDG04; For recent results, see: BES Collab., J. Z. Bai, *et al.*, Phys. Rev. Lett. **92**, 052001 (2004); Phys. Rev. **D69**, 012003 (2004), 072001 (2004); Phys. Lett. **B589**, 7 (2004); Phys. Rev. **D70**, 012005 (2004); M. Ablikim, *et al.*, Phys. Rev. **D70**, 112003 (2004), 112007 (2004); Phys. Lett. **B614**, 37 (2005).
4. BES Collab., M. Ablikim, *et al.*, Phys. Rev. **D71**, 072006 (2005).
5. BES Collab., M. Ablikim, *et al.*, Phys. Rev. **D72**, 012002 (2005).
6. BES Collab., M. Ablikim, *et al.*, Phys. Lett. **B619**, 247 (2005).
7. CLEO Collab., N. E. Adam, *et al.*, Phys. Rev. Lett. **94**, 012005 (2005); R. A. Briere *et al.*, hep-ex/0505101; T. K. Pedlar *et al.*, hep-ex/0505057.
8. W. S. Hou and A. Soni, Phys. Rev. Lett. **50**, 569 (1983); G. Karl and W. Roberts, Phys. Lett. **B144**, 263 (1984); S. J. Brodsky *et al.*, Phys. Rev. Lett. **59**, 621 (1987); M. Chaichian *et al.*, Nucl. Phys. **B323**, 75 (1989); S. S. Pinsky, Phys. Lett. **B236**, 479 (1990); X. Q. Li *et al.*, Phys. Rev. **D55**, 1421 (1997); S. J. Brodsky and M. Karliner, Phys. Rev. Lett. **78**, 4682 (1997); Yu-Qi Chen and Eric Braaten, Phys. Rev. Lett. **80**, 5060 (1998); T. Feldmann and P. Kroll, Phys. Rev. **D62**, 074006 (2000); M. Suzuki, Phys. Rev. **D63**, 05421 (2001); J. P. Ma, Phys. Rev. **D65**, 097506 (2002); M. Suzuki, Phys. Rev. **D65**, 097507 (2002).
9. J. L. Rosner, Phys. Rev. **D64**, 094002 (2001).
10. B. S. Zou, Nucl. Phys. **A675**, 167C (2000).
11. BES Collab., J. Z. Bai, *et al.*, Phys. Lett. **B510**, 75 (2001).
12. BES Collab., M. Ablikim, *et al.*, hep-ex/0405030.
13. BES Collab., M. Ablikim, *et al.*, Phys. Lett. **B598**,149 (2004).
14. X. H. Mo, *et al.*, HEP&NP **28**, 455 (2004).
15. S. P. Chi, *et al.*, HEP&NP **28**, 1135 (2004).
16. BES Collab., M. Ablikim, *et al.*, Phys. Lett. **B603**, 130 (2004).
17. BES Collab., J. Z. Bai, *et al.*, Nucl. Inst. Meth. **A458**, 627 (2001).
18. Particle Data Group, S. Eidelman, *et al.*, Phys. Lett. **B592**, 1 (2004).
19. BES Collab., M. Ablikim, *et al.*, hep-ex/0507092, Accepted by Phys. Rev. **D**.
20. P. Wang, *et al.*, Phys. Lett. **B574**, 41 (2003).
21. E. M. Aitala, *et al.*, Phys. Rev. Lett. **86**, 770 (2001).
22. H. Q. Zheng, *et al.*, Nucl. Phys. **A733**, 235 (2004).

REACTIONS

The Deck effect in $\pi N \to \pi\pi\pi N$

Jozef Dudek* and Adam Szczepaniak†

*Jefferson Laboratory MS 12H2, 12000 Jefferson Avenue, Newport News, VA 23606, USA
†Physics Department & Nuclear Theory Center, Indiana University, Bloomington, IN 47405, USA

Abstract. Recent high-statistics analyses of the reaction $\pi p \to \pi\pi\pi p$ have motivated a reconsideration of the Deck effect which was widely applied to such data in the 1960's and 70's. The Deck effect is essentially kinematic in origin yet considerable subtleties can arise when one attempts to model its contribution to the data. We will discuss previous Deck studies and propose how it may be constrained further than was possible before using the data now available.

Keywords: partial wave analysis, hadro-production, meson resonances, Deck effect
PACS: 11.80.Et, 12.40.Nn, 13.75.Gx, 13.75.Lb, 14.40.Cs

THE REACTION $\pi^- P \to \pi^+\pi^-\pi^- P$

This process is the source of much of our experimental knowledge of the $I = 1, G = -$ mesons such as the a_2, a_1, π_2 resonances, which have been extracted via Breit-Wigner fits to partial waves. The extraction of partial waves from the experimental angular distributions is usually performed within the framework of the *isobar model* where an assumption is made about the parentage of the three pions, specifically that a three-pion resonance is produced which then decays without memory of its production to a $\pi\pi$ isobar and a "bachelor" pion. This process is pictured in figure 1(a) and mathematically described by the equations below which show the angular basis of isobar partial waves labeled by $b = J^P LM^\eta$ [isobar].

$$I(W,t,s_1,\text{angles}) = \left|\sum_b C_b(W,t) A_b(s_1,\text{angles})\right|^2$$
$$A_b(s_1,\text{angles}) = \sum_\lambda D^{J*}_{M\lambda}(\Omega_1) D^{S*}_{\lambda 0}(\Omega_3) \langle L0; S\lambda | J\lambda \rangle F_L(p_1) F_S(p_3) \Delta^{\text{isobar}}_S(s_1)$$

The essential assumptions here are the factorisation of production and decay and the explicitly chosen dependence on s_1 present in the isobar parameterisation. This is usually a Breit-Wigner (for, say, the ρ meson) or an experimental phase shift (the $\pi\pi$ S-wave).

This is, however, not the only possible diagram which can contribute to this reaction. As early as 1964, it was realised [1] that a diagram as pictured in figure 1(b) could be present. In its simplified form this involves the diffractive dissociation of an incident high-energy pion into an on-shell $\pi\pi$ isobar and a near-on-shell pion which subsequently scatters quasi-elastically from the proton, being kicked on-shell in the process. It is this that we call the "Deck effect". Deck described this process by the product of a pion propagator and an empirical πN elastic scattering amplitude and the cross-section so obtained was observed to peak at low three-pion invariant mass, in excess of the phase

CP814 *Hadron Spectroscopy: Eleventh International Conference*
edited by A. Reis, C. Göbel, J. Sá Borges, and J. Magnin

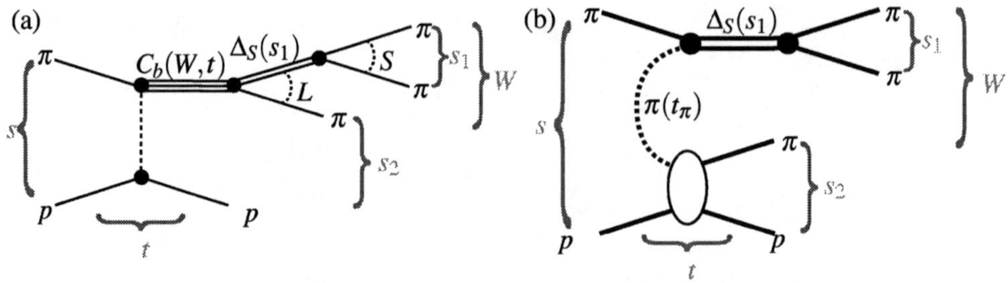

FIGURE 1. Diagrammatic description of (a) the isobar model, (b) the Deck model

space for such a process. This broad threshold enhancement below the narrow a_2 meson peak had already been seen in the data and was later dubbed the "A_1" region.

THE DECK EFFECT IN PARTIAL WAVES

Stodolsky [2] neatly demonstrated the kinematic origin of the Deck effect and at the same time obtained a basic partial wave expansion. We repeat here the essentials of his argument: We can describe the propagation of the off-shell-pion by a Feynman propagator $(t_\pi - m_\pi^2)^{-1}$, from which one would anticipate strength in many partial waves. The pion rescattering at high πp invariant mass will be dominantly diffractive (Pomeron exchange) and can be modeled by the phenomenological form $\mathscr{M}(\pi p \to \pi p) \propto i s_2 e^{at}$ which describes the elastic process well. The "diffractive peak" at small $t \to t_{\min}$ will dominate and in most experiments we are at sufficiently high incident pion energy that $s \gg W^2, m_p^2$. In this kinematic region it is found [3] that

$$s_2 \to s \frac{m_\pi^2 - t_\pi}{W^2 - m_\pi^2}. \tag{1}$$

The pole in the unphysical region of W^2 is often called the "Stodolsky pole" and is the basis of many, more complicated, models of the Deck effect which put to good use its simple structure in the complex energy plane. The zero in the numerator will tend to cancel much of the strength of the pion pole in the physical region; this cancellation leaves no dependence in this simplified Deck amplitude on the angle between the isobar and the rescattered pion and as such they must be in a relative S-wave. Thus one expects threshold enhancements of the Deck type in partial waves $0^-S0^+\varepsilon\pi$, $1^+S0^+\rho\pi$, $2^-S0^+f_2\pi$ ... The effect is likely to be washed out in the pseudoscalar wave by the broad nature of the σ enhancement, but the "A_1"(1^+) and "A_3"(2^-) regions warrant investigation given the relatively small widths of the ρ and f_2 isobars.

Deck amplitudes and the isobar model

At this stage it is worth pointing out a problem that may be encountered when comparing partial waves due to the Deck effect with those extracted from experiment using the isobar model. The isobar model's dependence on the isobar invariant mass s_1 is fixed as an input to the partial wave analysis to be that due to the isobar propagation and its decay (via a barrier factor). While this is undoubtedly correct for the diagram shown in figure 1(a), it is not sufficiently versatile to completely describe amplitudes of the Deck type. The upper vertex in figure 1(b) has the isobar propagation, but there is further dependence on s_1 in the lower vertex through the kinematic constraints on s_2. This dependence is mild compared to the rapidly varying isobar factor, but it will cause distortion of the true partial wave amplitudes in an isobar model fit. This distortion can be tested by putting events generated with a pure Deck model for which the exact partial wave decomposition is known through an isobar partial wave analysis code and observing in which partial waves the events actually end up. This was attempted in an unsystematic fashion in [4] where some distortions were observed - a more systematic study is underway.

THE "A_1" IMBROGLIO

The realisation of the existence of the Deck effect and the rise of the quark model were contemporaneous - this lead to a confused situation where parties argued whether the observed broad enhancement in $\rho\pi$ just above threshold was the axial $q\bar{q}$ resonance predicted by the quark model, or simply the kinematic enhancement due to the Deck effect. The Deck supporters pointed out the asymmetric structure of the bump, its large width and the lack of considerable phase motion were described in their model and while this was partially true, tricks had to be played in the model to get the high energy side of the bump to fall fast enough. One common "trick" was to "Reggeise" the pion, that is to model its propagator by a Regge form [3, 4],

$$\sim \frac{1}{t_\pi - m_\pi^2} e^{-i\frac{\pi}{2}\alpha_\pi(t_\pi)} (W^2)^{\alpha(t_\pi)}. \tag{2}$$

The extra t_π dependence mollifies the pole cancellation and distributes some of the higher W^2 amplitude into waves other than the S-wave, damping the S-wave more rapidly, and thus better describing the experimental data. A model of this type is described in [4] where it is seen that a Reggeised Deck model which has no free parameters (after the non-unique choice of how to Reggeise) can describe well the cross-section data in an energy-averaged way (see figure 5 in [5]).

The observation of a large enhancement in hadronic tau decays of the type $\tau \to \nu_\tau\pi\pi\pi$ independently demonstrated the need for a a_1 meson resonance. In light of this how was one to interpret the excellent description of the $\pi p \to \pi\pi\pi p$ data found by Ascoli *et al.* [4]? An answer lies in the strictly unjustified application of Regge parameterisation to this amplitude; Regge forms are analytic, unitary descriptions of scattering in the high energy limit, but here they have been applied in a case where the appropriate invariant mass, W, is barely above threshold. The energy-averaged description of the data might be explained in this context as being due to so called approximate Regge-duality which states that the Reggeon exchanged in the t-channel is somehow dual to all the possible resonances in the s-channel - including, in this case, the a_1 resonance. Thus if this explanation is correct, the Reggeised Deck supporters were unwittingly supporting the existence of the a_1 resonance. The interpretative problems of Reggeising the exchanged pion in this low W region render it a less useful phenomenological tool for simple fitting of experimental data, and we will not consider it further.

The marriage of Deck and a_1

Studies which took the middle ground of proposing both an a_1 meson and production via the Deck effect have also been considered with considerable phenomenological success. Examples include [6, 7, 8] where the essential structure of the model was that the Deck diagram with a Feynman pion (and possibly a t_π-dependent form-factor) should be considered as a Born term where the pions produced can be subsequently rescattered through the a_1 resonance. In addition, the a_1 can be produced directly, as in the isobar model diagram. Figure 1 in [8] shows this schematically. These models have the desirable feature of unitarising the $\rho\pi$ amplitude - a full three pion unitarisation being a considerably more involved task [9].

THE MOVING π_2 PEAK

The "A_3" region of the 1970's is what we would now call the π_2 meson. The 2^- waves containing this state are often used when interpreting isobar model PWA as reference waves, whose phase is considered known and against which unknown waves' phases are compared. That this is well motivated is not clear owing to a peculiar effect seen in $\pi p \to \pi\pi\pi p$.

With an f_2 isobar, one has a π_2 enhancement in both S and D waves. If the π_2 were simply an isolated resonance produced in the manner anticipated by the isobar model, the Breit-Wigner peak would have the same shape and position in both waves and the two waves would have zero relative phase. This is not what is observed [10, 11]; the enhancement in the D-wave is at a considerably higher mass and an $S-D$ phase shift of over π occurs between the two peaks. Simply fitting the two peaks independently by Breit-Wigner forms gives a mass for the S-wave state of 1660 MeV and 1840 MeV for the D-wave state. One possible explanation for the peak shifting might be that the S-wave feels a considerable amount of Deck amplitude that interferes with the true resonant π_2 shifting the peak down to a lower mass. Fits to the data using a simple but theoretically unjustified model are shown in figure 2. The model adds a direct meson production amplitude to a "Deck background" of the Ascoli type and hence runs the risk of double-counting or over-saturation of unitarity. It is seen that the addition of Deck amplitudes could help explain the effect with the π_2 meson constrained to have a consistent position in the two waves. It is not possible though to explain the phase motion in this picture without invoking a second π_2 resonance present in both S and D waves at a higher mass (see [10] who use a multi-partial wave fit involving two resonances and production via

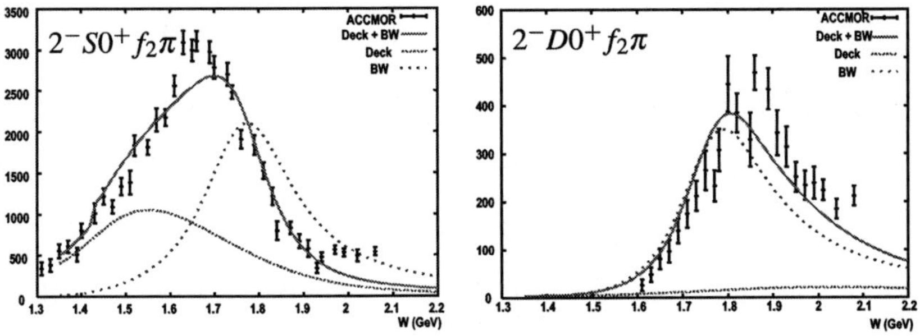

FIGURE 2. "Deck plus resonance" fits to ACCMOR data

rescattered Deck and extract a second resonance above 2 GeV which has a large width).

CONSTRAINING THE DECK EFFECT USING NEW DATA

We are reconsidering the Deck effect now owing to the availability of unprecedented high-statistics data. The recent completion [12] of a partial wave analysis of 20 million events from the E852 experiment opens up a number of new possibilities for constraining models of the Deck effect. This data set's large number of events has made it possible to perform independent partial-wave analyses of the data in different t-bins, allowing us to study the t-dependence of all partial waves. In addition we now have statistically significant signals in relatively minor waves such as $2^-S1^+f_2\pi$, which admits a test of the angular momentum projection properties of the Deck amplitudes. Furthermore, a large fraction of the data is in the alternative charge mode $\pi^0\pi^0\pi^-$ which gives an isospin-related cross-check of our models. Work is underway on the application of Deck-style models to this data.

We thank Alex Dzierba, Ryan Mitchell, Scott Tiege and Geoffrey Fox for discussions and assistance with PWA formalism. This work was supported by DOE contract DE-AC05-84ER40150 under which the Southeastern Universities Research Association operates the Thomas Jefferson National Accelerator Facility.

REFERENCES

1. R. T. Deck, Phys. Rev. Lett. **13**, 169 (1964).
2. L. Stodolsky Phys. Rev. Lett. **18**, 973 (1967).
3. E. L. Berger in *Proceedings of the Daresbury Study Weekend* DL/R34, pp. 35 (1975).
4. G. Ascoli, L. M. Jones, B. Weinstein and H. W. Wyld, Phys. Rev. D **8**, 3894 (1973).
5. G. Ascoli *et al.*, Phys. Rev. D **9**, 1963 (1974).
6. J. L. Basdevant and E. L. Berger, Phys. Rev. D **16**, 657 (1977).
7. I. J. R. Aitchison and M. G. Bowler, J. Phys. G **3**, 1503 (1977).
8. M. G. Bowler, M. A. V. Game, I. J. R. Aitchison and J. B. Dainton, Nucl. Phys. B **97**, 227 (1975).
9. I. J. R. Aitchison in *Proceedings of the Daresbury Study Weekend* DL/R34, pp. 81 (1975).
10. C. Daum *et al.* [ACCMOR Collaboration], Nucl. Phys. B **182**, 269 (1981).
11. S. U. Chung *et al.*, Phys. Rev. D **65**, 072001 (2002).
12. A. R. Dzierba, R. Mitchell, A. P. Szczepaniak, M. Swat and S. Teige, arXiv:hep-ex/0502022.

Semi-exclusive processes, grey track production and hadronization

C. Ciofi degli Atti

Department of Physics, University of Perugia & INFN, Sezione di Perugia, 06122 Perugia, Italy

Abstract. Semi-exclusive processes and grey track production in DIS off nuclear targets are analyzed as possible tools to investigate the space time evolution of hadronization.

Keywords: quarks, gluons, QCD, nuclei
PACS: 24.85.+p, 13.60.-r

INTRODUCTION

Nuclear targets serve as a natural and unique analyzer of the space-time development of strong interactions at high energies. Due to Lorentz time dilation, projectile partons may keep their coherence for some time, but once they become incoherent the cross section of final state interaction (FSI) increases. One needs observables sensitive to such modifications of FSI. One way for experimental testing of theoretical models is the measurement of the nuclear modification factor for inclusive production of leading hadrons [1], and various theoretical approaches have been advocated [2, 3, 4, 5] to explain recent data from HERMES experiment at HERA [6]. The production of hadrons with large fraction z_h of the initial parton momentum is a rare, nearly exclusive process, which has quite a specific time development. In the main bulk of events the jet energy is shared by many hadrons. It is a difficult task to find observables sensitive to the time development of hadronization in this case, and additional, complementary processes which could be sensitive to the space-time evolution of hadronization should be investigated. It is the aim of this paper to review two approaches which have been recently pursued along this line, namely the Semi- Exclusive DIS (SEDIS) processes $A(e,e'B)X$ [7, 8], where B is a detected recoiling heavy fragment (e.g a nucleus $A-1$) and X the unobserved jets from hadronization, and the grey track production in DIS off nuclei [9]. Unlike the rare process of leading hadron production, both processes, which are depicted in Fig. 1, depend upon the bulk of the FSI of the nucleon debris with the nuclear medium, and turn out to be very sensitive to the hadronization mechanism. Their theoretical treatment within a Glauber-like approach of FSI requires an *effective time-dependent nucleon debris-nucleon cross section*; a recent model of the latter [7] will be reviewed in the next Section and its use in the theoretical description of the two processes will be presented in Sections 3 and 4.

CP814 *Hadron Spectroscopy: Eleventh International Conference*
edited by A. Reis, C. Göbel, J. Sá Borges, and J. Magnin

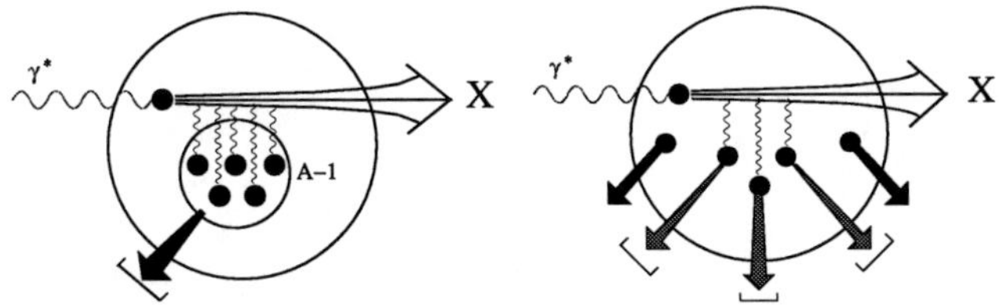

FIGURE 1. Cartoons of the process $A(e,e'B)X$ (with $B = A-1$) (left), and the grey track production (right) in DIS. *Left*: after $\gamma*$ interaction with a quark, the nucleon debris interacts with the spectator nucleus $A-1$, which coherently recoils and is detected in coincidence with the scattered lepton. *Right* the nucleon debris breaks apart the spectator nucleus $A-1$, and nucleons with momentum $200-600\,MeV/c$ are emitted and detected as grey tracks.

THE EFFECTIVE DEBRIS-NUCLEON CROSS SECTION

The effective debris-nucleon cross section obtained in [7], is based upon the hadronization model which combines the soft part of the hadronization dynamics, treated by means of the color string model, with the hard part one, described within perturbative QCD. In this model, which is inspired by Ref. [1], the formation of the final hadrons occurs during and after the propagation of the created debris through the nucleus, with a sequence of soft and hard production processes. Introducing a mass scale $\lambda = 0.65\,GeV$ [10], soft production occurs at $Q < \lambda$ and hard production at $Q > \lambda$; in the former case *npQCD* is taken care of by the color string model, whereas in the latter case pQCD is described within the gluon radiation model [1]. The details of the approach are given in Ref. [7], here only the basic elements of the model will be recalled. The ingredients of the treatment of hadron (mostly mesons (**M**)) formation from the string (**S**) decay, are the following ones: the probability $W(t)$ for a string to create no quark pairs since its origin; the time dependent length of the string $L(t)$ with $L_{max} = m_{qq}/\kappa$, where m_{qq} is the mass of the "diquark" and $\kappa \simeq 1\,GeV\,fm^{-1}$ the string tension; the creation of a baryon (**B**) and a shorter string after the first breaking of the latter (within $\Delta t \simeq 1\,fm$), and the sequence of decays according to the scheme $\mathbf{S} \Rightarrow \mathbf{B}+\mathbf{S} \Rightarrow \mathbf{B}+\mathbf{S}+\mathbf{M} \Rightarrow \mathbf{B}+\mathbf{S}+2\mathbf{M}+\ldots$ leading to the following multiplicity of mesons

$$n_M(t) = \ln(1+t/\Delta t)/\ln(2) \tag{1}$$

with $\Delta t \simeq 1\,fm$. The gluon radiation mechanism is governed by the coherence time $t_c = 2E_q\,\alpha\,(1-\alpha)/k_T^2$ which is the time which elapses from the creation of the leading quark and the emission of the gluon which lost coherence with the color field of the quark; here α, $k_T = |\mathbf{k}_T|$, and E_q are the fraction of the quark light-cone momentum carried by the radiated quantum, its transverse momentum, and the quark energy, respectively, and the time dependence of the gluon radiation, controlled by the parameter $t_0 = (m_N x_{Bj})^{-1} = 0.2fm/x_{Bj}$. The final result for n_G is ,

$$n_G(t) = \frac{16}{27}\left\{\ln\left(\frac{Q}{\lambda}\right) + \ln\left(\frac{t\,\Lambda_{QCD}}{2}\right)\ln\left[\frac{\ln(Q/\Lambda_{QCD})}{\ln(\lambda/\Lambda_{QCD}}\right]\right\}, \tag{2}$$

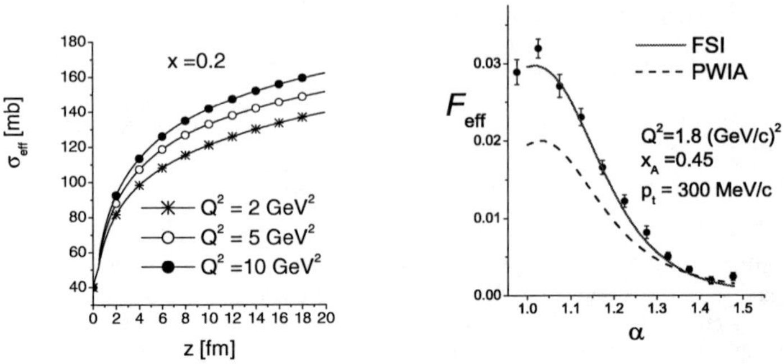

FIGURE 2. *Left*: The debris-nucleon effective cross section (Eq. (4)) plotted *vs* the distance z for a fixed value of the Bjorken scaling variable $x \equiv x_{Bj}$ and various values of the four-momentum transfer Q^2 (after Ref. [7]). *Right*: The effective semi exclusive deuteron structure function calculated including (FSI) and omitting (PWIA) the FSI of the debris, compared with preliminary experimental data from Jlab [14]. α and p_t are the light cone fraction and perpendicular momentum component of the detected proton and $x_A = x_{Bj}/(2-\alpha)$ (after [15]).

for $t < t_0$, and

$$n_G(t) = \frac{16}{27}\left\{\ln\left(\frac{Q}{\lambda}\frac{t_0}{t}\right) + \ln\left(\frac{t\,\Lambda_{QCD}}{2}\right)\ln\left[\frac{\ln(Q/\Lambda_{QCD}\sqrt{t_0/t})}{\ln(\lambda/\Lambda_{QCD})}\right]\right.$$
$$\left. + \ln\left(\frac{Q^2 t_0}{2\Lambda_{QCD}}\right)\ln\left[\frac{\ln(Q/\Lambda_{QCD})}{\ln(Q/\Lambda_{QCD}\sqrt{t_0/t})}\right]\right\}, \tag{3}$$

for $t > t_0$, with saturation at $t > t_0 Q^2/\lambda^2 = 2\nu/\lambda^2$. Using Eqs. (1), (2), and (3), one obtains the effective debris-nucleon cross section in the following form

$$\sigma_{eff}(t) = \sigma_{tot}^{NN} + \sigma_{tot}^{MN}\left[n_M(t) + n_G(t)\right] \tag{4}$$

The time (or z, for a particle moving with the speed of light) dependence of $\sigma_{eff}(t)$ is shown in Fig. 2 for a fixed value of x_{Bj} and various values of Q^2.

THE SEMI-EXCLUSIVE DIS PROCESS

This process has been discussed in [12, 13] within the Plane Wave Impulse Approximation (PWIA). If we consider, for ease of presentation, a deuteron (D) target, the PWIA process consists of the hard interaction of $\gamma*$ with a parton of, e.g. the neutron, the creation of a debris, with the spectator (s) proton recoiling and detected in coincidence with the scattered electron (cf. Fig. 1, left, disregarding the wavy lines that represent the FSI between the debris and the spectator nucleons). The PWIA cross sections reads as follows

$$\frac{d\sigma}{dxdQ^2\,d\alpha dp_T^2} = K(x_{Bj},Q^2,p_s)\,N_A^D(\alpha,\mathbf{p}_T)\,F_2^{N/D}(Q^2,x_{Bj},p_s), \tag{5}$$

is obtained, where $p_s = (p_s^0, \mathbf{p}_s)$ is the four-momentum of the recoiling detected nucleon, $\alpha = [p_s^0 - |\mathbf{p}_s|\cos\theta_s]/M$ (θ_s being the nucleon emission angle with respect to the direction of $\mathbf{q}$), $K(x_{Bj},Q^2,p_s)$ is a kinematical factor, $F_2^{N/A}(Q^2,x_{Bj},p_s)$ is the DIS structure function of the hit nucleon, N_A^D is the nucleon momentum distribution and

$$\Gamma^{NN}(\mathbf{b}_1 - \mathbf{b}_i) \longrightarrow \Gamma^{N^*N}(\mathbf{b}_1 - \mathbf{b}_i, z_1 - z_i) \tag{6}$$

takes care of the final state interaction between the debris and the spectator. The approach can readily be generalized to complex nuclei for which the experiments are difficult to carry out. However, preliminary experimental data for the deuteron have already been obtained at Jlab [14]. A limited set of these data is shown in Fig. 3, where the quantity $F_{eff}\left(\frac{x_{Bj}}{2-\alpha}, p_T, Q^2\right) \equiv \sigma^{exp}/K \cdot N_A^D(\alpha,\mathbf{p}_T)$ is compared with theoretical calculations [15]; it can be seen that the latter is extremely important.

HADRONIZATION AND GREY TRACK PRODUCTION

The dominant channels of DIS are the ones in which the recoiling nucleus B breaks apart to fragments; the investigation of the nature and the kinematical dependence of these fragments can shed light on the time evolution of the jet in the nuclear medium. This seems particularly true in the processes producing grey tracks, which are hadrons, predominantly protons, with momenta in the range of few hundred MeV/c. The basic mechanism of grey track production is the inelastic interaction of the jet with the spectator nucleons of the target, which recoil in the given momentum interval and are detected (cf. Fig. 1, right). The Q^2 and x_{Bj} dependence of the average number of grey tracks produced in the Fermilab E665 experiment [16] (protons with momentum $200-600\,MeV/c$ produced in $\mu - Xe$ and $\mu - D$ DIS at 490 GeV beam energy), have been recently analyzed [9] within the theoretical framework which employs the effective debris-nucleon cross section of [7]. The basic elements of the approach are the following ones:

- DIS on a bound nucleon occurs at coordinate $(\vec{b}, z)$ and the debris propagates through the nucleus interacting with spectator nucleons via $\sigma_{eff}(z-z')$. The mean number of collisions (plus the recoiling nucleon formed in the hard $\gamma* - N$ act) is

$$\langle \nu_c \rangle = \int d^2b \int_{-\infty}^{\infty} dz\, \rho_A(\vec{b},z) \int_{z}^{\infty} dz'\, \rho_A(\vec{b},z')\, \sigma_{eff}(z-z') + 1\,. \tag{7}$$

  where ρ_A is the nuclear density and $\sigma_{eff}(t) = \sigma_{in}^{MN}[n_M(t) + n_G(t)]$ with $\sigma_{in}^{MN} = \sigma_{tot}^{\pi N} - \sigma_{el}^{\pi N} - \sigma_{dif}^{\pi N} = 17.7\,mb$.
- the mean number of collisions has been calculated according to Eq. (7), and the mean number of grey tracks $< n_g >$ has been obtained using the empirical relation

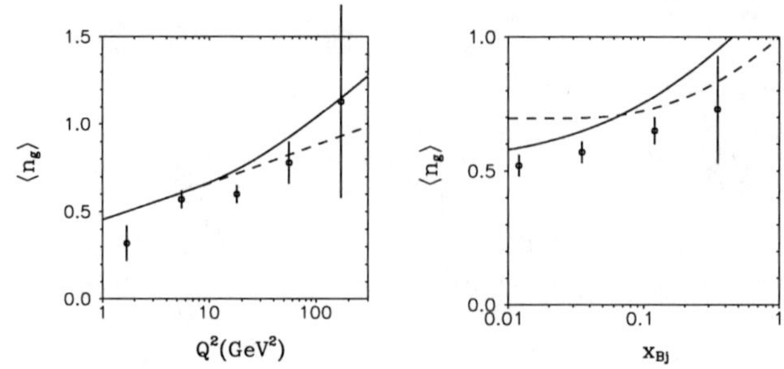

FIGURE 3. *Left*: mean number of grey tracks $< n_g >$ produced in the $\mu - Xe$ DIS experiment [16] *vs* Q^2 in the non-shadowing region with fixed $x_{Bj} = 0.07$ (dashed). *Right*: mean number of grey tracks $< n_g >$ *vs* x_{Bj} with fixed value of $Q^2 = 14.3\, GeV^2$ (dashed). In both Figures the solid curve includes the $Q^2 - x_{Bj}$ correlation found in the experiment (after Ref. [9]).

found in [16], *viz.*

$$< n_g >= \frac{\langle \nu_c \rangle - (2.08 \pm 0.13)}{(3.72 \pm 0.14)} \tag{8}$$

The results of the parameter-free calculations, which are shown in Fig. 4, exhibit a good agreement with the experimental data; worth being mentioned is the Q^2 dependence of the data, which is explained by the adopted hadronization model, namely by the gluon radiation mechanism. It turns out, therefore, that the observed Q^2 dependence of the average number of grey tracks is a very sensitive tool to discriminate between different models of hadronization.

CONCLUSIONS

The main conclusions of my talk can be summarized as follows:

- a time-dependent debris-nucleon cross section has been obtained; it incorporates both non perturbative (string) and Q^2-dependent perturbative (gluon radiation) effects; it accounts for the bulk of the FSI interaction of the debris created by the hard interaction of a bound nucleon with $\gamma *$; it should be used in all kinds of DIS off nuclei to describe the FSI of the hit nucleon debris with the spectator nucleons;
- the (parameter-free) calculation of the Semi Exclusive DIS off the deuteron $D(e,e'p)X$ exhibits good agreement with preliminary data from Jlab; a systematic investigation of the x_{Bj} and Q^2 dependence of these processes at Jlab and HERA energies would shed further light on the hadronization mechanism;
- the (parameter-free) calculation of the Q^2 and x_{Bj} dependence of the average number of grey tracks produced in Deep Inelastic $\mu - Xe$ scattering exhibits a very satisfactory agreement with the experimental data, thanks to the gluon radiation mechanism for hadron production.

ACKNOWLEDGMENTS

I am grateful to the Organizers of the Workshop for the invitation and to Leonid Kaptari and Boris Kopeliovich for a fruitful and stimulating collaboration.

REFERENCES

1. B.Z. Kopeliovich, J. Nemchik and E. Predazzi, in *Future Physics at HERA*, Proceedings of the Workshop 1995/96, edited by G. Ingelmsan, A. De Roeck and R. Klanner, DESY, 1995/1996, vol.2, p. 1038 (nucl-th/9607036);
   in Proceedings of the *ELFE Summer School on Confinement Physics*, edited by S.D. Bass and P.A.M. Guichon, Editions Frontieres, 1995, p. 391, Gif-sur-Yvette (hep-ph/9511214).
2. B.Z. Kopeliovich, J. Nemchik, E. Predazzi and A. Hayashigaki, Nucl. Phys. **A740** (2004)212.
3. E. Wang and X.-N. Wang, Phys. Rev. Lett. **89** (2002) 162301.
4. A. Accardi, V. Muccifora and H.J. Pirner, Nucl. Phys. **A720** (2003) 131.
5. T. Falter, W. Cassing, K. Gallmeister, U. Mosel, nucl-th/0303011; nucl-th/0406023.
6. HERMES Collaboration, A. Airapetian et al., Eur. Phys. J. **C20** (2001) 479; Phys. Lett. **B577** (2003) 37.
7. C. Ciofi degli Atti and B.Z. Kopeliovich, Eur. Phys. J. **A17** (2003) 133.
8. C. Ciofi degli Atti, L.P. Kaptari, and B.Z. Kopeliovich, Eur. Phys. J. **A19** (2004) 145.
9. C. Ciofi degli Atti and B.Z. Kopeliovich, Phys. Lett. **B606** (2005) 281.
10. B.Z. Kopeliovich, A. Schäfer and A.V. Tarasov, Phys. Rev. **D62** (2000) 054022 (hep-ph/9908245).
11. J.F. Gunion and G. Bertsch, Phys. Rev. **D25** (1982) 746.
12. W. Melnitchouk, M. Sargsian, M. I. Strikman, Z. Phys. **A359** (1997) 99.
    S. Simula, Phys. Lett. **B387** (1996) 245.
13. C. Ciofi degli Atti, L.P. Kaptari, and S. Scopetta, Eur. Phys. J. **A5** (1999) 191.
14. Jlab Experiment 94-102, S. E. Kuhn, K. A. Griffioen, co-spokespersons, *Inelastic electron scattering off a moving nucleon in deuterium*;
    S. E. Kuhn, *Private Communication.*
15. C. Ciofi degli Atti and L. P. Kaptari, *unpublished.*
16. E665 Collaboration, M.R. Adams et al., Z. Phys. **C65** (1995) 225.

Study of bound kaonic systems in nuclear matter with FINUDA at DAΦNE

A. Filippi
on behalf of the FINUDA Collaboration [1]

I.N.F.N. Torino, via P. Giuria, 1, 10125 Torino, Italy

Abstract. The possible existence of deeply bound kaonic states has been studied with the FINUDA spectrometer at DAΦNE, exploiting both the invariant mass spectroscopy technique, and the missing mass one. A reasonable evidence for the existence of a $(K^- pp)$ bound system has been found; the first hints for the existence of a strange tribaryon by other experiments are discussed, on account of FINUDA observations.

Keywords: Deeply bound kaon-nucleus states; invariant and missing mass spectroscopy
PACS: 21.45.+v, 21.30.Fe, 21.80.+a, 25.80.Nv

DEEPLY BOUND $\overline{K}$-NUCLEUS STATES: INTRODUCTION

The possible existence of deeply bound (anti)kaon-nucleus states has raised considerable interest quite recently, with the publication of some first possible experimental observations [1, 2, 3] in rough agreement with a few theoretical expectations based on the calculations of Yamazaki and Akaishi [4].

These states consist of few-body strange (S=-1) systems composed by nucleons strongly bound to a $\overline{K}$. The strength of the $\overline{K}N$ attractive interaction, in the $I = 0$ configuration, allows for the stability of the system, as well as for its compactness [5]. This feature is important as these aggregates could be possible constituents of high density forms of nuclear matter, such as the core of neutron stars.

Akaishi and Yamazaki predict for these states sizeable binding energies. In such a model the $\overline{K}N^{(I=0)}$ interaction is predicted to be so strong that the widths of these aggregates should be considerably narrow. In fact, if the potential well is deep enough, the main decay mode of the $\overline{K}N^{(I=0)}$ system, $\Sigma\pi$, is energetically closed, and moreover the $\Lambda\pi$ channel is suppressed because of isospin conservation. In this framework, these states are expected to be formed with larger probabilities in lighter nuclei.

---

[1] M. Agnello, G. Beer, L. Benussi, M. Bertani, S. Bianco, E. Botta, T. Bressani, L. Busso, D. Calvo, P. Camerini, P. Cerello, B. Dalena, F. De Mori, G. D'Erasmo, D. Di Santo, F.L. Fabbri, D. Faso, A. Feliciello, A. Filippi, V. Filippini, E. Fiore, H. Fujioka, P. Gianotti, N. Grion, V. Lucherini, S. Marcello, T. Maruta, N. Mirfakhrai, O. Morra, T. Nagae, A. Olin, H. Outa, E. Pace, M. Palomba, A. Pantaleo, A. Panzarasa, V. Paticchio, S. Piano, F. Pompili, R. Rui, G. Simonetti, H. So, S. Tomassini, A. Toyoda, R. Wheadon, A. Zenoni
Italy (Bari, Brescia, LNF, Pavia, Torino, Trieste), Canada (Victoria, TRIUMF), Japan (Tokyo, KEK, RIKEN), Iran (Teheran), S. Korea (Seoul)

CP814 *Hadron Spectroscopy: Eleventh International Conference*
edited by A. Reis, C. Göbel, J. Sá Borges, and J. Magnin

However, other models for the K^--nucleus interaction have been proposed, that require a shallower potential well ($50 \div 75$ MeV deep, allowing the existence of a bound system, though preventing its observation because of the large width) [6], and comply as well to the description of K-atoms and K-nucleon scattering experimental results. At the moment, no experimental indication has been found to favour one of the two possibilities. Recently, however, a critical revision of the calculations of the deep-potential model by Akaishi and Yamazaki has been made [7].

FIRST EXPERIMENTAL OBSERVATIONS

The observations of a possible $\overline{K}$-nucleus state are based on two methods: by reconstructing the invariant mass of the system starting from its decay products, namely a hyperon and a nucleon (*invariant mass spectroscopy*), or by measuring the momentum of the monochromatic recoiling nucleon in a $A(K^-,NX)$ reaction (*missing mass spectroscopy*). The first method is of course less ambiguous and allows for stricter selection criteria, such as cuts on angular correlations, but can be exploited only by performant apparatuses able to fully reconstruct the event topologies, as FINUDA is [8]. The second technique has been used so far and allowed to obtain the first observations.

The KEK-PS E471 experiment studied the inclusive spectra of a proton [1] or a neutron [2] emitted in the ^{4}He(K^-,NX) reaction. They quoted the existence of two different states: $S^0(3115)$, with $I=1$, produced in the reaction $K^-_{stop} + ^4\text{He} \rightarrow (K^-pnn) + p$ [1], and $S^+(3140)$, with $I=0$, produced in $K^-_{stop} + ^4\text{He} \rightarrow (K^-ppn) + n$ [2]. The two masses are slightly different, while the widths are comparable, with a quoted upper limit of 21 MeV/c^2. The preferred decay channel, for both, is suggested to be Σ^-NN. The two structures are suggested to be compatible with a deeply bound kaon-nucleus state with a binding energy of 169 and 194 MeV [9].

A third observation was performed by the BNL-AGS E930 Experiment, which measured the missing mass spectrum in the ^{16}O(K^-,n) reaction, with 930 MeV/c kaons. The distribution shows two slight enhancements, corresponding to a binding energy of the kaon of about 90 MeV (which can suggest the formation of the kaonic oxygen $^{16}_{\overline{K}}$O with the kaon in p-shell), and of 130 MeV ($^{16}_{\overline{K}}$O with the kaon in s-shell).

OBSERVATIONS WITH FINUDA

FINUDA is a multi-purpose magnetic spectrometer installed in the DAΦNE e^+e^- collider at Laboratori Nazionali di Frascati, Italy. Details of its set-up and performances may be found elsewhere [8]; its features are a high momentum resolution (from 0.5 to 1.5% FWHM depending on the mass and momentum of the particle), an angular acceptance close to 2π sr and a good particle identification. Moreover, the apparatus is equipped with targets of five nuclear species, in order to provide measurements of the same observables in several nuclear media at the same time.

Thanks to these features, the apparatus can be used not only to perform high resolution studies on hypernuclear spectroscopy and decay modes, but also to investigate new fields, such as the existence of deeply bound kaonic states, using effectively both the invariant mass and the missing mass spectroscopy methods.

The invariant mass method. Some indications of the existence of a $(K^- pp)$ system decaying into Λp have been found with FINUDA [10], thanks to its ability to fully reconstruct the particles produced in a $K^- A$ interaction. When a K^- coming from the decay of the ϕ meson stops in a nuclear target, it can interact with one or more nucleons, producing hyperons in quasi-free reactions, undergoing absorption processes or possibly forming deeply bound states. If a kaonic bound state is formed, one should be able to reconstruct its mass measuring a YN pair from its decay. Moreover, in the hypothesis of a bound state decaying at rest, a back-to-back correlation of these two particles can be required.

The Λ particle can be reconstructed in FINUDA by its decay products, a proton and a π^-. Because of the 1.0 T magnetic field, a kinematic cut on the minimum detectable pion momentum is set; its reconstruction is possible only down to 110 MeV/c. This of course imposes an acceptance cut on the Λ momentum, which can be detected only if exceeding 300 MeV/c. In a typical "deeply $\overline{K}$-bound" event, the back-to-back Λ and proton have a momentum of about 510 MeV/c, while the π from the Λ decay has a momentum in the range $(110 \div 220)$ MeV/c. The event may therefore be fully reconstructed by FINUDA.

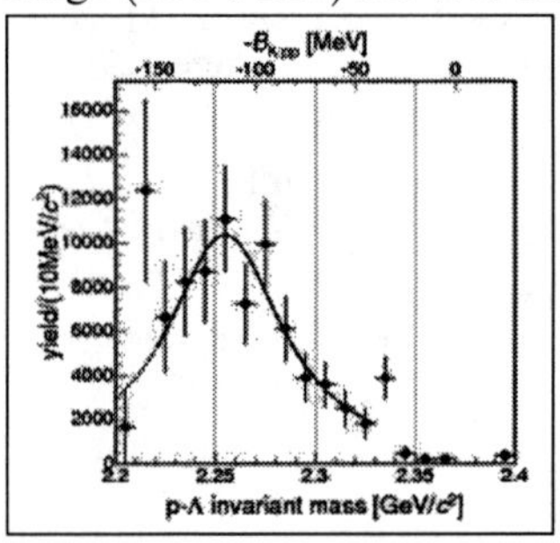

1: Acceptance corrected invariant mass of the (Λp) system (back-to-back particles). The peak has been fitted with a Lorenzian function smeared with the apparatus resolution (gaussian function with $\sigma = 4$ MeV).

Independently of the target, a clear back-to-back correlation is found for the events in which a Λ, reconstructed with a resolution of 6 MeV/c^2 FWHM, is produced in coincidence with a second proton (amounting to about the 5% of all events with a reconstructed Λ). In case of a two nucleon absorption reaction $K^- + (pp) \rightarrow Y^\star N \rightarrow \Lambda p$ with the production of a $Y^\star$ intermediate state, one would expect the invariant mass of the $(K^- pp)$ system to be peaked slightly below 2.37 GeV. On the contrary, the invariant mass of the back-to-back (Λp) system (selecting events from the lightest available targets only, ^{6}Li, ^{7}Li and ^{12}C), exhibits a much lower peak, shown in Fig. 1. This peak cannot simply be explained as due to the contribution of the $\Sigma^0 p$ channel, with the following electromagnetic decay $\Sigma^0 \rightarrow \Lambda\gamma$. In this case the peak is expected about 74 MeV/c^2 below the threshold. The observed peak, with proper acceptance corrections (which are mass dependent), shifts indeed further downwards. However, a large discrepancy comes out if we compare the observed ratio of Λp to $\Sigma^0 \pi$ production with old experiments' results [11]: this fact induces to discard the $\Sigma^0 p$ hypothesis.

Assuming a flat and negligible background contribution, one gets a peak at about 2250 MeV/c^2. Fitting it with a lorentzian function the binding energy and width of such a state can be obtained: $B = 115^{+6}_{-5}(\text{stat})^{+3}_{-4}(\text{sys})$ MeV, $\Gamma = 67^{+14}_{-11}(\text{stat})^{+2}_{-3}(\text{sys})$ MeV. Its tentative production yield is of the order of 0.1%/(stopped K^-).

This is the first observation of a bound $(K^- pp)$ system; recently, a similar behaviour has been observed by the FOPI Experiment at GSI, studying the (Λp) invariant mass system in Ni+Ni collision events at 1.093 GeV/A [12].

The missing mass method. The inclusive momentum distribution of a proton emitted after the capture at rest of a K^- on ^{6}Li is shown in Fig. 2. While the structures on the left side of the spectrum can be explained as due to the contribution of non mesonic decays of hyperfragments, the peak at about 500 MeV/c is similar to the signal observed by E471. The events in this peak are characterized by the simultaneous presence of a high momentum π^-, as shown in the scatter plot of the momenta of the two particles reported in Fig. 3. When a selection on the pion momentum is made ($p_\pi > 260$ MeV/c), the higher momentum region of the proton spectrum is enhanced. This enhancement could have a few possible explanations.

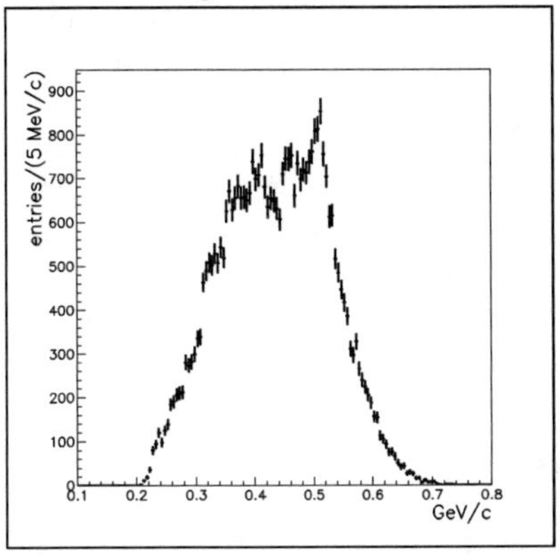

2: Inclusive momentum spectrum of a proton produced in a K^--^{6}Li absorption process in FINUDA, not acceptance corrected.

Following the analysis by E471, and accounting on the experimental observation that in several reactions a ^{6}Li nucleus behaves like a $(d+\alpha)$ cluster, with the two nuclei in relative s-wave [14], one might suggest the formation of a tribaryon on a "quasi"-^{4}He substructure of ^{6}Li. In this hypothesis, a fit of the missing mass spectrum, with proper background subtraction, delivers a mass and a width thoroughly compatible with the results from E471. However, the momentum of the negative pion seems to be too high to justify its production in the decay of the Σ^- from the $\Sigma^- NN$ or $\Sigma^- d$ decay channels of the tribaryon. Moreover, a back-to-back angular correlation between the proton and the pion can be observed, which can be hardly explained in the hypothesis of a deeply bound state decay. Therefore, this explanations seems questionable.

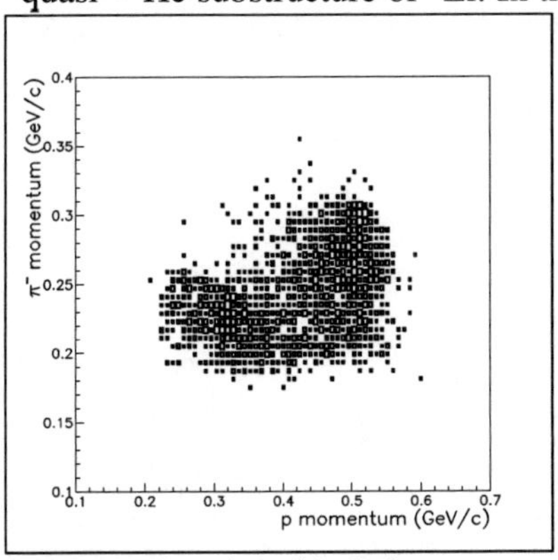

3: Scatter plot of semi-inclusive momenta spectra (p and π^- in coincidence: along the x axis, the distribution of the p momentum is plotted, along y, the distribution of the π^-. A clear correlation between the two can be observed, for higher π^- and p momenta.

Following the analysis of E930 [3], the production of a $(K^- ppnnn)$ cluster (corresponding to a $^{5}_{K^-}$He kaonic nucleus) recoiling against the measured proton can be hypothesized. The binding energy of the kaon in this hypothesis has a distribution peaked at about 160 MeV, value in reasonable agreement with the predictions of Mareš *et al.* [13], and of the same scale of those found by E930 for $^{16}_{K}$O. But again, this hypothesis does not comply to the observed angular correlation between the π^- and the proton.

A physical process producing 508 MeV/c protons is the "rare" decay of the $^4_\Lambda$He hypernucleus in two bodies: $^4_\Lambda\text{He} \to p + t$. This hyperfragment is produced from a

^{6}Li nucleus with the emission of a 255 MeV/c π^-. The authors on Ref. [1] discard this possibility studying the behaviour of the fast pions by means of scintillators light output. In FINUDA the specific energy-loss measurement provided by the vertex detector could help to identify the presence of tritons emitted from the ^{6}Li targets; however, no indications for the production of these nuclei in such reactions were found so far.

Without invoking the production of K^--nucleons aggregates, a simple explanation complying with the observed features could be the interaction of a K^- on a deuteron, or two nucleons, as in $K^- + d \rightarrow \Sigma^- p$ reaction. In this case, the momentum of the recoiling proton is 515 MeV/c, while the π^- momentum (from the Σ^- decay) is in the range $(200 \div 300)$ MeV/c. The expected production rate of this reaction is compatible with the observed yield, even taking into account that in the ^{6}Li nucleus, as previously mentioned, a sizeable content of d clusters is present.

CONCLUSIONS AND FUTURE PROSPECTS

With FINUDA, reasonable indications for the existence of a $(K^- pp)$ aggregate decaying in Λp have been found, using the invariant mass spectroscopy technique, while, from the analysis of the features of the inclusive and semi-inclusive proton spectra, no convincing evidence for the existence of a tribaryon emerged so far.

A new long FINUDA data taking is expected in 2006, which will allow the study, with large statistics, of the esistence of new possible deeply bound states like $(K^- pn)$ and $(K^- ppn)$, and of possible many-body deeply bound states decaying in Λd.

REFERENCES

1. T. Suzuki *et al.*, *Nucl. Phys.* **B597** (2004), 263
2. M. Iwasaki *et al.*, *Nucl. Instr. Meth.* **A473** (2001), 286
   T. Suzuki *et al.*, *Nucl. Phys.* **B754** (2005), 375c
3. T. Kishimoto *et al.*, **Nucl. Phys. A754** (2005), 383c
4. Y. Akaishi and T. Yamazaki, *Phys. Rev.* **C65** (2002), 044005
   T. Yamazaki and Y. Akaishi, *Phys. Lett.* **B535** (2002), 70
   T. Yamazaki *et al.*, *Phys. Lett.* **B598** (2004), 167
5. A. Doté *et al.*, *Phys. Lett.* **B590** (2004), 51
   A. Doté *et al.*, *Phys. Rev.* **C70** (2004), 044313
6. A. Cieplý *et al.*, *Phys. Rev.* **C60** (1999), 024314
   A. Ramos and E. Oset, *Nucl. Phys.* **A671** (2000), 481
7. E. Oset and H. Toki, nucl-th/0509048
8. FINUDA Collaboration, M. Agnello *et al.*, *Phys. Lett.* **B622** (2005), 35
9. Y. Akaishi, A. Dote, T. Yamazaki, Phys. Lett. **B606** (2005), 140
10. M. Agnello *et al.*, *Phys. Rev. Lett.* **94** (2005), 212303
11. P.A. Katz *et al.*, *Phys. Rev.* **D1** (1970), 1267
12. N. Herrmann for the FOPI Collaboration, talk presented at *Hadron Physics at COSY*, Bad Honnef, Germany, July 2005
13. J. Mareš *et al.*, *Phys. Lett.* **B606** (2005), 295
14. J. Favier *et al.*, *Nucl. Phys.* **A169** (1971), 540
    C. Cernigoi *et al.*, *Nucl. Phys.* **A456** (1986), 599

Study of the $^{7}_{\Lambda}$He hypernucleus in formalism of the Faddeev equations

Igor Filikhin*, Vladimir M. Suslov*,† and Branislav Vlahovic*,**

*North Carolina Central University, Durham, NC 27707, USA
†Fock Institute of Physics, Sankt-Petersburg State University, 198504, Petrodvorets, Russia
**Jefferson Lab, 12000 Jefferson Avenue, Newport News, VA 23606, USA

Abstract. P-shell $A = 7$ hypernuclei are considered in the cluster $^{5}_{\Lambda}\mathrm{He}+\mathrm{N}+\mathrm{N}$ model. The folding procedure are applied to construct the $^{5}_{\Lambda}\mathrm{He}-\mathrm{N}$ interaction. We use the OBE simulating NSC97f potential for ΛN interaction and various phenomenological potentials for $\alpha\Lambda$ interaction. Configuration space Faddeev calculations are performed for hyperon binding energy of the $^{7}_{\Lambda}\mathrm{He}(\frac{1}{2}^{+})$ and $^{7}_{\Lambda}\mathrm{Li}(\frac{1}{2}^{+}$ and $\frac{3}{2}^{+}$, T=0) nuclei. Predicted value for $B_{\Lambda}(^{7}_{\Lambda}\mathrm{He})$ is 5.35 MeV. This value is obtained when the $^{6}_{\Lambda}\mathrm{He}(2^{-})$ excitation energy is equal 0.26 MeV by the adjustment of pair $^{5}_{\Lambda}\mathrm{He}-\mathrm{N}$ effective potential to reproduce the experimental value of the $^{7}_{\Lambda}\mathrm{Li}(\frac{3}{2}^{+})$ excitation energy. Our results are compared with those of E. Hiyama et al. *Phys. Rev.* C **53**, 2075-2085 (1996).

Keywords: hypernuclei; cluster model; ΛN interaction; Faddeev equations; few-body systems
PACS: 21.80.+a, 11.80.Jy, 21.45.+v

INTRODUCTION

P-shell $A = 7$ hypernuclei are considered as appropriate objects for studying the ΛN interaction and its spin-dependence, particularly. The $^{7}_{\Lambda}$Li and $^{7}_{\Lambda}$Be nuclei have been experimentally studied quiet well [1, 2, 3], when there exists no experimental data for $^{7}_{\Lambda}$He [4] at the present time. Theoretical studies of $A = 7$ hypernuclei in the different approaches have been undertaken a long time ago by many authors [5, 6]. Recently the microscopic variational calculations [7] were performed on the basis of the three-body model $^{5}_{\Lambda}\mathrm{He}+\mathrm{N}+\mathrm{N}$, and within the framework of the four-body model ($\alpha+\Lambda+\mathrm{N}+\mathrm{N}$)[8, 9]. In this paper we report results of the study based on the configuration space Faddeev formalism [10] for the P-shell $A = 7$ hypernuclei which are considered as the cluster system $^{5}_{\Lambda}\mathrm{He}+\mathrm{N}+\mathrm{N}$. We calculate the hyperon binding energy of $^{7}_{\Lambda}$He in the $\frac{1}{2}^{+}$ state, and $^{7}_{\Lambda}$Li with T=0 in the $\frac{1}{2}^{+}$ and $\frac{3}{2}^{+}$ (excitation energy) states. In the framework of this model a new prediction of 5.35 MeV has been obtained for the hyperon energy of the ground state of $^{7}_{\Lambda}$He. This value corresponds to the excitation energy of the 2^{-} state of $^{6}_{\Lambda}$He of 0.26 Mev. The calculated hyperon energy of $^{7}_{\Lambda}$He is close to the one reported by E. Hiyama et al. [7], which was obtained using another value of the excitation energy equal to 0.36 MeV. This distinction disappears after an correction for method of the hyperon energy extracting from the calculated three-body energy of the $^{5}_{\Lambda}\mathrm{He}+n+n$ system. After such correction in the result of E. Hiyama et al. a new value 5.14 MeV is obtained which is in good agreement with the present calculation.

CP814 *Hadron Spectroscopy: Eleventh International Conference*
edited by A. Reis, C. Göbel, J. Sá Borges, and J. Magnin

MODEL DESCRIPTION

The system ${}^{5}_{\Lambda}\mathrm{He}+\mathrm{N}+\mathrm{N}$ can be studied in three-body model as bound cluster *Core*=$\alpha+\Lambda$ (the ${}^{5}_{\Lambda}$He nucleus). To construct the ${}^{5}_{\Lambda}\mathrm{He}-\mathrm{N}$ pair potential we take into account four-body nature of ${}^{7}_{\Lambda}$He and consider two subsystem which are ΛNN and αNN. Spin-isospin variables of our ${}^{5}_{\Lambda}\mathrm{He}+\mathrm{N}+\mathrm{N}$ system are defined by the ΛNN subsystem when angular variables are defined by the αNN subsystem. The $Core+\mathrm{N}+\mathrm{N}$ system is considered in the spin-isospin state with S=1/2 and T=0. The potential model used in our calculation for this system includes a semirealistic potential for NN interaction and an effective ${}^{5}_{\Lambda}\mathrm{He}-\mathrm{N}$ potential. The last being given by a sum of an αN potential and a ΛN potential averaged over the Λ density function of the αΛ cluster (folding procedure). A phenomenological local potential was used to describe the Nα interaction. For the ΛN potential an OBE simulating NSC97 model was applied. The ${}^{5}_{\Lambda}\mathrm{He}-\mathrm{N}$ folding potential that we thus obtain has s and p-wave components. The p-wave component is used for adjusting the parameters of the potential to reproduce the experimental values for the excitation energy of the 2^- states of the ${}^{6}_{\Lambda}$He hypernucleus and the $\frac{3}{2}^+$ state of ${}^{7}_{\Lambda}$Li.

POTENTIALS

The nuclear αN interaction is given by the phenomenological potential SBB [11] having s and p-wave components including $(\vec{l}\vec{s})$ coupling, as modified in [12]. We use the ΛN potential simulating the NSC97 model [13] for barion-barion interaction. This s-wave potential is the three-range Gaussian [14] corrected in [15]. Parameters of the potential was chosen to reproduce the scattering length and the effective range for a given model [13]. For the triplet and singlet NN interaction we used the semirealistic Malfliet-Tjon potential MT-I-III [16]. Both s and p-wave components of the ${}^{5}_{\Lambda}$He-N ($Core-$N) folding potential have the form

$$l=0:\;\; V^{l,t(s)}_{Core-\mathrm{N}} = V^{l}_{\alpha\mathrm{N}} + V^{f,t(s)}_{\Lambda\mathrm{N}}; \qquad l=1:\;\; V^{l,t(s)}_{Core-\mathrm{N}} = V^{l}_{\alpha\mathrm{N}} + \eta^{t(s)} V^{f,t(s)}_{\Lambda\mathrm{N}}, \quad (1)$$

where $\eta^{t(s)}$ is the spin-dependent coefficient. $V^{f,t(s)}_{\Lambda\mathrm{N}}$ is the triplet(singlet) component of the folding ΛN potential:

$$V^{f,t(s)}_{\Lambda\mathrm{N}}(|\vec{r}|) = \int d\vec{\xi}\rho(|\vec{\xi}|)V^{s(t)}_{\Lambda\mathrm{N}}(|\vec{r}-\vec{\xi}|), \quad (2)$$

where $\rho(|\vec{r}|)$ is the Λ-particle density function for the ${}^{5}_{\Lambda}$He hypernucleus ($\int d\vec{r}\rho(|\vec{r}|) = 1$). $V^{s(t)}_{\Lambda\mathrm{N}}$ and $V^{s(t)}_{\mathrm{NN}}$ are the s-wave ΛN and NN potentials, respectively. $V^{l}_{\alpha\mathrm{N}}$ is the partial component (s or p-wave) of the αN potential (spin-orbit coupling is neglected). The folding potentials averaged on spins of the $Core+\mathrm{N}$ pair are as follows:

$$V^{l}_{Core-\mathrm{N}} = \tfrac{3}{4}V^{l,t}_{Core-\mathrm{N}} + \tfrac{1}{4}V^{l,s}_{Core-\mathrm{N}}, \text{ for } {}^{7}_{\Lambda}\mathrm{He}(\tfrac{1}{2}^+, T=1),$$
$$V^{l}_{Core-\mathrm{N}} = \tfrac{3}{4}V^{l,s}_{Core-\mathrm{N}} + \tfrac{1}{4}V^{l,t}_{Core-\mathrm{N}}, \text{ for } {}^{7}_{\Lambda}\mathrm{Li}(\tfrac{1}{2}^+, T=0),$$
$$V^{l}_{Core-\mathrm{N}} = V^{l,t}_{Core-\mathrm{N}} \text{ for}{}^{7}_{\Lambda}, \qquad \text{for } \mathrm{Li}(\tfrac{3}{2}^+, T=0).$$

TABLE 1. Hyperon binding energy B_Λ (in MeV) of A=7 hypernuclei calculated with various $\Lambda\alpha$ models of the ${}^5_\Lambda\mathrm{He}-\mathrm{N}$ folding potential. The excitation energy E_x (in MeV) of the $(\frac{3}{2}^+)$ state of ${}^7_\Lambda\mathrm{Li}$ is also shown. In our calculation the Coulomb force is not taken into account and the excitation energy of ${}^6_\Lambda\mathrm{He}(2^-)$ was chosen equal to 0.36 MeV[7].

	$\alpha\Lambda$	${}^7_\Lambda\mathrm{He}(\frac{1}{2}^+)$	${}^7_\Lambda\mathrm{Li}(\frac{1}{2}^+, T{=}0)$	$E_x({}^7_\Lambda\mathrm{Li}(\frac{3}{2}^+, T{=}0))$
	TH	5.17	6.05	0.98
	Gibson	5.16	5.99	0.98
	MSA	5.10	5.88	0.98
	Isle	5.10	5.85	0.98
[7]	–	5.44	5.87	0.88
our correction of [7]	–	5.14	–	–
exp. [1],[3]		–	5.58	0.69

The hyperon density function $\rho(|\vec{r}|)$ of ${}^5_\Lambda\mathrm{He}$ used in equation (2) was calculated for a number of diverse phenomenological $\Lambda\alpha$ potentials (for instant see [17]).

RESULTS

The coupled configuration space Faddeev equations for a $Core\mathrm{N}+\mathrm{N}$ system was numerically solved, applying the method given in [18]. For the αNN system the following combinations of relative momenta $(\ell_{NN}, \lambda_{(\mathrm{NN})-\alpha}) = (0,0)$ and $(\ell_{\alpha\mathrm{N}}, \lambda_{(\alpha\mathrm{N})-\mathrm{N}}) = (0,0), (1,1)$, were taken into account. Using the potential model with the averaged total momentum of the pair we obtained the value -1.19 MeV for the binding energy of ^{6}He. This value is smaller than the value of -1.0 MeV from [12] where the spin-orbital interaction was taken into account. For better agreement with the experimental data we made a slight correction in dept parameter of the p-wave component of the αN potential. After this correction the calculated binding energy $E_B({}^6\mathrm{He})$ was -0.98 MeV, which coincides with the experimental value $E_B^{exp}({}^6\mathrm{He})$=-0.98 MeV [19]. For the ^{6}Li nucleus our calculation without the Coulomb interaction gives $E_B({}^6\mathrm{Li})$=-6.68 MeV, which is smaller than the Coulomb-adjusted experimental value -4.53 MeV [20, 21] by 2 MeV approximately. In Equation (1) there are two free parameters η^s and η^t to determinate the spin dependence of the effective folding ${}^5_\Lambda$He+N potential. The singlet coefficient η^s was chosen to reproduce the experimental value for the binding energy of the 1^- state of ${}^6_\Lambda$He which is located below the ${}^5_\Lambda$He+n threshold by -0.17 MeV [22]. The experimental data for the 2^- state of ${}^6_\Lambda$He does not still exist. Therefore one can use some theoretical prediction from [6, 7, 8] to adjust the triplet coefficient η^t. In initial calculation we chose an excitation energy for the 2^- state equal to 0.36 MeV as in [7] (in [6] the value 0.42 MeV was used). The hyperon binding energy B_Λ of ${}^7_\Lambda$He (${}^7_\Lambda$Li) calculated with diverse $\Lambda\alpha$ models in the ${}^5_\Lambda\mathrm{He}-\mathrm{N}$ folding potential is presented in Table 1. Note that the Coulomb force was not taken into account. The hyperon binding energy was calculated using the formula

$$B_\Lambda({}^7_\Lambda\mathrm{Z}) = E_B({}^6\mathrm{Z}) - (E_B({}^7_\Lambda\mathrm{Z}) + E_B({}^5_\Lambda\mathrm{He})). \quad (3)$$

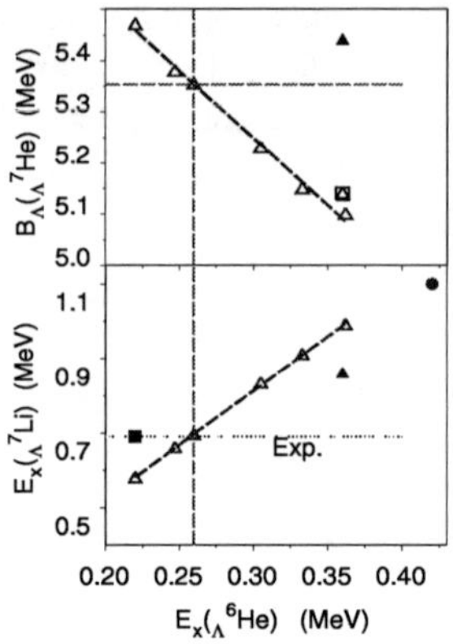

FIGURE 1. ${}^7_\Lambda\text{He}(\frac{1}{2}^-)$ hyperon binding energy B_Λ and the ${}^7_\Lambda\text{Li}(\frac{3}{2}^-,\text{T}=0)$ excitation energy E_x (open triangles) with variations of the ${}^6_\Lambda\text{He}(2^-)$ excitation energy. The Isle $\Lambda\alpha$ potential model has been used to construct the ${}^5_\Lambda\text{He}-\text{N}$ folding potential. In our calculations the Coulomb force is not taken into account. Results of the tree-body calculation from [7] and [6] are pointed by the solid triangles and circle, respectively. Our correction of $B_\Lambda({}^7_\Lambda\text{He})$ for calculation of [7] is denoted by the open "triangle in box". The solid square corresponds to the result of the Hiyama et al. four-body calculation [8].

For $E_B({}^5_\Lambda\text{He})$ the experimental value -3.12 MeV was chosen. The excitation energy of ${}^7_\Lambda\text{Li}(\frac{3}{2}^+)$ was computed as $E_x(\frac{3}{2}^+) = E_B(\frac{3}{2}^+) - E_B(\frac{1}{2}^+)$. One can see from Table 1 that some disagreement with results of [7] occurs for the ${}^7_\Lambda\text{He}$ hypernucleus while for ${}^7_\Lambda\text{Li}$ there is satisfactory agreement. The discrepancy for the ${}^7_\Lambda\text{He}$ hyperon energy does not comes from the three-body calculation. It comes from the way the hyperon energy $B_\Lambda({}^7_\Lambda\text{He})$ is extracted from the calculated tree-body energy. In [7] the value of -0.68 MeV for the ${}^6\text{He}$ binding energy, calculated in their $\alpha+\text{N}+\text{N}$ model, was used for the hyperon energy calculation with Equation (3). However, the three-body system ${}^5_\Lambda\text{He}+\text{N}+\text{N}$ does not include the αNN system as a subsystem explicitly. One has to take into account that in [7] an adjustment of the ΛN potential was also made. This adjustment is similar to our procedure for choosing the $\eta^{t(s)}$ coefficients for the ${}^5_\Lambda\text{He}$-N potential (1). But as follows from Equation (1) there exists the possibility of varying the parameters of the p-wave component of the αN potential within ${}^5_\Lambda\text{He}+\text{N}+\text{N}$, the three-body model, since any change of the p-wave parameters can be compensated by a adjustment of the $\eta^{t(s)}$ coefficients. After the correction of the $E_B({}^6\text{He})$ value from [7] by using the experimental data for $E_B({}^6\text{He})$ (-0.98 MeV) we obtain the value 5.14 MeV for $B_\Lambda({}^7_\Lambda\text{He})$ (see Table 1). This value is close to the results of our calculations with $E_x({}^6_\Lambda\text{He})$=0.36 MeV. We considered the dependence of the excitation energy of ${}^7_\Lambda\text{Li}$ on the excitation of ${}^6_\Lambda\text{He}$. To evaluate $B_\Lambda({}^7_\Lambda\text{He})$, we take into account this dependence together with the analogous one for $B_\Lambda({}^7_\Lambda\text{He})$. The illustration of this evaluation is presented in Fig. 1. The value 0.26 MeV for E_x of ${}^6_\Lambda\text{He}(2^-)$ corresponding to the observed excitation energy of ${}^7_\Lambda\text{Li}(\frac{3}{2}^+)$ (the horizontal dot-dashed line in Fig 1) was used to calculate the hyperon binding energy of ${}^{7}_{\Lambda}$He. From this figure follows that calculated value is equal to 5.35 MeV (the vertical dashed line). This is a new prediction for the hyperon energy of the ${}^{7}_{\Lambda}$He ground state obtained in the framework of a three-body model. One can see in Fig. 1 that the results of [6, 7, 8] do not contradict results obtained in our model. In presented model which does not include the Coulomb force the ${}^{7}_{\Lambda}$He and ${}^{7}_{\Lambda}$Be nuclei are identical, therefore the hyperon binding energy of ${}^{7}_{\Lambda}$Be is also equal to 5.35 MeV. The experimental value is 5.09±0.11 MeV [1]. The difference of $B_{\Lambda}^{cal.} > B_{\Lambda}^{exp.}$ can be caused by the Coulomb effect. It agrees with the calculations for the ${}^{7}_{\Lambda}$Li nucleus. The calculated value for the hyperon energy of ${}^{7}_{\Lambda}$Li is bigger than the experimental value when is clear from Table 1.

ACKNOWLEDGMENTS

This work was partially supported by the Department of Defense through grant No.DAAD 19-01-1-0795.

REFERENCES

1. M. Juric et al. *Nucl. Phys. B* **52**, 1 (1973).
2. M. May et al. *Phys. Rev. Lett.* **51**, 2085-2088 (1983).
3. H. Tamura et al. *Phys. Rev. Lett.* **84**, 5963-5966 (2004).
4. L. Tang, *Modern Phys. Lett. A* **18**, 112-114 (2003).
5. A. Gal, J. M. Soper and R. H. Dalitz, *Ann. Phys.* **63**, 53 (1971); *Ann. Phys.* **72**, 445 (1972); *Ann. Phys.* **113** 79 (1978); Richer O, Sotona M and J. Žofka, *Phys. Rev. C* **43**, 2753-2764 (1991); V. N. Fetisov, L. Majling, J. Žofka, R. A. Eramzhyan, *Z. Phys. A* **339** 399 (1991).
6. T. Motoba, H. Bando, K. Ikeda, *Prog. Teor. Phys.* **70**, 189-221 (1983).
7. E. Hiyama et al. *Phys. Rev. C* **53**, 2075-2085 (1996).
8. E. Hiyama et al. *Phys. Rev. C* **59**, 2351-2360 (1999).
9. E. Hiyama et al. *Nucl. Phys. A* **737**, 138-146 (2004).
10. L. D. Faddeev and S. P. Merkuriev *Quantum Scattering Theory for Several Particle Systems*, Kluwer Academic, Dordrecht, 1993.
11. S. Sack, L. C. Biedenharn, G. Bret, *Phys. Rev.* **93**, 321-325 (1954).
12. E. Garrido, D. V. Fedorov, A. S. Jensen, *Nucl. Phys. A* **617**, 153-75 (1997).
13. Th. A. Rijken, V. G. J. Stoks and Y. Yamamoto *Phys. Rev. C* **59**, 21-40 (1999).
14. E. Hiyama et al. *Prog. Theor. Phys.* **97**, 881-899 (1997).
15. Filikhin I N and Gal A *Phys. Rev. Lett.* **89**, 172502-1-172502-4 (2002); *Phys. Rev. C* **65**, 041001-1R 041001-4R (2002).
16. R. A. Malfliet and J. A. Tjon, *Nucl. Phys. A* **127**, 161-168 (1969); J. L. Friar et al. *Phys. Rev. C* **C42** 1838-1840 (1990).
17. V. M. Suslov, I. N. Filikhin and B. Valahovic, *J. Phys. G* **30**, 513-518 (2004).
18. J. Bernabeu et al. *Hyperfine Interaction*, **101/102**, 391-399 (1996).
19. F. Ajzenberg-Selove, *Nucl. Phys. A* **490**, 1-225 (1988); A. H. Wapstra, G. Audi and R. Hoekstra, *At. Data Nucl. Data Tables* **39**, 281-283 (988).
20. D. R. Lehman, M. Rai, A. Ghovanlou, *Phys. Rev. C* **17**, 744-754 (1979).
21. N. W. Shellingerhout, L. P. Kok, S. A. Coon, R. M. Adam, *Phys. Rev. C* **48**, 2714-2730 (1993).
22. Majling L 1994 *Proceedings of the National Conference on Physics of Few-Body and Quark-hadronic System*, edited by V. Bolyshev, V. Kotlyar and A. Shebeko, KFTI,Kharkov,Ukraine, 1992, pp. 345-348.

In-Medium modifications of the ρ^0 in light nuclei

George J. Lolos

Department of Physics, University of Regina, Regina, SK, S4S 0A2, Canada

Abstract. The results of sub-threshold production of ρ^0 via the $\gamma_t \to \pi^+\pi^-$ reaction on ^2H,^3He and ^{12}C are reported here with an emphasis on the analysis techniques, mainly the model independent identification of the ρ^0 among the competing reaction channels leading to $\pi^+\pi^-$ production. The data are consistent with the production of longitudinally polarized ρ^0 and this helicity signature has been a primary factor in the kinematical suppression of background reactions. The spectral shapes are extracted and compared to theoretical predictions.

Keywords: medium modifications, spectral shapes, ρ^0 meson
PACS: 14.40.Aq, 13.20.-v, 12.39.Fe, 12.40.Vv

INTRODUCTION

Of all particles, the ρ-meson has received the most attention with regard to medium modifications. Since the ρ^0 carries the quantum numbers of the conserved vector current, its properties are related to chiral symmetry and can be investigated with a variety of models. Most models predict a reduction of the renormalized vector meson mass in the nuclear medium. A review of the field, up to 1999, is given in Ref. [1]. One of the first models was "Brown and Rho scaling" [2] that was based on chiral symmetry and scale invariance and predicted that the mass of the ρ^0 should drop by $\sim 15\%$ from its free mass value at standard nuclear density. This is supported by Lattice QCD calculations [3], which suggest that chiral symmetry will be fully restored at $T_c \geq 150$ MeV and/or $\rho_c \geq 5\ \rho_{nuc}$. Two other well-known models are that of Rapp, Chanfray, and Wambach [4] and Saito and Thomas [5], respectively. However, the lack of experimental results makes comparisons with such theoretical predictions uncertain, since the experimental data carried a lot of uncertainties on their own.

MOTIVATION

Experimental evidence for in-medium ρ^0 mass modification has been widespread, but all interpretations suffer from significant model uncertainties. In this work, the uncertainties are minimized by isolating only kinematical signatures of the $\rho \to \pi^+\pi^-$ decay. The analysis and all the experimental information are reported in great detail in [6] and will be only briefly touched upon here.

The TAGX Collaboration has investigated the ^{2}H,^{3}He,^{12}C$(\gamma,\pi^+\pi^-)$ reaction in the sub-threshold region. In this case, diffractive ρ^0 production is suppressed, as the reaction must utilize Fermi momentum to produce a ρ^0. The resulting ρ^0's are produced with low boost with respect to the nuclear medium, so that they decay within 1 fm of their production point (i.e. still within even a small nucleus). These two properties of the sub-

CP814 *Hadron Spectroscopy: Eleventh International Conference*
edited by A. Reis, C. Göbel, J. Sá Borges, and J. Magnin

threshold region reaction enhance the nuclear-medium effect. The effects of πN final state interactions (FSI) are minimized by the choice of light nuclei, and, in addition, have been investigated through MC simulations and the effects of the cuts on them have been investigated thoroughly.

There are a number of reaction channels leading to $\pi^+\pi^-$ production, therefore, even kinematical signatures alone may not be able to isolate the two pion production coming from the $\rho^0 \to \pi^+\pi^-$ decay. The innovation of the analysis in this work has its foundation on the experimentally determined signature of helicity analysis, as reported in [7]. In this contribution, certain aspects of the analysis reported in [6] will be emphasized to address issues that have been brought forward and to address new developments in the field of medium modifications.

The $(\gamma,\pi^+\pi^-)$ experiments were carried out using the tagged photon beam of the 1.3-GeV Electron Synchrotron (ES) at the Institute of Nuclear Study at Tokyo (INS) and the TAGX magnetic spectrometer. The tagged photon energy region available to the experiment was between 600 and 1120 MeV. The data were divided into three tagged photon energy bins of 600-800 MeV, 800-960 MeV, and 960-1120 MeV. The choice of the limits of these bins was motivated by the photon energy thresholds for ρ^0 production via the quasi-free and non-quasi-free mechanisms in nuclei. An extended description of the TAGX spectrometer is found in Ref. [8]. Particle identification was provided by a combination of magnetic field, fast plastic scintillator counters and a system of tracking wire chambers. It is a well understood device with proven performance and reliable MC simulations of acceptance and detector thresholds.

The objective of the data analysis was the model-free identification of $\pi^+\pi^-$ pairs coming from ρ^0 decay, since at such low energies the latter does not exhibit the clear resonance shape above a featureless background one observes at higher beam energies. Therefore, the critical observables were *only* kinematical in nature. This removes the main objection that has been leveled at treatments where interference among the various other channels present may affect the conclusions [9]. Interference affects the amplitudes of the different channels *exclusively* and not their kinematical signatures that reflect *solely* energy-momentum and angular momentum considerations. Two key observables in this comparison are the lab frame pion-pion opening angle ($\theta_{\pi\pi}$) and the missing mass (m_{miss}).

The ρ^0 production generator was assumed to occur via a quasi-free $\gamma N_F \to \rho^0 N \to \pi^+\pi^- N$ mechanism, where N_F is the participating struck proton with initial Fermi momentum p_F and the remainder of the nucleus is a spectator. For the ρ^0 line-shape, we selected to use the line-shape of Benayoun et al. [10]. This line-shape is based on the $l = 1$ partial-wave analysis (PWA) of $e^+e^- \to \pi^+\pi^-$ data from $m_{\pi\pi} = 350$ to 1000 MeV/c^2, and so is the most authoritative description of the free ρ^0 low mass tail. The ρ^0 then proceeded to decay into a $\pi^+\pi^-$ pair. For the simulations assuming longitudinally polarized ($l = 1$, $m = 0$) ρ^0 production, the decay distribution was weighted by $\cos^2\theta_\pi$ in the ρ^0 rest frame; for unpolarized ρ^0 production the decay distribution was taken to be isotropic.

For the baryon resonances, we assumed quasi-free reactions leading to the production of $\Delta^{++}\pi^-$, $\Delta^-\pi^+$, $N^*(1520)\pi^-$, $N^*(1520) \to \Delta^{++}\pi^-$ and $N^*(1520) \to p\rho^0$. The line-shapes of the baryon resonances were assumed to be described by Breit-Wigner distributions with centroids and widths taken from Ref. [11], while the shape of the ρ^0

from $N^*(1520)$ decay uses the PWA shape from [10]. In addition, $\Delta^{++}\Delta^-$ production was simulated in a $\gamma A \rightarrow \Delta\Delta(A-2)$ mechanism, where $A-2$ was a spectator recoiling in some excited state with the appropriate Fermi momentum. As only quasi-free mechanisms have been simulated, m_{miss} cuts must be placed on both the experimental data and the MC simulated data to exclude regions populated by more complicated reaction mechanisms involving additional nucleons or those producing three or more pions.

For all processes leading to $\pi^+\pi^-$ production in the $600 \leq E_\gamma \leq 1120$ MeV energy range, the emitted pions are in the resonant energy region and pion re-scattering, as well as absorption, have large cross sections. Although FSI in the form of absorption may play a role in the extraction of the final ρ^0 invariant mass distribution, re-scattering is of particular concern because helicity signatures are used to extract the ρ^0_L content, to assign confidence levels and to define the actual background shapes and strengths to be subtracted from the data. For the sub-threshold region considered here, our simulations indicate a mean πN CM momentum of 230 MeV/c, which yields an approximate 50% probablity that one of two $\pi^\pm$ originating in the center of a uniform density nucleus with the radius of ^{12}C will undergo re-scattering. Therefore, a series of MC simulations were made in which a pion was randomly selected to scatter with a proton, $\pi + N_F \rightarrow \pi' + N'$, where the pion initial energy and angle are from the production and decay MC generator, and the nucleon has appropriate Fermi momentum for ^{12}C. The representative processes for which FSI were studied were quasi-free ρ^0 production, ρ^0 production via $N^*(1530) \rightarrow \rho^0 p$ decay, $\Delta^{++}\pi^-$ and $\Delta^{++}\Delta^-$. The πN amplitudes are taken from the phase shift analysis parameterization of [12].

Fig. 1 shows simulations, for the ^{12}C target, for three non-ρ^0 channels as well as quasi-free ρ^0 production, both with and without the effect of re-scattering FSI. The upper limit of the missing mass cut applied, $10970 < m_{miss} \leq 11305$ MeV/c^2, corresponds to 130 MeV excitation of the recoil system. This cut value was chosen to maximize the data statistics, while still remaining below 3π production.

In order to better understand the effect of FSI on the final conclusions, a question that has been raised in a number of occasions, one has to compare the two cases of ^3He and ^{12}C. The relative survival rates for the two nuclei are 34% and 19%, respectively [6]. This difference is mostly due to the effect of π re-scattering in ^{12}C, that, in combination with the two cuts and the co-planar nature of TAGX, eliminate essentially all the $\pi^+\pi^-$ events that had undergone FSI.

The ^{12}C data distributions are shown in Fig. 2. The reduction in the number of surviving events is dramatic and this is consistent with the corresponding MC distributions. There is a strong resemblance to the MC simulations of pure ρ^0_L contributions and no other $\pi^+\pi^-$ production process has been found that can account for such response to the cuts and the TAGX parameters. There is simply no combination of $\pi^+\pi^-$ producing reactions that can give such clear helicity signatures, either.

We now compare the data distributions with theoretical calculations and this is shown in Fig. 3. Ref. [13] was the only work to provide specific predictions for our experiment; their predicted result for ^{3}He was a 40 MeV reduction in the mass of the ρ^0. The mean of the 800-960 MeV data distribution is lower than the predicted $m^*_{\rho^0}$ value by approximately 20 MeV. This indicates that the reduction in the mass of the ρ^0 for this energy bin is approximately 60 MeV. As the model has no sensitivity to spectral shape

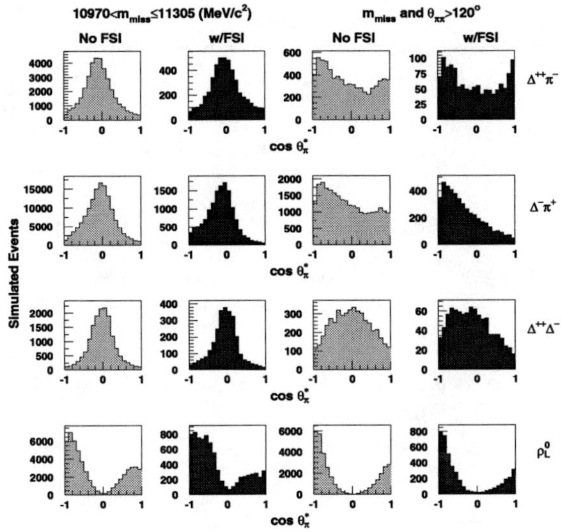

FIGURE 1. Simulated event distributions for four quasi-free reactions on ^{12}C, both with and without the effect of re-scattering FSI, as indicated. In the case with-FSI simulations, one of the two pions was randomly selected to undergo πp scattering, weighted according to the phase shift analysis parameterization of Ref. [12]. All distributions are as-detected in TAGX (i.e. detection thresholds and experimental acceptance included) and are for $800 < E_\gamma \leq 960$ MeV and $10970 < m_{miss} \leq 11305$ MeV/c^2. In addition, the opening angle cut $\theta_{\pi\pi} > 120^o$ has been applied to the two rightmost columns. The four simulations shown were each run for 10^6 simulated TAGX triggers over the tagged photon range, so the y-axis values indicate the survival probability after the various cuts.

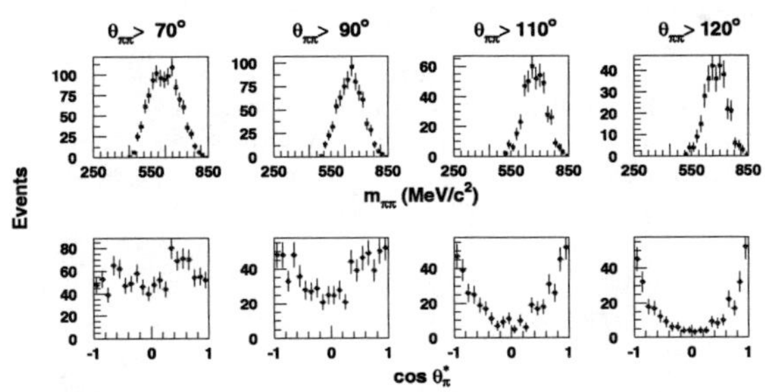

FIGURE 2. ^{12}C data distributions for $800 < E_\gamma \leq 960$ MeV and $10970 < m_{miss} \leq 11305$ MeV/c^2. In addition, a series of opening angle $\theta_{\pi\pi}$ cuts have been applied to the data, as indicated.

modifications, this agreement would imply that the ρ_L^0 line shape is largely unaffected by the nuclear medium, a fact consistent with the good agreement with our PWA-based kinematic models [6].

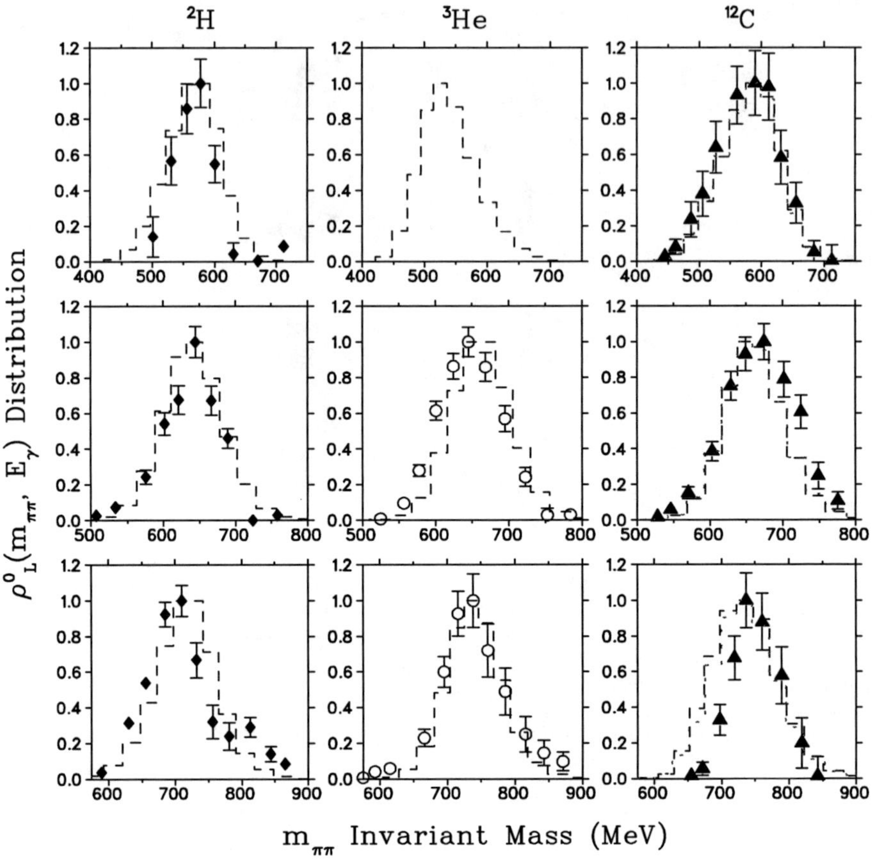

FIGURE 3. Invariant mass distributions of the ρ_L^0 strength extracted from the data [symbols] compared to quasi-free simulations [dashed lines] incorporating the STT model [13] in-medium ρ_L^0 mass. In the panels for ^{12}C, the dotted lines indicate the result when π absorption is added to the simulation. The data error bars reflect both the statistical precision of the data, as well as the uncertainty in the background subtraction.

CONCLUSIONS

Using only kinematical quantities, we have isolated the ρ^0 production in three light nuclei. The only model assumptions are the nature of production mechanisms (quasi-free) and the specific reactions leading to $\pi^+\pi^-$ production. As such, any reservations expressed based on the interference of amplitudes [9] are not even relevant in such analysis. The mass of the ρ_L^0, the only such contribution identified in the data, is approximately 700-730 MeV and its width is unmodified from that of the free ρ^0. Recent results in fact indicate medium modifications of the ω and ρ^0 mesons that are in agreement with our results [14, 15]. Our results also support several theoretical calculations that leave the width of the ρ_L^0 unmodified [4].

REFERENCES

1. G.Q. Li, Prog. Part. Nucl. Phys. **43** (1999) 619-682.
2. G.E. Brown and M. Rho, Phys. Rev. Lett. **66** (1991) 2720.
3. X. Jin and D.B. Leinweber, Phys. Rev. C **52** (1995) 3344.
4. R. Rapp, G. Chanfray and J. Wambach, Nucl. Phys. **A617** (1997) 472.
5. K. Saito, A.W. Thomas, Phys. Rev. C **51** (1995) 2757.
6. G.M. Huber et al., Phys. Rev. C **68** (2003) 065202.
7. G.J. Lolos *et al.*, Phys. Lett. **B528** (2002) 65.
8. K. Maruyama *et al.*, Nucl. Instr. and Meth. **A376** (1996) 335.
9. E. Oset, private communication.
10. M. Benayoun, *et al.*, Z. Phys. **C58** (1993) 31.
11. Particle Data Group, Eur. Phys. J. C **15** (2000) 1.
12. G. Rowe, M.Salomon, and R.H. Landau, Phys. Rev. C **18** (1978) 584.
13. K. Saito, K. Tsushima and A.W. Thomas, Phys. Rev. C **56** (1997) 566.
14. D. Trnka et al., Phys. rev. Lett. **94** (2005) 192303.
15. M. Naruki et al., arXiv:nucl-ex/0504016 v1 20 April 2005.

The Moscow-Jülich Frozen-Pellet Target[1]

M. Büscher, P. Fedorets*, V. Chernetsky, V. Chernyshev† and A. Bukharov, A. Semenov**

*Institut für Kernphysik, Forschungszentrum Jülich (FZJ), D-52425 Jülich, Germany
†Institute for Theoretical and Experimental Physics (ITEP), Cheremushkinskaya 25, 117259 Moscow, Russia
**Moscow Power and Engineering Institute (MPEI), Ul. Krasno-Kazarmennaja 13, 117218 Moscow, Russia

Abstract.
Frozen-pellet targets promise to be ideally suited for high luminosity storage-ring experiments with a 4π-detector geometry. Such a target has been set up at the FZJ by a group from ITEP and MPEI. It comprises a vibration-free cooling system based on liquid nitrogen and helium and, thus, allows one to use various liquefiable gases like H_2, D_2, N_2, Ne, Ar, Kr and Xe as target material.

During several test runs, stable production of pellets with diameters down to $\sim 20\,\mu$m has been achieved. Their spatial distributions and angular spreads have been determined with the help of fast CCD cameras and the pellet size and frequency stabilities have been measured.

Keywords: Internal target, frozen pellet
PACS: 29.25.Pj, 07.20.Mc

DESCRIPTION OF THE TARGET

Targets of frozen micro spheres ("pellets") have been proposed for high luminosity experiments ($L > 10^{32}\,\mathrm{cm}^{-2}\mathrm{s}^{-1}$) at internal accelerator beams. Pioneering work on pellet targets has been made at CELSIUS (Uppsala, Sweden) [1].

A pellet target with a different cooling concept has been developed at the *Institut für Kernphysik of the Forschungszentrum Jülich* by a group from the *Institute for Theoretical and Experimental Physics, Moscow* and the *Moscow Power and Engineering Institute*. In this device liquid nitrogen and cold helium gas are used for cooling, which provides the possibility for the liquefication and pellet production from different gases, *e.g.* H_2, D_2, N_2, Ar, Kr and Xe.

The layout of the target is shown in Fig. 1. Target consists of the target cryostat and the dumping system. There are two baths with liquid N_2 and liquid He in the upper part of the cryostat. The cooling is realized in three steps. In the first, hydrogen (for example) is flowing through the liquid nitrogen bath and cooled down to $\sim$100 K. The cold evaporated He gas is used for cooling in the second stage, the heat exchanger, where hydrogen is cooled down to 22 K. Finally, the hydrogen is liquefied in the condenser with cold He gas.

After condensation, the liquid hydrogen flows through a nozzle, mounted on the outlet

---

[1] Supported by BMBF, DFG, EC, ISTC and RFFI.

CP814 *Hadron Spectroscopy: Eleventh International Conference*
edited by A. Reis, C. Göbel, J. Sá Borges, and J. Magnin

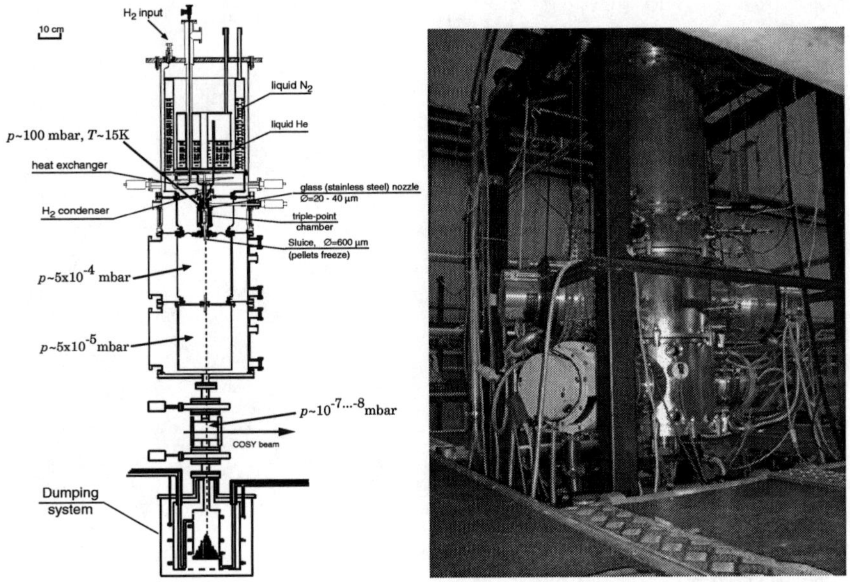

FIGURE 1. Layout of the pellet target and photo of the target cryostat in the test stand at COSY.

of the condenser, into the triple point chamber (TPC), where conditions close to the triple point for the particular target material (T_{TP}=14 K, $p_{\mathrm{TP}}\sim$100 mbar for hydrogen) are maintained.

The diameter of the liquid jet in the TPC can be as small as $\sim$10 μm, depending on the nozzle diameter. Inside the TPC the jet is broken into droplets with diameters down to $\sim$20 μm by a sonic generator with resonant frequency 3 kHz. The generator is mounted coaxially on the outer surface of the nozzle. Passing the sluice between TPC and first vacuum chamber the droplets freeze due to surface evaporation and a continuous flow of frozen pellets with diameters down to $\sim$20 μm is generated. Finally, the pellets pass through two vacuum chambers in order to decrease the pressure from 100 mbar (TPC) down to $10^{-7...-8}$ mbar (scattering chamber). In the last step, pellets enter the dumping system, where they are collected and evaporated and pumped away.

The diagnostic system is based on CCD cameras, which allows one to measure the velocities and sizes of individual droplets and pellets. Figure 2 displays photos in the TPC: a jet of liquid H_2 in the upper part of the chamber which splits into droplets in the bottom of the chamber just before entering the sluice to the first vacuum chamber. For these tests a steel nozzle with a diameter of 38 μm has been used, corresponding to a droplet diameter of 70 μm.

Figure 3 shows frozen N_2 and H_2 pellets at the sluice outlet. Due to the high pellet velocity ($\sim$70–80 m/s) and the sensitive time of the CCD ($\sim$5 μs) the pellets are seen as short tracks with lengths proportional to their velocities. Here the pellet diameters are 30 μm, which corresponds to a nozzle diameter of 18 μm. Systematic tests showed that the size and frequency stability of the pellets is not worse than 10%.

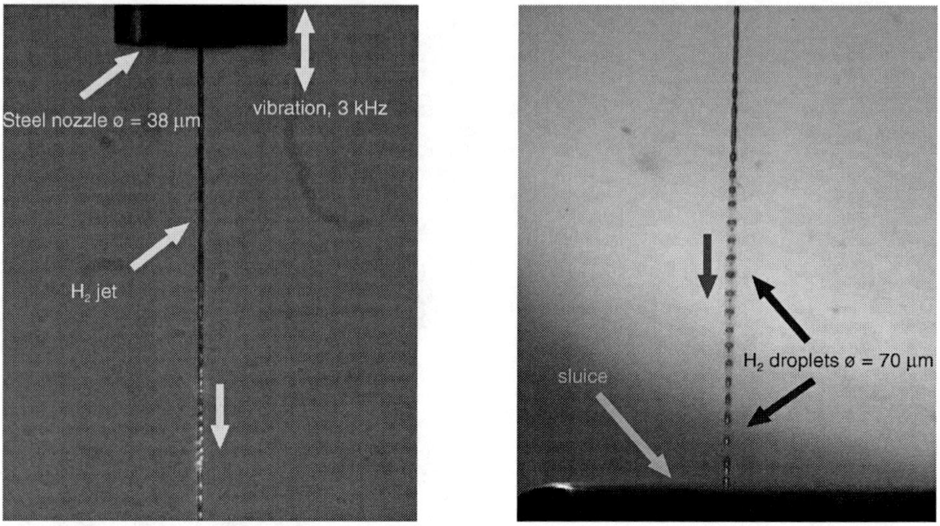

FIGURE 2. Photos in the TPC: jet of liquid H_2 at the top of the chamber and splitting of the H_2 jet into droplets in the bottom.

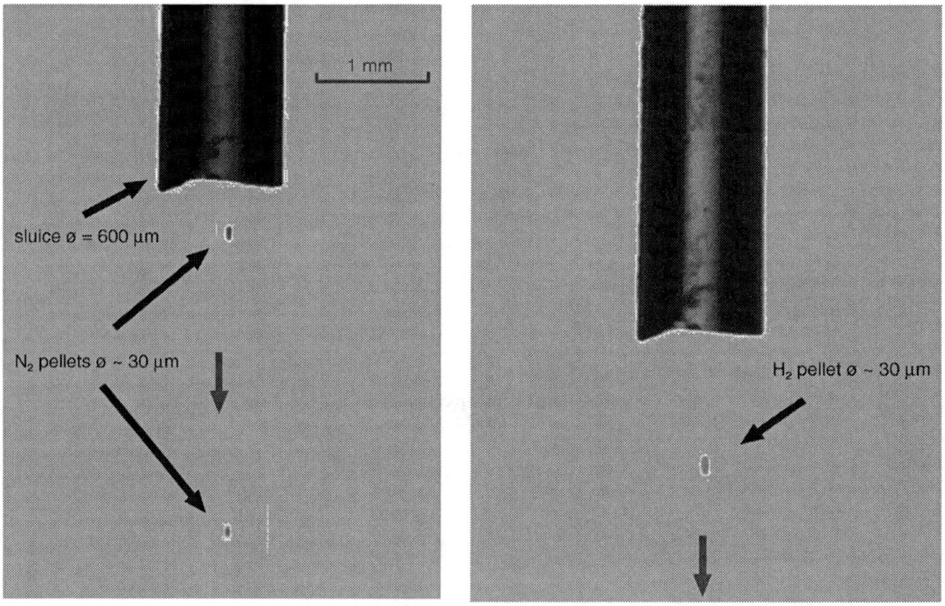

FIGURE 3. N_2 and H_2 frozen pellets at the sluice outlet in the first vacuum chamber.

FIRST TEST RESULTS

Two different nozzle types — stainless steel and glass nozzles glued into a brass housing — have been used in the target. During 2004 stable pellet production with different nozzle diameters has been achieved. The results are summarized in Table 1.

TABLE 1. Nozzle parameters with stable pellet production in the first vacuum chamber.

Nozzle type	Nozzle diameter	Target material	Pellet diameter
Steel	25 μm	H_2	50 μm
	38 μm	H_2	70 μm
	38 μm	N_2	70 μm
Glass	10 μm	H_2	$< 20\mu$m
	18 μm	H_2	30 μm
	18 μm	N_2	30 μm
	40 μm	H_2	70 μm

During a test in the spring 2005, after careful optimization of the nozzle/sluice adjustment, frozen hydrogen pellets (nozzle 15 μm, pellets 25 μm) have been observed in the scattering chamber, $\sim$ 120 cm downstream from the TPC. In Fig. 4 we show the measured velocity and spacial distributions of the pellets.

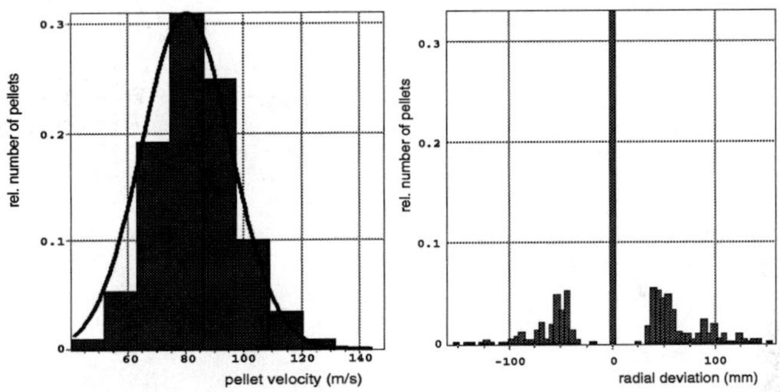

FIGURE 4. Velocity and spatial distributions of H_2 pellets in the scattering chamber.

The mean pellet velocity is about 80 m/s, thus significantly larger than the droplet velocity in the TPC. This is due to a strong gas flow from the TPC into the vacuum chamber which accelerates the pellets on their way through the sluice. In the scattering chamber the spatial distribution of the pellet beam exhibits a narrow central part (FWHM$<$500 μm) and two satellites at distances of about 4 cm. The latter are pellets which undergo scattering on the walls of the sluice to the vacuum chamber. If narrow beam profiles are required, these pellets can be removed by scrapers, *e.g.* between the first and second vacuum chamber, allowing about 6–10% of the initial droplets to reach the scattering chamber. We expect that this number can be increased to $\sim$30%.

REFERENCES

1. B. Trostell *et al.*, Nucl. Instr. and Meth. A **362**, 41 (1995); C. Ekström *et al.*, Nucl. Instr. and Meth. A **371**, 572 (1996)

RHIC

Improved Langevin Approach to Spinodal Decomposition in the Chiral Transition

Eduardo S. Fraga*, Gastão Krein† and Rudnei O. Ramos**

*Instituto de Física, Universidade Federal do Rio de Janeiro
C.P. 68528, 21941-972 Rio de Janeiro, RJ , Brazil
†Instituto de Física Teórica, Universidade Estadual Paulista
Rua Pamplona 145, 01405-900 São Paulo, SP, Brazil
**Departamento de Física Teórica, Universidade do Estado do Rio de Janeiro,
20550-013 Rio de Janeiro, RJ Brazil

Abstract. We use an improved Langevin description that incorporates both additive and multiplicative noise terms to study the dynamics of phase ordering. We perform real-time lattice simulations to investigate the role played by different contributions to the dissipation and noise. Lattice-size independence is assured by the use of appropriate lattice counterterms.

Keywords: QCD phase transition; Heavy-ion collisions; Dissipation in field theory
PACS: 25.75.-q, 11.30.Rd , 05.70.-a, 64.90.+b

The dynamics of spinodal decomposition after a phase transition in field theory plays a major role in physics, for instance in the nonequilibrium time evolution of order parameters [1]. The influence of the presence of a hot and dense medium on the dynamics of particles and fields is encoded "macroscopically" in attributes such as dissipation and noise, and can dramatically affect the relevant time scales for different stages of phase conversion. In particular, in high-energy heavy ion collisions, where one presumably forms a hot, dense, strongly interacting quark-gluon plasma [2], chiral fields evolve under extreme conditions of temperature and density during the QCD phase transition. To have a clear understanding of data coming from BNL-RHIC, and especially of data that will be produced at CERN-LHC, one needs a realistic description of the hierarchy of scales associated with dissipation, noise, radiation, expansion and finite size of the system.

Recently, some of us have considered the effects of dissipation in the scenario of explosive spinodal decomposition for hadron production [3, 4, 5, 6, 7] during the QCD transition after a high-energy heavy ion collision in the simplest fashion [8]. Using a phenomenological Langevin description for the time evolution of the order parameter in a chiral effective model [9], inspired by microscopic nonequilibrium field theory results [10, 11, 12], we performed real-time lattice simulations for the behavior of the inhomogeneous chiral fields. It was shown that the effects of dissipation could be dramatic in spite of very conservative assumptions: even if the system quickly reaches the unstable region there is still no guarantee that it will explode. The framework for the dynamics was assumed to be given by the following Langevin equation:

$$\frac{\partial^2 \phi}{\partial t^2} - \nabla^2 \phi + \eta \frac{\partial \phi}{\partial t} + \frac{dV_{\rm eff}(\phi)}{d\phi} = \xi(\vec{x},t)\,, \tag{1}$$

CP814 *Hadron Spectroscopy: Eleventh International Conference*
edited by A. Reis, C. Göbel, J. Sá Borges, and J. Magnin

where ϕ is a real scalar field, $V_{\rm eff}(\phi)$ is a Landau-Ginzburg effective potential and η, which can be seen as a response coefficient that defines time scales for the system and encodes the intensity of dissipation, is usually taken to be a function of temperature only, $\eta = \eta(T)$. The function $\xi(\vec{x},t)$ represents a stochastic (noise) force, assumed Gaussian and white, so that $\langle \xi(\vec{x},t) \rangle = 0$ and $\langle \xi(\vec{x},t)\xi(\vec{x}',t') \rangle = 2\eta T \delta(\vec{x}-\vec{x}')\delta(t-t')$, according to the fluctuation-dissipation theorem.

Although phenomenological equations of the form of Eq. (1) are widely used in the literature, formal derivations of effective equations of motion based on a more complete field-theoretic description of nonequilibrium dissipative dynamics [10, 11, 12, 13] show that much more complicated equations emerge. In general they are nonlocal equations of motion, with colored, non-Markovian dissipative and noise kernels, exhibiting both noise and dissipation terms that depend on the field amplitude. It is then important to analyze the possible outcome of dynamics emerging from these more realistic equations and compare them to the simple phenomenological equation (1). In fact, on very general physical grounds, one expects that dissipation effects should depend on the local density $\sim \phi^2 \dot{\phi}$ and, accordingly, the noise term should contain a multiplicative piece $\sim \phi$. These results emerge as appropriate approximations to the effective equations of motion derived *e.g.* in Refs. [10, 11, 13].

Motivated by previous results for effective equations of motion obtained from microscopic derivations [10, 11, 13], we consider an improved Langevin framework which includes the effects of multiplicative noise and density-dependent dissipation terms in the description of the time evolution of the order parameter, which should be closer to a more realistic physical realization than the one described by Eq. (1). To assess the effect of the new contributions on the relevant time scales for phase ordering, we consider a scalar $\lambda\phi^4$ theory in the broken phase. It was shown recently that the effects of multiplicative noise are rather significant in the Kibble-Zurek scenario of defect formation in one spatial dimension [14]. We perform $(3+1)$-dimensional real-time lattice simulations to study the behavior of the inhomogeneous scalar field, taking into account lattice counter-terms that guarantee lattice-size independence [15, 16, 17].

To model the mechanism of phase ordering we adopt a $\lambda\phi^4$ theory in the broken phase, given by the standard Lagrangian density

$$\mathscr{L} = \frac{1}{2}(\partial_\mu \phi)(\partial^\mu \phi) + \frac{1}{2} m^2 \phi^2 - \lambda \phi^4 \,, \tag{2}$$

where ϕ is a scalar field playing the role of the order parameter and $\lambda > 0$. In what follows we assume that the system was quenched from a high-temperature environment in which the sign of the mass term was the opposite, corresponding to a symmetric phase, down to a temperature below some critical temperature of symmetry restoration ($T < T_c$), in which the effective potential has the form shown in the Lagrangian density above.

In our analysis, the time evolution of the field $\phi(\vec{x},t)$ at each point in space and its approach to equilibrium will be dictated by an improved Langevin equation of the form

$$\left(\frac{\partial^2}{\partial t^2} - \nabla^2\right)\phi(\vec{x},t) + [\eta_1 + \eta_2 \phi^2(\vec{x},t)]\frac{\partial \phi(\vec{x},t)}{\partial t} + V'(\phi) = \xi_1(\vec{x},t) + \phi(\vec{x},t)\xi_2(\vec{x},t) \,, \tag{3}$$

which models the local approximation for the effective equation of motion derived e.g. in Ref. [10]. In Eq. (3), η_1 and η_2 will be taken to be functions of temperature only, $\eta_i = \eta_i(T)$. The functions $\xi_1(\vec{x},t)$ and $\xi_2(\vec{x},t)$ represent stochastic (noise) forces, assumed Gaussian and white.

Eq. (3) could, in principle, be obtained from a microscopic field-theoretic description of the real-time nonequilibrium dynamics of the chiral field at finite temperature [10, 12]. The noise and dissipation terms, which originate from quantum fluctuations, are engendered by either self-interactions of the chiral field or coupling to one or more different fields that play the role of a heat bath, provided one incorporates higher order terms in the computation of the effective equation of motion for $\phi(\vec{x},t)$. In fact, it is well-known that one has to go up to two-loop corrections in order to pick up imaginary parts in the self-energy associated with viscosity and dissipation [10, 11, 12, 18]. Self-interactions of the ϕ field, as well as possible interactions with other fields in the medium, justify the inclusion of dissipation and noise terms such as done in Eq. (3) in the framework adopted here.

In our numerical simulations we solve Eq. (3) on a cubic space-like lattice. We use the leap frog method (see *e.g.* Refs. [15, 16]) for the time evolution and treat the Laplacian using a Fast Fourier Transform for the spatial dependence [19]. To minimize lattice artifacts, we use counterterms calculated in lattice perturbation in the classical theory (tadpole plus setting-sun diagrams) [20]:

$$V_{CT} = \frac{1}{2}\left\{-3\times 0.252731\frac{T}{a} + \frac{6T^2}{(4\pi)^2}\left[\log\left(\frac{6}{\mu a}\right) + 0.009\right]\right\}\phi^2, \tag{4}$$

where T is the temperature, a is the lattice spacing and μ is an arbitrary renormalization scale. We note that the addition of this counterterm is supposed to guarantee independence of the results on a at *equilibrium* only. Initially we check the lattice independence of the results. We use units such that $m = 1$ in the Lagrangian of Eq. (2), *i.e.* all dimensional quantities are expressed in units of m. In Fig. 1, we present the volume average of ϕ as a function of time, for (a) additive noise only, and (b) for additive plus multiplicative noises, for three different values of lattice spacings, $a = 1/2, 1/3$ and $1/4$, and a time step of $\Delta t = 10^{-3}$. In these figures we fix the lattice extent to $L = 16$, and use $T = 1$, $\lambda = 1$, and $\mu = 1$. The initial configuration is taken to be a random distribution of the form $\phi(x,t) = \phi_{sp} + 0.001 * rand(seed)$, where $\phi_{sp} = 1/\sqrt{3}$ is spinodal point of the bare double-well potential, and $rand(seed)$ is a uniform random number distribution in the interval $[-1,1]$. We average the results over several runs for different realizations of the noises. Clearly, as seen in the figure, the long-time (equilibrium) value of the $\langle\phi\rangle$ is fairly lattice independent. We have used other sets of parameters and obtained the same lattice independence of the equilibrium values of $\langle\phi\rangle$.

In Figure 2 we present preliminary results for different values of the multiplicative dissipation parameter η_2. As η_2 increases, the effect is of delaying the equilibration of the order parameter, as expected. This is, of course, in the same direction as with purely additive noise [8]. In Fig. 3 we investigate the effect of increasing the temperature. Of course, the higher the temperature, the larger the thermal fluctuations. Clearly seen in the figure is the restoration of the symmetry for temperatures between $T = 3$ and $T = 5$. Here we should note that η_1 and η_2 are functions of T and so, increasing the

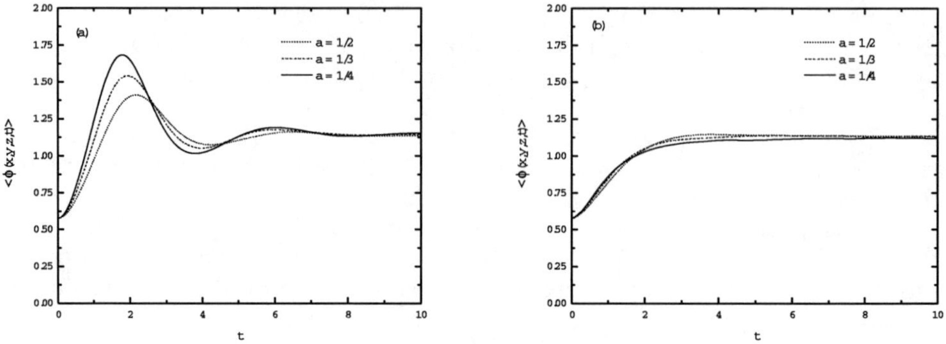

FIGURE 1. Volume average of $\phi(x,y,z,t)$ as a function of time for different lattice spacings, for temperature $T = 1$ and counterterms scale $\mu = 1$. (a) Additive noise only, $\eta_1 = 1, \eta_2 = 0$, (b) Additive plus multiplicative noises, $\eta_1 = \eta_2 = 1$. Dimensional quantities are in units of m.

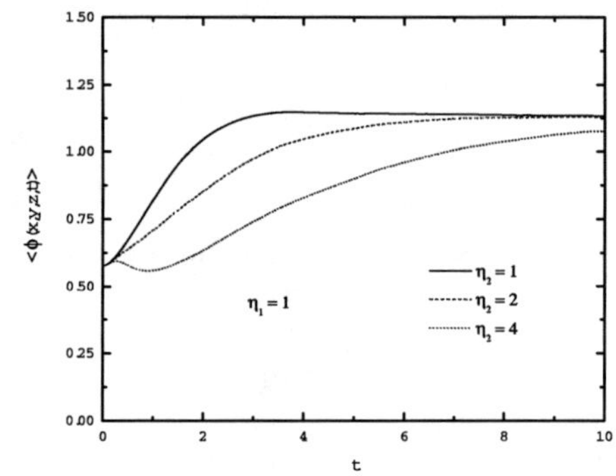

FIGURE 2. Volume average of $\phi(x,y,z,t)$ as a function of time for three different values of η_2, for $\eta_1 = 1$, temperature $T = 1$, and counterterms scale $\mu = 1$. Dimensional quantities are in units of m.

temperature and keeping these quantities fixed is not entirely consistent. Physically motivated parameterizations of $\eta_1(T)$ and $\eta_2(T)$ and a detailed study of the late-time behavior of the order parameter will be presented elsewhere [21].

ACKNOWLEDGMENTS

This work was partially supported by CAPES, CNPq, FAPERJ, FAPESP and FUJB/UFRJ.

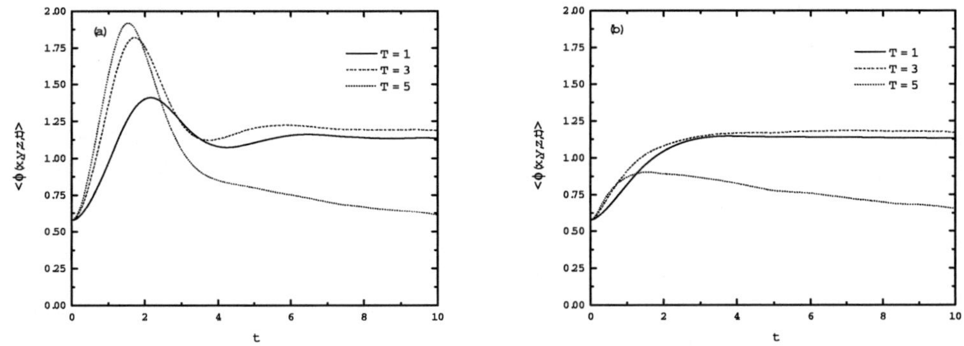

FIGURE 3. Volume average of $\phi(x,y,z,t)$ as a function of time for different lattice spacings, for three different temperatures, and counterterms scale $\mu = 1$. (a) Additive noise only, $\eta_1 = 1, \eta_2 = 0$, (b) Additive plus multiplicative noises, $\eta_1 = \eta_2 = 1$. Dimensional quantities are in units of m.

REFERENCES

1. J. D. Gunton, M. San Miguel and P. S. Sahni, in *Phase Transitions and Critical Phenomena* (Eds.: C. Domb and J. L. Lebowitz, Academic Press, London, 1983), vol. 8.
2. Proceedings of Quark Matter 2004, J. Phys. G **30** S633-S1425 (2004).
3. J. Rafelski and J. Letessier, Phys. Rev. Lett. **85**, 4695 (2000).
4. A. Dumitru and R. D. Pisarski, Nucl. Phys. A **698**, 444 (2002). O. Scavenius, A. Dumitru and A. D. Jackson, Phys. Rev. Lett. **87**, 182302 (2001).
5. A. Dumitru and R. D. Pisarski, Phys. Lett. B **504**, 282 (2001).
6. J. Randrup, Phys. Rev. Lett. **92**, 122301 (2004); arXiv:nucl-th/0406031.
7. E. S. Fraga and R. Venugopalan, Physica A **345**, 121 (2004).
8. E. S. Fraga and G. Krein, Phys. Lett. B **614**, 181 (2005).
9. O. Scavenius, A. Dumitru, E. S. Fraga, J. T. Lenaghan and A. D. Jackson, Phys. Rev. D **63**, 116003 (2001).
10. M. Gleiser and R. O. Ramos, Phys. Rev. D **50**, 2441 (1994).
11. A. Berera, M. Gleiser and R. O. Ramos, Phys. Rev. D**58**, 123508 (1998).
12. D. H. Rischke, Phys. Rev. C **58**, 2331 (1998).
13. R. O. Ramos and F. A. R. Navarro, Phys. Rev. D **62**, 085016 (2000).
14. N. D. Antunes, P. Gandra and R. J. Rivers, Phys. Rev. D **71**, 105006 (2005).
15. J. Borrill and M. Gleiser, Nucl. Phys. B **483**, 416 (1997).
16. C. J. Gagne and M. Gleiser, Phys. Rev. E **61**, 3483 (2000).
17. L. M. A. Bettencourt, K. Rajagopal and J. V. Steele, Nucl. Phys. A **693**, 825 (2001).
18. A. Mocsy, Phys. Rev. D **66**, 056010 (2002).
19. M. I. M. Copetti and C. M. Elliot, Mat. Sci. Tecn. **6**, 273 (1990).
20. K. Farakos, K. Kajantie, K. Rummukainen and M. E. Shaposhnikov, Nucl. Phys. B **425**, 67 (1994).
21. E. S. Fraga, G. Krein and R. O. Ramos, work in progress.

SCALARS

Effects of Mixing Between Low and High Mass Scalar Mesons for Radiative Decays Involving $f_0(980)$ and $a_0(980)$

T. Teshima, I. Kitamura and N. Morisita

Department of Applied Physics, Chubu University, Kasugai 487-8501, Japan

Abstract. Assuming $f_0(980)$ and $a_0(980)$ are mainly $qq\bar{q}\bar{q}$ low mass scalar mesons and mixed with $q\bar{q}$ high mass scalar mesons, we analyzed the mixing effects for $\phi \to f_0(980)\gamma$, $\phi \to a_0(980)\gamma$, $f_0(980) \to \gamma\gamma$ and $a_0(980) \to \gamma\gamma$. Used the vector meson dominance model (VDM), these decay amplitudes are expressed by coupling parameters B describing the $S(qq\bar{q}\bar{q}$ scalar meson)-V(vector meson)-V(vector meson) coupling and B' describing the $S'(q\bar{q}$ scalar meson)-V-V coupling. If the magnitudes for B and B' are $\sim 2.8\text{GeV}^{-1}$ and $\sim 12\text{GeV}^{-1}$, respectively, the mixing angle between $a_0(980)$ and $a_0(1450)$ are $\sim 9°$, and the mixing parameter λ_{01} causing the mixing between $I = 0$ $qq\bar{q}\bar{q}$ state and $q\bar{q}$ state is $\sim 0.24\text{GeV}^2$, the experimental data for these radiative decays are interpreted consistently.

Keywords: $f_0(980)$ and $a_0(980)$, radiative decays, scalar meson mixing
PACS: 12.39.Mk; 12.40.Vv; 12.40.Yx

INTRODUCTION

From the recent experimental and theoretical analyses for the $f_0(600)$ and $\kappa(900)$, these scalar mesons are considered as the light scalar nonet together with the $f_0(980)$ and $a_0(980)$, on the other hand high mass scalar mesons $f_0(1370)$, $a_0(1450)$, $K_0(1430)$, $f_0(1500)$, $f_0(1710)$ are considered as $L = 1q\bar{q}$ nonet + glueball. These low and high mass scalar mesons are considered to mix with appreciable mixing angle because of the fact that high mass scalar mesons are very high compared to the masses supposed from the $L = 1q\bar{q}$ 1^{++} and 2^{++} mesons mass relation $m^2(2^{++}) - m^2(1^{++}) = 2(m^2(1^{++}) - m^2(0^{++})$ [1, 2].

In previous work [3], we analyzed this mixing problem in the decay processes of low and high mass scalar mesons decaying to two pseudoscalar mesons, and obtained the result that mixing angle between $I = 0$ $a_0(980)$, $a_0(1450)$ and $I = 1/2$ $\kappa(900)$, $K_0^*(1430)$ are $\approx 10°$. In this work, we analyze the radiative decay processes, $\phi(1020) \to a_0(980)\gamma$, $\phi(1020) \to f_0(980)\gamma$, $a_0(980) \to \gamma\gamma$ and $f_0(980) \to \gamma\gamma$ assuming that the light scalar mesons have the $qq\bar{q}\bar{q}$ and $q\bar{q}$ component, and reveal the mixing ratio.

MIXING BETWEEN LOW AND HIGH MASS SCALAR MESONS

The $qq\bar{q}\bar{q}$ scalar $SU(3)$ nonet S^a_b are represented by the quark triplet as $S^a_b = \varepsilon^{acd} q_c q_d \varepsilon_{bef} \bar{q}^e \bar{q}^f$ and have the following flavor configuration:

CP814 *Hadron Spectroscopy: Eleventh International Conference*
edited by A. Reis, C. Göbel, J. Sá Borges, and J. Magnin

$$
\begin{array}{ccc}
\bar{s}\bar{d}us,\ \frac{1}{\sqrt{2}}(\bar{s}\bar{d}ds-\bar{s}\bar{u}us),\ \bar{s}\bar{u}ds & \Longleftrightarrow & a_0^+,\ a_0^0,\ a_0^- \\
\bar{s}\bar{d}ud,\ \bar{s}\bar{u}ud,\ \bar{u}\bar{d}us,\ \bar{u}\bar{d}ds & \Longleftrightarrow & \kappa^+,\ \kappa^0,\ \overline{\kappa^0},\ \kappa^- \\
\frac{1}{\sqrt{2}}(\bar{s}\bar{d}ds+\bar{s}\bar{u}us) & \Longleftrightarrow & f_{NS}\sim f_0(980) \\
\bar{u}\bar{d}ud & \Longleftrightarrow & f_{NN}\sim f_0(600)
\end{array}
$$

The high mass scalar mesons S'^a_b are ordinary $SU(3)$ nonet $S'^a_b=\bar{q}^a q_b$. The mixing between $qq\bar{q}\bar{q}$ and $q\bar{q}$ states is caused by the OZI rule allowed diagram then this mixing is assumed to be large. The transition interaction is represented as

$$
\begin{aligned}
L_{\text{int}} &= -\lambda_{01}\varepsilon^{abc}\varepsilon_{def}S^d_a S'^e_b \delta^f_c = \lambda_{01}\Big[a_0^+ a'^-_0 + a_0^- a'^+_0 + a_0^0 a'^0_0 + \kappa^+ K_0^{*-} \\
&\quad + \kappa^- K_0^{*+} + \kappa^0 K_0^{*0} + \bar{\kappa}^0 \bar{K}_0^{*0} - \sqrt{2} f_{NN} f'_N - f_{NS} f'_N - \sqrt{2} f_{NS} f'_S\Big].
\end{aligned}
$$

Inter-mixing between $I=1,1/2$ low and high mass scalar mesons

$I=1$ $a_0(980)$ and $a_0(1450)$ state mixing and the mixing matrix are represented as follows;

$$
\left\{\begin{array}{l}
a_0(980)=\cos\theta_a\overline{a_0(980)}-\sin\theta_a\overline{a_0(1450)} \\
a_0(1450)=\sin\theta_a\overline{a_0(980)}+\cos\theta_a\overline{a_0(1450)}
\end{array}\right., \quad
\begin{pmatrix}
m^2_{\overline{a_0(980)}} & \lambda^a_{01} \\
\lambda^a_{01} & m^2_{\overline{a_0(1450)}}
\end{pmatrix},
$$

where $\overline{a_0(980)}$ and $\overline{a_0(1450)}$ are the before mixing states. Similarly, $I=1/2$ $\kappa(900)$ and $K_0^*(1430)$ state mixing and the mixing matrix are represented as follows;

$$
\left\{\begin{array}{l}
\kappa(900)=\cos\theta_K\overline{\kappa(900)}-\sin\theta_K\overline{K_0^*(1430)} \\
K_0^*(1430)=\sin\theta_K\overline{\kappa(900)}+\cos\theta_K\overline{K_0^*(1430)}
\end{array}\right., \quad
\begin{pmatrix}
m^2_{\overline{\kappa(900)}} & \lambda^K_{01} \\
\lambda^K_{01} & m^2_{\overline{K_0^*(1430)}}
\end{pmatrix},
$$

where $\overline{\kappa(900)}$ and $\overline{K_0^*(1430)}$ are the before mixing states. In our previous work, we estimated the mixing angles analyzing the processes decaying to two pseudoscalar mesons and get the results;

$$
\begin{aligned}
&\theta_a=\theta_K=(9\pm4)^\circ,\quad \lambda^a_{01}=\lambda^K_{01}=0.19^{+0.07}_{-0.09}\text{GeV}^2, \\
&m_{\overline{a_0(980)}}=1.00^{+0.02}_{-0.01}\text{GeV},\quad m_{\overline{a_0(1450)}}=1.46\pm0.01\text{GeV}, \qquad (1) \\
&m_{\overline{K_0(900)}}=0.92\pm0.01\text{GeV},\quad m_{\overline{K_0^*(1430)}}=1.40\pm0.01\text{GeV}.
\end{aligned}
$$

Inter-mixing between $I=0$ low and high mass scalar mesons

The mass matrix for intra- and inter-mixing for $I=0$ low mass and high mass scalar mesons and glueball is expressed as

$$
\begin{array}{c} \\ f_{NN} \\ f_{NS} \\ f_{N'} \\ f_{S'} \\ f_G \end{array}
\begin{array}{c}
\begin{array}{ccccc} f_{NN} & f_{NS} & f_{N'} & f_{S'} & f_G \end{array} \\
\begin{pmatrix}
m_{NN}^2+\lambda_0 & \sqrt{2}\lambda_0 & \sqrt{2}\lambda_{01} & 0 & 0 \\
\sqrt{2}\lambda_0 & m_{NS}^2+2\lambda_0 & \lambda_{01} & \sqrt{2}\lambda_{01} & 0 \\
\sqrt{2}\lambda_{01} & \lambda_{01} & m_{N'}^2+2\lambda_1 & \sqrt{2}\lambda_1 & \sqrt{2}\lambda_G \\
0 & \sqrt{2}\lambda_{01} & \sqrt{2}\lambda_1 & m_{S'}^2+\lambda_1 & \lambda_G \\
0 & 0 & \sqrt{2}\lambda_G & \lambda_G & \lambda_{GG}
\end{pmatrix}
\end{array},
$$

where left-up 2×2 block represents the intra-mixing mass matrix for $I = 0$ low mass scalar mesons and right-down 3×3 block represents the intra-mixing mass matrix for $I = 0$ high mass scalar meson and glueball.

Using the result (1) and the relation $m_{NN}^2 = 2m_{\overline{K_0}}^2 - m_{\overline{a_0}}^2$, $m_{NS}^2 = m_{\overline{a_0}}^2$ and $m_{S'}^2 = 2m_{\overline{K_0'}}^2 - m_{\overline{a_0'}}^2$, $m_{N'}^2 = m_{\overline{a_0'}}^2$, we can obtain the mixing parameters for typical values of coupling parameters; $\lambda_0 = -0.03\text{GeV}^2, \lambda_1 = 0.04\text{GeV}^2, \lambda_G = 0.1\text{GeV}^2, \lambda_{GG} = 1.7^2\text{GeV}^2$,

$$
\begin{aligned}
&{}^t(f_0(600)\ \ f_0(980)\ \ f_0(1370)\ \ f_0(1500)\ \ f_0(1710)) \\
&\qquad = [R_{f_0(M)I}]\,{}^t(f_{NN}\ \ f_{NS}\ \ f_{N'}\ \ f_{S'}\ \ f_G),
\end{aligned} \tag{2}
$$

$$
[R_{f_0(M)I}] = \begin{pmatrix}
0.949 & 0.250 & -0.185 & -0.047 & 0.013 \\
0.270 & -0.925 & 0.067 & 0.257 & -0.017 \\
0.054 & -0.233 & 0.210 & -0.946 & 0.064 \\
0.150 & 0.161 & 0.932 & 0.160 & -0.239 \\
0.025 & 0.035 & 0.220 & 0.107 & 0.969
\end{pmatrix}.
$$

RADIATIVE DECAYS INVOLVING $f_0(980)$ AND $a_0(980)$

The decays $\phi \to f_0\gamma$, $\phi \to a_0\gamma$ have been analyzed for a long time by many authors assuming that the decays $\phi \to f_0(a_0)\gamma$ proceed through the K loop as $\phi \to K^+K^- \to f_0(a_0)\gamma$ [4]. The two γ decays of $f_0(980)$ and $a_0(980)$ are analyzed by various models, e. g., the linear sigma model, vector mesons resonance model and the quark-hadron duality model [5].

We assume the vector dominance model(VDM) and use the following interaction for SVV, $S'VV$ and GVV coupling;

$$
L_I = B\varepsilon^{abc}\varepsilon_{def}S_a^d V^{\mu\nu e}{}_b V_{\mu\nu}{}^f_c + B'S_a'^b\{V^{\mu\nu c}{}_b,\ V_{\mu\nu}{}^a_c\} + B''G\{V^{\mu\nu b}{}_a,\ V_{\mu\nu}{}^a_b\}. \tag{3}
$$

These interactions are represented graphically by the diagrams shown in Fig. 1.

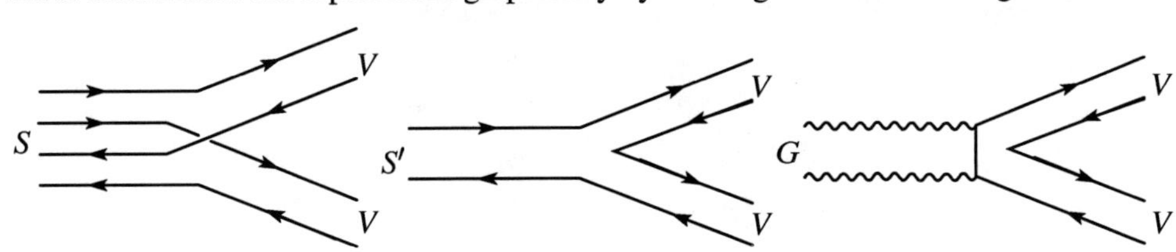

Fig.1. SVV, $S'VV$ and GVV coupling

We define the coupling constants g_{SVV} in the following expression,

$$\begin{aligned} L_I &= g_{a_0(980)\rho\omega}a_0\cdot\rho_{\mu\nu}\omega^{\mu\nu} + g_{a_0(980)\rho\phi}a_0\cdot\rho_{\mu\nu}\phi^{\mu\nu} + g_{f_0(M)\rho\rho}f_0(M)\rho_{\mu\nu}\cdot\rho^{\mu\nu} \\ &+ g_{f_0(M)\omega\omega}f_0(M)\omega_{\mu\nu}\omega^{\mu\nu} + g_{f_0(M)\omega\phi}f_0(M)\omega_{\mu\nu}\phi^{\mu\nu} + g_{f_0(M)\phi\phi}f_0(M)\phi_{\mu\nu}\phi^{\mu\nu}. \end{aligned}$$

Then the g_{SVV} are expressed by the definition (3) as

$$\begin{aligned} g_{a_0(980)\rho\phi} &= -2B\cos\theta_a,\ g_{a_0(980)\rho\omega} = -2\sqrt{2}B'\sin\theta_a, \\ g_{f_0(980)\phi\omega} &= 2BR_{f_0(980)NS},\ g_{f_0(980)\phi\phi} = 2B'R_{f_0(980)S'} + 2B''R_{f_0(980)G}, \\ g_{f_0(980)\rho\rho} &= -BR_{f_0(980)NN} + \sqrt{2}B'R_{f_0(980)N'} + 2B''R_{f_0(980)G}, \\ g_{f_0(980)\omega\omega} &= BR_{f_0(980)NN} + \sqrt{2}B'R_{f_0(980)N'} + 2B''R_{f_0(980)G}. \end{aligned}$$

Decay widths for radiative decays are expressed as

$$\begin{aligned} \Gamma(\phi\to a_0(980)\gamma) &= \frac{4\alpha}{3\gamma_\rho^2}(g_{a_0\rho\phi})^2|k_\gamma|^3, \\ \Gamma(\phi\to f_0(980)\gamma) &= \frac{4\alpha}{3\gamma_\rho^2}\left(\frac{g_{f_0\phi\omega}}{3} - \frac{\sqrt{2}g_{f_0\phi\phi}}{3}\right)^2|k_\gamma|^3, \\ \Gamma(a_0(980)\to\gamma\gamma) &= \frac{\pi\alpha^2}{\gamma_\rho^2}m_{a_0}^3\left(-\frac{\sqrt{2}}{3}g_{a_0\rho\phi} + \frac{1}{3}g_{a_0\rho\omega}\right)^2, \\ \Gamma(f_0(980)\to\gamma\gamma) &= \frac{\pi\alpha^2}{\gamma_\rho^2}m_{f_0}^3\left(-\frac{\sqrt{2}}{9}g_{f_0\phi\omega} + g_{f_0\rho\rho} + \frac{1}{9}g_{f_0\omega\omega} + \frac{2}{9}g_{f_0\phi\phi}\right)^2, \end{aligned}$$

where γ_ρ is the $\gamma-\rho$ coupling constant and $\frac{1}{\gamma_\rho} = 0.201\pm0.002$, and k_γ is the photon momentum in the V rest frame. We quote the experimental data results from PDG(2004) [6],

$$\begin{aligned} &\Gamma(\phi\to a_0(980)\gamma) = 0.323\pm0.029\ \text{keV},\quad \Gamma(\phi\to f_0(980)\gamma) = 1.87\pm0.11\ \text{keV}, \\ &\Gamma(a_0(980)\to\gamma\gamma) = 0.30\pm0.10\ \text{keV},\quad \Gamma(f_0(980)\to\gamma\gamma) = 0.39^{+0.10}_{-0.13}\ \text{keV}. \end{aligned}$$

From these results, we obtained the restriction on the parameters; Eq. (26) as

$$\begin{aligned} &|B\cos\theta_a| = 2.316\pm0.127\ \text{GeV}^{-1},\ \text{from}\ \phi\to a_0(980)\gamma\ \text{decay} \quad \text{(a)} \\ &|BR_{f_0NS} - \sqrt{2}B'R_{f_0S'} - \sqrt{2}B''R_{f_0G}| = 13.8\pm0.54\ \text{GeV}^{-1}, \\ &\qquad\qquad \text{from}\ \phi\to f_0(980)\gamma\ \text{decay} \quad \text{(b)} \\ &|B\cos\theta_a - B'\sin\theta_a| = 1.137\pm0.214\ \text{GeV}^{-1},\ \text{from}\ a_0(980)\to\gamma\gamma\ \text{decay} \quad \text{(c)} \\ &|B(R_{f_0NS} + 2\sqrt{2}R_{f_0NN}) - B'(5R_{f_0N'} + \sqrt{2}R_{f_0S'}) - 6\sqrt{2}B''R_{f_0G}| \\ &\qquad = 3.920^{+0.591}_{-0.789}\ \text{GeV}^{-1},\ \text{from}\ f_0(980)\to\gamma\gamma\ \text{decay} \quad \text{(d)} \end{aligned}$$

Seeing the result (a), we can say that because only the 4 quark component of $a_0(980)$ can contribute to the $\phi\to a_0(980)\gamma$ decay, one cannot explain this decay width without

TABLE 1.

Decay	Coupling strength	Experimental data (GeV^{-1})	Best-fit value (GeV^{-1})
$\phi \to a_0\gamma$	$\|B\cos\theta_a\|$	2.316 ± 0.127	2.76
$\phi \to f_0\gamma$	$\|BR_{f_0NS} - \sqrt{2}B'R_{f_0S'} - \sqrt{2}B''R_{f_0G}\|$	13.8 ± 0.54	7.15
$a_0 \to \gamma\gamma$	$\|B\cos\theta_a - B'\sin\theta_a\|$	1.137 ± 0.214	0.84
$f_0 \to \gamma\gamma$	$\|B(R_{f_0NS} + 2\sqrt{2}R_{f_0NN}) - B'(5R_{f_0N'} + \sqrt{2}R_{f_0S'}) - 6\sqrt{2}B''R_{f_0G}\|$	$3.920^{+0.591}_{-0.789}$	6.20

4 quark model. From the result (b), we can say that because of the opposite sign of R_{f_0NS} to that of $R_{f_0S'}$ recognized in Eq. (2), if we take the relative sign of B and B' as the same, $\sqrt{2}B'R_{f_0S'}$ factor contribute additively. In the result (c), the experimental result that the coupling strength for $a_0 \to \gamma\gamma$ is a half of that of $\phi \to f_0\gamma$ can be explained by the fact that B and B' have the same sign and θ_a is positive and then the term $B'\sin\theta_a$ contribute destructively. In the result (d), the term $BR_{f_0NS} - \sqrt{2}B'R_{f_0S'} - \sqrt{2}B''R_{f_0G}$ is the same as that in $\phi \to f_0\gamma$ decay. The term $2\sqrt{2}BR_{f_0NN}$ contributes destructively because of the opposite sign of R_{f_0NN} to that of R_{f_0NS}.

R_{a_0NS} etc. in coupling strengths vary with the variation of the values of λ_{01}, λ_0, λ_1, λ_G, λ_{GG}. Then, we take the χ^2 fit analysis of coupling strengths varying B, B', B'' and above λ_{01}'s. The best-fit values are given at the following values of coupling constants B, etc.,

$$B = 2.8\mathrm{GeV}^{-1},\ B' = 12.0\mathrm{GeV}^{-1},\ B'' = 7.2\mathrm{GeV}^{-1},\ \lambda_{01} = 0.24\mathrm{GeV}^2,$$
$$\lambda_0 = -0.03\mathrm{GeV}^2,\ \lambda_1 = 0.09\mathrm{GeV}^2,\ \lambda_G = 0.17\mathrm{GeV}^2,\ \lambda_{GG} = 1.66^2\mathrm{GeV}^2.$$

The best-fit values for coupling strengths are tabulated in Table 1. The best-fit values for mixing parameters are obtained as follows;

$$\begin{pmatrix} f_0(600) \\ f_0(980) \\ f_0(1370) \\ f_0(1500) \\ f_0(1710) \end{pmatrix} = \begin{pmatrix} 0.929 & 0.292 & -0.218 & -0.058 & 0.028 \\ 0.315 & -0.901 & 0.053 & 0.292 & -0.032 \\ 0.088 & -0.251 & 0.295 & -0.912 & 0.099 \\ 0.160 & 0.178 & 0.808 & 0.172 & -0.510 \\ 0.067 & 0.092 & 0.458 & 0.222 & 0.853 \end{pmatrix} \begin{pmatrix} f_{NN} \\ f_{NS} \\ f_{N'} \\ f_{S'} \\ f_G \end{pmatrix}.$$

REFERENCES

1. D. Black, A. H. Fariborz and J. Schechter, Phys. Rev. **D61**, 074001(2000).
2. T. Teshima, I. Kitamura and N. Morisita, J. Phys. **G28**, 1391(2002).
3. T. Teshima, I. Kitamura and N. Morisita, J. Phys. **G30**, 663(2004).
4. N. N. Achasov and V. N. Ivanchenko, Nucl. Phys. **B315**, 465(1989).
   F. E. Close and N. Isgur and S. Kumano, Nucl. Phys. **B389**, 513(1993).
   N. N. Achasov and V. V. Gubin, Phys. Rev. **D56**, 4084(1997).
   R. R. Akhmetshin, et al., Phys. Lett. **B462**, 380(1999).
5. R. Delbourgo, D. Liu and M. D. Scadron, Phys. Lett. **B446**, 332(1999).
   D. Black, M. Harada and J. Schechter, Phys. Rev. Lett. **88**, 181603(2002).
   S. B. Gerasimov, Nucl. Physics B (Proc. Suppl.) **126**, 210(2004).
6. S. Eidelman, et al., Particle Data Group, Phys. Lett. **B592**, 1(2004).

Unitary chiral dynamics in J/Ψ decay into VPP and the role of the scalar mesons[1]

L. Roca*, J. E. Palomar*, E. Oset* and H. C. Chiang†

*Departamento de Física Teórica and IFIC, Centro Mixto Universidad de Valencia-CSIC, Institutos de Investigación de Paterna, Aptdo. 22085, 46071 Valencia, Spain
†Institute of High Energy Physics, Chinese Academy of Sciences, Beijing 100039, China

Abstract. We make a theoretical study of the J/Ψ decays into $\omega\pi\pi$, $\phi\pi\pi$, $\omega K\bar{K}$ and $\phi K\bar{K}$ using the techniques of the chiral unitary approach stressing the important role of the scalar resonances dynamically generated through the final state interaction of the two pseudoscalar mesons. We also discuss the importance of new mechanisms with intermediate exchange of vector and axial-vector mesons and the role played by the OZI rule in the $J/\Psi\phi\pi\pi$ vertex, quantifying its effects. The results nicely reproduce the experimental data for the invariant mass distributions in all the channels considered.

PACS: 13.20.Gd; 12.39.Fe; 14.40.Cs

The J/Ψ decay into a pseudoscalar meson pair and a vector meson has been claimed to be one of the most suited reactions to study the long controversial nature of the scalar mesons, on which the interpretation as $q\bar{q}$ mesons or as meson-meson molecules has mainly centered the discussion. In the last years, a chiral unitary coupled channel approach [1] has proved to be successful in describing meson-meson interactions in all channels up to energies ~ 1.2 GeV, far beyond the natural limit of applicability of the standard Chiral Perturbation Theory (ChPT), which is ~ 500 MeV where the pole of the lightest resonance, the σ meson, appears. In this approach the scalar mesons rise up naturally as dynamically generated resonances, in the sense that, without being included as explicit degrees of freedom, they appear as poles in the s-wave meson-meson scattering amplitudes. The aim of the present work [2] is to make a consistent and comprehensive description of the $J/\Psi \to VPP$ decays, including all the mechanisms able to influence the region of pseudoscalar pair invariant masses up to ~ 1.1 GeV, addressing the role played by the scalar mesons and the OZI rule. Following the framework of the chiral unitary approach, we will implement the final meson-meson state interaction in order to generate dynamically the scalar resonances involved.

The first mechanisms considered are those involving a direct coupling of the J/Ψ to the two pseudoscalars and the vector, implementing the final state interaction of the pseudoscalars pair, as is depicted in Fig. 1 for the $J/\Psi \to \omega\pi^+\pi^-$ channel.

The thick dot in Fig. 1 means that one is considering the full $\pi\pi(K\bar{K}) \to \pi^+\pi^-$ t-matrix, involving the loop resummation of the Bethe-Salpeter equation and no just

---

[1] Contribution to the XI. INTERNATIONAL CONFERENCE ON HADRON SPECTROSCOPY (HADRON 05), 21-26 August 2005, Rio de Janeiro, Brazil
[2] All the details of the model can be found in [2], on which the present contribution is based

CP814 *Hadron Spectroscopy: Eleventh International Conference*
edited by A. Reis, C. Göbel, J. Sá Borges, and J. Magnin

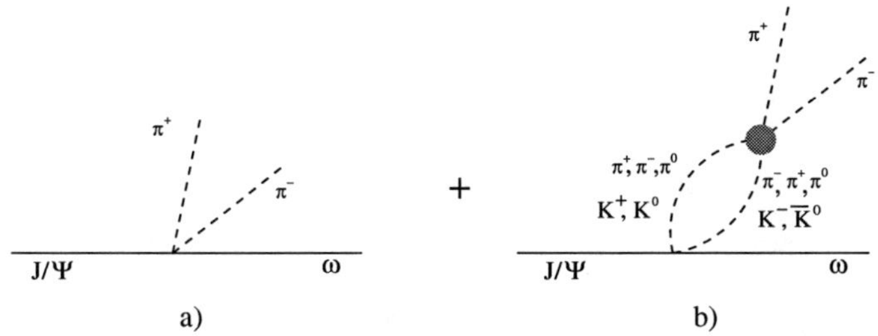

FIGURE 1. Diagrams with direct $J/\Psi VPP$ vertex at tree level, a), and with the iterated meson loops, b).

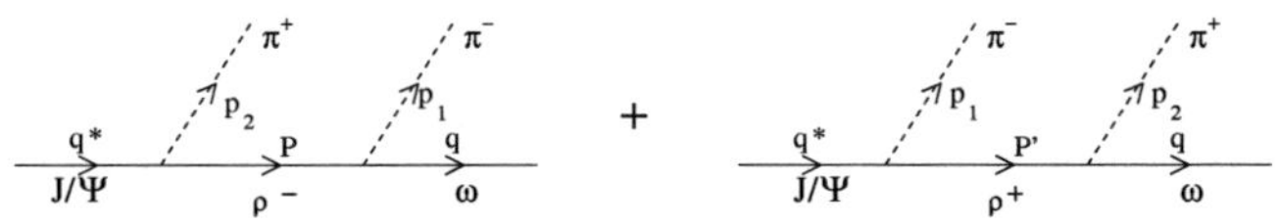

FIGURE 2. Diagrams for the tree level mechanisms with sequential vector meson exchange.

the lowest order $\pi\pi(K\bar{K}) \to \pi^+\pi^-$ amplitude. Actually this loop resummation is what dynamically generates the scalar resonances. The $J/\Psi VPP$ vertex can be constructed using $SU(3)$ arguments to relate the different isospin channels and parameterizing the amplitudes in a way which clearly manifest the role played by the OZI rule in this direct vertex, accounted for by means of a parameter λ_ϕ such that, if the OZI rule would be exact, then λ_ϕ would be zero.

In analogy with the approach in Ref. [3] (and in some sense to [4]), we consider next the mechanisms involving the sequential exchange of a vector or axial-vector meson as depicted in Fig. 2 for the vector meson case for the $\omega\pi\pi$ channel.

The vertices needed are obtained from suitable phenomenological Lagrangians (see [2, 3, 5]). In Figs. 3 and 4 we show the mechanisms with the implementation of the FSI of the pseudoscalar mesons.

The only two free parameters in our model are the coupling of the direct $J/\Psi VPP$ vertex and the OZI rule violation parameter, λ_ϕ. Fitting our model to the $J/\Psi \to \omega\pi^+\pi^-$ and $J/\Psi \to \phi\pi^+\pi^-$ experimental data we obtain values of λ_ϕ clearly different from zero ($\lambda_\phi = 0.12, 0.20$) and reasonably smaller than one, what manifests the OZI rule violation within reasonable values. In Fig. 5, left column, we show the results for $\omega\pi\pi$ and $\phi\pi\pi$ channels including an estimation of the theoretical error band. The experimental data shown in the figures have been obtained from [6, 7, 8, 9]. Concerning the scalar

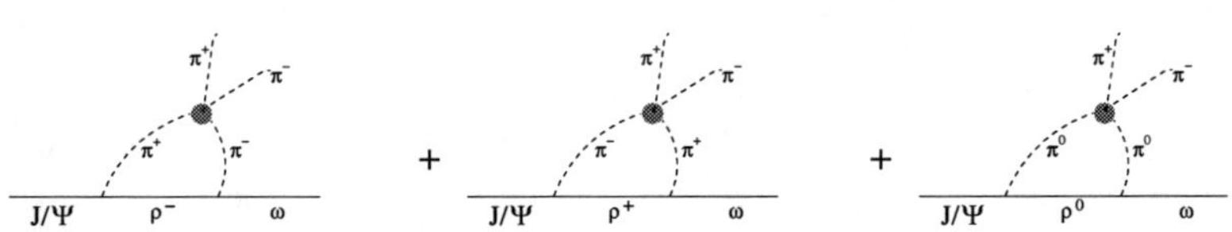

FIGURE 3. Sequential vector meson exchange diagrams with final state interaction of pions

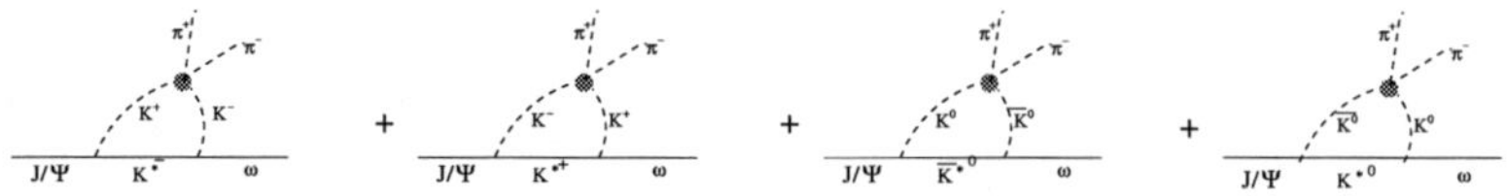

FIGURE 4. Sequential vector meson exchange diagrams with final state interaction of kaons

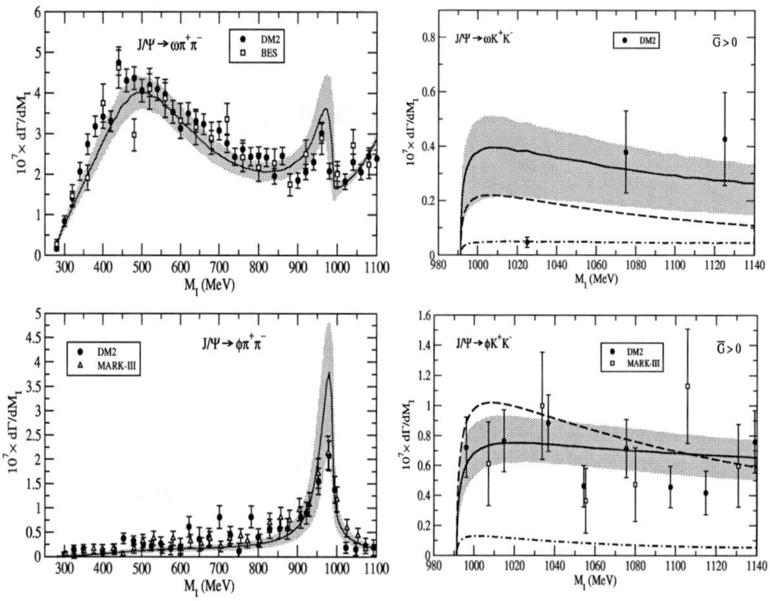

FIGURE 5.

mesons, it is important to stress first that the final shape and strength of the bump appearing in the $\pi\pi$ invariant mass distribution in the $\omega\pi\pi$ channel at ~ 500 MeV is determined by a subtle interference between the final state interaction in the direct $J/\Psi VPP$ mechanisms and the tree level direct $J/\Psi VPP$ decay. This means that the shape and position of the bump does not directly represent the physical properties of the σ meson, since it is distorted due to interferences with other terms not related to the σ meson. We will see in the following that the correct shape of the bump at lower energies comes mainly from a subtle interference of the $t^{I=0}_{\pi\pi,\pi\pi}$ amplitudes of Fig. 1b) with the contact term of Fig. 1a). In Fig. 6 we plot the modulus squared of $G_{KK}t^{I=0}_{KK,\pi\pi}$ (dashed-dotted line), $G_{\pi\pi}t^{I=0}_{\pi\pi,\pi\pi}$ (solid line) and $1+G_{\pi\pi}t^{I=0}_{\pi\pi,\pi\pi}$ (dashed line). In the unitary chiral models, the scalar mesons appear as poles in the second Riemann sheet of the $t_{MM,M'M'}$ scattering amplitude in the scalar-isoscalar channel. Actually the $f_0(980)$ appears in the $987-i14$ MeV position and the σ meson in $445-i221$ MeV, but its physical mass, given by the distribution in the real axis, is around 600 MeV because of the large width and the complicated distribution in the complex plane of the σ meson pole. Therefore the σ bump would be seen in around 600 MeV if only the term containing the $t^{I=0}_{\pi\pi,\pi\pi}$

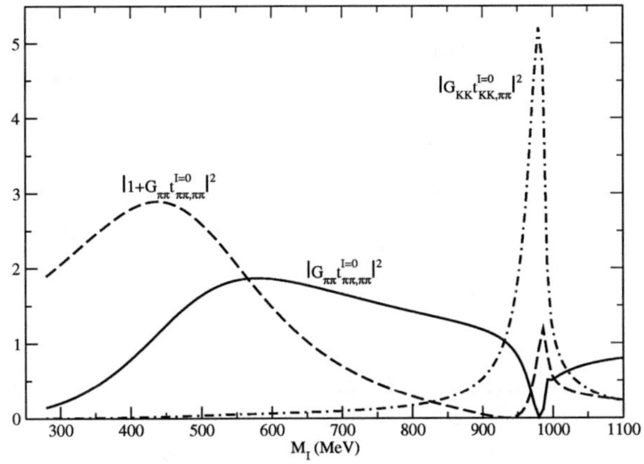

FIGURE 6. Modulus square of the different pieces of the amplitude for the mechanisms in Fig. 1 involving the two meson loop and the meson-meson unitarized amplitude.

amplitude would be present, this is, with a shape similar to the solid line in Fig. 6. When the $t^{I=0}_{\pi\pi,\pi\pi}$ term interferes with the contact term of Fig. 1a), represented by the "1" in the formulae, the shape in the σ region is strongly modified to something much more similar to the final shape of the $d\Gamma/dM_I$ curve. Therefore one can conclude that when doing an analysis of the experimental data to extract the σ mass and width one has to allow the σ meson amplitude to interfere with a polynomial containing at least the constant term. It is also interesting to present a different interpretation of this peak. Since in the σ region the $K\bar{K}$ channel is not important we can use the Bethe-Salpeter equation with only one channel, $\pi\pi$, and then we have

$$t^{I=0}_{\pi\pi,\pi\pi} = V + VG_{\pi\pi}t^{I=0}_{\pi\pi,\pi\pi} = V(1+G_{\pi\pi}t^{I=0}_{\pi\pi,\pi\pi}). \tag{1}$$

Since in the amplitude of the mechanisms of Fig. 1, neglecting the $K\bar{K}$ channel, one has the contribution $1+G_{\pi\pi}t^{I=0}_{\pi\pi,\pi\pi}$, then one can make

$$1+G_{\pi\pi}t^{I=0}_{\pi\pi,\pi\pi} = \frac{t^{I=0}_{\pi\pi,\pi\pi}}{V}. \tag{2}$$

Hence, the shape in the $J/\Psi \to \omega\pi^+\pi^-$ channel around $M_I = 500$ MeV is given roughly by $|t^{I=0}_{\pi\pi,\pi\pi}/V|^2$ and it happens that V ($V = -(M_I^2 - m_\pi^2/2)/f^2$) (which provides the Adler zero) is more strongly dependent on M_I than $t^{I=0}_{\pi\pi,\pi\pi}$, it grows faster as a function of M_I and the ratio $t^{I=0}_{\pi\pi,\pi\pi}/V$ decreases faster as a function of M_I than $t^{I=0}_{\pi\pi,\pi\pi}$, producing this apparent narrower peak, which does not reflect the M_I dependence of the $t^{I=0}_{\pi\pi,\pi\pi}$ matrix but rather the M_I dependence of the kernel V. In order to see the $t^{I=0}_{\pi\pi,\pi\pi}$ amplitude one has to resort to reactions where the Born term is forbidden like in the $\gamma\gamma \to \pi^0\pi^0$ reaction. In conclusion one has to be extremely careful when using the experimental data to extract the physical σ meson properties by fitting Breit-Wigner-like shapes.

On the other hand, the relative weights of the $f_0(980)$ and the σ meson are well reproduced in both the $\omega\pi\pi$ and $\phi\pi\pi$ channels in spite of their large difference in these channels. This relative weight is mainly determined by the OZI rule violation parameter and the interferences of the "direct" terms with the other mechanisms, specially in the $f_0(980)$ region. In our model, since the scalar mesons are dynamically generated through the resummed meson-meson amplitude, the relative weight between the $f_0(980)$ and the σ mesons is related to the relative weight between the $K\bar{K} \to \pi\pi$ and $\pi\pi \to \pi\pi$, in $I = 0$, scattering amplitudes. Specially remarkable is the fair agreement in the $f_0(980)$ region of the $\omega\pi\pi$ channel despite the smallness of the bump and the fact that many mechanisms contribute in this region.

Finally, in Figs.5, right column, we apply our results to the $\omega K\bar{K}$ and $\phi K\bar{K}$ decay channels, obtaining a fair agreement without introducing any extra freedom in the model. This is a nice test of the present model, both reproducing the absolute strength, and also the shape, which shows much strength close to threshold as a reflection of the proximity of the $f_0(980)$ resonance below threshold.

In conclusion, we have obtained a good description of these interesting J/Ψ decays combining phenomenological Lagrangians and the techniques of the chiral unitary approach to implement the final state rescattering of the pseudoscalar pairs, quantifying the controversial non-trivial role of the scalar mesons and the violation of the OZI rule. The fact that once more one is able to reproduce the shape and strength of the $f_0(980)$ and the σ resonances without the need to introduce them as explicit degrees of freedom provides an extra support to the idea of the nature of these resonances as dynamically generated from the interaction of the mesons.

REFERENCES

1. J. A. Oller, E. Oset and A. Ramos, Prog. Part. Nucl. Phys. **45** (2000) 157.
2. L. Roca, J. E. Palomar, E. Oset and H. C. Chiang, arXiv:hep-ph/0405228.
3. J. E. Palomar, L. Roca, E. Oset and M. J. Vicente Vacas, Nucl. Phys. A **729** (2003) 743 [arXiv:hep-ph/0306249].
4. E. Oset, J. R. Pelaez and L. Roca, Phys. Rev. D **67** (2003) 073013 [arXiv:hep-ph/0210282].
5. L. Roca, J. E. Palomar and E. Oset, arXiv:hep-ph/0306188.
6. N. Wu, arXiv:hep-ex/0104050.
7. J. E. Augustin *et al.* [DM2 Collaboration], Nucl. Phys. B **320** (1989) 1.
8. A. Falvard *et al.* [DM2 Collaboration], Phys. Rev. D **38** (1988) 2706.
9. W. S. Lockman [MARK-III Collaboration], SLAC-PUB-5139 *Presented at 3rd Int. Conf. on Hadron Spectroscopy, Ajaccio, France, Sep 23-27, 1989*

The Scalar Resonances $a_0/f_0(980)$ at COSY

M. Büscher for the ANKE collaboration

Institut für Kernphysik, Forschungszentrum Jülich, D-52425 Jülich, Germany

Abstract. Fundamental properties of the scalar resonances $a_0/f_0(980)$, like their masses, widths and couplings to $K\bar{K}$, are poorly known. In particular, precise knowledge of the latter quantity would be of great importance since it can be related to the $K\bar{K}$ content of these resonances.

An experimental program is under way at COSY-Jülich aiming at the extraction of the isospin violating a_0/f_0 mixing amplitude Λ which is in leading order proportional to the product of the coupling constants of the a_0 and f_0 to kaons. a_0/f_0 production is studied in pp, pn and dd interactions, both for the $K\bar{K}$ and the $\pi\eta/\pi\pi$ decays, using the ANKE and WASA spectrometers. The latter will be available for measurements at COSY in 2007.

As a first step, isovector $K\bar{K}$ production has been measured in the reaction $pp \to dK^+\bar{K}^0$. The data reveal dominance of the a_0^+ channel, thus demonstrating the feasibility of scalar meson studies at COSY. Analyses of $K\bar{K}$- and $\bar{K}d$-FSI effects yield the corresponding scattering lengths, $a(K\bar{K})_{I=1} = -(0.02\pm0.03) - i(0.61\pm0.05)$ fm and $|\text{Re } a(\bar{K}d)|\leq 1.3$ fm, Im $a(\bar{K}d)\leq 1.3$ fm.

Keywords: Scalar mesons, isospin violation
PACS: 14.40.Gx; 13.75.Cs; 13.75.Lb

THE LIGHT SCALAR RESONANCES

QCD is the fundamental theory of the strong interactions. How quarks and gluons are bound into hadrons is a yet unsolved strong coupling problem. QCD can be treated explicitly in this regime using lattice techniques [1], which are, however, not yet in the position to make quantitative statements about the light scalar mesons. Alternatively, QCD-inspired models, which employ effective degrees of freedom, can be used. The constituent quark model is one of the most successful in this respect (see *e.g.* [2]). However, more states with quantum numbers $J^P{=}0^+$ have been identified experimentally than would fit into a single SU(3) scalar nonet: the $f_0(600)$ (or σ), $f_0(980)$, $f_0(1370)$, $f_0(1500)$ and $f_0(1710)$ with $I{=}0$, the $\kappa(800)$ and the $K^*(1430)$ ($I{=}1/2$), as well as the $a_0(980)$ and $a_0(1450)$ ($I{=}1$) [3]. Consequently, the $a_0/f_0(980)$ have been associated with crypto-exotic states like $K\bar{K}$ molecules [4] or compact qq-$\bar{q}\bar{q}$ states [5].

In the sixties Weinberg suggested how to decide whether a (stable) particle is composite or elementary [6]. The formalism was applied to the deuteron and it was shown that it is not an elementary particle but made of a proton and a neutron. Baru *et al.* [7] have extended Weinberg's approach to the (unstable) $a_0/f_0(980)$ and find evidence for significant mesonic components in these states. From the formalism in Ref. [7] it follows that precise knowledge of the resonance parameters, in particular the coupling constants of the a_0 and f_0 to kaons, would allow one to model independently determine the $K\bar{K}$ content of the a_0/f_0. However, the values for $g_{a_0K\bar{K}}$ and $g_{f_0K\bar{K}}$ are still poorly known and range from 0.224 to 0.834 and from 1.31 to 2.84, respectively [8] — depending on the fitted Flatté parameterizations of measured $\pi\eta/\pi\pi$ and $K\bar{K}$ spectra.

CP814 *Hadron Spectroscopy: Eleventh International Conference*
edited by A. Reis, C. Göbel, J. Sá Borges, and J. Magnin

EXPERIMENTAL PROGRAM AT COSY

An experimental program is under way at COSY-Jülich aiming at the extraction of the isospin-violating (IV) a_0/f_0 mixing amplitude Λ which is in leading order proportional to the product of $g_{a_0K\bar{K}}$ and $g_{f_0K\bar{K}}$ [9, 10]. a_0/f_0 production has been or will be studied in pp, pn and dd interactions, both for the $K\bar{K}$ and the $\pi\eta/\pi\pi$ decay channels [11]. While near-threshold decay channels with at least one charged kaon can well be investigated with the magnetic spectrometer ANKE ("Apparatus for the detection of Nucleonic and Kaon Ejectiles"), the $\pi\eta$ and $\pi\pi$ final states will be measured with WASA ("Wide Angle Shower Apparatus") which will be available for measurements at COSY in 2007 [12].

Isospin filters and violation

A $pp \to dX$ reaction must lead to a_0^+ (I=1) production, a $pn \to dX$ interaction is not isospin selective, whereas the $dd \to \alpha X$ reaction — neglecting the small IV contributions which are the final goal of the proposed experimental program — is a filter for the f_0 (I=0) resonance, because the initial deuterons and the α particle in the final state have isospin I=0 ("isospin filter").

Since at COSY it is possible to fix the initial isospin one can selectively produce the a_0 or f_0 resonances and can identify observables that vanish in the absence of IV. The idea behind the proposed experiments is the similar to recent measurements of IV effects in the reaction $dd \to \alpha\pi^0$ [13]. However, the interpretation of the signal from the scalar mesons is much simpler as compared to the pion case. Since the a_0 and the f_0 are rather narrow overlapping resonances, the a_0-f_0 mixing in the final state is enhanced by more than an order of magnitude compared to IV in the production operator (*i.e.* "direct" IV $dd \to \alpha a_0$ production) and should give the dominant contribution to the IV effect via the reaction chain $dd \to \alpha f_0(I{=}0) \to \alpha a_0^0(I{=}1) \to \alpha(\pi^0\eta)$ [10, 14].

Any observation of $\pi^0\eta$ production in the $dd \to \alpha X$ reaction is a direct indication of IV and the cross section $\mathrm{d}\sigma/\mathrm{d}m_{\pi\eta}$ is given by the product of the mixing amplitude $\Lambda(m)$ and the $dd \to \alpha f_0$ production operator. It is therefore compulsory to determine the latter in an independent measurement in order to extract the mixing amplitude. A corresponding measurement of the $dd \to \alpha(K^+K^-)_{I=l=0}$ cross section is foreseen for winter 2005/06 at ANKE. These data, together with the information on the $dd \to \alpha(\pi^0\eta)$ reaction from WASA will allow one to determine the a_0-f_0 mixing model independently.

First measurements with ANKE: isovector $K\bar{K}$ production

As a first measurement, the production of $K\bar{K}_{I=1}$ pairs has been investigated in the reaction $pp \to dK^+\bar{K}^0$ at excitation energies $Q_{K\bar{K}} = 48$ and 105 MeV. This reaction has the advantage of lower background as compared to dd scattering, a $K\bar{K}$ production cross section which is expected to be ~100 times larger, and the fact that admixtures of $I = 0$ components are forbidden by charge conservation. So far, in pp collisions the $a_0(980)$ production has been measured at $p_p = 450$ GeV/c *via* $f_1(1285) \to a_0^{\pm}\pi^{\mp}$

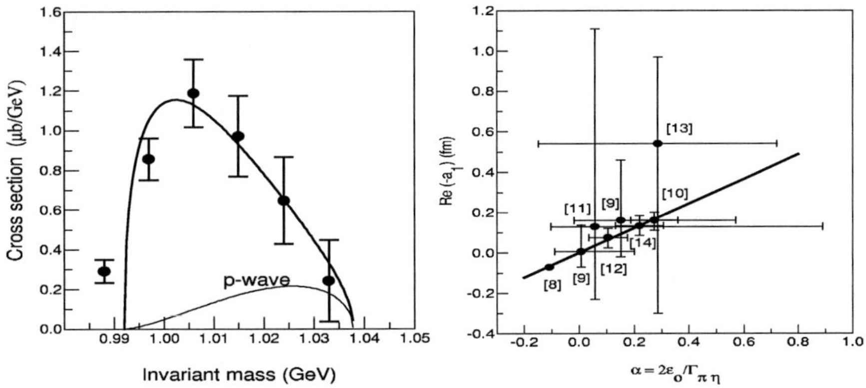

FIGURE 1. Left: $K^+\bar{K}^0$ mass distribution from the reaction $pp \to dK^+\bar{K}^0$ measured at an excitation energy of $Q = 48$ MeV with ANKE [18]. The solid line shows a fit to the data from which the imaginary part of the isovector $K\bar{K}$ scattering length has been determined [22]. Right: Plot of $Re(a_1)$ (filled circles) *vs.* the dimensionless parameter $\alpha = 2\varepsilon_0/\Gamma_{\pi\eta}$. The error bars show the errors introduced into α and $Re(a_1)$ by the uncertainties in the associated $a_0(980)$ mass values.

decays [15] and in inclusive measurements of the $pp \to dX^+$ reaction at $p_p = 3.8$, 4.5, and 6.3 GeV/c [16]. The best data so far on the $K\bar{K}_{I=1}$ decay mode of the a_0 have been obtained with Crystal Barrel at LEAR, with a mass resolution of $\delta(m_{K\bar{K}}) \approx$ 20 MeV [17].

A model-independent analysis of the ANKE angular distributions at $Q = 48$ MeV shows dominance of an *s*-wave between the two kaons accompanied by a *p*-wave deuteron with respect to the mesons. This has been interpreted as dominant production via the $a_0(980)$ channel near threshold [10, 18]. Model calculations of the reaction $pp \to dK^+\bar{K}^0$ performed by Grishina *et al.* [19] confirm this conclusion. In particular, the non-resonant (*p*-wave) $K^+\bar{K}^0$ production is shown to be small for excess energies Q less than $\sim$ 100 MeV.

In principle, the $dK^+\bar{K}^0$ final state is sensitive to three FSIs, dK^+, $d\bar{K}^0$ and $K^+\bar{K}^0$. The former is known to be small and can be neglected for the extraction of the latter two. The FSI in the $\bar{K}d$ system reduces the cross section, but leaves the invariant mass distributions virtually unaffected [19]. Sibirtsev *et al.* [20] have extracted limits for the $\bar{K}d$ scattering length from the ANKE data [18] and found $|\text{Re } a(\bar{K}d)| \leq 1.3$ fm, $\text{Im } a(\bar{K}d) \leq 1.3$ fm. A recent study [21] based on this formalism finds an optimal fit to the COSY data for a purely imaginary $d\bar{K}$ scattering length of $a_{d\bar{K}} = (0.0 + i1.0)$ fm.

The ANKE experiment has achieved a resolution of $1-8$ MeV for the invariant $K\bar{K}$ mass-spectrum (see Fig. 1) which is considerably better than in previous measurements. The relatively small influence of the antikaon-deuteron FSI on the shape of the measured invariant mass distributions as indicated in Refs. [19, 20, 21] allows one to determine the $K\bar{K}$ scattering length by a fit to the measured mass spectrum assuming that the deuteron's role is that of a spectator [22].

In Ref. [22] it has been assumed that $K^+\bar{K}^0$ production in $pp \to dK^+\bar{K}^0$ is dominated by FSI in the meson-meson channels only via the resonant intermediate mode $\pi^+\eta \to$

$a_0^+ \to K^+\bar{K}^0$. Then the cross section is proportional to the $I = 1$, $I_3 = +1$ cross section $\sigma(\pi^+\eta \to K^+\bar{K}^0)$ and the s- (p-)wave contributions to the cross section can be expressed as

$$\frac{d\sigma^{(s)}}{dm_{K\bar{K}}} = C_s \frac{k\Delta t}{p^2}[1 + 2Im(a_1)k] \quad \text{and} \quad \frac{d\sigma^{(p)}}{dm_{K\bar{K}}} \approx C_p \frac{k^3\Delta t}{p^2} \tag{1}$$

with overall normalizations $C_{s,p}$, and factors k, Δt and p which are given by the reaction kinematics. A fit to the data (solid lines in Fig. 1) yields for the imaginary part of the scattering length $Im(a_1) = -(0.61 \pm 0.05)$ fm at $\chi^2_{\min} = 0.81$.

The spectator model only fixes the imaginary part of the scattering length. In order to determine $Re(a_1)$ one has to introduce further assumptions regarding the structure of the $K\bar{K}$ scattering amplitude in the isovector channel. If one expresses the $a_0(980)$ mass distribution by the Flatté formula, there is a linear relation between the real and imaginary parts of the scattering lengths, $Re(a_1) = Im(a_1)\alpha$ with $\alpha = 2\varepsilon_{a_0}/\Gamma_{\pi\eta}$. ε_{a_0} is the binding energy M_{a_0} relative to the $K\bar{K}$ threshold and $\Gamma_{\pi\eta}$ the $a_0(980)$ partial decay width. If one takes $Im(a_1) = -0.61$ fm from the spectator model fit, this fixes the slope of the straight line in a $Re(a_1)$-*vs.*-α plot, see right hand side of Fig. 1. The result can then be compared with the values of $Re(a_1)$ and α extracted from data obtained at other experiments [8, 22]. This comparison is also shown in Fig. 1. The spectator model fit is seen to be consistent with all of the data once the uncertainties in α and hence $Re(a_1)$ due to the experimental error in the $a_0(980)$ mass are taken into account.

By assuming scale invariance of Flatté distributions [8], one can infer a value for α by calculating the mean of the α's shown in Fig. 1, $\alpha_c = 0.031 \pm 0.054$. Inserting this estimate one obtains $Re(a_1) = -(0.02 \pm 0.03)$ fm. The final result for the extracted isovector scattering length thus reads

$$a_1 = -[(0.02 \pm 0.03) + i(0.61 \pm 0.05)]\text{fm}\,. \tag{2}$$

ACKNOWLEDGMENTS

This work has been supported by: Deutsche Forschungsgemeinschaft (DFG), Russian Academy of Science, Russian Fund for Basic Research, European Community, and the Ernest Oppenheimer Memorial Trust for research support.

REFERENCES

1. T. Kunihiro, S. Muroya, A. Nakamura, C. Nonaka, M. Sekiguchi and H. Wada [SCALAR Collaboration], Phys. Rev. D **70**, 034504 (2004).
2. D. Morgan, Phys. Lett. B **51**, 71 (1974); K.L. Au, D. Morgan, and M.R. Pennington, Phys. Rev. D **35**, 1633 (1987); D. Morgan and M.R. Pennington, Phys. Lett. B **258**, 444 (1991); D. Morgan and M.R. Pennington, Phys. Rev. D **48**, 1185 (1993); A.V. Anisovich *et al.*, Eur. Phys. J. A **12**, 103 (2001); S. Narison, hep-ph/0012235.
3. S. Eidelman *et al.* (Particle Data Group); Phys. Lett. B **592**, 1 (2004).
4. J. Weinstein and N. Isgur, Phys. Rev. Lett. **48**, 659 (1982); Phys. Rev. D **27**, 588 (1983); Phys. Rev. D **41**, 2236 (1990); G. Janssen *et al.*, Phys. Rev. D **52**, 2690 (1995); J.A. Oller and E. Oset, Nucl. Phys. A **620**, 438 (1997) [Erratum-ibid. A **652**, 407 (1999)].

5. N.N. Achasov, hep-ph/0201299; R.J. Jaffe, Phys. Rev. D **15**, 267 (1977); J. Vijande *et al.*, Proc. Int. Workshop MESON 2002, May 24–28, 2002, Cracow, Poland, World Scientific Publishing, ISBN 981-238-160-0, p.501, hep-ph/0206263.
6. S. Weinberg, Phys. Rev. **130** (1963) 776.
7. V. Baru *et al.*, Phys. Lett. B **586** (2004) 53 [arXiv:hep-ph/0308129].
8. V. Baru *et al.*, Eur. Phys. J. A **23** (2005) 523 [arXiv:nucl-th/0410099].
9. N. N. Achasov *et al.*, Phys. Lett. B **88** (1979) 367.
10. C. Hanhart, Phys. Rept. **397** (2004) 155 [arXiv:hep-ph/0311341].
11. M. Büscher, arXiv:nucl-ex/0401010.
12. H. H. Adam *et al.* [WASA-at-COSY Collaboration], arXiv:nucl-ex/0411038.
13. E. J. Stephenson *et al.*, Phys. Rev. Lett. **91** (2003) 142302 [arXiv:nucl-ex/0305032].
14. V. Y. Grishina *et al.*, Phys. Lett. B **521** (2001) 217 [arXiv:nucl-th/0103081].
15. D. Barberis *et al.*, Phys. Lett. B **440** (1998) 225; Phys. Lett. B **488** (2000) 225.
16. M. A. Abolins *et al.*, Phys. Rev. Lett. **25** (1970) 469.
17. A. Abele *et al.*, Phys. Rev. D **57** (1998) 3860.
18. V. Kleber *et al.*, Phys. Rev. Lett. **91** (2003) 172304 [arXiv:nucl-ex/0304020].
19. V. Y. Grishina *et al.*, Eur. Phys. J. A **21** (2004) 507 [arXiv:nucl-th/0402093].
20. A. Sibirtsev *et al.*, Phys. Lett. B **601** (2004) 132 [arXiv:nucl-th/0406061].
21. F.P. Sassen, PhD dissertation, University of Bonn (2004); to be published.
22. M. Büscher *et al.*, arXiv:hep-ph/0508118.

Study of the radiative decays $\phi \to f_0(980)\gamma$ and $\phi \to a_0(980)\gamma$ with the KLOE detector

The KLOE collaboration[1]
presented by C.Bini

Universitá "La Sapienza" and INFN, Roma

Abstract. The KLOE experiment at the electron-positron collider DAPHNE has collected in the 2001-2002 running period an integrated luminosity of 450 pb^{-1}. Through the radiative ϕ decays, large samples of the light scalar mesons $f_0(980)$ and $a_0(980)$ are available. The f_0 meson is clearly observed in $\pi^+\pi^-$ and $\pi^0\pi^0$ final states, the a_0 in the $\eta\pi^0$ final states. Invariant mass spectra and Dalitz plot densities are analyzed to investigate the controversial structure of these particels.

Keywords: e^+e^- collisions, ϕ radiative decays, scalar mesons
PACS: PACS:13.66.Bc,14.40.-n

INTRODUCTION

At the ϕ factory DAPHNE, the lightest scalar mesons of isospin 0 and 1 are accessible through the radiative decay $\phi \to$scalar+γ. The study of these decays is motivated by the following considerations. Since the ϕ is a pure $s\bar{s}$ state, the coupling of any meson to the ϕ is related to the s-quark content of the meson. This is an important ingredient to discriminate between the quark-antiquark and the tetraquark interpretation of these particles [1]. Moreover a comparison between the f_0 and a_0 properties, allows to study possible effects of isospin mixing in the scalar sector [2]. Finally an high statistics study of invariant mass spectra and Dalitz plot densities allows to check the existence of the $f_0(600)$.

The $f_0(980)$ is studied looking at the 2 pion decays, $\pi^+\pi^-$ and $\pi^0\pi^0$ since the $f_0 \to$KK decay is strongly disfavoured. In both cases a huge unreducible background partly interfering with the signal is present: initial and final state radiation in the case of $\pi^+\pi^-$, $e^+e^- \to \omega\pi^0$ with $\omega \to \pi^0\gamma$ in the case of $\pi^0\pi^0$.

---

[1] F. Ambrosino, A. Antonelli, M. Antonelli, C. Bacci, P. Beltrame, G. Bencivenni, S. Bertolucci, C. Bini, C. Bloise, V. Bocci, F. Bossi, D. Bowring, P. Branchini, R. Caloi, P. Campana, G. Capon, T. Capussela, G. Carboni, F. Ceradini, S. Chi, G. Chiefari, P. Ciambrone, S. Conetti, E. De Lucia, P. De Simone, G. De Zorzi, S. Dell'Agnello, A. Denig, A. Di Domenico, C. Di Donato, S. Di Falco, B. Di Micco, A. Doria, M. Dreucci, G. Felici, A. Ferrari, M. L. Ferrer, G. Finocchiaro, C. Forti, P. Franzini, C. Gatti, P. Gauzzi, S. Giovannella, E. Gorini, E. Graziani, M. Incagli, W. Kluge, V. Kulikov, F. Lacava, G. Lanfranchi, J. Lee-Franzini, D. Leone, M. Martini, P. Massarotti, W. Mei, S. Meola, S. Miscetti, M. Moulson, S. Müller, F. Murtas, M. Napolitano, F. Nguyen, M. Palutan, E. Pasqualucci, A. Passeri, V. Patera, F. Perfetto, L. Pontecorvo, M. Primavera, P. Santangelo, E. Santovetti, G. Saracino, B. Sciascia, A. Sciubba, F. Scuri, I. Sfiligoi, T. Spadaro, M. Testa, L. Tortora, P. Valente, B. Valeriani, G. Venanzoni, S. Veneziano, A. Ventura, S. Ventura, R. Versaci, G. Xu

CP814 *Hadron Spectroscopy: Eleventh International Conference*
edited by A. Reis, C. Göbel, J. Sá Borges, and J. Magnin

The $a_0(980)$ is studied looking at $\eta\pi^0$ final states using two different samples: a fully neutral one due to $\eta \rightarrow \gamma\gamma$ and a charged and neutral one due to $\eta \rightarrow \pi^+\pi^-\pi^0$. The unreducible background is in this case much lower.
DAFNE is an e^+e^--collider with a center of mass energy around 1 GeV, providing a peak luminosity of 10^{32}cm^{-2}s^{-1}. The KLOE detector consists of a large-volume cylindrical drift chamber [3], surrounded by a sampling calorimeter [4] made of lead and scintillating fibers. The entire detector is located within a magnetic field B=0.52 T. The trigger [5] is based on the detection of at least two energy deposits in the calorimeter. In the following we present the analysis of the mass spectra using two distint models.

DATA ANALYSIS AND RESULTS

We search the $f_0(980)$ signal in the $\pi\pi$ invariant mass (m in the following) spectrum of $\pi^+\pi^-\gamma$ events with a photon at large polar angle ($\theta_\gamma > 45°$) to reduce the initial state radiation background. We require a vertex with two tracks at the e^+e^- interaction region and a photon matching the missing momentum of the tracks pair. The kinematics of the event, together with the time of flight signature, allows the rejection of $e^+e^-\gamma$, $\mu^+\mu^-\gamma$ and $\pi^+\pi^-\pi^0$ background events.
Fig.1 shows the m spectrum of the selected events. The distribution is dominated by the initial state radiation events characterized by the ρ^0 peak and by the ρ-ω interference pattern. In the region around m=980 MeV we observe a small bump due to the $f_0(980)$. Two fits of the full spectrum are shown according to two different approaches for the description of the f_0 amplitude: the first is based on the kaon-loop model [1] the second (called no-structure model in the following [6]) assumes the scalar amplitude as the sum of a Breit-Wigner and a polynomial continuum. A good fit is obtained in both cases by introducing one scalar meson only with a mass compatible with the $f_0(980)$ mass. Preliminary values of the parameters are given in Tab.1 for the kaon-loop fit.
We study also the charge asimmetry $A = (N(\theta^+ > 90°) - N(\theta^+ < 90°))/(N(\theta^+ > 90°) + N(\theta^+ < 90°))$ with θ^+ the polar angle of the positive pion, that is expected to be different from zero due to the presence of C-odd and C-even terms in the cross-section. The m dependence of A is shown in Fig.2 compared to the Montecarlo prediction with and without a scalar amplitude based on the kaon-loop results of the invariant mass fit shown in Fig.1. A negative bump is observed in the region around m=980 MeV in this case also. Notice that in order to have a good description of the spectrum, the inclusion of the scalar amplitude is necessary.
$\pi^0\pi^0\gamma$ events are selected by requiring no tracks and 5 photons at an angle larger than 21°. Two kinematic fits are applied: the first requiring energy-momentum conservation, and the second associating the photons and imposing the π^0 masses. The analysis strategy is partly changed with respect to the previous one [7]: $\omega\pi^0$ events are not rejected anymore since they have in principle a sizeable interference term with the signal events; moreoever in order to appreciate better the signal vs. background composition of the selected sample, a 2-dimensional analysis $m^2_{\pi^0\gamma}$ vs. $m^2_{\pi^0\pi^0}$ is performed. The resulting Dalitz plot is shown in Fig.3. The two "bands" on the left side of the diagram are due to the $\omega\pi^0$ background, the cluster of events on the right corner, is due to the f_0 signal.

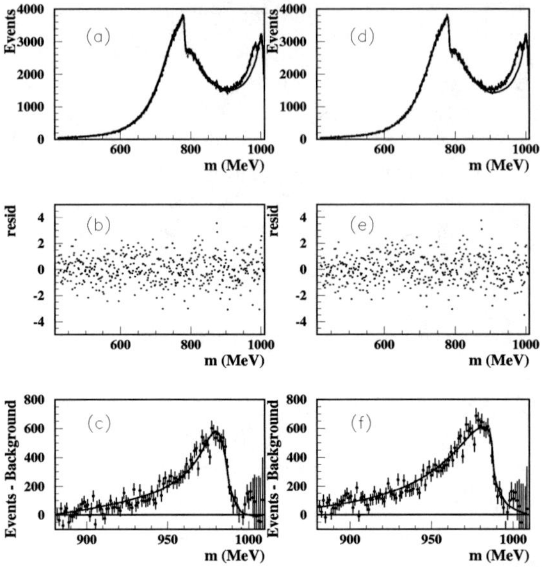

FIGURE 1. Result of the no-structure fit (a)-(b)-(c) and of the kaon-loop fit (d)-(e)-(f). (a)-(d) Data spectrum compared with the fitting function (upper curve following the data points) and with the estimated non-scalar part of the function (lower curve); (b)-(e) fit residuals as a function of *m*; (c)-(f) the fitting function is compared to the signal events obtained subtracting to the measured data spectrum the non-scalar part of the function in the f_0 region.

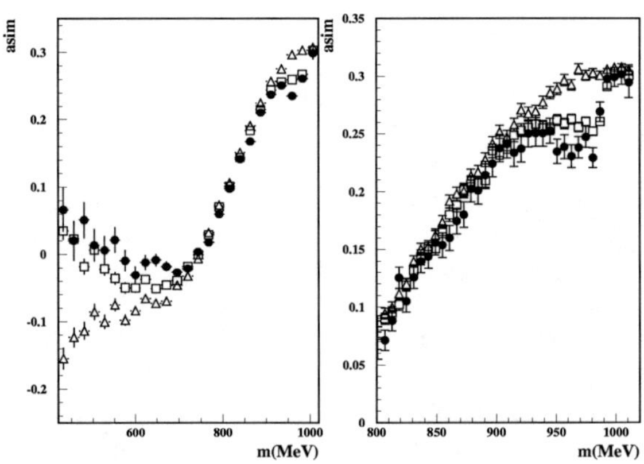

FIGURE 2. Charge asymmetry data (full circles) compared to a simulation based on the non-scalar part of the spectrum only (open triangles), and on the non-scalar plus the f_0 part obtained from the kaon-loop amplitude(open squares). The second plot shows the detail of the comparison in the f_0 region.

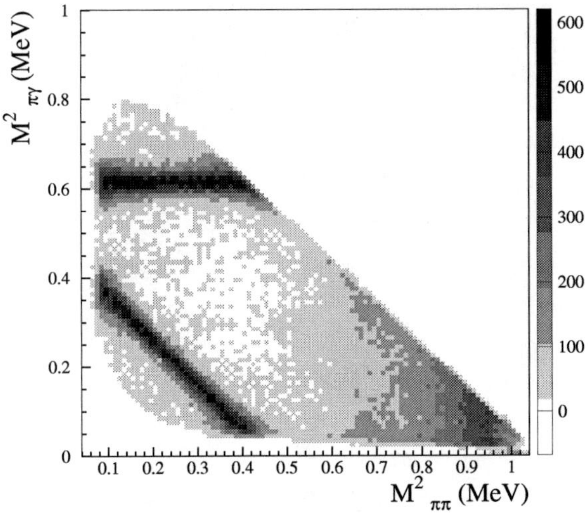

FIGURE 3. Dalitz plot distribution ($m^2_{\pi^0\gamma}$ vs. $m^2_{\pi^0\pi^0}$) of the $\pi^0\pi^0\gamma$ data sample.

TABLE 1. Preliminary results for the masses and the couplings of the f_0 and a_0 to KK, $\pi^+\pi^-$ and $\eta\pi^0$, within the kaon-loop framework.

m_{f_0} (MeV)	983	m_{a_0} (MeV)	987
g_{f_0KK} (GeV)	5.9	g_{a_0KK} (GeV)	2.3
$g_{f_0\pi^+\pi^-}$ (GeV)	3.6	$g_{a_0\eta\pi}$ (GeV)	2.6
$R=g^2_{f_0KK}/g^2_{f_0\pi^+\pi^-}$	2.7	$R=g^2_{a_0KK}/g^2_{a_0\eta\pi}$	0.8

An acceptable description of the Dalitz-plot distribution is obtained using the kaon-loop model to describe the scalar amplitude and the vector dominance model to describe the $\omega\pi^0$ background. Evaluation of the parameters is in progress.
Finally, the $\eta\pi^0$ invariant mass spectra are shown in Fig.4 for the two selected samples. A good fit is obtained using the kaon-loop model as in Ref.[8]. Preliminary values of the parameters are given in Tab.1.

SUMMARY AND OUTLOOK

The kaon-loop frame describes our entire data set. Looking at the couplings reported in Tab.1, we observe that the $f_0(980)$ is strongly coupled to kaons (R=2.7) the a_0 is less strongly coupled to kaons (R=0.8). The previous KLOE results [7, 8] are confirmed in both cases.

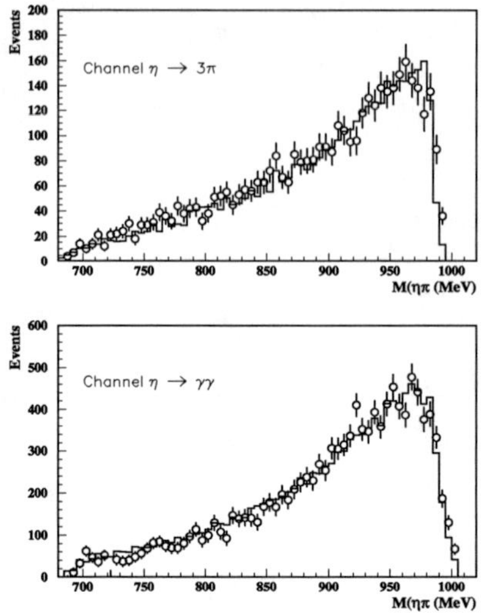

FIGURE 4. $\eta\pi^0$ invariant mass distributions (open points) from the two $\eta\pi^0\gamma$ selected samples. The histograms are the results of a combined fit of the two distributions.

The no-structure approach still requires theoretical effort. By the way it also describes in an acceptable way the observed spectra.
Finally, the $\pi^+\pi^-\gamma$ fits do not require any $f_0(600)$ to describe the data.
By the end of the year, KLOE will have collected a data sample of $\sim$2 fb^{-1} that is a factor $\sim$5 more than the data presented here. With such a statistics it will be possible to look for f_0, a_0 going to KK pairs also.

ACKNOWLEDGMENTS

We are indebted to N.N.Achasov for providing us all the required information concerning the kaon-loop model, G.Isidori and L.Maiani for many useful discussions concerning the no-structure model.

REFERENCES

1. N.N. Achasov and V.N. Ivanchenko, *Nucl. Phys.* **B315** 465 (1989).
2. F.E. Close and A. Kirk, *Phys. Lett.* **B515** 13 (2001).
3. M. Adinolfi et al., *Nucl. Instr. and Meth.* **A488** 1 (2002).
4. M. Adinolfi et al., *Nucl. Instr. and Meth.* **A482** 364 (2002).
5. M. Adinolfi et al., *Nucl. Instr. and Meth.* **A492** 134 (2002).
6. G. Isidori and L. Maiani, private comunication.
7. A. Aloisio et al., KLOE Collaboration, *Phys. Lett.* **B537** 21 (2002).
8. A. Aloisio et al., KLOE Collaboration, *Phys. Lett.* **B536** 209 (2002).

Role of $\pi\pi$ S-wave in CLEO D 3-body decays

S.A. Dytman - for the CLEO Collaboration

Department of Physics and Astronomy
University of Pittsburgh
Pittsburgh, PA 15260

Abstract. D meson 3-body hadronic decay data in the past were analyzed with Breit-Wigner amplitude fits. However, the complicated dynamics of the final states is known from previous studies to be poorly described by Breit-Wigner models. This short paper discusses work within CLEO to use coupled channel techniques to describe these decays.

Keywords: charm,Dalitz,coupled channel
PACS: 12.20.Fc,12.39.Mk

INTRODUCTION

Historically, interest in D 3-body decays comes from the desire to search for CP violation and mixing in the $D^0\bar{D}^0$ system. Most recently, the emphasis has switched to developing ways to understand B meson decays as a way to measure the CKM unitarity angle γ or ϕ_3 [1]. The most straightforward way to measure γ is presently based on the decay $B^{\pm} \to DK^{\pm}$, where the D (actually a mixture of D^0 and $\bar{D}^0$) decays to $K_s\pi^+\pi^-$. This analysis requires a model for the D decay; since strong interactions play a strong role in this decay, phenomenology is required. Recent results [1] estimate an error of 10-11° due to uncertainties in this model. Additionally, these decays provide a way to preferentially produce light scalar mesons (e.g. $f_0(600)$).

These final states are complicated mixtures of resonance+meson states where light resonances such as ρ, K^*, and K_0 contribute. The various final states very easily couple to each other. In addition, overlapping resonances such as $f_0(1370)$ and $f_0(1500)$ interfere. Medium energy experiments such as Crystal Barrel [2] have developed ways to use coupled channel techniques for these final states. They used a K-matrix technique that provides a unitary description of a large number of reactions. On the other hand, D 3-body decays are traditionally analyzed with Breit-Wigner models even though the resonances are not isolated or narrow. The FOCUS collaboration used a truncated version of the K-matrix model to analyze D and D_s 3-body decays [3]. They applied a K-matrix description of the $\pi\pi$ component of the final state using the work of Anisovich and Sarantsev [4]. This allowed them to accurately describe the $\pi\pi\pi$ final state without an explicit $f_0(600)$.

Data from the CLEO detector at Cornell have been used to study D 3-body decays. They are tagged through the $D^* \to D\pi$ decays where the soft pion can be detected. This also provides a tag of the D flavor. We have recently published papers studying $D^0 \to \pi^+\pi^-\pi^0$ [5] and $D^0 \to K_S\pi^+\pi^-$ [6]. Both contain the first full study of that particular final state. Here, we report model studies of the final state.

CP814 *Hadron Spectroscopy: Eleventh International Conference*
edited by A. Reis, C. Göbel, J. Sá Borges, and J. Magnin

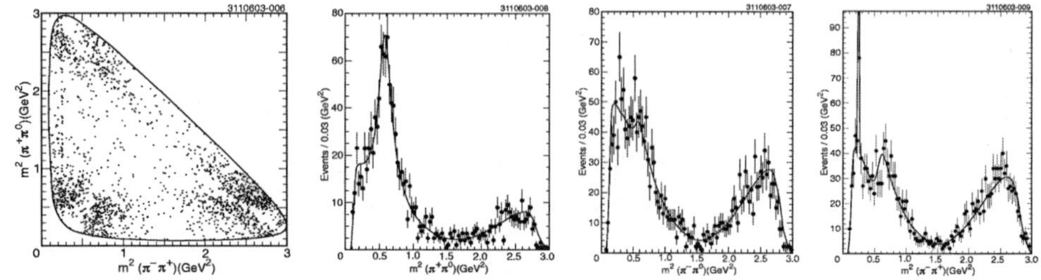

FIGURE 1. Dalitz plot and 3 projections for $D \to \pi^+\pi^-\pi^0$. The assumes PDG Breit-Wigner parameters for all ρ resonances and a flat nonresonant contribution.

$D^0 \to \pi^+\pi^-\pi^0$

These data were taken with the CLEO-III detector during data runs at $\sqrt{s} \sim 10$ GeV. The silicon detector had particular importance for maximizing resolution. A moderately clean sample of 1917 events were identified in a 9.0 fb^{-1} data sample. The data and the Breit-Wigner fit are shown in Fig. 1. When reflections are carefully considered, all features in the Dalitz plot can be described by ρ resonances of all 3 charges.

This standard fit assumes no scalar resonances. They would be seen as sharp structures in the $\pi^+\pi^-$ projection or as deviations between the fit and the data. As a test, we then included various models for scalar resonances. In each case, the evidence is weak. The 95% confidence limits for the fit fraction were flat nonresonant (6.4%), $f_0(600)$ [7] (0.2%), and K-matrix (1.9%). The K-matrix model was very similar to that employed by FOCUS, but the $\pi\pi$ model of Au, Morgan, and Pennington [9] was used here.

Although there is no strong evidence for scalar resonances here, the FOCUS $D^+ \to \pi^+\pi^-\pi^+$ [3] fit determined a fit fraction of 56±4%. Because of the charge differences, the ratio of the branching fractions for $D^+ \to f_0\pi^+$ to $D^0 \to f_0\pi^0$ should 18 due to tree level factors. We see a ratio of 60^{+51}_{-19} where the large errors are due to the large CLEO estimated error. A quark model for D decay of Bucella, Lusignoli, and Pugliese [8] used the scalar meson wave functions of Isgur and Weinstein, fitted model parameters to existing decay data and predicted this ratio to be 46.7, good agreement with the new data although the data point has a significant error estimate. We see that these decays can be used to test wave functions for scalar mesons.

$D^0 \to K_S\pi^+\pi^-$

This decay was studied in the same CLEO-III data set as for the $D^0 \to \pi^+\pi^-\pi^0$ decay described in the previous section. In both cases, we have the most complete analysis to date. With a large branching fraction and a final state with all charged particles, the data for $D^0 \to K_S\pi^+\pi^-$ is very clean. The data sample is 5299 events with an estimated background of 3.4%. The 3 published papers for this reaction [6] explore branching fractions, CP violations, and D^0-$\bar{D}^0$ mixing. using a Breit-Wigner model to characterize the final state. The best fit to the full data set is shown in Fig. 2.

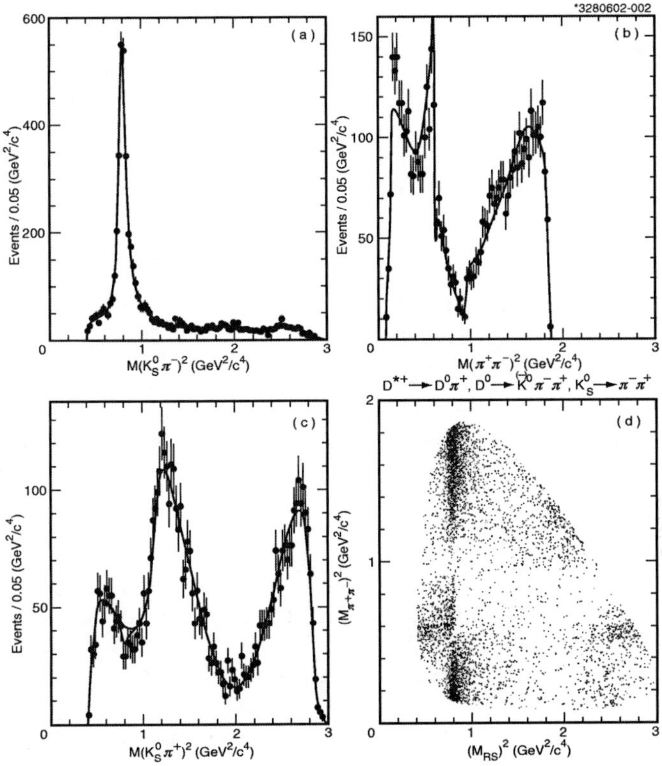

FIGURE 2. Dalitz plot and 3 projections for $D \to K_S\pi^+\pi^-$. The fit assumes PDG Breit-Wigner parameters for many resonances and a flat nonresonant contribution.

This fit has 10 resonances, including K^*, ρ, K_0, f_0, f_2, and ω, and a flat background representing all unknown sources. The $\mathscr{M}^2_{K_S\pi^-}$ distribution is dominated by $K^*(892)$ and the $\mathscr{M}^2_{K_S\pi^+}$ distribution is mostly populated by reflections of sharp states in from the other 2 combinations. Nevertheless, there are notable problems in $\mathscr{M}^2_{\pi^+\pi^-}$ in the same mass region where scalar states have strong influence. A $f_0(600)$ was not preferred; $f_0(980)$ and $f_0(1370)$ were included using PDG values for mass and width. ρ and $f_0(980)$ states are prominent.

Thus, we were anxious to see if the $\pi\pi$ S-wave desciption could be improved with a more modern treatment. As in the previous section, we chose the Au, Morgan, and Pennington [9] model. Although it is older than Anisovich and Sarantsev [4], it has fewer parameters and was updated by the Crystal Barrel group [2] for interpreting their high statistics $p\bar{p}$ experiments, e.g. $\pi^0\pi^0\pi^0$ final state. Comparing the raw $\pi\pi$ S-wave T matrix, we see the 2 formulations are very similar in the energy region sampled by these data.

This fit included the truncated K-matrix model used by FOCUS [3]. The 2 f_0 states

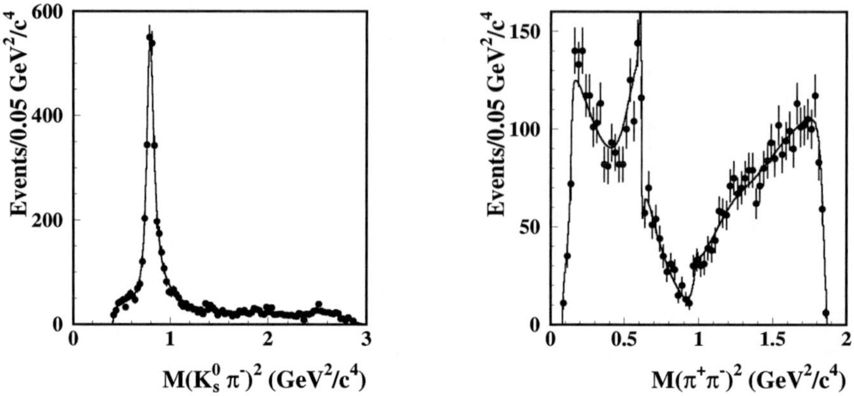

FIGURE 3. Projections for $D \to K_S\pi^+\pi^-$. The fit assumes PDG Breit-Wigner parameters for all resonances but those that couple to the $\pi\pi$ S-wave. For those states, the Au, Morgan, Pennington model was substituted.

and the flat background from the Breit-Wigner fit were taken out and replaced by the Crystal Barrel $\pi\pi$ S-wave model. The new parameters are coupling strengths for D decay into particular final states, β, often called production vectors. These are complex numbers. For $p\bar{p}$ experiments, the reactions are too complicated to ever expect a calculation; however, that is an intriguing possibility for D decay.

The results of this new fit are shown in Fig. 3. The improvements to the $\pi^+\pi^-$ invariant mass plot (compare to Fig. 2) are noticeable. The region $\mathscr{M}^2_{\pi^+\pi^-}$>1 (GeV/c)^2 is described very well and the fit to the region below 0.5 (GeV/c)^2 is much improved. Our tentative conclusion is that the more sophisticated treatment of the $\pi\pi$ S-wave used here provides a significant improvement in fit quality.

CLEO-c is presently taking data at the ψ(3770) resonance which decays strongly to correlated D^0-$\bar{D}^0$ pairs. At this time (accumulated luminosity of 281 pb^{-1}) statistics in the data sample accumulated is roughly a factor of 4 larger than that used in the CLEO-III analyses. Many decay channels are clearly seen and analyses are expected to provide significant information.

For extraction of γ or ϕ_3, lack of knowledge of D decay will have an impact on the systematic error. Until recently, both BABAR and Belle used Breit-Wigner analyses with data samples significantly larger than those of CLEO. The features seen in the CLEO also show up in the newer data. However, the Breit-Wigner model wasn't able to fit all features in the new data and 2 'extra' scalar resonances were needed corresponding a significant fit fraction. The uncertainty of these unphysical states led to a significant systematic error, $\sim 11^o$ in the final result. The recent BABAR fit used the FOCUS truncated K-matrix model and no extra states were required, more evidence that the improved dynamics of the final state are significant. In addition, the estimated systematic error with this new model alone is much smaller.

Our conclusion is that the experience in previous studies of the $\pi\pi$ S-wave can have important consequences for the extraction of the CKM unitarity angle γ. We should also note that none of these recent efforts use the full K-matrix model and that the process by which systematic errors are determined don't properly account for modeling errors. Work on these issues is in progress.

ACKNOWLEDGMENTS

We gratefully acknowledge the effort of the CESR staff in providing us with excellent luminosity and running conditions. This work was supported by the National Science Foundation and the U.S. Department of Energy.

REFERENCES

1. B. Aubert, *et al.* [BaBar Collaboration], submitted to *Phys. Rev. Lett.*, arXiv:hep-ex/0504039; A. Poluketov, *et al.* [Belle Collaboration], *Phys. Rev. D***70**,072003 (2004).
2. C. Amsler, *Rev. Mod. Phys.* **70**, 1293 (1998).
3. J. M. Link *et al.* [FOCUS Collaboration], *Phys. Lett. B***585**, 200 (2004).
4. V. V. Anisovich and A. V. Sarantsesv, *Eur. Phys. A***16**, 229 (2003).
5. D. Cronin-Hennessy *et al.* [CLEO Collaboration], *Phys. Rev. D***72**, 031102 (2005).
6. H. Muramatsu *et al.* [CLEO Collaboration], *Phys. Rev. Lett.* **89**, 251802 (2002) [Erratum-ibid. **90**, 059901 (2003)], D. M. Asner *et al.* [CLEO Collaboration], *Phys. Rev. D***70**, 091101 (2004). D. M. Asner *et al.* [CLEO Collaboration], *Phys. Rev. D***72**, 012001 (2005).
7. E. M. Aitala *et al.*, [E791 Collaboration], *Phys. Rev. Lett.***86**, 765 (2001).
8. F. Buccella, M. Lusignoli, and A. Pugliese, *Phys. Lett. B***379**, 249 (1996).
9. K. L. Au, D. Morgan, and M. R. Pennington, *Phys. Rev. D***35**, 1633 (1987).

Scalar Meson $\sigma(500)$ Phase Motion at $D^+ \to \pi^-\pi^+\pi^+$ Decay

Jussara M. de Miranda* and Ignacio Bediaga*

*Centro Brasileiro de Pesquisas Físicas,
Rua Xavier Sigaud 150, 22290-180 – Rio de Janeiro, RJ, Brazil

Abstract. Applying the Amplitude Difference Method to Fermilab experiment E791 $D^+ \to \pi^-\pi^+\pi^+$ data, we measure the low mass $\pi^+\pi^-$ phase motion. Our results suggest a significant phase variation, compatible with the existence of an isoscalar $\sigma(500)$ meson, as previously reported using an isobar model fit to the full Dalitz-plot density.

Keywords: Heavy Meson Decays; Dalitz-plot; Scalar Mesons; Resonances.
PACS: 13.25.Ft, 14.40.Ev, 14.40.Cs,11.80.Et.

Recently we proposed the Amplitude Difference (AD) Method to extract the phase motion of a complex amplitude in three body heavy meson decays [1]. With this method, the phase variation of a generic complex amplitude can be directly revealed through interference in the Dalitz-plot region where it crosses a well established resonant state, used as a probe. As a test, this method was successfully applied to data [2] to extract the well known phase motion of the scalar amplitude $f_0(980)$ observed in $D_s^+ \to \pi^-\pi^+\pi^+$ decay. In the present paper we intend to use the same method to study the low $\pi^+\pi^-$ mass region of the $D^+ \to \pi^-\pi^+\pi^+$ decay where Fermilab experiment E791 showed evidence for the existence of a light and broad scalar resonance [3].

The present study is a reanalysis of the Fermilab experiment E791 data. Here we investigate a subset of the total phase space used by the experiment in their full Dalitz plot analysis. A description of the experiment, data selection criteria, background parametrization and detector acceptance are found in references [3, 4]. The final $\pi^-\pi^+\pi^+$ invariant mass distribution is shown in ref. [3]. There are 1686 candidates events with invariant mass between 1.85 and 1.89 GeV$/c^2$. The integrated signal-to-background ratio in this range is about 2:1. Fig. 1 shows the folded Dalitz plot. The horizontal and vertical axes are the squares of the $\pi^+\pi^-$ invariant mass high (s_{12}) and low (s_{13}) combinations. The present analysis, use the hatched area of Fig.2. Based in the background study done for the previous analysis, we find that its distribution shows no dependence in the s_{12} variable in this region. We estimated 60 background events from the 197 total analyzed in this sample.

The detector acceptance in this region is almost constant. There is a very mild slope in the s_{13} and no variation in s_{12}. Nevertheless the acceptance shape is taken into account to correct the events distributions showed later on.

CP814 *Hadron Spectroscopy: Eleventh International Conference*
edited by A. Reis, C. Göbel, J. Sá Borges, and J. Magnin

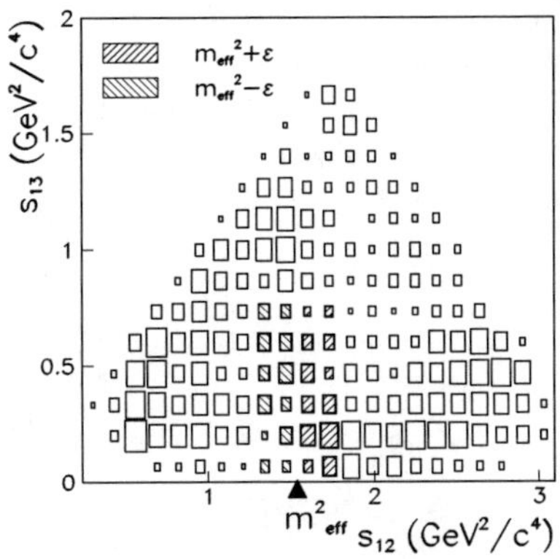

FIGURE 1. $D^+ \to \pi^-\pi^+\pi^+$ Dalitz plot folded distribution. The events used by the AD analysis are in the hatched region. The size of the area of each bin in the plot corresponds to the number of events in that bin.

There are two conditions necessary to extract the phase motion of a generic amplitude with the AD Method:

- A crossing region between the amplitude under study and a probe resonance has to be dominated by these two contributions. Other amplitudes must be negligible.
- The integrated amplitude of the probe resonance must be symmetric with respect to an reference effective mass squared (m^2_{eff}) [1].

Assuming that the best contributions in this region are the $f_2(1270)$ amplitude [2] in s_{12} and the $\pi\pi$ complex amplitude under study in s_{13}, we can write:

$$\mathscr{A}(s_{12},s_{13}) \sim a_R \;\; \mathscr{BW}_{f_2(1270)}(s_{12})^{J=2}\mathscr{M}_{f_2(1270)}(s_{12},s_{13}) + \\ a_s/(p^*/\sqrt{s_{13}}) \;\; sin\delta(s_{13}) \;\; e^{i(\delta(s_{13})+\gamma)}, \tag{1}$$

where γ is the overall relative phase-difference between the two amplitudes (assumed to be constant and arising from both production and final-state-interaction (FSI) in the di-pion system); $sin\delta(s_{13})e^{i\delta(s_{13})}$ represents a general amplitude for a two-body elastic scattering, $p^*/\sqrt{s}_{13}$ is a phase space factor to make this description compatible with $\pi\pi$ scattering; and a_R and a_s are the real magnitudes related to the two contributions. In the AD Method, a_R and a_s must be taken as independent of s, it assumes that the $\pi^+\pi^-$ production is constant over the effective mass range considered. Finally, the resonance $f_2(1270)$ is parameterized by a relativistic Breit-Wigner distribution with a mass dependent width, $\mathscr{BW}$, multiplied by an angular distribution function, $^{J=2}\mathscr{M}_{f_2(1270)}(s_{12},s_{13})$, both defined in reference [2].

[1] Analysis with more amplitudes in a limited phase space could be done, but in a model dependent way.

Both the width and the angular function of the $f_2(1270)$ resonance produces asymmetries in the its s_{12} distribution . We use a $D^+ \to f_2(1270)\pi^+$ Monte Carlo simulation to study the behavior of the probe resonance distribution in s_{12} and s_{13} variables.

To be brief, from here on, we use ${}^{J=2}\bar{\mathscr{M}}_{f_2(1270)}(s_{13}) = \bar{\mathscr{M}}$ and $p^*/\sqrt{s}_{13} = p'$. The main equation of the AD method applied to the integrated amplitude-square difference is [1, 2]:

$$p'/\bar{\mathscr{M}}\Delta \int \mathscr{A}^2 = p'/\bar{\mathscr{M}}(\int_{m^2_{eff}}^{m^2_{eff}+\varepsilon} \mid \mathscr{A}(s_{12},s_{13}) \mid^2 ds_{12} - \int_{m^2_{eff}-\varepsilon}^{m^2_{eff}} \mid \mathscr{A}(s_{12},s_{13}) \mid^2 ds_{12})$$

$$\sim -\mathscr{C}(sin(2\delta(s_{13})+\gamma) - sin\gamma), \qquad (2)$$

where $\mathscr{C}$ is an overall constant. Clear from Eq.2, the function $\Delta \int \mathscr{A}^2 p'/\bar{\mathscr{M}}$ directly reflects the behavior of $\delta(s_{13})$. A constant $\Delta \int \mathscr{A}^2 p'/\bar{\mathscr{M}}$ would implies constant $\delta(s_{13})$. This would be the case for a constant non-resonant contribution. In the same way, a slow phase motion will produce a slowly varying $\Delta \int \mathscr{A}^2 p'/\bar{\mathscr{M}}$ distribution, and a full resonance phase motion produces a clear signature in $\Delta \int \mathscr{A}^2 p'/\bar{\mathscr{M}}$ with the presence of zero, maximum and minimum values.

As mentioned previously, the background has no dependence in s_{12} and its contributions drops out of the $\Delta \int \mathscr{A}^2$ distribution.

To extract the phase motion, we divide $\Delta \int \mathscr{A}^2$ by $\bar{\mathscr{M}}$, and multiply by p', both known functions of s_{13}. By doing this, the only s_{13} dependence of the right hand side of 2 is in the phase motion $\delta(s_{13})$. The angular function has zero at $s_{13} \sim 0.48$ GeV$^2/c^4$ which implies a singularity around this value in $\Delta \int \mathscr{A}^2 p'/\bar{\mathscr{M}}$ distribution. In Fig. 2 we show the $\Delta \int \mathscr{A}^2 p'/\ \bar{\mathscr{M}}$ distribution. To treat the effect of the singularity, we have used a binning such that the singularity is placed in the middle of one bin. Doing this, we isolate the singularity in a single bin (bin 6) and discard it in the analysis. We point out that the location of this singularity can only affect the exact position of the minimum of $\Delta \int \mathscr{A}^2 p'/\bar{\mathscr{M}}$. It does not change the general feature observed, in figure 2, that this quantity starts at zero, has maximum and minimum values, and comes back to zero, the signature for a strong phase variation. We can see that the 6th bin, has a huge error, which corresponds to the bin of the singularity.

Assuming that $\delta(s_{13})$ is an analytical function of s_{13}, Eq. 2 allows us to set the two following conditions at the maximum and minimum values of $\Delta \int \mathscr{A}^2 p'/\bar{\mathscr{M}}$ respectively:

$$\Delta \int \mathscr{A}^2 p'/\bar{\mathscr{M}}_{max} \to sin(2\delta(s_{13})+\gamma) = -1 \qquad (3)$$

and

$$\Delta \int \mathscr{A}^2 p'/\bar{\mathscr{M}}_{min} \to sin(2\delta(s_{13})+\gamma) = 1. \qquad (4)$$

With these two conditions, we obtain γ and $\mathscr{C}$ and with these values, we can extract directly $\delta(s_{13})$ from Fig. 2 by inverting Eq. 2.

To propagate the data statistical errors to the values of the γ and $\delta(s_{13})$, we "produce" statistically compatible "experiments" by allowing each bin fluctuate randomly following a Poisson law. We then solve the problem for each set. The statistical error in each bin of $\delta(s_{13})$ will be the RMS of the distribution of $\delta(s_{13})$ solutions for the Monte Carlo "experiments". For the systematic errors, we change the ε parameter ($\varepsilon = 0.22GeV^2/c^4$

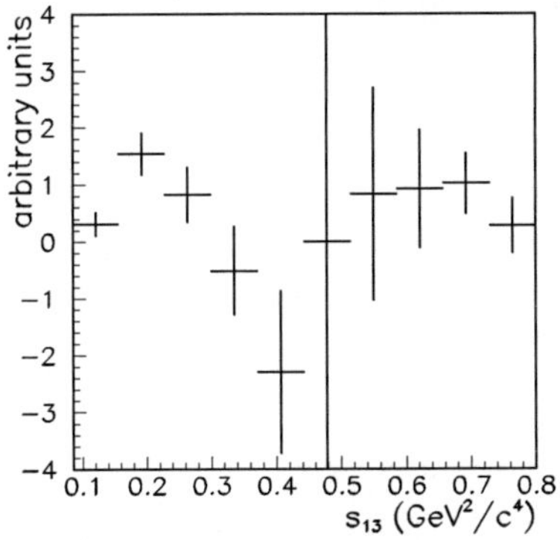

FIGURE 2. s_{13} distribution for $\Delta \int \mathscr{A}^2\ p'/\bar{\mathscr{M}}$. Note that the scale is arbitrary once the data, $\Delta \int \mathscr{A}^2$, is multiplied by an unnormalized function, $p'/\bar{\mathscr{M}}$. Statistical error only

and $\varepsilon = 0.30 GeV^2/c^4$); we examine the possible influence of other neglected amplitudes contributing in this region of the phase space (based on E791 amplitudes measurement for non-resonant, $f_0(1370)\pi^+$ and $\rho^0(1450)\pi^+$ [3]); and finally, we study the effect of averaging ${}^{J=2}\bar{\mathscr{M}}_{f_2(1270)}(s_{13})$ distributions. The three systematic errors, while treated separately bin-by-bin, are found to be of an average size, relative to the statistical uncertainty, of 1, 0.6, and 0.4 respectively before addition in quadrature.

For the FSI phase we measure $\gamma = 3.31 \pm 0.33 \pm 0.49$ (with the first error statistical and the second systematic). Here, in addition to the systematic error sources mentioned above, we have also incorporated one related to the binning choice that we made to avoid the singularity in $\bar{\mathscr{M}}$. If we change the binning such that the singularity is placed between bins we measure $\gamma = 3.52 \pm 0.53$. The value is somewhat larger than the E791 full Dalitz-plot analysis value ($\gamma_{Dalitz} = 2.59 \pm 0.19$) [3]. The consequence of the use of an effective mass-squared for the $f_2(1270)$ = 1.535 GeV^2/c^4 instead of the nominal mass is responsible for the shift observed in the relative phase. To evaluate the magnitude of this effect, we generated 1000 samples of fast-MC, with only two amplitudes, $f_2(1270)$ and $\sigma(500)$. For both, we used Breit-Wigner functions with the E791 parameters, including the phase-difference of 2.59 rad. Then we extract γ from these 1000 samples, using the method presented here. The result has a mean value of 3.07 ± 0.10 rad, instead the 2.59 rad used as input value. To correct for this systematic effect, we used the difference between the generated and measured γ in Monte Carlo to estimate an offset correction factor (2.59 - 3.07 = -0.48). After applying this correction to the above γ measured, our best estimate become $\gamma_{corr} = 2.83 \pm 0.38 \pm 0.49$. So the observed FSI phase-difference between the $f_2(1270)\pi^+$ and the $\sigma(500)\pi^+$ states in D^+ decay in the Dalitz analysis and the γ_{corr} in AD Method are in good agreement.

With γ and $\mathscr{C}$ values we solve Eq. 2 for $\delta(s_{13})$ for each s_{13} bin, but there are ambiguities that arise due to the sin^{-1} operations To resolve the ambiguities we use the assumption that the phase starts at zero at threshold and is an increasing, monotonic,

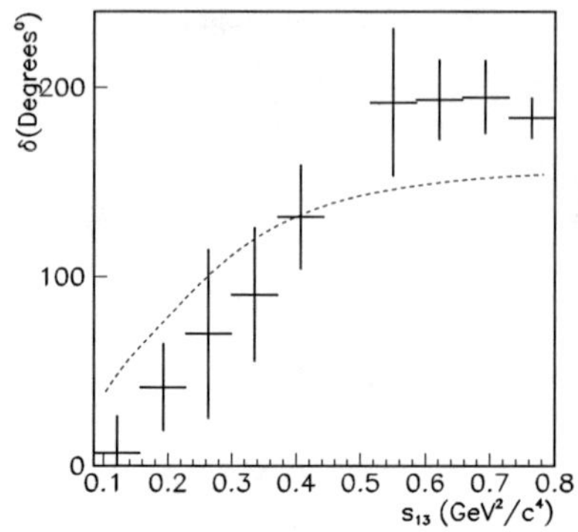

FIGURE 3. Phase motion vs s_{13}, with errors shown (systematic and statistical in quadrature). The continuous line is the Breit-Wigner phase motion, with the E791 parameters for the $\sigma(500)$.

and smooth function of s_{13}.

Using the above criteria, the best solution for $\delta(s_{13})$, including systematic and statistical errors, is shown in Fig. 3. We see a strong phase variation of about 180^0, starting at threshold and saturating around $s_{13} = 0.6\ \text{GeV}^2/c^4$. Our limited statistics implies the use of a few bins to describe a wide phase variation. It does not allow us to accurately assign the mass and width parameters to a BW resonance fitted to this plot. We can only say that, at least for the preferred solution, the observed phase variation is compatible with the complete phase motion through 180^0 expected for a resonance. This result is in agreement with the evidence for a broad and low mass scalar resonance suggested by the previous E791 result using the full Dalitz-plot analysis [3].

The Breit-Wigner phase motion for the E791 mass and width parameters, combined with other scalar contributions obtained from that fit [3], is shown as the continuous line of Fig. 3. We can see a two standard deviation difference at low and also at high mass, in opposite directions, but in overall agreement with the mass region where a scalar meson σ must have its strong phase variation.

We have presented a direct measurement of the phase motion of the low mass $\pi^+\pi^-$ scalar amplitude using the well known $f_2(1270)$ tensor meson in the crossing channel acting as an interferometer.. The result in Fig. 2 come from a simple event counting procedure in a region of phase space dominated by a D-wave $f_2(1270)$ interfering with the S-wave. The derivation of the phase motion, Fig. 3, relies heavily on the assumption that the maximum and minimum bins in Fig. 2 correspond to the quantity $S = sin(2\delta+\gamma) = +1$ and -1. The clear presence of a maximum and a minimum indicates this is a reasonable assumption. Given this caveat, the most likely solution for $\delta(s_{13})$ has a variation of about 180^0, consistent with a resonant $\sigma(500)$ contribution. We also obtain good agreement between the FSI γ_{corr} observed with the Amplitude Difference Method and the γ observed in the full Dalitz-plot analysis using an isobar model [3]. These results support the previous evidence for an important contribution of the isoscalar $\sigma(500)$ meson in $D^+ \to \pi^- \pi^+ \pi^+$ decay [3].

ACKNOWLEDGMENTS

We would like to acknowledge the E791 Collaboration for allowing us to use their data for this paper and for the valuable discussions.

REFERENCES

1. I. Bediaga and J. Miranda, *Phys. Lett.*, **B550** 135,(2002).
2. I. Bediaga for E791 Collaboration, *Scalar Meson Workshop*, May 2003, SUNYIT, Utica, NY and hep-ex/0307008.
3. E791 Collaboration, E.M. Aitala *et al.*, *Phys. Rev. Lett.*, **86**,770 (2001).
4. E791 Collaboration, E.M. Aitala *et al.*, *Phys. Rev. Lett.*, **86**,765 (2001).

Structure of $\pi^+\pi^-$ S-wave in Partial Wave Analysis of the $\pi^-A \to A'\pi^+\pi^-\pi^-$ reaction

Yu. Khokhlov for the VES Experiment [1]

IHEP, 142281 Protvino, Russia

Abstract. The $\pi^+\pi^-$ S-wave model is constructed in the course of the Partial Wave Analysis of the data on the reaction $\pi^-A \to A'\pi^+\pi^-\pi^-$. Its structure, including the narrow resonances $f_0(980)$ and $f_0(1500)$ as well as broad "objects" $\varepsilon(1300)$ and $\sigma(500)$, is discussed.

Keywords: Scalar meson, Resonance, PWA
PACS: 13.85.Hd, 14.40.Cs, 11.80.Et

1. INTRODUCTION

In spite of a long history of its study, both theoretical and experimental, the situation in the scalar sector of light mesons remains controversial. The number and assignment of scalars to multiplet(s), their "composition" and role in theory – all are unknown.

For the experimental study the scalars are accessible almost exclusively through an S- wave state of two pseudo-scalars P. Probably the most convenient from this point of view are the scattering processes of the type $a+b \to PP$, such as $\pi\pi \to PP$, $\gamma\gamma \to PP$, $\mathscr{P}\mathscr{P} \to PP$ ($\mathscr{P}$ stands for a pomeron), 3-particles decays of well determined states like $\psi\prime \to \psi\pi\pi$, $D \to PPP$, $B \to PPP$ (see [1]) and, in a lesser extent, annihilation $p\bar{p} \to PPP$.

In the present paper the structure of the S-wave in the $\pi^+\pi^-$ system accessing the isoscalar scalar mesons is studied less directly, in a course of the detailed Partial Wave Analysis (PWA) of the $\pi^+\pi^-\pi^-$ system produced diffractively by pions on nuclei.

2. DATA SAMPLE AND THE PWA MODEL

A sample of about 8 billion events of the exclusive reaction $\pi^-A \to A\pi^+\pi^-\pi^-$ collected with the VES setup at 37 GeV/c beam momentum is used for the analysis. The description of the spectrometer can be found elsewhere [2]. The general formalism of the PWA of 3 pseudoscalars and its implementation in our experiment are considered in [3] and [4] respectively. Here we give shortly the notation and specific features of the PWA model used.

---

[1] V. Dorofeev, R. Dzhelyadin, A. Fenyuk, Yu. Gavrilov, Yu. Gouz, A. Ivashin, V. Kabachenko, I. Kachaev, A. Karyukhin, Yu. Khokhlov, A. Konoplyannikov, V.F. Konstantinov, V. Matveev, V. Nikolaenko, A. Ostankov, B. Polyakov, D. Ryabchikov, A.A. Solodkov, A.V. Solodkov, O. Solovianov, E. Starchenko, A. Zaitsev, A. Zenin

CP814 *Hadron Spectroscopy: Eleventh International Conference*
edited by A. Reis, C. Göbel, J. Sá Borges, and J. Magnin

The production and decay of the 3 pion system with is characterized with a density matrix ρ of a set of final states with quantum numbers $J^P M \eta L X$, where J is a spin of the system, P is a parity, M is a projection quantum number w.r.t. z axis in the (Gottfried-Jackson) coordinate system, L is an angular momentum between an *isobar* X and the *bachelor* pion. The matrix has two non interfering blocks according to the value of the "reflectivity" η w.r.t. the reaction plane.

The task of the mass independent PWA is to determine elements of the density matrix as a function of the invariant mass $M_{3\pi}$. It is achieved by a fitting procedure with the use of the method of the event-by-event log-likelihood. The change of its fitted maximum (further on *LogLk*) w.r.t. variations of the model serves as a criteria of the relative change in the quality of data description by the PWA solution.

The final set of our model consists of some 40 waves with J up to 4, L up to 3. The density matrix of an arbitrary rank which amounts about 600 fitting parameters, is supplemented with a non-interfering block of a single structureless and isotropic upon all variables "wave" (so called "FLAT"). Note, however, that the "wave" "flat" proved out to be practically consistent with zero.

The well known ([5]) resonances in the $\pi^+\pi^-$ system $\rho(770)$, $f_2(1270)$, $\rho_3(1690)$ prescribed with the relativistic Breit-Wigner formula constitute the starting set of isobars. However, it proved out that to describe data this set must be complemented by a structure in the scalar channel. The trial of constructing reasonably economical but still informative model of the S-wave in a wide range of the $M_{\pi^+\pi^-}$ constitutes an integral part of the analysis. We present some PWA results and discuss its features relevant only to the S-wave study.

3. THE S-WAVE AMPLITUDE CONSTRUCTION

As a basis for the S-wave amplitude the AMP M-solution ([6]) was chosen:

$$M_{ij} = \frac{a_{ij}}{s - s_0} - \sum_p \frac{f_i^p f_j^p}{s_p - s} + \sum_{n=0}^{4} c_{ij}^n \left[\frac{s}{4m_K^2} - 1\right]^n, \tag{1}$$

where the first term accounts for Adler zero, i and j enumerate channels $(\pi\pi, KK)$, p enumerates poles. (To reproduce the amplitude behavior as it is in the figure in [6] we replaced the misprinted plus sign at the second term with "minus"). Starting from (1) and referencing to our data we modify it significantly. Prior to describe our parameterization we should make some comments.

In the original work the second term of (1) has a single pole corresponding to the narrow resonance $f_0(980)$ which exhibits itself in the amplitude squared as a sharp dip on a broad bump. In the 2→2 processes this pole can be compensated by a pole in the production amplitude, if required by the experimental spectrum. In the reaction under study, without the dip in the $M_{\pi\pi}$ and within the given PWA formalism with a density matrix, this receipt is not applicable directly.

Further on, outside the relatively low mass region to which authors were addressed, two peculiarities are observed in the original amplitude due to the behavior of the 4-th order polynomial in (1). Around $M_{\pi\pi} = 1.6$ GeV/c^2 the intensity shows a shoulder, and

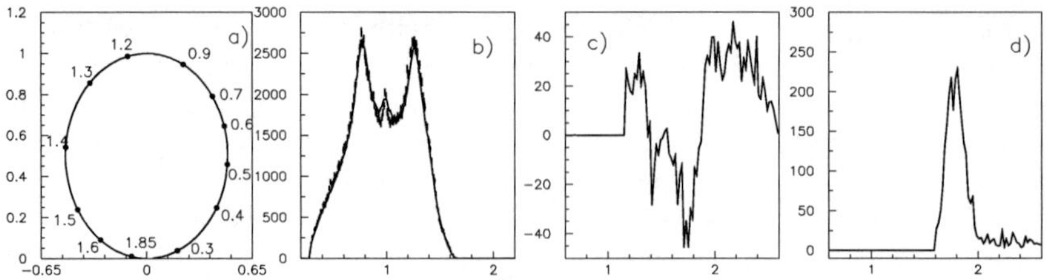

FIGURE 1. (a) The Argand plot for the "$\varepsilon(1300)$"; (b) the $M_{\pi\pi}$ (in GeV/c^2) spectrum for central part of Euler angles and $M_{3\pi}$=1.6–1.8 GeV/c^2, for data (points with errors) and the model (line) ; (c) the difference $LogLk_F$-$LogLk_{BW}$ vs.$M_{3\pi}$ for two models of the $f_0(980)$; (d) the gain in the *LogLk* vs. $M_{3\pi}$ when the $f_0(1500)$ is included

at higher masses the trajectory of the complex amplitude moving to the $(Re, Im) = (0,0)$ of the Argand plot has an unphysical loop with the reversed sign of the curvature. We relate the appearance of such rather high order of the polynomial in [6] with a need to account for effectively a contribution of the $f_0(1500)$ meson which is not resolved explicitly.

We model our parameterization in following way: 1) in the M-matrix null the couplings with the pole and cancel the highest term in the polynomial. The Argand plot for the residual amplitude, which we call "$\varepsilon(1300)$", is shown in fig.1(a). 2) Add outside the M-matrix "traditional" amplitudes for narrow resonance(s).

4. THE $F_0(1500)$ AND $F_0(980)$ CASES

The $M_{\pi\pi}$ "raw" spectrum is dominated by signals of the $\rho(770)$, the $f_2(1270)$ and broad bump related to $\varepsilon(1300)$. The $f_0(980)$ resonance which is practically indistinguishable on it, can be "visualized" as a peak (see fig.1(b)) selecting an S-wave enriched central region of the two Euler angles in the phase space of 3π .

The replacement of the AMP-M amplitude with the modified one, "$\varepsilon(1300)$" plus the Breit-Wigner (*BW*) $f_0(980)$, leads to a huge gain of the total (over $M_{3\pi}$) *LogLk*. The "best" $M = 975$ MeV/c^2 and $\Gamma = 78$ MeV/c^2 at which the integral win is maximal are found using many PWA runs scanning over these parameters.

Despite of so significant improvement, the model still does not describe fully the complicated shape of the $f_0(980)$ changing rapidly with the $M_{3\pi}$. Along with the *BW* we have probed the (*F*) Flatte model ([7]) for it, with the *best* parameters $M = 966$ MeV/c^2, $\Gamma_{\pi\pi} = 120$ MeV/c^2, $\Gamma_{KK} = 160$ MeV/c^2. The corresponding change of the *LogLk* is a sign-alternating function of the $M_{3\pi}$ (see fig.1(c)), though the *F* has some advantage over the *BW* in integral. We are not able to make a unique choice in favor of either of them. There are also other parameterizations (see e.g. [8]) for the line shape of this object with a complicated dynamics.

The $f_0(1500)$ displays itself as statistically very significant distortion of the spectrum $M_{2\pi}$. The fig.1d shows the increase of the *LogLk* vs. $M_{3\pi}$ when introducing the *BW* isobar with the best parameters $M = 1420$ MeV/c^2, $\Gamma = 140$ MeV/c^2.

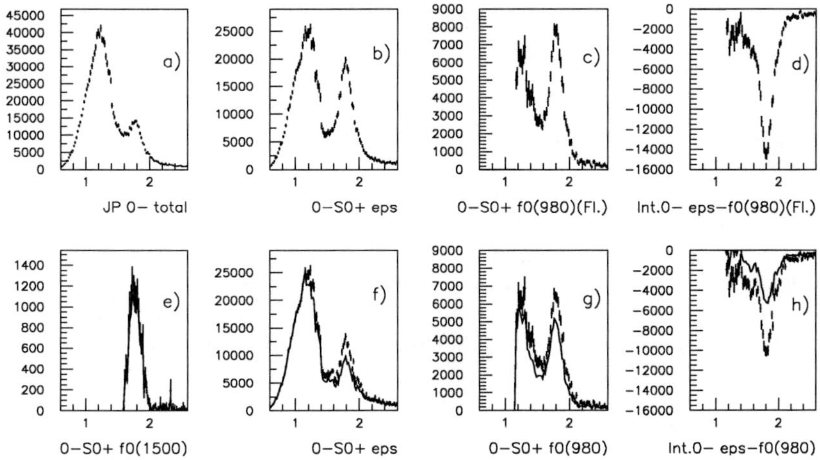

FIGURE 2. Upper row: selected PWA results without the $f_0(1500)$. The total wave 0^- (a); its channels into "ε" (b) and $f_0(980)$ in Flatte parameterization (c) and interference between them (d). Lower row: the $0^- S0 + f_0(1500)$ wave (e); (f) – (h) correspond to (b)–(d) but with the $f_0(1500)$ included. Solutions with two models of the $f_0(980)$ are shown in (e) to (h) : Flatte (points with errors) and Breit-Wigner (line). Everywhere the x axis is for the $M_{3\pi}$ in Gev/c^2.

The fig.2 shows some results of the PWA, relating to the $0^- S$ wave, which "absorbs" most of the salars. In general, the splitting of this wave to channels with "$\varepsilon(1300)$" and narrow f_0's depends strongly on their parameterization, contrary to its total intensity.

In a wide bump peaking at about 1200 MeV/c^2 (Fig.2a), which can be (at least partly) associated with the $\pi(1300)$, the contribution of the "$\varepsilon(1300)$" dominates. We will focus the attention to the signal of the well established resonance $\pi(1800)$.

The very large negative interference between the $0^- f_0(980)\pi$ and $0^- \varepsilon(1300)\pi$ waves is prominent in it. The inclusion of the $f_0(1500)$ along with the F-parameterized $f_0(980)$ results in almost double damping of the $\pi(1800)$ decay mode into the "$\varepsilon(1300)$" and the comparable suppression of the interference with it. Even stronger suppression of the "$\varepsilon(1300)$" contribution in intensity and interference is observed when using the *BW*-parameterized $f_0(980)$, which is an argument for *BW* against the Flatte model.

The wave $0^- S0 + f_0(980)$ is more stable w.r.t. the variations of the model, the inclusion of the $f_0(1500)$ among them. The variations of the respective intensities determine the overall uncertainty in the branching fraction measured:

$$\frac{Br\big(\pi^-(1800) \to f_0(1500)\pi^-\big) \times Br\big(f_0(1500) \to \pi^+\pi^-\big)}{Br\big(\pi^-(1800) \to f_0(980)\pi^-\big) \times Br\big(f_0(980) \to \pi^+\pi^-\big)} = 0.25 \pm 0.10 \qquad (2)$$

5. THE $\sigma(500)$

The most prominent discrepancy of the data with its modeling according to the aforesaid S-wave - parameterization is observed in the region $m_{\pi\pi} < 1$ GeV. A significant improvement of the description can be obtained with the supplement of the scalar $\sigma(500)$, with Breit-Wigner parameters $M = 500$ MeV, $\Gamma = 550$ MeV.

However we have to qualify the use of very broad "resonances". The ability of the non-unitary isobar model without final state interaction to incorporate them correctly is questionable. The behavior of waves "decaying" to them indicates likely that only a total $J^P M\eta$ state is meaningful, and not its splitting to different modes.

In this sense the $\sigma(500)$ should be thought of as a mean to loose too "rigid" effective "ε" model to fit better the broad part of the S-wave to the data in the wide range of $M_{3\pi}$.

6. CONCLUSION

The $\pi^+\pi^-$ S-wave parameterization is built in the framework of the 3- pions PWA.

The narrow scalar resonances $f_0(980)$ and $f_0(1500)$ are reliably observed. They dominantly contribute to the decay of the $\pi(1800)$ which is remarkable by itself by the suppression of the decays into "usual" modes such as $\rho(770)\pi$.

The branching ratio of the $\pi(1800)$ decays to these two scalars is measured, the major contribution to its uncertainty being the variations of their parameterization.

The suggested "reduced" AMP-like solution named "$\varepsilon(1300)$" together with the two narrow resonances yields a significant gain in a quality of the data description against the original AMP-solution. Its further improvement is obtained by inclusion of the broad $\sigma(500)$ resonance. Taking into account reservations on the use of wide "isobars" in the PWA model, we regard the "ε+σ" structure like an effective tool to describe the data rather than a measurement of parameters of real resonance(s).

ACKNOWLEDGMENTS

The work is supported by RFBR grants 05-02-08082 and 05-02-17664 and by Presidential grant NSh-1695.2003.2. The participation of Yu.Kh. in the Conference is supported by RFBR grant 05-02-26884.

REFERENCES

1. E. Klempt, *HADRON-05 Summary*, this Conference.
2. S. Bityukov et al., Phys.Lett. **B 268** (1991) 137
3. J. Hansen et al., Nucl.Phys. **B 81** (1974) 403
4. D.I. Amelin et al., Phys.Lett. **B 356** (1995) 595
5. PDG Authors, S.Eidelman et al., Phys.Lett. **B 592** (2004) 520
6. K.L. Au, D. Morgan, M.R. Pennington, Phys.Rev. **D 35** (1987) 1633
7. S.M. Flatte, Phys.Lett. **63 B** (1976) 224, 228
8. N. Achasov, G. Shestakov, hep-ph/0506046

Scalar Mesons As A Mixing Of Two And Four Quark States

B. Silvestre-Brac[+], J. Vijande[++], F. Fernandez[++], A. Valcarce[++]

[+] *LPSC, 53 Avenue des martyrs ,F38026 Grenoble Cedex, France*
[++] *Grupo de Física Nuclear, Universidad de Salamanca, E37008 Salamanca, Spain*

Abstract. The scalar mesons are a puzzling problem in meson spectroscopy: they appear to be too numerous and with a mass often incompatible with usual quark-quark potentials. In this paper, we study the possibility to describe them as a mixing of states composed of one and two quark-antiquark pairs. A potential containing confinement, gluon exchange and boson exchange, as expected from chiral symmetry, is used in a consistent way to calculate the two and four quark states separately. Then, a coupling between these states is introduced as a constant term depending only on the flavour of the created pair. The description is largely improved. To refine the treatment, a coupling with a glueball is also considered. All the experimental resonances seem to fit correctly in this scheme.

Keywords: Chiral symmetry, tetraquarks, glueball, $q\bar{q}$ pairs, scalar mesons.
PACS: 12.39.-x, 12.38.-t .

THE SCALAR MESON PUZZLE

It appears that, with a sufficiently realistic interquark potential, the spectroscopy of pseudoscalar, vector and tensor resonances can be understood quite well in a framework limited to quark-antiquark pairs. But scalar mesons (spin 0, parity + and C parity + when it makes sense) raise a number of problems.

First, we observe experimentally too many resonances as compared to what can be accommodated in such a simple model. Indeed we do know presently (at least below 2.2 GeV) 15 such states: 2 isovectors $a_0(980)$, $a_0(1450)$, 3 with isospin ½ $K_0(800)$, $K_0(1430)$, $K_0(1950)$, and 10 isoscalars $f_0(600)$, $f_0(980)$, $f_0(1200\text{-}1600)$, $f_0(1370)$, $f_0(1500)$, $f_0(1710)$, $f_0(1790)$, $f_0(2020)$, $f_0(2100)$, $f_0(2200)$. As we will see, a naïve model based on a quark-antiquark pair can justify only 11 scalar resonances. Moreover, even when a description in terms of $q\bar{q}$ pair can be invoked, the calculated mass is generally quite different to the experimental value. Lastly, the strong and electromagnetic transitions concerning scalar resonances are hardly compatible with such a simple $q\bar{q}$ structure.

Possible explanations for this puzzling behaviour have been proposed; in particular, tetraquarks, glueballs and hybrids may play an important role. We examine this subject in the following.

CP814 *Hadron Spectroscopy: Eleventh International Conference*
edited by A. Reis, C. Göbel, J. Sá Borges, and J. Magnin

THE INTERQUARK POTENTIAL

The first thing to do is to chose a quark-quark potential that can provide, in a consistent way, a good description of the underlying physics in every sector. It is phenomenological but sticks, as close as it can, to the basic properties of QCD dynamics. It contains 3 main different sources [1]:

$$V_{ij}(r) = V_{ij}^{conf}(r) + V_{ij}^{oge}(r) + V_{ij}^{ch}(r)$$

First, the confinement $V_{ij}^{conf}(r)$ is a basic ingredient of QCD. It is essentially non perturbative, acts at long range and concerns all types of quarks. It is composed of a central part and a spin-orbit contribution. The central part is a screened potential, linear at intermediate range but which saturates to a constant at large distance.

The second source $V_{ij}^{oge}(r)$ is due to one gluon exchange. It is perturbative and concerns all types of quarks. It contains central, spin-orbit and tensor contributions. It is well known; let us just mention that the Fermi-Breit prescription, leading to a collapse, has been regularized by a Yukawa form factor that is mass dependent. Moreover, we use an effective strong coupling constant $\alpha_s(\mu_{ij})$ which is scale dependent through the reduced mass of the interacting pair.

The last source of potential $V_{ij}^{ch}(r)$ comes from the spontaneous chiral symmetry breaking at some momentum scale. The resulting boson exchange potentials are essentially non perturbative and concerns only light quarks. They include a scalar σ exchange, and a pseudoscalar exchange, which after symmetry breaking, reduces to π, K and η exchanges. The scalar potential gives rise to a relativistic spin-orbit contribution, whereas the pseudoscalar exchanges give rise to a tensor contribution.

The parameters have been determined from N-N interaction (for the chiral part), and on meson spectroscopy (for the rest). They have quite reasonable values.

QUARK-ANTIQUARK STATES

With the previous potential, one can calculate the states composed of one $q\bar{q}$ pair, that are called diquonia. For heavy quarks, one has just confinement and one gluon exchange interactions. For light quarks, one has in addition boson exchange terms.
The kinetic energy operator is chosen to be non relativistic, and the corresponding Schrödinger equation is solved by the Noumerov algorithm, which is very accurate.
The resulting spectra for pseudoscalar, vector and tensor resonances are quite good (error less than 100 MeV), whatever the flavour sector considered [1].
But the situation deteriorates a lot, passing to the scalar sector. For the a_0 family, the result could be considered as correct. Although the right number of K_0 resonances is obtained, the calculated values are dramatically too large, since the discrepancy can reach 500 MeV. Below 2.2 GeV, one sees 10 isoscalar resonances, while in the same region there exist only 6 theoretical states. In particular, one observe 3 large gaps in which diquonia are absent: between 0.5 GeV and 1.3 GeV, where lie $f_0(600)$ and $f_0(980)$, between 1.4 GeV and 1.7 GeV, where lies $f_0(1500)$, and between 1.9 GeV and 2.2 GeV where lie $f_0(2020)$ and $f_0(2100)$.

Thus, it is absolutely clear that diquonia alone cannot accommodate the experimental data and one must search elsewhere explanations to this feature.

TETRAQUARK STATES

A possibility to explain the missing states is to appeal for tetraquark systems. A tetraquark is a compact object made of two quarks and two antiquarks. Depending upon the coupling, we have two different types of structure. The diquark-antidiquark structure is obtained by grouping the quarks together and the antiquarks together $(qq)(\overline{qq})$; the most natural Jacobi coordinates are denoted x (relative distance between quarks), y (relative distance between antiquarks) and z (diquark-antidiquark distance). This coupling is convenient for treating the Pauli principle. The meson-meson structure is obtained by grouping a quark and an antiquark $(q\overline{q})(q\overline{q})$; it is best suited to study the thresholds.

We solve the dynamical problem as a variational one [2], based on a wave function expanded on basis states $\psi_k(\alpha)$ depending on non linear parameters α. The resulting equation of motion is known as a Hill-Wheeler equation, which contains two kernels : the norm kernel (overlap between basis states due to non orthogonality) and the energy kernel (corresponding to the matrix elements of the Hamiltonian). The eigenvalues depend on the parameters α; these are determined by a usual minimization procedure based on energy $\partial E(\alpha)/\partial\alpha = 0$.

The basis states are a tensor product of colour, isospin, spin and space functions. There exist 2 possibilities for the colour to be a singlet : either the diquark is in an antitriplet state or in a sextet state. The isospin and spin functions result from the usual SU(2) couplings; their number depends on each peculiar case. The space function $\phi_k(\alpha)$ is of gaussian type, depending on 6 free parameters : $\alpha = \{a, b, c, d, e, f\}$.

$$\phi_k(\mathbf{x},\mathbf{y},\mathbf{z};\alpha) = \exp(-a_k\mathbf{x}^2 - b_k\mathbf{y}^2 - c_k\mathbf{z}^2 \pm d_k\mathbf{x}.\mathbf{y} \pm e_k\mathbf{x}.\mathbf{z} \pm f_k\mathbf{y}.\mathbf{z})$$

The mixed terms $\mathbf{x}.\mathbf{y}$, $\mathbf{x}.\mathbf{z}$, $\mathbf{y}.\mathbf{z}$ are usually neglected; however we proved that they are very important. We kept them in our calculations, although they lead to serious complications. Solving the tetraquark problem with the presented interquark potential is time consuming due to the large number of free parameters (several hundreds), but one gets reliable solutions. The interesting point is that tetraquark states fit nicely in the scheme, since they appear precisely in the energy gaps where diquonia are absent. However the situation is not yet satisfactory because the calculated energy values do not match the experimental data.

A FULL DESCRIPTION

The scalar mesons are particles with $B = 0$ baryonic number. The Fock space is in principle infinite, composed of one $q\overline{q}$ pair, two $q\overline{q}$ pairs, three $q\overline{q}$ pairs and so on so for. In the following, we restrict it to the diquonia plus the tetraquark sectors and consider the coupling between them. Thus the Hamiltonian $H = H_0 + H_1$ has a

diagonal part H_0 which represents the diquonia and tetraquark sectors separately and can be treated as explained in the two preceeding chapters, and a term H_1 which allows the coupling between both sectors. Our prescription for this coupling is very simple: we take a constant term whose value depends only on the flavour of the created pair, C_n associated to the creation of an ordinary quark $n\bar{n}$ pair, and C_s associated to the creation of the strange quark $s\bar{s}$ pair. After a fit to the data, the resulting values C_n = 165 MeV and C_s = 70 MeV show that the coupling can be treated perturbatively.
Indeed the mixing has nevertheless a crucial importance, and the physical states are most of the time a complicated mixing of the various components. For example the $f_0(600)$ is 84% a diquonium and 16% a tetraquark, while the $f_0(980)$ is 83% a tetraquark and 17% a diquonium. The coupling improves the situation, especially in the kaon sector and in the lowest part of the f_0 family, but in the isoscalar sector there is still a missing state.
It has been suggested [3] that, in the energy range 1.5-1.7 GeV, there exists a scalar glueball. This conclusion is supported by lattice calculations. We suppose that such a glueball exists and take its unperturbed mass equal to 1643 MeV, a value derived from lattice QCD. Since gluons do not carry flavour, we consider a coupling which is flavour blind and we have the relationship $V_{G\to n\bar{n}} = \sqrt{2} r V_{G\to s\bar{s}}$. For a pure SU(3) symmetry, the parameter r should equal 1, but, due to a number of corrections, QCD prefers a value a bit larger; we chose r = 1.2. The value for the coupling $V_{G\to s\bar{s}}$ has been proposed to be 64 MeV, and we adopted this value. We suppose that the coupling of a glueball to tetraquarks is negligeable.
Owing to all the couplings taken into consideration, we redid all our calculations and we found improved spectra for scalar resonances. We present these results in the figure. The influence of the coupling plays a major role, and improves appreciably the situation. Most of the resonances are described now within an error less than 100 MeV, and, in the isoscalar sector, each resonance can be assigned for.

CONCLUSIONS

To get a correct description of scalar mesons, the tetraquark states are an unavoidable ingredient. They appear in the energy gaps where diquonia are absent.
The coupling between diquonia and tetraquarks that, for the moment, are considered as pure phenomenological constants, plays an important role; not only in the value of the resulting masses, but also in the mixing of the various components of the wave function. Thus, the scalar resonances are complex states built with a large number of non negligeable components. This is certainly a crucial point to improve the strong and electromagnetic transitions concerning those resonances.
It seems also that a glueball state is likely in the isoscalar sector to reproduce correctly the data.
We want to stress that we do not pretend that our theory reflects completely the physical reality, but it offers a possible satisfactory explanation, at least as a first step,

of this long standing scalar puzzle. The spectra are largely improved, but it is also quite clear that problems remain and need for further studies and developments.

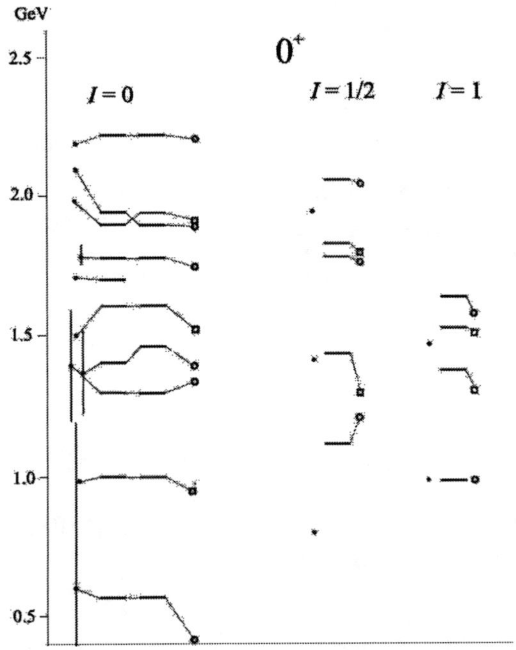

FIGURE. Spectra of scalar mesons. In the right hand part is the a_0 family, in the middle the K_0 family and in the left the f_0 family. The experimental data are denoted by full circles with error bars. The open circles and open squares stand for the diquonia and tetraquark states respectively. The horizontal bars represent the values when the coupling between them is switched on. In the very left column, one has in addition the coupling to the glueball.

ACKNOWLEDGMENTS

This work has been partially funded by Ministerio de Ciencia y Tecnologia under Contract No FPA2004-05616, by Junta de Castilla y Leon under Contract No SA104/04 and by the IN2P3-CICYT agreement PTh05-1.

REFERENCES

1. J. Vijande, F. Fernandez and A. Valcarce, *J. Phys. G* **31**, 481-506 (2005).
2. J. Vijande, F. Fernandez, A. Valcarce and B. Silvestre-Brac, *Euro. Phys. J A,* **19**, 383-389 (2004).
3. G.B West, *Phys. Rev. Lett,* **77**, 2622-2625 (1996).

Light scalar and vector mesons in the large N_c limit from unitarized chiral perturbation theory

José R. Pelaez

Dept. Física Teórica II. Universidad Complutense, 28040-Madrid, Spain

Abstract. We review the large N_c behavior of light scalar and vector resonances generated from unitarized meson-meson scattering amplitudes at one loop in Chiral Perturbation Theory. The vectors nicely follow the behavior expected for $\bar{q}q$ states, whereas scalar mesons do not. This suggests that the main component of light scalars is not $\bar{q}q$. We also comment on t-channel vector exchange as well as on the large N_c behavior of the mass splittings between the vectors generated from the inverse amplitude method.

Keywords: mesons, scalars, large N_c, chiral perturbation theory, unitarizatiom
PACS: 12.39.Fe, 11.15.Pg, 12.39.Mk, 13.75.Lb

INTRODUCTION

Chiral Perturbation Theory (ChPT) is the QCD low energy Effective Lagrangian built as the most general derivative expansion respecting SU(3) symmetry and containing only π, K and η mesons [1]. These particles are the QCD low energy degrees of freedom since they are Goldstone bosons of the QCD spontaneous chiral symmetry breaking. For meson-meson scattering, ChPT is an expansion in even powers of momenta, $O(p^2), O(p^4)$..., over a scale $\Lambda_\chi \sim 4\pi f_0 \simeq 1\,\mathrm{GeV}$. Since the u, d and s quark masses are so small compared with Λ_χ they are introduced as perturbations, giving rise to the π, K and η masses, counted as $O(p^2)$. At each order, ChPT is the sum of *all terms* compatible with the symmetries, multiplied by "chiral parameters", that absorb loop divergences order by order, yielding finite results. The leading order is universal, containing only one parameter f_0, that sets the scale of spontaneous symmetry breaking. Different underlying dynamics correspond to different values of the higher order parameters, called L_i, that, once renormalized, depend on a regularization scale as $L_i(\mu_2) = L_i(\mu_1) + \Gamma_i \log(\mu_1/\mu_2)/16\pi^2$, where Γ_i are constants[1]. In physical observables the μ dependence is canceled with that of the loop integrals.

We will also make use of the large N_c expansion [2], which is the only analytic approximation to QCD in the whole energy region and provides a clear definition of $\bar{q}q$ states, that become bound, and whose masses and widths behave as $O(1)$ and $O(1/N_c)$, respectively. In fact, the π, K, η masses scale as $O(1)$ and f_0 as $O(\sqrt{N_c})$. The L_i parameters that determine meson-meson scattering up to $O(p^4)$ and their N_c scaling [1, 3] is given in Table 1. In order to apply the large N_c expansion, the μ scale, a dependence suppressed by $1/N_c$, has to be chosen[1] between μ =0.5 and 1 GeV.

In recent years ChPT has been extended to higher energies by means of unitarization [4, 5, 6, 7]. The main point is that the partial waves, t_{IJ}, of definite angular momentum

CP814 *Hadron Spectroscopy: Eleventh International Conference*
edited by A. Reis, C. Göbel, J. Sá Borges, and J. Magnin

TABLE 1. $O(p^4)$ chiral parameters ($\times 10^3$) and their N_c scaling. In the ChPT column, L_1, L_2, L_3 come from [8] and the rest from [1]. The IAM columns correspond to different fits [4].

$O(p^4)$ Parameter	N_c Scaling	ChPT $\mu = 770\,\text{MeV}$	IAM I $\mu = 770\,\text{MeV}$	IAM II $\mu = 770\,\text{MeV}$	IAM III $\mu = 770\,\text{MeV}$
L_1	$O(N_c)$	0.4 ± 0.3	0.56 ± 0.10	0.59 ± 0.08	0.60 ± 0.09
L_2	$O(N_c)$	1.35 ± 0.3	1.21 ± 0.10	1.18 ± 0.10	1.22 ± 0.08
L_3	$O(N_c)$	-3.5 ± 1.1	-2.79 ± 0.14	-2.93 ± 0.10	-3.02 ± 0.06
L_4	$O(1)$	-0.3 ± 0.5	-0.36 ± 0.17	0.2 ± 0.004	0 (fixed)
L_5	$O(N_c)$	1.4 ± 0.5	1.4 ± 0.5	1.8 ± 0.08	1.9 ± 0.03
L_6	$O(1)$	-0.2 ± 0.3	0.07 ± 0.08	0 ± 0.5	-0.07 ± 0.20
L_7	$O(1)$	-0.4 ± 0.2	-0.44 ± 0.15	-0.12 ± 0.16	-0.25 ± 0.18
L_8	$O(N_c)$	0.9 ± 0.3	0.78 ± 0.18	0.78 ± 0.7	0.84 ± 0.23
$2L_1 - L_2$	$O(1)$	-0.55 ± 0.7	0.09 ± 0.10	0.0 ± 0.1	-0.02 ± 0.10

J and isospin I, in the elastic regime satisfy the unitarity condition:

$$\text{Im}\, t_{IJ} = \sigma |t_{IJ}|^2, \quad \text{where } \sigma = \frac{2q}{\sqrt{s}} \quad \Rightarrow \quad \text{Im}\frac{1}{t_{IJ}} = -\sigma \quad \Rightarrow \quad t_{IJ} = \frac{1}{\text{Re}\, t_{IJ}^{-1} - i\sigma}, \tag{1}$$

where q is the meson CM momentum. In order to have a unitary amplitude we only need $\text{Re}\,t^{-1}$, that can be obtained from ChPT: this is the Inverse Amplitude Method (IAM) [5, 4]. In this way, the IAM generates the ρ, K^*, σ and κ resonances not initially present in ChPT, ensures unitarity in the elastic region and respects the ChPT expansion.

When inelastic two-meson processes occur, the IAM can be generalized [4, 6] to $T \simeq (\text{Re}\,T^{-1} - i\Sigma)^{-1}$, within a coupled channel formalism, where T is a matrix containing all partial waves between all physically accessible two-body states, whereas Σ is a diagonal matrix with their corresponding phase spaces. Using one-loop ChPT calculations, the coupled channel IAM provides a remarkable description [4] of two-body π, K or η scattering up to 1.2 GeV. In addition, it generates the ρ, K^*, σ, κ, $a_0(980)$, $f_0(980)$ and the octet ϕ. Such states are not included in the ChPT Lagrangian, but each one has an associated pole in the second Riemann sheet of its corresponding partial wave. These poles appear already with the L_i set used for standard ChPT, which is compatible with the L_i sets in Table 1, obtained from fits to data. For narrow, Breit-Wigner like, resonances, their mass and width is roughly given by $\sqrt{s_{pole}} \sim M_R - i\Gamma_R/2$. Furthermore, the IAM respects the $O(p^4)$ correct low energy expansion, with chiral parameters compatible with standard ChPT. Different IAM fits[4] are mostly due to different ChPT truncation schemes, equivalent up to $O(p^4)$, and to the estimation of systematic errors in the data.

Since ChPT amplitudes are renormalized, and therefore scale independent, there are no cutoffs or subtraction constants where a spurious N_c dependence could hide. All the QCD N_c dependence appears correctly through the L_i, f_0 and the π, K, η masses.

Recently[7], by rescaling the ChPT parameters, we have studied how the resonances generated from unitarization behave in the large N_c expansion, around "real life", $N_c = 3$. Thus, in Fig.1 we see that the modulus of partial waves associated to the $\rho(770)$ and $K^*(892)$ vector mesons presents a peak, obtained from a fit to data, that becomes narrower as N_c increases, whereas the mass remains almost the same. This is exactly the behavior expected for a $\bar{q}q$ state, namely, $M \sim O(1)$, $\Gamma \sim O(1/N_c)$.

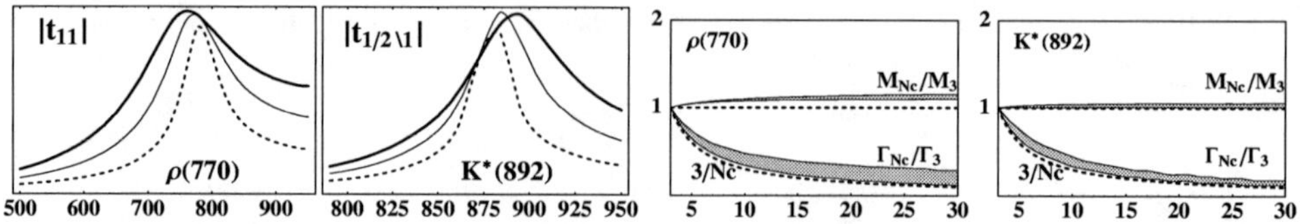

FIGURE 1. Left: Modulus of $\pi\pi$ and πK elastic amplitudes versus $\sqrt{s}$ for $(I,J)=(1,1),(1/2,1)$: $N_c=3$ (thick line), $N_c=5$ (thin line) and $N_c=10$ (dotted line), scaled at $\mu=770\,\mathrm{MeV}$. Right: $\rho(770)$ and $K^*(892)$ pole positions: $\sqrt{s_{pole}} \equiv M - i\Gamma/2$ versus N_c. The gray areas cover the uncertainty $\mu = 0.5-1\,\mathrm{GeV}$. The dotted lines show the large N_c scaling expected for a $\bar{q}q$ state.

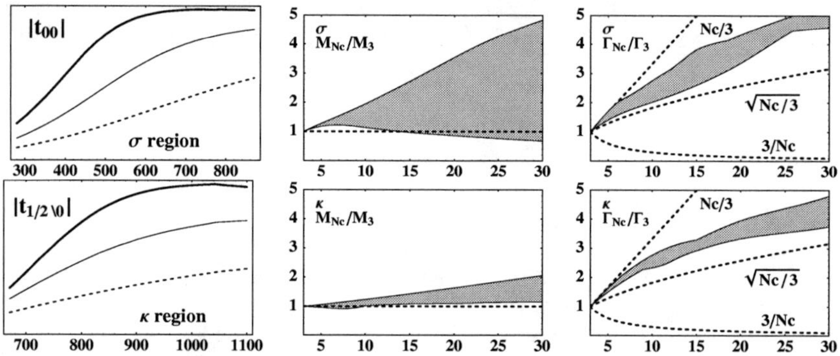

FIGURE 2. (Top) Left: Modulus of the $(I,J)=(0,0)$ scattering amplitude, versus $\sqrt{s}$ for $N_c=3$ (thick line), $N_c=5$ (thin line) and $N_c=10$ (dotted line), scaled at $\mu=770\,\mathrm{MeV}$. Center: N_c evolution of the σ mass. Right: N_c evolution of the σ width. (Bottom): The same but for the $(1/2,0)$ amplitude and the κ.

In contrast, in Figure 2 we see the corresponding behavior for the σ (or $f_0(600)$) and the κ. The results for the $f_0(980)$ and $a_0(980)$ (the latter, except in a corner of parameter space) are roughly similar, but more subtle [7]. It is evident that these *scalars behave completely different to $\bar{q}q$: The modulus of their partial waves in the resonance region vanish and their widths grow as N_c increases from $N_c=3$*, as $O(N_c^{1/2}) < \Gamma < O(N_c)$. These results have been confirmed later [9]. Thus, we can conclude the following

- *The dominant component of the σ and κ in meson-meson scattering does not behave as a $\bar{q}q$.*
  - Why "dominant"? Because, most likely, scalars are a mixture of different kind of states. If the $\bar{q}q$ was *dominant*, they would behave as the ρ or the K^* in Figure 1. *But it cannot be excluded that there is some smaller fraction of $\bar{q}q$.*
  - Also, since scalars could be an admixture of states with different nature and wave functions, the small $\bar{q}q$ component could be concentrated in the core and better seen in other reactions, whereas in scattering we are seeing mostly the outer region.
- *Two-meson and some tetraquark states [10] have a consistent "qualitative" behavior*, i.e., both disappear in the continuum of the meson-meson scattering amplitude as N_c increases (also the glueballs for the σ case, but not for the κ). Waiting for more

quantitative results, we have not been able to establish yet the nature of that dominant component, but two-meson states or some kind of tetraquarks are, qualitatively, candidates to form that dominant component.

Next we will address comments concerning the role of vector mesons in the IAM.

T-CHANNEL VECTOR MESON EXCHANGE

A common concern for people who have modeled meson-meson scattering including explicitly resonance fields, is the t-channel vector meson exchange, since it should contribute sizably to $\pi\pi$ or πK scattering. Since the ChPT Lagrangian does not have an explicit ρ or K^* field, and the usual resummations in the literature involve only s-channel loops, it may seem that we are not taking into account this contribution in the IAM. However, the resonance saturation mechanism [11] explains the L_i values as the contact terms that remain once resonances heavier than pions, kaons and etas, are integrated out from a chiral Lagrangian including all resonances. In other words, resonance propagators are reduced to constants when $M_V^2 >> s$. From vector resonances some L_i get contributions of the form

$$L_i^V \sim 1/M_V^2, \tag{2}$$

where M_V is the typical mass of the ρ and K^* multiplet.

Hence, a t-channel vector exchange, after integration of the heavy fields, becomes a combination of L_i parameters in the effective theory, and is thus *effectively* included in the one loop amplitudes later used in the IAM [7]. This is another important reason to consider the completely renormalized $O(p^4)$ ChPT amplitudes in the IAM, because we can then use L_i parameters compatible with standard ChPT, which also ensure the correct *low energy* expansion and the *effective* crossed channel resonance exchange.

ρ AND K^* MASS SPLITTINGS FROM THE LARGE N_C IAM

As shown in [7], using L_i parameters compatible with those of standard ChPT, the IAM is able to fit the experimental data on many meson-meson channels below 1.2 GeV, and in particular the different physical masses of the ρ and K^* resonances. However, it has been remarked [12] that the vector mass splittings are

$$M_{K^*}^2 = M_\rho^2 + O(M_{\pi,K}^2) \simeq M_V^2 + O(M_{\pi,K}^2) \tag{3}$$

and that the $O(p^4)$ IAM does not yield this dependence in the large N_c limit. Certainly, it does not, but if we took into account the vector mass splittings, Eq.(3), in the resonance saturation hypothesis, Eq.(2), we would find,

$$L_i^{V+split} \sim \frac{1}{(M_V^2 + O(M_{\pi,K}^2))} \sim \frac{1}{M_V^2}(1 + O(M_{\pi,K}^2)) \sim L_i^V\,(1 + O(M_{\pi,K}^2)).$$

But in ChPT, $O(M_{\pi,K}^2)$ counts as $O(p^2)$. Since L_i^V is multiplied by an $O(p^4)$ operator, the splitting term contributes, at least, at $O(p^6)$. Obviously, just with $O(p^4)$ ChPT one does not have to get the $O(p^6)$ splitting terms. In particular, this does *not* mean that "a systematic expansion in powers of quark masses has not been performed" [12], but that it has only been performed up to $O(p^4)$ in ChPT. Therefore, *nothing is to blame on the IAM, but just on the truncation at* $O(p^4)$. Any other unitarization scheme matching ChPT only up to $O(p^4)$ could also fail at $O(p^6)$ in some observable. In particular, even using $O(p^4)$ ChPT *without unitarization* to fit the low energy $\pi\pi$ and πK scattering in the vector channels, the resulting L_i will not correspond to splittings including $O(p^6)$ corrections. Of course, when dealing with a truncated expansion to describe data, the L_i absorb the effect of higher orders, and contain information on the physical splittings.

In summary, the observation that the $M^2_{\pi,K}$ dependence of the splittings is not obtained at $O(p^4)$ is interesting, but the IAM is not to blame for it, but just the $O(p^4)$ truncation. In particular, as we see in Fig.1, it cannot be concluded [12] that the IAM at $O(p^4)$ *does not yield parametrically reasonable large* N_c *dependence for the vector masses*, $O(1)$, which is the only relevant issue for our discussion on the nature of resonances. For mass *splittings*, which is another issue, one should consider the $O(p^6)$ *at least* [13].

ACKNOWLEDGMENTS

The author thanks the HADRON05 organizers for creating such a nice working atmosphere during the Conference and S. Leupold and R. Kaminski for comments and discussions. Work partially supported by the DGICYT contracts BFM2002-01868, and the EURIDICE network contract HPRN-CT-2002-00311 as well as the EU Hadron Physics Project, contract number RII3-CT-2004-506078.

REFERENCES

1. J. Gasser and H. Leutwyler, Annals Phys. **158**, 142 (1984). Nucl. Phys. B **250** (1985) 465.
2. G. 't Hooft, Nucl. Phys. B **72** (1974) 461. E. Witten, Annals Phys. **128** (1980) 363.
3. A. A. Andrianov, Phys. Lett. B **157**, 425 (1985). A. A. Andrianov and L. Bonora, Nucl. Phys. B **233**, 232 (1984). D. Espriu, E. de Rafael and J. Taron, Nucl. Phys. B **345** (1990) 22 S. Peris and E. de Rafael, Phys. Lett. B **348** (1995) 539
4. A. Gomez Nicola and J. R. Pelaez, Phys. Rev. D **65**, 054009 (2002) J. R. Pelaez, Mod. Phys. Lett. A **19**, 2879 (2004)
5. T. N. Truong, Phys. Rev. Lett. **61** (1988) 2526. Phys. Rev. Lett. **67**, (1991) 2260; A. Dobado, M.J.Herrero and T.N. Truong, Phys. Lett. **B235** (1990) 134. A. Dobado and J. R. Pelaez, Phys. Rev. D **47** (1993) 4883. Phys. Rev. D **56** (1997) 3057.
6. J. A. Oller, E. Oset and J. R. Pelaez, Phys. Rev. Lett. **80** (1998) 3452; Phys. Rev. D **59** (1999) 074001 J. A. Oller and E. Oset, Nucl. Phys. A **620**, 438 (1997) [Erratum-ibid. A **652**, 407 (1999)] and Phys. Rev. D **62** (2000) 114017. F. Guerrero and J. A. Oller, Nucl. Phys. B **537**, 459 (1999) [Erratum-ibid. B **602**, 641 (2001)] M. Uehara, [arXiv:hep-ph/0204020].
7. J. R. Pelaez, Phys. Rev. Lett. **92**, 102001 (2004). Also [arXiv:hep-ph/0306063] and [arXiv:hep-ph/0307018].
8. J. Bijnens, G. Colangelo and J. Gasser, Nucl. Phys. **B427** (1994) 427.
9. M. Uehara, [arXiv:hep-ph/0308241]. [arXiv:hep-ph/0401037]. [arXiv:hep-ph/0404221].
10. R. L. Jaffe, Proc. of the Intl. Symposium on Lepton and Photon Interactions at High Energies. Physikalisches Institut, U. of Bonn (1981). ISBN: 3-9800625-0-3. (Scanned version available at KEK)
11. G. Ecker, J. Gasser, A. Pich and E. de Rafael, Nucl. Phys. B **321**, 311 (1989).
12. S. Leupold, arXiv:hep-ph/0506220v1. (version 1)
13. J.R. Peláez, in preparation.

Measurement of the $K^- \pi^+$ S-wave System in $D^+ \to K^- \pi^+ \pi^+$ Decays from Fermilab E791 [1]

B. Meadows

University of Cincinnati, Cincinnati, OH, 45221, USA
Representing the Fermilab E791 Collaboration

Abstract. A new approach to the analysis of three body decays is presented. Measurements of the S-wave $K\pi$ amplitude are made in independent ranges of invariant mass from threshold up to the upper kinematic limit in $D^+ \to K^- \pi^+ \pi^+$ decays. These are compared with results obtained from a fit where the S-wave is assumed to have κ and $K_0^*(1430)$ resonances. Results are also compared with measurements of $K^- \pi^+$ elastic scattering. Contributions from $I = \frac{1}{2}$ and $I = \frac{3}{2}$ are not resolved in this study. If $I = \frac{1}{2}$ dominates, however, the Watson theorem prediction, that the phase behavour below $K\eta'$ threshold should match that in elastic scattering, is not well supported by these data. Production of $K^- \pi^+$ from these D decays is also studied.

Keywords: charm, 3-body decays, hadronic decays, $K\pi$ S-wave
PACS: 13.25.Ft, 13.25.Es, 13.75.Lb, 14.40.Aq, 14.40.Lb

INTRODUCTION

Decays of D and B mesons show promise as a source of information on the light-quark mesons they produce. Their decays to S-wave systems in three pseudo-scalar final states, may help to improve our knowledge of the particularly confusing scalar meson ($J^P = 0^+$) spectrum. Until now, extracting such information has been done in model-dependent ways that make assumptions about the scalar states observed that can influence the results. With large, clean samples of such decays, anticipated to be coming from the B factories and the Tevatron collider, the need for new approaches is a priority.

Knowledge of strange scalar mesons has relied on measurements of S-wave $K^- \pi^+$ scattering. These come principally from SLAC experiment E135 (LASS) [1] and cover the invariant mass range above 825 MeV$/c^2$. Data from other experiments below this range exist, but with less precision [2]. More measurements in the low mass region are required if the possibility of the existence of a κ state is to be properly evaluated.

In this paper, a study of the decays [Note] $D^+ \to K^- \pi^+ \pi^+$ observed in data from Fermilab experiment E791 is presented. In an earlier analysis [3] the $K^- \pi^+$ S-wave was modelled with Breit-Wigner (BW) amplitudes for κ and $K_0^*(1430)$ resonances. A less model-dependent analysis is presented here. S-wave amplitudes are measured without assuming any specific dependence on $K^- \pi^+$ invariant mass, or on the presence of scalar resonances. Results are compared with the amplitude from our earlier analysis, and also with measurements from LASS.

---

[1] Preprint numbers: FERMILAB-CONF-05-467-E, UCHEP-05-06

CP814 *Hadron Spectroscopy: Eleventh International Conference*
edited by A. Reis, C. Göbel, J. Sá Borges, and J. Magnin

DATA SAMPLE

The selection process for events used in this paper is described in Ref. [3]. A signal consisting of 15,079 $D^+ \to K^-\pi_a^+\pi_b^+$ decays, with a purity of $\sim 94\%$, is obtained. Fig. 1 shows the Dalitz plot with the $K^-\pi_a^+$ invariant mass squared plotted vs. that for the $K^-\pi_b^+$ system. Horizontal (and symmetrized vertical) bands corresponding to the $K^*(892)$ resonance are clearly seen. A complex pattern of both constructive and destructive interference is seen near 2 $(\text{GeV}/c^2)^2$ due to presence of $K_0^*(1430)$, $K_1^*(1410)$ and $K_2^*(1430)$. Evidence for $K_1^*(1680)$, difficult to see due to smearing of the Dalitz plot boundary resulting from the finite resolution in the three-body D^+ mass, may also exist.

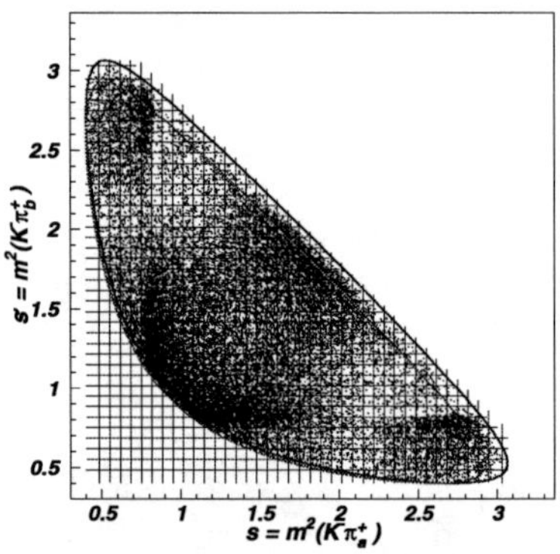

FIGURE 1. Dalitz plot for $D^+ \to K^-\pi_A^+\pi_B^+$ decays. The squared invariant mass of one $K^-\pi^+$ combination is plotted against the other combination. The plot is symmetrized, each event appearing twice. Lines in both directions indicate values, equally spaced in squared effective mass, at each of which the S-wave amplitude is determined by the method described in the text. Kinematic boundaries for the Dalitz plot are drawn for three-body mass values $M = 1.810$ and $M = 1.890$ GeV$/c^2$, between which data are selected.

An asymmetry along the $K^*(892)$ bands, most easily described by interference with a significant S-wave component to these decays, is also observed. Information on this S-wave amplitude is obtained from its interference with the $K^*(892)$, and also the other well-established resonances in the Dalitz plot.

METHOD

In Ref. [3], as in most earlier analyses of D decays to three pseudo-scalar particles ijk, the isobar model, with BW resonance forms, is used. The decay amplitude $\mathscr{A}$ is described by a sum of quasi two-body terms $D \to R + k$, $R \to i + j$, in each of the three channels $k = 1, 2, 3$:

$$\mathscr{A} \;=\; d_0 e^{i\delta_0} + \sum_{n=1}^{N} d_n e^{i\delta_n} \frac{F_R(p, r_R, J)}{m_{R_n}^2 - s - i m_{R_n}\Gamma_{R_n}(s)} \times F_D(q, r_D, J)\, M_J(p, q) \qquad (1)$$

The squared invariant mass of the ij system is s, J is the spin, m_{R_n} the mass and $\Gamma_{R_n}(s)$ the width of each of the N resonances R_n seen to be contributing to the decay. For $J > 0$, F_R and F_D are Blatt-Weisskopf form factors [4], with effective radius parameters r_R and r_D, for all R_n and for the parent D meson, respectively. For $J = 0$, a Gaussian form suggested by Tornqvist [5] is used for the D. The momenta, $\vec{p}$ and $\vec{q}$ for i and k, respectively, are defined in the ij rest frame, and $M_J(p,q) = (-2pq)P_J(\hat{p}\cdot\hat{q})$ is introduced to describe spin conservation in the decay. The complex coefficients $d_n e^{i\delta_n}$ $(n = 0,N)$ depend on the D decay and are determined by a fit to the data. The first term describes non-resonant (NR) decay to $i+j+k$ with no intermediate resonance, and is assumed to be independent of s. For $D^+ \to K^-\pi_a^+\pi_b^+$ decays we Bose-symmetrize $\mathscr{A}$ with respect to interchange of π_a^+ and π_b^+.

In Ref. [3], an excellent fit to the data is obtained with κ and $K_0^*(1430)$ resonant terms and the NR term comprising the $J = 0$ part of Eq. (1).

In this paper, the $K^-\pi^+$ S-wave is examined in a less model-dependent way, outlined here. A more detailed account is given in [6].

Terms appearing in Eq. (1) are grouped according to the value of J. The S-wave part (all terms with $J = 0$, including the NR term) is factored

$$\mathscr{S} \;=\; \mathrm{S}(s)\times M_0(p,q)F_D(q,r_D,0) \tag{2}$$

into a partial wave $\mathrm{S}(s)$, describing $K^-\pi^+$ scattering, and the product $M_0(p,q)F_D(q,r_D)$ describing the D decay. The P- and D- ($J = 1,2$, respectively) waves are factored in the same way:

$$\mathscr{P} \;=\; \mathrm{P}(s)\times M_1(p,q)F_D(q,r_D,1)\;;\;\mathscr{D} \;=\; \mathrm{D}(s)\times M_2(p,q)F_D(q,r_D,1), \tag{3}$$

with partial waves $\mathrm{P}(s)$ and $\mathrm{D}(s)$ consisting of resonant terms given as in Eq. (1). The S-wave $\mathrm{S}(s)$, however, is replaced by a set of values $c_k e^{i\gamma_k}$ defined at 40 discrete squared invariant masses $s = s_k$. These are indicated by the lines in Fig. 1. The c_k and γ_k values are regarded as independent parameters to be determined by the data.

An unbinned likelihood fit is made to the data, in a similar way to that described in Ref. [3]. An incoherent function describing the 6% background in the sample from events that are not true D decays is added in proper proportion at each three-body $K^-\pi^+\pi^+$ mass to a signal distribution proportional to $|\mathscr{A}|^2$. There are 86 parameters - all (c_k,γ_k) and the coefficients $d_k e^{i\delta_k}$ for $K^*(892)$, $K_1^*(1680)$ in the P-wave and $K_2^*(1430)$ in the D-wave. For the $K^*(892)$, $d_k e^{i\delta_k} = 1$ is used to provide the reference amplitude.

RESULTS AND COMPARISON WITH MODEL-DEPENDENT FIT

This fit also results in an excellent description of the data. Comparison of the observed and predicted population of the Dalitz plot gives a χ^2 probability of 50% for 363 bins. The S-, P- and D-waves resulting from the fit are shown in Fig. 2.

These results are compared with the model-dependent fit from Ref. [3] which provided an excellent description of the observed Dalitz plot distribution. Partial waves for this fit are also shown in Fig. 2. The main S-wave features of both fits agree well. Some differences, particularly at the highest and lowest ends of the invariant mass range, result from shifts in the P- and D-waves. Resonant fractions and the total S-wave fraction (about 75%) also agree well within statistical limits.

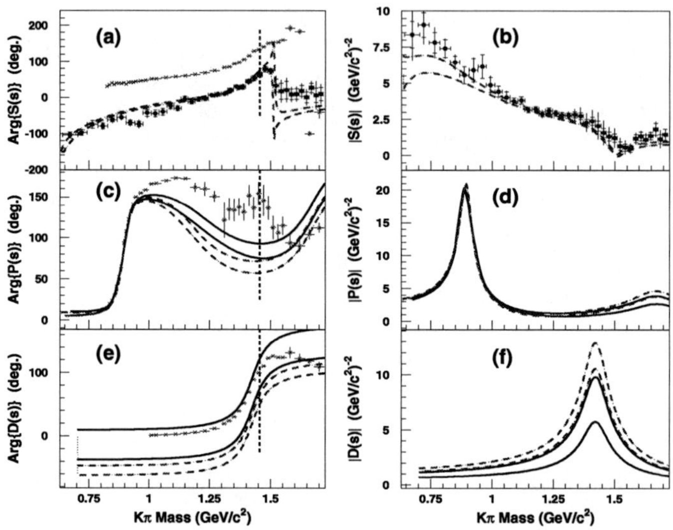

FIGURE 2. (a) Phases $\gamma_k = \arg S(s_k)$ and (b) magnitudes $c_k = |S(s_k)|$ of S-wave amplitudes for $K^-\pi^+$ systems from $D^+ \to K^-\pi^+\pi^+$ decays with the amplitude and phase of the $K^*(892)$ as reference. Solid circles, with error bars, show the values obtained from the model-independent fit described in the text. The effect of adding systematic uncertainties in quadrature is indicated by extensions on the error bars. The magnitudes plotted include a D^+ form-factor $F_D(q, r_D, 0) = e^{-25q^2/12}$. The phase and magnitude of P(s) are shown in (c) and (d), respectively. In each case, the parameters and error matrix from the fit are used in Eq. (3) to produce the solid curves shown one standard deviation above and below the central values. Similar plots for the D-wave amplitude D(s) are shown in (e) and (f). In all plots, the dashed curves show similar, one standard deviation limits for the amplitudes obtained from the isobar model fit in Ref. [3]. In (a), (c) and (e), $I = \frac{1}{2}$ phases for $K^-\pi^+$ scattering measured in the LASS experiment are shown as ×'s with error bars indicating statistical uncertainties.

It can be concluded that, with the present sample size, no significant distinction between the model-dependent and model-independent parametrizations of the S-wave can be made.

A comparison of the S-wave amplitudes $\mathrm{S}(s)$ measured here with the amplitudes $T(s)$ obtained in $K^-\pi^+$ scattering is now made. For each partial wave J (for each isospin I) it is expected that $\mathrm{S}(s) = Q(s)\mathrm{T}(s)/F_D(q, r_D, J))\sqrt{s}/p^{(J+1)}$ where $Q(s)$ describes the s-dependence of $K^-\pi^+$ production in D decays. The Watson theorem [7] requires, provided there is no re-scattering of the $K^-\pi^+_a$ from π^+_b, that $Q(s)$ is a real function, so that phases found in D decay should match those in $K^-\pi^+$ elastic scattering data.

$I = 1/2$ phases measured by LASS are plotted in Fig. 2. There is a large offset in the S-wave, about 75°, not seen in P- or D-waves. The shapes of S- and P-waves are also not the same. Unless significant admixture of $I = 3/2$ $K^-\pi^+$ production occurs, these results suggest that the conditions for the Watson theorem are not met in these data.

The production function $Q(s)$ is shown in Fig. 3. The scattering amplitude $T(s)$ is assumed to be elastic, $T = \sin(\gamma - \gamma_0)$. The phase offset γ_0 is required to account for the difference between elastic scattering and D decay. There appears to be significant s-dependence above $\sim$ 1.2 GeV$/c^2$ that grows in the region near 1.4 GeV$/c^2$ where $\gamma \sim \gamma_0$.

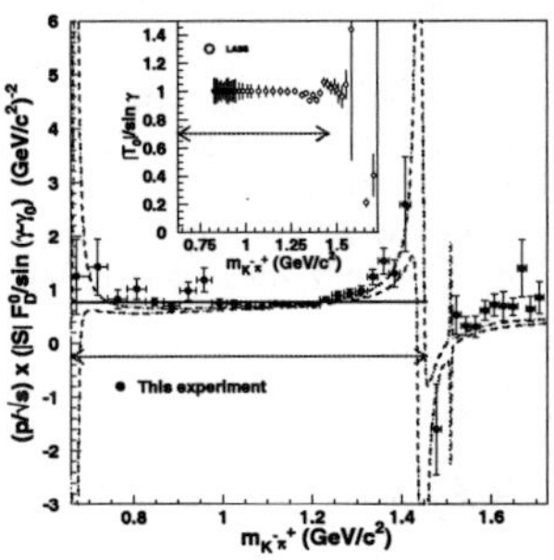

FIGURE 3. The quantities $p/\sqrt{(s_k)} \times |S(s_k)| \times F_D(q, r_D, 0)/\sin(\gamma_k - \gamma_0)$ plotted as solid circles for each point obtained for the S-wave amplitude in the model-independent fit described in the text. Three points between 1400 and 1450 MeV$/c^2$ are omitted from the plot as their values for $\sin(\gamma_k - \gamma_0)$ are very small, making their errors extremely large. The region between the dashed lines shows the one standard deviation limits of this quantity for the S-wave amplitude obtained from the model-dependent fit. The inset shows, as small open circles, the quantities $|T_0(s)|/\sin(\gamma)$ taken from the LASS experiment. The elastic range, up to $K\eta'$ threshold, is indicated by double-headed arrows in both plots.

ACKNOWLEDGMENTS

We thank the LASS collaboration for making their data available to us. We gratefully acknowledge the assistance of the staffs of Fermilab and of all the participating institutions. This research was supported by the Brazilian Conselho Nacional de Desenvolvimento Científico e Tecnológico, CONACyT (Mexico), FAPEMIG (Brazil), the Israeli Academy of Sciences and Humanities, the U.S. Department of Energy, the U.S.-Israel Binational Science Foundation, and the U.S. National Science Foundation. Fermilab is operated by the Universities Research Association for the U.S. Department of Energy.

REFERENCES

1. D. Aston, et al., *Nucl. Phys.* **B296**, 493 (1988).
2. P. Estabrooks, et al., *Nucl. Phys.* **B133**, 490 (1978).

Note. Note, Charge conjugate states are always implied unless explicitly stated otherwise.

3. E. M. Aitala, et al., *Phys. Rev. Lett.* **89**, 121801 (2002), `hep-ex/0204018`.
4. J. M. Blatt, and V. F. Weisskopf, *Wiley, New York* p. 361 (1952).
5. N. A. Tornqvist, *Z. Phys.* **C68**, 647–660 (1995), `hep-ph/9504372`.
6. E. M. Aitala, et al. (2005), Submitted to Phys. Rev. D., `hep-ex/0507099`.
7. K. M. Watson, *Phys. Rev.* **88**, 1163–1171 (1952).

Dalitz Analysis of $B \to Khh$ Decays at Belle

A.Garmash

Princeton University, Princeton, NJ 08544, USA

Abstract. We report results on the Dalitz analysis of three-body charmless $B^+ \to K^+\pi^+\pi^-$, $B^0 \to K^0\pi^+\pi^-$ and $B^+ \to K^+K^+K^-$ decays including searches for direct CP violation in $B^+ \to K^+\pi^+\pi^-$ mode. Branching fractions for a number of quasi-two-body intermediate states are reported. We also observe 3.9σ evidence for a large direct CP violation in $B^\pm \to \rho(770)^0K^\pm$ channel. This is the first evidence for CP violation in a charged meson decay. The results are obtained using Dalitz analysis technique with a large data sample of $B\bar{B}$ pairs collected with the Belle detector operating at the KEKB asymmetric energy e^+e^- collider.

Keywords: charmless B decays, CP violation

INTRODUCTION

Decays of B meson to three-body charmless hadronic final states is a rich laboratory for studying B meson decay dynamics and provide new possibilities for CP violation searches. In decays to two-body final states ($B \to K\pi$, $\pi\pi$, etc.) direct CP violation can only be observed as a difference in B and $\bar{B}$ decay rates. In decay to three-body final state which often is a set of quasi-two-body channels, direct CP violation can also manifest itself as a difference in relative phases between two quasi-two-body channels. Large direct CP violation is expected in charged B decays to some quasi-two-body charmless hadronic modes [1].

APPARATUS, DATA SAMPLE & EVENT SELECTION

The Dalitz analysis of $B^+ \to K^+\pi^+\pi^-$ and $B^+ \to K^+K^+K^-$ decays is performed with $140\,\mathrm{fb}^{-1}$ data sample; for Dalitz analysis of the $B^0 \to K^0\pi^+\pi^-$ decay and for CP violation searches in decay $B^+ \to K^+\pi^+\pi^-$, we use a data sample of $357\,\mathrm{fb}^{-1}$. The data are collected with the Belle detector [2] operating at the KEKB asymmetric-energy e^+e^- collider with a center-of-mass (c.m.) energy at the $\Upsilon(4S)$ resonance.

We identify B candidates using two almost independent kinematic variables: $\Delta E = (\sum_i \sqrt{c^2\mathbf{p}_i^2 + c^4m_i^2}) - E^*_{\text{beam}}$ and $M_{bc} = \frac{1}{c^2}\sqrt{E^{*2}_{\text{beam}} - c^2(\sum_i \mathbf{p}_i)^2}$, where the summation is over all particles from a B candidate; $\mathbf{p}_i$ and m_i are their c.m. three-momenta and masses, respectively. The dominant background to studied processes is due to $e^+e^- \to q\bar{q}$ ($q = u,d,s$ and c quarks) continuum events. This background is suppressed using variables that characterize the event topology. A detailed description of the continuum suppression technique can be found in Ref. [3] and references therein. From MC study we find the dominant background originating from other B decays that peaks in the signal region is due to $B \to Dh$, where h stands for a charged pion or kaon and due to

CP814 *Hadron Spectroscopy: Eleventh International Conference*
edited by A. Reis, C. Göbel, J. Sá Borges, and J. Magnin

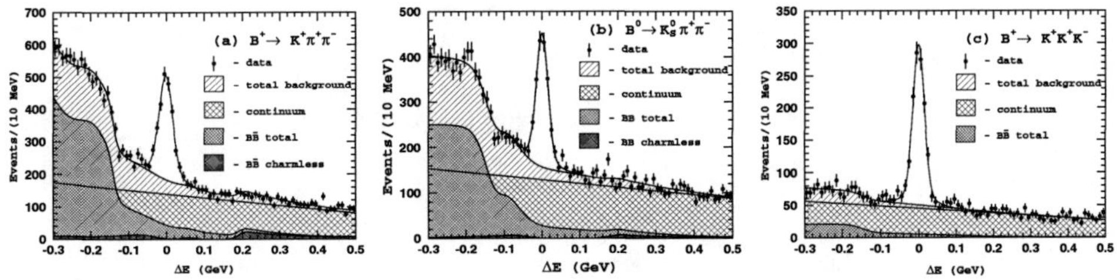

FIGURE 1. ΔE distribution for the (a) $B^+ \to K^+\pi^+\pi^-$, (b) $B^0 \to K^0\pi^+\pi^-$ and (c) $B^+ \to K^+K^+K^-$ candidate events with $|M_{bc} - M_B| < 7.5$ MeV/c^2. Points with error bars are data; the upper curve is the fit result; the hatched histograms are various background components.

$B \to J/\psi(\psi(2S))[\mu^+\mu^-]K$ decays, where muons are misidentified as pions. We veto these backgrounds by applying requirements on invariant mass of the appropriate two-particle combination. The most significant background from charmless B decays to $B \to K\pi\pi$ channels originates from $B \to \eta'[\gamma\pi^+\pi^-]K$, $B^+ \to \pi^+\pi^+\pi^-$, where one of the two same charge pions is misidentified as a kaon, and from $B \to K\pi$ processes. These backgrounds cannot be removed and are taken into account when fitting the data. We find no charmless B decay modes that produce a significant background to the $K^+K^+K^-$ final state. The ΔE distributions for $K^+\pi^+\pi^-$, $K^0_S\pi^+\pi^-$ and $K^+K^+K^-$ candidates that pass all the selection requirements are shown in Fig. 1.

DALITZ ANALYSIS RESULTS

For the amplitude analysis we select events in the B signal region defined as an ellipse around the M_{bc} and ΔE mean values: $\left[\frac{M_{bc}-M_B}{7.5\ \mathrm{MeV}/c^2}\right]^2 + \left[\frac{\Delta E}{40\ \mathrm{MeV}}\right]^2 < 1$. We find that $B^+ \to K^+\pi^+\pi^-$ signal is well described by a coherent sum of the $K^*(892)^0\pi^+$, $K^*_0(1430)^0\pi^+$, $\rho(770)^0K^+$, $f_0(980)K^+$, $f_X(1300)K^+$, $\chi_{c0}K^+$ quasi-two-body channels and a non-resonant component. The channel $f_X(1300)K^+$ (with mass and width of $f_X(1300)$ to be determined from the fit) is added to account for an excess of signal events visible in $M(\pi^+\pi^-)$ spectrum near 1.3 GeV/c^2. Results of the best fit are shown in Figs. 2 (a,b). The mass and width of the $f_X(1300)$ state obtained from the fit are consistent with those for the $f_0(1370)$, however more data are required for more definite conclusion. To test for the contribution of other possible quasi-two-body intermediate states such as $K^*(1410)^0\pi^+$, $K^*(1680)^0\pi^+$, $K^*_2(1430)^0\pi^+$ or $f_2(1270)K^+$, we include an additional amplitude of each of these channels in the decay amplitude one by one and repeat the fit to data. None of these channels have a statistically significant signal. Branching fraction and upper limit results are summarized in Table 1. For more details see Ref. [4].

In the fit to $K^0_S\pi^+\pi^-$ events we use decay amplitude $\mathcal{M}$ similar to those constructed in the analysis of $B^+ \to K^+\pi^+\pi^-$ decay. The masses and widths of all resonances are fixed at either their world average values from PDG or at values determined from the analysis of $B^+ \to K^+\pi^+\pi^-$ decay (for $f_0(980)$ and $f_X(1300)$). Since in this analysis

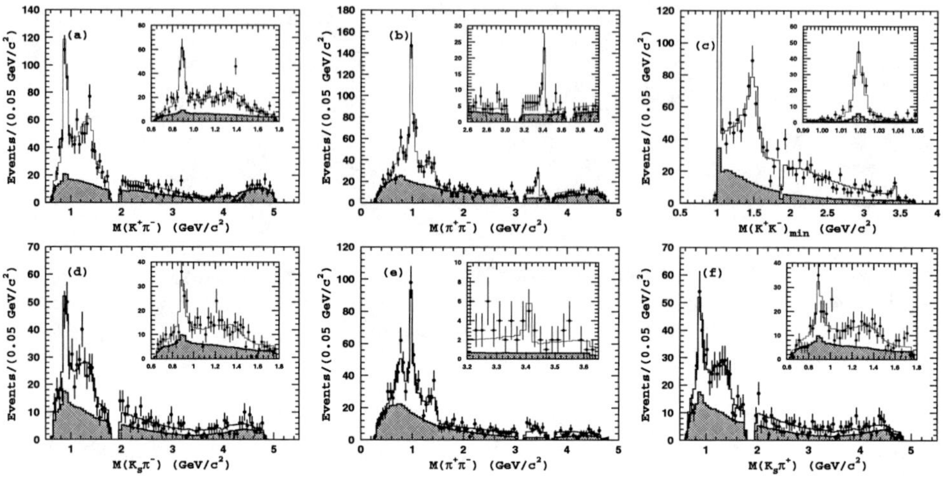

FIGURE 2. Results of the fit to events in the B signal region for the (a,b) $K^+\pi^+\pi^-$, (c) $K^+K^+K^-$, and (d,e,f) $K_S^0\pi^+\pi^-$ final state. Points with error bars are data, the open histogram is the fit result and hatched histogram is the background component.

we do not distinguish between B or $\bar{B}$ decays the signal PDF is a non-coherent sum $S(K_S^0\pi^\pm\pi^\mp) = |\mathcal{M}(K_S^0\pi^+\pi^-)|^2 + |\mathcal{M}(K_S^0\pi^-\pi^+)|^2$. Results of the best fit are shown in Figs. 2 (d-f). All plots demonstrate good agreement between data and the fit. Branching fraction results are summarized in Table 1. For more details see Ref. [5].

We find that $K^\pm K^+K^-$ signal is well described by an amplitude that is a coherent sum of the ϕK^+, $f_X(1500)K^+$, $\chi_{c0}K^+$ quasi-two-body channels and a non-resonant component, where the $f_X(1500)K^+$ (with mass and width of $f_X(1300)$ to be determined from the fit) channel is added to describe the excess of signal events visible in K^+K^- mass spectrum near 1.5 GeV/c^2. As there are two identical kaons in the final state, the decay amplitude is symmetrized with respect to interchange of two kaons of the same charge $S(K^+K^+K^-) = |\mathcal{M}(K_1^+K_2^+K^-) + \mathcal{M}(K_2^+K_1^+K^-)|^2$. Results of the fit are shown in Fig. 2 (c) and summarized in Table 1. For more details see Ref. [4].

SEARCH FOR DIRECT *CP* VIOLATION IN $B^\pm \to K^\pm\pi^\pm\pi^\mp$

For *CP* violation studies the amplitude for each quasi-two-body channel is modified from $ae^{i\delta}$ to $ae^{i\delta}(1 \pm be^{i\varphi})$, where the plus (minus) sign corresponds to B^+ (B^-) decay. With such a parameterization charge asymmetry, A_{CP}, for a particular quasi-two-body $B \to f$ channel can be calculated as

$$A_{CP}(f) = \frac{N^- - N^+}{N^- + N^+} = -\frac{2b\cos\varphi}{1+b^2}. \quad (1)$$

Results of the fit are given in Table 2. The statistical significance of the asymmetry

TABLE 1. Summary of branching fraction results. The first quoted error is statistical, the second is systematic and the third is the model error.

Mode	$\mathcal{B}(B^+ \to Rh^+) \times \mathcal{B}(R \to h^+h^-) \times 10^6$	$\mathcal{B}(B^+ \to Rh^+) \times 10^6$
$K^+\pi^+\pi^-$ charmless total	–	$46.6 \pm 2.1 \pm 4.3$
$K^*(892)^0[K^+\pi^-]\pi^+$	$6.55 \pm 0.60 \pm 0.60^{+0.38}_{-0.57}$	$9.83 \pm 0.90 \pm 0.90^{+0.57}_{-0.86}$
$K_0^*(1430)^0[K^+\pi^-]\pi^+$	$27.9 \pm 1.8 \pm 2.6^{+8.5}_{-5.4}$	$45.0 \pm 2.9 \pm 6.2^{+13.7}_{-8.7}$
$K^*(1410)^0[K^+\pi^-]\pi^+$	< 2.0	–
$K^*(1680)^0[K^+\pi^-]\pi^+$	< 3.1	–
$K_2^*(1430)^0[K^+\pi^-]\pi^+$	< 2.3	–
$\rho(770)^0[\pi^+\pi^-]K^+$	$4.78 \pm 0.75 \pm 0.44^{+0.91}_{-0.87}$	$4.78 \pm 0.75 \pm 0.44^{+0.91}_{-0.87}$
$f_0(980)[\pi^+\pi^-]K^+$	$7.55 \pm 1.24 \pm 0.69^{+1.48}_{-0.96}$	–
$f_2(1270)[\pi^+\pi^-]K^+$	< 1.3	–
Non-resonant	–	$17.3 \pm 1.7 \pm 1.6^{+17.1}_{-7.8}$
$K_S^0\pi^+\pi^-$ charmless	–	$47.5 \pm 2.4 \pm 3.7$
$K^*(892)^+[K^0\pi^+]\pi^-$	$5.61 \pm 0.72 \pm 0.43^{+0.43}_{-0.29}$	$8.42 \pm 1.08 \pm 0.65^{+0.64}_{-0.43}$
$K_0^*(1430)^+[K^0\pi^+]\pi^-$	$30.8 \pm 2.4 \pm 2.4^{+0.8}_{-3.0}$	$49.7 \pm 3.8 \pm 3.8^{+1.2}_{-4.8}$
$K^*(1410)^+[K^0\pi^+]\pi^-$	< 3.8	–
$K^*(1680)^+[K^0\pi^+]\pi^-$	< 2.6	–
$K_2^*(1430)^+[K^0\pi^+]\pi^-$	< 2.1	–
$\rho(770)^0[\pi^+\pi^-]K^0$	$6.13 \pm 0.95 \pm 0.47^{+1.00}_{-1.05}$	$6.13 \pm 0.95 \pm 0.47^{+1.00}_{-1.05}$
$f_0(980)[\pi^+\pi^-]K^0$	$7.60 \pm 1.66 \pm 0.59^{+0.48}_{-0.67}$	–
$f_2(1270)[\pi^+\pi^-]K^0$	< 1.4	–
Non-resonant	–	$19.9 \pm 2.5 \pm 1.5^{+0.7}_{-1.2}$
$K^\pm K^+K^-$ charmless total	–	$30.6 \pm 1.2 \pm 2.3$
$\phi[K^+K^-]K^+$	$4.72 \pm 0.45 \pm 0.35^{+0.39}_{-0.22}$	$9.60 \pm 0.92 \pm 0.71^{+0.78}_{-0.46}$
$\phi(1680)[K^+K^-]K^+$	< 0.8	–
$f_0(980)[K^+K^-]K^+$	< 2.9	–
$f_2'(1525)[K^+K^-]K^+$	< 4.9	–
$a_2(1320)[K^+K^-]K^+$	< 1.1	–
Non-resonant	–	$24.0 \pm 1.5 \pm 1.8^{+1.9}_{-5.7}$
$\chi_{c0}[\pi^+\pi^-]K^+$	$1.37 \pm 0.28 \pm 0.12^{+0.34}_{-0.35}$	–
$\chi_{c0}[K^+K^-]K^+$	$0.86 \pm 0.26 \pm 0.06^{+0.20}_{-0.05}$	–
$\chi_{c0}K^+$ combined	–	$196 \pm 35 \pm 33^{+197}_{-26}$
$\chi_{c0}[\pi^+\pi^-]K^0$	< 0.56	< 113

TABLE 2. Results of the best fit to $K^\pm\pi^\pm\pi^\mp$ events in the B signal region. The first quoted error is statistical and the second is the model dependent uncertainty. The quoted significance is statistical only.

Channel	b	φ, (°)	A_{CP}, (%)	Significance, (σ)
$K^*(892)\pi^\pm$	$0.078 \pm 0.033^{+0.012}_{-0.003}$	$-18 \pm 44^{+5}_{-13}$	$-14.9 \pm 6.4 \pm 3.0^{+0.8}_{-0.8}$	2.6
$K_0(1430)\pi^\pm$	$0.069 \pm 0.031^{+0.010}_{-0.008}$	$-123 \pm 16^{+4}_{-5}$	$+7.6 \pm 3.8 \pm 3.0^{+2.0}_{-0.9}$	2.7
$\rho(770)^0K^\pm$	$0.28 \pm 0.11^{+0.07}_{-0.09}$	$-125 \pm 32^{+10}_{-85}$	$+30 \pm 11 \pm 3.0^{+11}_{-4}$	3.9
$f_0(980)K^\pm$	$0.30 \pm 0.19^{+0.05}_{-0.10}$	$-82 \pm 8^{+2}_{-2}$	$-7.7 \pm 6.5 \pm 3.0^{+4.1}_{-1.6}$	1.6
$f_2(1270)K^\pm$	$0.37 \pm 0.17^{+0.11}_{-0.04}$	$-24 \pm 29^{+14}_{-20}$	$-59 \pm 22 \pm 3.0^{+3}_{-3}$	2.7
$\chi_{c0}K^\pm$	$0.15 \pm 0.35^{+0.08}_{-0.07}$	$-77 \pm 94^{+154}_{-11}$	$-6.5 \pm 19.6 \pm 3.0^{+2.9}_{-1.4}$	0.7

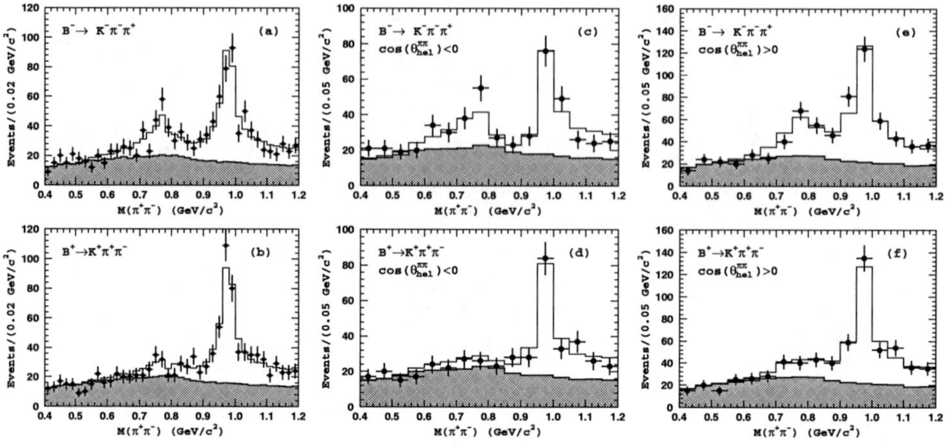

FIGURE 3. $\pi^+\pi^-$ mass spectra for B^- (top row) and B^+ (bottom row) events for different helicity regions: (a,b) no helicity cuts; (c,d) $\cos\theta_H^{\pi\pi} < 0$; (e,f) $\cos\theta_H^{\pi\pi} > 0$; Points with error bars are data, the open histogram is the fit result and the hatched histogram is the background component.

quoted in Table 2 is calculated as $\sqrt{-2\ln(\mathscr{L}_0/\mathscr{L}_{\max})}$, where $\mathscr{L}_{\max}$ and $\mathscr{L}_0$ denote the maximum likelihood with the nominal fit and with the asymmetry fixed at zero, respectively. The only channel where the statistical significance of the asymmetry exceeds the 3σ level is $B^\pm \to \rho(770)^0 K^\pm$, where we find a 3.9σ effect. Figures 3(a,b) show the $M(\pi^+\pi^-)$ distributions for the $\rho(770)^0 - f_0(980)$ mass region separately for B^- and B^+ events. The effect is more apparent when $M(\pi^+\pi^-)$ spectrum for the two helicity angle regions shown in Figs. 3(c-f) are compared. Results on the A_{CP} measurement are summarized in Table 2. To cross check the asymmetry observed in $B^\pm \to \rho(770)^0 K^\pm$ we make an independent fit to B^- and B^+ subsamples. We also confirm the significance of the asymmetry observed in $B^\pm \to \rho(770)^0 K^\pm$ channels with MC pseudo-experiments. For more details see Ref. [6]. Large value of A_{CP} measured in $B^\pm \to \rho(770)^0 K^\pm$ is in agreement with a recent update by BaBar [7] and some theoretical predictions [1]. The statistical significance of the asymmetry is varies from 3.7σ to 4σ depending on the model used to fit the data. This is the first evidence for *CP* violation in the decay of a charged meson.

REFERENCES

1. See for example: M. Beneke and M. Neubert, Nucl. Phys. **B675**, 333 (2003); C.-W. Chiang, M. Gronau, Z. Luo, J. Rosner, and D. Suprun, Phys. Rev. **D69**, 034001 (2004) and references therein.
2. A. Abashian *et al.*, Nucl. Instr. and Meth. **A479**, 117 (2002).
3. A. Garmash *et al.* (Belle Collaboration), Phys. Rev. **D69**, 012001 (2004).
4. A. Garmash *et al.* (Belle Collaboration), Phys. Rev. **D71**, 092003 (2005).
5. K. Abe *et al.* (Belle Collaboration), BELLE-CONF-577, hep-ex/0509047.
6. K. Abe *et al.* (Belle Collaboration), BELLE-CONF-528, hep-ex/0509001.
7. B. Aubert *et al.* (BaBar Collaboration), hep-ex/0507004. Submitted to Phys. Rev. D.

Chiral Breit-Wigner

L. O. Arantes and M. R. Robilotta

Instituto de Física, Universidade de São Paulo,
C.P. 66318, 05315-970, São Paulo, SP, Brazil.

Abstract. Chiral symmetry and unitarization are combined into generalized Breit-Wigner expressions describing scalar resonances, which contain free parameters and allow flexible descriptions of masses, widths and pole positions. This theoretical tool is especially designed to be used in analyses of low-energy data.

Keywords: scalar resonances, low-energy, chiral symmetry
PACS: 14.40 Cs, 13.75 Lb

The vacuum of quantum chromodynamics (QCD) is non-trivial and, at low-energies, it can be excited in the form of pions, kaons or etas. These particles can be clearly perceived by experiment because their quantum numbers allow a neat contrast with the vacuum. The same kind of contrast is not possible in the case of scalar mesons and their identification is much more difficult.

Evidences for resonances with low masses and large widths produced by the E791 Fermilab experiment in D decays[1] and confirmed in a number of other reactions[2] have motivated a renewal in the interest on the scalar sector. The information extracted from these experiments relies on theoretical expressions for resonance widths and amplitudes, which eventually do play major roles in final results. As a consequence, "empirical" values for masses and coupling constants depend heavily on the theoretical input used in data analyses.

CHIRAL SYMMETRY - The best theoretical framework available for describing hadronic decays of heavy mesons is low-energy QCD, which can be reliably represented by means of effective theories with approximate chiral symmetry. This symmetry is especially suited for describing processes involving pions and has been successfully applied to a rather large sample of physical problems.

About ten years ago, Törnqvist[3] employed chiral symmetry in the description of scalar resonances and obtained a width for the decay into S-wave pions of the form

$$\Gamma_T(s) = \alpha\, (2s - \mu^2)\, \frac{\sqrt{s - 4\mu^2}}{\sqrt{s}}\, e^{-(s-M^2)/4\beta^2}\,, \tag{1}$$

where α and β are free parameters and μ and M are the pion and resonance masses. This result was derived from a quark model and has the merit of including the chiral factor $(2s - \mu^2)$, which makes the width small at low-energies. As we discuss in the sequence, effective lagrangians incorporating chiral symmetry give rise to more general expressions, which include other forms of s-dependence. Another advantage of the lagrangian framework is that it allows the unambiguous separation between resonance and background contributions and avoids the risk of double counting.

CP814 *Hadron Spectroscopy: Eleventh International Conference*
edited by A. Reis, C. Göbel, J. Sá Borges, and J. Magnin

Modern chiral perturbation theory employs non-linear realizations of the symmetry and gives rise to results which are very general and incorporate all the possible freedom compatible with the symmetry. The old and well known linear $\sigma-$model, which has been much used in the study of scalar resonances, can be recovered as a special case.

RESONANCES AND $\pi\pi$ AMPLITUDE - The scalar resonance propagator at low energies is closely related with the elastic $\pi\pi$ scattering amplitude, which is the main locus of chiral symmetry. Quite generally, the amplitude $T_{\pi\pi}$ for the process $\pi(p)\,\pi(q) \to \pi(p')\,\pi(q')$ can be written as

$$T_{\pi\pi} = T_0(s,t,u)\,P_0 + T_1(s,t,u)\,P_1 + T_2(s,t,u)\,P_2\,, \tag{2}$$

where $s = (p+q)^2$, $t = (p-p')^2$, $u = (p-q')^2$ and the operator P_I is the projector into channel with isospin I.

A chiral theorem[4] ensures that the leading order amplitude T_0 at low energies is given by

$$T_0 = (2s-\mu^2)/f_\pi^2 + \cdots\,, \tag{3}$$

where μ and f_π are the pion mass and decay constant, whereas the ellipsis indicates higher order contributions. When resonances are included, it is very important to ensure consistency with this theorem. In the case of a scalar-isoscalar resonance, the tree level amplitude for $\pi\pi$ scattering is given by the four diagrams

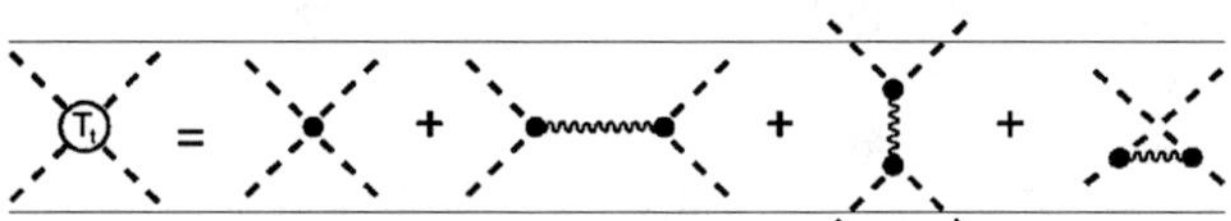

In the non-linear approach, the resonance field f is assumed to be a chiral scalar and couples to pion fields ϕ, which behave non-linearly under chiral transformations[5]. The effective lagrangian for this system is written as

$$\begin{aligned}\mathcal{L} &= \frac{1}{2}\left(\partial_\mu f\,\partial^\mu f - M^2 f^2\right) + \frac{1}{2}\left(1+c_s f/f_\pi\right)\left(\partial_\mu\phi\cdot\partial^\mu\phi + \partial_\mu\sqrt{f_\pi^2-\phi^2}\;\partial^\mu\sqrt{f_\pi^2-\phi^2}\right)\\ &+ \mu^2 f_\pi\left(1+c_b f/f_\pi\right)\left(\sqrt{f_\pi^2-\phi^2} - f_\pi\right)\,,\end{aligned} \tag{4}$$

where the dimensionless constants c_s and c_b represent, respectively, the scalar-pion couplings that preserve and break chiral symmetry. The explicit evaluation of the diagrams yields the following scalar-isoscalar tree level amplitude

$$\begin{aligned}T_t &= -\frac{\gamma^2}{s-M_\sigma^2}\,,\\ \gamma^2(s) &= \left\{(2s-\mu^2)(M_\sigma^2-s) + 3\left[c_s s/2 - (c_s-c_b)\,\mu^2\right]^2 + B(M_\sigma^2-s)\right\}/f_\pi^2\,,\end{aligned} \tag{5}$$

where the function $\gamma(s)$ incorporates chiral dynamics and implements effective couplings at the vertices. Its first term reproduces the low-energy theorem, given by eq.(3), the second one describes the resonance couplings, and the last function represents a non-resonating background, which receives contributions from t and u channels only. The particular choice $(c_s=2, c_b=1)$ corresponds to the linear σ−model.

UNITARIZATION - The range of validity of the field theoretical $\pi\pi$ amplitude can be extended by means of unitarization. We consider only iterated contributions from a single loop, without form factors. The advantage of this procedure is that there is little model dependence and loop contributions are given by a compact analytical expression, which is real below threshold and complex afterwards. The dressed propagator is determined by the three diagrams shown in figure (a) below, whereas the full amplitude T, described in figure (b), corresponds to a composite Dyson series and includes all possible iterations of the $\pi\pi$ tree amplitude.

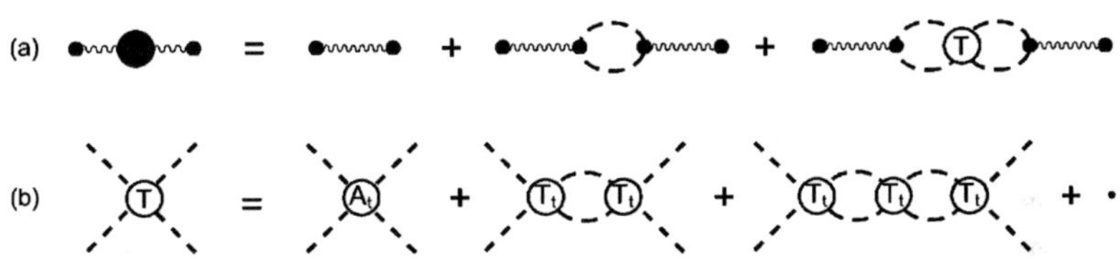

This geometrical series can be summed and one finds

$$\bar{T}(s) = -\frac{\gamma^2}{s-\mathscr{M}^2+iM\Gamma_\chi} \quad \leftrightarrow \quad \Gamma_\chi(s) = \gamma^2\,\frac{\sqrt{s-4\mu^2}}{32\pi M\sqrt{s}}\,\Theta(s-4\mu^2)\,, \qquad (6)$$

where $\mathscr{M}^2(s)$ is a running mass such that $\mathscr{M}^2(M^2)=M^2$. This result corresponds to the most general expression possible for the width dictated by chiral symmetry and should be compared with eq.(1). The signature of the symmetry is the factor $(2s-\mu^2)/f_\pi^2$ in the the function $\gamma^2(s)$ [eq.(5)], which ensures that the low energy theorem for $\pi\pi$ scattering is satisfied.

RESULTS - In the figure below, we explore the interplay between chiral symmetry and resonance in the function $|T(s)|^2$, for a scalar mass fixed at $M=\sqrt{12}\,\mu\sim 485$ MeV. The first term of eq.(5) yields the *leading order* curve, an unbound parabola which blows up at large energies. The inclusion of the remaining two terms with $(c_s=2, c_b=1)$ in that amplitude gives rise to the *tree* curve. The *unitarized σ-model* curve corresponds to eq.(6) and is obtained by iterating the *tree* amplitude by means of two-pion loops. Finally, the *resonance* curve is derived by iterating just the second term in eq.(5) with $s=M^2$ and corresponds to the width normally used in data analyses,which is given by

$$\Gamma_R(s) = \frac{3\,(M_\sigma^2-\mu^2)^2}{f_\pi^2}\,\frac{\sqrt{s-4\mu^2}}{32\pi M\sqrt{s}}\,\Theta(s-4\mu^2)\,. \qquad (7)$$

Inspecting the figure, one learns that the proper implementation of chiral symmetry forces the *leading order*, *tree* and *unitarized σ-model* curves to stay very close together at low energies. This kind of behavior also holds for other choices of the parameters (c_s, c_b). In the *resonance* curve, on the other hand, the symmetry is badly violated, since it does not tend to the *leading order* one when $s \to 0$, as predicted by the low-energy theorems.

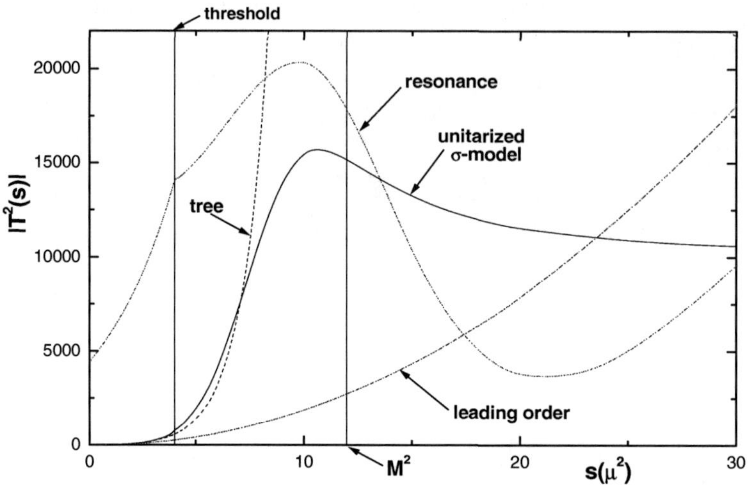

Resonances manifest themselves as poles in the second Riemann sheet of scattering amplitudes and another problem with the use of eq.(7) is that it does not allow one to explain how a scalar mass around 850 MeV, as observed in phase shifts, can coexist with a pole at much lower energy. The strong s-dependence of the function γ in the chiral width [eq.(6)] is essential to explain[6] how scalar parameters $(M_\sigma, \Gamma_\sigma/2) \simeq (846, 265)$ MeV can be compatible with a pole position at $\sqrt{s} \simeq (455 - i290)$ MeV.

CHIRAL BREIT-WIGNER - Our chiral expression describing a single scalar resonance has the following properties:

- at low energies, it reproduces the theorem derived by Weinberg[4] for $\pi\pi$ scattering;
- it is compatible with a rather general non-linear realization of chiral symmetry and involves two free parameters (c_s, c_b), which allow a rather flexible description of masses, widths and pole positions as shown in the figure in the next page;
- results from the usual K-matrix unitarization can be recovered by identifying the running mass in eq.(6) with the resonance mass and correspond to the expression

$$\bar{T}(s) = -\frac{\gamma^2}{s - M^2 + i\,M\,\Gamma}, \tag{8}$$

with

$$\Gamma_\chi(s) = \gamma^2 \, \frac{\sqrt{s-4\mu^2}}{32\pi M \sqrt{s}} \, \Theta(s-4\mu^2), \tag{9}$$

$$\gamma(s) = \left\{ (2s-\mu^2)(M^2-s) + 3\left[c_s s/2 - (c_s-c_b)\,\mu^2\right]^2 \right\} / f_\pi^2 \,. \tag{10}$$

These expressions represent a compromise between simplicity and chiral low-energy dynamics, which can be used in data analyses by treating the parameters M, c_s and c_b as being adjustable.

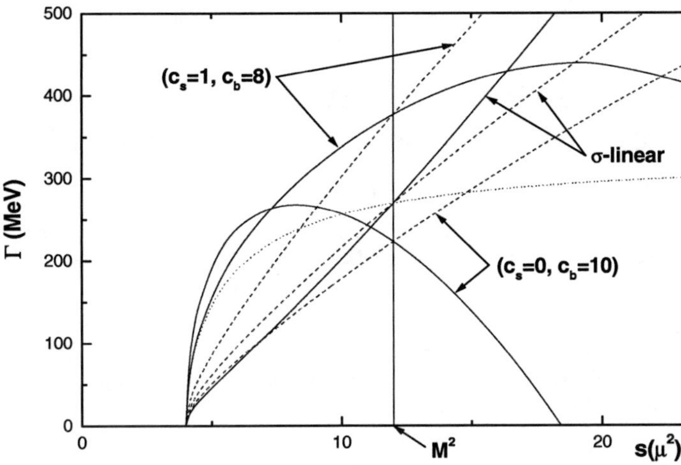

• the inclusion of a second resonance with the same quantum numbers is straightforward. Dynamical unitarization by means of quantum field theory techniques corresponds to first combining tree-level amplitudes and subsequently evaluating loop corrections. This leads to the following trial function:

$$\bar{T}(s) = -\frac{\gamma_\alpha^2\,(s-M_\beta^2)+\gamma_\beta^2\,(s-M_\alpha^2)}{(s-M_\alpha^2)\,(s-M_\beta^2)+i[(s-M_\beta^2)M_\alpha\Gamma_\alpha+(s-M_\alpha^2)M_\beta\Gamma_\beta]}\,, \tag{11}$$

where $\Gamma_{\alpha,\beta}$ and $\gamma_{\alpha,\beta}$ are obtained by using the parameters suited for each resonance in eqs.(9) and (10).

• There is no simple relationship between the unitarized amplitude for the coupled system and Breit-Wigner expressions for individual resonances. The price to be paid for representing each resonance by a Breit-Wigner is a very high one, since one will have to deal with coefficients and phases that depend strongly on the variable s.

An extended version of this work has been submitted for publication elsewhere.

REFERENCES

1. E791 Collaboration, E. M. Aitala et al., Phys. Rev. Lett. **86**, 770 (2001); E791 Collaboration, E. M. Aitala et al., Phys. Rev. Lett. **89**, 121801 (2002).
2. CLEO Collabotration, H. Muramatsu et al., Phys. Rev. Lett. **89**, 251802 (2002); Belle Collaboration, K. Abe et al., preprint hep-ex/0308043; BaBar Collaboration, B. Aubert et al. preprint hep-ex/0408088; KLOE Collaboration, A. Aloisio et al. Phys. Lett. B**537**, 21 (2002); BES Collaboration, J-Z. Bai et al., preprint hep-ex/0404016; BaBar Collaboration, B. Aubert et al. preprint hep-ex/0408032.
3. N. A. Törnqvist, Z. Phys. C**68**, 647 (1995).
4. S. Weinberg, Phys. Rev. Lett. **17**, 616 (1966).
5. S. Weinberg, Phys. Rev. **166**, 1568 (1968).
6. G. Colangelo, J. Gasser and H. Leutwyler, Nucl. Phys. B **603**, 125 (2001).

On Scalar Meson Dominance in Semi- and Non-Leptonic Weak Pseudoscalar Meson Decays

Frieder Kleefeld [1]

Centro de Física das Interacções Fundamentais (CFIF), Instituto Superior Técnico, Edifício Ciência, Piso 3, Av. Rovisco Pais, P-1049-001 LISBOA, Portugal

Abstract. Aspects of Scalar Meson Dominance in semi- and non-leptonic weak pseudoscalar meson decays are shortly discussed on the basis of the Quark-Level Linear Sigma Model.

Keywords: scalar meson dominance, weak meson decays, linear sigma model, penguins
PACS: 13.20.-v, 13.25.-k, 14.40.Aq

Introduction. The theoretical description of leptonic, semi- and non-leptonic weak decays of pseudoscalar mesons has been known to be extremly rewarding since at least 5 decades [1] in particular due to technical and numerical challenges in the description of strong interactions on the basis Quantum Chromodynamics (QCD) [2]. The experimental and theoretical study of non-leptonic K-meson decays has for instance a long tradition [1, 3, 4, 5, 6], yet the theoretical issues concerning these decays are far from settled [7, 8, 9]. To use the words of G. Colangelo [9]: "… *The calculation of the $K \to \pi\pi$ amplitude in the Standard Model still remains one of the most difficult and yet unsolved problems of today's particle physics, despite many years of efforts and progress in our understanding of various related physics aspects. Indeed we lack yet a satisfactory explanation of the $\Delta I = 1/2$ rule, and do not yet have a calculation of ε'/ε in the Standard Model which would make a comparison to the measured value somewhat useful — the typical size of the theoretical uncertainties attached to any of the calculations available in the literature is around 50 % …. A recurring disussion in the literature on calculations of $K \to \pi\pi$ concerns the role of final state interactions (FSI) …*". In view of the outlined situation we want to recall that a growing amount of meson production data available from charm/bottom quark factories allows for the first time a direct view [10] to the nature and properties of light scalar mesons which up now were only hardly accessable in elastic meson-meson scattering due to the unfortunate presence of Adler-zeros [11] and were taken into account in QCD-based formalisms to meson weak decays only indirectly as hardly quantifiable non-perturbative effects. The very existence of light scalar mesons draws renewed attention to the strongly neglected *Partial* Conservation of flavour-changing Vector Current (PCVC) which has coexisted with often celebrated Partial Conservation of Axialvector Current (PCAC) for about 4 decades [12]. We shall here report shortly on first results of a "Scalar Meson Dominance" (SMD) approach to pseudoscalar meson weak decays taking into account PCVC manifestly by including scalar mesons, rather than guessing PCVC non-perturbatively like in QCD.

---

[1] e-mail: kleefeld@cfif.ist.utl.pt, URL: http://cfif.ist.utl.pt/~kleefeld/

CP814 *Hadron Spectroscopy: Eleventh International Conference*
edited by A. Reis, C. Göbel, J. Sá Borges, and J. Magnin

The Linear Sigma Model (LσM) and its dynamical generation. Facing [13, 14] the yet lacking conclusive experimental evidence for the existence of gluons, new developments in mathematical physics and puzzling experimental data we will follow here Ref. [15] and model strong interactions on the basis of an alternative approach to QCD known under the name "Quark-Level Linear Sigma Model" (QLLσM) [16], which makes use of degrees of freedom being found in nature, i.e. mesons and (anti)quarks. In other words we dynamically generate as in Ref. [15] our strong interaction Lagrangean on the basis of the following Yukawa-like interaction Lagrangean relating (anti)quark fields (denoted by $q_\pm(z)$) and scalar ($S(z)$), pseudoscalar ($P(z)$), vector ($V(z)$), and axialvector ($Y(z)$) $U(6)\times U(6)$ meson field matrices in flavour space (see also [14]) (The undetermined signs s_s, s_p, s_v, $s_y \in \{-1,+1\}$ are here irrelevant!):

$$\begin{aligned}\mathscr{L}^{strong}_{int}(z) &= \\ &= \sqrt{2}g\,\overline{q^c_+}(z)\left(s_s S(z)+s_p iP(z)\gamma_5+\frac{e^{-i\alpha}}{2}\left(s_v \not{V}(z)+s_y \not{Y}(z)\gamma_5\right)\right)q_-(z)\,, \quad (1)\end{aligned}$$

with $g=|g|\exp(i\alpha)$ being the eventually complex strong interaction coupling constant. The dynamical generation of an effective meson-meson interaction Lagrangean (like e.g. the well known $U(3)\times U(3)$ LσM [16, 17]) makes then use — to avoid double counting — of the idea that meson-meson interaction proceeds through quark-loops only and *not on the basis of any further extra direct meson-meson interaction terms* in the Lagrangean. $\mathscr{L}^{strong}_{int}(z)$ is now added — as done by Cabibbo & Maiani [16] — to the Lagrangean describing (in Fermi's limit) weak interactions among leptons (denoted here as $\ell_\pm(x)$) and (anti)quarks, i.e. (G_F = Fermi's constant, θ_W = Weinberg's angle, T_3 = isospin matrix, Q_q, Q_ℓ = charge matrices, V_{CKM} = CKM-matrix, $P_L \equiv (1-\gamma_5)/2$):

$$\begin{aligned}\mathscr{L}^{weak}_{int}(x) &= -4\frac{G_F}{\sqrt{2}}\times \\ \times & \left[\left(\overline{\ell^c_+}(x)\,\gamma^\mu\,P_L\begin{pmatrix}0_3 & 1_3\\ 0_3 & 0_3\end{pmatrix}\ell_-(x)+\overline{q^c_+}(x)\,\gamma^\mu\,P_L\begin{pmatrix}0_3 & V_{CKM}\\ 0_3 & 0_3\end{pmatrix}q_-(x)\right)\right. \\ & \left(\overline{\ell^c_+}(x)\,\gamma_\mu\,P_L\begin{pmatrix}0_3 & 0_3\\ 1_3 & 0_3\end{pmatrix}\ell_-(x)+\overline{q^c_+}(x)\,\gamma_\mu\,P_L\begin{pmatrix}0_3 & 0_3\\ \overline{V}_{CKM} & 0_3\end{pmatrix}q_-(x)\right) \\ & +\left(\overline{\ell^c_+}(x)\,\gamma^\mu(T_3P_L-Q_\ell\sin^2\theta_W)\,\ell_-(x)+\overline{q^c_+}(x)\,\gamma^\mu(T_3P_L-Q_q\sin^2\theta_W)\,q_-(x)\right) \\ & \left.\left(\overline{\ell^c_+}(x)\,\gamma_\mu(T_3P_L-Q_\ell\sin^2\theta_W)\,\ell_-(x)+\overline{q^c_+}(x)\,\gamma_\mu(T_3P_L-Q_q\sin^2\theta_W)\,q_-(x)\right)\right] \\ & + \text{ h.c.}\,. \quad (2)\end{aligned}$$

Leptonic weak meson decays and pseudoscalar decay constants. As a first step we study leptonic decays of pseudoscalar mesons to extract pseudoscalar decay constants f_P. The relevant part $S_{eff}[P\bar{\ell}\ell]$ of the effective action in the local limit resulting by dynamical generation is given in Ref. [15]. By inspection of this effective action we conclude as in Ref. [15] that the finite decay constant $f_{\eta_{q_1\bar{q}_2}}$ of a pseudoscalar meson

$\eta_{q_1\bar{q}_2}$ corresponds in the considered formalism to a log.-divergent integral, i.e. [15][2] [3]:

$$f_{\eta_{q_1\bar{q}_2}} \longleftrightarrow -4\,iN_c\,|g| \int \frac{d^4p}{(2\pi)^4}\,\frac{(m_{q_1}+m_{\bar{q}_2})/2}{(p^2-m_{q_1}^2)(p^2-m_{\bar{q}_2}^2)}\,. \tag{3}$$

Semi-leptonic weak meson decays and SMD. As a first instructive example we consider in Ref. [15] the semi-leptonic decay $K^+ \to \pi^0\, e^+\, \nu_e$. Ingredient for our calculation has been the dynamical generation of the effective action $S_{eff}[SPP]$ between one scalar and two pseudoscalar mesons displayed in the footnote [4]. It is then straight forward to dynamically generate the (local) effective action for the process $K^+ \to \pi^0\, e^+\, \nu_e$ [15]:

$$\begin{aligned}
S_{eff} &= \int d^4z\; (-i\,e^{2i\alpha}) \left(-\frac{G_F}{\sqrt{2}}\right) \overline{V}_{us}\; \overline{e^c_+}(z)\; \gamma_\mu\, (1-\gamma_5)\; \nu_{e-}(z) \\
&\times\; \frac{1}{\sqrt{2}} \Big\{ \pi^0(z) \left(\frac{2\,|g|\, f_{K^+}}{m_u+m_s}\; \partial^\mu\, K^+(z) \right) - K^+(z) \left(\frac{2\,|g|\, f_{\eta_{u\bar{u}}}}{m_u+m_u}\; \partial^\mu\, \pi^0(z) \right) \\
&\quad +4iN_c\,|g|^2\,(m_s-m_u)^2\; K^+(z)\, (\partial^\mu\, \pi^0(z)) \int \frac{d^4p}{(2\,\pi)^4}\, \frac{1}{(p^2-m_s^2)(p^2-m_u^2)^2} \\
&\quad + \frac{\lambda}{g^2}\, m_s\, \frac{(m_s-m_u)}{m^2_{K^+}}\, \frac{2\,|g|\, f_{K^+}}{m_u+m_s} \Big(\pi^0(z)\,(\partial^\mu\, K^+(z)) + K^+(z)\,(\partial^\mu\, \pi^0(z)) \Big) \Big\} \\
&+\; K^*\text{-exchange} + \ldots\,.
\end{aligned} \tag{5}$$

For $\lambda \simeq 2g^2$ the weak transition formfactors to relevant order in the scale $\delta = (m_s/m_u) - 1 \simeq 0.44$ according to the nonrenormalization theorem of Ademollo & Gatto [18] are determined as [15] $f_+^{K^+\pi^0}(0) = 1 + O(\delta^2)$ and $f_-^{K^+\pi^0}(0) = 4\,|g|\,e^{2i\alpha}\,\delta\,(1+\delta)\,(2+\delta)^{-1}|m_u||f_{K^+}|m_{K^+}^{-2} + O(\delta^2)$ displaying SMD. Because of [19] $f_-^{K^+\pi^0}(0)/f_+^{K^+\pi^0}(0) \simeq -0.125 \pm 0.023$, $|f_{K^+}| \simeq 159\,\text{MeV}/\sqrt{2}$, $m_{K^+} \simeq 797$ MeV, and $|m_u| \simeq 337$ MeV there has to hold $0 > e^{2i\alpha} \simeq -1$ due to the 1-loop dynamically generated value $|g| = 2\pi/\sqrt{3}$ [15]. In conclusion experiment is suggesting [15] an *asymptotic free* QLLσM with Yukawa coupling $g \simeq -i2\pi/\sqrt{3}$ and quartic coupling $\lambda \simeq 2g^2 \simeq -8\pi^2/3 < 0$, respectively.

---

[2] The colour factor N_c is displayed here for traditional reasons and can absorbed by rescaling g.

[3] This replacement of logarithmic divergences by finite experimental numbers renormalizes the formalism without the need of introducing unpleasant regulators like cutoffs.

[4] In the considered local limit we obtained [15] ($M_q \equiv \text{diag}[m_u, m_c, m_t, m_d, m_s, m_b]$, "$\text{tr}_F$"= flavour trace):

$$\begin{aligned}
S_{eff}[SPP] &= \int d^4z\; \sqrt{2}\; g^2\; e^{i\alpha}\; s_s\; (-4\,i\,N_c\,|g|) \int \frac{d^4p}{(2\pi)^4} \\
&\times\; \Big\{ \text{tr}_F \Big[S(z)\, \frac{1}{(p^2-M_q^2)}\, \{P^2(z), M_q\}\, \frac{1}{(p^2-M_q^2)} \Big] \\
&\quad + \text{tr}_F \Big[[S(z), P(z)]\, \frac{1}{(p^2-M_q^2)}\, [P(z), M_q]\, \frac{1}{(p^2-M_q^2)} \Big] \\
&\quad - \text{tr}_F \Big[\{S(z), M_q\}\, \frac{1}{(p^2-M_q^2)} [P(z), M_q]\, \frac{1}{(p^2-M_q^2)}\, [P(z), M_q]\, \frac{1}{(p^2-M_q^2)} \Big] \Big\} + \ldots\,.
\end{aligned} \tag{4}$$

.

Non-leptonic weak meson decays and SMD. For the theoretical explanation of two longstanding problems based on non-leptonic meson weak decays, i.e. the quantitative calculation of $|M(K^+ \to \pi^+\pi^0)|$ and the so-called $\Delta I = 1/2$ rule, there might exist an attractive solution on the basis of SMD [5]. Unfortunately the solution of both problems requires a sizable yet unexplained mechanism for *S-P* mixing [4]. In the QLLσM it is tempting to construct e.g. effective actions $S^{semi}_{eff}[S\bar{q}q]$ or $S^{semi}_{eff}[P\bar{q}q]$ for semi-strong $S\bar{q}q$- or $P\bar{q}q$-penguins [6] in the hope that they contribute to explain *S-P* mixing. Unfortunately it turns out that relevant meson tadpoles constructed from these actions vanish, while non-tadpole graphs constructed from these actions don't contribute to the solution of the problems under consideration[7]. We conclude that explanation of *S-P* mixing [4, 6, 22] being probably of similar nature as π^0-η or η-η' mixing [17] requires further research.

---

[5] The observation that the "vacuum saturation" (VS) W-emission contribution to the non-leptonic decay $K^+ \to \pi^+\pi^0$ predicts an amplitude *two times* larger than the experimental value $|M_{K^+\to\pi^+\pi^0}(exp)| = m_{K^+}(8\pi\Gamma/p_{cm})^{1/2} = (1.832\pm 0.007)\cdot 10^{-8}$ GeV led SVZ [6] to decrease the discrepancy between theory and experiment on the basis of so-called "penguin-diagrams" [6] which are now constraint strongly by the fact that *"polychromatic* [i.e. QCD] *penguins don't fly"* [20]. 1987 Scadron proposed a VS being *half* of the VS of SVZ and agreeing to experiment, as $|M_{K^+\to\pi^+\pi^0}(Scadron)| = |G_F V_{ud}\overline{V}_{us}(f_{\pi^-}(m^2_{K^+} - m^2_{\pi^0}) - 2 f_{K^+}(m^2_{\pi^+} - m^2_{\pi^0}))/(2\sqrt{2})| = (1.886\ldots - 0.026\ldots)\cdot 10^{-8}\,\mathrm{GeV} = 1.860\ldots\cdot 10^{-8}\,\mathrm{GeV}$. Scadron's mysterious factor 1/2 might be explained by adding to the SVZ W-emission contribution SMD meson-exchange diagrams like in Fig. 1 of Ref. [4]. Unfortunately these extra contributions require a yet unexplained mechanism to mix scalar and pseudoscalar mesons (*S-P* mixing). Another testing ground for SMD in non-leptonic meson weak decays is the so-called "$\Delta I = 1/2$ rule" [1, 5, 6, 9]. A probable scenario was summarized by M.D. Scadron *et al.* [21] in 2002: "... *The well-known* $\Delta I = 1/2$ *rule* $\Gamma(K_S \to \pi^+\pi^-)/\Gamma(K^+ \to \pi^+\pi^0) \approx 450$ *for nonleptonic weak* $K_{2\pi}$ *decays suggests that the parity-violating (PV) amplitude* $\langle 2\pi|H^{pv}_w|K_S\rangle$ *could be dominated by the* $\Delta I = 1/2$ *weak transition* $\langle\sigma|H^{pv}_w|K_S\rangle$ ...". Although supported by other work [4, 6, 22] also this conjecture lacks yet to explain *S-P* mixing.

[6] Dynamical generation yields the following semi-strong "penguin" effective actions in the local limit:

$$S^{semi}_{eff}[S\bar{q}q] = \int d^4z\, G_F\, s_s\, e^{i\alpha}\frac{1}{N_c}\times$$

$$\times\ \left(\overline{q^c_+}(z)\begin{pmatrix} 0_3 & V_{CKM}\\ 0_3 & 0_3\end{pmatrix}[\partial_\mu\left[\frac{fS}{M}\right](z), M_q]\,\gamma^\mu P_L\begin{pmatrix} 0_3 & 0_3\\ \overline{V}_{CKM} & 0_3\end{pmatrix} q_-(z)\right.$$

$$\left. + \overline{q^c_+}(z)\begin{pmatrix} 0_3 & 0_3\\ \overline{V}_{CKM} & 0_3\end{pmatrix}[\partial_\mu\left[\frac{fS}{M}\right](z), M_q]\,\gamma^\mu P_L\begin{pmatrix} 0_3 & V_{CKM}\\ 0_3 & 0_3\end{pmatrix} q_-(z)\right) + \ldots, \qquad (6)$$

$$S^{semi}_{eff}[P\bar{q}q] = \int d^4z\, G_F\, s_p\, e^{i\alpha}\frac{2i}{N_c}\times$$

$$\times\ \left(\overline{q^c_+}(z)\begin{pmatrix} 0_3 & V_{CKM}\\ 0_3 & 0_3\end{pmatrix}\partial_\mu[fP](z)\,\gamma^\mu P_L\begin{pmatrix} 0_3 & 0_3\\ \overline{V}_{CKM} & 0_3\end{pmatrix} q_-(z)\right.$$

$$\left. + \overline{q^c_+}(z)\begin{pmatrix} 0_3 & 0_3\\ \overline{V}_{CKM} & 0_3\end{pmatrix}\partial_\mu[fP](z)\,\gamma^\mu P_L\begin{pmatrix} 0_3 & V_{CKM}\\ 0_3 & 0_3\end{pmatrix} q_-(z)\right) + \ldots, \qquad (7)$$

with $\left[\frac{fS}{M}\right] \equiv (-4iN_c|g|)\int\frac{d^4p}{(2\pi)^4}\frac{1}{p^2-M_q^2}S\frac{1}{p^2-M_q^2}$ and $[fP] \equiv (-4iN_c|g|)\int\frac{d^4p}{(2\pi)^4}\frac{1}{p^2-M_q^2}\frac{1}{2}\{P, M_q\}\frac{1}{p^2-M_q^2}$.

[7] $S^{semi}_{eff}[S\bar{q}q]$ does not contribute to the cascade $K_S \to \sigma \to \pi^+\pi^-$, as PCVC enforces $S^{semi}_{eff}[S\bar{q}q] = 0$ for the "flavour-neutral" σ-meson, while $S^{semi}_{eff}[P\bar{q}q]$ does not lead to an I=3/2 final state in $K^+ \to \pi^+\pi^0$.

Acknowledgments. This work has been supported by the FCT of the *Ministério da Ciência, Tecnologia e Ensino Superior* of Portugal, under Grants no. SFRH/BDP/9480/2002, POCTI/FNU/49555/2002, and POCTI/FP/FNU/50328/2003.

REFERENCES

1. A.J. Buras, *"Weak decays and QCD: 18 years later"*, pp. 591-633 in Proc. Workshop on "QCD: 20 Years Later", Aachen, Germany, 9-13.6.1992, River Edge, N.J., World Scientific, 1993; W. Bernreuther *et al.*, *Z. Phys. C* **4** (1980) 257; M.D. Scadron, *Rept. Prog. Phys.* **44** (1981) 213; R. Fleischer, hep-ph/0507156; I.I.Y. Bigi, A.I. Sanda, *Camb. Monogr. Part. Phys. Nucl. Phys. Cosmol.* **9** (2000) 1.
2. R.C. Verma, *Indian J. Phys. A* **72A** (1998) 579; A.J. Buras, L. Silvestrini, *Nucl. Phys. B* **569** (2000) 3; G. Buchalla *et al.*, *Rev. Mod. Phys.* **68** (1996) 1125; R. Rückl, hep-ph/9810338; A.J. Buras *et al.*, *Nucl. Phys. B* **268** (1986) 16; H.Y. Cheng *et al.*, hep-ph/0409317, *Phys. Rev. D* **59** (1999) 092004; M. Neubert, B. Stech, *Adv. Ser. Direct. High Energy Phys.* **15** (1998) 294 [hep-ph/9705292]; M. Beneke, M. Neubert, *Nucl. Phys. B* **675** (2003) 333; Z. Xiao, W. Zou, *Phys. Rev. D* **70** (2004) 094008; R. Casalbuoni *et al.*, *Phys. Rept.* **281** (1997) 145; D. Delepine *et al.*, hep-ph/0501022.
3. N. Cabibbo, *Phys. Rev. Lett.* **12** (1964) 62; J. Schwinger, *Phys. Rev. Lett.* **12** (1964) 630; R.M. Delaney *et al.*, *Phys. Rev.* **176** (1968) 1841; D.G. Hill *et al.*, *Phys. Rev. D* **8** (1973) 1290; P.C. McNamee *et al.*, *Phys. Rev. D* **10** (1974) 2280; T.J. Devlin *et al.*, *Rev. Mod. Phys.* **51** (1979) 237; M. Milosevic *et al.*, *Nucl. Phys. B* **187** (1981) 514; H. Wahl, *Phys. Rept.* **403-404** (2004) 19.
4. D.N. Goswami *et al.*, *Phys. Rev. D* **5** (1972) 2276; P. Minkowski *et al.*, *Eur. Phys. J. C* **39** (2005) 71.
5. Fayyazuddin *et al.*, *Nucl. Phys. B* **76** (1974) 125; B.H.J. McKellar *et al.*, *Phys. Rev. D* **27** (1983) 157.
6. M.A. Shifman, A.I. Vainshtein, V.I. Zakharov, *Sov. Phys. JETP* **45** (1977) 670 [*Zh. Eksp. Teor. Fiz.* **72** (1977) 1275], Nucl. Phys. B **120** (1977) 316; Y. Tosa, *"Penguins In Weak Interactions"* (UR-746), talk at Joint Annual McGill-Rochester-Syracuse-Toronto Meeting, Syracuse, N.Y., 1-2.5.1980.
7. M.D. Scadron, S.R. Choudhury, *"Simple Theory Of Nonleptonic Kaon Decays"*, Abdus Salam Int. Centre for Theor. Physics Miramare-Trieste preprint IC/87/180, KEK-library preprint 200033325.
8. E. Pallante, A. Pich, *Nucl. Phys. B* **592** (2001) 294; M.D. Scadron, *Mod. Phys. Lett. A* **14** (1999) 1273; N.N. Trofimenkoff, *Prog. Theor. Phys.* **85** (1991) 859; R.E. Karlsen, M.D. Scadron, *Phys. Rev. D* **44** (1991) 2192; H.Y. Cheng, *Phys. Rev. D* **36** (1987) 2056 [Erratum **37** (1988) 1338]; A.N. Kamal, *Phys. Rev. D* **33** (1986) 1344; Riazuddin, Fayyazuddin, *Phys. Rev. D* **32** (1985) 1720.
9. H.Y. Cheng, *Z. Phys. C* **32** (1986) 237; J.O. Eeg, *Acta Phys. Polon. B* **18** (1987) 639; M.D. Scadron *et al.*, *Mod. Phys. Lett. A* **19** (2004) 2267, *Phys. Rev. D* **53** (1996) 2421; J.M. Gérard *et al.*, hep-ph/0508189; P. Minkowski, *Phys. Lett. B* **88** (1979)138; G. Colangelo, *AIP Conf. Proc.* **660** (2003)25.
10. D.V. Bugg, hep-ex/0510019, *Phys. Rept.* **397** (2004) 257; E. van Beveren *et al.*, hep-ph/0510120, hep-ph/0509351, *Phys. Rev. Lett.* **91** (2003) 012003; H.Y. Cheng *et al.*, hep-ph/0508104.
11. G. Rupp *et al.*, *AIP Conf. Proc.* **756** (2005) 360; F. Kleefeld, *AIP Conf. Proc.* **717** (2004) 332 [hep-ph/0310320]; J.R. Ellis, *Nucl. Phys. B* **21** (1970) 217; S.L. Adler, *Phys. Rev.* **137** (1965) B1022.
12. I.S. Gerstein *et al.*, *Phys. Rev.* **175** (1968) 1876; P.R. Auvil *et al.*, *Phys. Rev.* **183** (1969) 1463.
13. F. Kleefeld, hep-th/0506142; X. Chen, Y. Mao, B.Q. Ma, *Nucl. Phys. A* **759** (2005) 188.
14. F. Kleefeld, *AIP Conf. Proc.* **660** (2003) 325 [hep-ph/0211460]; hep-th/0408028, hep-th/0408097.
15. F. Kleefeld, hep-th/0506140 (accepted for publication in *Czech J. Phys.* **55** (2005)).
16. M. Lévy, *Nuovo Cim.* **LII A** (1967) 23; N. Cabibbo *et al.*, *Phys. Rev. D* **1** (1970) 707; R. Delbourgo, M.D. Scadron, *Mod. Phys. Lett. A* **10** (1995) 251, *Int. J. Mod. Phys. A* **13** (1998) 657.
17. F. Kleefeld, nucl-th/0510017; S. Gasiorowicz, D.A. Geffen, *Rev. Mod. Phys.* **41** (1969) 531.
18. M. Ademollo, R. Gatto, *Phys. Rev. Lett.* **13** (1964) 264.
19. S. Eidelman *et al.* [PDG], *"Review of particle physics"*, *Phys. Lett. B* **592** (2004) 1.
20. R.S. Chivukula, J.M. Flynn, H. Georgi, *Phys. Lett. B* **171** (1986) 453.
21. M.D. Scadron *et al.*, *AIP Conf. Proc.* **660** (2003) 311 (see also *Nucl. Phys. A* **724** (2003) 391).
22. E. P. Shabalin, *Sov. J. Nucl. Phys.* **48** (1988) 172 [*Yad. Fiz.* **48** (1988) 272], *Sov. J. Nucl. Phys.* **48** (1988) 864 [*Yad. Fiz.* **48** (1988) 1357]; T. Morozumi *et al.*, *Prog. Theor. Phys.* **87** (1992) 185, *Phys. Rev. Lett.* **65** (1990) 404, *Phys. Rev. D* **46** (1992) 2240; M. Harada *et al.*, *Phys. Rev. D* **62** (2000) 014002; Y. Tosa, *Phys. Lett. B* **93** (1980) 443; S. Narison, *Nucl. Phys. B* **593** (2001) 3; J.M. Gérard *et al.*, *Phys. Lett. B* **503** (2001) 99; S. Trine, hep-ph/0305048.

$K^*(892)^0$ lineshape and $K\pi$ scalar contribution from $D^+ \to K^-\pi^+\mu^+\nu$ at FOCUS

Carla Göbel

Depto. de Física, Pontifícia Universidade Católica, Rio de Janeiro, Brazil
On behalf of the FOCUS Collaboration

Abstract.
FOCUS (Fermilab) is a photoproduction experiment designed to collect and reconstruct charm particles. Using a large and clean sample of semileptonic $D^+ \to K^-\pi^+\mu^+\nu$ decays, we provide a characterization of the $K^*(892)^0$ resonance, measuring its mass and width. We also investigate contributions from other structures which couple to the $K\pi$ final state. A small scalar component is needed to describe the data. We compare results obtained here to those from $K\pi$ elastic scattering and hadronic $D^+ \to K^-\pi^+\pi^+$ decays.

Keywords: charm meson, semileptonic decays, $K^*(892)^0$ resonance, $K\pi$ S-wave
PACS: 13.20.Fc, 14.40.Ev, 13.25.Ft, 13.75.Lb

INTRODUCTION

Semileptonic decays of charm meson offer an interesting place to study some hadronic processes, due to the factorization of the leptonic and hadronic currents. The decay $D^+ \to K^-\pi^+\mu^+\nu$ [1] can be used to investigate the $K\pi$ system in the absence of interactions with other hadrons. Due to Watson's theorem [1], one expects that the observed $K\pi$ phase shifts in $D^+ \to K^-\pi^+\mu^+\nu$ are the same as those measured in $K\pi$ elastic scattering.

The study of the $K\pi$ system is particularly relevant for the scalar sector. The LASS experiment [2] on elastic $K\pi$ scattering has shown that the S-wave amplitude can be well represented by the sum of a $K_0^*(1430)^0$ resonance and a slowly-varying background. On the other hand, the scalar $K^-\pi^+$ system from a D^+ decay has been recently investigated through the hadronic decay mode $D^+ \to K^-\pi^+\pi^+$ from Fermilab E791 [3, 4]. The phase shift for the S-wave with respect to $K^*(892)^0$ obtained from $D^+ \to K^-\pi^+\pi^+$ [5] is substantially different from the one obtained by LASS.

Here we present results from Fermilab experiment FOCUS for $D^+ \to K^-\pi^+\mu^+\nu$. Since this decay is known to be dominated by the $K^*(892)^0$ resonance, our large and clean sample provides an excellent environment to measure the $K^*(892)^0$ mass and width, as well as the effective Blatt-Weisskopf [6] damping factor parameter. Besides the $K^*(892)^0$ characterization, we present results on the extra (small) scalar component needed to describe the $D^+ \to K^-\pi^+\mu^+\nu$ decay. This component has been reported earlier by FOCUS [7] and here we present a measurement for its fraction [8] and discuss

[1] Charge conjugate states are implied throughout this paper.

CP814 *Hadron Spectroscopy: Eleventh International Conference*
edited by A. Reis, C. Göbel, J. Sá Borges, and J. Magnin

the interference pattern of this channel with the $D^+ \to \overline{K}^*(892)^0\mu^+\nu$ channel. We also present upper limits for contributions from heavier states. We then compare the results obtained here with those from LASS and E791.

THE FOCUS EXPERIMENT AND DATA SAMPLE SELECTION

The FOCUS experiment collected data during the Fermilab 1996-97 fixed-target run. A forward multi-particle detector is used to measure the interactions of photons with an average energy of 175 GeV with a segmented BeO target. The FOCUS beamline and detector has been described elsewhere [9]. For this analysis, requirements were made on tracking, vertexing and particle identification and are described in detail in [8]. A total of 18245 $D^+ \to K^-\pi^+\mu^+\nu$ candidates remain after the selection criteria. We estimate the charm and non-charm background contributions to be, respectively, 17.8% and 3.2% of the total number of events within our signal region.

THE $D^+ \to K^-\pi^+\mu^+\nu$ ANALYSIS FORMALISM

Four-body decays of spinless particles are described by five kinematical variables. In this analysis, the variables chosen are the $K\pi$ invariant mass ($m_{K\pi}$), the square of the $\mu^+\nu$ mass (q^2), and three decay angles: the angle between the π^+ and the D^+ direction in the $K^-\pi^+$ rest frame (θ_v), which defines one decay plane, the angle between the ν and the D^+ direction in the $\mu^+\nu$ rest frame (θ_ℓ), which defines a second decay plane, and the acoplanarity angle (χ) between these two planes. The differential decay rate can be represented by a coherent sum of resonant and non-resonant contributions to the angular momentum eigenstates of the $K^-\pi^+$ system,

$$\frac{d\Gamma}{dm_{K\pi}} = \int |\sum_J \sum_R a_{J(R)} \mathscr{M}_J \mathscr{A}_{J(R)}|^2 \varepsilon \phi d\Omega \tag{1}$$

where $d\Omega \equiv dq^2 d\cos\theta_v d\cos\theta_\ell d\chi$, ϕ is the phase space density, ε is the efficiency/acceptance function due to our detector, $\mathscr{M}_J$ is the weak matrix element for a transition with angular momentum J, and $\mathscr{A}_{J(R)}$ is the hadronic amplitude for resonance R, or non-resonant (NR). The strengths $a_{J(R)}$ are parameters to be determined by the fit.

The weak matrix elements for the vector and scalar processes, $\mathscr{M}_1$ and $\mathscr{M}_0$ respectively, are written as a function of helicity amplitudes and are described in [8].

The resonant hadronic amplitudes $\mathscr{A}_{J(R)}$ with mass m_0 and width Γ_0 are parametrized as

$$\mathscr{A}_{J(R)} = \frac{m_0\Gamma_0}{m_{K\pi}^2 - m_0^2 + im_0\Gamma(m_{K\pi})} \mathscr{F}_J(m_{K\pi}) \tag{2}$$

where $\Gamma(m_{K\pi}) = \Gamma_0 \mathscr{F}_J^2 \frac{p^*}{p_0^*} \frac{m_0}{m_{K\pi}}$, p^* is the decay momentum in $K\pi$ rest frame, $p_0^* = p^*(m_0)$, $\mathscr{F}_0 = 1$ and $\mathscr{F}_1 = \frac{p^*}{p_0^*}\frac{B(p^*)}{B(p_0^*)}$. B is the Blatt-Weisskopf damping factor, which is $1/\sqrt{1+r_0^2 p^{*2}}$ for $J=1$. The effective radius r_0 is an additional fit parameter for the parametrization of the $K^*(892)^0$ resonance, along with its mass and width.

Besides the dominant $K^*(892)^0$ contribution, the hadronic amplitudes tested in this analysis are: a NR amplitude, a light scalar resonance κ, the heavier scalar $K_0^*(1430)^0$ and the heavier vector $K^*(1680)$.

The parametrization used for the NR decay is the effective-range form used by LASS [2]. By removing the phase space factor (included in Eq. 1) it reads:

$$\mathscr{A}_{NR} = \frac{m_{K\pi}}{p^*} \frac{1}{\cot(\delta_{NR}) - i} \tag{3}$$

where $\cot\delta_{NR} = \frac{1}{a p^*} + \frac{1}{2} b\, p^*$ with $a = 4.03 \pm 1.72 \pm 0.06$ GeV^{-1} and $b = 1.29 \pm 0.63 \pm 0.67$ GeV^{-1} [2]. The shape used for κ follows the parametrization used by E791 [3].

$K\pi$ MASS SPECTRUM ANALYSIS AND RESULTS

With only scalar and vector amplitudes, the differential decay rate given in Eq.1 can be rewritten as

$$\frac{d\Gamma}{dm_{K\pi}} = |\mathscr{V}|^2 F_{11} + |\mathscr{S}|^2 F_{00} + 2\Re(\mathscr{V}^* \mathscr{S}) F_{01} \tag{4}$$

where $F_{JJ'} \equiv \int \varepsilon \mathscr{M}_J^* \mathscr{M}_{J'} \phi d\Omega$ are real functions that depend only on $m_{K\pi}$, and are obtained from a full simulation of $D^+ \to K^- \pi^+ \mu^+ \nu$ events (generated according to phase space and weighted by $\mathscr{M}_J^* \mathscr{M}_{J'}$). The vector $\mathscr{V}$ and scalar $\mathscr{S}$ amplitudes are given by the sum of the states under study, thus $|\mathscr{V}|^2$, $|\mathscr{S}|^2$ and $\Re(\mathscr{V}^* \mathscr{S})$ depend on $m_{K\pi}$ as well as on the fit parameters.

Altogether, the fit parameters are the strengths $a_{J(R)}$ of the individual amplitudes, the mass and width of the $K^*(892)^0$, and the parameter r_0 appearing in B. Masses and widths of all other resonances are fixed to PDG values [10].

The $K\pi$ mass spectrum is fitted by an unbinned maximum-likelihood method, minimizing the quantity $-2\ln\sum_{\text{events}}\{(1-f_B)\mathscr{L}_S + f_B\mathscr{L}_B\}$, where $\mathscr{L}_S$ and $\mathscr{L}_B$ are the probability density functions for signal and background, respectively, and f_B is the background fraction, fixed in the estimated value. The $K^*(892)^0$ amplitude is taken as the reference channel. Decay fractions f_R are obtained by integrating each individual amplitude squared over the phase space and dividing by the integral of the overall amplitude squared over the phase space.

To account for resolution effects on the $K^*(892)^0$ parameters, we use a probability density function convoluted with a Gaussian distribution with $\sigma = 5.88$ MeV/c^2, the value obtained from our Monte Carlo simulation.

We first fit the data assuming only a contribution from the $K^*(892)^0$ channel. This fit does not adequately reproduce the data with a fit confidence level of only 0.21%. This confirms previous FOCUS results regarding the necessity of other contributions in this decay [7].

If an additional scalar amplitude is included, the fit improves significantly. With only the mass spectrum information, this contribution could be well represented by a constant

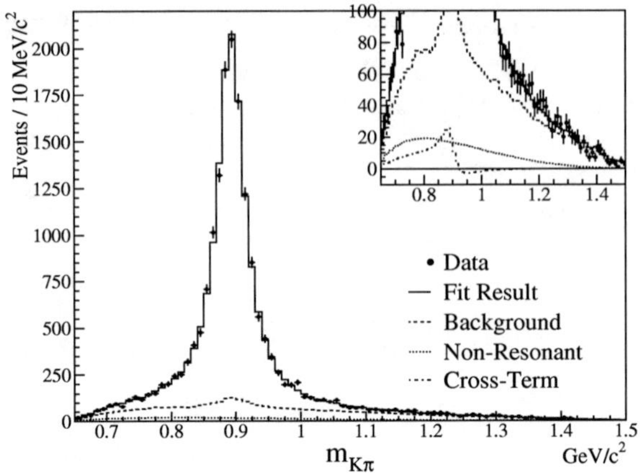

FIGURE 1. $K\pi$ Mass spectrum fit. Error bars correspond to data, solid line to the fit. Individual contributions to the fit are indicated in the figure.

TABLE 1. $K\pi$ Mass Spectrum Fit Results

$K^*(892)^0$ Mass	$895.41 \pm 0.32^{+0.35}_{-0.43}$ MeV/c^2
$K^*(892)^0$ Width	$47.79 \pm 0.86^{+1.32}_{-1.06}$ MeV/c^2
r_0	$3.96 \pm 0.54^{+1.31}_{-0.90}$ GeV^{-1}
f_{NR}	$(5.30 \pm 0.74^{+0.99}_{-0.96})$ %
$f_{K_0^*(1430)^0}$	$< 4.0\%$ @ 90% CL
$f_{K^*(1680)^0}$	$< 0.64\%$ @ 90% CL

NR decay, or an effective-range shape as in LASS [2], or a κ Breit-Wigner shape. The mass spectrum fit is not sensitive to the interference pattern of the $D^+ \to \overline{K}^*(892)^0\mu^+\nu$ channel and this additional scalar state. From an angular distribution analysis, however, this information can be determined. As discussed in the following section, a Breit-Wigner κ amplitude does not represent well the interference pattern observed between the $K^*(892)^0$ channel and the scalar channel in the absence of additional phase shifts (as imposed by the Watson theorem). The results we present here assume an effective-range form for the NR decay (Eq. 3), which also gives good results for the angular analysis (see next section). The confidence level of this fit is 66%. The results from this fit are displayed in Table 1, where systematic errors are already included [2]. In Fig. 1 we present our fitted $K\pi$ mass spectrum. Including contributions from the heavier resonant states $K_0^*(1430)^0$ and $K^*(1680)^0$ gave no improvement in the fit. We quote upper limits for the decay fraction of these modes, as shown in Table 1.

[2] For details about the systematic studies performed, see [8].

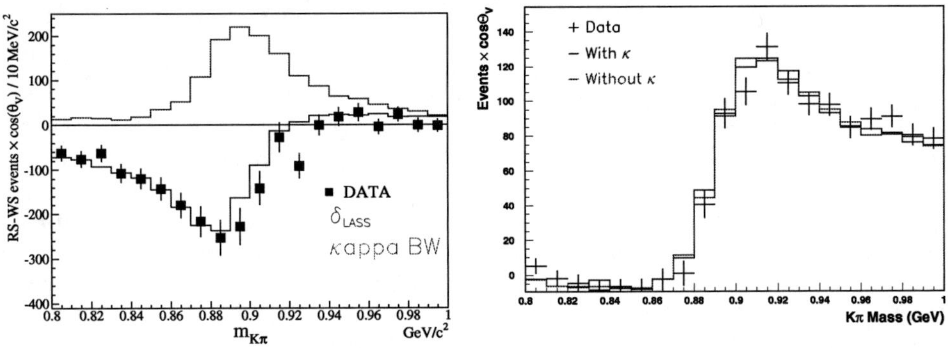

FIGURE 2. (a) Distribution of $m_{K\pi}$ weighted by $\cos\theta_v$ for $D^+ \to K^-\pi^+\mu^+\nu$ events at FOCUS. The data (points) are consistent with a NR decay parametrized as in LASS experiment (solid histogram), but not with a κ model (dashed line). (b) Distribution of $m_{K\pi}$ weighted by $\cos\theta_v$ for $D^+ \to K^-\pi^+\pi^+$ events from E791 [11].

ANGULAR DISTRIBUTION RESULTS. COMPARISONS.

As presented in the previous section, the $K\pi$ mass spectrum is dominated by the formation of the $K^*(892)^0$ resonance, although a small scalar contribution is also required. Angular distributions of the $D^+ \to K^-\pi^+\mu^+\nu$ decay give information on the interference pattern between the vector and scalar components. In fact, it has been shown [7] that a nearly constant amplitude and phase contribution to the helicity zero amplitude of the virtual W^+ was required to adequately fit the observed angular distributions. In particular, the $m_{K\pi}$ distribution weighted by $\cos\theta_v$ in the region of the $K^*(892)^0$ mass provides information on the phase of the additional structure relative to that of the $K^*(892)^0$. It can be used to discriminate between possible low mass states. Fig. 2(a) compares the distribution observed from data to those of a NR model (fit described in the previous section) and a κ Breit-Wigner model (in the absence of additional phase shifts). The NR model (using LASS parametrization) is consistent with the observed distribution.

In Fig. 2(b) we show the $m_{K\pi}$ distribution weighted by $\cos\theta_v$ for the hadronic decay $D^+ \to K^-\pi^+\pi^+$ as obtained by E791 [11]. One can see that it is very different from that of $D^+ \to K^-\pi^+\mu^+\nu$ process. Thus the interfering patterns between the $K^*(892)^0$ resonance and the scalar low-mass structure in these two $K\pi$ systems differ. The same conclusion comes from reading the phase of the S-wave relative to that of $K^*(892)^0$ near its mass. In $D^+ \to K^-\pi^+\pi^+$ [3, 4, 11] this relative phase is about $-45°$ at the $K^*(892)^0$ mass while in $D^+ \to K^-\pi^+\mu^+\nu$ it is about 45° (consistent with LASS elastic scattering).

SUMMARY AND DISCUSSIONS

The semileptonic sample of $D^+ \to K^-\pi^+\mu^+\nu$ decays from FOCUS was used to extract the $K^*(892)^0$ resonance parameters as well as to study the S-wave $K\pi$ system. The values obtained for the mass and width of the $K^*(892)^0$ are below the current PDG values by about 1σ and 2σ, respectively. The small scalar component observed to contribute to the final state can be well described by the effective-range parametrization used by LASS in elastic $K\pi$ scattering, including the phase behavior, as expected by Watson's theorem. Nevertheless, this phase behavior is different from that of $K\pi$ S-wave from $D^+ \to K^-\pi^+\pi^+$ decays. An important difference between $D^+ \to K^-\pi^+\mu^+\nu$ and $D^+ \to K^-\pi^+\pi^+$ environments, with respect to the $K\pi$ system, is the absence of final state interactions (FSI) - other than 2-body $K\pi$ itself - in the former process. It suggests that the FSI of the bachelor pion with the $K\pi$ pair in the $D \to (K\pi)\pi$ system plays an important role. This could be a clue to understanding the observed difference of the scalar phase shifts in $D^+ \to K^-\pi^+\pi^+$ and "isolated" $K\pi$ systems such as $D^+ \to K^-\pi^+\mu^+\nu$ and elastic $K\pi$ scattering.

ACKNOWLEDGMENTS

We wish to acknowledge the assistance of the staffs of Fermi National Accelerator Laboratory, the INFN of Italy, and the physics departments of the collaborating institutions. This research was supported in part by the U.S. National Science Foundation, the U.S. Department of Energy, the Italian Instituto Nazionale di Fisica Nucleare and Ministero dell'Instruzione dell'Università e della Ricerca, the Brazilian Conselho Nacional de Desenvolvimento Científico e Tecnológico, CONACyT-México, the Korean Ministry of Education, and the Korean Science and Energy Foundation.

REFERENCES

1. K. M. Watson, Phys. Rev. **88**, 1163 (1952).
2. D. Aston *et al.* (LASS Collaboration), Nucl. Phys. **B296**, 493 (1988).
3. E. M. Aitala *et al.* (E791 Collaboration), Phys. Rev. Lett. **89**, 121801 (2002).
4. E. M. Aitala *et al.* (E791 Collaboration), hep-ex/0507099.
5. See also the contribution from B. Meadows in these proceedings.
6. J. M. Blatt and V. F. Weisskopf, *Theoretical Nuclear Physics*, John Wiley & Sons, New York, p. 361 (1952).
7. J. M. Link *et al.* (FOCUS Collaboration), Phys. Lett. B **535**, 43 (2002).
8. J. M. Link *et al.* (FOCUS Collaboration), Phys. Lett. B **621**, 72 (2005).
9. P. L. Frabetti *et al.*, Nucl. Instrum. Methods A **329** 62 (1993); P. L. Frabetti *et al.* (E687 Collaboration), Nucl. Instrum. Methods A **320** 519 (1992); J. M. Link *et al.* (FOCUS Collaboration), Nucl. Instrum. Methods A **484** 270 (2002); J. M. Link *et al.* (FOCUS Collaboration), Nucl. Instrum. Methods A **516** 364 (2002);
10. S. Eidelman *et al.* (Particle Data Group), Phys. Lett. B **592**, 1 (2004).
11. C. Göbel (for E791 Collaboration), AIP Conf. Proc. **688**, 266 (2004), hep-ex/0307003.

Recent J/ψ Results from BESII

Xiaoyan SHEN (for the BES Collaboration)

Institute of High Energy Physics, Chinese Academy of Science, Beijing 100049, P. R. China

Abstract. The studies on the light scalar mesons and excited baryon states at BES are presented, based on 5.8×10^7 J/ψ data collected with BESII detector. The measurements of some J/ψ and η_c decays are also presented in this report.

Keywords: J/ψ decay; scalar meson; branching ratio; baryon
PACS: 14.40.Gx, 12.38.Qk, 13.25.Gv

INTRODUCTION

BES, as described in ref. [1], is a large general purpose solenoidal detector at the Beijing Electron Positron Collider (BEPC). The 5.8×10^7 J/ψ events were accumulated with BESII, which provides a good laboratory for the study of the hadron spectroscopy. All the results presented in this report are based on this data sample.

LIGHT SCALAR MESONS

The σ and κ

There have been hot debates on the existence of σ and κ. The decay of $J/\psi \to \omega\pi^+\pi^-$, with the ω decaying to $\pi^+\pi^-\pi^0$, is studied for σ [2].

Figure 1 (left) shows the $\pi^+\pi^-$ invariant mass spectrum recoiling against the ω for the selected $J/\psi \to \omega\pi^+\pi^-$ events. The Dalitz plot of this channel is shown in Fig. 1 (right). In addition to the well known $f_2(1270)$, a broad bump which is due to the σ pole in the low mass region is clearly seen. This peak is evident as a strong band along the upper right-hand edge of the Dalitz plot.

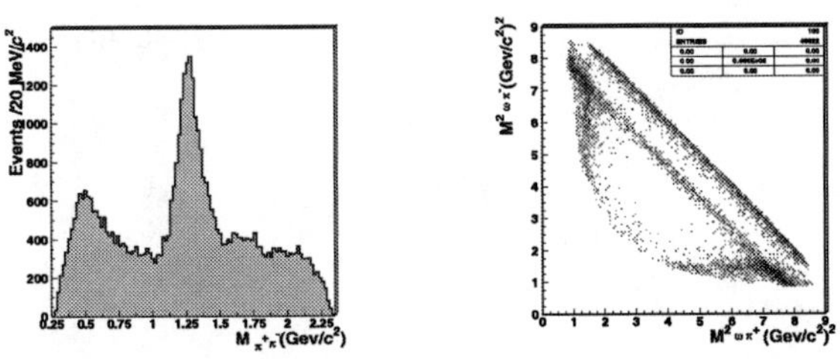

FIGURE 1. Left: distribution of $\pi^+\pi^-\pi^0$ mass. Right: Dalitz plot.

Two independent partial wave analyses (PWA) have been performed on this channel. Different analysis methods and four parametrizations of the σ amplitude give

CP814 *Hadron Spectroscopy: Eleventh International Conference*
edited by A. Reis, C. Göbel, J. Sá Borges, and J. Magnin

consistent results for the σ pole. The average σ pole position is determined to be $(541\pm39)-i(252\pm42)$ MeV/c^2 [2].

The κ is studied from $J/\psi \to \bar{K}^*(892)K^+\pi^-$[3] and $K^+K^-\pi^+\pi^-$ decays through partial wave analyses. The κ pole are needed in the fits. Figure 2 is the $K\pi$ invariant mass spectrum which recoils against $K^*(892)$. The shaded area shows the κ contribution. The preliminary pole position of κ from $J/\psi \to K^*K\pi$ is determined to be $(841\pm30^{+81}_{-73})-i(309\pm45^{+48}_{-72})$ MeV/c^2 [3].

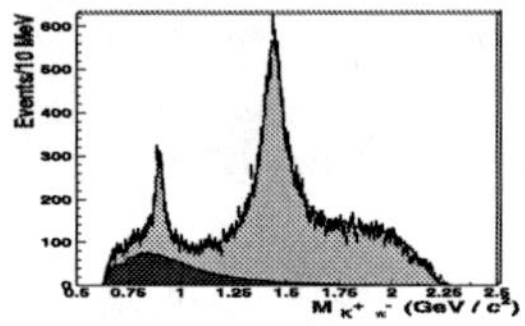

FIGURE 2. The $K\pi$ invariant mass recoiling against K^*. The crosses are data and histograms represent the PWA fit projection. The shaded area shows the κ contribution.

$f_0(980)$ and $f_0(1370)$

The $f_0(980)$ is observed clearly in both $\pi^+\pi^-$ and K^+K^- modes in the $J/\psi \to \phi\pi^+\pi^-$ and ϕK^+K^- decays at BESII. The crosses in Fig. 3(left) and 3(right) show the $\pi^+\pi^-$ and K^+K^- invariant mass spectra from $J/\psi \to \phi\pi^+\pi^-$ and ϕK^+K^-, respectively. The $\phi\pi^+\pi^-$ and ϕK^+K^- data are fitted simultaneously by using the partial wave analysis, constraining the resonances' masses and widthes to be the same in both sets of data [4]. The Flatté form:

$$f = \frac{1}{M^2 - s - i(g_1\rho_{\pi\pi} + g_2\rho_{K\bar{K}})}$$

is used to fit the $f_0(980)$ amplitude. The full histograms in Fig. 3 show the maximum likelihood fits.

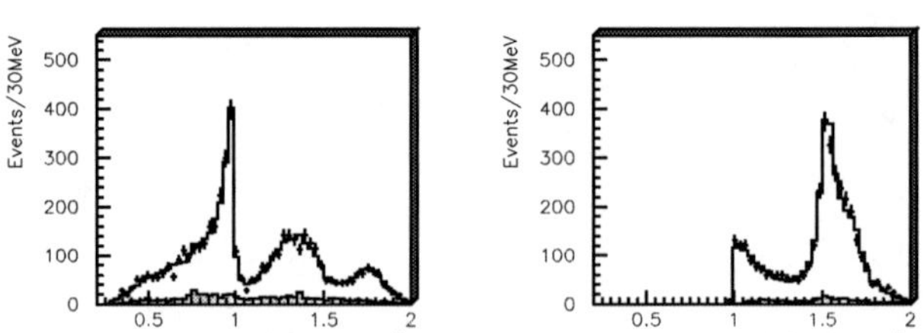

FIGURE 3. Left: The $\pi^+\pi^-$ invariant mass distribution from $J/\psi \to \phi\pi^+\pi^-$ (crosses). Right: The K^+K^- invariant mass distribution from $J/\psi \to \phi K^+K^-$ (crosses). The full histogram shows the maximum likelihood fit and the shaded histogram the background estimated from ϕ sidebins.

The present data offer the opportunity to determine the parameters of $f_0(980)$ accurately: $M = 965 \pm 8(sta) \pm 6(sys)$ MeV/c^2, $g_1 = 165 \pm 10(sta) \pm 15(sys)$ MeV/c^2, $g_2/g_1 = 4.21 \pm 0.25(sta) \pm 0.21(sys)$.

The $\phi\pi^+\pi^-$ data also exhibit a strong peak at around 1335 MeV/c^2. It is dominated by $f_0(1370)$ with the mass and width of $M = 1350 \pm 50$ MeV/c^2 and $\Gamma = 265 \pm 40$ MeV/c^2.

$f_0(1710)$ and $f_0(1790)$

A peak at around 1.7 GeV/c^2 is clearly seen in the K^+K^- invariant mass distribution from $J/\psi \to \omega K^+K^-$ decays, shown as crosses in Fig. 4. A partial wave analysis [5] shows that the J^P of this structure favors 0^+ and the mass and width are optimized at $M = 1740 \pm 30$ MeV/c^2 and $\Gamma = 125 \pm 20$ MeV/c^2. They are consistent with $f_0(1710)$ obtained in $J/\psi \to \gamma K^+K^-$ [6]. However, there is no $f_0(1710)$ in the $\pi^+\pi^-$ mass in $J/\psi \to \omega\pi^+\pi^-$. The upper limit at the 95% C.L. on the relative production rate of $f_0(1710)$ is:

$$\frac{BR(f_0(1710) \to \pi\pi)}{BR(f_0(1710) \to K\bar{K})} < 0.13$$

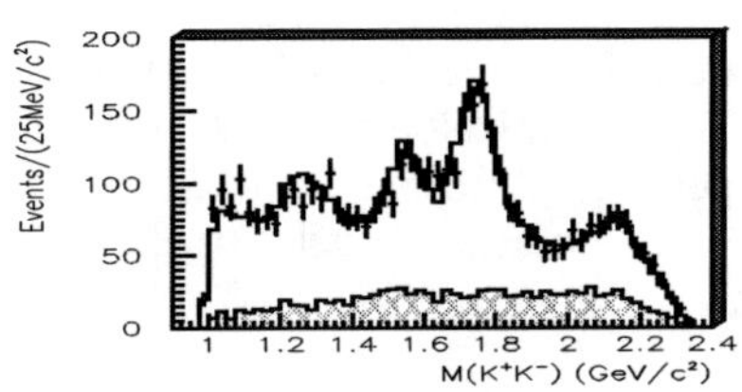

FIGURE 4. The K^+K^- invariant mass distribution from $J/\psi \to \omega K^+K^-$ (crosses). The full histogram shows the maximum likelihood fit and the shaded histogram the background estimated from ω sidebands.

The PWA of $\phi\pi^+\pi^-$ data also show a strong 0^+ peak (denoted as $f_0(1790)$) with the mass and width as $M = 1790^{+40}_{-30}$ MeV/c^2 and $\Gamma = 270^{+60}_{-30}$ MeV/c^2. It can not arise from $f_0(1710)$, since the branching fraction ratio $\pi\pi/K\bar{K}$ for $f_0(1790)$ is about one order larger than that mentioned above. The large discrepancy in branching fractions might imply the possible existence of two distinct states $f_0(1790)$ and $f_0(1710)$. Further study is needed to clarify it.

$J/\psi \to \gamma\pi\pi$ (preliminary)

The $\pi^+\pi^-$ and $\pi^0\pi^0$ invariant mass distributions from $J/\psi \to \gamma\pi^+\pi^-$ and $\gamma\pi^0\pi^0$ are shown in Fig. 5 (left) and (right). The partial wave analyses are carried out in the 1.0-2.3 GeV/c^2 $\pi\pi$ mass range. Two 0^{++} states exist in the mass lower than 1.8 GeV/c^2. The first one peaks at $1466 \pm 6 \pm 16$ MeV/c^2 with a width of $108^{+14}_{-11} \pm 21$ MeV/c^2, which is approximately consistent with $f_0(1500)$. Due to the large interference between S-wave states, a possible contribution from $f_0(1370)$ can not be excluded. The second 0^{++} peaks at around 1.75 GeV/c^2. If it is the same state with that observed in $J/\psi \to \gamma K\bar{K}$ [6], we obtain the ratio of decaying to $\pi\pi$ and $K\bar{K}$ as $0.41^{+0.08}_{-0.15}$.

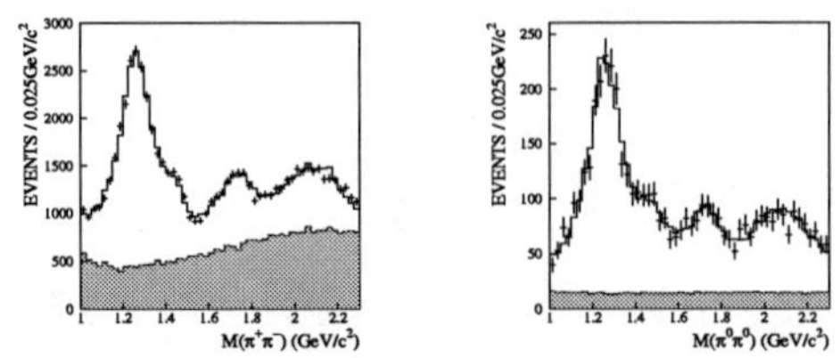

FIGURE 5. Left: the mass spectrum of $\pi^+\pi^-$ in $J/\psi \to \gamma\pi^+\pi^-$. Right: the mass spectrum of $\pi^0\pi^0$ in $J/\psi \to \gamma\pi^0\pi^0$. Crosses are data and histograms are PWA fit projections.

STUDY OF THE EXCITED BARYON STATES FROM $J/\psi \to P\bar{N}\pi^-$

The πN system in decays of $J/\psi \to \bar{N}N\pi$ is limited to be isospin 1/2 by isospin conservation. This provides a big advantage in studying $N^* \to \pi N$. Figure 6 shows the πN invariant mass spectrum from $J/\psi \to p\bar{n}\pi^-$. Besides two well known N^* peaks at 1500 MeV/c^2 and 1670 MeV/c^2, there are two new, clear N^* peaks in the $p\pi$ invariant mass spectrum around 1360 MeV/c^2 and 2030 MeV/c^2. They are the first direct observation of the $N^*(1440)$ peak and a long-sought 'missing" N^* peak above 2 GeV/c^2 in the πN invariant mass spectrum. A simple Breit-Wigner fit gives the mass and width for the $N^*(1440)$ as $1358 \pm 6 \pm 16$ MeV/c^2 and $179 \pm 26 \pm 50$ MeV/c^2, and for the new N^* peak as $2068 \pm 3^{+15}_{-40}$ MeV/c^2 and $165 \pm 14 \pm 40$ MeV/c^2, respectively [7].

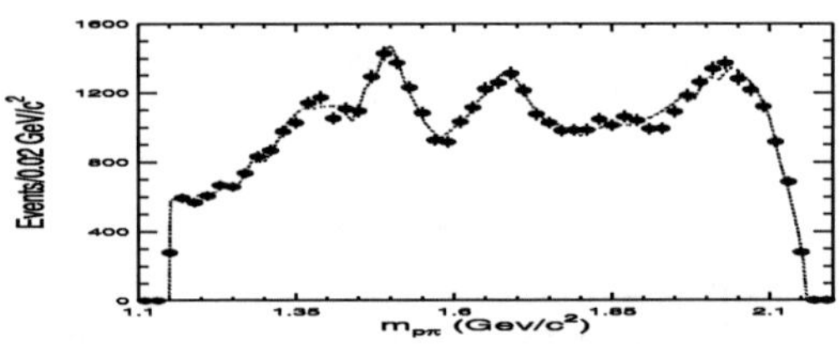

FIGURE 6. The invariant mass spectrum of πN.

MEASUREMENTS OF J/ψ AND η_C DECAY BRANCHING RATIOS

The largest J/ψ sample, 58 million J/ψ events, offers a unique opportunity to precisely measure the branching fractions of J/ψ and η_c decays. We measured the branching fractions of $J/\psi \to \phi\pi^0$, $\phi\eta$ and $\phi\eta'$ in different η and η' decay modes. The results are listed in Table 1.

TABLE 1. Branching fractions of $J/\psi \to \phi\pi^0$, $\phi\eta$, and $\phi\eta'$.

$J/\psi \to$	Final states	BF($\times 10^{-4}$)	Average BF($\times 10^{-4}$)	PDG ($\times 10^{-4}$)
$\phi\pi^0$	$K^+K^-\gamma\gamma$	< 0.064 (C.L. 90%)		
$\phi\eta$	$K^+K^-\gamma\gamma$	8.67±0.19±0.93	8.99±0.18±0.89	6.5±0.7
	$K^+K^-\pi^+\pi^-\gamma$	9.79±1.02±1.17		
	$K^+K^-\pi^+\pi^-\gamma\gamma$	9.41±0.30±1.19		
$\phi\eta'$	$K^+K^-\gamma\gamma$	6.10±1.34±0.73	5.40±0.25±0.56	3.3±0.4
	$K^+K^-\pi^+\pi^-\gamma$	5.64±0.35±0.70		
	$K^+K^-\pi^+\pi^-\gamma\gamma$	5.11±0.31±0.65		

The decays of $J/\psi \to \eta 2(\pi^+\pi^-)$ and $\eta 3(\pi^+\pi^-)$ are observed and measured for the first time [9]. They are:

$$Br(J/\psi \to \eta 2(\pi^+\pi^-)) = (1.88 \pm 0.06 \pm 0.25) \times 10^{-3},$$
$$Br(J/\psi \to \eta 3(\pi^+\pi^-)) = (7.24 \pm 0.96 \pm 1.23) \times 10^{-4}.$$

New decay modes of $\eta_c \to K^+K^-2(\pi^+\pi^-)$ and $3(\pi^+\pi^-)$ are observed [10]. The preliminary product branching ratios are:

$$Br(J/\psi \to \gamma\eta_c) \cdot Br(\eta_c \to K^+K^-2(\pi^+\pi^-)) = (1.21 \pm 0.32 \pm 0.23) \times 10^{-4},$$
$$Br(J/\psi \to \gamma\eta_c) \cdot Br(\eta_c \to 3(\pi^+\pi^-)) = (2.59 \pm 0.32 \pm 0.48) \times 10^{-4}.$$

SUMMARY

Based on 5.8×10^7 J/ψ events accumulated at the BESII detector, the studies on the light scalar mesons and excited baryon states are reviewed. We also report the measurements of J/ψ and η_c decays, of them some are the first measurements and some improve the precision from previous measurements.

ACKNOWLEDGMENTS

I would like to acknowledge my colleagues for providing me the results presented in this report and suggestions and comments for the improvement of my report.

REFERENCES

1. BES Collaboration, Nucl. Instr. Meth., **A344** 319 (1994), **A458** 627 (2001).
2. BES Collaboration, Phys. Lett. **B598**, (2004) 149
3. BES Collaboration, hep-ex/0506055
4. BES Collaboration, Phys. Lett. **B607**, (2005) 243
5. BES Collaboration, Phys. Lett. **B603**, (2004) 138
6. BES Collaboration, Phys. Rev. **D68**, 052003 (2003).
7. BES Collaboration, hep-ex/0405030
8. BES Collaboration, Phys. Rev. **D71**, 032003 (2005)
9. BES Collaboration, Phys. Lett. **B610**, (2005) 192
10. BES Collaboration, hep-ex/0505093

Consistency checks of different $\pi\pi$ scattering data sets using forward dispersion relations

José R. Peláez

Dept. Física Teórica II. Universidad Complutense, 28040-Madrid, Spain

Abstract. We review our evaluation of forward dispersion relations for direct fits to the different, and often conflicting, $\pi\pi$ scattering experimental analyses. We find that some of the most commonly used data sets do not satisfy these constraints by several standard deviations. We also provide a consistent $\pi\pi$ amplitude by improving a global fit to data with these dispersion relations.

Keywords: mesons, pions, scattering, dispersion relations
PACS: 12.39.Fe, 13.75.Lb, 11.55.Jy, 11.80.Et

A precise knowledge of the $\pi\pi$ scattering amplitude provides crucial tests for Chiral Perturbation Theory (ChPT), as well as information on light meson spectroscopy, pionic atom decays and CP violation in kaons. However, there are several $\pi\pi$ scattering data sets in the literature, in conflict among themselves, even within the same experiment. The reason is that the data are extracted from other reactions and thus with large theoretical and systematic uncertainties. Here we review our recent works [1, 2, 3] where we checked dispersion relations on the different sets of data and provided simple parameterizations of $\pi\pi$ scattering amplitudes consistent with such requirements.

Fits to different sets of data

We first consider *fits to data* [3] for the S0, S2, P waves, below $s^{1/2} \lesssim 1\,\text{GeV}$. We parameterize the phase shifts, $\delta(s)$, taking into account the analytic properties, zeros and poles of the amplitude, in a conformal expansion of $\cot\delta(s)$.

The P wave, up to $\simeq 1\,\text{GeV}$, comes from the following fit to the pion form factor [4]

$$\cot\delta_1(s) = \frac{s^{1/2}}{2k^3}(M_\rho^2 - s)\left\{B_0 + B_1\frac{\sqrt{s}-\sqrt{s_0-s}}{\sqrt{s}+\sqrt{s_0-s}}\right\}; \quad s_0^{1/2} = 1.05\text{ GeV}. \quad (1)$$

$$B_0 = 1.069 \pm 0.011, \quad B_1 = 0.13 \pm 0.05, \quad M_\rho = 773.6 \pm 0.9$$

(s_0 is the point where inelasticity begins to be nonnegligible). The fit is seen in Fig.1a.

For the S2 wave at low energies, we first fix the Adler zero at $z_2 = M_\pi$ and fit only the low energy data, $s^{1/2} < 1.0\,\text{GeV}$; later on we allow z_2 to vary. We have

$$\cot\delta_0^{(2)}(s) = \frac{s^{1/2}}{2k}\frac{M_\pi^2}{s-2z_2^2}\left\{B_0 + B_1\frac{\sqrt{s}-\sqrt{s_0-s}}{\sqrt{s}+\sqrt{s_0-s}}\right\}, \quad {s_0}^{1/2} = 1.05\text{ GeV}, \quad (2)$$

$$B_0 = -80.4 \pm 2.8, \quad B_1 = -73.6 \pm 12.6; \quad (3)$$

CP814 *Hadron Spectroscopy: Eleventh International Conference*
edited by A. Reis, C. Göbel, J. Sá Borges, and J. Magnin

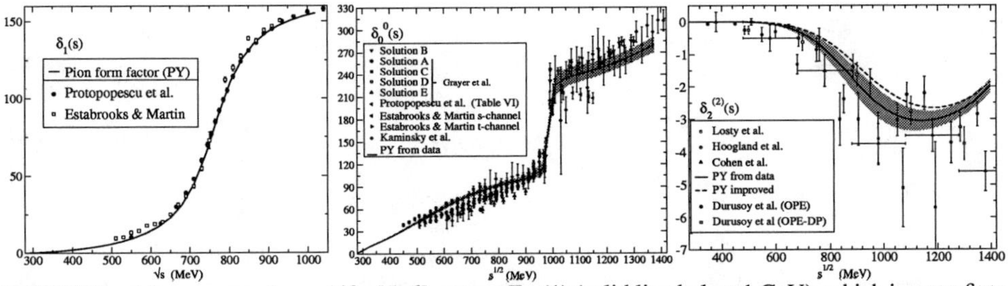

FIGURE 1. a) P wave $\pi\pi$ phase shifts [5, 6], versus Eq.(1) (solid line below 1 GeV), which is *not* a fit to these data, but to the pion form factor [4]. The uncertainty is the line thickness. Above 1 GeV, the dotted line and error are as in [3]. b) S0 phase shifts and error band as given by Eq.(4) below 1 GeV, and from [3] above. The K_{l4} and $K_{2\pi}$ decay data are not shown. (see our [3] for details). c) Continuous line: The fit to $I = 2$, D-wave phase shift data. Broken line: fit improved with dispersion relations. The experimental points are from [8].

Except for the very reliable K_{l4} and $K \to 2\pi$ decay experiments [7], that we always fit, the S0 data is very confusing and thus we have adopted two approaches. *In the first method*, the "*global fit*", we fit averaged phase shift data between $0.81\,\text{GeV} \leq s^{1/2} \leq 0.97\,\text{GeV}$, (where the experiments agree within $1.5\,\sigma$), composing their errors carefully. The fit, shown in Fig.1.b, is valid for $s^{1/2} \leq 0.95$ GeV, and with the Adler zero fixed at $z_0 = M_\pi$, corresponds to

$$\cot\delta_0^{(0)}(s) = \frac{s^{1/2}}{2k}\frac{M_\pi^2}{s-\frac{1}{2}z_0^2}\frac{M_\mu^2-s}{M_\mu^2}\left\{B_0 + B_1\frac{\sqrt{s}-\sqrt{s_0-s}}{\sqrt{s}+\sqrt{s_0-s}}\right\}, \tag{4}$$

$$B_0 = y - x; \quad B_1 = 6.62 - 2.59x; \quad y = 21.04 \pm 0.70, \quad x = 0 \pm 2.6.$$

The second method is to fit only K_{l4} and $K \to 2\pi$ data, or to add to them the data from the various experimental analyses separately. The results can be found in Table 1.

The D2 inelasticity is negligible below $4M_\rho^2$. A pole term is needed since the data [8] gives a small negative phase above ~ 500 MeV, but the Froissart–Gribov representation, yields [9] a positive scattering length, $a_2^{(2)} = (2.72 \pm 0.36) \times 10^{-4} M_\pi^{-5}$, which is included in the fit. Also, the inflection seen in data around 1 GeV asks for a third order conformal expansion. So we write

$$\cot\delta_2^{(2)}(s) = \frac{s^{1/2}}{2k^5}\left\{B_0 + B_1 w(s) + B_2 w(s)^2\right\}\frac{M_\pi{}^4 s}{4(M_\pi{}^2+\Delta^2)-s}, \; w(s) = \frac{\sqrt{s}-\sqrt{s_0-s}}{\sqrt{s}+\sqrt{s_0-s}}.$$

And we find $B_0 = (2.4 \pm 0.3) \times 10^3$, $B_1 = (7.8 \pm 0.8) \times 10^3$, $B_2 = (23.7 \pm 3.8) \times 10^3$, $\Delta = 196 \pm 20$ MeV. The fit is shown in Fig.1.c.

For brevity, we do not discuss here the D0 and F waves as they do not present special features, nor the intermediate energy region $1\,\text{GeV} \geq s^{1/2} \geq 1.42$ GeV, but a detailed account of their parameterizations can be found in [3].

We also need the imaginary part of the scattering amplitude at $s^{1/2} \geq 1.42$ GeV, that we take from a Regge fit to experimental data (see [2] for details and the slightly improved rho residue of [3]), shown in Fig.2. As discussed in [2, 3], standard Regge factorization describes experiment [10, 2] and is consistent with crossing sum rules.

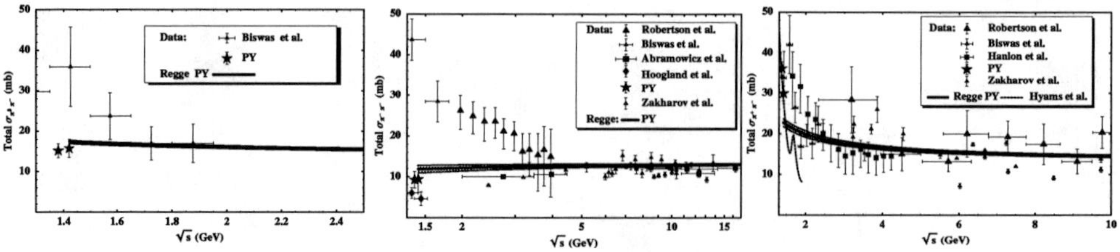

FIGURE 2 The $\pi\pi$ cross sections. Experimental points from [10]. The stars at 1.38 and 1.42 GeV (PY) are from the phase shift analysis of experimental data given in [3]. Continuous lines, from 1.42 GeV (PY): Regge formula, with parameters as in [2] (the gray bands cover the uncertainties). Below 2 GeV, the dotted line corresponds to the $\pi^+\pi^-$ cross section from the Cern–Munich analysis of Hyams et al.[6]

	B_0	B_1	M_μ (MeV)	I_t=1 χ^2/d.o.f.	$\pi^0\pi^0$ $\chi^2/d.o.f.$
PY, Eq.(4)	21.04	6.62	782 ± 24	0.3	3.5
K decay only	18.5 ± 1.7	$\equiv 0$	766 ± 95	0.2	1.8
K decay + Grayer B [6]	22.7 ± 1.6	12.3 ± 3.7	858 ± 15	1.0	2.7
K decay +Grayer C [6]	16.8 ± 0.85	-0.34 ± 2.34	787 ± 9	0.4	1.0
K decay + Grayer E [6]	21.5 ± 3.6	12.5 ± 7.6	1084 ± 110	2.1	0.5
K decay + Kamisnki [6]	27.5 ± 3.0	21.5 ± 7.4	789 ± 18	0.3	5.0
K decay + Grayer A [6]	28.1 ± 1.1	26.4 ± 2.8	866 ± 6	2.0	7.9
K decay+ EM, s-channel [6]	29.8 ± 1.3	25.1 ± 3.3	811 ± 7	1.0	9.1
K decay+ EM, t-channel [6]	29.3 ± 1.4	26.9 ± 3.4	829 ± 6	1.2	10.1
K decay+Protopopescu VI [5]	27.0 ± 1.7	22.0 ± 4.1	855 ± 10	1.2	5.8
K decay+Protopopescu XII [5]	25.5 ± 1.7	18.5 ± 4.1	866 ± 14	1.2	6.3
K decay+ Protopopescu 3 [5]	27.1 ± 2.3	23.8 ± 5.0	913 ± 18	1.8	4.2

TABLE 1 Fits to K decays and different data sets of S0 phase shifts. PY is our global fit, Eq.(4), and its B_0 and B_1 uncorrelated uncertainties can be obtained from Eq.(4). Note that many of these sets have very large $\chi^2/d.o.f.$ for the $I_t = 1$ and $\pi^0\pi^0$ dispersion relations below 950 MeV. Let us also remark that some of them (like that of Sol. E) have a relatively small $\chi^2/d.o.f.$ just because it has huge uncertainties in its parameters. Ideally one would require a low $\chi^2/d.o.f.$ with small uncertainties in the parameters.

Checking and improving amplitudes with forward dispersion relations

Let us study how well the previous fits to data satisfy three independent scattering amplitudes. We choose the t-symmetric or antisymmetric combinations, that form a complete set: $F_{00} \equiv F(\pi^0\pi^0 \to \pi^0\pi^0)$, $F_{0+} \equiv F(\pi^0\pi^+ \to \pi^0\pi^+)$, and the t channel isospin one amplitude, $F^{(I_t=1)}$. Final uncertainties are small for the two first, since they depend only on two isospin states, and their imaginary parts are sums of positive terms. Thus, we find two dispersion relations by choosing either $F = F_{00}$ or $F = F_{0+}$ in

$$\operatorname{Re} F(s) - F(4M_\pi^2) = \frac{s(s-4M_\pi^2)}{\pi} \,\text{P.P.} \int_{4M_\pi^2}^{\infty} \mathrm{d}s' \, \frac{(2s'-4M_\pi^2)\operatorname{Im} F(s')}{s'(s'-s)(s'-4M_\pi^2)(s'+s-4M_\pi^2)}. \quad (5)$$

By setting $s = 2M_\pi^2$, and $F = F_{00}$, we find a sum rule important to fix the Adler zeros.

$$F_{00}(4M_\pi^2) = F_{00}(2M_\pi^2) + \frac{8M_\pi^4}{\pi}\int_{4M_\pi^2}^{\infty} ds \frac{\mathrm{Im}\, F_{00}(s)}{s(s-2M_\pi^2)(s-4M_\pi^2)}. \quad (6)$$

Finally, for isospin unit exchange, which does not require subtractions,

$$\mathrm{Re}\, F^{(I_t=1)}(s,0) = \frac{2s-4M_\pi^2}{\pi} \text{ P.P.} \int_{4M_\pi^2}^{\infty} ds' \frac{\mathrm{Im}\, F^{(I_t=1)}(s',0)}{(s'-s)(s'+s-4M_\pi^2)}, \quad (7)$$

at threshold this is known as the Olsson sum rule.

Depending on the method we use to fit the S0 wave we find the results in Table 1, where, we have separated on top those fits to data with a total $\chi^2/d.o.f. < 6$ for the $\pi^0\pi^0$ and $I_t = 1$ dispersion relations up to 0.925 GeV, a fairly reasonable $\chi^2/d.o.f.$ since these fits were obtained independently of the dispersive approach.

However, in Table 1 we also list the very frequently used t and s-channel solutions of Estabrooks and Martin [6], those of Protopopescu *et al.*[5], from Table VI, VIII and table XII, as well as the solution A of Grayer *et al.* [6]. Their $I_t = 1$ plus $\pi^0\pi^0$ dispersion relation total $\chi^2/d.o.f.$ is surprisingly poor: 11.3, 10.1, 7, 6, 7.5, 9.9, respectively. *Therefore, any result that relies heavily on these sets should be taken very cautiously.*

We have also improved the previous low energy fits parameters by fitting also the dispersion relations up to 0.925 GeV, thus obtaining parameterizations more compatible with analyticity and $s-u$ crossing. Improving from Eq.(4), we find, in M_π units,

$$\begin{aligned}
&\text{S0};\ s^{1/2} \leq 2m_K: && B_0 = 17.4 \pm 0.5;\ B_1 = 4.3 \pm 1.4;\\
& && M_\mu = 790 \pm 21 \text{ MeV};\ z_0 = 195 \text{ MeV [Fixed]};\\
&\text{S2};\ s^{1/2} \leq 1.0: && B_0 = -80.8 \pm 1.7;\ B_1 = -77 \pm 5;\ z_2 = 147 \text{ MeV [Fixed]};\\
&\text{P};\ s^{1/2} \leq 1.05: && B_0 = 1.064 \pm 0.11;\ B_1 = 0.170 \pm 0.040;\ M_\rho = 773.6 \pm 0.9 \text{ MeV};\\
&\text{D2};\ s^{1/2} \leq 1.42: && B_0 = (2.9 \pm 0.2) \times 10^3;\ B_1 = (7.3 \pm 0.8) \times 10^3;\\
& && B_2 = (25.4 \pm 3.6) \times 10^3;\ \Delta = 212 \pm 19. \qquad (8)
\end{aligned}$$

The D0 and F waves do not change appreciably and we just refer to [3] for details. In Fig.3 we show the improved curves for S0 and S2, and that of D2 in Fig.1.c.

Concerning the improved fits to individual data sets, we get somewhat different results for S0, listed in Table 2 with the $\chi^2/d.o.f.$ of each forward dispersion relation and the standard deviations for the sum rule in Eq.(6) (which are more than four for K decay plus the Grayer B or E or Kaminski improved solutions). For other waves, no matter what S0 fit is used, we find very similar values to those in Eq.(8). This can be checked in Fig.3.b, where we show the improved "K decay + Grayer Sol. B" S2 wave. Even though it is the one for which we obtained the most different central values for the S0 wave compared with those given in Eq.(8), it falls within the uncertainty of our improved solution.

In summary, with forward dispersion relations, we have checked the consistency of different phase shift analyses available in the literature. Surprisingly, some of the most frequently used phase shift sets do not satisfy these dispersive constraints and sum rules, and should therefore be used cautiously. We have provided a simple parameterization of $\pi\pi$ scattering consistent simultaneously with some data sets and all three forward dispersion relations, that we hope could be of use for future studies of $\pi\pi$ scattering.

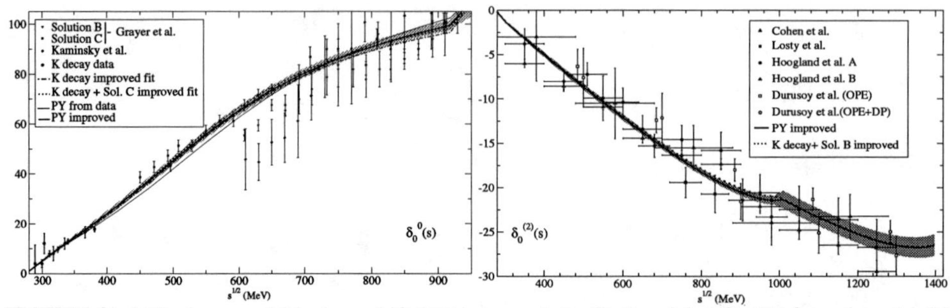

FIGURE 3. a) The improved S0 phase shift (PY improved, Eq.8), the global fit (PY from data, the S0 in Eq. (4)), and the *improved* solutions "*K* decay only" and "Grayer C" of Table 2 (almost on top of PY improved). b) S2 improved Phase shift (PY improved, Eq. (8)); global fit (PY from data, Eq. (4)) and the improved parameterization with K decays and So. B of Grayer et al.[6].

TABLE 2. Improved fits (only those already with $\chi^2/dof < 6$ in Table 1). Although errors are given for the Adler zero, we fix it when evaluating other errors, to break the otherwise very large correlations.

Improved fits:	Improved PY, Eq.8	*K* decay only	*K* decay+ Grayer C	*K* decay+ Grayer B	*K* decay+ Grayer E	*K* decay+ Kamiński
B_0	17.4 ± 0.5	16.4 ± 0.9	16.2 ± 0.7	20.7 ± 1.0	20.2 ± 2.2	20.8 ± 1.4
B_1	4.3 ± 1.4	$\equiv 0$	0.5 ± 1.8	11.6 ± 2.6	8.4 ± 5.2	13.6 ± 43.7
M_μ (MeV)	790 ± 30	809 ± 53	788 ± 9	861 ± 14	982 ± 95	798 ± 17
z_0 (MeV)	195 ± 30	182 ± 34	182 ± 39	233 ± 30	272 ± 50	245 ± 39
$I_t = 1$, $\chi^2/d.o.f.$	0.40	0.30	0.37	0.37	0.60	0.43
$\pi^0\pi^0$, $\chi^2/d.o.f.$	0.66	0.29	0.32	0.83	0.09	1.08
$\pi^+\pi^-$, $\chi^2/d.o.f.$	1.62	1.77	1.74	1.60	1.40	1.36
Eq.(6)	1.6σ	1.5σ	1.5σ	4.0σ	6.0σ	4.5σ

REFERENCES

1. Peláez, J. R., and Yndураín, F. J., *Phys. Rev.* **D68**, 074005 (2003).
2. Peláez, J. R., and Yndurain, F. J., *Phys. Rev.* **D69**, 114001 (2004).J. R. Pelaez, Int. J. Mod. Phys. A **20**, 628 (2005)
3. Peláez, J. R., and Yndurain, F. J.,*Phys. Rev.* **D71**, 074016 (2005).
4. de Trocóniz, J. F., and Yndurain, F. J., *Phys. Rev.*, **D65**, 093001, (2002), *Phys. Rev.* **D71**, 073008 (2005).
5. Protopopescu, S. D., et al., *Phys Rev.* **D7**, 1279, (1973).
6. Cern-Munich experiment: Hyams, B., et al., *Nucl. Phys.* **B64**, 134, (1973); Grayer, G., et al., *Nucl. Phys.* **B75**, 189, (1974). Estabrooks, P., and Martin, A. D., *Nucl. Physics*, **B79**, 301, (1974); Kamiński, R., Lesniak, L, and Rybicki, K., *Z. Phys.* **C74**, 79 (1997) and *Eur. Phys. J.* **C4**, 4 (2002);
7. Rosselet, L., et al. *Phys. Rev.* **D15**, 574 (1977); Pislak, S., et al. *Phys. Rev. Lett.*, **87**, 221801 (2001).
8. Losty, M. J., et al. *Nucl. Phys.*, **B69**, 185 (1974); Hoogland, W., et al. *Nucl. Phys.*, **B126**, 109 (1977); Cohen, D. et al., *Phys. Rev.* **D7**, 661 (1973); Durusoy, N. B., et al., *Phys. Lett.* **B45**, 517 (19730.
9. Palou, F. P., and Yndurain, F. J., *Nuovo Cimento*, **19A**, 245, (1974); Palou, F. P., Sánchez-Gómez, J. L., and Yndurain, F. J., *Z. Phys.*, **A274**, 161, (1975).
10. Biswas, N. N., et al., *Phys. Rev. Lett.*, **18**, 273 (1967) Cohen, D. et al., *Phys. Rev.* **D7**, 661 (1973). Robertson, W. J., Walker, W. D., and Davis, J. L., *Phys. Rev.* **D7**, 2554 (1973). Hanlon, J., et al, *Phys. Rev. Lett.*, **37**, 967 (1976); Abramowicz, H., et al. *Nucl. Phys.*, **B166**, 62 (1980). B.G. Zakharov and V.N. Sergeev, *Sov. J. Nucl. Phys.* **39**, 448 (1984) also in Yad. Fiz. **39**, 707 (1984).

Improved Pion Pion Scattering Amplitude from Dispersion Relation Formalism

I. P. Cavalcante*, Y. A. Coutinho† and J. Sá Borges**

*Universidade Federal de Mato Grosso do Sul
†Universidade Federal do Rio de Janeiro
**Universidade do Estado do Rio de Janeiro

Abstract. Pion-pion scattering amplitude is obtained from Chiral Perturbation Theory at one- and two-loop approximations. Dispersion relation formalism provides a more economic method, which was proved to reproduce the analytical structure of that amplitude at both approximation levels. This work extends the use of the formalism in order to compute further unitarity corrections to partial waves, including the D-wave amplitude.

Keywords: pion-pion scattering, dispersion relation, chiral symmetry
PACS: 11.40.Ex, 11.55.Fv, 13.75.Lb, 12.39.Fe

INTRODUCTION

In the absence of the underlying theory for hadron interactions, in the early sixties, many results have been derived from the assumptions of a local $SU(2) \times SU(2)$ chiral algebra of vector and axial-vector current densities together with the partial conservation hypothesis relating the divergence of the axial-vector current to the π-meson field. In particular, a technique was developed for explicitly displaying all the information that current algebra implies for three- and four-point functions of currents[1]. The hard meson method assumes that the ρ, A_1, and π mesons dominate the 1^-, 1^+, and 0^- channels, respectively. In order to implement this assumption one defines diagonalized n-point functions in such way that, on mass shell, a channel associated with an axial current receives contributions only from 1^+ states, by subtracting the off-shell 0^- contribution.

This method was applied, long ago, to pion-pion scattering. In that case, the result is a crossing symmetric amplitude written as a combination of scalar and vector pion form factors and propagators[2]. It can be shown that its low energy limit reproduces the soft-pion Weinberg result[3]. In this context, an unitarization program of current algebra (UPCA) was developed using the fact that the soft-pion limit gives a real amplitude so that the nearby discontinuities across the right-hand cut are small as compared with current algebra result[2]. By this unitarization procedure one computes corrected form factor and propagators using the consequences of partial wave elastic unitarity.

The hard meson method based on the Ward identities technique was abandoned for several reasons, summarized in Weinberg's sentence[4] " ... Unfortunately, such calculations are tedious, especially when three or more Goldstone bosons are involved ... For this reason a different and more physical calculational technique was introduced, based on the use of effective Lagrangian: We simply calculate the Goldstone boson amplitudes by the methods of perturbation theory ...".

CP814 *Hadron Spectroscopy: Eleventh International Conference*
edited by A. Reis, C. Göbel, J. Sá Borges, and J. Magnin

As a consequence of this kind of reasoning[5], new techniques have been developed in order to extend the results of the current algebra days and *systematically* explore corrections to the soft-pion predictions based on symmetry properties of QCD Green functions. The method is called chiral perturbation theory (ChPT) [6] and describes the dynamics of Goldstone bosons in the framework of an effective field theory.

As soon as the $\mathscr{O}(p^4)$ ChPT pion-pion scattering result was produced[7], one of us compared it with UPCA first-order corrected pion-pion amplitude and concluded that they coincide. That indication motivates him to present the tools for constructing second order corrections to the soft-pion amplitude[8]. We emphasize that only ten years later the $\mathscr{O}(p^6)$ ChPT was derived[9], and applied to pion-pion scattering[10]. It was proved that this result has the same analytical structure as UPCA second order corrected amplitude as well[11].

In the present work we claim that UPCA is an economic method to introduce corrections to the low energy theorem for pion-pion scattering. In the past, the main difficulties were to handle dispersive integrals of the model. At present, modern algebraic computational packages does the job[12], allowing one to improve the approximations leading to reliable amplitude expressions. Another advantage of UPCA, discussed here, relies on the number and meaning of involved arbitrary constants. While ChPT low energy constants (LECs), at each order of approximation, are new couplings, essential to a complete chiral invariant Lagrangian, UPCA free parameters are subtraction constants related to different orders of derivatives of form factors at some energy value. This way, eliminating ChPT LECs corresponds to modifying the Lagrangian but, on the other hand, some UPCA subtraction constants can be assumed to be zero without changing the energy dependence of the amplitude. As a result, we show the resulting second order corrected D-wave phase and also present the tools to construct higher order corrections.

DISPERSION RELATION METHOD AND IMPROVEMENTS

Let us recall the main points of the current algebra unitarization program (UPCA). The starting point in our derivation was an exact hard-pion expression for the correlation function of four currents, with the quantum numbers of the pion, written in terms of three- and two-point functions. The hard-meson method leads to the following tree structure

$$T^{abcd} = \bar{h}_0(s)P_0^{ab,cd} + \bar{h}_2(s)P_2^{ab,cd} + (t-u)\bar{h}_1(s)P_1^{ab,cd} + \text{crossed terms}$$

where $P_I^{ab,cd}$ are the isospin I s-channel projectors and $\bar{h}_I$ are the following combinations of form-factors F_I and propagators Δ_I:

$$\bar{h}_I(x) = \frac{(F_I(x)-1)^2}{\Delta_I(x)} \quad \text{for I=0,2 and} \quad \bar{h}_1(x) = \frac{(F_1(x)-K)^2}{\Delta_V(x)} - \frac{C_V}{4f_\pi^4} \tag{1}$$

>From this expression, and using very plausible form factors and propagators estimates at $x \simeq m_\pi^2$:

$$F_I(x) \simeq 1 + f_I^{(1)}(x), \;\; \Delta_V(x) \simeq C_V + \delta_V^{(1)}, \;\; \text{and} \;\; \Delta_I(x) \simeq \delta_I^{(1)}(x) \;\; \text{for I} = 0{,}2,$$

we obtained the so-called soft-pion Weinberg crossing symmetric amplitude, namely,

$$A^{ca}(s,t,u) = \frac{1}{f_\pi^2}\left(s - m_\pi^2\right).$$

We have introduced the superscript (1) to characterize first-order corrections of order m_π/m_ρ to the relevant functions. The remainder of the amplitude reflects the difference between soft- and hard-meson methods. In order to explore its content, one of us used the implications for form factors and propagators discontinuities of elastic unitarity to the partial wave, namely:

$$\mathrm{Im}\, F_I(x) = \frac{1}{32\pi}\rho(x)\, T_I^*(x)\, F_I(x) \quad \text{and} \quad \mathrm{Im}\,\Delta_I(x) = \frac{1}{32\pi}\rho(x)\, |F_I(x)|^2, \tag{2}$$

where

$$\rho(s) = \sqrt{\frac{s-4m^2}{s}} \quad \text{for } s \geq 4m^2$$

is the phase-space factor which plays an important role in our derivation.

The first order correct amplitude follows from the expressions of $h_I^{(1)}$. To compute these functions one takes their discontinuities on the right-hand side cut from Eq. (2). In such a way, we calculate, for example

$$h_{11}^{(1)}(s) = s\lambda_3 + \frac{s^2}{32\pi^2}\int_{4m^2}^{\infty} \frac{\rho(x)\, T_1^{ca}(x)}{x^2(x-s)}\, dx,$$

where $T_1^{ca}(s) = (s-4m^2)/3f_\pi^2$ is the P-wave projection of the Weinberg amplitude and λ_3 is a subtraction constant related to the pion electromagnetic charge radius. We thus obtain,

$$A^{(1)}(s,t,u) = \frac{1}{f^4}(s-4m^2)^2 G_1(s) + \frac{2}{3f^4}\left\{\left(t^2 - 10m^2 t - 4m^2 s + 14m^4\right) G_1(t) + \right.$$
$$\left.(t \leftrightarrow u)\right\} + P_{\lambda_1,\lambda_2,\lambda_3}(s,t,u),$$

where P is a polynomial depending of the three subtraction constants, and

$$32\pi^2\, G_1(s) = \rho(s) L(s), \quad \text{with} \quad L(s) = \ln\frac{\rho(s)-1}{\rho(s)+1}. \tag{3}$$

The unitarity corrected first order partial-wave amplitudes was computed from $A^{(1)}$ and the free parameters have been adjusted to fit the model to the experimental phases[2].

The next order UPCA approximation corresponds to two loops calculations in ChPT. In this case, the expressions of right-hand discontinuities for the functions $h_{\ell I}^{(2)}$ follows from the discontinuities of corrected form factors and propagators and are written as:

$$\mathrm{Im}\; h_{\ell I}^{(2)}(s) = \frac{\rho(s)}{16\pi^2}\;\mathrm{Re}\; T_{\ell I}^{(1)}(s), \quad \text{for } s \geq 4m^2$$

where one has to consider the real part of the first order corrected amplitudes. For example, to calculate the function $h_{11}^{(2)}$, we have to consider

$$\begin{aligned} f_\pi^4 T_{11}^{(1)}(s) &= \frac{1}{9}\left(s-4m^2\right)G_1(s) - \frac{1}{288\pi^2}\frac{s^3-16s^2m^2+72sm^4-36m^6}{(s-4m^2)}\frac{L(s)}{\rho(s)} - \\ & \frac{m^4}{48\pi^2}\frac{3s^2-13m^2s-6m^4}{(s-4m^2)^2}L(s)^2 + \frac{1}{288\pi^2}\left(s-4m^2\right)(\lambda_1 s+\lambda_2) \\ & -\frac{1}{864\pi^2}\frac{s^3+37s^2m^2-149sm^2+120m^6}{(s-4m^2)}, \end{aligned}$$

consequently, we have to calculate integrals like

$$G_2(s) = \frac{s-4m^2}{32\pi^2}\int_{4m^2}^{\infty}\frac{\rho(x)\mathrm{Re}\,G_1(x)}{(x-4m^2)(x-s)} \text{ or } Z(s) = \frac{s-4m^2}{32\pi^2}\int_{4m^2}^{\infty}\frac{L(x)}{(x-4m^2)(x-s)}dx.$$

The results are combination of polylogarithm functions,

$$\left(32\pi^2\right)^2 G_2(s) = -\rho(s)^2\left\{\frac{\pi^2}{6} + \mathrm{dilog}\left(2\frac{\rho(s)}{\rho(s)-1}\right) + \mathrm{dilog}\left(2\frac{\rho(s)}{\rho(s)+1}\right)\right\},$$

but, as in the physical region, $s > 4m^2$, and $(\rho(s)+1)/(\rho(s)-1) < -1$, we obtain, by using Maple's *combine polylog* function[12],

$$\left(32\pi^2\right)^2 G_2(s) = \frac{\rho^2}{2}L(s)^2 \quad \text{and} \quad 64\pi^2 Z(s) = L(s)^2 - \pi^2.$$

Similar procedure can be employed to solve the integral involving $L(s)^2$ in Eq. (4) and also one can easily gets algebraic expressions for angular momentum projections of these functions.

Our strategy was to keep just three arbitrary parameters for all partial-waves, when computing the functions $h_{\ell I}^{(2)}$. Then we fit, to the experimental P and D_0 phase-shifts[13], the phases of the sum of contributions to h_{11} and to h_{20} respectively. The resulting fits are shown in Fig. 1. Having fixed all model free parameters, we obtained the h_{00} phase, present in the Fig. 2 together with experimental results for S_0[13]. We do not include the experimental errors of the experimental phase-shits and the resulting parameters values are: $\lambda_1 = 5.88$, $\lambda_2 = 11.75$ and $\lambda_3 = -.156$. Finally we present the tools for next order calculation. Expanding the equation (2), we obtain

$$h_{\ell I}^{(3)}(s) = \frac{\rho(s)}{32\pi}\left\{T_I^{ca}(s)\,\mathrm{Re}h_{\ell I}^{(2)}(s) + \mathrm{Re}T_{\ell I}^{(1)}(s)\,\mathrm{Re}h_{\ell I}^{(2)}(s) + \mathrm{Re}T_{\ell I}^{(2)}(s) + \rho^2 T_I^{ca3}(s)\right\},$$

consequently, next order calculation *via* the UPCA method involves other integrals leading to combinations of polylogarithms with different arguments, for example

$$G_3(s) = \frac{s-4m^2}{32\pi^2}\int_{4m^2}^{\infty}\frac{\rho(x)\mathrm{Re}G_2(x)}{(x-4m^2)(x-s)} = \frac{\rho^2}{6(32\pi^2)^3}\left\{\rho L(s)^3 - \pi^2\rho L(s) - 8\pi^2\right\}.$$

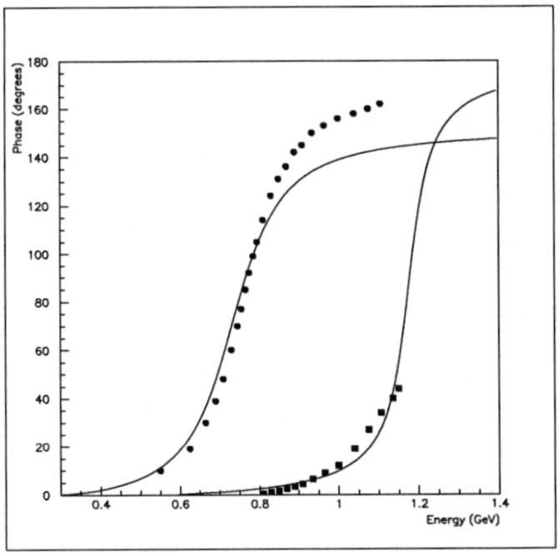

FIGURE 1. Three parameters fit of P and D_0 phases. The experimental[13] error bars are not shown.

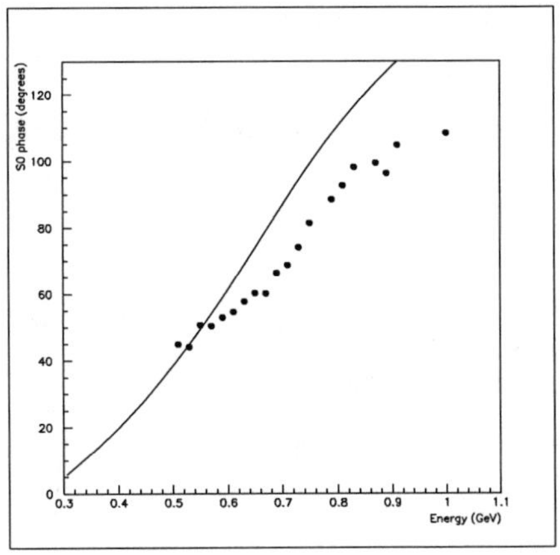

FIGURE 2. The resulting S_0 phase. The error bars are not shown[13].

CONCLUSIONS

In this work we stress that UPCA, when combined with modern algebraic computacional packages, is an economical method to construct higher order approximations to the low energy theorem of current algebra for $\pi\pi$ scattering, satisfying analitycity, crossing symmetry and approximate unitarity. We have also observed that one can safely reduce the number of model free parameters. In particular we have presented a three parameters simultaneous fit of $P-$, D_0 and S_0 phases. It turns out that the h_{20}- phase reaches 90^0 but do not fit the $f_2(1270)$ resonance, while it results a much broader P-wave and a rapid rising S_0-phase. We believe that better fits can be obtained by increasing the number of model free parameters.

REFERENCES

1. I. S. Gerstein, J. H. Schnitzer, and S. Weinberg, *Phys. Rev.* , **170** 1638 (1968).
2. J. Borges, *Nucl. Phys.*, **B51** 189 (1973).
3. S. Weinberg, *Phys. Rev. Lett.*, **17** 616 (1966).
4. S. Weinberg,"Spontaneous Broken Global Symmetries", in *The Quantum Theory of Fields*, Press Syndicate of the University of Cambrigde, Cambridge, 1996, pp. 192.
5. S. Weinberg, *Physica A*, **96** 327 (1979).
6. J. Gasser and H. Leutwyler, *Annals Phys.*,**158** 142 (1984).
7. J. Gasser and H. Leutwyler, *Phys. Lett.* , **125B** 325(1983).
8. J. Sá Borges, *Phys. Lett.*, **149B** 21 (1984).
9. H. W. Fearing and S. Scherer, *Phys.Rev.* **D53** 315-348 (1996).
10. J. Bijnens, G. Colangelo, G. Ecker, J. Gasser and M. E. Sainio, *Phys. Lett.*, **374B** 210 (1996).
11. J. Sá Borges, J. Soares Barbosa and M. D. Tonasse, *Phys. Rev.*, **D57** 4108 (1998).
12. MapleV, http://www.maplesoft.com/
13. S. D. Protopopescu et al, *Phys. Rev*, **D7** 1279 (1973).

Glueball-$Q\bar{Q}$ mixings in the scalars

Qiang Zhao

Department of Physics, University of Surrey, Guildford, GU2 7XH, United Kingdom
Institute of High Energy Physics, Chinese Academy of Sciences, Beijing, 100049, P.R. China

Abstract. Possibly significant Okuba-Zweig-Iizuka (OZI) rule violation is found important for the scalar meson production in J/ψ hadronic decays. Its correlation with the glueball-$Q\bar{Q}$ mixings provides a natural explanation for the "scalar puzzle" arising from the data for $J/\psi \to V f_0$ at BES. Nevertheless, such a correlation can be examined at different energy scales and other heavy quarkonium hadronic decays, where insights into the scalar meson structures can be gained.

Keywords: Scalar glueball, OZI rule, charmonium hadronic decay
PACS: 13.25.Gv,12.39.Mk,12.39.St

The idea of glueball-$Q\bar{Q}$ mixings based on the old-fashion perturbation theory leads to interesting implications about the scalar meson properties [1, 2]. Although the detailed dynamics is still not well understood, such a prescription provides an empirical treatment for the scalar meson decays into pseudoscalar meson pairs, and manages to accommodate the scalars in a self-consistent framework for the meson spectrum [3].

In a recent work [4], we reported a coherent study of the production of f_0^i ($i = 1, 2, 3$ corresponding to $f_0(1710)$, $f_0(1500)$, and $f_0(1370)$) in $J/\psi \to V f_0 \to VPP$, where $V = \phi,\ \omega$. There are advantages of studying isoscalar vector meson ϕ and ω in association with the f_0 states: i) Since ω and ϕ are approximately ideally mixed in the flavor space, the process $J/\psi \to \phi f_0(s\bar{s})$ and $J/\psi \to \omega f_0(n\bar{n})$ may probe the $s\bar{s}$ and $n\bar{n}$ components separately to leading order via the singly OZI disconnected (SOZI) processes; ii) If there exist glueball components in the f_0 states, those two processes also probe the glueball components at the same order via $J/\psi \to VG$; iii) As discussed in Ref. [4], the ratio of these two decay channels also highlights the importance of the doubly OZI disconnected (DOZI) process in the interpretation of the BES data for the production of $f_0(1370)$ and $f_0(1500)$ in the J/ψ decays, which is contrary to expectations based on naive OZI rule [5]. Therefore, by studying the correlation between the glueball-$Q\bar{Q}$ mixing in the scalar meson wavefunctions and OZI-rule violations in the scalar production we can gain more insights into the nature of those scalar mesons.

In Ref. [4], we showed that such a correlation played an important role in the understanding of the "scalar puzzles" arising from the recent BES data [6, 7, 8, 9]. For $J/\psi \to \omega f_0^i$ and ϕf_0^i, the ratio between these two partial decay widths becomes

$$R_i^{OZI} = \frac{\Gamma_{J/\psi\to\phi f_0^i\to\phi K\bar{K}}}{\Gamma_{J/\psi\to\omega f_0^i\to\omega K\bar{K}}} = \frac{|\mathbf{p}_{\phi i}|}{|\mathbf{p}_{\omega i}|}\frac{[x_i+y_i+\sqrt{2}rz_i]^2}{2[x_i+ry_i+\sqrt{2}z_i]^2}\,, \tag{1}$$

where the correlation between the scalar meson structure and the OZI-rule violation is highlighted. As illustrated by this relation, apart from the kinematic factors ($\mathbf{p}_{\phi i}$ and $\mathbf{p}_{\omega i}$), the OZI-rule violation factor r is the only quantity dependendent on energy scale.

CP814 *Hadron Spectroscopy: Eleventh International Conference*
edited by A. Reis, C. Göbel, J. Sá Borges, and J. Magnin

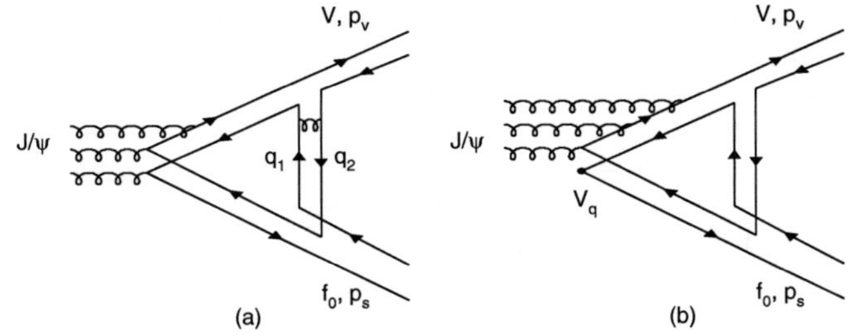

FIGURE 1. The DOZI processes in terms of intermediate virtual meson exchanges.

In other words, a measurement of R_i^{OZI} at different energies will reflect the energy evolution of r factor, which should be an important evidence for the contributions of DOZI processes in $J/\psi \to V f_0^i$. In this proceeding, we will explore the role played by the DOZI processes and look for additional evidence for the glueball component in the scalars.

Theoretically, the energy evolution of r can be investigated by calculating the relative strengths between the DOZI and SOZI processes. However, due to the break-down of the perturbative character of the DOZI processes, a pQCD calculation of r becomes unreliable. We hence introduce a dynamical assumption based on sucessful phenomenological treatments available in the literature. We assume that the DOZI processes are dual with the intermediate meson exchange processes [10, 11, 12]. It was pointed out by Lipkin before that the OZI-rule violations can always proceed by a two-step process involving intermediate virtual mesons [10]. In this case, the DOZI processes can occur via $J/\psi \to K^*\bar{K} + c.c. \to V f_0^i$, or $J/\psi \to \rho\pi + c.c. \to V f_0^i$. As we know that $J/\psi \to K^*\bar{K} + c.c.$ and $\rho\pi + c.c.$ are two of the largest decay channels of J/ψ [13], the virtual meson mediated transitions as the dominant contributions to the DOZI processes may not be small.

In the picture of intermediate virtual meson exchanges, the DOZI process can be illustrated by two topological diagrams, Fig. 1(a) and (b). Qualitatively, such processes will be suppressed relative to the SOZI ones at higher energies. In Fig. 1(a), the two colored quark-antiquark pairs must exchange a colored gluon to keep overall color-singlet. In the case that the created quark pairs scatter out to form the final state V and f_0^i, there must need large momentum transfers between quark line q_1 and q_2. Nevertheless, the momentum difference between q_1 and q_2 will increase with the increasing masses of the initial quarkonia since they must be recoiled back to the directions of their original quark (antiquark) partners to form the mesons in the final state. As a consequence, the exchanged gluon will become very hard and the transition amplitude will be suppressed at higher masses. One can also understand this as the suppression from the wavefunction distribution of the virtual meson formed by q_1 and q_2. Higher energies will correspond to the wavefunction tail at shorter distances, which is far away from the large wavefunction density determined by the typical size of meson.

In Fig. 1(b), the color-singlet quark-antiquark pair of the scalar meson is created and recoiled to the backward direction. In the virtual meson exchange model, as shown by Fig. 1(b), the quark-antiquark pairs must correlate with each other to form the virtual mesons. Therefore, at least at one vertex, e.g. V_q, which is the $^3P_0$ production vertex, the quark line must be recoiled back approximately to the momentum direction of the final antiquark. Because of such a large momentum transfer at vertex V_q, the $^3P_0$ mechanism is suppressed, which means that the transition via Fig. 1(b) will decrease with the increasing energies, i.e. in the decay of heavier quarkonia.

In principle, by summing over all the possible meson exchange loops, the meson exchange mechanisms would provide the same results for the DOZI processes. However, difficulty will arise due to possible contributions from a large number of intermediate processes. A further reasonable assumption then is to consider the dominant processes based on the available experimental information. As shown by PDG [13], J/ψ has the largest partial decay widths to $\rho\pi$, and ω meson has also strong couplings to $\rho\pi$. Meanwhile, for these f_0 states, the decay of $f_0^i \to \pi\pi$ is one of the most important channels, e.g. $f_0(1370) \to \pi\pi$. This makes the $\rho\pi$ intermediate exchange a leading contribution to $J/\psi \to \omega f_0^3$. Similarly, the intermediate $K^*\bar{K} + c.c.$ exchange should be also a dominant contribution to $J/\psi \to \phi f_0^1$: the partial decay width of $J/\psi \to K^*\bar{K} + c.c.$ is about one order of magnitude larger than $J/\psi \to \phi f_0^1$; the ϕ meson couples strongly to K^*K; and $K\bar{K}$ is the dominant decay channel for the $f_0(1710)$. Furthermore, note that the other hadronic decay channels of J/ψ are almost the same order of magnitude as $J/\psi \to \omega f_0(1710)$, or just one magnitude larger than $J/\psi \to \phi f_0(1710)$. Their rescatterings to final state $V f_0^i$ would be strongly suppressed due to their weak couplings to $V f_0^i$ and energy suppressions at the rescattering vertices.

Following the meson exchange mechanisms, we calculate the meson loops in Fig. 2 as a leading contribution to the doubly disconnected process. Supposing the OZI violation effects are from the intermediate meson exchange processes, we can estimate the meson exchange contribution and compare it with the "tree" diagram for $J/\psi \to V\bar{P} + c.c.$, e.g. $J/\psi \to K^*\bar{K} + c.c.$ and $J/\psi \to \rho\pi + c.c.$ Since the $J/\psi V\bar{P}$ vertices are the same between the tree and loop transitions, the ratio between the meson-exchange loop and the tree process of $J/\psi \to V\bar{P} + c.c.$ will then highlight the OZI violations in this dynamical process. For the purpose of gaining information about the rough behavior of the energy-evolution of the doubly OZI processes, the energy-evolution of this ratio is sufficient to account for the energy-dependence of the doubly OZI violation transitions.

The transition amplitude can be written as

$$M_{fi} = -i\int \frac{d^4p_2}{(2\pi)^4} T_v^\beta \left(g_\beta^\lambda - \frac{p_{1\beta}p_1^\lambda}{p_1^2} \right) T_{0\lambda} T_s \frac{F(p_2)}{a_1a_2a_3} \delta^4(p_0 - p_v - P_s) \,, \tag{2}$$

where the vertex functions are

$$T_v^\beta = i\frac{g_v}{M_v}\varepsilon^{\mu\nu\alpha\beta} p_{\nu\mu}\varepsilon_{f\nu}p_{2\alpha}, \;\; T_\lambda = i\frac{g_0}{M_0}\varepsilon_{\lambda\sigma\tau\delta}p_0^\sigma\varepsilon_i^\tau p_3^\sigma, \;\; T_0 = ig_sM_s \,, \tag{3}$$

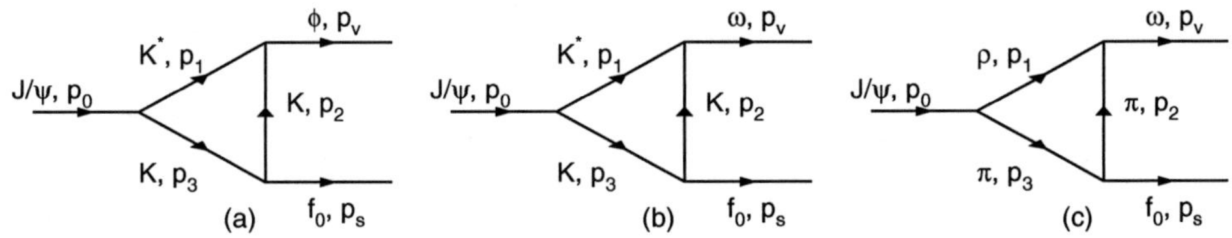

FIGURE 2. The OZI violations via intermediate K^*K and $\rho\pi$ exchanges as the dominant contributions to the DOZI processes.

and $a_1 = p_1^2 - m_1^2 + i\varepsilon$, $a_2 = p_2^2 - m_2^2 + i\varepsilon$, and $a_3 = p_3^2 - m_3^2 + i\varepsilon$ are the denominators of the meson propagators. The coupling constants g_v, g_s and g_0 can be determined independently in meson decays [14]. For the J/ψ hadronic decays, large branching ratios for $J/\psi \to K^*\bar{K} + c.c.$ and $\rho\pi + c.c.$ imply that the meson loop transitions may have significant contributions to those decays where the final state mesons have also large couplings to the exchanged mesons. This is indeed the case for $J/\psi \to V f_0(1710)$, where the couplings for ϕK^*K, ωK^*K, and $f_0(1710)K\bar{K}$ are significantly large. Therefore, this simple argument will allow us to consider only the dominant meson exchange loops in the calculations, and estimate the OZI violation factor r at the Υ energies, i.e. in $\Upsilon \to V f_0^i$.

We refer the readers to Ref. [14] for detailed calculations, but proceed to the results for the ratio $R_i^{OZI}(\Upsilon)$ at $r(\Upsilon)$, which is illustrated by Fig. 3. It shows that the ratio R_i^{OZI} has a nontrivial energy-dependence via r. The contributions from the loop transitions will decrease much faster than the tree transitions for the Υ decays. The values of R_i^{OZI} for Υ and J/ψ are indicated by different values of r, which are dramatically different in $J/\psi \to V f_0^i$ and $\Upsilon \to V f_0^i$. Therefore, comparing the values of R_i^{OZI} for Υ and J/ψ, the role played by the DOZI processes can be highlighted. The dramatic change of the relative positions of R_i^{OZI} in Fig. 3 for these three f_0 states also depend on the mixing matrix elements. For example, in addition to R_1^{OZI} for the $f_0(1710)$ in $\Upsilon \to V f_0(1710) \to VK\bar{K}$, it is also interesting to measure R_3^{OZI} for the $f_0(1370)$ in $\Upsilon \to V f_0(1370) \to V\pi\pi$. The values for R_1^{OZI} and R_3^{OZI} if match the pattern of Fig. 3, would be a strong support for the glueball-$Q\bar{Q}$ mixings.

In brief, the measurement of R_i^{OZI} in the Υ decays examines the glueball-$Q\bar{Q}$ mixing scheme proposed for the scalar mesons. We also note, the correlation between the OZI-rule violation and glueball-$Q\bar{Q}$ mixing in the scalars should be a general feature in scalar meson production via heavy quarkonium decays. A systematic study of $\chi_{c0,2} \to VV$, PP and SS also supports such an idea [15]. In particular, the presence of significant DOZI

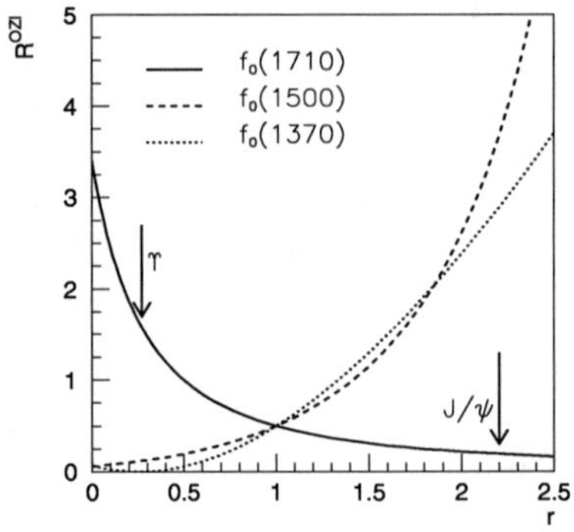

FIGURE 3. Energy evolution of R_i^{OZI} in terms of parameter r.

contributions predicts unique decay patterns for $J/\psi \to f_0^i f_0^j$, which are consistent with the new data from BES [9].

ACKNOWLEDGMENTS

The author thanks Dr. A. C. dos Reis for the warm invitation. This work is supported, in part, by grants from the U.K. Engineering and Physical Sciences Research Council Advanced Fellowship (Grant No. GR/S99433/01), and the University of Surrey.

REFERENCES

1. F.E. Close and C. Amsler, Phys. Lett. B **353**, 385 (1995); Phys. Rev. **D53**, 295 (1996).
2. F.E. Close and A. Kirk, Phys. Lett. B **483**, 345 (2000).
3. F.E. Close and N.A. Tornqvist, J. Phys. **G 28**, R249 (2002).
4. F.E. Close and Q. Zhao, Phys. Rev. D **71**, 094022 (2005) [arXiv:hep-ph/0504043];
5. A. Seiden, H. F.-W. Sadrozinski, and H. E. Haber, Phys. Rev. D **38**, 824 (1988).
6. M. Ablikim *et al.* [BES Collaboration], Phys. Lett. B **607**, 243 (2005) [arXiv:hep-ex/0411001].
7. M. Ablikim *et al.* [BES Collaboration], Phys. Lett. B **603**, 138 (2004) [arXiv:hep-ex/0409007].
8. S. Jin, Plenary talk at 32nd International Conference on High-Energy Physics (ICHEP2004), Beijing (2004).
9. C.Z. Yuan, Plenary talk at this conference. See also parallel talk by X.Y. Shen.
10. H.J. Lipkin, Phys. Rev. Lett. **13**, 590 (1964); **14**, 513 (1965); Phys. Rep. **8C**, 173 (1973); Nucl. Phys. **B244**, 147(1984); **B291**, 720 (1987); Phys. Lett. **B179**, 278 (1986).
11. P. Geiger and N. Isgur, Phys. Rev. D **47**, 5050 (1993).
12. H.J. Lipkin and B.S. Zou, Phys. Rev. D **53**, 6693 (1996).
13. S. Eidelman *et al.* (Particle Data Group), Phys. Lett. **B 592**, 1 (2004).
14. Q. Zhao, B.S. Zou and Z.B. Ma, Phys. Lett. **B** in press; hep-ph/0508088.
15. Q. Zhao, Phys. Rev. D **72**, 074001 (2005).

SUMMARY TALKS

Glueballs, Hybrids, Baryons and Pentaquarks: Hadron05 summary on light-quark spectroscopy

Eberhard Klempt

Helmholtz–Institut für Strahlen– und Kernphysik, Nussallee 14-16, 53115 Bonn

Abstract. Results on light–quark spectroscopy presented at Hadron05 are reviewed. Particular emphasis is given to the status of pentaquarks and of glueballs and exotic mesons. Experiments are proposed to decide on open issues.

Keywords: Scalar and pseudoscalar glueballs, J^{PC} exotics, baryon structure and spectroscopy, pentaquarks, dynamically generated resonances
PACS: 12.38Qk,13.25Jx,112.39-x,3.30Eg,13.75-n,13.85-t,14.20c, 14.40-n,13.60-r

INTRODUCTION

New and exciting results were presented at Hadron05. In this first part of the conference summary, selected results on light quark spectroscopy are discussed while the field of heavy quarks is reviewed by Ted Barnes. Quark masses vary over a wide range, from a few MeV to the top quark mass, but their interactions do not depend on flavour, so we deal with the same physics. This is visualized in figure 1 showing the mass gap for the lowest angular momentum excitation and the hyperfine splitting between ρ and π, J/ψ and η_c, between $\Delta(1232)$ and the nucleon, and other baryon splittings.

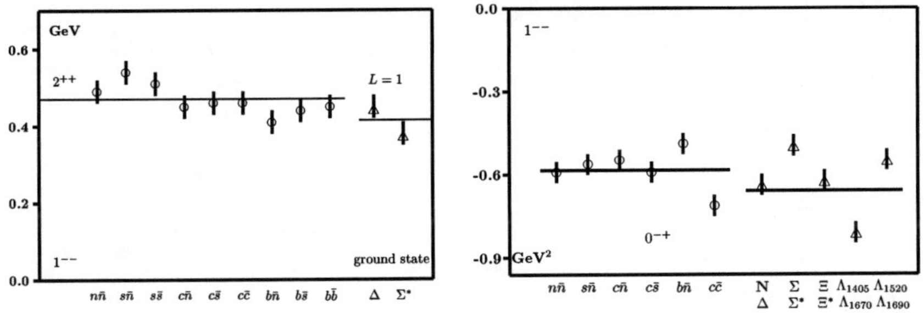

FIGURE 1. Flavor independence of the strong forces: Left panel: Mass differences for $L=1$ excitations of mesons and baryons (for systems with aligned spin). Right panel: Mass square differences between PS and V mesons for $L=0$ octet-decuplet and $L=1$ singlet-octet baryons.

The report begins with a section on aspects of baryon structure and spectroscopy. The subsequent section is devoted to the status of pentaquarks. Scalar mesons and the search for the lowest mass glueball, pseudoscalar mesons and exotic mesons are covered in three separate sections. Dynamical generation of resonances is reviewed, and a few remarks are made concerning the quark–gluon plasma. The summary ends with an outlook on results expected from ongoing experiments and from forthcoming facilities.

CP814 *Hadron Spectroscopy: Eleventh International Conference*
edited by A. Reis, C. Göbel, J. Sá Borges, and J. Magnin

BARYON STRUCTURE AND SPECTROSCOPY

"Why N*'s are important" was the title of the Baryon98 summary by N. Isgur, and he answered: "baryons are sufficiently complex to reveal physics which may be hidden from us in mesons". Also today, there is a gap in our understanding of the partonic structure of nucleons (see Haas [1]) and static properties of nucleons and their excitation spectrum. A first approach to bridge this gap was presented by Brodsky [1] who gave a fascinating 30 min talk presenting 125 transparencies.

Already neutrino scattering experiments had revealed a sizeable contribution ($\sim 15\%$) of strange quarks to the structure of nucleons [2]. Unknown so far is the effect of strange quarks to the electric and magnetic formfactors. Smith [1] reported measurements of the small parity violating contribution to electron nucleon scattering as a function of Q^2 from which rather small formfactors $\mathrm{G_E^s} = -0.01 \pm 0.03, \mathrm{G_M^s} = +0.55 \pm 0.28$ were determined. The puzzling observation that, in the infinite momentum frame, the largest fraction of proton spin is not carried by quarks is now accompanied by measurements by the COMPASS collaboration of the gluon spin contribution. The statistics of 'golden' events with open charm production is still too limited to allow for a significant constraint; from high-$\mathrm{P_T}$ events, Paul and Marchand [1] concluded that at medium x values gluons contribute to the proton spin at the 10% level. The results are shown in Fig. 2. Obviously, the orbital angular momentum wins the race to provide the most significant contribution to the proton spin; the old–fashioned idea of the magnetic moment being generated by a virtual pion orbiting (part time) around a neutron seems to find further experimental support (in addition to the neutron electromagnetic formfactor or to charge asymmetries in the n–p structure functions). The GDH sum rule connects the static magnetic moment of nucleons with their excitation spectrum (more precise: the helicity dependent photo absorption cross sections). Results were reported by Pedroni [1]).

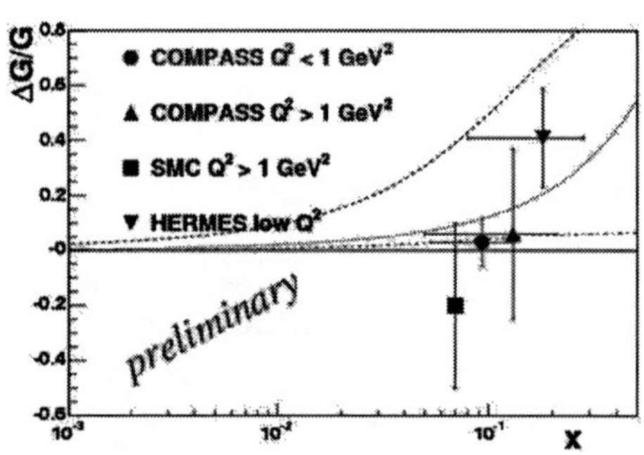

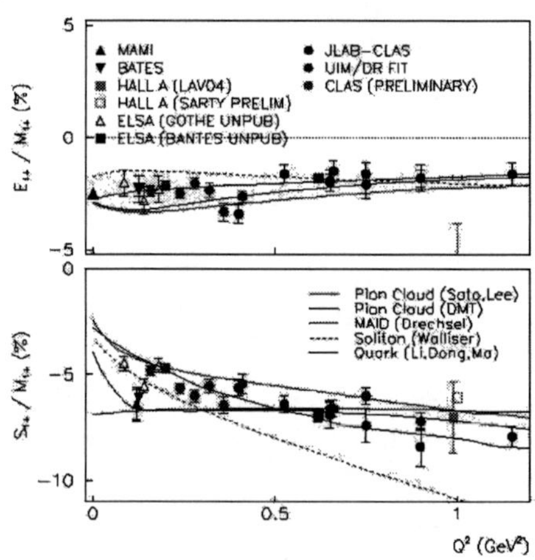

FIGURE 2. Left: $\Delta G/G$ for large p_T from COMPASS. Only curve (1) assigns the proton spin to the spin of quarks (from Paul [1]). Right: $N \rightarrow \Delta(1232)$ quadrupole transitions (from Smith [1]).

The $N \rightarrow \Delta(1232)$ quadrupole transitions (E_1^+ and S_1^+) are in the focus of interest since a few decades. It should reveal how the quark magnetic moments organise themselves as a function of their spin direction. Is there a preference for configurations in which quark spins align themselves with colour magnetic flux lines (like normal dipole magnets)? Results reported by DeVita, Gothe, Joo and Smith [1] are shown in Fig. 2b. Dynamical models reproducing the data attribute most of the deformation to the proton's pion cloud even though a small contribution from the expected oblate or prolate shape of the static nucleon or $\Delta(1223)$ is not excluded. With increasing Q^2 (for Q^2 up to 5 GeV^2), the data show no trend for a transition to perturbative QCD.

It is interesting to note that photoproduction of $\Delta(1232)$ by the 2.7 Kelvin background radiation limits the range of energies of cosmic particles to about 10^{20} eV. The Agasa collaboration reported a sizeable yield above this energy [3], the Hires experiment is compatible with a suppression due to photoproduction of $\Delta(1232)$'s [4]. The final answer will comes from the Auger observatory, status and first results were presented by Mello Neto [1].

The Roper resonance, suspected to be of hybrid or pentaquark nature, was studied at Jlab in electroproduction; the couplings and their Q^2 dependence are compatible with quark model calculations and the interpretation of the Roper as ordinary qqq resonance. Baryons as three-particle systems support a large number of excitations. The mass spectrum below 1.8 GeV is reasonably well understood even though certainly not yet in all aspects. Above 1.8 GeV the predicted level density becomes increasingly complex. It is unclear at present how the levels organise: is chiral symmetry restored [5] (see however [6]), is the level density reduced due to diquark effects [7], or should we expect the full richness of states as predicted in symmetric quark models [8]? Personally, I expect substantial new insight from an exploration of the high–mass baryon spectrum. Brodsky [1] presented an important step in relating the parton structure of baryons to the baryon excitation spectrum. Sarantsev [1] reported on a multichannel partial wave analysis of data on photoproduction of pseudoscalar mesons evidencing several new states. The cross section for η photoproduction is dominated by three resonances, the well known $S_{11}(1535)$, the $P_{13}(1720)$ (for which the large $N\eta$ coupling was not yet known), and a newly suggested $D_{15}(2070)$. These resonances have similar structure: their angular momenta couple with $\vec{J}=\vec{L}+\vec{S}; L=1,2,3; S=1/2; J=L-1/2$. The reason for the pattern is unknown. Inconsistencies in electroproduction of data on $p\pi^+\pi^-$ at JLab (reported by Joo [1]) may indicate the presence of a NP_{13} doublet at 1720 MeV. More results on baryon spectroscopy are to be expected for the near future.

PENTAQUARKS

The new high–statistics data on photoproduction of pentaquarks from CLAS presented by Smith and Tedeschi [1] were awaited for impatiently. A large number of experiments had reported evidence for narrow peaks at about 1540 MeV with significance of 3-5 σ. The peaks were interpreted as a new baryon resonance and called $\Theta^+(1540)$. Its valence structure requires a minimum of five quarks, $udud\bar{s}$. Due to non-observation in KN scattering, tight bounds were set on the pentaquark width, $\Gamma_{\Theta^+(1540)} < 1$ MeV. Such a small width was exciting or, depending on the point of view, intriguing. Pentaquarks were predicted within the framework of chiral soliton models [9] but also quark models were formulated [10, 11] which account for the new type of hadrons. Difficult to understand is the extremely narrow width.

A common feature of the data was their low statistics. High–statistics data, mostly from hadronisation studies in high–energy experiments, did not show evidence for pentaquarks, except for a few cases where a peak due to the $\Theta^+(1540)$ was observed. A recent review can be found in [12]. In addition, new pentaquarks were suggested, the $\Phi^{--}(1860)$ [13] and the $\Theta_c(3100)$ [14].

The absence of the $\Theta^+(1540)$ in hadronisation was a point of concern, upper limits from the DELPHI detector were presented by Gavillet [1]. However, the production rates of pentaquarks in hadronisation are unknown. Hadronisation is a fascinating topic in itself (see Bigi and Atti [1]) with a well–understood phenomenology but there is no ab–initio understanding of the fundamental processes and the non-observation of $\Theta^+(1540)$ in hadronisation does not rule out its existence. The final answer has to come from the "discovery" experiments, from low–energy photoproduction. The new high–statistic CLAS data seem to overrule the first claims for a peak in the reactions $\gamma d \to \Theta^+(1540)pK^-$ and in $\gamma p \to \Theta^+(1540)K^0$; the new upper limits are incompatible with the yields determined from the first round of experiments. Hansen [1] reported on an experiment by Hall A. They presented null results in the search for a Θ^{++} (isotensor partner of the Θ^+) and the $\Sigma^0_{\overline{10}}$ (anti-decuplet member of the Θ^+ with stangeness -1). Stringent upper limits from Babar for hadro- and electroproduction of the $\Theta^+(1540)$ and the $\Phi^{--}(1860)$ were presented by Hryt'nova and Muller [1]. The limits were incompatible with the findings of Hermes and Zeus. No $\Theta_c(3100)$ (see Eisenberg, Kluge [1]) was observed at ZEUS; the upper limit is in conflict with the H1 result. The existence of the ZEUS $\Theta^+(1540)$ was however confirmed in the ZEUS reanalysis.

Seth [1] asked: are pentaquarks on life support? The evidence for their existence is certainly shaken. Nevertheless, a few observations need to be clarified in further analyses. A 7σ $\Theta^+(1540)$ peak in $\gamma p \to \Theta^+(1540)K^-\pi^+$ was reported by CLAS [15]. Spring8 has taken higher statistics, and supporting evidence for the $\Theta^+(1540)$ was observed in a new reaction, in $\gamma d \to \Lambda(1520)\Theta^+(1540)$ [16]. At COSY, a seemingly convincing structure was seen in the reaction $pp \to \Sigma^+\Theta^+(1540)$ [17]. In all these cases, data are on tape, and the final results should be evaluated. The expectations of the community have certainly changed: so far, we all had hoped that the new data from CLAS would confirm the perhaps premature claims; now we have to hope for a miracle.

It is of course impossible that so many experiments have observed a statistically significant but nonexisting peak at the same mass. Cumulat [1] calculated the statistical probabilty that all these observations were unrelated statistical fluctuations, the probabilty is extremely small. There must be a physical or technical reason. A reflection, as advocated by the Indiana group [18], seems excluded by the new Jlab results. If the $\Theta^+(1540)$ is not resurrected, the most likely fault which may have happened is an incorrect pattern recognition of overlapping tracks. This may have accidently created a peak at 1540 MeV depleting the invariant mass distribution at both sides of the peak.

The $\Theta^+(1540)$ poses also new questions to theory: how stable are the claims for the $\Theta^+(1540)$ stability against falling apart in different models ? I think both, experimentalists and theoreticians have to go back to their data and equations to find out what went wrong. Hence the study of the $\Theta^+(1540)$ must (and will) continue, both experimentally and theoretically (see talks by Eidemueller, Gavillet, Goldman, Hansen, Hry'nova, S. Ishida, Juge, Lee, Lesiak, Ma, Navarra, Noya, Reyes, Smith, and Tedeschi [1]).

SCALAR MESONS AND THE SCALAR GLUEBALL

The scalar mass spectrum and its interpretation remained one of the most controversial issues at this conference, see the contributions by Afanasev, Beveren, Bini, Borges, Bugg, Büscher, Dytman, Garmash, Göbel, Jin, Kleefeld, Khokhlov, Meadows, Maeda, Miranda, Ochs, Pelaez, Polosa, Polycarpo, Robilotta, Roca, Shen, Silvestre-Brac, Teshima, Yuan, Zhao, Zheng [1] and possibly others who all have discussed important aspects of scalar meson spectroscopy from threshold to 2 GeV.

Two–pion interactions at the lowest energy are explored in the Dirac experiment. In matter, $\pi^+\pi^-$ atoms have the chance to annihilate into $\pi^0\pi^0$ or to be broken apart in atomic collisions. The break–up can be detected and thus the natural life–time of $\pi^+\pi^-$ atoms be measured. The result of Dirac, $\tau = 2.9\,\mathrm{ps}$, is in excellent agreement with predictions from chiral perturbation theory (see Afanasev [1]).

The first hadronic scalar isoscalar meson, often called σ, is quoted as $f_0(600)$ by the Particle Data Group, with a mass between 400 and 1200 MeV and a width in the 600 and 1000 MeV range. Bugg [1] emphasized the need to constrain production data on the low–energy $\pi\pi$ interactions by removing the Adler–Weinberg zero present in $\pi\pi$ scattering. In his view, the correct treatment of the Adler zero enforces a low–mass pole in the scalar isoscalar S–wave, the $f_0(600)$. It is accompanied by the $K_0^*(900)$, $a_0(980)$ and $f_0(980)$, forming a natural nonet of dynamically generated resonances. The low–energy behaviour of the $\pi\pi$ S–wave is constrained by chiral symmetry and by dispersion relations (see talks by Pelaez and Borges [1]). The scalar isoscalar S–wave structure at 1300 MeV observed in several experiments is interpreted by Bugg as $f_0(1370)$; the 3 states $f_0(1370)$, $f_0(1500)$, and $f_0(1710)$ are the observable signatures of a scalar glueball intruding the scalar meson spectrum and mixing with two quarkonium states. The additional $f_0(1790)$ observed in BES data on $\phi\pi\pi$ (see Jin and Yuan [1]) is interpreted as $f_0(1370)$ radial excitation.

Ochs [1] presented a different view. He underlined the model dependence of the σ mass and suggested that the σ and the $f_0(1370)$ may very well be separate regions of a common object, called red dragon or $f_0(1000)$, having a width of 1000 MeV or more.

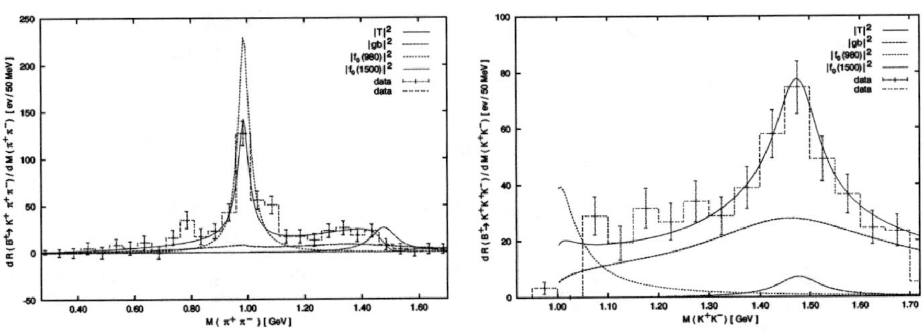

FIGURE 3. $\pi^+\pi^-$ mass spectrum in B-decays in comparison with a model including $f_0(980), f_0(1500)$ and a broad glueball. The individual resonance contributions are also shown. In $\pi\pi$ the glueball interferes destructively, in $K\bar{K}$ constructively with both $f_0(980)$ and $f_0(1500)$ (from Ochs [1]).

Its actual mass is suggested to depend on the parameterisation and mass range which is fitted. He reported on an analysis of B decays and demonstrated that the data can be interpreted naturally adopting his view. The scalar glueball is identified with the wide scalar background intensity.

Alternatively, the wide scalar $\pi\pi$ background amplitude can be understood quantitatively as dynamically generated by (mainly) ρ exchange, and there is no need to introduce additional poles due to s–channel resonances [19]. The $f_0(1370)$ is supposed to have strong couplings to 4π. The question arises if $f_0(1370)$ in 4π is a s–channel resonance or if it represents a background amplitude describing meson–meson interaction dynamics with a slowly rising phase. In central production, the $f_0(1370)$ decays only into $\rho\rho$ and not to $4\pi^0$ (see Fig. 4 (nor to $\eta\eta$) [20, 21]. The dip in the $\rho\rho$ scalar mass distribution is due to $f_0(1500)$ and unitarity constraints in the same way as unitarity produces a dip at the $f_0(980)$ in the $\pi\pi$ mass distribution. The Crystal Barrel collaboration observed both mesons, $f_0(1370)$ and $f_0(1500)$, in their $\rho\rho$, $\eta\eta$ and $\sigma\sigma$ decays. These discrepancies can be interpreted as evidence that (likely non–resonant) scalar isoscalar intensity is generated dynamically by isovector t–channel exchange (e.g. ρ exchange to yield $\pi\pi$, π exchange to yield $\rho\rho$) between two Pomerons leading to scalar intensity in $\rho\rho$ but not in two isoscalar particles like $\eta\eta$ or $\sigma\sigma$.

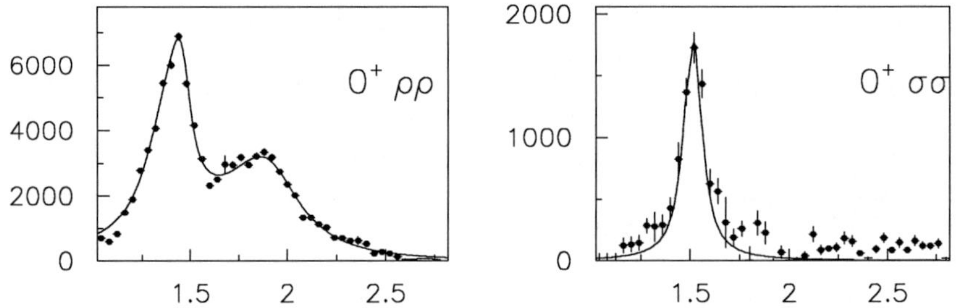

FIGURE 4. The $\rho\rho$ and $\sigma\sigma$ scalar intensity in WA102 [20]. The absence of $\sigma\sigma$ can be understood when the intensity is due to Pomeron–Pomeron interactions via isovector (e.p. pion) exchange.

These different views lead to experimental consequences. If the $f_0(1370)$ is a real particle, it will be observed as a dip in $\pi\pi \to 4\pi$ scattering [22]. Otherwise, it is part of the $f_0(1000)$. The glueball nature of the latter can be probed by radiative J/ψ decays into $\eta\eta$ and into $4\pi^0$. There is no reason for scalar mesons not to decay into theses two modes while background amplitudes for $\eta\eta$ or $4\pi^0$ in radiative J/ψ decays should be very small compared to the production of $f_0(1500)$. These are two crucial experiments, in particular the study of J/$\psi \to \gamma\eta\eta$ or $\gamma 4\pi^0$. It will yield the final answer to the question: does a scalar glueball exist as intruder mixing with scalar quarkonia, does it exist as a the broad object seen e.g. in central production or has nature chosen a different scenario. The search for glueballs has accompanied us for about 25 years. The answer could come in the very near future from CLEO. If they decide to continue with the $\psi(2S)$ program, the community will likely have to wait for the BESIII upgrade.

PSEUDOSCALAR MESONS

Masoni [1] suggested the pseudoscalar glueball should be identified with the low–mass component of the $\eta(1440)$ called $\eta(1405)$ by the Particle data group [23]. In his interpretation, the $\eta(1295)$ is the radial excitation of the η. Its mass is degenerate with the $\pi(1300)$, hence the pseudoscalar radial excitations seem to be ideally mixed. The $\bar{s}s$ partner should then have a 240 MeV higher mass. The high–mass component of $\eta(1440)$, called $\eta(1475)$, decays dominantly into $K\bar{K}^* + cc$ and is assigned to be the $\bar{s}s$ partner of the $\eta(1295)$. The $\eta(1405)$ does not find a slot in the spectrum of $\bar{q}q$ mesons; it is interpreted as a glueball.

This view could be wrong. The $\eta(1295)$ is observed only in πN scattering and in no other process. Radiative J/ψ decays into $\gamma K\bar{K}\pi$ show an asymmetric peak in the $\eta(1440)$ region [24]: both, $\eta(1405)$ and $\eta(1475)$, contribute to the process. Hence not only glueballs but also radial excitations are produced in radiative J/ψ decays. The $\eta(1295)$ should therefore be produced as well, but it is not - at least not with the expected yield. At BES, $\eta(1295)$ and $\eta(1440)$ were studied in J/$\psi \to (\rho\gamma)\gamma$ and $\to (\phi\gamma)\gamma$ [25]. The $\eta(1440)$ (seen at 1424 MeV) is seen to decay strongly into $\rho\gamma$ and not into $\phi\gamma$. This is inconsistent with the peak being a glueball (which should not decay radiatively) or with its interpretation as $s\bar{s}$ state (which should decay into $\phi\gamma$ and not into $\rho\gamma$). The $\eta(1295)$ is missing again, a peak below 1300 MeV is assigned to $f_1(1285)$ (even though a small contribution from $\eta(1295)$ cannot be excluded). Finally, photons couple to charges; in $\gamma\gamma$ fusion a radial excitation is hence expected to be produced more frequently than a glueball. The L3 collaboration studied $\gamma\gamma \to K_s^0 K^{\pm}\pi^{\mp}$ [26]. At low q^2, a peak at 1440 MeV is seen but no peak due to $\eta(1295)$. Neither a glueball nor a $s\bar{s}$ state can have stronger two–photon couplings than a $n\bar{n}$ state. Hence $\eta(1295)$ should have been seen if it were the η radial excitation. The latter argument is however weakened since CLEO$_c$ does not see neither $\eta(1295)$ nor $\eta(1440)$ and reports an inconsistency with L3 with more than two standard deviations [27]. The Crystal Barrel collaboration searched for $\eta(1295)$ and $\eta(1440)$ in the reaction $p\bar{p} \to \pi^+\pi^-\eta(xxx)$, $\eta(xxx) \to \eta\pi^+\pi^-$ [30]. A pseudoscalar resonance signal was observed at 1405 MeV decaying into $a_0(980)\pi$ and $\eta\sigma$. A scan for an additional 0^+0^{-+} resonance gave no evidence for $\eta(1295)$ but for a second resonance at 1480 MeV. The phase of the $a_0(980)\pi$ or $\sigma\eta$ isobar changed by π indicating the presence of only one resonance in the 1250 to 1500 MeV region instead of 3. The $\eta(1295)$ does not have the properties of a $q\bar{q}$ resonance; it might be faked by a combination of Deck effect and feed-through from the $f_1(1285)$. The splitting of $\eta(1440)$ into two separate peaks can be understood assuming that it is a radial excitation. Its node in the wave function has an impact on the decay matrix elements calculated by Barnes *et al.* [31] within the $^3P_0$ model. The node occurs at different momenta for $\eta(1440) \to a_0(980)\pi$ and K*K decays. The K*K distribution is shifted towards higher masses, the $a_0(980)\pi$ and $\sigma\eta$ distribution are split into a low-mass and a weak high–mass peak. One $\eta(1420)$ and the assumption that it is a radial excitation are sufficient to describe the data. There is no need to introduce two independent states.

What is the radial excitation of the η'? BES may have observed up to seven pseudoscalar states in radiative J/ψ decays, $\eta(1440)$ in K$\bar{\text{K}}\pi$ and $\eta\pi\pi$, $\eta(1760)$ in $\eta\pi\pi$, possibly a $\eta(1650)$ in K$\bar{\text{K}}\pi$, X(1835) (likely $\eta(1835)$) in $\pi\pi\eta'$, $\eta(1860)$ in $p\bar{p}$ and $\eta(2040)$ in K$\bar{\text{K}}\pi$ and a broad $\eta(1800)$ background amplitude which is interpreted as

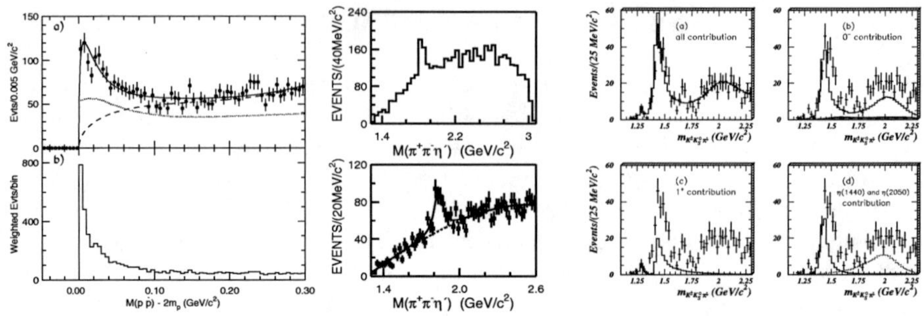

FIGURE 5. Left: The $p\bar{p}$ threshold enhancement from $J/\psi \to \gamma p\bar{p}$ after correction for phase space. Center: The $\eta'\pi^+\pi^-$ spectrum from radiative J/ψ decays. No partial wave analysis has been performed. (from Jin [1]). Right: In $J/\psi \to \gamma K\bar{K}\pi$ the $\eta(1440)$ is seen, and a (disregarded) $\eta(1650)$. The pwa gives additionally evidence for a $\eta(2040)$.

glueball. In [28], $\eta(1650)$ was discarded because of its narrow widths of 30 MeV. However, the $\eta(1835)$ has about the same width. The $\eta(1835)$ and $\eta(1860)$, shown in Fig. 5, are suggested to represent two decay modes of one resonance which is interpreted as $p\bar{p}$ bound state (see talks by Yuan, Jin, and Shen [1]). I prefer to keep them separate and to link the $\eta(1860)$ to $N\bar{N}$ physics. The $\eta(1760)$ is seen with little evidence by BES, the original claim by MARK3 and DM2 [23] was later shown to be wrong [29]. From the masses one may speculate and attempt an identification as

1^1S_0	$\pi(138)$	$\eta(958)$	$\eta(548)$	K(495)
$\delta M^2 =$	1.9	1.8	1.8	1.9 GeV^2
2^1S_0	$\pi(1375)$	$\eta(1650)$	$\eta(1440)$	K(1460)
$\delta M^2 =$	1.4	1.4	1.3	1.2 GeV^2
3^1S_0	$\pi(1800)$	$\eta(\sim 2040)$	X(1835)	K(1830)

EXOTIC MESONS

Adams [1] reported on the status of exotic mesons with quantum numbers $J^{PC} = 1^{-+}$. There are three regions in which exotic mesons were reported, at 1400, 1600 and 2000 MeV. A $\pi_1(1400)$ seen in $\eta\pi$ and $\rho\pi$, a $\pi_1(1600) \to \eta'\pi, \rho\pi, f_1(1285)\pi$ and $b_1(1230)\pi$, and a $\pi_1(2000)$ in its $f_1(1285)\pi$ and $b_1(1230)\pi$ decay modes. For the $\pi_1(1400)$ and $\pi_1(1600)$, the dynamics of the production process of the final states $\rho\pi$ and $b_1(1230)\pi$ (produced by both natural and unnatural parity exchange) is different from that of the $\eta\pi$, $\eta'\pi$, and $f_1(1285)\pi$ decay modes (which proceed only via natural parity exchange). Hence there must be four exotic states in the 200 MeV mass region. It is unlikely that these are hybrids, and also their mass is too low. Hence these four resonances must be four–quark states or generated by meson–meson interactions. The conjecture that a state doubling occurs is supported for the $\pi_1(1400)$ by the LEAR results: the Crystal Barrel $\pi_1(1400)$ in $\pi\eta$ is produced from the $^3S_1$ initial state of the

$p\bar{p}$ atom; the Obelix $\pi_1(1400)$ decays into $\rho\pi$ but is produced from the $^1S_0$ state.

If we accept the $\pi_1(1400)$ and $\pi_1(1600)$ to be split and (likely) generated by meson–meson interactions, you expect (due to $3 \otimes 3 \otimes 3 \otimes 3 = 1 + 8 + 8 + 8 + 8 + 10 + \bar{10} + 27$) several π_1 resonances. This is seen experimentally. Only the highest–mass resonance $\pi_1(2000)$ has the properties as expected for hybrid mesons. But how can one be shure of its hybrid nature given such an abundancy of four–quark exotic states ? A convincing case for observation of a hybrid can only be made when the observations at 1400 and 1600 MeV can be discredited.

Mitchel [1] reported on a reanalysis of the Indiana group of E852 data on the 3π final states with the aim to scrutinize the existence of the $\pi_1(1600)$. The older E852 analysis used 20 waves plus background to describe the data; the Indiana group 35. In this way they succeeded in reducing the intensity of the $\pi_1(1600)$ but the phase motion between the π_1 and π_2 partial waves (not shown in their paper) remained stable. The criticism of the Indinana group was refused by Adams speaking for the E852 collaboration. He presented a fit to the Indiana partial wave analysis result, and found (from their analysis !) the π_1 partial wave to demand a resonance with (M, Γ=1550,320) MeV, fully compatible with the published result ($1593 \pm 8^{+29}_{-47}$, $168 \pm 20^{+150}_{-12}$). The $\pi_1(1600) \to \pi\eta'$ is fitted by both groups with a Breit Wigner amplitude, hence in this decay mode the $\pi_1(1600)$ is not controversial. Adams also commented on another analysis of the Indiana group. The Indiana group constructed $\pi\eta$ interactions which allows them to generate the $\pi_1(1400)$ dynamically from final state interactions. They argue that the $\pi_1(1400)$ is not a 'QCD pole', intead, the physical origin responsible for intensity and phase motion are meson–meson interactions. Both groups agree however, at least qualitatively, on amplitude and phase. The $\pi_1(1400) \to \rho\pi$ decay has not been reported by BNL or VES since the $\rho\pi$ low–mass region is dominated by the Deck effect. Progress in the treatment of this background amplitude was reported by Dudek [1].

In summary, exotic mesons have been identified in different experiments. Their identification as hybrids is difficult since most of the observations are certainly not hybrids but rather tetraquarks or originating from meson–meson dynamics.

I would like to mention here the open issues in the spectrum of vector radial excitations and their interpretation, and refer to the talks by Sibidanov and Ishida [1].

DYNAMICALLY GENERATED RESONANCES

At low energies, QCD can be developed into an effective field theory which can be expanded systematically with q/Λ_χ and m_π/Λ_χ ($\Lambda_\chi \sim 1$ GeV) as small expansion parameters. Hadrons coupling to an S–wave decay mode which is just closed due to phase space limitations can be generated from an effective chiral invariant Lagrangian in which pseudoscalar (and vector) mesons and octet (and decuplet) baryons are the 'fundamental' particles. Spectral functions can be calculated which can be confronted with experimental data. Several contributions to Hadron05 demonstrated the wide range of phenomena to which chiral perturbation theory can be applied.

Hofmann [1] reminded us of the rich history of dynamically generated resonances which goes back to Dalitz. The Λ(1405) is one example, other well known cases are the N*(1535) coupling to KΣ and Nη, the Λ(1520) coupling to K*N. The question arises:

are dynamically generated resonances distinct from $q\bar{q}$ states, do these resonances exist independently and in addition to quark model states? Or can chiral dynamics predict properties of resonances like mass shifts due to hadronic interactions and couplings to different final states even though the resonances still have a $q\bar{q}$ seed? This question has been posed since many years, I remind you of the $f_0(980)$ and $a_0(980)$ mesons which are interpreted as dynamically generated resonances, $qq\bar{q}\bar{q}$ states, as chiralons, or as 1^3P_0 $q\bar{q}$ states. Likely, they are a bit of all but need the $q\bar{q}$ seed. This has important consequences: if they have a $q\bar{q}$ seed attracting molecular components due to final–state interactions, they should not be disregarded when forming meson multiplets.

In some cases, dynamically generated resonances may have no overlap with quark model states. The doubling of the $\Lambda(1405)$ could be such an example. The $\Sigma\pi$ and NK pairs in S–wave have thresholds at 1332 and 1435 MeV. Their interaction is fully fixed by an effective chiral invariant Lagrangian; Oset [1] showed that the interaction leads to three poles, two are found just below the NK threshold, and he conjectured that the $\Lambda(1405)$ has a two–pole structure. The (low–statistics) data on $\pi^-p \to \Sigma^+\pi^-$ and $\pi^-p \to \Sigma^-\pi^+$ are consistent with this picture. Of course, the $\Lambda(1405)$ plays a role (and occupies a slot) in the spectrum of (qqq) baryon resonances. Its low mass may pose a problem for quark modelists, however it is known that strong couplings to S–wave decays may introduce considerable mass shifts. The two $\Lambda(1405)$ poles are predicted to have different flavour structures, one being dominantly a flavour singlet, the other dominantly a flavour octet state. In the quark model, the $\Lambda(1405)$ has a single pole (in the second Rieman sheet), and it is a flavour singlet. This is a prediction which can be tested experimentally at BES: J/ψ decays into a ground–state Λ plus $\Lambda(1405)$ are allowed only when the $\Lambda(1405)$ has a large octet component. The quark model predicts $J/\psi \to \Lambda\Lambda(1405)$ to be small compared to $J/\psi \to \Lambda\Lambda(1520)$ while Oset's model predicts the two reactions to have similar strengths. A second example are the axial vector mesons. The $f_1(1285)$, $h_1(1170)$, and $h_1(1380)$ can be understood as bound state of pseudoscalar and vector mesons generated dynamically, the $f_1(1420)$ (or $f_1(1510)$) however not (see Roca [1]). In quark models, all four states are predicted and indeed, all four are seen experimentally. A second point is the wave function. The generated $f_1(1285)$ is an octet state and has a $s\bar{s}$ component. In the quark model, $f_1(1285)$ is a $(u\bar{u}+d\bar{d})/\sqrt{2}$ state and should not be produced by $J/\psi \to \varphi f_1(1285)$. The sizable branching ratio of $J/\psi \to \varphi f_1(1285)$ can only be understood if $f_1(1285)$ is dynamically generated. In this case, the quark model gives the correct pattern of states, their properties are better understood from final–state interactions of their decay products. More work on the relation between dynamically generated resonances and quark model states is certainly needed [33].

THE QUARK–GLUON PLASMA

At high density and temperature, nuclear matter is predicted and possibly observed to undergo a phase transition. Takahashi [1] reported on results of the STAR collaboration at RHIC which has the goal to reconstruct from the final–state particles the dynamics of the quark–gluon plasma and its expansion in a hadronic gas from which the final state is reached via freeze out at 165 MeV. Munhoz [1] (see also Fraga [1]) showed impressive distributions of shower devolpments in Au Au collisions. The compound system is shown to dissipate the energy of a jet almost completely. Hence a thermal equilibrium seems to be reached. This is of course of considerable importance for the interpretation $c\bar{c}$ states and their interaction in a dense and hot environment.

Nuclear medium effects can be observed already in normal nuclei as Lolos [1] reported. The ρ spectral function from the reaction $\gamma^{12}C \rightarrow \pi^+\pi^- + X$ can still be understood without medium modifications but at larger densities (and heavier nuclei) in–medium mass shifts have been observed [34, 35].

OUTLOOK

The beautiful results on heavy quark spectroscopy are reviewed by Ted Barnes. The discovery potential of J/ψ decays for light quark spectroscopy and the search for gluonic degrees of freedom in spectroscopy is known since the pioneering experiments at Stanford and Orsay in the 70's and 80's. BESII has taken up the field providing a multitude of interesting results. Luminosity and detector performance will undergo a substantial upgrade to reach BESIII; CLEO$_c$ has the potential for substantial contributions also to quark–gluon spectroscopy. It was an unexpected surprise that B and D mesons contribute substantially to light–quark spectroscopy. Initial state radiation (see Aston, Berger [1]) is a further novel tool to study the spectrum of vector mesons from the ρ and ω up the X(4260) - suggested to be a $c\bar{c}$ hybrid [32] - and beyond, and to complement the work of DAΦNE and VEPP2. BNL has a rich data set on light mesons and will produce more results.

The Thomas Jefferson Lab is producing a wealth of beautiful data on the structure of nucleons. COMPASS will provide a solution to the proton spin puzzle and will yield new results on meson spectroscopy (see Colantoni, Paul [1]). And also FERMILAB has an active hadron program (see Appel and Maciel [1]). Several laboratories like ELSA, Spring8, MAMI, GRAAL and of course Jlab again are producing results on the baryon spectrum. At the low energy end, chiral perturbation theory provides a frame in which low-energy concepts of QCD can be tested with high precision. Active programs are persued at DAΦNE, MAMI and COSY and were presented at Hadron05 by Ambrosini, Filippi, Miscetti, Papandreou, Pedroni and Machner [1].

Two major new facilites are planned, GlueX and FAIR (see contributions of Carman, Peters and Boca [1]). GlueX is designed to optimise the identication of hybrid mesons. This task requires high luminosity to find small signals in the presence of a large background of conventional $q\bar{q}$ resonances and full solid angle coverage for both charged and neutral particles to fully reconstruct the complex decay chains predicted for hybrids. The use of a photon beam with linear polarisation constrains the naturality of the particle exchange and thus facilitates the analysis. FAIR is an umbrella which houses a broad spectrum of experimental facilities. Nuclear structure is studied using beams of radioactive isotopes; relativistic nuclear collisions will explore compressed hadronic matter; plasma physics, the physics of highly charged ions and the study of low–energy antiprotons are part of the scientific program. Of special interest for this conference are the prospects of a continuation of the successful Fermilab E835 experiment, with greatly improved luminosity and detector, a special pellet target (see Büscher's talk) and a much broader scientific program. The spin–dependence of antiprotons is proposed to be studied with two different techniques and aims. I believe that an eventual understanding of confinement will not come from heavy quark spectroscopy but from a study of high–mass excitations of light mesons and baryons. FAIR has the potential to scan the 2 to 3 GeV mass range in $p\bar{p}$ annihilation with a polarised target and possibly even with a polarised antiproton beam. Such experiments would be unique, with GSI being the only place in the world where such experiments could be carried out.

In summary, there are excellent facilities supporting intense studies of one of the most challenging issues of hadron physics: what is the nature of confinement, and what is the relation between the partonic degrees of freedom and the nucleon as a (qqq) bound state.

At the end, I would like to thank the organisers, in particular Alberto Corrêa dos Reis, for setting up this interesting conference and for their warm hospitality.

REFERENCES

1. See contribution to these proceedings.
2. M. Abramowitz et al., Z. Phys. **C15** (1982) 19.
3. M. Takeda et al., PRL **81** (1998) 1163.
4. HiRes Collab., arXiv:astro-ph/0501317.
5. L. Y. Glozman, Phys. Lett. B **475** (2000) 329.
6. E. Klempt, Phys. Lett. B **559** (2003) 144.
7. E. Klempt, arXiv:nucl-ex/0203002.
8. S. Capstick and W. Roberts, Prog. Part. Nucl. Phys. **45** (2000) S241.
   U. Löring, B. C. Metsch, and H. R. Petry, Eur. Phys. J. A **10** (2001) 395, 447.
9. D. Diakonov, V. Petrov and M. V. Polyakov, Z. Phys. A **359** (1997) 305.
10. R. L. Jaffe and F. Wilczek, Phys. Rev. Lett. **91** (2003) 232003.
11. M. Karliner and H. J. Lipkin, arXiv:hep-ph/0307243.
12. K. H. Hicks, Prog. Part. Nucl. Phys. **55** (2005) 647.
13. C. Alt *et al.* [NA49 Collaboration], Phys. Rev. Lett. **92** (2004) 042003.
14. A. Aktas *et al.* [H1 Collaboration], Phys. Lett. B **588** (2004) 17.
15. V. Kubarovsky *et al.* [CLAS Collab.], Phys. Rev. Lett. **92** (2004) 032001 [E.: **92** (2004) 049902].
16. T. Nakano, Int. Conf. on QCD and Hadronic Physics June 2005, Beijing, China, www.phy.pku.edu.cn/ qcd/.
17. M. Abdel-Bary *et al.* [COSY-TOF Collaboration], Phys. Lett. B **595** (2004) 127.
18. A. R. Dzierba, D. Krop, M. Swat, S. Teige and A. P. Szczepaniak, Phys. Rev. D **69** (2004) 051901.
19. B. S. Zou and D. V. Bugg, Phys. Rev. D **50** (1994) 591.
20. D. Barberis *et al.* [WA102 Collaboration], Phys. Lett. **B474** (2000) 423.
21. D. Barberis *et al.* [WA102 Collaboration], Phys. Lett. B **479** (2000) 59.
22. A. Sarantsev, private comminication, 2005.
23. S. Eidelman *et al.* [Particle Data Group Collaboration], Phys. Lett. B **592** (2004) 1.
24. L. Köpke and N. Wermes, Phys. Rept. **174** (1989) 67.
25. J. Z. Bai *et al.* [BES Collaboration], Phys. Lett. B **594**, 47 (2004).
26. M. Acciarri *et al.* [L3 Collaboration], Phys. Lett. B **501**, 1 (2001).
27. R. Ahohe *et al.* [CLEO Collaboration], Phys. Rev. D **71**, 072001 (2005).
28. J. Z. Bai *et al.* [BES Collaboration], Phys. Lett. B **476** (2000) 25.
29. D. V. Bugg *et al.*, Phys. Lett. B **353** (1995) 378.
30. E. Klempt, arXiv:hep-ph/0409148.
31. T. Barnes, F. E. Close, P. R. Page and E. S. Swanson, Phys. Rev. D **55** (1997) 4157.
32. F. E. Close and P. R. Page, arXiv:hep-ph/0507199.
33. U. G. Meissner, Int. J. Mod. Phys. A **20** (2005) 514.
34. D. Trnka *et al.* [CBELSA/TAPS Collaboration], Phys. Rev. Lett. **94** (2005) 192303.
35. M. Naruki *et al.*, arXiv:nucl-ex/0504016.

HADRON05 Summary: Heavy Quark Hadrons and Theory

T.Barnes

Physics Division, ORNL
and
Department of Physics and Astronomy,
University of Tennessee

Abstract. This HADRON05 summary covers the topics of (1) mesons containing heavy quarks, and (2) theory. The new material discussed here is taken mainly from plenary presentations. We specifically emphasize new or recent results in spectroscopy that are likely to appear in future editions of the PDG. An exception is made for the pentaquark, which was withdrawn at this meeting. We undoubtedly have something important to (re)learn about multiquarks from the pentaquark saga, and this merits a phrase in Portuguese. The three general areas we consider are: I. $Q\bar{Q}$ spectroscopy, II. $Q\bar{q}$ spectroscopy, and III. lessons from the pentaquark. Finally, in Section IIIb. we conclude with "Our moment of Zen".

Keywords: hadrons
PACS: 12.38.-t

$Q\bar{Q}$ SPECTROSCOPY

Introduction

The most exciting new results presented at HADRON05 (excepting the withdrawal of the pentaquark) were the new experimental developments in charmonium spectroscopy. After a hiatus of approximately 20 years, there have been many exciting new results in the charmonium sector, due largely to the rather surprising effectiveness of B decays in producing $c\bar{c}$ resonances [2]. This was discussed here by Trabelsi [1], who noted that the CKM-favored B meson weak decay process leads to a relatively large (*ca.* 1%) inclusive $c\bar{c}$ branching fraction, due to the favorable V_{bc} and V_{cs} CKM matrix elements in the quark-level process $b \to c\bar{c}s$. The $c\bar{c}$ mesons identified in B decays to date include not only the S-wave J/ψ, η_c, ψ' and η_c', but also the P-wave states χ_0 and χ_1 and the dominantly D-wave $\psi(3770)$. This is a surprising result, since the W propagator in the diagram might *a priori* have been expected to couple preferentially to a final $c\bar{c}$ system at small separations of $\approx 1/M_w \approx 0.01 fm$. Production of P-wave and dominantly D-wave $c\bar{c}$ states with branching fractions comparable to those of S-wave states has nonetheless been reported experimentally. This gives us an effective method for producing higher-mass radially- and orbitally-excited $c\bar{c}$ states, and perhaps other states in this flavor sector, such as $c\bar{c}$ hybrids or charm molecules.

Four new states which might be charmonia were discussed at this meeting in plenary presentations; these are the Z(3931), Y(3943), X(3943) and Y(4260). (The earlier dis-

CP814 *Hadron Spectroscopy: Eleventh International Conference*
edited by A. Reis, C. Göbel, J. Sá Borges, and J. Magnin

covery of the X(3872) has apparently led to a preference for the far end of the Latin alphabet in naming these new states.) We will discuss the new states individually, and review the possibilities discussed for assignments in the conventional $c\bar{c}$ spectrum.

Z(3931)

The new states Z(3931), Y(3943) and X(3943) reported by Belle all have possible assignments as conventional 2P $c\bar{c}$ states; to misquote Hamlet, "2P or not 2P, that is the question". The Z(3931), discussed here by Trabelsi [1] was reported in $\gamma\gamma \to Z \to DD$ (both charge states [3]; see figure 1). It has a reported mass and width of

$$M = 3931 \pm 4 \pm 2 \text{ MeV}, \tag{1}$$

$$\Gamma = 20 \pm 8 \pm 3 \text{ MeV}. \tag{2}$$

Belle suggests a 2P χ_2' assignment for the Z(3931). This remarkable state has a surprisingly small total width of about 20 MeV, which is about 1/2 of what is expected for a 2P χ_2' $c\bar{c}$ meson at this mass [4]. The reported two-photon width is

$$\Gamma_{\gamma\gamma} \cdot B_{DD} = 0.23 \pm 0.06 \pm 0.04 \text{ keV}, \tag{3}$$

which is about 1/2 the theoretical expectation for a χ_2' $c\bar{c}$ state [5] (neither the total width nor the two-photon width is a disaster, given the typical accuracy of these calculations).

Another possibility is that the Z(3931) might be a 2P χ_0' $c\bar{c}$ scalar; decay calculations using the $^3P_0$ model show that this state should be surprisingly narrow, due to a node in the DD decay mode which is accidentally near 3.95 GeV. One can differentiate between these options by searching for DD and DD* modes; the scalar χ_0' will only produce DD, whereas the tensor should decay to both DD and DD*, with an expected branching fraction ratio of DD*/DD $\approx 1/3$.

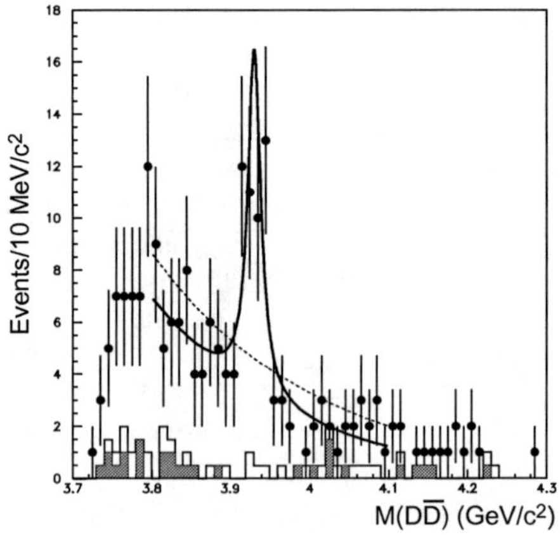

FIGURE 1. The Z(3931) reported in $\gamma\gamma \to$Z$\to$DD by Belle [1].

Y(3943)

The Y(3943) was reported in B decays to KY, Y$\to \omega J/\psi$ [6], with a mass and width of

$$M = 3943 \pm 11 \pm 13 \text{ MeV}, \tag{4}$$

$$\Gamma = 87 \pm 22 \pm 26 \text{ MeV}. \tag{5}$$

The combined Y(3943) formation and decay branching fraction found by Belle is

$$B_{\mathrm{B}\to\mathrm{KY}(3943)} \cdot B_{\mathrm{Y}(3943)\to\omega J/\psi} = 7.1 \pm 1.3 \pm 3.1 \cdot 10^{-5}. \tag{6}$$

Since a typical total branching fraction for production of conventional charmonium states in B decays is 10^{-3} to 10^{-4}, this suggests that the Y(3943) has a large branching fraction to the closed charm channel $\omega J/\psi$, if this enhancement is indeed due to a real resonance.

The remarkable closed-charm $\omega J/\psi$ mode led Belle to suggest that the Y(3943) might be a charmonium hybrid [6]. They note however that the mass of the Y(3943) is 500 MeV below LGT estimates; this makes the hybrid assignment appear implausible. In view of the mass of the Y(3943), another possibility which should be considered is that this is another conventional 2P $c\bar{c}$ state; this would lead to a total width comparable to what is reported for the Y(3943), with DD* as the dominant decay model. A search for this DD* mode is underway [6], and if found, this may lead to identification of this rather wide enhancement with the 2P χ_1' $c\bar{c}$ state.

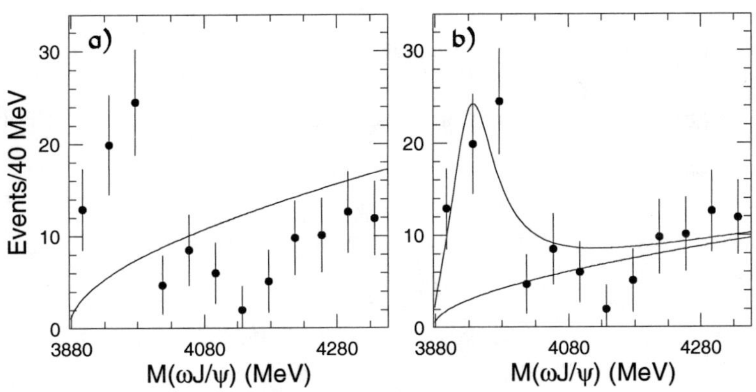

FIGURE 2. The Y(3943) reported by Belle in the closed-charm mode B$\to$ KY, Y$\to \omega J/\psi$ (fitted in right panel) [1, 6].

X(3943)

The X(3943) was reported by Belle in the double charmonium reaction $e^+e^- \to J/\psi X$ (X is mainly DD*) [7]. The fitted parameters are

$$\mathrm{M} = 3943 \pm 6 \pm 6\ \mathrm{MeV}, \tag{7}$$

$$\Gamma = 15.1 \pm 10.1\ \mathrm{MeV}. \tag{8}$$

The X(3943) has an estimated total width of only about 15 MeV, however the missing mass bump in the 3940 region may have a more complicated structure (see figure 3); this small width estimate evidently carries considerable uncertainty.

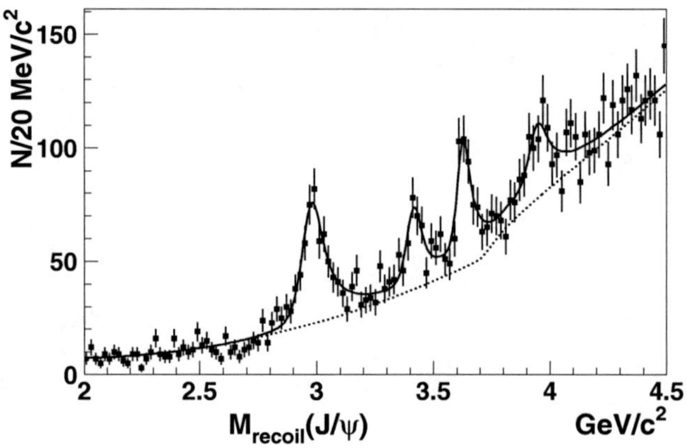

FIGURE 3. The X(3943) reported by Belle in the double charmonium production reaction $e^+e^- \to J/\psi X$. The η_c, χ_0 and η_c' are also evident [1, 7].

Belle report that DD* is the dominant X(3943) decay mode, with a branching fraction of

$$B_{\mathrm{DD}^*} = 96 {}^{+45}_{-32} \pm 22\%. \tag{9}$$

The other available open-charm mode is DD, which is only known as an upper limit,

$$B_{\mathrm{DD}} < 41\%,\ \ 90\%\ c.l. \tag{10}$$

Similarly the closed-charm mode $\omega J/\psi$ (of interest since this mode is apparently preferred by the Y(3943)) is also known as an upper limit,

$$B_{\omega J/\psi} < 26\%,\ \ 90\%\ c.l. \tag{11}$$

Eichten has suggested a $3^1\mathrm{S}_0$ η_c $c\bar{c}$ radial excitation assignment for this state; given the $3^3\mathrm{S}_1$ $\psi(4040)$, the mass is roughly correct, and the observation of the η_c and η_c' with comparable strength in double charmonium production supports this identification. The other possibility with a dominant DD* mode in this mass region is the $2^3\mathrm{P}_1$ $c\bar{c}$, but the absence of a strong signal for the $1^3\mathrm{P}_1$ $\chi_1(3510)$ in double charmonium production argues against this 2P assignment. The $3^1\mathrm{S}_0$ η_c assignment could be tested by searching for evidence of this state in $\gamma\gamma \to$DD*.

Y(4260): A charmonium hybrid?

The most remarkable new state discussed at HADRON05 was the Y(4260), which was reported by BaBar in initial state radiation (ISR) in the reaction $e^+e^- \to \gamma_{ISR} J/\psi\pi^+\pi^-$ [8]. There may also be evidence for an enhancement in $J/\psi\pi^+\pi^-$ states near 4.26 GeV in the decay B$\to$K$J/\psi\pi^+\pi^-$ (in both neutral and negative B/K charge states) [9]. The mass and width reported for the Y(4260) [8] are

$$\mathrm{M} = 4259 \pm 8 \pm {}^{+2}_{-6} \ \mathrm{MeV}, \tag{12}$$

$$\Gamma = 88 \pm 23 \pm {}^{+6}_{-4} \ \mathrm{MeV}. \tag{13}$$

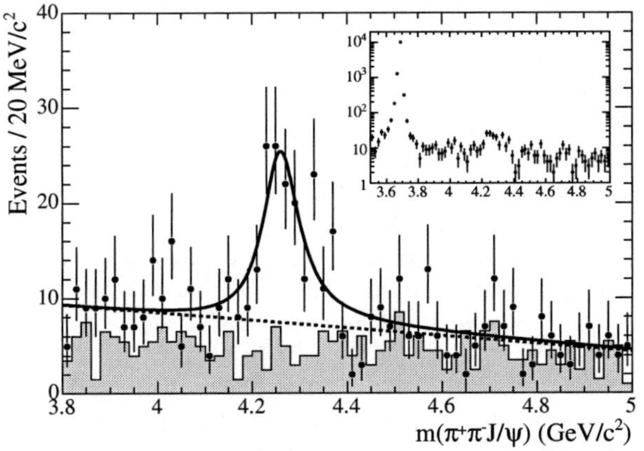

FIGURE 4. The Y(4260) reported by BaBar in the ISR process $e^+e^- \to \gamma_{ISR} J/\psi\pi^+\pi^-$ [8].

The ISR production mechanism tells us that this state must be 1^{--}, but it cannot be a conventional $c\bar{c}$ state because the 1^{--} states in this mass region are well established from earlier e^+e^- annihilation experiments. (The Y(4260) is bracketed by the $2^3\mathrm{D}_1$ $\psi(4160)$ and the $4^3\mathrm{S}_1$ $\psi(4415)$, which have masses that are in excellent agreement with the expectations of $c\bar{c}$ potential models.)

The Y(4260) (if a real resonance) evidently represents "overpopulation" of the expected quark potential model spectrum of 1^{--} $c\bar{c}$ states. This observation, combined with the fact that there is no enhancement visible in R near this mass [10] has led to suggestions that this state may be a charmonium hybrid [11]. The lightest hybrid multiplet in both the flux tube and bag models contains a 1^{--} state. In the flux tube model this lightest $c\bar{c}$-hybrid multiplet is predicted to lie near 4.2 GeV [12], and in LGT studies the lightest 1^{-+} exotic hybrid (expected to be near the 1^{--} hybrid in mass) is found at a rather similar mass of about 4.4 GeV [13]. In these models hybrids have a vanishing $c\bar{c}$ wavefunction at contact, so it has long been speculated that they would have small e^+e^- widths, and thus make rather weak contributions to R. A LGT study by the UKQCD group [14] of the strong decay couplings of exotic $b\bar{b}$ hybrids found strikingly large couplings to closed flavor modes (specifically to $\chi_b S$, where S is a light scalar isoscalar meson that would decay to $\pi\pi$). This is sufficiently similar to the BaBar observation of the Y(4260) in the single closed-charm mode $J/\psi\pi^+\pi^-$ to be cited as additional possible evidence for a hybrid interpretation.

The unusual $J/\psi\pi^+\pi^-$ mode and the UKQCD study suggest searches of any other accessible closed-charm modes with 1^{--} quantum numbers, such as $J/\psi\eta$, $J/\psi\eta'$, $\chi_J\omega$ and so forth. Ideally the light system should have quantum numbers thought to couple strongly to pure glue, such as 0^{++} and 0^{-+}.

In specific decay models (flux tube and constituent glue), theorists have anticipated that the dominant open-charm decay modes of charmonium hybrids would be a meson pair with one internal S-wave (D, D*, D_s, D_s^*) and one internal P-wave (such as D_J and D_{sJ}). In the case of the Y(4260) this suggests dominance of the decay mode $DD_1(2430)$. This broad D_1 has a width of *ca.* 400 MeV, and decays to $D^*\pi$, so this suggests a search for evidence of the Y(4260) in $DD^*\pi$. Since this is a prediction of a decay model in an untested regime (hybrids), one should be cautious and search the more familiar two-body modes DD, DD*, D*D*, D_sD_s, $D_sD_s^*$ and $D_s^*D_s^*$ for evidence of the Y(4260) as well. If there is evidence of a large $DD_1(2430)$ signal, the Y(4260) would then be quite convincing as a hybrid having properties predicted by the flux tube model. If it appears in some of these open charm modes such as DD* and $D_sD_s^*$ at rates comparable to or larger than $J/\psi\pi^+\pi^-$, one might claim a hybrid but speculate that the flux tube decay model was inaccurate in predicting hybrid decay modes. Finally, if the Y(4260) does not appear in any other mode, one might be skeptical about whether the $J/\psi\pi^+\pi^-$ signal is due to a resonance at all; there are nonresonant possibilities, such as production of DD_1 in $e^+e^- \to DD_1$ followed by an inelastic FSI that produces a broad $J/\psi\pi^+\pi^-$ enhancement due to the (very broadened) onset of $DD_1(2430)$ threshold events (which would appear near 4.3 GeV).

In any case the evidence for the Y(4260) from BaBar is not strong, with an estimated 125 ± 23 events [8]. Since it is clear from the reported mass that the Y(4260) is not a conventional $c\bar{c}$ state, it will be very important to test the existence of this state through the accumulation of better statistics.

Other developments in the $Q\bar{Q}$ sector: X(3872)

Developments in other sectors of $Q\bar{Q}$ were rather less striking, although there have been some interesting developments. The unusual and very narrow X(3872) was reviewed here by Trabelsi [1] and by Swanson [15]. There is now clear evidence that this state is 1^{++}, which was a prediction of the DD* molecule picture promoted in particular by Tornqvist [16] and Swanson [17]. New results from CDF II discussed here by Maciel [18] show that the $\pi^+\pi^-$ mass distribution in the decay mode X(3872)$\to J/\psi\pi^+\pi^-$ is well described by the assumption that it is due to $J/\psi\rho^0$. One especially striking prediction of the (neutral D) D^0D^{*0} molecule model is that one should observe comparable strength $J/\psi\omega$ and $J/\psi\rho^0$ decay modes [17], due to the maximal isospin breaking present in the initial state. This prediction appears to have been confirmed by the evidence from Belle for the ω in the 3π mode X(3872)$\to J/\psi\pi^+\pi^-\pi^0$ [19]. The 3π invariant mass peaks at the highest mass, as expected for a virtual ω, and the ratio of 2π

to 3π branching fractions is close to unity,

$$\frac{\Gamma(X(3872) \to J/\psi\pi^+\pi^-\pi^0)}{\Gamma(X(3872) \to J/\psi\pi^+\pi^-)} = 1.0 \pm 0.4 \pm 0.3, \tag{14}$$

as expected in the DD* model. There is also evidence from Belle for the radiative transition X(3872)→ $\gamma J/\psi$ [19], with the width ratio

$$\frac{\Gamma(X(3872) \to \gamma J/\psi)}{\Gamma(X(3872) \to J/\psi\pi^+\pi^-)} = 0.14 \pm 0.05, \tag{15}$$

which should be useful in testing the details of different models of the X(3872).

HEAVY-LIGHT MESON SPECTROSCOPY

Most of the studies of heavy-light mesons reported at HADRON05 considered decays of known $Q\bar{q}$ states, and these have provided very extensive new results on light meson spectroscopy (see for example the talk by Asner [20]). There has been rather less recent activity on the spectroscopy of heavy-light mesons themselves. Several new results on $Q\bar{q}$ spectroscopy were presented at HADRON05, and I will specialize to a few of these in the B and D sectors. There are also some new results on heavy-quark baryons, such as the identification by SELEX of the first doubly-charmed baryon, a Ξ_{cc}^{+} at 3518.7 ± 1.7 MeV; baryon spectroscopy was discussed here by Cumulat [21] and in the summary by Klempt [22].

Maciel [18] discussed results from CDF on the $B_c^{\pm}$ pseudoscalar meson, which is reported in decays to $J/\psi\pi^{\pm}$ [23]. The fitted mass determined from the *ca.* 19 events was

$$M_{B_c} = 6287.0 \pm 4.8 \pm 1.1 \text{ MeV}, \tag{16}$$

which is roughly consistent with or somewhat below theoretical expectations.

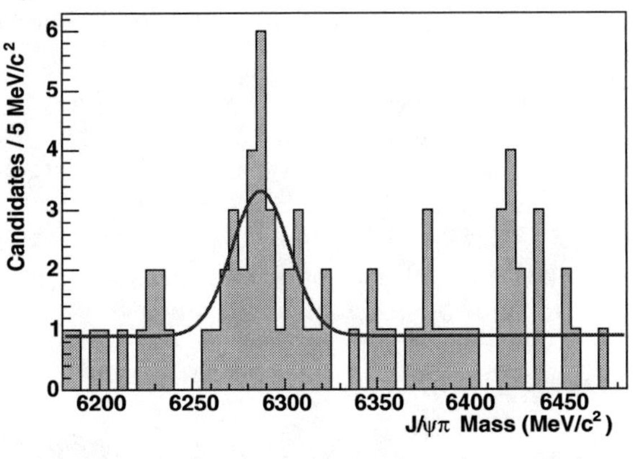

FIGURE 5. The B_c reported by CDF in $B_c^{\pm} \to J/\psi\pi^{\pm}$ [18, 23].

Results from CDF and D0 on orbitally excited B mesons were also presented; for the first time we have evidence for the separate states in the P-wave B meson multiplet, rather than seeing an unresolved superposition of states. Both experiments identify B mesons in J/ψK events, which are combined with another π and studied for evidence of excited B mesons. In both experiments the results are preliminary, but do show evidence of separate B_2^* and B_1 contributions. The D0 mass results are

$$M_{B_1} = 5724 \pm 4 \pm 7 \text{ MeV}, \tag{17}$$

$$M_{B_2^*} - M_{B_1} = 23.6 \pm 7.7 \pm 3.9 \text{ MeV}. \tag{18}$$

Their Bπ mass plot is shown in figure 6. The total widths, constrained in the fit to be equal (and predicted to be comparable theoretically) are

$$\Gamma_{tot.} = 23 \pm 12 \text{ MeV}, \tag{19}$$

which is consistent with theoretical expectations for the two narrow P-wave B mesons. The experimental resolution is estimated to be about 10 MeV.

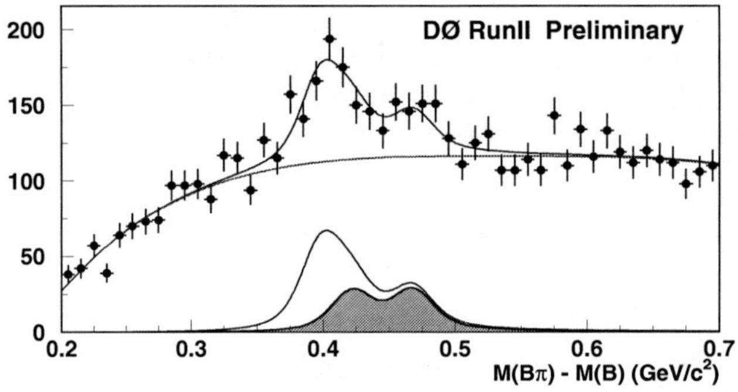

FIGURE 6. Orbitally-excited P-wave B mesons, as reported by D0. Fitted contributions from $B_1 \rightarrow B\pi$, $B_2^* \rightarrow B\pi$ and $B_2^* \rightarrow B^*\pi$ are shown [18, 24].

At the previous meeting HADRON03, the spectroscopy of the D_s sector was a topic of great interest, due to the discovery of the surprisingly light and narrow states $D_{s0}^*(2317)$ and $D_{s1}(2460)$. Here we have heard relatively few new experimental results on these states. The most interesting new result may be their observation in B decays by Belle, reported here by Bianco [25]. The $D_{s0}^*(2317)$ is seen in the process $B^0 \rightarrow D_{s0}^{*+}(2317)K^-$, $D_{s0}^{*+}(2317) \rightarrow D_s^+\pi^0$, and the $D_{s1}^+(2460)$ may have been seen (at a much lower rate) in a similar chain to $D_s^+\gamma$. The reported branching fractions are

$$B_{B^0 \rightarrow D_{s0}^{*+}(2317)K^-} \cdot B_{D_{s0}^{*+}(2317) \rightarrow D_s^+\pi^0} = 4.4 \pm 0.8 \pm 0.6 \pm 1.1 \cdot 10^{-5}, \tag{20}$$

$$B_{B^0 \rightarrow D_{s1}^{*+}(2460)K^-} \cdot B_{D_{s1}^{*+}(2460) \rightarrow D_s^+\gamma} = 0.53 \pm 0.20 \pm 0.6 \pm \begin{matrix} +0.16 \\ -0.15 \end{matrix} \cdot 10^{-5}. \tag{21}$$

The latter branching fraction produce is also quoted as $< 0.86 \cdot 10^{-5}$, 90% *c.l.*

Although it is very interesting to see the narrow D_{sJ} mesons in B decays, uncertainty regarding the production mechanism makes the interpretation of these numbers in terms of the internal structure of the D_{sJ} states somewhat problematic.

THEORY AT HADRON05

There were just four plenary talks at HADRON05 by theorists specifically devoted to theory. Since these contributions appear in the proceedings, here I will simply cite the topics discussed, and will consider one of these talks, which addresses a long-standing problem in hadron physics, in more detail. The theory plenary talks were given by Brambilla (effective field theories), Brodsky (anti deSitter space and conformal field theory), Fodor (recent developments in lattice QCD), and van Beveren ("complex meson spectroscopy").

Brambilla [26] discussed recent work on the derivation of effective Lagrangians for heavy-quark mesons (treated as fields) and gluons. This approach was cited as having many potential applications in the description of the physics of heavy quark hadrons, including meson and baryon spectroscopy, radiative widths, determination of m_c, m_b and α_s, and (in the pure glue sector) the spectrum of glueballs.

Brodsky [27] gave a remarkable talk on connections between string theory, Maldecena duality, anti deSitter space, conformal field theory and hadrons. These connections suggested a generalization of the MIT bag model, which leads to light cone wavefunctions that allowed the evaluation of many properties of hadrons in the hard scattering regime. Predictions of this "template for QCD" included results for the spectrum of mesons, baryons and gg glueballs, and the absence of ggg C-negative "odderon" glueballs (in a classical approximation to this approach).

Fodor [28] reviewed aspects of lattice QCD, and concentrated on recent developments in the simulation of QCD at nonzero temperature. Impressive new results have been reported for properties of the QCD phase diagram and equation of state, as a result of developments in algorithms for treating systems at nonzero chemical potential. These results have potential applications to an unusually broad set of topics, including for example heavy ion collisions at RHIC and the internal structure of neutron stars.

Finally, van Beveren [29] discussed the effect of coupling the discrete, stable hadron basis states of the naive quark model to the two-hadron continuum. This program, often referred to as "unquenching the quark model", is an extremely important topic in hadron physics. It is clear that coupling naive quark model states (or quenched lattice QCD states) to the continuum of real and virtual decay channels will induce many important effects, such as mass shifts, nonzero decay widths, and modified electromagnetic couplings due to continuum components in the state vector.

The narrow D_{sJ} mesons may be examples of the importance of the coupling of quark model states to the continuum. These states lie *ca.* 100 MeV below the expectations of the naive quark model, which does not include loop effects. The fact that these states are near DK and D*K thresholds and have very large S-wave couplings to these channels suggests that their anomalously low masses may be due to mass shifts that result from virtual meson loops [30].

The effects of virtual hadron loops on hadron properties have been studied in detail only in a few specific cases, usually with a severely truncated set of continuum channels. Evaluation of these loop effects is complicated by the fact that the general three-hadron decay vertex is not well established, and several different models of this coupling have been proposed. Two specific forms that have seen wide application are the $^3P_0$ model of Micu [31] (which was used here by van Beveren, and by most studies of loop effects) and a linear vector potential pair-production model, which was used by the Cornell group in their study of the effect of hadron loops on charmonium states [32]. Application of the $^3P_0$ decay model to loop effects in $c\bar{c}$ suggests that the mass shifts are typically 100s of MeV [33, 34], much larger than the *ca.* 10 MeV quoted for the Cornell decay model [35]. Distinguishing between these apparently very different predictions is complicated by the fact that the $^3P_0$ model predicts large *overall* mass shifts, but much smaller observable *differences* in mass shifts between states.

These $^3P_0$ and Cornell decay models give quite different predictions for some observables, such as the famous D/S ratio in $b_1 \to \omega\pi$. Comparison of the different predictions with experimentally well established decay amplitudes should allow the development of a realistic working model of the crucial three-hadron vertices in future. The predicted mass shifts and other loop effects will clearly depend strongly on the model used to describe this coupling.

In their contribution to HADRON05, van Beveren *et al.* describe a T-matrix formalism for iterating the effect of these meson loops, and apply it to the calculation of $2 \to 2$ scattering amplitudes. Their results appear to be successful in describing many puzzling aspects of meson spectroscopy, such as the low-mass "σ" and "κ" scalar enhancements and the masses of the narrow D_{sJ} states [36]. This type of model shows considerable promise in attempts to understand the importance of hadron loops in the physics of hadrons, and in answering such long-standing questions as why the valence quark model works as well as it does, and under what circumstances it should fail.

O PENTAQUARK MORREU!

In Elton Smith's presentation [37] we learned that Jefferson Lab has now accumulated considerable data on photoproduction of the kaon nucleon system (from both proton and deuteron targets), so that they now have much better statistics than in the data originally cited as evidence for a narrow pentaquark [38, 39]. There is no evidence for any such state in this new high-statistics data. One may compare the nK^+ mass distribution originally reported by the CLAS Collaboration in the reaction $\gamma d \to K^+K^-pn$ in Fig.4 of Ref.[39] (our figure 7) to the new, high-statistics distribution shown here by Smith (our figure 8). The previously claimed 5σ peak at 1.542 ± 0.005 GeV has disappeared. Thus, "The pentaquark is dead!"

Since we have heard many experimental reports of observations of this or other related pentaquark states over the previous two years, and many theorists have claimed to understand this now vanished signal, it will certainly be of interest to revisit the pentaquark saga in future and ask just what mistakes were made, and what lessons we have learned as a result. Here I will discuss one important issue in the theory of multiquark systems generally; experimental problems with the pentaquark in particular

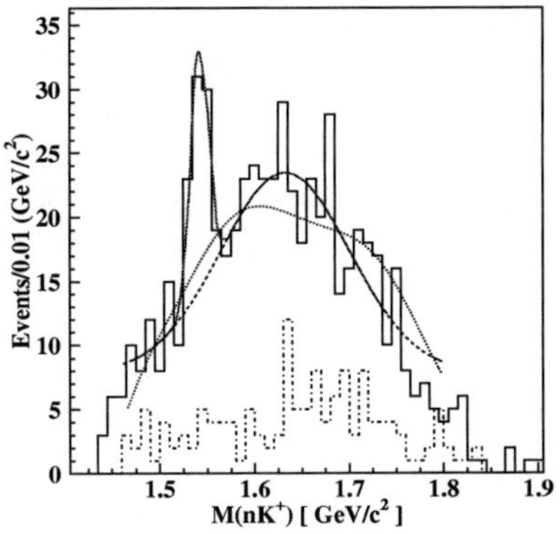

FIGURE 7. The flavor-exotic pentaquark state previously reported by the CLAS Collaboration in the nK^+ invariant mass distribution near 1540 MeV, in the reaction $\gamma d \rightarrow K^+K^-pn$ [39].

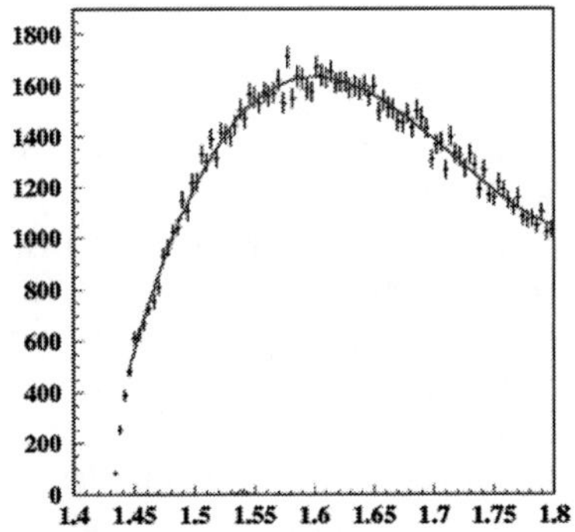

FIGURE 8. The corresponding nK^+ invariant mass distribution from CLAS for photoproduction on protons, presented at HADRON05 by Smith [37]. Neither this nor the deuteron data show evidence of a narrow pentaquark.

were reviewed recently by Dzierba *et al.* [40].

It is both interesting and unsettling that we have traveled this dead-end multiquark road many times before. The repeated claims of narrow multiquark resonances in low-statistics data sets accompanied by spurious theoretical support was such a problem in the 1970s that Isgur in 1985 wrote of "The Multiquark Fiasco" in a series of Schladming lectures [41]. The discussion of multiquarks in this reference should be required reading for anyone working on the pentaquark, or on any multiquark resonance.

The conceptual problem with multiquarks was primarily one of understanding strong decays. Most models of multiquark systems simply excluded open decay channels from consideration, and assumed that the entire multiquark system was confined into a single hadron. Since there are many degrees of freedom present, this type of model predicts a very rich spectrum of discrete levels, many of which have exotic quantum numbers. One famous example is the I=2 $u^2\bar{d}^2$ 0^{++} flavor-exotic scalar, which was predicted by the MIT bag model to have a mass of about 1.2 GeV. A resonant state with these quantum numbers should be clearly visible in the I=2 $\pi\pi$ S-wave phase shift, but experimentally there is no indication of any such resonance; one sees only repulsive scattering. Although the bag model predicts a flavor-exotic multiquark scalar meson at 1.2 GeV, none is observed. What has gone wrong?

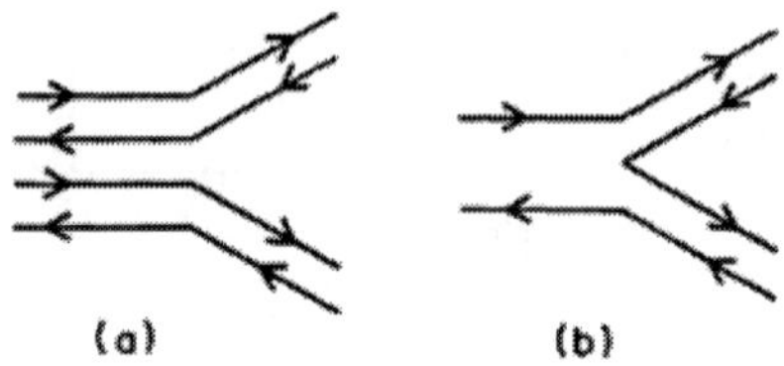

FIGURE 9. Quark line diagrams showing (a) the spontaneous dissociation "fall-apart decay" of a multiquark meson into two color singlets, versus (b) the decay of a $q\bar{q}$ meson through $q\bar{q}$ pair production [42].

The crucial mistake appears to have been the neglect of strong decay couplings, specifically the assumption by most models of a fully confined multiquark system. In a (perhaps oversimplified) quark model picture one can see that the decays of multiquark states are fundamentally different from those of conventional $q\bar{q}$ or qqq hadrons, which are the minimal color singlets. Decay of a $q\bar{q}$ resonance into two $q\bar{q}$ mesons (for example) requires a decay interaction, specifically $q\bar{q}$ pair production. In contrast, a multiquark system such as a $q^2\bar{q}^2$ meson or a $q^4\bar{q}$ pentaquark can simply spontaneously dissociate "fall apart" into two separate minimal color singlets, without a decay interaction. These two very different decay processes are shown in figure 9.

Although the hadron community has absorbed this difference as the "folklore" that multiquark states are expected to be very wide, the difference in decays can be much more drastic; unless there is some form of fission barrier, a multiquark system above its strong fall-apart decay threshold will not exist as a resonance at all. This was appreciated in the late 1970s, and is the reason for Jaffe's statements [43] regarding his table of bag model dibaryon (q^6) multiquarks: "Most of the "states" listed in the table probably do not correspond to particles or resonances." The problem is that "The object is classically unstable against small deformations leading to fission into separate n and p bags. [...] Quantizing about the six-quark object would lead to instabilities analogous to those in a field theory with negative squared mass." In other words, these are not resonant states; the fall-apart effect implies that they do not exist as resonances.

Since the proposed pentaquark was above the mass of a kaon and a nucleon, it too has a fall-apart mode, and should not exist as a resonance barring exceptional circumstances such as a fission barrier. Alas this lesson appears to have been forgotten, and multiquark models that did not consider fall-apart into the KN continuum were accepted as justification for the existence of such states with undue credulity.

FIGURE 10. Jorge Agustín Nicolás Ruiz de Santayana y Borrás. Professor of Philosophy, Harvard University, 1889-1912.

Santayana's familiar quote unfortunately appears relevant to the pentaquark saga: "Those who cannot remember the past are condemned to repeat it." [44]. A less well-known observation by Santayana, which is also relevant to the pentaquark story, is that "Scepticism is the chastity of the intellect, and it is shameful to surrender it too soon or to the first comer." [45]

Our moment of Zen

"The story of the pentaquark shows how poorly we understand QCD." [46]

ACKNOWLEDGMENTS

It is a pleasure to thank Alberto Reis and colleagues at CBPF for their kind invitation to present this summary, and for the opportunity to discuss recent developments in hadron physics with my fellow participants. I also gratefully acknowledge useful communications from S.Bianco, S.Godfrey, J.Pursley and E.Swanson regarding aspects of the material discussed here. This research was supported in part by the U.S. National Science Foundation through grant NSF-PHY-0244786 at the University of Tennessee, and the U.S. Department of Energy under contract DE-AC05-00OR22725 at Oak Ridge National Laboratory.

REFERENCES

1. K.Trabelsi, "Hadron Physics at Belle", these proceedings.
2. E. J. Eichten, K. Lane and C. Quigg, Phys. Rev. Lett. **89**, 162002 (2002) [arXiv:hep-ph/0206018].
3. K. Abe *et al.*, arXiv:hep-ex/0507033.
4. T. Barnes, S. Godfrey and E. S. Swanson, Phys. Rev. D **72**, 054026 (2005) [arXiv:hep-ph/0505002].
5. The two photon width of a $\chi_2'(3900)$ 2^3P_2 $c\bar{c}$ state was estimated to be 0.64 keV; see T. Barnes, in Proc. IXth Internatl. Conf. on Photon-Photon Collisions, (LaJolla, CA, 22-26 March 1992).
6. K. Abe *et al.* [Belle Collaboration], Phys. Rev. Lett. **94**, 182002 (2005) [arXiv:hep-ex/0408126].
7. K. Abe *et al.*, arXiv:hep-ex/0507019.
8. D. Muller, "Hadron Physics at BaBar", these proceedings; B. Aubert *et al.* [BABAR Collaboration], Phys. Rev. Lett. **95**, 142001 (2005) [arXiv:hep-ex/0506081].
9. B. Aubert *et al.* [BABAR Collaboration], arXiv:hep-ex/0507090.
10. K.Seth, these proceedings.
11. F. E. Close and P. R. Page, arXiv:hep-ph/0507199.
12. T. Barnes, F. E. Close and E. S. Swanson, Phys. Rev. D **52**, 5242 (1995) [arXiv:hep-ph/9501405].
13. X. Liao and T. Manke, arXiv:hep-lat/0210030.
14. C. McNeile, C. Michael and P. Pennanen [UKQCD Collaboration], Phys. Rev. D **65**, 094505 (2002) [arXiv:hep-lat/0201006].
15. E.S.Swanson, these proceedings.
16. N. A. Tornqvist, arXiv:hep-ph/0308277.
17. E. S. Swanson, Phys. Lett. B **588**, 189 (2004) [arXiv:hep-ph/0311229].
18. A.Maciel, " Hadron Physics at the Tevatron", these proceedings.
19. K. Abe *et al.*, arXiv:hep-ex/0505037.
20. D.Asner, these proceedings.
21. J.Cumulat, these proceedings.
22. E.Klempt, these proceedings.
23. D. Acosta *et al.* [CDF Collaboration], arXiv:hep-ex/0505076.
24. [D0 Collaboration], "Study of excited B mesons", D0note 4517 (1 Aug 2004).
25. S.Bianco, these proceedings.
26. N.Brambilla, these proceedings.
27. S.J.Brodsky, these proceedings.
28. Z.Fodor, these proceedings.
29. E.van Beveren, these proceedings.
30. D. S. Hwang and D. W. Kim, Phys. Lett. B **601**, 137 (2004) [arXiv:hep-ph/0408154].
31. L. Micu, Nucl. Phys. B **10**, 521 (1969).
32. E. Eichten, K. Gottfried, T. Kinoshita, K. D. Lane and T. M. Yan, Phys. Rev. D **21**, 203 (1980).
33. T. Barnes, J. Phys. Conf. Ser. **9**, 127 (2005) [arXiv:hep-ph/0412057].
34. Y. S. Kalashnikova, Phys. Rev. D **72**, 034010 (2005) [arXiv:hep-ph/0506270].
35. E. J. Eichten, K. Lane and C. Quigg, Phys. Rev. D **69**, 094019 (2004) [arXiv:hep-ph/0401210].
36. E. van Beveren, J. E. G. Costa, F. Kleefeld and G. Rupp, arXiv:hep-ph/0509351.
37. E. Smith, "Review of Recent Jlab Results", these proceedings.
38. T. Nakano *et al.* [LEPS Collaboration], Phys. Rev. Lett. **91**, 012002 (2003) [arXiv:hep-ex/0301020].
39. S. Stepanyan *et al.* [CLAS Collaboration], Phys. Rev. Lett. **91**, 252001 (2003) [arXiv:hep-ex/0307018].
40. A. R. Dzierba, C. A. Meyer and A. P. Szczepaniak, J. Phys. Conf. Ser. **9**, 192 (2005) [arXiv:hep-ex/0412077].
41. N. Isgur, Acta Physica Austriaca Suppl. XXVII, 177 (1985). Published in Schladming School 1985:0177 (QCD161:I8:1985), also in Los Alamos Wkshp. 1985:0619 (QCD161:W675:1985).
42. R. L. Jaffe and K. Johnson, Phys. Lett. B **60**, 201 (1976).
43. R. L. Jaffe, Phys. Rev. Lett. **38**, 179 (1977).
44. J.Santayana, *Life of Reason, Reason in Common Sense* (1905).
45. *ibid.*, *Scepticism and Animal Faith* (1923).
46. F.Wilczek (attributed).

LIST OF PARTICIPANTS

Adams, Garry	Rensselaer Polytechnic Institute
Afanasev, Leonid	JINR - Dubna
Alves, Gilvan	CBPF
Almeida, Katia	Universidade Federal do Rio de Janeiro
Ambrosino, F	Universita di Napoli
Amsler, Claude	University of Zurich
Appel, Jeffrey	Fermilab
Arantes, Lecio	Universidade de São Paulo
Araujo, Wilson	Instituto Tecnológico da Aeronáutica
Asner, David	Carleton University
Aston, David	SLAC
Barbi, Mauricio	University of Regina
Barnes, Ted	Tennessee University
Bediaga, Ignacio	CBPF
Bessa, Andre	Universidade Federal do Rio de Janeiro
Bianco, Stefano	Laboratori Nazionali di Frascati dell'INFN
Bigi, Ikaros	Notredame University
Bini, Cesare	Universita di Roma I
Boca, Gianluigi	University of Pavia and INFN
Boito, Diogo R.	Universidade de São Paulo
Borges, Jose Sa	Universidade do Estado do Rio de Janeiro
Boschi-Filho, Henrique	Universidade Federal do Rio de Janeiro
Braghin, Fabio	Universidade de São Paulo
Brambilla, Nora	Universita di Milano
Brito, Thiago	Universidade de São Paulo
Brodsky, Stanley	SLAC/Stanfrod University
Büscher, Markus	Forschunhszentrum Juelich
Bugg, David	Queen Mary, University of London
Carman, Daniel	Ohio University
Cassol, Nadiane	Instituto de Física Teórica
Cavalcante, Isabela	Universidade Fed. M. Grosso do Sul

Ciofi degli Atti, C.	INFN – Perugia
Chen, Jiangchuan	Institute of High Energy Physics, Beijing
Chung, Suh-Urk	BNL
Colantoni, Marialaura	INFN – Torino
Coutinho, Yara	Universidade Federal do Rio de Janeiro
Credé, Volker	Florida State University
Cumalat, John	Colorado University
De Vita, Raffaella	INFN - Genova
Dexheimer, Veronica	Univ. Federal do Rio Grande do Sul
Dudek, Jozef	Jefferson Lab
Dumm, Daniel G.	Universidad de La Plata
Dunne, James	Jefferson Lab
Dytman, Steven	University of Pittsburgh
Eisenberg, Yehuda	Weizmann Institute
Faccioli, Pietro	Bologna University & INFN
Fadeeva, Ekaterina	ITEP – Moscow
Favart, Laurent	Universite Libre de Bruxelles
Ferreira, Erasmo	Universidade Federal do Rio de Janeiro
Filippi, Alessandra	INFN - Torino
Fodor, Zoltan	University of Wuppertal
Fogaça, David A.	Universidade de São Paulo
Fraga, Eduardo	Universidade Federal do Rio de Janeiro
Frederico, Tobias	Instituto Tecnológico da Aeronáutica
Gambin, Ediana	Instituto Tecnológico da Aeronáutica
Gandelman, Miriam :	Universidade Federal do Rio de Janeiro
Garmash, Alexei	Princeton University
Gavillet, Philippe	CERN
Giovannella, Simona	Laboratori Nazionali di Frascati dell'INFN
Göbel, Carla	PUC-Rio
Goldman, Terry	Los Alamos National Laboratory
Gothe, Ralf W	University of South Carolina
Grunfeld, Ana G.	Comision Nacional de Energia Atomica
Haas, Tobias	DESY
Hadjimichef, Dimiter	Universidade Federal de Pelotas
Hansen, Ole	Jefferson Lab
Hofmann, Julian	GSI
Hrynova, Tetiana	SLAC

Ishida, Muneyuki	Meisei Universityu
Ishida, Shin	Nihon University
Jacobsen, Rafael	Univ. Federal do Rio Grande do Sul
Jin, Shan	Institute of High Energy Physics, Beijing
Joo, Kyungseon	University of Connecticut
Juge, Keisuke J.	Trinity College
Khokhlov, Yury	IHEP, Protvino
Kichimi, Hiromichi	KEK
Kleefeld, Frieder	CFIF/Instituto Superior Técnico - Lisboa
Klempt, Eberhard	University Bonn
Kodama, Takeshi	Universidade Federal do Rio de Janeiro
Komada, Toshihiko	Nihon University
Korpar, Samo	Ljubljiana
Krein, Gastão	Instituto de Física Teórica
Lee, Su Houng	Yonsey University
Lesiak, Tadeusz	Institute of Nuclear Physics - Krakow
Linares, Luis P.	Universidade Federal Fluminense
Lolos, George	University of Regina
Luna, Emerson	Instituto de Física Teórica
Ma, Bo-qiang	Peking University
Machner, Hartmut	Forschunhszentrum Juelich
Maciel, Arthur	CBPF
Maeda, Tomohito	Nihon University
Magnin, Javier	CBPF
Machado, Ana A.	CBPF
Malheiro, Manuel	Instituto Tecnológico da Aeronáutica
Marchand, Claude	CEA - Saclay
Masoni, Alberto	INFN - Cagilari
Matheus, Ricardo	Universidade de São Paulo
Matsuki, Takayuki	Tokyo Kasei University
Meadows, Brian	Univesrity of Cincinnati
Melo, Joao P. B.	CETEC/Instituto de Física Teórica
Miranda, Jussara	CBPF
Miscetti, Stefano	Laboratori Nazionali di Frascati dell'INFN
Mitchell, Ryan	University of Indiana
Mizher, Ana J.	Universidade Federal do Rio de Janeiro
Mo, Xiaohu	Institute of High Energy Physics, Beijing

Mocsy, Agnes	Frankfurt University
Moloney, Glenn	University of Melbourne
Montalvo, Jorge C.	Universidade do Estado do Rio de Janeiro
Motta, Helio	CBPF
Munhoz, Marcelo	Universidade de São Paulo
Muller, David	SLAC
Navarra, Fernando	Universidade de São Paulo
Negrini, Matteo	Universita di Ferrara
Neto, João R. T. M.	Universidade Federal do Rio de Janeiro
Nielsen, Marina	Universidade de São Paulo
Noya, Hiroshi	Hosei University at Tama
Ochs, Wolfgang	Max Planck Institut Munich
Oller, Jose	Universidad de Murcia
Orth, Herbert	GSI
Oset, Eulogio	Universidad de Valencia
Papandreou, Zisis	University of Regina
Palhares, Leticia	Universidade Federal do Rio de Janeiro
Paul, Stephan	University of Munich
Paula, Leandro	Universidade Federal do Rio de Janeiro
Pedroni, Paolo	INFN- Sezione di Pavia
Peaslee, David	Hope College
Pelaez, Jose	Universidad Complutense - Madrid
Peters, Klaus	GSI
Poireau, Vincent	Laboratoire de Physique de Particules – Annecy-le-Vieux
Polosa, Antonio	Dip. Fisica, Bari
Polycarpo, Erica	Universidade Federal do Rio de Janeiro
Portugal, Licinio	Universidade Federal do Rio de Janeiro
Predazzi, Enrico	Universita di Torino
Razeira, Moises	Univ. Federal do Rio Grande do Sul
Reis, Arthur V. R.	Universidade de São Paulo
Reis, Alberto	CBPF
Reyes, Marco	University of Guanajuato
Rezende, Fabio	CBPF
Rodriguez, Marcos	Universidade Federal do Rio Grande
Robilotta, Manoel	Universidade de São Paulo
Rocha, Carlos A.	Universidade São Judas Tadeu
Rong, Gang	Institute of High Energy Physics, Beijing
Roca, Luis	Universidad de Valencia

Rodrigues, Fernando	Universidade Federal do Rio de Janeiro
Sarantsev, Andrey	HISKP Uni-Bonn, PNPI St. Petersburg
Shan, Jin	Institute of High Energy Physics - Beijing
Shellard, Ronald	CBPF
Scoccola, Norberto	Comision Nacional de Energia Atomica
Seth, Kamal	Northwestern University
Sibidanov, Alexei	Budker Insitute of Nuclear Physics
Silva, Daniel T.	Univ. Federal do Rio Grande do Sul
Silva, Mario L.L.	Univ. Federal do Rio Grande do Sul
Silvestre-Brac, Bernard	CNRS - Grenoble
Shen, Pengnian	Institute of High Energy Physics - Beijing
Shen, Xiaoyan	Institute of High Energy Physics - Beijing
Smith, Elton	Jefferson Lab
Souza, Maria S.K.	Universidade de São Paulo
Suslov, Vladimir	North Carolina Central University
Swanson, Eric	University of Pittsburgh
Takahashi, Jun	Universidade de São Paulo
Takamatsu, Kunio	KEK
Taketani, Bruno	Universidade Federal do Rio de Janeiro
Tavares, Bernardo	Universidade Federal do Rio de Janeiro
Teshima, Tadayuki	Chubu University
Tedeschi, David	University of South Carolina
Trabelsi, Karim	University of Hawaii
Vairo, Antonio	Universita' di Milano
Valcarce, Alfredo	University of Salamanca
van Beveren, Eef	University of Coimbra
Vasconcellos, César	Univ. Federal do Rio Grande do Sul
Vijande, Javier	Universidad de Salamanca
Vlahovic, Branislav	North Carolina Central University
Wahlberg, Hernan	Utrecht University/NIKHEF
Yamada, Kenji	Nihon University
Yuan, Changzheng	Institute of High Energy Physics, Beijing
Zarnauskas, Gabriel	Universidade de São Paulo
Zhao, Qiang	University of Surrey
Zheng, Hanqing	Peking University
Zhu, Yongsheng	Institute of High Energy Physics, Beijing
Ziegler, Veronique	University of Iowa
Zisis, Papandreou	University of Regina
Zou, Bingsong	Institute of High Energy Physics, Beijing

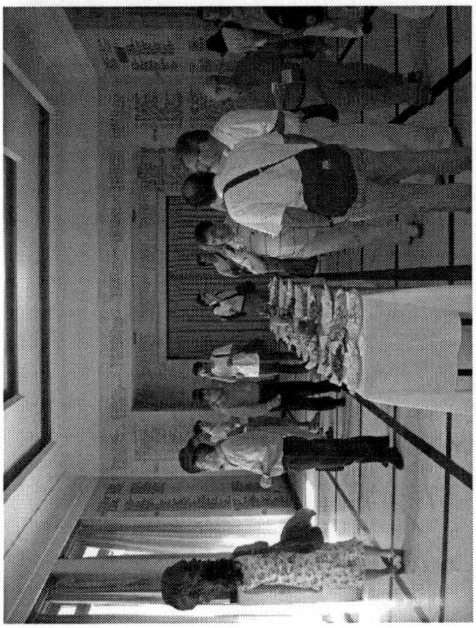

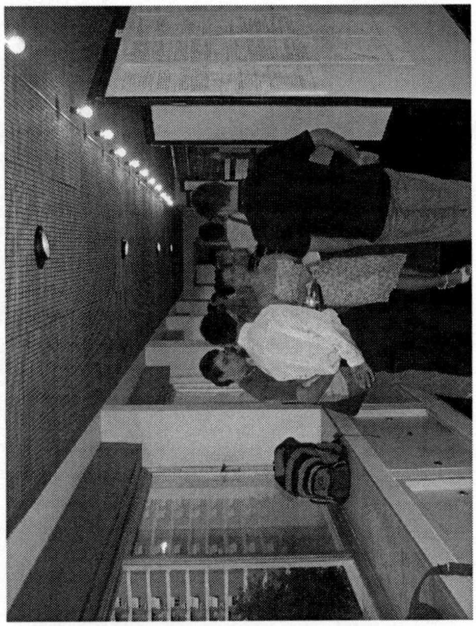

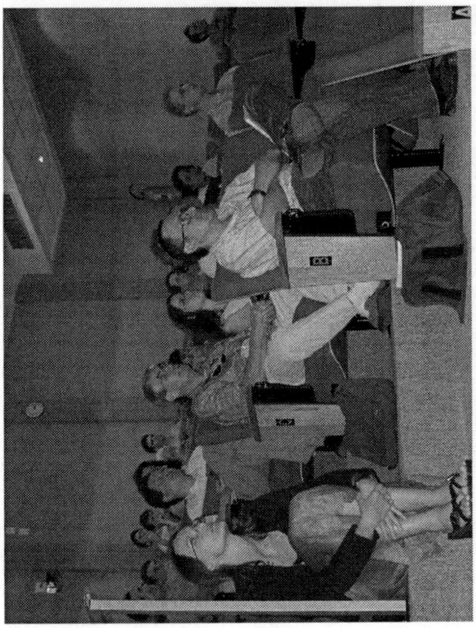

Author Index

S

T

V

W

Y

Z